8th International Scientific Conference on Engineering, Technologies and Systems (TechSys 2019)

IOP Conference Series: Materials Science and Engineering
Volume 618

Plovidv, Bulgaria
16 - 18 May 2019

Editors:

Valyo Nikolov
Nikola Shakev

ISBN: 978-1-5108-9825-7
ISSN: 1757-8981

Printed from e-media with permission by:

Curran Associates, Inc.
57 Morehouse Lane
Red Hook, NY 12571

Some format issues inherent in the e-media version may also appear in this print version.

Institute of Physics
Dirac House, Temple Back
Bristol BS1 6BE UK

Phone: 44 1 17 929 7481
Fax: 44 1 17 920 0979

techtracking@iop.org

Additional copies of this publication are available from:

Curran Associates, Inc.
57 Morehouse Lane
Red Hook, NY 12571 USA
Phone: 845-758-0400
Fax: 845-758-2633
Email: curran@proceedings.com
Web: www.proceedings.com

Table of contents

Volume 618

8th International Scientific Conference on Engineering, Technologies and Systems (TechSys 2019)

16–18 May 2019, Plovdiv, Bulgaria

Accepted papers received: 21 August 2019
Published online: 29 October 2019

Preface

Papers

Automation, Control Systems and Robotics

Electrical Engineering and Electronics

Automotive and Aeronautical Engineering

IOP Conf. Series: Materials Science and Engineering **618** (2019) 011001 doi:10.1088/1757-899X/618/1/011001

Preface

8th International Scientific Conference "TechSys 2019" – Engineering, Technologies and Systems is organized by Technical University of Sofia, Plovdiv Branch. It was held on 16-18 May 2019 in Plovdiv, Bulgaria. The main goal of TechSys is to bring together researchers from academia and industry, scientists and experts in the field of engineering and technologies for sharing the ideas and innovations. Selected articles that represent the significant results of the conference have been collected to publish in this volume.

The conference is organized under the frame of "Science Days of Technical University of Sofia" and supported by the Scientific and Research Sector.

We would like to thank all the authors who have trusted the conference. To thank all those researchers who recognized the TechSys as an appropriate forum to share their recent developments and to get the new trends in the field of technical sciences and engineering.

We must also thank all the reviewers and topic editors for their quiet but very important work, which ensured high quality of scientific research.

Thanks to the plenary speakers Prof. Zezulka and Prof. Bradac for valuable presentation.

Last but not least, we sincerely thank the whole team involved in organizing the TechSys 2019 conference.

Editors: Prof. Valyo Nikolov and Assoc. Prof. Nikola Shakev

IOP Conf. Series: Materials Science and Engineering **618** (2019) 011002 doi:10.1088/1757-899X/618/1/011002

TECHNICAL UNIVERSITY OF SOFIA, PLOVDIV BRANCH

8-th INTERNATIONAL SCIENTIFIC CONFERENCE
"ENGINEERING, TECHNOLOGIES AND SYSTEMS"
TECHSYS 2019

Organizing Committee:
Chairman: Prof. Dr. Valyo Nikolov
Vice Chairman: Assoc. Prof. Dr. Nikola Shakev
Members:

Prof. Dr. Grisha Spasov
Assoc. Prof. Dr. Krum Kutryansky
Assoc. Prof. Dr. Anton Lechkov
Assoc. Prof. Dr. Ivan Kostov
Assoc. Prof. Dr. Nikola Georgiev
Assoc. Prof. Dr. Nikolay Kakanakov
Assoc. Prof. Dr. Valentin Vladimirov
Assoc. Prof. Dr. Pepo Yordanov
Assoc. Prof. Dr. Hristian Panayotov
Assoc. Prof. Dr. Angel Lengerov
Assoc. Prof. Dr. Dimitar Petrov
Assoc. Prof. Dr. Krasimir Ambarev
Assoc. Prof. Dr. Silvina Ilieva
Assoc. Prof. Dr. Toni Mihova
Assoc. Prof. Dr. Valentina Proycheva

International Programme Committee:
Honorary Chairman: Prof. DSc Georgi Mihov, Rector of the Technical University of Sofia
Chairman: Prof. Dr. Michail Petrov
Members:

Prof. DSc Andre Barraco, France
Prof. DSc Okyay Kaynak, Turkey
Prof. DSc. Venelin Jivkov, Bulgaria
Prof. DSc. Vesko Panov, Bulgaria
Prof. DSc Marin Nenchev, Bulgaria
Prof. DSc Emil Nikolov, Bulgaria
Prof. DSc Radi Romanski, Bulgaria
Prof. DSc Todor Stoilov, Bulgaria
Prof. DSc Edmunds Teirumnieks, Latvia
Prof. DSc Mark Himbert, France
Prof. DSc Ivan Jachev, Bulgaria
Prof. DSc. Chavdar Damianov, Bulgaria
Prof. Dr. Nikolay Ganev, Czech Republic
Prof. Dr. Ivaylo Banov, Bulgaria
Prof. Dr. Veselka Boeva, Bulgaria
Prof. Dr. Mladen Velev, Bulgaria
Prof. Dr. Daniela Goceva, Bulgaria
Prof. Dr. Petar Gecov, Bulgaria
Prof. Dr. Lubomir Dimitrov, Bulgaria
Prof. Dr. Frantisek Zezulka, Czech Republic
Prof. Dr. Ilia Iliev, Bulgaria
Prof. Dr. Ernst Wintner, Austria
Prof. Dr. Nikola Kasabov, New Zealand
Prof. Dr. Ivan Kralov, Bulgaria
Prof. Dr. Petr Louda, Czech Republic
Prof. Dr. Valeri Mladenov, Bulgaria
Prof. Dr. Ognyan Nakov, Bulgaria
Prof. Dr. Galidia Petrova, Bulgaria
Prof. Dr. Marsel Popa, Romania
Prof. Dr. Vladimir Pulkov, Bulgaraia
Prof. Dr. Georgi Todorov, Bulgaria
Prof. Dr. Andon Topalov, Bulgaria
Prof. Dr. Sedat Sunter, Turkey
Prof. Dr. Ahmed Hafaifa, Algeria
Prof. Dr. Abdellah Kouzou, Algeria
Prof. Dr. Zdenek Bradach, Czech Republic

Scientific Secretary: Dr. Sevil Ahmed
Technical Secretary: Eng. Tsvetan Petrov, Eng. Christo Christev, Eng. Lalka Boteva

IOP Conf. Series: Materials Science and Engineering **618** (2019) 011003 doi:10.1088/1757-899X/618/1/011003

Peer review statement

All papers published in this volume of *IOP Conference Series: Materials Science and Engineering* have been peer reviewed through processes administered by the proceedings Editors. Reviews were conducted by expert referees to the professional and scientific standards expected of a proceedings journal published by IOP Publishing.

Digital Twin and AAS in the Industry 4.0 Framework

Z Bradac, P Marcon, F Zezulka, J Arm, T Benesl

Brno University of Technology, Faculty of Electrical Engineering and Communication, Brno 61600, Czech Republic

zezulka@feec.vutbr.cz

Abstract. The paper goes out from the state of the art in the area of Industry 4.0 initiative of the most developed countries. There are also discussed reasons of Industry 4.0 development and implementation into existing still developed technologies. According to the last development, the term digital twin became a very often-used word. Authors explain differences in the content of the digital twin and recommend to use for the Industry 4.0 area the more appropriate and precise specified term – the Asset Administration Shell (AAS). Next parts of the contribution therefore deal with the Asset Administration Shell (AAS) in details to enable engineers, technicians and informatics from the praxis much more simple introduction in the Industry 4.0 problems and their solution. There is introduced and explained the structure of the AAS, the model and sub-models architecture, parameters and features in the contribution.

1. Introduction

The contribution deals with Industry 4.0 (I4.0) as with a phenomenon, with its properties and procedures of its solution and implementation. Authors introduce readers in the state of the art after some years of the I4.0 initiative has been started in the most developed countries. Authors catch attention on what has been already done and how to utilize these outputs from the research and development for implementation for practical use. They point out also on the importance of standardization in each step of the value chain of industrial production systems, they present a list of them, and show an outlook of the next standards. One part of the contribution deals with a comprehensive list of requirements, which led to the specification of the most important idea and the most important component – the Asset Administration Shell – the electronic rucksack or digital twin of components in the I4.0 smart production systems.

As it has been told in many forums by different opportunities, the I4.0 phenomenon is something like the 4th Industrial revolution. But as has been also said, the I4.0 is more then revolution a rapid and complex evolution of existing automation and automated production with sometimes still a significant level of robotics, complex automation, global communication, digitization, standardization, and openness. The existing industrial production, which can be titled as production in the intention of the Industry 3.0, still shows attributes, which are a challenge and respectable steps towards the creation of a higher type of organization, architecture, and realization of industrial production and even next social aspects of human activities. It can lead to too optimistic evaluation (self-evaluation) of managements of firms, that they already implemented I4.0 principles into their production systems. However, in the most case, there is missing a lot of very important features, characterizing the really I4.0 ideas, principles and technologies. For example, the digitization is used for many years, but not systematically. Authors are the opinion, that only systemically utilized digitalization can grow int the Industry 4.0 world period.

The truth is, that still many features of existing Industry 3.0 (I3.0) production support a statement that we are on the boarder towards a new, higher level of industrial production, tell the 4th industrial revolution This statement is supported by following phenomenones.

Accelerated chain from idea via research, development, and realization of a new product, next a high value of digitization of information from production, next utilization of this digitized information in the control infrastructure of machines, production lines and technological processes. In addition, the high level of intelligence in process instrumentation, control, and monitoring systems, in MES and ERP systems supports the chance of a possibility of transformation existing production systems from the I3.0 to the I4.0 world. Also digitization of all attributes of physical components, marketing methods and procedures and technical and technological development periods, big data flow in the production process, storing and processing of measurement and control values in the cloud supports idea of the coming Industry 4.0 period. This opinion is supported as well as by activities of standardization organizations in highly developed countries to standardize interfaces, communication protocols, production procedures, requirements on the functional safety and security of the all value chain play a significant role.

One of the most promising phenomenon of this historical period, which still influences significantly all the human society is Internet. Internet influences the production processes - the Internet of things (IoT) and the Industrial Internet of things (IIoT) represent the highest development phenomenon horizontal and vertical integration for Industry 4.0 purposes.

Authors are an opinion, that still the slow progress in the implementation of I4.0 principles is caused by non- systematic digitization, non- optimal data acquisition and data processing and low level of application of standards, an unwillingness of specialists in control, measurement and informatics to apply still existing standards and in missing some of needed standards. It appears particularly in communication subsystem, in communication interfaces and protocols. On the other hand the I4.0, its popularity, broaden up, much work which already has been done during last years, and due to competences of many standardization working groups, mutually cooperated across Germany, France, Italy, but also with support and cooperation with the USA, China, Japan bring and open an excellence opportunity and challenge in technical development, that was not possible still 5-7 years ago. I4.0 activities bring the following opportunities:

- An opportunity for unified communication in the all control pyramid.
- Communication by a unified communication protocol from the shop to the highest levels of the office floors of the all value chain.
- Interconnection of all activities in the value chain of industrial production thanks to unified interfaces of HW as well as SW interfaces, protocols, production procedures, production documentation, quality control, safety, and security.
- A significantly higher degree of cooperation among producers and consumers.
- Support of the informatics branch in research, development, standardization, and implementation of unified real-time communication not only in time-critical production processes but also inside the all value chain.
- Acceleration of the activities design – development – production, increasing of digitization and the use of it, decreasing of redundancy in data acquisition processes.

These opportunities will be utilized only, when following preconditions will be to provided systematically and in a standardized way:

- Solid, systematic and standardized digitalization of information from the all value chain.
- Virtualization of production, process modeling in a virtual environment and after tuning all the process to transmit optimized control algorithms into the physical production, output quality checking, marketing and service.
- A higher level of the horizontal integration of the all value chain.

IOP Conf. Series: Materials Science and Engineering **618** (2019) 012001 doi:10.1088/1757-899X/618/1/012001

- Intelligence until the component of the production.
- Creation of a standardized architecture of a digital twin of physical components (machines, components of machines), components of a transport system, supermarkets, control SW, technical documentation, product, and market documentation.
- Production components will be designed, developed and implemented as the Industry 4,0 components, specified by the ZVEI, VDE/VDI, and cooperating organizations [1], [2], [5], [8].

The contribution goal is to help technical experts from praxis to understand the importance of the I4.0 component model and to win skills in working with it. This I4.0 component model is described in the next chapter. The official term of the production component is the "I4.0 component" and its electronic form is oft titled "the digital twin".

2. Digital twin alias Asset Administration Shell (AAS)

The term digital twin was for the first time used by NASA (National Agency of Space and Aeronautics of the U.S.A.) for approximately 60 years to name an electronic version of the physical model of physical comics systems (space shifts, and other systems to fly in the space. Such a very precious mathematical description and consequently very precise digital realization (digital twin) enables monitoring, control and maintenance of the American space system on very long cosmic distances. Because of similar functionality and an appropriate and with the most important element of the I4.0 activity, the Asset Administration Shell – the electronic rucksack is the AAS oft in the last time named digital twin. It was seen also during the last Hannover Fair that the term digital twin very oft used the title. However, in the sense of the NASA, the digital twin is a virtual representation, an embodiment of an asset of any type, material or non-material – including everything from power turbines to services and maintenance. The digital twin is described by the structure and behavior of connected "things" generating real-time data [3].

In comparison, the asset administration Shell (AAS) is the crucial item in the all I4.0 idea. It creates an interface between the physical and virtual production steps. AAS is a virtual digital and active representation of an I4.0 component in the I4.0 system [2]. Any component of production in the I4.0 environment has to have an administrative shell [2].

Fig. 1 shows the structure and connection of the physical thing and the administration shell (AS). The component of the I4.0 is a unity of an asset and the electronic model – the corresponding AS.

The AAS in Fig. 1 is composed of a body and a header. The header contains identifying details regarding the AAS and the represented asset. The body contains a certain number of submodels for an asset-specific characterization of the AAS [4-7].

The co-author of this contribution lived such a misunderstanding of term digital twin during one oral presentation of one tutor by a tutorial in the ZVEI Forum I4.0 in this April in Hannover. The tutor needed several digital twins for one I4.0 component, but from the AAS definition would be fully acceptable and recommended only one AAS with several sub-models.

Much wrong is, that from commercial reasons, the term digital twin is used also for the 3D model, e.g. of a production unit, machine, or car, including simulation. This interpretation is currently state of the art and used by a broad industrial community. However, the 3D model is an I3.0 technology only [3].

Maybe, that existing digital twins in the commercial interpretation will step by step grow to cover all useful information which is relevant across the lifetime of the related asset, from the initial idea to the engineering, logistics, operation, maintenance, reuse, and destruction. They could become a future digital twin that will contain a simulation model, the 3D model, a lot of other properties, historical data, handbooks, installation guidelines, property function blocks, interlockings, state models, alarms, event definition, etc. On the other hand – a static asset will not include in its digital twin any simulation model [3].

In this context, the term AAS is for purposes of the I4.0 more appropriate title then the term digital twin. Therefore authors emphatically recommend to preferably use the term AAS in the I4.0

environment. How great is the difference between the AAS and a pure digital twin in the commercial interpretation can be seen from the following chapters, dealing with the structure of the AAS.

3. Asset Administration Shell Advanced Topic

Much work has been done by working groups of the ZVEI, VDI/VDE, BITCOM exactly in the structure and its components in the Asset Administrative Shell specification. A very comprehensive material of the detail of the AAS was prepared for publication at the end of 2018.

This material is a result of the new situation in Europe. The initial I4.0 idea of the German state institutions ZVEI, VDI/VDE, BITCOM, and some private companies and organizations became newly a larger European background when it has been started initiatives to keep up and improve three the most developed industrial European countries in the manufacturing industry. Alliance Industrie du Future in France, Platform Industrie 4.0 in Germany and Piano Industria 4.0 in Italy have agreed to join forces working on a shared action plan towards internationalization as an end to end digital continuity and global standardization are of crucial importance for a digitized economy [4].

Let us explain more comprehensively, because of high importance, the term AAS:

- The I4.0 component is the combination of the asset and its logical representation, the AAS.
- The AAS is the standardized digital representation of the asset, cornerstone of the interoperability between the applications managing the manufacturing systems.
- The AAS may be the logical representation of a simple component, a machine or a plant at any level of the equipment hierarchy.
- From the manufacturer point of view, the asset is a product. The manufacturer manages different types that have a history with different versions. In parallel, he produces instances of these different types and versions.
- The manufacturer provides the standardized digital representation to his customers, creating both an AAS for the asset type and for asset instance. The system designers, the asset users, the applications, the processes and the asset itself update the information of AAS during the life of the asset until its disposal [4].

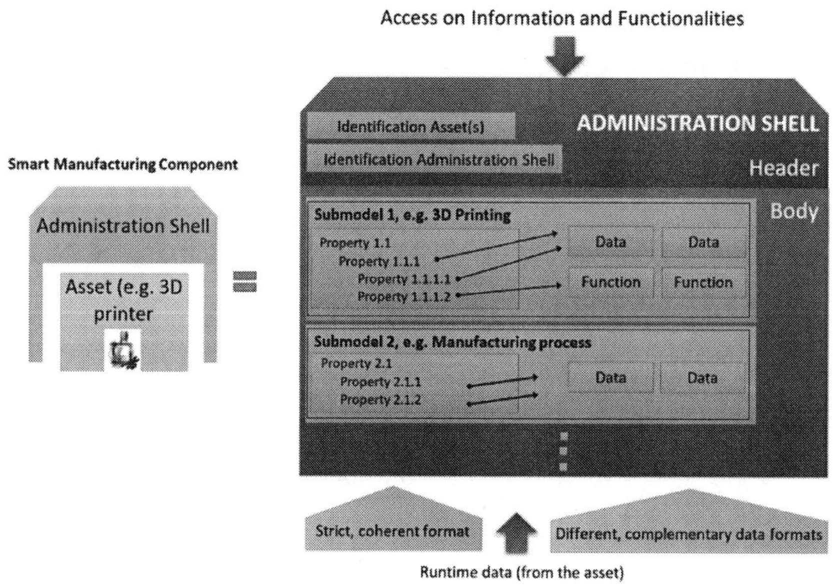

Figure 1. AAS Structure in details (inspired by [4,8]).

Figure 1 shows the AAS structure associated with some specific use cases and much more specifies the contents of the AAS towards the body's part, hence towards submodels.

What is a submodel? Submodels represent different aspects of an asset. Possible aspects and therefore a possible submodel could be: Identification, Communication, Engineering, Configuration, Safety, Security, Lifecycle status, Energy Efficiency, Condition Monitoring, etc.

Each submodel contains a structured quantity of properties that can refer to data and functions. Properties can be specified in accordance with the standard IEC 61360, but data and functions can be specified in various formats [4]. The following example stemming from the project RACAS shows how a specific communication process (bidding process or "interaction pattern") is directed towards the domain-specific submodels in the AAS, Fig. 1 [8]. The bidding between two assets in an industrial production line with 3D printers is described in the top of Fig. 2: an asset (e.g. semi-finished product) asks another asset (3D printer) situated in the production line, if its capacity, functionality, availability are able to provide the specified operation (printing on the semi-finished product of its dimensions <150x200x50 mm from material PLA by filament density 50 % printing operation in color RAL1003, a quality 0.2 mm taking time shorter than 4 hours.

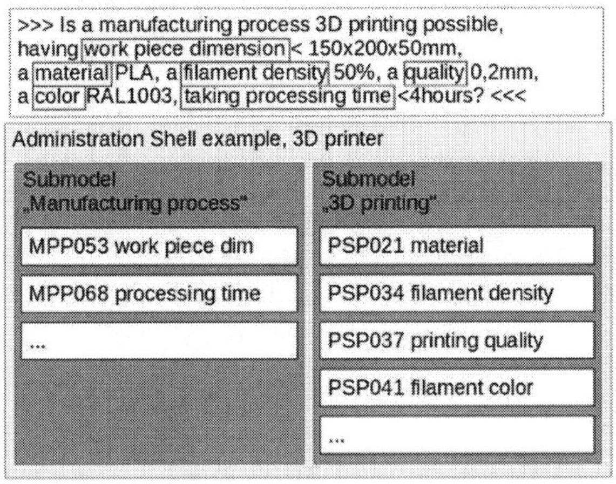

Figure 2. Bidding process directed towards the specific submodels of the AAS.

For such purposes, there have to be implemented in the AAS of the 3D printer at least two submodels ("Manufacturing process" submodel with parameters of the printer such as MPP053 work-piece dimension, MPP068 processing time and, and. There should be also implemented the second submodel the "3D printing" with parameters of the printing – PSP021 material, PSP034 filament density, PSP041 filament color.

This example recovers a principle of decentralized control in the I4.0 environment. The I4.0 components of the I4.0 production would control self their live cycle by negotiation with other I4.0 components. By this way, many problems with fronts in narrow places in production processes will be solved dynamically without a central control system. This architecture also enables more rapid and more flexible reaction of production processes in malfunction of machines, production lines, control systems, transport system and other parts of enterprises technologies.

An important pre-condition for such a production process is, that each I.0 component will be equipped by its standardized AAS.

4. Requirements regarding the AAS

The structure, properties, content of submodels, parameters and other features the AAS have been specified, developed and implemented into the AAS on requirements which have been collected, sorted and specified in detail by working groups of ZVEI, VDI/VDE, GMA, and others and will be openly published in the 2019 year. Contribution contents a proposal of them.

Requirements regarding the AAS sorted into three groups:

1. General requirements (R#1 – R#5)
2. Requirements regarding identifiers (R#6 - R#7)

3. Requirements regarding the AAS self (R#8 – R#22)

Particularly the Requirements regarding the AAS self have significantly influenced the existing model of the AAS [8,9]. All requirements are listed in the following summary [4].

Requirement # 1
The AS shall accept properties from different technical domains in mutually distinct submodels that can be version-controlled and maintained independently of each other.

Requirement # 2
The AS should be capable of including properties from a wide range of technical domains and of identify which domain they derive from.

Requirement # 3
For finding definitions within each relevant technical domain, different procedural models should be allowed that respectively meet the requirements of standards, consortium specifications, and manufacturer specifications sets.

Requirement # 4
Different ASs in respect of an asset must be capable of referencing each other. In particular, elements of an AS should be able to play the role of a "copy" of the corresponding components from another AS. E.g., one or more assets can be portrayed in an AS - mechanical axis, motor, servo amplifier, and additional assets constitute an "encapsulate-capable" Smart Manufacturing Component. The ASs of several individual assets that a manufacturer brings into the market individually is consolidated into one AS, if this manufacturer also sells a whole axis system.

Requirement # 5
Individual ASs should, while retaining their structure, be combined into an overall AS

Requirement # 6
Identification of assets, ASs, properties, and relationships shall be achieved using a limited set of identifiers (IRDI, URI, and GUID), providing as far as possible offer global uniqueness.

Requirement # 7
The AS should allow retrieval of alternative identifiers such as a GS1 and GTIN identifier in return to asset ID (referencing).

Requirement # 8
The AS consists of header and body, see Fig. 1.

Requirement # 9
The header contains information about the identification, Fig. 1. The header contains minimal information about identification. It uniquely identifies the AS. This identification can therefore also serve as a root entry point for an application programming interface (API) to browse for information and functionalities. The header contains also the identification of one or multiple assets that are described by the AS. The header also indicates if these assets are asset types or asset instances.

Requirement # 10
The body contains information about the respective asset(s).
The body contains information about the asset(s) and describes functionalities that are associated with the asset(s) or the AS. The information can concern asset type(s) and/ or asset instance(s). Thus, the body serves as the actual carrier of information and functionality.

Requirement # 11
The information and functionality in the AS are accessible by means of a standardized application programming interface (API).

IOP Conf. Series: Materials Science and Engineering **618** (2019) 012001 doi:10.1088/1757-899X/618/1/012001

Requirement # 12

The Administration Shell has a unique ID.

Requirement # 13

The asset has a unique ID. It should be ensured that the link between assets and ASs does not break, even if they are saved in digital repositories or saved in a manner that spans all value-added partners.

Requirement # 14

An industrial facility is also an asset; it has an AS and is accessible by means of ID. The concepts of the AS shall be applicable on all hierarchy levels of an industrial facility, such as factories/plants, production lines, stations, controls and field devices.

Requirement # 15

Types and instances must be identified as such. ASs can be formulated for both types and instances of assets. It must be possible to differentiate between these. Ideally, an information relationship will also be established between component producers and the system integrator that, where required, allows updated developments regarding asset types to be communicated to the system integrator and conversely feedback to be transmitted to the component producer about the component use.

Requirement # 16

The AS can include references to other ASs or Smart Manufacturing information. For the cross-linking of information to knowledge, it is important that this can also take place on an over-arching basis. Thus, for example, a component can model the dependencies on other components or can contain a circuit diagram, which refers to other components.

Requirement # 17

Additional properties, e.g. manufacturer specific must be possible. The Smart Manufacturing component can only meet future requirements if, in addition to the information content stipulated by standards, consortia and manufacturer properties can also be quickly agreed and processed. The AS should, therefore, support this consortia and proprietary information content and, associated accordingly, necessary collaboration processes.

Requirement # 18

A reliable minimum number of properties must be defined for each AS. ASs shall be a reliable source of information to other ASs or other systems. To do so, it shall be possible to define for each asset class a minimum set of properties and value statements that can be relied upon. The following requirements are applicable to the properties of an AS; the properties are structured by submodels. Standardized submodels types can require the presence of properties in submodel instances.

Requirement # 19

The properties and other elements of information in the AS must be suitable for types and instances. ASs can be formulated for both types and instances of assets; thus, properties need to be able to describe particularities of on asset type and, maybe, in addition, the asset instance. An AS of an asset instance shall also feature the properties of the AS of the respective asset type, as long as these properties were not overridden. NOTE: This can for example also mean, that the descriptions of an asset type are extended over the lifetime or, for an asset instance, properties are added, amended or deleted depending on (maintenance) activities of the respective asset.

Requirement # 20

There must be a capability of hierarchical and countable structuring of the properties. The volume of properties to be organized is rather large and it is anticipated that it will steadily increase in the progress of Smart Manufacturing. This means that these quantities should remain manageable for humans and machines. It is thus necessary to be able to organize properties using combinations of structures and arrays.

Requirement # 21

Properties shall be able to reference other properties, even in other ASs. Properties referring to other properties allow expressing dependencies on values contained in other ASs. In addition, knowledge can be modeled by interrelating two properties by a predicated relationship.

Requirement # 22

Properties must be able to reference the information and functions of the AS. The structure of submodels and properties serves as a clearly defined "table of contents" for all information and functions within the AS. Properties are of uniform structure, they are standardized and they are thus providing a very stable source of information. Complex data (digital models) and functions, on the other hand, can have a large variance and can be very complex in structure. Therefore, properties shall be able to refer to these complex data and functions in order to provide an anchor point for these entities in the above "table of contents". NOTE: This concept relies on an extended understanding of an IEC 61360 property concept.

5. Conclusion

Contribution deals with the most important term of the Industry 4.0 theory and application – the AAS. This fundamental term has been already specified in details and can create a reliable basis for the realization of not only I4.0 case studies and test beds, but for the real design, development and implementation of I4.0 principles in reality of factories of the future. Authors repeat terminology and associate technical as well as theoretical basis of the I4.0 idea based on the AAS.

Acknowledgment

The authors acknowledge financial support from the Technology Agency of the Czech Republic (TF04000074 – Digital Representation of Assets as a Configurable AAS for OT and IT Production Systems).

References

[1] Die Verwaltungsschale im detail von der Idee zum implementierbaren Konzept. Platform I4.0.
[2] ZVEI (April 2017). Examples of Assets Administration Shell for Industrie 4.0 Components-Basic Part.
[3] Industrial IoT – Digital Twin, Fraunhofer IOSB, Ovtchvarova J., Grethler M.: Beyond the Twin- making analytics come alive, IMI, Karlsruhe, Germany, 2018, ISSN 1616-8240.
[4] The structure of the Administration Shell, Trilateral Perspectives from France, Italy and Germany, Piano, Industria, I4.0. Platform Industrie 4.0, International paper 2018.
[5] Special Issue to the Hannover Industriemesse 2016, VDE Verlag, pp. 2 – 4.
[6] Marcon, P.; Arm, J.; Benesl, T.; Zezulka, F.; Diedrich, C.; Schröder, T.; Belyaev, A.; Dohnal, P.; Kriz, T.; Bradac, Z. New Approaches to Implementing the SmartJacket into Industry 4.0 ‡. *Sensors* 2019, *19*, 1592.
[7] Bradac Z.; Zezulka F.; Marcon P.: Technical and Theoretical basis of Industry 4.0 implementation. *Proc. of the TechSys 2018*, Technical University-Sofia, Plovdiv branch, Bulgaria, ISSN 1310-8271.
[8] ZVEI. Details of the Asset Administration Shell. Available online: https://www.zvei.org/en/press-media/publications/details-of-the-asset-administration-shell/ (accessed on 27.11.2018).
[9] Marcon, P et al. The Asset Administration Shell of Operator in the Platform of Industry 4.0. *In Proceedings of the 18th International Conference on Mechatronics-Mechatronika (ME)*, Brno, Czech Republic, 5–7 December 2018.

IOP Conf. Series: Materials Science and Engineering **618** (2019) 012002 doi:10.1088/1757-899X/618/1/012002

Application of Auto_Threshold plugin for cut surface evaluation of white cheese in brine

Atanaska Bosakova-Ardenska[1*], Peter Panayotov[2], Petya Boyanova[2]

[1]Department of Computer Systems and Technologies, University of Food Technologies, Plovdiv, Bulgaria

[2]Department of Milk and Dairy products Technology, University of Food Technologies, Plovdiv, Bulgaria

[*]corresponding author's e-mail address: a_bosakova@uft-plovdiv.bg

Abstract. This paper presents a research over application of global thresholding methods for evaluation of cut surface of white cheese in brine. Six trademarks Bulgarian white cheese in brine produced of cow, sheep and goat milk are used for cut surface evaluation. Examined cheese samples are evaluated by experts and the results are used as referent assessment for results of image processing trough thresholding algorithms. A Java program is developed for image segmentation with specific thresholding algorithm using two or three colors in different modes to produce segmented image. The images of cheese samples (their cut surfaces) are processed with developed program using all of the seventeen thresholding algorithms which are implemented in Auto_Threshold plugin for open source software ImageJ in different modes. A correlation analysis shows that some of thresholding algorithms which are implemented in Auto_Threshold plugin could be used for cut surface evaluation of white cheese in brine.

1. Introduction

Milk and milk products are essential foods for human and thus they are very popular food in everyday meal. A lot of milk products there are known such as cheese, yoghurt, curd and etc. Cheese has a high energy, nutritional and biological value for human body. Various human-useful vitamins are included in the composition of the cheese, such as: group A and D vitamins which are fat-soluble; water-soluble vitamins of group B (B2 and B6) and others [1]. The consumption of mature cheese, such as the Bulgarian white cheese in brine, give to the human necessary quantities of essential amino acids. From ancient to now the technologies for dairy processing are constantly evolving. As the analysis shows European countries, and particularly Bulgaria, have well-functioning dairy industry [2]. The cow, sheep and goat milk is most popular raw materials for dairy industry in Bulgaria [3]. According to information in statistical report of Bulgarian Ministry of Agriculture, Food and Forestry, white cheese in brine, yoghurt and packed milk are the most produced milk products in Bulgaria. The quality of milk and milk products are strictly controlled by standards. In Bulgaria there are a lot of standards for quality control of milk products. For white cheese in brine quality control standards are about fourteen (BDS 15:2010).

Some parameters of food quality could be evaluated visually. In the field of food quality evaluation, computer vision systems are popular and effective technique for quality investigation [4, 5, 6, 7, 8]. A lot of researches are done about application of computer vison for cheese quality evaluation. Area occupied of gas holes on Emental, Ragusano and Cheddar cheese are successfully measured using image processing [9]. Some functional features which are related to thermal processing (cooking) of cheese of type Mozzarella are also evaluated through image processing [10, 11]. Computer vision is applied also in inspection of distribution and amount of ingredients in pasteurized cheese [12]. Cut surface of Bulgarian white cheese in brine is inspected through processing of microscopic images (SEM) which are specifically colored [13, 14], through image processing with NI VisionBuilder [15] and through image processing in HSV color space [16].

The aim of current research is to propose fast and effective approach for cut surface evaluation of white cheese in brine using image segmentation based on methods which are implemented in Auto_Threshold plugin.

2. Materials and methods

2.1. Image segmentation

The main purpose of image processing is to extract information about foreground (objects) and background of images [17]. In the beginning every pixel has to be associated with a specific group i.e. the pixel has to be grouped with other pixels according to its color and relation with neighbours. This process is popular as image segmentation. The segmented image gives significant information about objects on the image. Thus the image segmentation is a very important step in process of automatic object recognition because the result of image segmentation is an image with pixels which are grouped in regions by their similarity. Algorithms for image segmentation are based on different techniques such as region growing, edge detection, thresholding and etc. The algorithms for image segmentation are widely used for biomedical images processing [18, 19, 20, 21], food quality evaluation [4, 22, 23], microbiological analysis [24, 25, 26] and etc.

2.2. Software

It is developed a Java application (ImgSegJ) for image segmentation using seventeen thresholding algorithms which are implemented in Auto_Threshold plugin for open source software ImageJ. These algorithms are: Default (private implementation of IsoData method), Huang, Huang2 (alternative implementation of Huang's method by J. Schindelin), Intermodes, IsoData, Li, MaxEntropy, Mean, MinError, Minimum, Moments, Otsu, Percentile, RenyiEntropy, Shanbhag, Triangle, Yen. On figure 1 it is presented a model of application ImgSegJ. Developed application works with BMP images and uses global thresholding algorithms in two basic modes: binarization and segmentation.

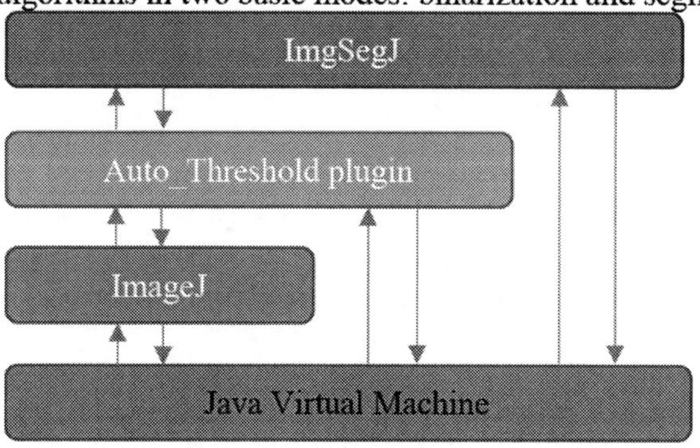

Figure 1. Application model of ImgSegJ.

When the images are processed in binarization mode then the results are binary images which contain pixels with black or white color. When the images are processed in segmentation mode then the results are images which contain pixels with black, gray or white color. The segmentation mode has two options: "Separate in Black area" and "Separate in white area". When the option "Separate in black area" is selected then image processing consists of next steps:

1) Major threshold (Th_1) value has to be found with selected algorithm;

2) A histogram of image has to be modified so that all pixels which have color which is bigger than threshold (i.e. the pixels in the result image have to be black color) are removed from histogram;

3) Second threshold (Th_2) value has to be found using selected thresholding algorithm and based on modified histogram;

4) The result image contains three colors (white, gray and black) and it is generated using major and second thresholds by formulae:

$$p_{out} = \begin{Bmatrix} white, p_{in} \leq Th_1 \\ gray, Th_1 < p_{in} \leq Th_2 \\ black, p_{in} > Th_2 \end{Bmatrix} \quad (1)$$

where p_{in} is color of pixel in input image, pout is color of pixel in resulting image (after segmentation), Th_1 is major threshold and Th_2 is second threshold (based on modified histogram).

When the option "Separate in white area" is selected then image processing consists of next steps:

1) Major threshold (Th_1) value has to be found with selected algorithm;

2) A histogram of image has to be modified so that all pixels which have color which is smaller than threshold (i.e. the pixels in the result image have to be white color) are removed from histogram;

3) Second threshold (Th_2) value has to be found using selected thresholding algorithm and based on modified histogram;

4) The result image contains three colors (white, gray and black) and it is generated using major and second thresholds by formulae:

$$p_{out} = \begin{Bmatrix} white, p_{in} \leq Th_2 \\ gray, Th_2 < p_{in} \leq Th_1 \\ black, p_{in} > Th_1 \end{Bmatrix} \quad (2)$$

where p_{in} is color of pixel in input image, pout is color of pixel in resulting image (after segmentation), Th_1 is major threshold and Th_2 is second threshold (based on modified histogram). On figure 2 is presented GUI (Graphical User Interface) of application ImgSegJ.

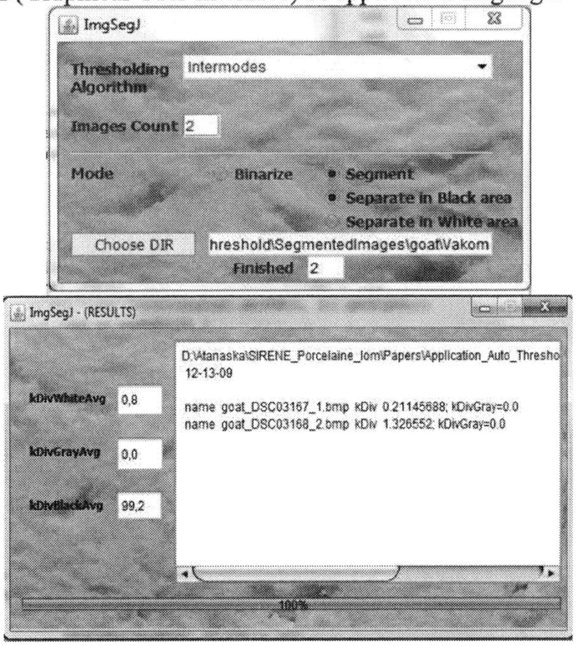

Figure 2. ImgSegJ – GUI (Performance with two images in segmentation mode with option "Separate in black area" and thresholding algorithm Intermodes).

2.3. White cheese in brine

Six trademarks of Bulgarian white cheese in brine are bought from marketplace. They are produced from cow, sheep and goat milk. On figure 3 are presented three samples of examined trademarks.

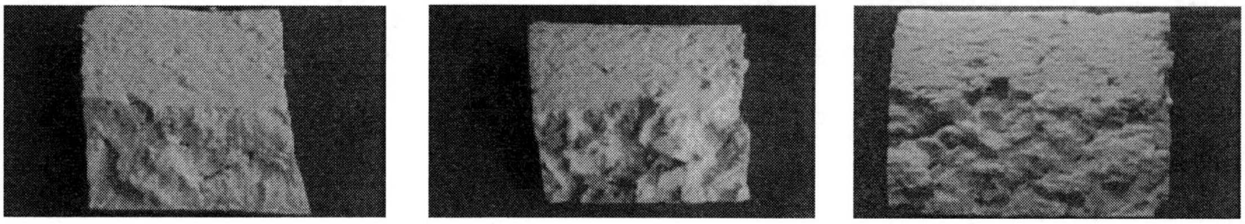

(a) (b) (c)

Figure 3. (a) Cheese of cow milk; (b) Cheese of sheep milk; (c) Cheese of goat milk.

Two samples of every trademark are used for experiments. All examined samples of white cheese in brine are cut and their cut surface are captured with camera Sony DSC-HX300. On table 1 are presented cut surfaces of examined trademarks of cheese.

Table 1. Cut surface of six trademarks Bulgarian white cheese in brine.

	Cow milk		Sheep milk		Goat milk
Trademark 1	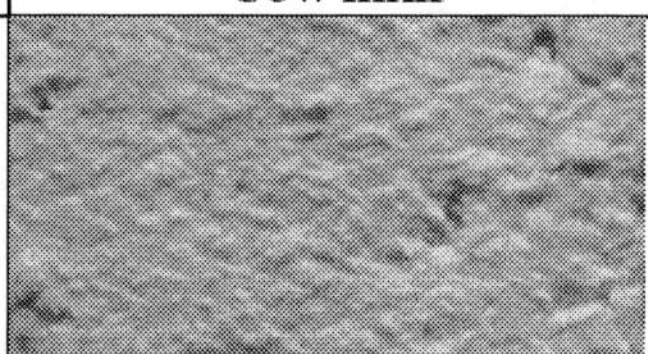	Trademark 3	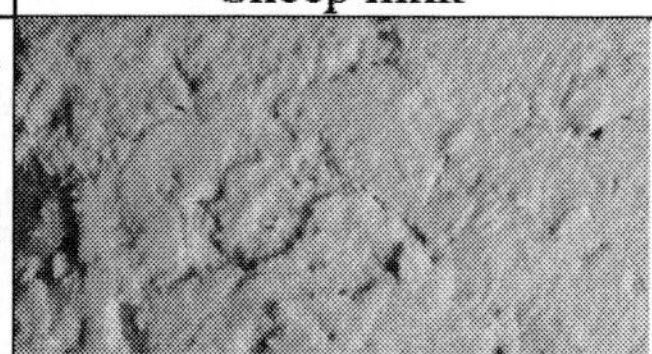	Trademark 5	
Trademark 2	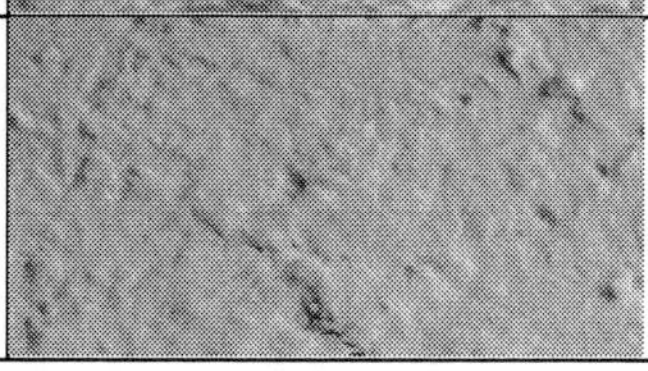	Trademark 4	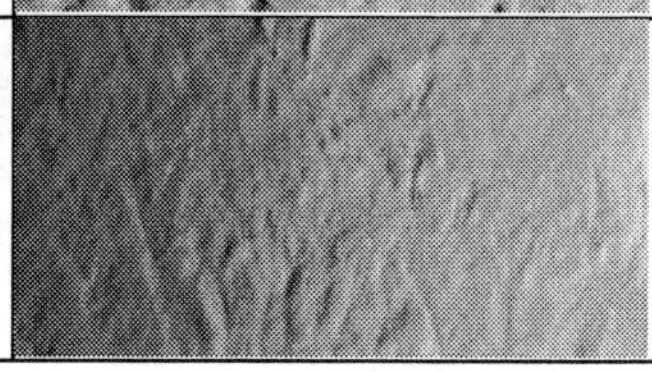	Trademark 6	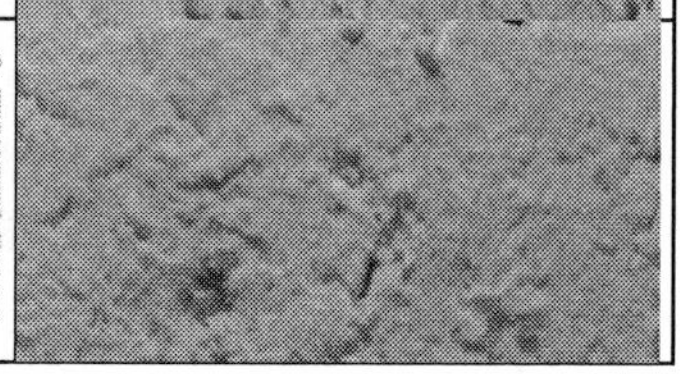

3. Results and discussion

The samples of white cheese in brine are evaluated by experts and results for quality of cut surface are presented on figure 4.

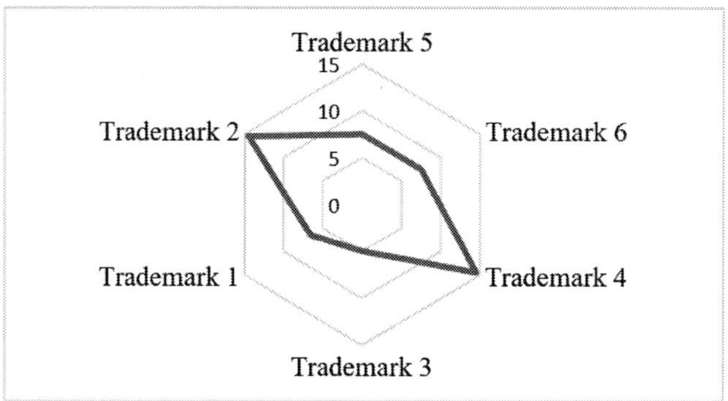

Figure 4. Cut surface – expert's evaluation

It is obviously that the best result for quality of cut surface have cheese "Trademark2" (produced of cow milk) and cheese "Trademark 4" (produced of sheep milk). Cut surface of these two trademarks is smooth, without big technical pores thus they receive high score from experts (near to maximal value which is 15 according to Bulgarian national standard for sensory evaluation of milk and milk

IOP Conf. Series: Materials Science and Engineering **618** (2019) 012002 doi:10.1088/1757-899X/618/1/012002

products). The cheese which have low score for cut surface of expert evaluation are "Trademark 1" (produced of cow milk) and "Trademark 3" (produced of sheep milk) because their surface is grainy with technical pores. Cheese of goat milk (trademarks "Trademark 5" and "Trademark 6") have score for cut surface near to the middle of the scale (about 8) because they have small number of big pores and structure with visually highlighted layers. Images of evaluated trademarks are processed with ImgSegJ and ratio of number of white pixels to all pixels in segmented images is presented in table 2 as mean value of the samples.

Table 2. Ratio of white pixels to all pixels in segmented images of Bulgarian white cheese in brine. (ImgSegJ, mode "Binarize").

Trademark / Algorithm	Cow milk		Sheep milk		Goat milk	
	Tm 1	**Tm 2**	**Tm 3**	**Tm 4**	**Tm 5**	**Tm 6**
Default	45,27	36,03	29,71	44,33	32,49	29,88
Huang	53,72	50,73	49,45	47,62	46,27	53,37
Huang2	56,03	50,94	49,05	48,53	46,73	56,42
Intermodes	27,99	0,084	1,66	27,36	2,57	0,769
IsoData	47,68	34,06	30,54	45,15	33,85	34,55
Li	51,32	50,73	39,74	45,93	44,44	48,21
MaxEntropy	3,42	0,05	0,31	0,154	1,13	0,912
Mean	48,93	48,37	44,89	47,57	46,27	46,66
MinError	55,96	23,99	44,89	47,57	46,27	23,93
Minimum	28,93	0,012	0,189	14,89	0,485	0,066
Moments	45,27	45,24	26,57	45,91	31,23	30,40
Otsu	48,93	45,24	35,53	45,15	39,63	40,37
Percentile	50,09	48,37	50,38	50,10	49,61	50,16
RenyiEntropy	3,57	0,059	0,408	0,232	1,09	1,02
Shanbhag	51,32	0,02	17,89	45,15	42,71	48,21
Triangle	2,66	1,61	4,05	0,39	2,27	4,22
Yen	3,29	0,046	0,261	0,054	1,05	0,78

Table 3. Ratio of white pixels to all pixels, gray pixels to all and black pixels to all in segmented images of Bulgarian white cheese in brine. (ImgSegJ, mode "Separate in black area").

Trademark / Algorithm	Cow milk						Sheep milk						Goat milk					
	Trademark 1			**Trademark 2**			**Trademark 3**			**Trademark 4**			**Trademark 5**			**Trademark 6**		
	w/all	**g/all**	**b/all**	**w/all**	**g/all**	**b/all**	**w/all**	**g/all**	**b/all**	**w/all**	**g/all**	**b/all**	**w/all**	**g/all**	**b/all**	**w/all**	**g/all**	**b/all**
Default	45,27	33,52	21,20	36,03	39,68	24,29	29,71	38,42	31,86	44,33	26,19	29,47	32,49	40,14	27,37	29,88	42,53	27,59
Huang	53,72	24,99	21,29	50,73	30,39	18,88	49,45	24,47	26,08	47,62	22,06	30,32	46,27	26,78	26,95	53,37	26,13	20,50
Huang2	56,03	23,59	20,38	50,94	26,99	22,07	49,05	26,46	24,49	48,52	20,14	31,34	46,73	28,13	25,14	56,42	24,19	19,39
Intermodes	27,99	0	72,01	0,08	49,01	50,91	1,66	0	98,34	27,36	32,86	39,77	2,57	0	97,43	0,77	70,91	28,32
IsoData	47,68	33,51	18,80	34,06	45,74	20,20	30,54	39,05	30,41	45,15	26,39	28,46	33,85	40,48	25,67	34,55	42,75	22,69
Li	51,32	31,41	17,27	50,73	38,49	10,78	39,74	36,98	23,28	45,93	28,76	25,31	44,44	37,99	17,56	48,21	35,22	16,57
MaxEntropy	3,42	65,95	30,63	0,05	49,19	50,75	0,31	5,76	93,93	0,15	42,53	57,31	1,13	27,03	71,84	0,91	6,13	92,96
Mean	48,93	27,98	23,09	48,37	32,75	18,88	44,89	30,45	24,65	47,57	26,04	26,39	46,27	29,65	24,08	46,66	30,64	22,69
MinError	55,96	21,92	22,12	23,99	26,00	50,00	44,89	30,46	24,65	47,57	19,65	32,78	46,27	29,65	24,08	23,93	14,91	61,15
Minimum	28,93	0,01	71,06	0,01	0,02	99,97	0,19	1,01	98,79	14,89	30,21	54,89	0,48	0,24	99,28	0,07	0,43	99,50
Moments	45,27	36,71	18,02	45,24	45,09	9,68	26,57	46,79	26,64	45,91	33,24	20,86	31,23	47,64	21,12	30,40	48,82	20,77
Otsu	48,93	33,05	18,02	45,24	41,12	13,64	35,53	37,83	26,64	45,15	27,42	27,43	39,63	40,39	19,98	40,37	40,06	19,57
Percentile	50,09	25,06	24,85	48,37	25,68	25,96	50,38	25,13	24,49	50,10	24,58	25,32	49,61	26,31	24,08	50,16	24,80	25,04
RenyiEntropy	3,57	65,80	30,63	0,06	49,56	50,38	0,41	6,56	93,03	0,23	54,13	45,64	1,09	28,43	70,48	1,02	7,89	91,08
Shanbhag	51,32	32,84	15,84	0,02	48,35	51,63	17,89	42,53	39,58	45,15	27,42	27,43	42,71	38,37	18,92	48,21	36,04	15,75
Triangle	2,66	96,49	0,85	1,61	95,88	2,51	4,05	49,98	45,97	0,39	80,95	18,66	2,27	95,78	1,95	4,22	94,04	1,74
Yen	3,29	69,79	26,91	0,05	49,50	50,45	0,26	5,14	94,60	0,05	42,63	57,31	1,05	26,27	72,68	0,78	4,49	94,72

* w/all – ratio of count of white pixels to all pixels in segmented image in percent; g/all – ratio of count of gray pixels to all pixels in segmented image in percent; b/all – ratio of count of black pixels to all pixels in segmented image in percent;

On this table (table 2) Trademark 1 to Trademark 6 are named "Tm 1" to "Tm 6".Table 3 and table 4 present ratio of white to all pixels, gray to all pixels and black to all pixels for segmented images in modes "Separate in black area" and "Separate in white area". The results are calculated as a mean value for samples of every trademark of white cheese in brine.

Table 4. Ratio of white pixels to all pixels, gray pixels to all and black pixels to all in segmented images of Bulgarian white cheese in brine. (ImgSegJ, mode "Separate in white area").

Trademark	Cow milk						Sheep milk						Goat milk					
	Trademark 1			Trademark 2			Trademark 3			Trademark 4			Trademark 5			Trademark 6		
Algorithm	w/all	g/all	b/all	w/all	g/all	b/all	w/all	g/all	b/all	w/all	g/all	b/all	w/all	g/all	b/all	w/all	g/all	b/all
Default	9,08	36,19	54,73	4,57	31,45	63,97	4,60	25,11	70,29	17,87	26,46	55,67	2,08	30,41	67,51	3,97	25,90	70,12
Huang	22,84	30,87	46,28	16,46	34,27	49,27	18,08	31,37	50,55	23,92	23,70	52,38	18,90	27,37	53,73	19,54	33,84	46,62
Huang2	21,00	35,03	43,97	21,13	29,82	49,06	17,06	31,99	50,95	23,92	24,61	51,47	19,75	26,98	53,27	20,58	35,84	43,58
Intermodes	22,04	5,95	72,01	0,02	0,07	99,91	0,33	1,32	98,34	0,11	27,25	72,64	0,72	1,85	97,43	0,14	0,63	99,23
IsoData	9,48	38,21	52,31	4,75	29,31	65,94	4,69	25,84	69,46	18,46	26,69	54,85	2,17	31,68	66,15	4,51	30,04	65,45
Li	13,87	37,45	48,68	14,10	36,63	49,27	7,59	32,15	60,26	20,61	25,33	54,06	11,27	33,17	55,56	7,33	40,87	51,79
MaxEntropy	0,15	3,27	96,58	0,01	0,04	99,95	0,08	0,23	99,69	0,03	0,12	99,85	0,11	1,02	98,87	0,13	0,78	99,09
Mean	18,39	30,53	51,07	16,46	31,90	51,63	15,55	29,35	55,10	21,87	25,70	52,43	16,87	29,40	53,73	15,24	31,42	53,34
MinError	32,56	23,40	44,04	8,15	15,85	76,00	23,15	21,75	55,10	26,06	21,51	52,43	16,87	29,40	53,73	7,76	16,17	76,07
Minimum	20,89	8,05	71,06	0	0,01	99,99	0,08	0,11	99,81	0	14,89	85,10	0,02	0,47	99,52	0	0,07	99,93
Moments	9,08	36,19	54,73	4,57	40,66	54,76	2,60	23,97	73,43	17,52	28,38	54,09	2,84	28,39	68,77	3,59	26,81	69,60
Otsu	10,93	37,99	51,07	8,11	37,13	54,76	5,65	29,88	64,47	19,24	25,90	54,85	5,02	34,61	60,37	5,37	34,99	59,63
Percentile	23,77	26,32	49,91	22,39	25,98	51,63	23,42	26,96	49,62	23,86	26,25	49,89	23,94	25,67	50,39	23,73	26,43	49,84
RenyiEntropy	0,22	3,35	96,43	0,02	0,04	99,94	0,09	0,32	99,59	0,04	0,19	99,77	0,12	0,98	98,90	0,15	0,88	98,98
Shanbhag	11,08	40,23	48,68	0,01	0,01	99,98	0,78	17,11	82,11	17,87	27,28	54,85	1,25	41,46	57,29	1,87	46,34	51,79
Triangle	0,34	2,32	97,34	0,12	1,49	98,39	0,35	3,70	95,95	0,06	0,33	99,61	0,72	1,54	97,73	0,73	3,49	95,78
Yen	0,17	3,12	96,71	0,01	0,04	99,95	0,08	0,18	99,74	0,004	0,05	99,95	0,13	0,91	98,95	0,12	0,67	99,21

* w/all – ratio of count of white pixels to all pixels in segmented image in percent; g/all – ratio of count of gray pixels to all pixels in segmented image in percent; b/all – ratio of count of black pixels to all pixels in segmented image in percent;

On figures 5, 6 and 7 are presented correlation coefficients for expert evaluation and calculated ratios of segmented images in modes "Binarization", "Separate in black area" and "Separate in white area".

On figure 5 are presented absolute values of correlation coefficients which are calculated using MS Excel. The results show that algorithms Triangle and Moments have significant correlation with expert evaluation for cut surface of white cheese in brine (their correlation coefficients are near to 1). Because of this, these two algorithms could be used for automatic evaluation of cut surface of cheese using ImgSegJ in mode "Binarization".

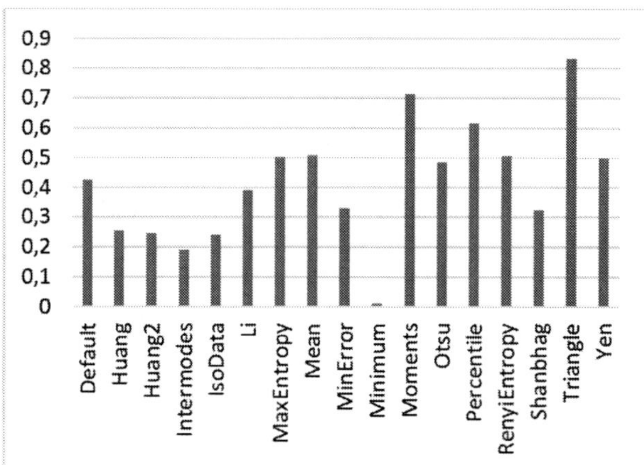

Figure 5. Correlation coefficients for expert evaluation and ratio of white pixels to all (ImgSegJ, mode "Binarization").

On figure 6 are presented absolute values of correlation coefficients which are calculated using MS Excel. The results show that algorithms Triangle, Percentile and Moments have significant correlation with expert evaluation for cut surface of white cheese in brine (their correlation coefficients are near to 1). Algorithms Triangle and Moments have correlation coefficients above 0,7 for ratio of white pixels to all. Only algorithm Percentile has correlation coefficients above 0,7 for ratio of black pixels to all.

On figure 7 are presented absolute values of correlation coefficients which are calculated using MS Excel. The results show that algorithms Huang2, Li, MaxEntropy, Moments, RenyiEntropy and Triangle have significant correlation with expert evaluation for cut surface of white cheese in brine (their correlation coefficients are near to 1). Algorithms Huang2, Li, MaxEntropy and RenyiEntropy have correlation coefficients above 0,7 for ratio of white pixels to all. Only algorithms Moments and Triangle have correlation coefficients above 0,7 for ratio of black pixels to all (algorithm Triangle has also high correlation coefficient for ratio of gray pixels to all). The highest correlation coefficient is achieved for ratio of black pixels to all using algorithm Triangle in mode segmentation- "Separate in white area". The same algorithm (Triangle) gives significant correlation with expert evaluation using "Binarization" mode too. On figure 8 are presented images of white cheese in brine before and after processing with Triangle algorithm in different modes.

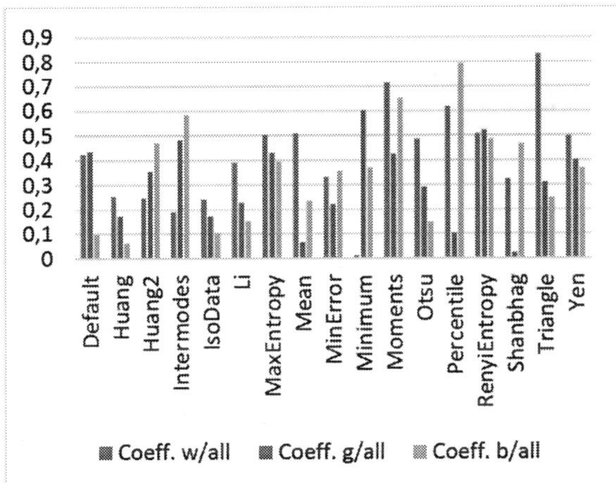

Figure 6. Correlation coefficients for expert evaluation and ratio of white, gray and black pixels to all (ImgSegJ, mode segmentation- "Separate in black area").

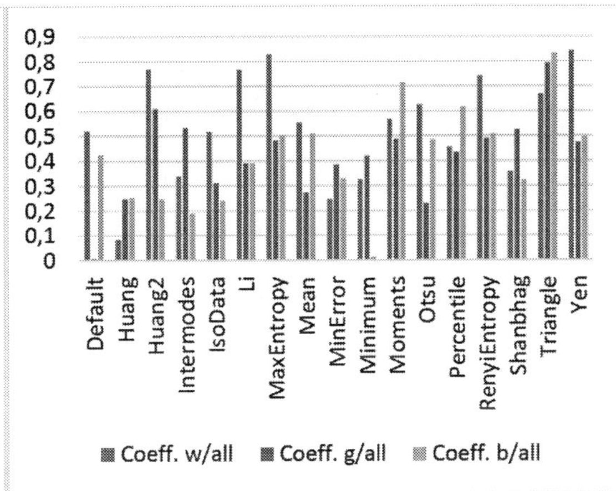

Figure 7. Correlation coefficients for expert evaluation and ratio of white, gray and black pixels to all (ImgSegJ, mode segmentation- "Separate in white area").

Processing with Triangle algorithm emphasize on technical pores and some borders of areas in cheese structure. The next equations could be used to calculate the score for cut surface of white cheese in brine based on ratio of white or black pixels to all for segmented image (ImgSegJ, mode "Binarization" or ImgSegJ, mode Segmentation- "Separate in white area"). With c in formulae 3 and 4 it is noted a constant.

$$score = \left\{ \begin{matrix} 5, ratio_{white/all} + c > 15 \\ ratio_{white/all} + c \end{matrix} \right\} \quad (3)$$

$$score = \left\{ \begin{matrix} 15, ratio_{black/all} + c > 8 \\ ratio_{black/all} + c \end{matrix} \right\} \quad (4)$$

On figure 9 are presented expert evaluation and calculated scores by formulae 3 and 4 for cut surface of tested trademarks of white cheese in brine (for score calculation are used data of table 2 and table 4 for algorithm Triangle and constant value 13,8 for formula 3 and -90 for formula 4).

The score for cut surface of white cheese in brine is near to their expert evaluation (fig. 9) and average relevant error by evaluation is about 16% using formula 3 and about 10% using formula 4.

This is the reason that developed software (ImgSegJ) could be applied for objective evaluation of cut surface of white cheese in brine. Developed software could support experts in their work.

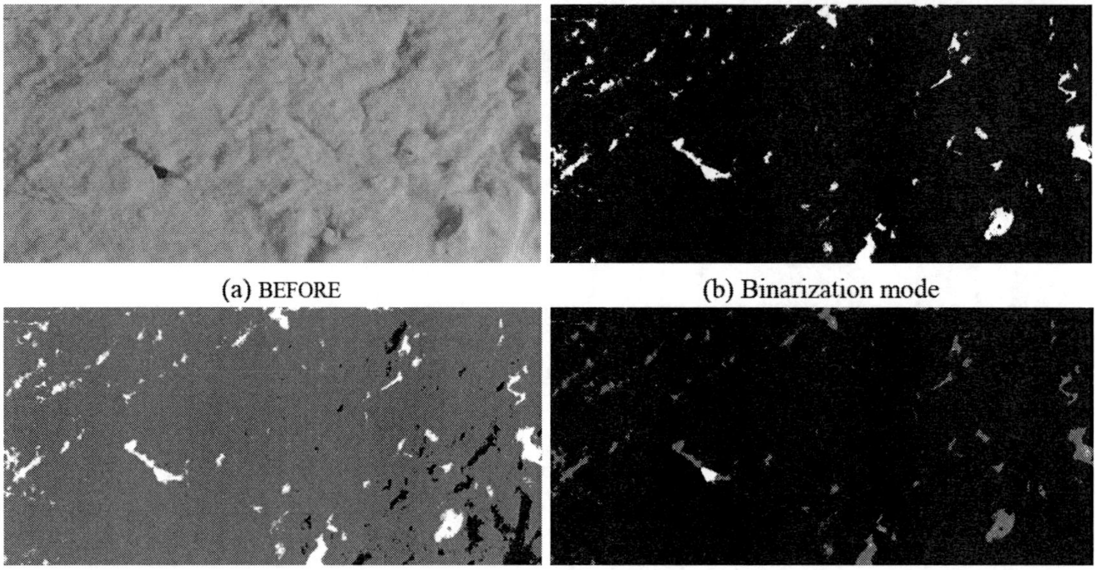

(a) BEFORE (b) Binarization mode

(c) Segmentation, "Separate in black area" (d) Segmentation, "Separate in white area"

Figure 8. Cut surface of sample white cheese in brine (trademark "Trademark 3" produced of sheep milk) before and after processing with ImgSegJ.

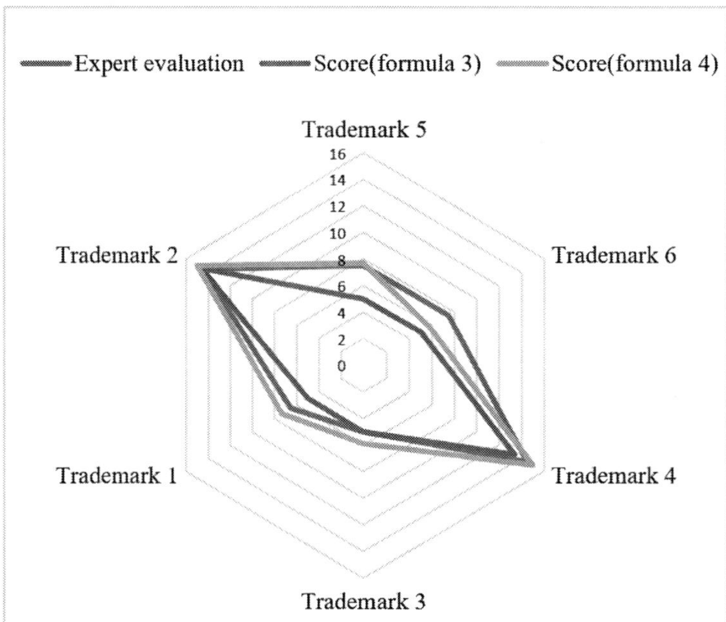

Figure 9. Expert evaluation and score for cut surface of white cheese in brine (ImgSegJ, thresholding algorithm Triangle).

4. Conclusion

Six trademarks white cheese in brine produced of cow, sheep and goat milk are used for evaluation of cut surface. Evaluation of cut surface is performed using two methods: expert evaluation according to standard BDS 15612-83 (Dairy products. Sensory analysis.) and computer analysis using image segmentation through proposed software ImgSegJ. There are proposed equations to calculate the score for cut surface based on ratio of white or black pixels to all in segmented images and statistical analysis (correlation). The results show that developed software (ImgSegJ) could be applied for objective evaluation of cut surface of white cheese in brine.

Future work will continue with research over larger set of trademarks and with developing automatic classifier for white cheese in brine based on quality of cut surface.

5. References

[1] Walther B. Schmid A. Sieber R. Wehrmuller K. 2008 Cheese in nutrition and health *Dairy Sci. Technol.* **88** pp 389–405

[2] Weichselbaum E. Benelam B. Costa H. 2009 Synthesis report No 6: Traditional Foods in Europe *In EuroFIR Project Management Office/British Nutrition Foundation United Kingdom*

[3] Ministry of Agriculture Food and Forestry *State and Trends in the Production of Milk and Dairy Products in Bulgaria* [http://www.mzh.government.bg/media/ filer_public/ 2018/03/01/mliako.pdf, accessed on 10.01.2019] (in Bulgarian)

[4] Sun DW 2008 *Computer vision technology for food quality evaluation* Food Sc. and Tech.

[5] Wu Di Sun DW 2013 Colour measurements by computer vision for food quality control- A review *Trends in Food Sc. & Tech.* **29** pp 5-20

[6] Zhang B. Huang W. Li J. Zhao Ch. Fan Sh. Wu J. Liu Ch. 2014 Principles, developments and applications of computer vision for external quality inspection of fruits and vegetables: A review *Food Research Int.* **62** pp 326–343

[7] Kostadinova L. Shopov N. Madjarov N. Jelezarov I. 2015 New Computer Aided Ultrasonic Technology for Classification of Industrial Produced Bulgarian Yogurt *Proc. of the Bulgarian Academy of Sciences* V. **68** No **1** pp 89-94 ISSN 2367-5535

[8] Georgiev, A. Kostadinova L. Gabrova R. 2007 Automatic Quality Classifiers of Food Products – Rate of the Speed Problems *Proc. of XII International Conference on Informatiom, Communication and Energy Systems and Technologies -ICEST 2007, Ohrid, Macedonia* **2** pp 803-805 ISBN: 9989-786-06-2

[9] Caccamo M. Melilli D. Barbano M. Licitra G. 2004 Measurement of Gas Holes and Mechanical Openness in Cheese by Image Analysis *Journal of Dairy Sc.* **87(3)** pp 739–48 DOI: 10.3168/jds.S0022-0302(04)73217-8

[10] Wang Hai-Hong Sun DW 2003 Assessment of cheese browning affected by baking conditions using computer vision *Journal of Food Engineering* **56** pp 339–345

[11] Ma Xixiu 2013 Evaluation of Functional Properties and Microstructure of Mozzarella Cheese and their Correlation *PhD thesis The University of Auckland*

[12] Jelinski T. Du Cheng-Jin Sun DW Fornal J. 2007 Inspection of the distribution and amount of ingredients in pasteurized cheese by computer vision *Journal of Food Engineering* **83** pp 3–9

[13] Bosakova–Ardenska A. Boyanova P. Panayotov P. 2014 Application of Global Thresholding Technique in White Brined Cheese Evaluation (IJIRSE) *Int. Journal of Innovative Research in Sc. & Eng.* **2(12)** pp 839-846 ISSN (Online) 2347-3207

[14] Bosakova-Ardenska A. Andreeva H. Boyanova P. Panayotov P. 2017 CutSurf– a software tool for cut surface evaluation of white cheese *Int. Conference on Computer Systems and Technologies - CompSysTech'17* pp 241-248 ISBN: 978-1-4503-5234-5

[15] Ganchovska V. 2014 Objectively determination the characteristics of milk and dairy products in real time *PhD thesis UFT-Plovdiv* (in Bulgarian)

IOP Conf. Series: Materials Science and Engineering **618** (2019) 012002 doi:10.1088/1757-899X/618/1/012002

[16] Vasilev M. 2016 Evaluation of surface of dairy products with optical techniques *Int. Sc. Conf. for Young Scientists "Management and quality"* pp 91-97 ISBN 978-619-160-679-5 (in Bulgarian)
[17] Gonzalez R. Woods R. 2008 *Digital Image Processing (Third Edition)* Prentice-Hall ISBN 978-93-325-7032-0
[18] Gamarra M. Zurek E. San-Juan H. 2017 A Study of Image Analysis Algorithms for Segmentation, Feature Extraction and Classification of Cells *Journal of Information Systems Engineering & Management* **2(4)** DOI: https://doi.org/10.20897/jisem.201720
[19] Bozek J. Mustra M. Delac K. Grgic M. 2009 A Survey of Image Processing Algorithms in Digital Mammography *In: Grgic M., Delac K., Ghanbari M. (eds) Recent Advances in Multimedia Signal Processing and Communications Studies in Computational Intelligence* **231** (Springer, Berlin, Heidelberg)
[20] Topor A. Malin S. Babarada F. Ravariu C. Salageanu A. Caras I. 2017 Applications of images processing algorithms for bacterial meningitis diagnosis: Bacterial image processing in Mathlab *9th Int. Conf. on Electronics, Computers and Artificial Intelligence (ECAI, Targoviste)* pp. 1-4 DOI: 10.1109/ECAI.2017.8166447
[21] Prodanov D. Verstreken K. 2012 Automated Segmentation and Morphometry of Cell and Tissue Structures. Selected Algorithms in ImageJ *Molecular Imaging, Bernhard Schaller, IntechOpen* DOI: 10.5772/36729
[22] Nandhini P. Jaya J. 2014 Image Segmentation for Food Quality Evaluation Using Computer Vision System *Int. Journal of Engineering Research and Applications* **4(2)** pp 01-03 ISSN: 2248-9622
[23] Lukinac J. Jukić M. Mastanjević K. Lučan M. 2018 Application of computer vision and image analysis method in cheese-quality evaluation: a review *Ukrainian Food Journal* **7(2)** pp 192-214 DOI: 10.24263/2304-974x-2018-7-2-4
[24] Danev A. Gabrova R. Yaneva-Marinova T. Angelov A. 2018 Application possibilities of open-source software for microbiological analyses *Bulgarian Chemical Communications* **50** pp 239-245 ISSN: 08619808
[25] Alonso C. Domínguez C. Heras J. Mata E. Pascual V. Torres C. Zarazaga M. 2017 Antibiogramj: A tool for analysing images from disk diffusion tests *Computer Methods and Programs in Biomedicine* **143** pp. 159-169 DOI: 10.1016/j.cmpb.2017.03.010
[26] Puchkov E. 2016 Image Analysis in Microbiology: A Review *Journal of Computer and Communications* **4** pp 8 -32 ISSN Online: 2327-5227 ISSN Print: 2327-5219 DOI: 10.4236/jcc.2016.415002

Acknowledgements
The authors would like to thank to University of Food Technologies for its financial support. This research was supported by University of Food Technologies according to contract number 16/18-H UFT Plovdiv.

IOP Conf. Series: Materials Science and Engineering **618** (2019) 012003 doi:10.1088/1757-899X/618/1/012003

Optimal selection of feed electric drives for boring machines

M Zhilevski[1] **and M Mikhov**[1]
[1] Faculty of Automatics, Technical University of Sofia, Bulgaria

E-mail: mzhilevski@tu-sofia.bg

Abstract. In this paper are analyzed and formulated the main requirements for the drive systems of a class of modernized boring machines with digital program control. On this basis, a methodology for optimal selection of feed drives is offered. The respective algorithm takes into account the specific features of the technological process, the treated material, the tools used and their wear, the mechanical gears types and the guides. Concrete examples with DC and AC electric drives are presented, illustrating the practical application of the methodology. The research held as well as the results obtained can be used in the development of such electric drives for the studied class of machine tools with digital program control.

1. Introduction

The technical potential of modern machine tools with digital program control depends on both their control systems and the functionality of the respective drive systems.

To meet the constantly growing requirements, modern electric drives should possess a number of features such as: a wide range of speed regulation; high precision of the respective coordinate control; good dynamics; reliability; economic operation; good communication abilities, etc. The role of the drives in machine tools increases constantly and currently they affect even the structures of the driven mechanisms.

Boring machines are used to mill, drill, bore, cut threads or face turn. They can be horizontal and vertical, which refers to the axis orientation of the machine spindle rotation that provides the primary motion [1 - 4].

A drive system for boring machines includes feed drives, spindle drive and auxiliary drives. For modernization of a class of boring machines with digital program control, two additional controlled axes and a device for angular position of the workpieces are introduced. The goal is to extend the capabilities of these machines to process more complex machine parts and to enhance their performance. In this way, the machines under consideration can be attributed to the class of those with multi- coordinate electric driving systems.

Feed electric drives in boring machines are used to position the workpieces and the cutting tools at the desired location and participate in the machining process. For this reason, their positional accuracy and speed significantly influence the quality and performance of the machine tools with digital program control [5 - 8].

In this paper, the main requirements for the respective drives of a type of boring machines are formulated and on this basis, a methodology for selection of appropriate feed electric drives is developed. In choosing suitable feed drives, a number of essential factors were taken into account, namely: the technological process features, the treated materials, the tools used, the mechanical gears and the respective guides. Examples for selection of feed drives with DC and AC motors are shown, illustrating the practical application of the offered methodology. Some experimental results are presented and discussed.

2. Requirements for the drive system

The block diagram of a drive system for these machines is shown in Fig. 1, where the following notations are used: DPC – digital program control device; ED1 - ED10 – electric drives; G1 - G8 – mechanical gears; L1 - L8 – loads. Each of the presented drive subsystems has specific features and requirements that need to be taken into account in the process of design and study.

The feed drives of the *x, y, z, a,* and *c* coordinate axes are used for positioning of the tool and the workpiece at the desired locations and they participate in the machining process. The main requirements for them can be formulated as follows: smooth speed regulation in a wide range; good dynamics; high position accuracy; formation of the necessary position cycles; providing the required torque; reversible speed and torque control; compensation of the disturbances; easy maintenance; reliability; economy.

The spindle drive is involved in the machining process and it should meet the following requirements: dual-zone speed regulation (by constant torque and constant power respectively); high maximum speed; sufficient torque; oriented braking with high accuracy; reversible speed control.

The main requirements for the auxiliary drives can be formulated as follows: angular positioning of the workpiece with high precision; instrument subsystem providing choice of the desired tool with high precision; dosing lubrication of the machine coordinate axes; cooling subsystem with possibility of automatic and manual braking and starting from the control panel.

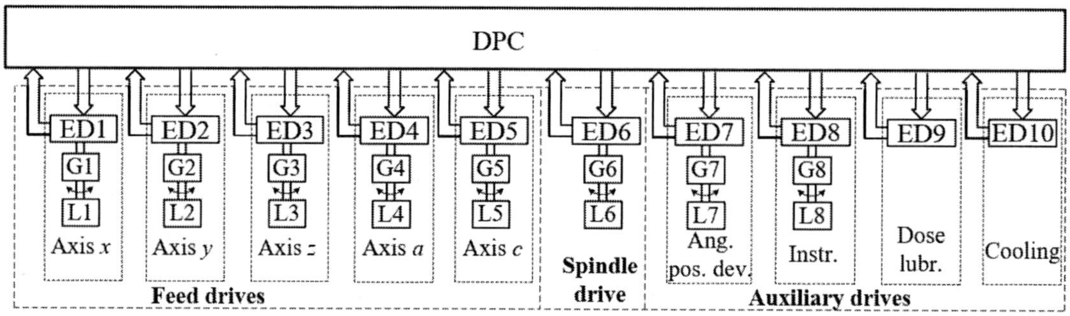

Figure 1. Block diagram of the drive system.

The choice of appropriate feed electric drives for the studied class of machines includes selection of motors, power converters, controllers, sensors and mechanical gears. This problem is extremely important in terms of performance, techno-economic parameters, reliability, etc.

3. Selection of feed drives

A unified approach for the designing process of feed drives for boring machines is applied, including the following basic stages: development of a methodology for optimal choice of these drives; calculations, according to the respective procedures of the methodology; technical and economic analysis of the possible variants of electric drives; modeling and computer simulation of the relevant dynamic and static modes of operation; experimental study to evaluate the actual performance of the drive system.

3.1. Features of the methodology

When choosing an electric drive, it is essential to define the heaviest processing mode of operation, the mass of the coordinate axis and the guides of the machine to optimize the power and the price.

Fig. 2 shows the simplified block diagram of the developed algorithm for selection of feed drives. The following notations are used in this section: D_{gb} – nominal diameter of a ball screw; a_b – acceleration of the ball screw; $D_{mb\max}$ – maximum diameter of the boring machine; $D_{cb\max}$ – maximum diameter of boring instrument; H_B – Brinell hardness of the processed material; $m_{m\max}$ –

maximum mass of the driven mechanism; $V_{b\max}$ – maximum speed of the driven mechanism; V_{cb} – cutting speed; ω_b – spindle speed; V_{fb} – feed speed; f_{rb} – feed per radian; a_{pb} – cutting depth, which is the difference between the uncut and the cut hole radius; K_{cfzb} – specific cutting force, when feeding the tool tooth; η_b – efficiency of the boring machine; $P_{cb\max}$, P_{cbw} – maximum powers needed to perform cutting distributed between feed and spindle drives, respectively without and with taking into account the tool wear; P_{fb} – power required only for the feed drive; F_{fb} – force required only for the feed drive; F_g – force required for the guides; μ_g – friction coefficient of the guides; P_{fbgc} – power required for the feed drive with the required friction power; $P_{fbg\max}$ – maximum power required for the feed drive in maximum cutting speed of the mechanism; h_{bi} – nominal steps of the ball screw; ω_{fbi} – speeds of the motor for different nominal steps of the ball screw; M_{fbi} – motor torques for different nominal steps of the ball screw; $i = 1 \div n$, where n is the variant number.

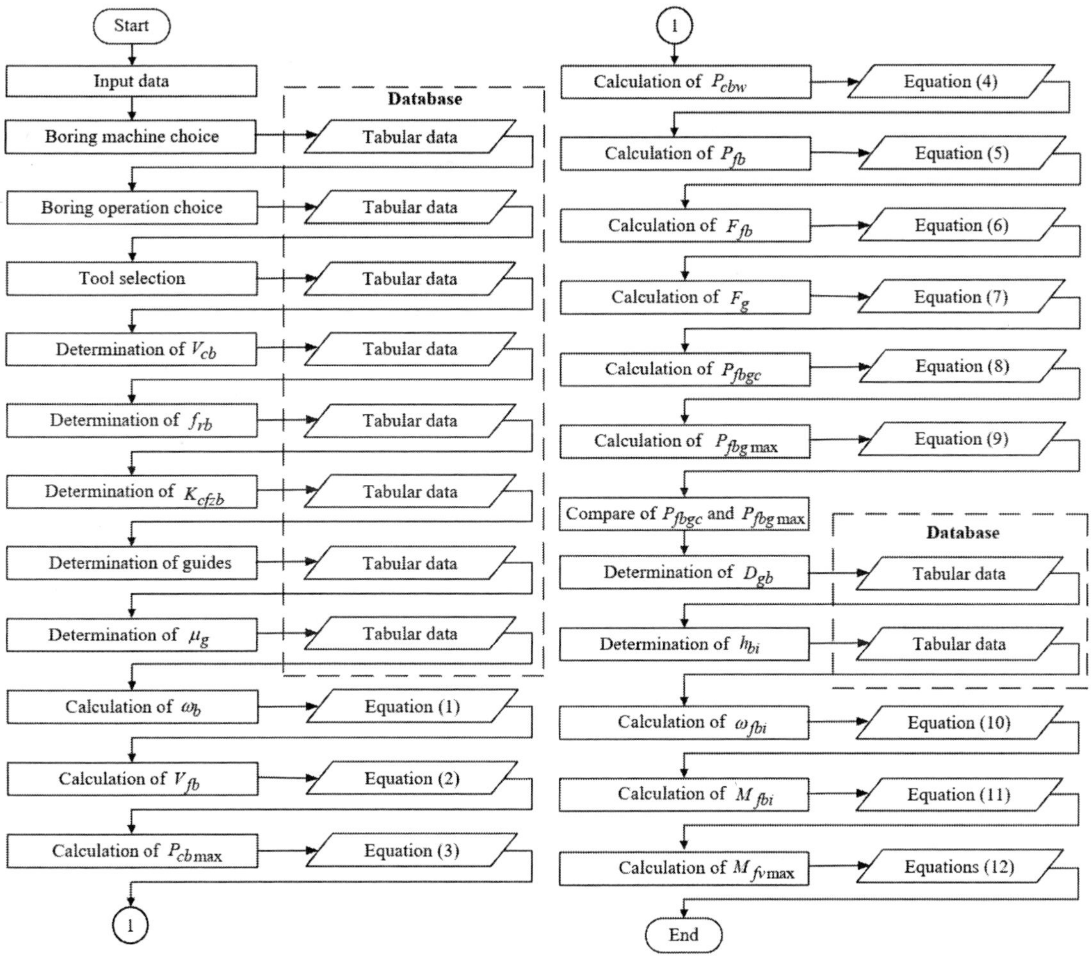

Figure 2. Block diagram of the algorithm for drive selection.

In the proposed methodology, a ball screw couple was used as a mechanical gear with a specified screw diameter. The design and calculation of ball screws is described in ISO/DIN standards [9 - 11].

The tabular data used in this methodology are taken from [9] and [12 - 14]. The choice of boring

IOP Conf. Series: Materials Science and Engineering **618** (2019) 012003 doi:10.1088/1757-899X/618/1/012003

tools goes through the following several stages: definition of the operation type; selection of a boring system; definition of the boring diameter and hole requirements; choosing of the entering angle; selection of an adaptor and inserts for the tools.

The input data are as follows: the heaviest cutting regime, H_B, $D_{cb\max}$, $D_{mb\max}$, $a_{pb\max}$, $m_{m\max}$, $a_{b\max}$, η_b and $V_{b\max}$.

The spindle speed in the boring process is determined by the expression [12]:

$$\omega_b = \frac{V_{cb} \times 2}{D_{mb\max}} \tag{1}$$

The boring feed speed is calculated using the following equation [12]:

$$V_{fb} = f_{rb} \times \omega_b \tag{2}$$

The maximum power needed to perform cutting, distributed between both feed drive and spindle drive without taking into account the tool wear, is calculated by the next expression [12]:

$$P_{cb\max} = \frac{V_{cb} \times a_{pb\max} \times f_{rb} \times K_{cfzb} \times 2\pi \times 10^6}{\eta_b} \times (1 - \frac{a_{pb\max}}{D_{cb\max}}) \tag{3}$$

The maximum power needed to perform cutting, distributed between both feed drive and spindle drive with taking into account the tool wear, is determined by the equation [12]:

$$P_{cbw} = (1.1 \div 1.3) \times P_{cb\max} \tag{4}$$

The power required only for the feed drive is determined as follows [15]:

$$P_{fb} = (1 \div 5)\% \times P_{cbw} \tag{5}$$

The force required for the feed drive is determined as follows [16]:

$$F_{fb} = \frac{P_{fb}}{V_{fb}} \tag{6}$$

The force required for guides of the feed drive is determined by the expression [9]:

$$F_g = m_{m\max} a_{b\max} \mu_b \tag{7}$$

The power required for the feed drive taking into the guides, is determined as follows [16]:

$$P_{fbgc} = (F_g + F_{fb}) V_{fb} \tag{8}$$

The power required for the feed drive in rapid travel for positioning, is determined as follows [16]:

$$P_{fbg\max} = F_g V_{fb\max} \tag{9}$$

The motor speeds for different nominal steps of the ball screw are calculated as follows [16]:

$$\omega_{fbi} = \frac{V_{fb\max} \times 2 \times \pi}{h_{bi}} \tag{10}$$

The torques of the motor for different nominal steps of the ball screw in the cutting process are determined by the following equation [16]:

$$M_{fbi} = \frac{P_{fbgc}}{\omega_{fbi}} \tag{11}$$

The torques of the motor for different nominal steps of the ball screw in maximum feed speed are determined by the following equation [16]:

$$M_{fv\max i} = \frac{P_{fbg\max}}{\omega_{fbi}} \tag{12}$$

3.2. Practical application

The offered methodology for drives selection takes into account the specific features of the technological processes, the processed materials, the tools used and their wear, the guides, as well as the mechanical gear types. It gives opportunity for choice of motors, power converters, mechanical gears and sensors.

Fig. 3 shows a diagram, illustrating the elements used for spindle and axis z drives of the studied machine. The notations used are as follows: DPC – digital program control device; 1 – position sensor for z axis; 2 – motor for z axis; 3 – coupling between the motor and the ball screw for z axis; 4 – ball screw for z axis; 5 – guides for z axis; 6 – speed sensor for the spindle drive; 7 – spindle motor; 8 – coupling between the spindle motor and gearbox; 9 – gearbox for the spindle drive; 10 – boring instrument; 11 – driven mechanism (work desk for the workpiece); 12 – processed workpiece; h_{bz} – nominal steps of the ball screw for z axis; D_{gbz} – nominal diameter of the mechanical gear made by a ball screw for z axis; V_{bz} – speed of the z axis; F_{gz} – force required for the guides of the z axis; F_{fbz} – force for the feed drive in the cutting process of z axis.

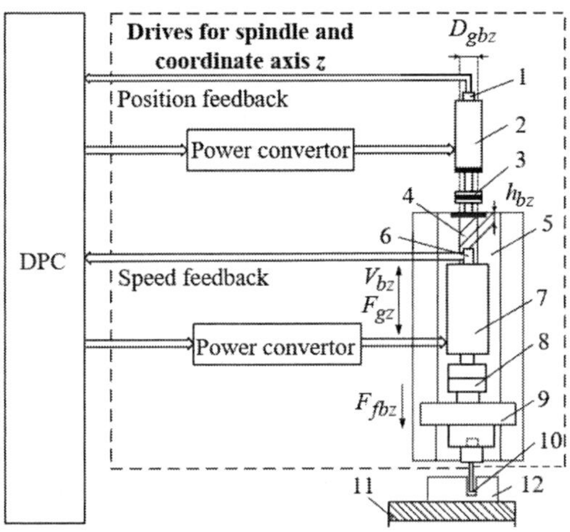

Figure 3. Diagram of the elements used for spindle and axis z drives.

Some examples of using this methodology for selection of feed electric drive for cutting materials with different hardness are presented below.

The input data are as follows: the heaviest cutting regime of operation – at unalloyed steel with $H_B = 170$ and cast iron with $H_B = 130$; $D_{mb\max} = 0.05$ m, $D_{cb\max} = 0.04$ m, $a_{pb\max} = 0.002$ m, $m_{m\max} = 500$ kg, $a_{b\max} = 10$ m/s, $\eta_b = 0.85$ and $V_{b\max} = 0.5$ m/s. The results are given in Table 1.

Table 1. Results from the calculations.

Step	Operation	Unalloyed steel	Cast iron
1.	Boring machine choice	Vertical	Vertical
2.	Boring operation choice	Roughing	Roughing
3.	Tool selection	Core 820	Core 820

Table 1. Results from the calculations (Cont.).

4. Determination of V_{cb} [m/s]	≈ 3.17	≈ 5
5. Determination of f_{rb} [m/rad]	3.1×10^{-5}	6.3×10^{-5}
6. Determination of K_{cfzb} [MPa]	2200	940
7. Determination of guides	Rolling	Rolling
8. Determination of μ_g	0.003	0.003
9. Calculation of ω_b [rad/s]	126.8	200
10. Calculation of V_{fb} [m/s]	3.93×10^{-3}	12.6×10^{-3}
11. Calculation of $P_{cb\max}$ [W]	3034.86	4156.55
12. Calculation of P_{cbw} [W]	3338.34	4572.2
13. Calculation of P_{fb} [W]	166.9	228.6
14. Calculation of F_{fb} [N]	42472.5	18142.9
15. Calculation of F_g [N]	15	15
16. Calculation of P_{fbgc} [W]	166.98	228.8
17. Calculation of $P_{fbg\max}$ [W]	7.5	7.5
18. Compare of P_{fbgc} and $P_{fbg\max}$	$P_{fbgc} > P_{fbg\max}$	$P_{fbgc} > P_{fbg\max}$
19. Determination of D_{gb} [m]	0.04	0.025
20. Determination of h_{bi} [m]	$h_{b1} = 0.005$ $h_{b2} = 0.01$ $h_{b3} = 0.02$ $h_{b4} = 0.04$	$h_{b1} = 0.005$ $h_{b2} = 0.01$ $h_{b3} = 0.02$
21. Calculation of ω_{fb_i} [rad/s]	$\omega_{fb_1} \approx 628$; $\omega_{fb_2} \approx 314$; $\omega_{fb3} \approx 157$; $\omega_{fb4} \approx 78.5$	$\omega_{fb_1} \approx 628$; $\omega_{fb_2} \approx 314$; $\omega_{fb3} \approx 157$
22. Calculation of M_{fb_i} [Nm]	$M_{fb_1} \approx 0.27$; $M_{fb_2} \approx 0.53$; $M_{fb_3} \approx 1.06$; $M_{fb_4} \approx 2.12$	$M_{fb_1} \approx 0.36$; $M_{fb_2} \approx 0.72$; $M_{fb_3} \approx 1.53$
23. Calculation of $M_{fv\max_i}$ [Nm]	$M_{fv\max_1} \approx 0.01$; $M_{fv\max_2} \approx 0.02$; $M_{fv\max_3} \approx 0.05$; $M_{fv\max_4} \approx 0.01$	$M_{fv\max_1} \approx 0.01$; $M_{fv\max_2} \approx 0.02$; $M_{fv\max_3} \approx 0.05$

The selected motors must have nominal torques about 10% greater than the calculated ones, in order to compensate for the allowable wear over time. These obtained values are used for the motor choice from the respective technical catalogues.

Table 2. Data of the selected feed drives.

Elements	In machining of unalloyed steel	In machining of cast iron
Ball screw	diameter 0.04 m with step 0.02 m	diameter 0.025 m with step 0.01 m
Motor	DC motor PI10.06: $M_{nom} = 1.6$ Nm $\omega_{nom} = 314$ rad/s	AC motor DS4: $M_{nom} = 1.6$ Nm, $\omega_{nom} = 439.64$ rad/s
Power convertor	model SA-12	model KW2
Position sensor	encoder	resolver

As a result of the calculations made for these two materials, appropriate DC and AC electric drives were chosen from [17 - 21]. Some of their basic parameters are presented in Table 2.

4. Experimental studies

On the basis of the formulated requirements, the developed methodology and the selected DC and AC electric drives for a machine of the considered class, a detailed experimental study was carried out at different settings of the controllers and operating modes.

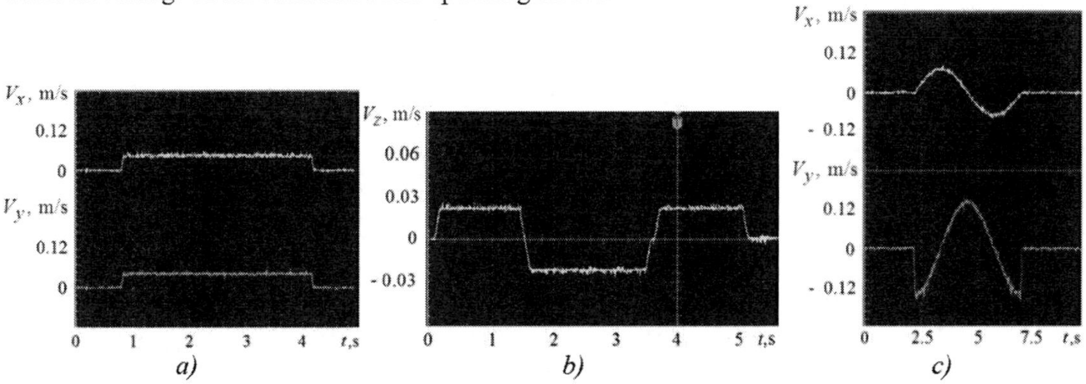

Fig. 4. Time diagrams for feed drives with DC motors.

Some experimentally obtained oscillograms with feed DC and AC electric drives are presented respectively in Fig. 4 and Fig. 5.

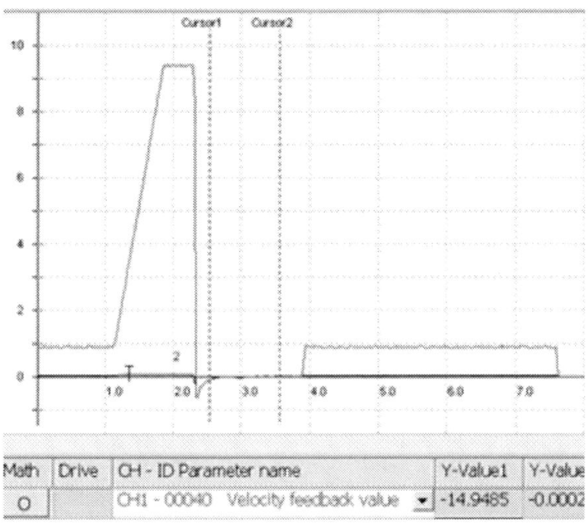

Figure 5. Experimentally obtained time diagram for an implemented feed electric drive with AC motor.

Fig. 4*a* and 4*b* show speed trajectories along the two-coordinate drive *x-y* and the axis z. Fig. 4*c* presents circular interpolation with the two-coordinate drive *x-y*. The selected mechanical gears are of ball-screw type with the following coefficients: $K_{gx} = K_{gy} = K_{gz} = 10$ mm/rev $\approx 1.6\text{x}10^{-3}$ m/rad.

Fig. 5 presents a time diagram, obtained experimentally for an implemented feed AC electric drive with vector control. In this case the respective velocity feedback signal [V] is displayed as a function of the time [s].

Based on the studies carried out with DC and AC electric drives the following practical inferences can be drawn: the implemented DC drives have good tuning qualities and provide the necessary static and dynamic characteristics, however a disadvantage is the presence of the brush collector; electric drives with AC motors have easier operational maintenance but their price is relatively higher.

IOP Conf. Series: Materials Science and Engineering **618** (2019) 012003 doi:10.1088/1757-899X/618/1/012003

5. Conclusion

The basic requirements for the drive systems of a class of boring machines with digital program control are formulated. A methodology for optimal selection of feed drives is offered and some examples with DC and AC electric drives are described.

The developed algorithm takes into account the technological process, the tools used, the mechanical gears and the guides.

Experimental results with implemented feed electric drives are presented and analyzed.

The research held and the results obtained can be used in the choice of such drives for the studied class of machine tools.

Acknowledgments

This work is supported by the Technical University of Sofia, Bulgaria under Research Project No. 191PR0002-08/2019.

References

[1] Popov G, *Machine Tools, Part I: Applicability, Device and Control*, Technical of University of Sofia, Sofia, 2010.

[2] Tata McGraw-Hill Education, *Manufacturing Technology: Metal Cutting and Machine Tools*, Vol. 2, 2013.

[3] Youssef H, H El-Hofy, *Machining Technology: Machine Tools and Operations*, CRC Press, 2008.

[4] Girsang I, J Dhupia, Jaspreet, *Machine Tools for Machining,* DOI: 10.1007/978-1-4471-4670-4_4, 2015.

[5] Mikhov M, *Electric Drive Systems*, Technical University of Sofia, Sofia, 2011.

[6] Altintas Y, A Verl, C Brecher, L Uriarte and G Pritschow, Machine Tool Feed Drives, *CIRP Annals - Manufacturing Technology*, Vol. 60, No. 2, pp. 779 -796, 2011.

[7] Mikhov M and M Zhilevski, Analysis of a Multi-Coordinate Drive System Aiming at Performance Improvement, *Proceedings of the International Conference "Research and Development in Mechanical Industry"*, Vol. 2, pp. 1102-1107, Vrnjacka Banja, Serbia, 2012.

[8] Zhilevski M and M Mikhov, Performance Improvement of a Type of Turning Machines, *European Journal of Engineering Research and Science*, Vol. 3, No. 8, pp. 44-49, 2018.

[9] *Ball screw*, THK, General Catalogue.

[10] *Ball Screw and Nuts*, ServoMech, Catalogue.

[11] DIN 69051-3: *Werkzeugmaschinen; Kugelgewindetriebe; Abnahmebedingungen und Abnahmeprüfungen*, 1998.

[12] Sandvik Coromant, *Metalcutting Technical Guide: Turning, Milling, Drilling, Boring, Toolholding*, Sandvik, 2005.

[13] *Rolling bearings*, SKF Group, Catalogue, 2018.

[14] *Precision Rolled Ball Screw*, SKF Group, Catalogue, 2013.

[15] Andonov I, *Cutting of Metals*, Softtrade, Sofia, 2001.

[16] Braitinger H, *Elektrische Antriebstechnik*, AMK Arnold Müller, 2004.

[17] *Precision Ball Screw Assemblies*, Rexroth Bosch Group, Catalogue, 2009.

[18] SERVOMOTORS, *Gama Motors Catalogue*, 2016.

[19] Graphical Peripheral Devices, *Catalogue Sensors*.

[20] AMKASYN, *Servo Drives KE/KW*, AMK Catalogue, 2014.

[21] DYNASYN, *Servo Motors DT and DP*, AMK Catalogue, 2014.

IOP Conf. Series: Materials Science and Engineering **618** (2019) 012004 doi:10.1088/1757-899X/618/1/012004

Embedded Walking Algorithm for Biped Humanoid Robot with 17 Degrees-of-Freedom

Sherif Sherif [1], Jordan Kralev[1], Tsonyo Slavov[1] and Veselin Kunchev[1]
[1]Technical University of Sofia, Faculty of Automatics, Department of Systems and Control, Sofia, Bulgaria

E-mail: mr.shark94@abv.bg; jkralev@yahoo.com; ts_slavov@tu-sofia.bg; v.kanchev1993@gmail.com

Abstract. The realization of the control algorithm consists of two main stages: planning of a joint coordination sequence and implementation of an embedded walking algorithm in the Arduino Mega 2560 hardware platform. The joint coordination sequence consists of 13 steps in order for the humanoid robot to move forward without falling. The embedded walking algorithm is realized around the operating points of the servo motors. The software is developed in Arduino IDE in C language.

1. Introduction

Biped walking machines have been developed since 1970. Today popular biped robots are ASIMO of HONDA [1], WABIAN-2 of Waseda University [2], and HRP-3 of AIST [3]. The walking algorithm of a biped robot can be either static [4] or dynamic [5]. In static walking the robot stops its walking motion at each step in order to dampen the oscillations from the previous step. Dynamic walking is smoother because each parameter of the walking cycle is continuously varied such that centre of gravity (CG) translation speed is close to constant. The realization of a stable biped robot walking is a considerable challenge even in deterministic environment [6]. The key to successful walking requires the projection of CG onto the ground to lie within the supporting polygon of the two feet. In the case of dynamic walking the same condition is required for the zero moment point.

Considered system is laboratory robot with 17 degrees-of-freedom (DOFs) provided by servo motors LD-2015 mounted at the key joints. The microcontroller which controls the walking is ATmega2560 which is programmed with the Arduino IDE.

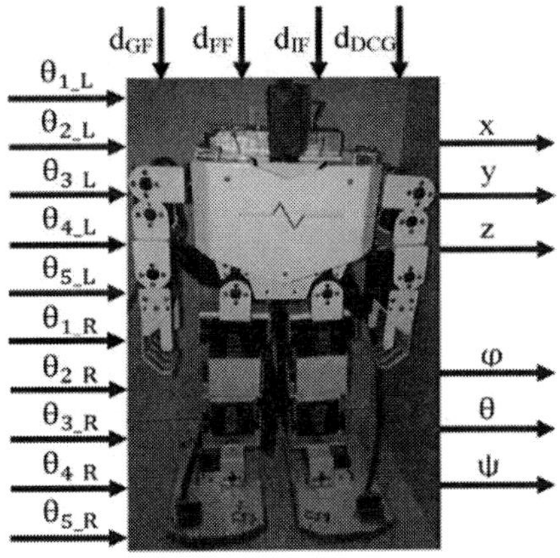

Figure 1. Laboratory model of biped robot with 17 degrees-of-freedom with 10 inputs and 6 outputs

The aim of the open-loop control for biped walking robots is to attenuate external disturbances acting on the mechanical system. In order to achieve such control the external disturbances have to be from deterministic origin. For the present laboratory model of humanoid robot the key external disturbances are: **gravitational force, friction forces, inertial forces** and **CG motion.**

The control actions are $\boldsymbol{\theta}_{i,L}$ and $\boldsymbol{\theta}_{i,R}$ are the angular offset in the left and right key joints, with $i = 1 \dots 5$. The output signals $\boldsymbol{x}, \boldsymbol{y}, \boldsymbol{z}$ represent the position for the robot CG and Euler angles $\boldsymbol{\varphi}, \boldsymbol{\theta}, \boldsymbol{\psi}$ represent the robot orientation.

2. External disturbances

The external disturbance forces acting on the mechanical construction of the robot during walking can be pointed:

- **Gravitational force - $\mathbf{d_{GF}}$**

The gravitation is disturbing the stability on the robot when the incidence angle between the body and the ground surface is increased. If that angle goes beyond certain limit the robot cannot be stabilized with the control action from the servo motors alone.

- **Friction - $\mathbf{d_{FF}}$**

Successful walking requires a certain amount of contact force between the ground surface and robot feet which prevents foot slipping. However when a foot is translated over the surface developed friction force can disturb the orientation of the robot and consequently its trajectory.

- **Inertial forces - $\mathbf{d_{IF}}$**

These forces are fictitious forces representing the effect of kinematic variables upon the orientation of the body. If the frequency of the step cycle is increased beyond certain limit the abundant energy from uncompensated inertial forces cannot be dampen fast enough and leads to instability.

- **CG motion - $\mathbf{d_{DCG}}$**

Generally if the center of gravity is not projected upon the feet base then the mechanical construction becomes less stable and small uncompensated disturbance force would cause the robot to fall down.

The location of the robot CG is the most important factor which prevents the robot from falling during the execution of the walking step cycle. Location of the CG can be expressed with **(1)** by accounting for the mass action of the various components of the robot.

$$\vec{X}_{CG} = \frac{m_B}{M}\vec{X}_{CG,B} + \frac{m_{HL}}{M}\vec{X}_{CG,HL} + \frac{m_{HR}}{M}\vec{X}_{CG,HR} + \frac{m_{LL}}{M}\vec{X}_{CG,LL} + \frac{m_{LR}}{M}\vec{X}_{CG,LR} \qquad (1)$$

where $\vec{\boldsymbol{X}}_{\boldsymbol{CG}}$ is the location vector of the CG with components

$$\vec{X}_{CG} = \begin{pmatrix} x_{CG} \\ y_{CG} \\ z_{CG} \end{pmatrix} \qquad (2)$$

and $\vec{\boldsymbol{x}}_{\boldsymbol{CG,i}}$ is the location of CG of respective body element $i \in \{B, HL, HR, LL, LR\}$, $\boldsymbol{M}$ is the total mass of the robot, $\boldsymbol{m_B}$ is the mass of the trunk, $\boldsymbol{m_{HL}}$ is the left hand mass, $\boldsymbol{m_{HR}}$ is the right hand mass, $\boldsymbol{m_{LL}}$ is the left leg mass and $\boldsymbol{m_{LR}}$ is the right leg mass such that

$$M = m_B + m_{HL} + m_{HR} + m_{LL} + m_{LR} \qquad (3)$$

The CG of each component of the robot at time instant $\boldsymbol{k}$ is shifted with respect to CG location at time instant $\boldsymbol{k-1}$

$$\vec{X}_{CG,i}(k) = \vec{X}_{CG,i}(k-1) + \vec{\Delta}_i \tag{4}$$

with $\vec{\boldsymbol{\Delta}}_i$ representing the CG displacement caused by joint reconfiguration during walking. Then the displacement of the robot CG $\vec{\boldsymbol{\Delta}}_{CG}$ will be a weighted sum over the displacements $\vec{\boldsymbol{\Delta}}_i$ of CGs of its components with weights $\boldsymbol{m_i/M}$. Hence by rotating a joint in particular direction the robot CG is translated in this direction. In order to keep the upper stance the projected location of the robot CG onto the ground surface should be kept within the supporting foot base.

3. Walking algorithm

At initial position the robot joint angles are set to their trim values. They are experimentally tuned to guarantee its upper standing position and are summarized in **Table 1**.

Table 1. Absolute values of servo motors

Joint	Notation	Value [deg]
Left ankle roll	$\theta_{\_1\_L\_Trim}$	97
Left ankle pitch	$\theta_{\_2\_L\_Trim}$	74
Left knee pitch	$\theta_{\_3\_L\_Trim}$	6
Left hip pitch	$\theta_{\_4\_L\_Trim}$	85
Left sholder pitch	$\theta_{\_5\_L\_Trim}$	155
Right ankle roll	$\theta_{\_1\_R\_Trim}$	79
Right ankle pitch	$\theta_{\_2\_R\_Trim}$	82
Right knee pitch	$\theta_{\_3\_R\_Trim}$	142
Right hip pitch	$\theta_{\_4\_R\_Trim}$	70
Right sholder pitch	$\theta_{\_5\_R\_Trim}$	87

The joint coordination sequence consists of 13 steps in order for the humanoid to move forward without falling. The starting position of the walking robot is denoted as **"A"** when its center of gravity is distributed on both legs as in figure 1.

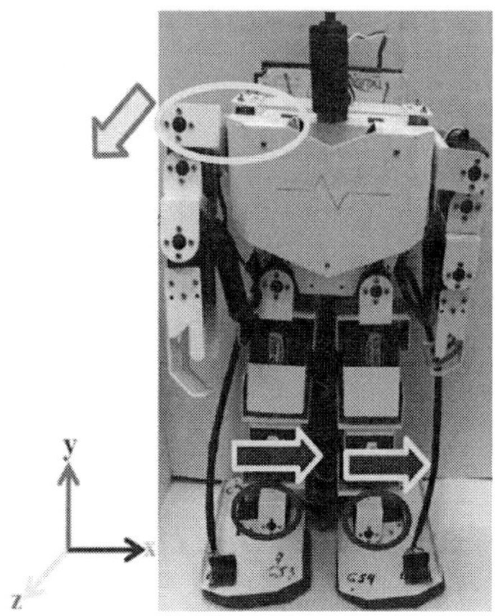

Figure 2. Stage "B"

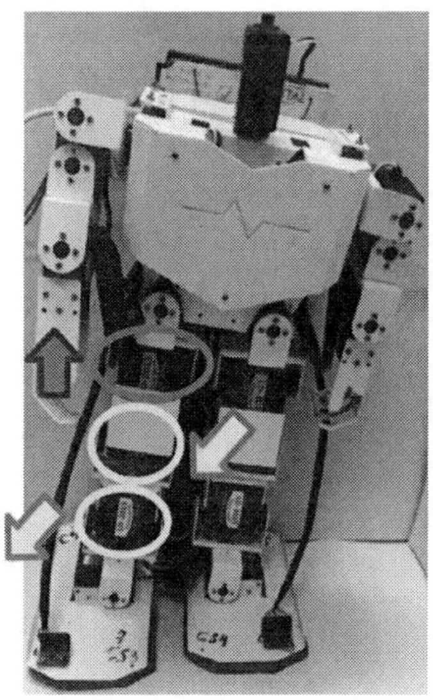

Figure 3. Stage "C"

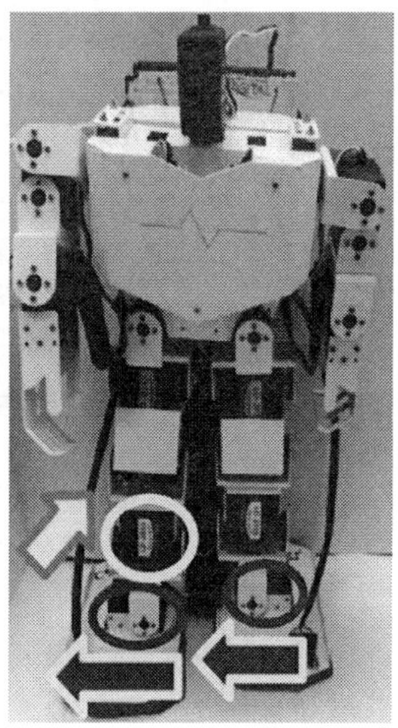

Figure 4. Stage "D"

In **stage "B"** as figure 2 shows the center of gravity projection is shifted to the left leg by applying ankle roll and right shoulder pitch.

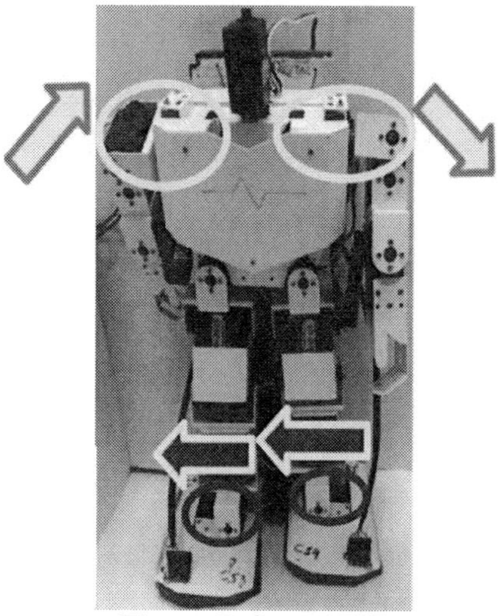

Figure 5. Stage "E"

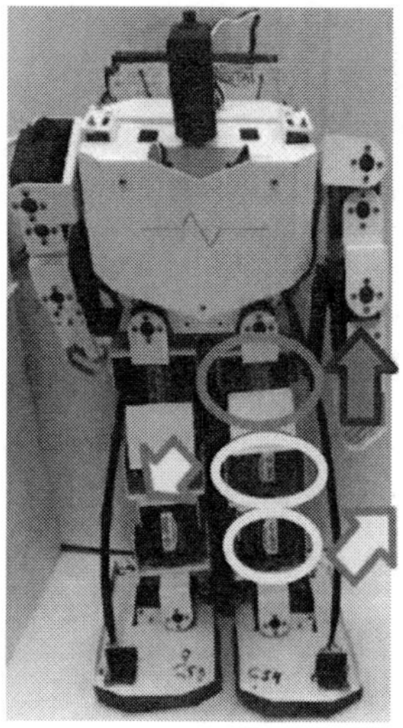

Figure 6. Stage "F"

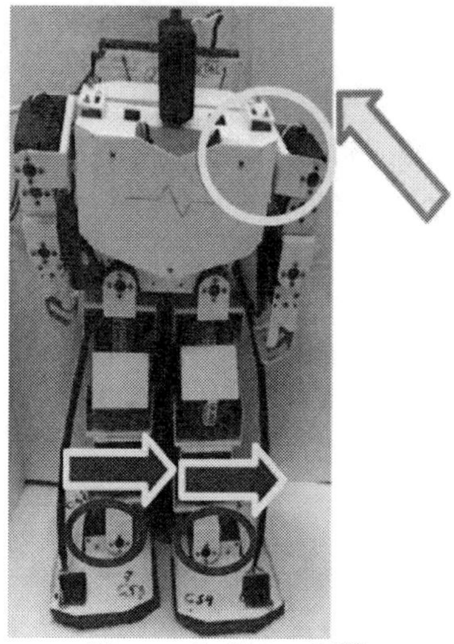

Figure 7. Stage "G"

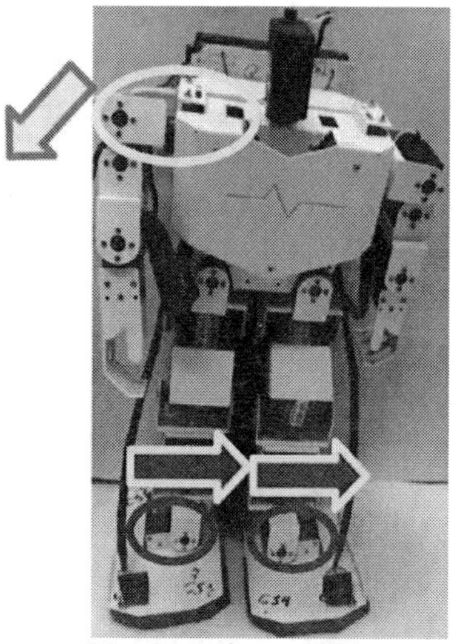

Figure 8. Stage "H"

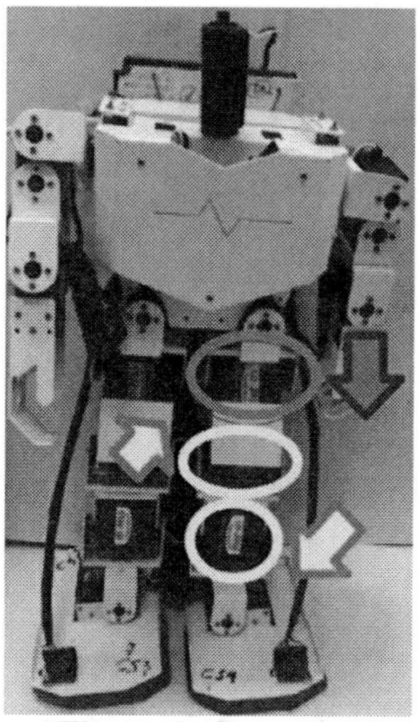

Figure 9. Stage "J"

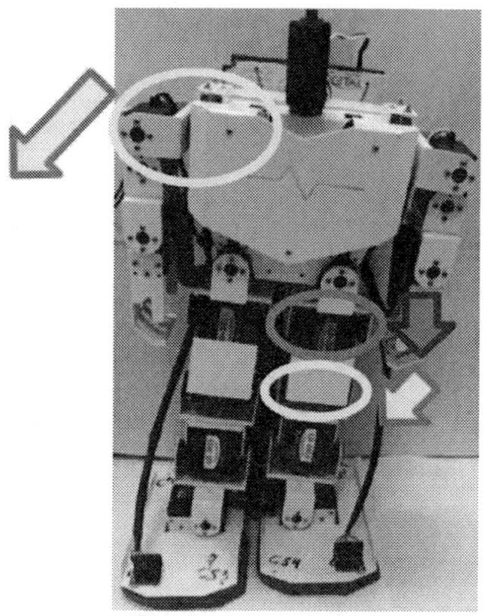

Figure 10. Stage "K"

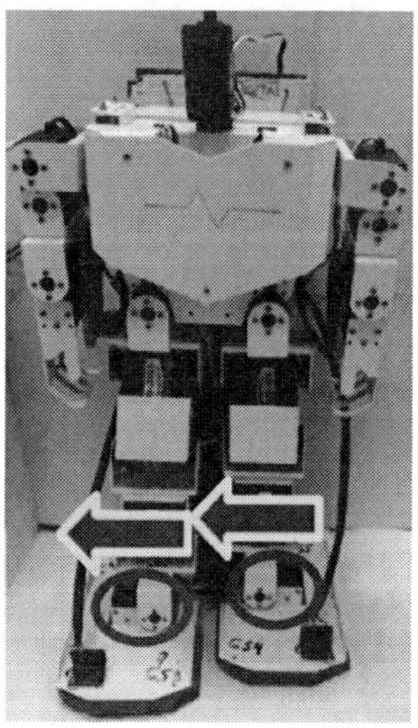

Figure 11. Stage "L"

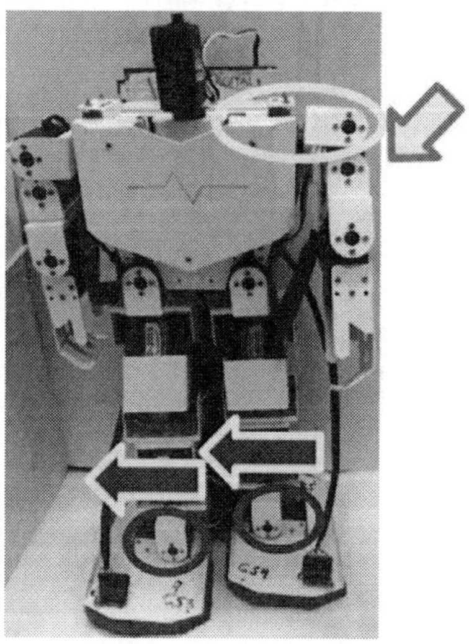

Figure 12. Stage "M"

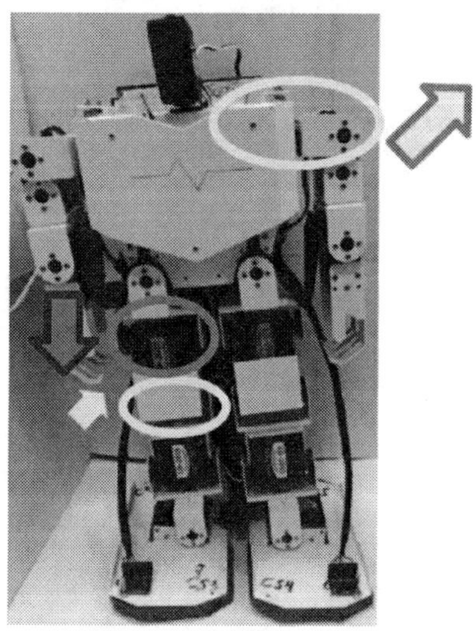

Figure 13. Stage "N"

In **step "C" (figure 3.)** the right leg is moved forward but the center of gravity projection remains on the left foot and keeps the stability of the robot. In this stage the right leg pitch is performed in ankle, knee and hip joints.

In **step "D"** given in **figure 4.** it can be noticed that the right leg is exerted a little further from the left, which is due to the changes in the previous stage (Stage "C" – figure 3.). Also in **step "D"** center of gravity projection is again distributed on both legs. To accomplish this step, three changes are made to the current values of the servo motors – roll of both ankles and pitch of the right ankle.

In the next **step "E" (figure 5.)** the center of gravity projection is shifted to the right foot by rotating four of the joints – ankle's roll and shoulder's pitch.

During the sixth **stage "F"** shown in **figure 6.** the left foot of the robot is moved forward by performing ankle, knee and hip pitch. The center of gravity projection then lies on a right leg which is static to ensure the stability of the robot. The difference between this stage compared to stage "C" is in supporting leg where the center of gravity projection is located.

In **step "G" (figure 7.)** the weight of the robot is distributed over the both legs, but it is not in the starting position because the right knee joint, the right shoulder joint and some of the left leg joints are still displaced. To achieve this step roll of both ankles is applied and also a pitch of the left shoulder.

The **step "H"** from **figure 8.** shifts the center of gravity projection to the left leg. The difference here with respect to stage "B" is that the two servos in the knee of the right leg are not in their initial position and also the servo in the shoulder joint of the right hand. The other difference is in the left leg and in particular in the servo motor for ankle pitch and the two servo motors in the knee, which in the current stage ("H") are not in their initial position.

In **step "J" (figure 9.)** the left foot of the humanoid begins its return to its starting position. To prevent the robot from falling this is done in two stages ("J" and "K"). The need for two steps is due to the fact that the current angle of the left knee joint is twice as large as that of the right knee. The difference in this stage with respect to "F" stage is the foot on which supports the center of gravity projection. In order to accomplish this step, three changes were made to the current values of the joints. The ankle and knee servos are still not returned to their starting position. Their current angels are set to an intermediate values with respect to stages "F" and "K".

In the tenth **stage "K" (figure 10.)** the process of returning of the ankle and knee joints of the left leg to initial position is completed. Also, at this stage, the right shoulder joint is set to its trim value.

In **step "L"** shown in **figure 11.** the center of gravity projection lies on both robot's legs, but it is still not in the starting position because the right foot is exerted slightly further than the left. The right leg is ahead of the left due to its ankle and knee joints. In order to realize this stage, there are changes in the current values of two servo motors, which are located in the feet joints.

In **stage "M" (figure 12.)** the center of gravity projection is shifted to the right leg. In order to accomplish this step, changes are made to the current values of three servo motors pointed in the figure.

In next **step "N" (figure 13.)** the robot's right foot returns to the starting position. For the completion of this stage right hip and knee joint are rotated. Also the left shoulder joint is rotated. After the rotations the servo motors in the knee of the right leg and the servo motor in the left shoulder are set to their trim values.

A sub-step before returning to **stage "A"** after the last stage "N" is executed in order to return all joints to their trim values. In this sub-step, the weight of the robot is distributed over both legs. Then the robot completes the joint coordination sequence and its location is one step ahead.

4. Implementation in Arduino

The robot control algorithm is embedded in Arduino Mega 2560 hardware platform which is attached at the back of the body element together with a signal distribution board (**figure 14.**). The software is developed in Arduino IDE in C language. Also created special shield, which plays a roll on the distribution board. The shield allowed easily installation of servo motors and the connection of components to the battery.

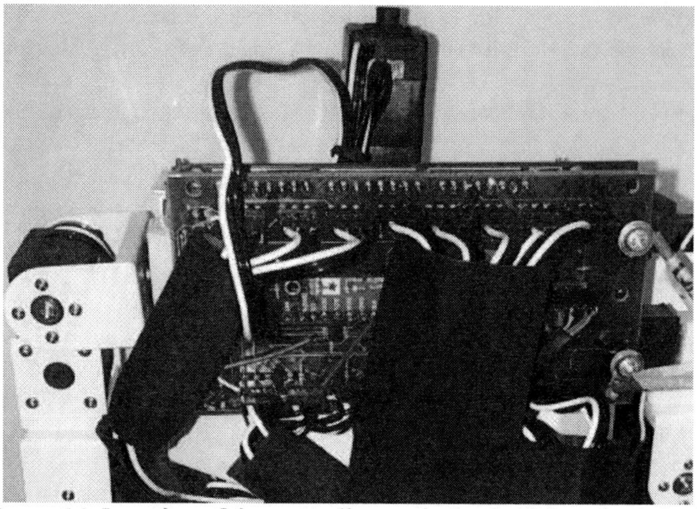

Figure 14. Location of the controller on the back of the robot's body

The signal line of each servo motor is connected to a digital output pin of the Arduino platform. The trim values of the joints $\boldsymbol{\theta_{i,j,TRIM}}$ are declared as constants. And the current offset values $\boldsymbol{\theta_{i,j}}$ around the trim constants are calculated according to the presented step sequence from the previous section. The angular position of the joint is commanded to the servo motors as a pulse width modulated signal, with duration $\boldsymbol{u_{i,j}}$ of the pulse between **1 [ms]** to **2 [ms]** corresponding angles from **0** to **180 degrees**.

The progress through the step sequence is monitored by a dedicated temporal progress variable $\boldsymbol{\tau} \in [\boldsymbol{0}, \boldsymbol{T_{step}}]$, where $\boldsymbol{T_{step}}$ determines the necessary time for performing a single step. The value of $\boldsymbol{T_{step}}$ is tuned experimentally and it is set to **2760 [ms]**. If a shorter period is used the humanoid robot becomes unstable. The current time is obtained through the processor internal clock $\boldsymbol{t}$ and the progress variable $\boldsymbol{\tau}$ is calculated as $\boldsymbol{\tau} = \boldsymbol{t} - \boldsymbol{T_{start}}$, where $\boldsymbol{T_{start}}$ is the variable for storing the initial timing of the current step. The progress variable $\boldsymbol{\tau}$ can be represented in percentage scale for convenience so when **100 [%]** are reached the step sequence is completed the start time $\boldsymbol{T_{start}}$ is set to the current value of the clock $\boldsymbol{t}$.

The programming of each stage of a single step form the walking sequence requires a particular configuration of the robot joints. For example let focus on the **stage "B"** from **figure 2.**. This stage occurs when $\mathbf{0.83}\ [\%] \leq \boldsymbol{\tau} \leq \mathbf{5}\ [\%]$. After checking that condition the respective servo motors are rotated by setting a new value to $\boldsymbol{\theta_{i,j}}$. The pulse width is then calculated as (**eq. (5)**)

$$u_{i,j}(t) = 0.001\left(1 + \frac{\theta_{i,j,TRIM} + \theta_{i,j}(t)}{180}\right) \tag{5}$$

Figure 15. and **figure 16.** shows the control signal send to the servo motor in order to achieve the desired joint angular position. It is a pulse width modulated signal with varying pulse duration between **1** to **2 [ms]**. The walking algorithm requires all 17 joint servo motors to receive appropriate signal. According to the servo communication standard the period of the command signal should be $\boldsymbol{T_{pwm}} = \mathbf{20}\ [\boldsymbol{ms}]$. Since the correctness of the servo commands depend critically on timing it have to be precisely synchronized with the rest of the tasks within the Arduino **loop()** function. The algorithm for synchronization is schematized in **figure 17**.

In the beginning of the **loop()** function all of the servo pins are set to high level by the **digitalWrite()** routine. Then the starting time of this 'high' event is stored in the variable **start_time**. Then while all servo command pins are help high the joint configuration is updated if required with a function **Calculate_Servo_Commands()**.The updated servo pulse width are stored in variables $\boldsymbol{u_{i,j}}$. They are consequently filled in the **array_pulse[]** array. Another array named **array_pins[]** will held the corresponding to $u_{i,j}$ values pin numbers. These two array are then sorted in an ascending order of the pulse width by the function **Sord_Pins_By_Pulse_Width(pins, angles)**. Then the pin with the shortest pulse width is driven to low state first. The duration of the pause for the consequent pins is calculated as a temporal difference between the pulse widths with the function **First_Difference()**.

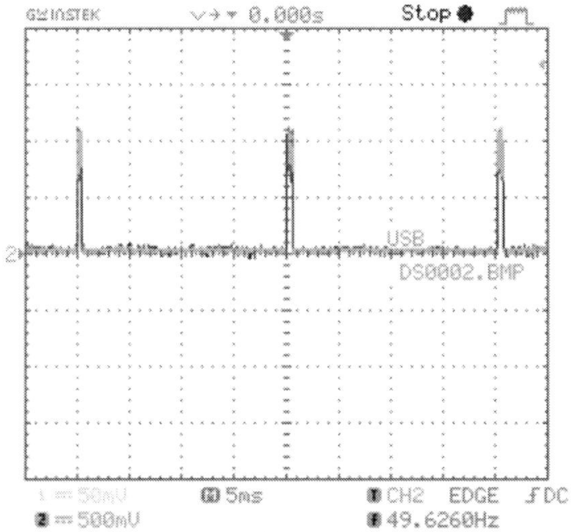

Figure 15. Servo control signal with pulse width of **1 [ms]** corresponding to rotation of **0 [degress]**

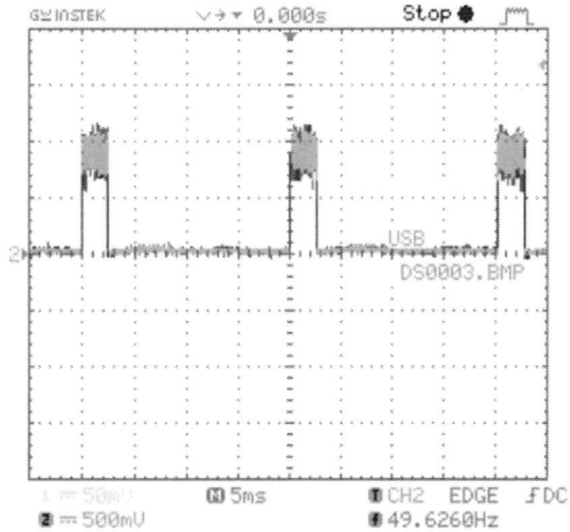

Figure 16. Servo control signal with pulse width of **2 [ms]** corresponding to rotation of **180 [degress]**

```
loop()
{
  digitalWrite(pin_1L,HIGH);
            ...
  digitalWrite(pin_5R,HIGH);
  start_time = Get_Time();
  Calculate_Servo_Commands();
  array_pins[0] = pin_1L; array_pulse[0] = u_1L;
            ...
  array_pins[9] = pin_5R; array_pulse[9] = u_1R;
  Sort_Pins_By_Pulse_Width(array_pins,array_pulse);
  array_pulse_d = First_Difference(array_pulse);
  while (Get_Time() - start_time < array_pulse_d[0]);
  digitalWrite(array_pins[0],LOW);
            ...
  while (Get_Time() - start_time < array_pulse_d[9]);
  digitalWrite(array_pins[9],LOW);
  while (Get_Time() - start_time < 20ms);
}
```

Figure 17. Servo command synchronization algorithm

On **figure 18.** and **figure 19.** the servo control actions from both sides of the robot are presented. From these figures it can be determined how many steps the humanoid robot is making forward and the distance that travels for **50 [s]** can be estimated. Noting that one step ahead is **2.76 [s]** in duration, and the one-step distance is **2.14 [cm]** the distance covered for period of **50 [s]** by the robot which performs **18 steps** is approximately **39 [cm]**.

IOP Conf. Series: Materials Science and Engineering **618** (2019) 012004 doi:10.1088/1757-899X/618/1/012004

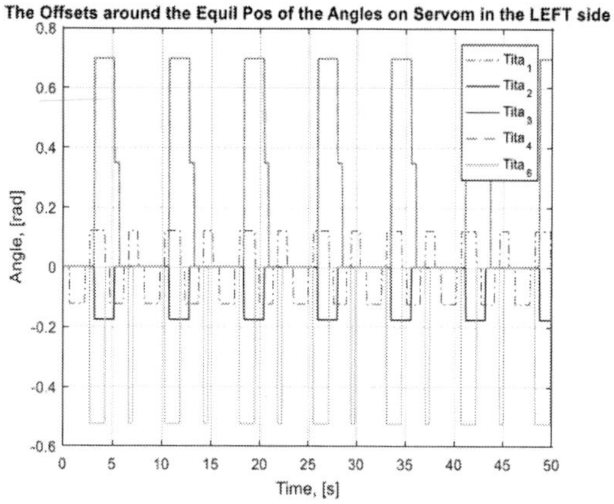

Figure 18. Joint angles in radians of the **left side** of the robot during the walking algorithm

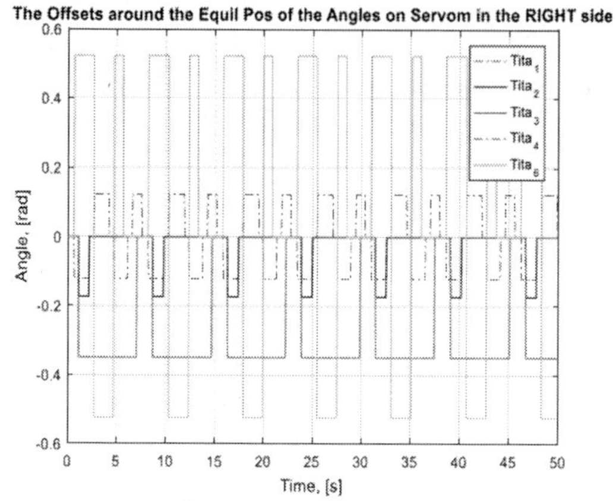

Figure 19. Joint angles in radians of the **right side** of the robot during the walking algorithm

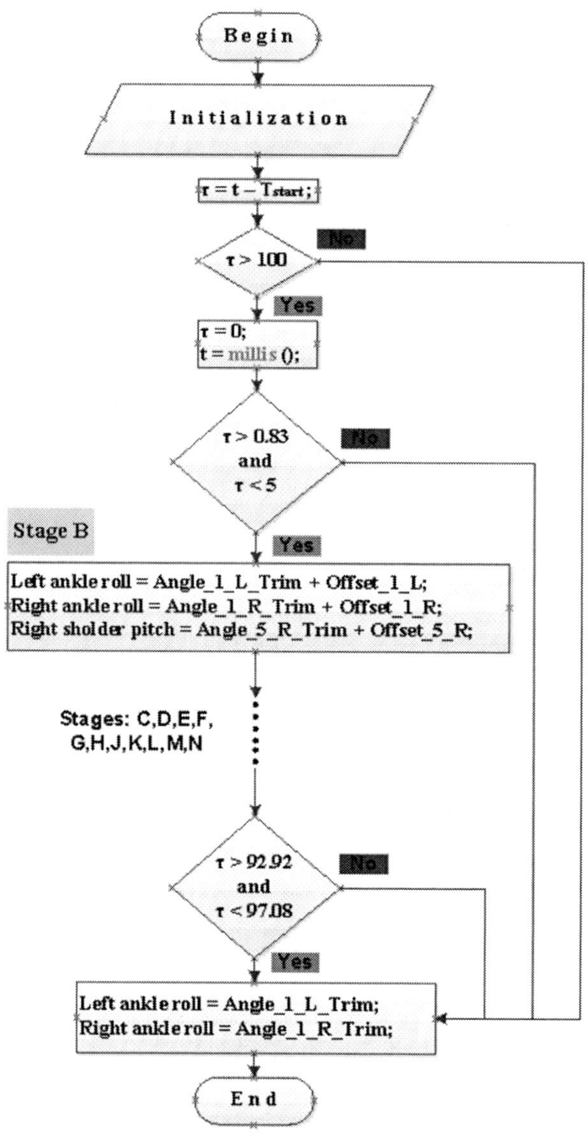

Figure 20. Block scheme of Embedded Walking Algorithm

On **figure 20.** is shown block scheme of part from embedded walking algorithm for biped humanoid robot. As shown, the program starts with initialization of the low level drivers of the servo motors, the input-output ports for pulse width modulated (PWM) signals, the absolute (Trim) angular values, the current values and the offset values over the operating point, the time period for one step $\boldsymbol{\tau}$, and measured time for one step $\boldsymbol{T_{start}}$ and the clock value $\boldsymbol{t}$.

After initializing starts a procedure of comparing $\boldsymbol{\tau}$ with the percentage of the time elapsed for one single step. If the value is bigger than **100 [%]** it is necessary to assign zero values to $\boldsymbol{\tau}$ and $\boldsymbol{t}$. We check then if $\boldsymbol{\tau}$ is between **0.83 [%] AND 5 [%]**. If the compare return flag is **TRUE**, then the Left ankle roll, Right ankle roll and the Right shoulder pitch are rolled with the accordingly offsets. This procedure is according to **Stage "B"**. The other stages are analogic, excluding the percentages for compare, which are different for every stage. Last part of the cycle is to check $\boldsymbol{\tau}$ and if the condition is fulfilled, Left ankle roll and Left ankle roll are the only variables with current values, so it is necessary to assign the absolute values to them.

5. Conclusions

Biped walking robot will have an important part in many future activities of humanity. The paper presented a laboratory model of a humanoid biped platform with 17 degrees of freedom. However, only 10 of the servo motors are necessary for achieving stable process of walking. The paper presented in detail the developed walking algorithm together with its implementation in a commonly used in hobby robotics Arduino platform.

Acknowledge

The authors would like to thank the Research and Development Sector at the Technical University of Sofia for the financial support.

References

[1] J. Chestnutt, M. Lau, G. Cheung, et al., *Footstep Planning for the Honda ASIMO Humanoid*, International Conference on Robotics and Automation (ICRA), 2005 Barcelona, Spain, doi: 10.1109/ROBOT.2005.1570188

[2] Yu Ogura, H. Aikawa, K. Shimomura, et al., *Development of a new humanoid robot WABIAN-2*, International Conference on Robotics and Automation (ICRA), 2006 Orlando, USA, doi: 10.1109/ROBOT.2006.1641164

[3] K. Kaneko, K. Harada, F. Kanehiro, et al., *Humanoid robot HRP-3*, International Conference on Robotics and Automation (ICRA), 2008 Nice, France, doi: 10.1109/IROS.2008.4650604

[4] F. Plestan, J.W. Grizzle, E.R. Westervelt, et al. *Stable walking of a 7-DOF biped robot*, IEEE Trans on Robotics and Automation, vol 14(4), 2013, doi: 10.1109/TRA.2003.814514

[5] M. Gienger, K. Loffler, F. Pfeiffer, *Towards the design of a biped jogging robot*, Int. Conf. on Rob. and Autom. (ICRA), 2001 Seoul, South Korea, doi: 10.1109/ROBOT.2001.933265

[6] D. K. Pratihar, L. C. Jain, *Intelligent Autonomous Systems: Foundations and Applications*, Springer 2010, ISBN 9783642116759

IOP Conf. Series: Materials Science and Engineering **618** (2019) 012005 doi:10.1088/1757-899X/618/1/012005

Assessment of bias removal techniques used in case of sensor parametric modelling

Pl Nikovski

University of food technologies – Plovdiv, Technical faculty, bul. Mariza 26, Plovdiv, Bulgaria

E-mail: plmnn@abv.bg

Abstract. The study assesses how two techniques for bias removing influence the parameterization of a humidity sensor model with known structure. Experimental data consist of SHT 31-DIS sensor responses to a step change of humidity from 12 %RH to 65 %RH. The test performed after parameter estimation shows that the time-series subtraction unbiased method produces a model with a better coefficient of determination. This and other results obtained are useful when a data pre-processing for black-box modelling of humidity sensors is used.

1. Introduction

Environmental sensors have been applied to a variety of fields such as industrial and home automation (figure1), environmental monitoring, energy management etc. [1-6]. The increasing advances in CMOS technology have resulted in availability of high accuracy and small size digital sensors used to track temperature, humidity, particulate matter, VOC and CO_2 [7-10]. However, the precise sensor construction is not always a decisive factor when smart devices are designed. Particularly applicable to data mining and machine learning, sophisticated algorithms for analysis or decision making need an adequate mathematical model describing the real-time data processing in the sensor.

Available from the Swiss company Sensirion SHT 31-DIS is a humidity sensor that integrates in the same chip a polymer-based capacitive sensing element and an electronic circuit [11]. Under different conditions polymer material adsorbs or desorbs some amount of water vapor molecules

Figure1. Sensirion SHT31 Smart Gadget [5] for humidity and temperature measurements.

depending on the relative humidity of the surrounding environment. As a result, the relative dielectric permittivity of the polymer and the capacitance of the sensing element are changed. An electronic circuit measures this change in capacitance and after correction and linearization, sampled values of relative humidity can be read through the I²C digital interface. SHT 31-DIS does not require any user calibration and is applicable for measuring humidity from 0 %RH to 100 %RH. In most cases, its error does not exceed ±2 %RH [11].

Due to the complexity of the processes in SHT 31-DIS, the lack of sufficient information about the polymer properties and the electronic circuit, it is impossible to obtain a physics-based sensor model. In this case, it is more appropriate to implement black box modelling [12]. Black box modelling uses experimental input-output data and does not require any knowledge about the internal structure and processes in the sensor. However, its effectiveness strongly depends on the data and their pre-processing. The main purpose of this study is to assess how two techniques for bias removing influence the parameterization of a humidity sensor model.

2. Materials and methods

By experiments, several step responses of the sensor SHT 31-DIS were obtained (output was sampled every 0,3 s): four of them when the input humidity rises rom φ_A to φ_B and four when it drops from φ_B to φ_A. The relative humidities φ_A and φ_B, approximately equal to 12 %RH and 59 %RH, were reproduced by saturated salt solutions. The collected data were used to estimate a sensor model of the following type:

$$\begin{aligned} y[k] = &-a_1y[k-1] - a_2y[k-2] - a_3y[k-3] \\ &+b_1u[k-d-1] + b_2u[k-d-2] + b_3u[k-d-3] \\ &+e[k] + a_1e[k-1] + a_2e[k-2] + a_3e[k-3] \end{aligned} \tag{1}$$

where backward shift operator is denoted by q^{-1}, $u[k]$ and $y[k]$, k=0,1,2,... are samples of the sensor input and output, $\boldsymbol{\theta} = [b_1, b_2, b_3, a_1, a_2, a_3]^T$ is the parameter vector, $e[k]$ is an additive white noise and $d = 1$ is the input-output delay in samples. The order of the difference equation (1) and time delay are based on results of the preliminary data analysis.

Usually immediate use of the acquired data to build the model is not recommended. In addition to the noise, many factors such as outliers, missing values, offsets, drifts etc. degrade the quality of the data and may have a significant negative effect on the estimation process [13, 14]. Pre-processing of the raw data is an important step that allows many difficulties to be avoided later. The visual inspection of the data records shows that the number of gross errors like outliers and missing values is negligible (less than 10^{-5} %) and therefore a simple strategy for their correction is applied: a destroyed value is replaced by the previous neighboring value. The situation with the other impacts is considerably more complicated. It can be seen in figure 2a that under the repeatability experimental conditions the initial values of separate realization are different (this is also true for the final values in figure 3a). These values involve not only the true steady state sensor response but also the low-frequency disturbances, drifts and offsets from both the sensor and the experimental setup. By removing the bias from the experimental data, most of these unwanted components can be eliminated. Data, treated in this way, will no longer contain information about the actual values of the sensor steady states. But in most cases this does not affect seriously the estimation procedures.

3. Bias removing and sensor model estimation

Many approaches are used to remove the bias from the experimental data but two of them are fundamental [13, 14]. When the input bias u_B and the output bias y_B are known, they can just be subtracted from the experimental raw data $u[k]$ and $y[k]$. Now model parameters are determined by using deviations:

$$\begin{aligned} &u[k] - u_B \\ &y[k] - y_B \end{aligned} \tag{2}$$

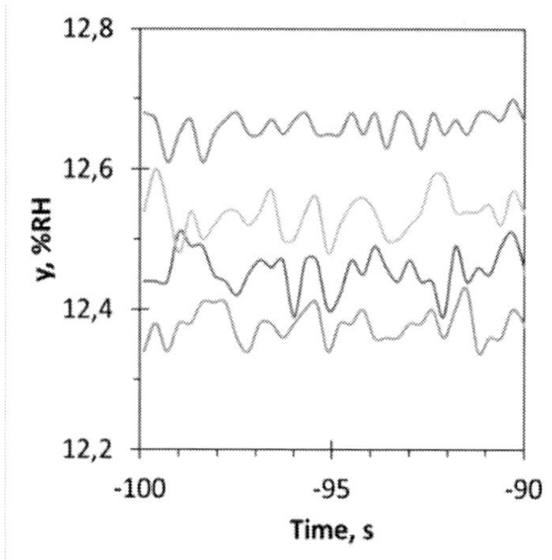

(a)

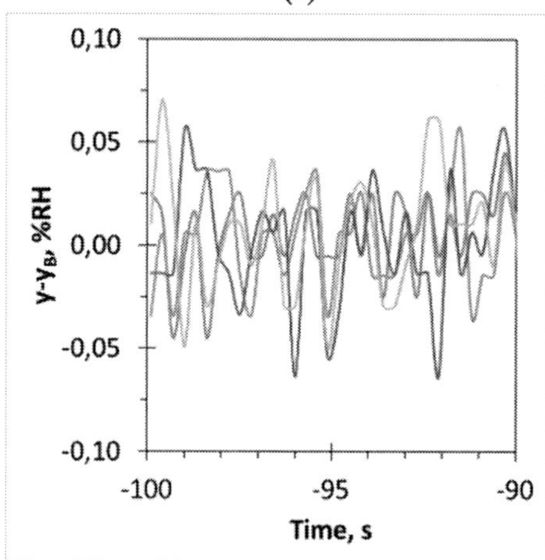

(b)

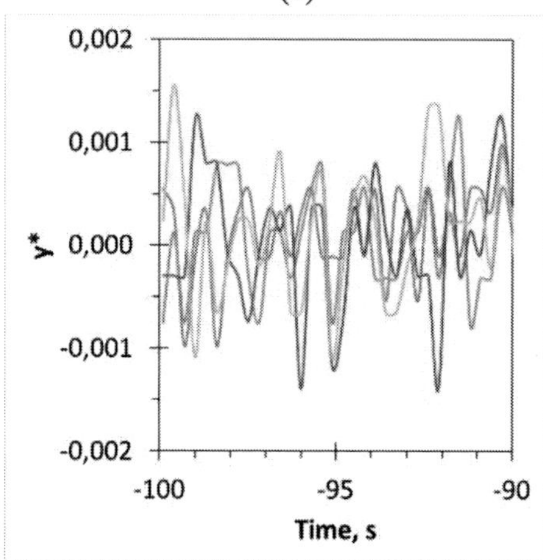

(c)

Figure 2. Raw (a), unbiased (b) and scale unbiased (c) output experimental data before input rise.

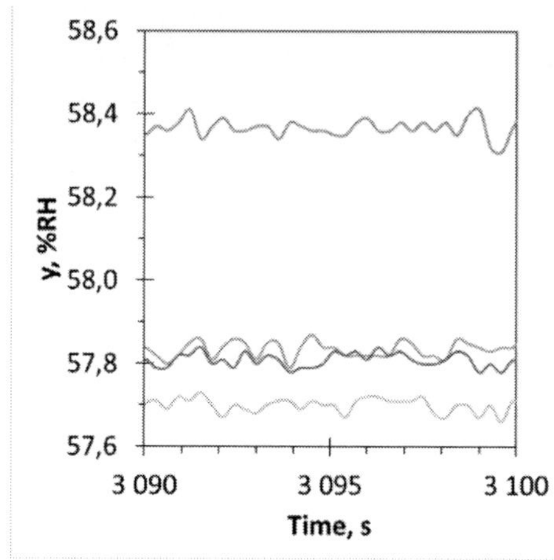

(a)

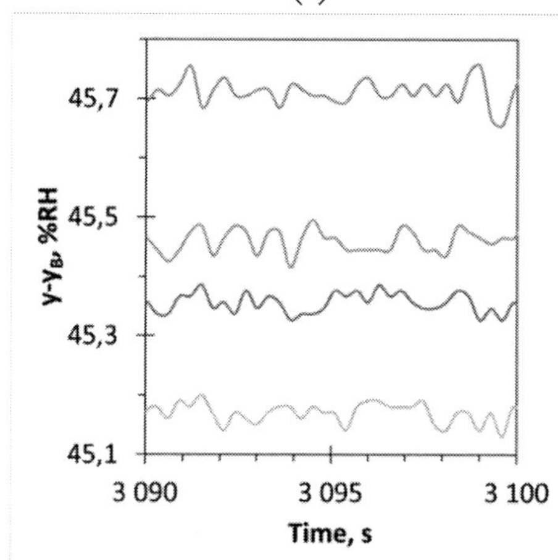

(b)

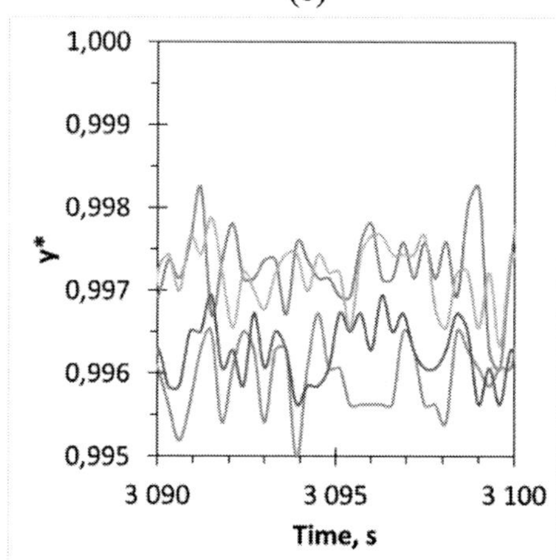

(c)

Figure 3. Raw (a), unbiased (b) and scale unbiased (c) output experimental data after input rise.

For a linear time-invariant object the use of absolute value $u[k]$ and $y[k]$ or deviations (2) will lead to the same parameter vector $\boldsymbol{\theta}$. Removing the bias as described above zeroing the initial data values.The final values also change. As can be seen from figure 3b they continue to be different from each other and the unbiased step responses will have varying heights. When the model parameters are found by minimizing the mean squared error (i.e. using the prediction error method) this is not desirable because the individual responses will have a different weight in the estimation. To solve this problem, unbiased responses are scaled:

$$u^*[k] = \frac{u[k] - u_b}{u[\infty] - u_b} \qquad y^*[k] = \frac{y[k] - y_b}{y[\infty] - y_b} \tag{3}$$

where $u[\infty]$ and $y[\infty]$ are the final values after a sufficiently long time. In the case being considered here, the biases and the final values are calculated as sample means from groups of 60 observations at the beginning and the end of each raw step response. Taking into account equations (2) and (3), the model (1) can be further re-written as

$$\begin{aligned} y^*[k] = &-a_1 y^*[k-1] - a_2 y^*[k-2] - a_3 y^*[k-3] \\ &+b_{1s} u^*[k-d-1] + b_{2s} u^*[k-d-2] + b_{3s} u^*[k-d-3] \\ &+C(e[k] + a_1 e[k-1] + a_2 e[k-2] + a_3 e[k-3]) \end{aligned} \tag{4}$$

where

$$C = \frac{u[\infty] - u_b}{y[\infty] - y_b} \tag{5}$$

$$b_{is} = C b_i, \; i = 1,2,3. \tag{6}$$

The unbiased and scaled sensor data are shown in figure 2c and figure 3c.

The implementation of the second approach does not require prior knowledge of the bias. Static components are removed by a special case of high-pass filtering realized by using the differences

$$\begin{aligned} u[k] - u[k-1] &= (1 - q^{-1}) u[k] \\ y[k] - y[k-1] &= (1 - q^{-1}) y[k] \end{aligned} \tag{7}$$

instead of the variables $u[k]$ and $y[k]$. Scaling the new variables for the reasons previously discussed,

$$u^{**}[k] = \frac{u[k] - u[k-1]}{u[\infty] - u_b} \qquad y^{**}[k] = \frac{y[k] - y[k-1]}{y[\infty] - y_b} \tag{8}$$

and having in mind the linearity of the difference equation (1), we obtain a model with the same structure as (4):

$$\begin{aligned} y^{**}[k] = &-a_1 y^{**}[k-1] - a_2 y^{**}[k-2] - a_3 y^{**}[k-3] + b_{1s} u^{**}[k-d-1] \\ &+ b_{2s} u^{**}[k-d-2] + b_{3s} u^{**}[k-d-3] \\ &+ C(e[k] + a_1 e[k-1] + a_2 e[k-2] + a_3 e[k-3]) \end{aligned} \tag{9}$$

Parameterization of the model can be done in several different ways [13, 14]. According to the Prediction error method (PEM), the "best" parameters will be those, by means of which the output can be accurately predicted at any moment of time. Most often, an estimate of $\boldsymbol{\theta}$ (denoted by $\widehat{\boldsymbol{\theta}}$) is determined by minimizing the sum of the error squared between the pre-processed $y^*[k]$ and one-step ahead predicted value

$$\hat{y}^*[k|k-1] = -a_1 y^*[k-1] - a_2 y^*[k-2] - a_3 y^*[k-3] + b_1 u^*[k-d-1] + b_2 u^*[k-d-2] + b_3 u^*[k-d-3] \tag{10}$$

over a set of N samples:

$$\hat{\theta}^* = \underset{\theta}{argmin} \sum_{k=0}^{N-1} (y^*[k] - \hat{y}^*[k|k-1])^2 \tag{11}$$

The difference

$$\varepsilon^* = y^*[k] - \hat{y}^*[k|k-1] \tag{12}$$

is known as a prediction error. An advantage of PEM is its ability to find good approximation when there is a mismatch between the user defined model and the real dynamics of the studied system [13, 14]. The solving equation (11) with $N-1 = 10000$ yields the model

$$y^*[k] = 2{,}917y^*[k-1] - 2{,}834y^*[k-2] + 0{,}918y^*[k-3] + 0{,}066u^*[k-2] - 0{,}131u^*[k-3] + 0{,}065u^*[k-4] + e[k] - 2{,}917e[k-1] + 2{,}834e[k-2] - 0{,}918e[k-3] \tag{13}$$

Using the same approach and taking into account equations (7) - (9), the second model was also evaluated:

$$y^{**}[k] = 2{,}930y^{**}[k-1] - 2{,}859y^{**}[k-2] + 0{,}930y^{**}[k-3] + 0{,}060u^{**}[k-2] - 0{,}120u^{**}[k-3] + 0{,}060u^{**}[k-4] + \frac{1}{1-q^{-1}}(e[k] - 2{,}930e[k-1] + 2{,}859e[k-2] - 0{,}930e[k-3]) \tag{14}$$

or after some transformation

$$y^{**}[k] = 3{,}930y^{**}[k-1] - 5{,}789y^{**}[k-2] + 3{,}789y^{**}[k-3] - 0{,}930y^{**}[k-4] + 0{,}060u^{**}[k-2] - 0{,}180u^{**}[k-3] + 0{,}180u^{**}[k-4] - 0{,}060u^{**}[k-5] + e[k] - 3{,}930e[k-1] + 5{,}789e[k-2] - 3{,}789e[k-3] + 0{,}930e[k-4] \tag{15}$$

The impact of the compared methods for bias removing on the quality of the sensor model can be assessed in different ways [13, 15, 16]. The figure 4 and figure 5 show the prediction error in both cases, calculated on fresh data not used in the identification process.

When data are preprocessed by the time-series subtraction method, the maximum prediction error does not exceed 0,028. The model, obtained by the time-series differencing method has approximately the same magnitude of error: -0,031. Another widely used metric in model estimation is the coefficient of determination R^2:

$$R^2 = 1 - \frac{\sum_{k=0}^{N-1} \varepsilon^2[k]}{\sum_{k=0}^{N-1} (y[k] - \bar{y})^2} \tag{16}$$

where $\bar{y}$ is the mean value of $y[k]$. R^2 is a number, usually between 0 and 1. In an ideal case, i.e. when $\hat{y}[k] = y[k]$, $R^2 = 1$ and it decreases as the discrepancy between the original system and the model output increases. In the first case the coefficient of determination is $R^2 = 0{,}99$ and in the second case $R^2 = 0{,}81$: the first model fits the data better than the second one. In addition, its structure is simpler (see equations (13) and (15)).

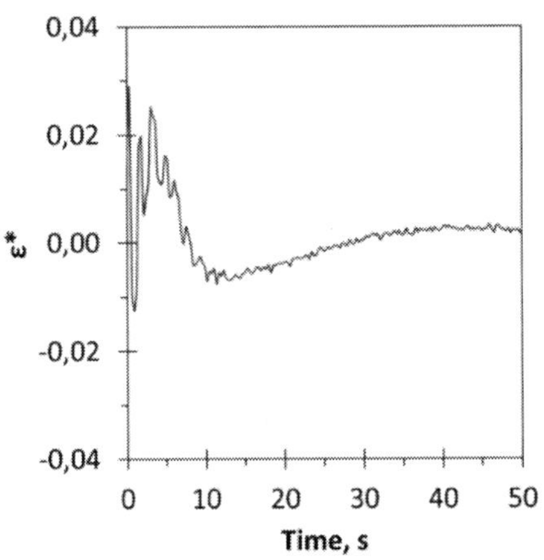

Figure 4. Prediction error of time-series subtraction method.

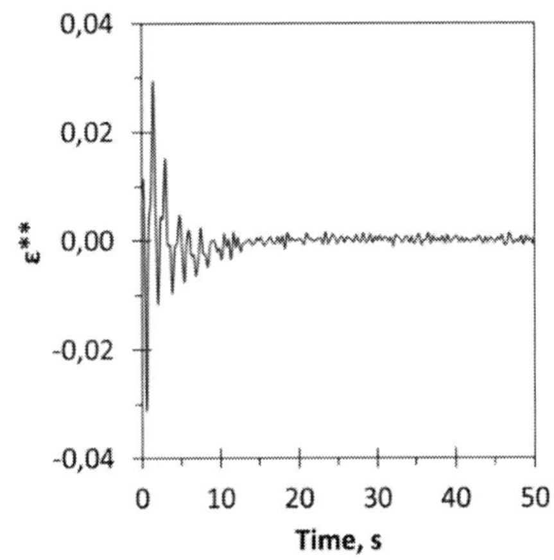

Figure 5. Prediction error of time-series differencing method.

4. Conclusion

This paper discusses a case where the parameters of a humidity sensor SHT 31-DIS discrete model with a known structure are estimated applying two different unbiasing techniques. Experimental data consist of sensor responses to a step change of humidity from 12 %RH to 65 %RH. Separate realizations have some bias and two techniques have been applied to remove it. In the first case the bias stated in the raw data before input changing is subtracted and after that the obtained variation is scaled. In the second case, the raw data are first differentiated and after that they are scaled. Replacing the original variables with pre-processed variables yields the new model with unbiased input and output. Its parameterization is done by the prediction error method. After estimation, a test was conducted with fresh data not used in the identification process. The coefficients of determination for the time-series subtraction and time-series differencing unbiased methods were $R^2 = 0{,}99$ and $R^2 = 0{,}81$ respectively. Therefore, in cases when step response data are used to parametrize a model with the considered structure it is more appropriate to apply the first approach for bias removing.

References

[1] Hancke G, Silva B and Hancke Jr G 2013 The role of advanced sensing in smart cities. *Sensors* **13**, 393-425
[2] Kim G, Son Y, Lee J et al. 2016 Air Pollution Monitoring and Control System for Subway Stations Using Environmental Sensors. *Journal of Sensors* **2016**
[3] Nedev N, Nenova Z and Ivanov S 2014 Virtual instruments for humidity and temperature measurements. *Information Technology Based Higher Education and Training*, York pp 1-6
[4] Kumar A and Hancke G 2014 An Energy-Efficient Smart Comfort Sensing System Based on the IEEE 1451 Standard for Green Buildings. *IEEE Sensors Journal* **14** 4245-52
[5] Sensirion 2015 SHT31 Smart Gadget User Guide. (Revision 1) p11
[6] Nenova Z and Dimchev G 2013 Compensation of the Impact of Disturbing Factors on Gas Sensor Characteristics. *Acta. Polytech. Hung.* **10** 97–111
[7] Ueno K, Hirose K, Asai T and Amemiya Y 2006 Ultralow-Power Smart Temperature Sensor with Subthreshold CMOS Circuits. *International Symposium on Intelligent Signal Processing and Communications* (Tottori) pp 546-549
[8] Passeri D et al. 2003 High-resolution CMOS particle detectors: design and test issues. *IEEE Nuclear Science Symposium* (Portland) pp 331-334

[9] Maeng S, Guha P, Udrea F et al. 2008 SOI CMOS-based smart gas sensor system for ubiquitous sensor networks. *ETRI Journal* **30** 516–525
[10] Nizhnik O, Higuchi K and Maenaka K 2012 Self-calibrated humidity sensor in CMOS without post-processing. *Sensors* **12** 226–232
[11] Sensirion 2019 *Datasheet SHT3x-DIS Humidity and Temperature Sensor*. (Version 6) p 22
[12] Ljung L 2001 Black-box models from input-output measurements. *IMTC 2001* (Budapest) pp 138-146
[13] Ljung L 2014 *System identification: Theory for the user*. (Prentice Hall) p 640
[14] Tangirala A 2014 *Principles of system identification: Theory and practice*. (NW: CRC Press) p 908
[15] Botchkarev A 2018 Performance Metrics (Error Measures) in Machine Learning Regression, Forecasting and Prognostics: Properties and Typology. (*Available online:* https://arxiv.org/abs/1809.03006 accessed on 9 September 2018)
[16] Tian Y, Nearing G, Peters-Lidard C et al. 2016 Performance Metrics, Error Modeling, and Uncertainty Quantification. *Monthly Weather Review* **144** 607-613

Evolution of entanglement in quantum neural network

Ufuk KORKMAZ, Deniz TÜRKPENÇE, Tahir Çetin AKINCI*, Serhat ŞEKER

Department of Electrical Engineering, Faculty of Engineering, Istanbul Technical University, Istanbul, Turkey

*Correspondence: akincitc@itu.edu.tr

Abstract. This study concerns with the evolution of entanglement in a quantum neural network (QNN) model that is locally in contact with data environments. As a valuable resource, duration of entanglement in quantum systems is extremely important. Therefore, the effect of various initial states on the occurrence or decay of entanglement are examined in the presence of information reservoirs. In this study, central spin model was investigated as a quantum version of neural networks inspired by biological models. The architecture of the model is based on a central spin system with two nodes where the nodes are coupled to independent spin baths. Numerical results show that initial state preparation has a profound effect on the fate of entanglement. The results show that the entanglement lifetime can be adjusted by engineering the reservoir states as well as the initial states of the system of interest. The results can be used to improve the performance of the formation or distribution of entanglement in realistic communication network states.

1. Introduction

In popular discussions on artificial intelligence, especially artificial neural networks (ANN) and machine learning have been the focus of interest. Machine learning, which is a computer science area followed by the beginning of the ANN, provides the ability to learn to computers without being programmed explicitly [1-3]. Artificial neural networks are interconnected computing structures composed of a set of layers that process and yield output for corresponding input information. The basic unit of a neural network is a perceptron [3] with a single output node connected to several inputs. The mathematical structure of a perceptron was developed by Rosenblatt inspired by the work of McCulloch and Pitts presents a model [2] mimics the functionality of a biological neuron. The Hebb's learning rule was also developed regarding the biological context in which specifies that how much weight of a connection between connected units should be adjusted in proportion to their activations [4]. Classical learning rules govern the dynamics in a statistical information environment defined in terms of probability density functions [5]. The formulations and constraints of learning laws therefore depend on the relationships between the global and local information environments of each transaction item.

Networks formed of quantum systems are known as quantum networks that are important for quantum communication and the distribution of quantum resources [6]. Therefore, the idea of the use of quantum systems as neural networks [7, 8] has triggered various studies [9-14], including attempts to implement QNN suggestions in the context of quantum computations [14-18]. In particular, quantum resources make the scheme more attractive over classical communication tasks. In this context, entanglement is one of the most interesting features of quantum mechanics [19-21].

Therefore, understanding their characteristics and dynamics is very important for a number of applications in quantum mechanics. In addition, quantum cryptography, quantum information processing, secure communication, the transmission of unknown quantum data and quantum measurement are the examples in which the entanglement has the central role [20-24]. Because of the fragile nature of the entanglement, it is a challenge to maintain and distribute entanglement for a sufficient amount of time.

When the system units carrying entanglement content get in contact to environmental degrees of freedom, entanglement decays to zero in finite time. This situation is generally known as entanglement sudden death (ESD) [20, 25-27] and has also been observed experimentally [28]. As we mentioned earlier [21], in the case of Markovian reservoirs, the dynamics of entanglement in the common and independent reservoirs varies dramatically depending on the system initial conditions. In the presence of a common reservoir, the entanglement can be lost for a finite time, and then it may revive again. This is because the common reservoirs tend to create entanglement rather than completely destroy it, since they indirectly combine the qubits. Even if qubits are initially prepared in a factorized situation, the correlation created by the environment may lead to a phenomenon known as entanglement sudden birth (ESB) [20, 29].

This study focuses on entanglement dynamics based on a small-scale QNN architecture [30]. The dynamic behaviour of quantum entanglement under certain assumptions in the QNN architecture should be reassessed for the sake of exploiting the quantum resources, although the dynamics of the entanglement is well studied for a few qubits scale. In addition, due to the rapid progress of technology, the production of entanglement on demand was achieved faster than the entanglement decay [31]. More specifically, we investigate the dynamics of a small QNN unit that contacts a quantum information reservoir. The concept of the information reservoir was originally introduced in a classical style and was studied for quantum systems [21]. We simulate the proposed quantum system up to three input nodes with different states and examine the open system dynamics. The present study is related to the numerical simulation of a dynamically evolution of an open QNN that is initially entangled and initially separable. The basic assumption of the manuscript is that the QNNs are open quantum systems and they exist in the information environments. Therefore, the aim of the study is to examine the dynamics of entanglement based on quantum data of QNNs. Open system dynamics are modelled through repeated interactions [32, 33] between the local nodes of the network and the units representing the reservoir. This type of repeated interactions has been shown to be dynamic maps equivalent to the Markovian master equation approach due to the divisibility of quantum channels [34]. We have found that the dynamics of entanglement is strongly related to the initial state of the quantum network. Both asymptotic decay and the decay of entanglement in finite time were observed due to the diversity in environmental states. The following sections are devoted to describe the model and the results of the dynamics.

2. Model and system dynamics

Studies aimed at the application of advanced cognitive tasks revealed neural networks [5]. The simplest neural network unit is a perceptron which is a data classifier with $x_1, x_2, \cdots, x_N$ input nodes connected to an output node with corresponding adjustable weights $w_1, w_2, \cdots, w_N$. The output node experience the weighted linear summation of the input information as $y = \sum_i x_i w_i$ modulated by an activation function $f(y)$ which returns an output depending on the value of y.

The QNN model we investigated is equivalent to the central spin model (also known as the spin-star network) with various applications such as of quantum coherence [35] or quantum communication [36]. Previously, we investigated the dynamic evolution of the central spin quantum coherence based on different spin coupling types under Markov Dynamics [37]. And also, we demonstrated that the non-equilibrium nature of the model may have advantages on a quantum scale for the harvesting of thermodynamic work [38]. In this study, central spin model was investigated as a quantum version of neural networks inspired by biological models (Figure 1 (a)). Figure 1 (b) summaries the basic ideas of the proposed model dynamics provided that the system is an interacting spin system coupled to two independent spin baths locally. Therefore, the system of interest is an open quantum system and consists three interacting qubits with a flip-flop type Hamiltonian. The time independent Hamiltonian representing the system dynamics as,

$$H = \frac{\omega}{2}\left(\sum_{i}^{N} \sigma_i^z + \sum_{i=1}^{N-1} \sigma_{u_i}^z\right) + \left(\sum_{i=1}^{N-1} J_i \sigma_i^+ \sigma_{out}^- + \sum_{i=1}^{N-1} J_i \sigma_{u_i}^+ \sigma_i^- + H.c.\right) \quad (1)$$

which is plausible for quantum effects also in biological systems [39]. Here, $\sigma^z, \sigma^-$and σ^+ are the Pauli-z, lowering and raising operators for the qubits, respectively. Also, σ_{out}^- is the respective Pauli operator for output qubit and J_i is the coupling coefficient between input units and the output node, σ_i is the Pauli operator representing an individual node in contact with the information reservoir and σ_{u_i} is the Pauli operator of the individual unit representing the information reservoir. The Bohr frequency ω of each spin and single qubit information unit has been taken equal for simplicity. We employ density matrix formalism in our study to represent the quantum states of the system in question. The quantum neurons are initially assumed to be in a product state as $\rho(0) = \rho_{\text{Res}} \otimes \rho_{\text{Sys}}(0)$, where $\rho_{\text{Res}} = \rho_{\text{Res1}} \otimes \rho_{\text{Res2}}$ and $\rho_s(0) = \rho_1(0) \otimes \rho_2(0) \otimes \rho_{\text{out}}(0)$ are Reservoir units and System units, respectively. Individual qubit states were chosen as $\rho(0) = |\uparrow\rangle\langle\uparrow|$ in order to initially provide a blank memory. Here, $\rho_{up} = |\uparrow\rangle\langle\uparrow|$, $\rho_{dn} = |\downarrow\rangle\langle\downarrow|$, $|\mp\rangle = \frac{1}{\sqrt{2}}\big((|\uparrow\rangle \mp |\downarrow\rangle)\big)$ are the cat states where $|\uparrow\rangle$, $\langle\downarrow|$ are orthogonal spin states known as computational basis in quantum computing language. We define maximally entangled Bell states in the generic form as $|\Psi^{\mp}\rangle = \frac{1}{\sqrt{2}}\big((|\uparrow\rangle \otimes |\downarrow\rangle) \mp (|\downarrow\rangle \otimes |\uparrow\rangle)\big)$ or $|\phi^{\mp}\rangle = \frac{1}{\sqrt{2}}\big((|\uparrow\rangle \otimes |\uparrow\rangle) \mp (|\downarrow\rangle \otimes |\downarrow\rangle)\big)$.

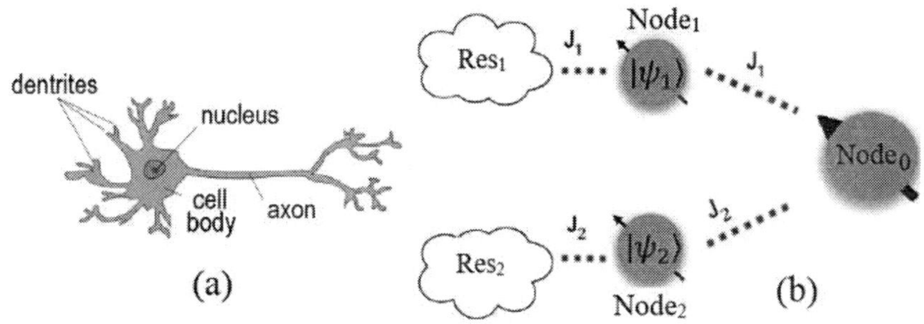

Figure 1. Comparison of a biological neuron (a) and the open quantum system we investigate (b). Dendrites of a cell receive signals from the connected cells to the cell body. On the other hand, in our model, the open quantum system receives information from the reservoirs in which it is connected. The open quantum system is composed of three qubits with two inputs nodes ($Node_1$ and $Node_2$) and an output node ($Node_0$). For simplicity, each input node is coupled to the corresponding reservoir and to the output node with equal coupling strengths J_1 and J_2. All the nodes have an initial quantum state with an entanglement between any two of them and the evaluation of entanglement in time domain is the goal of the current study.

As mentioned above, the open system dynamics are discussed by adopting a repetitive interaction process, also known as a collision model [40]. Figure 1 (b) simulates the open dynamics of the network to illustrate the specific use of the collisional model. In our model, the initially and identically arranged qubit states interact in turn with the nodes of the neural network. The dynamical evolution of the QNN would be obtained by tracing out the reservoir (environment) degrees of freedom as

$$\rho_s(t+\tau) = \text{Tr}_{u_i}\left[U(\tau)\rho_s(t) \otimes \rho_{u_i} U(\tau)^\dagger\right]. \quad (2)$$

Here, $U(\tau) = e^{-iH\tau}$ is the unitary operator representing the system plus environment dynamics for time independent Hamiltonians, τ is the duration of each unit-node interaction and Tr_{u_i} stands for partial trace over environmental degrees of freedom. Thus, the open quantum evolution was simulated by the discrete steps of the quantum channels represented by Equation 2. The calculations were performed by exact diagonalization. In this model, $\rho_{\mathrm{Res}} = \rho_{u_i}$ represents the quantum state of each single qubit environment unit and ρ_s the state of the system of interest (QNN). Also, the state vector and the density matrix formalism were used interchangeably to represent the quantum states of the respective systems.

The aim of our study is to examine the fate of entanglement for various initial states. Therefore, we investigate two situations. The first is that there is no initial entangled between the readout node and one of the input nodes in contact with the environment. In the second case, the system starts in one of the maximally entangled Bell states.

After the system is interacted with the information reservoir, the environment is expected to send information to the system, i.e. the system is equilibrated by the environment carrying quantum information content.

The concurrence is a scalar function that is used to measure the entanglement of density matrices that initially define the mixed states of a system [41]. In this study concurrence is accepted as a measure of entanglement and calculated as $C = \max[0, \lambda_1 - \lambda_2 - \lambda_3 - \lambda_4]$ where λ_i are the square roots of non-Hermitian matrix $\rho\tilde{\rho}$, ρ is the density matrix to be calculated and $\tilde{\rho}$ is its spin flipped from such as $\tilde{\rho} = (\sigma_y \otimes \sigma_y)\rho^*(\sigma_y \otimes \sigma_y)$ where ρ^* is the complex conjugated from of ρ, the descending order [42].

3. Results

To understand the field of open quantum systems, quantum control [43,44] is important to protect the system from unwanted environmental noise [45] or engineering system environmental interactions [46]. Especially small quantum systems may be monitored by any quantum observable, given the mild assumptions on initial conditions of the system that relaxes to a steady value in the long term limit [47,48]. In this section, we analyze the QNN as a system that interacts weakly with the information reservoir in the Markov approach. In these interactions, the quantum state of the system is lost irreversibly through reservoir degrees of freedom, and the temporal evolution of the system depends only on the current state of the system.

In our case, a quantum information reservoir interacts weakly with the local subsystems (input nodes) of the QNN unit. Particularly, time-dependent entanglement dynamics of the discussed QNN is investigated in an information environment. Information environments can be represented by $|\uparrow\rangle\langle\uparrow|$ and $|\downarrow\rangle\langle\downarrow|$ that are single qubit pure states.

As mentioned above, the initial quantum state of the environment plus system is represented by $\rho(0) = \rho_{\mathrm{env}} \otimes \rho_{\mathrm{sys}}(0)$ where$\rho_{\mathrm{sys}}(0) = \rho_1(0) \otimes \rho_2(0) \otimes \rho_0(0)$. Here, ρ_1, ρ_2 are the quantum states of the input nodes ($Node_1$, $Node_2$) and ρ_0 stands for the output node ($Node_0$). On the other hand, quantum state of the environmental qubit states are $\rho_{\mathrm{env}} = \rho_{\mathrm{Res1}} \otimes \rho_{\mathrm{Res2}}$ and fixed. In the following, the time evolution of concurrence between two specific qubits of the three-qubit QNN system will be analysed for various initial states.

Figure 2 represents the evolution of concurrence for the absence and the presence of the initial entangled of the quantum neuron depending on the number of collisions (nc). Figure 2 a) depicts the evolution of concurrence for separable initial states. More specifically, the initial system plus environment state is $\rho(0) = \rho_{up} \otimes \rho_{\mathrm{up}}(0) \otimes \rho_{\mathrm{dn}}(0) \otimes \rho_{\mathrm{dn}}(0) \otimes \rho_{|+\rangle}(0)$ for Figure 2 a). Despite there is no initial entangled between the nodes of the QNN, during the early evolution of concurrence a rapid formation of entanglement between the nodes is evident. The observed entanglement rapidly decays as the system evolution continues. More interestingly in the later evolution, a sudden birth of entanglement between $Node_1$ and $Node_2$ was observed where the nodes do not directly interact.

Conversely, the evalution in Figure 2 b) corresponds to the initial $\rho_{sys}(0) = \rho_{\Psi^-} \otimes \rho_2$ where ρ_{Ψ^-} is the initially entagled state between $Node_1$ and $Node_0$. Here, $\rho_2(0) = |+\rangle\langle+|$ and both the reservoir states are given $\rho_1,\ \rho_2 = |\uparrow\rangle\langle\uparrow|$. As obvious in Figure 2 b), initial entangled decays exponentially and oscillatory. Also formation of entanglement between $Node_1$ and $Node_2$ was observed from the beginning of the evolution this time. Plots with smaller timesclaes can be found as the insets of Figure 2 a)-b). In Figure 2 c) the initial states of the system is given as $\rho_{sys}(0) = \rho_{\phi^+} \otimes \rho_2(0)$ where $\rho_2(0) = |\downarrow\rangle\langle\downarrow|$ and ρ_{ϕ^+} is one of the maximally entangled states described in the introduction. The reservoir states in this case are gives as ρ_{up} and ρ_{dn} respectively. In this case, initial entangled decays fast and end up in finite time [49].

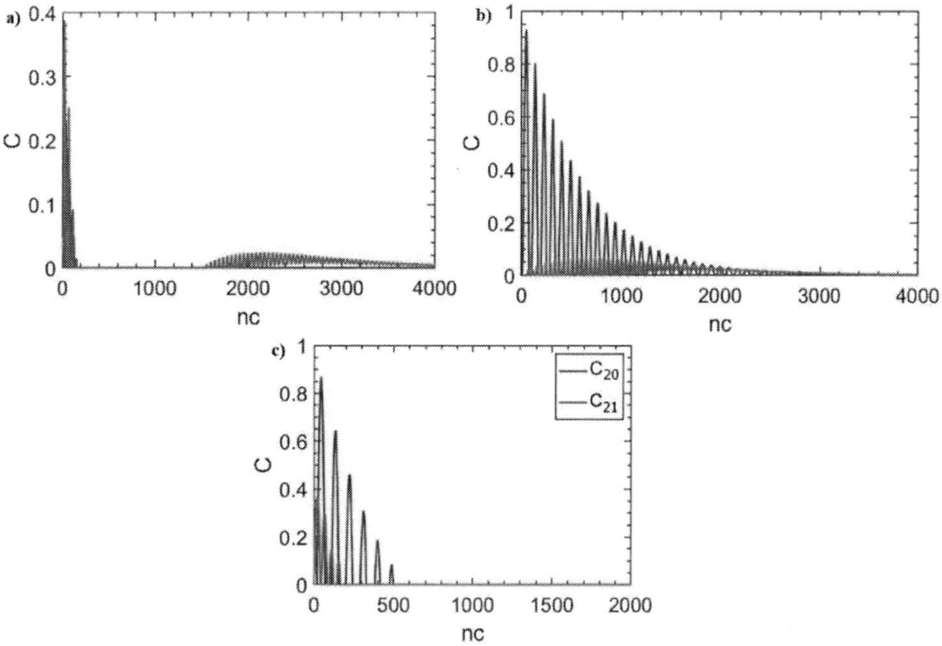

Figure 2. Evolution of concurrence for absence and presence of the initial entangled of the quantum neuron depending on the number of collisions (nc). a) The quantum state of the quantum neuron initially separable as $Node_1 = \rho_{\mathrm{dn}}$, $Node_2 = \rho_{\mathrm{dn}}$ and $Node_0 = \rho_{|+\rangle}$ states. The state of the units representing the environment were set as ρ_{up} for both of them. b) The quantum state of the quantum neuron was $\rho_{|+\rangle}$ state for $Node_2$ while $Node_1$ and $Node_0$ were initially entangled as ρ_{Ψ^-} state. The state of the units representing the environment were set as ρ_{up} for both of them. c) The quantum state of the quantum neuron was ρ_{dn} state for $Node_2$ while $Node_1$ and $Node_0$ *were* initially entangled as ρ_{ϕ^+} state. The state of the units representing the environment were set as $Res_1 = \rho_{\mathrm{dn}}$ and $Res_2 = \rho_{\mathrm{up}}$. The coupling between the environment unit and the input node is equal to the coupling between the input and the readout node $J_1 = J_2 = J = 0.1$. The duration of the each unit interaction τ between the units and the input node is $\tau = 5 \times 10^{-2}/J$.

As in Figure 3, this time we present the disentanglement dynamics of the proposed model for initially maximally entangled $|\phi^{\mp}\rangle$ and $|\Psi^{\mp}\rangle$ states. This time the only examined reservoir states are ρ_{up} pure states for both of the reservoir states in contact with the local nodes of the system. With the same parameters used in Figure 2, entanglement lifetime is longer so that it is not negligible for the $\Psi^{\mp}$ case compared with the $\phi^{\mp}$ case even though the initial states are the same. A rapid formation of entanglement in the very beginning of the evolution between $Node_1$ and $Node_2$ (C_{12}) in which they are initially disentangled is evident. This formation of entanglement rapidly decays while the decay of entanglement between $Node_2$ and $Node_0$ last longer compared to C_{12}. Again a revival scheme is observed in the later evolution for C_{12} but this is much weaker than observed in Figure 2. The most interesting result of the model for the Figure 3 parameters is that a dynamical entanglement swapping occurs between $Nodes_{1-0}$ and $Nodes_{2-0}$. As is clear in Figures 3 b) and d), a swap mechanism between three nodes of the neural network is apparent while the decay of entanglement occurs.

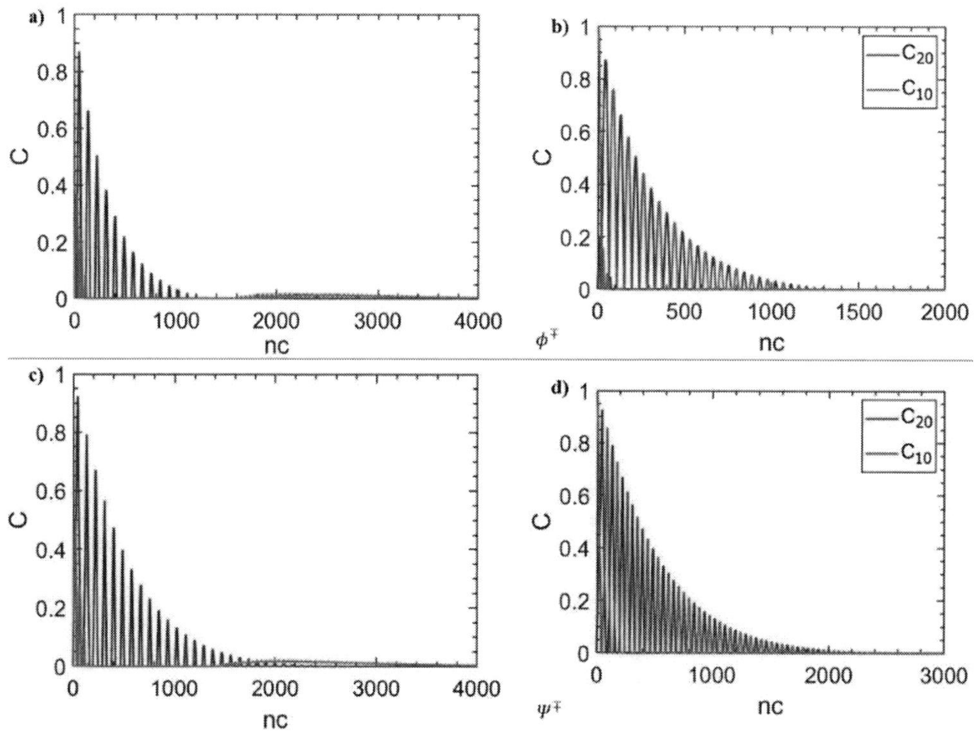

Figure 3. Evolution of concurrence for presence of the initial entanglement of the quantum neuron depending on the number of collisions (nc). a) and b) The quantum state of the quantum neuron was ρ_{dn} state for $Node_2$ while $Node_1$ and $Node_0$ were initially entangled as $\rho_{\phi^{\mp}}$ states. c) and d) The quantum state of the quantum neuron was ρ_{dn} state for $Node_2$ while $Node_1$ and $Node_0$ were initially entangled as $\rho_{\psi^{\mp}}$ states. The state of the units representing the environment were set as ρ_{up} for both of them. The coupling between the environment unit and the input node is equal to the coupling between the input and the readout node $J_1 = J_2 = J = 0.1$. The duration of the each unit interaction τ between the units and the input node is $\tau = 5 \times 10^{-2}/J$.

In conclusion, we have presented the dynamics of entanglement of a simple QNN, locally in contact with quantum data environments, in the time domain. We showed that initial state preparation plays a significant role in the time evolution of the entangled states as well as the reservoir states. Note that as depicted in Figure 2 c) entanglement lifetime is too short compared to the other cases. The reason for this is that two orthogonal reservoir states (spin-up and spin-down) are in contact with the corresponding nodes with equal coupling rates. In Ref. [40] it's shown that, in the steady state limit, a small quantum system in contact with orthogonal reservoir states results with a mixed quantum state causes entanglement sudden death. Though it's not surprising to observe the entanglement decay in finite time in the presence of the environments with mixed states, i.e., thermal environments, we showed that when the entangled nodes experience the linear combination of pure environmental states this affects equivalent to the mixed environmental states. For this reason Figure 2 c) exhibits the worst performance for the corresponding environmental states. We also analyse the dynamical entanglement swapping between the nodes of QNN and observe the generation of entanglement between the nodes in which they are initially not entangled. Figure 3 b)-d) clearly depict the exchange of entangled states during the dynamical evolution. As another result, we observed that a rapid entanglement generation occurs during the dynamical evolution where the nodes are initially separable (see Figure 2 a)-b)). Though this generated entanglement rapidly decays, a revival (or rebirth) of entanglement is evident (Figure 2 a)) only in the case of separable initial states. This finding is important to the cases in which initial entanglement generation is difficult or costly.

There have been proposals aiming efficient distribution of entanglement over quantum networks [50-53]. Optical implementations are particularly promising due to the already existing architecture built for efficient communication, i.e., optical fibers promote flying qubits [53]. Most recently, it's been experimentally shown that the generation of entanglement is possible faster than the decay of entanglement in quantum networks [54]. In that work, it's been pointed out that deterministic distribution of entanglement, generated between distinct NV centres, is possible due to their findings. At this point, our results could contribute deterministic generation of entanglement in terms of preparation initial entangled states as well as coupling or decoupling the system from the reservoirs. Our results could also be used to obtain rebirth of entanglement in case of non-entangled separable initial states.

4. Conclusions

This study, we examined a QNN unit for open quantum system dynamics. More specifically, we considered an interacting spin model as a quantum version of a neural network unit. We examined the QNN unit for open system dynamics and then connect to two reservoirs by a conventional flip-flop Hamiltonian. We examined the open system dynamics for Markov regimes for two input nodes, one output node and two different states of information reservoirs. We adopt a collisional model to simulate open system quantum dynamics. The study also examines the generation of entanglement due to initially separable states and initially entangled states. According to the results obtained from numerical calculations, the dynamic evolution of disentanglement varies dramatically according to the initial situation.

The obtained results show that disentanglement is slower in the presence of pure information reservoirs. We also show that entanglement revival occurs in the separable initial states. Another significant result is that there exist a commutative and periodic swapping of entanglement during the relaxation of the system through reservoir degrees of freedom. The results could be used to improve the schemes to distribute or connect the entanglement resource through quantum versions of artificial networks.

References

[1] Russell S and Norvig P 2009 *Artificial Intelligence: A Modern Approach* (USA: Pearson, Upper Saddle River)
[2] McCulloch W S and Pitts W 1943 A logical calculus of the ideas immanent in nervous activity *Bull. Math. Biophys.* **5** 115-133
[3] Rosenblatt F 1958 The perceptron: A probabilistic model for information storage and organization in the brain *Psychological Review* **65** 386-408
[4] Hebb DO 2002 *The Organization of Behavior: A Neuropsychological Theory* (Mahwah NJ: Psychology Press)
[5] Hecht-Nielsen R1990 *Neurocomputing* (Boston, MA, USA: Addison-Wesley)
[6] Cirac J I, Zoller P, Kimble H J and Mabuchi H 1997 Quantum state transfer and entanglement distribution among distant nodes in a quantum network *Phys. Rev. Lett.* **78** 3221
[7] Lewenstein M and Olko M 1991 'Quantum' neural networks *Network: Computation in Neural Systems* **2** 207-230
[8] Kak S 1995 On quantum neural computing *Information Sciences* **83** 143-160
[9] Lewenstein M 1994 Quantum Perceptrons *J. Mod. Opt.* **41** 2491-2501

[10] Lagaris I E, Likas A and Fotiadis D I 1997 Artificial neural network methods in quantum mechanics *Comput. Phys. Commun.* **104** 1-14
[11] Zak M and Williams C P Quantum Neural Nets *Int. J. Theor. Phys.* **37** 651-684
[12] Narayanan A and Menneer T 2000 Quantum artificial neural network architectures and components. *Information Sciences* **128** 231-255
[13] Ventura D and Martinez T 2000 Quantum associative memory *Information Sciences* **124** 273-296
[14] Gupta S and Zia R K P 2001 Quantum Neural Networks. *J. Comput. Syst. Sci.* **63** 355-383
[15] Panella M and Martinelli G 2011 Neural networks with quantum architecture and quantum learning *Int. Journal of Circuit Theory and Applications* **39** 61-77
[16] Zhou R, Wang H, Wu Q and Shi Y 2012 Quantum Associative Neural Network with Nonlinear Search Algorithm *Int. J. Theor. Phys.* **51** 705-723
[17] Schuld M, Sinayskiy I and Petruccione F 2015 Simulating a perceptron on a quantum computer. *Phys. Lett. A* **379** 660-663
[18] Banchi L, Pancotti N and Bose S 2016 Quantum gate learning in qubit networks: Toffoli gate without time-dependent control *npj Quantum Information* **2** 16019
[19] Horodecki R, Horodecki P, Horodecki M and Horodecki K 2009 Quantum entanglement *Rev. Mod. Phys.* **81** 865
[20] Mazzola L, Maniscalco S, Piilo J, Suominen K A and Garraway B M 2009 Sudden death and sudden birth of entanglement in common structured reservoirs *Phys. Rev. A* **79** 042302
[21] Türkpençe D 2018 Disentanglement Dynamics of a Data Driven Quantum Neural Network *NeuroQuantology* **16** 14-19
[22] Bennett C H, Brassard G, Crépeau C, Jozsa R, Peres A and Wootters W K 1993 Teleporting an unknown quantum state via dual classical and Einstein-Podolsky-Rosen channels *Phys. Rev. Lett.* **70** 1895-1899
[23] Scarani V, Bechmann-Pasquinucci H, Cerf N J, Dušek M, Lütkenhaus N and Peev M 2009 The security of practical quantum key distribution *Rev. Mod. Phys.* **81** 1301
[24] Galindo A and Martin-Delgado M A 2002 Information and computation: Classical and quantum aspects *Rev. Mod. Phys.* **74** 347
[25] Al-Qasimi A and James D F V 2008 Sudden death of entanglement at finite temperature *Phys. Rev. A* **77** 012117
[26] Yu T and Eberly J H 2009 Sudden Death of Entanglement *Science* **323** 598-601
[27] León J and Sabín C 2009 Photon exchange and correlation transfer in atom-atom entanglement Dynamics *Phys. Rev. A* **79** 012301
[28] Almeida M P, de Melo F, Hor-Meyll M, Salles A, Walborn S P and et al 2007 Environment-Induced Sudden Death of Entanglement, *Science* **316** 579-582
[29] Ficek Z and Tanaś R 2008 Delayed sudden birth of entanglement, *Phys. Rev. A* **77** 054301
[30] Schuld M, Sinayskiy I and Petruccione F 2014 The quest for a quantum neural network *Quantum Inf. Process.* **13** (11): 2567-2586
[31] Humphreys P C, Kalb N, Morits J P, Schouten R N, Vermeulen R F and et al 2018 Deterministic delivery of remote entanglement on a quantum network *Nature* **558** 268
[32] Ziman M, Štelmachovič P, Bužek V, Hillery M, Scarani V and et al 2002 Diluting quantum information: An analysis of information transfer in system-reservoir interactions. *Phys. Rev. A* **65** 042105
[33] Scarani V, Ziman M, Štelmachovič P, Gisin N and Bužek V 2002 Thermalizing Quantum Machines: Dissipation and Entanglement *Phys. Rev. Lett.* **88** 097905
[34] Wolf M M and Cirac J I 2008 Dividing quantum channels *Commun. Math. Phys.* **279** 147-168
[35] Zurek W H 2003 Decoherence, einselection, and the quantum origins of the classical *Rev. Mod. Phys.* **75** 715
[36] Chen Q, Cheng J, Wang K L and Du J 2006 Optimal quantum cloning via spin networks *Phys. Rev. A* **74** 034303

[37] Türkpençe D, Akıncı T C and Şeker S 2018 Decoherence in a quantumneural network. *NeuroQuantology* **16** 1-5
[38] Türkpençe D, Altintas F, Paternostro M and Müstecaplıoğlu O E 2017 A photonic Carnot engine powered by a spin-star network *EPL (Europhysics Letters)* **117** (5) 50002
[39] O'Reilly E J and Olaya-Castro A 2014 Non-classicality of the molecular vibrations assisting exciton energy transfer at room temperature *Nature Communications* **5** (3012)
[40] Türkpençe D, Akıncı T C and Şeker S 2019 A steady state quantum classifier *Phys. Lett. A* **13** 1410-1418
[41] Hildebranda R 2007 Concurrence revisited *J. Math. Phys.* **48** 102108
[42] Hill S and Wootters W K 1997 Entanglement of a Pair of Quantum Bits *Phys. Rev. Lett.* **78** 5022
[43] Leibfried D, Monroe C and Wineland D 2003 Quantum dynamics of single trapped ions *Rev. Mod. Phys.* **75** 281-324
[44] Raimond J M, Brune M and Haroche S 2001 Manipulating quantum entanglement with atoms and photons in a cavity *Rev. Mod. Phys.* **73** 565-582
[45] Suter D and Álvarez G A 2016 Colloquium: Protecting quantum information against environmental noise *Rev. Mod. Phys.* **88** 041001
[46] Verstraete F, Wolf M M and Cirac J I 2009 Quantum computation and quantum-state engineering driven by dissipation *N. Phys.* **5** 633–636
[47] Reimann P 2008 Foundation of Statistical Mechanics under Experimentally Realistic Conditions. *Phys. Rev.iew Lett.* **101** 190403
[48] Linden N, Popescu S, Short A J and Winter A 2009 Quantum mechanical evolution towards thermal equilibrium *Phys. Rev. E* **79** 061103
[49] Yu T and Eberly J H 2004 Finite-time disentanglement via spontaneous emission *Phys. Rev. Lett.* **93** 140404
[50] Maître X et. al. 1997 Quantum memory with a single photon in a cavity *Phys. Rev. Lett.* **79** 769–772
[51] Sillanpää M A, Park J I and Simmonds R W 2007 Coherent quantum state storage and transfer between two phase qubits via a resonant cavity *Nature* **449** 438-442
[52] Stephan R et. al. 2012 An elementary quantum network of single atoms in optical cavities *Nature* **484** 195–200
[53] Haroche S and Raimond J M 2006 *Exploring the Quantum: Atoms, Cavities, and Photons* 1st edn, Ch. 5 (Oxford Univ. Press, Oxford)
[54] Humphreys P C, Kalb N, Morits J P, Schouten R N, Vermeulen R F, Twitchen D J, Markham M and Hanson R 2018 Deterministic delivery of remote entanglement on a quantum network *Nature* **558** 268-273

Algorithm development with LoRaWAN for monitoring system

Stoitcho Penkov, Albena Taneva

TU Sofia-Branch Plovdiv, Control Systems Department, 4000, Plovdiv, Bulgaria

stoitcho@abv.bg, altaneva@tu-plovdiv.bg

Abstract. This article presents the developed algorithm with LoRa specification. Based on it is achieved LoRaWAN for monitoring application in industrial systems. There is no universal solution for using LoRa Media Access, in such cases. The combination of low power hardware and radio communication is an advantage of the work. The developed algorithm ensures a new and modern network structure for monitoring in control systems. The real time observation is conducted in TTN. The experiments verified the algorithm applicability in the frame of systems of new digital industry.

1. Introduction

Nowadays are widely used networked devices, to achieve information in variety of applications. There is growing interest in such solutions in almost each area. Many systems need low power consumption, task distributions and collect data remotely using media access. There is no universal solution for monitoring and control using low energy and low cost networked devices. The aim of this work is to present the developed algorithm for monitoring in networked system using LoRa specification [1]. Appling network in industrial applications is widely popular, but complex, responsible, and sometimes dangerous task. More difficult case is when there is no power supply or electricity and a strict requirement for low power consumption, and long-range operation and protection. The work is devoted on algorithm development for industrial system purpose in the case of missing 220V supply using LoRaWAN.

LoRa means Long Range modulation which is suitable for the purpose of recent case, figure 1. LoRa is a proprietary spread spectrum modulation scheme that is derivative of Chirp Spread Spectrum modulation (CSS), which was developed for radar applications in the 1940's.

Traditionally is applied in a number of military and secure communications applications; over past years this technique has seen increased adoption in a number of data communications applications. Due to its relatively low transmission power requirements and inherent robustness from channel. It trades data rate for sensitivity within a fixed channel bandwidth. It implements a variable data rate, utilizing orthogonal spreading factors, which allows the system designer to trade data rate for range or power, so as to optimize network performance in a constant bandwidth [3]. It is evident that a PHY layer of LoRa is modulation to higher-layer implementations, figure 2. This allows coexisting and interoperating with available network architectures.

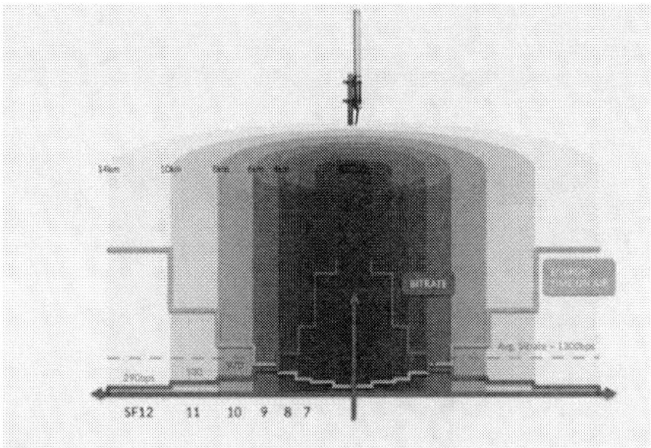

Figure 1. LoRa Adaptive Data Rate

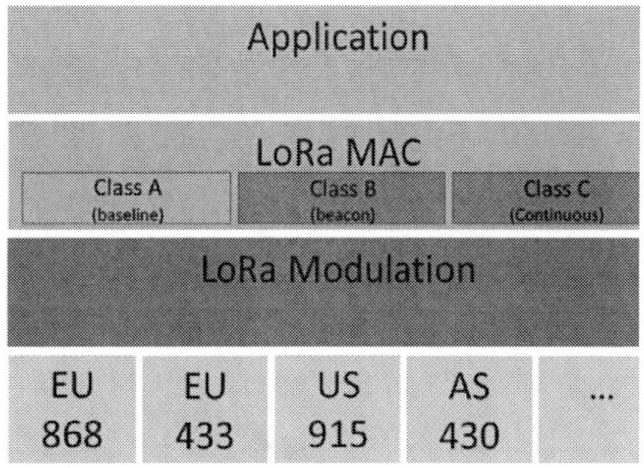

Figure 2. LoRa in OSI model structure

Thus can be noted some of the basic concepts of LoRa modulation and the advantages. The modulation can provide when deploying both fixed and mobile low-power real-world communication networks.

2. System configuration

Classical control system consists: plant (target), connected with feedback to the controller and predefine reference. Figure 3 shows the overall control scheme with single loop. The difference of recent case is in the use of network connection between plant, controller and variety of devices for monitoring and reference set up. The Media could be used between: control device and plant, reference value and control, between each element, as is depicted with dashed circle line on the draw.

It is evident that the data and another type of information in the system loop can be transferred in between, marked with ellipse using media, figure 4. The developed solution covers the connections between End Nodes (in recent case it is BME 680, figure 7) and the developed LoRa Gateway, figure 7.

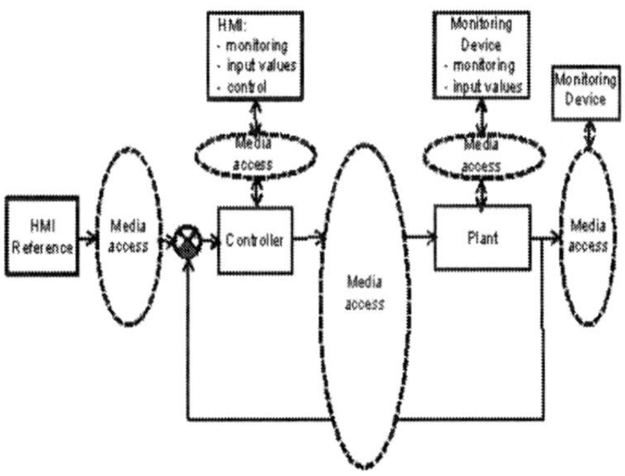

Figure 3. The classical system with LoRa

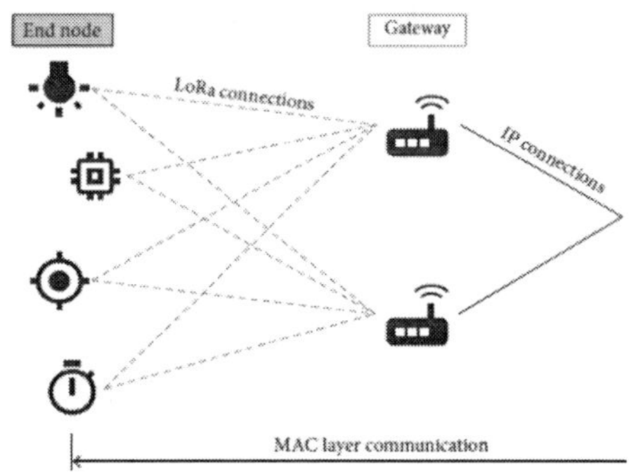

Figure 4. Overall scheme for monitoring with LoRa.

The figure shows that there are a variety of capabilities and tasks, on the one hand, a variety of different input devices are observed, with LoRa via Gateway being transmitted to many custom-defined applications. The network can contain thousands of nodes. If a new extension or addition is required for LoRaWAN, it is necessary to add the Gateway where there is no network coverage of an existing one. In the presented work, such a gateway has been designed and implemented with a freeze-dried spectrum. For Europe this is 868MHz frequency. This is an advantage and a contribution to this development, figure 7, LoRaWAN ™.

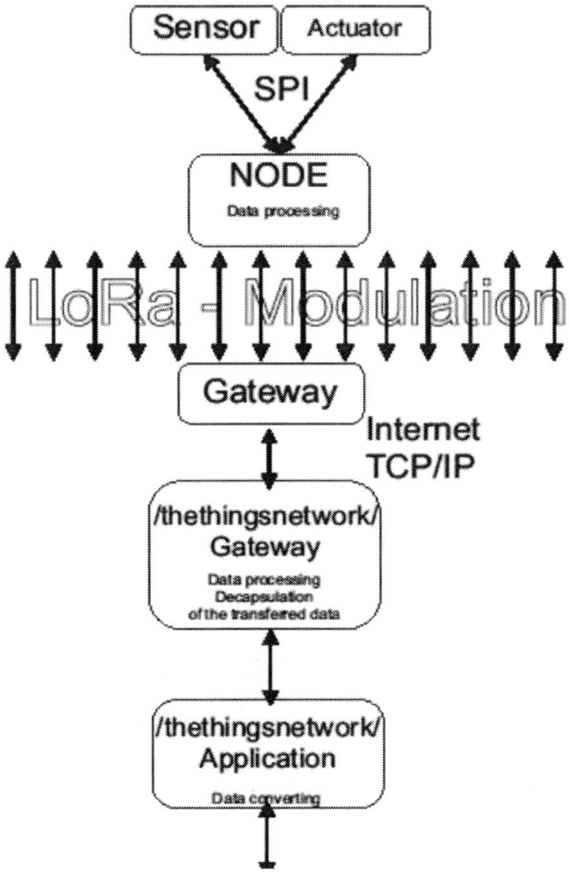

Figure 5. The developed Application with LoRa and shifting to the LoRaWAN

2.1. Gateway design

The developed Gateway supports the links, depicted on figure 4 with dashed lines. It relates and corresponds to the feedback in control scheme on figure 3. Furthermore it can cover and apply to all positions, denoted with dashed circle at the same figure. The data from (so called) end devices are transferred through CMWX1ZZABZ-091 LoRa® module made by MuRata to the Gateway or many Gates. It forwards received from Air packets to a public LoRa server, in our case *TheThingsNetwork* using Semtech packet-forwarding protocol. Hence it is created LoRaWAN for industrial system purpose. The developed Application with LoRa and shifting to the LoRaWAN is given on figure 5.

3. Algorithm development

On the figure 6 is given the structure of the developed algorithm for system operating. It consists of 6 sub algorithms, depicted with ALGORITHM 1-5. The advantage of the presented solution is the programming in free text editor. It is used Arduino environment 1.8.7.

3.1. Algorithm

As it is shown on the algorithm block structure, figure 6, it has 5 main parts.

3.2. Programming of the algorithm

For algorithm programming is used ST-Link programator STM32L0. Part of the program code for the

IOP Conf. Series: Materials Science and Engineering **618** (2019) 012007 doi:10.1088/1757-899X/618/1/012007

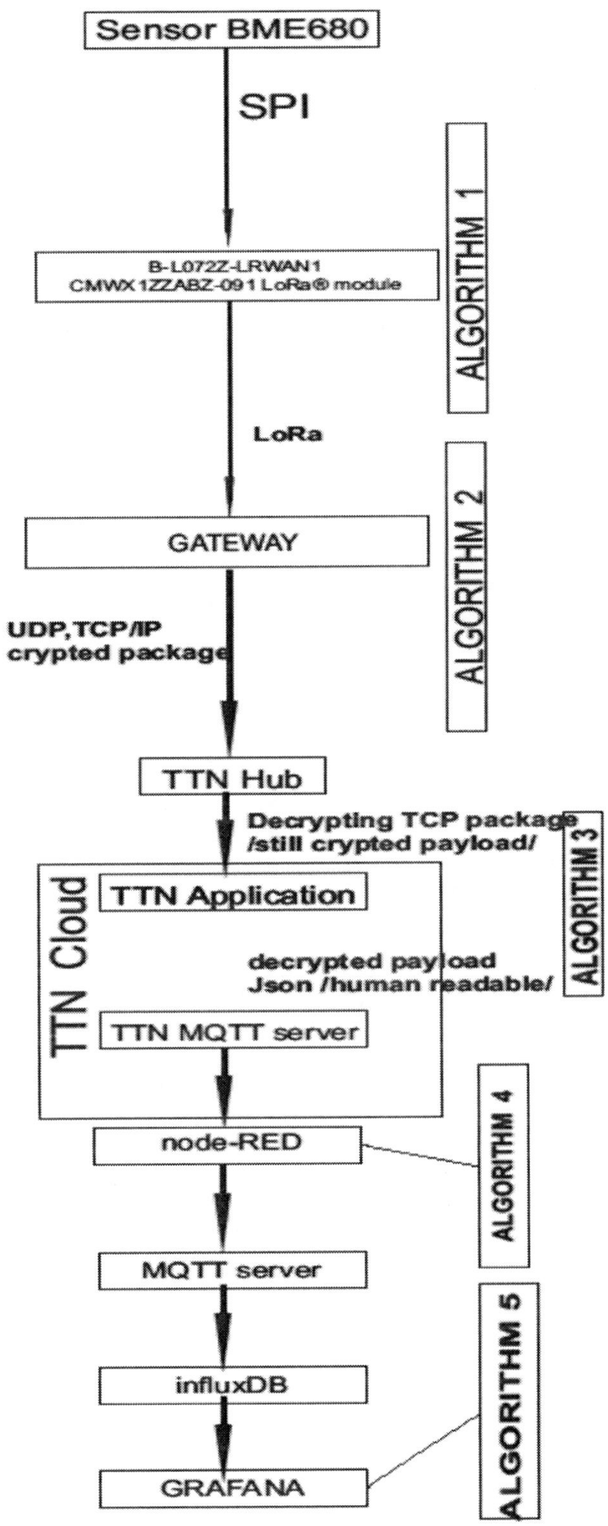

Figure 6. The Flow chart of the developed algorithm

ALGORITHMs 1 and 2,
follows:

```
#include <bme680.h>
#include <bme680_defs.h>
// I2C scanner by Nick
Gammon.  Thanks Nick.
#include <Wire.h>
void setup() {
  Serial.begin (115200);
//*****  make sure serial
monitor baud matches *****
  // Leonardo: wait for serial
port to connect
  while (!Serial)
    {
    }
  Serial.println ();
  Serial.println ("I2C scanner.
Scanning ...");
  byte count = 0;
    Wire.begin();
  for (byte i = 1; i < 120; i++)
  {
    Wire.beginTransmission (i);
    if (Wire.endTransmission ()
== 0)
      {
      Serial.print ("Found
address: ");
      Serial.print (i, DEC);
      Serial.print (" (0x");
      Serial.print (i, HEX);
      Serial.println (")");
      count++;
      delay (1);  // maybe
unneeded?
      } // end of good response
  } // end of for loop
  Serial.println ("Done.");
  Serial.print ("Found ");
  Serial.print (count, DEC);
  Serial.println (" device(s).");
}  // end of setup
void loop() {}
```

IOP Conf. Series: Materials Science and Engineering **618** (2019) 012007 doi:10.1088/1757-899X/618/1/012007

Programming of the ALGORITHM 3

```
function kur(d,c,b,a) {
  var bytes = (a<<24) + (b<<16) + (c<<8) + d;
  var sign = (bytes & 0x80000000) ? -1 : 1;
  var exponent = ((bytes >> 23) & 0xFF) - 127;
  var significand = (bytes & 0x7FFFFF);
  if (exponent == 128)
    return sign * ((significand) ? Number.NaN : Number.POSITIVE_INFINITY);
  if (exponent == -127) {
    if (significand === 0) return sign * 0.0;
    exponent = -126;
    significand /= (1 << 22);
  } else significand = (significand | (1 << 23)) / (1 << 23);
 return sign * significand * Math.pow(2, exponent);
}
function Decoder(a) {
 var decoded = {}
 decoded.tmp = kur(a[0 ],a[1 ],a[2 ],a[3 ])/1;
 decoded.hum = kur(a[4 ],a[5 ],a[6 ],a[7 ])/1;
 decoded.pre = kur(a[8 ],a[9 ],a[10],a[11])/100;
 decoded.gas = kur(a[12],a[13],a[14],a[15])/1000;
 decoded.alt = kur(a[16],a[17],a[18],a[19])/1;
 return decoded
}
```

4. Equipment

In this system is used as an End device, BME680, figure 8. It is an open - source hardware, made by Adafruit Industries. The library adafuit/Adafruit_BME, [4], provides gas, humidity, temperature and pressure measurements. The used Arduino library for BME680 version is 1.0.7.

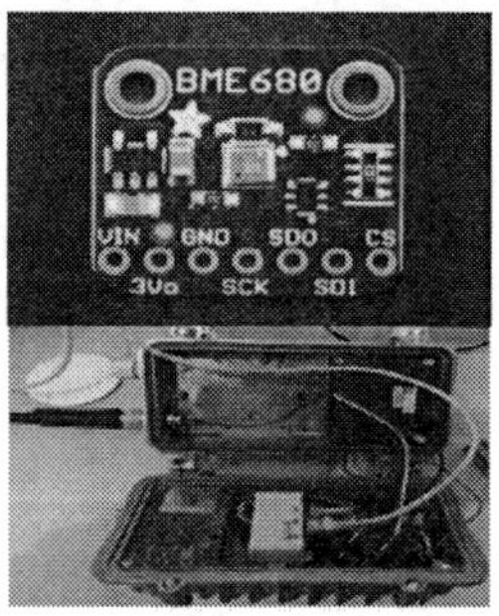

Figure 7. The developed and assembled Gateway for LoRaWAN, and the End node device.

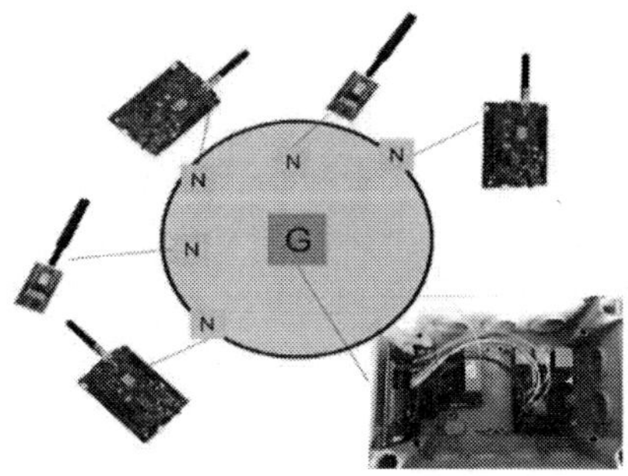

Figure 8. The developed system with LoRaWAN

5. Experiments and results

The programmed algorithm was ran and proved in real time experiment. On the next figure 9, the project in TTN is presented. On the figure 10 is shown the payload that is received from device test-device-bme680-02. The snapped picture shows only encrypted data in the frame of payload. This payload has to be decoded, as it explained in ALGORITHM 3. On figure 11 is shown the result of the applied ALGORITHM 3 for data decoding transferred via LoRa. The obtained information is presented on figure 12. The payload is presented in Json format. This is widely used but not so user-friendly data format.

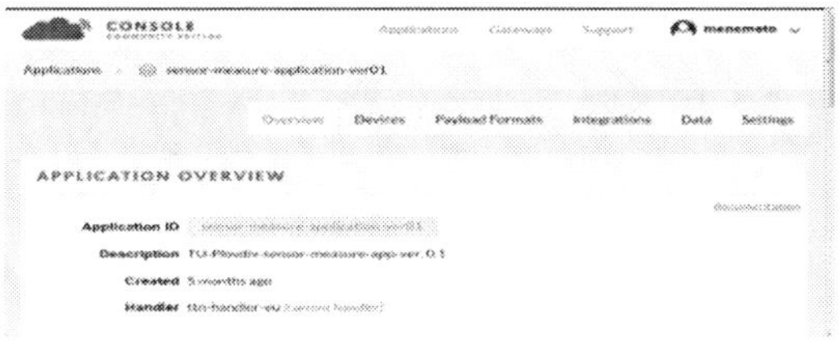

Figure 9. The developed project In TTN with LoRaWAN

Figure 10. Visualisation of the received payload from device

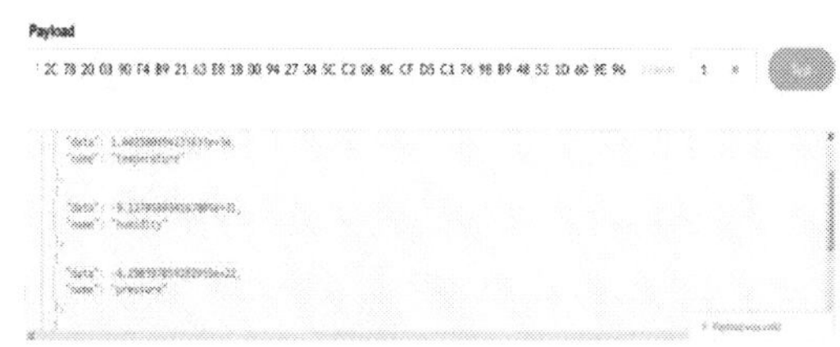

Figure 11. The detailed view of the payload.

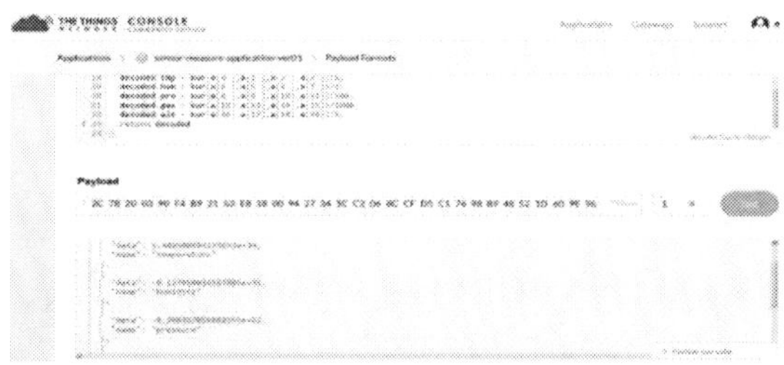

Figure 12. The decoded payload from the device via LoRa

6. Conclusion

The work is devoted to the industrial system application in the case of missing 220V supply using LoRaWAN. The main goal is the developed and programmed algorithm, ensures data monitoring in industrial systems with LoRa specification. An advantage is that it can be easy programmed in text environment. The presented solution with LoRa specification is a basement in LoRaWAN for monitoring and furthermore application in industrial environment. Due to the lack of universal approach for using such Media Access in industrial cases, this can be viewed as an effort to achieve new standard for LoRaWAN. From different point of view this solution can be assumed as an effort for future standardization to the new modern digital industry. The achieved results, security transferred, crypted and decrypted (figures 10-12) show the promising future work. It could be devoted on program development to ensure the Algorithm 4 and 5 for proper user interface, regarding to the process control tasks. Future works will improve the ALGORITHM 5, to ensure the control action with classical control low.

Acknowledgments

This work was supported by the European Regional Development Fund within the OP "Science and Education for Smart Growth 2014-2020", Project Competence Centre "Smart Mechatronic, Eco-And Energy Saving Systems And Technologies", № BG05M2OP001-1.002-0023.

The authors would like to thank the Research and Development Sector at the Technical University of Sofia for the financial support.

References

[1] LoRa Alliance, LoRaWAN™ 1.1 Specification Kim J., J. Song, A Dual Key-Based Activation Scheme for Secure LoRaWAN. Wireless Communications and Mobile Computing Volume 2017, Article ID 6590713, https://doi.org/10.1155/2017/659071

[2] Semtech , Application Note, Revision 2, May 2015© , Corporation

[3] http://www.adafruit.com/products/3660

[4] Marais J., R.Malekian, A.Abu-Mahfouz, LoRa and LoRaWAN Testbeds: a Review, , Conference: IEEE Africon 2017, South Africa, DOI: 10.1109/AFRCON.2017.8095703

[5] Durkop L., B. Czybik, J.Performance evaluation of M2M protocols over cellular networks in a lab environment, 18th International Conference on Intelligence in Next Generation Networks (ICIN), 2015, Paris, France DOI: 10.1109/ICIN.2015.7073809

[6] https://www.st.com/content/dam/technology-tour-2017/session-1_track-2_lora-enabling-low-power.pdf

[7] https://fenix.tecnico.ulisboa.pt/downloadFile/1689468335603030/LoRaWAN%20Introduction.pdf

A Stable Control Algorithm for Multi Robot Formation

Nikola Shakev, Andon V Topalov, Sevil A Ahmed and Vasil L Popov

Control Systems Department, Technical University of Sofia, Branch Plovdiv;
25 Tsanko Dyustabanov Str, 4000 Plovdiv, Bulgaria

shakev@tu-plovdiv.bg

Abstract. This paper presents the developed trajectory tracking controller for a formation of nonholonomic robots, which combines features from the leader-follower and virtual-structure approaches. The implemented decentralized control strategy allows the robots to be relatively independent and to switch easily between the executed individual tasks and the collective tasks. Convergence is thoroughly analyzed and guarantied using the Lyapunov approach.

1. Introduction

In the recent years, the cooperation of multiple robots like unmanned aerial vehicles (UAVs), autonomous underwater vehicles (AUVs), mobile robots systems (MRSs) is gaining significant importance in the control and robotics communities. Using formation of multiple robots to accomplish a mission or objective offers has obvious advantages such as increased feasibility, scalability, accuracy, flexibility, cost and energy efficiency, and so on. Potential applications include geophysical survey, search and rescue, farm aerial spraying, forest fire suppression, pipeline and power line patrol, and also many military applications such as general marine mine sweeping , exploration, surveillance and tracking, etc.

Several approaches have been proposed to solve the formation control problem. Behavior-based approach, leader-follower approach, virtual-structure, artificial potential, and graph theory are classified as the main formation control methods.

In the behavior-based approach [1], [2] numerous desired behaviors are supported for each robot, and the effective formation control is obtained as a result from a weighted summation of individual behavioral outputs.

A powerful technique to design control laws comes from the use of artificial potential functions where the formation control and collision avoidance can be solved through a combination of attractive and repulsive vector fields [3].

Other formation schemes are based on formation graphs and the so-called consensus problem. Some tools of graph theory and linear systems are used to prove the convergence to a desired formation [4].

In the virtual structure approach [5], the entire formation is managed as one unit. The desired posture for every robot is calculated and each robot has to track its own trajectory.

In the leader-following approach some agents are designated as leaders, where-as others are designated as followers [6], [7]. Each follower robot has been controlled to track the state that is obtained from the states of one or two leaders. The leader-following method has a simple structure and if it is necessary, it can rely on local sensor data only.

In this paper an original trajectory tracking controller for multi-robot formation, which combines features from leader-follower and virtual-structure approaches is proposed and its performance is evaluated. The designed control law is obtained as a generalization of an earlier presented [8] trajectory tracking controller for a single nonholonomic mobile robot.

2. The control problem statement

The task of the control algorithm is to drive a group of robots to follow a de-sired trajectory while keeping the predefined space structure of the formation. The adopted approach concerns that the leader of the formation has to track the reference trajectory, while the other robots have to keep their positions in the formation relative to the position of the leader. So this approach can be classified as a leader following one. It is possible to modify the control problem discussed above if we assume that the leader of the formation is not a physical, but a virtual robot following perfectly the reference trajectory. In this case, the approach can be classified as a virtual structure approach. In this investigation, we will assume the case where the leader is a physical robot.

The structure of the formation is defined by specifying the position and orientation of each robot as a position and orientation of a leader with added offsets by constant values. In this way, there are two possibilities: (i) the offsets of the coordinates relative to the leader's coordinates can be considered in the inertial coordinate frame; (ii) the offsets of coordinates can be considered in the moving coordinate frame attached to the leader. We will consider the first case where the offsets are defined in the inertial coordinate frame.

In order to implement the above strategy a kinematic controller, initially pro-posed for trajectory tracking for a single robot [8] will be applied. The usage of this controller for the multi-robot system case ensures an asymptotic convergence of the trajectories of the controlled robots to the leader's trajectory shifted by some constant displacement.

2.1. Robot Movement Representation

For the description of the motion of each robot on a 2D plane two coordinate frames have to be used: a global inertial frame *(O, x, y)* and a local frame fixed to the center of mass G_i of the *i*-th robot (G_i, x_i, y_i).

The configuration,

$$p_i = \begin{bmatrix} x_i \\ y_i \\ \theta_i \end{bmatrix} \tag{1}$$

of the *i*-th robot represents the position and orientation of the local coordinate system in the global frame, where the orientation θ_i is taken counterclockwise from the global X-axis. The motion of the mobile robot is controlled by its linear and angular velocities - $\upsilon(t)$ and $\omega(t)$ respectively. The relation between $[\upsilon(t) \quad \omega(t)]^T$ and the velocities in Cartesian frame is given by equation (2), where $J(\theta)$ denotes the Jacobian matrix.

$$\dot{p}_i = \begin{bmatrix} \dot{x}_i \\ \dot{y}_i \\ \dot{\theta}_i \end{bmatrix} = Jq_i = \begin{bmatrix} \cos(\theta_i) & 0 \\ \sin(\theta_i) & 0 \\ 0 & 1 \end{bmatrix} \begin{bmatrix} \upsilon_i \\ \omega_i \end{bmatrix} \tag{2}$$

2.2. The Controller Design

The control task for a leader following behavior is to ensure that the trajectory of the *i*-th robot $p_i(t)$ will follow the trajectory of the leading robot $p_L(t)$ conceivably shifted to some predefined constant displacement c_i. The displacement $c_i = \begin{bmatrix} c_{ix} & c_{iy} & c_{i\theta} \end{bmatrix}^T$ represents the desired distance between the configuration of i-th robot and the leading robot in the global frame.

Then the tracking error in global reference frame $E_i(t)$ is defined as follows:

$$E_i(t) = \begin{bmatrix} E_{ix} \\ E_{iy} \\ E_{i\theta} \end{bmatrix} = p_L(t) - p_i(t) - c_i = \begin{bmatrix} x_L(t) - x_i(t) - c_{ix} \\ y_L(t) - y_i(t) - c_{iy} \\ \theta_L(t) - \theta_i(t) - c_{i\theta} \end{bmatrix} \quad (3)$$

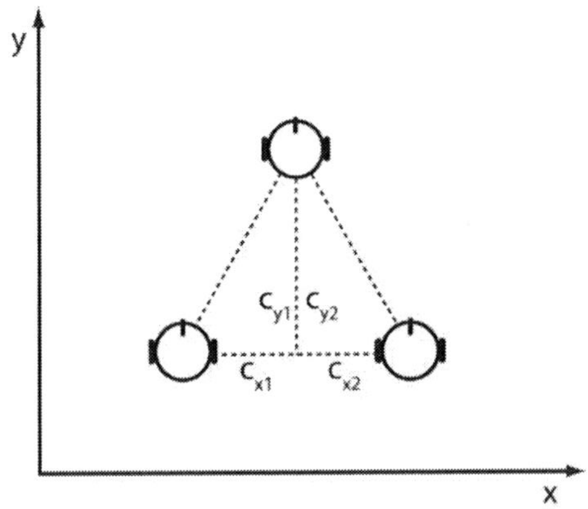

Figure 1. Formation consisting of a leading robot and two following robots.

Let $e_i(t)$ denotes the expression of $E_i(t)$ in the i-th robot local reference frame:

$$e_i(t) = \begin{bmatrix} e_{i1} \\ e_{i2} \\ e_{i3} \end{bmatrix} = T_i E_i(t) = \begin{bmatrix} \cos(\theta_i) & \sin(\theta_i) & 0 \\ -\sin(\theta_i) & \cos(\theta_i) & 0 \\ 0 & 0 & 1 \end{bmatrix} \begin{bmatrix} E_{ix} \\ E_{iy} \\ E_{i\theta} \end{bmatrix} \quad (4)$$

where $T(\theta)$ is a transformation matrix. From equation (2) it is possible to obtain the following two equations:

$$\upsilon_i = \dot{x}_i \cos(\theta_i) + \dot{y}_i \sin(\theta_i) \quad (5)$$

$$0 = \dot{x}_i \sin(\theta_i) - \dot{y}_i \cos(\theta_i) \quad (6)$$

The time derivative vector $\dot{e}_i(t)$ can be calculated from (4) and using equations (5) and (6) the following expressions for its elements can be obtained:

$$\begin{aligned}\dot{e}_{i1}(t) &= (\dot{x}_L - \dot{x}_i)\cos\theta_i + (\dot{y}_L - \dot{y}_i)\sin\theta_i - (x_L - x_i - c_{ix})\dot{\theta}_i \sin\theta_i + (y_L - y_i - c_{iy})\dot{\theta}_i \cos\theta_i = \\ &= \dot{x}_L \cos\theta_i + \dot{y}_L \sin\theta_i - \upsilon_i + e_{i2}\dot{\theta}_i = \\ &= \dot{x}_L \cos(\theta_L - (c_{i\theta} + e_{i3})) + \dot{y}_L \sin(\theta_L - (c_{i\theta} + e_{i3})) - \upsilon_i + e_{i2}\omega_i = \\ &= \dot{x}_L\left[\cos(\theta_L)\cos(c_{i\theta} + e_{i3}) + \sin(\theta_L)\sin(c_{i\theta} + e_{i3})\right] + \\ &+ \dot{y}_L\left[\sin(\theta_L)\cos(c_{i\theta} + e_{i3}) - \cos(\theta_L)\sin(c_{i\theta} + e_{i3})\right] - \upsilon_i + e_{i2}\omega_i = \\ &= (\dot{x}_L \cos(\theta_L) + \dot{y}_L \sin(\theta_L))\cos(c_{i\theta} + e_{i3}) + \\ &+ (\dot{x}_L \sin(\theta_L) - \dot{y}_L \cos(\theta_L))\sin(c_{i\theta} + e_{i3}) - \upsilon_i + e_{i2}\omega_i = \\ &= \upsilon_L \cos(c_{i\theta} + e_{i3}) - \upsilon_i + e_{i2}\omega_i\end{aligned} \tag{7}$$

$$\begin{aligned}\dot{e}_{i2}(t) &= -(\dot{x}_L - \dot{x}_i)\sin\theta_i + (\dot{y}_L - \dot{y}_i)\cos\theta_i - (x_L - x_i - c_{ix})\dot{\theta}_i \cos\theta_i - (y_L - y_i - c_{iy})\dot{\theta}_i \sin\theta_i = \\ &= -\dot{x}_L \sin\theta_i + \dot{y}_L \cos\theta_i + \dot{x}_i \sin\theta - \dot{y}_i \cos\theta_i - e_{i1}\dot{\theta}_i = \\ &= -\dot{x}_L \sin(\theta_L - (c_{i\theta} + e_{i3})) + \dot{y}_L \cos(\theta_L - (c_{i\theta} + e_{i3})) - e_{i1}\omega_i = \\ &= -\dot{x}_L\left[\sin(\theta_L)\cos(c_{i\theta} + e_{i3}) - \cos(\theta_L)\sin(c_{i\theta} + e_{i3})\right] + \\ &+ \dot{y}_L\left[\cos(\theta_L)\cos(c_{i\theta} + e_{i3}) + \sin(\theta_L)\sin(c_{i\theta} + e_{i3})\right] - e_{i1}\omega_i = \\ &= (\dot{x}_L \cos(\theta_L) + \dot{y}_L \sin(\theta_L))\sin(c_{i\theta} + e_{i3}) + \\ &+ (-\dot{x}_L \sin(\theta_L) + \dot{y}_L \cos(\theta_L))\cos(c_{i\theta} + e_{i3}) - e_{i1}\omega_i = \\ &= \upsilon_L \sin(c_{i\theta} + e_{i3}) - e_{i1}\omega_i\end{aligned} \tag{8}$$

$$\dot{e}_{i3}(t) = \dot{\theta}_L - \dot{\theta}_i = \omega_L - \omega_i \tag{9}$$

Then the linear and angular velocities of the i-th robot are calculated according to the following control law:

$$q_i(t) = \begin{bmatrix} \upsilon_i \\ \omega_i \end{bmatrix} = \begin{bmatrix} \upsilon_L \cos(c_{i\theta} + e_{i3}) + k_1 e_{i1} \\ \omega_L + k_2 \upsilon_L e_{i2} + k_3 \upsilon_L \sin(c_{i\theta} + e_{i3}) \end{bmatrix} \tag{10}$$

The analysis of stability can be implemented by defining the Lyapunov function candidate $V(t)$:

$$V_i(t) = \frac{1}{2}(e_{i1}^2 + e_{i2}^2) + \frac{1 - \cos(c_{i\theta} + e_{i3})}{k_{i2}} \tag{11}$$

The time derivative $\dot{V}_i(t)$ has to be calculated:

$$\dot{V}_i = \dot{e}_{i1}e_{i1} + \dot{e}_{i2}e_{i2} + \dot{e}_{i3}\frac{\sin(c_{i\theta} + e_{i3})}{k_2} \tag{12}$$

Substituting equations (7), (8), and (9) into (12) and applying after that the control law (10) for $\upsilon_i(t)$ and $\omega_i(t)$ leads to the following result:

$$\begin{aligned}\dot{V}_i &= \left[\upsilon_L \cos\left(c_{i\theta}+e_{i3}\right) - \upsilon_i + e_{i2}\omega_i\right] e_{i1} + \\ &+ \left[\upsilon_L \sin\left(c_{i\theta}+e_{i3}\right) - e_{i1}\omega_i\right] e_{i2} + \left[\omega_L - \omega_i\right] \frac{\sin(c_{i\theta}+e_{i3})}{k_2} = \\ &= -k_1 e_{i1}^2 - k_3\upsilon_L \frac{\sin^2(c_{i\theta}+e_{i3})}{k_2} \le 0\end{aligned} \tag{13}$$

If the linear velocity $\upsilon_L(t)$ of the leading robot has nonnegative values and the controller parameters k_1, k_2, and k_3 are also positive, then the inequality (13) is satisfied and the system is uniformly asymptotically stable around $e_i(t) = 0$. This means that the trajectory of the i-th robot will converge to the trajectory of the leader shifted by some constant displacement $c_i = \left[c_{ix} \quad c_{iy} \quad c_{i\theta}\right]^T$.

3. Simulation experiments

In order to investigate more systematically the dynamics of the collaboration between the robots in the formation, we have conducted a simulation experiment in the Webots environment, a 3D simulator for autonomous robots (figure2). Webots simulator computes trajectories and sensory inputs of the robots in an area corresponding to the physical set-up. The simulation is sufficiently faithful for the controllers to be transferred to the real robots without changes, and for the robot behaviors to be very similar to those of the real robots.

The functionality of the proposed trajectory tracking controller has been tested on a scenario, where three robots have to perform a cooperative mission. A formation consisting of one leading robot and two following robots has been considered. The spatial structure of the formation has been defined by the following offsets:

$$c_1 = [c_{1x}, c_{1y}, c_{1\theta}]^T = [-20\,cm, 10\,cm, 0\,rad]^T$$
$$c_2 = [c_{2x}, c_{2y}, c_{2\theta}]^T = [-20\,cm, -10\,cm, 0\,rad]^T$$

which represents an isosceles triangle.

The desired trajectory of the leading robot has been selected to be a square with a side length of 100 cm and rounded corners (figure 3). The linear velocity on the straight-line parts of the path has to be 1.246 cm/s, while on the curves the linear velocity has to be 1.09 cm/s and the angular velocity has to be 0.12 rad/s.

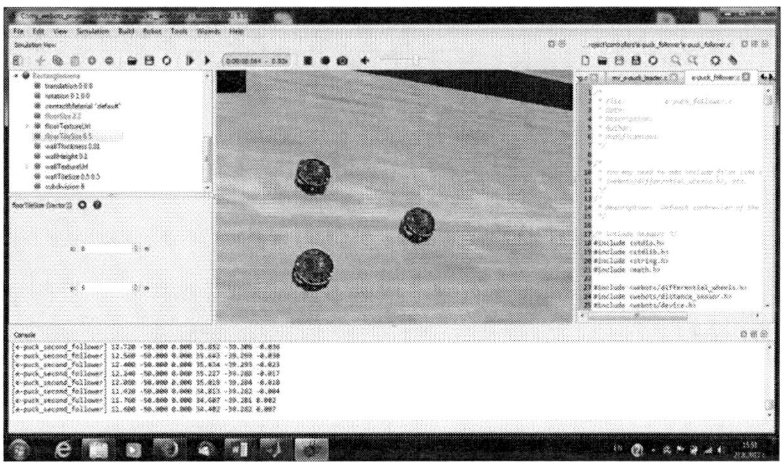

Figure 2. The simulation setup using software Webots.

IOP Conf. Series: Materials Science and Engineering **618** (2019) 012008 doi:10.1088/1757-899X/618/1/012008

In addition, we have introduced initial tracking errors for the two followers:

$$E_1(0) = \begin{bmatrix} E_{1x}(0) \\ E_{1y}(0) \\ E_{1\theta}(0) \end{bmatrix} = \begin{bmatrix} x_L(0) - x_1(0) - c_{1x} \\ y_L(0) - y_1(0) - c_{1y} \\ \theta_L(0) - \theta_1(0) - c_{1\theta} \end{bmatrix} = \begin{bmatrix} 0\,cm \\ -10\,cm \\ 0\,rad \end{bmatrix}$$

$$E_2(0) = \begin{bmatrix} E_{2x}(0) \\ E_{2y}(0) \\ E_{2\theta}(0) \end{bmatrix} = \begin{bmatrix} x_L(0) - x_1(0) - c_{1x} \\ y_L(0) - y_1(0) - c_{1y} \\ \theta_L(0) - \theta_1(0) - c_{1\theta} \end{bmatrix} = \begin{bmatrix} -8.4\ cm \\ 15\ cm \\ 0\ rad \end{bmatrix}$$

The trajectories of the leading robot and the two following robots are shown on figure 3. It can be seen that by using the proposed control law, the two following robots keep the desired displacement with respect to the leader and the initially introduced tracking error tends asymptotically to zero. The time varying errors of both followers are shown on figure 4 and figure 5 respectively.

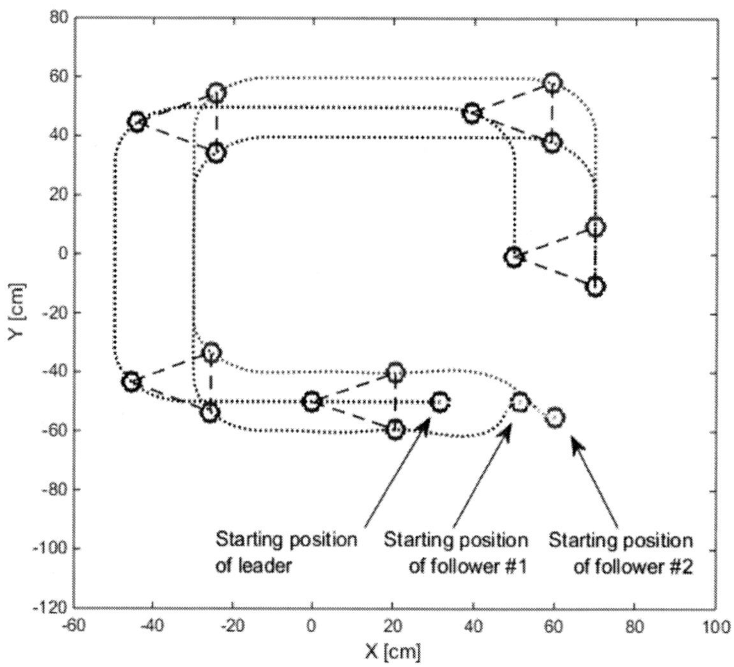

Figure 3. Reference trajectory of the leading robot and the obtained trajectories for the following robots.

IOP Conf. Series: Materials Science and Engineering **618** (2019) 012008 doi:10.1088/1757-899X/618/1/012008

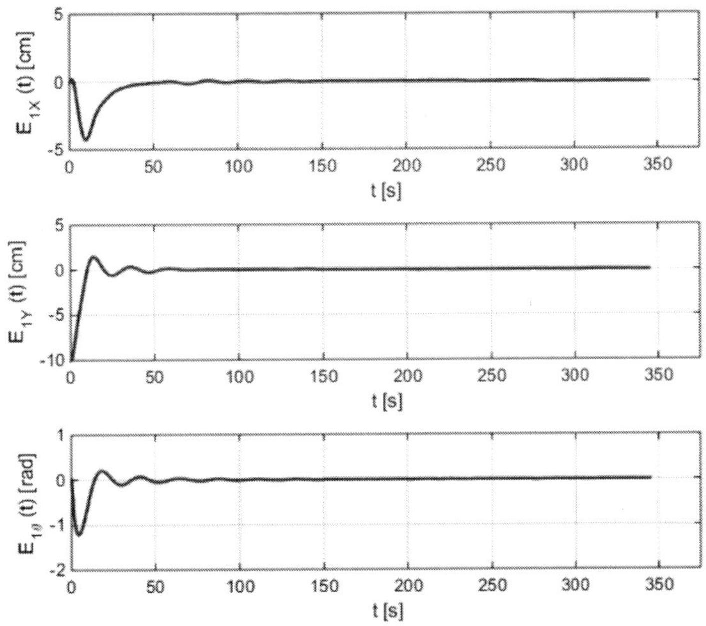

Figure 4. The errors E_{1X}, E_{1Y}, and $E_{1\theta}$ of the first following robot along the trajectory.

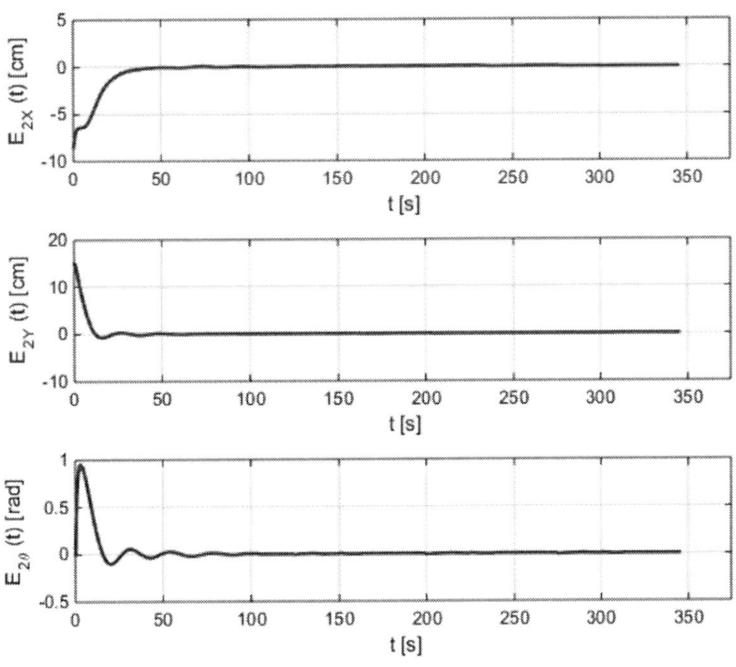

Figure 5. The errors E_{2X}, E_{2Y}, and $E_{2\theta}$ of the second following robot along the trajectory.

4. Conclusion

In this paper, a control algorithm for a multi robot formation is proposed. The aim of the algorithm is to provide a reliable tracking control of the mobile robots. The formation movement strategy consists of two tasks: 1) the leader of the formation has to track the desired trajectory and 2) the robots in the formation have to track the leader by keeping at the same time the predefined structure of the formation. The presented control law is an extension of the control law proposed by Kanayama et al. (1991) and allows the accomplishment of both tasks mentioned above. The asymptotic convergence of the error is proved for the case of formation trajectory tracking. The presented simulation experiment confirm the convergence and the stability of the system.

Acknowledgments

The authors gratefully acknowledge to the Scientific and Research Sector NIS of TU-Sofia for the support under the project 182PD0026-19.

The work of V. Popov is supported by National program "Young scientists and Postdoctoral candidates" of Ministry of Education and Science.

References

[1] Arkin R 1987 Motor schema based navigation for a mobile robot: an approach to programming by behavior *Proc. of the 1987 IEEE Int. Conf. on Robotics and Automation* **4** pp. 264–271

[2] Brunete A, Hernando M, Gambao E and Torres J E 2012 A behaviour-based control architecture for heterogeneous modular, multiconfigurable, chained micro-robots *J. Robotics and Autonomous Systems* **60** 1607–1624

[3] Hernandez-Martinez E, Aranda-Bricaire E 2011 Convergence and collision avoidance in formation control: A survey of the artificial potential functions approach. *In Multi-Agent Systems—Modeling, Control, Programming, Simulations and Applications* (INTECH, Austria) ed Alkhateeb, Maghayreh E A, Doush I A (eds.) pp. 103–126

[4] Desai J P 2002 A Graph Theoretic Approach for Modeling Mobile Robot Team Form *ations J. of Robotic Systems* **19** 11 pp. 511-525

[5] Chen Y and Wang Z 2005 Formation control: a review and a new consideration *Proc. 2005 IEEE/RSJ Int. Conf. on Intelligent Robots and Systems* pp. 3181–3186.

[6] Tanner H G, Pappas G J, Kumar V 2004 Leader-to-formation stability *IEEE Transactions on Robotics and Automation* **20** p. 443

[7] Lewis M A, Tan K-H 1997 High Precision Formation Control of Mobile Robots Using Virtual Structures *J. Autonomous Robots* **4** p. 387

[8] Kanayama Y, Kimura Y, Miyazaki F and Noguchi T 1991 A stable tracking control method for a non-holonomic Mobile Robot *In Proc. of IEEE International Workshop IROS'91* (Osaka, Japan) pp. 1236-1241

IOP Conf. Series: Materials Science and Engineering **618** (2019) 012009 doi:10.1088/1757-899X/618/1/012009

Process model as a basis for intellectual control in real time by the processes of the production enterprise

O Novikova[1], A Kotsyubinsky[2], N Yudina[2], S Livshits[2], T Dunaeva[2]

[1]Peter the Great St. Petersburg Polytechnic University, St. Petersburg, Russia
[2]Kazan State Power Engineering University KSPEU, Kazan, Russia

novikova-olga1970@yandex.ru

Abstract. Modern enterprises are working in a complex and volatile business environment. In order to adapt to this situation, enterprises are trying to realize horizontal and vertical integration of the enterprise processes (managerial and production processes) leading to the creation of a real-time enterprise. In such an enterprise, integral data is used with different time horizons, starting from the millisecond, generated by automatic devices, and ending with daily, generated business applications. However, these integral data are rarely used for operational management of time by enterprise processes of the enterprise. To overcome this limitation, a process model for identifying and assimilating knowledge is proposed that encompasses four steps: (i) analyzing and (re-designing the enterprise's controlled processes, (ii) developing a data model and enterprise data flow diagrams for automatic devices and business applications , (iii) identification of knowledge based on the discovery of knowledge in working with databases, (iv) continuous monitoring and management of enterprise processes, using complex event processing.

1. Introduction

The difficult, changeable and uncertain business environment of a modern enterprise exerts on the last (enterprise) increased pressure to provide quality services, produce high-quality components, shorten the time for services and production, and reduce costs. Due to this pressure, the business and production processes of creating value added must be flexible, adaptable and constantly monitored. In this connection, vertical and horizontal integration of enterprise processes becomes necessary for survival. For business process management (supply chain management), customer relationship management (CRM) can be available in the market of the system of enterprise applications integration (EAI) [1]. Along with horizontal integration, vertical integration of various levels of the enterprise is a necessary condition for continuous monitoring of enterprise processes and creation of a real-time enterprise (RTE - Real-Time Enterprise) [2].

The enterprises of production, gas and electric transport - can be divided into three levels of manufacturing execution systems (MES): the level of enterprise management, the level of production management and the production level, as shown in figure 1.

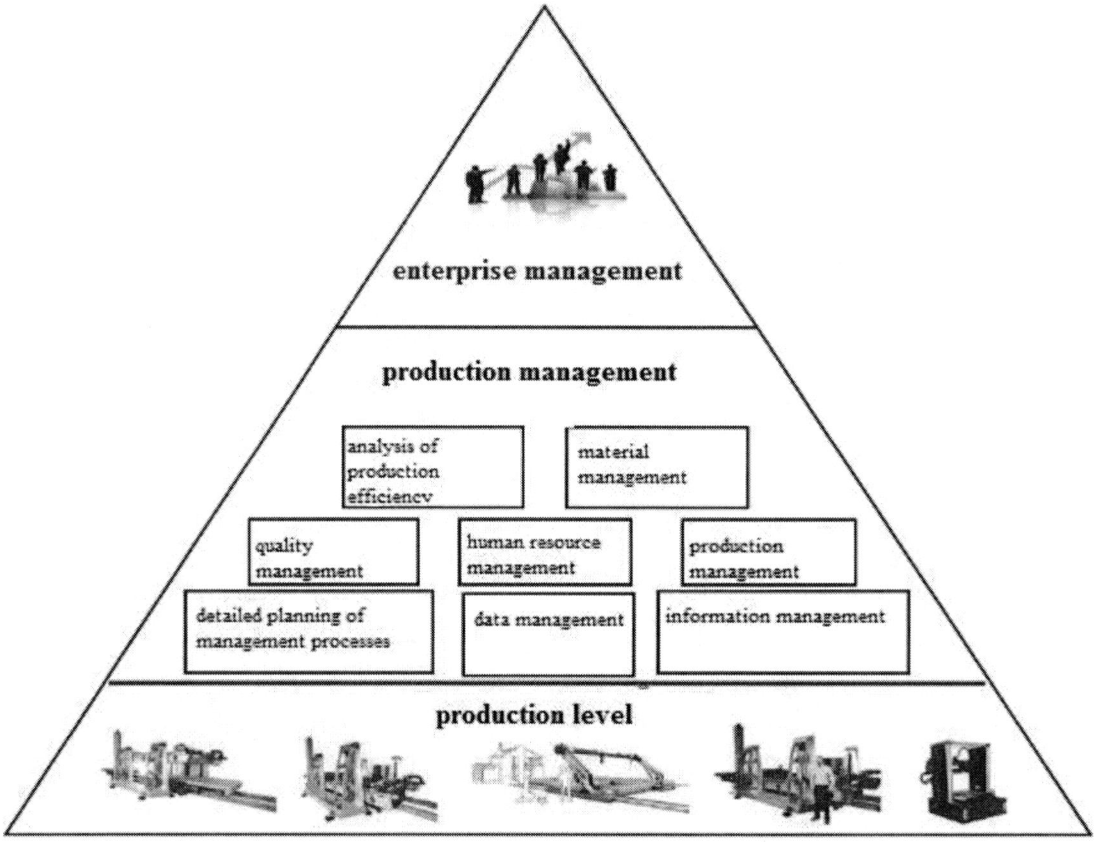

Figure 1. Levels of manufacturing systems of the enterprise.

At the enterprise management level, it is business processes that allow to achieve strategic targets, so business processes must be developed, launched and analyzed in accordance with the four steps of the BPM (BPM - Business Process Management): design, configuration, execution, diagnostics (monitoring and optimization). During the configuration of the business process at the enterprise management level, are laid transactional planned performance values (i.e., meanings "as expected" - TO-BE) based on the strategic objectives of the enterprise and with a measurement period of weeks and months.

Production processes are used to achieve the goals and indicators set at the enterprise management level and use automatic devices of the production level. A huge amount of data (so-called sensory data) is generated by these devices in real time (i.e., seconds and millisecond intervals). In addition, operators define data related to automatic devices, such as predetermined causes of equipment failure, instructions defining the order of supply of parts at the time the operation is started, and so on. In general, these values (i.e., the value "as is" - AS-IS) reflect the real productivity at the production level. MES solutions and process applications allow the creation of computerized and automated vertical integration. However, the problem of interaction between the enterprise management level and the production level remains unresolved [3]: implementation of multi-loop control within the enterprise and through all levels from the level of enterprise management to production level. Insufficient vertical integration prevents the accumulation of knowledge of the enterprise, the creation of training cycles [4] and the use of accumulated knowledge. First, data from various levels of MES are not integrated and, therefore, cannot be used to gain new knowledge. Secondly, if new knowledge has been received, it is not used in the continuous management of enterprise processes. As a consequence, the concepts of RTE: feel-and-react and learn-and-adapt are little used in enterprise processes.

2. Current status

Since the mid-1990s, several reference enterprise integration architectures (CIMOSA, PERA, ARIS and GRAI / GIM) have become available to guide the design and implementation of the integrated enterprise. However, these reference architectures differed in theoretical basis, understanding of the enterprise, approaches to modeling and goals. To overcome these problems, a generalized enterprise reference architecture and methodology (GERAM-Generalized Enterprise Reference Architecture and Methodology) was developed [5]. The mentioned reference architectures contributed to the development of GERAM, which was later standardized as ISO 15704 - requirements for the enterprise reference architecture and methodologies. It is still largely taken into account in the new standards (for example, ISO 19439: 2006 "Enterprise integration Framework for enterprise modeling").

The GERAM scheme provided for four groups of aspects of the enterprise architecture, called Views - the types of models ("functions", "data", "resources", "organization"), purpose, implementation and "physical representations" (hardware), but also the possibility identify additional aspects. It included a description of all aspects or some part of them in each of the seven or eight phases of the architecture and operation of the enterprise, and also included the specification of the architecture model at three levels - generalized, level of partial models and specific models. GERAM defines common adjacent enterprise concepts recommended for application in enterprise engineering and implementation of integration projects. Such concepts can be classified as:

1) concepts that are human-centered.

2) concepts aimed at describing the business process of the enterprise;

3) concepts focused on the description of technology to support the business process, involved in the work process and strengthening the enterprise engineering (modeling and application of the model).

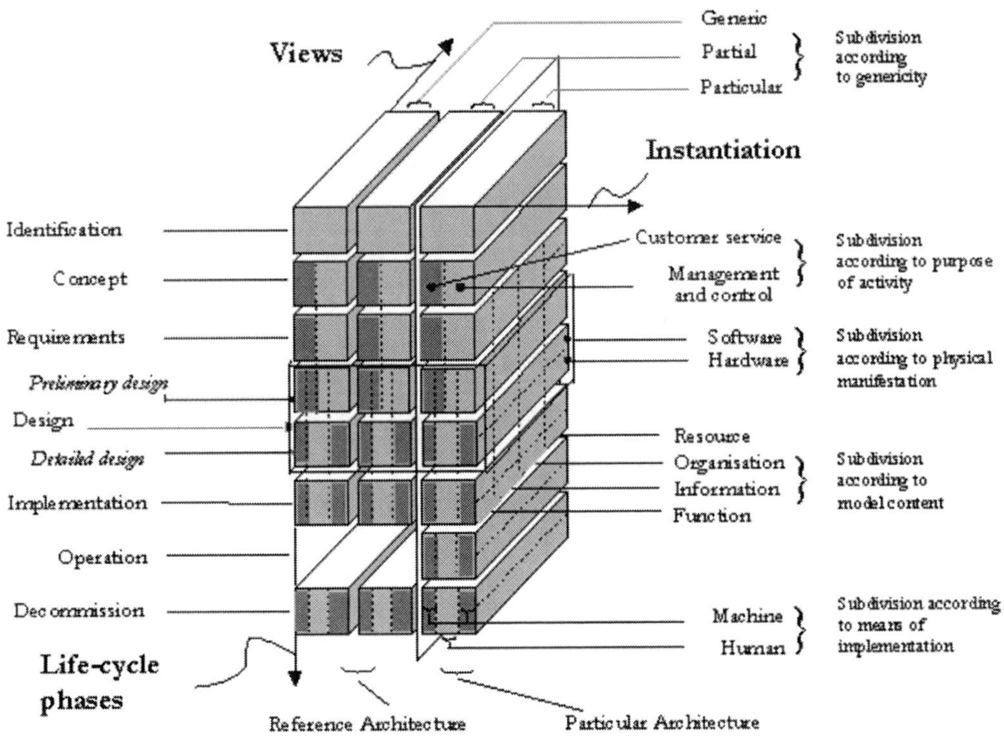

Figure 2. Levels of manufacturing systems of the enterprise.

The company's reference architectures do not specify how to implement them in terms of specific technologies, i.e. are platform-independent. In addition to the company's reference architectures, some software vendors have developed solutions to overcome bottlenecks of vertical integration between MES-levels, for example, MES-HYDRA. But data exchange between MES-levels continues to be done manually or, at best, semi-automatically due to inflexible and closed interfaces [3].

An agent-oriented production monitoring and control system (PMC) was developed and based on the JADE (JADE framework) [6] PMC Provis.Agent, which integrates various information systems and machine control devices, determines the use of information between different systems (for example, for visualization). The NIIP-SMART architecture provides horizontal and vertical integration and interoperability, using workflow, enterprise rules, agents and STEP.

In the service-oriented architecture (SOA) used to integrate enterprise applications, business applications offer their functionality as loosely coupled services (services) that are combined into complex work flows. Due to the loosely coupled structure, the implemented IT architecture is flexible and adaptable. European projects SIRENA and SOCRADES [7] use the SOA paradigm for seamless integration of heterogeneous production-level resources with business applications at the enterprise management level. Unlike the prevailing "request-response" communication approach, an enterprise must react to current events, which requires the implementation of the "publication-subscription" mechanism. SOA and event handling are complementary concepts that provide modularity, weak connectivity and flexibility.

The enterprise data model is a necessary part of the EI for correlating the TO-BE and AS-IS criteria of the various MES levels. IEC 62264-2 [8] describes models and terminology, which allows to realize interaction between the level of enterprise management and the level of production management. For system integration, the structure of the MES database with respect to resources, system plans, system statuses and system configuration was analyzed [9]. The production data model is defined for represent strategic objectives, the organization structure and the enterprise's behavior. This data model consists of the classes "strategy", "object", "process", "resource", "token" and "flow". The "token" class is a physical flow of material (for example, a workpiece, an account). The "flow" class is a reference to the "token" and "process" classes. For example, The Data Reference Model (DRM) is one of the five major reference models in Enterprise Architecture, which gave rise to a comprehensive description and planning of the development of the business device of the enterprise. It is necessary for any leader, at least in order to know that his organization is a whole, to maintain an expedient order of its arrangement, and then to proceed to its current development or transformation taking into account all important recommendations. It is a structure whose main purpose is to ensure the exchange of information and its reuse through the federal government through a standard description and detection of common data and the promotion of common data management methods. In other words, it is a model for describing information.

Not explicit and context-specific knowledge is embedded in the processes of the enterprise by employees of the enterprise, i.e., "know-who" in the form of “know-what”, “know- why” and "know-how". This knowledge is specified by the employees of the enterprise and can be expressed using the data (attributes) of the enterprise process (i.e., TO-BE values). This data as a process feedback is constantly updated during each process execution (i.e., AS-IS values). TO-BE and AS-IS values from different MES levels are integrated for the operational management of enterprise processes [4]. These integrated values are stored in the relational database as historical data that are periodically used in the processes of calculating key performance indicators (KPIs), overall equipment effectiveness (OEE) and the like.

The process of decision-making related to the management of enterprise processes covering different MES levels is a complex task, depending on the quantity and quality of the information. A framework for the organization and application of knowledge in decision-making in the field of production was developed [10] in which the decision-making process was supported by knowledge obtained using data mining algorithms.

The European project K-NET (sub-project of the FInES project cluster - Future Internet Enterprise Systems) presented a conceptual approach to improving, monitoring and reusing knowledge in a networked enterprise. In addition, the modeling and enterprise integration framework based on Knowledge Discovery in Databases (KDD) with the extension of CIMOSA views. In most enterprises, explicit knowledge is codified as rules in managed systems based on rules (RBS - Rule-Based Systems).

RBS Drools Expert, for example, often lacks time and cause-effect relationships between events, therefore, studies were conducted on the use of explicit knowledge to create patterns of events used in the complex event processing (CEP) program. The concept of unified event management architecture for the simplest and complex events for monitoring and control of production processes was developed [11]. The architecture is positioned at the enterprise management level and integrates data in real time from production level, but neglects the integration of transaction data from the enterprise management level. Later, the architecture for the expandable event-driven production system was refined. This architecture is built on a MES solution with a tight integration of enterprise-level management, and uses the CEP to manage events generated by the production level. However, the approach presented does not take into account the knowledge identification required to define the event templates.

The Insights system is an open source framework for operational monitoring and analysis of manufacturing enterprises [12], which contains components for the delivery, collection and analysis of data. Data delivery from various devices is achieved through the MTConnect standard and the MTConnect data bus. Data is stored in high-speed databases. To monitor enterprise processes, data analysis is performed online, using the services of Esper Tech or Drools Fusion [13] CEP. Saved data is used to calculate various indicators.

3. Methodology

For digital integration of the enterprise, an IT framework was developed (figure 3), which uses tracking objects along with RBS.

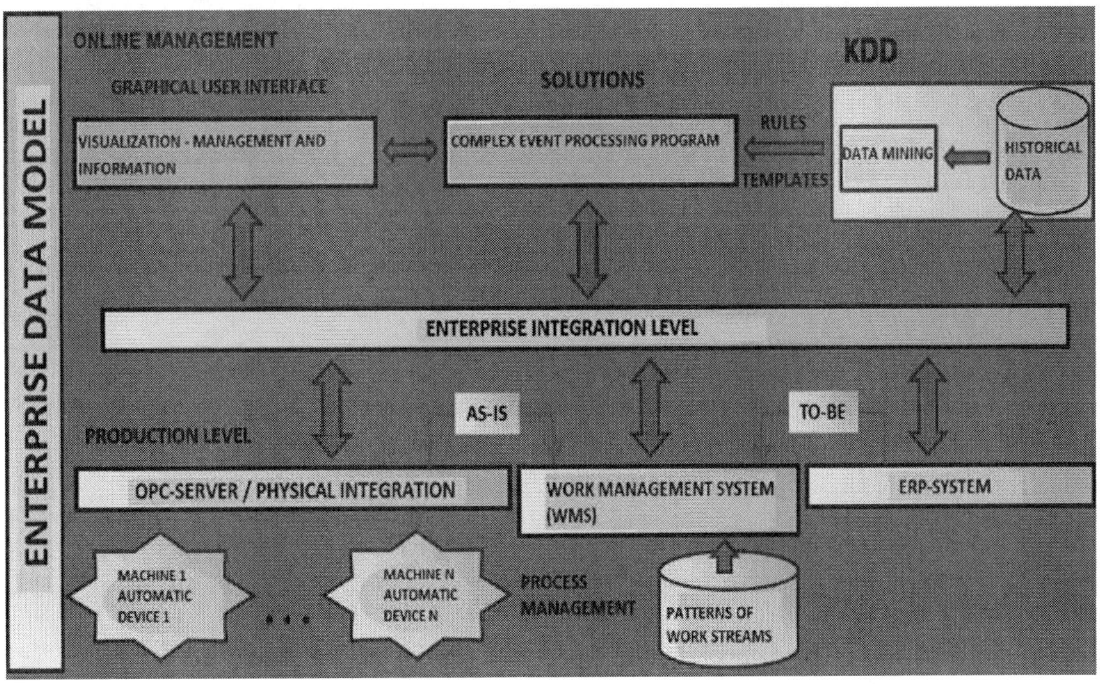

Figure 3. IT-framework for digital integration of the enterprise.

To take into account temporal and cause-effect relationships between events launched at different levels of the enterprise, it is necessary to supplement RBS with modern CEP. This framework is based on the developed methodology for the identification and use of knowledge for the operational control of enterprise processes, including the components of the analysis of enterprise processes, enterprise data models and diagrams of data flows, knowledge identification, and assimilation of knowledge.

3.1. Overview of the IT framework for digital enterprise integration

The architecture of the IT framework digital enterprise integration is shown in the figure 2. The flow of process data and control data between the various components of this IT framework is shown in figure 4.

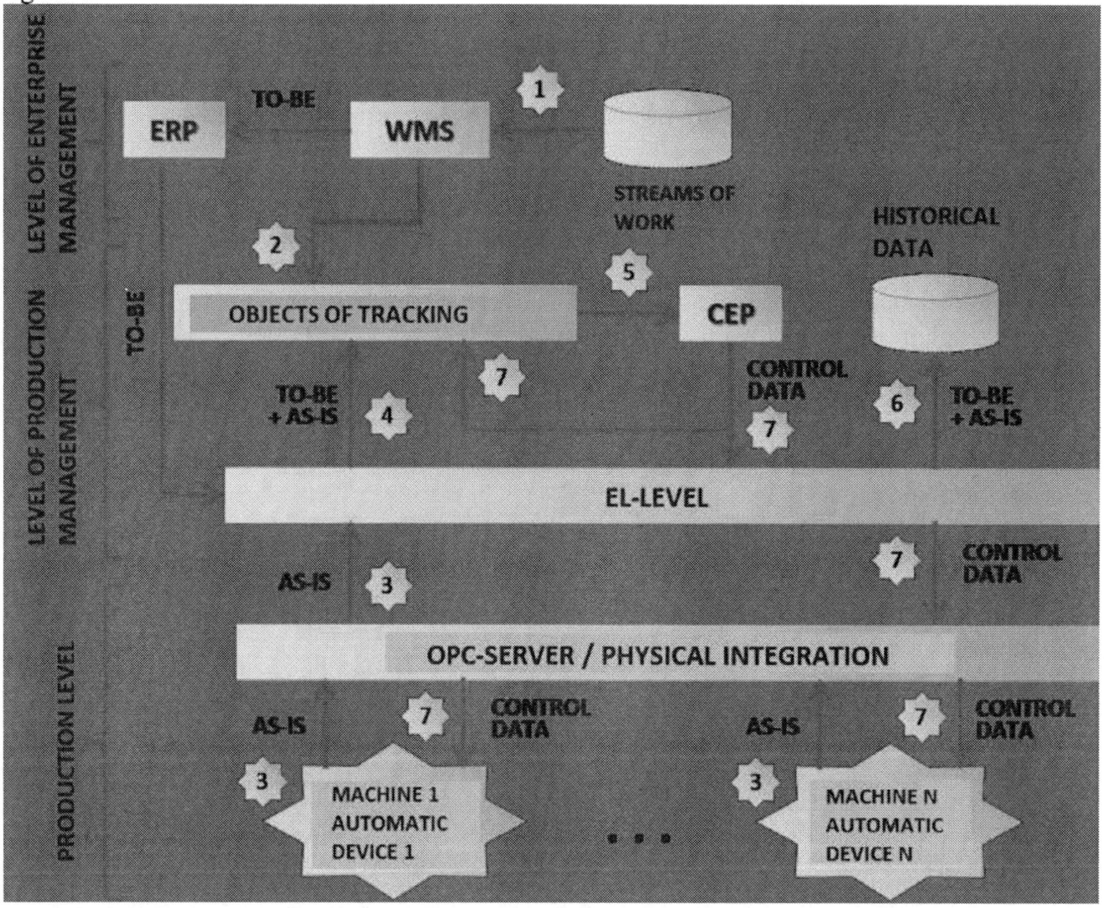

Figure 4. Process data flow and control data IT enterprise integration digital enterprise.

The IT framework is based on current standards (ISO15704, IEC 62264), technologies and paradigms (SOA paradigm) and includes various IT systems. Enterprise processes are defined earlier by the created workflow templates (step 1 in figure 4) and are supported with the required planned performance or process criteria (TO-BE values) such as the ERP system (figure 3 and Step 2 in figure 4). Business applications or their critical functionality in a specific context (access to planned performance values) are available as services (services) within the SOA.

For continuous monitoring of enterprise processes, as one of the goals of the IT framework, a publishing-subscription mechanism used in event-driven architectures (EDA) is used. It is a software architecture template that allows the creation, definition, consumption and response to events.

This architectural template can be used to develop and implement applications and systems that transfer events among loosely coupled software components and services. An event-driven system typically contains sources of events (or agents) and event consumers (or sinks). The drains are responsible for the response as soon as the event has occurred. The reaction can be fully or not completely created by runoff. For example, the runoff can only be responsible for filtering, converting and delivering an event to another component, or it can create its own response to this event. The first category of waste can be based on traditional components, such as middleware for messaging, and the second category of wastewater (which forms its own reaction in the process of operation) may require a more suitable transaction execution platform. Event-driven architecture corresponds to a service-oriented architecture (SOA), because services can be activated by triggers triggered by incoming events.

Service-oriented architecture, driven by events, develops SOA and EDA architectures to provide a deeper and more reliable service level by using previously unknown cause-effect relationships to form a new event model. This new business intelligence template leads to further automation of the processing, adding to the enterprise previously unachievable performance by adding valuable information to the recognized pattern of activity.

The EI layer subscribes to events generated by production resources (automatic devices) at the production level. To collect data from these production resources and forward them to all subscribers, a three-level physical integration architecture is used (step 3 in figure 4). The received real-time data from the production level contain the actual performance (AS-IS values). The enterprise management level plan values together with the actual production level performance values are combined according to the enterprise data model and stored as a historical data in the relational database (step 6 in figure 4).

The IT-framework for digital integration of the enterprise is expanded by the operational monitoring and control of the enterprise processes using the tracking objects, which are objects such as orders, products and resources that are the entities of the corresponding workflow in the WMS (step 2 in figure 4). These objects change during execution of the process with the values of different MES levels (step 4 in figure 4). Changes in the state of the tracking objects are analyzed in a constant mode (online) using RBS (for example, Drools Expert). The use of RBS implies the absence of time and cause-effect relationships between events. It should be noted that only a few WMSs support the collection and interpretation of real-time data, hence, it becomes necessary to use CEP (for example, Esper Tech) instead of RBS. SERE-continuously analyzes the tracking objects and directs the control data to the required levels (step 7 in figure 3).

In addition to monitoring enterprise processes using CEP, historical data are periodically used to calculate various indicators, including key performance indicators. However, historical data are rarely used to acquire new knowledge, and in the future this knowledge is not used for operational monitoring and management of enterprise processes. To overcome these shortcomings, a methodology for identifying and assimilating knowledge for operational monitoring and management of enterprise processes has been developed.

3.2. Methodology of identification and assimilation of knowledge

The model of the process of implementing the digital integration of the enterprise is shown in figure 5.

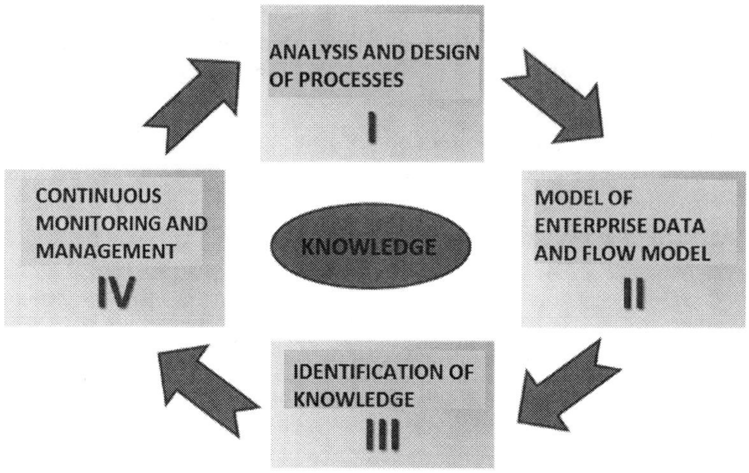

Figure 5. Model of the process of realizing the operational control of enterprise processes.

This process model consists of four stages, each of which is unique to a particular enterprise and they need not necessarily be performed sequentially; moreover, individual stages of the process can be performed from time to time to improve the processes of the enterprise.

Before the implementation of the IT infrastructure of enterprise digital integration, it is necessary to analyze and redesign the enterprise processes, similar to the life cycle of BPM (Phase I in figure 5). The company data model based on industry standards (IEC 62264) is extended by technical models (DIN EN 61512-2) and relates the corresponding AS-IS and TO-BE values of different MES levels (step II in figure 5). Data link diagrams (DFDs) should be created to identify the relationships between business applications, resources, and associated events generated at different MES levels.

The knowledge (pressure, temperature) generated in the data of the enterprise's processes generated before and during the execution of these processes are TO-BE and AS-IS values. These process data are mapped to the mentioned enterprise data model and stored in the relational database as historical data. Subsequently, the standalone KDD process can be run on historical data to determine new knowledge (phase III in figure 5). Reuse of time-consuming queries to the database, data is not optimal for the on-line monitoring of enterprise processes. Event streams (TO-BE and AS-IS values) created during the execution of the process must be analyzed and processed operatively (online) using CEP (step IV in figure 5). Explicit knowledge must be codified as event templates and their corresponding rules, and these event patterns should be used to detect complex events. Along with the detection of complex events, event processing should be used for operational control of enterprise processes.

3.2.1. Step 1. Analysis and (re-) design of enterprise processes

The analysis and redesign of enterprise processes is part of the enterprise reference architecture and business process reengineering (BPR) [14]. BPR can be described in four main steps: (i) identifying critical enterprise processes, (ii) reviewing, updating and analyzing enterprise processes (AS-IS analysis), (iii) (re-designing enterprise processes based on AS IS- analysis and (iv) implementation (re-engineered enterprise processes).

Business processes are very diverse, but there are more definite requirements that all of them must meet. One can single out the following principles of organization of business processes formed during the reengineering:

1. Integration of business processes. The most characteristic property of the redesigned processes is the lack of assembly pipelines as a way of coordinating the work of personnel with relatively simple labor functions. When performing complex labor functions, a different organization

of work is required. In practice, of course, it is not always possible to reduce all the stages of the process to the work performed by one person. In this case, a team is created that is responsible for this process. There may be malfunctions and errors, but the losses will be much less than with the traditional organization of work.

2. Horizontal compression of business processes. Comparative evaluations performed by companies that reengineered show that the transition from the traditional organization of work to the implementation of the process by one person allows to reduce the number of personnel and accelerate the process approximately 10-fold. Reduces the number of errors and eliminates the need to keep professionals to eliminate these errors. By reducing the number of employees and a clear distribution of responsibilities between them, manageability is improved.

3. Decentralization of responsibility (vertical compression of business processes). Performers make independent decisions in cases in which they have traditionally had to turn to management.

4. The logic of the implementation of business processes. Linear execution of work is replaced by a logical order (that is, often the work is carried out in parallel). This saves time, which was spent on interconnection of works on different sites.

5. Diversification of business processes. There are different versions of the execution process. A traditional process, oriented to the production of mass production, must be performed equally for all inputs, leading to concerted outputs. Traditional processes are usually very complex, as they are very detailed and in many respects are calculated for exceptions and special cases.

6. Development of various versions of business processes in an ever-changing market is necessary for processes to have different options depending on situations, entrances and market conditions. New processes that have different versions begin with a check step, which determines which version of the process is most appropriate for the current situation. Therefore, new processes, unlike traditional ones, are simpler and more understandable, since each option is oriented only on one, the corresponding situation.

7. Rationalization of management impact. It is a matter of reducing the number of inspections and reducing the degree of managerial influence that does not lead directly to the acquisition of material values. Therefore, the task of reengineering is to implement them only to the extent that it is economically feasible.

8. Culture of the solution of the problem. It is assumed that the harmonization is minimized, since they too have no material value. The task of reengineering is to minimize reconciliations in the course of execution of the process by reducing external contacts.

9. Rationalization of "company-customer" ties. Improvement of the company's organizational structure should create conditions under which the authorized manager provides a single channel of communication.

10. Preservation of positive moments of centralization of management. In practice, this is achieved by improving the information support of the divisional management organization. Modern IT enables the company's units to operate autonomously, while retaining the ability to use centralized data. Thus, the company can eliminate the bureaucratic regional structures necessary to serve a geographically dispersed clientele, and simultaneously improve the quality of service.

11. Rationalization of horizontal links. Creation of linear functional units. Work is carried out in the place where it is most appropriate. Previously, in companies, the work was organized according to the "thematic" principle in the relevant divisions: the settlement department, the transport department, the supply department, etc., so if the accounting department needed pencils, then he applied to the supply department with the application. This department found the manufacturer, negotiated the price, placed the order, inspected the goods, paid for it, and transferred it to the settlement department. This process is long and uneconomical. An analysis conducted in one of the companies showed that the cost of buying a battery for $ 3. amounted to 100 dollars. During reengineering, horizontal management connections between departments are most often created. This eliminates unnecessary integration.

IOP Conf. Series: Materials Science and Engineering **618** (2019) 012009 doi:10.1088/1757-899X/618/1/012009

12. Authorized leader. This principle is applied in those cases when the steps of the process are either complex or distributed in such a way that their integration by the forces of a small team is impossible. The authorized manager is a buffer between the complex process and the customer. The manager in the relationship with the customer is responsible for the entire process. To play this role, the manager should be able to answer customer questions and solve his problems. The content of the task makes it necessary to provide the manager with access to all the information systems used in this process, as well as to his executors.

Understanding the processes of the enterprise and their integration within the organizational structure of the enterprise is crucial for the implementation of the strategy of operational (online) management of enterprise processes. Thus, the analysis and redesign of processes involves a corporate organizational structure at the same time as the process-oriented organization itself.

Activities and functions of the enterprise process are performed by various resources (IT systems) responsible for ensuring the operation of the organizational units of the enterprise, which are modeled by organizational charts. Functions and activities of the enterprise process accept several types of inputs, this can be data (printed documents) and intangible resources, as implicit knowledge of the participants of the enterprise. Each function of the enterprise process can be associated with the organizational unit with various inputs and outputs. Functions are combined into enterprise processes using logical connectors "and", "or", "exclusive or".

Several modeling languages and methodologies are used to model enterprise processes, both on the basis of an event-driven process chain (EPC), and models and notations of business processes (BPMN - Business Process Model and Notation). In addition to these modeling languages that focus on material inputs and outputs (data and documents), the descriptive knowledge management languages (KMDL - Knowledge Management Description Language) can be used to describe "knowledge" processes with the creation and use of knowledge in enterprise processes.

3.2.2. Step 2. Enterprise data model and data flow diagram

Today's business applications and automatic devices are complex and require a lot of data to determine the processes of the enterprise, in real time automatic devices produce a huge amount of data, such as feedback, product position, alerts. For operational monitoring and control of these processes, it is necessary to analyze business applications, automatic devices and their corresponding processes to identify important parameters. In this regard, the modeling of enterprise data is an important step and affects the quality of information needed to perform the processes, achieve EI, and improve operational monitoring and management of enterprise processes.

In addition to the static structure of the enterprise data model, DFDs identifies the interdependencies between process systems and production resources, either alone or in combination. In addition, DFDs identifies the flow of data describing the dynamic behavior of enterprise processes. Data is generated and managed by several resources at runtime. DFDs can be used to show the relationships between different information systems and resources. As the number of systems and resources increases, DFDs are organized in the DFDs hierarchy. Large-scale DFDs can display, for example, a trading floor and its resources, while small-scale DFDs is created for a particular trading room sub-process.

3.2.3. Step 3. Identification of knowledge

In an environment where information and knowledge - these entities that do not receive adequate objectification beyond their own power - become the main production factor, the problem loses its former economic character and to some extent becomes a sociological and even psychological problem. This is confirmed, in particular, by the fact that it identifies and develops problems.

Knowledge can be defined from different points of view. In the current context, the following definition is used: "Data is raw numbers and facts, information is processed data, and knowledge is verified information".

IOP Conf. Series: Materials Science and Engineering **618** (2019) 012009 doi:10.1088/1757-899X/618/1/012009

From the historical data, new knowledge is used, used in online monitoring and process management of the enterprise. Knowledge identification should be performed using a KDD process defined as "the process of mapping low-level data to other forms that can be more compact or abstract, or useful." It is the process of finding useful knowledge in "raw" data. KDD includes questions: preparing data, selecting informative features, cleaning data, applying Data Mining (DM) methods, post-processing data, and interpreting the results. Of course, the "core" of this whole process is the DM methods, which allow us to discover knowledge. These knowledge can be rules that describe the relationships between data properties (decision trees), frequently encountered patterns (associative rules), and classification results (neural networks), data clustering, etc. The KDD process consists of the following steps:

1) Prepare the original data set. This stage consists in creating a data set, including from various sources, the selection of the training sample, and so on. To do this, there must be developed tools for accessing various data sources. It is desirable to have support for working with data warehouses and the existence of a semantic layer that allows you to use business concepts to prepare the initial data, not technical terms. The system includes Deductor Warehouse - a multidimensional data warehouse, focused on solving problems of consolidating information from heterogeneous sources and quickly retrieving the data set of interest.

2) Preprocessing of data. In order to effectively apply the methods of Data Mining, you should pay attention to the issues of data preprocessing. Data may contain omissions, noise, abnormal values, etc. In addition, the data may be redundant, insufficient, etc. In some problems it is required to supplement the data with some a priori information. It is naive to assume that if you submit data to the input of the system in its existing form, then at the output you will get useful knowledge. The data must be qualitative and correct in terms of the DM method used. Therefore, the first stage of KDD is the preprocessing of data. Moreover, sometimes the dimension of the original space can be very large, and then it is desirable to apply special algorithms for reducing the dimension. This is both the selection of significant features, and the mapping of data into a space of smaller dimensions. It also includes a Deductor that contains a large set of mechanisms for preprocessing and cleaning data: filling in blanks, editing anomalies, removing noise, smoothing, filtering and many others with the ability to combine preprocessing methods.

3) Transformation, normalization of data. This step is necessary to bring information to a form suitable for later analysis. For what it is necessary to do such operations as casting, quantization, and reduction to a "sliding window" and so on. In addition, some analysis methods that require that the source data be in some specific form. Neural networks, for example, work only with numerical data, and they should be normalized. Here Deductor includes a large set of data transformation mechanisms that allow all the preparatory work for the subsequent analysis. In addition, the system contains a wide range of normalization mechanisms for all types of data: numeric, string, date / time and logical.

4) Data Mining. At this step, various algorithms are used to find knowledge. These are neural networks, decision trees, clustering algorithms, associations, etc. The package includes algorithms that implement popular and effective Data Mining methods: neural networks, decision trees, Kohonen self-organizing maps, associative rules and so on.

5) Post processing the data. Interpretation of results and application of acquired knowledge in business applications. The results of any processing can be displayed using a large set of visualization mechanisms: OLAP, tables, diagrams, trees and many others. For some mechanisms, specialized visualizers are provided, which make it easy to interpret the results. Results can be exported for later processing with other applications. This makes it possible to effectively use the acquired knowledge or models on other data.

The KDD process is depicted in Fig. 6. The KDD process receives historical data, and at the output we get templates with some definite quality, known as a measure of interest. This measure of interest can be objectively measured by statistical methods, based on the properties found by patterns, can be based on subjective indicators that are derived from user beliefs or expectations. A template is an abstract representation of a subset of data and should be evaluated by domain experts to identify knowledge.

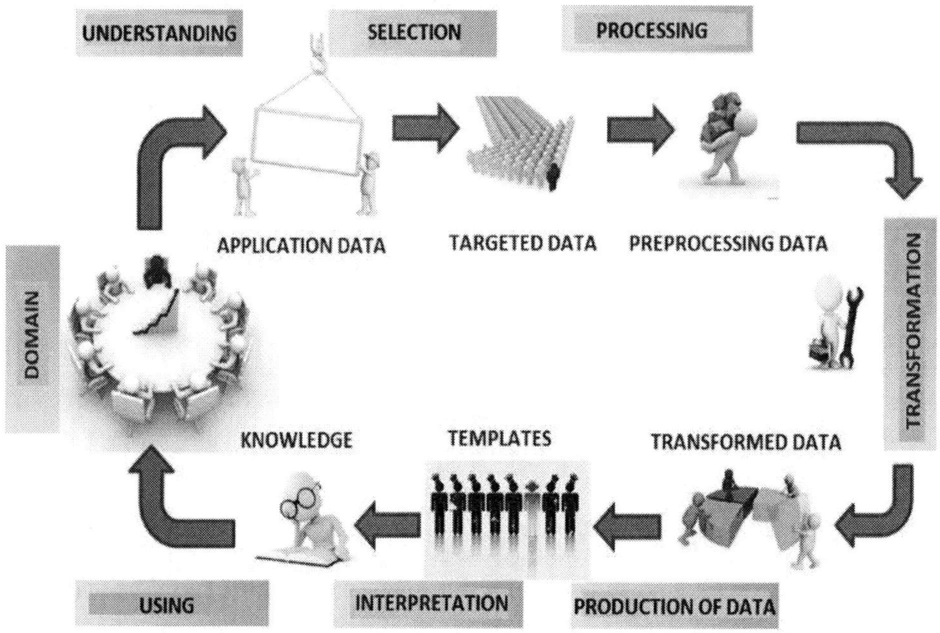

Figure 6. Data discovery process in databases (KDD process)

Understanding the domain is necessary for the successful implementation of the KDD process. This should be done as described in the analysis and redesign of the process, enterprise data models and DFDs (steps I and II in Fig. 5). Based on the main activities of manufacturing and gas distribution companies (production (transportation), maintenance, quality management and production preparation), it is necessary to determine the objectives of the KDD process. Depending on the goals of the KDD process, the specific data (target data) from which historical templates are searched will be extracted from the historical data.

For various reasons, historical data may be inaccurate, influencing the identified knowledge. During data collection, the latter come from operators through consoles, analog equipment and digital meters with programming logic controllers (PLCs - Programming Logic Controllers) (Fig. 3). In the AS-IS IT framework, the values become available for the Enterprise Integration layer through Object Linking and Embedding for OLE for Process Control (Fig. 3 and Step 3 in Fig. 4). The collected data consists of noise, inaccuracies, missed values, which complicates the search for templates. This can be caused by limitations of measuring instruments, operator misprints or errors in PLCs logic. To overcome these inaccuracies, it is necessary to study and understand the subject domain (stage I in Fig. 5). Understanding the domain along with the enterprise data model will help determine appropriate statistical methods for removing noise, a strategy to fill in the missing values and remove duplicate data. Data collection should be supplemented by verification of collected data from production resources in coordination with operators. In general, cleaning and pre-treatment should be performed on the selected data for further processing.

To achieve the above goals, the KDD process requires a subset of pre-processed data. Data conversion causes a reduction in the number of parameters in the target data or the presentation of the target data in a more general or acceptable format. To reduce the number of parameters, filtering and data convolution should be used [15]. Understanding production processes, operations and constraints will help in the transformation process supported by the enterprise data model and DFDs.

Data mining is a specific sub-process of the KDD process and is based on proven techniques of machine learning, pattern recognition, statistics, artificial intelligence, knowledge collection, data visualization and high-performance computing. Data mining consists of three stages: selecting a method for searching for data, determining the appropriate algorithms for searching for data, and using these algorithms to find a template.

Data mining methods should be selected in accordance with the objectives of the KDD process, as they determine the type of knowledge that will be extracted, i. E. description of the concept, classification, association, clustering and forecasting. The knowledge necessary to support decision-making processes must be identified. To determine new knowledge, it is necessary to use methods of classification and regression. Classification is a way of categorizing new instances of data into one of several predefined classes. It consists of two steps: constructing a model (classification) based on the analysis of tuples of a database (training set) described by attributes, and using this constructed (classification) model to classify new instances of data. Regression is a function that maps data to a predictable real variable.

KDD includes multi-disciplinary events. This includes storing and accessing data, scaling algorithms to massive data sets, and interpreting the results. The process of data cleaning and data access included in the data warehouse. Artificial intelligence supports KDD, revealing the empirical laws of experimentation and observation. Patterns recognized in the data must be valid with new data and have some degree of certainty. These templates are considered new knowledge.

In accordance with the previously selected methods of data extraction (for example, classification), data mining algorithms should be chosen to search for templates; Decision trees, decision rules, programmable inductive logic can be used to define rules [10]. Selected methods and data mining algorithms are used to detect templates.

The detected patterns can be large enough or require a subset of the detected patterns. To improve the quality of the detected patterns, the measure of interest should include subjective and objective approaches [15]. The detected template can be obtained using structured interviews of experts in the subject area. If necessary, all or individual steps of the KDD process must be repeated in order to obtain more accurate knowledge. If significant integrated data is not available for the KDD process, structured interviews with domain experts should be conducted to identify the primary knowledge. Later, the collected historical data can be used in the KDD process to increase and expand the knowledge base.

3.2.4. Step 4. Knowledge-Based Operational Control of Enterprise

Repeated and time-consuming requests to the database (for example, the use of online analytical processing requests, OLAP - OnLine Analytical Processing) result in non-operative (offline) monitoring of enterprise processes. Consequently, real-time data must be processed in real time-online (or almost real-time) to monitor enterprise processes, using pre-identified knowledge, tracking objects, and CEP, as shown in Fig. 3 and 4.

Tracking objects represent the entities (products, orders, resources) of a particular enterprise process that are monitored by the workflow of the WMS process route (step 1 in Fig. 3) and the associated performance targets (TO-BE values). During the execution of enterprise processes, AS-IS production-level values are available in the OPC servers and simultaneously directed to the EI layer using the publish-subscribe mechanism (step 3 in Fig. 4). These AS-IS values are integrated with the corresponding TO-BE values from the ERP system obtained using the request-response mechanism.

The EI layer manages the integrated process data simultaneously in several ways. First, tracking objects are updated with the corresponding integrated process data (step 4 in Fig. 4) and therefore contain the latest status information about the entities of the particular enterprise in its process. Secondly, integrated process data is delivered by subscription to all subscribed clients with a graphical user interface (GUI) for operational monitoring of enterprise processes (Fig. 3). Finally, integrated process data is stored in the relational database as historical data for offline analysis (Fig. 3 and Step 6 in Fig. 4). Tracking objects are constantly analyzed at the level of production management and the result is used for the on-line monitoring of enterprise processes using the CEP to improve the main lines of business activity (step 5 in Fig. 4). The CEP is responsible for scheduling the control data to the production resources (step 7 in Fig. 4), for updating the tracking objects with the control data.

The event is characterized by its source (a certain automation device), type, attribute (data) and time stamp or time interval and, in addition, the receiver (operator, enterprise manager). Events are initiated at various levels of the MES during the execution of enterprise processes and form an event cloud [12]. Events can be classified as simple and complex (composite), described by their levels of abstraction. Simple events are run at different levels of the MES and have no abstraction. Consequently, a simple event does not provide enough information for the on-line monitoring of enterprise processes [11]. For example, a simple event is triggered each time a document is printed on the printer.

Conversely, a complex event with a higher abstraction can be described by an event model based on simple events [11]. For example, a complex event is triggered whenever the total number of printed pages of documents exceeds the corresponding article of the department's monthly budget. Higher abstraction events can be obtained from composite events, as shown in Fig. 7. A composite event can be created by combining basic events using a specific set of constructors, such as disjunction, join, sequence, and so on. An event flow is a linearly ordered sequence of events that is ordered by arrival time or is limited to a certain time interval. Event streams consist of simple events that indicate the creation or update of tracking objects at runtime.

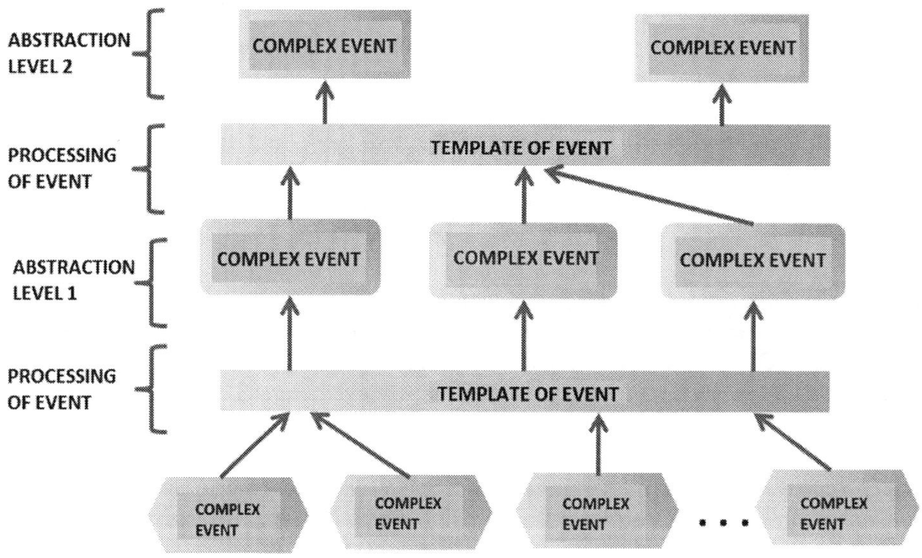

Figure 7. Abstraction of a hierarchical event.

An event template is a template that contains event patterns, relational operators, and variables. The relationships between various (simple and composite) events can be basic, (temporal and spatial) [12], logical, (temporal and causal) [11, 16]. Simple events are created at a certain time period and have timestamps associated with them. The participants of the enterprise from the level of enterprise management and the level of production management are interested in aggregated events for the on-line monitoring and management of products, orders or resources. To make event aggregation available, it is necessary to use temporal event patterns with time slots to support slippery time boundaries [11, 16], as shown in Fig. 8. Temporal event patterns include overlap, match, content and event patterns before or after [12]; and also defined the operators (concatenation, sequence) associated with templates of temporal events.

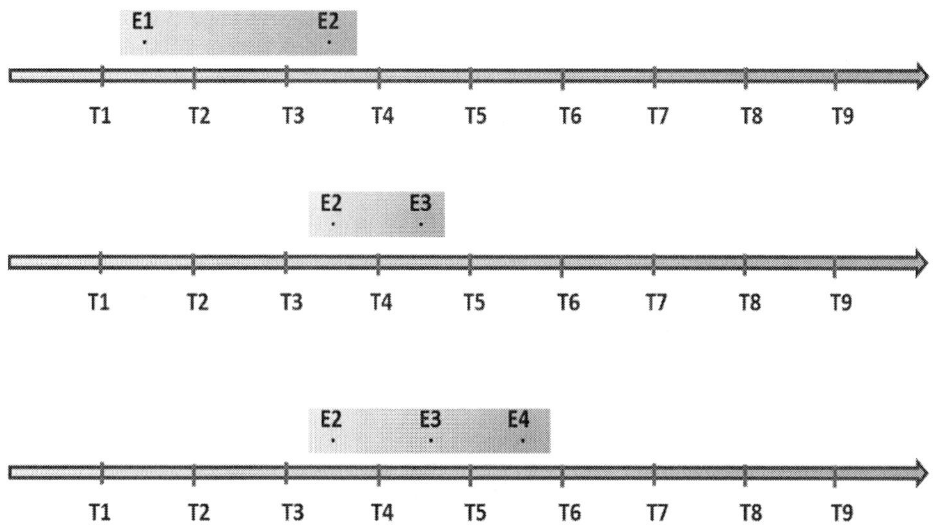

Figure 8. Processing of time events using sliding time boundaries.

CEP performs operations on complex events, including reading, creating, converting, or abstracting [16]. The main task of CEP - continuously analyzing the flow of events (tracking objects), monitor the processes of the enterprise. Tracking objects contain the current status information of entities of enterprise processes. When updating new incoming integral data, the tracking objects are analyzed by CEP (step 5 in Fig. 4). Inside the CEP, capable of analyzing logical, temporal and cause-effect event templates, event templates written in the Event Processing Language (EPL) are used. Rules of event templates (reactive rules) determine how the CER reacts to the occurrence of a specific event template. Incoming events are analyzed using event template rules and the necessary control data. Events with higher abstractions are created and added to the already existing event cloud for further processing.

For effective monitoring and management of enterprise processes, it is necessary to identify and characterize events. Previously, certain knowledge is assimilated by creating event templates and rules of event templates with EPL programming. Structured interviews with domain experts are used to refine and improve event templates and corresponding rules of event templates. Event templates and rules for event templates are stored in a centralized database. Participants in the enterprise or decision-makers do not need all-events. Event receivers (or event users) can be conFig.d in accordance with the roles of the enterprise participants (manager, plant manager) and the appropriate privileges (defined in the LDAP server protocol, LDAP - Lightweight Directory Access Protocol). The event configuration provides the necessary functionality for configuring event templates.

There are two realizations of how CEP influences or controls enterprise processes. First. CEP uses the interfaces and services provided by the EI layer (Fig. 3), for automatic dispatch control commands. The second. Before manipulating the processes of the enterprise, CEP exposes the envisaged solution as a proposal for customers with a GUI. The enterprise participant accepts or rejects the offer. Human interaction is used in cases when the participants of the enterprise must take responsibility. Access to the above functionality of dispatching control data depends on the role of members of a particular enterprise and their respective privileges.

SOA software vendors will extend their proposed CEP features, enabling them to support an event-based architecture as an alternative or complement to SOA. More and more manufacturers will implement CEP support in their BAM tools for BPM environments in order to be able to react almost instantly to changing business conditions. Corporate bus manufacturers will invest in CEP to offer user-friendly means of combining events, correlating and visualizing them for their publishing and subscription environments. But the CEP market does not realize its full potential until the producer community prepares a coherent platform for interoperability that uses open SOA standards. In this respect, the creation of the Event Processing Technical Society (EPTS) community, a group of manufacturers and other interested parties at the end of 2006, is very encouraging in order to popularize CEP applications, clarify the relevant terminology and define a reference platform to support interoperability.

However, EPTS representatives made it clear that they do not intend to become a standards-setting organization, although they do not deny the possibility of subsequently collaborating with groups dealing with standards. EPTS has not yet submitted open proposals for the reference platform, nor has it accessed other community groups, such as the Organization for Advancement of Structured Information Standards (OASIS), which is developing a standard for CEP. The WS-Notifications standard supports event-driven SOA interaction templates and provides notification exchange.

With the advent of open standards in the next few years, the CEP market will begin to develop very quickly. This will also be facilitated by the creation of freely available alternatives and the acquisition by the leading producers of SOA, ESB, BI and EII of the most promising companies specializing in CEP. By the end of this decade, the situation in the CEP market will radically change, and enterprises will be able to implement standards-based CEP infrastructure, calculated for different applications and not depending on the manufacturer.

4. Practice

The mentioned IT framework and the corresponding model of the process of digital integration of the enterprise and achievement of operational control of the enterprise processes are developed and adapted to manage the innovative activity of the gas transportation and gas distribution enterprise. Expected results are applicable to grid companies in the energy sector.

Acknowledgment

The research carried out with the financial support of the grant from the Program Competitiveness Enhancement of Peter the Great St.Petersburg Polytechnic University, Project 5-100-2020

References

[1] D. Linthicum, Enterprise Application Integration, Addison-WesleyLongman, Amsterdam, 2000.

[2] S. Karnouskos, D. Guinard, D. Savio, P. Spiess, O. Baecker, V. Trifa, and L. de Souza, "Towards the Real-Time Enterprise: Service-basedIntegration of Heterogeneous SOA-ready Industrial Devices with Enterprise Applications," Proc. 13th IFAC Symposium on Information Control Problems in Manufacturing (INCOM '09), June 2009, pp. 2127-2132.

[3] H. Panetto and A. Molina, "Enterprise Integration and Interoperability in Manufacturing Systems: Trends and Issues," Computers in Industry, vol. 59, no. 7, Sept. 2008, pp. 641-646.

[4] M. Grauer, D. Metz, S. Karadgi, W. Schäfer, and J. W. Reichwald, "Towards an IT-Framework for Digital Enterprise Integration", Proc. 6th International Conference on Digital Enterprise Technology (DET 2009), AISC, vol. 66, Springer, Berlin, Dec. 2009, pp. 1467-1482.

[5] P. Bernus and L. Nemes, "The Contribution of the Generalised Enterprise Reference Architecture to Consensus in the Area of Enterprise Integration," Proc. of ICEIMT97, K. Kosanke and J. Nell,Eds., Springer, 1997, pp. 175-189.

[6] O. Sauer and G. Sutschet, "Agent-Based Control", Computing & Control Eng. J., vol. 17, no. 3, June 2006, pp. 32-37.

[7] L. M. S. de Souza, P. Spiess, D. Guinard, M. Köhler, S. Karnouskos, and D. Savio, "SOCRADES: A Web Service Based Shop Floor Integration Infrastructure," Internet of Things, C. Floerkemeier, M. Langheinrich, E. Fleisch, F. Mattern, and S. Sarma, Eds., Springer, Berlin, LNCS 4952, 2008, pp. 50-67.

[8] IEC 62264, Enterprise-control system integration.
[9] B. Zhou, S. Wang, and L. Xi, "Data Model Design for ManufacturingExecution System," J. of Manufacturing Technology Management,vol. 16, no. 8, 2005, pp 909-935.
[10] [A. Kusiak, "Data Mining: Manufacturing and Service Applications,"International J. of Production Research, vol. 44, nos. 18-19, Sept./Oct. 2006, pp. 4175-4191.
[11] K. Walzer, J. Rode, D. Wünsch, and M. Groch, "Event-Driven Manufacturing: Unified management of Primitive and Complex Events for Manufacturing and Control", IEEE International Workshop on Factory Communication Systems, 2008.
[12] A. Vijayaraghavan, "MTConnect for Realtime Monitoring andAnalysis of Manufacturing Enterprises," Dec. 2009; http://www.systeminsights.com/, 02.04.2016.
[13] Drools Fusion, http://www.drools.org/, 02.04.2016.
[14] M. Hammer and J. Champy, Reengineering the Corporation: A Manifesto for Business Revolution, Harper Business, 1994.
[15] A. A. Freitas, "A Survey of Evolutionary Algorithms for Data Mining and Knowledge Discovery," Advances in Evolutionary Computing: Theory and Applications, Springer-Verlag, New York, 2003, pp. 819-845.
[16] K. Walzer, A. Schill, and A. Löser, "Temporal Constraints for RuleBased Event Processing", Proc. ACM first Ph.D. workshop in CIKM,Nov. 2007, pp. 93-100.
[17] George L. Bagiev, Olga A. Pogrebova, Evgenii A. Konnikov "Marketing platform of organization of system-spatial complex «PSE» (Production — Science — Education)" Published in: "Quality Management,Transport and Information Security, Information Technologies" (IT&QM&IS). 2017. DOI: 10.1109/ITMQIS.2017.8085864

IOP Conf. Series: Materials Science and Engineering **618** (2019) 012010 doi:10.1088/1757-899X/618/1/012010

Development of Algorithms and Software for Automatic Pedestrian Recognition for Intelligent Situation Control Systems

N Abramov[1], V Rakova[1], D Barinov[1], A Leksashov[2] and E Pridanova[1]

[1] Peter the Great St. Petersburg Polytechnic University St. Petersburg, Russia
nikolay.abramov@spbpu.com; valeriya.rakova@spbpu.com;
dmitriy.barinov@spbpu.com; ekaterina.pridanova@spbpu.com

[2]Tetracube LLC St. Petersburg, Russia
leksashov@mail.ru

Abstract. The article is devoted to the development of algorithms and software that automatically recognize road users - pedestrians, build their trajectory according to the analysis of the video stream. The architecture of the system for pedestrian recognition and inter-frame tracking is proposed. The developed algorithms can be used to estimate road situation with participation of pedestrians for the traffic monitor systems.

1. Introduction

Image processing algorithms have been in the focus of research in the field of intelligent transport systems [1, 2, 3,13] for a long time. Pedestrian detection and tracking tasks are those kinds of problems that can be solved using image processing algorithms. One of the active consumers of these algorithms are traffic monitor systems. These systems make it possible to observe whether drivers let pedestrians cross a road on regulated and unregulated pedestrian crossings equipped with such systems. Fig. 1 represents a typical image from one of those systems. However, this is not the only function of such systems, which also have to detect and track cars, determine their speed, form violation protocols, etc. That's why in this case pedestrian detection algorithms are not only bound by strict requirements for accuracy of detection, but also have strict limitations on the quota of computing resources on the built-in computing platform, which already performs a range of tasks at the same time. Pedestrian detection is a specialized subfield of object detection, which is currently in demand in areas such as automotive, surveillance and security. Caltech Dataset [4] and Kitti [5] are up to date benchmarks for comparing the accuracy and processing time of the algorithms.

During last years diverse efforts have been made to improve the performance of pedestrian detection. Following the success of integral channel feature detector (ICF) [14], many variants [15, 16, 17] were proposed and showed significant improvement. From some recent works one can see that improved features have been driving performance and are likely to continue doing so. These features are often complemented by additional sources of information, such as optical flow [18], depth information [7], context information [19] and many others to even further boost detection accuracy.

Another class of classification algorithms – neural networks – have vastly advanced during recent years. By fine-tuning a model pre-trained on external data convolution neural networks (convolutional networks) have also reached state-of-the-art performance [20, 21].

Currently, neural networks occupy a leading position in recognition accuracy. However, the processing time of these algorithms takes from 0.1 s to several seconds [5] when executed on a graphics accelerator, which is unacceptable in our case. That's why we focus on algorithms suitable for execution on embedded platforms with small computing capabilities.

Figure 1. Typical frame from the traffic monitor system's camera.

Among such algorithms are Fastest pedestrian detection in the west [6], Pedestrian detection at 100 frames per second [7], C4: A Real-Time Object Detection Framework [8,9].

The authors of the Fastest Pedestrian Detection in the West present a method to significantly reduce execution time of multi-scale object detectors that use several types of features, including gradient histograms, with a slight decrease in the accuracy of pedestrian detection.

Two new algorithmic accelerations are presented in [7], one of which is based on improved scale processing (on monocular images), and the other on better use of depth information (on stereo images).

Article [9] makes two major contributions. First, the authors show that the outline defining the silhouette of a person is an important information for detection. Secondly, it is proposed to detect people using contour signals, using CENTRIST descriptor [10] for this purpose. Higher classification speed can be achieved due to the fact that one does not need to explicitly calculate CENTRIST descriptor, because it seamlessly embedded in the classifier estimation.

A distinctive feature of the first two approaches is that they offer more efficient methods of classifier response estimation in a multi-scale detection environment, i.e. when pedestrian size

in image may vary greatly. However, in our case, the camera of a traffic monitor system is located at a distance of 10+ meters from the observed region (pedestrian crossing) and there is no need to use a large number of scales to detect pedestrians. Therefore, as a basis, we decided to use C4 pedestrian detection algorithm [9]. The following drawbacks were identified in this approach: it only uses feature that cover local region structure. In our work, we propose extensions to eliminate this drawback.

Thus, in this article, pedestrian detection algorithm for traffic monitor systems is proposed. In "Baseline classifier C4", we describe C4 pedestrian detection algorithm, taken as a basis in our work. "Multi-scale CENTRIST descriptor" suggests classifier improvement, in the form of a multi-scale approach. Approach to training is presented in "Detection framework". The testing conditions and the evaluation of the results are presented in "Results".

2. Pedestrians detection algorithm for traffic monitor systems

2.1 Baseline classifier C4

C4 pedestrian detector feature calculation process is based on Census transform (CT) [11] which was originally intended to establish similarities between the local image regions. To calculate it one needs to find image gradient at each point of image first. To achieve this, Sobel operator is often used (1). Gradient values at pixel positions are calculated along the X and Y axis, and then combined together to obtain a single value. Thus, G_x is the image with gradient values along the X axis, G_y is the image with gradient values along the Y axis, G is the image with combined gradient values and I is the original image.

$$G_x = \begin{bmatrix} -1 & 0 & +1 \\ -2 & 0 & +2 \\ -1 & 0 & +1 \end{bmatrix} * I$$

$$G_y = \begin{bmatrix} -1 & -2 & -1 \\ 0 & 0 & 0 \\ +1 & +2 & +1 \end{bmatrix} * I \qquad (1)$$

$$G = |G_x| + |G_y|$$

Next, the gradient value of the center pixel is compared to the gradient values of eight neighboring pixels, as shown in Fig.2. If the central pixel value is larger (or equal) to one of its neighbors, bit 1 is set at the appropriate location. Otherwise, bit 0 is set.

$$\begin{array}{c|c|c} 32 & 64 & 96 \\ \hline 32 & \mathbf{64} & 96 \\ \hline 32 & 32 & 96 \end{array} \Rightarrow \begin{array}{ccc} 1 & 1 & 0 \\ 1 & & 0 \\ 1 & 1 & 0 \end{array} \Rightarrow (11010110)_2 \Rightarrow \mathrm{CT} = 214$$

Figure 2. Example calculation of Census transform.

Eight bits generated from the intensity comparisons can be combined in any order to obtain the base 10 number. This is the CT value for the center pixel. The CENTRIST descriptor is a histogram of these CT values [10]. As shown in Fig. 2, the CT values briefly encode the signs of comparison between neighboring pixels. However, the CENTRIST descriptor lacks the ability to describe global (or larger) structures and contours beyond a small 3×3 range.

CENTRIST has some close relationship with local binary pattern, another feature used for human detection. One important difference is how the LBP values are utilized. Pedestrian detection methods use “uniform LBP”, in which certain LBP values that are called “non-uniform” are lumped together. These procedures make their descriptors to only encode a blurred version of the most important information, i.e., signs of neighboring pixel comparisons.

2.2 Multi-scale CENTRIST descriptor

As described above, to obtain the CENTRIST descriptor, a Census transform histogram is computed within the super block with a size of 24 x 16 pixels. Then these histograms are concatenated to form a final vector of features for the classifier. However, these histograms describe only their local area and are not able to represent a larger scale structure [28-31].

To overcome this disadvantage, we propose to use a multiscale approach to calculate CENTRIST descriptor. To do this, one can take four neighboring superblocks, calculate the CENTRIST descriptor for each of them, and obtain a larger-scale descriptor by applying the max pooling operation. Let’s call CENTRIST of a single superblock a 1-st level descriptor. Second level CENTRIST descriptor combines adjacent lower level descriptors of the first level, and so on. Thus, the *i*-th element of the $k + 1$ level CENTRIST descriptor located at the (x, y) – $C_{k+1}^{(x,y)}$ is obtained by applying maximum function to the *i*-th elements of *k* level descriptors – $C_{k+1}^{(x,y)}$.

$$C_{k+1}^{(x,y))}(i) = max(C_k^{(x*2,y*2))}(i), C_k^{(x*2+1,y*2))}(i), C_k^{(x*2,y*2+1))}(i), C_k^{(x*2+1,y*2+1))}(i)) \quad (2)$$

Computation continues while candidate image region contains enough *k* level descriptors to form $k + 1$ level descriptor. Visually, this principle is represented in Fig. 3. After all descriptors are calculated we concatenate them together to form a feature vector for the classifier.

It’s worth to note that Wu et al. already proposed a multi-scale approach for CENTRIST calculation [8]. However, it requires to use image pyramid to extract corresponding CENTRIST descriptors on each pyramid level. It means that we would need to calculate multiple integral images, which is acceptable when pedestrians may greatly vary in size, since we may reuse them from other detection scales. In our approach we expect to have small variation in pedestrian sizes and don’t plan to build integral images for several scales, because of complexity constraints. For the calculation of our multi-scale CENTRIST descriptor no additional image scales are required, since we can get all the required information from level 1 CENTRIST histograms.

Our approach allows to capture the structure of objects on a larger scale and improves the accuracy of pedestrian detection, which is presented in the next section [32, 33].

3. Detection framework

In the training phase, we have a set of 96×32 positive training image patches P and a set of larger negative images N that do not contain any pedestrians. We use block size of 12 x 8 pixels, which we found to show best detection results.

During training we follow the methodology of C4 original authors [9].

We first randomly choose a small set of patches from the images in N to form a negative training set N_1. Using P and N_1 we train a linear SVM classifier H_1.

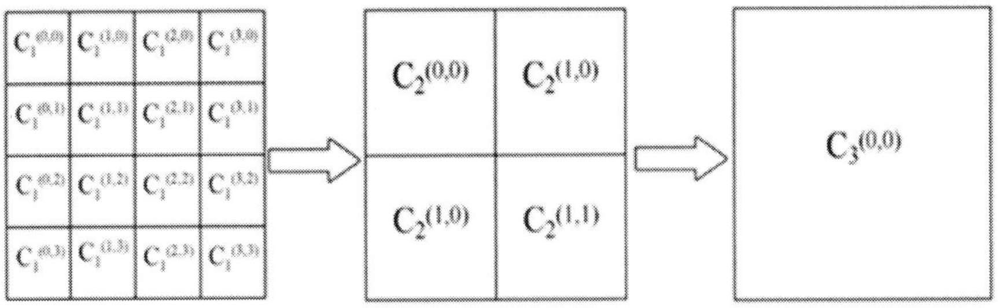

Figure 3. A bootstrap process is used to generate a new negative training set N_2. H_1 is applied to all patches in the images in N. Multi-scale CENTRIST approach.

We then train H_2 using P and N_2. This process is repeated until all patches in N are classified as negative by at least one of previous level classifiers. We then train a linear SVM classifier using P and the combined negative set, which we call H_{lin}.

To goal of the linear classifier is to preserve as much pedestrians as possible (i.e. have high recall) and to filter out considerable amount image regions that don't contain pedestrians. To obtain higher detection accuracy we train a second H_{hik} classifier. We use H_{lin} on $N$to bootstrap a new negative training set N_{final} and train an SVM classifier using the libHIK HIK SVM solver [8]. In the testing and detection phase, a cascade with two nodes H_{lin} and H_{hik} is used.

4. Results

The classifier is written in C ++ language, the learning was carried out according to the procedure described in Sect. III, using the Liblinear [12] and libHIK [8] libraries. For the learning and evaluation of results, the sampling of records from the traffic monitor system cameras was used, containing more than 200,000 positive examples with pedestrians [25-27].

The testing was performed on the Intel Nano-6060 hardware platform with a 4-core Atom E-3845 processor, clocked at 1.95 GHz. Classifier evaluation was carried out on a CPU, using singe core to simulate real-world situation, where most resources would be occupied by other tasks. As an input data we used grayscale images with a resolution of 1280x960 pixels with a region of interest specifying the position of expected pedestrian crossing.

Table 1. Classifier evaluation metrics.

Algorithm	True positive rate	False positive rate	Time per frame, ms
C4 baseline	0.974	0.085	25
C4 ours	0.985	0.054	28

Test results show that the proposed approach slightly increases the recall of pedestrians and reduced the number of false positives by 37%. Due to the use of the hierarchical multi-scale approach, execution time slightly increased by 12%.

After analysis of Table 1 results, we came the following conclusion - the majority of false positives and false negatives in detection results related to the difficult visibility conditions such as night environment, heavy rain, fog, snow and other. In the future, we plan to adapt the algorithm to improve its efficiency in such conditions [22, 23].

5. Conclusion

In this article we presented a way to improve the accuracy of the C4 pedestrian recognition algorithm by using a multi-scale approach. We demonstrated that this modification allowed to increase the coverage of this classifier and reduced the number of false positives. This modification allows using this algorithm to recognize pedestrians in automatic traffic fixing systems built on the basis of low-power embedded systems. A further development direction is to increase the accuracy of the classifier in poor visibility conditions and to develop a tracking system for pedestrians.

Acknowledgements
We are grateful for the support of the "Olvia" LLC for providing numerous pedestrian samples in real-world city environment. The research carried out with the financial support of the grant from the Program Competitiveness Enhancement of Peter the Great St.Petersburg Polytechnic University, Project 5-100-2020.

References

[1] P. Dollar, C. Wojek, B. Schiele, and P. Perona, "Pedestrian Detection: An Evaluation of the State of the Art," IEEE Transactions on Pattern Analysis and Machine Intelligence, vol. 34, no. 4, pp. 743–761, Apr. 2012.
[2] R. Benenson, M. Omran, J. Hosang, and B. Schiele, "Ten Years of Pedestrian Detection, What Have We Learned?," in Computer Vision - ECCV 2014 Workshops, vol. 8926, L. Agapito, M. M. Bronstein, and C. Rother, Eds. Cham: Springer International Publishing, 2015, pp. 613–627.
[3] S. Zhang, R. Benenson, M. Omran, J. Hosang, and B. Schiele, "How Far are We from Solving Pedestrian Detection?," 2016, pp. 1259–1267.
[4] Caltech pedestrian dataset [Electronic resource: http://www.vision.caltech.edu/Image_Datasets/CaltechPedestrians/] (Date of the application: 24.04.2018)
[5] Kitti object detection benchmark [Electronic resource: http://www.cvlibs.net/datasets/kitti/eval_object.php] (Date of the application: 24.04.2018)
[6] P. Dollar, S. Belongie, and P. Perona, "The Fastest Pedestrian Detector in the West," 2010, pp. 68.1-68.11.
[7] R. Benenson, M. Mathias, R. Timofte, and L. Van Gool, "Pedestrian detection at 100 frames per second," 2012, pp. 2903–2910.
[8] Jianxin Wu, Nini Liu, C. Geyer, and J. M. Rehg, "C4: A Real-Time Object Detection Framework," IEEE Transactions on Image Processing, vol. 22, no. 10, pp. 4096–4107, Oct. 2013.
[9] J. Wu, C. Geyer, and J. M. Rehg, "Real-time human detection using contour cues," 2011, pp. 860–867.
[10] Jianxin Wu and J. M. Rehg, "CENTRIST: A Visual Descriptor for Scene Categorization," IEEE Transactions on Pattern Analysis and Machine Intelligence, vol. 33, no. 8, pp. 1489–1501, Aug. 2011.
[11] R. Zabih and J. Woodfill, "Non-parametric local transforms for computing visual correspondence," in Computer Vision — ECCV '94, vol. 801, J.-O. Eklundh, Ed. Berlin/Heidelberg: Springer-Verlag, 1994, pp. 151–158.
[12] R.-E. Fan, K.-W. Chang, C.-J. Hsieh, X.-R. Wang, and C.-J. Lin, "LIBLINEAR: A Library for Large Linear Classification," p. 30.
[13] M. Enzweiler and D. M. Gavrila, "Monocular Pedestrian Detection: Survey and Experiments," IEEE Transactions on Pattern Analysis and Machine Intelligence, vol. 31, no. 12, pp. 2179–2195, Dec. 2009.
[14] P. Dollar, Z. Tu, P. Perona, and S. Belongie, "Integral Channel Features," 2009, pp. 91.1-91.11.
[15] W. Nam, P. Dollar, and J. H. Han, "Local Decorrelation For Improved Pedestrian Detection," p. 9.
[16] S. Zhang, R. Benenson, and B. Schiele. Filtered channel features for pedestrian detection. In CVPR, 2015.
[17] S. Zhang, C. Bauckhage, and A. B. Cremers. Informed haar-like features improve pedestrian detection. In CVPR, 2014.
[18] D. Park, C. L. Zitnick, D. Ramanan, and P. Dollár. Exploring weak stabilization for motion feature extraction. In CVPR, 2013.
[19] W. Ouyang and X. Wang. Single-pedestrian detection aided by multi-pedestrian detection. In CVPR, 2013.
[20] S. Ren, K. He, R. Girshick, and J. Sun, "Faster R-CNN: Towards Real-Time Object Detection

with Region Proposal Networks," IEEE Transactions on Pattern Analysis and Machine Intelligence, vol. 39, no. 6, pp. 1137–1149, Jun. 2017.
[21] F. Yang, W. Choi, and Y. Lin, "Exploit All the Layers: Fast and Accurate CNN Object Detector with Scale Dependent Pooling and Cascaded Rejection Classifiers," 2016, pp. 2129–2137.
[22] N.A. Abramov, M.V. Bolsunovskaya, A.V. Leksashow, , D.S.Barinov, "Algorithms for detecting and tracking of objects with optical markers in 3D space", Proceedings of the 19th International Conference on Soft Computing and Measurements, SCM 2016; Saint Petersburg Electrotechnical University (SPbETU) "LETI", Saint Petersburg, Russian Federation; 25-26 May 2016; Editor Shaposhnikov S.; Institute of Electrical and Electronics Engineers Inc., 2016. Article # 7519782, PP. 362-364, DOI: 10.1109/SCM.2016.7519782)
[23] A.S. Makarov, M.V.Bolsunovskaya, "The 360° around view system for large vehicles, the methods of calibration & removal of barrel distortion for omnidirectional cameras", CEUR Workshop Proceedings. 5th International Conference on Analysis of Images, Social Networks and Texts, AIST 2016; Yekaterinburg; Russian Federation; 6-8 April 2016;. CEUR-WS, 2016. Volume 1710. PP.182-190 (ISSN: 16130073)
[24] V.N. Volkova, A.V. Loginova, S.V. Shirokova, E.A. Kozlovskay, "Development of the innovative IT-project and managing project human resources", Proceedings of the 19th International Conference on Soft Computing and Measurements, SCM 2016, Saint Petersburg Electrotechnical University (SPbETU) "LETI", Saint Petersburg; Russian Federation; 25-27 May 2016; 2016. Article # 7519816, PP. 470-473, DOI: 10.1109/SCM.2016.7519816)
[25] Fisher, W. D., Camp, T. K., & Krzhizhanovskaya, V. V. (2016). Crack detection in earth dam and levee passive seismic data using support vector machines. Paper presented at the Procedia Computer Science, , 80 577-586. doi:10.1016/j.procs.2016.05.339
[26] Gelgor, A., Gorlov, A., & Nguyen, V. P. (2018). Performance analysis of SEFDM with optimal subcarriers spectrum shapes. Paper presented at the 2017 IEEE International Black Sea Conference on Communications and Networking, BlackSeaCom 2017, , 2018-January 1-5. doi:10.1109/BlackSeaCom.2017.8277680
[27] Gelgor, A., Gorlov, A., & Nguyen, V. P. (2016). The design and performance of SEFDM with the sinc-to-RRC modification of subcarriers spectrums. Paper presented at the International Conference on Advanced Technologies for Communications, 65-69. doi:10.1109/ATC.2016.7764831
[28] Inochkin, F. M., Kruglov, S. K., Bronshtein, I. G., Kompan, T. A., Kondratiev, S. V., Korenev, A. S., & Puhov, N. F. (2018). Superresolution contour reconstruction approach to a linear thermal expansion measurement. Paper presented at the Proceedings - International Conference on Image Processing, ICIP, 3843-3847. doi:10.1109/ICIP.2018.8451081
[29] Liokumovich, L., Medvedev, A., Muravyov, K., Skliarov, P., & Ushakov, N. (2017). Signal detection algorithms for interferometric sensors with harmonic phase modulation: Distortion analysis and suppression. Applied Optics, 56(28), 7960-7968. doi:10.1364/AO.56.007960
[30] Lutin, V. I., Desyatirikova, E. N., Kuripta, O. V., Akamsina, N. V., & Mager, V. E. (2018). The processing of signals from sensors to observe objects in various physical fields. Paper presented at the Proceedings of the 2018 IEEE Conference of Russian Young Researchers in Electrical and Electronic Engineering, ElConRus 2018, , 2018-January 1113-1116. doi:10.1109/EIConRus.2018.8317285
[31] Plotnikov, A., & Gelgor, A. (2018). Spectral efficiency comparison between FTN signaling and optimal PR signaling for low complexity detection algorithm doi:10.1007/978-3-030-01168-0_18
[32] Tsikin, I., & Melikhova, A. (2016). Angle-of-arrival GPS integrity monitoring insensitive to satellite constellation geometry doi:10.1007/978-3-319-46301-8_49

IOP Conf. Series: Materials Science and Engineering **618** (2019) 012011 doi:10.1088/1757-899X/618/1/012011

Face Validation Based Anomaly Detection Using Variational Autoencoder

B Zeno[1], Yu Matveev[2], B Alkhatib[3]

[1,2] ITMO University
49 Kronverksky Pr., St. Petersburg, 197101,Russia
[3] Damascus University

[1]basilzeno@gmail.ruline, [2]matveev@mail.ifmo.ru, [3]t_balkhatib@svuonline.org

Abstract. In unconstrained facial images, large visual variations concerning pose, scale, the presence of occlusions, expressions and lighting usually cause difficulties in discriminating faces from the background accurately. As a result, some non-face regions are recognized as faces (false positive) and that influences the effectiveness of face detection algorithms which is characterized by low false positive (FP) rate, high detection rate and high speed of processing. In order to reduce these non-face regions, they are considered as anomalies and then try to detect them. In this paper, we propose an anomaly detection method using reconstruction error from variational autoencoder (VAE), which is a generative machine learning model. We train VAE to learn reconstructing faces that are close to its original input faces using FDDB dataset, then the difference between the original input face and the reconstructed output is measured to obtain the reconstruction error which can be used as an anomaly score. Consequently, the regions resulting from faces detection algorithm with high reconstruction error are defined as anomalies or false positives.

1. Introduction

Face detection is one of the mostly studied problems in vision, it has been actively researched for over two decades [1]. Face detection is considered as a problem of single-class object detection, and it is an important field of research in computer vision, because it forms a necessary first step for many face processing systems such as face recognition, face tracking, face verification and identification or facial expression analysis. Most famous of face detection algorithms have been based on cascading approach [2,3,4,5], Deformable Parts Models (DPMs) [6,7,8,9] and deep convolutional neural networks [10,11,12,13,14], detect faces within the rectangular search region specified by ROI (region of interest), which may contain many non-face regions or false positives that must be rejected.

This work was financially supported by Government of Russian Federation (Grant 08-08).

Many approaches have been proposed to reject non-face regions, that lead to improve the overall detection precision. Some of them based on eyes detection using support vector machine [15,16]. In [17,18,19] skin color methods are proposed to filter non-faces, where YCbCr color space, statistical data in different color spaces and some morphologic operations are used to achieve that.

The nature of non-face data makes its detection very challenging, where non-face data is far greater than that of face and the distribution of non-face pattern is very widespread. Moreover, non-face data are rare and, in most cases, non-existent. This makes the non-face detection a semi-supervised learning problem.

In this study, we try to solve this problem using generative machine learning methods, which recently have become popular due to their ability to model input data distributions and generate new examples from those distributions.
In section II, we briefly present some background about the anomaly detection and generative learning models, and in section III, we present the proposed anomaly detection method. Implementation is presented in section IV. Then we present the conclusion.

2. Ease of Use BACKGROUND

2.1. Anomaly Detection
Anomaly detection refers to the problem of finding patterns in data significantly deviated from expected normal behaviour. It is applied in many areas such as credit card fraud detection, network intrusion detection, sensor network fault detection, medical diagnosis and video surveillance [21].
Many anomaly detection algorithms have been proposed, and they can be further categorized as generative or discriminative approaches. A generative approach such as a Gaussian-mixture model, Variation Autoencoder, or Adversarial Autoencoders (e.g., [22]) builds a model solely based on normal training data and evaluates each testing case to see how well it fits the model. A discriminative approach (e.g., [23]), on the other hand, utilizes the data from normal and abnormal classes to create an explicit decision boundary separating them. The output of an anomaly detection algorithm can be one of two types:

- Anomaly scores: It is computed by evaluating the quality of the fit between the data point and the model. In other words, it quantifies the level of anomaly of each data point.
- Binary labels: A second type of output is a binary label indicating whether a data point is an abnormal or not. Although some algorithms might directly return binary labels, anomaly scores can also be converted into binary labels. This is typically achieved by imposing thresholds on anomaly scores, and the threshold is chosen based on the statistical distribution of the scores. A binary labeling contains less information than a scoring mechanism, but it is the final result that is often needed for decision making in practical applications.

In [22] anomalous event detection approach in images was proposed. It is Adversarial Autoencoders method, which is based on autoencoders in combination with Generative Adversarial Networks. The adversarial error of the learned autoencoder was used as anomaly score, since it is low for regular events and high for irregular events, such as walking on grass, bike, cart, etc. In [24] an anomaly detection method using variational autoencoders (VAE) was proposed. It used the reconstruction probability as anomaly score. Experimental results show that this method outperformed autoencoder based and PCA based methods.

2.2. Variational Autoencoder
Variational Autoencoder (VAE) [27] is generative model that belongs to the field of representation learning in Artificial Intelligence where an input is mapped into hidden representations. In a VAE we have examples X that are distributed according to some unknown distribution Pgen(X), which is specified over latent variables, and the goal is to learn a model P, such that P is as similar as possible to Pgen . Then data is generated or reconstructed from input by mapping the latent variables through a non-linear function implemented by a neural network. In the VAE, the highest layer of the directed graphical model z is treated as the latent variable where the generative process starts. g(z) represents the complex process of data generation that results in the data x, which is modeled in the structure of a neural network. The objective function of a VAE is the variational lower bound of the marginal likelihood of data, since the marginal likelihood is intractable. The marginal likelihood is the sum over the marginal likelihood of individual data points:

$$\log p_\theta(x^{(1)},\ldots,x^{(N)}) = \sum_{i=1}^{N} \log p_\theta(x^{(i)}) \tag{1}$$

where the marginal likelihood of individual data points can be rewritten as follows.

$$\log p_\theta(x^{(i)}) = D_{KL}(q_\phi(z \mid x) \parallel p_\theta(z)) + L(\theta, \phi; x^{(i)}) \tag{2}$$

$q_{\phi}(z \mid x)$ is the approximate posterior and $p_{\theta}(z)$ is the prior distribution of the latent variable z. The first term of the right-hand side of equation (2) is the KL divergence of the approximate posterior and the prior. The second term of the right-hand side of equation (2) is the variational lower bound on the marginal likelihood of the data point i. Since the KL divergence term is always bigger than 0, equation (2) can be rewritten as follows.

$$\log p_{\theta}(x^{(i)}) = -D_{KL}(q_{\phi}(z \mid x) \| p_{\theta}(z)) + E_{q_{\phi}(z|x^{(i)})}[\log p_{\theta}(x \mid z)] \tag{3}$$

$p_{\theta}(x \mid z)$ is the likelihood of the data x given the latent variable z. The first term of equation (3) is the KL divergence between the approximate posterior and the prior of the latent variable z. This term forces the posterior distribution to be similar to the prior distribution, working as a regularization term. The second term of equation (3) can be understood in terms of the reconstruction of x through the posterior distribution $q_{\phi}(z \mid x)$ and the likelihood $p_{\theta}(x \mid z)$. The VAE models the parameters of the approximate posterior $q_{\phi}(z \mid x)$ and by using a neural network. This is where the VAE can relate to the autoencoder. As shown in figure 1.

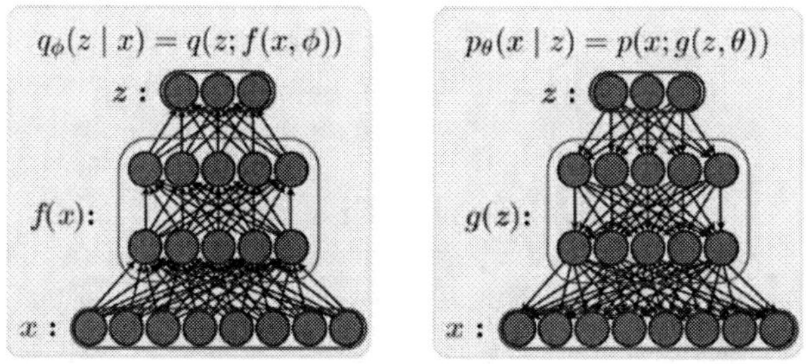

Figure 1. Neural networks in variational autoencoder

3. THE PROPOSED METHOD

We begin by presenting the problem statement, then the proposed method to detect anomalies.

3.1. Problem statement

Let $X = \{x_1, x_2, ..., x_n\}, x_i \in X$, set of detected faces from a face detection algorithm, which contains false-positive detections or non-face regions that effects on False Alarm Rate. To decrease this rate, we consider the non-faces as abnormalities and faces as normal data, then we apply anomaly detection algorithm f after face detection. See figure 2.

$$f(x_i) = \begin{cases} face(normal) & score \geq \varphi \\ non-face(abnormal) & score < \varphi \end{cases} \tag{4}$$

where, f anomaly detection algorithm, x_i: detected face, score: anomaly score, φ: the anomaly threshold, determined empirically to achieve best false alarm rate.

3.2. The proposed method

Our system consists of two components as shown in figure 3, an autoencoder network including an encoder network E(x) and decoder network D(z), and an anomaly detection. An input image x is encoded as a latent vector z = E(x), which will be decoded back to image space $\bar{x} = D(z)$. In order to train a VAE, we need two losses, one is KL divergence, which is used to make sure that the latent vector z is an independent unit. The other is reconstruction loss, which is that directly comparing the input image and the generated image in the pixel space. We choose mean square error (MSE) as reconstruction loss and we consider the reconstruction loss as anomaly score. Finally,

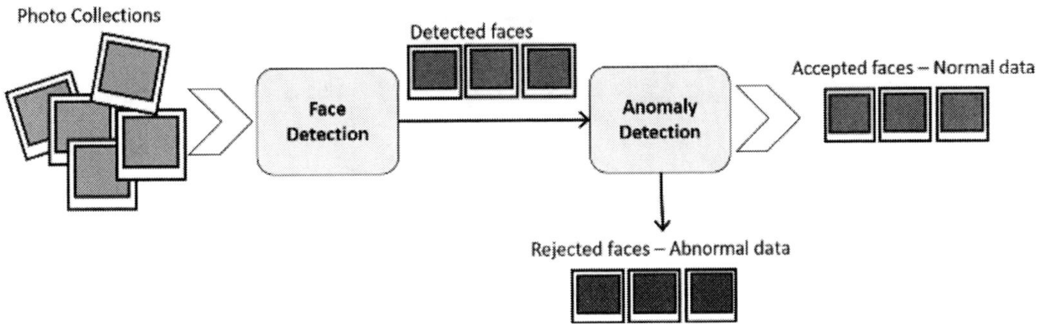

Figure 2. Anomaly detection system

the VAE model can be trained by optimizing the sum of the reconstruction loss (L_{rec}) and KL divergence loss (L_{kl}) by stochastic gradient descent.

$$\text{anomaly score} = L_{rec} = -E_{q_\phi(z|x^{(i)})}[\log p_\theta(x \mid z)] \tag{5}$$

$$L_{kl} = D_{KL}(q_\phi(z \mid x) \,\|\, p_\theta(z)) \tag{6}$$

$$L_{vae} = L_{kl} + L_{rec} \tag{7}$$

Both encoder and decoder network are based on deep CNN like AlexNet [28] and VGGNet [29]. We used the architecture that proposed in [30].

The algorithm of the proposed method is in Algorithm 1. The anomaly detection task is conducted a semi-supervised framework, using only data of normal instances for training the VAE. We use the autoencoder architecture that is proposed in [30].

Algorithm 1 Variational autoencoder based anomaly detection algorithm

INPUT: Normal dataset X, Anomalous dataset $x^{(i)} i = 1,\ldots,N$, threshold α

OUTPUT: reconstruction loss L_{rec}

$\phi, \theta \leftarrow$ train a variational autoencoder using the normal dataset X

for i = 1 **to** N **do**

$\mu_z(i), \sigma_z(i) = f_\theta(z \mid x^{(i)})$

draw one sample $\bar{x}$ from z ~ N ($\mu_z(i), \sigma_z(i)$)

$L_{rec}{}^{(i)} = MSE(x^{(i)}, \bar{x}^{(i)}) = \frac{1}{L}\sum_{i=1}^{L}(x^{(i)} - \bar{x}^{(i)})^2$,

L: number of pixels.

if $L_{rec}{}^{(i)} > \alpha$ **then**

$x^{(i)}$ is an anomaly

else

$x^{(i)}$ is not an anomaly

end if

end for

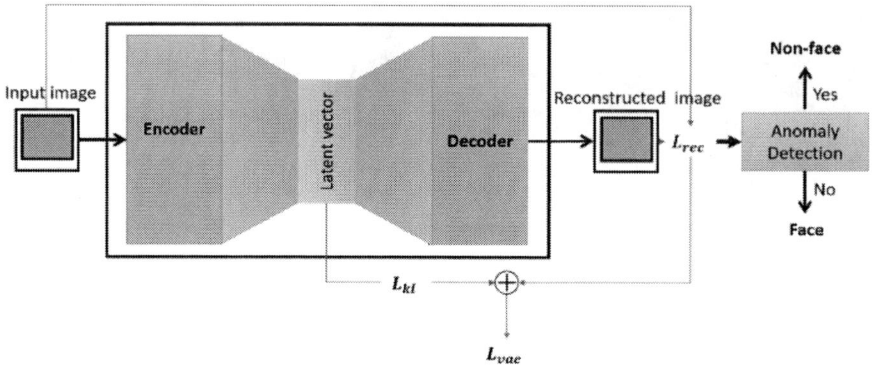

Figure 3. Variational autoencoder based anomaly detection model

4. IMPLEMENTATION

We conduct two experiments, in the first one, our system is trained on LWFcrop dataset, and the second one on Pubfig83-lfw dataset, while FDDB dataset is chosen for testing.

4.1. Baseline Methods

Viola Jones Face Detector: Viola and Jones proposed an efficient cascade framework [25] that rapidly discards negatives and spends more time in positive candidates. The cascade framework is one of the most successful practical products of vision research. This algorithm minimizes computation time while achieving high detection accuracy. It is 15 times quicker than any technique at the time of release with 95% accuracy at around 17 fps. Viola Jones algorithm has four stages: Haar feature selection, creation of an integral image, Adaboost training, and cascaded classification. We use this face detector as a first step before anomaly detection.

Hog-SVM Classifier using MMOD method as an anomaly detection algorithm: To detect abnormalities in set of detected faces that are obtained from the Haar-Cascade Face Detector, we use a fast detector based on Maximum Margin Object Detection (MMOD) method [26]. This method does not perform any sub-sampling, but instead optimizes over all candidate windows. The scoring function that used in MMOD's parameters outputs anomaly score, which is converted to anomaly labels by thresholding empirically.

4.2. Datasets

LFWcrop: It is a cropped version of the Labeled Faces in the Wild (LFW) [31], contains 3694 colored cropped faces with size 64x64 pixels.

Pubfig83-lfw: It is a combination of PubFig83 and the LFW dataset for open-universe face identification. Faces are pre-aligned by the eye positions as reported by PittPatt-reported fiducials [32].

We evaluate our system on the Face Detection Data Set and Benchmark (FDDB) challenge [33]. This challenging dataset contains images of human faces in multiple poses captured in indoor and outdoor settings. There are 2845 images with a total of 5171 faces contain a wide range of difficulties including occlusions, difficult poses, and low resolution and out-of-focus faces.

a) Reconstructed faces (Right) of faces from LFWcrop

b) Reconstructed faces (Right) of faces from Pubfig83-lfw

Figure 4. Examples of reconstruction faces from validation dataset

We preprocess the images by applying standardization technique as the following:
new_image = (image – db_mean) / db_stddev

4.3. Hardware Specifications
We used virtual machine instance in Google Cloud with the following details: operating system is Ubuntu 16.04, 6 cores CPU, 10GB RAM and GPU is Tesla K80, 12GB.

4.4. Training on LFWCrop and Pubfig83-lfw
In the first experiment, VAE model is trained on LFWcrop with 2000 images for training and 500 images for validation, while in the second experiment, is trained on Pubfig83-lfw with 5000 images and 2000 images for validation. In both experiments, we scale images to 64x64 and use batch size equals to 16. The size of latent vector z is set to 100 also. Figure 4. shows examples of reconstructed faces from validation datasets.

4.5. Experimental Results on FDDB

After training the models for 5000 epochs, we test them on FDDB dataset. figure 5. shows examples of reconstructed faces. Table. 1 presents the evaluation results of our VAE model-based anomaly detection in comparison with results of Hog-SVM results. We note that the first experiment, when the VAE model is trained on LFWcrop dataset gives better results than the second experiment when trained on Pubfig83-lfw dataset. In addition, the performance of our system achieves the performance of Hog-SVM anomaly detection model. Since, the false positives decreased from 421 to 128 and 135 for Hog-SVM classifier and out method respectively.

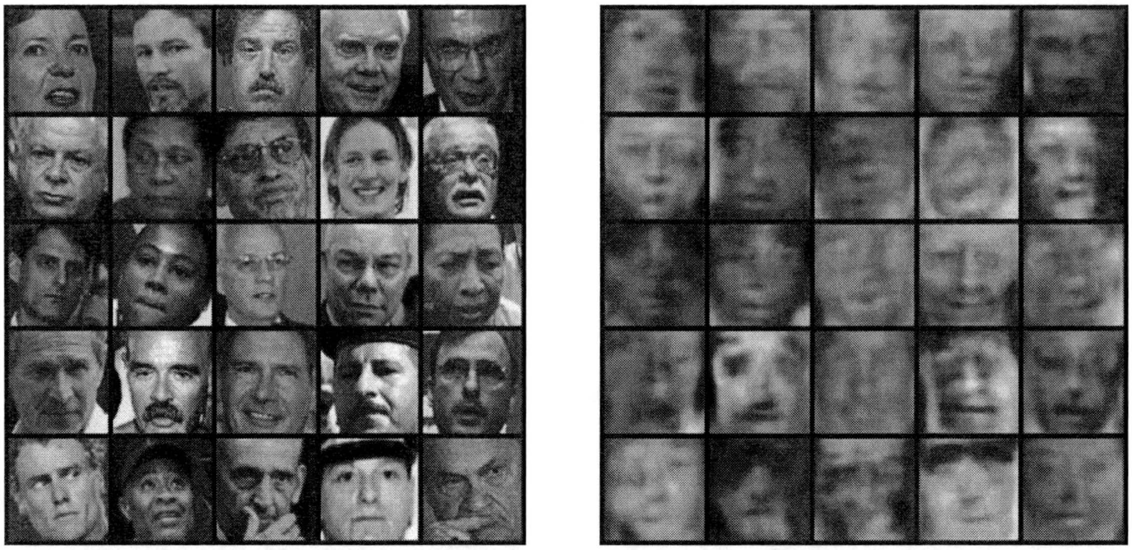

c) Reconstructed faces (Right) of faces from LFWcrop after 1st experiment

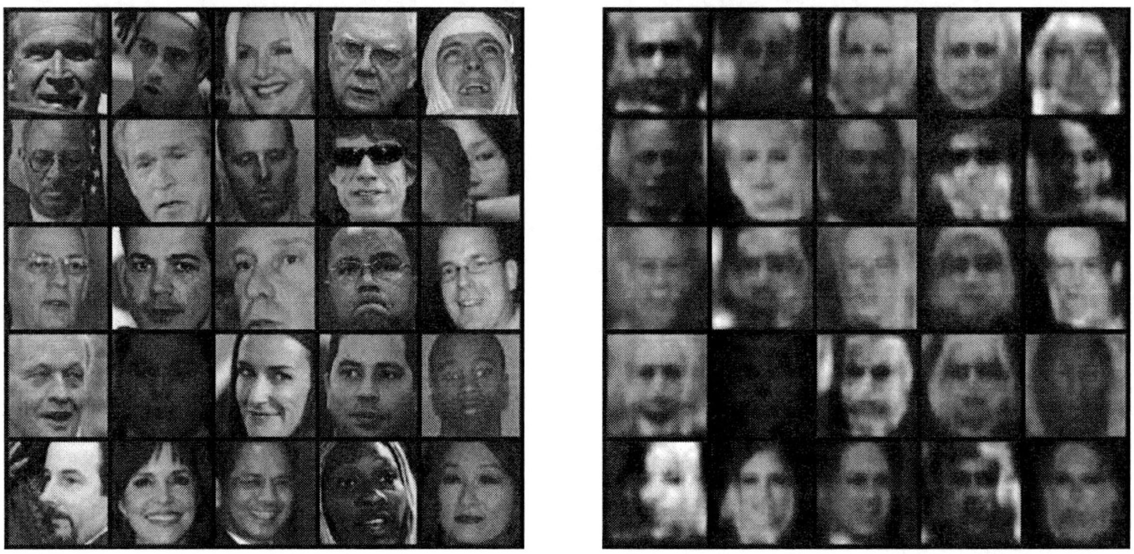

d) Reconstructed faces (Right) of faces from FDDB after 2nd experiment

Figure 5. Examples of reconstruction faces from FDDB dataset

Table 1. Evaluation Results of Proposed Method

Face detection step			
	Min window size	***Recall***	***FP***
Haar-Cascading OpenCV	80 x 80	64%	421
Anomaly detection step			
	Threshold φ	***Recall***	***FP***
Hog-SVM	>= 1	29 %	86
Hog-SVM	>= 0	55 %	121
Hog-SVM	>= -1	**55 %**	**128**
Ours (Exp1)	<=3500,000	34 %	94
Ours (Exp1)	<=4500,000	**54 %**	**135**
Ours (Exp2)	<=2500,000	18 %	102
Ours (Exp2)	<=3500,000	41 %	188

Acknowledgments
This work was financially supported by the Government of the Russian Federation (Grant 08-08).

References

[1] Zhang and Z. Zhang. A survey of recent advances in face detection. Microsoft Research, Technical Report 2010, no. MSR-TR-2010-66, 1 p.

[2] Viola P and Jones MJ. "Robust real-time face detection", Int J Comp Vision, 2004, vol. 57, no. 2, pp. 137–154.

[3] H. Yang, X.Wang. "Cascade classifier for face detection", Journal of Algorithms & Computational Technology, 2016, vol. 10, no. 3, pp. 187 – 197.

[4] R. Mohan, N. Sudha, "Fast face detection using boosted eigenfaces", IEEE Symposium on Industrial Electronics & Applications, 2009, vol. 2, pp. 1002-1006.

[5] Chen , S. Ren, Y. Wei, X. Cao, J. Sun. "Joint cascade face detection and alignment", In: Proc. Eur. Conf. Comput. Vis. (ECCV), 2014, pp. 109–122.

[6] X. Zhu and D. Ramanan. "Face detection, pose estimation, and landmark localization in the wild", Computer Vision and Pattern Recognition (CVPR), 2012, pp. 2879–2886.

[7] J. Yan, X. Zhang, Z. Lei, and S. Z. Li. "Face detection by structural models", Image and Vision Computing, 2014, vol. 32, no. 10, pp. 790–799.

[8] M. Mathias, R. Benenson, M. Pedersoli, and L. Van Gool. "Face detection without bells and whistles", Lecture Notes in Computer Science, 2014, vol. 8692, pp. 720–735.

[9] T. Alafif, Z. Hailat, M. Aslan, and X. Chen. "On detecting partially occluded faces with pose variations, in proceedings of the 14th Inter- national Symposium on Pervasive Systems, Algorithms, and Networks (I-SPAN)", IEEE Computer Society Conference Publishing Services (CPS), 2017.

[10] S. S. Farfade, M. J. Saberian, L. Li. Multi-view Face Detection Using Deep Convolutional Neural Networks. ICMR '15 Proceedings of the 5th ACM on International Conference on Multimedia Retrieval, 2015, vol. 2, pp. 643-650.

[11] H. Jiang, E. Learned-Miller. Face Detection with the Faster R-CNN. ArXiv e-prints:1606.03473. 2016.

[12] S. Sarkar, V. M. Patel, and R. Chellappa. Deep feature-based face detection on mobile devices. ArXiv e-prints, abs/1602.04868, 2016.

[13] R. Ranjan, V. M. Patel, and R. Chellappa. A deep pyramid deformable part model for face detection, In Biometrics Theory, Applications and Systems (BTAS), 2015, pp 1–8.

[14] S. Yang, P. Luo, C. C. Loy, X. Tang. From facial parts responses to face detection: A deep learning approach. 2015, pp. 3676–3684.

[15] L. Nanni, A. Lumini, Sh. Brahnam. Ensemble of Face/Eye Detectors For Accurate Automatic Face Detection, International Journal of Latest Research in Science and Technology, 2015, vol. 4, no. 3, pp.8-18.
[16] L. Nanni, A. Lumini. Combining Face and Eye Detectors in a High- Performance Face-Detection System, 2012, vol. 19, no. 4, 3 p.
[17] J. Prinosil, J. Vlach. Face detection in image with complex background, IFIP — The International Federation for Information Processing, 2007, vol. 245, pp 533-544.
[18] Nefian. Georgia Tech face database, URL: http://www.anefian.com/research/face_reco.htm
[19] Md. Hafizur, J. Afrin. Human Face Detection in Color Images with Complex Background using Triangular Approach, Global Journal of Computer Science and Technology Graphics & Vision, 2013, vol. 13, no. 4. 2 p.
[20] F. Alizadeh, S. Nalousi, C. Savari. Face Detection in Color Images using Color Features of Skin, 2011, vol. 5, no. 4, pp. 368-369.
[21] Varun Chandola, Arindam Banerjee, and Vipin Kumar. Anomaly detection: A survey. ACM computing surveys (CSUR), 41(3):15, 2009.
[22] Dimokranitou A. Adversarial Autoencoders For Anomalous Event Detection In Images. Thesis Submitted to the Faculty of Purdue University. 2017.
[23] Qian J., Root J., Saligrama V., and Chen Y. A Rank-SVM Approach to Anomaly Detection, 2014. https://arxiv.org/abs/1405.0530.
[24] An, J. and Cho, S. Variational Autoencoder based Anomaly Detection using Reconstruction Probability. 2015.
[25] Viola, P and Jones, M.J. Robust real-time face detection. Int J Comp Vision, 2004, vol. 57, no. 2, pp. 137–154.
[26] E. King. Max-Margin Object Detection. in Computer Vision and Pattern Recognition, 2015, arXiv:1502.00046.
[27] Kingma, M. Welling. Auto-encoding variational bayes. 12 2013.
[28] Krizhevsky, I. Sutskever, and G. E. Hinton. Imagenet classification with deep convolutional neural networks. In Advances in neural information processing systems, pages 1097–1105, 2012.
[29] K. Simonyan and A. Zisserman. Very deep convolutional networks for large-scale image recognition. arXiv preprint arXiv:1409.1556, 2014
[30] X. Hou, L. Shen, K. Sun, G. Qiu. Deep Feature Consistent Variational Autoencoder. 2016.
[31] B. Huang, M. Marwan, L. Honglak, and L. Erik. Learning to align from scratch. In NIPS, 2012.
[32] http://www.briancbecker.com/blog/research/pubfig83-lfw-dataset/
[33] V. Jain and E. Learned-Miller. Fddb: A benchmark for face detection in unconstrained settings. Technical Report UM-CS-2010-009, University of Massachusetts, Amherst, Tech. Rep. UM-CS-2010-009, 2010.

Comparative Review of Cross-Domain Generative Adversarial Networks

B Zeno[1], I Kalinovskiy[2] and Yu Matveev[1]

[1]ITMO University, St. Petersburg, Russia
basilzeno@gmail.com; matveev@mail.ifmo.ru

[2]STC-innovation Ltd s, St. Petersburg, Russia
kalinovskiy@speechpro.com

Abstract. This paper provides the comparative analysis between two recent image-to-image translation models that based on Generative Adversarial Networks. The first one is UNIT which consists of coupled GANs and variational autoencoders (VAEs) with shared-latent space, and the second one is Star-GAN which contains a single GAN model. Given training data from two different domains from dataset CelebA, these two models learn translation task in two directions. The term domain denotes as a set of images sharing the same attribute value. So, the attributes that are prepared: eye glasses, blond hair, beard, smiling and age. Five UNIT models are trained separately, while only one Star-GAN model is trained. For evaluation, we conduct some experiments and provide a quantitative comparison using direct metric GAM (Generative Adversarial Metric) to quantify the ability of generalization and the ability of generating photorealistic photos. The experimental results show the superiority of cross-model UNIT over multi-model StarGAN on generating age and eye glasses attributes, and the equivalent performance to synthesize other attributes.

1. Introduction

Many problems in computer vision can be formulated as a translation problem, mapping an input image in one domain to a corresponding output image in another domain by changing a particular aspect or more in the given image. For example, changing the seasons of scenery images; changing the resolution of images from low to high; translating cat or dog images between different species. And more, this translation can handle details of the face such as adding expressions like a smile and adding attributes like beard or sunglass. For many years of research, this problem is studied in supervised settings, where paired input-output examples in different domains are available [1,2]. Due to the difficulty and cost of collecting this data, little a number of unsupervised solutions are proposed [3,4,5,6], where there are two independent sets of images, the first one consists of images in one domain, while the other consists of images in another domain.

Traditionally, many of image-to-image translation tasks have been tackled for special purpose. For example, image colorization [7]; removing the effects of camera shake from blurred images [8]; composing realistic picture from a simple freehand sketch is presented [9,10]; synthesizing a plausible image at a different time of day from an input image [11]. But Lately with the remarkable progress in generative machine learning models, a quite a number of research dedicated to the use of Generative Adversarial Networks (GANs) to address translation problem. Some of them are general to solve all these tasks. For instance, pix2pix [2] for synthesizing photos from label maps, reconstructing objects from edge maps; and colorizing images; CycleGAN [5] for style transfer, season transfer, photo enhancement and generating photos from paintings; icGAN [6], StarGAN [4] and UNIT [3] for adding facial expressions and attributes; UNIT also tested for other purposes such as changing street scene image and cat species conversion.

The two-recent image-to-image translation models, StarGAN and UNIT have achieved impressive results, but there no experiments to determine which one of them is better. In UNIT, the domain adaption experiment is conducted between the Street View House Number (SVHN) dataset, MNIST and USPS datasets. Then the performance of adapted classifiers is evaluated using classification accuracy and compared to performance of other domain adaptation algorithms. In StarGAN, facial attributes transfer task is evaluated using Amazon Mechanical Turk (AMT), while facial expression synthesis task is evaluated using external facial expression classifier and the results are compared to [5,6].

In this paper, we conduct some experiments and provide a comparative analysis between StarGAN and UNIT to measure the ability of generalization and ability of generating photorealistic photos. The main contribution is demonstrating that the cross-model UNIT is better than the multi-model StarGAN. Despite, the superiority of StarGAN over its baseline cross-domain models.

2. Background

We begin by reviewing the relevant background material on Variational Autoencoders [22] and Generative Adversarial Networks [23].

2.1 Variational Autoencoder

Variational Autoencoder (VAE) [22] as a deep generative model which is based on a regularized version of the standard autoencoder. It modifies the autoencoder architecture by learning the representation as a latent variable z in a probabilistic generative model (e.g. Gaussian distribution) instead of the deterministic function. VAE is comprised of two multilayer perceptrons: encoder and decoder. The encoder maps an observation xi into its latent representation zi (or latent code) and a decoder (the generative network) decodes an observation from a latent code. The generative process can be written mathematically:

$$z_i \sim p(z), x_i \sim p_\theta(x \mid z_i) \quad (1)$$

where p(z) is the prior distribution over the latent variables and $p_\theta(x \mid z_i)$ is probabilistic decoder network with parameters θ. The approximate posterior distribution of this generative process, call it, $q_\phi(z \mid x_i)$ (with parameters φ).

To admit efficient inference, the variational Bayes approach simultaneously learns both the parameters of $p_\theta(x \mid z_i)$ as well as those of a posterior approximation $q_\phi(z \mid x_i)$. This is achieved by maximizing the evidence lower bound (ELBO):

$$L(\phi,\theta,x) = \mathrm{E}_{q_\phi(z|x)}[\log p_\theta(x,z) - \log q_\phi(z \mid x)], \quad (2)$$

with $L(\phi,\theta,x) \le \log p_\theta(x)$.

2.2 Generative Adversarial Networks

As one of the most significant improvements on the research of deep generative models, GAN framework was introduced by Goodfellow et al. [23] for generative modeling of data through learning a transformation from points belonging to a simple prior distribution (z ~ pz) to those from the data distribution (x ~ pdata). A typical GAN model consists of two modules that play an adversarial game: a discriminator and a generator. while the generator learns to generate fake samples G(z) that are indistinguishable from real samples, the discriminator learns to distinguish between these fake samples (G(z) ~ pG) and real ones (x ~ pdata), thus giving a scalar output (y ∈ {0, 1}). The goal of the generator is to fool the discriminator by generating photorealistic samples that resemble those from the real data while that of the discriminator is to accurately distinguish between real and generated data. The two models, typically designed as neural networks, play a min max game with the objective function as shown in Equation (1):

$$\min_G \max_D V(D,G) = \mathrm{E}_{x \sim p_{data}(x)}[\log D(x)] + \mathrm{E}_{z \sim p_z(z)}[\log(1 - D(G(z)))] \quad (3)$$

IOP Conf. Series: Materials Science and Engineering **618** (2019) 012012 doi:10.1088/1757-899X/618/1/012012

Image to Image Translation Models

In this section, we present overview for both image-to- image translation models, StarGAN and UNIT:

2.3 UNIT:Unsupervised Image-to-Image Translation Networks

Liu et al. [3] proposed unsupervised image-to-image translation framework based on Coupled VAE-GANs. They analysed the problem from a probabilistic modeling perspective, where the goal is to learn a joint distribution of images in different domains. But in the unsupervised settings, where the two sets consist of images from two marginal distributions in two different domains, make inferring the joint distribution from these marginal distributions is a highly ill-posed problem, due to existence an infinite set of joint distributions. To address this ill-posed problem, they make a shared-latent space assumption [13], which assumes a pair of corresponding images in different domains can be mapped to a same latent representation in a shared-latent space. Based in this assumption, as shown figure 1, the system consists of 6 subnetworks: including two encoders E1 and E2 for mapping images to shared-latent codes z, two generators G1 and G2 for mapping latent codes z to images, and two adversarial discriminators D1 and D2 for the respective domains, in charge of evaluating whether the translated images are fake or realistic.

In UNIT, the input image can be reconstructed from translating back the translated input image. Since the assumption mentioned above implies the cycle consistency constraint [14,15], which means an image in the source domain can be mapped to an image in the target domain and this translated image in the target domain can be mapped back to the original image in the source domain.

To learn UNIT, learning problems of coupled VAE-GANs are solved jointly for the image reconstruction streams, the image translation streams, and the cycle-reconstruction streams:

$$\begin{aligned}\min_{E_1,E_2,G_1,G_2}\max_{D_1,D_2} & L_{VAE_1}(E_1,G_1)+L_{GAN_1}(E_1,G_1,D_1)+\\ & L_{CC_1}(E_1,G_1,E_2,G_2)+L_{VAE_2}(E_2,G_2)+\\ & L_{GAN_2}(E_2,G_2,D_2)+L_{CC_2}(E_2,G_2,E_1,G_1)\end{aligned} \tag{4}$$

Where, L_{VAE_1} and L_{VAE_2} are objective functions of VAEs; L_{GAN_1} and L_{GAN_2} are objective functions of GANs; L_{CC_1} and L_{CC_1} are objective functions to model the cycle-consistency constraint.

UNIT learns translation between only two domains. So, as demonstrated in figure 2(a), in order to learn mapping between k domains, (k(k-1))/2 UNIT model have to be trained, where each one consists from two coupled generators.

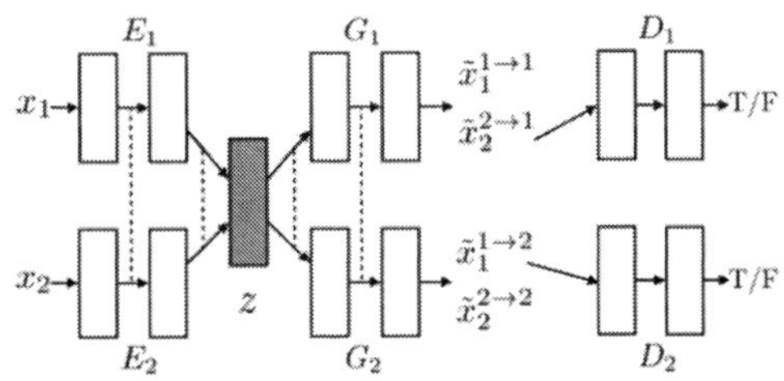

Figure 1. UNIT framework. E1, E2, G1 and G2 are presented using CNNs. Here, Here, $\tilde{x}_1^{1\to1}$ and $\tilde{x}_2^{2\to2}$ are self-reconstructed images, and $\tilde{x}_1^{1\to2}$ and $\tilde{x}_2^{2\to1}$ are domain-translated images, z is shared-latent space assumption using a weight sharing constraint (illustrated using dashed lines)

2.4 StarGAN: Unified Generative Adversarial Networks for Multi-Domain Image-to-Image Translation

In contrast to UNIT framework, Choi et al. [4] proposed StarGAN, a novel generative approach that can perform image-to-image translations for multiple domains using only a single GAN model. As demonstrated in figure 2(b), StarGAN takes training images of multiple domains, and learns the mappings between all available domains using only one generator. To achieve this mapping, the idea is simple, instead of learning fixed translation (e.g., young-to-old face), the model takes both input image and target domain information (x,c) and learns to flexibly translate the input image into an output image y on the corresponding domain. The domain information is represented by a label (binary or one hot vector for categorical attributes). And during training, target domain labels c are generated randomly so that the generator G learns to flexibly translate the input image: c, G(x, c) → y. Auxiliary classifier [15] is introduced on top of the discriminator to allow him to control multiple domains. So, two probability distributions are produced by discriminator over both sources and domain labels, D: x → {Dsrc(x), Dcls(x)}.

To learn StarGAN, the following objective functions are used to optimize both discriminator D and generator G, respectively:

$$L_D = -L_{adv} + \lambda_{cls} L^r_{cls}, \tag{5}$$

$$L_G = L_{adv} + \lambda_{cls} L^f_{cls} + \lambda_{rec} L_{rec}, \tag{6}$$

where, L_{adv} is adversarial loss, to make the generated images indistinguishable from real images; L_{rec} is reconstruction loss to make the generated images are realistic and classified to its correct target domain with preserving the content of its input images; L^r_{cls} is domain classification loss for real images, while L^f_{cls} is domain classification loss for fake images; λ_{cls} and λ_{rec} are hyper-parameters.

In StarGAN, the original image can be reconstructed from translated image due to applying the cycle consistency loss [14,15] to the generator.

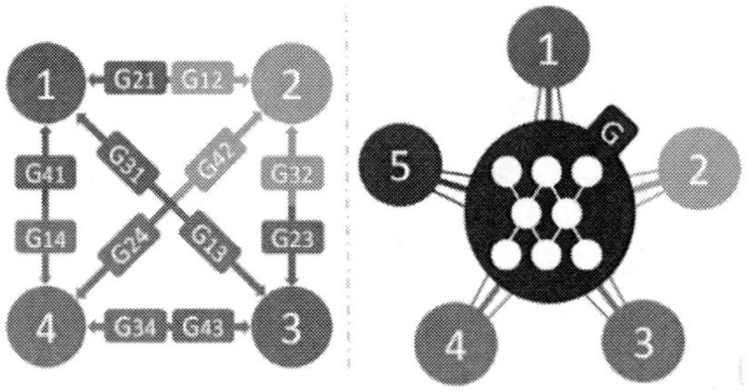

(a) UNIT (b) StarGAN

Figure 2. Comparison between UNIT and StarGAN frameworks [4]. a) Six UNIT models are required to learn translation task between 4 attributes. Every UNIT model contains two generators. b) One StarGAN is required only to learn any number of attributes.

3. The Proposed Evaluation Method

Several measures have been introduced for evaluating and comparing GANs, but there is no consensus as to which measure best captures strengths and limitations of models. In comparative analysis between two image-to-image translation models, we must determine which model generates photorealistic images with desired attribute and which model preserves the identity. So, in this section, we review firstly some quantitative and qualitative measures for evaluating GAN models. Then we present our evaluation method to compare StarGAN and UNIT models.

3.1 Qualitative Measures

Visual examination of samples by humans is one of the common and most intuitive ways to evaluate GANs. For instance, in [4] to evaluate StarGAN two user studies are performed in a survey format using Amazon Mechanical Turk (AMT), where the Turkers were asked to judge whether given images with single and multiple transferred attribute are real from the dataset or generated from StarGAN model. In that case, the discriminator can be viewed as a human, while the generator is a trained GAN. The problems with this approach are that Turkers may have high variance, which makes it necessary to average over a large number of them, and the experimental setup is expensive, time-consuming, difficult to reproduce, and does not fully reflect the capacity of models.

3.2 Quantitative Measures

Quantitative measures are less subjective may not directly correspond to how humans perceive and judge generated images. We dived them into two categories: indirect and direct measures. Indirect measures when the evaluation approaches rely on trained external model. For example, FaceNet convolutional neural network in [17] is used to measure the identity-similarity between each reconstructed face image and the source face image; facial expression classifier is used in [4] to evaluate the generator of StarGAN model in synthesizing realistic facial expression, where the classifier is trained, then the model is evaluated by computing the classification error of a facial expression on reconstructed images. It achieves the lowest classification error in comparison with CycleGAN [5] and IcGAN [6]. Inception Score (IS) [18] is used as indirect metric to measure the quality and diversity of the generated images. The second category is direct measures when the evaluation of the model depends on its own structure without any external model. For example, classification accuracy of UNIT discriminator [3] is used to evaluate its performance in domain adaptation task.

Generative Adversarial Metric (GAM) [19] is the direct measure to quantify the ability of generalization and the ability of generating photorealistic images. It is used to determine which GANs generator is better, considering the performance of discriminators are equivalent.

GAM compares two generative adversarial models by pitting each generator against the opponent's discriminator. Consider two generative adversarial models M1={(G1,D1)} and M2={(G2,D2)}, in the training phase, G1 competes with D1. Likewise, for G2 and D2. In the test phase, G1 engages in a battle against D2 while G2 against D1. The ratios of their classification errors ε on real test dataset and on generated samples are then calculated as rtest and rsamples as shown in Equations 7 and 8.

$$r_{test} = \frac{\varepsilon(D_1(x_{test}))}{\varepsilon(D_2(x_{test}))} \tag{7}$$

$$r_{samples} = \frac{\varepsilon(D_1(G_2(z)))}{\varepsilon(D_2(G_1(z)))} \tag{8}$$

The test ratio, rtest, tells us which model generalizes better since it is based on discriminating the test data. While, the sample ratio, rsamples, tells us which model can fool the other model more easily, since it is based on generating photorealistic photos. Therefore, for G1 to win against G2, both rsamples < 1 and rtest ≈ 1 must be satisfied.

In [20] GAM is used in order to compare between Original GAN and CapsuleGAN, which have different discriminators, but the ratios of classification accuracies A are calculated instead of errors to avoid numerical problems. So, in order to compare StarGAN and UNIT models, and Following [3,4,19,20], the new rtest and rtranslating are shown in Equations 9 and 10.

$$r_{test} = \frac{A(D_{StarGAN}(x_{test-on}))}{A(D_{UNIT}(x_{test-on}))} \tag{9}$$

$$r_{translating} = \frac{A(D_{StarGAN}(G_{UNIT:OFF\to ON}(x_{test-off})))}{A(D_{UNIT}(G_{StarGAN:OFF\to ON}(x_{test-off})))} \tag{10}$$

where DStarGAN and DUNIT are discriminators of StarGAN and UNIT respectively. GStarGAN and GUNIT are generators of StarGAN and UNIT respectively. xtest-off is a dataset of images, which has not an attribute that the generators will add. The winning model is defined as follows:

$$Winner = \begin{cases} Unit, & if: r_{translating} > 1 \quad and \quad r_{test} \cong 1 \\ StarGAN, & if: r_{translating} < 1 \quad and \quad r_{test} \cong 1 \\ Tie, & otherwise \end{cases} \tag{11}$$

4. Experiments

In this section, we first review the training and testing datasets. Next, we perform facial attribute transfer using generators of both StarGAN and UNIT models on unseen images during the training phase. lastly, we calculate the evaluation metric GAM by conducting a classification experiment on the translated images using the discriminators of models.

IOP Conf. Series: Materials Science and Engineering **618** (2019) 012012 doi:10.1088/1757-899X/618/1/012012

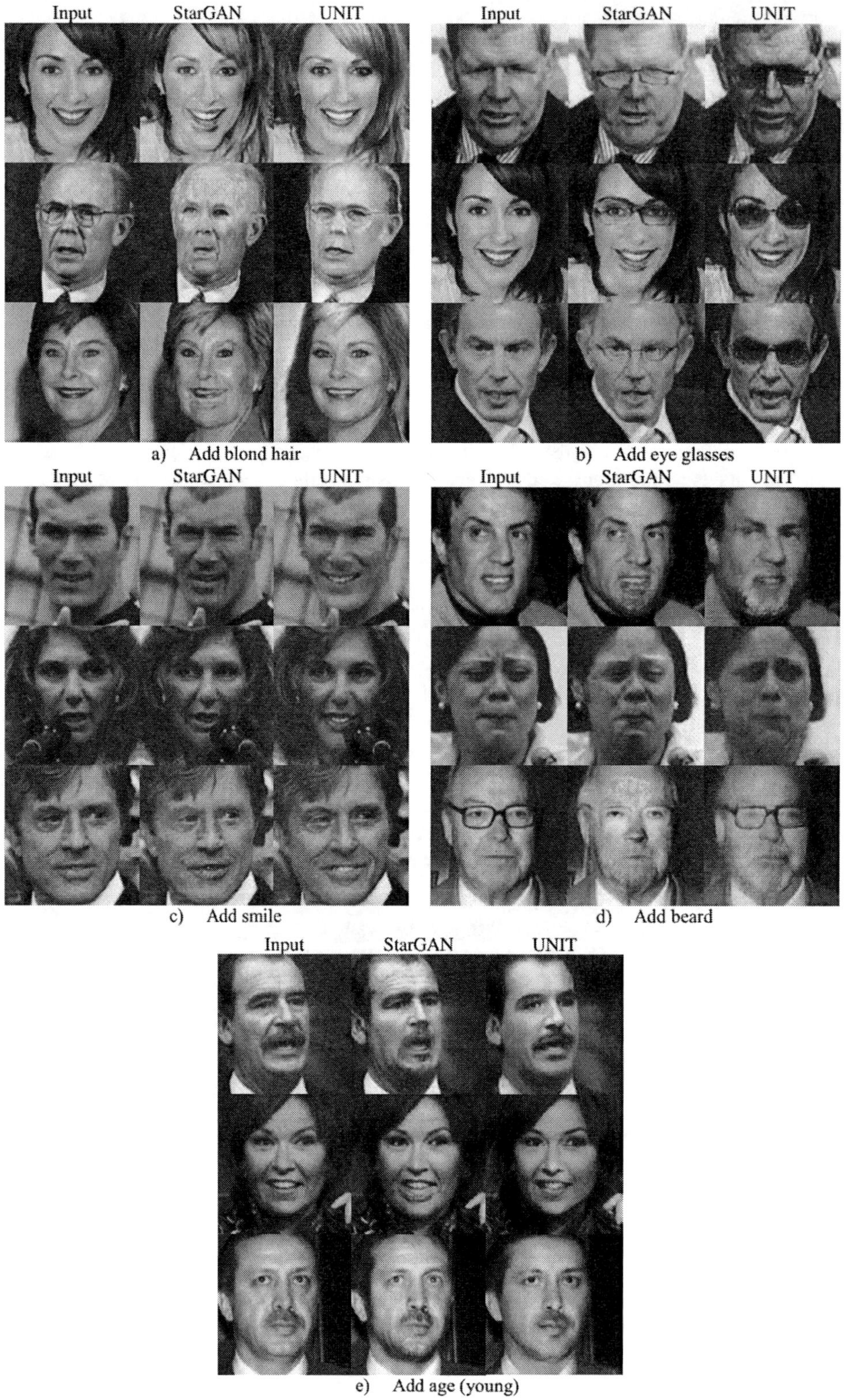

Figure 3. Examples of facial attributes transfer task using StarGAN and UNIT

4.1 Datasets

Several image datasets come with a number of labeled attributes. For instance, the CelebA [17] dataset contains 40 labels related to facial attributes such as hair color, gender, and age.

- **CelebA**. The CelebFaces Attributes (CelebA) dataset [21] contains 202,599 face images of celebrities, each annotated with 40 binary attributes. We use the MTCNN [24] detector to align the face images and scale them to a size of 128x128. We select all images for training step. Five attributes are chosen as follows: eye glasses, blond hair, beard, smiling and young. We construct ten domains using the following attributes: eye glasses (yes/no), blond hair (yes/no), beard (yes/no), smiling (yes/no) and age (young/old).
- **LFW+**. The extended LFW database (LFW+) [25] contains 15,699 unconstrained face images of about 8,000 subjects, each image is annotated with 3 categorical attributes (age, gender, race). We use LFW+ version with extended annotations, where there are more 15 categorical attributes. We perform preprocessing similarly to CelebA. Then, we choose images with the following attributes: eye glasses, blond hair, beard, smiling and age. Next, we delete images with attribute value (age=baby, age=children) to ensure that the model correctly learns the remaining values. Lastly, we convert the attributes to binary form.

4.2 Training on CelebA

Both models StarGAN and UNIT are trained on NVIDIA GeForce GTX 1080Ti using all images in CelebA. In more detail, one StarGAN model is trained for 2 days, while 5 models of UNIT are trained for 4 days, one UNIT model for each attribute. For data augmentation we flip the images horizontally with a probability of 0.5. The batch size is set to 16 for training StarGAN and 1 for training UNIT.

4.3 Experiminal Results on LFW+

During training UNIT model, we notice that the training could be unstable. This problem is mentioned in the original work [3]. So, after more than one million iterations, we choose the best models that give high classification accuracy. Next, to perform facial attribute transfer task, we select from LFW+ images with no targeted attribute, in order to add them by generators. Ten experiments are conducted to add 5 attributes for both models.

- **Qualitative evaluation**. Figure 3, shows the facial attribute transfer results on LFW+ dataset. In contrast to results of StarGAN work, we observed that UNIT method provides a higher visual quality of translation results on test data compared to the multi-domain model StarGAN. We noticed also disadvantage in StarGAN, where we can not turn off the translation of the existed attributes. For example, if the source image is labeled with two attributes: age (old) and eye glasses (yes), and we want to change age from old to young. The translated image will be with eye glasses surly, but in the fact, eye glasses will be added also into the translated image. Whilst, in UNIT, only the targeted attribute will be added. See figure 4.

Input StarGAN UNIT

Figure 4. Translating image example with attribute age (young)

- **Quantitative evaluation.** In order to evaluate the ability of generators to synthesize photorealistic images, GAM is used as discussed in 5th section. StarGAN discriminator is used to classify translated images by UNIT, while UNIT discriminator is used to classify the translated images by StarGAN. Sigmoid function is used to obtain a probability for sample classification. Table 1, shows the results of classification testing images that have the targeted attribute. The reported numbers are classification accuracies.

Table 1. Classification accuracy of testing images (XTEST-ON)

Facial attribute	StarGAN	UNIT	Number of images	r_{test}
Eye glasses	93 %	80 %	2348	1.16
Blond hair	91 %	41 %	1064	2.22
Beard	8 %	72 %	3486	0.11
Smiling	5 %	90 %	5064	0.05
Age (young)	18 %	25 %	3336	0.72

Translate images with eye glasses Translate images with age (young)

Input

StarGAN

UNIT

Figure 5. Examples of facial attributes transfer task using StarGAN and UNIT, the left column for adding eye glasses attribute and the right column to add age (young) attribute

We notice that the discriminator of StarGAN is superior to discriminator of UNIT in classifying blond hair attribute. While the discriminator of UNIT achieves better performance in classifying attributes: beard and smiling. But, it does not mean the generator of StarGAN or UNIT is better. So, according to our metric, both models, StarGAN and UNIT on these three attributes are equivalent, because the rtest is not equal approximately to one. While we can continue comparing on attributes: eye glasses and age (young). Table. 2, shows the classification accuracies of translated images and its ratio $r_{translating}$.

Table 2. Classification accuracy of translated images (XTEST-OFF→ON)

Facial attribute	$D_{StarGAN}(G_{UNIT})$	$D_{UNIT}(G_{StarGAN})$	$r_{translating}$
Eye glasses	89.40 %	39 %	2.30
Age (young)	54.32 %	7.43 %	7.31

So, from Table. II, we notice high classification accuracy in StarGAN discriminator, while low value in UNIT discriminator. It means that the generator of UNIT superior to StarGAN generator, where the 89.4% of total testing images have eye glasses, while only 22.20% that generated by StarGAN have eye glasses. We refer to the translated xtest-off with targeted attributes as xtest-off→on. The number of xtest-off images is 9263 and 8275 for eye glasses attribute and age attribute respectively. Figure 5, shows examples of generated images for age (young) and eye glasses attributes.

5. Conclusion

In this paper, the comparative analysis between two recent image-to-image translation models that based on Generative Adversarial Networks, multi-domain model StarGAN and cross-domain model UNIT provided. To quantify the ability of generating photorealistic photos and the ability to generalization, GAM metric is used to compare these models by having them engage in a "battle" against each other. Since, both StarGAN and UNIT models consist of a discriminator and a generator in pairs, firstly we trained them on CelebA dataset to learn facial attribute transfer task on five attributes: eye glasses, blond hair, beard, smiling and age. Then we exchanged the pairs of generators and discriminators and have them play the generative adversarial game against each other by generating images from testing dataset LFW+ with targeted attributes. The experimental results show the superiority of UNIT model over StarGAN, although they are equivalents in generating: beard, blond hair, smiling, but generator of UNIT performs better than StarGAN generator in translating images with age and eye glasses attributes. Next, we will conduct experiments to measure the quality and diversity of the generated images using Inception Score (IS), and to measure the identity-preserving using FaceNet classifier.

Acknowledgments
This work was financially supported by the Government of the Russian Federation (Grant 08-08).

References

[1] Z. Liu, P. Luo, X. Wang, and X. Tang, "Deep learning face attributes in the wild," International Conference on Computer Vision, 2015.
[2] P. Isola, J.-Y. Zhu, T. Zhou, and A. A. Efros, "Image-to-image translation with conditional adversarial networks," Conference on Computer Vision and Pattern Recognition, 2017.
[3] M.-Y. Liu, T. Breuel, and J. Kautz, "Unsupervised image-to-image translation networks," arXiv preprint arXiv:1703.00848, 2017.
[4] Y. Choi, M. Choi, M. Kim, J. Ha, S. Kim, and J. Choo, "StarGAN: Unified Generative Adversarial Networks for Multi-Domain Image-to-Image Translation," IEEE Conference on Computer Vision and Pattern Recognition (CVPR), 2018.
[5] J.-Y. Zhu, P. Isola, T. Zhou, and A. A. Efros, "Image-to-image translation with conditional adversarial networks," Conference on Computer Vision and Pattern Recognition, 2017.
[6] icGAN: G. Perarnau, J. van de Weijer, B. Raducanu, and J. M. Alvarez, "Invertible conditional gans for image editing," arXiv preprint arXiv:1611.06355, 2016.
[7] R. Zhang, P. Isola, and A. A. Efros, "Colorful image colorization," ECCV, 2016.
[8] R. Fergus, B. Singh, A. Hertzmann, S. T. Roweis, and W. T. Freeman, "Removing camera shake from a single photograph. ACM Transactions on Graphics (TOG), 2006.
[9] T. Chen, M.-M. Cheng, P. Tan, A. Shamir, and S.-M. Hu,. "Sketch2photo: internet image montage," ACM Transactions on Graphics (TOG), 28(5):124, 2009.
[10] G.A. Kukharev, Yu.N. .Matveev, A.L. Oleinik, "Mutual image transformation algorithms for visual information processing and retrieval," Scientific and Technical Journal of Information Technologies, Mechanics and Optics, 2017, vol. 17, no. 1, pp. 62–74. doi: 10.17586/2226-1494-2017-17-1-62-74 (in Russian).
[11] Y. Shih, S. Paris, F. Durand, and W. T. Freeman, "Data-driven hallucination of different times of day from a single outdoor photo," ACM Transactions on Graphics (TOG), 32(6):200, 2013.
[12] Y. Netzer, T. Wang, A. Coates, A. Bissacco, B. Wu, and A. Y. Ng, "Reading digits in natural images with unsupervised feature learning," Advances in Neural Information Processing Systems workshop, 2011.
[13] M.-Y. Liu and O. Tuzel, "Coupled generative adversarial networks," Advances in Neural Information Processing Systems, 2016.
[14] J.-Y. Zhu, T. Park, P. Isola, and A. A. Efros, "Unpaired image-to-image translation using cycle-consistent adversarial networks," International Conference on Computer Vision, 2017.
[15] T. Kim, M. Cha, H. Kim, J. Lee, and J. Kim, "Learning to discover cross-domain relations with generative adversarial networks," International Conference on Machine Learning, 2017.
[16] A. Odena, C. Olah, and J. Shlens, "Conditional image synthesis with auxiliary classifier gans,"

arXiv preprint arXiv:1610.09585, 2016.

[17] L. Zhigang, and L. Yupin, "Generate Identity-Preserving Faces by Generative Adversarial Networks," arXiv preprint arXiv: 1706.03227. 2017

[18] T. Salimans, I. Goodfellow, W. Zaremba, V. Cheung, A. Radford, and X. Chen, "Improved techniques for training GANs," In Advances in Neural Information Processing Systems, pp. 2226–2234, 2016.

[19] D. J. Im, C. D. Kim, H. Jiang, and R. Memisevic, "Generative adversarial metric," 2016.

[20] A. Jaiswal, W. AbdAlmageed, Y. Wu, and P. Natarajan, "CapsuleGAN: Generative Adversarial Capsule Network," arXiv: 1802.06167, 2018

[21] Z. Liu, P. Luo, X. Wang, and X. Tang, "Deep learning face attributes in the wild," In Proceedings of the IEEE International Conference on Computer Vision (ICCV), 2015.

[22] D. Kingma, and M. Welling. "Auto-encoding variational bayes," 12 2013.

[23] I.J. Goodfellow, J. Pouget-Abadie, M. Mirza, B. Xu, D. Warde-Farley, S. Ozair, A. Courville, and Y. Bengio, "Generative adversarial nets," In Advances in Neural Information Processing Systems (NIPS), pages 2672–2680, 2014.

[24] K. Zhang, Zhang, Z. Zhang, Z. Li and Y. Qiao, (2016), "Joint face detection and alignment using multitask cascaded convolutional networks," IEEE Signal Processing Letters, 23(10):1499–1503.

[25] H. Han, A. K. Jain, F. Wang, S. Shan, and X. Chen, "Heterogeneous Face Attribute Estimation: A Deep Multi-Task Learning Approach," To appear in IEEE Transactions on Pattern Analysis and Machine Intelligence (T-PAMI), 2017, DOI: 10.1109/TPAMI.2017.2738004.

IOP Conf. Series: Materials Science and Engineering **618** (2019) 012013 doi:10.1088/1757-899X/618/1/012013

Examination of a repetitive process control system

Nina G. Nikolova

Technical University-Sofia; FA, DIA; 8, Kliment Ohridski str., Sofia-1000, Bulgaria
ninan@tu-sofia.bg

Abstract. In the present work, the application of repetitive filters in the robust process control systems is examined. The functionality of the proposed system and the improved performance, robust performance and filtering properties has been proven.

1. Introduction.

By increasing the performance requirements of the modern technological process control systems, there is considerable interest in the classes of non-stationary control systems in the conditions of a priori uncertainty and unmodulated dynamics, some of which are the following known classes of systems: robust systems; systems with parametric compensation; systems with interference absorption; systems using fractal algorithms; adaptive systems; systems with reduced sensitivity; systems with variable structure; extreme systems; repetitive systems; intelligent control systems (using neural networks, fuzzy multi-theory, genetic algorithms, chaos theory, etc.).

There are differences between them in structural and parametric terms. They are predominantly governed by the class of control process, the type of control and interfering performance, and the controlled objective. In the case where the parametric disturbances in the processes are dominant and there are periodic external disturbances whose nominal parameters are known in advance but which fluctuate operationally around the nominal ones and one of the main requirements for the performance of the system is the performance and the robust performance of the system, application of robust repetitive systems is recommended as an effective.

In this context, **the aim** of the current work is to study the properties and applicability of the repetitive filters in robust systems and systems with robust properties. [1].

2. Robust systems and systems with robust properties.

In literature are known systems with internal model RMM [1-3], systems with interference absorption DAS [1], and gain scheduled control system GSC [1], as well as the methods and the algorithms for their synthesis.

➢ The free parameter method solves the task of analytical synthesis of robust systems RMM (figure 1) with an internal model for process control in the conditions of a priori uncertainty, under the criteria: (1): robust stability and robust performance of the system for a predetermined in the synthesis parameter set Π, satisfying the requirements (1.a); local criteria for quality LCQ (1.b) – process with minimal integral squared error, and initial conditions for synthesis – a priori know G^*, Π, σ, where: $Q^*(p)$ classic regulator (2) with filter $F_\lambda(p)$ (4) [2], corresponding to a nominal model of the controlled process G^* by the set LCQ; $R_M(p)$ (3) – robust controller with an internal model of the process G^*. Every l element от $f_l(p)$ (4) should meet the requirements (5), where π_l are the poles G^* in the right half-space.

IOP Conf. Series: Materials Science and Engineering **618** (2019) 012013 doi:10.1088/1757-899X/618/1/012013

$$\left\{\begin{array}{l} a)\text{->>}\ \Pi=\left\{\begin{array}{c} G:\left|\dfrac{G(j\omega)-G^*(j\omega)}{G^*(j\omega)}\right|\le\bar{\ell}_m(\omega)=G^{\square} \\ \bar{\ell}_m(\omega)<|\eta|^{-1},\forall\omega \\ |\eta\bar{\ell}_m|+|e\,y^0|<1,\forall\omega \end{array}\right\} \\ b)\text{->>}\ \min_{\lambda}\|\varepsilon\|_2^2=\min\dfrac{1}{2\pi}\displaystyle\int_{-\infty}^{\infty}|\varepsilon(j\omega)|^2\,d\omega \end{array}\right\} \quad (1)$$

$Q=Q(Q^*,F_\lambda)$ (2.a) $\quad Q(p)=Q^*(p)F_\lambda(p)$ (2.b) $\quad R_{(M)}(p)=\dfrac{Q(p)}{1-G^*(p)Q(p)}$ (3)

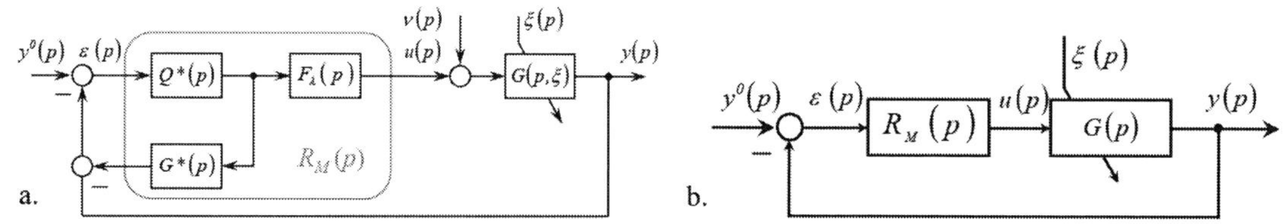

Figure 1. Robust control system with internal model

$$F_\lambda(p)=diag\{f_1(p),\ldots,f_n(p)\} \quad (4)$$

$f_i(\pi_i)=1,\ (i=0,1,\ldots,k)$ (5.a) $\quad \dfrac{d^j}{dp^j}f_i(p)\Big|_{s=\pi_i}=0,\ (j=1,\ldots,m)$ (5.b)

$$\left\{\begin{array}{ll} a)\text{->>} & \{\|Eu_d+F\xi\|=min,\ (u=u_r+u_d)\} \\ b)\text{->>} & \left\{\begin{array}{l} \left(\left|\mathcal{R}^*(j\omega_\pi^{A^0})\dfrac{k_{opt}^A\,G^{\square}(j\omega_\pi^{A^0})}{Q_2(j\omega_\pi^{A^0})}\right|= \; =|R^*(j\omega_\pi)G^*(j\omega_\pi)|\right)^{*} \\ \left(20\log_{10}\left|\dfrac{1}{1+\mathcal{R}^A(k^A,j\omega)G^{\square}(j\omega)}\right|\le 0,\ \forall\omega\ \ge 0.1\,\omega^{\square}\right)^{**} \end{array}\right\} \\ c)\text{->>} & \sigma=const \end{array}\right\} \quad (6)$$

➢ **The method** of the equilibrium partial absorption equation [2] solves the task for analytical synthesis of systems (figure 2), absorbing the influence of wave structured disturbances on the adjustable magnetite, for controlled process under a priori uncertainty, **under the criteria** (6): minimum Euclidean norm (6.a) of the members of the equilibrium balanced absorption equation on the adjustable magnitude; equality stock sustainability (6.b) (or optimal module); *LCQ* (6.c) (process with reordering set $\sigma=const$) etc., and **initial conditions** for synthesis: a priori known G^*, Π, σ, a representative trend of the system error in operating (simulation) conditions. The control u (6) of the systems (figure 3) has two additive constituents: u_r - on the basis controller R^*, realizing the aim of control of the respective *LCQ* and u_d - to absorb the impact of interference ξ. The matrix E, F и C (7) reflect the dependence of the vector y from the controlled vector u, the disturbance ξ and the condition of the system x. The "nominal" system corresponding to the absorption system is a hypothetical parametric, non-intrusive linear system synthesized at the given *LCQ* (6.c) to the nominal model G^*, determined with the "nominal" controller R^* (8).

$y=Cx+E(u_r+u_d)+F\xi$ (7) $\quad R^*(p)\Leftrightarrow G^*(p)$ (8)

Corresponding to the disturbance absorbing system (figure 2) "the classical" parametric disturbed system is with controller R^* and process control, modeled using the $G(p,\xi)$ in real operational conditions (or with adequate computer simulation). y° , u are signal external interference aggregated on the assignment channel y° and/or the loading u , and ξ are aggregated internal additive or multiplicative parametric and/or structural disturbances with a wave structure in process control. Analytical design of the absorber A in the structure (figure 2) of the system is according (9), where the fluent $Q_\Sigma^{-1}(p)$ is determined by a known methodology [1], and the optimum gain value k^A is determined based on (10) or (11). ω_π^{CLASS} e ω_π - frequency of the closed classical system, $\omega_\pi^{A^*}$ e ω_π - frequency of the closed, disturbed, interference-absorbing system; G^D - a model of disruption at the highest level process control; $\Phi_{y^*\varepsilon}^{CLASS^*}$ - transmitting fluent of the closed, top-class classical system.

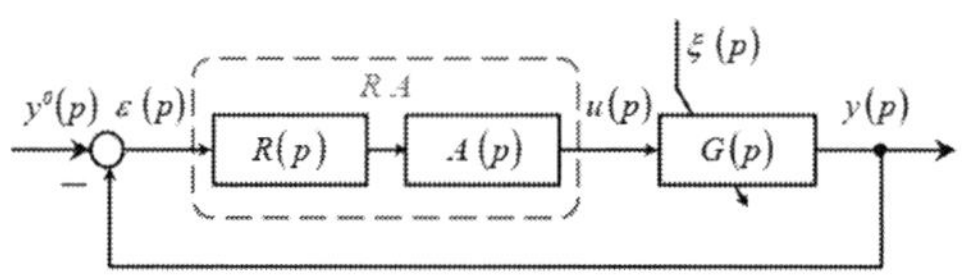

Figure 2. Disturbance absorbing system

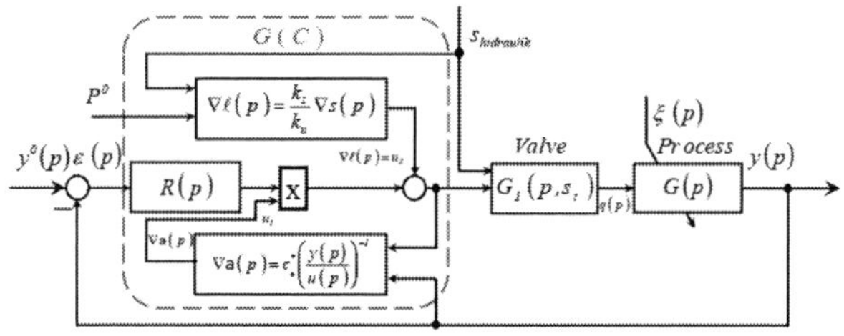

Figure 3. Gain scheduling control system

$$A(p)=k^A Q_\Sigma^{-1}(p) \quad (9) \qquad k^A_{opt}=\frac{\left|R^*\left(j\omega_\pi^{CLASS}\right)G^*\left(j\omega_\pi^{CLASS}\right)Q_\Sigma\left(j\omega_\pi^{A^*}\right)\right|}{\left|\mathcal{R}^*\left(j\omega_\pi^{A^*}\right)G^D\left(j\omega_\pi^{A^*}\right)\right|} \quad (10)$$

$$\left\{\begin{array}{l} k^A_{opt}\Leftrightarrow\left|\left[1+\left(\mathcal{R}^*(j\Omega)G^D(j\Omega)\dfrac{k^A_{opt}}{Q_\Sigma(j\Omega)}\right)\right]^{-1}\right|\le 1 \\ \left\{\Omega=0.1\omega^D;\left(\begin{array}{l}\omega^D\Leftrightarrow\dfrac{d}{d\omega}\left|\dfrac{1}{1+\mathcal{R}^*(j\omega)G^D(j\omega)}\right|= \\ =\dfrac{d}{d\omega}\left|\Phi_{y^*\varepsilon}^{CLASS^*}(j\omega)\right|=0\end{array}\right)\right\} \end{array}\right. \quad (11)$$

➢ **The method** of the compensation equation of the parametric balance [1] solves the task of analytical synthesis of gain scheduled control system with combined $G(C)$ (figure 3) parametric compensation for controlled process under a priori uncertainty, **under the criteria** (12): constant value (corresponding to optimal system setup under σ) of the transmission coefficient k_{rl} of the regulatory body (12.a); constant value (corresponding to optimal system setup under σ) of the transmission coefficient $k_{SYSTEM\,l}$ of the system (12.b); *LCQ* (12.c) (process with reordering set $\sigma=const$ etc.),

$$\left\{\begin{array}{ll} a)->> & \left\{\begin{array}{l}k_{rl}(\nabla s_l,t)=k_{rl}=const\\ \Leftrightarrow\left(k_p,T_i,T_d\right)_{optimal}\end{array}\right\} \\ b)->> & \left\{\begin{array}{c}k_{rl}(\nabla s_l,t)=k_{rl}=const\\ \Leftrightarrow\left(k_p,T_i,T_d\right)_{optimal}\\ k_{SYSTEM\,l}(\nabla s_l,\xi_l,t)=k_{SYSTEM\,l}=const\\ \Leftrightarrow\left(k_p,T_i,T_d\right)_{optimal}\end{array}\right\} \\ c)->> & \sigma=const \end{array}\right\} \quad (12)$$

and **initial conditions** for synthesis: a priori known σ , type of regulator, where: $R^*(p)\Leftrightarrow G^*(p)$, $(l=l_0^*,\ s=s_0^*)$; y° , s are signal external interference aggregated on the assignment channel y° and/or loading s , and ξ are aggregated internal additive or multiplicative parametric and/or structural interferences in process control; l_t, y_t, q_t - position of the valve control system CV; adjustable magnitude and cost of output at CV; $\nabla \ell_t^{G(A)}$, $\nabla \ell_t^{G(C)}$, $\nabla a_t^{G(C)}$ - compensation variables; $\nabla \ell_{t,romp}$, $\nabla \ell_{t,rolin}$ (13) - compensation variables for CV with equal and linear flaw characteristics; ∇s_t (13) - ∇ - loading changes.

$$\begin{cases} \nabla \ell_t^{G(C)} = -\dfrac{k_{z;}}{k_{l;}} \nabla s_t, & \begin{pmatrix} \nabla \ell_{t,romp} = \dfrac{0.5\left(e^{2n(1-l)}-1\right)}{n\, s\, e^{2n(1-l)}} \nabla s_t & ; \\ \nabla \ell_{t,rolin} = \dfrac{0,5\left(l^{-2}-1\right)l^{3}}{s} \nabla s_t \end{pmatrix} \\ \nabla a_t^{G(C)} = c_0^{*}\left(\dfrac{d\left(y_t\right)}{d\left(l_t\right)}\right)^{-1}, & \begin{pmatrix} \nabla a_t^{G(C)} \approx c_0^{*}\left(\dfrac{\Delta\left(y_t\right)}{\Delta\left(l_t\right)}\right)^{-1^{-1}}, \\ c_0^{*} = k_{SYSTEM1}^{G(C)}\ k_p^{-1} = const \end{pmatrix} \end{cases} \tag{13}$$

3. Repetitive control and *ML* filters.

The repetitive control is effectively applied in systems for controlled technological objects characterized by periodic signal interference [4,5]. In the structure of the repetitive regulator, the internal model in the *ML* (***Memory Loop***) [6-11] $M_L(p)$ (figure 4 and figure 6) is used, which consists of a periodic signal generator allowing compensation of periodic disturbances with a predetermined and fixed value of the period T_p. The base *ML* has the ability to "memorize" the frequency of "cut-off" and effectively to counteract it, through its specific structure as a dynamic system. There are dependency bridges for it (14)-(16):

$$\varepsilon^*(p) = \varepsilon(p) + \varepsilon^{0}(p) \tag{14}$$

$$M_L(p) = \frac{\varepsilon^{0}(p)}{\varepsilon(p)} = \frac{e^{-pT_p}}{1-e^{-pT_p}} \tag{15}$$

$$\varepsilon^{0}(p) = M_L(p)\varepsilon(p) = \left(\frac{e^{-pT_p}}{1-e^{-pT_p}}\right)\varepsilon(p) \tag{16}$$

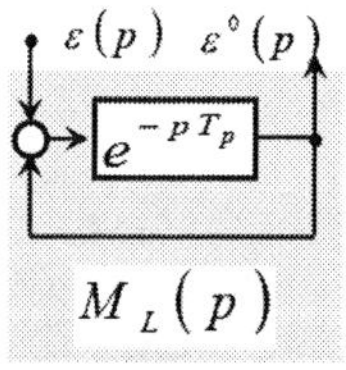

Figure 4. Base ML filter

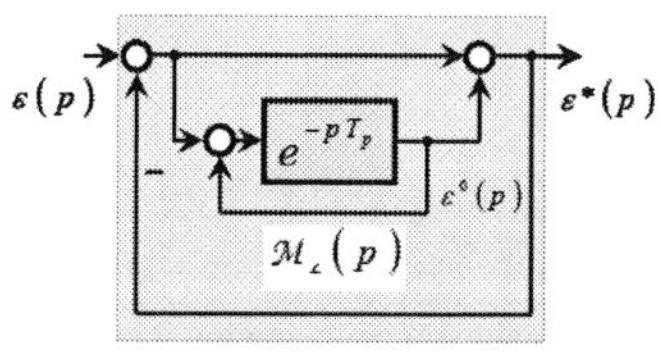

Figure 5. Improved *ML* filter

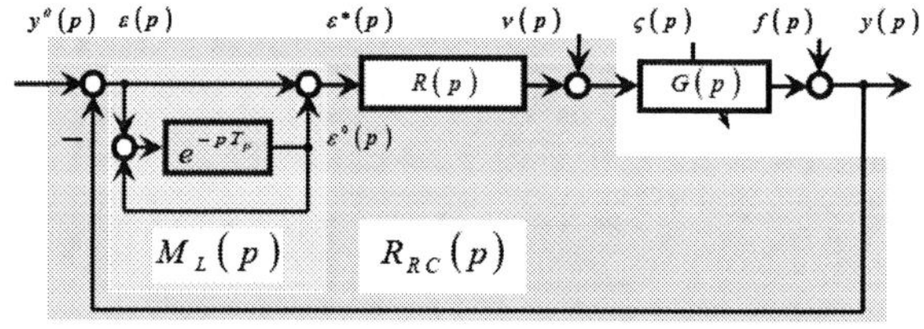

Figure 6. Repetitive control system

In the literature is well-known and improved [9] *ML* (figure 5). It differs from the basic one [6-11] because it uses a model of delay, but in a different structure. Unlike basic one (17), the improved *ML*-filter (19) is a sustainable dynamic system.

$$M_L(p) = \frac{\varepsilon^*(p)}{\varepsilon(p)} = \frac{1}{1-e^{-pT_p}} \tag{17}$$

$$\frac{\varepsilon^{0}(p)}{\varepsilon(p)} = \frac{e^{-pT_p}}{2-e^{-pT_p}} \tag{18}$$

$$\mathcal{M}_L(p) = \frac{\varepsilon^*(p)}{\varepsilon(p)} = \frac{1}{2-e^{-pT_p}} \tag{19}$$

The values of the controller setting parameters in the controlled technological process systems are subject to the aim and the criteria presented to the system in its synthesis. They are not a function of the respective *ML*-filter, which is added to the controller. In this sense, the design of the *ML*-filter in the control systems is autonomous and is not related to the synthesis of the controller. For designing $\mathcal{ML}$-filters with memory is used "*rejecting module bandpass filter criterion*". The essence of this method is a *sequential frequency correction.* The design filter criterion is defined by (20).

$$|\mathcal{M}_{L,i}(j\omega)| \equiv \begin{cases} 0, & \forall\omega \in \Delta\omega_i, (\omega_{b,i} < \omega_p < \omega_{h,i}) \\ 1, & \forall\omega \in [0,\omega_{b,i}], \\ & \forall\omega \in [\omega_{h,i},\infty), (\omega_{b,i} < \omega_p < \omega_{h,i}) \end{cases} \tag{20}$$

4. Robust repetitive control systems with internal model. Synthesis.

1. They are set below (chart 1): nominal model $G^*(p)$ of a minimal-phase object $G(p)$; an embarrassed object (model) at the top $G^{\Box}(p)$; synthesis criteria (1.a); period T_p of constant intermittent disturbance $d = sin(\omega_p t) = const$.

2. It is necessary to synthesize: robust R_M controller (free parameter method); repetitive $\mathcal{ML}$-filter (19), (20); compensation variables $\underline{\nabla\ell}_i^{PCS}$, ∇a_i^{PCS} (13) (method of the parametric balance equation); absorber $A^{daz}(p)$ of the wave structure disturbances (equilibrium partial absorption of the disturbances method).

Based on the synthesis methods and the advanced $\mathcal{ML}$-filter described above, the configuration of robust repetitive systems is based on development in the structure (figure 1) using the principle of consistent correction of R_M with $\mathcal{ML}$-filter, compensation variables $\underline{\nabla\ell}_i^{PCS}$, ∇a_i^{PCS} and/or absorber $A^{daz}(p)$. In fact, four completely new systems are illustrated respectively:

- in figure 7 – robust repetitive internal model control *RMREP*, where the controlled algorithm $R_M^{\mathcal{ML}}(p)$ is presented analytically with (21), and the results of the synthesis are plotted in table 1;
- in figure 8 – robust stabilization repetitive *RMGSCREP*, where the controlled algorithm $R_M^{GSC\mathcal{ML}}(p)$ is presented analytically with (22), and the results of the synthesis are plotted in table 1;
- in figure 9 – robust repetitive disturbance absorbing *RMDASREP*, where the controlled algorithm $R_M^{GSC\mathcal{ML}}(p)$ is presented analytically with (23), and the results of the synthesis are plotted in table 1;
- in figure 10 – robust stabilization repetitive disturbance absorbing *RMGSCDASREP*, where the controlled algorithm $R_M^{DASGSC\mathcal{ML}}(p)$ is presented analytically with (24), and the results of the synthesis are plotted in table 1.

Table 1. Results of the synthesis

initial conditions of synthesis	$G^*(p)$	$G^{\Box}(p)$	control valves CV	
	$\frac{1,25e^{-2p}}{10p+1}$	$\frac{2,1e^{-4p}}{10p+1}$	equal percent control valves ECV	
permanent intermittent disturbance	$T_p = 100s$ $d = sin(\omega_p t) = sin(0,0628t) = const$		$(\omega_p = \omega_{pf} = 0,0628 rad/s ; T_{pf} = 100s)$	
system synthesis results				
$R_M(p)$	$1,25\left(1+0,909p+\frac{1}{11p}\right)\frac{1}{(0,71p+1)}$			
$\mathcal{M}_L(p)$	$(2-e^{-pT_p})^{-1}, (T_p \equiv 2\pi(\omega_p)^{-1} = 2\pi(0,0628)^{-1})$			
compensation variable $\underline{\nabla\ell}_i^{PCS}$	$\frac{0.5(e^{2\pi(1-l)}-1)}{nse^{2\pi(1-l)}}\nabla s_i$	compensation variable ∇a_i^{PCS}	$c_0^*(d(y_i)/d(l_i))^{-1}$	
$A^{daz}(p)$	$\frac{0,08666}{p^2+5p+1}$			

$$R_M^{\mathcal{ML}}(p) = \mathcal{ML}\, R_M, \left(R_M \underset{\{min\|\epsilon\|_i^*\}}{\Leftrightarrow} G^*\right) \tag{21}$$

$$R_M^{GSC\mathcal{ML}}(p) = \mathcal{ML}\, R_M^{GSC}, \left(R_M \underset{\{min\|\epsilon\|_i^*\}}{\Leftrightarrow} G^*\right) \tag{22}$$

$$R_M^{DASML}(p) = \mathcal{ML}\, R_M^{DAS} = \mathcal{ML}\, R_M\, A_i\left(R_{M\,\{\min\|\varepsilon\|_i\}} \Leftrightarrow G^* \right) \quad (23)$$

$$R_M^{DASGSCML}(p) = \mathcal{ML}\, R_M^{GSC}\, A_i\left(R_{M\,\{\min\|\varepsilon\|_i\}} \Leftrightarrow G^* \right) \quad (24)$$

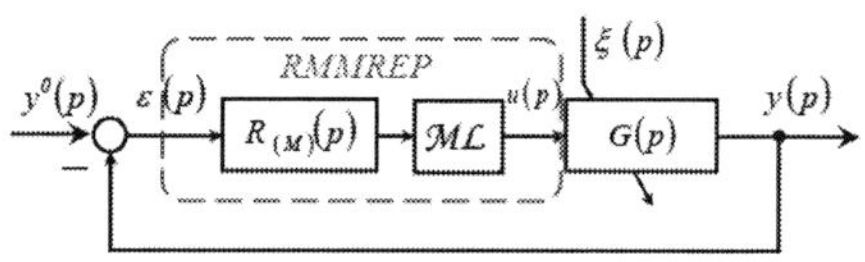

Figure 7. Robust repetitive internal model control system

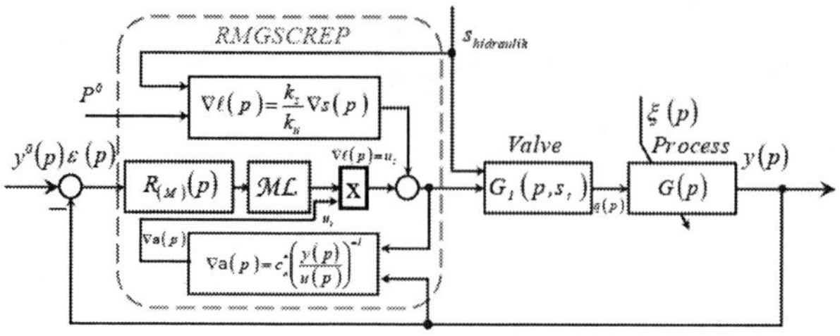

Figure 8. Robust stabilization repetitive control system

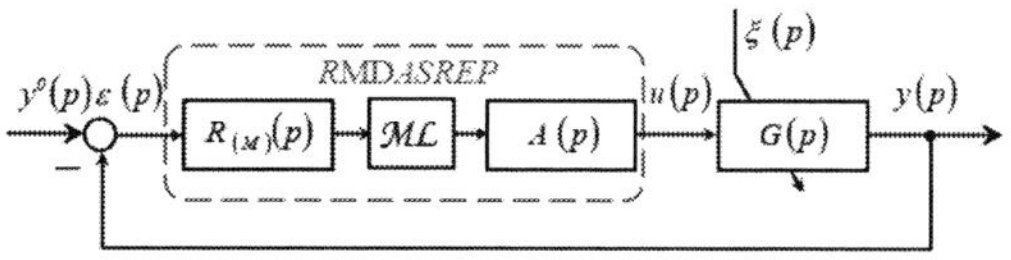

Figure 9. Robust repetitive disturbance absorbing system

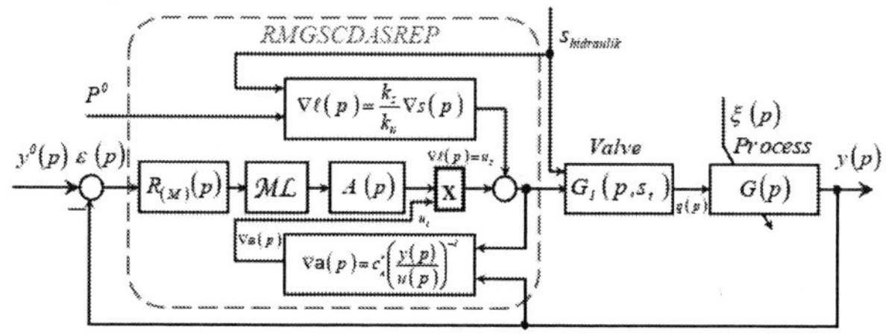

Figure 10. Robust stabilization repetitive disturbance absorbing system

5. Analysis and quality assessment. Comparative analysis.

The four synthesized systems are modelled and simulated under the same conditions. The simulation model results - time and frequency characteristics for closed Φ_i and for open W_i control systems are shown in figure 11 and figure 12. The system features are visualized *in nominal parametric mode* ($G \hat{=} G^*$, $d \equiv 0$, figure 11) and in *perturbance mode on the upper limit* ($G \hat{=} G^{U}$, $d \equiv 0$, figure 12) without the presence of a permanent intermittent external signal disturbance d. The following symbols are used: h_{Φ_i} - transient functions, $\Phi(j\omega)$, $W(j\omega)$- frequency characteristics of closed and open systems.

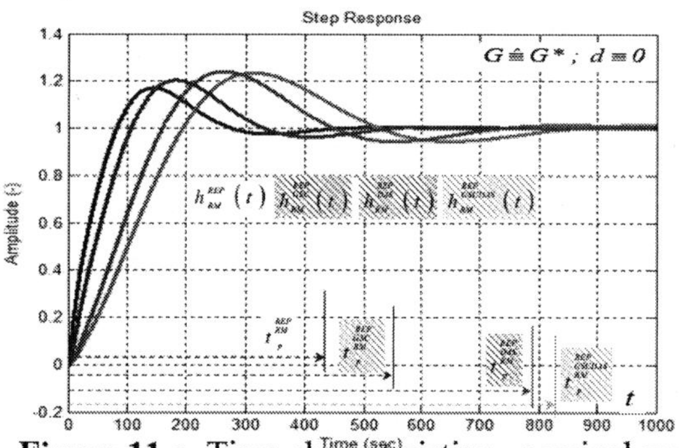

Figure 11.a. Time characteristics - nominal systems

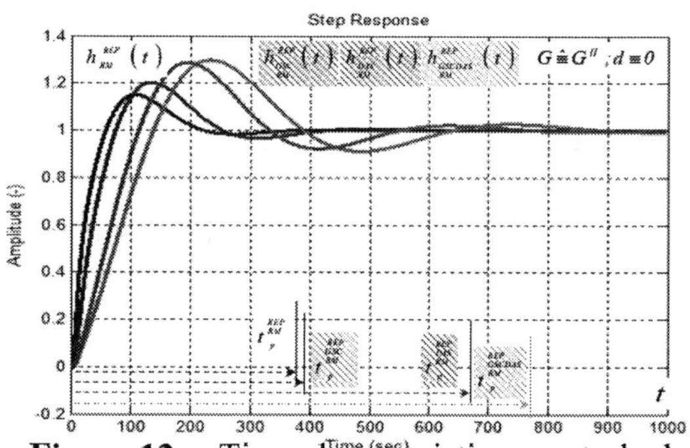

Figure 12.a. Time characteristics - perturbed systems

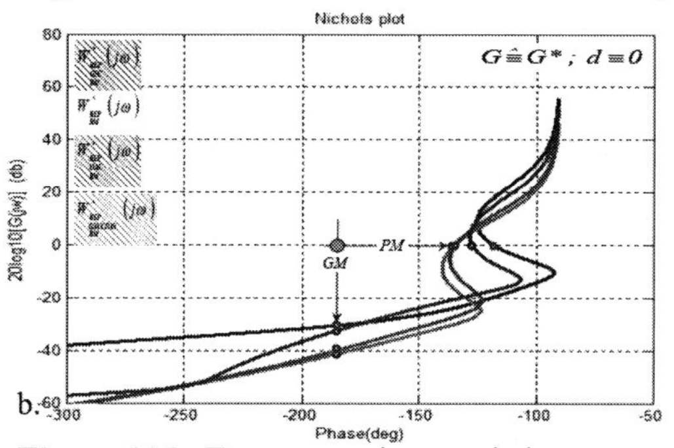

Figure 11.b. Frequency characteristics - nominal systems

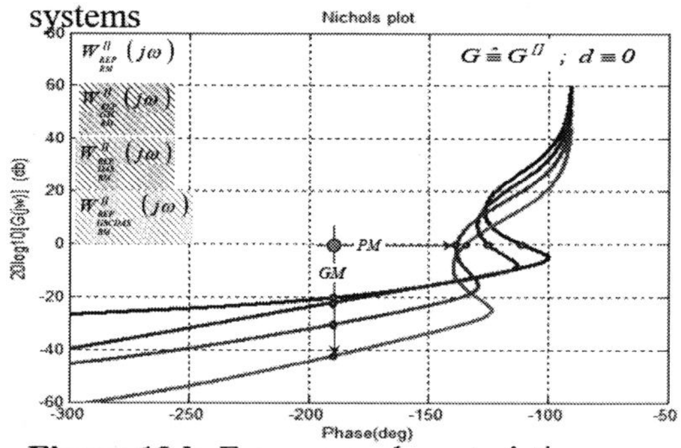

Figure 12.b. Frequency characteristics - perturbed systems

Quality analysis [1,12] *in nominal parametric mode of the systems* (figure 11) confirms that all four systems meet *LCQ*, reflecting the adjustment times (figure 11.a) and their stocks by module and phase (figure 11.b).

Quality analysis in perturbance mode on the upper limit systems (figure 12) *confirms that*: systems remain stable, reflecting the times regulation (figure 12.a) and their gain margin and phase margin (figure 12.b).

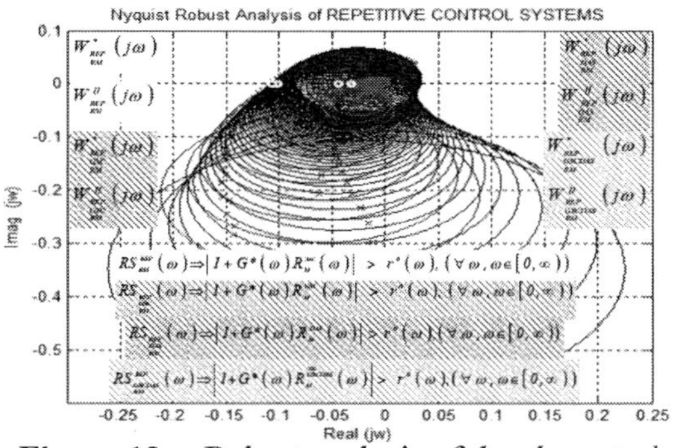

Figure 13.a. Robust analysis of the characteristics of the open loop systems

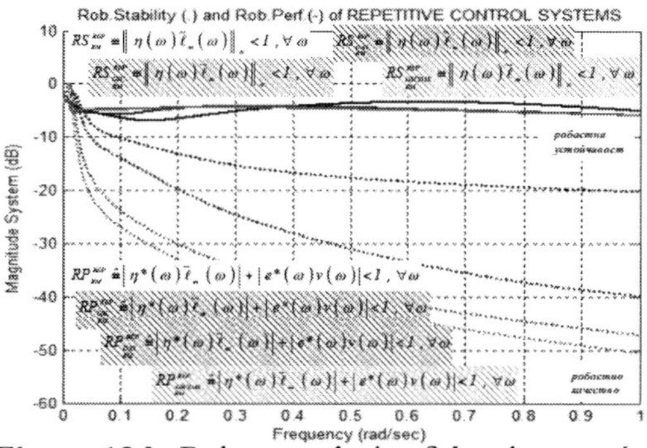

Figure 13.b. Robust analysis of the characteristics of the closed loop systems

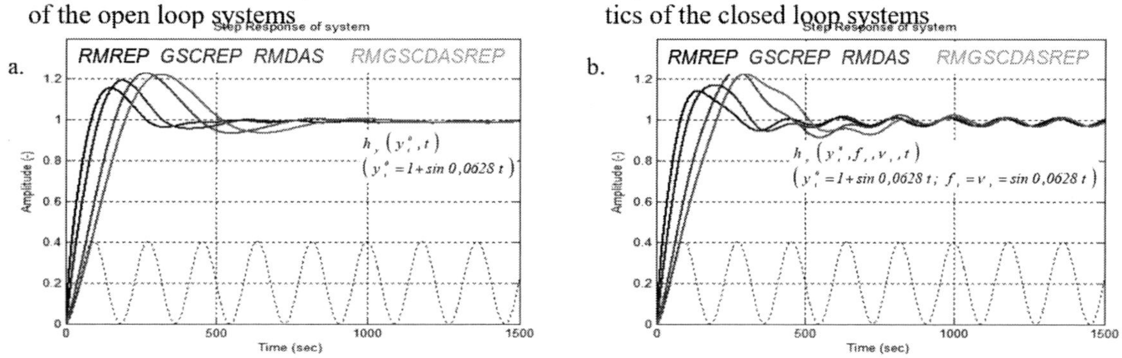

Figure 14. Time analysis of the filtering properties systems

6. Robust analysis.

By definition, the system analyzed is: robust stability and robust performance [1] for the set Π (1.a), if it meets the requirements (29) и (31) for the entire frequency range. If the nominal W_i^* and the perturbed $W_i^{\square}$ are open systems (25), so the circle π (26) with radius r° (27) and with centers in points ω_i, depicting the uncertainty in the process control, so the requirements for the robust stability and robust performance can also be expressed in the characteristics of an open system with (28) and respectively with (30).

$$W_i^* = R_i G^* ; \quad W_i^{\square} = R_i G^{\square} \ (25) \quad \pi(j\omega) \in W(j\omega), (\forall \omega, \omega \in [0, \infty)) \tag{26}$$

$$r^{\circ}(\omega_i) = | \ l_m(\omega_i) R(\omega_i) G^*(\omega_i) \ | \tag{27}$$

$$RS(\omega) \Rightarrow | \ 1 + G^*(\omega) R(\omega) \ | \ > \ r^{\circ}(\omega), (\forall \omega, \omega \in [0, \infty)) \tag{28}$$

$$RS(\omega) \Rightarrow \| \ \eta^*(\omega) \bar{\ell}_m(\omega) \ \|_{\infty} < 1, (\forall \omega, \omega \in [0, \infty)) \tag{29}$$

$$RP(\omega) \Rightarrow | \ 1 + G(\omega) R(\omega) | \geq | \ 1 + G^*(\omega) R(\omega) | - r^{\circ}(\omega), (\forall \omega, \omega \in [0, \infty)) \tag{30}$$

$$RP(\omega) \Rightarrow | \eta^*(\omega) \bar{\ell}_m(\omega) | + | e^*(\omega) \upsilon(\omega) | \ < \ 1, (\forall \omega, \omega \in [0, \infty)) \tag{31}$$

The results (figure 13) from the robust analysis of the characteristics of the open loop $W_i(j\omega)$ (figure 13.a) and close loop $\Phi_i(j\omega)$ (figure 13.b-c) systems, *analytically prove* that under uncertainty ξ, $\xi^{\square}$ (table 1): the repetitive systems (figure 7-figure 10) meet the requirements (28)-(31) both for robust stability and robust performance.

7. Time analysis.

A *time analysis of the filtering properties* of the four (figure 7÷figure 10) synthesized systems with "test" periodic effects was also carried out [11], by modeling the systems and evaluating the simulation results by time criteria. The four systems were modeled with nominal parametric mode. Their transitional functions $h_i(t)$ of periodic impact d (table 1) are simulated.

The time characteristics of the systems as a result of their parallel simulation for a "test" frequency $\omega_p = 0,0628 rad/s$, are illustrated in figure 14 – under disturbance $d(t)$ on the assignment channel y^o (figure 14.a), as well as in parallel along the three channels y^o, v и f (figure 14.b).

The results of the study of the four systems with a *time analysis of the robust filtering properties* with respect to the quality evaluation for performance of the process confirm that the repetitive systems (figure 7, figure 8, figure 9, figure 10) have a reduced sensitivity to external harmonic impact $d(t)$ with frequency $\omega_p = 0,0628 rad/s$.

Conclusion

The present study examines and analyzes the results of the research on the peculiarities and properties of the repetitive robust systems. Results of research are presented, which are confirmation and proof of their quality. The stability, performance, accuracy, robust performance and stability, filtering properties of external periodic effects of this class of systems are all proven. In comparative terms, the properties of robust repetitive (figure 7), robust stabilization repetitive (figure 8), robust repetitive disturbance absorbing (figure 9) and robust stabilization repetitive disturbance absorbing (figure 10) control systems of the process G, described in table 1, using the methods for: • ***Performance analysis in nominal and parametric mode*** *of the systems* (figure 11, figure 12); • ***time analysis of the filtering properties*** *of systems with "test" periodic effects* (figure 14); • ***robust analysis*** of the behavior of uncertainty systems Π, G, G^*, $G^{\square}$, ξ, $\xi^{\square}$ of the object (figure 13).

Acknowledgements: The author would like to thank the Research and Development Sector at the Technical University of Sofia for the financial support.

References

[1] E. Nikolov, D. Jolly, N. Nikolova, B. Benova, 2005, *Commande Robuste*, Sofia, Ed de l'Université Technique de Sofia, ISBN 954-438-500-2, **216**

[2] M. Morari, E. Zafiriou, 1989, *Robust Process Control*, Prentice-Hall Int., Inc. Englewood Cliffs, NJ, **479**

[3] V.Karlova-Sergieva, 2016, *Control systems with conditional feedback*, International Conference Automatics'2016, FA, June 03-05, Sozopol, Publishing House of Technical University of Sofia, ISSN 0374-342X, ISSN 1311-0829, Vol. 66 (2), pp. **181-187**

[4] S. Guergov, 2018, *A review and analysis of the historical development of machine tools into complex intelligent mechatronic systems*, Journal of Machine Engineering, ISSN 1895-7595 (Print), ISSN 2391-80-71 (Online), Vol. 18, No. 1, **107-119**

[5] S. Guergov, 2017, *A modular multifunctional system for complex physiotherapy*, Series on Biomechanics, Peer-reviewed Journal Edited by Bulgarian Academy of Science, Vol.31, No.2, ISSN 1313-2458, **41-44**

[6] R. Gricy, R. Cardoner, R. Costa-Castelló, E. Fossas, 2007, *Digital repetitive control of a three-phase four-wire shunt active filter*, IEEE Transactions on Industrial Electronics 54 (3), **1495-1503**

[7] M. Steinbuch, 2002, *Repetitive control for systems with uncertain period-time,* Automatica 38, **2103-2109**

[8] R. W. Longman, 2010, *On the Theory and Design of Linear Repetitive Control Systems*, European Journal of Control, vol. 5, DOI: 10.3166/EJC.16, **447-496**

[9] N. Nikolova, E. Nikolov, 2007, *ML- Structures In The Repetitive Robust Control Systems*, Cybernetics and Information Technologies Journal, Vol. 8, No 2, BAS, **15-28**

[10] N. G. Nikolova, 2013, *Filters in Repetitive Control Systems*, Proceedings of the Technical University Sofia, Publishing House of Technical University of Sofia, ISSN 0374-342X, ISSN 1311-0829, Vol 63, (2), pp. **45-52**

[11] N. G. Nikolova, E. Nikolov, 2014, *Predictive and Repetitive Control (Applied Methods for Process Control - part III)*, Sofia, Publishing House of Technical University of Sofia, ISBN 978-619-167-136-6, **258**

[12] V. Karlova-Sergieva, 2013, *Design of control systems with guaranteed performance*, Publishing House "Radix" OOD, ISBN 978-619-7140-01-9, **164**

IOP Conf. Series: Materials Science and Engineering **618** (2019) 012014 doi:10.1088/1757-899X/618/1/012014

Sputtering of Ga-doped ZnO nanocoatings on silicon for piezoelectric transducers

M Aleksandrova[1], T Tsanev[1], G Dobrikov[1], G Kolev[1], M Sophocleous[2], J Georgiou[2] and K Denishev[1]

[1]Technical University of Sofia, Department of Microelectronics, 8 Kliment Ohridski blvd, bl. 1, 1000 Sofia, Bulgaria

[2] University of Cyprus, Department of Electrical and Computer Engineering, Holistic Electronics Research Laboratory & EMPHASIS Research Centre, 75 Kallipoleos Str., 1678 Nicosia, Cyprus

E-mail: m_aleksandrova@tu-sofia.bg

Abstract. In this study, lead-free, piezoelectric devices were fabricated by vacuum radiofrequency (RF) sputtering of Ga–doped, ZnO (GZO) nanocoatings at different deposition modes on silicon substrates. Several deposition rates were varied (2 to 16 nm/min) by tuning the sputtering voltage and pressure, in order to control the microstructure and surface morphology of the films. The dependences of the piezoelectric properties of GZO thin films on the different deposition conditions were investigated. Significant dependency of the films' microstructure and surface roughness was observed to the deposition modes used for fabrication. The results showed that after optimization, the surface roughness of the films grown using sputtering voltage 1.1 kV and Ar=$2.5.10^{-2}$ Torr, is 6 nm. Films with such microstructures are characterized with crystallites orientation (002) and exhibit the highest piezoelectric coefficient of 96.3 pC/m, which is superior compared to other known, lead-free, piezoelectric materials. A simple, membrane–based, vibration sensor was fabricated at the optimal set of conditions to demonstrate the functionality of the coating in a real microelectromechanical (MEMS) device. The working range of the device is between 2 and 800 g and the piezoelectric voltage generated after minimum deformation of 5.4 nm (2g), was 9.66 mV.

1. Introduction

Silicon-based, pressure/vibration sensors have a broad applicability, because of their favorable technological properties, such as ease of control, flexibility by anisotropic etching, lack of mechanical hysteresis and compatibility with the integrated circuits fabrication technology [1]. However, problems with their sensitivity and response time still exist. For this reason, the researchers combined the purely capacitive behavior of such membrane-type, bulk micromachined sensors with piezoelectric response by growing different piezoelectric films (mostly based on lead zirconium titanate - PZT), by variety of methods like sputtering [2], epitaxy [3, 4], pulsed laser deposition, etc. [5]. Recently, material science allowed production of lead-free, piezoelectric materials with high piezoelectric

IOP Conf. Series: Materials Science and Engineering **618** (2019) 012014 doi:10.1088/1757-899X/618/1/012014

coefficients, such as $Ba_{1-x}Ca_xTi_{1-y}Zr_yO_3$, $(K_{0.5}Na_{0.5})NbO_3$,$(Bi_{0.5}Na_{0.5})TiO_3$, ZnO doped by boron (B), gallium (Ga), vanadium (V) and aluminum (Al) [6-9]. The doping approach is applied to increase the conductivity of the piezoelectric films, typically these materials are dielectric in nature, and thus enhance the power produced. The most promising results for achieving controllable and reproducible structures and morphology of the films seem to exist for Ga-doped ZnO. However, the focus of the known studies was on the optical characteristics for application in solar cells and photoluminescent devices (on glass substrates), and less on the piezoelectric response of the films, which require different features of the crystallites as compared to these necessary for the optoelectronic elements [10,11].

In this paper silicon wafer with crystallography orientation (100) was used as a substrate and ZnO was co-sputtered with Ga by RF sputtering for films with different thicknesses, produced at different growth rates. Analysis of the samples by X-ray diffraction (XRD), X-ray Photoelectron Spectroscopy (XPS) and Atomic Force Microscopy (AFM) gave information about the dependences of the structural, elemental and morphological properties of the GZO thin films at different deposition conditions on the piezoelectric behavior and sensing properties.

2. Experimental section

In this study, radio frequency (RF) (13.56 MHz) sputtering was applied to deposit the GZO thin films. ZnO (50 wt %) and Ga (50 wt %) were compounded in target with a 3 inch diameter. The area of the Si (100) substrate was about 3 cm x 3 cm. The substrates were preliminary cleaned in hydrofluoric acid to strip the native oxide film on the top of silicon. The chamber was evacuated to vacuum level of 10^{-5} Torr, and then the deposition parameters were controlled at pressures between 10^{-4} Torr and 10^{-1} Torr and voltages between 0.7 kV and 1.3 kV (plasma power in the range 35 W-234 W) in order to achieve different crystalline parameters and films compositions. Two different growth times were set for each combination of deposition modes to produce different films thicknesses. The thicknesses of the GZO thin films were measured using α-step Tencor profiler. The deposition rate was between 2 and 16 nm/min and it was calculated based on averaging 3 samples thickness measurements data. The roughness (or flatness) was monitored using atomic force microscopy (AFM) MFP-3D, Asylum Research, Oxford Instruments in non-contact mode, and the crystalline structure was investigated using Philips PW-1724 X-ray diffraction (XRD) patterns with CuKα radiation. The surface chemical composition and bonding of the GZO thin films were analyzed using the X-ray photoelectron spectroscope (XPS) ULVAC-PHI 5000. Piezoelectric coefficients were calculated based on capacitance and voltage generated measurements, following standard charge excitation procedure, described elsewhere [12]. Capacitance was measured by LCR-819 - GW Instek RLC Impedance meter and the shape of the voltage generated was recorded with digital oscilloscope DQ2042CN. The test setup for the piezoelectric coefficient was chosen, considering elasticity coefficient of the additional vibration system. It was proven that in this way the width of the frequency range at constant vibration could be broaden, as well as the amount of generated charge is greater [13]. The simple, membrane-based, sensing structure was obtained following standard bulk micromachining procedure (thermal oxidation of silicon – photolithography patterning of SiO_2 – wet anisotropy etching of uncoated with SiO_2 wafer in 40% KOH solution at 30°C). For the sensor device, back side and top membrane electrodes were produced by thermal evaporation of aluminum (240 nm). Intermediate piezoelectric film was grown by using the optimal deposition parameters, namely plasma power of 136 W and working pressure of $2.5.10^{-3}$ Torr, because the deposited GZO thin films had the most uniform and regular surface, and acceptable growth rate. Sensor element loading ability and working range was detected by laboratory-made testers for dynamic loading with variable magnitude and frequency of the vibrations generated.

3. Results and discussion

As figure 1 shows, the films deposited at different sputtering voltage showed different peak intensities (002), typically appearing between 28° and 35° depending on the substrate used and film thickness. These results suggest that the Ga-doped ZnO thin films deposited at higher plasma power, hence higher temperatures, have a stronger asymmetry in the crystalline structure. The rest of the peaks are suppressed in the range $2\theta = \sim 50\text{-}60°$.

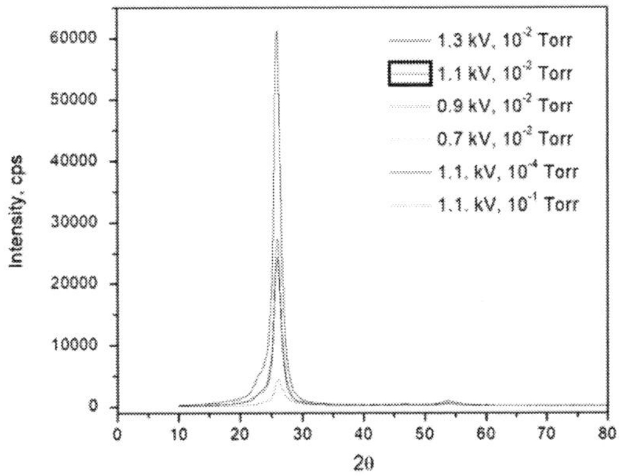

Figure 1. XRD patterns of Ga-doped ZnO thin films on silicon as a function of the deposition voltage and pressure.

The grown films exhibited surface states with increasing average roughness when the sputtering voltage and pressure increased, as shown in the AFM 3D topographies (figure 2).

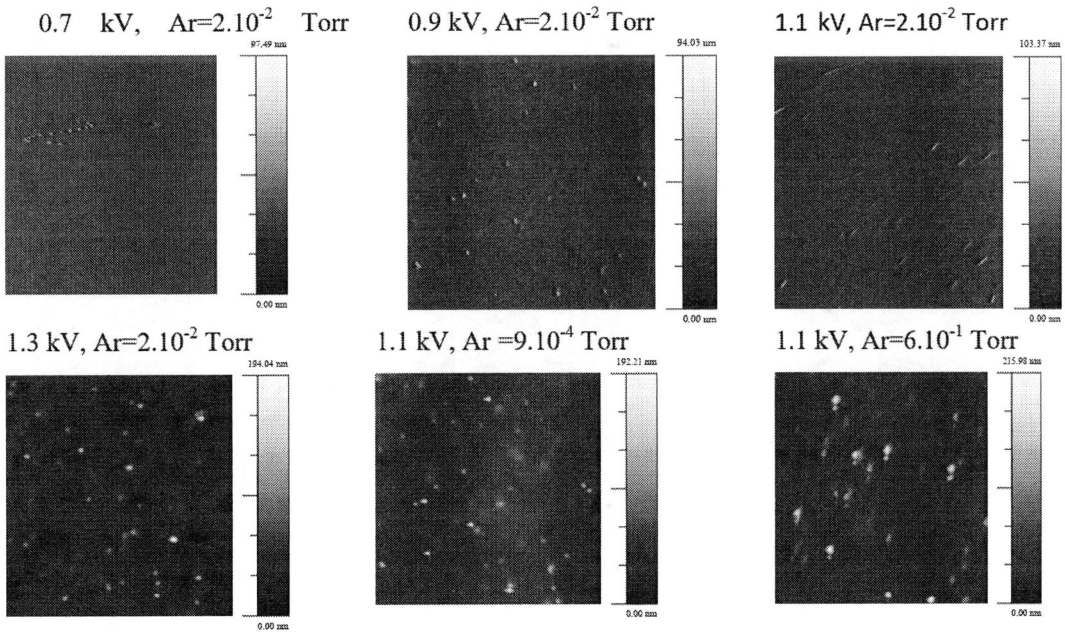

Figure 2. AFM images of the Ga-doped ZnO nanocoatings sputtered on silicon wafers at different conditions.

18 nm roughness was observed for film thickness of 194 nm, grown at sputtering voltage of 1.3 kV and pressure Ar= 2.10^{-2} Torr. The coating, grown at 0.7 kV and Ar=2.10^{-2} Torr shows high roughness of almost 14 nm. It is in contrast to the normal trend that at low deposition rates the films consist of small, uniform in size grains. Similar inconsistence is reported by Qui et al. [14] for evaporated film produced of low density vapor flux. By similarity, the possible reason for this reverse behavior at the lowest sputtering voltage is the insufficient energy of the particles, forming columnar grains and voids between them without possibility for particle diffusion and homogenization. Increasing the sputtering pressure at constant voltage results in decrease of the mean free path of the particles and due to overcrowding in the gap of the vacuum deposition chamber, the roughness also increased. It could be seen that rms roughness can reach 21 nm for film with thickness of 216 nm at Ar=6.10^{-1} Torr and U=1.1 kV. Acceptable surface roughness of ~6 nm, considering reliable contact with the next electrode film, exhibited the coatings produced at 0.9kV and 1.1kV at Ar = 2.10^{-2} Torr, which is in agreement with the results in [15]. From both layers, the XRD results showed better crystallinity for those, grown at 1.1 kV revealing four additional peaks in its diffractogram between 33° and 36.24° that could be ascribed to the substance under study (figure 3 up). This is the reason to select these particular deposition conditions for the further simple sensor element design. To completely investigate the relationship between the deposition conditions and chemical composition, the chemical structure of Ga-doped ZnO thin film at the optimum deposition conditions was investigated by XPS (figure 3 down).

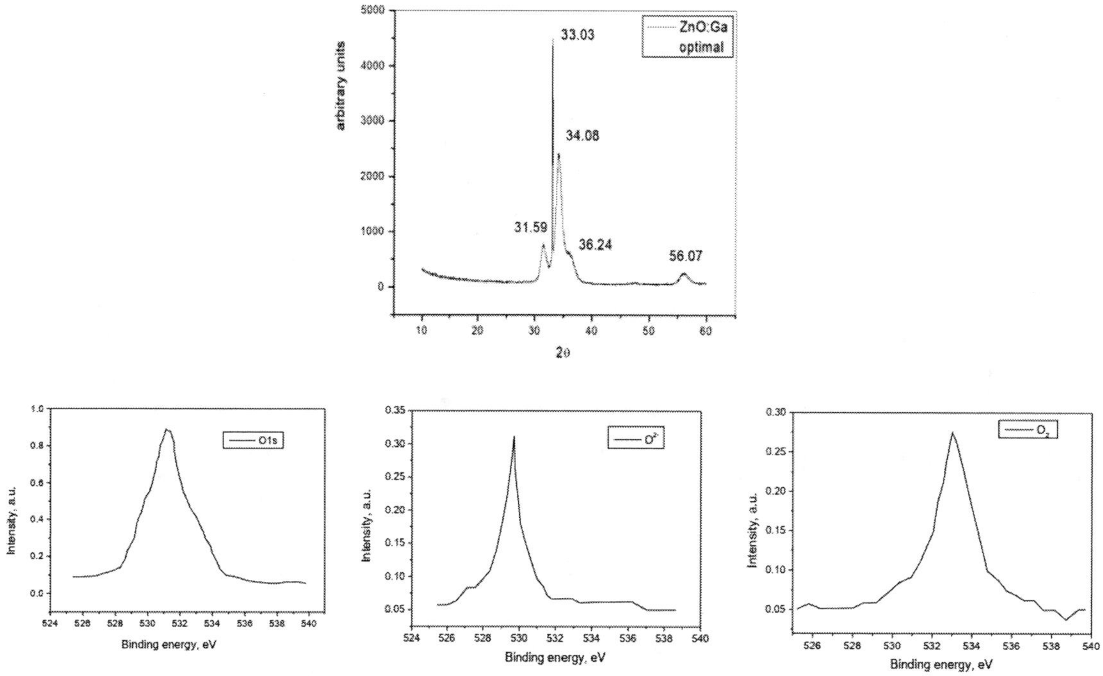

Figure 3. XRD of selected film grown at optimal deposition conditions (up) and XPS spectroscopy of the same film (down).

The measurement showed that no apparent peak is observed in the spectra of the Ga and ZnO constituents of the film, which suggest that binding state of new formation of Ga_2O_3 appears and starts dominating. Similar results are reported, where GZO films are sputtered on glass [15]. This could be a possible reason for enhancement of the piezoelectric properties. Different bond existance was detected by the spectrophotometer for Ga-doped ZnO thin films deposited on silicon (100) substrates, as can be seen from XPS spectrum (figure 3 down). The state corresponding to the O1s spectrum was divided into three components, centered in the range ~530 eV-533 eV for all characteristic peaks. The peak with higher binding energy is associated with O^{2-} ions in oxygen vacancies within substrates that have poor crystallization, as the figure shows. Based on the results, it can be concluded that there are more oxygen vacancies existing in GZO thin films. Regarding the low-energy peak, it was due to poor adhesive strength of Ga-doped ZnO thin films to the silicon wafer. Decrease of oxygen vacancies and the increase of oxygen absorption, may improve the piezoelectric properties of the films [16,17].

Piezoelectric coefficients were obtained, following the measurement procedure validated for silicon cantilever with defined geometry and size [18]. The cantilever was fixed at one of the ends, while the opposite side was free standing. By supplying vibration with certain frequency (related to ω) and magnitude (A), producing deflection and due to suitable electrode patterning and placement, it was possible to determine the transvers piezoelectric coefficient d_{33} (equation 1). In this way, the measured voltage from peak to peak (U_{p-p}) is related to the strain applied, the area activated between the electrodes, resonant frequency of the beam, due to its geometry and some parameters of the measurement equipment (resistance and capacitance R and C of the oscilloscope).

$$U_{p-p} = \frac{A.S.e_{33}}{C} \cdot \frac{R^2C^2\omega^2}{1 + R^2C^2\omega^2\left[1 + exp^{-\frac{t}{RC}}\right]} \quad [18] \qquad (1)$$

The formula was used to determine the electrical displacement coefficients e_{33}. They are related with the piezoelectric coefficients through the elastic compliances for stress under steady electric field (in reverse direction to the Young's modulus). This method is simple, reliable, provides high resolution and accuracy better than ± 0.5 pC/N [19]. The data obtained after this measurement for films, grown at different sputtering voltages and pressures, as a function of the vibration frequency are shown in figure 4.

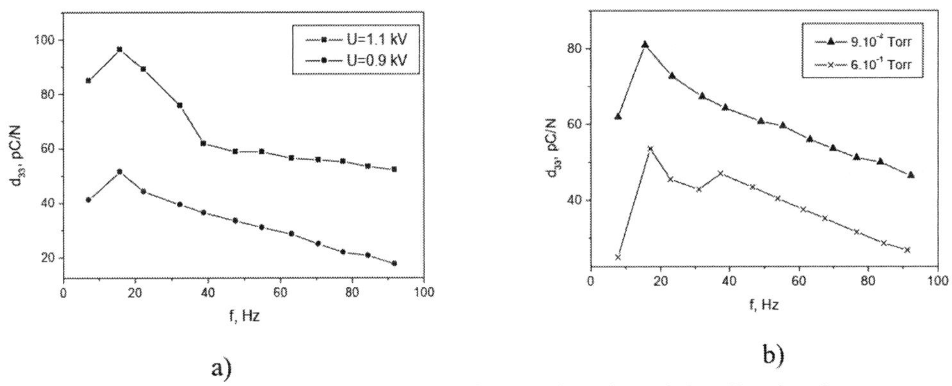

Figure 4. a) Piezoelectric coefficient d_{33} as a function of the vibration frequency for GZO sputtered at: a) 0.9 kV and 1.1 kV (constant pressure of $2.5x10^{-2}$ Torr); b) 9.10^{-4} Torr and 6.10^{-1} Torr (at constant sputtering voltage of 1.1 kV).

The results show enhanced conversion efficiency of the coating at lower frequencies with maximum at 16-17 Hz, respectively, depending on the deposition parameter and decreasing with further increase of the frequency. Overall, the results suggest that the modification of the deposition mode by increasing the sputtering voltage, results in enhancement of the asymmetry of the crystal lattice, as it can be seen from the XRD measurements. Although irregularities on the surface still exist, it was possible to enhance the piezoelectric behavior of Ga-doped ZnO nanocoatings. On the contrary, the increase of the sputtering pressure resulted in degradation of the properties. Two basic factors, i.e. the irregularities and crystallinity of the material, have contributed to the variation of the piezoelectric coefficients. Grain boundaries are considered as non-polar type traps for the charge carriers, therefore, films, deposited at voltage higher than 0.8 kV, which are characterized with larger grains, have affinity

IOP Conf. Series: Materials Science and Engineering **618** (2019) 012014 doi:10.1088/1757-899X/618/1/012014

to generate greater piezoelectric response due to the lower number of such boundaries. The behavior is similar at higher argon partial pressure. Such relations have been reported in [20] for piezoelectric ceramic with ZnO content. The observed peaks in the 3D AFM scanned images, contributed to the formation of defects, influencing the domain zones uniformity and inducing polarization quenching, which results in the weaker piezoelectric properties [21].

In order to estimate the sensing characteristic, membrane-based, vibration sensor was fabricated using Ga-doped ZnO film that exhibit maximal piezoelectric coefficient (figure 5).

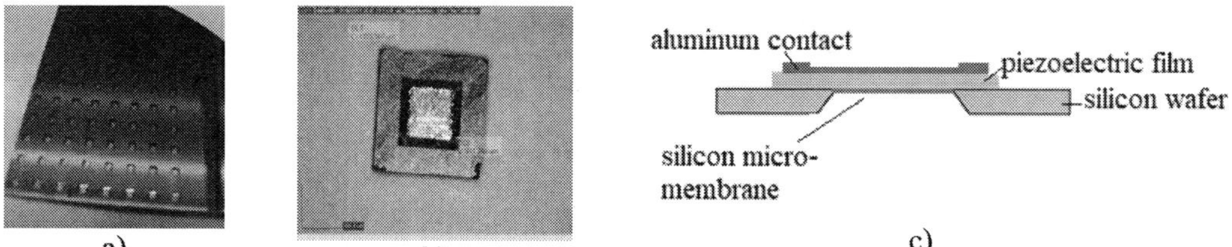

a) b) c)

Figure 5. a) Multiple membranes, bulk micromachined in silicon (100); b) single membrane, scribed and separated from the wafer having coated inner part with piezoelectric – face side; c) cross-sectional view of piezoelectric coated silicon micromembrane.

The backside was coated with patterned aluminum electrode. The element was tested at constant frequency of 16 Hz and varying vibration strength, creating in this way mechanical load in the range of 2 – 800 g. Due to the excellent mechanical properties of the silicon membrane without hysteresis, the output voltage, generated under strong wave, was tested at high load up to 800 g. The sensing element worked properly, without damage, even just before the strongest excitation and then breaks.

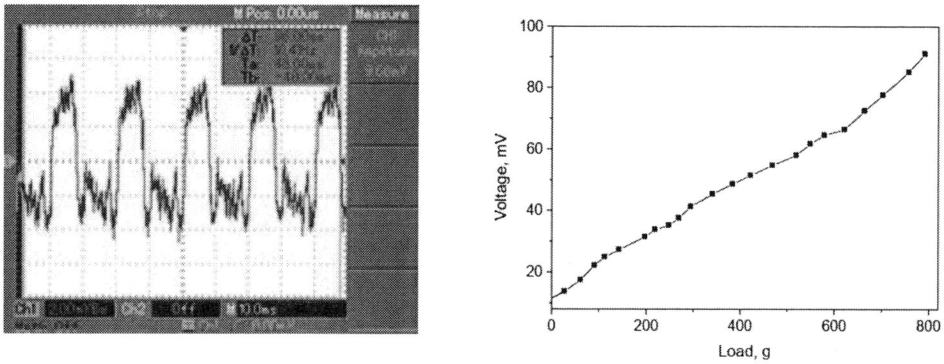

Figure 6. a) Voltage vs time (oscilloscope view) for cyclic repeating compression – tension of the structure at relatively low load of 2 g, demonstrating the sensor sensitivity; b) voltage versus load weight in the range of 2-800 g.

Figure 6a shows the variation of the output voltage with the time under cyclic loading with repeated pressure and tension with the sinusoidal excitation wave, equivalent to load of 2 g. Figure 6b shows the effective (rms) value of the piezoelectric voltage, produced during the membrane deflection. Although the linear response of the sensor, the signal should be further filtered by a high-order filter with enhanced selectivity and frequency tuning [22]. Nevertheless, at minimum film deformation of 5.4 nm, calculated according to Hook's law [23], the piezoelectric voltage was 9.66 mV. After applying greater load than 800 g, the membrane started tearing apart and the signal is abruptly interrupted, therefore the limiting factor in this case is not related to the GZO films capabilities. By the best of the author knowledge, the results are superior as compared to the sensor response of similar constructive design, but using different lead-free materials.

4. Conclusions

In this study, nanosized films of Ga-doped ZnO were successfully RF sputtered on silicon wafer with crystallographic orientation (100) suitable for bulk micromachining and membrane-based sensor fabrication. It was found that film with rms roughness ~6 nm (thickness ~103 nm) and dominant orientation of crystallites (002) grown at sputtering voltage 1.1 kV and Ar=2.10^{-2} Torr, exhibited the highest piezoelectric coefficient of 96.3 pC/m for single frequency of 16 Hz. The sensor was tested at loading at this frequency and different vibration strengths whilst the piezoelectric response was analyzed. The sensor is characterized by high sensitivity and broad operational range, as compared to similar sensors using other lead-free piezoelectric materials, due to the optimized sputtering conditions of GZO films. Future work will be related to filtration of the produced signal and study of the pyroelectric behavior of this material for fabricating multifunctional sensor "2 in 1", detecting pyro- and piezoelectric effect simultaneously for variety of transducers and energy harvesting applications.

Acknowledgement: The authors acknowledge funding support of BNSF's grant КП06-Н27/1. Part of this study is related to PhD project for thesis development 192ПД0006-03 at NIS, TU-Sofia. The authors are thankful to all colleagues from the Bulgarian Academy of Sciences participating in the spectroscopic and microscopic measurements.

References

[1] Bogue R 2013 *Sensor Review* **33** 300
[2] Setter N 2006 *Electroceramic-Based MEMS: Fabrication-Technology and Applications* (Springer Science & Business Media)
[3] Yang E H, Hishinuma Y, Cheng J G, Trolier-Mckinstry S, Bloemhof E and Levine BM 2006 *Journal of Microelectromechanical Systems* **15** 1214
[4] Isarakorn D *et al.* 2010 *Journal of Micromechanics and Microengineering* **20** 055008
[5] Eason R 2007 *Pulsed Laser Deposition of Thin Films: Applications-Led Growth of Functional Materials* (John Wiley & Sons)
[6] Villafuerte-Castrejón M E, Morán E, Reyes-Montero A, Vivar-Ocampo R, Peña-Jiménez J A, Rea-López S O and Pardo L 2016 *Materials* **91**
[7] Wang W, Ai T and Yu Q 2017 *Scientific Reports* **7** 42615
[8] Laurenti M, Castellino M, Perrone D, Asvarov A, Canavese G and Chiolerio A 2017 *Scientific Reports* **7** 41957.
[9] Blagoev B S *et al.* 2016 *J. Phys.: Conf. Ser.***700** 012040
[10] Duan J, Xiong Q, Hu J and Wang H 2015 *Journal of Power and Energy Engineering* **3** 62059
[11] Mishra A, Mohapatra S, Gouda H S and Singh U P 2016 *AIP Conference Proceedings* **1728** 020495.
[12] Liu J M, Pan B, Chan H L W, Zhu S N, Zhu Y Y and Liu ZG 2002 *Materials Chemistry and Physics* **75** 12
[13] Nedelchev K and Kralov I 2017 *AIP Conference Proceedings* **1910** 020015
[14] Qui H, Wang F, Wu P, Pan L, Li L, Xiong L and Tian Y 2002 *Thin Solid Films* **414** 150
[15] Wang F H, Chen K N, Hsu C M, Liu M C and Yang C F 2016 *Nanomaterials* **6** 88
[16] Yu C F, Chen S H, Sun S J and Chou H 2011 *Appl. Surf. Sci.* **257** 6498
[17] Rao L K and Vinni V 1993 Appl. Phys. Lett. **63** 608
[18] Liu J M, Pan B, Chan H L W, Zhu S N, Zhu Y Y and Z G Liu 2002 Materials Chemistry and Physics **75** 12

[19] Verardi P, Cracium F and Dinescu M 1997 *IEEE Proceedings of the Ultrasonics Symposium* vol. 1 (New Jersey, USA) p. 569
[20] M Promsawat, Z GYe and Watcharapasorn A 2017 *Materials Letters* **205** 126
[21] Promsawat M, Watcharapasorn A, Ye Z G and Jiansirisomboon S 2015 *J. Am. Ceram. Soc.* **98** 848
[22] Pandiev I 2018 *IEEE 23rd International Symposium for Design and Technology in Electronic Packaging – Proceedings* (SIITME 2017 Constanta, Romania) p. 165
[23] Jalili N 2009 *Piezoelectric-Based Vibration Control: From Macro to Micro/Nano Scale Systems* (Springer Science & Business Media)

Optical diagnostics of the condition of the cardio vascular system on the basis of optoelectronic methods

S S Yesman[a], D V Veligotsky[a], S O Mamilov[a], A I Gisbrecht[b], V S Mircheva *[b,c] and L B Zaharieva[b,c]

[a] Institute of applied problems of physics and biophysics NAS of Ukraine
[b] Institute of Electronics, Bulgarian Academy of Sciences
[c] Faculty of Physics, Sofia University

e-mail: victoriamircheva@gmail.com

Abstract. In the present work we have investigated the parameters of peripheral blood flow by using a non-invasive method for processing photoplethysmograms (PPGs). The developed original technique allows us to register a photoplethysmogram, isolate the pulse wave and determine the duration of cardio intervals, the duration of the anacrotic and dicrotic phases, the duration of fast and slow blood filling, the expulsion period, the dicrotic and diastolic indices, temporal and amplitude parameters of vascular tone, etc. We observe three types of photoplethysmograms after taking medicaments Corvitol-50 and nitroglycerin. The analysis of the amplitude-temporal characteristics of PPGs allows the determination of key diagnostic parameters.

1.Introduction

Optical diagnostics of biological objects has gained wide application in recent years. The advantages of optical diagnostic methods consist primarily in noninvasiveness, rapidness and high accuracy. Unlike radiological and ultrasonic methods, optical irradiation is natural for the organism and does not cause any side effects with the powers used. Determination of the relative saturation of arterial blood with oxygen (SaO_2) - pulse oximetry has become the gold standard in resuscitation and intensive care. The spectral range between 600 and 1000 nm (near IR) is well suited for optical tissue oximetry because at these wavelengths, the absorption spectrum of hemoglobin is closely related to its oxygenation state. Moreover, the penetration depth of biological tissues is high in this spectral range.

Today, cardiovascular diseases are the leading cause of death and disability. The most important tasks are the improvement of methods for diagnosing and monitoring, the development of effective methods for prevention. Doctors need non-invasive research methods that could record the main indicators of central and peripheral hemodynamics and would characterize the contractile activity of the heart. At the present stage of the development of medicine, it is no longer enough to take only electrocardiographic and blood pressure measurements to make a comprehensive diagnosis. In assessing the adaptive abilities of a person (physiology of labor and sports medicine, etc.), determination of basic hemodynamic parameters is necessary to establish rational modes of work and rest, optimal and maximum levels of physical activity.

IOP Conf. Series: Materials Science and Engineering **618** (2019) 012015 doi:10.1088/1757-899X/618/1/012015

There are many methods of measuring CO, both invasively and non-invasively; each has advantages and drawbacks. Invasive methods are well accepted, but these methods are quite complicated, traumatic and often lead to changes in heart rate. Consequently, the focus on development of non-invasive methods is growing [1].

Ultrasound Dopplerography is most used by these methods. Being accurate and inexpensive, Doppler ultrasound is a routine part of clinical ultrasound; it has high levels of reliability and reproducibility, but it does not measure blood volume and is complicated to use. This method is most suitable for large clinics [2].

Electrical impedance plethysmography and Electrical cardiometry are similar methods based on the model of Electrical Velocimetry, and non-invasively measures stroke volume, cardiac output and other hemodynamic parameters through the use of 4 surface ECG electrodes. The disadvantage of rheoplethysmography to determine the parameters of peripheral blood flow is the placement of current and measuring electrodes at a great distance. At the same time, the signal is influenced by the properties of vessels throughout the pathway.

Photoplethysmography (PPG) is a simple and low-cost optical technique that can be used to detect blood volume changes in the microvascular bed of tissue. It is often used non-invasively to make measurements at the skin surface. A PPG is often obtained by using a pulse oximeter which illuminates the skin and measures changes in light absorption. The PPG waveform comprises a pulsatile physiological waveform attributed to cardiac synchronous changes in the blood volume with each heart beat [3-5]. PPG offers significant potential for data mining and a range of innovative pulse wave analysis techniques.

The aim of our study is to determine the parameters of peripheral blood flow using an optoelectronic method for processing photoplethysmograms.

2. Materials and Methods

The photoplethysmogram is a change in the transmittance of light by a biological tissue under the influence of a pulse wave. The sensor for recording photoplethysmography is a light-emitting diode and a photodetector, between which biological tissue is placed. Taking into account the absorption spectra of blood and tissue, the use of near-infrared emitters is most appropriate. The use of photoplethysmogram allows local research of peripheral blood flow.

Similar to the analysis of the rheoplethysmogram proposed by Gurevich [6], the information parameters of the photoplethysmographic pulse curves can be divided into primary (amplitude and time) and derivatives of the primary, obtained by simple mathematical operations. The pulse curve depends on the systolic ejection, blood flow intensity, blood viscosity, condition of the vascular walls, the ratio of precapillary and postcapillary pressure, etc.

The pulse wave has the following main components (Figure 1): a steep systolic ascent - the anacrotic phase and the descending part, which corresponds to the catacrotic phase. On the anacrot there are two sections - the period of rapid blood filling from the beginning of the anacrot to the point of maximum steep rise (maximum of the first derivative of the pulse wave) and the period of slow blood filling. On the catacrot is the so-called dicrotic wave.

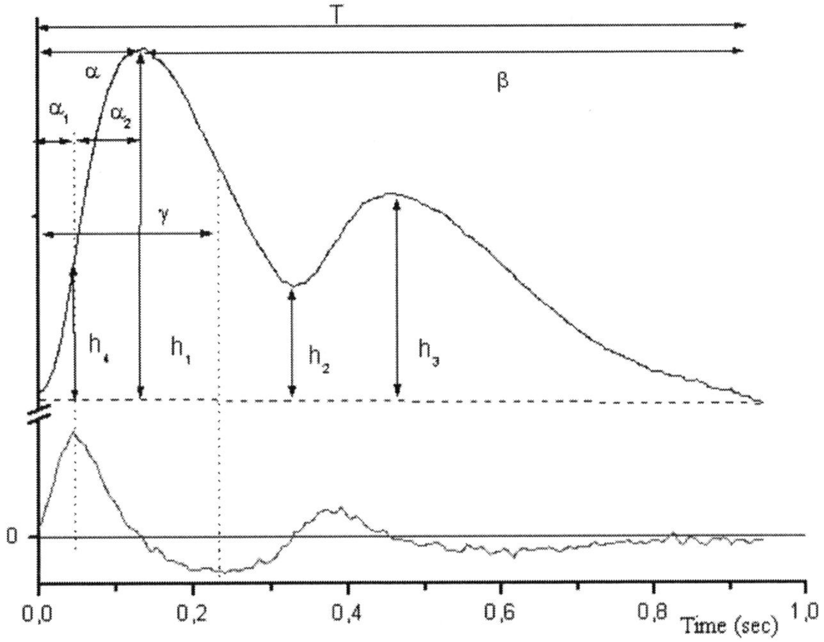

Figure 1. Pulse wave (upper curve) and its first derivative (lower curve).

Primary characteristics of the pulse wave:
T - the duration of the cardiac cycle.
α - the duration of the ascending part of the wave (anacrotic phase)
α_1 - the duration of rapid blood filling.
α_2 - the duration of slow blood filling .
β - the duration of the catacrotic phase
γ - the ejection period.
h_1 - the maximum amplitude of the pulse wave, the indicator of the magnitude of the pulse of blood.
h_2 - the amplitude of the pulse wave at the level of the incisura.
h_3 - amplitude of the pulse wave at the level of a dicrotic wave.
h_4 - amplitude of the pulse wave at the level of the projection of the maximum of the first derivative.

Secondary parameters:

$\frac{h_2}{h_1}$ – dicrotic index , $\frac{h_3}{h_1}$ – diastolic index, $\frac{h_4}{h_1}$ – amplitude index of vascular tone.

There are other secondary indices that are left out of our study.

In medical practice, the averaging of the SaO_2 value over 8 or 16 heartbeats is used, which simplifies the requirements for recording a photoplethysmographic signal. However, when we developed methods for determining changes in the relative volume of peripheral blood [7], the relative concentration of carboxyhemoglobin in arterial blood [8], and the relative oxygen concentration in venous blood [9], to reduce the error we need a lot of points, so it is necessary to record a photoplethysmogram with a frequency of 200-250 Hz [10]. This system enables us to implement the diagnosis of the cardiovascular system by analyzing the volume pulse wave.

The developed original technique allows us to register a photoplethysmogram, isolate the pulse wave and determine the duration of cardio intervals, the duration of the anacrotic and dicrotic phases, the duration of fast and slow blood filling, the expulsion period, the dicrotic and diastolic indices, temporal and amplitude parameters of vascular tone, etc.

All results are given as mean ± (SD). The Student's test for connected sampling was used for the statistical calculations. Statistical results with a value of $p < 0.05$ were viewed as being significant. The evaluation was carried out using the program "Origin 7.5".

3. Results and discussion

Measurements of these parameters were carried out with a medicament impact on a 51-year-old volunteer with coronary heart disease. The registration was carried out in a calm state with a satisfactory state of health. The measurements were carried out 30 minutes after taking "Corvitol-50" and 2-3 minutes after taking nitroglycerin.

The analysis revealed photoplethysmograms of several types (Figure 2). The y-axis are arbitrary units, because we are only interested in the shape of the signal. We attributed the first type of photoplethysmogram (Figure 2a) to a curve with a well-expressed dicrotic wave, the maximum of which is located approximately in the middle of the catacrotic phase. The difference between the amplitudes of the wave at the level of incisura and dicrotic wave is small. Photoplethysmograms of the first type were observed in a calm state and after taking Corvitol-50.

The photoplethysmogram shown in Figure 2b was attributed to the second type. In this type the dicrotic wave is also well expressed, but is shifted closer to the end of the catacrot. Incisura has a much smaller amplitude than in the previous case and the difference between the amplitudes of the wave at the level of incisura and the dicrotic tooth is very significant. Such photoplethysmograms were observed after taking nitroglycerin. The third type of photoplethysmogram (Figure 3b) is characterized by the absence of a expressed dicrotic wave. This leads to the fact that it is impossible to isolate the incisura and dicrotic prong and measure their amplitudes. This type of photoplethysmogram was also observed after taking nitroglycerin.

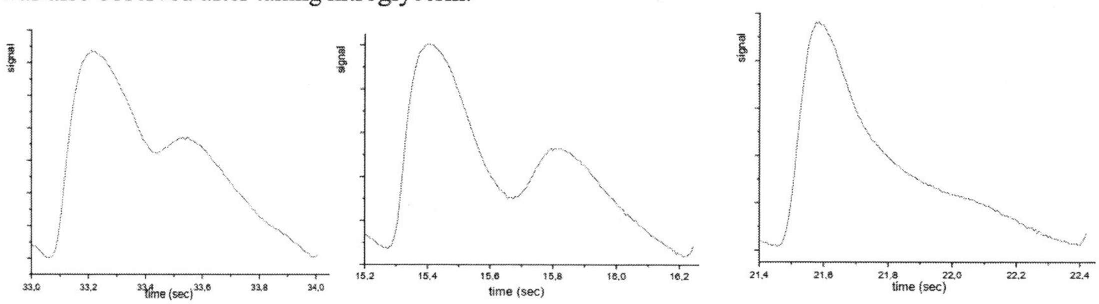

a) b) c)

Figure 2. Photoplethysmograms: a) first, b) second, c) third type.

After taking nitroglycerin, the second type of photoplethysmogram mainly occurs at the beginning of the recording, and the third is closer to the end.

3.1. Temporal characteristics

During the experiments, the following primary properties were determined: the duration of the cardio intervals, the duration of the anacrotic phase, the duration of the fast and slow blood filling, the ejection period. In a calm state, the temporal characteristics were: $\alpha_1 = 0.053 \pm 0.006$ sec., $\alpha_2 = 0.089 \pm 0.007$ sec., $\alpha = 0.142 \pm 0.008$ sec., $\gamma = 0.279 \pm 0.018$ sec., $T = 0.945 \pm 0.023$ sec.

Changes in temporal characteristics after taking medications are shown in Figure 3 and Figure 4.

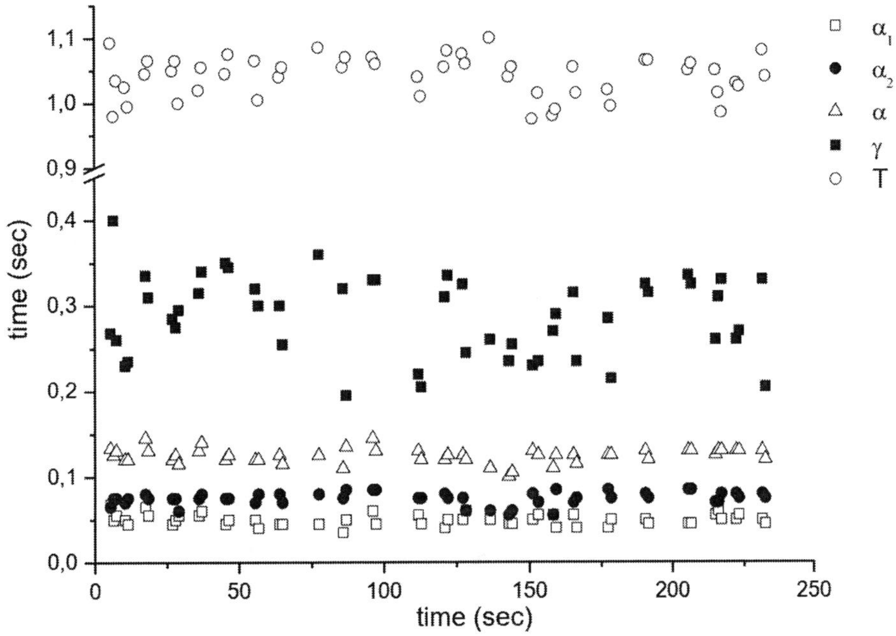

Figure 3. Temporal characteristics after taking "Corvitol-50".

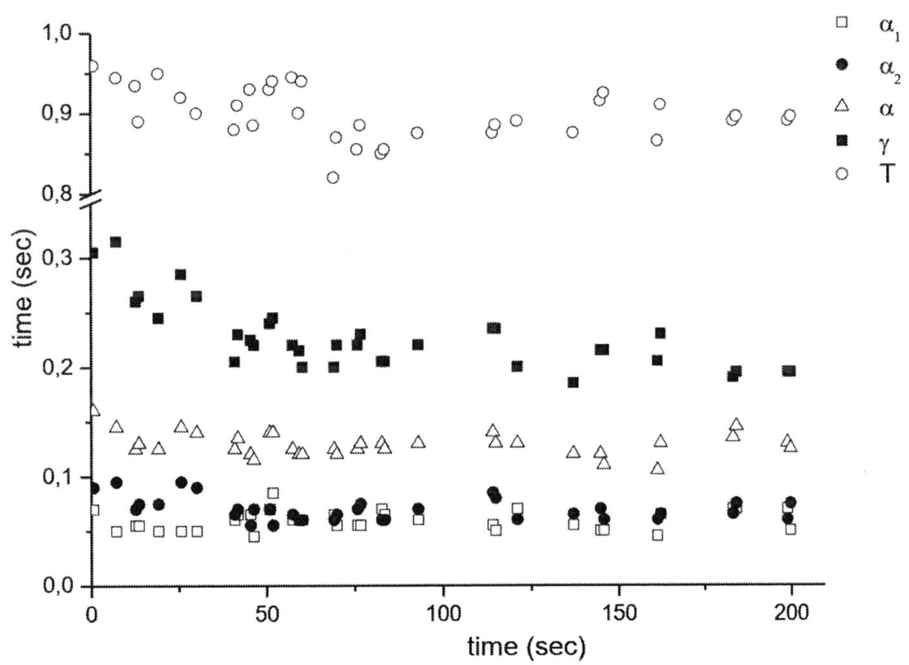

Figure 4. Temporal characteristics after taking nitroglycerin.

After taking Corvitol, the duration of cardio intervals increased (T = 1.040 ± 0.031 sec), which corresponds to a decrease in heart rate. The duration of the anacrot decreased, the durations of the fast and slow blood filling were respectively $\alpha_1 = 0.050 \pm 0.007$ sec., $\alpha_2 = 0.075 \pm 0.007$ sec., $\alpha = 0.124 \pm$

0.009 sec. The ejection period increased slightly - $\gamma = 0.287 \pm 0.040$ sec., the spread of its values also increased.

The correlation characteristics between these parameters were determined. High correlation is observed between α_1 and α and between α_2 and α (0.53 and 0.67, respectively). The correlation of these values with the ejection period was 0.02 for α_1, 0.31 for α_2 and 0.29 for α. With the duration of the cardio interval, the correlation was 0.13 for $\alpha 1$, 0.01 for α_2, 0.11 for α and 0.23 for γ.

After taking nitroglycerin, the duration of cardio intervals decreased ($T = 0.899 \pm 0.026$ sec.), which corresponds to an increase in heart rate. The duration of anacrotic and the duration of slow blood supply decreased, but the duration of fast blood filling increased. They amounted to, respectively, $\alpha_1 = 0.059 \pm 0.007$ seconds, $\alpha_2 = 0.070 \pm 0.008$ seconds, $\alpha = 0.129 \pm 0.008$ seconds. It is noted, unlike other experiments, the excess of the duration of fast blood filling over the duration of slow blood filling. The ejection period significantly decreased - $\gamma = 0.227 \pm 0.023$ sec.

The correlation between $\alpha 1$ and α and between α_2 and α is slightly lower than in the previous case, but quite high (0.43 and 0.64, respectively). The correlation of the duration of fast blood filling with the ejection period became negative and amounted to -0.17, while the correlation between the duration of slow blood filling and the length of anacrot with the ejection period was high - 0.74 for α_2 and 0.59 for α. With the duration of the cardio interval, the correlation was 0.06 for α_1, 0.26 for α_2, 0.31 for α and 0.55 for γ.

The secondary temporal characteristics were determined: the temporal index of the vascular tone and the specific time of the anacrot. The temporal index of vascular tone was 0.60 ± 0.10 for a calm state, 0.68 ± 0.11 after taking Corvitol and $0.88 + 0.18$ after taking nitroglycerin, and the specific anacrot time was 0.15 ± 0.01 for a calm state, 0.12 ± 0.006 after taking Corvitol and 0.14 ± 0.01 after taking nitroglycerin.

3.2Amplitude characteristics

The amplitudes h_1, h_2, h_3, h_4 and secondary characteristics were determined: dicrotic index, diastolic index, amplitude index of vascular tone. At calm state, the dicrotic index was 0.48 ± 0.06, the diastolic index was 0.59 ± 0.03, and the amplitude index of vascular tone was 0.40 ± 0.04.

After taking Corvitol all parameters slightly increased and amounted to a dicrotic index - 0.53 ± 0.10, diastolic index - 0.67 ± 0.05, amplitude index of vascular tone - 0.43 ± 0.03. Taking nitroglycerin leads to a decrease in the dicrotic and diastolic indices and is characterized by the appearance of a pulsogram without a dicrotic wave. The mean values were respectively: dicrotic index - 0.20 ± 0.06, diastolic index - 0.35 ± 0.11, amplitude index of vascular tone - 0.45 ± 0.04.

4. Conclusion

The method of optoelectronic determination of the characteristics of the cardiovascular system allows the analysis of the amplitude-temporal characteristics of the photoplethysmographic curve to determine such interesting diagnostic parameters as the duration of the anacrotic and dicrotic phases, the duration of fast and slow blood filling, dicrotic and diastolic indices, temporal and amplitude vascular tone. In contrast to the other methods, the use of photoplethysmography allows us to conduct local research, to determine changes in the state along the vessels, excludes electrical effects on the patient when conducting research.

Acknowledgments

This work is partially supported by the Bulgarian Science Fund Grant H28/11/2018 "Novel bio-optical methods for determination of chemical and conformational reorganization of collagen - based tissue structures in vitro and in vivo".

References

[1] Chiaki I, Yoshiaki T, Kazunori Y, Makoto F, Masafumi T and Koji S 2008 Comparison of oxygen consumption calculated by Fick's principle and measured by indirect calorimetry

Journal of Anesthesia **22**(2) 163-166
[2] Hennerici M G. 2008 *Ultrasound in Medicine & Biology* **34**(4), 691 DOI: 10.1016/j.ultrasmedbio.2007.10.003
[3] Allen J 2007 Photoplethysmography and its application in clinical physiological measurement *Physiological Measurement* **28** (3) DOI: 10.1088/0967-3334/28/3/R01
[4] Kowal M, Kubal S, Piotrowski P and Staniec K 2017 Practical remarks on the heart rate and saturation measurement methodology *Meas. Sci. Technol.* 28 054003 https://doi.org/10.1088/1361-6501/aa5d8f
[5] Merkepci M, Ozyazici M and Dogru N 2018 Photoplethysmography based instant remote monitoring of non-invasive blood pressure and O2 saturation by using zigbee network *Biomedical Research* **29**(11) 2401-04 doi: 10.4066/biomedicalresearch.60-17-1900
[6] Gurevich M 1982 Impedance reopletizmografiya (Kiev: Naukova Dumka) p 103
[7] Kravchenko V, Mamilov S, Plaksiy Yu, Yesman S, Kazak N and Ryzhevich A 2005 Noninvasive optical method for determining the scattering coefficients and specific volume of the blood in biological tissue in vivo *Jour. of Appl. Spectrosc.* **72**(1), 126-131
[8] Mamilov S, Yesman S, Veligodski D, Asimov M and Gisbrecht A 2014 Wavelength Dependence of the Efficiency of in vivo Photolysis of Carboxyhemoglobin *Jour. of Appl. Spectrosc.* **81**(3) 432-436
[9] Yesman S, Mamilov S, Asimov M and Gisbrecht A 2011 Noninvasive methods of measuring oxygen saturation in venous blood *Jour. of Appl. Spectrosc.* **78**(3) 406-413
[10] Asimov M, Asimov R, Rubinov A, Mamilov S and Plaksiy Yu 2006 High sensitive pulseoximeter-spectrophotometer for laser optical dosimetry in biology and medicine *Proc. of SPIE* **6251** 147-154

Interference wedged structures as high efficient optical beam splitters – development and simple practical realization

Margarita Deneva[1,6], Pepa Usunova[2], Nadejda Kaimakanova[3,1], Dimitar Atanasov[4], Nikolay Ivanov[4], Marin Nenchev[1], Elena Stoykova[5] and Petar Petrov[4]

[1] Technical University of Sofia, R&D Division, Sci.Lab. Quantum and Optoelectronics , Branch Plovdiv, 25 Tcanko Djustabanov Str., Plovdiv, Bulgaria; E-mail: mar.deneva@abv.bg
[2] Medical University of Sofia, Sofia 1431, 15 "Acad. Ivan Geshov" Blvd., Bulgaria
[3] University of Plovdiv, "Tzar Assen" str. 24, Plovdiv, Bulgaria
[4] Institute of Electronics, Bulgarian Academy of Sciences, 72, Tzarigradsko chaussee Blvd, 1784-Sofia, Bulgaria
[5] Institute of Optical Materials and Technologies, Bulgarian Academy of Sciences, Sofia 1113, 109, "Acad. G. Bontchev" Str., Bulgaria

Abstract. Having essential experience and results, concerning Interference Wedged Structures (IWS), we have developed and practically realized optical beam splitter elements as suitably built variants of IWS. The principle of action, theoretical treatment, including computer simulation for optimization is briefly discussed. As an essential point, the technological approach for practical realization of simplest applicable Interference Wedged Based Splitters (IWBS), are carried out. The splitters are built as compact thin flat and plane, practically – sheet-like, elements with dimensions e.g. ~ 4 x 3 x 0.2 cm. They present advantages as: smooth and linear varying the power ratio between the formed beams by simple sliding of the splitter in its plane; keeping the direction of propagation of the formed beams during the control of their variations; very small losses in splitting (theoretically ~5 %); work with high power incident beams (kW/cm^2 and more for CW and MW/cm^2 for ns pulses beams).

1. Introduction

The conventional wedged type interferometers, realized as a single wedged structure of the type of Interference Wedge (IW) - solid version of the Fizeau Interferometer [1-3], present in one hand, research interest as spectroanalyser elements [1,2,4]. Such applications are due to ability of these elements to provide simplest linear dispersion for a wide spectrum light beam. The applications of the IW arrangement include also measurement of plate flattnes and refractive index of substances [5]. In other hand, that is less discussed, promising type of applicability is based on the IW property for wavelength selection from illuminating, small diameter, multi-wavelengths beam, in combination with the wavelength tuning. The tuning, is by simple sliding of the IW along the plane of its surface and as results the transmitted and reflected beams keep their direction of propagation This property in transmission [1,3] and in reflection [6] permits sucssesively application in the laser technology as methods for laser spectral control, including efective solutions of lasers for two-wavelength generation [6,7] and for high selective and contiuously tunable single-mode lasers [8,9]. Recently, we have introduced a new type of interference wedged structure – Composed Tunable Interference Wedged Structure (CTIWS), which represents suitable composition of wedged layers [10]. Such structure has essentially increased selectivity (~0.01 nm and better) from a multi-wavelength beam, combined with widened tuning range (over ~40 nm and more). In the last our works, we have presented starting results, concerning investigation of the possibility the IWS to be used as splitter, which divides a laser beam with diameter permitting fully entering and translation in IWS spatial resonance (physical base and analysis) [11]. Here, we present for such type of splitter (IWBS) new results, including also the technology for practical fabrication of simple working IWBS's.

The light beam splitters are widely used elements in the range of light investigations and technology. In the literature are presented solutions [12] that are mainly of practical interest. Each of them has advantages and limitations. Shortly, for the group of compact and simples splitters, based on transparent substance doped with absorption, the limitations are constant ratio of separation and losses from the absorption, leading to heating the splitter and its cracking for high power incident beam. For the polarizing based splitting and control, limitations are work with polarized light and non-compactness of the control. For the compact group, based on variants of Fabry-Perot Interferometer (FPI) type elements (filters, reflectors), the drawbacks of typical simplest control by rotation of the elements, are change of the propagation direction of the formed beams and varying of the reflectivity of multi-layer mirrors and components. The change of power ratios for the formed beams by pieso-electric control of FPI thickness is complicated, not-compact and sensitive to parasitic influences of electrical and magnetic fields. For the compact splitters, based on separation, using optical fiber, the problem is the absence of control of power ratios and polarizing requirement for the light.

The proposed by us splitters assure precisely and variably controlled ratio of the reflected and transmitted powers by simple sliding in their planes and division with negligible energy losses. In addition to the simplicity and compactness of the construction and the control, important property of such sheet-like splitters, is power ratio control keeping the propagation direction of the formed beams, that is of essential interest for applications in optical schemes with complex. No polarization requirements are related with the splitter work. For splitter built from optical layers with high light damage resistivity, the work with the beams of power densities of kW/cm^2 and more for CW, and MW/cm^2 for pulses with ns duration is possible. In general, any solution of splitter element presents advantages and limitations. For specific application in given problematic, a given solution of the splitter can be of competitive interest. Thus, the development of new solutions of optical splitting elements with new properties or improvement, are constantly actual problematic [13].

2. IW as physical base of IWBS. Discussion of creation IWBS with application of IW

The IW that is simplest interference wedged structure is a single wedge gap with reflective sides (figure 1a). In the pictures are schematically given the obtained transmitted and reflected beams after division of the incident beam. Here R_1, R_2 mark the corresponding reflection sides of the wedge layer and gives their reflectivity; n_1 is the refractive indexes inside the layer-gap, n_0, n_3 of outside and of the support; α is the wedge apex angle and $e_1.n_1= e_1^*$ is the optical thickness of the gap in the marked resonance (e_1 is the geometrical thickness). The figure represents part of the wedge around the given resonance. Following the theory of IW, along the line, perpendicular to the wedge apex, is disposed a sequence of the transmission resonances of the wedge. Each resonance is relatively narrow spatial region with high transmission in comparison to the distance between the resonances. The resonance condition for the position of the resonance line of maximum transmission is $k(\lambda/2)=e.n.\cos\theta$, where $e.n$ is optical thickness in the considered point, θ - incident angle, λ - wavelength of the illuminating beam (monochromatic) and k is the integer k=1,2,...20. With some losses, mainly due to the absorption and scattering in the layer and mirrors, the non transmitted part of the incident beam, is reflected. Theoretically, for ideal gap and mirror, all power of the incident light is divided between the transmitted and reflected beams. The graphs of the transmission and reflection curves for ideal IW (without absorption and scattering), computed on the discussed below FPI approach for treatment of IW are plotted in figure 1(b). The discussed division of the power between the reflected (R) and transmitted beams (T) can be seen. Two sequent resonances are shown.

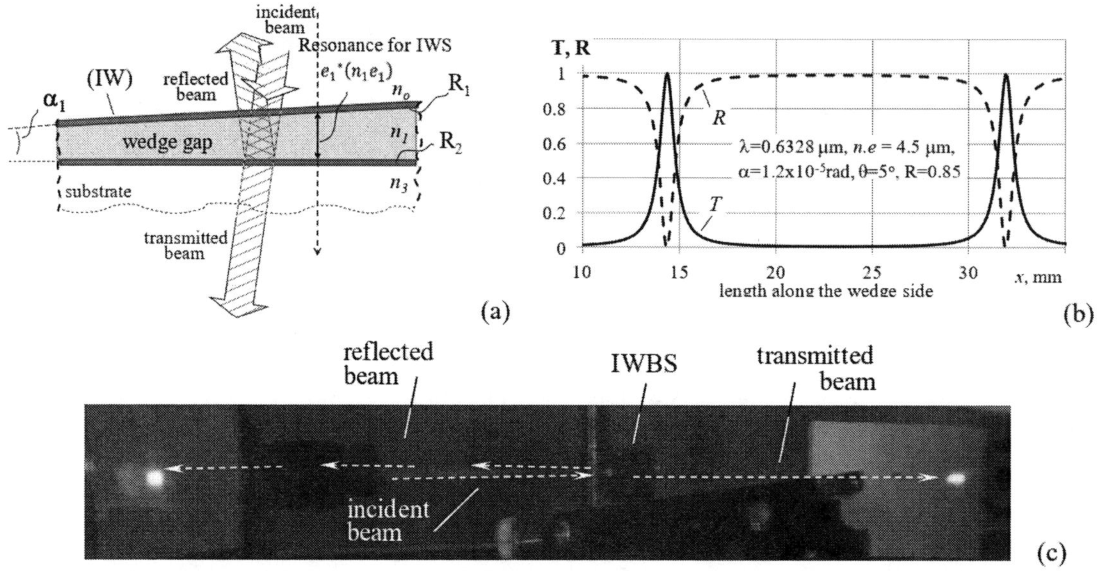

Figure 1. (a) Schematic of IW. R_1, R_2 mark the reflection sides of the wedge layer and gives their reflectivity; n_1 - refractive index inside the layer-gap, n_0 and n_3 - of outside and of the support, respectively; α is the wedge apex angle and $e_1.n_1= e_1$* is the optical thickness of the gap in the marked resonance (e_1 is the geometrical thickness). The figure represents part of the wedge around the given resonance. (b) –Typical transmission and reflection curves for ideal IW; (c) – division of the incident light beam (from He-Ne laser, 0.6328 μm line)

The transmission resonance is not single thin line - it has, around the maximum transmission, some spatial distribution, presented by a resonance profile curve and described by the IW theory [1]. The precise theoretical treatment of IW is relatively complicated [3,8] and no so simply adaptable for all treated cases. As we have shown in [14] there is the possibility for simplified treatment of IW with acceptable for practical evaluations precision. We have developed simplified model to describe mathematically the action of the IW (single layer-gap IWS) and combination of two layers IWS on the base of FPI theory [1]. The basic idea is to consider the IWS as sequence of FPIs with linearly increasing (or decreasing) thickness located one to other along the line, perpendicular to the wedge's apex – axis x. The given comparison in [14] between the results for particular cases, obtained using the discussed approach and exact theory of IW [3,8] and the experiment, shows that such approach permits evaluation with acceptable precision in practical interesting cases [14]. As illustration, some calculated graphs, following the discussed FPI approach (a) and the exact theoretical treatment (b)-(d) are plotted in figure 2.

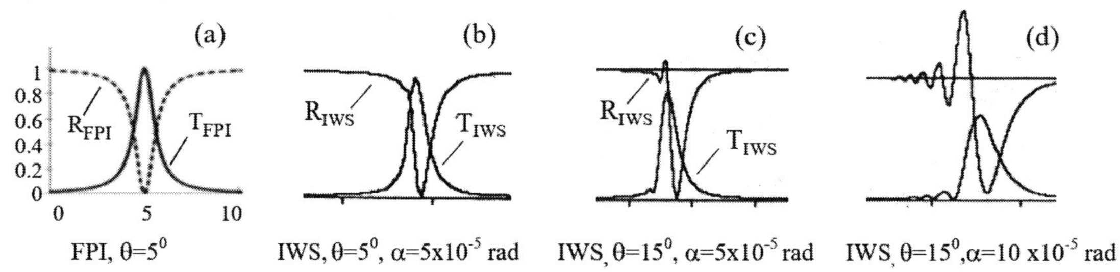

Figure 2. Computed transmission T and reflectivity R around the spatial resonance of IWS, calculated by use: (a) the approximation with sequences of continuously variable thickness FPI and (b) – (d) - the exact theory, given in [3,8]; for all graphs λ=0.6328 μm, e=20μm, R=0.9.

The approach is convenient for use of layers with relatively small thicknesses (of ~1 to 50 μm) and angles (of few to ten μrad, beam incident angles of~ to 10 angular degrees). The comparison between (a) and (b)–(d) leads to the noted above range of acceptable correctness for the simplified FPI approach. In the noted limits, this contributes essential part in important for the practice applications of wedged interference structure, including also the analysis of the proposed by us beam splitters. Thus, for analysis of resonance of IW we will apply the computer simulation using combination by the formula (1) [1] for FPI transmission and the property of IW to have linear variation of the thickness $e_1(p)$ with distance along line perpendicular to the wedge apex – i.e. axis x- formula (2). Here p is distance from starting thickness e_0 – the last being taken the thickness of given chosen resonance – expression (2).

$$T = (1 - R^2) / (1 - 2.R.\cos\delta + R^2), \tag{1}$$

where T is the FPI transmission, R reflectivity of the mirrors and $\delta = (2\pi / \lambda).2.e_1.n.\cos\theta$ [1], with λ - wavelength, θ- incident angle and n - refractive index.

$$e_1(p) = e_0 + p.\tan\alpha \tag{2}$$

From (1) and (2) we obtain

$$\delta=(2\pi/\lambda).2.(e_0+p.\tan\alpha).n.\cos\theta \tag{3}$$

Programming (3) we can compute the corresponding resonance transmission curve as function of λ, p, θ for given e_0, α and n (or can work with optical thickness of $e^* = e.n$, taken $\tan\alpha = \lambda/2.n$). Note that from (3) directly follows that the resonant thicknesses $e = e_{0k}$ for maximal transmission, wavelength λ, α, n and θ are related with the expression $k.(\lambda/2) = e.n.\cos\theta$ with k being integer 1,2,3 ... (from δ to be equal of $k.2\pi$, i.e. $\cos\delta$ to be 1).

We will consider the case of beam splitting of the illuminating beam with smaller diameter with respect to the width of resonance lines, and incident point that can move inside the resonance line along the structure (figure 7, below). In this case, sliding the wedge in its plane, the transmission can vary – if the incident point moves away from the maximum, the transmission will decrease and vise-versa. The IWS itself - if it is illuminated by larger laser beam than the width of the resonance, it can transmit part of the incident light that coincides with the resonance (resonances) and will reflect the other part that is outside of the resonance, thus to split the beam. Such simplest splitting in practice can deform strongly the reflected and transmitted beams compared with the incident beam. This will have limited use to give information about power variation in the incident beam, using the observation of only one part of it [15].

We will follow the idea of splitting by moving the small size beam inside the large resonance. To obtain good splitting, the splitter must have large resonance width for splitting the large diameter incident beams and linear decreasing of the transmission of the wing – the last will permit linear decreasing of the transmitted power (and increasing of the reflected). Thus, it is necessary the IW to have such type resonances – with long and linear wings. The computed resonance curves of IW of the dependence of reflection R, using the expression (1) and (3) are given in figure 3 (with used parameters in it).

The wedge thickness e^* ($e.n$) practically has no influence on the width of resonance curves that show from the analysis based on the expressions (1) and (3). The wedge angle α is chosen realistic for technological realization (2.1x10^{-5}rad). The incident angle θ is enough smaller, less than 10°, permitting separation of both formed reflected and incident beams. One IWS structure can be used as beam splitter for many wavelengths because different wavelength resonances enter in the length of the IWBS (of few cm) for different order k. For calculations, realistic parameters of $e_0.n$, α, λ (He-Ne laser) and θ are chosen. The reflectivity of the mirrors is parameter, given as part of the maximal value of 1.The curves show that, the needed resonance behavior of IW for good splitting can be obtained for relatively low reflectivity (R~0.15 - 0.35; the transmissivity is also given as part of 1 (1 corresponds to 100%). Note that the maximum transmission for light incidence of 0° is 1 in the used approach for the treated IW(FPI-theory). For exact theoretical analysis [3,8] it is less – of ~ 0.96. In practice, due to the absorption and scattering in the real layers and Fresnel reflection the transmission is lower. In our investigation the maximal obtained transmission is value of 83% for IWS that we have, with full losses of ~10 %.

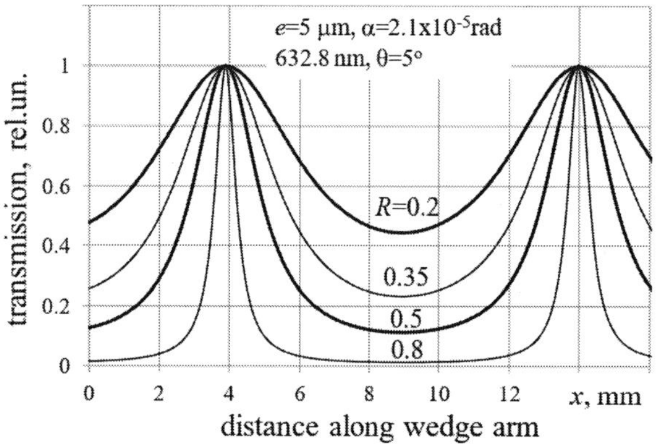

Figure 3. The computed resonance curves of IW: the dependence of reflection R, using the expression (1) and (3).

The considered (developed and investigated) IWBS uses the specific properties of the IWS: to provide by simple sliding in its plane variable ratio of the reflected and transmitted power for produced beams by division of the incident beam (with negligible energy losses, theoretically - without); to present important property, being plane, sheet-like, optical element, to provide the power ratio control without causing change of the propagation directions of the formed beams. No polarization requirements are related with such splitter work. The splitter has high compactness and small dimensions. Such structure, with conveniently chosen parameters, assures: long and linear range of variation of the transmission, permitting division of relatively large diameter incident beam; acceptable slope of the line of decreasing/increasing of the transmission that provide lower or negligible deformation of the spatial intensity distribution in formed beams; difference of variation of the transmission at the start and the end of this line to be sufficient for the practical use of splitter.

3. IWBS fabrication as solid, flat sheet-like and compact construction Experimental Investigations of laboratory realized working IWBS.

3.1. Dielectric layers fabrications

Following the discussion in point 2, we have fabricated the laboratory working samples of IW with discussed parameters that are most suitable, as it is possible, in our conditions to work as good IWBS. The described already models, that approach this requirement and the component of which is realistic for production, was used for experimental investigations.

To compose the IW we use as wedge gaps air-gaps with thicknesses of ~5 μm, closed between two TiO (or TaO) layers of thickness ~ 0.22μm. We specially creates the wedged structures with noted characteristics to have low reflectivity (~ 0.15 -0.5) and thus to be with a large spatial resonance, that was discussed already in section 2. The TaO and TiO, fabricated by DC reactive magnetron sputtering in vacuum were chosen as very resistive of internal influences layers. The produced layers, using such technique, are with good mechanical properties and resistivity (especially compared to these ones, producing by vaporization or chemical layering).

The technology was the following. The sputtered cathode Ti with diameter 100 mm and thickness $d = 5$ mm was made of pure Ti (99,8 % brand BT 1). The sputtered cathode Ta with diameter 100 mm and thickness $d = 6$ mm was made of pure Ta (99.95 % brand TVCH). Depending on the specific requirements of the analytical techniques used, the layers were deposited on substrates of microscope polish glass (DELTALAB – Spain) with dimensions 40×26×1 mm. Prior to the deposition, the substrates were sputter cleaned for 10 minutes - working pressure p_{Ar} = 8 Pa; discharge voltage U = 1050 V; current I = 0.1 A. The films were produced in the Titan 22 (Bulgaria) high-vacuum installation under the next regime parameters. For TiO: lowest pressure value p_1=1,5×10^{-3} Pa; working pressure - reactive gas (O_2), p_{O2} = 5,5×10^{-2} Pa; discharge voltage U = 420V; discharge current I = 1A; substrate temperature T = 180 C°; cathode – sample distance H=120 mm; deposition time t = 60 min.

IOP Conf. Series: Materials Science and Engineering **618** (2019) 012016 doi:10.1088/1757-899X/618/1/012016

For TaO: lowest pressure value p_l = $1{,}5\times10^{-3}$Pa; partial pressure (Ar), $p_{Ar}=1{,}0\times10^{-2}$ Pa; working pressure (Ar $+O_2$)n p_{tot} = $5{,}6\times10^{-2}$ Pa; discharge voltages U = 535 V; discharge current I = 1 A; substrate temperature T = 20 C°; cathode – sample distance H=120 mm; deposition time t = 60 min.

The layers was layered on the flat glass supports (~ 4.5 x 3 cm) with thicknesses of 0.1 cm The reflectivity of single layer on the glass support was ~ 0.20, measured for wavelengths – red (0.6328 μm), yellow (0.5941μm) and green (0.532 μm).

3.2. The IWBS construction, operation, useful properties and limitations.

The described layers with reflectivity of ≈ 0.2 are specially producing to be very convenient as layer-mirrors R_1=R_2 for creating the IWBS with desired long line of transmissivity variation. The schematic of the simplest constructions used was shown in figure 4(a). The two supports (see end of subsection *3.1*), turned with the layers to be one to other, were separated at the one end by ~ 3 μm special foil and in the other - at 4.5 cm – with 4.5 μm thick foil, thus the wedge angle to be ~3×10^{-5} rad. The composition of plate with the layers and spacer was adjusted by suitably pressing in homemade mechanism (figure 4b) that permits easy variation of the wedge angle, or and glued variant to represent solid, flat sheet-like and compact construction (figure 4c), with the given above dimensions of the supports. In figure 5 real photograph of experimental installation using the described simples of IWBS is presented.

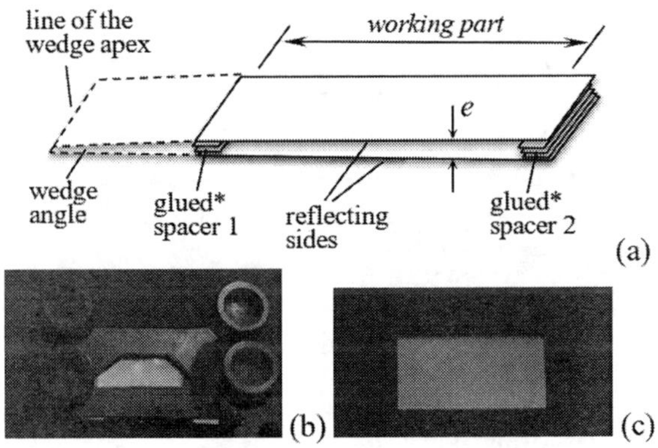

Figure 4. (a) Schematic construction of the IWBS; (b), (c) basic realizations, described in the text.

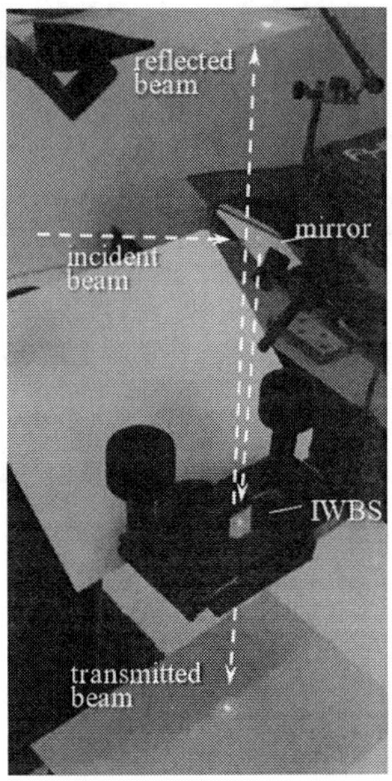

Figure 5. Photograph of real operation of the IWBS.

From the graphs in figure 3 we can estimate that for R~0.2 (0.15-0.25), the range of linear variation of transmission is ≈ 4.5 mm, the last permits to control splitting for the beams with diameters Ø to 2.5 mm (net range 4.5– ø mm, i.e. 3.5÷2 mm). The variation – decreasing/increasing of the transmitted power is smooth and linear between maximal of T ≈ 0.67 and minimal T ≈ 0.33 and with the beam diameter width (i.e. for the He-Ne Gaussian laser beam with Gaussian diameter, of 1.4 mm,). Note that, measuring the reflection, transmitted and incident power, we have estimated the losses in the prepared IWBS. For incident beam power of 1.6 mw, transmission power was measured to be 0.672 mW and reflected of ~ 0.66 mW. The corresponding losses are estimated to be 0.18 (18%). Thus good quality of the layers is important for IWBS work.

The main series of experimental investigation was made with the compact IWBS with vertical disposition of IWBS, as it is shown in figure 5. In figure 6(a) is plotted photographs of the TaO layer on the glass supports used to created the IWBS with fixed mirrors (dispositive in figure 4c).The

IOP Conf. Series: Materials Science and Engineering **618** (2019) 012016 doi:10.1088/1757-899X/618/1/012016

formed large Fizeau fringes are shown in photograph in figure 6(b). In general, they have some distortion, however the aim of the IWBS to be used with small beam diameter (~ 1-1.5 mm) permits to be selected good part for IWBS work – the last shown with dashed rectangle in figure 6(b). This spatial range permits good work of IWBS. In figure 6(c) is given the graph of the working range, where in the bottom – the selected resonance in expanded scale for beam-splitting. The series of photographs of splitting of incidence beam (IB) by sliding of the IWBS are shown in figure 7(a)-(c) that illustrate the change of the position of point of incidence of the beam with respect to change the position of the resonance line – line of maximal transmission. The variation of the transmission is well evident from the figure - the spots in increased scale are shown in figure 7 (a',b',c'). The graph (d) shows variation of the transmission when the incident beam (diameter of 1mm) moves toward the maximum of the transmission in selected resonance. The unit of 1 mm for the scale is marked. Finally, in figure 7(e) are plotted real graphs for controlled beam splitting of the incident beams of 1 mW – for the transmitted beam – transmission *T* and for reflected – *R*. The spot for the smaller slope of transmission of curve RL (figure 7d) reproduce well the intensity distribution of the incident beam.

The graphs show real applicability and potential of the developed technique for controlled beam splitting and opportunity for easy organizing fabrication of such IWBS.

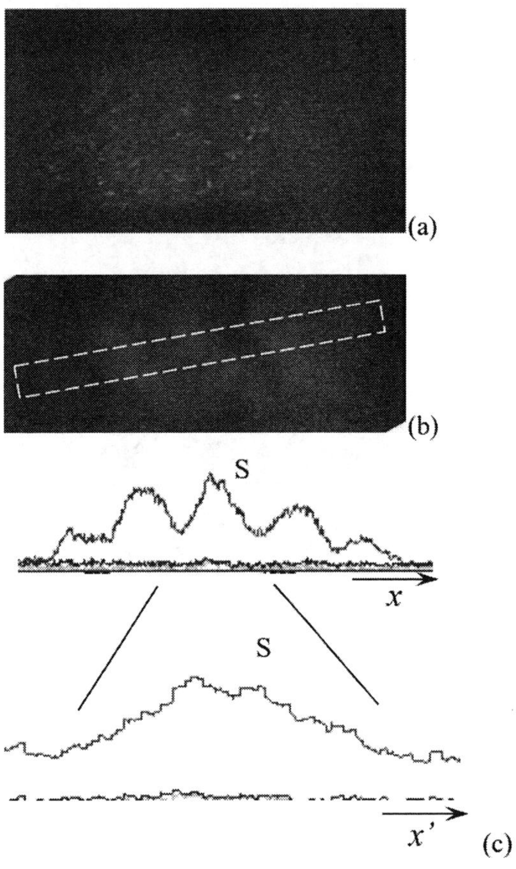

Figure 6. Composition component of IWBS and large interference resonances for the described IWBS realization; (a) layers; (b) composition and resonances; (c) – expanded resonances.

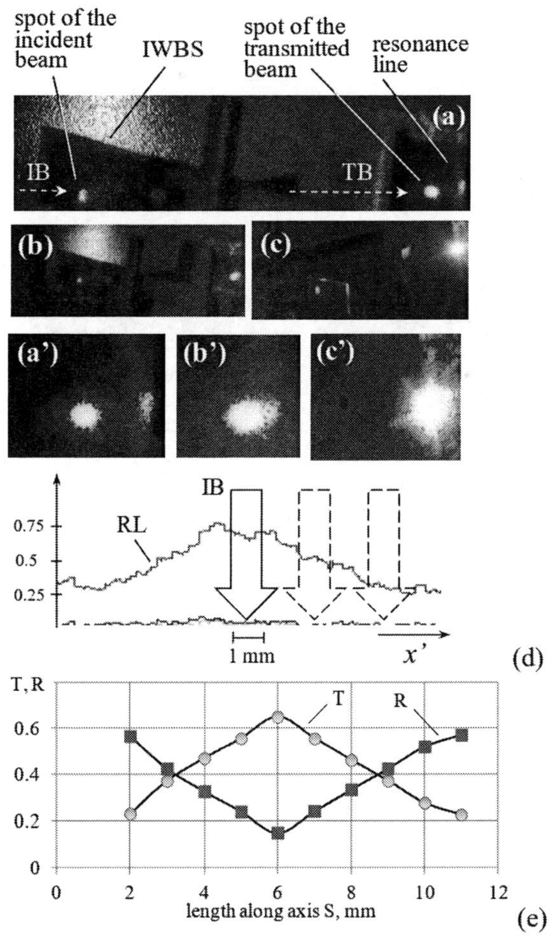

Figure 7. Beam Splitting with developed IWBS. Schematic of splitting and Graphs of controlled powers in the transmitted and reflected beams, obtained by separation of the incident beam.

4. Conclusion

In the work we have presented the developed and practically realized by us new beam splitting element, based on interference wedged structure (IWBS). The IWBS present advantages such as: the splitters are built as compact thin flat and plane, practically – sheet-like, elements .They assure smooth and linear varying the power ratio between the formed beams by simple sliding of the splitter in its plane; keeping the direction of propagation of the formed beams during the control; very small losses in splitting (theoretically zero); work with high power incident beams (kW/cm^2 CV, MW/cm^2 for nanosecond pulses). The developed of new solutions of optical splitting elements with new properties or improvement, enriches the variety of beam splitters and the choice for consummators'.

Acknowledgment

This work was supported by Bulgarian Science Fund, project DN-17/7 "Elements and devices in optics and quantum electronics on the base of wedged interferential structures for laser technology, optoelectronics, optical communications, metrology and spectrum analysis."

References

[1] Demtreder W., 2008 *Laser spectroscopy*, Springer, 4th ed.
[2] Kajava T, Lauranto H and Saloma R 1993 Fizeau interferometer in spectral measurements, *J. Opt. Soc. Am.* **B10**, pp 1980 – 1989
[3] Stoykova, E and Nenchev M 2010, Gaussian Beam Interaction with Air-gap Fizeau Wedge, *J. Opt. Soc. Am.* **27**, pp 58-68 and the literature cited therein
[4] Siddique M, Yang S, Li Z, Li P 2007 Fizeau interferometery for THz-waves' frequency and intensity measurement, *J. Beijing Institute of Technology* **16**(3), pp 326-329
[5] Kumar Y and Chatterjee S 2009 Simultaneous measurement of refractive index and wedge angle of optical windows using Fizeau interferometry and a cyclic path optical configuration *Applied Optics* **48**(24), pp 4756-4761
[6] Nenchev M. N and Meyer Y H 1981 Two-wavelength dye-laser operation using a reflecting Fizeau interferometer *Applied Physics* **24**, pp 7-9; 1984 Laser tuning with combined system of Interference wedges, *Proc. SPIE* 473, p 181
[7] Deneva M, Uzunova P, Nenchev M 2007 Tunable subnanosecond laser pulse generation using an active mirror concept *Optcal and Quantum Electronics* **27**(39), pp 193-212
[8] Stoykova E and Nenchev M 2001 Fizeau wedge with unequal mirrors for spectral control and coupling in a linear laser oscillator-amplifier system *Aplied Optics* **40** pp275402-5411 and the literature cited therein
[9] Nenchev M N and Meyer Y H Continuous-scanning system for single-mode wedge dye lasers, *Optics Letters* **7**(5):199-200 , DOI: 10.1364/OL.7.000199
[10] Nenchev M, Stoykova E and Deneva M 2018 Composite wavelength tunable wedged interference structures with increased free spectral range *Optical and Quantum Electronics* **50**(12) DOI: 10.1007/s11082-018-1668-9
[11] Deneva M, Nenchve M and Stoykova E 2019 Interference wedged structures as light beam splitting elements *Proc. of SPIE* Vol. 11047, 110471G, CCC code: 0277-786X/19/$18, doi: 10.1117/12.2516696
[12] RP Photonics Encyclopedia, https://www.rp-photonics.com/beam_splitters.html
[13] Turner M D, Saba M, Zhang Q, Cumming B P, Schröder-Turk G E and Gu M 2013 Miniature chiral beamsplitter based on gyroid photonic crystals *Nature Photonics* **7**, pp 801–805
[14] Nenchev M, Deneva M and Stoykova E 2019 Competitive light wavelength division multiplexing element based on tunable interference wedged structures *Proc. of SPIE* Vol. 11047, 110471F, CCC code: 0277-786X/19/$18, doi: 10.1117/12.2516698
[15] Beers Y 1974 The Theory of the Optical Wedge Beam Splitter, *U.S. Government Printing Office Washington, Nat. Bur. Stand.* (U.S.), NBS *Monogr.* 146, 31 pages

IOP Conf. Series: Materials Science and Engineering **618** (2019) 012017 doi:10.1088/1757-899X/618/1/012017

Design of phased antenna array for UHF RFID applications

I Nachev[1] and P Z Petkov[1]

[1]Technical University of Sofia, Faculty of Telecommunications, RCVT

ivaylonachev@yahoo.com, pjpetkov@tu-sofia.bg

Abstract. The paper present design of low-cost antenna array for RFID indentification in the frequency range 865 to 868 Mhz. The design antenna has enhanced coverage (pattern) and beam-steering capabilities. It can serve as a base element for more sophisticated RFID readers with direction finding capabilities and improved distance of tag reading.

1. Introduction

In the recent years, Radio Frequency Identification (RFID) systems have been gaining strong technological development. RFID technology is a technology that reads remotely (several centimeters to several meters) a tag in order to tracking it, and this tag can be sticked to a person, vehicle, cargo, etc in order to track it. A reader is basically a radio frequency (RF) transceiver, controlled by a microprocessor or another type of a digital controller. An RFID reader emits electromagnetic signals where an RFID tag draws power and retranslate the response. The reader antenna should have a circular polarization (CP) characteristic since the tag antenna can be arbitrarily positioned on the target. In RF design, patch array can be used because of their advantages in this field: such as light weight, small volume, low fabrication cost, high gain, the possibility of integrating a scanning systems, et cetera. In this paper, an improved RFID reader design is demonstrated, with employment of scanning phased antenna array, which will increase reader's technical characteristics [1].

2. Patch array design approach.

The employed substrate for the proposed design is Rogers FR4 with thickness h=2mm and Dissipation factor tanδ = 0.03. In order to improve the bandwidth and efficiency of antennas, a design method called "suspended substrate" is chosen, where the thickness of the gap between ground plane and antenna is Δ=3mm. In this method we use the equivalent Dielectric constant (6). Distribution of the elements of the simulated model, as with radars frequency fc = 866.5MHz. For calculation of antenna dimensions a straightforward basic algorithm is proposed:

Step 1: Calculation of width (W): Width of the Micro strip antenna is given by the common known equation:

$$W = \frac{c}{2fc}\left(\frac{\varepsilon_r + 1}{2}\right)^{-\frac{1}{2}}, \frac{W}{h} > 1 \tag{1}$$

Where c is the speed of the light c=3*10^8, ε_r is the relative dielectric constant of the substrate, f_c is radar frequency and h is substrate thickness.

Step 2: Calculation of Effective Length (L): Length of the microstrip radiator is given by:

$$L = \frac{c}{2f_c\sqrt{\varepsilon_{reff}}} \tag{2}$$

Where fc is radar frequency and ε_{eff} is effectively dielectric constant of substrate

Step 3. Calculation of the Length extension (L_{ext}): The actual length is obtained by the equation (accounting for the fringing fields):

$$L_{ext} = 0.412h\frac{(\varepsilon_{eff}+0.3)(\frac{W}{h}+0.264)}{(\varepsilon_{eff}-0.258)(\frac{W}{h}+8)} \tag{3}$$

Follow these instructions as carefully as possible so all articles within a conference have the same style to the title page. This paragraph follows a section title so it should not be indented.

Step 4: Calculation of the actual length of the Patch (L) and the rest of the design parameters: The actual length of the patch is given by:

$$\varepsilon_{eff} = \frac{\varepsilon_r+1}{2}+\frac{\varepsilon_r-1}{2}\left(1+10\frac{h}{W}\right)^{0.5} \tag{4}$$

The length of the guided wave and the equivalent permittivity accounted for the air gap:

$$\lambda_g = \frac{\lambda}{\sqrt{\varepsilon_{eff}}} = \frac{c}{fc\sqrt{\varepsilon_{eff}}} \tag{5}$$

$$\varepsilon_{r\,eq} = \frac{\varepsilon_r*(h+\Delta)}{\varepsilon_r+h*\Delta} \tag{6}$$

Where ε_r is a dielectric constant of substrate and Δ is thickness of the air gap. The dielectric constant in accounted for the gap between ground and the suspended substrate is marked with $\varepsilon_r eq$.

In table 1 are presented the values obtained for dimensions of the antenna components. Fig.1 shows simulation model of antenna system - consists of 4 array each one has 4 elements with series fed patch array, which are relevant by using of corporate feed. Using this model, scanning is available only in azimuth. The rounded edges of the square model of patch antenna provides circular polarization.[2][3][4]

Table 1. *Values obtained for dimensions of the antenna components.*

	W, cm	L_{eff} ,mm
Calculated	12.2	13.6
Adjusted after simulation	12.5	13.25

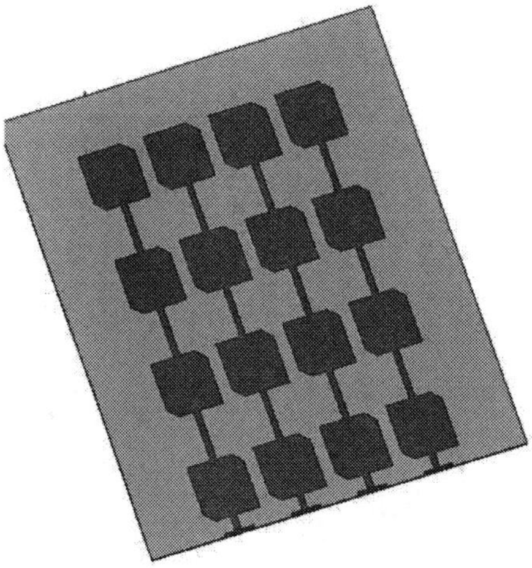

Figure 1. *Patch array – the simulation model*

3. Transmission line phase shifter.

The Butler matrix employed in this designis a type of beam-forming network. Depending on which input is excited, the antenna beam is steered in a specific direction in one plane. It performs a similar function to a Rotman lens and is a simple form of phased array antenna system. The Butler matrix consists N inputs and N output port. The design of Butler matrix is consisted of combination of hybrid couplers, phase shifters and crossovers. Values of the typical Butler matrix is show on table 2.

Described beamformer of the phased antenna array uses a hybrid T junction system that divides the input power in 4 equal parts. In this paper is described how - by changing electric length of microstrip lines (ML), changes the phases of the RF signals, thus steering the beam, for this purpose use the second and third row of Butler matrix (Fig. 2). This arrangement gives the opportunity to steer the beam in three different positions: -45; 0; +45 deg. Similar problems are discussed also in [5],[6],[7].

Table 2. *Values of the Butler Matrix*

Patch array	I	II	III	IV
ectrical length extensions	45°	90 °	135 °	190 °
electrical length extensions	135 °	0 °	225 °	90 °
electrical length extensions	190 °	135 °	90 °	45 °
electrical length extensions	90 °	225 °	0 °	135 °

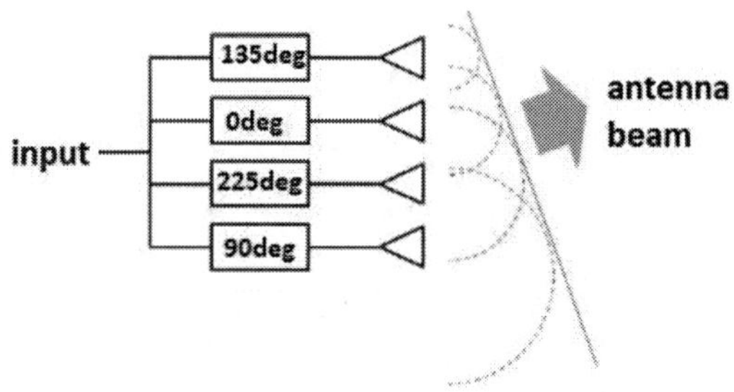

Figure 2. *Switched line phase shifter*

Table 3. Relation between beam streering and radiator's phases

Beam direction	I	II	III	IV
0 deg	0°	0°	0°	0°
+45 deg	135°	0°	225°	90°
-45 deg	90°	225°	0°	135°

4. Phase delay line control system

On Figure 3 a graphical relation between radiating antenna, feed, phase delay lines and hybrid T junctions is shown. Beam steering is accomplished by switching between microstrip lines with different lengths, using PIN diodes or different models RF switches, in this case we can use PE4259, Peregrine Semiconductors. This leads to required beam scanning (Tabl. 3).

5. Simulations results

In this section results obtained by simulation are shown. On Fig. 4 is displayed the return loss of the single segment (vertical seral array) of antenna, with the presence of the adjacent elements. In other words, the coupling between elements is included. Figures 5 and 6 show S_{11} parameters of the entire array with feed network for different scan angles. It is evident, that the operational impedance bandwidth of the array exceeds the required bandwidth for RFID applications with a great margin. The next figures, Fig. 7 and 8 show the antenna radiation pattern in azimuth for 0, +45 and -45deg beam steering. The beam shape is steered according to calculations and performs as described in the theory. The results show that even a small array with significant phase step between elements can bring a satisfactory results for such type of applications.

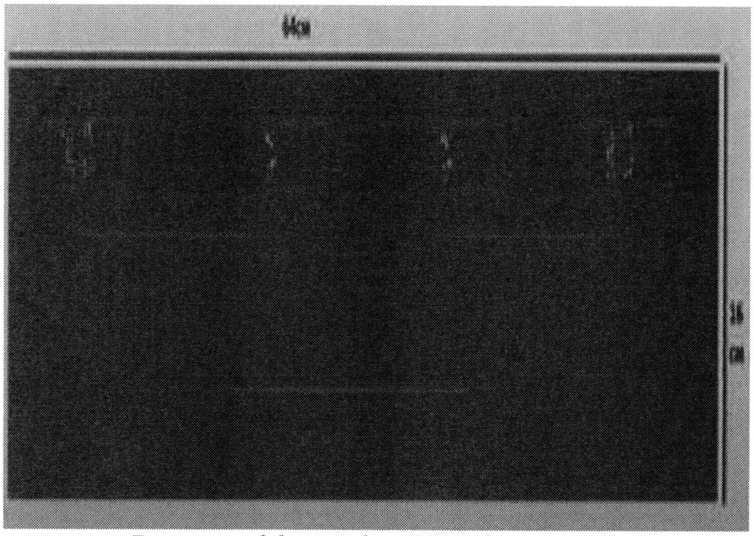

Figure 3. *Design model printed on PCB of antenna phasing section*

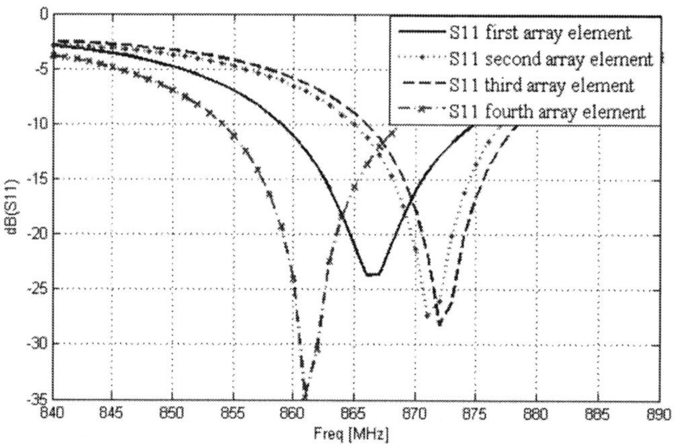

Figure 4. *Antenna array bandwidth (vertical segments), dB*

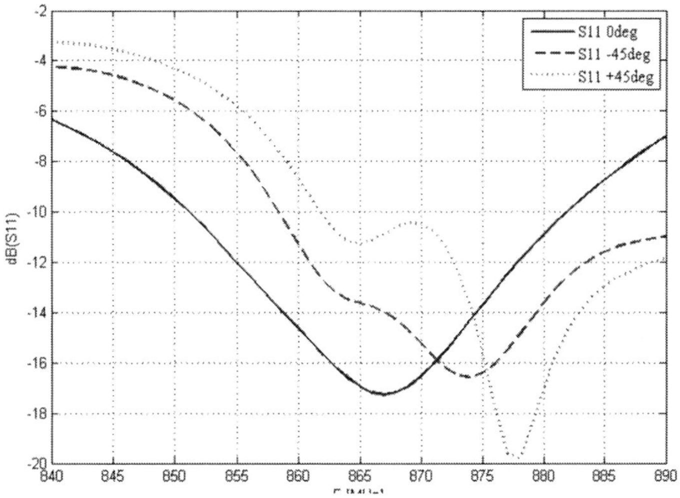

Figure 5. *Frequency bandwidth of the feed network inputs, dB*

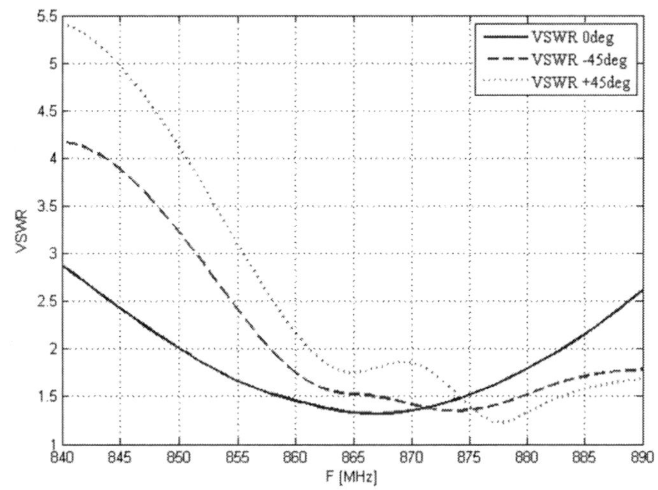

Figure 6. *Antenna array bandwidth for the 3 cases of scan angle*

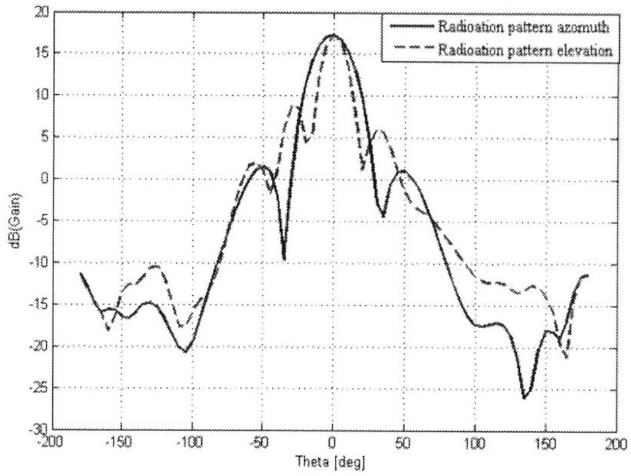

Figure 7. *Radiation Pattern at 0 deg mode*

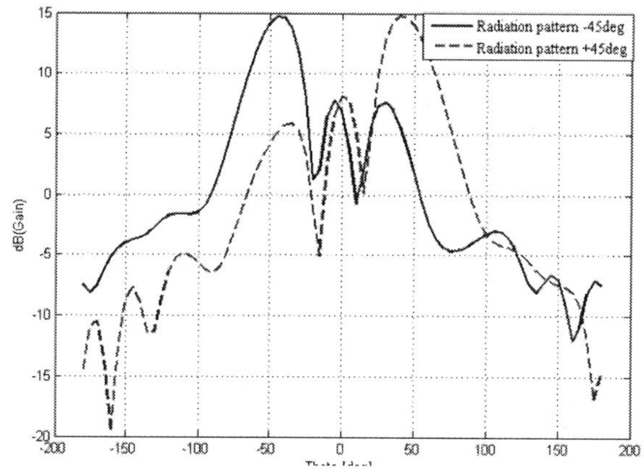

Figure 8. *Radiation Pattern at +-45 deg mode*

In table 4, a comparison of gains in beam`s different state. Figure 10 and 11 shows return loss and VSWR on antenna array inputs, after connecting the array and the fed network, depending on the mode of operation.

Table 4. Simulated Antenna array gain

Beam direction	Antenna array gain, dBi
0 deg	16.23
+45deg.	14.83
-45deg	14.86

6. Sample algorithm for beamforming control code

As addition to the beam steering antenna array, authors propose a simple algorithm for programmable microprocessors/controllers to perform scanning for RFID tags (targets) and control the phase shifters. Fig. 9 shows algorithm block diagram of the code for beamforming control.

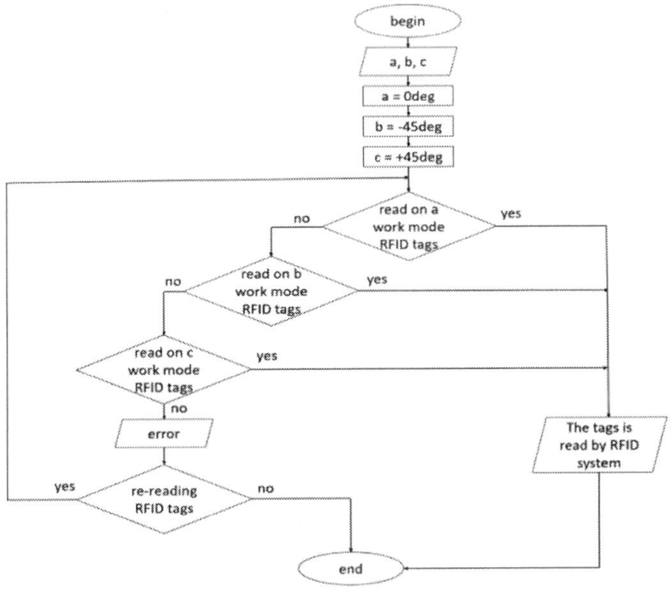

Figure 9. *Proposed aalgorithm for beamforming control.*

7. Conclusion

A simulation design RFID antenna array can be integrated into advanced, controllable RFID reader system, which work on EU-UHF license for Radio-frequency identification. The proposed antenna array has many advantages over typical antennas for that technology – simplicity, cost effectiveness and beam steering capability, as a disadvantage it can be pointed out the increased size of the entire antenna array.

Acknowledgments

This material is based upon work supported by the Bulgarian Science Fund under Grant No DN07/19/15.12.2016

References

[1] Klaus Finkenzeller, "RFID HANDBOOK", *Wiley*, 2010
[2] Chang, Kai; "RF and Microwave Wireless Systems",*Wiley*, 2000
[3] FR4 specification: www.rogerscorp.com
[4] J R James & P S Hall, "Handbook of microstrip antennas ", *Peregrinus*, 1993
[5] Akhtar A., Mateen Alahi H., Sehnan M., "Simulation of Phased A rrays with Rectangular Microstrip Patches on Photonic Crystal Substrates", MSc thesis, *Linaeus University*
[6] J. Ehmouda, Z. Briqech, and A. Amer, "Steered Microstrip Phased Array Antennas", *World Academy of Science, Engineering and Technology*, issue: 49, 2009, p. 321 - p. 323
[7] Ilia Iliev, Stanio Kolev, Stoicho Manev „ Reduction of the frequency offset influence in OFDM using an optimal PRS coding and MLSD „ *Internat Journal of Research in Computer and Communication Technology*, Vol 3, Issue 1

Primary analytical model for determining the electron temperature of a CuBr laser

S G Gocheva-Ilieva[1,], I P Iliev[2] and D V Fidanov[1]

[1] Faculty of Mathematics and Informatics, University of Plovdiv Paisii Hilendarski 24 Tzar Assen Street, 4000 Plovdiv, Bulgaria
[2] Technical University Sofia – Branch Plovdiv, 25 Tzanko Diusstabanov Street, 4000 Plovdiv, Bulgaria

E-mail: snegocheva@gmail.com

Abstract. A family of copper bromide metal vapor lasers used as brightness amplifiers in recently developed optical systems is considered. One of the main problems of this area is to investigate the basic amplification characteristics of the laser medium. In this study, a simplified mathematical model is developed based on the one-dimensional stationary equation for determining the electron temperature of the discharge. The proposed model allows to calculate the radial distribution of the electron temperature of the discharge by taking into account the radial distribution of the supplied electric power. Computer simulations are run to investigate the dependence of electron energy as a function of the electrical power supplied to the gas discharge tube. The natural limits of this process are specified.

1. Introduction

It is thought that copper bromide and copper compound metal vapor lasers are well-known and studied. These have numerous unique properties – still being the most powerful sources in the visible spectrum, with 516.6 nm and 578.2 nm wavelengths and high laser beam coherence and convergence. They are sources of ultraviolet radiation (248.6 nm, 259.2 nm, 260.0 nm and 270.3 nm).

Copper bromide vapor lasers have a wide range of applications in practice and scientific studies [1-6].

- *Applications in medicine and medical research*

In medicine, copper bromide vapor lasers are mainly used in dermatology and for photocoagulation.

- *Industrial applications*

Copper bromide lasers and laser systems are used for the microprocessing of various types of materials: drilling, cutting, marking, etching, etc.

- *In scientific research*

The laser is often used in scientific studies: for isotope isolation of various chemical elements, to study the magnetic properties of materials, etc. Copper bromide vapor lasers are used to pump titanium-sapphire lasers, dye lasers, etc.

- *Other applications*

CuBr lasers have numerous other applications: in high-speed photography, show business and advertising, laser microscopy, laser displays and nanotechnology, in the defense industry for aerial and underwater navigation, as well as studies of air and ocean pollution, etc.

- *Brightness amplifiers*

Copper bromide vapor lasers have numerous unique characteristics, which make them ideal for use as brightness amplifiers in optical systems. A brightness amplifier operating in ASE (Amplified spontaneous emission) mode lights the observed object, amplifies the reflected signal, which carries the image and serves as a narrow band filter with a passband in picometers. Recent results in this field are reported in [7, 8] and cited therein literature. It is constructed hardware-software complex called "laser monitor" including brightness amplifier of active CuBr medium, optical circuit speed digital video camera system for registration of motion images, and a computer program for processing the resulting images. This type of laser monitor achieves a time resolution of 50 microseconds and in space - up to 1 micrometer in tests and in real high-speed processes such as coherent beam combining (CBC), corona discharge in air, the process for obtaining nanopowder by the laser ablation method, etc.

One of the methods for studying amplification characteristics is kinetic modeling of processes in the plasma of metal vapor lasers' active mediums. At present, there are numerous papers about investigations of energy processes within the plasma of metal vapors and active mediums, which use kinetic modeling. In order to solve the set problem, around 240 kinetic equations are prepared and solved using custom-designed software PLASER [9].

The concentration of copper and neon energy states, ions and electrons is described by the one-dimensional reaction-diffusion system [8-9], which is solved with the following electron temperature equation:

$$\begin{aligned}\frac{\partial}{\partial t}\left(\frac{3}{2}N_eT_e\right)&=\frac{E_S^2\sigma}{e}-2N_e\left(T_e-T_g\right)\frac{m_e}{m_{Ne}}\nu_{e-Ne}+Q_r-\frac{1}{r}\frac{\partial}{\partial r}\left(r\lambda_e\frac{\partial}{\partial r}T_e\right)+\\&+\frac{3}{2}\frac{1}{r}\frac{\partial}{\partial r}\left(r\left(-T_eN_e\mu_eE_r+D_e\frac{\partial}{\partial r}\left(T_eN_e\right)\right)\right)\end{aligned}\tag{1}$$

where T_e is the electron temperature, N_e is the concentration of the buffer gas (here neon), E_S is the longitudinal electric field strength, λ_e is the heat conductivity of electrons, D_e is the electron diffusion coefficient, μ_e is the coefficient of electron mobility, E_r is the radial strength of the electric field, which maintains plasma quasineutrality, Q_r is the contribution of reactions to electron temperature, ν_{e-Ne} is the electron-neon elastic scattering rate, m_e is the electron mass, m_{Ne} is the neon atom mass, e is the electron charge, T_g is the gas temperature, which is determined by the stationary equation of the heat conductivity of the neutral gas, and σ is the conductivity of the plasma.

The energy of the electrons is one of the most important characteristics of the gas discharge. It defines the kinetics of operating levels of copper and neon energy states and has a defining role in laser generation. One of the crucial parameters, which determines electron temperature is the electric power supplied to the gas discharge. In order to examine the dependence of electron temperature only as a function of supplied electric power following [9], it is necessary to solve the compiled system of equations every time, which impedes such evaluations.

This paper proposes a simplified analytical model, which allows for the performance of such evaluations without the need to follow the approach developed in [9].

2. Problem rationale

To derive the model for determining the electron temperature it is necessary to solve the problem

$$\begin{gathered}\nabla\lambda_e\nabla T_e=-\xi q_v,\\\left.\frac{\partial T_e}{\partial r}\right|_{r=0}=0,\quad T_e(R)=T_{e,w}\end{gathered}\tag{2}$$

where ξ indicates the portion of the total supplied power that goes to heating the electrons, q_v is the power per unit volume, $T_{e,w}$ is the known electron temperature on the wall, R is the radius of the discharge tube, and $0 \le r \le R$.

There are two main difficulties in solving problem (2):

- unlike the heat conductivity coefficient of the neutral gas λ_g, for the coefficient of electron heat conductivity λ_e, there are practically no table of experimental values.
- it is difficult to determine in advance what part of the total supplied power ξ goes to direct heating of electrons (the predominant part of the energy goes to heating the neutral gas neon).

In order to solve the first problem, λ_e needs to be determined directly. To this end, it is most adequate to use the Wiedemann-Franz law for the electron gas:

$$\lambda_e / \sigma = C_c T_e$$

where the coefficient C_c assumes the value $C_c = \pi^2 (k/e)^2 / 3$. In this case, σ is the electron conductivity of the gas, k is the Boltzmann constant, e is the electron discharge.

Equation (1) assumes the form

$$\nabla C_c \sigma T_e \nabla T_e = -\xi q_v \tag{3}$$

In order to simplify it, the mean value of the power per unit volume q_v and the mean value of conductivity $\bar{\sigma}$ are introduced as quantities averaged along the cross-section of the tube which therefore do not depend on the coordinates.

After these assumptions, equation (3) takes the following form:

$$\nabla T_e \nabla T_e = -\frac{\xi q_v}{C_c \bar{\sigma}} \tag{4}$$

Equation (3) is of type

$$\operatorname{div}\left(\lambda_e \operatorname{grad} T_e\right) + q_v = 0 \tag{5}$$

where $\lambda_e = \lambda_0 T_e^m$. Here m and λ_0 are the constants to be determined.

Further the case of a radial distribution of power per unit volume q_v is considered. For this reason, some qualitative theoretical dependencies are used. From $q_v = jE$ and $j = \sigma E$ the result is $q_v = jE^2$. Here j is the density of electric current and E is the electric field intensity. According to [10], the intensity distribution of the electric field in the cross-section of the tube follows the law $E(r) = E_0 J_0(2.4r/R)$, where E_0 is a constant and J_0 is the zero order Bessel function of first kind. In this way, the result is

$$q_v(r) = c\left[J_0\left(\frac{2.4r}{R}\right)\right]^2 \tag{6}$$

where c is an unknown constant subject to defining. The function $J_0(2.4r/R)$ is sufficiently well known and tabulated, for example in [11]. Since in the general case, it is not convenient for engineering calculations, the volume power density is approximated by a second order polynomial of type [12]

$$q_v(r) = Kq_0\left(a + br^2\right), \tag{7}$$

where K, a and b are constants subject to defining, and q_0 is the mean volume power density (supposing that Q is the supplied electric power in the laser medium and V is the volume of the discharge), i.e. the power, obtained under the assumption that it is distributed equally along the tube's cross-section. Using the least squares method, the following values of the unknown constants are obtained: $K = 1.4383;\ a = 1.0183471;\ b = -0.001077$.

The solution to equation of type (5) under assumption (7) has the form [12]

$$T_e(r)=\left(T_{e,w}^{m+1}-\frac{(m+1)K\dfrac{\xi q_v}{C_c\bar{\sigma}}(r^2-R^2)(4a+br^2+bR^2)}{16\lambda_0}\right)^{\frac{1}{m+1}} \tag{8}$$

In the considered case we take $\lambda_0=1,\ m=1,\ T_{e,w}=0$.

The unknown quantity in equation (8) is ξ. In order to find it, the results from [9] are used. Since T_e in [9] is averaged along the tube's cross-section, the following integral needs to be solved:

$$\bar{T}_{e,\,ex}=\frac{1}{R}\int_0^R\sqrt{\frac{2K\dfrac{\xi q_v}{C_c\bar{\sigma}}(r^2-R^2)(4a+br^2+bR^2)}{16\lambda_0}}\,dr \tag{9}$$

This integral does not have an exact solution. For this reason, it is necessary to use approximate numerical methods for its solution. If in (9) the respective values are replaced by those from [9]: $R=3.5mm,\ \bar{T}_{e,ex}=1.5eV,\ \sigma=4.10^{-5}\Omega^{-1}$, for ξ the result is $\xi=4.1\times10^{-5}$.

3. Subject of investigation

Further, the derived model is applied for a real operational CuBr laser, described in [9]. The principle scheme and the geometric characteristics are given in figure 1 and Table 1.

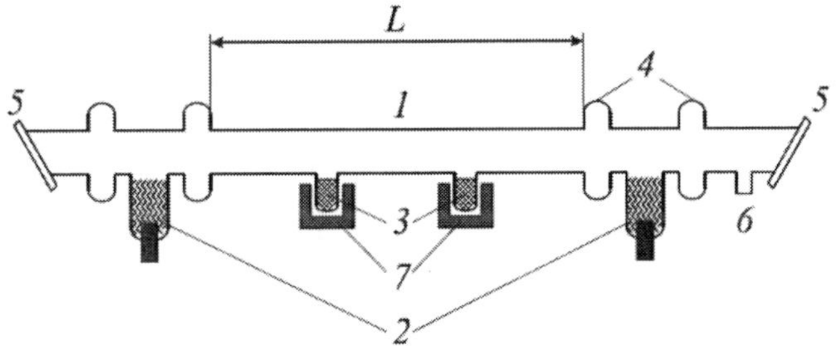

Figure 1. Gas discharge tube (GDT): (1) working channel, (2) electrodes, (3) CuBr containers, (4) traps, (5) exit windows, (6) HBr generator, (7) CuBr container heaters; *L* is the active zone length [9].

Table 1. GDT parameters.

Parameter	Value
GDT diameter ($d=2R$), cm	0.7
Active zone length (L), cm	14
Maximum pump PRF, MHz	1.1
Buffer gas (Ne) pressure, Torr	25

4. Application of the model and discussion

The solution and computer simulations are carried out by using *Wolfram Mathematica* software system. The distribution of the electron temperature in the tube's cross-section for three values of the supplied electric power ($Q_1=800W,\ Q_2=970W,\ Q_3=1080W$) is shown in figure 2.

The next figure 3 shows the changes in the electron temperature at the center of the tube as a function of the electric power supplied in the GDT. The dependence is given in relative units, taking as a basis the point $Q=800W,\ T_e=1.91eV$.

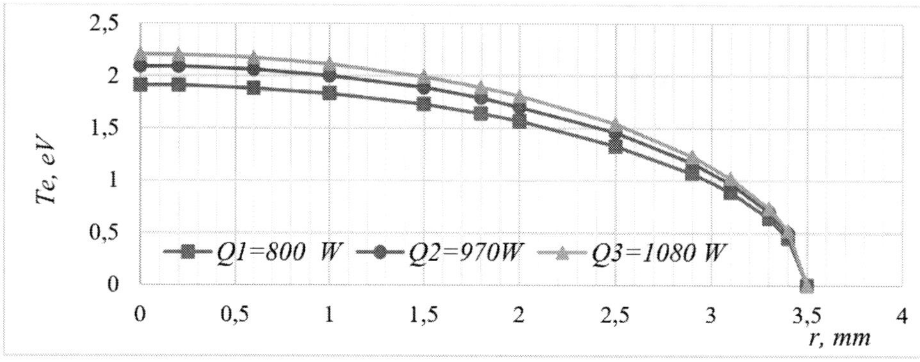

Figure 2. Calculated distribution of the electron temperature $T_e(r)$ depending on the radius for three values of the supplied electric power Q .

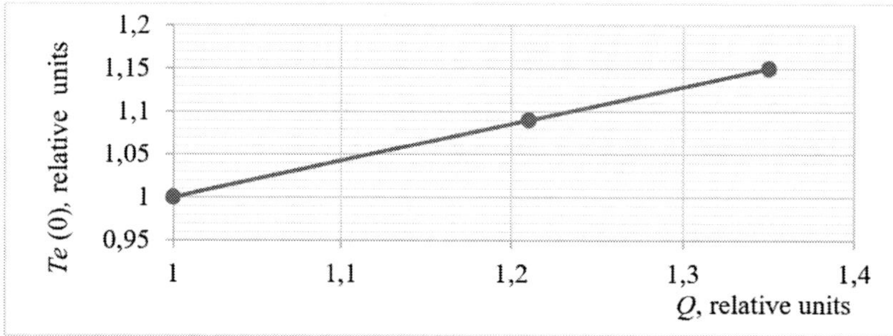

Figure 3. Dependence of the electron temperature $T_e(0)$ in tube center with respect to supplied electric power Q in relative units.

This figure shows that an increase in the energy of the electrons is significantly slower than the supplied electric power. For example, when Q is increased by 21%, the electron temperature T_e goes up just 9%, and when Q is changed by 35%, T_e increases by just 15%.

Figure 4 shows a comparison of the simulation results for T_e adopted from [8] with the calculated by our model at $Q = 800W$. Our model has the advantage of calculating the distribution of $T_e(r)$ by taking into account the radial distribution of the supplied electric power.

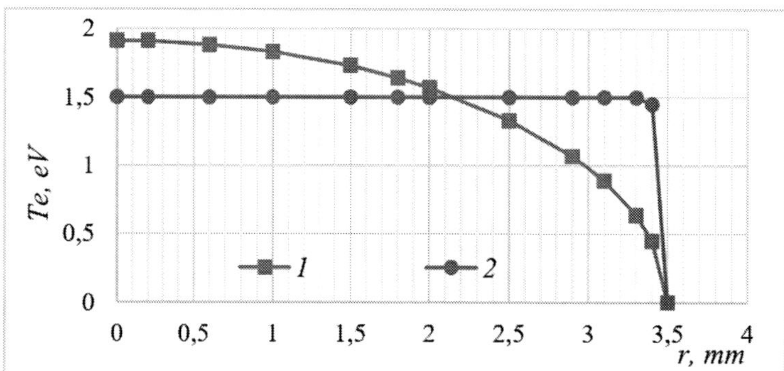

Figure 4. Comparison of the calculated distribution (curve 1) of the electron temperature $T_e(r)$ with the results from [8] (curve 2) for the case of supplied electric power $Q = 800W$.

The obtained results have important operational significance when studying laser generation. The supplied electric power is mainly significant for the increase of laser generation. This process has its natural limits. Together with the increase in electron temperature, the temperature of the neutral gas T_g also goes up. When critical values are reached for T_g, the gas may overheat and laser generation may cease. Other possibilities also need to be sought to increase laser generation: structural, energetic, thermodynamic and optical.

5. Conclusion

Based on experimental data and an actual existing CuBr laser, a primary mathematical model for defining the energy of the electrons T_e has been developed. The degree of influence of the supplied electrical power Q to the gas discharge tube is investigated. It has been found that Q has a limited influence on T_e from the positions on the stable operation of the laser.

Acknowledgement

This work was partial supported by the Grant No BG05M2OP001-1.001-0003, financed by the Science and Education for Smart Growth Operational Program (2014-2020), co-financed by the European Union through the European structural and Investment funds.

References

[1] Sabotinov N V 2006 Metal vapor lasers *Gas Lasers* eds M Endo and R F Walter (Boca Raton: CRC Press) pp 449–494
[2] McCoy S, Hanna M, Anderson P, McLennan G and Repacholi M 1996 An evaluation of the copper–bromide laser for treating telangiectasia *Dermatol Surg.* **22**(6) 551–557
[3] Foster P G 2005 *Industrial applications of copper bromide laser technology* Thesis (Ph.D.) (Adelaide Australia: University of Adelaide, School of Chemistry and Physics)
[4] Warner B 1998 Atomic vapour laser isotope separation *Proc. SPIE* **737** pp 2–6
[5] Little C E 1999 *Metal Vapour Lasers: Physics, Engineering and Applications* (Chichester: John Wiley & Sons)
[6] Carman R J 1996 Computer modelling of longitudinal excited elemental copper vapour lasers *Pulsed Metal Vapor Lasers – Physics and Emerging Applications in Industry, Medicine and Science* eds C E Little and N V Sabotinov (Dordreht: Kluwer Academic Publishers) pp 203–214
[7] Trigub M V, Evtushenko G S, Torgaev S N, Shiyanov D V and Evtushenko T G 2016 Copper bromide vapor brightness amplifiers with 100 kHz pulse repetition frequency *Optics Communications* **376** 81–85
[8] Torgaev S N, Kulagin A E, Evtushenko T G and Evtushenko G S 2019 Kinetic modeling of spatio-temporal evolution of the gain in copper vapor active media *Optics Communications* **440** 146–149
[9] Boichenko A M, Evtushenko G S, Nekhoroshev V O, Shiyanov D V and Torgaev S N 2015 CuBr–Ne–HBr laser with a high repetition frequency of the lasing pulses at a reduced energy deposition in the discharge *Physics of Wave Phenomena* **23**(1) 1–13
[10] Blau P 1996 Impedance matching and electric field penetration in metal vapour lasers *Pulsed Metal Vapour Lasers, NATO Science Partnership* (*Sub-Series 1* vol 5) eds C E Little and N V Sabotinov (Springer, 1996) pp. 215–220
[11] Janke E, Emde F and Lösch F 1977 *Special Functions* (Moscow: Nauka, 1977)
[12] Iliev I P and Gocheva-Ilieva S G 2010 Model of the radial gas temperature distribution in a copper bromide vapour laser *Quantum Electronics* **40**(6) 479–483

IOP Conf. Series: Materials Science and Engineering **618** (2019) 012019 doi:10.1088/1757-899X/618/1/012019

Energy efficiency of atmospheric pressure plasma-aided porous media surface finishing

D N Gospodinova, P D Dineff

Department of Electrical Engineering, Technical University of Sofia, 8 Kliment Ohridski Blvd, 1756 Sofia, Bulgaria

Abstract. Atmospheric pressure plasma pretreatment based on dielectric barrier discharge technology at industrial frequency (50 Hz) has been used for surface activation (oxidation, functionalization) of four species of European and tropical wood samples in a multi-step plasma enhanced capillary process of flame retardancy. Plasma-chemical activation is an effective surface alteration by attaching polar or functional groups to wood and cellulosic surfaces. Oxygen containing functional groups enhance effectively capillary impregnation processes for flame retardation of wood as porous media finishing. Oxidation level was studied by wettability in order to follow the increased surface free energy. For each species, a critical active power surface density was necessary to obtain the best impregnation results more quickly. The relationship between the penetration-spreading parameter and the active power density has been investigated experimentally as process quality parameter.

1. Introduction

In general, the surface free energy (*SFE*) of wood, wooden and cellulosic porous media is not sufficiently high to permit good capillary impregnation by flame retardant containing water solution. In case of finishing such as gluing, painting, varnishing, coating, it prevents better bonding between two or more materials. Hence, for better quality of finishing processes, *SFE* of the low energy wooden surfaces should be increased. The "*open air*", "*non-pressure*" or "*non-vacuum*" atmospheric pressure plasma (*APP-*) pretreatment founded on dielectric barrier discharge (*DBD*) is widely used nowadays for surface activation (oxidation, functionalization) to achieve a higher *SFE* of porous wooden and cellulosic surfaces for enhancement of finishing such as impregnation, gluing, painting, varnishing and coating processes. It is a well-known by the used common three-step impregnation processes to ensure flame retardancy of wood, wooden and cellulosic materials – lumber, timber, paneling, floor, plywood, *MDF*, *OSB*, paper, corrugated board, cardboard [1, 2].

The purpose of this paper is to provide an energy efficiency assessment not only for the *APP*-pretreatment process, but also for the overall plasma enhanced finishing. The ultimate goal of every oxidative *APP*-pretreatment technique is to increase porous media *SFE* or so-called "dyne level", [3, 4, 5], their wettability and capillary activity, as well as to enhance surface finishing by increasing the penetration-spreading parameter (*PSP*) of common process: both *SFE* and liquid *surface tension* (*SFT*) are essential in porous media finishing phenomena. The addition of anionic micelle-forming and trisiloxane-ethoxylate surfactants can reduce the surface tension of a phosphorous flame retardant (*FR*) containing impregnating water solution figure 1.

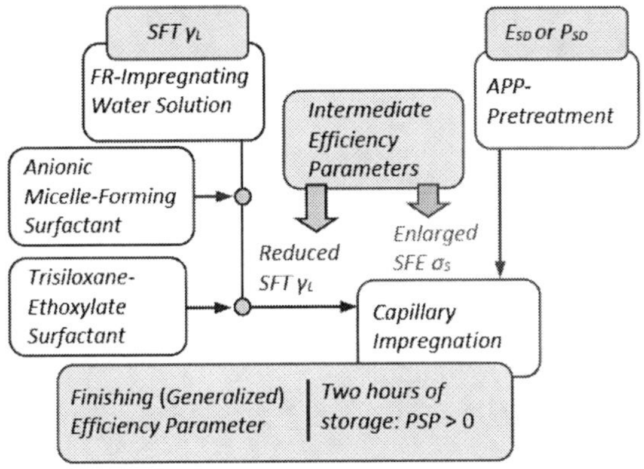

Figure 1. Finishing or generalized efficiency parameter – *penetration-spreading parameter* (*PSP*), and intermediate efficiency parameters expressing an enlarged *SFE* σ_S of the *APP*-pretreated wood surface, and a reduced *SFT* γ_L of impregnating water solution, in a multi-steep plasma-enhanced capillary impregnation process. The relationship between the finishing parameter *PSP* and the active power density P_{SD} (or the corresponding energy dosage E_{SD}) predicts and displays the efficiency of the used plasma enhanced process of capillary impregnation with flame retardant (*FR*) containing water solutions.

Ultimately the energy dosage (or power factor) is known to get a particular wood material to a certain *SFE* or common plasma-aided finishing to a certain *PSP*, it can be used to predict the results if any of the conditional parameters change such as operative speed, duration of treatment or electrode capacity to distribute to necessary level of applied energy (power) equation (1) [3, 5].

2. Factor and parameter of atmospheric pressure plasma pretreatment

The *APP* pretreatment of a wood side in operative gap is done schematically between earthed ferromagnetic (carbon steel) electrode and dielectric barrier (alkali glass) as shown in figure 2. The operating gap between earthed electrode and dielectric barrier was changed from 3 to 15 mm. The high-voltage ferromagnetic electrode was connected to *AC* power supply having high voltage (below 30 kV peak value) and industrial frequency (50 Hz) [6].

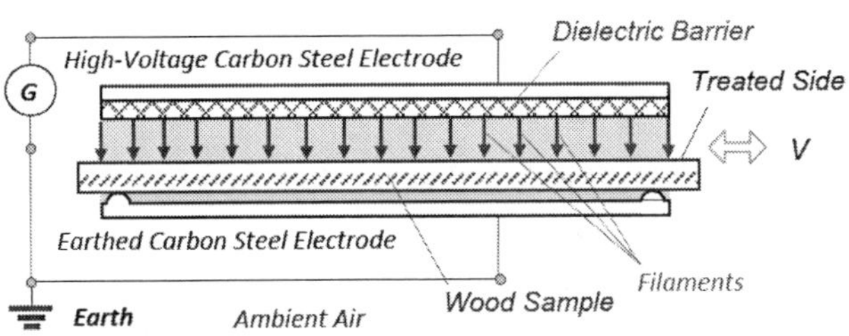

Figure 2. The atmospheric-pressure plasma (*APP*-) pretreatment of a wood surface is done by passing the wood sample through dielectric barrier discharge in air at atmospheric pressure and industrial frequency (50 Hz) created between earthed electrode and dielectric glass barrier in a coplanar electrode system with two structural carbon steel (BDS 2592) electrodes.

2.1. Atmospheric pressure plasma pretreatment terms

As its name suggests "current–voltage" characteristic of a burning *DBD* is a relationship, represented as a chart or graph, between the discharge current flowing through an operating gap and the applied voltage across electrodes. The typical current–voltage characteristic of a burning *DBD* is a relationship between the average current value I_{avg} and the root mean square value of applied voltage V_{rms}: $I_{avg} = \varphi\ (V_{rms})$ [3]. These current and voltage values can be measured directly in an experimental investigation. According to a well-known methodology (*Fillipov-Emelyanov*, 1957) the current-voltage characteristic can be used to establish a new mathematical model of *DBD* behavior – "*active power-voltage*": $P = \varphi\ (V_{rms})$, or "*active power density-voltage*": $P_{SD} = P/S = \varphi\ (V_{rms})$ characteristic, where $S = l\ w$, is the active area of electrode system; l is the length and w is the width of electrodes figures 2 and 4a.

2.1.1. Energy dosage, energy factor or specific energy

The *energy dosage* (E_{SD}) is the specific quantity of *DBD* energy applied to the treated wood surface. This measurement identifies the energy dose E_{SD} as the surface density of discharge energy or specific

IOP Conf. Series: Materials Science and Engineering **618** (2019) 012019 doi:10.1088/1757-899X/618/1/012019

energy (per unit surface) to which a one side or two sides of wood samples is subjected. The being applied active (real, true) power *P*, W; production line speed *V*, m/min; number of side to be treated *N* – one or two; treating surface or electrode width *w* in m, and material dependent D_M, are included in it, figure 2 [3, 4, 5]:

$$E_{SD} = D_M P / (N V w), \text{W min/m}^2. \quad (1)$$

The specific energy E_{SD}, identifies too the *DBD*-energy dosage to which a stationary ($V = 0$) wood surface equals to the electrode surface ($S = l\,w$) is subjected. The plasma pretreatment time (duration) *t* and the length *l* of treated wood sample (or electrodes) are included in it:

$$E_{SD} = D_M P\,t / (N\,l\,w), (V = 0), \text{W min/m}^2. \quad (2)$$

Sometimes, the energy dosage E_{SD} is called "*watt density*" [3, 4, 5].

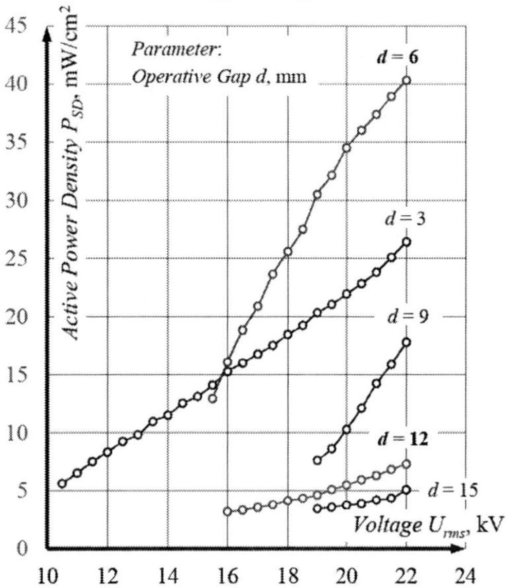

Figure 3. Relationship between *DBD* active power density P_{SD} and applied voltage V_{rms} for different operative gap *d* in the preferred *APP pretreatment* operative area. Fitting the best trendlines/regression types to experimental data of active power density for two air gap (*d*):

i) polynomial trendline illustrates the relationship between active power surface density (P_{SD}) and voltage (V_{rms}) at air gap $d = 6$ mm:

$$P_{SD} = -0.2292 V_{rms}^2 + 12.803 V_{rms} - 130.31;$$

ii) polynomial trendline illustrates the relationship between real power surface density (P_D) and voltage (V_{rms}) at air gap $d = 12$ mm:

$$P_{SD} = 0.06\, V_{rms}^2 - 1.5869 V_{rms} + 13.191.$$

Notice that the *R*-squared values (R^2) are respectively 0.9990 and 0.9991, which are good fits of the trendlines to the data.

Using this *DBD*-treatment (impact) factor, it is possible to calculate the increase of the *SFE* or to correctly match a *DBD*-treatment device for a particular application.

Once the energy dosage E_{SD} to get a defined wood surface to a certain *SFE* (dyne level) is known, it can be used to predict the results if any of the process parameters change. For example, let's say that the line speed increases from V_1 to V_2 ($V_2 > V_1$):

$$E_{SD} = D_M P_1 / (N.V_1.w) = D_M P_2 / (T.V_2.w);$$

$$P_1 V_2 = P_2 V_1 \rightarrow P_2 = P_1 (V_2 / V_1), \text{ or } P_2 > P_1. \quad (3)$$

If we use a power supply with active power *P*, knowing the energy dosage E_{SD} which is able to get the intended *SFE*, the fastest line speed (*max V*) we can achieve under these conditions well be:

$$max\ V = D_M P / (E_{SD} N w). \quad (4)$$

2.1.2. Power factor or active power density

The active (real, true) power P discloses the ability of a DBD to carry through an effective plasma activation and decides how much faster this activation can be done. The power factor or active power density PSD specifies the power per unit surface of sample subjected to the plasma treatment or per unit length in direction along the production line. The time t defines the length L of the plasma activated surface at a set line speed V ($L = V t$):

$$P_{SD} = E_{SD}/t = D_M P / (N V t w) = D_M P / (N L w), \text{W/m}^2. \quad (5)$$

It is possible to define a specific power for a certain length L, for example, 1 cm.

The specific active power P_{SD} identifies also the *DBD*-power density to which a stationary ($V = 0$) treated wood surface is subjected. The plasma pretreatment time (duration) t and the length l of treated wood sample are included in it also:

$$P_{SD} = D_M P / (T\, l\, w), \text{ W/cm}^2. \quad (6)$$

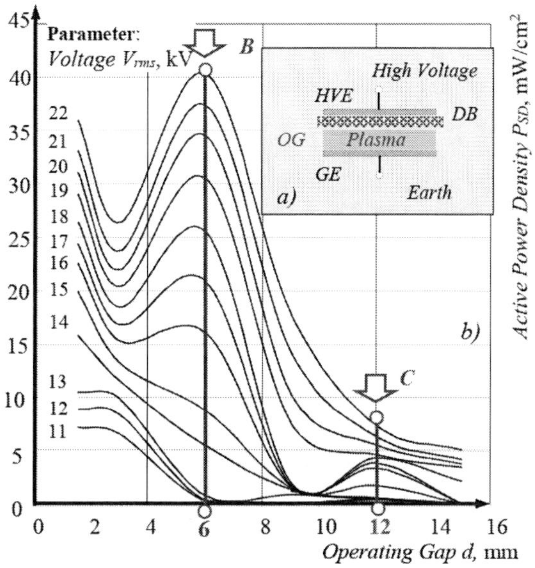

Figure 4. Dielectric barrier air discharge at atmospheric pressure (*DBD*) in asymmetric coplanar electrode system with one glass barrier (***a***), technological *DBD* characteristic "*active power density* (P_{SD}) *–operating gap size* (d)", parameter – applied voltage (V_{rms}), and operational streamer regimes: *regime B* - cathode directed streamers; *regime C* - anode directed streamers, of atmospheric pressure plasma pretreatment at industrial frequency – 50 Hz (***b***).

HVE – high voltage electrode; *GE* – earthed (grounded) electrode; *DB* – dielectric (alkali glass: $\varepsilon_r = 10$; $\rho_V = 10^9\ \Omega$m; $tg\delta = 25$ at 20°C) barrier with $b = 3$ mm; *OG* – operating gap.

The wood samples dimensions (length and width) should be the same as those of the flat electrodes. That is so called power factor of a *DBD* coplanar electrode system.

The effectiveness of a *DBD* can be assessed using that impact factor, or a *DBD*-treatment station can be dimensioned for a specific treatment application. It is very important in order to assess the saturation of plasma activation effect. Also, impact factor can be accepted as an assessment criterion referred to the degree of thermal shock applied to the wood surface by an electrical discharge.

2.1.3. Wood surface treatability

Not all wooden or cellulosic materials can be treated equally well. One of them is very hard to DBD-plasma treatment and requires more treatment energy or power ($D_1 \neq 0$). Another one is less difficult to treatment compared to the previous one ($D_{M,2} < D_{M,1}$). Also, there may be such material that does not require plasma treatment for finishing as their initial surface free energy is sufficiently high ($D_{M,3} = 0$). This experimental study was carried out on four kind of wood samples: European white pine (Pinus Sylvestris L) from Bulgaria and three woods from Mexican rain forest: Tzalam (Lysiloma Bahamensis L), Mexican white cedar (Cupressus Lusitanica L), and Mahogany caoba (Swietenia Mahogany L), according a well-known methodology of experiment used by us [1, 2].

The selected wood species form the following range of decreasing density, in kg/m³: Tzalam - 780; Mahogany caoba - 650; Mexican white cedar – 470; European white pine – 371. There are two hard wood species with high density, D > 540 kg/m³: Tzalam and Mahogany caoba (Mahogany), and two soft wood species with low density, D < 540 kg/m³: Mexican white cedar (Cedar) and European white pine (Pine). On the basis of our own former experience in plasma-aided capillary impregnation of wood and wooden materials, an oxidative APP-pretreatment has been applied on the test samples for 60 sec figure 2.

The storage (aging, open) time means a time period after APP-pretreatment which enables an effective plasma-aided finishing processes such as impregnating, gluing, bonding, coating and painting, to be achieved. Plasma pre-activation most often takes place "in-line" immediately prior to the onset of the finishing: the storage time is zero. The storage time should be different from zero when the plasma-aided capillary impregnation is applied in the finishing "out of the line". The substrates with the lowest initial SFE after storage time also have the lowest SFE. This process is commonly described as "hydrophobic recovery" [5]. Comparing the rate of hydrophobic recovery, there is an exception for the mahogany: the mahogany sample with the lowest SFE after treatment has the slowest decrease in surface energy figure 5.

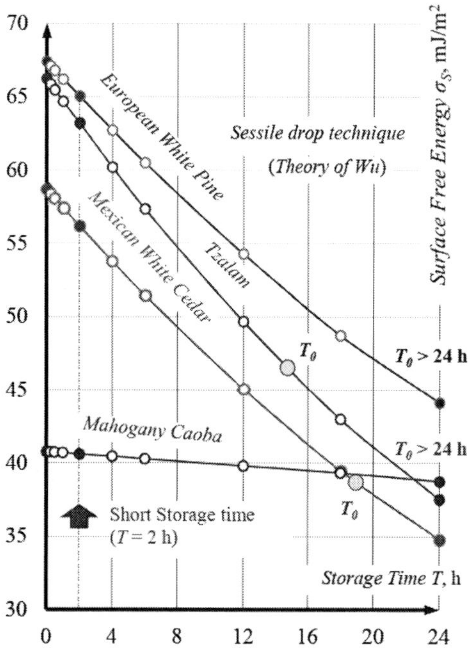

Figure 5. Alteration of the *surface free energy* (*SFE*) σ_S, mJ/m^2, measured by *Sessile drop technique* (*Theory of Wu*) as a function of storage (aging, open) time *T*, h, after *APP-pretreatment* at 18 kV rms (25.6 kV peak value) and 50 Hz. The measured initial *SFE* were, mJ/m^2: *European White Pine* – 29.86; *Tzalam* – 46.50; *Mexican White Cedar* – 38.59; *and Mahogany Caoba* – 27.87. The short storage time below two hours is suitable for "out of the production line" plasma-aided capillary impregnation process. The *storage time leading to zero effect* (T_0), h, is this storage time that eliminates the positive effect of an *APP-pretreatment.*

Table 1. Storage (aging) time (*T*) after *APP*-pretreatment and storage time to zero effect (T_0), h.

	Aging time T	0	2	24	T_0
FE, mJ/m^2	Pine	67.37	65.04	44.13	45.19
	Tzalam	66.24	63.18	37.54	15.44
	Cedar	58.70	56.20	34.71	19.04
	Mahogany	40.79	40.62	38.77	190.24

3. Factor and parameter of plasma-aided capillary impregnation

Plasma-aided capillary impregnation finishing will be more successful and the porous media will be more susceptible as the *PSP* >> 0. This is the essence of the rule for obtaining effective plasma-enhanced capillary impregnation finishing. The *PS*-parameter depends only on two thermodynamic factors of the finishing process - the *SFE* σ_S of the wood sample increased by the *APP*-pretreatment; and the *SFT* γ_L of impregnating water solution reduced by the added surfactants. These two factors can be quantified individually at two well-known methods of contact angle measurement - the sessile drop and the pendant drop techniques: $PSP = (\sigma_S - \gamma_L)$.

Studying the relationship between the *PSP* after two hours of storage and the active power density P_{SD} it becomes clear that two effective "*out of the line*" finishing processes are applicable – with *FR-A5* and *FR-A5-S* impregnating solutions: *PSP* >> 0. An effective common process can be realized also with *FR* impregnating solution after an *APP*-pretreatment at $P_{SD} > 8.8$ mW/cm^2 (15 kV rms; AC): *PSP* >> 0 figure 6.

However, the two modified impregnating solutions *FR-A5* and *FR-A5-S* with low *SFT* provide high *PSP* values (over 15 mJ/m^2) and good impregnation process quality at too low power levels - below 1 mW/cm^2, figure 6.

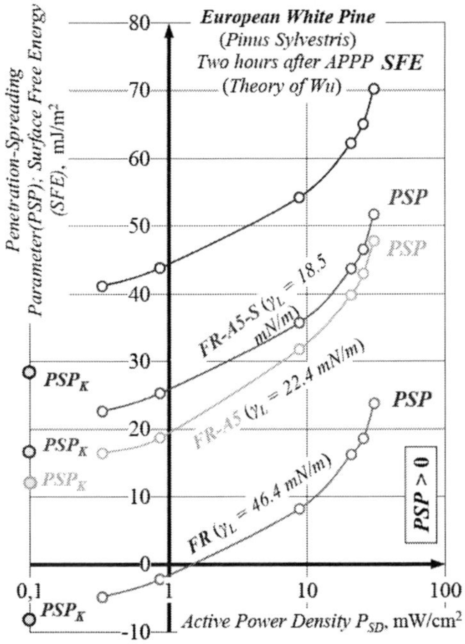

Figure 6. Relationship between the *penetration-spreading parameter* (*PSP*), mJ/m^2, and *surface free energy* (*SFE*), mJ/m^2, measured by *Sessile drop technique* (*Theory of Wu*) and the *active power density* P_{SD} after two hours of storage for *European White* (*Pinus Sylvestris L.*) heart wood samples.

Penetration-spreading parameter (*PSP*) is equal to the minimum *interfacial free energy* (*IFE*) value at $\cos\theta = 1$ ($\theta = 0$ deg): $PSP = \sigma_{SL}$. An *effective finishing* is characterized by a positive value of *PSP*: $PSP > 0$; a non-effective finishing has a *PSP*-value equal or less to zero: $PSP \leq 0$.
Plasma-aided surface impregnation was accomplished by tree types of water solutions of phosphorous flame retardant (*FR*) with different surface tension γ_L: *FR* – bare 30 mass % solution of *FR*; *FR-A5* – *FR*-water solution with micelle-forming anionic surfactant at 5 vol. %; *FR-A5-S* – *FR*-water solution with micelle-forming anionic surfactant at 5 vol. % and trisiloxane surfactant at 0.01 vol. %. PSP_K – *PSPs* of bare or non-treated wood samples.

4. Conclusion

It is often recommended that the active power density applied to the treated side of wood sample during *APP* pre-treatment to be as high as necessary to reach the maximum *SFE* but to be as low as possible in order to save energy – an energy efficiency restriction.

In the present study, contrary to expectations, it has been shown that the "*excessive treatment*" increases *PSP*, although *SFE* is not further increased at higher active power densities, for example over 20 mW/cm^2. The use of reduced *SFT* impregnating solutions allows the use of energy-efficient low active power levels even below 0.33 mW/cm^2.

Acknowledgments

The authors would like to thank the Research and Development Sector at the Technical University of Sofia for the financial support.

References

[1] Dineff P, Gospodinova D and Ivanov I 2017 Efficiency assessment of plasma-aided porous media surface finishing *Adv. Sci. Technol. Eng. Syst. J.* (Special issue on Advancement in Engineering Technology) **2** (6) p. 242-251.

[2] Dineff P, Gospodinova D and Ivanov I 2016 Study on processing efficiency of plasma-aided capillary impregnation of porous media through a penetration-spreading parameter analysis *Proc. Int. Conf. on Advanced Manufacturing Technologies* 20-th 26-29 June Sozopol Bulgaria (Publishing house "St. Ivan Rilski") p. 54-65.

[3] Dineff P 2009 Homogeneous dielectric barrier discharge (Sofia: Technical University of Sofia Press) **59** (2) p. 146-155. (in Bulgarian).

[4] Eisby J 2016 Corona treatment (Vetaphone Corona & Plasma A/S, Denmark) www.vetaphone.com.

[5] Lindner M, Rodler N, Jesdinszki M, Schmid M and Sängerlaub S 2018 Surface energy of corona treated PP, PE and PET films, its alteration as function of storage time and the effect of various corona dosages on their bond strength after lamination J. *Appl. Polym. Sci.* 10.1002/APP.45842.

[6] Benelli C 2011 Surface Treatment – Necessity or Luxury? *Coating* **8** p. 18-20.

Study of the short-circuit currents in branches of distribution networks with trilateral power supply

M I Matsankov

Faculty of Engineering and Pedagogy, Sliven, Technical University of Sofia, Bulgaria

misho.ivanov@abv.bg

Abstract. The article presents the determination of the short-circuit currents in distribution networks with trilaterally power supplied branches – form a supplying substation and two decentralized energy sources. After connecting decentralized energy sources, branches with bilateral and trilateral power supply appear. This changes the network configuration and, respectively, its mode parameters, and also requires adjustments to the settings of the relay protection.

1. Introduction

Calculation of the short-circuit currents in the distribution networks with decentralized energy sources (DES) is necessary for the purpose of choosing apparatuses and current-carrying parts, as well as for correct setting of the relay protection and the means of automation [1÷3].

Following the connection of DES, sections with bilateral or trilateral supply appear in certain branches of the distribution networks, which changes the structure of the network and its mode parameters.

From the literature review it has been established that the classical methodologies for calculation of short-circuit currents in distribution networks do not take into account the presence of three power supplies in the connection branches.

The selective turning of the short circuits off requires a precise setting procedure both by the observed operational parameter and by time of action in case a number of supplying sources is available.

The aim of the present article is to describe the possibilities and the results from the implementation of an especially developed software program for calculation of the short-circuit currents in distribution networks and in branches with trilateral supply: from the power supplying substation and from two DESs. The program facilitates the correct adjustment of the protective devices in the trilaterally power supplied branches and helps to avoid unnecessary and deceptive exploits.

2. Algorithm for calculating the short-circuit currents

The compiled algorithm should be applicable to both medium-voltage and low-voltage distribution networks.

The calculation of the short-circuit currents is done taking into account the different ways of grounding of the star center of the medium voltage distribution networks in the Republic of Bulgaria [2 ÷ 5]:

- Insulated in the networks with small current of grounding connection;
- Grounded through active resistance in cable and air networks;
- Grounded through an arc-extinguishing coil in air networks;
- Combined with an arc-extinguishing coil and an active resistor, which is automatically switched on by a single-phase short-circuit breaker in air and cable lines.

The low voltage distribution network is a four-wire one with a directly grounded star center, and when compiling the zero sequence diagram for calculating unsymmetrical modes, the presence of a neutral (zero) conductor with grounding devices must be taken into account.

The size of the connected DES, included in a common connection node, affects the short-circuit currents that may exceed the permissible values. The power transformers in the transformer stations of the DES connection node in the distribution networks can be with the following connection groups: Yo/Δ, Y/Δ or Yo/Yo, and the currents for the different types of short circuits on both sides of the transformers have different values [3] [4].

In electrical networks, operating with an insulated star center, a double ground connection may occur. For the schemes for switching DES to the distribution network, the locations of the two successive ground connections are, respectively, in the connection node and in the supply node of the distribution network.

The compiled algorithm is part of the computer simulation system, described in detail in [6], and aiming at simulating the mode parameters under the following three modes: a normal mode, an emergency and tracking modes.

The input data for the compiled algorithm are actually the ones which influence the magnitude of the short-circuit currents: rated voltage, kV; effective grounding resistance of the electric pole, Ω; impedance of the first branch $R_1 + jX_1;\Omega$; impedance of the second branch $R_2 + jX_2;\Omega$; impedance of the third branch $R_3 + jX_3;\Omega$; source 1 – the power supplying sub-station; source 2 – a DES with power S_2 and $\cos\varphi_2$; source 3 – a DES with power S_3 and $\cos\varphi_3$ (Figure1).

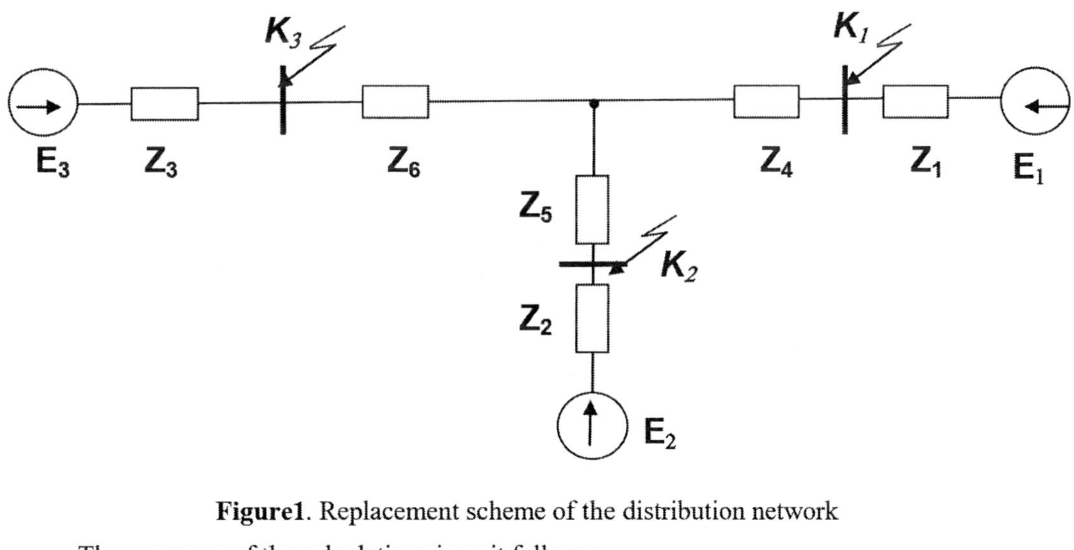

Figure1. Replacement scheme of the distribution network

The sequence of the calculations is as it follows:

Step 1 – the input data are introduced and the place of the short circuit is chosen.

Step 2 – the impedance of the connected in short circuit is calculated.

The following three cases are considered at this step, depending on the place, where the short circuit occurs:

- *Case 1 – short circuit in point K1 (Figure 1).*

Calculation unit 1:

IOP Conf. Series: Materials Science and Engineering **618** (2019) 012020 doi:10.1088/1757-899X/618/1/012020

$$Z_7 = Z_2 + Z_5; Z_8 = Z_3 + Z_6; \tag{1}$$

$$E_4 = E_2 // E_3 = \frac{E_3.Z_7 + E_2.Z_8}{Z_7 + Z_8}; Z_9 = \frac{Z_7 Z_8}{Z_7 + Z_8}; \tag{2}$$

$$Z_{10} = Z_9 + Z_4; \quad E_{екв1} = E_4 // E_1 = \frac{E_4.Z_1 + E_1.Z_{10}}{Z_1 + Z_{10}}; \quad Z_{екв1} = Z_1 // Z_{10}. \tag{3}$$

where Z_1 and Z_2 are the impedances of DES 1 and 2; Z_3 – the impedance for connection to the supplying network; Z4, Z5, Z6 – the impedances of the power lines; E1, E2, E3 – the e.m.f. of the generating sources.

- *Case 2: short circuit in point K2 (Figure1):*

Calculation unit 2:

$$Z_8 = Z_3 + Z_6; \; Z_{11} = Z_4 + Z_1; \tag{4}$$

$$E_5 = E_3 // E_1 = \frac{E_3.Z_{11} + E_1.Z_8}{Z_{11} + Z_8}; \; Z_{12} = Z_8 // Z_{11}; \tag{5}$$

$$Z_{13} = Z_{12} + Z_5; \quad E_{екв2} = E_5 // E_2 = \frac{E_5.Z_2 + E_2.Z_{13}}{Z_2 + Z_{13}}; \quad Z_{екв2} = Z_{13} // Z_2. \tag{6}$$

- *Case 3: short circuit in point K3 (Figure1):*

Calculation unit 3:

$$Z_{11} = Z_4 + Z_1; Z_7 = Z_2 + Z_5; \tag{7}$$

$$Z_{14} = Z_2 // Z_{11}; E_6 = E_1 // E_2 = \frac{E_1.Z_2 + E_2.Z_{11}}{Z_2 + Z_{11}}; \tag{8}$$

$$Z_{15} = Z_6 + Z_{14}; \quad E_{екв3} = E_3 // E_6 = \frac{E_3.Z_{15} + E_6.Z_3}{Z_3 + Z_{15}}; \quad Z_{екв3} = Z_3 // Z_{15}. \tag{9}$$

Step 3 – the short-circuit currents (three-phase, two-phase, and single-phase short-circuit, ground connection, double coupling) are calculated using the traditional approach and the classical methods [1,7].

Step 4 – the procedure is repeated after changing some of the influencing factors and the obtained results are visualized.

Following the developed algorithm, a software program, based on the calculating capabilities of Excel, has been compiled, which performs the described procedures and is part of the computerized system for calculating the mode parameters in the electrical networks of EPF-Sliven [6].

3. Numerical experiment

The calculations are performed at the following input data: rated voltage 20 kV; effective grounding resistance of the electric pole 4Ω; impedance of the first branch $R_1 + jX_1; \Omega$; impedance of the second branch $R_2 + jX_2; \Omega$; impedance of the third branch $R_3 + jX_3; \Omega$.

The voltage of the supplying sub-station (source 1) is 110/20 kV; source 2 – a DES with power S_2 and $\cos\varphi_2$; source 3 – DES with power S_3 and $\cos\varphi_3$.

The results from the conducted experiments are presented in Figures 2, 3, 4 and 5.

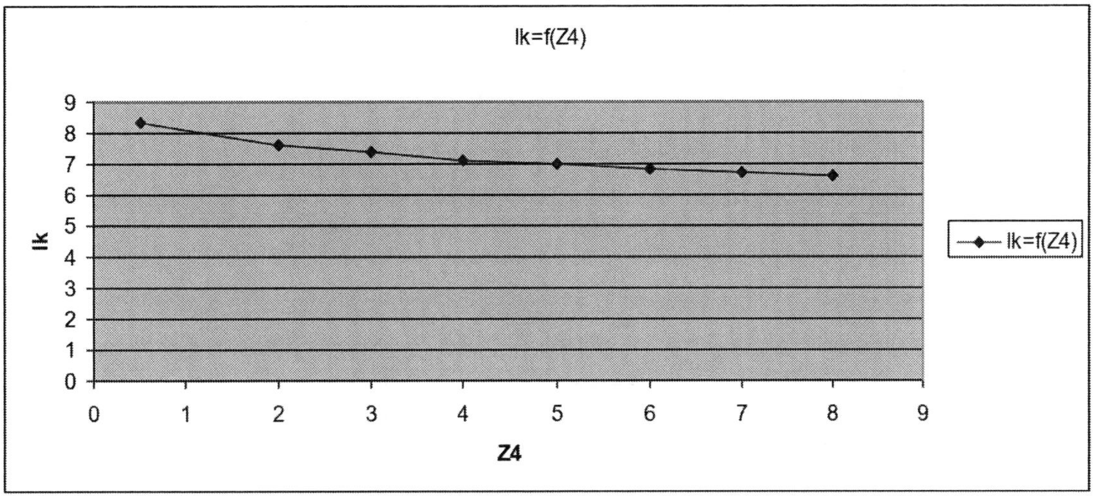

Figure 2. Short-circuit current at changing Z4

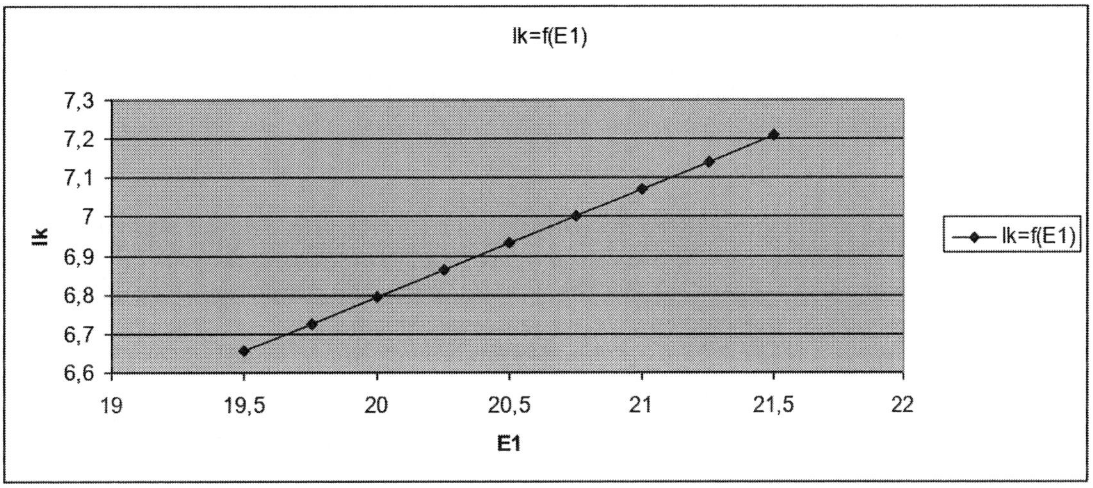

Figure 3. Short-circuit current at changing E1

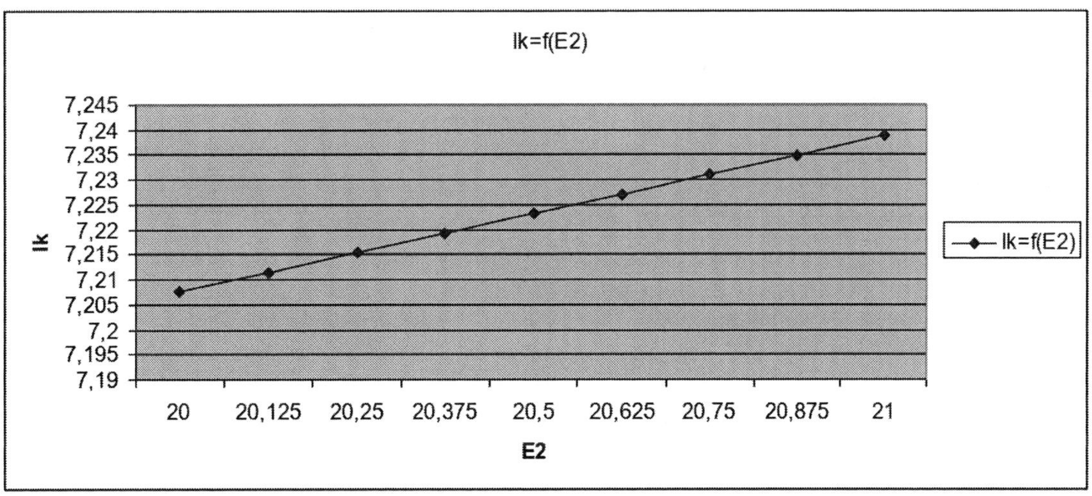

Figure 4. Short-circuit current at changing E2

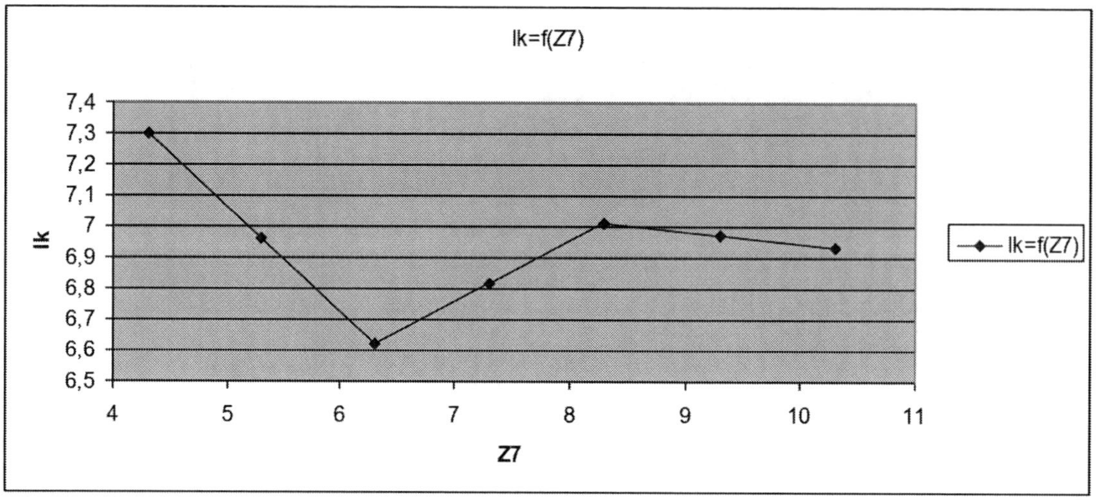

Figure 5. Short-circuit current at changing Z7

4. Summary

By means of the possibilities, offered by the algorithm and by the performed calculations, the magnitude of the short-circuit currents in the branches of distribution networks with power supply from three sources can be found.

5. Conclusions

- The compiled algorithm allows for defining the short-circuit currents in different points of the branches in distribution networks in case they are power supplied by three sources.
- The algorithm allows to change the factors, influencing the magnitude of the short-circuit currents and obtain results for their most unfavorable combinations.
- The obtained results are useful for the relay protection settings, which makes it possible to avoid unnecessary and false outages and interruptions in the power supply, which, in turn, results in more reliable operation of the electrical network.
- The results from the calculations are useful for selecting the apparatuses and the current-carrying parts both at the design stage and during the exploitation of electrical networks with associated DESs.

6. References

[1] Notov P. St., Nedeltcheva, Electroenergetics, Volume 2, TU – Sofia Publishing house, Sofia, 2009.

[2] Ivanov A. M., Limitation of short-circuit currents in electrical networks with decentralized sources, dissertation, TU-Sofia 2014

[3] Nedeltcheva St., Electrical networks and systems with decentralized generating sources, TU – Sofia Publishing house, Sofia, 2018, ISBN 978-619-167-348-3

[4] Nedeltcheva St., Methodology for analysis of unsymmetrical modes in distribution networks, Notices of TU-Sliven № 3, 2006, pp 11 – 19, ISSN 1312-3920

[5] Nedeltcheva St., N.Nedeltchev, Analysis of short circuit currents in distribution networks with dispersed generation, First international symposium on environment identities and Mediterranean area, ISEIM-2006, IEEE France section, ISBN 1-4244-0232-8, IEEE Catalog Number 06EX13229C, Paper V2s437, Corte-Ajaccio, 10-13.07.2006, http://ieeexplore.ieee.org/Xplore/l

[6] Nedeltcheva St., Simulator for modeling the modes in distribution networks with decentralized sources, Notices of TU-Sliven № 4, 2005, pp 3 – 17, ISSN 1312-3920

[7] Notov P., Short-circuits in power systems, TU – Sofia Publishing house, Sofia, 2000.

Study of a two-spring electromagnetic harvester

N P Georgiev[1] and R P Raichev[2]

[1]Department of Electrical Engineering, Technical University of Sofia, Branch Plovdiv, 25Tsanko Dystabanov Street, Plovdiv, Bulgaria
[2]Department of Mechanical Engineering, Technical University of Sofia, Branch Plovdiv, 25Tsanko Dystabanov Street, Plovdiv, Bulgaria

[1]nikola.georgiev@tu-plovdiv.bg
[2]rpraichev @tu-plovdiv.bg

Abstract. The paper studies an electromagnetic harvester, consisting of a Π-shaped plate with 4 magnets, placed on two springs, rigidly fixed at their low end. A coil lies between the magnets. Modeling by means of ANSYS has been done and the horizontal deviation of the mechanical system has been obtained. The magnetic field of the electromagnetic harvester has been modeled by Femm and its electromotive force has been defined. Experimental studies have been conducted for validating the developed model.

1. Introduction

Magnetic harvesters are more and more frequently used for converting mechanical energy into electrical and thus they replace batteries and low-consumption power-supplying electronic devices [1, 2]. Two types of motion of the magnet with respect to the coil are normally used. In the first case the magnet, which is most frequently placed in the air gap of the coil, moves in parallel with it [3], while in the second its movement is vertical in relation to the coil [4].

The present paper studies a two-spring electromagnetic harvester with two different types of mass in the form of two Π-shaped steel plates, with 4 magnets each. The harvester has a rigidly fixed coil, around which two pairs of rare-earth magnets move in parallel. Modeling by means of ANSYS has been done and the horizontal deviation of the mechanical system has been obtained. With the help of Femm 4.2 the magnetic field has been modeled and the magnetic induction has been determined. The magnetic fluxes through the winding and the induced e.m.f. of the harvester have been calculated. The voltages and the powers have been both calculated and measured at direct-current mode.

2. Exposition

Figure 1 presents the construction of the two-spring electromagnetic harvester. The following notations have been used: 1 – base; 2- steel springs; 3 – Π-shaped steel plate; 4 - rare-earth magnets; and 5 – a coil. The springs are with wire diameter d_s=0,8 mm, diameter of the spring D_s=6,7 mm and length of the spring l_s=20 mm. The Π-shaped steel plates are two in number, with the same dimensions and with 4 permanent magnets fixed to each of them. The magnets on the first plate are with dimensions 20x10x2 mm and weight m_1=15 grams, while the second plate is with magnets, measuring 20x10x3 mm and weighing m_2=22 grams. The diameter of the coil is $D`_c$=20 mm, its thickness - l_c=10 mm, number of turns N=800, and wire diameter d_c=0,1 mm.

The trajectory of the plates with the magnets in the considered electromagnetic harvester is an arch and thus a non-linear mechanical oscillating system is obtained.

When a force, changing in a sinusoidal way, with a resonant frequency is applied, the mechanical system, consisting of a mass (Π-shaped plate with permanent magnets) and two springs starts oscillating in parallel with the coil. Thus the magnetic flux goes differently through the coil and creates an alternating magnetic field, whereas alternating e.m.f. is induced.

The harvesters under consideration are non-linear mechanical oscillating systems. During their simulations by ANSYS R19.1 the following parameters have been taken into account: the way of fixing, the influence of the force of weight on the Π-shaped plates with permanent magnets, as well as the mechanical characteristics of the used springs. During the process of modelling the horizontal deviation x of the mechanical system, consisting of a mass (Π-shaped plate with permanent magnets) and two springs, has been obtained, Figure 2.

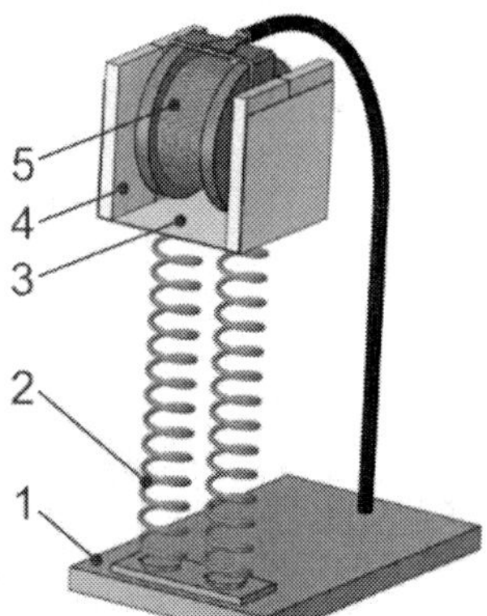

Figure 1. Two-spring electromagnetic harvester

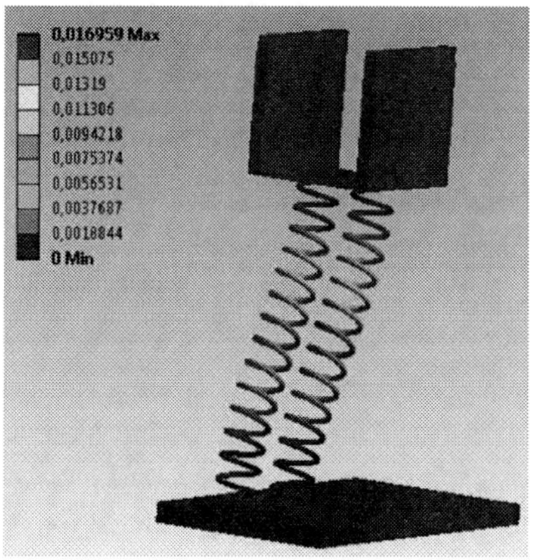

Figure 2. Simulation of the mechanical deviation of a harvester

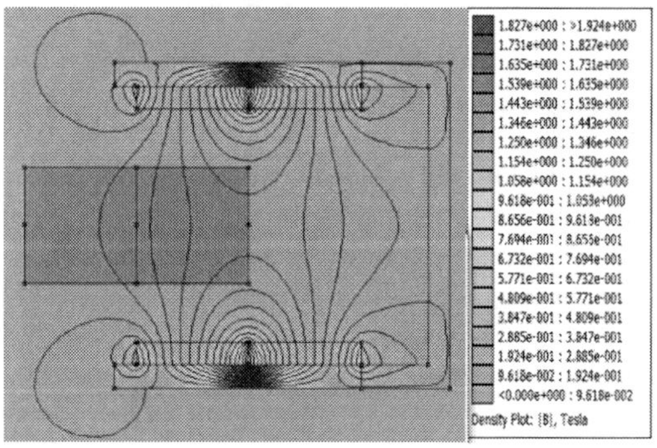

Figure 3. Distribution of the magnetic field for a small-mass harvester

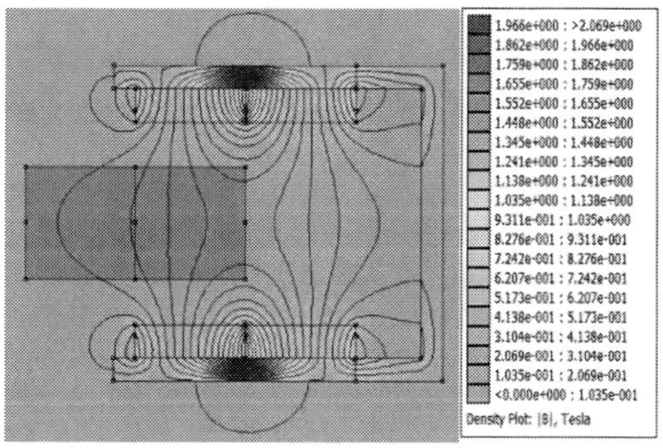

Figure 4. Distribution of the magnetic field for a big-mass harvester

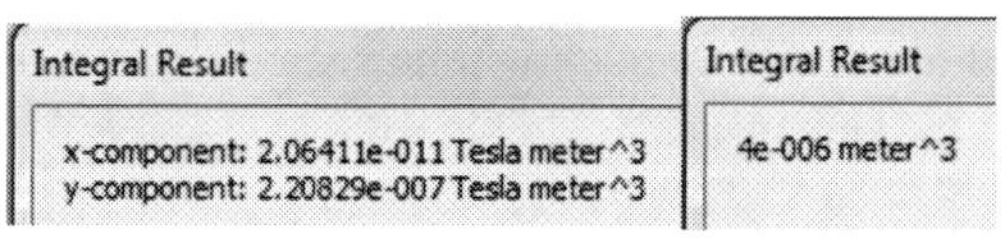

Figure 5. Densities of the magnetic induction for a harvester with small springs

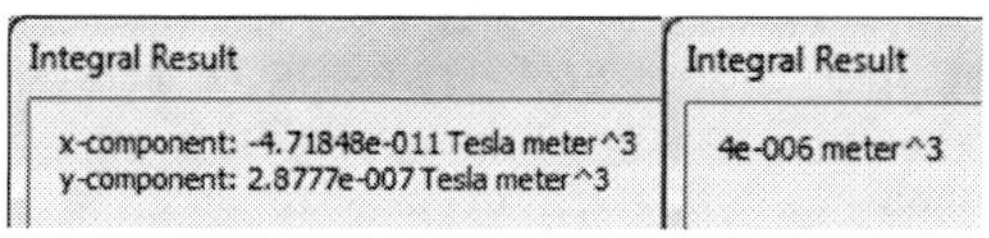

Figure 6. Densities of the magnetic inductions for a harvester with big springs

The distribution of the magnetic field in the electromagnetic harvester is obtained by means of Femm 4.2. Figure 3 illustrates the distribution of the magnetic induction at maximum horizontal deviation for a small-mass harvester, while Figure 4 gives the picture for a harvester with a big mass. Figure 5 shows the densities of the magnetic induction along the horizontal x - B_{cx} and the vertical y - B_{cy}, as well as the volume of the winding V_c for a small-mass harvester, and Figure 6 shows the same picture for the harvester with a big mass. The maximum and the minimum magnetic induction ($B'_{cy\max}$ and $B'_{cy\min}$) along y are calculated with the help of B_{cy} and V_c for the two types of mass.

$$B'_{cy\max} = \frac{B_{cy\max}}{V_c} \qquad B'_{cy\min} = \frac{B_{cy\min}}{V_c} \tag{1}$$

Thus the difference between the maximum and the minimum magnetic induction ΔB through the coil can be obtained

$$\Delta B = B'_{cy\max} - B'_{cy\min} \tag{2}$$

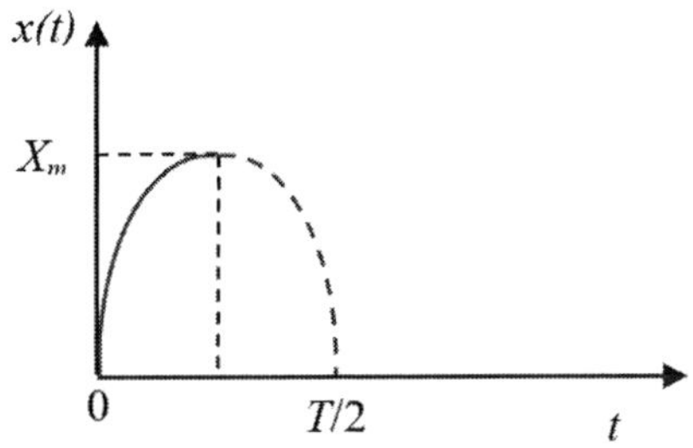

Figure 7. Horizontal deviation at a sinusoidal force

The horizontal deviation at a force, changing in a sinusoidal way, with a resonant frequency ω, is also sinusoidal, Figure 7

$$x(t) = X_m \sin \omega t \tag{3}$$

The horizontal oscillation velocity is equal to

$$v(t) = \frac{d\,x(t)}{d\,t} = \omega X_m \sin\left(\omega t + \frac{\pi}{2}\right) \tag{4}$$

The magnetic flux can be expressed by the change in the cross-section of the coil A, through which the difference between the maximum and the minimum magnetic induction ΔB goes

$$\Phi(t) = \Delta B \frac{dA(t)}{dt} \tag{5}$$

The diameter of the coil $D`_c$ is multiplied by a coefficient 0,785 due to the transformation from a circle to a square because of Femm 4.2

$$D_c = 0{,}785 D'_c \tag{6}$$

The change of the cross-section A in the course of time is equal to the changing horizontal deviation $x(t)$ multiplied by the diameter D_c

$$\frac{dA(t)}{dt} = D_c\, x(t) \tag{7}$$

From (5) and (7) for the magnetic flux through the coil it is obtained

$$\Phi(t) = \Delta B\, D_c\, x(t) \tag{8}$$

The induced e.m.f. in the winding of the electromagnetic harvester is

$$e(t) = -N\frac{d\Phi(t)}{dt} \tag{9}$$

From (8) and (9) for the induced e.m.f. at idle run mode it is obtained

$$e(t) = -N\,\Delta B\, D_c \frac{d\,x(t)}{dt} \tag{10}$$

The amplitude of the induced e.m.f. in the winding is in function of the amplitude of the oscillation velocity ωX_m

$$E_m = N\,\Delta B\, D_c\, \omega X_m \tag{11}$$

The circular frequency of the forced oscillations can be expressed by means of the frequency f, Figure 7

$$\omega = 2\pi f \tag{12}$$

From (11) and (12) for the amplitude of the induced e.m.f. in the winding at resonance it is obtained

$$E_m = N\,\Delta B\, D_c\, 2\pi f\, X_m \tag{13}$$

Thus with the help of the calculated by Femm 4.2 magnetic induction and the given resonant frequency, number of turns, diameter of the coil and horizontal deviation, the amplitude of the induced e.m.f. in the winding can be calculated at idle run for the corresponding resonant frequency.

Figure 8 shows the equivalent circuit of the electromagnetic harvester with a doubling voltage amplifier at active load. By R_c and L_c the active resistance and the inductance of the winding of the electromagnetic harvester are denoted, by $e_c(t)$ - the induced e.m.f., and by R_L – the active load resistance. The rectified voltage on the load resistance is denoted by U_L

The active power at direct-current mode is calculated using the amplitude of the induced e.m.f. in the winding at idle run and at resonance, as well as with the help of the voltage on the germanium diode U_D and the parameters of the equivalent circuit (14).

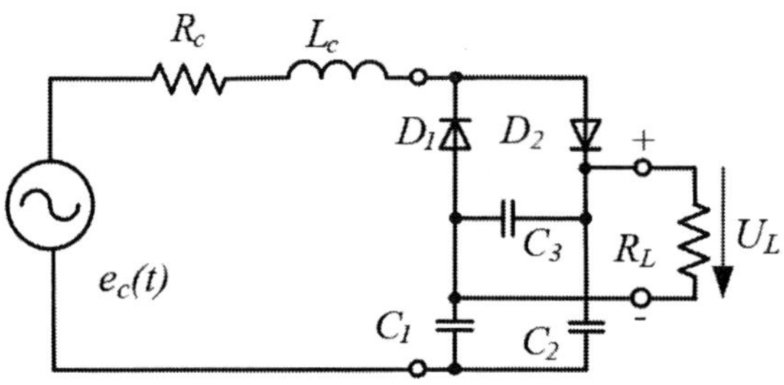

Figure 8.

$$P_L = \left(1{,}41\left(\frac{E_m R_L}{\sqrt{\left(\frac{R_C}{b}+R_L\right)^2+(\omega L_C)^2}} - U_D\right)\right)^2 \cdot \frac{1}{R_L} \tag{14}$$

In the obtained expression b is the coefficient of attenuation, defined by means of the logarithmic decrement of attenuation δ [5], where T is the period of oscillation and $E_m(t)$ - the amplitude of the measured e.m.f.

$$b = 2m\delta \quad , \tag{15}$$

$$\delta = \frac{1}{T}\ln\frac{E_m(t)}{E_m(t+T)} \tag{16}$$

3. Experimental studies

Figure 9 presents the characteristics of the relationship between the rectified voltage at idle run and the frequency of the small mass harvester (m_1=15 grams), while Figure10 gives them for a big mass (m_2=22 grams).

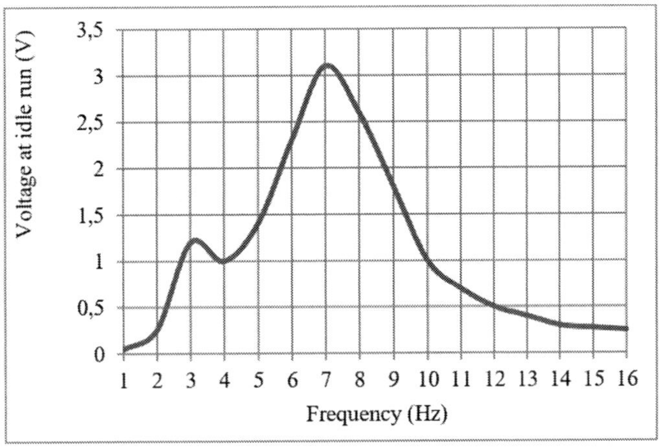

Figure 9. Rectified voltage at idle run as a function of the frequency of the small mass harvester

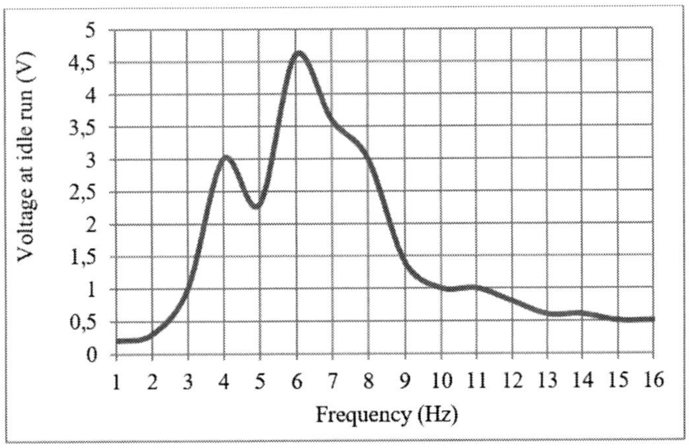

Figure 10. Rectified voltage at idle run as a function of the frequency of a big mass harvester

It can be seen from the characteristics that with the increase in the mass, the resonant frequency of the harvesters decreases, while the maximum voltage at idle run increases. Figure 11 shows the characteristics of both the measured and the obtained during modeling direct-current resistances of a small-mass harvester, while Figure 12 gives the same characteristic for the big-mass harvester at load resistances from 0 до 100 kΩ. The voltages of a big-mass harvester are three times greater than the ones of a small-mass harvester for the entire studied range of resistances.

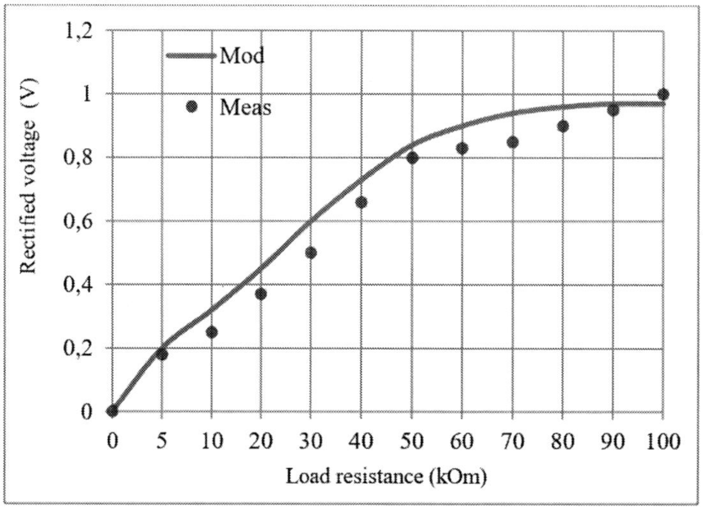

Figure 11. Rectified voltage as a function of the load resistance for a harvester with a small mass.

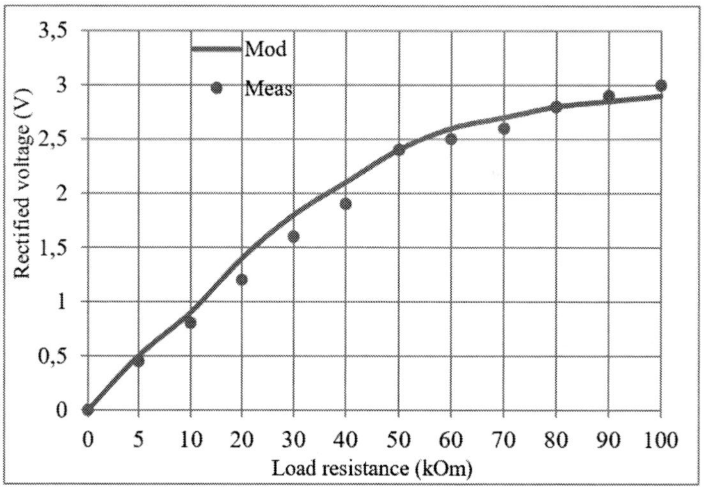

Figure 12. Rectified voltage as a function of the load resistance for a harvester with a big mass

Figure 13 shows the characteristics of the measured and the obtained by modeling direct-current powers for a small-mass harvester, while Figure 14 gives the characteristics for a harvester with a big mass. It can be seen from the characteristics that with the increase in the active load the output powers increase and reach a maximum at 50 kΩ. With the increase in the mass both the measured and the obtained by modeling powers considerably increase. The powers of a big-mass harvester are ten times greater than the ones, of a harvester with a small mass for the entire studied range of load resistances.

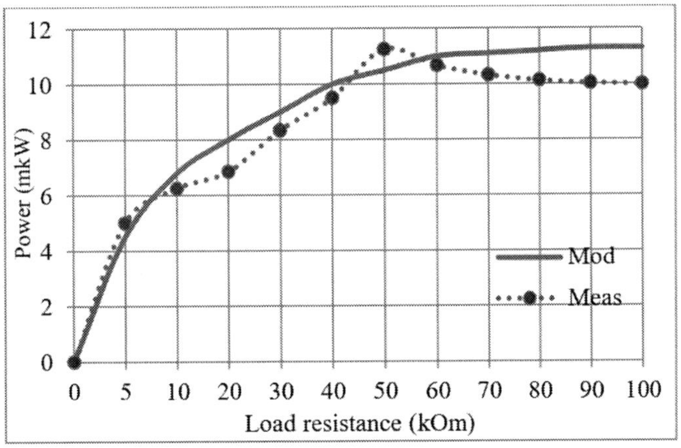

Figure 13. Active power in function of the load resistance of a harvester with a small mass

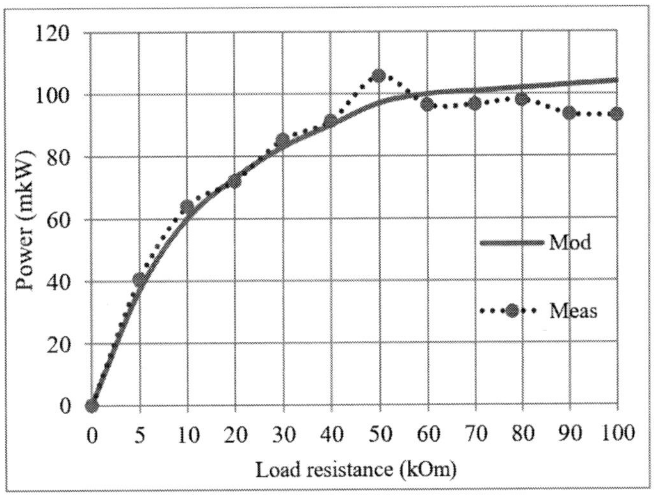

Figure 14. Active power in function of the load resistance of a harvester with a big mass

Table 1 presents the maximum voltages and powers for the corresponding resonant frequencies of the considered electromagnetic harvester with the two types of mass.

Table 1. Measured voltages, powers and resonant frequencies for a harvester at two types of mass

Mass (gr)	Maximum voltage (V)	Maximum power (μW)	Resonant frequency (Hz)
15	0.99	11.3	7
22	3.12	105.8	6

4. Conclusions

Theoretical derivations, programming simulations and experimental studies of an electromagnetic harvester with a mechanical system, consisting of a mass (Π-shaped plate with permanent magnets), two springs and a fixed coil, have been carried out. The harvester under consideration has two types of mass – a smaller and a bigger one.

Modeling of the two-spring electromagnetic harvester for the two types of mass has been done by ANSYS R19.1 and the maximum horizontal deviation X_m of the mechanical system has been obtained. The distribution of the magnetic field in the electromagnetic harvester has been defined by means of Femm 4.2. Modeling has been performed, by means of which an expression for the induced e.m.f. has been derived. An equivalent electric circuit of the electromagnetic harvester has been drawn and an expression for the direct-current power at active load has been found. From the frequency characteristics it can be seen that with the increase in the concentrated mass the resonant frequencies of the harvester decrease, while the rectified voltages increase. Both the measured and the obtained by modeling maximum powers increase with the increase in the mass. When the active load increases, the output powers increase and reach a maximum at 50 kΩ.

The voltages of the big-mass harvester are three times greater than the ones, of the harvester with a small mass for the entire studied range of load resistances. The powers of a big-mass harvester are ten times greater than the powers of the harvester with a small mass for the entire range of load resistances. Consequently, a harvester with a big mass has a considerably better output electrical parameters and is more appropriate for practical use.

Acknowledgements
The authors would like to thank the Research and Development Sector at the Technical University of Sofia for the financial support.

References

[1] Williams C, Shearwood M, Mellor P, Birch S and Yates R 2001 *IEE Proc. Circuits Devices Systems*. **148**

[2] Stephen N 2006 *Journal of Sound and Vibration*. **293**

[3] Saha C2011 *Score Project, School of Electrical & Electronic Engineering University of Nottingham, NG7 2RD*

[4] Zhu D 2011 *Sustainable Energy Harvesting Technologies - Past, Present and Future. INTECH*

[5] www.fs.vsb.cz/export/sites/fs/330/.content/files/technical_vibration_1.pdf

Study of two-spring piezoelectric harvesters

N P Georgiev[1] and R P Raichev[2]

[1]Department of Electrical Engineering Technical University of Sofia,
Branch Plovdiv, 25Tsanko Dystabanov Street, Plovdiv, Bulgaria
[2]Department of Mechanical Engineering Technical University of Sofia,
Branch Plovdiv, 25Tsanko Dystabanov Street, Plovdiv, Bulgaria

[1]nikola.georgiev@tu-plovdiv.bg
[2]rpraichev @tu-plovdiv.bg

Abstract. The paper considers two piezoelectric harvesters, consisting of two springs, attached to a piezoelectric transformer, and having a common concentrated mass. Modeling has been done by means of ANSYS, whereas the horizontal deviation of the mechanical system has been obtained, and the charge, accumulated on the electrodes of the piezo plate has been calculated. An equivalent circuit has been drawn and experimental studies have been conducted in order to validate the developed model.

1. Introduction

More and more frequently harvesters replace in practice batteries and accumulating power-supplying low-consumption electronic devices. [1]. The main advantage of these devices is that they do not require maintenance and can be used in large-scale infrastructure projects, where they convert mechanical vibrations into electrical energy [2]. Piezoelectric harvesters are among the most widely used type of them, as their construction is simple and reliable, on the one hand, and due to their small size and low price, on the other hand [3].

This paper studies and models two piezoelectric harvesters, consisting of piezoelectric transformers, rigidly fixed at their low end and having two springs with a common concentrated mass, fixed at the top. One of the harvesters has smaller springs and a smaller concentrated mass than the other. Modelling of the piezoelectric two-spring harvesters has been done by means of ANSYS R19.1 and experimental tests have been carried out to validate the models.

2. Exposition

Figure 1 presents the construction scheme of the two-spring piezoelectric harvester with the smaller springs, while Figure 2 illustrates the construction of the one with the bigger springs. The following notations are used in the schemes: 1–piezo plate; 2-brass base; 3-steel springs; 4-concentrated mass.

The piezo plates are disk-type, made of piezoelectric ceramics PZT 5A, with a diameter D_p=20mm and thickness T_p=0,3 mm; the brass base is rectangular in shape and measures 28x20x0,3 mm. The springs are two types. For the first harvester they are smaller, with wire diameter d_s=0,4 mm, spring diameter D_s=4,3 mm and length l_s=25 mm, while for the second harvester they are bigger, with wire diameter d_s=0,8 mm, spring diameter D_s=7,2 mm and length l_s=25 mm. The concentrated mass for the first harvester is made of steel and weighs 2, 4 and 6 grams, and for the second it weighs 3, 5, 7 and 10.5 grams, respectively.

IOP Conf. Series: Materials Science and Engineering **618** (2019) 012022 doi:10.1088/1757-899X/618/1/012022

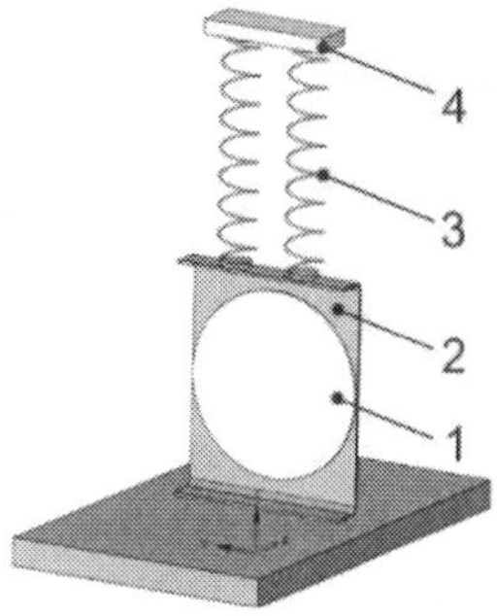

Figure 1. Two-spring piezoelectric harvester with small springs

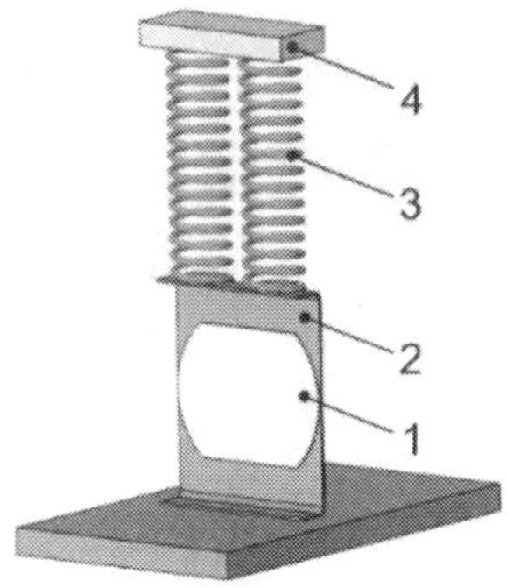

Figure 2. Two-spring piezoelectric harvester with big springs.

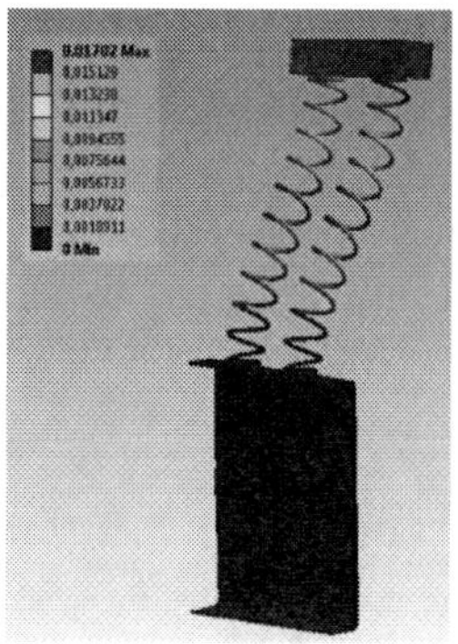

Figure 3. Simulation of the mechanical deviation for the small-spring harvester

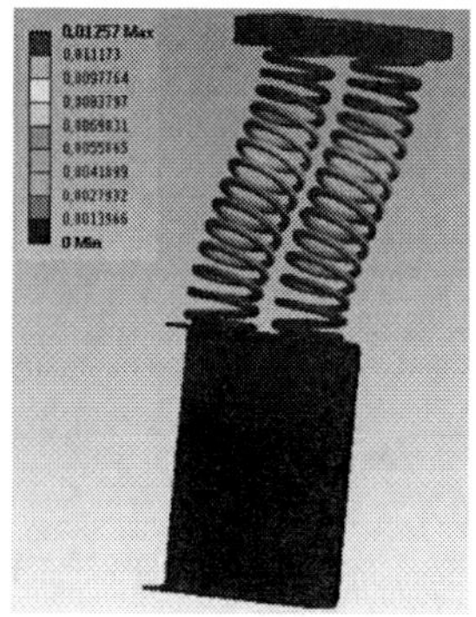

Figure 4. Simulation of the mechanical deviation for the big-spring harvester

Both studied harvesters are non-linear mechanical oscillating systems. During their simulation by means of ANSYS R19.1 the following parameters have been taken into account: the way of fixing, the influence of the force of weight of the concentrated mass, as well as the mechanical characteristics of the used springs, piezo plate and brass base. The horizontal deviation x of the mechanical system, consisting of the concentrated mass, the springs, and the piezo-plate, glued to a brass base, has been obtained and presented for the harvester with the smaller springs in Figure 3, for the harvester with the bigger springs - in Fig.4. The horizontal deviation of the piezo plate, obtained during the simulation, is given in Fig. 5, while Figure 6 shows the minimum, average and maximum horizontal deviations of the piezo plate, calculated by means of ANSYS for the two-spring harvester with the smaller springs.

IOP Conf. Series: Materials Science and Engineering **618** (2019) 012022 doi:10.1088/1757-899X/618/1/012022

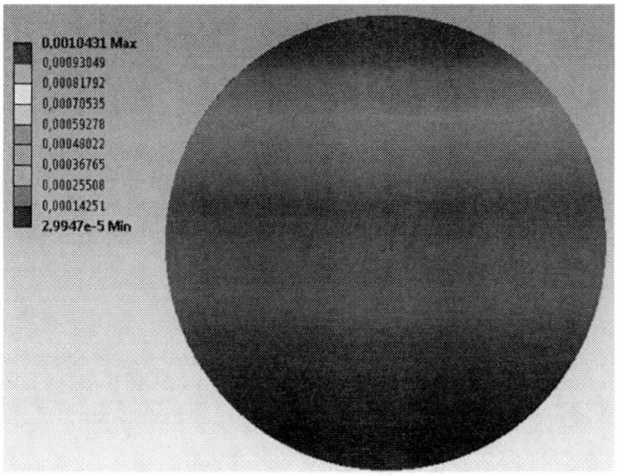

Figure 5. Simulation of the horizontal mechanical deviations of the piezo plate for the harvester with the smaller springs.

Tabular Data

	Time [s]	✔ Minimum [m]	✔ Maximum [m]	✔ Average [m]
1	1,	2,9947e-005	1,0431e-003	4,6797e-004

Figure 6. Horizontal deviations of the piezo plate for the harvester with the smaller springs.

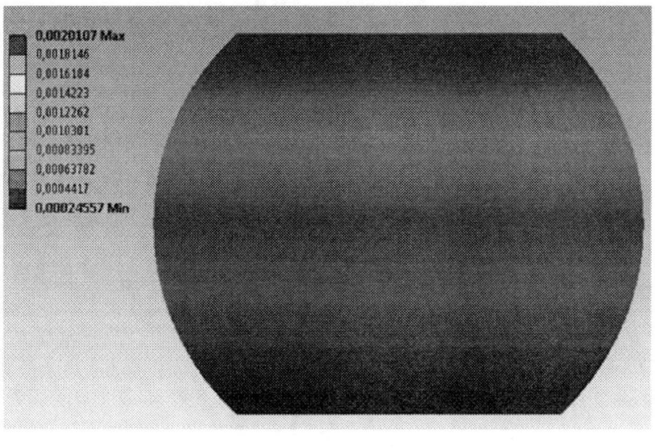

Figure 7. Simulation of the horizontal mechanical deviations of the piezo plate for the harvester with the bigger springs.

Tabular Data

	Time [s]	✔ Minimum [m]	✔ Maximum [m]	✔ Average [m]
1	1,	2,4557e-004	2,0107e-003	1,1098e-003

Figure 8. Horizontal deviations of the piezo plate for the harvester with the bigger springs.

Figure 7 shows the horizontal deviation of the piezo plate for the harvester with the big springs, obtained during the simulation, while Figure 8 illustrates the minimum, average and maximum deviations of the piezo plate for the same harvester, calculated by means of ANSYS.

The force, acting on the middle area of the piezo plate is proportional to the average horizontal deviation $x_{avr}(t)$, calculated by means of ANSYS, where r_p is the radius of the piezo plate, Y is the Young modulus and J is the moment of inertia [4]

$$F(t) = \frac{3YJ}{r_p^3} x_{avr}(t) \quad \textbf{(1)}$$

The average horizontal deviation $x_{avr}(t)$ is a variable, changing in a sinusoidal way, with an amplitude x_m and circular frequency ω

$$x_{avr}(t) = x_m \sin \omega t \quad \textbf{(2)}$$

With the help of the force $F(t)$, acting on the middle area of the piezo plate, the idle run voltage on the electrodes is defined [5], where g_{31} is the transverse voltage constant

$$u(t) = \frac{g_{31} F(t)}{r_p} \quad \textbf{(3)}$$

By means of (1), (2) and (3) the voltage at idle run of the piezo plate is found

$$u(t) = \frac{3g_{31}YJ}{r_p^4} x_m \sin \omega t \quad \textbf{(4)}$$

The amplitude of the voltage at idle run U_m can be calculated with the help of the amplitude of the average horizontal deviation x_m

$$U_m = \frac{3g_{31}YJ}{r_p^4} x_m \quad \textbf{(5)}$$

The maximum charge, accumulated on the electrodes of the piezo plate with capacity C_p, is equal to

$$Q_m = C_p U_m = \frac{3g_{31}YJC_p}{r_p^4} x_m \quad \textbf{(6)}$$

The equivalent electric circuit of the piezo plate is presented in Figure 9, where $\dot{J}_p$ denotes the source of electricity, proportional to the accumulated charge, C_p is the capacity of the piezo plate, and R_L is the active load, connected to the harvester.

$$\dot{J}_p = j\omega Q_m \quad \textbf{(7)}$$

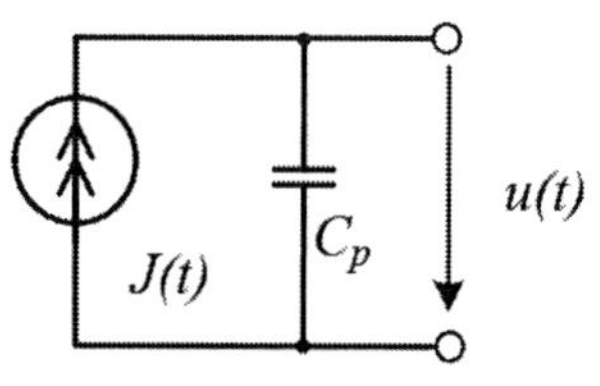

Figure 9.

Figure 10 presents the equivalent electric circuit of the piezoelectric harvesters with a voltage doubling amplifier at active load. By R_L here the active load resistance is denoted, and U is the rectified voltage.

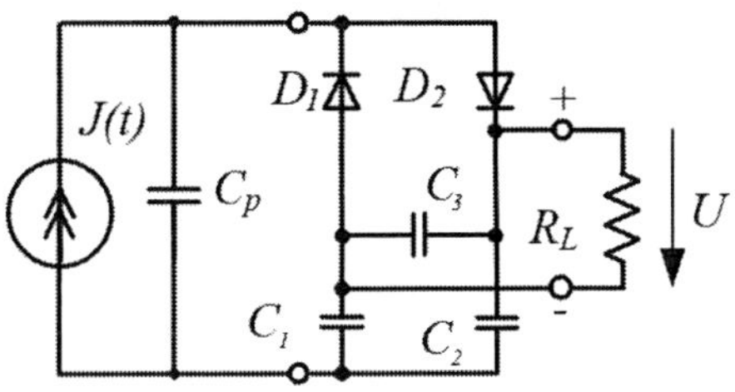

Figure 10.

The active power at direct-current mode is calculated with the help of the charge on the electrodes of the piezo plate Q, the voltage on the germanium diode U_D and the parameters of the equivalent circuit.

$$P_L = \left(1{,}41\left(\frac{\omega Q R_L}{\sqrt{1+\left(\omega C_p R_L\right)^2}} - U_D\right)\right)^2 \cdot \frac{1}{R_L} \qquad \textbf{(8)}$$

3. Experimental studies

Figure 11 presents the characteristics of the relationship between the rectified voltage at idle run and the frequency for the harvester with the smaller springs at concentrated masses of 2, 4, and 6 grams. It can be seen from the characteristics that with the increase in the concentrated mass the resonant frequencies of the harvesters decrease, and the maximum effective voltages at idle run also decrease. The harvester with a concentrated mass of 2 grams has the highest voltage and two resonant frequencies (of 18 and 6 Hz) are obtained for it.

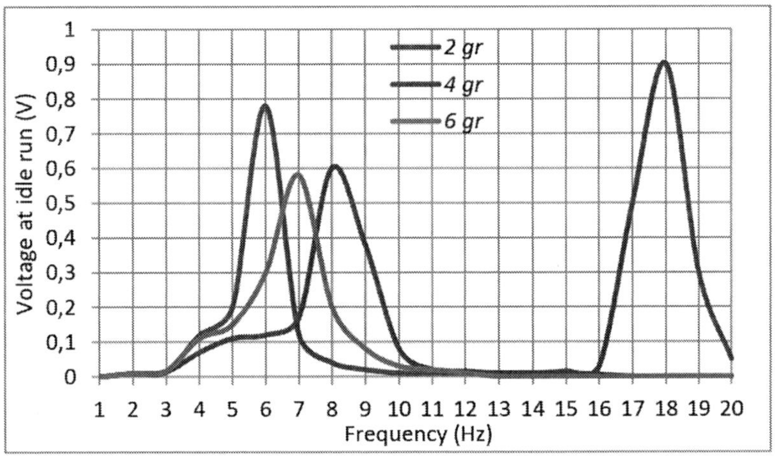

Figure 11.

Figure 12 presents the characteristics of the relationship between the rectified voltage at idle run and the frequency of the harvester with the bigger springs at a concentrated mass of 3, 5, 7 and 10.5grams, respectively. It can easily be seen from the graph that when the concentrated mass increases, the basic resonant frequencies, for which there are maximum voltages at idle run, decrease. For these frequencies together with the increase in the mass, the voltages at idle run go down. Second resonant frequencies appear for harvesters with big springs, which are lower than the basic ones and have lower voltages at idle run. A harvester with a concentrated mass of 3.5 grams at a frequency of 16 Hz has the highest voltage.

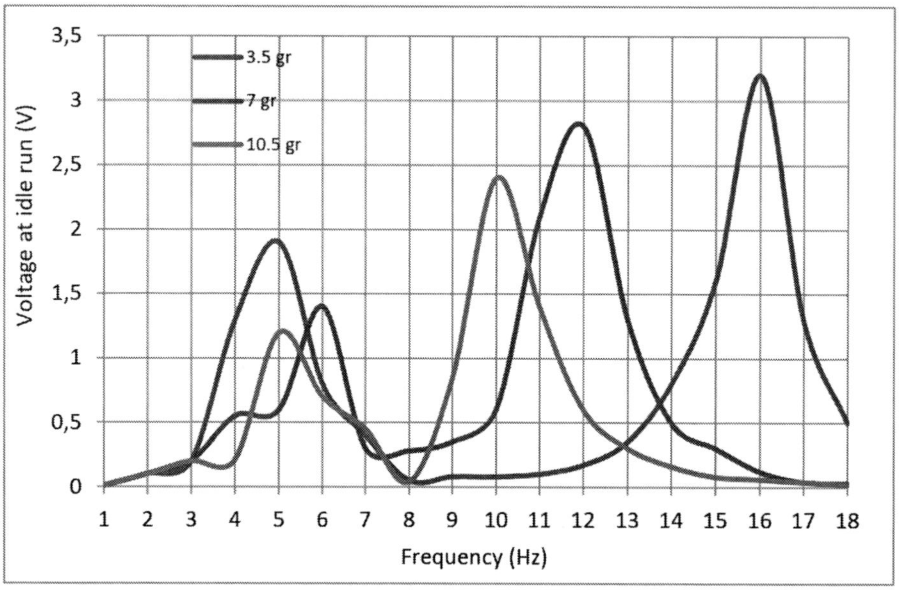

Figure 12.

Figure 13 shows the characteristics of both the measured and the calculated in the process of modeling direct-current powers for the harvester with the smaller springs at the three different weights of the concentrated mass: 2 grams and resonant frequency f=18 Hz; 4 grams and resonant frequency f=8 Hz; and 6 grams and resonant frequency f=7 Hz. It can be seen from the characteristics that together with the increase in the concentrated mass both the measured and the obtained by modeling powers decrease. When the active load increases, the output powers also go up.

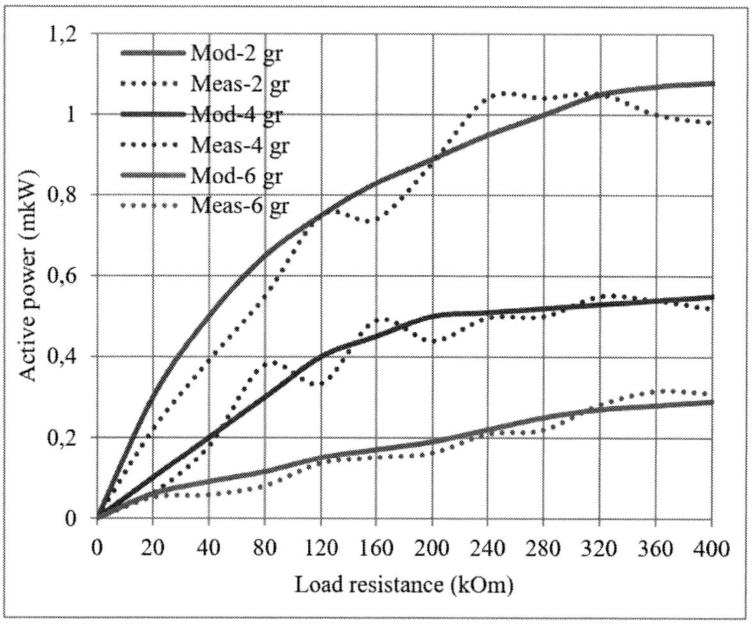

Figure 13.

The characteristics of both the measured and the obtained by modeling direct-current powers of the harvester with the bigger springs for a concentrated mass of 3.5 grams and resonant frequency *f*=16 Hz, for a mass of 7 grams and *f*=12 Hz, as well as for a mass of 10.5 grams and *f*=10 Hz, are shown in Figure 14. It can be seen that with the increase in the concentrated mass both the measured and the obtained by modeling powers decrease. When the active load increases, the output powers increase.

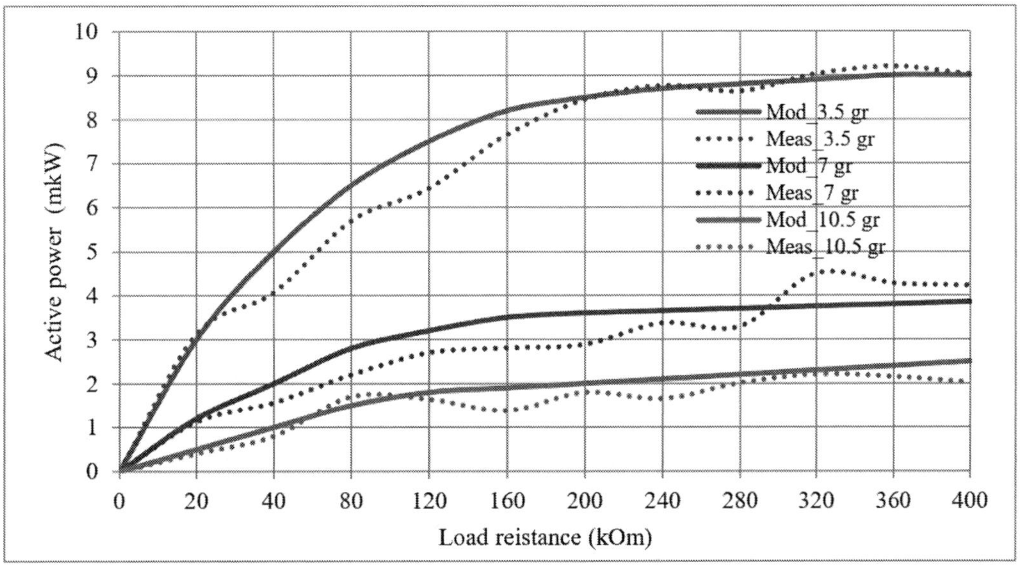

Figure 14.

Table 1 presents the measured maximum direct-current voltages and powers and the resonant frequencies for the harvesters with small springs at their three types of concentrated mass. A harvester with a concentrated mass of 2 grams has the highest measured voltage and power, while one with a mass of 6 grams has the lowest measured voltage and power.

Table 2 gives the measured maximum direct-current voltages and powers, as well as the resonant frequencies for the harvesters with big springs at their three types of concentrated mass. The highest measured voltage and power characterize a harvester with a concentrated mass of 3.5 grams, while the lowest voltage and power are measured for a harvester with a concentrated mass of 10.5 grams.

Table 1. Measured voltages, power and resonant frequencies for harvesters with small springs.

Mass (gr)	Maximum voltage (V)	Maximum power (μW)	Resonant frequency (Hz)
2	0.62	1.04	18
4	0.43	0.46	8
6	0.352	0.31	7

Table 2. Measured voltages, power and resonant frequencies for harvesters with big springs

Mass (gr)	Maximum voltage (V)	Maximum power (μW)	Resonant frequency (Hz)
3.5	1.91	9.57	16
7	1.32	4.49	12
10.5	0.91	2.21	10

4. Conclusions

Theoretical derivations, programming simulations and experimental studies of two harvesters, consisting of a transformer, two springs and a concentrated mass, have been done. One of the harvesters is with smaller springs and a smaller concentrated mass, the other is with bigger ones.

Modeling of the two-spring piezoelectric harvesters has been done by means of ANSYS R19.1 and the maximum horizontal deviations x_m of the mechanical systems have been found. An equivalent electric circuit of the piezo plate has been drawn and an expression for the maximum power at active load has been obtained. From the frequency characteristics it can be seen that with the increase in the concentrated mass the resonant frequencies of both types of harvesters decrease and so do the rectified voltages at idle run. Both the measured and the obtained by modeling maximum powers decrease together with the increase of the concentrated mass for the two types of harvesters. When the active load increases, the voltages and the powers increase for the two types of harvesters again.

The maximum voltages for the harvester with big springs are nine times higher than for the harvester with smaller springs. The powers of the harvester with big springs are nine times higher than the measured values for a harvester with small springs. Therefore a harvester with big springs is more appropriate for use in practice.

Acknowledgements
The authors would like to thank the Research and Development Sector at the Technical University of Sofia for the financial support.

References

[1] Kubba C and Jiang K 2014 *Sensors.* **14**

[2] Lu F, Lee H and Lim S 2004 *Smart Mater. Struct.* **13**

[3] Dhakara L, Liua H, Tayb F and Leea Ch 2013 *Sensors and Actuators*. **A199**

[4] https://www.fs.vsb.cz/export/sites/fs/330/.content/files/technical_vibration_2.pdf

[5] Zizys D 2018 *Doctoral Thesis Engineering Sciences, Mechanical Engineering. Kaunas university of technology*

Determining the parameters of induction motors by Genetic Algorithms

Stela Stoykova[1] and Vasil Spasov[2]

[1]Industrial Engineering speciality, Technical University - Sofia, Branch Plovdiv
[2]Department of Electrical Engineering, Technical University - Sofia, Branch Plovdiv

E-mail: vasilspasov@yahoo.com

Abstract. The paper presents an approach for determining the equivalent circuit parameters of two widely used induction motors types by Genetic Algorithms. The T-shaped equivalent circuit without considering the steel losses is utilized. Genetic Algorithms are applied and the results are compared by using one, two and three data sets. The accuracy of the proposed approach is analyzed by determining the relative error in the parameters, obtained by Genetic Algorithms, with regard to analytical values. The different computation times for one, two and three data sets are discussed. The results indicate that to obtain low relative error, the proposed approach using the conventional parameters settings and fitness function needs three points from the motors load curves.

1. Introduction

Induction motors are the most widely used in domestic, commercial and various industrial applications. Particularly, the squirrel cage type is characterized by its simplicity, robustness and low cost and for these reasons took the leading place in industrial sectors.

Two types of induction motors are analysed in this paper - the T80B4 type and the AO-90S-4 type. The former is manufactured by Elmatech Shareholders Company and the latter - by the M&C Electric Power CO. These motors are widely used in various industrial and household applications such as cooling fans and blowers, machine tools, die stamping, crushers, conveyors and compressors.

The necessity of knowing the equivalent circuit parameters of IM is constantly growing due to the following reasons:

- students should get deeper, up-to-date and accurate knowledge in the physical processes occurring in IM;
- increasingly adequate models of IM are needed for their research and improvement.

The conventional method for estimation of IM equivalent circuit parameters is based on the no-load and blocked rotor tests which are a time-consuming task, especially if the motor is already coupled to driving equipment.

This paper presents a more sophisticated approach for determining the equivalent circuit parameters of induction motors. The approach is based on only a few sets of measured or analytically obtained data such as stator voltage, current, power factor and slip. A genetic algorithm model for determining the parameters is synthesized that enables the simultaneous satisfaction of steady-state stator current and power factor for various operating points by defining an objective function. The

IOP Conf. Series: Materials Science and Engineering **618** (2019) 012023 doi:10.1088/1757-899X/618/1/012023

results obtained by the synthesized model are compared with analytical data. An analysis is performed that proves the validity and adequacy of the proposed approach.

2. Methods for IM equivalent circuit parameters estimation

The methods for IM equivalent circuit parameters estimation can be classified in two main groups - experimental and computational.

The classical experimental methods are a good alternative to the methods using nameplate data. They are performed by two tests – no-load test, blocked rotor test and measurement of the stator winding resistance [1]. The no-load test is used to determine the core loss resistance. The blocked rotor test enables to calculate the rotor resistance, the magnetizing reactance and the sum of the stator and rotor leakage reactances. By this approach, however, it is not possible to know how the leakage reactances are shared between the rotor and stator. This deteriorates the accuracy when predicting the dynamic performance of the motor. Moreover, in order to perform these tests in practice, several difficulties are faced. First, it is difficult to block the rotor when the motor is incorporated in a drive system. Second, the no-load test is often hard to perform since IM usually rotate with load such as fan or gear. Third, IEEE Standard 112 requires performing the motor tests with a voltage unbalance not exceeding 0.5% [1]. Field conditions, however, may exceed this limit significantly. Thus when evaluating motor performance in the field a more accurate and reliable approach is needed.

The modern experimental methods include all methods using tests different from the classical no-load and blocked rotor tests. Such methods include the use of transients in the motor equivalent circuit when supplied from direct voltage and/or direct current [2]. These methods have the following advantages - they are of short duration (only a few seconds) and the motor is not separated from the driving mechanism. Their disadvantage is the necessity of converter to have additional functions in order to perform the tests and to be provided with software to analyze the motor response to these tests. These functions are comparatively easy to realize. Recently electric drives appeared that perform auto adjustment by no-load and standstill tests.

An efficient modern experimental method is proposed in [3]. It determines the equivalent circuit parameters based on the recorded time variations of voltage, current, power and speed from start-up till no-load. The method is accurate but has several disadvantages. It needs expensive equipment to record the time variations of the above electrical and mechanical quantities. The method is intrusive since all loads should be decoupled from the motor during the test. Finally, it is applicable only to high-voltage induction machines rated 1 MW and above.

Some of the computational methods are based on the motor nameplate and catalog data. Three methods for determining the equivalent circuit parameters when taking into account the steel losses are described in [4]. The methods are based on several assumptions such as:

- equal leakage inductances of stator and rotor windings;
- zero value of the referred leakage inductance of rotor winding when determining copper losses;
- zero value of the referred leakage inductance of stator winding when determining steel losses, etc.

All three methods require knowing the rated supply voltage, stator current, rated power, power factor, rated speed and efficiency. It is also necessary to know the stator resistance which is easy to measure. The errors of the different parameters vary from $(4\div60)\%$ [4].

When catalog data for motors is available, it is easy to develop procedures for changing one type of equivalent circuit with another, as well as to relate the obtained results with synchronous speed, rated power, rated, breakdown and starting torque. One of the main disadvantages of the methods based on the motor nameplate and catalog data is the fact that the three most frequently used standards (NEMA, JEC, IEC) are not in agreement which may results in different efficiencies for a given motor. Secondly, the motor may have been rewound and the nameplate or catalog data may no longer be valid. On the other hand, the field voltage unbalance and harmonics content may be different from that for which the nameplate or catalog data is derived. In this way when estimating the equivalent circuit parameters a great percentage of statistical error may be introduced. Another problem is the fact that due to various reasons, most manufacturers usually do not publish detailed data about their production.

Other computational methods for equivalent circuit parameters estimation are the analytical methods using analytical expressions, developed decades ago [5]. They aim to obtain the steady-state performance of a motor for a given set of dimensions. The solutions by these methods are obtained very quickly, typically in seconds, on modern computers. The analytical methods, however, make quite a number of approximations, as IM operation involves 3D phenomena, saturation, eddy currents, etc. Some important details of geometry are also overlooked. These approximations deteriorate the accuracy of analytical methods.

The fast improvement of computer performance, combined with the development of the finite element method (FEM), lead to another important class of computational methods – the numerical methods. The numerical methods predict IM parameters using the magnetic field numerical solutions [6]. A number of professional software packages using FEM are now available that provide two or three dimensional magnetic field solutions. The 3D solutions are accurate but need long preprocessing and solution times. Therefore mostly 2D models of IM are analyzed. The 2D FEM analysis of IM yields reliable results, but has several disadvantages. First, the good software packages are commercial and expensive. Second, the finite element method requires detailed information about the stator and rotor geometry, number of turns, wire diameter, reluctivity curve of steel, etc. Third, it is necessary to compute analytically the stator end turn leakage reactance, the rotor end ring reactance and resistance [7].

The last class of computational methods are the methods based on genetic algorithms, applied in the present paper.

The survey of the methods for equivalent circuit parameters estimation shows that intrusiveness, cost and accuracy are the major considerations when selecting a method for determining IM parameters. Users prefer a cheap and low intrusive method providing a good accuracy.

3. Main operators of Genetic Algorithms

The performance of GA depends on three main operators [8]:

Selection - the process of choosing individuals in the population that will create fitter offspring. This is a method that randomly chooses chromosomes from the population according to their evaluation function. The higher the fitness, the bigger chance an individual has to be selected.

The selection process must balance the selection pressure and population diversity. The degree to which the better individuals are preferred is called selection pressure. It is one of the main factors affecting the improvement of population fitness over successive generations.

The most common types of selection are roulette wheel and tournament selection. The roulette wheel selection only guarantees that fitter individuals have greater opportunity to be selected. Their selection, however, is not guaranteed. The opportunity of selection is proportional to the individual's fitness. The roulette wheel is easy to implement but inconsistent in terms of promoted individual's fitness.

For this reason in the present analysis the tournament selection is used. Selective pressure is applied by holding a tournament competition among the individuals of the population. The winner of that tournament competition is the individual with the highest fitness thus consequently to be introduced to the mating pool. The tournament is repeated until the mating pool is filled. Since the mating pool is comprised of tournament winners, it has higher average fitness.

The tournament selection is very efficient and usually leads to an optimal solution.

Crossover - the process of taking two parent solutions and producing a child. Selection reproduces the fit individuals but does not create new ones. The crossover operator is applied to the mating pool. It aims at creating a better offspring through recombination of the best characteristics (genes) of the individuals in the population. Thus crossover actively supports the population diversity.

In the present work scattered crossover is preferred. A random binary vector called a mask is utilized. It selects the genes where the vector is 1 from the first parent and the genes where the vector is 0 from the other parent and introduces them into the offspring. A recombination of the characteristics of the parent individuals generates the child.

Mutation - it prevents the algorithm from being trapped in a local minimum. Mutation introduces new genetic structures in the population through random modification of some of its building blocks. GA is allowed to explore wider search space and genetic diversity is instituted in the future offspring.

Genetic algorithms have several advantages over the other optimization methods. They find the global minimum, instead of a local minimum. GA do not require the use of the derivative of the function, which is not easily obtained or may not even exist.

4. Implementation of the GA model and objective function definition

In order the natural selection to be used for estimating the parameters of induction motor equivalent circuit, an objective function should be defined. This function is based on the equations of the T-shaped equivalent circuit without considering the steel losses (figure 1).

There are five unknowns in the circuit in figure 1, namely: stator resistance R_1 and stator leakage reactance X_1, rotor resistance R_2 and rotor leakage reactance X_2 (both referred to stator), and magnetizing reactance X_m. The known quantities from measurement are the input voltage V_1 that equals the rated voltage, the input current I_1, the power factor $\cos\varphi$ and slip s.

Based on the T-shaped induction motor equivalent circuit in figure 1, the stator current can be calculated:

$$\dot{I}_1 = \frac{\dot{V}_1}{Z_{eq}} = \frac{\dot{V}_1}{R_{eq} + jX_{eq}}, \tag{1}$$

where Z_{eq} is the equivalent circuit impedance.

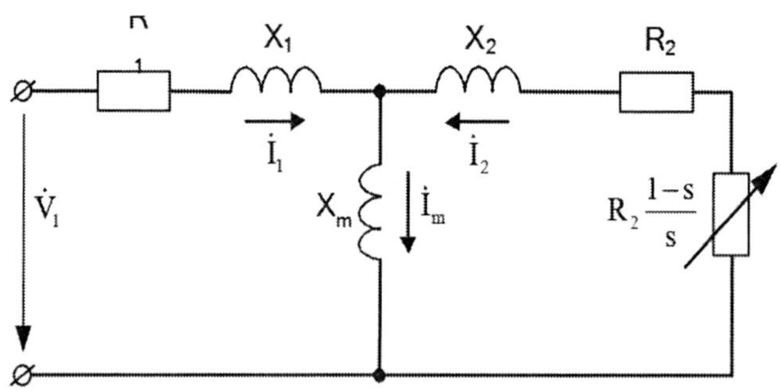

Figure 1. T-shaped equivalent circuit.

The equivalent circuit resistance R_{eq} is equal to:

$$R_{eq} = R_1 + \frac{\frac{X_m^2 R_2}{s}}{\left(\frac{R_2}{s}\right)^2 + (X_2 + X_m)^2}. \tag{2}$$

The equivalent circuit reactance X_{eq} is computed as follows:

$$X_{eq} = X_1 + \frac{\left(\frac{R_2}{s}\right)^2 X_m + X_2 X_m (X_2 + X_m)}{\left(\frac{R_2}{s}\right)^2 + (X_2 + X_m)^2}. \tag{3}$$

Now the power factor can be calculated:

$$\cos\varphi = \cos\left(\tan^{-1}\frac{X_{eq}}{R_{eq}}\right). \tag{4}$$

Based on formulae (1) and (4), we shall develop the objective function [9]:

$$F_{obj} = \sum_{i=1}^{n}\left(\frac{\cos\varphi_{c,i}}{\cos\varphi_{m,i}} - 1\right)^2 + \sum_{i=1}^{n}\left(\frac{I_{1c,i}}{I_{1m,i}} - 1\right)^2. \tag{5}$$

In (5) $I_{1c,i}$ and $\cos\varphi_{c,i}$ are the values computed by (1) and (4). $I_{1m,i}$ and $\cos\varphi_{m,i}$ are the measured values [10]. The variable n varies from 1 to 3 in our case.

The analysis of (5) shows that the developed objective function reflects the error in the measured motor data and the results obtained with GA optimization. This means that the lower the error in the computed and the measured data, the smaller the values of the objective function. Thus, the fitness of the individuals rises because the objective function is reciprocal of the fitness function.

The aim of GA is to minimize the error of the objective function defined by (5). In the present work the value of F_{obj} is set to zero and in this way the global minimum of the objective function is found.

5. IM equivalent circuit parameters estimation

The equivalent circuit parameters of a T80B-4 and an AO-90S-4 type squirrel cage induction motors are estimated. The motors have 0.75 kW and 1,5 kW output power, respectively. The following nameplate data applies to both motors: 380 V phase-to-phase voltage, 50 Hz frequency and 2 poles.

The obtained GA results for the above mentioned motors are compared to the analytical data sets shown in Table 1 and Table 2.

Table 1. Data sets used in GA for the AO-90S-4 motor.

Stator current [A]	Slip	Power factor
2.58	0.06	0.63
3.36	0.10	0.72
4.09	0.14	0.74

Table 2. Data sets used in GA for the T80B-4 motor.

Stator current [A]	Slip	Power factor
2.06	0.06	0.59
2.56	0.10	0.70
3.07	0.14	0.74

The performance of the GA optimization greatly depends on the proper choice of the GA parameters such as selection, crossover, mutation, population size, generations, etc. According to the parameters setup, different results are obtained [11].

Their accuracy is measured by the relative error:

$$\varepsilon = (X_{GA} - X_{an}) / X_{an}. \tag{6}$$

In (6) X_{GA} is the value of the parameter computed by the GA optimization and X_{an} is the analytical value of the same parameter.

The population size value is significant for GA performance. If the population size is too small, accurate solutions may not be reached. A too large population size requires additional computation time and renders the configuration inefficient.

The default setting for population size is 20 to 100 individuals [11]. In order to guarantee good accuracy, however, population size higher than the default (500) is chosen, although it leads to slightly higher computation time. Otherwise the population would lack diversity, the algorithm would explore only a small part of the search space and not find global optimal solutions.

The conventional configuration of the GA parameters is given in Table 3.

Table 3. Conventional Genetic Algorithm parameters.

Population size	500
Selection function	Tournament
Tournament size	4
Elite count	2
Mutation function	Adaptive Feasible
Crossover function	Scattered
Crossover fraction	0.8

It must be noted that an elite count is set up. Elitism guarantees that the first best chromosome or the few best chromosomes (two in our case as evident from the elite count) are copied to the new population.

These settings are chosen by analogy, namely using past experience that has proved successful for similar problems [11, 12].

Tables 4 and 5 show the equivalent circuit parameters estimated by the genetic algorithm as well as the relative error in the estimated parameters with regard to the analytical values.

Table 4. Equivalent circuit parameters and relative error for the AO-90S-4 motor.

Parameter		R_1	R_2	X_1	X_2	X_m
Analytical value [Ω]		8,12	5,26	22,42	1,05	91,95
1 data set	GA [Ω]	42,17	6,11	51,64	31,97	45,13
	ε [%]	419,33	16,16	130,33	294,47	-50,91
2 data sets	GA [Ω]	8,38	5,30	22,50	1,054	91,51
	ε [%]	3,1	0,76	0,39	0,38	-0,47
3 data sets	GA [Ω]	8,21	5,24	22,37	1,048	91,76
	ε [%]	1,01	-0,35	-0,21	-0,22	-0,21

Due to the random nature of GA, every computed parameter value in Tables 4 and 5 is an average of the best values from GA obtained in 10 runs. The results in Tables 4 and 5 show that the proposed GA approach is very sensitive to the number of the data sets used.

Table 5. Equivalent circuit parameters and relative error for the T80B-4 motor.

Parameter		R_1	R_2	X_1	X_2	X_m
Analytical value [Ω]		12,20	7,33	26,37	1,46	103,53
1 data set	GA [Ω]	63,11	8,23	51,79	8,13	52,76
	ε [%]	417,29	12,28	96,40	456,84	-49,04
2 data sets	GA [Ω]	12,60	7,29	29,61	1,47	102,78
	ε [%]	3,24	-0,51	0,89	0,89	-0,72
3 data sets	GA [Ω]	12,36	7,31	26,49	1,47	103,27
	ε [%]	1,3	-0,23	0,45	0,45	-0,25

When one data set is used, there is a great discrepancy between analytical and estimated values. In this case the maximum error exceeds 400%, because one single point can not define the nonlinear IM current-slip curve accurately.

The use of two data sets greatly improves the accuracy, the maximum relative error being slightly more than 3%.

With three data sets the maximum relative error is within 1%. Thus, it can be concluded that when more data sets are used, the accuracy of the results is enhanced.

The average computation (CPU) time for one data set is 11 minutes, for 2 data sets - 8 and for three sets - 5 minutes. The computations are performed on a quad core Intel Pentium Processor N3710, 1.6 GHz and 4GB RAM.

The greater the number of points from the motor load curves we use for the GA, the lower the CPU time. This is due to the fact that the GA has more reference points on which to base probable solutions.

The settings of the GA have a strong effect on the computation time. One of the main factors, impacting the CPU time in GA, is the population size. The greater the population size, however, the more CPU time is required.

The choice of the fitness function also greatly impacts the computation time and accuracy. This choice will be discussed in another paper.

6. Conclusion

A genetic algorithm approach for estimating the equivalent circuit parameters of two types of induction motors is proposed. The approach is based on a widely used fitness function and the conventional parameter settings of genetic algorithms.

When one data set is used, there is a great discrepancy between analytical and estimated values, because one single point can not uniquely define the nonlinear IM current-slip curve.

The increase in the number of data sets used improves the accuracy and decreases the CPU time. To obtain relative error within 1%, three sets of electrical input data (voltage, current, power factor and slip) of the motor are needed.

The proposed approach has several advantages over the similar methods for estimating the equivalent circuit parameters of induction motors. The approach is simple, accurate, faster, cheaper and less intrusive than the other experimental or computational methods.

In our future work we shall propose ways for reducing the number of required data sets while preserving the accuracy of the proposed approach.

Acknowledgements

The authors would like to thank the Research and Development Sector at the Technical University of Sofia for the financial support.

References

[1] IEEE Standard Test Procedure for Polyphase Induction Motors and Generators 1996 IEEE Std. 112-1996 IEEE Power Engineering Society (New York)

[2] Barrass P 2005 Automatic characterization of induction motors for optimum performance *TechOPS*

[3] Babau R and Boldea I 2001 Parameter identification for large induction machines using direct online startup test *Workshop on Electrical Machines Parameters* (Technical University of Cluj-Napoca, Romania) pp 47-52

[4] Ozyurt C and Ertan B 2005 Prediction of induction motor parameters from manufacturer's data *XIV-th International Symposium on Electrical Apparatus and Technologies SIELA 2005* (Plovdiv, Bulgaria) vol II pp 96-105

[5] Boldea I and Nasar S 2002 *The induction machine handbook* (CRC Press)

[6] Zhou P et al. 1998 Finite element analysis of induction motors based on computing detailed

equivalent circuit parameters *IEEE Transactions on Magnetics* vol 34 no 5 pp 3499-3502

[7] Ertan B et al. 2003 Identification of parameters of an induction motor from field solution *The Sixth International Conference on Electrical Machines and Systems ICEMS 2003* vol 1, pp 1-6

[8] Petrowski A and Ben-Hamida S 2017 *Computer Engineering Series: Evolutionary Algorithms* vol 9 (Wiley-ISTE)

[9] Spasov V, Rangelova V, Kostov I and Drambalov V 2014 An efficient approach for determining induction motors parameters *Acta Technika Corviniensis - Bulletin of Engineering* Tome VII pp 117-122

[10] Kostov I and Rangelova V 2010 Power factor determination of induction motor frequency controlled drives *International Journal of Engineering* (Faculty of Engineering Hunedoara) Tome VIII issue 2 pp 67-72

[11] Kramer O 2017 *Studies in Computational Intelligence* vol 679: *Genetic Algorithm Essentials* (Springer International Publishing)

[12] Iba H 2018 *Evolutionary approach to Machine Learning and Deep Neural Networks: Neuro-Evolution and Gene Regulatory Networks* (Springer Singapore)

IOP Conf. Series: Materials Science and Engineering **618** (2019) 012024 doi:10.1088/1757-899X/618/1/012024

Choice of fitness functions and parameter settings in Genetic Algorithms for analysis of induction motors

Stela Stoykova[1] and Vasil Spasov[2]

[1]Industrial Engineering speciality, Technical University - Sofia, Branch Plovdiv
[2]Department of Electrical Engineering, Technical University - Sofia, Branch Plovdiv

E-mail: vasilspasov@yahoo.com

Abstract. The paper proposes methods for enhancing the accuracy and speed of Genetic Algorithms when determining the parameters of induction motors. A systematic study of the impact of genetic operators and algorithm parameters on optimization process performance is done. New fitness functions are developed containing additional quantities such as the input power of motors. Various genetic operator types and constraints are analysed providing an insight into a specific settings choice. Special attention is given to the factors generating stochastic noise and the ways to eliminate it. To evaluate the effectiveness of the proposed fitness functions and parameter settings, ten combinatorial optimization experiments are conducted on two types of induction motors. The results show that adequate fitness functions, combined with proper genetic operators setups, can significantly enhance the accuracy and speed of Genetic Algorithms while reducing the required input data.

1. Introduction

The notion of Genetic Algorithms (GA) optimization dates back to the principles of natural evolution introduced by Charles Darwin in his work The Origin of Species. The foundation of most modern GA optimization approaches is the basic principle of survival of the fittest in biological evolution.

The smallest operational unit of a GA is a chromosome or an individual that consists of a set of binary parameters called genes. The aim of GA is to optimize the individual solutions guided by a fitness function. Typically Genetic Algorithms incorporate the following genetic operators: initialization, selection, recombination, mutation and replacement [1]. A successive iteration through these processes allows for better fitness of individual solutions to be attained.

Initialization most commonly occurs with an initial population of randomly generated solutions. The optimization of solutions is performed in generations. A set of solutions is selected based on their fitness in each generation. The selected solutions are then modified to form a new population of fitter solutions. The new population is used in the next iteration. Termination of the algorithm occurs when either a maximum number of generations is reached, or a satisfactory fitness value is attained [2].

This paper presents a systematic study of the influence of the genetic operators and the algorithm parameters (fitness function and population size) on the optimization process accuracy and speed. For this purpose two new fitness functions are developed and various combinations of genetic operators and algorithm parameters are compared. The comparison is based on the following criteria: accuracy, computation time and convergence.

To verify the efficacy of the proposed new fitness functions and parameter settings, two types of induction motors are analysed in the paper. The results show that the new fitness functions, implemented with appropriate parameter settings, greatly improve the accuracy and speed of the analysis. In addition, less input data is needed as compared to the conventional GA optimization.

2. GA model and development of new fitness functions

To determine the parameters of induction motors, the T-shaped equivalent circuit is used [3, 4]. Two squirrel cage induction motors are analysed- the T80B4 type and the AO-90S-4 type. These motors are widely used in various industrial and household applications. Details of the motors and the nameplate data are given in [3].

The fitness function is an evaluation function that is used to summarize how close a given solution is to achieving the set aim. The fitness not only indicates how good the solution is, but also shows how close the chromosome is to the optimal one.

Since the fitness function is always problem dependent, its appropriate design for the specific aim is a major concern. Problems occurring due to inappropriately designed fitness function include difficulty in algorithm convergence or finding local and not global solutions.

The fitness function has to be concise and fastly computable. Since a typical Genetic Algorithm must be iterated many times in order to produce a viable solution, the execution speed is very important. In our case of multiobjective optimization the fitness function is even more difficult to determine as it should seek optimal solutions corresponding to all objectives.

A widely used fitness function (called also objective function) for determining the parameters of induction motors is [3, 5]:

$$F_{obj} = \sum_{i=1}^{n}\left(\frac{\cos\varphi_{c,i}}{\cos\varphi_{m,i}} - 1\right)^2 + \sum_{i=1}^{n}\left(\frac{I_{1c,i}}{I_{1m,i}} - 1\right)^2 . \tag{1}$$

In (1) $I_{1c,i}$ and $\cos\varphi_{c,i}$ are the computed values of current and power factor. $I_{1m,i}$ and $\cos\varphi_{m,i}$ are the measured values [6] and n designates the number of electrical input data sets of the motor.

The aim of GA is to minimize the error of F_{obj}. In the present work the value of the fitness function is set to zero and in this way its global minimum is found. To obtain relative error within 1%, the conventional fitness function (1) requires at least three data sets, i.e. three points from the motors load curves. With two data sets the maximum relative error is about 3% [3].

To improve the accuracy and reduce the number of required data sets, two new fitness functions are developed in this paper. The first developed fitness function FF1 is the reciprocal of the sum of modules of the type:

$$FF_{s1} = \frac{1}{F_1^2 + F_2^2} . \tag{2}$$

For each data set a new module of the type FF_{s1} is added.

F_1 and F_2 in (2) are functions of the computed and measured stator current and input power:

$$F_1 = \frac{I_{1c} - I_{1m}}{I_{1m}}; \; F_2 = \frac{P_{1c} - P_{1m}}{P_{1m}} . \tag{3}$$

The second developed fitness function FF2 is the reciprocal of the sum of modules of the type:

$$FF_{s2} = \frac{1}{F_1^2 + F_2^2 + F_3^2} . \tag{4}$$

For each data set a new module of the type FF_{s2} is added.

F_3 in (4) is defined as a function of the power factor:

$$F_3 = (\cos\varphi_c - \cos\varphi_m) / \cos\varphi_m . \tag{5}$$

It should be noticed that the module F_3 in the second fitness function FF2 supplies a supporting data input that balances out the selection-produced stochastic noise (see Section 4). The new data points, obtained by the inclusion of F_3, facilitate more accurate determination of globally optimal solutions.

3. Choice of parameter settings and convergence of GA

3.1 Initial population size

The quality of a solution is directly dependent on the initial population size [7]. A large population size guarantees higher diversity in the population and exploration of a more encompassing search space. Thus Genetic Algorithms have a greater opportunity to converge to a global optimum.

Large initial population sizes, however, require a considerable amount of computation time for evaluation of the fitness of the possible solutions. Thus in selecting appropriate population size, a trade-off must exist between efficient exploration of the search space and the computation resources available [7].

If the information for resuming a run of GA has been saved for a specific set of parameters, the initial population of a new run can be replaced by the final population of the previously saved run. This approach allows search to be localized in ranges of the search space that are known to produce optimal solutions. It, however, limits the range of possible solutions and can cause premature convergence to a locally optimal solution instead of a globally optimal one.

The presented GA setting combinations, detailed in Section 4, use randomly generated initial population to allow for diversity of possible solutions.

3.2 Selection and crossover

Four types of selection are used in the present study: tournament, roulette wheel, stochastic uniform sampling and stochastic remainder. Next their main characteristics, advantages and disadvantages are discussed.

Tournament selection has several benefits: it is efficient to code, applicable to parallel architectures and allows easy adjustment of the selection pressure through modification of tournament size. It, however, requires considerably more computational time than other selection techniques. GA achieve low error rates but are slow to converge on optimal solutions.

Roulette-wheel (or fitness proportionate) *selection* allows for weaker solutions to survive the selection process. This, however, can be considered an advantage, as though a solution may be weak, it may include some genes which could prove useful following the recombination process. The lack of guarantee that weaker solutions will be eliminated in this selection process generates stochastic noise and optimal solutions for some ranges of the search space may not be attained.

Stochastic uniform sampling is similar to roulette-wheel selection as individuals are assigned contiguous segments of a line, where the size of the segment corresponds to the individuals fitness. Pointers are placed over the line at equal distances from each other. The number of individuals to be selected equals the number of pointers.

The number m of individuals to be selected determines the distance between the pointers as 1/m. The position of the first pointer is a randomly generated number in the range [0, 1/m].

While roulette-wheel selection chooses solutions from the population by repeated random sampling, stochastic uniform sampling utilizes a single random value for sampling of all the solutions by choosing them at evenly spaced intervals. This approach ensures that stochastic uniform sampling produces solutions closer to the globally optimal and eliminates stochastic noise.

This selection technique is applicable to any number of selections but is preferably used in situations where more than one sample is required.

Stochastic remainder selection chooses chromosomes from the population based on their relative fitness [8]. The relative fitness λ of a chromosome is the ratio of the fitness associated with the individual fitness f_i and the average fitness f_{avg} of the population:

$$\lambda = f_i / f_{avg} . \qquad (6)$$

In the first stage of the selection solutions are chosen deterministically based on the integer part of λ. In the second stage the free places in the mating pool are filled up by the roulette-wheel selection on the remaining fractional part. If the value of λ for an individual is 1,4 that individual is listed once in the mating pool because the integer part is 1. After all the individuals have been assessed and assigned based on the integer part of their relative fitness, the roulette-wheel selection commences. The probability that an individual is chosen in this stage is proportional to the fractional part of λ.

Two types of crossover (recombination) are implemented in the present study: scattered and constraint dependent.

Scattered crossover uses a random binary vector. It selects the genes where the vector is 1 from the first parent and the genes where the vector is 0 from the other parent and introduces them into the offspring. A recombination of the characteristics of the parent individuals generates the child. This approach produces better results when applied to problems without linear constraints.

As shown in Section 4, scattered crossover produces low error rates with minimal computational time when combined with roulette-wheel selection. Due to the scattered crossover principle of generating offspring as a mixture of the genes of parent solutions, the stochastic noise effect of roulette selection is minimized.

Constraint dependent crossover is the default function when solving linear constraint problems. It generates offspring by taking a weighted average of the parents [9].

3.3 Mutation and constraints

The mutation operator is critical to the success of genetic algorithms since it determines the search directions and avoids convergence to local optima. A gene is selected randomly from the chromosome and undergoes a change to diversify the solution pool. Two types of constraint tolerant mutation operators are implemented in this study- adaptive feasible and constraint dependent.

Adaptive feasible mutation is the default with constraint bound problems. The mechanism of operation is based on randomly generated directions adaptive with respect to the last successful or unsuccessful generation [10]. Problem specific linear constraints set the direction and mutation probability.

Mutation probability controls the ratio of new individuals to the population size. Generally, it determines the rate of occurrence of mutation with consideration of the above-mentioned ratio.

This algorithm parameter allows for individuals with high fitness that are lost during optimization to be returned in the population. With too high mutation rate some good properties of the individuals are lost due to frequent random changes to the chromosomes. Too low mutation rates cause a decrease in the population diversity and possibility of converging on local optimum is high. The default mutation rate of 0,1 is used in the presented GA combinations.

Constraint dependent mutation is applied in our study in a combination with non-problem specific constraints: function and tolerance constraint.

Function constraint can be utilized in customizing the range of the stopping criteria of the algorithm with respect to the relative change in the best fitness function value over the stall generations.

Tolerance constraint is not implemented as a stopping criterion. It determines the general feasibility of the solutions generated by the algorithm in the last iterations with respect to linear constraints and exacts a change in direction if necessary.

3.4 Convergence of GA

The main disadvantage of most optimization algorithms is that after they find a local minimum, they are trapped and do not converge further to a better globally optimal solution. Genetic Algorithms possess the ability to locate and converge to globally optimal solutions if the appropriate parameter settings are applied.

To assess the ability of the algorithm to provide globally optimal solutions, the coefficient of variation C_V is introduced for every genetic algorithm parameter combination at the end of each run. The coefficient is obtained by the formula $C_V = \sigma/\overline{x}$, where σ is the standard deviation of the given parameter and $\overline{x}$ is its mean value [11, 12]. The mean value and σ are computed as follows:

$$\bar{x} = \frac{1}{n}\sum_{i=1}^{n} X_i\,; \quad \sigma = \sqrt{\frac{\sum_{i=1}^{n}(X_i - \bar{x})^2}{(n-1)}}\,. \tag{7}$$

Here X_i is the set of end configurations for the run and n is the number of runs.

An average value of C_V for each combination is obtained over multiple runs. Higher C_V means more diversity in individual genes, hence poor convergence. C_V is not affected by gaps in parameter ranges at boundary conditions since the affected σ and $\bar{x}$ cancel each other out in the ratio.

4. Results and discussion

The equivalent circuit parameters of the T80B-4 and the AO-90S-4 type squirrel cage induction motors are estimated using the new fitness functions and various combinations of genetic operators discussed in Section 3. The obtained GA results for the two motor types are compared to the analytical data sets shown in Table 1. The accuracy is measured by the relative error:

$$\varepsilon = (X_{GA} - X_{an}) / X_{an}\,. \tag{8}$$

In (8) X_{GA} is the value of the parameter computed by the GA optimization and X_{an} is the analytical value of the same parameter from Table 1.

Table 1. Data sets used in GA for the motors.

Motor type	Stator current [A]	Slip	Power factor
AO-90S-4	2,58	0,06	0,63
	3,36	0,10	0,72
T80B-4	2,06	0,06	0,59
	2,56	0,10	0,70

The two new fitness functions FF1 and FF2 are applied to the induction motors with different parameter settings. Ten combinatorial optimization experiments, designated as GA1 to GA10, are presented in Table 2.

Table 2 shows the induction motors equivalent circuit parameters estimated by GA as well as the relative errors with regard to the analytical values. The relative errors are for tests with two data sets, shown in Table 1.

The induction motors parameters are: stator resistance R_1 and stator leakage reactance X_1, rotor resistance R_2 and leakage reactance X_2 (both referred to stator), and magnetizing reactance X_m.

First the new fitness functions are discussed. Proof of the positive effect of appropriately chosen auxiliary data modules in the fitness functions on stochastic noise occurrences is the performance of GA7. Even though this GA parameter combination utilizes roulette wheel selection that produces stochastic noise, it yields one of the lowest coefficients of variation (figure 1). Thus GA7 has high convergence, high speed performance, i.e. low computation (CPU) time and low errors (Table 2). This is due to the use of the fitness function FF2 with the balancing auxiliary data module F_3.

The computations in this paper are performed on a quad core Intel Pentium Processor N3710, 1.6 GHz and 4GB RAM.

To evaluate the effect of the population size, simulations with various population sizes are carried out (figure 2). Population size selection is problem specific and as a result fitness function dependent. The combinations used (GA1 and GA6) correspond to the two fitness functions FF1 and FF2.

A scale of the best fitness of solutions is used. The actual fitness values for GA are not given and the results are scaled by a factor of 1E6. Best fitness near zero indicates poor performance whereas best fitness of one indicates excellent convergence.

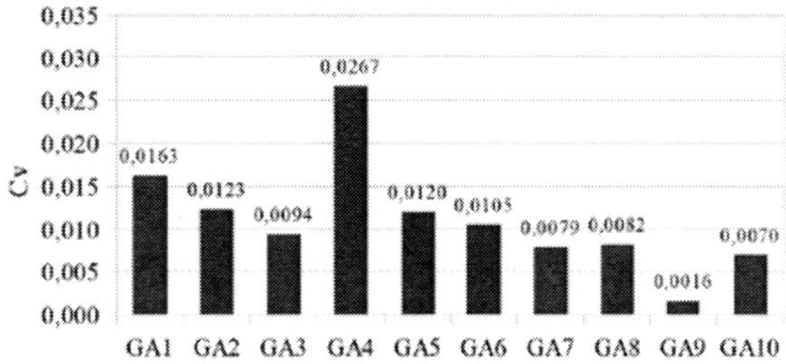

Figure 1. Values of the coefficient of variation.

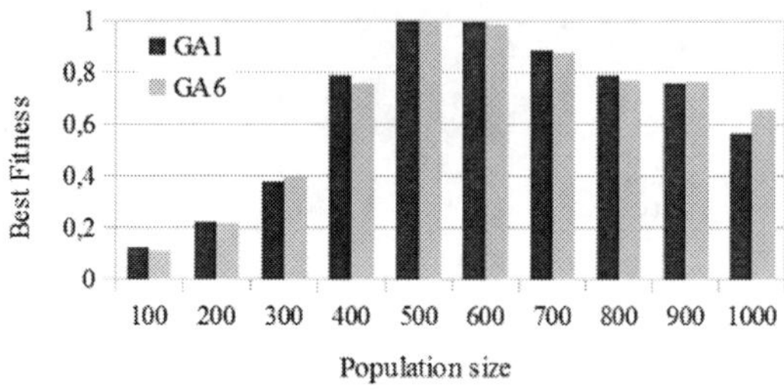

Figure 2. Effect of the population size.

Figure 2 shows that standard population sizes of 250 and below result in premature termination of the GA. Thus optimal solutions are not reached for the given fitness functions FF1 and FF2. Initial population sizes in the range of 350 to 750 result in termination due to the satisfactory value of fitness function attained.

An indication of good exploration of the search space is the high success rate of reaching the global optimum with population sizes above 500. Population sizes larger than 600 do not yield noticeable improvement of the solution. Hence, a mean value of population size 550 that allows converging to an optimal solution without excessive computation time is used further in this paper.

Next the parameter settings and convergence of GA are analysed. GA2 and GA5 that implement tournament selection have the highest CPU times (Table 2). Combinations GA2 and GA5 achieve low error rates but are slow to converge.

The roulette-wheel selection in GA3 does not eliminate weaker solutions. As a result this selection generates stochastic noise (figure 3). Hence the search space may not be fully explored.

There is an obvious difference in the effectiveness of GA3 and GA7, both of which use roulette-wheel selection. GA3 uses FF1 and GA7 uses FF2 with the auxiliary module F3. It is evident that GA7 outperforms GA3. GA7 converges faster (figure 1), has significantly lower CPU time and is more accurate (Table 2). Figure 4 shows the convergence history of GA7 which lacks stochastic noise due to the auxiliary input data in F3.

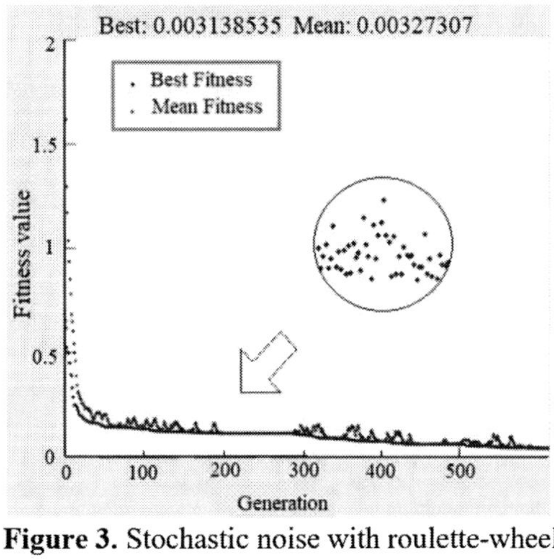

Figure 3. Stochastic noise with roulette-wheel.

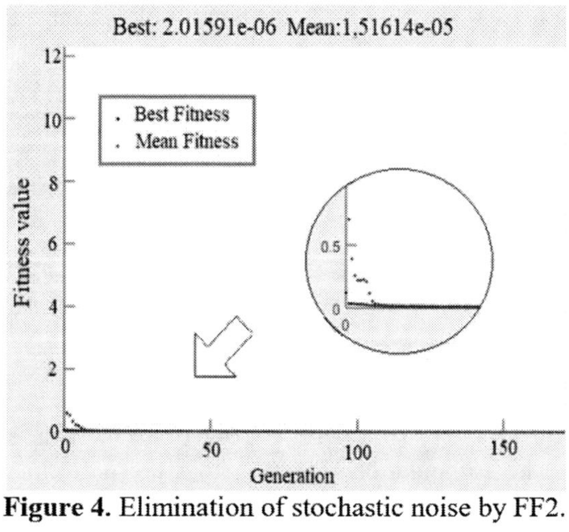

Figure 4. Elimination of stochastic noise by FF2.

The comparison of GA1 to GA3, utilizing FF1, and GA6 to GA7, based on FF2, shows that GA with stochastic uniform sampling have slower rate of convergence than those with roulette-wheel. The final error assessment, however, indicates that the stochastic uniform sampling based GA produce considerably lower errors than the roulette-wheel based GA.

Table 2. Combinatorial optimization experiments for the two induction motors.

		GA1	GA2	GA3	GA4	GA5	GA6	GA7	GA8	GA9	GA10
Motor type	AO-90S-4	+		+		+		+			+
	T80B-4		+		+		+		+	+	
Fitness function	FF1	+	+	+	+						+
	FF2					+	+	+	+	+	
Selection	Tournament		+			+					
	Roulette			+				+			
	Stochastic uniform	+					+			+	+
	Remainder				+				+		
Crossover	Scattered			+	+			+	+		
	Constraint dependent	+	+			+	+			+	+
Mutation	Constraint dependent			+	+	+			+		
	Adaptive feasible	+	+				+	+		+	+
Function constraint		1E-6	1E-6	1E-6	1E-6	1E-6	1E-6	1E-6	1E-6	1E-8	1E-8
Tolerance constraint		1E-6	1E-6	1E-6	1E-6	1E-6	1E-6	1E-6	1E-6	1E-7	1E-7
Relative error [%] of:	R_1	0,06	0,06	0,24	-0,18	0,81	0,14	0,22	-0,06	0,05	0,12
	R_2	0,34	-0,34	-0,16	-0,67	-0,01	-0,14	-0,04	-0,01	0,05	0,03
	X_1	-0,01	0,09	0,04	0,06	0,09	-0,02	0,05	-0,05	0,01	-0,004
	X_2	-0,02	0,09	0,04	0,06	0,08	-0,02	0,04	-0,05	0,01	0,04
	X_m	-0,02	0,01	0,07	0,03	-0,10	-0,001	-0,03	0,01	-0,003	-0,001
CPU time [s]		201	369	234	236	337	215	202	250	281	299

The stochastic remainder selection in GA8 results in moderate speed, good accuracy and fast convergence.

As evident from GA9 and GA10, constraint dependent crossover achieves outstanding results in linear constraint problems. Both algorithms exhibit fast convergence and very good accuracy.

The combination of scattered crossover and stochastic remainder selection in GA4 slows down the convergence rate (Figure 1), though the CPU time is hardly affected and the accuracy remains good (Table 2). This is due to the deterministic nature of the two approaches.

When combined with roulette-wheel selection, scattered crossover in GA3 and GA7 is very fast and produces low errors (Table 2). Since scattered crossover generates offspring solutions as a mixture of the genes of parent solutions, the stochastic noise effect of roulette-wheel selection is eliminated.

GA9 and GA10 use constraint dependent mutation combined with function and tolerance constraint. Since double precision of data is default in MATLAB, the allocation of constraints is in the range of [1E-7, 1E-8]. The constraints added to GA9 and GA10 result in high rates of convergence (figure 1) and minimal errors with an insignificant increase in the CPU time (Table 2).

5. Conclusion

The paper analyzes the choice of fitness functions and parameter settings for enhancing the accuracy and speed of Genetic Algorithms. To validate the analysis, the equivalent circuit parameters of two induction motors are evaluated.

Two new fitness functions (FF1 and FF2) are developed and compared. Ten combinatorial optimization experiments are conducted including various genetic operator types and constraints. To assess the ability of Genetic Algorithms to provide globally optimal solutions, the coefficient of variation is introduced for every combination.

It is found out that the inclusion of auxiliary data modules in FF2 balances out the selection-produced stochastic noise. The impact of the population size on the convergence is analyzed. It is shown that a mean value of population size 550 allows converging to an optimal solution without excessive computation time.

Four types of selection are compared: tournament, roulette-wheel, stochastic uniform sampling and stochastic remainder. The accuracy assessment indicates that the stochastic uniform sampling produces considerably lower errors than the other selection types.

Two types of crossover are implemented: scattered and constraint dependent. The constraint dependent crossover achieves outstanding results. When combined with roulette-wheel selection, scattered crossover is fast, produces low errors and eliminates the stochastic noise.

Two mutation operators are implemented: adaptive feasible and constraint dependent mutation. The constraint dependent mutation results in high rates of convergence and minimal errors with only a slight increase in the computation time.

To obtain relative error within 1%, the conventional fitness function and parameter settings require at least three data sets, i.e. three points from the motors load curves. With two data sets the maximum relative error is about 3%. In contrast, the proposed new fitness functions and GA parameter settings enable to obtain relative error less than 1% with only two data sets.

Based on the results it can be concluded, that adequate fitness functions, combined with proper genetic operators setups, can significantly enhance the accuracy and speed of Genetic Algorithms. In addition, the number of required input data sets can be reduced.

Acknowledgements
The authors would like to thank the Research and Development Sector at the Technical University of Sofia for the financial support.

References

[1] Wagner S et al. 2018 *Genetic Algorithms and Genetic Programming: Modern Concepts and Practical Applications* (Chapman and Hall / CRC)
[2] Haupt R and Haupt S 2004 *Practical genetic algorithms* (Wiley & Sons)
[3] Stoykova S and Spasov V 2019 Determining the parameters of induction motors by Genetic AlgorIthms *8th International Scientific Conference TECHSYS 2019* (Technical University – Sofia, Plovdiv branch) (to be published)
[4] Ozyurt C and Ertan B 2005 Prediction of induction motor parameters from manufacturer's data *XIV-th International Symposium on Electrical Apparatus and Technologies SIELA 2005* (Plovdiv, Bulgaria) vol II pp 96-105
[5] Spasov V, Rangelova V, Kostov I and Drambalov V 2014 An efficient approach for determining induction motors parameters *Acta Technika Corviniensis - Bulletin of Engineering* Tome VII pp 117-122
[6] Kostov I and Rangelova V 2010 Power factor determination of induction motor frequency controlled drives *International Journal of Engineering* (Faculty of Engineering Hunedoara) Tome VIII issue 2 pp 67-72
[7] Lobo F et al. 2007 *Parameter setting in evolutionary algorithms* (Springer-Verlag)
[8] Kramer O 2017 *Studies in computational intelligence* vol 679: *Genetic algorithm essentials* (Springer International Publishing)
[9] Petrowski A and Ben-Hamida S 2017 *Computer Engineering Series: Evolutionary algorithms* vol 9 (Wiley-ISTE)
[10] Merjalili S 2018 *Evolutionary algorithms and Neural Networks: Theory and Applications*
[11] Iba H 2018 *Evolutionary approach to Machine Learning and Deep Neural Networks: Neuro-Evolution and Gene Regulatory Networks* (Springer Singapore)
[12] Mikhailov K and Sclocco A 2018 The Apertif monitor for bursts encountered in real-time auto-tuning optimization with genetic algorithms *Astronomy and Computing* vol 25 pp 139-148

IOP Conf. Series: Materials Science and Engineering **618** (2019) 012025 doi:10.1088/1757-899X/618/1/012025

Analysis of the influence of some factors on the contact resistance of bolted busbar connections

Ivan Hadzhiev[1], Dian Malamov[2], Iosko Balabozov[3] and Ivan Yatchev[3]

[1]Department of Electrical Engineering, Technical University of Sofia, Plovdiv Branch, 25 Tsanko Dyustabanov Str., 4000 Plovdiv, Bulgaria
[2]Department of Communications and Power Engineering, University of Plovdiv Paisii Hilendarski, 24 Tzar Asen Str., 4000 Plovdiv, Bulgaria
[3]Department of Electrical Apparatus, Technical University of Sofia, 8 Kliment Ohridski Blvd, 1000 Sofia, Bulgaria

E-mail: hadzhiev_tu@abv.bg, deanmalamov@abv.bg, i.balabozov@tu-sofia.bg, yatchev@tu-sofia.bg

Abstract. The paper presents an experimental setup and a method for determining the contact resistance between busbars at a given contact force. Experiments have been conducted by means of an experimental setup, developed for the purpose of studying the influence of the contact force on the contact resistance between busbars with cross-sectional area of 60x6 mm. The study has been performed for two types of bolted connections, complying with DIN 43673, which have been studied for both copper and aluminum busbars.

1. Introduction

A typical functional element in low voltage switchboards are the busbar systems, characterized by their complex configurations and relationships. This results in a significant number of contact connections in these systems. The presence of contact connections is a prerequisite for increase in the resistance, as well as for overheating of the electrical equipment. It is known [1], [2] that the contact resistance depends on complex physical processes, big part of which random in character, and therefore its analytic calculation is related to big inaccuracies. Papers [3] and [4] propose formulas, in which the influence of micro and macro geometry and contact surface hardness have been taken into account and expressed by means of empirical coefficients. The range of change of these coefficients is quite big, which additionally leads to insufficient accuracy of the obtained results. Besides, the proposed formulas do not reflect any specific constructive peculiarities of the contact surfaces. The analysis of planar contact connections, taking into account geometric and physical characteristics of the contact surfaces, proposed in [5], [6] and [7], is of great interest. The method, proposed here, is used for analysis of planar contact connections with specific construction, which is not possible by the methods, given in [1], [2], [3] and [4]. Despite the progress in theoretical research, however, the experimental studies of contact resistance are still valid at this stage.

This paper presents a study of the influence of the contact force, the material of the busbars and the construction of the contact connection between the busbars on the contact resistance, realized by means of a specially developed experimental setup.

IOP Conf. Series: Materials Science and Engineering **618** (2019) 012025 doi:10.1088/1757-899X/618/1/012025

2. Description of the method of studying and the experimental setup

A comparative method was chosen for studying the contact resistance between the busbars. It consists in comparing the resistance in sections of the busbar without and with a contact connection, as shown in Fig. 1. The sections have the same length. Under this condition:

$$R_{12} = R_0 \tag{1}$$

$$R_{23} = R_0 + R_K \tag{2}$$

$$R_K = R_{23} - R_{12} \tag{3}$$

where:

R_0 - resistance between the points 1 and 2, measured in ohms;

R_K - contact resistance;

R_{23} - resistance in the section between the points 2 and 3.

The resistance can be defined by the voltage drops and the current through the busbar:

$$R_K = \frac{U_{23} - U_{12}}{I} \tag{4}$$

where:

U_{12} - voltage drop in the section between the points 1 and 2;

U_{23} - voltage drop in the section between the points 2 and 3.

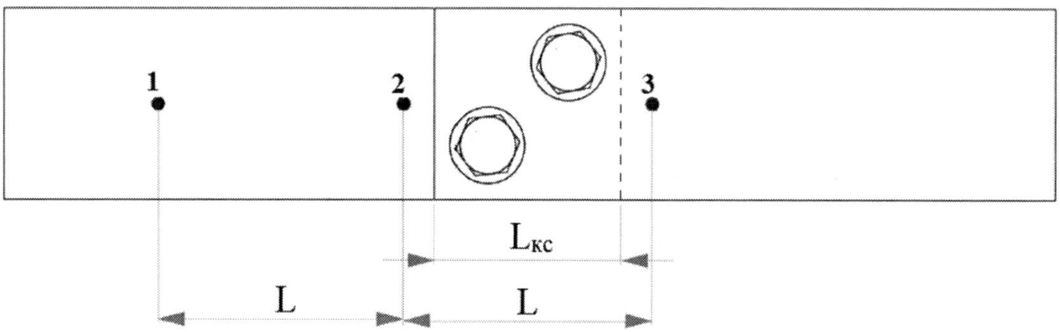

Figure 1. Location of the points 1, 2 and 3 for determining the contact resistance by the comparative method.

The experimental setup is presented in Fig. 2. The testing current is set by means of both an autotransformer and a load transformer. Voltmeters with high input resistance are used to measure the voltage drops in the sections mentioned above.

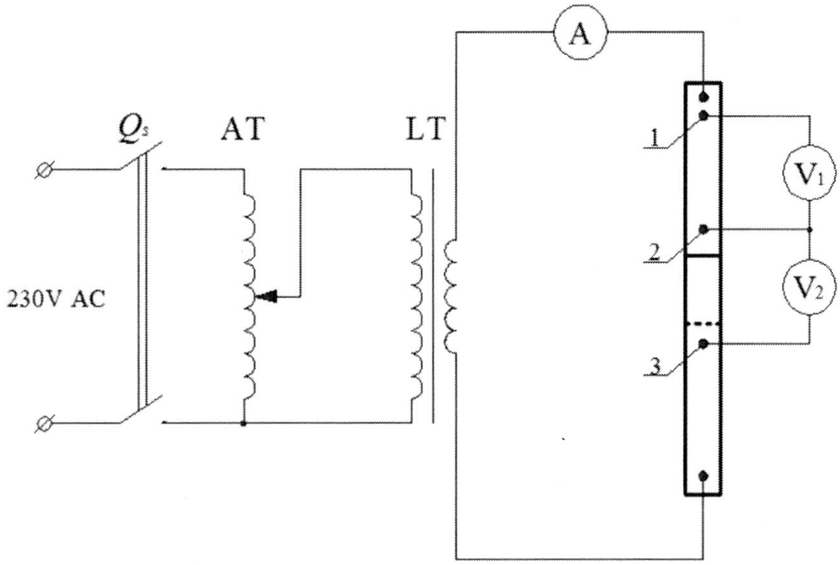

Figure 2. Electric circuit of the experimental setup: *Qs* – circuit breaker; AT – autotransformer; LT – load transformer; A – ammeter; V_1, V_2 – voltmeters.

The contact force is set by means of a screw drive. Fig. 3 presents the device for measuring the force. It is done by measuring the pressure of the hydraulic fluid. The construction of the hydraulic system is shown in Fig. 3. When the piston is loaded by means of a given force, the force is converted into pressure on the liquid, from which the value of the force is defined. The compressive force for the developed system can be determined by the following formula:

$$F[kg] = p.S = k.p \tag{5}$$

where:
k – constant of the measuring device;
p – pressure.

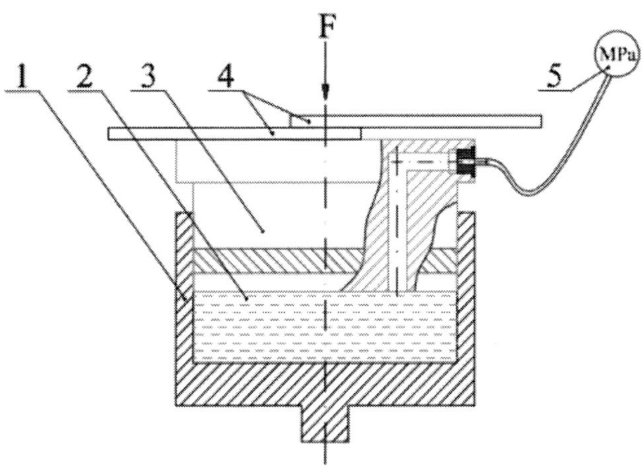

Figure 3. Device for setting the contact force: 1 - cylinder; 2 – hydraulic oil; 3 - piston; 4 - busbars; 5 - manometer.

Fig. 4 presents a picture of the experimental setup for studying the contact resistance.

Figure 4. Appearance of the experimental setup.

3. Construction of the studied contact connections

Two types of contact connections between busbars without additional treatment of the contact surfaces were the object of the presented study (as shown in Fig. 5 and Fig. 6).

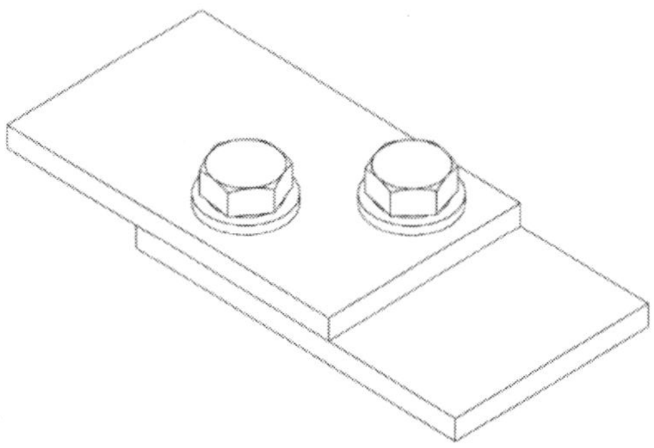

Figure 5. Contact connection between busbars (60x6) mm in accordance with [8].

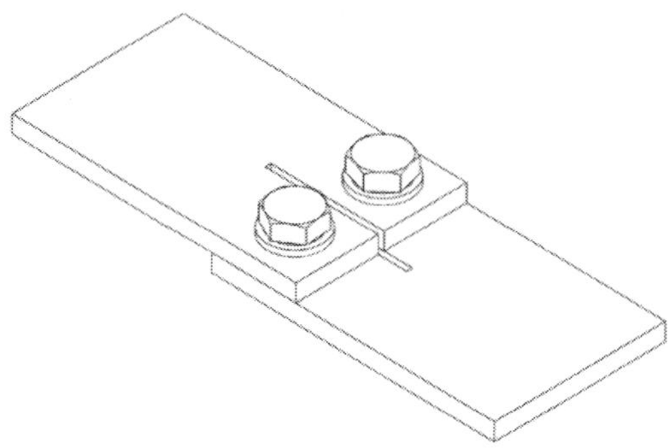

Figure 6. Contact connection between busbars (60x6) mm in accordance with [8].

All tests for measuring the contact resistance were conducted for the following cases:
- contact connection type 1 as in Fig. 5;
- contact connection type 2 as in Fig. 6.

The contact connections were studied for both copper and aluminum busbars, measuring (60x6) mm.

4. Experimental results

By means of the experimental setup, described above, studies of the dependence between the contact resistance and the contact force, the construction of the contact connection and the material of the contact connection were realized. Four samples were produced for each type of a contact connection. Twelve independent experiments for measuring the contact resistance were conducted for each sample.

4.1. Influence of the contact force on the contact resistance

Fig. 7÷10 illustrate the change in the contact resistance depending on the contact force for both copper and aluminum busbars for the two types of contact connections (type 1 and 2).

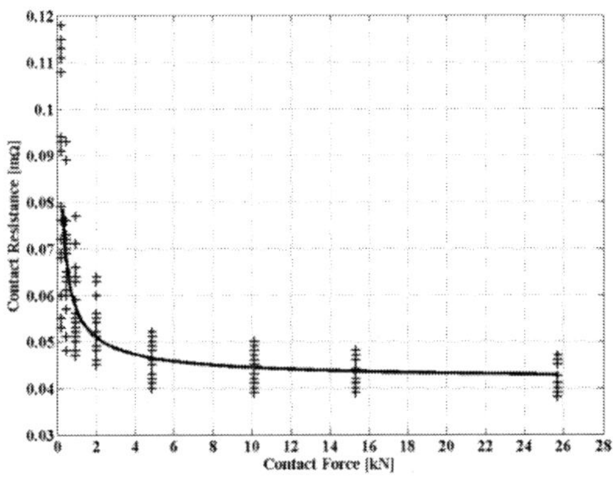

Figure 7. Experimental data and a curve, describing the dependence of the contact resistance on the contact force for type 1 bolted connection between copper busbars (60x6) mm.

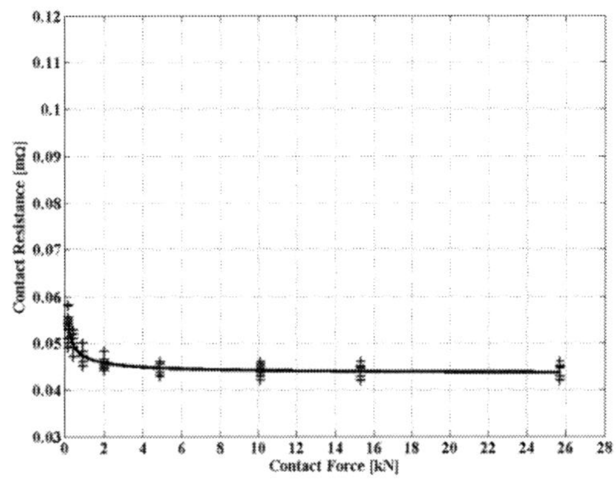

Figure 8. Experimental data and a curve, describing the dependence of the contact resistance on the contact force for type 2 bolted connection between copper busbars (60x6) mm.

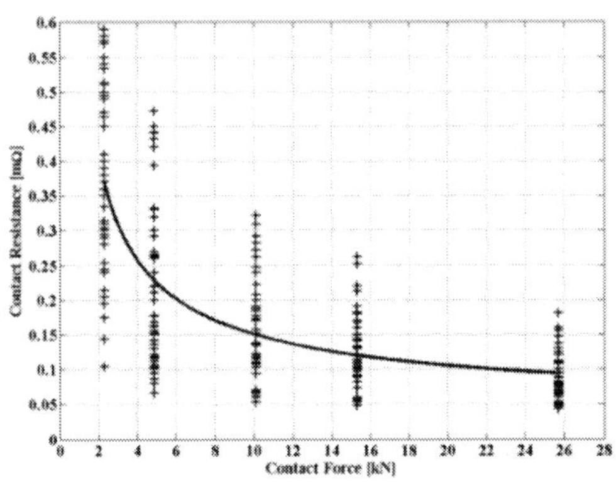

Figure 9. Experimental data and a curve, describing the dependence of the contact resistance on the contact force for type 1 bolted connection between aluminum busbars (60x6) mm.

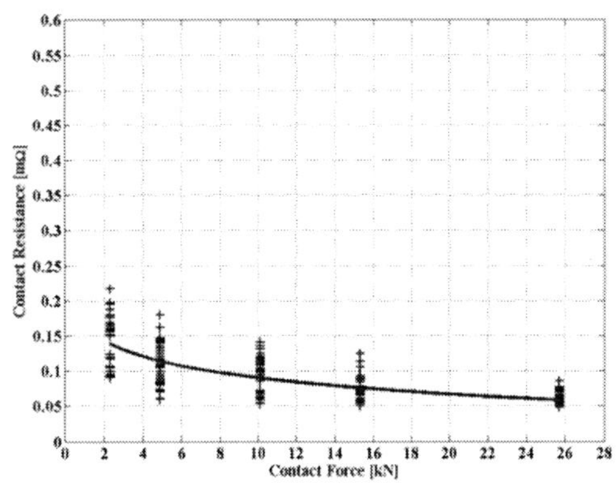

Figure 10. Experimental data and a curve, describing the dependence of the contact resistance on the contact force for type 2 bolted connection between aluminum busbars (60x6) mm.

4.2. Influence of the contact connection construction on the contact resistance

Fig. 11 and Fig. 12 draw a comparison between the contact resistance for type 1 and type 2 construction for copper and aluminum busbars.

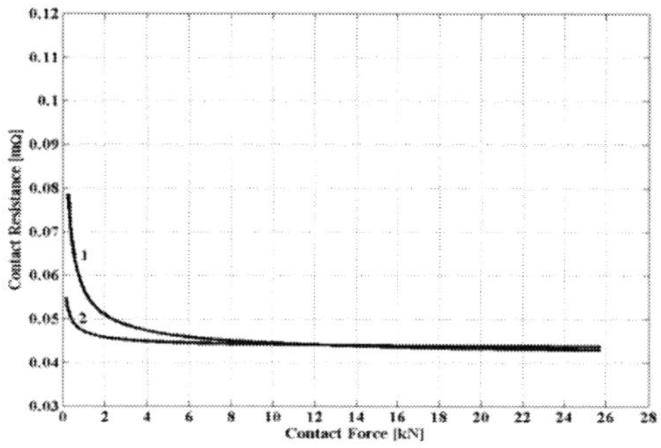

Figure 11. Dependence of the contact resistance on the force for type 1 connection construction (curve 1) and type 2 (curve 2) in case of using copper busbars.

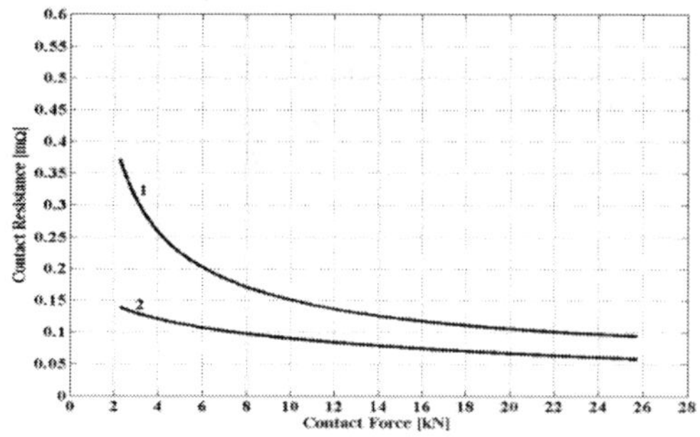

Figure 12. Dependence of the contact resistance on the force for type 1 connection construction (curve 1) and type 2 (curve 2) in case of using aluminum busbars.

5. Conclusion

The following conclusions can be drawn up based on the obtained results:

- When considering contact connections between copper busbars, the average values of the contact resistance for contact force of over 5 kN are practically the same for both types (type 1 and type 2) of contact connections. When the contact force is below 5 kN, the average value of the contact resistance for type 1 increases by 20 %, compared to the value for type 2.
- In case of bolted connections between aluminum busbars the average value of the contact resistance for type 2 is less than the one for type 1. This difference is quite significant at a minimum value of the contact force (about three times) and it decreases at a maximum force (about 30%).
- The dissipation of the contact resistance values for type 2 connections at a low force is significantly less than it is for type 1.

Acknowledgments

The authors would like to thank the Research and Development Sector at the Technical University of Sofia for the financial support. The work of I. Hadzhiev is supported by National program "Young scientists and Postdoctoral candidates" of Ministry of Education and Science.

References

[1] Alexandrov Al and Yatchev I 2010 *Electrical apparatus – part II,* Avangard Prima Sofia (in Bulgarian).
[2] Tzeneva R 2004 *Electric contact* Sofia (in Bulgarian).
[3] Dzekcer S 1987 *Multi-ampere contact connections* Leningrad Energoatom Publishing Leningrad Department.
[4] Nouri H, Keavey M and Williams A *Mathematical Modelling and Related Interfacial Phenomena in Contactor Contacts* Faculty of Engineering University of the West of England.
[5] Tzeneva R 2007 *Electric field distribution in perforated bolted busbar connections* XVth International Symposium on Electrical Apparatus and Technologies, SIELA 2007, Proceedings, vol. I, pp. 191-196.
[6] Tzeneva R 2009 *New perforation design of aluminum bolted busbar connections*, XVIth International Symposium on Electrical Apparatus and Technologies, SIELA 2009, Proceedings, vol. I, pp. 320-326.
[7] Tzeneva R 2018 *3D Coupled electric-thermal analyis of bolted busbar connections* 20th International Symposium on Electrical Apparatus and Technologies, SIELA 2018 – Proceedings, pp. 1-4.
[8] Standart DIN 43673 Parts 1 and 2.

IOP Conf. Series: Materials Science and Engineering **618** (2019) 012026 doi:10.1088/1757-899X/618/1/012026

Analysis of the active and reactive power of a wind power plant with SCIG generators, connected to a high power electrical grid

Ivan Hadzhiev[1], **Dian Malamov**[2] **and Aleksandar Georgiev**[1]

[1]Department of Electrical Engineering, Technical University of Sofia, Plovdiv Branch, 25 Tsanko Dyustabanov Str., 4000 Plovdiv, Bulgaria
[2]Department of Communications and Power Engineering, University of Plovdiv Paisii Hilendarski, 24 Tzar Asen Str., 4000 Plovdiv, Bulgaria

E-mail: hadzhiev_tu@abv.bg, deanmalamov@abv.bg, alex_georgiev94@abv.bg

Abstract. The paper presents the results from studies of both the active and reactive power of a wind power plant, connected in parallel with a high power electrical grid. The wind power plant is with SCIG asynchronous generators and a step-up substation 21/115 kV, realized by means of cable lines. The analyses have been made with the help of a computer model, synthesized in MatLab environment. The dependencies of the power, the power factor and the phase angle on the mechanical moment of the generator shaft in the cases of only one generator operating and all generators operating have been obtained. The inductive reactive power for compensating the capacitive power of the wind plant has been calculated at turned off generators. The parameters of the inductive compensating reactor have been determined. The capacitive power for compensating the inductive reactive power has been defined at loaded generators.

1. Introduction

Wind power plants have a significant share in the volume of electricity production from renewable energy sources. What makes them specific with respect to the other types of electrical power stations is the significant fluctuation of the primary energy, provided by the wind. Mostly asynchronous generators SCIG or DFIG type are used in the wind power plants. Compared to synchronous generators, they have certain advantages in the great fluctuations of the parameters (velocity, momentum) of rotation of the generators [1].

The paper presents the results from studies of the active and reactive power of a wind power plant, connected in parallel to a high power electrical grid. A single-line diagram of the considered wind power plant is given in Fig. 1. It consists of eight identical asynchronous DFIG generators, complemented with 0.69/21 kV transformers, whose outputs are connected to a step-up transformer 21/115 kV. The substation is realized by means of cable lines, made of NA2XS(F)2Y cables. The wind power plant is connected in parallel with the electrical power grid through a cable line W_1. A cable line for the future expansion of the wind power plant is envisaged.

IOP Conf. Series: Materials Science and Engineering **618** (2019) 012026 doi:10.1088/1757-899X/618/1/012026

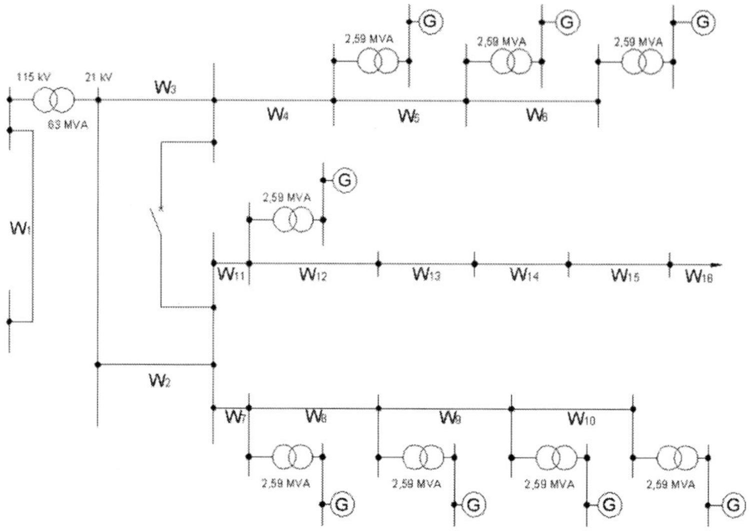

Figure 1. One-line diagram of the wind power plant.

Fig. 2 presents a time segment of 24 hours from the load schedule of the active and reactive power of the wind plant.

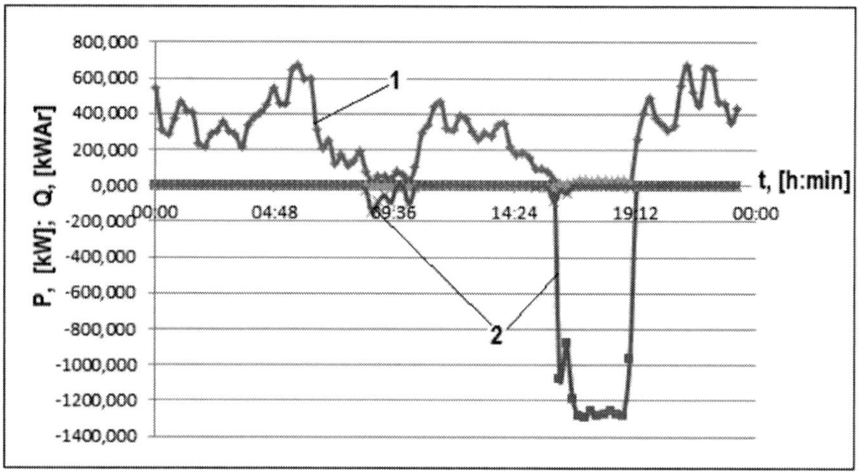

Figure 2. Load schedule of the active and reactive power of the wind power plant for a period of 24 hours: 1 – generated active power; 2 – reactive power.

The typical for this type of electrical power stations fluctuations of the generated active power are observed, as well as consumption of considerable reactive capacitive power at turned off generators or generators with a small load. The load schedule is for the wind plant operation with DFIG generators, allowing for automatic compensation of the reactive inductive power at working generators by means of controlling the rotor chain of the generator [1]. This paper presents an analysis of the wind plant operation with SCIG generators. The other parameters of the wind power plant are not changed. The main reason for the presence of a significant reactive capacitive power at turned off generators are the capacitive conductivities of the cables, by means of which the substation, connecting the wind plant with the electrical power grid is realized. The reactive capacitive power is one of the factors, degrading the quality of electricity [2], [3] and therefore the operation of the wind power plant in this mode is not recommended. One of the basic ways of solving this problem without turning off the power plant from the electrical power grid is to connect an inductive compensating reactor to it. Some specific problems, related to the influence of the reactive power at parallel operation of wind power plans have been considered in [3], [4], [5], [6].

The paper presents the results from studies of the active and reactive power of the described wind power plant. The studies were conducted by means of especially developed computer models in MatLab/Simulink. The computer models were synthesized taking into account: the network topology in accordance with 1; the length of the cable lines and their parameters in accordance with the catalogue data of the cables, transformers and generators.

2. Results, obtained by means of the computer model at turned off generators

Both the reactive capacitive power Q_C and the active power P_{nx} were defined at turned off generators. The active power is defined by the losses in the active resistances of the cables and by the losses in the windings and the magnetic system of the transformers.

$$Q_C = 1407\,kVAr;\quad P_{nx} = 29.82\,kW \tag{1}$$

The defined parameters of the phase shift angle between the idle run powers are:

$$tg\varphi_{nx} = -47.203;\ cos\,\varphi_{nx} = 0.212;\ \varphi_{nx} = -88.8^{\circ} \tag{2}$$

Based on the above data at a normed power factor and after compensating by $cos\,\varphi_1 = 0,94$ (inductive), the value of the inductive compensating power is obtained.

$$Q_\kappa = P_{nx.}.\left(tg\varphi_{nx} - tg\varphi_1\right) = 1418\ kVAr \tag{3}$$

where: Q_κ is the value of the inductive compensating power; φ_1 is the phase shift angle after compensation.

The compensation is realized by means of two identical compensating reactors, connected at the beginning of the 21 kV line (at the input of the 21/115 kV transformer). The reactors are oil type with discrete regulation of the reactive power under a five-degree load.

Table 1. Parameters of the compensating reactor.

Position of the regulator	Rated power $[kVAr]$	Rated current $[A]$	Ind. resistance $[\Omega]$
1	640	17.6	690.41
2	680	18.7	651.16
3	720	19.8	615.92
4	760	20.9	583.98
5	800	22	554.91

3. Results, obtained by means of the computer model at loaded generators

The dependencies of the active power P, the reactive power Q, the power factor $cos(FI)$ and the phase angle FI on the shaft moment of the generators M were studied by means of the computer model. The values of the active power, the reactive power and the shaft moment of the generators M are presented in relative units.

$$P^* = \frac{P}{S_n}; \quad Q^* = \frac{Q}{S_n}; \quad M^* = \frac{M}{M_n} \tag{4}$$

where: P and Q are the measured values of the active and reactive power; S_n is the rated power of the wind power plant.

The studies were conducted for the following cases:

- load on one generator in the range up to a maximum moment at a compensating reactor, turned on at maximum power, as well as at a turned off compensating reactor;
- equal load on all generators in the range up to a maximum moment at a compensating reactor, turned on at maximum power or at a turned off compensating reactor.

The results are graphically presented in Fig. 3÷10 with the following location of the capacities in the complex plane:

- the generated active power is along the positive part of the real axis, while the active power at operating generators in motion mode is along the negative part of the real axis;
- the capacitive reactive power is along the negative part of the imaginary axis, and the inductive reactive power is along the positive part of the imaginary axis.

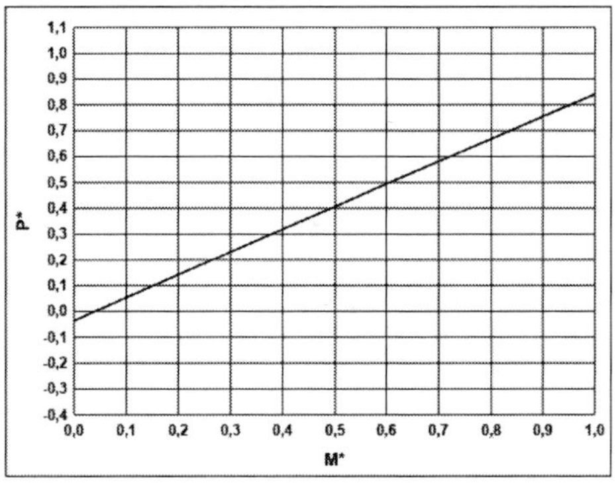

Figure 3. Active power at one loaded generator with turned on or turned off compensating reactor.

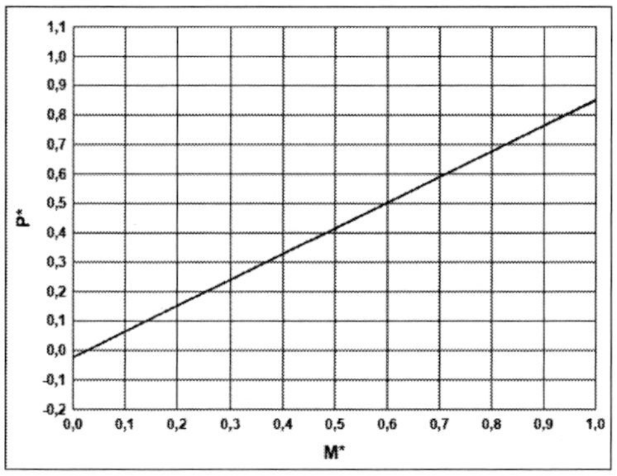

Figure 4. Active power at load on all generators with turned on or turned off compensating reactor.

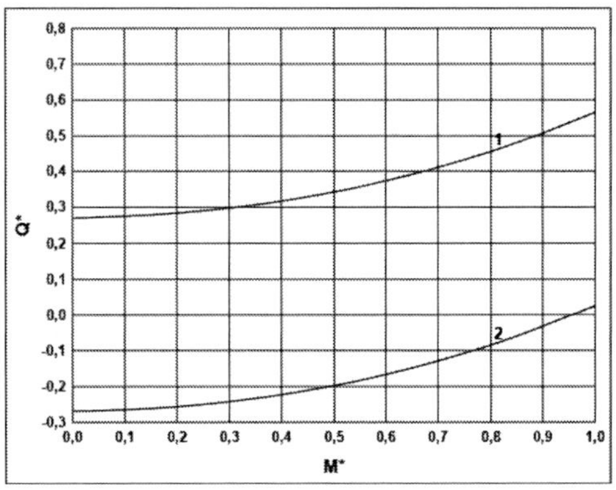

Figure 5. Reactive power at one loaded generator with: 1 – turned on compensating reactor; 2 – turned off compensating reactor.

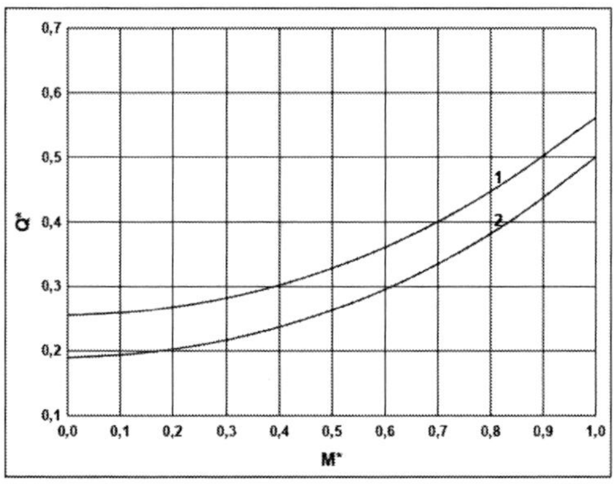

Figure 6. Reactive power at load on all generators with: 1 – turned on compensating reactor; 2 – turned off compensating reactor.

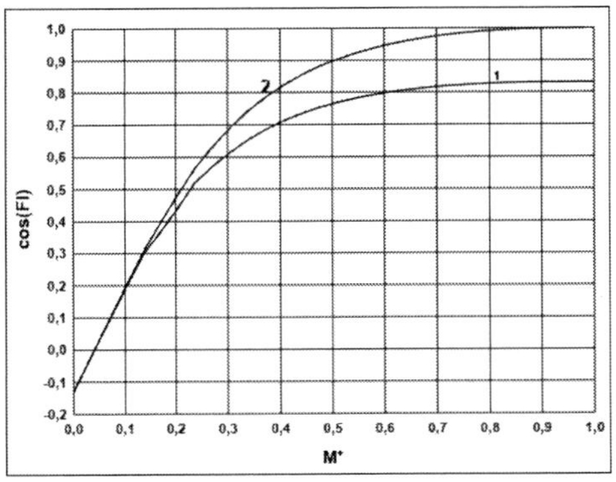

Figure 7. Power factor at one loaded generator with: 1 – turned on compensating reactor; 2 – turned off compensating reactor.

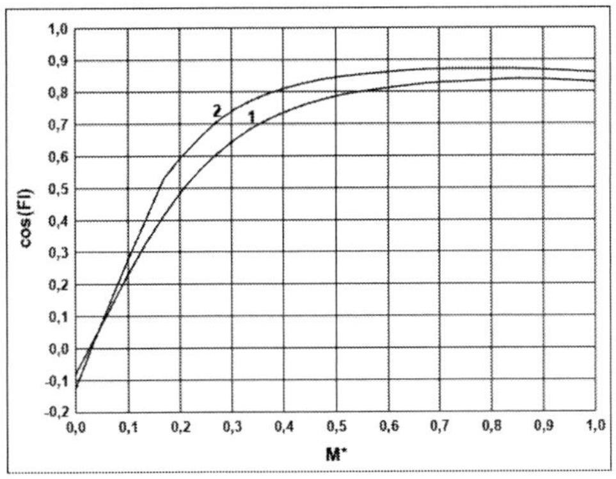

Figure 8. Power factor at load on all generators with: 1 – turned on compensating reactor; 2 – turned off compensating reactor.

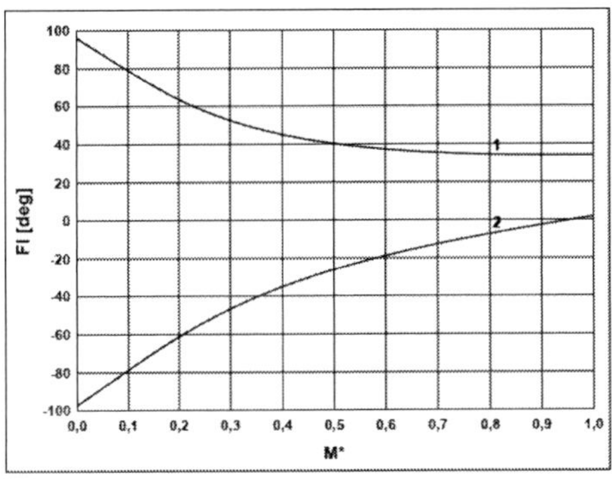

Figure 9. Phase angle at one loaded generator with: 1 – turned on compensating reactor; 2 – turned off compensating reactor.

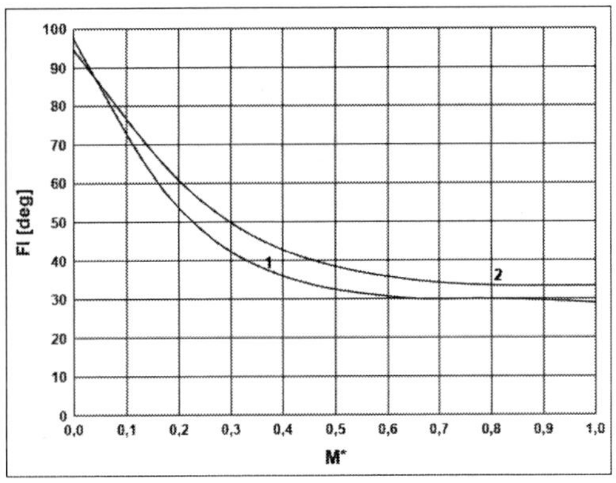

Figure 10. Phase angle at load on all generators with: 1 – turned on compensating reactor; 2 – turned off compensating reactor.

4. Conclusion

Based on the obtained results from the conducted studies the following conclusions can be drawn up:

- The required inductive compensating power at turned off generators has been determined. Two inductive controllable reactors have been chosen. The parameters of the reactors provide a reserve for the future expansion of the wind power plant.
- The inductive reactors do not influence the value of the generated active power, regardless of the number of working generators.
- When one generator is loaded without a compensating reactor, the capacitive power is compensated by the inductive power of the generator at reaching the rated load of the generator. This imposes the need to connect a compensating reactor at smaller loads, defining its power by choosing the corresponding degree.
- At load on all generators without a compensating reactor in the interval from idle run to summed rated power of all generators, the reactive power is inductive. If connected, the compensating reactors additionally degrade the power factor.
- The maximum power of the source of capacitive compensating power can be determined by (3) and by the data about the active power and the power factor at load on all generators, as well as by the

IOP Conf. Series: Materials Science and Engineering **618** (2019) 012026 doi:10.1088/1757-899X/618/1/012026

power factor after compensation. The required compensating power depends on the generated active power, which imposes the use of a source, capable of smooth and stepwise adjustment in the entire range, determined by its maximum power.

Acknowledgments
The authors would like to thank the Research and Development Sector at the Technical University of Sofia for the financial support. The work of I. Hadzhiev is supported by National program "Young scientists and Postdoctoral candidates" of Ministry of Education and Science.

References

[1] Ackerman T 2005 *Wind power in power systems* John Wiley & Sons ChicesterU.K.
[2] Nikolov D A 1994 *Electrical grids and systems* Techniques Sofia.
[3] Niiranen J 2008 *About the active and reactive power measurements in unsymmetrical voltage dip ride-through Testing* Wind Energy 11 121–131 doi:10.1002/we.254.
[4] Manusov V, Udalov S, Chivenkov A, Achitaev A and Suyakov S 2013 *The technical complex of increasing energy efficiency of wind power station* European Science and Technology: materials of the V international research and practice conference (issue 1) Vela Verlag Waldkraiburg Vol. 1pp. 432–440.
[5] Manusov V Z 2012 *Peculiarities of the parallel operation of wind power plants and electrical power grids* Siberian federal university http://conf.sfu-kras.ru/sites/mn2012/section06.html.
[6] Milovanova K A 2011 *Integration of a wind power plant into the operation of an electrical power grid* Abstract for obtaining a qualification degreeMoscow MEI.

Winding distribution influence of some current limiting devices certain parameters

Dinyo Ivanov Kostov

Addres:Yane Sandanski str. №8, Radnevo
E-mail:snapon_891@abv.bg

Abstract: This report presents a study of connection between winding distribution and some current limiting devices parameters. Influence of separation coefficient on the variation of the mass parameters and the active resistance of an air coil is examined. The results of the simulation study are presented and a critical analysis is performed.

1. Introduction

In the technical literature, the design issues for current-limiting devices (CLD) are well known [1], [2], [3]. In the last years, the group of CLD equipment also includes devices of a new type - the so-called "Fault Current Limiters" (FCL) [4], [5]. FCLs are an alternative to traditional CLD - current-limiting reactors, chokes, dampers. In general, they can be divided into two groups [4]: FCL using superconducting materials [5], [6] and FCLs made of conventional materials [6], [7], [8]. In the second group are included semiconductor type [6], FCL with submerging [7], transformer type FCL [8], electromagnetic type FCL [9], resistive type FCL [9]

It is necessary to know the distribution of the magnetic field created by the coil sequentially incorporated into the protected circuit in order to design an FCL with conventional materials.

There are some assumptions and simplifications in the engineering methodology designing current-limiting devices [1], [3], [10]. When designing a new type of CLD, these assumptions and simplifications may lead to a deterioration in their characteristics - increased losses, insufficient efficiency, higher end product costs. One of the main parameters for CLD improving efficiency is the coefficient of winding distribution [1].

The report aim is to investigate influence of the coil distribution coefficient on CLDs basic parameters - active resistance and change of the magnetic induction in the coil bright hole. Some dependencies have been used in the engineering methodology for calculating CLDs designed to operate in a particular power system (PPS) [10], [11]. A simulation study was conducted using the Finite Element Method (FEM).

2.CLD input parameters

In the design of the CLD, input parameters are assumed to be those that depend on the placement of the device and the type of PPS [12]. One of the input parameters is the admissible value of the paid-in inductance from the CLD [12].

In order to evaluate effect of the winding distribution coefficient on the bright hole winding volume field, the parameters of a specific PPS [13] are given in Table 1.

Table 1

Xbr [Ω]	Lbp [H]	I2n[A]	Q[mm^2]	σ[A/mm^2]	k3
0,519	0,01	577,35	300	2,3	0,25

With Xbr the induced resistance in the system is indicated, while residual network voltage is maintained in short circuit (SC) - 70%. Lbp is the inductance of the winding; I2n is the nominal operating current in the system; q is the cross section of the conductor with which the winding of the current limiter is wound; torque density; kz - winding load coefficient [12].

3.CLD output parameters

When calculating the output parameters of the device, the mean radius of the winding Rbp is determined by the expressions [3], [10], [14]:

$$R_{bp} = \left[\frac{L_{bp}.(6+19.k).I_{2n}{}^{2}}{32.10^{-6}.k1^{2}.k_{3}{}^{2}.\gamma^{2}}\right]^{0.2}, m \tag{1}$$

In (1) the distribution coefficient denoted by k1 is unknown. Winding distribution coefficient choice depends on the constructor's considerations. In this study it is assumed that its value is in the range of k1 = 0.33 to k1 = 0.71. The values of k1 are shown in Table 2.

The geometric dimensions a (winding package height) and b (winding package width) are calculated from the dependencies [3], [10]:

$$b = k.R_{bp}, m \tag{2}$$

$$a = \frac{2.R_{bp} + b}{2} a, m \tag{3}$$

The number of Wbp windings, outer diameter D2bp and inner diameter D1bp are calculated with the following dependencies:

$$D_{1bp} = 2.R_{bp} - a, m \tag{4}$$

$$D_{2bp} = 2.R_{bp} + a, m \tag{5}$$

$$Wbp = k_z.k^2.\frac{R_{bp}{}^2}{q.10^{-6}} \tag{6}$$

The results of the expressions (1), (2), (3), (4), (5), (6) are given in Table 2

Table 2

K1	Rbp[m]	a[m]	b[m]	D1bp[m]	D2bp[m]	Wbp[num.]
0,33	0,517	0,602	0,171	0,863	1,205	25
0,42	0,482	0,583	0,202	0,761	1,166	34
0,54	0,449	0,570	0,243	0,656	1,141	49
0,66	0,426	0,566	0,281	0,570	1,132	66
0,71	0,417	0,566	0,296	0,539	1,131	73

The variation of the coil distribution coefficient k1 results in a change of average radius Rbp (Fig. 1) and the other output parameters (figure 2) – the coil windings number Wbp, the bright hole coil diameter D1bp. Increasing the coefficient k1 reduces the geometric dimensions – mean coil radius

Rbp, winding height a, internal and external diameters (figure 1); the winding width b and the number of windings are increas. (figure 2).

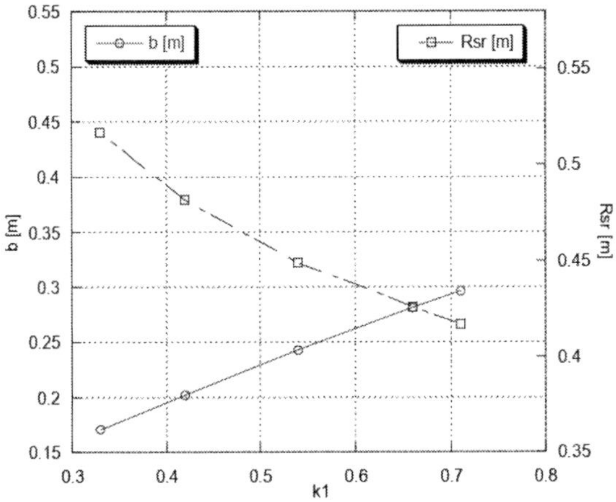

Figure 1. Winding height and average radius changing at different k1 values

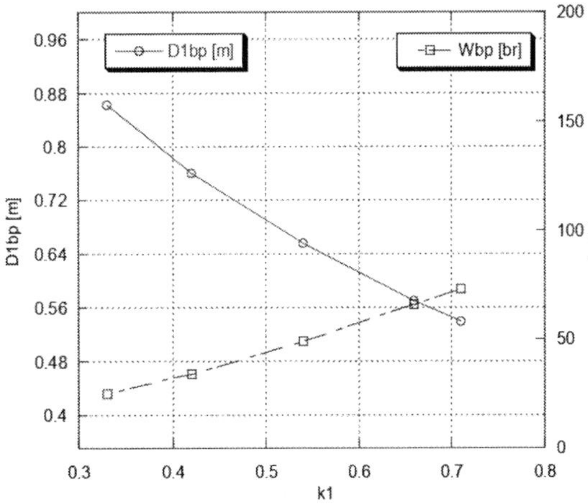

Figure 2. Bright hole diameter and number of windings changing at different k1 values

The three-dimensional simulation model [15], [16], has been formed from the CLD coil parameters. The software product COMSOL [17], [18], [19] was used. The simulation is performed in both modes - nominal mode and SC mode. Some of the results are shown in figure 3 and figure 4

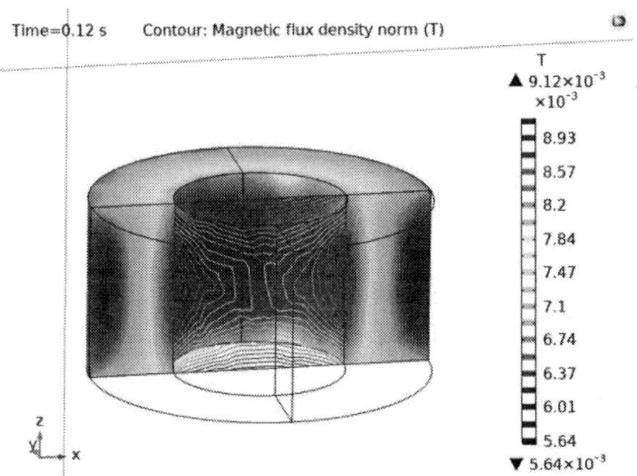

Figure 3. Nominal mode magnetic induction change

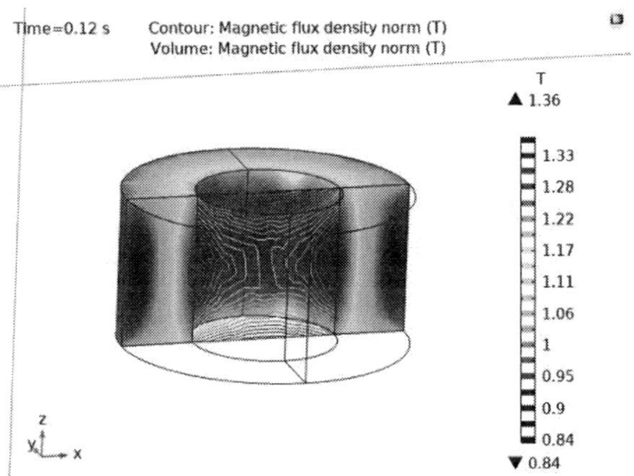

Figure 4. SC mode mode magnetic induction change

To verify the simulation model, value of the active resistance of CLDs winding was compared (Table 3). With Re the active resistance obtained by the engineering methodology is indicated, and with Rc - the resistance obtained in the simulation test. The minimum and maximum values of the winding brigt hole magnetic induction are determined. The magnitude of magnetic induction change in the air gap Kp is a disproportionate coefficient. It is accepted to evaluate the uniformity of the field [1]:

$$K_p = \frac{B_{max}}{Bmin} \tag{7}$$

Table 3 shows the influence of winding distribution coefficient k1 on the minimum and maximum induction in nominal operating mode and in SC mode. The active resistance of the winding Re determined by the simulation is not different from the value calculated by the engineering methodology Rc. At kl = 0,66 the multiplicity of air gap magnetic induction change is independent of the operating mode of the CLD - nominal or SC.

Table 3

			Nominal mode		SC mode			
kl	**Re**	**Rc**	Bminn	Bmaxn	Bminf	Bmaxf	Kpn	Kpf
	Ω	Ω	mT	mT	T	T		
0.33	0.0048	0.0048	0.433	0.741	0.97	1.68	1.74	1.73
0.42	0.0061	0.0061	1	1.68	0.95	1.62	1.68	1.71
0.54	0.0082	0.0082	2.6	4.3	0.91	1.52	1.65	1.67
0.66	0.0105	0.0105	5.64	9.12	0.84	1.36	1.62	1.62
0.71	0.0114	0.0114	7.32	11.36	0.8	1.27	1.55	1.59

Figure 5 shows the geometric size variation and the coil distribution coefficient variation kl. It also shows the active resistance change. By increasing the coil distribution coefficient value kl, the dimension a decreases and the resistance of the coil increases. The intersection in the characteristics of the active resistance and the characteristic of the geometric dimensions is found at winding distribution coefficient kl = 0,66.

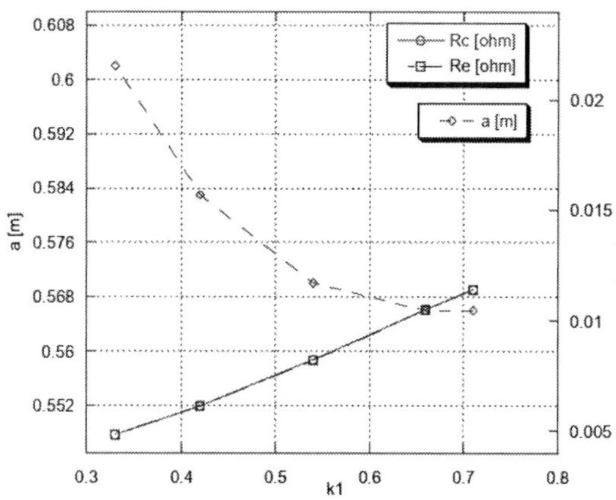

Figure 5. Physical parameters change and winding active resistance depending on the distribution coefficient

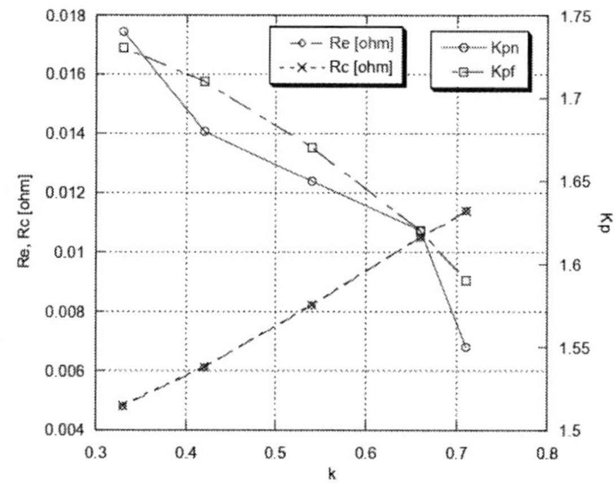

Figure 6. Multiplicity magnetic induction change in the winding bright hole depending on the coil distribution coefficient

Figure 6 gives the dependence of the magnetic inductance multiplication variation and the resistance variation depending on the coil distribution coefficient. In both the geometric dimensions and the winding bright hole induction multiplicity, the intersection of the characteristics is found at kl = 0,66.

4. Conclusions

From the results presented in figure 5 and figure 6, the following conclusions can be made:

- The winding output current of the CLD depends on the coil distribution coefficient.
- The increase in the coil distribution coefficient results in the designed coil geometric dimensions change.
- For values of kl greater than 0.5 the relative dimensional change decreases.

- The increase in coil distribution coefficient results in a more even magnetic field distribution in the designed coil bright hole. This is essential in the case of conventional materials made FCLs coil.
- The increase in coil distribution coefficient leads to an increase in the active resistance, hence the designed coil losses.
- For coil distribution coefficient k1 = 0,66 the multiplication factor of the magnetic induction in the winding bright hole does not depend on the operating mode.
- At coil distribution coefficient k1 = 0,66 good weight-to- gauge ratio is achieved with comparatively low added active resistance;

5.Inference

Winding distribution coefficient has a significant effect on current limiter output parameters.

The report demonstrates that if the coil distribution coefficient is k1 = 0,66 the weight-to- gauge ratio and coil losses will be acceptable; multiplicity magnetic induction coefficient in the bright hole coil bore does not depend on the mode of operation of the CLD.

This will allow for a even distribution of the magnet field of the coil when designing conventional materials FCL and hence limiting the inductance of the scattering of the CLD.

References

[1] Georgieva Z., Vliqnie na konstrukciqta vyrhu razpredelenieto na nagnitnite potoci i induktivnostta pri reaktori s magnitoprowod, Avangard Prima, Sofia, 2012 (in bulgarian)
[2] J.Schlabbach, Short – circuit currents, IET Power and Energy series 51, 2008
[3] Maslarov I., Electricheski aparati, Sofia, 2015 (in bulgarian).
[4] Kostov D., Srawnitelen analiz na ogranichiteli na tok na kyso syedinenie, TECHSYS, Plovdiv, 2018 (in bulgarian)
[5] Fault current limiters in electrical medium and high voltage systems, CIGRE WG A3.10, august, 2003
[6] Janos Kosaa. Novel 3-phase inductive fault current limiter.2005
[7] E.S.Ibrahim. Electromagnetic fault current limiter, Electric Power System Research, 42, 1997,pp.189
[8] T.Hoshino, K.Ohkubo, M.Etoh. Recovery time of superconducting non-inductive reactor type fault current limiter, IEEE Trans. on Magnetics, vol.32.no.4, july, 1996, pp.2403-2406.
[9] Fault current limiters – application, principles and experience, CIGRE WG A3.16, Tokyo, 2005.
[10] Aleksandrov A., Georgieva Z., Rykovodstvo za seminarni upravneniq po elektricheski aparati, Avangard Prima, Sofia, 2000 (in bulgarian).
[11] Strnin V.G., Suhie tokoogranichavaiushtie reaktoryi, 1965 (in russian)
[12] Stefanov I., Stoqnov S., Spravochnik na energetika, t.1, Tehnika, Sofia, 1981 (in bulgarian)
[13] Aleksandrov K., Specialen kurs po elektricheski aparati Avangard Prima, Sofia, 2015 (in bulgarian)
[14] Aleksandrov A., Electricheski aparati, Avangard Prima, Sofia, 2004 (in bulgarian)
[15] Qchev I., Marinova I., Chisleni metodi i modelirane na verigi i poleta chast 1, Sofia, 2011 (in bulgaian)
[16] N.Watson, Power System Electromagnetic Transient Simulation, IET Power and Energy series 39, 2007
[17] Aleksandrov A., Kompiutyrno proektirane na electricheski aparati, Avangard Prima, Sofia, 2004 (in bulgarian)
[18] Pryor, Roger-Multiphysics Modeling Using COMSOL-2011
[19] Praht V.A., Modelirowanie teplowyih i elektromagnitnyih processov Prograna COMSOL, 2011 (in russian)

Application of silver antibacterial nanolayers for hard contact lenses coating

Krassimir Koev*[1,2], Nikolay Donkov[1], Nadya Stankova[1], Hristo Najdenski[3], Timerfayaz Nurgaliev[1], Rumen Nikov[1], Latchezar Avramov[1]

[1]Institute of Electronics, Tzarigradsko shosse 72 blvd, Sofia 1784, Bulgaria,
[2]Department of ophthalmology, Medical University Sofia, 8 Byalo More str., Sofia, Bulgaria,
[3]The Stefan Angeloff Institute of Microbiology, Bulgarian Academy of Sciences, 26 Georgi Bonchev str., 1113 Sofia, Bulgaria

e-mail address: k00007@abv.bg

Abstract. The antibacterial action of Ag-doped Al_2O_3 nanolayers, deposited by RF reactive magnetron sputtering on hard contact lenses, was established. Synthesis of Ag/Al_2O_3 protective coatings is necessary for suppressing the infections caused by pathogenic microorganisms following the placement of hard contact lenses. The chemical composition, morphology and optical properties of the coatings were studied. Microbiological studies conducted of the nanocomposite Ag/Al_2O_3 layers against Gram-positive and Gram-negative bacteria (Pseudomonas aeruginosa, Escherichia coli, and Staphylococcus aureus) proved their antibacterial effects. Intense action of the layers was found against Pseudomonas aeruginosa - full inactivation after 2 hours; Escherichia coli - full inactivation after 5 hours; Staphylococcus aureus – full inactivation after 24 hours. Our experimental findings suggest a very promising application of such antibacterial Ag/Al_2O_3 nanolayers regarding the reduction of eye infections when the hard contact lenses are used.

1. Introduction

Contact lenses are widely used to treat myopia, hyperopia, astigmatism, and presbyopia. Different types of lenses have been applied in the practice: rigid or "hard" contacts - made from polymethyl methacrylate (PMMA); rigid gas-permeable lenses (RGP's) - made from plastics combined with other materials, which allow oxygen in the air to pass directly through the lens; soft contact lenses - made from hydrated polymer, hydroxyl-ethyl methacrylate (HEMA) [1].

Developing eye infections is a potential danger for people wearing contact lenses [2]. Concerning the eye infections, the contact lens are associated with several risk factors such as microbial keratitis and conjunctivitis [2].

Previous studies suggested that patients who do not follow the contact lens hygiene recommendations run greater risk for corneal inflammatory events [3-5].

Although, the hard contact lenses have been used less than the soft lenses, they are the most appropriate choice for some eye diseases like astigmatism and keratokonus. While, the rigid gas permeability lenses have been used to perform ortho-k, a non-surgical procedure designed to improve vision.

It is well known that silver (Ag) particles possess antibacterial properties [5-8]. They act as antioxidants, blocking cellular processes and leading to bacterial cell death. Already, the antibacterial activity of Ag

IOP Conf. Series: Materials Science and Engineering **618** (2019) 012028 doi:10.1088/1757-899X/618/1/012028

nanoparticles has been determined against Gram-positive Staphylococcus aureus and Gram-negative Escherichia coli and P. aeruginosa bacteria and Candida albicans by many authors [9-11].

The silver nanocomposite layers are a promising tools for preventing eye infection. As a broad-spectrum agent, it kills a wide range of microorganisms in low levels without associated toxicity [11,12]. The antibacterial power of silver nanoparticles has been widely documented [13-20], although the exact mechanism causing it is not known well. It is believed that the silver provokes inactivation of enzymes essential for the respiratory chain of the pathogens, or generates hydroxyl radicals [15,16] that, in turn, would cause damage to the pathogen.

The purpose of this study was to synthesize Ag/Al_2O_3 nanocomposite layers on hard contact lenses with a broad-spectrum antibacterial effect on Gram-positive and Gram-negative bacteria; and to investigate the chemical composition, optical and morphological properties of the layers as well as their antibacterial activity.

2. Experimental

2.1. Material synthesis and Analytical methods

The nanocomposite thin layers of Ag/Al_2O_3 were synthesized on hard contact lenses and on glass substrates (to facilitate the analytical analyses) by RF magnetron co-sputtering of Ag and Al under the following conditions: frequency 14 MHz; working pressure in the vacuum system was fixed at ~ 5.5 Pa with optimal mixture of Ar and O_2 gasses. The optimal concentration of Ag in the Al_2O_3 matrix was obtained by precise control of the process parameters during the deposition.

The surface elemental composition of the nanolayers on the glass samples was investigated by X-ray photoelectron spectroscopy (XPS) on an ESCALAB MkII (VG Scientific) electron spectrometer at a base pressure in the analysis chamber of 5×10^{-8} Pa (during the measurement 1×10^{-6} Pa), using an AlKα X-ray source (excitation energy of 1486.6 eV). The thickness and the optical transmission of the layers in UV-Visible spectrum were determined by using a Woollam M2000D rotating compensator spectroscopic ellipsometer and an Ocean Optics HR 4000 spectrophotometer, respectively. Scanning electron microscopy (SEM) (SEM/FIB Lyra/Tescan dual beam system) was used to investigate the coatings' surface morphology.

2.2. Antibacterial activity assay

Microbiological assay was carried out to examine the antibacterial properties of the coatings in regard to deactivation of Gram-positive and Gram-negative bacteria. The strains used in these experiments were: Staphylococcus aureus strain 29213 and Escherichia coli strain 35218 of the American Collection of Cell Cultures (ATCC); and Pseudomonas aeruginosa strain 1390 of the collection of "Stefan Angeloff" Institute of Microbiology, Bulgarian Academy of Sciences.

500 μL of microbiological suspension of the respective microorganisms in saline with concentration of 1×10^6 CFU/mL (colony-forming units per mL) was plated in 24 wells plate for cell culture. Uncoated hard contact lenses were placed in the control wells and the hard contacts lenses covered with the test coating of Ag/Al_2O_3 were placed in the sample wells.

Samples and controls were incubated at a temperature of 37 °C and shaken continuously. At pre-defined intervals of 0, 2, 5, and 24 hours, the samples were taken to determine the number of viable microorganisms. This was performed by seeding ten-fold diluted solutions of the incubation mixtures and plating in 25 μL nutrient medium of Tryptic Soy Agar for microbiology (Sigma-Aldrich). After 24 hours of incubation at 37 ° C, the number of bacterial colonies emerged. All the results were expressed in CFU/mL.

2.2.1. Statistics

Each experiment was performed in triplicate and the data were presented as a mean ± standard deviation (SD). The difference between two means was compared by a two-tailed unpaired Student's test. The values of $P<0.05$ were considered as significant.

3. Results

3.1 Analytical characterisations

The thickness measured of the Ag/Al_2O_3 layers synthesized and deposited on the substrates was about 18 nm was. All layers were highly transparent in the visible light spectrum with transmittance of about 95%. The optical transparency is great advantage, since it makes the as-obtained layers suitable for application like transparent coatings on eye lenses. The SEM imaging (figure 1) showed a uniform surface morphology without cracks, which determines the possibility for successful applications as coatings on eye lenses. Quasi spherical nanoparticles with average size distribution about 100 nm were formed. This leads to an increase of the active surface area of the layers thus further promoting their antibacterial function as protective coating.

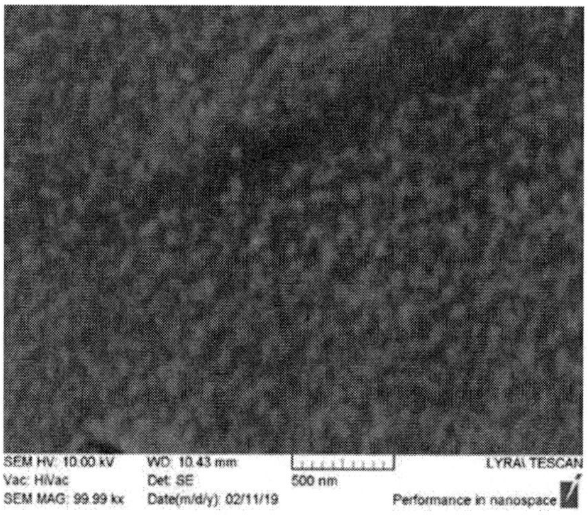

Figure 1. SEM image of the surface morphology of the Ag/Al_2O_3 nanocomposite layer

XPS measurements were performed to assess the surface chemical composition and the corresponding concentration of the Ag in the layers synthesized. The binding energies of peaks in the high resolution XPS spectra are assigned as Ag3d, Al2p, O1s and Si2p shown in Table 1 [21,22]. Based on these values the atomic concentrations are calculated and shown in the Table 1 as well.

Table 1. Surface atomic concentrations of the elements measured on the surface of the Ag/Al_2O_3 layers.

Photoelectron peaks	O1s	Al2p	Si2p	Ag3d
Concentration [at.%]	33.20	29.29	28.13	9.38
Binding Energy [eV]	531.7	74.4	103.6	368.7

3.2 Antibacterial test

The antibacterial activity of the layers deposited on hard contact lenses was examined by microbiological tests as described above.

The most effective activity of the layers was found against *Pseudomonas aeruginosa*. Complete elimination of pseudomonads occurred within two hours (figure 2). Reduction of 4 log in bacterial count of *Escherichia coli* was observed at the 2th hour. The bacteria were eliminated completely after the fifth hour (figure 3). A ten-fold reduction of the *Staphylococcus aureus* bacterial count was observed at the fifth hour, and complete inactivation of the bacteria was established at the 24th hour (figure 4).

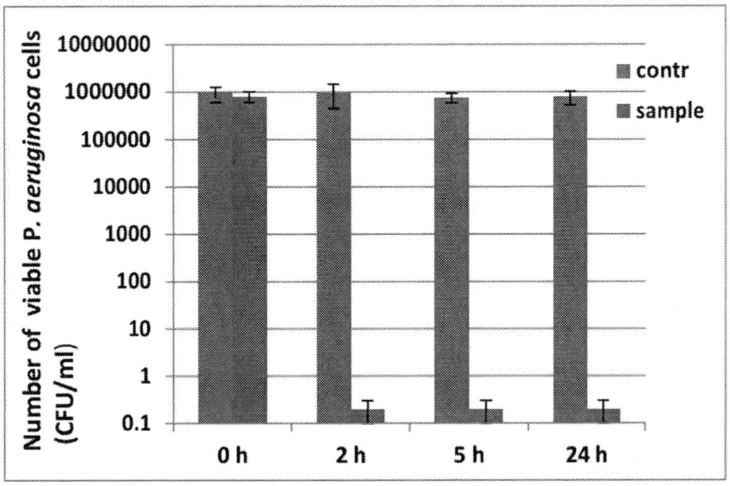

Figure 2. Number of viable *Pseudomonas aeruginosa* cells incubated for 24 hours at temperature of 37 °C with pristine eye lenses (control) and Ag/Al_2O_3 layers deposited on eye lenses (sample). Complete inactivation at the 2th hour

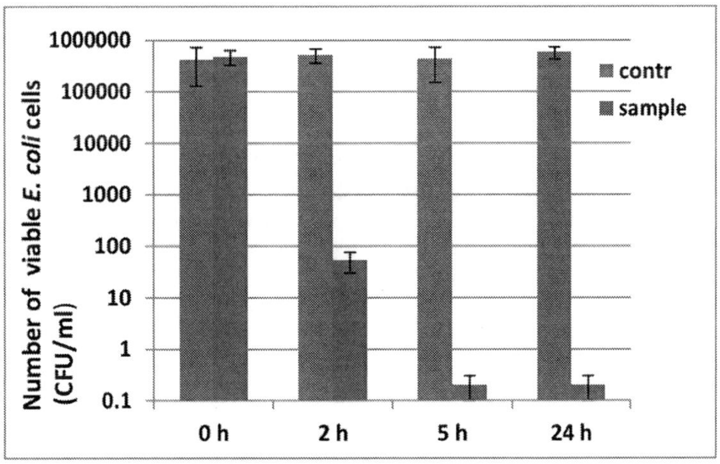

Figure 3. Number of viable *Escherichia coli* cells incubated for 24 hours at temperature of 37 °C with pristine eye lenses (control) and Ag/Al_2O_3 layers deposited on eye lenses (sample). Complete inactivation at the 5th hour

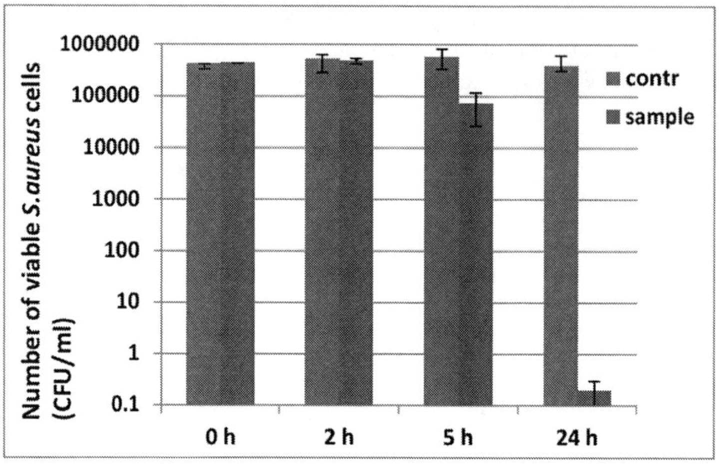

Figure 4. Number of viable *Staphylococcus aureus* cells incubated for 24 hours at temperature of 37 °C with pristine eye lenses (control) and Ag/Al_2O_3 layers deposited on eye lenses (sample). Complete inactivation at the 24th hour

4. Discussion

The antimicrobial efficiency of silver nanoparticles was reported by Salomoni R. et al. [17,18] on microbial keratitis causing microorganisms. Authors Maryam Shayani et al. [19] explored the effect of monomer composition on silver nanoparticles' (SNPs) binding capacity of hydrogels and evaluated their antibacterial efficacy. They established that SNP-loaded hydrogels demonstrate excellent antimicrobial effects against P. aeruginosa and S. epidermidis after soaking in 10 and 20 ppm SNP suspensions. In their work the SEM images revealed inhibitory effect of SNPs against biofilm formation on the surface of the hydrogels. We obtained similar results for Ag/Al_2O_3 coatings on hard lens surface.

Our research found that the Ag/Al_2O_3 coatings exhibited antibacterial action against Gram-positive and Gram-negative bacteria (*Pseudomonas aeruginosa, Escherichia coli, Staphylococcus aureus*).

5. Conclusions

Optically transparent antibakterial Ag/Al_2O_3 nanocomposite layers were synthesized. The Ag/Al_2O_3 coating were deposited by using RF magnetron sputtering on the hard contact lenses. Antibacterial effects against Pseudomonus aerogginosa, E. Coli, and Staphylococus aeroginosa were established. The antimicrobial properties of the surface of Ag doped Al_2O_3 synthesized nanocompisite products are very promising for many biomedical applications.

6. Acknowledgements

The work was supported in part by the Bulgarian National Science Fund under project DFNIB 02/9 "Development of bio-photonics as the basis of oncological theranostics".

References

[1] Findik F 2011 A case study on the selection of materials for eye lenses *ISRN Mechanical Engineering* **2011** Article ID 160671

[2] Cope JR, Collier SA, Rao MM, Chalmers R, Mitchell GL, Richdale K, Wagner H, Kinoshita BT, Lam DY, Sorbara L, Zimmerman A, Yoder JS and Beach MJ 2015 Contact lens wearer

demographics and risk behaviors for contact lens-related eye infections - United States, 2014 *MMWR* 64 (32) 865–70

[3] Stapleton F, Edwards K, Keay L, Naduvilath Th, Dart JKG, Brian G and Holden B 2012 Risk factors for moderate and severe microbial keratitis in daily wear contact lens users *Ophthalmology* **119** (8) 1516–21

[4] Zimmerman A, Richdale K, Mitchell G, Kinoshita B, Lam D, Wagner H, Sorbara L, Chalmers R, Collier S, Cope J, Rao M, Beach M and Yoder J 2017 Water exposure is a common risk behavior among soft and gas-permeable contact lens wearers *Cornea* **36** (8) 995–1001

[5] Dart JK, Radford CF, Minassian D, Verma S and Stapleton F 2008 Risk factors for microbial keratitis with contemporary contact lenses: a case-control study *Ophthalmology* **115** (10) 1647–54

[6] Durán N, Durán M, De Jesus MB, Seabra AB, Fávaro WJ and Nakazato N 2016 Silver nanoparticles: A new view on mechanistic aspects on antimicrobial activity *Nanomedicine: Nanotechnology, Biology and Medicine* **12** (3) 789-99

[7] Kascatan-Nebioglu A, Panzner MJ, Tessier CA, Cannon CL and Youngs WJ 2007 N-Heterocyclic carbene–silver complexes: A new class of antibiotics *Coord. Chem. Rev.* **251** 884-95

[8] Le Ouay B and Stellacci B 2015 Antibacterial activity of silver nanoparticles: A surface science insight *Nano Today* **10** 339-54

[9] Polívková M, Hubácek T, Staszek M, Švorcík V and Siegel J 2017 Antimicrobial Treatment of Polymeric Medical Devices by Silver Nanomaterials and Related Technology *Int. J. Mol. Sci.* **18** (2) 419

[10] Siegel J, Polivkova M, Staszek M, Kolarova K, Rimpelova S and Svorcik V 2015 Nanostructured silver coatings on polyimide and their antibacterial response *Mater. Lett.* **145** 87-90

[11] Polivkova M, Štrublová V, Hubácek T, Rimpelová S, Švorcík V and Siegel J 2017 Surface characterization and antibacterial response of silver nanowire arrays supported on laser-treated polyethylene naphthalate *Mater. Sci. Eng.* C **72** 512-18

[12] Chernousova S and Epple M 2013 Silver as antibacterial agent: ion, nanoparticle, and metal *Angew. Chem. Int. Ed.* **52** (6) 1636-53

[13] Gordon O, Slenters TV, Brunetto PS, Villaruz AE, Sturdevant DE, Otto M, Landmann R and Fromm KM 2010 Silver coordination polymers for prevention of implant infection: thiol interaction, impact on respiratory chain enzymes, and hydroxyl radical induction *Antimicrob Agents Chemother.* **54** (10) 4208-18

[14] Ferraris M, Balagna C, Perero S, Moila M, Ferraris S, Baino F, Battiato A, Manfredotti C, Vittone E and Verne E 2012 Silver nanocluster/silica composite coatings obtained by sputtering for antibacterial applications *IOP Conf. Ser.: Mater. Sci. Eng.* **40** 012037

[15] Hachicho N, Hoffmann P, Ahlert K and Heipieper HJ 2014 Effect of silver nanoparticles and silver ions on growth and adaptive response mechanisms of Pseudomonas putida mt-2 *FEMS Microbiol Lett.* **355** 71-77

[16] Sotiriou GA and Pratsinis SE 2011 Engineering nanosilver as an antibacterial, biosensor and bioimaging material *Curr. Opin. Chem. Eng.* **1** (1) 3-10

[17] Salomoni R, Léo P and Rodrigues MFA 2015 Antibacterial activity of silver nanoparticles (AgNPs) in Staphylococcus aureus and cytotoxicity effect in mammalian cells *The Battle Against Microbial Pathogens: Basic Science, Technological Advances and Educational Programs* Edition: 1ª Editors: Formatex Research Centre Formatex Microbiol. Series 5 p 851–857

[18] Salomoni R, Léo P, Montemor AF, Rinaldi BG and Rodrigues MFA 2017 Antibacterial effect of silver nanoparticles in Pseudomonas aeruginosa *Nanotechnol. Sci. Appl.* **10** 115-21.

[19] Rad MS, Khameneh B, Sabeti Z, Mohajeri SA and Bazzaz BSF 2016 Antibacterial activity of silver nanoparticle-loaded soft contact lens materials: the effect of monomer composition *Current Eye Research* **41** (10) 1286-93

[20] Siegel J,Staszek M, Polivkova M, Reznickova A, Rimpelova S, Svorcik V 2016 Green synthesized noble metals for biological applications *Mater. Today Proc.* **3** 608-16
[21] Pargar F, Kolev H, Koleva DA, and Van Breugel K 2018 Potentiometric response of Ag/AgCl chloride sensors in model alkaline medium *Advances in Materials Science and Engineering* (Hindawi Publishing Corporation) **2018** Article ID 8135492
[22] V. K. Kaushik VK 1991 XPS core level spectra and Auger parameters for some silver compounds *J. Electron Spectros. Relat. Phenomena* **56** (3) 273-77

IOP Conf. Series: Materials Science and Engineering **618** (2019) 012029 doi:10.1088/1757-899X/618/1/012029

Development of decision algorithms in the driver fatigue monitoring and prediction system based on a wireless multisensor device

A Vasiliev[1], N Martiusheva[2], A Boykov[3], N Vyatkin[4], K Nikitin[5]

[1] Peter the Great St. Petersburg Polytechnic University St. Petersburg, Russia
aleksandr.vasilyev@spbpu.com

[2] Peter the Great St. Petersburg Polytechnic University St. Petersburg, Russia
nadezhda.martiusheva@spbpu.com

[3] Peter the Great St. Petersburg Polytechnic University St. Petersburg, Russia
alexey.boykov@spbpu.com

[4]Neuromatics LLC St. Petersburg, Russia
vnv.ru@mail.ru

[5] Peter the Great St. Petersburg Polytechnic University St. Petersburg, Russia
nikitin@kspt.ftk.spbstu.ru

Abstract. This paper presents the system for the continuous monitoring of the physiological state of the vehicle driver. The developed system allows to make adequate decisions and forecast the situation on the basis of the driver fatigue monitoring data. Identification of drowsiness, state if carried out on the basis of blinking duration and frequency analysis. The neural network is used to detect blinking patterns. Signal acquisition and processing are performed on the wearable multisensor microcontroller-based device.

1. Introduction

According to the Global status report of the World Health Organization [1], over 1.2 million people die on the roads each year, and up to 50 million suffer non-fatal injuries. AAA Foundation reports that about 9% of all road accidents and 11% of serious crashes were caused by driver's loss of concentration induced by fatigue or drowsiness [2]. Because of this, development of the automated system performing continuous monitoring of driver's physiological state and alerting before a potentially dangerous situation occurs is of immediate interest.

This paper is devoted to designing a system for real-time detection of transitional state, when the driver is losing control of a situation. It`s called a lapse of responsiveness [3]. This state is characterized as temporal sensorimotor and cognitive systems dysfunction for a period of 0.5 to 15 seconds. Drowsiness state is associated with a set of physiological and behavioral signs, which can be automatically detected by a monitoring system [4]. The rest of the paper covers the development of the algorithms for such system and presents a wearable device designed for real-time driver state monitoring.

SleepAlert is a wearable device for driver's drowsiness detection. It measures EEG (electroencephalogram) signal using a set of three active dry-contact electrodes. In addition, this device obtains pulse wave signal with a pair of non-contact active electrodes. Also the device contains an inertial sensor, tracking user's head movements. Since individual signal may suffer from temporal quality degradation, the variety of sensors may potentially enable reliable detection of drowsiness.

The drowsiness detection system requires a set of sophisticated algorithms, designed for processing input signals and making decisions based on processing results. Multisensor nature of the system entails potentially high efficiency of simultaneous processing of data from different sensors, like EEG and head movement. The proposed operation scheme of the system is as follows:

- EEG signal contains EMG (electromyographic) artifacts mostly associated with eyes movement and blinking, that used as primary source of information about the driver's state.
- EEG spectrum is analyzed in order to determine a relation between high-frequency and low-frequency components. Prevalence of lower band signals indicates a drowsy state [5].
- Pulse wave signal is used to identify user's heart rate
- Head movement data is used to improve error rate of previous algorithms and as an additional source of information, such as presence of monotonic movements.

In this paper we primarily discuss the blinking analysis algorithm. In section 2 we provide a review of existing approaches towards EMG analysis. Section 3 presents the algorithm development of the drowsiness detection system. Section 4 contains a description of training data collection and evaluation process. Some details about hardware implementation are given in section 5, and section 6 gives a conclusion.

2. Review

The task of detection drowsiness stage is studied in various ways. Subjective evaluation techniques for drowsiness states detection are based on self-assessment of the driver's condition and are formalized in the form of a certain scale. These technics allow to describe drowsiness states in detail, but it can`t covered transition between the states of alertness and sleep.

The driver`s state can be determined by a number of typical facial expressions: frequent or constant blinking, nodding and rocking the head and frequent yawning. PERCLOS [8] - percentage eyelid closure. Despite the fact that some researchers claim to have 100% and 98% accuracy when using PERCLOS indicators and blink frequency [7], the situation changes greatly when the driver wears glasses and the lighting changes (by remote monitoring).

The other type of indicators, physiological indicators, are more suitable for the solution of this article`s problem, as they begin to change right in the early stages of sleep. Physiological indicators are EOG, ECG, EMG and EEG signals [6, 9].

The EOG signal, the field created by the potential difference between the cornea and the retina, is sometimes used to determine sleep by the movements of the eyes [7, 10]. The ECG and Heart Rate Variability (HRV) are also utilized [8, 10, 11]: this signs are the basis for calculating the indicator characterizing the transition from wakefulness to sleep [12].

The most commonly used physiological indicator for drowsiness detection is the EEG [13, 14, 15] - the recording of the oscillatory process of brain activity using electrodes placed on the head. In the classical EEG technique, up to 32 sensors are used, but as practice shows [6, 14], in some cases, several sensors are sufficient. Wearable applications, compared to standard clinical EEG, require special dry-contact or non-contact electrodes with strict positioning limitations induced by ergonomics considerations. Number of studies propose to calculate the EEG and EMG simultaneously to improve accuracy [6]. Electromyogram (EMG) allows determining the movement of muscles and localizing them to the EEG.

In research we decided to use physiological indicators - blink parameters, calculated by EMG, because it`s reliable detecting indicators and suitable for the applied task of wearable device development.

3. Methods

Drowsiness detection algorithm is currently based on detection of blinking patterns (figure 1) and evaluating statistical parameters of these signals. The preliminary research has shown that the artificial neural network (ANN) model has the best performance in terms of detection accuracy and computational complexity, since the signal processing has to be implemented in a wearable device.

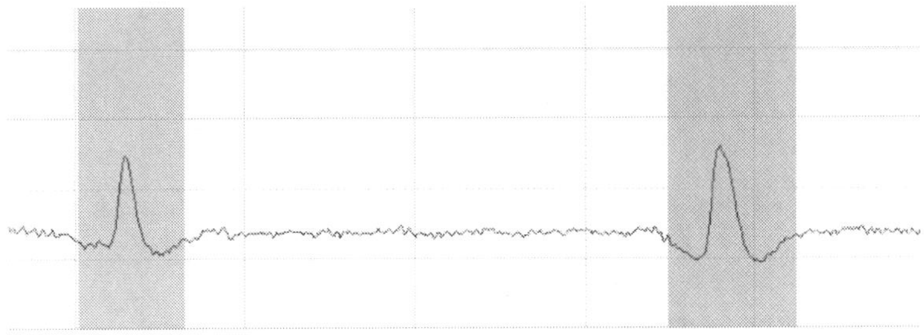

Figure 1. An example of blinking signal on EEG recording of SleepAlert device

We use a focused time-delay ANN model. Two NNs are created to separately detect the beginning and end moments of a blink. Input signal, containing EMG component, is fed to the NN in a way that NN input windows width would be equal to the expected blink signal duration. In order to reduce the amount of memory required for storing the delayed input signal, it is decimated with a factor of 5. With input sampling frequency equal to 250 Hz and amount of inputs being 51, the time step between neural network input samples is 20 ms and total input window is 1 s. Moreover, the input signal is filtered and normalized.

NN has one hidden layer and one output layer. The hidden layer contains non-linear sigmoidal transfer functions. Best performance was achieved with 15 hidden neurons. The output layer is linear. Network output is rounded to values 0 and 1, where 1 indicates detection of blinking start or end and 0 represents absence of these events.

The post-processing of neural network output involves merging together events which occurred within a short interval of 100 ms (NN may output several similar events after one blink, so they will be treated as one). After that, blinking beginning and end from separate NNs are joined into one blinking event, if the end event occurred within 2 second interval of the corresponding start event. In addition, we measure blinking duration which can be used as a parameter for drowsiness detection.

Two statistical parameters of EMG signal are evaluated: mean amount of detected blinks per minute and mean duration of a blink per minute. We use the following procedure to determine driver's condition based on given parameters. The lapse in responciveness is detected if one of the conditions is met: blinking period is more than 600 ms; mean blinking frequency exceeds the average value by 70% or mean blinking frequency is 60% below the average.

4. Training and evaluation

In order to collect the required dataset for training and evaluating the blinking analyzer model, we have built up a test setup to imitate driving process and to continuously record data acquired by the SleepAlert device. This setup is based on a PC with a steering wheel controller (fig 2.). The SleepAlert device is connected to the PC using the Bluetooth 4.0 adapter. The PC monitor displays a track which is slowly scrolled down. The cursor is positioned in vertical center of a screen and a subject is asked to follow the track by the cursor, rotating the wheel and moving the cursor left and right in a way to position it as closely to the track as possible. The track was previously computed as a set of overlapping sinusoidal signals with frequencies from 0.02 to 0.2 Hz.

Figure 2. Test setup for driving process simulation and SleepAlert device.

Assessment of subject's physiological state, which was used as training and test data, was made based on the difference between the displayed track and the user's cursor at every moment of sampling. In addition, the experiment was recorded on a video camera for subsequent manual analysis [16,17].

Experiments were conducted in the evening in a quiet room with ambient lighting. Duration of each experiment was one hour. All the data from SleepAlert sensors were recorded with timestamps, as well as the displayed track and the trajectory of user's cursor. 15 subjects took part in data collection, 12 of them reported feeling drowsy and 3 fell asleep.

The presented models were evaluated on the test data acquired in the driving imitation experiment. The blinking detector training and testing additionally required manually marking input signal, indicating blinking moments as seen on figure 1. The performance of the blinking detector is shown in table 1. Results are presented for several individual subjects as well as for whole dataset. Significant difference in error rate is due to inherently different signal quality on different subjects.

Table 1. Error rate of blinking detector.

Test data	False positive error,%	False negative error,%
Subject 1	12.2	8.5
Subject 2	14.2	0
Subject 3	12.9	19.8
Overall	16.9	7.4

Figure 3 presents an example of successful responsiveness lapse detection. Driver's loss of concentration, indicated by a major difference in road and steering trajectories starting, is preceded by a rapid growth in blinking periods. The system reacts 15 seconds before actual loss of control.

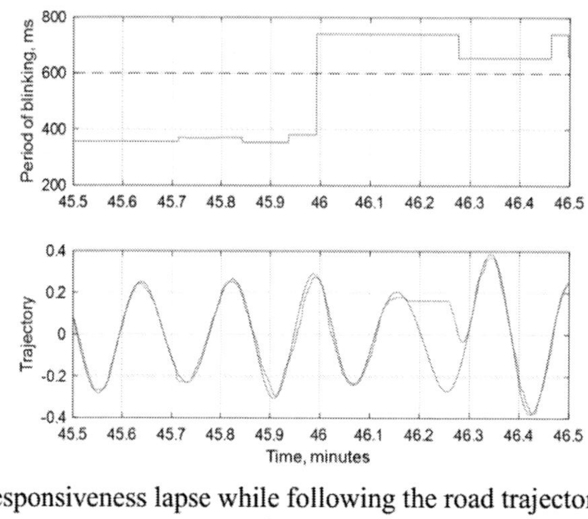

Figure 3. An example of responsiveness lapse while following the road trajectory and it's detection by proposed algorithm

Figure 4 is the time series plot of average blinking period (top) and relative trajectory tracking error (bottom) recorded during one hour driving imitation experiment. Dashed line shows the drowsiness state threshold. Generally, the correlation between high values of blinking period and major tracking errors is clearly visible. At 38 and 54 minutes there are the largest tracking deviations detected. In these moments the algorithm issued the warning about possible responsiveness lapse to the driver.

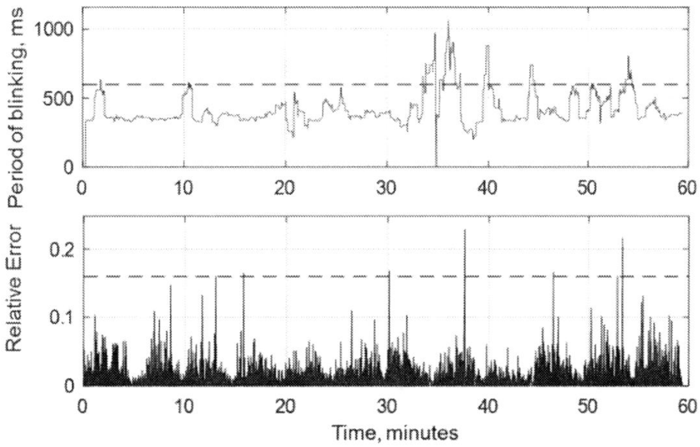

Figure 4. One-hour experiment recording with drowsiness detection.

5. Sleepalert device

The described signal processing for the drowsiness detection system is performed on a microcontroller-based wearable device. The form factor is a head band with five frontal dry-contact and non-contact active electrodes [18, 19]. The device acquires following signals:

- EEG signal using dry-contact electrodes with sampling rate of 1000 Hz. This signal also contains elecromyograhic artifacts such as blinking and eye movement.
- Pulse wave using non-contact electrodes with sampling rate of 1000 Hz
- Driver's head movement signals using accelerometer and gyroscope IC with 125 Hz sampling rate

The mentioned signals are acquired and processed by microcontroller-based control system [20]. The board is equipped with STM32L431 MCU with ARM Cortex-M4 core and FPU coprocessor. Its advanced digital signal processing capabilities are used for on-board data processing, which involves digital filtering, vector and matrix operations and Fourier transform. The device also transmits raw and processed data via the Bluetooth 4.0 interface.

The microcontroller is running FreeRTOS operating system, and the signal processing is implemented as a task in this OS, taking data from input queue, processing it and putting results in output queue. Input sample rate is 250 Hz, since raw EEG signal is preliminary decimated and filtered using a FIR filter with 2-50 Hz bandpass.

Input signal is smoothed out with the 9th order Savitzky-Golay filter. This filter requires four previous samples and four following samples of input signal for every output point. Because of this, the output is delayed by 4 samples. In addition, to reduce the amount of RAM required by smoothing and neural network algorithms, input signal is decimated by a factor of 5. We employ a certain technique to optimize the filtering and decimation operation. It is illustrated on figure 5. Columns represent memory array cells and rows represent steps. Vertical arrow show the windows used by filter. The first smoothing filter window is processed on step 9, it's center is 4th point. The next window is processed on 14th step, centered at 9th point. The next window, taken on 19th step, again is centered in 4th point. To make a new filter window ready by this moment, steps 11th to 14th require an additional copy to cells 0 to 3.

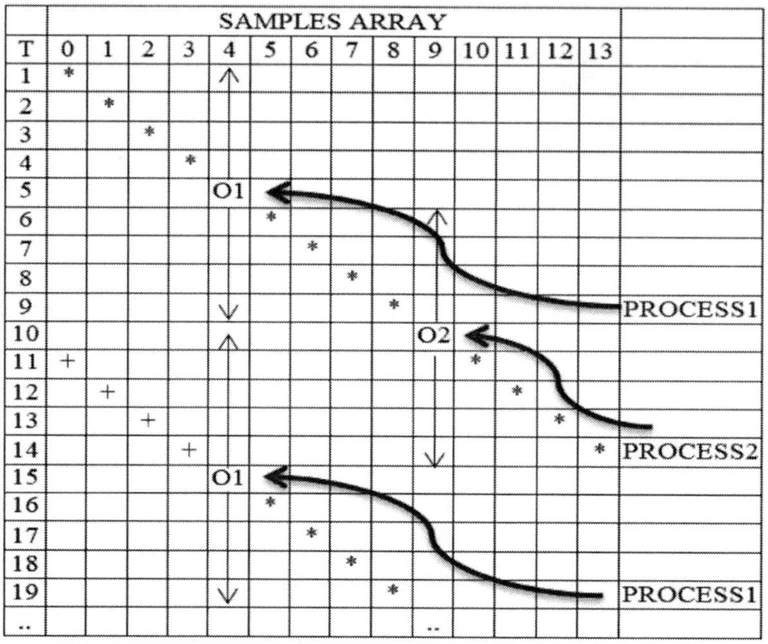

Figure 5. Buffer utilization scheme used by smoothing filte.

Decimated and smoothed signal is then transferred to the ANN input. Network output evaluation is implemented as matrix multiplication and addition operations. Microcontroller implementation of algorithm was tested to provide the same output on the same test data compared to modelling version. In addition, blinking detector was tested in natural environment.

6. Conclusion

In this paper we described the development of the driver drowsiness detection system. This system's operation is based on the blinking detector, employing artificial neural network algorithm to detect blinks in the EEG signal and measure their period and duration. These parameters allowed us to estimate subject's current physiological condition during the experiments with driving process simulation.

Embedded versions of these algorithms are designed for the wearable microcontroller-based device. System performance was tested to be suitable for real-time monitoring.

On the next stages of this research we are going to implement other algorithms utilizing all of the signals acquired by the device to improve the accuracy of drowsiness detection.

Acknowledgment

The work was carried out with the financial support of the FASIE, project №53GRNTIS5/26082 dated 22.12.2017 and grant from the Program Competitiveness Enhancement of Peter the Great St.Petersburg Polytechnic University, Project 5-100-2020.

References

[1] Global status report on road safety. WHO, 2015. http://www.who.int/violence_injury_prevention/road_safety_status/2015/en/.

[2] Prevalence of drowsy driving crashes: estimates from a large-scale naturalistic driving study. AAA Foundation, 2018. http://aaafoundation.org/prevalence-drowsy-driving-crashes-estimates-large-scale-naturalistic-driving-study/

[3] M.T.R. Peiris, P.R. Davidson, P.J. Bones, and R.D. Jones, "Detection of lapses in responsiveness from the EEG," J. Neural Eng. – 2011. – Vol. 8, № 1.

[4] I. Ilin, S. Shirokova, and A. Lepekhin, "IT Solution concept development for tracking and analyzing the labor effectiveness of employees," E3S Web of Conferences 2018. Paper number 33,03007

[5] P. Y. Tsai, W. Hu, T. B. J. Kuo, and L. Y. Shyu, "A portable device for real time drowsiness detection using novel active dry electrode system," 2009 Annual International Conference of the IEEE Engineering in Medicine and Biology Society, Minneapolis, MN, 2009, pp. 3775-3778.

[6] M.T.R. Peiris "Lapses in responsiveness: Characteristics and detection from the EEG : Doctor Thesis," Christchurch, 2008. – 192 p.

[7] A. Sahayadhas, K. Sundaraj, and M. Murugappan, "Detecting Driver Drowsiness Based on Sensors: A Review," Sensors. – 2012. Vol. 12. – P. 16937-16953

[8] D.F. Dinges, M.M. Mallis, G. Maislin, and J.W. Powell, "Final Report: Evaluation of Techniques for Ocular Measurement as an Index of Fatigue and as the Basis for Alertness Management," Report for NHTSA: Washington, DC, USA, 1998.

[9] A. Kokonozi, I. Chouvarda, and N. Maglaveras, "EEG and HRV markers of sleepiness and loss of control during car driving," IEEE Engineering in Medicine and Biology Society. Conference February 2008

[10] K. Sugiyama, T. Nakano, S. Yamamoto, T. Ishihara, H. Fujii, and E. Akutsu, "Method of detecting drowsiness level by utilizing blinking duration," JSAE Review. – 1996. – Vol. 17. – P. 159-163.

[11] M. Miyaji, "Method of Drowsy State Detection for Driver Monitoring Function," Int. Journal of Inform. and Electr. Eng. – 2014. – Vol. 4, № 4.

[12] J. Vicente, P. Laguna, A. Bartra, and R. Bailyn, "Drowsiness detection using heart rate variability," Med Biol Eng Comput. – 2016. Vol. 54. – P. 927–937.

[13] P.R. Davidson, R.D. Jones, and M.T.R. Peiris, "EEG-Based Lapse Detection With High Temporal Resolution," IEEE Trans. on Biomedical Eng. – 2007. Vol. 54, № 5.

[14] R.S. Huang, T.P. Jung, A. Delorme, and S. Makeig, "Tonic and phasic electroencephalographic dynamics during continuous compensatory tracking," Neuroimage. – 2008. – Vol. 39. – P. 1896-1909.

[15] J.I.S. Paiva "Predicting lapses in attention: a study of brain oscillations, neural synchrony and eye measures : Master's Thesis," Coimbra, 2014. – 154 p.

[16] N.A. Abramov, M.V. Bolsunovskaya, A.V. Leksashow, and D.S. Barinov, "Algorithms for detecting and tracking of objects with optical markers in 3D space (Conference Paper)," Proceedings of the 19th International Conference on Soft Computing and Measurements, SCM 2016; Saint Petersburg Electrotechnical University (SPbETU) "LETI", Saint Petersburg, Russian Federation; 25-26 May 2016; Editor Shaposhnikov S.; Institute of Electrical and Electronics Engineers Inc., 2016. Article # 7519782, PP. 362-364 (ISBN: 978-146738919-8, DOI: 10.1109/SCM.2016.7519782)

[17] A.S. Makarov, and M.V. Bolsunovskaya, "The method of obtaining orthographic top view for around view system at vehicles (Conference Paper)," Proceedings of the 2017 IEEE Russia Section Young Researchers in Electrical and Electronic Engineering Conference, ElConRus 2017, Saint-Petersburg, 01-03 February 2017, Institute of Electrical and Electronics Engineers Inc., 2017. Article # 7910652, PP. 697-699. (ISBN: 978-150904865-6, DOI: 10.1109/EIConRus.2017.7910652)

[18] Y. M. Chi, P. Ng, E. Kang, J. Kang, J. Fang, and G. "Cauwenberghs, Wireless non-contact cardiac and neural monitoring," Wireless Health 2010. – ACM, 2010. – P. 15-23.

[19] E. Valchinov, A. Rutkovskis, and N. Pallikarakis, "Wireless dry-contact biopotential electrode," E-Health and Bioengineering Conference (EHB), 2011. – IEEE, 2011. – P. 1-4.

[20] A.A. Efremov, A.V. Loginova, B.D. Mikeladze, and S.V. Shirokova, "The models and technologies for supporting decision making in design of information-control complexes," Proceedings of 2017 20th IEEE International Conference on Soft Computing and Measurements, SCM 2017. 6 July 2017, Paper number 7970742, Pages 846-848

IOP Conf. Series: Materials Science and Engineering **618** (2019) 012030 doi:10.1088/1757-899X/618/1/012030

Financial investments by Portfolio optimization

Todor Stoilov [1,*], **Krasimira Stoilova**[1] **and Miroslav Vladimirov** [2]

[1] Institute of Information and Communication Technologies – Bulgarian Academy of Sciences, Sofia, Bulgaria
[2] Varna University of Economics, Varna, Bulgaria

todor@hsi.iccs.bas.bg

Abstract. Application of the Portfolio theory to appreciation of the Bulgarian stock exchange following the most popular financial indices is presented. The data of these indices corresponds to the same watching periods an year before. Comparison of the optimization problem's solutions and analysis of the financial activities are commented. Modification of the portfolio optimization problem is proposed for calculation facilitation.

1. The first section in your paper

The resource allocation is done by investing capital in financial assets (or goods) which give a return to the investor after a period of time. The theoretical and formalized background of the investment process is based on the portfolio theory. For the investor the target is to maximize the return and to minimize the investment risk [1, 2, 3].

2. Description of the Portfolio Problem

The classical portfolio problem is analytically given by relation [1]

$$\min_x [\frac{1}{2}(1-\sigma)x^T \operatorname{cov}(.)x - \sigma E^T x] \qquad (1)$$
$$x^T.\mathbf{1} = 1,$$

where

cov(.) – is a symmetrical positively defined square matrix $n \ x \ n$ which presents the covariance between returns of n assets, participating in the portfolio. This matrix gives also numerical values of asset' risks;

E - is a $(n \ x \ 1)$ vector of the average profits of the assets for the period of time $T = [1, 2, ..., N]$;

σ - is a parameter of the investor's preferences to undertake risk in the investment process, $0 \leq \sigma \leq \infty$,

$\mathbf{1} = \begin{vmatrix} 1 \\ \vdots \\ 1 \end{vmatrix}$, is a unity vector, $n \times 1$;

The constraint of the optimization problem is the equation $x_1 + x_2 + \cdots + x_n = 1$, which is a formalization of the assumption that the full amount of the investment resources will be used for the portfolio. The solutions x_i , $i=1,n$ give the relative values of the investments, which are allocated for the assets i, $i=1,n$.

The first component $V_p = x^T \text{cov}(.)x$ of the goal function is the quantitative assessment of the portfolio risk. The second component $E_p = E^T x$ is the quantitative value of the portfolio return. The goal function of problem (1) targets the minimization of the portfolio risk V_p and also the maximization of its return E_p. The parameter σ formalizes the investor's ability to undertake risk and it has numerical value in the range [0, ∞]. When σ=0 the investor's main goal is to decrease the risk of the investment, $\min_x[x^T \text{cov}()x]$. For $\sigma=\infty$ the investor's target is to obtain a maximal return from the investment $\min_x[-\sigma E^T x] \equiv \max_x[E^T x]$.

For different values of the parameter σ, the portfolio problem (1) gives different solutions. The set of solutions for given set of σ graphically represents the "efficient frontier", which is drown in the space "risk-return", $V_p(E_p)$. The MPT recommends that the investor has to choose such an investment, which portfolio is a point, lying on the "efficiency frontier". An example of a graphical representation of the "efficiency frontier" is given in figure 1.

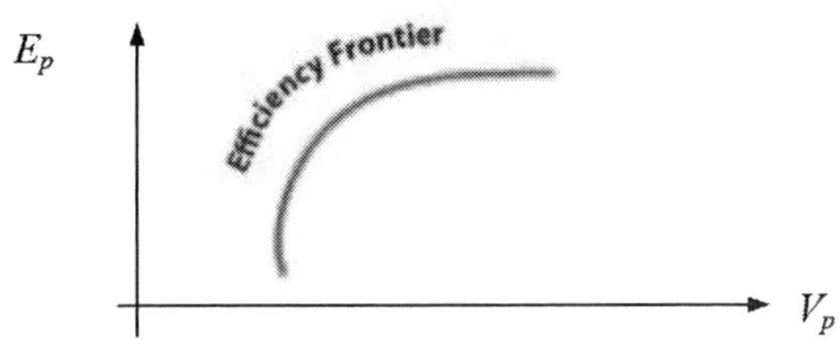

Figure 1. Efficiency frontier of the portfolio optimization.

The main characteristics of the Portfolio optimization problem are presented in [4]. The Portfolio theory was applied there for assessment of four indices of the Bulgarian stock exchange. In this paper we would like to see the changes of the same indices a year later. The data of their behaviour have been taken from the public data [5] from the same dates but a year later.

3. Application of the Portfolio Problem for Appreciation of the Bulgarian Financial Market

The Bulgarian stock exchange is presented of four indices: SOFIX, BGBX40, BGTR30 and BGREIT. Here we assume that each index represents particular security. According to the Portfolio theory we solve optimization problem - problem (1), which solutions show the preferable securities (here indices) for investments. Practically, each of the indices represents the behaviour of the included different Bulgarian companies (corresponding securities). For instance, the index SOFIX includes 16 Bulgarian companies (OEA, ALB, ATERA, BREF, CCB, CENHL, CHIM, EUBG, FIB, HVAR, MCH, MONBAT, SFARM, SKK, SKKW2, TRACE).
The values of the financial data are given in Table 1.

The volume of the financial data is in million leva. In order to have relative relations and to calculate the covariance matrix we transform Table 1 to relative units where the values are calculated according to the relation in percentage:

*[(New value – Old value)/Old value]*100*

The transformed data are presented in Table 2.

Table 1. Data of the four Bulgarian indices

Date	SOFIX	BGBX40	BGTR30	BGREIT
26.1.2019	573.65	113.3	487.28	120.45
29.12.2018	595.65	116.1	496.59	121.19
24.11.2018	588.93	114.9	488.74	117.35
29.10.2018	599.65	117.46	499.03	116.45
29.9.2018	624.39	121.93	520.34	117.79
25.8.2018	636.32	123.07	525.98	116.47
28.7.2018	632.31	122.52	523.56	116.44
30.6.2018	634.26	124.9	525.39	116.21
26.5.2018	642.09	127.44	531.11	114.77
28.4.2018	658.2	130.34	541.7	115.53
31.3.2018	649.17	128.54	536.29	114.93
24.2.2018	687.26	130.71	557.92	114.65
27.1.2018	709.62	137.91	567.67	116.09

Table 2. Relative data of the four Bulgarian indices in %

SOFIX	BGBX40	BGTR30	BGREIT
-3.69	-2.41	-1.87	-0.61
1.14	1.04	1.61	3.27
-1.79	-2.18	-2.06	0.77
-3.96	-3.67	-4.10	-1.14
-1.87	-0.93	-1.07	1.13
0.63	0.45	0.46	0.03
-0.31	-1.91	-0.35	0.20
-1.22	-1.99	-1.08	1.25
-2.45	-2.22	-1.95	-0.66
1.39	1.40	1.01	0.52
-5.54	-1.66	-3.88	0.24
-3.15	-5.22	-1.72	-1.24

Using data in Table 2 the covariance matrix is in the form:

Cov() =

	SOFIX	BGBX40	BGTR30	BGREIT
SOFIX	**4.315599**	2.760744	3.287769	1.404938
BGBX40	2.760744	**3.305114**	2.230184	1.483142
BGTR30	3.287769	2.230184	**2.819019**	1.213343
BGREIT	1.404938	1.483142	1.213343	**1.415463**

IOP Conf. Series: Materials Science and Engineering **618** (2019) 012030 doi:10.1088/1757-899X/618/1/012030

This matrix is obligatory for the evaluation of the goal function of the Portfolio Optimization Problem (1). For different values of the coefficient σ which represents the investor's behaviour for taking risk in investments we solve the optimization problem (1). For different σ we obtain the solutions of the optimization problem, the portfolio risk, the portfolio return and the goal function F, Table 3. The solutions of the optimization problem are the unknown variables x which here are the four indices of the Bulgarian Stock exchange. The columns SOFIX, BGBX40, BGTR30, BGREIT are x1, x2, x3, x4 and these solutions are the relative values of the investments, which are allocated for the SOFIX, BGBX40, BGTR30, BGREIT.

For $\sigma=0$ the investor should invest 0.89 units in BGREIT, 0.11 units in BGTR30 and zero in SOFIX and BGBX40.

According to the Portfolio optimization problem the best investment policy for $\sigma=0.15$ and more (here σ is changed to 3.5 but the tendency is the same for larger values) to invest all capital in BGREIT.

The Efficiency frontier for the presented data is shown in figure 2.

Table 3. Data of the four Bulgarian indices

sigma	SOFIX	BGBX40	BGTR30	BGREIT	Risk	Return	*F*
0	0	0	0.111805	0.888195	1.392865	0.139757	0.696433
0.001	0	0	0.110939	0.889061	1.392867	0.141111	0.696292
0.01	0	0	0.103149	0.896851	1.393001	0.153303	0.694967
0.02	0	0	0.094492	0.905508	1.393407	0.166849	0.693367
0.03	0	0	0.085836	0.914164	1.394084	0.180395	0.69163
0.05	0	0	0.068523	0.931477	1.396252	0.207487	0.687752
0.1	0	0	0.025242	0.974758	1.406411	0.275218	0.675684
0.15	0	0	0	1	1.415463	0.314719	0.660524
0.2	0	0	0	1	1.415463	0.314719	0.644788
0.25	0	0	0	1	1.415463	0.314719	0.629052
0.3	0	0	0	1	1.415463	0.314719	0.613316
0.35	0	0	0	1	1.415463	0.314719	0.59758
0.4	0	0	0	1	1.415463	0.314719	0.581844
0.45	0	0	0	1	1.415463	0.314719	0.566108
0.5	0	0	0	1	1.415463	0.314719	0.550372
0.6	0	0	0	1	1.415463	0.314719	0.5189
0.8	0	0	0	1	1.415463	0.314719	0.455957
1	0	0	0	1	1.415463	0.314719	0.393013
1.5	0	0	0	1	1.415463	0.314719	0.235654
2	0	0	0	1	1.415463	0.314719	0.078294
2.5	0	0	0	1	1.415463	0.314719	-0.07907
3	0	0	0	1	1.415463	0.314719	-0.23642
3.5	0	0	0	1	1.415463	0.314719	-0.39378

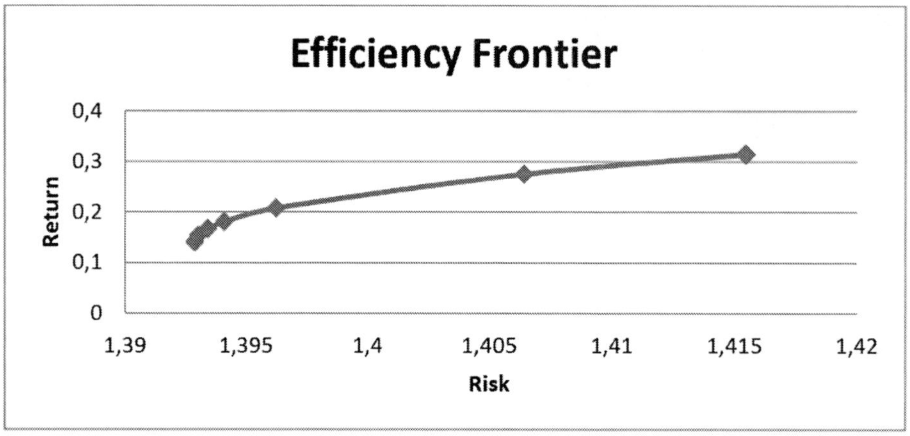

Figure 2. The efficiency frontier of the current stock exchange data.

The variations of BGTR30 and BGREIT according to σ are given in figure 3 and figure 4. The other two indices – SOFIX and BGBX40 have solutions zero that means that the investments are not recommended.

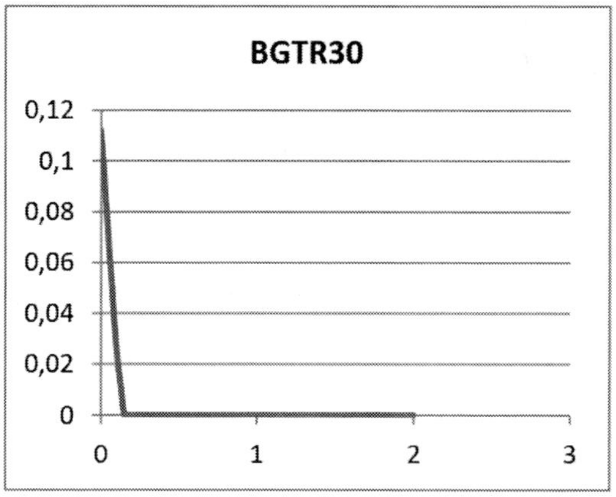

Figure 3. Variation of BGTR30 according to σ.

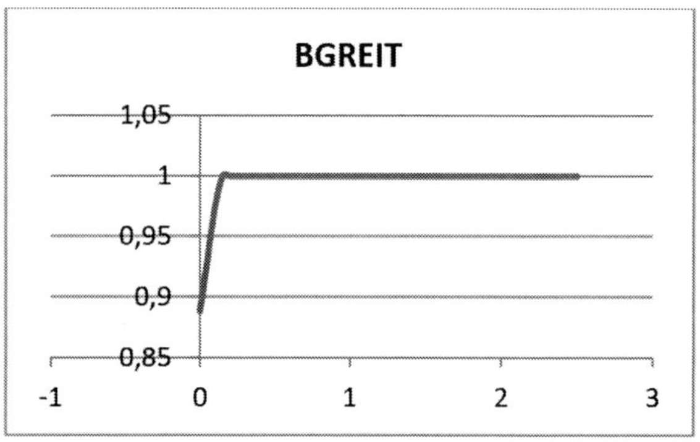

Figure 4. Variation of BGREIT according to σ.

The variations of Risk and Return to sigma are given in figure 5 and figure 6

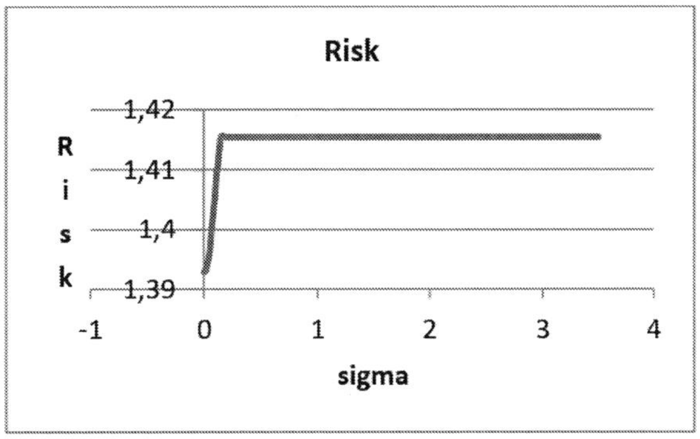

Figure 5. Variation of Risk according to σ.

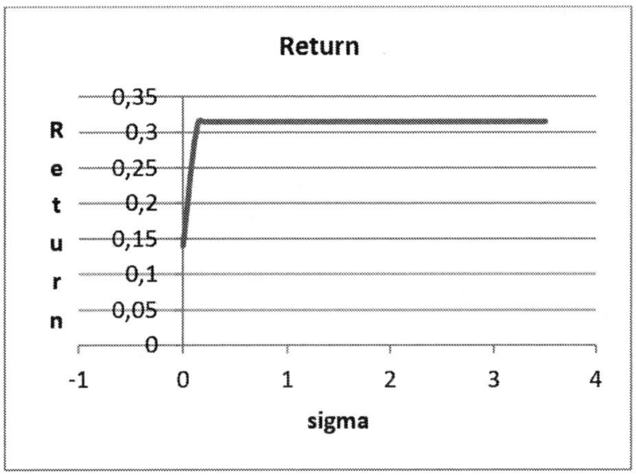

Figure 6. Variation of Return according to σ.

The results from the application of the Portfolio theory to real data of Bulgarian stock exchange show the following. The variations of problem's solutions are in changing σ from 0 to 0.15. When the coefficient of investor's preferences of taking risk σ is greater than 0.15 the problem's solutions, the Risk and Return obtain constant values.

4. Modification of the Portfolio optimization problem

The solution of the portfolio optimization problem (1) gives the optimal values of x_i that represent the best investment activities. Each solution of this problem gives one point of the Efficiency Frontier of the plane Return/Risk. By giving different values of σ, $0 \le \sigma \le \infty$ the Portfolio optimization problem is solved and accordingly one point of the Efficiency Frontier is received. The variations of σ are determined by the person, making decision. We propose more universal approach for decision making by normalization of the Portfolio optimization problem by proposing new variable ψ, $0 \le \psi \le 1$. The Portfolio optimization problem (1) has the following modification

$$\min_x [\frac{1}{2}(1-\psi)x^T \operatorname{cov}(.)x - \psi E^T x]$$

$$x^T.\mathbf{1} = 1,$$

$$0 \le \psi \le 1 \qquad (2)$$

Problem (2) is solved for different values of ψ, $0 \le \psi \le 1$.

IOP Conf. Series: Materials Science and Engineering **618** (2019) 012030 doi:10.1088/1757-899X/618/1/012030

The results are shown in Table 4.

The efficiency frontier of the initial (1) and modified (2) portfolio optimization problem is the same for both cases according to figures 2 and 7. The similarity can be seen also from the solutions, presented in Table 3 and Table 4. This equivalence can be used for changing ψ in smaller interval, $0\leq \psi \leq 1$ instead of changing σ in bigger interval, $0\leq \sigma \leq\infty$.

Table 4. Solutions of the modified Portfolio optimization problem.

ψ	SOFIX	BGBX40	BGTR30	BGREIT	Risk	Return
0	0	0	0.107216	0.892784	1.392903	0.146938
0.01	0	0	0.107216	0.892784	1.392903	0.146938
0.05	0	0	0.066245	0.933755	1.396618	0.211052
0.1	0	0	0.015624	0.984376	1.409589	0.290269
0.15	0	0	0.015624	0.984376	1.409589	0.290269
0.2	0	0	0	1	1.415463	0.314719
0.25	0	0	0	1	1.415463	0.314719
0.3	0	0	0	1	1.415463	0.314719
0.35	0	0	0	1	1.415463	0.314719
0.4	0	0	0	1	1.415463	0.314719
0.45	0	0	0	1	1.415463	0.314719
0.5	0	0	0	1	1.415463	0.314719
0.55	0	0	0	1	1.415463	0.314719
0.6	0	0	0	1	1.415463	0.314719
0.65	0	0	0	1	1.415463	0.314719
0.7	0	0	0	1	1.415463	0.314719
0.75	0	0	0	1	1.415463	0.314719
0.8	0	0	0	1	1.415463	0.314719
0.85	0	0	0	1	1.415463	0.314719
0.9	0	0	0	1	1.415463	0.314719
0.95	0	0	0	1	1.415463	0.314719
1	0	0	0	1	1.415463	0.314719

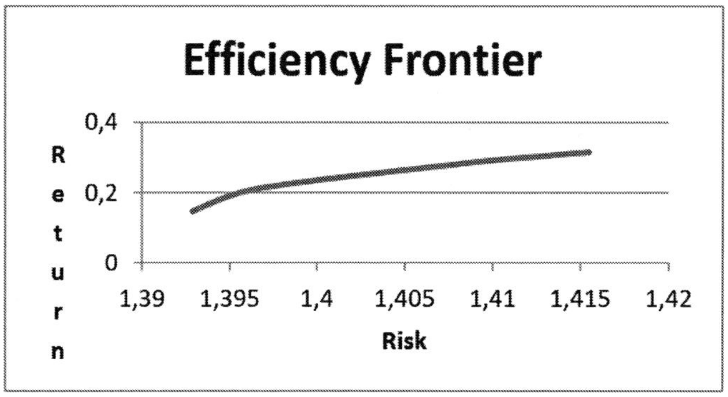

Figure 7. Efficiency Frontier of the modified Portfolio optimization problem.

5. Comparison of Portfolio problem's solutions

In [4] the Portfolio theory was applied for making decision of financial investments of the same financial indices like in this research. The data here of the four indices is taken at the same period like in [4]. We want to know how one year period reflects to the problem's solutions or how the financial investments are changed. Here are presented the previous graphical results from [4].

The optimal solutions of the portfolio optimization problem from the both groups of data are quite different. The first solution x_1 is about SOFIX. The current results show that it is not recommended the investments in SOFIX because the solutions are zero (see Table 3). A year before this index is preferable for investments for $\sigma>2$ (see figure 8).

The second solution x_2 is about BGBX40. For both periods this index is not suitable for investments according to the portfolio optimization problem (see Table 3 for current results and figure 9 about previous results).

The third solution x_3 is about BGTR30. This index is preferable for both periods however for different changes of σ. For the current results BGTR30 is suitable for investments for $0\leq\sigma\leq0.15$ (see figure 3). The previous variation of BGTR30 shows that preferable investments are for $1\leq\sigma\leq2.5$ (see figure 10).

The fourth solution x_4 is about BGREIT. This is the preferable index about current solution for $\sigma\geq0.15$ (see figure 4). In the previous calculations BGREIT has decreasing character for $0\leq\sigma\leq2$ and for $\sigma\geq2$ it is not recommended for investments, fig.11.

The efficiency frontier has similar variation for both cases. The variations of Risk according to σ and the variations of Return according to σ are similar for the both experiments.

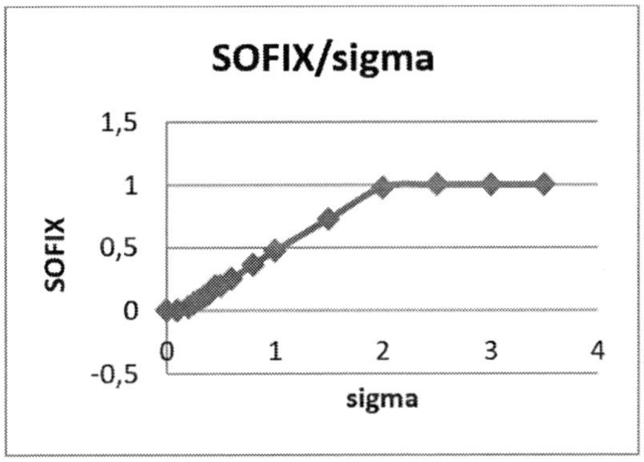

Figure 8. Optimal solutions changes SOFIX /sigma [4].

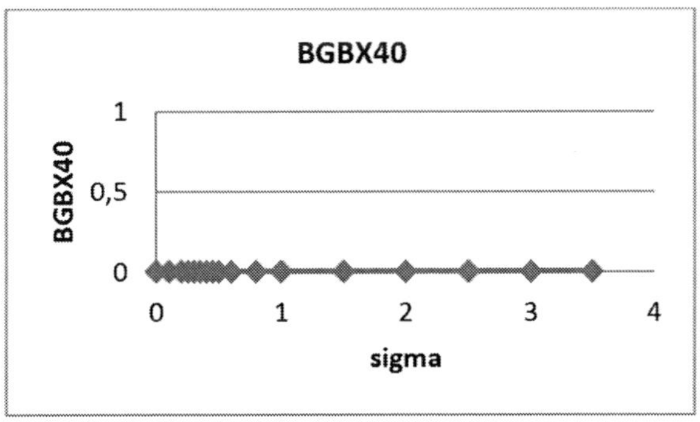

Figure 9. Optimal solutions changes BGBX40/sigma [4].

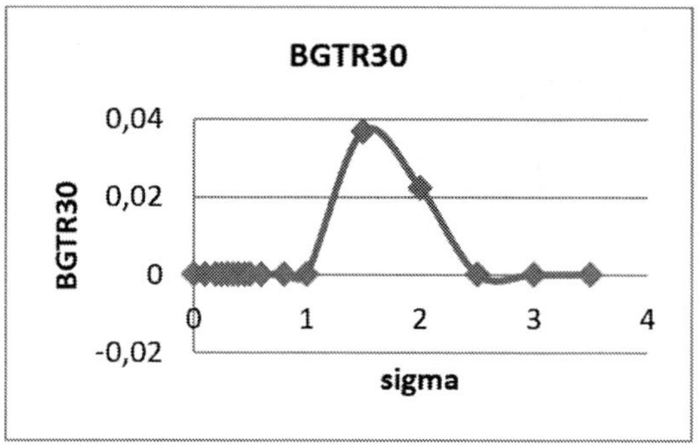

Figure 10. Optimal solutions changes BGTR30/sigma [4].

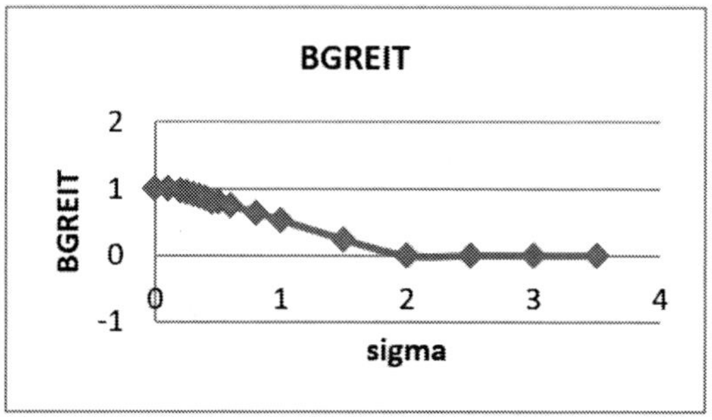

Figure 11. Optimal solutions changes BGREIT/sigma [4].

Nevertheless the above similarities, the received calculations show the differences about the preferable financial investments. It means that one year period is quite different for making decisions for financial investments at Bulgarian stock exchange. One year is too long period for using the previous results of the same indices and for better results the Portfolio optimization problem has to be more frequently solved for determining the best investments.

6. Conclusions

The portfolio optimization is applied for optimal resource investments. Four typical financial indices of the Bulgarian stock exchange are analysed and compared with previous results. The financial market in our country is very dynamic and the obtained results an year before are not valid today. The authors' recommendation is to apply more frequently the portfolio optimization for better financial investments.

Acknowledgements

This work has been partly supported by project H12/8, 14.07.2017 of the Bulgarian National Science fund: Integrated bi-level optimization in information service for portfolio optimization contract ДН12/10, 20.12.2017

References

[1] W. Sharpe 2010 Adaptive asset allocation polices. *J. Financial Analysts*, **66**, 5, pp.45-49 (2010)

[2] P. Magiera, A. Karbowski 2001 Dynamic portfolio optimization with expected value-variance criteria. *Preprints of the 9th IFAC Symposium on LSS'01*, Bucharest, Romania, pp. 308-313,

[3] J. Campbell, G. Chacko, J. Rodriguez, L. Viceira 2002 Strategic asset allocation in a continuous-time VAR model, Harvard Institute of Economic Research, Harvard University Cambridge, Massachusetts, pp.1-21

[4] K. Stoilova, T. Stoilov, M. Vladimirov 2018 Resource allocation by portfolio optimization. *Journal of the Technical University - Sofia Plovdiv branch, Bulgaria, "Fundamental Sciences and Applications"* **24** pp.19-24

[5] http://infostock.bg

A Study of Wi-Fi Security in City Environment

H Valchanov[1], J Edikyan[2], V Aleksieva[3]

[1] Assoc. Professor, Head of Department of Computer Science, Department of Computer Science, Technical University of Varna, Varna, Bulgaria
[2] Programmer, Varna, Bulgaria
[3] Assoc. Professor, Department of Computer Science, Technical University of Varna, Varna, Bulgaria

E-mail: hristo@tu-varna.bg

Abstract. Wireless networks are widely used with increasing number of users. Typically, they are based on the IEEE 802.11 standard (Wi-Fi). Because of its broadcast nature, they are very susceptible to security attacks. It is very important to study the risks and potential vulnerabilities of Wi-Fi networks, which will allow defining recommendations for users to enhance their security. This paper presents a methodology and research of the security of wireless networks in Varna city based on war-driving technique. The obtained results are analyzed and compared with those from previous research. The recommendations are given.

1. Introduction

Wireless networks allow easy to build small-sized home and corporate networks. The disappearance of the need to build cableways, additional networking devices and the possibility of mobility are the basis for their increasing use in recent years [1]. The spreading of Internet of Things is a prerequisite for increasing the number of user worldwide. Typically, the wireless networks are based on the IEEE 802.11 standard (Wi-Fi) [2]. Technologies have evolved continuously, giving faster speeds for users. But because of the nature of radio signals, the wireless networks are easily susceptible to security attacks. This may include both disconnect attacks and unauthorized access to resources on a network, stealing information or hiding behind a legitimate network for attacks against other networks. One of the main tasks of network security is to study the risks and potential vulnerabilities of Wi-Fi networks, which will allow defining recommendations for users to enhance their security [3].

This paper presents a methodology and research of the security of wireless networks in Varna city. Information was gathered using the use of the war-driving technique. The obtained results are analyzed and compared with those from previous studies.

2. Background

The wireless networks are primarily built on the 802.11 standard. It includes a number of standards that define the use of various wireless access technologies, starting with 802.11a and 54Mbs up to today's 802.11ah at 347Mbps [4]. The development of the 802.11 standard is tied up with the development of wireless network security protocols. There are 3 main protocols: WEP, WPA and WPA2 [5].

Wired Equivalent Privacy (WEP) – this is historically the first protocol to encrypt the connection between wireless devices based on a pre-shared key. This is also its main drawback, allowing the hacker to get this key after capturing a large number of packets. Its use is not recommended.

Wi-Fi Protected Access (WPA) – it improves security compared to WEP. It provides two ways for access control. For personally use it is recommended WPA Pre-Shared Key with a 256-bit key length. For enterprise solutions, WAP Enterprise (802.1x) and authentication server are used. To ensure the integrity of the data, Temporal Key Integrity Protocol (TKIP) is used to dynamically generate 128bit keys for every packet. Despite improvements, it is recommended to use WPA2.

Wi-Fi Protected Access 2 (WPA2) – it is an enhanced version of WPA, more resistant to attacks. It uses a new asymmetric Advanced Encryption Standard (AES) encryption algorithm and a new protocol replacing TKIP – Counter Mode with Cipher Block Chaining Message Authentication Code Protocol (CCMP).

Despite the improved security of WPA and WPA2, both have the common problem – a possible attack through Wireless Protected Setup (WPS) [6]. This is a standard that makes it easy to configure wireless devices. The vulnerability is expressed in the ability of an attacker to break the used password within 2-4 hours and to get unauthorized access to the network.

3. Methodology

The purpose of the study is to answer the main question: *What is the current state of the security of wireless networks in Varna?*

Achieving of the goal requires the following tasks:

- Obtain information about a large number of wireless networks in Varna;
- Determine the percentage of encrypted networks;
- Identify the types of manufacturers and networks SSID;
- Determine the percentage of use of WPS.

To realize the tasks, the study is carried out in following phases: design of a system for scanning, collecting of data, data analysis and comparison of the results with ones of previous research.

4. Design of the system

Wireless security research has been a manual process since its creation, but due to the much available information, the process has not been effective for wider use. The first automated wireless network security research was carried out at the end of 1999 in California by Pete Shipley and first introduced at the DEFCON 9 conference in Las Vegas in 2001 [7]. The basic principle is to visit the area to be explored and walk around it (on foot, by bicycle or car) with hardware capable of collecting and recording information transmitted from wireless access points, as well as GPS information about their approximate position.

This study uses the war-driving technique to crawl larger areas for analysis [8, 9, 10]. A passive study has been performed and the conclusions about security being based on only the publicly available information broadcasted from each wireless access point [11].

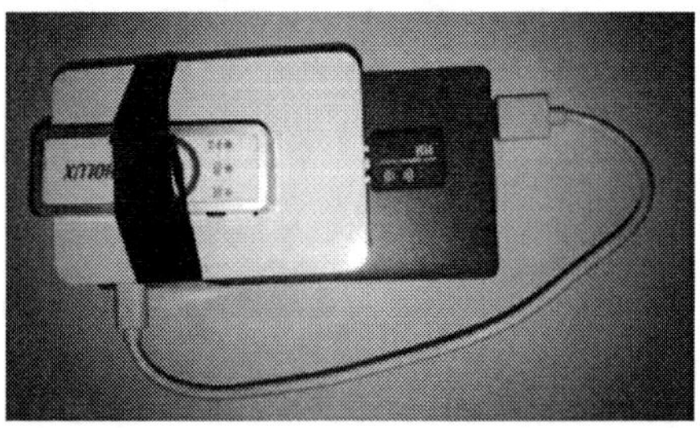

Figure 1. Wi-Fi scanning system.

The data collection system is based on a single-board computer Raspberry Pi 3 Model B. It has an ARM Cortex-A53, 1.2GHz processor, built-in Wi-Fi and Bluetooth functionality, as well as increased input current up to 2.5A. For the purpose of the conversion, it is necessary to record the position of each wireless access point. The selected GPS module is Holux M-1200E due to its support for the standard NMEA 0183, long-life battery and large memory support. To scan the wireless networks, the Wi-Fi module CanaKit is selected. To ensure uninterrupted power supply to the Raspberry Pi, a Canyon CNS-TPBP5DG portable battery with a 5000mAh capacity is used (figure 1). Such as embedded system makes it possible to use it also in various geographical locations where there is no possibility of driving by car (such as pedestrian, rural, village or industrial areas).

Wireless network scanning is done using Kismet open source software. The software has been compiled and installed under the Raspbian operating system. The Kismet data is recorded in *netxml* format then they are converted using a Python script in *csv* format. This is necessary for data to be presented in tabular format for easier processing and analysis with Microsoft Excel.

5. Data collection

The selected area of analysis includes the central part of Varna, because there are many offices and it includes a large part of the residents. Also, the area coincides with a similar study [12] to compare the obtained results (figure 2a).

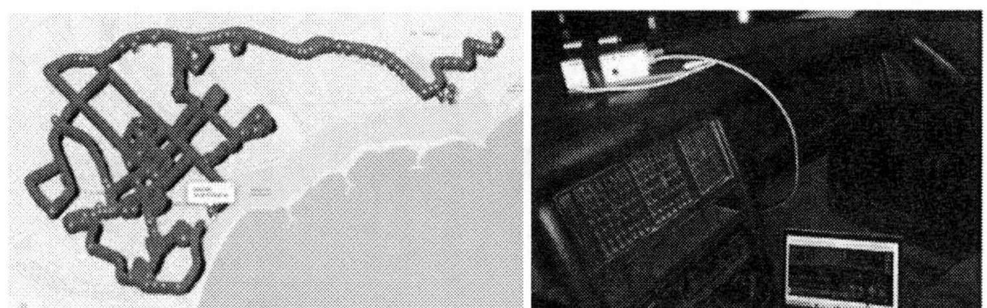

(a) (b)

Figure 2. Scanned areas of the city (a) and war-driving in action (b).

Driving the area was done by car, with a sustained speed of 0-35km/h. In order to monitor and control the process in real time, the Raspberry Pi Ethernet port connects to a laptop where, thanks to a VNC client, the state and information presented by Kismet are monitored (figure 2b).

6. Data analysis

The collected data includes information for a total of 19136 networks. For the purpose of the analysis, all networks which use channel 0 are excluded as invalid because of such a channel is incorrect. The reason for their occurrence is probably that there was not enough time to perform a full scan of such as networks. Thus, the number of networks to be analyzed is reduced to 11534.

6.1. Wi-Fi security

Figure 3 shows the percentage of encryption methods used. These data lead to summarized results showing that the percentage of networks with WEP protection is negligible small. From the figure is shown that 7% use no encryption. It is more likely that these are public networks with open access than they are not configured. This is typical for restaurants and shops as for the public city areas.

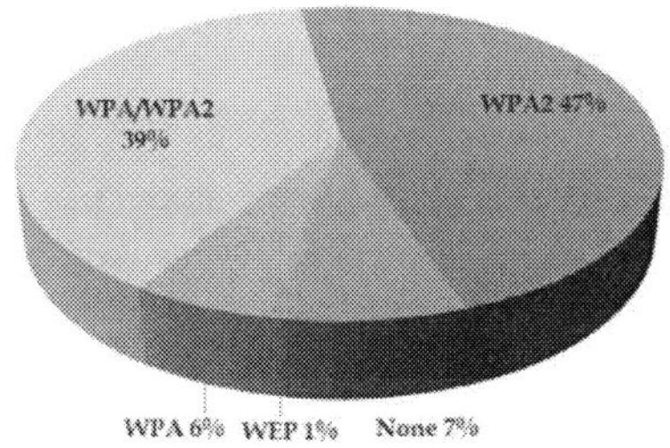

Figure 3. Use of Wi-Fi encryption.

Only 6% of the networks use WPA, which is positive, but unfortunately, another 39% also support it because, besides the AES CCMP protocol (this mean that they support WPA2), they also support TKIP. In this way these networks, which appear to be secure (as they are WPA2 networks), are in fact unsecure because they actually work in mixed mode. In many cases, however, mixed mode is needed for old client devices that have not support for WPA2.

6.2. SSID use

Focusing only on WPA2 security networks and their SSIDs, it can be concluded that at least 13.3% of these devices use the default network name (SSID), this may mean that they use default password too which is known for the vendors (figure 4). This is a potential vulnerability for penetration.

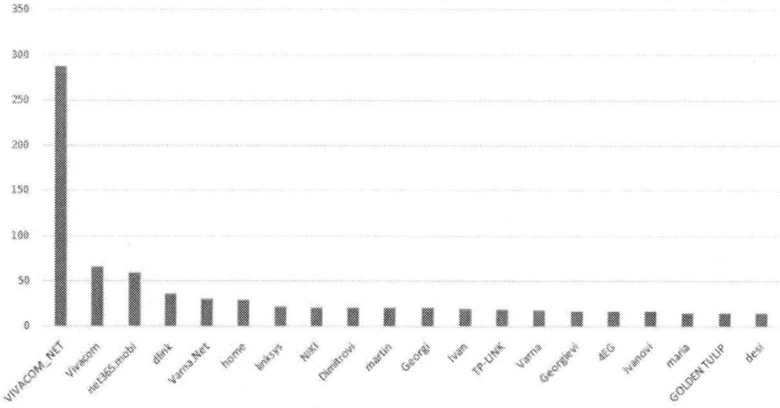

Figure 4. Commonly used SSIDs.

The analysis of the SSIDs shows that only 2% of the SSID identifiers are hidden. This method for security is good because the network does not appear on a Windows/Mac OS/Linux scan, but this is not a problem for any more serious hacker. A particular problem is that 1/3 of these networks only rely on the fact that their SSID is hidden – they do not use any method of protection.

6.3. Wi-Fi manufacturers

The most commonly used brands for Wi-Fi devices in the city are shown on figure 5.

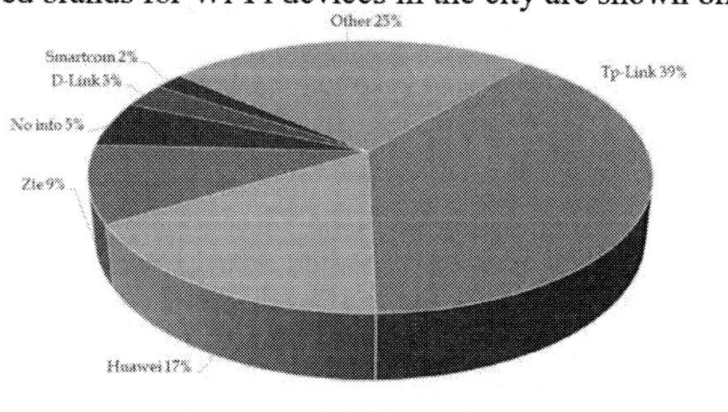

Figure 5. Wi-Fi vendors.

The TP-Link brand is on the first place – the reasons for this are both the low price and the fact that one of the largest suppliers in the city (TCV) provides models of this brand for its customers free of charge. For the second and third, Huawei and Zte, the situation is similar. Mtel provides Huawei devices, Vivacom (previously provided ZTE) in recent years also provides Huawei. For both providers, due to the type of infrastructure, the wireless router settings can be made remotely and in group. However, thanks to better default settings, only for TP-Link models the percent for WPA2-protected networks is larger than combined WPA/WPA2 method (figure 6).

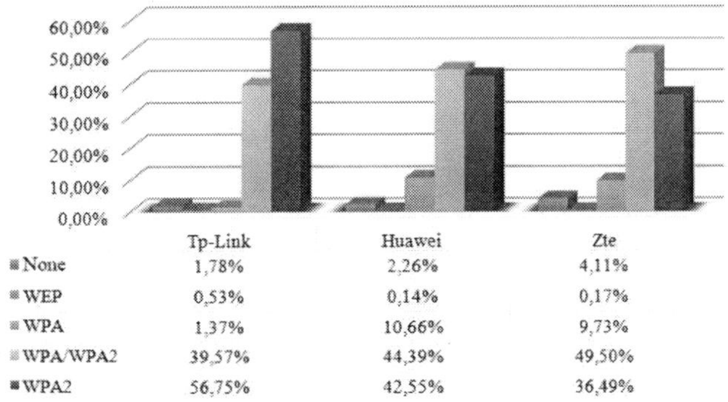

Figure 6. Wi-Fi security by vendors.

6.4. WPS support

The figure 7 shows how many of the networks in Varna offer the WPS function. It is a very important factor for security as mentioned before. It is supposed that many users are not familiar with the problems in the technology.

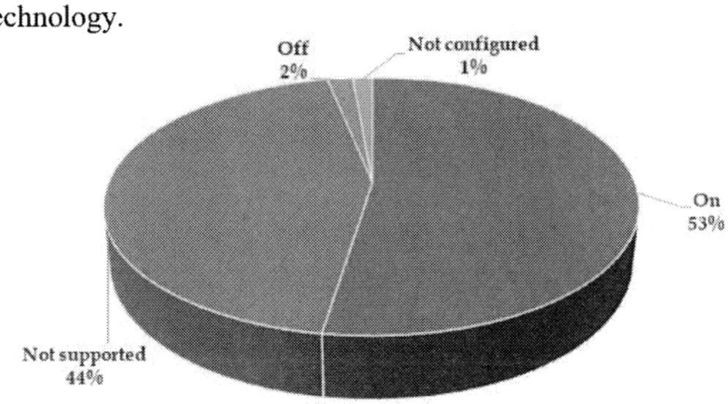

Figure 7. WPS support.

When analyzing WPS support for top brands, it is noticed that for TP-Link 77% of the devices allow WPS to be used. For Mtel and Huawei devices, the default WPS is off. For Vivacom and ZTE, the situation is the opposite - WPS is implicitly switched on (figure 8).

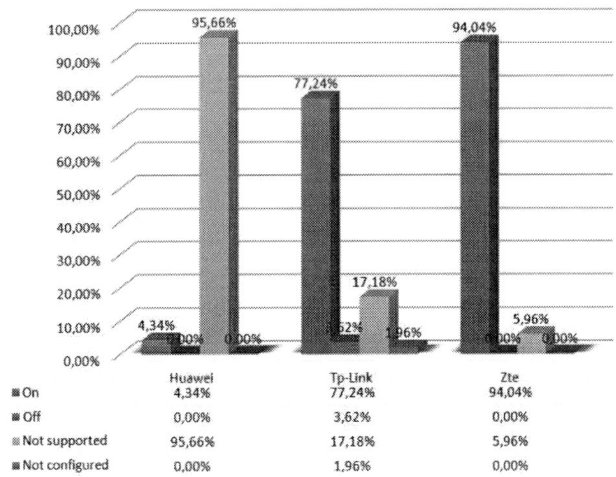

Figure 8. WPS support by vendors.

7. Comparison with previous research

This study aims to compare the results obtained with similar ones obtained in 2008 [12]. The survey collected data for 688 networks. Compared with the current number of detected networks, this means that for 10 years in the city the number of networks has increased more than 16 times. The real percentage may be lower as the network card with which the scanning was performed, detected networks with the highest standard of 802.11g. But summary, the results show that for this period there was a significant increase in the use of Wi-Fi networks in Varna.

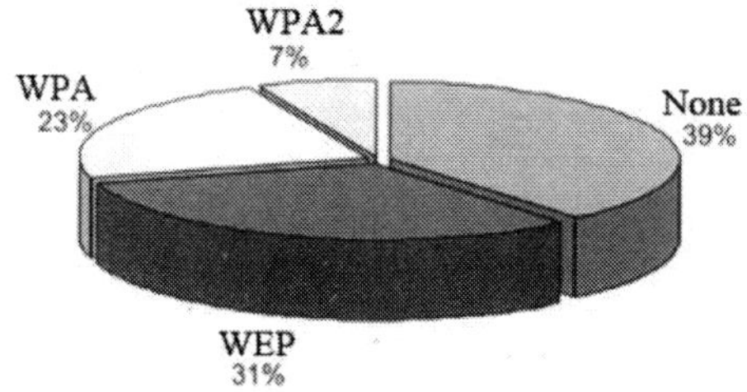

Figure 9. Use of Wi-Fi encryption in 2008 [12].

In addition to a significant increase in the number, there is also a major improvement in network security as shown on figure 9. The percentage of networks without any protection has decreased almost 4 times, from 39% to 7%, and networks with WEP protection level are virtually not found (as opposed to over 30% in 2008). The WPA level has also decreed to a small number of 6%, and the increase is directed to WPA2 (40% growth) and WPA/WPA2 mixed devices.

With regard to manufacturers, the only device brand retained and even increased its share is ZTE. All other major players in 2008 have a market share of less than 5% (now Cisco-Linksys has 2% and D-Link - 3%). The previously obtained results are shown on figure 10.

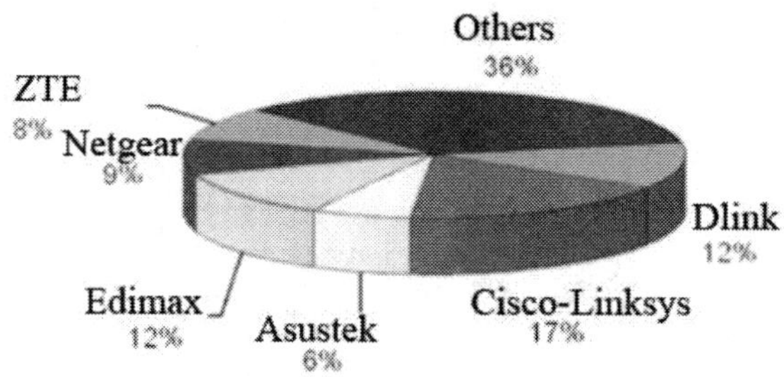

Figure 10. Wi-Fi vendors in 2008 [12].

Comparing the results from previous research can be summarized that the security level has been greatly increased. But in the same time due to the detected shortcomings in the wireless router security over the years, the number of real users with vulnerable security has also actually increased (the total number of unprotected networks and those using WEP or WPA is 1634 – 60% more than analyzed in 2008.

In Varna 47% of the networks use the current WPA2 standard and thanks to this, the wireless networks in the city have a better relative security level.

An extremely worrying fact is that 59% of the WPA2 protection networks in Varna use the WPS function, significantly reducing their security. In addition, at least 13.3% of these networks use default network names. This means that they also use a default password with a minimum length susceptible to dictionary attacks.

The budget required for implementing an attack on WEP, WPA, WPA/WPA2 mixed mode or WPS is extremely low - up to 50 EUR are needed to buy a wireless adapter that allows active mode of operation. There are a myriad of methods and programs that make it easy to implement an attack by a few mouse clicks.

Although there is a significant improvement in security since 2008, there are still many steps to be taken to ensure good security for wireless networks in Varna. This is critical, as the growth in the number of wireless access points and devices will continue to grow in the future.

8. Conclusions

This paper presents a wireless security research carried out in Varna city. A war-driving technique was used based on a developed Raspberry Pi scanning system. The obtained results are compared with similar ones made in 2008.

The answer to the main question of the survey can be formulated as follows: *The results show a significant increase in the security of the Wi-Fi networks in the city, but there is a need for further improvement.*

The main recommendations can be in the following directions:

1. Increase awareness for the security issues and the effect they may have on the users.
2. Use only WPA2 method. If there is a need to support older devices, build a separate, limited network that supports mixed WPA/WPA2 mode.
3. Disable WPS for all devices.
4. Select a password of at least 12 characters with uppercase and lowercase letters and numbers.
5. Update the firmware used by the wireless router to the latest available version. Another option is the use of an alternative firmware such as OpenWRT/DD-WRT, in which security issues are generally removed much faster.

As guidelines for the future work may be the analysis of newest 802.11ac and 802.11ah networks, and study and comparison of the security level in the wide districts of Varna and region.

References

[1] Colbach G 2017 *Wireless Networking: Introduction to Bluetooth and WiFi* (Independently published)
[2] IEEE 802.11 *Wireles local area networks http://www.ieee802.org/11/* (last visited 03.2019)
[3] Gopalakrishnan S 2014 *A survey of wireless network security IJCSMC* 1 **3** pp 53–68
[4] IEEE 802.11-2019 standards https://standards.ieee.org/standard/802_11-2016.html (last visited 03.2019)
[5] NetSpot *Wireless security protocols: WEP, WPA, WPA2, and WPA3 https://www.netspotapp.com/wifi-encryption-and-security.html (last visited 03.2019)*
[6] Viehböck S *Brute forcing Wi-Fi Protected Setup https://sviehb.files.wordpress.com/2011/12/viehboeck_wps.pdf* (last visited 03.2019)
[7] DEF CON Nine Schedule *https://www.defcon.org/html/defcon-9/defcon-9-speakers.html#Peter%20Shipley* (last visited 03.2019)
[8] Dobrilovic D 2016 *A method for comparing and analyzing wireless security situations in two capital cities Acta Polytechnica Hungarica* 6 **13** pp 67-86
[9] Sebbar S B 2016 *An empirical study of WIFI security and performance in Morocco - wardriving in Rabat Int. Conf. on Electrical and Information Technologies*
[10] Kyaw K A, Tian Z and Cisak B 2016 *Wi-Pi: a study of WLAN security in Auckland city IJCSNS* 8 **16** pp 68-80
[11] Priya S U 2013 *The Impact of War Driving On Wireless Networks IJCSET* 6 **3** pp 230-235
[12] Hristo Valchanov H, Ruskov I and Varbanov A 2009 *A Study of the Wireless Network Security Proc of 5th Conference "Computer Science 2009" (Sofia)* pp 273-278

IOP Conf. Series: Materials Science and Engineering **618** (2019) 012032 doi:10.1088/1757-899X/618/1/012032

Object-oriented pseudo-random data generator realizations

Tony Karavasilev[1]

[1]Computer informatics, Plovdiv University "Paisii Hilendarski", Bulgaria

E-mail: tony.karavasilev@gmail.com

Abstract. With the development of computational science and computer simulations, the randomness source has become a necessary part of every complex system and testing environment. Most programming languages, software frameworks and application libraries do not provide all different types of randomness generators and do not even implement high-quality algorithms for secure data generation. This paper presents an object-oriented solution for providing all the main classifications of randomness via random data generator classes that offer frequently used formats in information technology and in statistical hypothesis testing. It also presents usages for secure data generation and confirms the importance of true randomness in cryptography. The purpose of the developed practical experiments is to evaluate and show the differences in the sample generation quality of the sub-classifications of software randomness generators using the developed object-oriented components.

1. Introduction

The general language definition of randomness is the absence of a pattern or any predictability in an event. The random sequence of events, symbols or steps must also be unpredictable or without any visible combination order. Frequently, when testing with a huge amount of unpredictable events, also known as trials, patterns may still appear. An example of this may be the throwing of a dice or a coin multiple times until revealing some probability model or repeated behavior. [1]

It is important to mention that there is a link between randomness and many philosophical discussions about existence, natural laws and free will. For example, the indeterminism theory completely relies on randomness via the assumption that all events in nature are a cause of pure chance or even luck. While on the other hand, the determinism theory implies that everything happens for a necessary reason and has an underlying cause, which completely opposes the existence of true randomness itself. [2]

In mathematics and statistics, a random variable is an assignment of a numerical value to each possible outcome of an event scope or domain. This makes it possible to calculate a probability value for each valid event. When applying this scientific view, randomness becomes a valid measure of uncertainty or even applies a concept of chance in an outcome. There are many applications, such as the evaluation of the information entropy of a message by assigning a higher information rate to lower-probability events and finding the low-level building blocks in huge data samples. This way we can use it also as a measure of disorder in a system or a type of uncertainty associated with a random variable in a certain domain. [3]

The realization of practical randomness sources in computational science and software technologies relies completely on number theory algorithms and the building of random enough looking number sequences. Most of them rely on the current system time with a combination of binary operations between numbers in a certain valid domain. These algorithms have the ability to produce the same output by setting a custom value on initialization, also known as an initial seed or just seed. As a term in computer science, randomness is considered as noise, which represents some irrelevant or meaningless data. Having this said, some high-quality algorithms are initialized with values generated by radio waves caused from magnetic reconnections of the computer's hardware devices, such as the central processing unit or the hard disk drive, etc. Operating systems make such low-level data accessible via the kernel's software drivers as a source of noise that has higher entropy and can make the used mathematical algorithms far more unpredictable. Even though the generated noise is still limited to the byte event space, it is a lot harder to guess it and is more immune to logical attacks. [4]

Having this said, even if the used noise sources seem to be truly unpredictable, with time some of them will be scientifically explained and will no longer be looked upon as random. This can make our systems vulnerable to new zero-day attacks in the near future but also provides an easy mitigation method by just switching to another source or generation algorithm. Philosophically said, everything is random until we find an explanation and can reproduce the exact logic of the given process.

In computer systems, three main branches of the software random data generator realizations exist:

- Quasi-random number generator (QRNG);
- Pseudo-random number generator (PRNG);
- Cryptography secure pseudo-random number generator (CSPRNG).

The quasi-random type, also known as sub-random, is a low-discrepancy sequence. This type of sequence is uniformly distributed and the number of points in it is usually close to the measure of the whole count of the used set. Although they may fail most statistical tests for randomness, sub-random numbers have an advantage over pseudo-random numbers because they cover or fill the domain more evenly and quite faster. They tend to be very useful in financial mathematics and data mining algorithms. They also have the ability to repeat the same result via an initialization value. The most famous algorithms for this type are the Halton sequence and the Sobol sequence. [5]

On the other hand, the pseudo-random type appears to be truly random but in fact, it is not. The whole point of the generated sequence is to exhibit statistical randomness, which means to be close enough to true randomness on the first look by not having any recognizable patterns or visible order. This type is also controllable via an initialization value and is very useful in the realization of computer simulations, statistical analysis and software testing. Most software developers face this type of computer randomness realization first and may even use it for years before understanding the concept of true randomness. The most famous algorithms for this type are the linear congruential generator (LCG) and the Mersenne Twister generator (MTG). [3]

The most important type of randomness realization is the cryptographically secure pseudo-randomness. Its output sequence is harder to guess and is not controllable via seed values. The crypto-randomness algorithms used may vary but are far more complex than what the previous two types use internally. They pass statistical tests and are immune to attacks, even if the attacker knows the initial state passed to the algorithm or the consumed entropy pool location. They harvest the full force of high-quality sources of noise via the operating system's kernel. This type of randomness realization is suitable for cryptography purposes and network protocol algorithms. The best cyber security practices completely rely on the quality of the unique identifiers, tokens, keys, salts and nonces. The most famous algorithms for this type are the Yarrow generator and the Fortuna generator. [6]

Most programming languages, software frameworks and application libraries do not provide a lot of different types of randomness sources, which creates many problems with the overall application security status. In most cases, there is only a common pseudo-randomness algorithm and some basic cryptographically secure pseudo-randomness realization, both providing only basic data formats like positive numbers and raw bytes. This creates many disadvantages, for example:

- Not suitable for all scientific use cases;
- Not easily injected into other class services;
- Not enough data generation output formats available;
- No object-oriented class hierarchy exists or only global functions exist;
- Security, performance and thread safety may vary across different language versions.

The main purpose of this article is to provide an object-oriented solution for the implementation of all the main classifications of randomness via random data generator classes that support frequently used formats in information technology and computational science. This paper includes a practical implementation of an object-oriented class hierarchy for randomness realizations and extended sample generation experiments to illustrate the main differences between the three sub-classifications of software randomness. This research is also a part of ongoing work in developing an advanced PHP object-oriented software framework for cryptographic services.

2. The theoretical object-oriented solution

In most programming languages, the provided functions or classes only support basic pseudo-randomness implementations and in most cases at least one cryptography secure pseudo-randomness algorithm. No matter the case, the first step of creating an object-oriented solution is to build sophisticated wrapper classes and create low-level native realizations for all missing algorithms.

2.1. Building a suitable class hierarchy

Have in mind that even if we create classes as simple wrappers of other language provided classes or functions this would not be enough for an object-oriented solution to work. We must stay away from the anti-pattern known as God object, which is a class object that knows too much or does too much. The realization must carefully split reusable code into an abstract base class and force the desired implementations via interfaces or abstract methods. The abstract representation of randomness must force all derived objects to implement at least integer and byte generation methods. It must provide a constructor method specification for forcing the same type of instance creation for all derived classes.

Note that not every sub-classification of randomness can support initialization seeding, like the cryptography secure pseudo-randomness. It is a good idea to keep this functionality away from the base class and move it to an optional interface specification. Even if you leave it as an abstract method at the base class, you must never implement any logic for it when creating a cryptography secure generator and can force it to throw an exception on invocation. Figure 1 shows an example of a class hierarchy as a Unified Modelling Language (UML) class diagram.

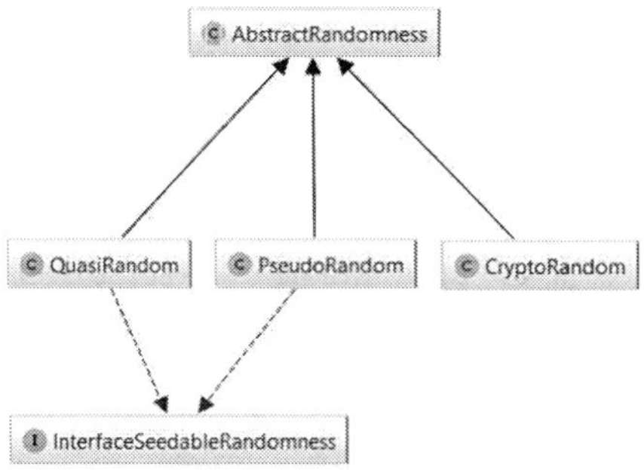

Figure 1. UML Diagram for an Example Randomness Class Hierarchy

This hierarchy can become more sophisticated as a realization, for example moving the generation or validation methods into interfaces and implementing them directly into the abstract base class. In some cases, you may even want to split methods into multiple abstract classes that inherit each other.

2.2. Creating native language realizations

Most programming languages have a limited amount of features in them for several reasons like compatibility, test coverage, security, performance and paradigm specific features. When a researcher or a developer needs some specific computational algorithm, in most cases, the implementation is a complete language native one. In the case of algorithmic information theory studies, the native code simulation of randomness sources is done via number theory specific realizations based on binary operations and mathematical formulas. For cryptography purposes, the realization must also always consume a high entropy physical source of noise provided via the operating system's kernel.

Typically, programming languages do not provide a low-discrepancy sequence generator. When creating one, we must keep in mind that the implementation of this type of space-filling algorithm is resource hungry. This comes from the need to keep and access a huge set of numbers in memory for manipulations that can truly slow things down. The most common optimizations are using a faster but worse shuffling algorithm or a smaller supported domain size. Some approaches even use the same hardcoded sequence for speed reasons, which is not truly acceptable for research purposes.

Pseudo-randomness is a part of every widely used language. Native implementations are important only when the built-in algorithm is old or not good enough. In many cases, the seeding algorithm implied is time-based and may need access to more precise sources of noise. Note that supplying certain seeding values based on the used algorithm may also create a poorer output sequence than expected. The realization of cryptography secure pseudo-random generators is mostly necessary when the built-in language functionalities do not provide a secure enough implementation. Most known approaches modify existing pseudo-random algorithms or use 3rd party implementations. As a strict rule, when writing in native code we must never insert a higher predictability rate into our algorithms.

2.3. Adding optional output formats

As discussed before, bundled language features mostly provide the generation of only positive numbers and raw bytes. The first problem every developer faces is the need for a wider range of output generation formats. When providing a software solution for random data generation, the first step is to implement negative number ranges. A simple algorithm for achieving this is extracting a number with at least half of the supported range on every generated number to provide both negative and positive numbers. Another approach may be pooling a number between zero and one, to simulate a Boolean expression, to choose if the number is positive or negative.

Another missing feature is the floating number output format that uses a certain rounding precision. This is very useful for statistical or financial simulations. A well-known algorithm is generating a number from a range and dividing it by the highest supported one in the same range. This way we achieve a probability floating number format between zero and one (as 0.0 – 1.0), widely used for statistical representation. Having this, we can easily multiply it to generated integer numbers to provide floating precision numbers in a bigger range or use it directly for some chance simulations.

A huge missing feature for every professional developer is the random string generation. This may include hexadecimal, Base64, ASCII, Unicode and other string formats. Every complex system needs the ability to provide secure passwords, strong authentication tokens, encryption initialization vectors and globally unique identifiers. A good realization would be creating a custom mapping of certain characters to certain positive numbers and just concatenating printable output characters based on generated random integer values in the range of the collection size.

2.4. Using the dependency injection principle

When creating a hierarchy of objects, we may come to the need for writing higher logic in container-based classes that reuse some of our objects as services. The approach of building container classes that can inject different objects of some type and use them as services that follow the same specification is the general definition of the dependency injection technique. [7]

Advanced examples for the usage of the dependency injection principle in random data generators is defining container objects for data shuffling and secure token generation. This way it is possible to supply different quality generators for the internal processing logic of the container. The usage of dependency injection increases the code reuse and decreases the coupling between a certain class and its dependencies. This specification allows all abstract randomness representations to be injectable into any class that inherits the container abstraction or implements some kind of service setter interface.

IOP Conf. Series: Materials Science and Engineering **618** (2019) 012032 doi:10.1088/1757-899X/618/1/012032

2.5. Backward compatibility and thread safety precautions

When creating an object-oriented solution, two main problems seem to always arise. The first one is backward compatibility and the second one is thread safety. When dealing with backward compatibility issues, a developer is attempting to keep the same behavior between language versions. To achieve this, the object creation requires native implementations, parameter type checks, various integer allocation optimizations and the execution of different logic for each used language version.

Thread safety issues exist even if the language does not have any asynchronous capabilities. The most frequent example is when the run-time execution fetches the same numerical sequence on each dynamic call in a certain scope or even causes process isolation failures by calling the same sequence for parallel requests. To mitigate these failure cases, a developer must always implement the algorithm logic in static memory and force the dynamic object creation to consume it via internal static method calls only. This will decrease the scope problems to a minimum and defines a single unified data pool for all object instances. When we want to deal with parallel request processing issues, which is a problem in both asynchronous and request-response architectures, we need to create a high-quality initialization logic that seeds unique values for every coexisting request. To cope with these problems, we must create shared memory locks for used resources at the object creation logic and consume only high-entropy sources of noise. A simplified method for non-cryptography secure generators would be using the current time to a microsecond as the initial value for the sequence generation algorithm.

3. The software solution implementation

With the wide case usage of pseudo-random data generators, the need for a practical implementation of the discussed theoretical solution is tremendous. The created software solution for all three main subtypes of abstract randomness is introduced as a part of a PHP object-oriented software framework for cryptographic services. The next sections provide a detailed practical analysis of the implementation and completely rely on the previously discussed realization approaches. The complete source is available and published online at the framework's GitHub repository. [8]

3.1. Quasi-randomness implementation

As discussed before, the realization of quasi-randomness algorithms is always resource-demanding because of the uniform distribution and the huge amount of points used in a set. Any simple optimization may end up breaking the distribution and converting it to a pseudo-random sequence instead. This is why many operating systems reuse the same hardcoded quasi-random sequence into their kernel internal resources, which may be a great optimization but lowers the dynamic computational use cases. [5]

The basic algorithm for producing a low-discrepancy sequence has three main steps – generation (and scrambling), initial skipping and between calls leaping. The generation of a set of any amount of numbers is simple in any programming language. The shuffling or scrambling, on the other hand, may create different performance problems or if not used properly can even convert the sequence into a pseudo-random one instead. The solution for the implementation is to shuffle the set of points only once or one in a dozen of times for each program execution. This way we ensure that the sequence seems scrambled enough and will not hurt the uniform distribution of the randomness generator. The skipping step is a simple way of ignoring the first amount of numbers in a scrambled uniform set of numbers. This part can be skipped to use the whole set. It is a good idea to use a different skipping amount depending on the seed value for the generator or to hardcode a smaller value.

Finally, we must use a leap step to specify the number of points to ignore on every call to the generator. The leap step may be applied either on every call or be included in the initial sequence generation to reduce the overall memory usage. Defining a leap step of value one will make every call to the generator skip exactly one point and may lead to the decreasing of the usable sequence size to exactly half of the initial set. We may need to set this at zero if we want to use the generated set for domain filling purposes. Note that there may be a need for more than one internal sequence, because fetching in closer ranges (like bytes – 2^8) will loop or scramble the set too often, thus cause extra load.

IOP Conf. Series: Materials Science and Engineering **618** (2019) 012032 doi:10.1088/1757-899X/618/1/012032

3.2. Pseudo-randomness implementation

The realization of pseudo-randomness algorithms in software platforms is simpler and more resource friendlier than the one for quasi-random generation. There is a huge amount of different methods for simulating statistical randomness in computer systems. As discussed before, most programming languages have at least one realization available internally.

If your language does not have any implemented, you can code one yourself or use one of the many open source implementations online. In most cases, when creating an implementation for your system it would be enough to create a wrapper object for the language provided functionality and manipulate values in a certain way to provide more features without breaking the pseudo-randomness characteristics. If the generated sequences are not good enough, then you can provide a high-entropy source for the algorithm's initialization seeding value to improve its overall quality and security.

3.3. Cryptography secure pseudo-randomness implementation

As discussed before, the realization of a cryptographically secure pseudo-random generator is always tricky. When there is an algorithm available in your language, you must definitely use it directly. The best solution is to create a wrapper object for choosing and switching secure algorithms in the correct backward compatible manner for different language versions. Your code may also need to measure and evaluate all entropy sources for the current system and chose the best one available. Note that most programming languages update their cryptography secure pseudo-random algorithms and their internal cryptography libraries quite frequently, so make sure to keep your system up to date.

If your language does not support a secure enough algorithm, then always use only official high-grade open source implementations and consume the highest entropy source available. In some rare cases, you can even disable the seeding feature of a high-quality pseudo-random algorithm that has proven to be prone to security attacks and supply initialization values directly from the operating system's high-entropy source or even via a 3rd party hardware chip [6]. It is important to note, that you must never create your own homemade algorithms and always stick to proven cryptography standards.

4. A practical comparison of different randomness realization types

After the complete implementation of all three main generator subtypes, there is a need for a practical comparison of their theoretical specification. The next experiments evaluate the correct source code realization and produce a visual correlation measurement between the outputs of the created data generator classes. The next sections describe the executed experiments and show their results.

4.1. The testing environment specification

All the practical experiments are run under a virtual machine environment created with Oracle VM VirtualBox version 6.0.6 hypervisor installed under Windows 10 Home Version 1809 x64 (Build 17763.437) as a host operating system. The setup of the guest virtual machine consists of running Apache 2.4.39 with PHP 7.3.5-FPM (the FastCGI Process Manager implementation). The hardware specification of the virtual machine and the used guest operating system is shown in Table 1.

Table 1. Virtual Machine Hardware Specification

Component	**Details**
CPU	Intel i7-6700HQ, 2 cores, 2.59 GHz, 6 MB L3
RAM	DDR4, SODIMM, 4096 MB, 2.40 MHz
GPU	Intel HD Graphics 530, 16 MB, 2.40 MHz
HDD	42 GB, 7200 RPM, 32 MB cache, 2GB swap
OS	Ubuntu Server 18.04.2 LTS x64, Linux Kernel 5.0.10

The installation of the virtual machine has all existing updates, kernel drivers and virtualization-specific packages. The PHP and Apache configuration uses all default installation settings. The next

IOP Conf. Series: Materials Science and Engineering **618** (2019) 012032 doi:10.1088/1757-899X/618/1/012032

experiments generate images to create a visual representation of each randomness generator type. The generation of all images is done with the PHP extension GD2, also known as the GD library. The colour scheme used for the image generation is the common RGB representation, which is an additive model that uses a combination of red, green and blue as 8-bit integers to produce 16 777 216 (24-bits, 256^3) unique colours. The output images are in the famous Video Graphics Array (VGA) format with a resolution of 640x480 pixels, also known as the 480p standard, which uses a 4:3 aspect ratio.

4.2. Generation of a black and white 480p image

The first test for our randomness generator objects is to represent each of the output numerical sequences as a black and white output images to measure the quality of the produces statistical noise. The implementation of the experiment includes generating numbers in a Boolean format to map the integer number zero to the black colour and the integer number one to the white colour. Each of the three randomness visual representations can be seen in Figure 2.

Figure 2. Black & White Visual Product – QRNG vs. PRNG vs. CSPRNG

As we can see from the results, the quasi-randomness implementation correctly differs from the other two types because of its uniformly distributed numerical set. The chaotic output of both the pseudo-randomness and the secure pseudo-randomness is evidently random enough, with more image artifacts for the secure one. It is important to note that running the same test with all RGB colours via the generation of a set of three 8-bit numbers (of 0-255 range) did produce the same output patterns.

4.3. Drawing of a polygon figure at a 480p image

The next experiment includes the drawing of a polygon using 10 000 generated points of white colour over a completely black background. The implementation generates points via pairs of two integer numbers to represent the X and the Y coordinates. Each of the generated points is only in the visible realm of the output image. The experiment's result can be seen in Figure 3.

Figure 3. Polygon Visual Product – QRNG vs. PRNG vs. CSPRNG

The quasi-randomness visual product confirms the uniform distribution of the internally used set and the algorithm's space-filling characteristics. The secure pseudo-randomness implementation produces a higher point density and has less visible patterns than the pseudo-randomness one does.

4.4. Drawing of multiple ellipses at a 480p image

The last test includes the drawing of 10 000 ellipses using random colours, shapes and coordinates. The initial background colour of the image is completely black. The width and the height of the ellipses will be between 3.00% and 80.00% of the picture's resolution to produce different shapes. The implementation generates all central points via pairs of two integer numbers to represent the X and the Y location coordinates. The generated points are only in the visible realm of the output image, but the figure may not be completely. The visual product for each of the generators is shown in Figure 4.

Figure 4. Ellipses Visual Product – QRNG vs. PRNG vs. CSPRNG

The quasi-randomness visual product is evidently not statistically random. The secure pseudo-randomness generator produces more different ellipses than the pseudo-randomness one does.

5. Conclusion

This paper has created an object-oriented solution for providing all the main classifications of randomness via pseudo-random data generator classes that can output a vast range of information formats. It maps the best uses of each of the randomness subtypes and demonstrates their characteristics via visual representations.

This article successfully confirms the importance of software randomness in the fields of cryptography, mathematics, statistics and computer science.

Acknowledgments
University of Plovdiv Paisii Hilendarski, 24 Tzar Asen, 4000 Plovdiv, Bulgaria

References

[1] Menezes A, van Oorschot P and Vanstone S 1996 *Handbook of Applied Cryptography* ISBN: 9781439821916
[2] Svozil K 2018 *Physical (A) Causality: Determinism, Randomness and Uncaused Events* ISBN: 9783319708157
[3] Goldreich O 1998 *Modern Cryptography, Probabilistic Proofs and Pseudorandomness (Algorithms and Combinatorics)* ISBN: 9783540647669
[4] Matsumoto M, Saito M, Haramoto H and Nishimura T 2006 *Pseudorandom Number Generation: Impossibility and Compromise* DOI: 10.3217/jucs-012-06-0672
[5] Alon N, Coja-Oghlan A, Hàn H, Kang M, Roedl V and Schacht M 2007 *Quasi-Randomness and Algorithmic Regularity for Graphs with General Degree Distributions* DOI: 10.1137/070709529
[6] Stipčević M and Kaya Koç Ç 2014 *True Random Number Generators* DOI: 10.1007/978-3-319-10683-0_12
[7] Hudli S and Hudli R 2013 *A Verification Strategy for Dependency Injection* DOI: 10.7763/LNSE.2013.V1.16
[8] Tony Karavasilev (Тони Каравасилев) 2019 *TonyKaravasilev/CryptoManana* DOI: 10.5281/zenodo.2604693

IOP Conf. Series: Materials Science and Engineering **618** (2019) 012033 doi:10.1088/1757-899X/618/1/012033

Introduction to Health Information Technologies

Donka Nesheva

Technical University of Sofia – branch Plovdiv

dnesheva@tu-plovdiv.bg

Abstract. Health information technologies have the potential to advance healthcare delivery and assist patients and providers in achieving better outcomes. While development and adoption of electronic health records (EHR) already have reached high levels, health information exchange is still a challenge. This paper presents some of the basic terms and concepts in the domain. The lack of standards and security concerns are discussed as the main technical barriers for reaching integration and interoperability. Mobile technologies are described as a new source of data and promising tool for providing personalized care. Latest trends and solutions are presented, as well as the need of advancement.

1. Health Information Technology (HIT)

1.1. Introduction and definitions

The term "health information technology" refers to the electronic systems that create, transmit, store and manage individuals' health data [1]. Adoption of electronic records and digital systems in the area of health care delivery occurred later compared to other industries. The current use of health information technologies is widespread among providers, patients and other stakeholders. That makes possible aggregation of data of millions patients, their analysis in order to gain new knowledge, and using it to make care delivery better quality and more efficient at a lower cost. According to another definition [2], health information technology is defined as any clinical information technology system that captures patient data in an electronic record.

Electronic records can be divided into two groups: The first group includes those records created and utilized by licensed professionals (health providers, physicians, etc). They are considered institution-centered records, this kind of records are Electronic Medical Records (EMR) and Electronic Health Records (EHR). The second group includes Personal Health Records (PHR), which are created and managed by patients. The records in this group are considered patient-centered [3].

1.2. Electronic medical records

An electronic record of health-related information on an individual, that can be created, gathered, managed, and consulted by authorized clinicians and staff within one health care organization [4]. Garets and Davis in [3] define EMR as an application environment composed of clinical data repository, clinical decision support, controlled medical vocabulary, order entry, computerized provider order entry, pharmacy, and clinical documentation applications. That environment is used by healthcare practitioners to document, monitor, and manage health care delivery within a care delivery organization (CDO). EMR contains medical and treatment history of the patients in one CDO. This is the reason it is considered an internal, organizational system.

1.3. Electronic health records

An electronic record of health-related information on an individual that conforms to nationally recognized interoperability standards and that can be created, managed, and consulted by authorized clinicians and staff across more than one health care organization [4]. EHR is a subset of each care delivery organization's EMR. It is owned by the patient and has patient input and access that includes episodes of care across multiple CDOs within a community, region, country or several countries [3]. EHR summaries patient's health data from multiple sources into one record (e.g., electronic prescription, emergency information, medical tests orders, decision-support systems, digital images, and telemedicine) and this makes it inter-organizational system. Making providers accessible to patient health data, EHR improves clinical decision support and achieves care coordination between different providers, which leads to fewer medical errors and duplicated lab tests. The effect of reducing errors is prominent in using electronic prescriptions. A study from 2010 shows that medication errors decreased nearly sevenfold, from 42.5 per 100 prescriptions (95% confidence interval (CI), 36.7-49.3) to 6.6 per 100 (95% CI, 5.1-8.3) prescriptions within a year of adopting e-prescribing [5].

1.4. Personal health record
PHR is an individual's electronic record of health-related information, that can be obtained from multiple sources. It is managed, shared and controlled by the individual [4]. PHR can include clinical health records shared by EHR/EMR systems or personal's daily activity data gathered from sensors, mobile or wearable devices. The main benefit of PHR is to leverage patient's engagement, involvement and interest, that has the potential to increase the effectiveness of care delivery and therapeutic processes.

2. Health Information Exchange (HIE)
Health information exchange (HIE) includes a variety of technologies to improve the provider's access to patient information collected and maintained by other organizations [4]. By providing access to timely and comprehensive patient information, HIE is a solution intended to address the threats to quality, safety, and efficiency caused by inaccessible, incomplete or missing information at the point of care. A patient survey from 2016 shows that 19% of patients in U.S. and 24% in France say that their medical records or test results had not been available at the time of an appointment or that duplicate tests had been ordered in the past two years [6].

HIE is essential for care coordination especially in chronic diseases management. According to the World Health Organization, 60% of all deaths in our days are due to chronic diseases [7]. Chronic diseases that are not well controlled cause other chronic diseases and complications. They cause 70% of all deaths and account for 78% of the healthcare expenses in the United States [8]. As a result, those patients yearly encounter multiple specialists, which need to coordinate the care among them. Using paper records is not efficient, it is error prone and often providers need to make decisions without the needed medical information as the survey shows. HIE has the potential to improve care coordination and reduce costs through avoidance of unnecessary testing, duplicate medications and procedures.

Sharing data with patients improves patient engagement in their own care. Aggregated health and behavioral data are suitable for big data analysis and medical research.

2.1. Privacy and security
Health information exchange requires core technologies to assure privacy, security and trust. Privacy means that only individuals or entities authorized by the patient may access their data. Security is preventing access to data by unauthorized entities. Trust ensures the identity of entities with which the data is being shared.

Since health data protection has been a serious concern for users for a long time, governments put in place laws and regulations to protect personal data. In USA this is Health Insurance Portability and Accountability Act of 1996 (HIPAA) [9]. HIPAA defines administrative, organizational and technical requirements for storing and transmitting personal health information. The General Data Protection Regulation (GDPR) [10] was enacted by the European Union in 2018 to establish personal data protection regulations. It is a comprehensive set of guidelines that acknowledges that sensitive data such health, biometrics or genetics require highest level of protection.

Technical safeguards for achieving HIPAA and GDPR compliant architecture include: setting up private cloud infrastructure, data encryption at rest (AES-256), data encryption at transit (HTTPS, SSL/TLS), data de-identification, data backups, implementation of auditing tools and other.

2.2. Interoperability

The major technical issue for HIE is interoperability. Interoperability is the ability of different information systems, devices or applications to connect in a coordinated manner, within and across organizational boundaries to access, exchange and cooperatively use data amongst stakeholders, with the goal of optimizing the health of individuals and populations [11].

Structural interoperability defines the structure or format of data exchange. It ensures that the clinical or operational purpose and meaning of the data is preserved and unaltered. Structural interoperability defines the syntax of data exchange. It ensures that data exchanges between information technology systems can be interpreted at the data field level.

Semantic interoperability is the ability of two or more systems to exchange information and to interpret and use that information. Semantic interoperability requires both structuring of the data exchange and codification of the data, including standard, publicly available vocabulary, so that the receiving information management systems can interpret the data [12][11]. Achieving semantic interoperability is still an ongoing challenge.

In the current situation there are multitude of EHR systems, mobile health applications, clinical decision support systems, etc., that were not designed to record data in a standard format or to share data with other systems using common protocols. Different vendors distribute EHR systems with different data models (e.g., HL7 v2, RIM v3, FHIR, CDA, OpenEHR, ISO/IEEE 11073, CEN EN13606, and DICOM). They use different encoding terminologies (e.g., SNOMED CT, LOINC, UMLS, RxNorm, ICD, MeSH), different datatypes, different schemas, different vendors, different formats, etc.

An example illustrating the lack of semantic interoperability is the need to integrate data from mobile health applications with EHRs and clinical decision support systems (CDSS). An EHR contains comprehensive patient medical and administrative data. Mobile health applications provide real-time patient generated data, captured from wearables, sensors, devices (such as blood glucose meters, blood pressure meters, etc.) or manually entered by the patient. The collected data can be supervised and studied by physicians. CDSS assists physicians with personalized decision support based on patient data. Integrating all provides end-to-end comprehensive care that supports patients and providers in achieving better outcomes. In addition, these data can be integrated with other patient data and used by big data and machine learning tools to discover model patterns. Linking all of these heterogeneous systems requires solving the problem of semantic interoperability, because currently there is no universal standard or terminology to ensure this [15].

Different data and interoperability standards have been developed for achieving health information exchange. Some of the most popular currently in use are:

2.3. Data standards

International Classification of Diseases (ICD) is a coding system of diseases and signs, symptoms, abnormal findings, complaints, social circumstances and external causes of injury or diseases, as classified by the World Health Organization (WHO). It is used to classify diseases and other health problems recorded on many types of health and vital records including death certificates and health records. Current revision of ICD is 11.

Logical Observation Identifiers Names and Codes (LOINC) is a universal code system for identifying laboratory and clinical observations. LOINC has standardized terms for all kinds of observations and measurements that enable exchange and aggregation of electronic health data from many independent systems.

Systematized Nomenclature for Medicine (SNOMED) is a universal standard for identifying laboratory observations and clinical results. It includes not just medical and laboratory code names, but also nursing diagnosis, nursing interventions, outcomes classification, and patient care data set.

RxNorm provides normalized names for clinical drugs and links its names to many of the drug vocabularies commonly used in pharmacy management and drug interaction software.

Digital Imaging and Communications in Medicine (DICOM) is the international standard to transmit, store, retrieve, print, process, and display medical imaging information.

2.4. Interoperability standards

Interoperability standards define the way health data is packaged, transported and shared.

Health Level Seven (HL7) is a not-for-profit, ANSI-accredited standards developing organization dedicated to providing a comprehensive framework and related standards for the exchange, integration, sharing, and retrieval of electronic health information that supports clinical practice and the management, delivery and evaluation of health services [16]. *HL7's Version 2.x (V2)* messaging standard is one of the most widely implemented standard, released in 1987. The standard is based on EDI/X12 technology. However, V2 requires significant site customization, which leads to semantic inconsistencies across implementations [17]. The HL7 *Reference Implementation Model (RIM)* defines the semantics of a common set of administrative, financial and clinical concepts in order to foster interoperability. It consists of five abstract concepts: act, relationship, participation, roles and entities. RIM offers a model for expressing clinical statements in a semantically consistent way, but implementation complexity leads to incompatible documents [18]. The *Clinical Document Architecture (CDA)* defines HL7 Version3 RIM-based documents that are assembled from elements, including administrative and clinical data, for particular purposes [19]. The current version of CDA is called *Consolidated CDA(CCDA)*. CCDA documents are assembled from templates of XML components. The unsuccess of HL7 v2 and v3 standards to ensure semantically consistent approach for sharing granular clinical data forces the development of *Fast Healthcare Interoperability Resources (FHIR)*. FHIR represents clinical data as resources that are subset of RIM. Each resource is an expression of meaning stated in terms of well-defined fields and data types. Data are queried through RESTful API and serialized in XML or JSON formats. FHIR's clinical resource definitions are concepts such as Patient, MedicationPrescription, Observation, and Device. These resources constitute a graph of clinical data by explicit inter-resource references. For example, a MedicationPrescription resource explicitly references its prescriber (a FHIR Practitioner), its patient (a FHIR Patient), and the drug prescribed (a FHIR Medication). FHIR does not include detailed models for every aspect of a clinical record, but provides a built-in extensibility mechanism to enrich existing resource definitions. FHIR resource definitions do not directly promise semantically consistent data out-of-the-box. Instead, to serve distinct contexts (e.g., EHRs, public health reporting workflows, wearable devices), FHIR resources might use specific data payloads with distinct terminologies. Semantic consistency relies on an abstraction layer called FHIR profiles that constrain and extend resource definitions in particular contexts [20][18].

SMART on FHIR is a set of open specifications aimed to integrate third party apps with EHRs and other health information systems. SMART acronym stands for "Substitutable Medical Apps, Reusable Technology" [21]. It is based on FHIR models and API, including authorization (OAuth2), authentication (OpenID Connect) and UI integration. A SMART on a FHIR app (application, service) runs against a SMART on a FHIR system, extending its functionality through the use of clinical and contextual data[18].

The *openEHR* approach for achieving interoperability is based on multi-level, single source modeling within a service-oriented software architecture. The openEHR design specification is based on reference model (RM), archetypes and templates. The RM defines the logical structures of EHR and demographic data. All EHR data in any openEHR system obey this RM. Archetypes are data points or data groups, independent of particular use. These data-points and data-groups are assembled into context-specific data sets called templates. The international library of openEHR archetypes (CKM) currently contains about 500 archetypes, or 6,500 data points. The benefits of this approach according to [22] are systems and tools built around the same data model which solve the problem with semantic interoperability.

Another approach for HIE is *Direct Project* created in 2010 to specify a simple and secure way to send authenticated, encrypted health information directly to known and trusted recipients using secure email. DirectTrust is a non-profit association of health care providers that authenticate the identity of the participants and issues email addresses. To assure secure transport via email Direct uses Simple Mail Transfer Protocol (SMTP) with special Direct email addresses, associated with X.509 certificates. It uses Secure/Multipurpose Internet Mail Extensions (S/MIME), a public key standard for encrypting email attachments that can be in any format (image, ODF, HL7 CCDA document, etc.). The system supports both provider-to-provider and bi-directional patient-provider exchange[23][12].

Blockchain is the newest technology addressing health information exchange issues. In [24], *John D. Halamka* discusses the potential of blockchain to transform electronic health records. *Azaria et al.* [25][26] provides a proof-of-concept that uses blockchain as a mediator to health information, called MedRec. For MedRec, the block content represents data ownership and viewership permissions shared by members of a private peer-to-peer network. The prototype uses smart contracts on an Ethereum blockchain to log patient-provider relationships that associate a medical record with viewing permissions and data retrieval instructions (essentially data pointers) for execution on external databases. Health insurance companies are running pilot project to use blockchain as a way to manage provider's data inputs [27]. In [12] *Mark Braunstein* mentioned a possible use of blockchain to create a uniform patient ID and storing patients' specific permissions for data sharing.

3. Mobile health technologies

I already mentioned the growing prevalence of chronic diseases nowadays and how they are the greatest cause of raising healthcare costs. Poor lifestyle choices as smoking, overuse of alcohol, poor diet, lack of physical activity are the main reasons for development of chronic diseases, including obesity, type 2 diabetes mellitus, hypertension, cardiovascular disease and other. Behavior change and continuous monitoring are the key drivers for positive outcome. The daily routines made by patients with chronic diseases (e.g. tracking vital signs as blood glucose, blood pressure, weight, pulse; tracking physical activity; medications adherence; keeping a healthy diet) are very important for disease management.

Mobile health technologies (mHealth) are capable to assist patients and providers to perform chronic disease management in a dynamic and more effective manner. The World Health Organization (WHO) says that "The use of mobile and wireless technologies to support the achievement of health objectives (mHealth) has the potential to transform the face of health service delivery across the globe. A powerful combination of factors is driving this change. These include rapid advances in mobile technologies and applications, a rise in new opportunities for the integration of mobile health into existing eHealth services, and the continued growth in coverage of mobile cellular networks."[28]. mHealth facilitates patient-provider connection, creating a smart environment and engaging the patients into their own care. Data from mobile applications, wearable devices and sensors are used by health care teams to make accurate decisions and provide the most adequate treatment according to the patient's state.

A good example of application of mHealth technologies are the AliveCor's devices used to screen for Atrial fibrillation (AF) [29]. AF is the most common cardiac arrhythmia affecting millions of peoples and an important risk factor for stroke. Early diagnosing of AF for patients in risk can prevent the occurrence of stroke. AliveCor's Kardia Mobile device detects electrocardiogram (ECG) heart trace and sends it to a mobile smartphone app using inaudible ultrasonic signals and phone's microphone.
AliveCor's KardiaBand is a special band for Apple watch. The SmartRythm software takes data from the watch's heart rate sensor and accelerometer and uses neural network to recognize rates that may indicate atrial fibrillation. Then the system notifies the user to take an ECG using the band. The user sees the ECG on the watch and may send it to their physician. There are studies that reports the positive results of AF screening using the devices [30].

Another area where mobile technologies already have taken significant place is diabetes management. Regarding diabetes guidelines physical activity, healthy eating, medication adherence (e.g., insulin dosing), monitoring (e.g., blood glucose and weight tracking), education, and problem solving are the six essential behaviors for improving diabetes self-management [15]. There are multitude of applications, helping patients in these routines. Usually they come integrated with wireless or Bluetooth-compatible devices, such as blood glucose meters, continuous glucose monitoring devices, insulin dosing devices, etc.

A new approach for diabetes management emerged in 2015 as a research collaboration between Emory University/Healthcare, the Atlanta VA Medical Center, Georgia Tech and Grady Memorial Hospital and later released as a remote patient monitoring system called Diasyst [31]. Diasyst is dealing with the problem of *clinical inertia*, defined as a failure of health care providers to initiate or intensify therapy when indicated [32]. The cloud-based system incorporates 4 strategies for improving glycemic control: personalized management based on individual patients' clinical markers and medications; facilitated transmission of patient data to clinicians; patient education and engagement; implementation of diabetes management guidelines and algorithms. Patient A1C levels are managed by personalized plan change recommendations based on patient real time data. The early results published in [33] show improving of A1C levels within a month of use with little hypoglycemia occurrence.

An article from 2018 reviewed the current state of the art in mobile health in general, and concentrated on the diabetes domain in particular. The study reviews mobile health from various points of view, including relationships between mobile health and EHRs, semantic interoperability, the clinical decision support systems, the wireless body area networks, the IoT, cloud computing, and big data analytics. At the end authors say "According to the results, mobile health applications still have plenty of room for improvement in order to take full advantage of unique mobile platform features and to truly fulfill their potential."[15]

4. Conclusion

This paper presented the basic concepts and challenges in the area of health information technologies in general. The current situation in the domain is the presence of multiple EMR/EHR systems and tools, adopted by providers and patients, that cannot communicate and share data among them. The key challenges now are health information exchange and receiving useful information of storing and analyzing vast quantities of digital health data. Interoperability is the major issue in order to achieve HIE. That need is especially important in the present state of increasing number of chronic diseases worldwide. The lack of standards and security concerns are determined as the main technical barriers to HIE. Organizations as HL7 have facilitated development of standards but widespread interoperability has not been achieved yet. Except data from traditional medical records, new sources of health data have appeared with the spread of mobile and wearable devices. Access to medical and behavioral data facilitates personalized care but still greater level of enhancement is needed.

References

[1] https://www.healthit.gov/sites/default/files/pdf/health-information-technology-fact-sheet.pdf
[2] Heart, T., Ben-Assuli, O., & Shabtai, I. (2017). A review of PHR, EMR and EHR integration_ A more personalized healthcare and public health policy. Health Policy and Technology, 6(1), 20–25. http://doi.org/10.1016/j.hlpt.2016.08.002
[3] Garets D, Davis M. Electronic medical records vs. electronic health records: yes, there is a difference. Policy white paper Chicago, HIMSS Analytics [Internet]. 2006 October 24, 2015: [1–14 pp.]
[4] The National Alliance for Health Information Technology. Report to the Office of the National Coordinator for Health Information Technology on Defining Key Health Information Technology Terms. 2008; https://s3.amazonaws.com/rdcms-himss/files/production/public/HIMSSorg/Content/files/Code%205%20Defining%20Key %20Health%20Information%20Technology%20Terms.pdf
[5] Kaushal R et al 2010. Electronic Prescribing Improves Medication Safety in Community-Based Office Practices. J Gen Intern Med. Jun 25(6): 530–536.
[6] http://www.commonwealthfund.org/interactives-and-data/surveys/international-health-policy-surveys/2016/2016-international-survey
[7] https://www.who.int/chp/chronic_disease_report/part1/en/index1.html
[8] G. Anderson, J. Horvath, "The growing burden of chronic disease in America," Public Health Rep. vol. 119, no. 21, pp. 263–270, 2004.
[9] https://aspe.hhs.gov/report/health-insurance-portability-and-accountability-act-1996
[10] https://eugdpr.org
[11] https://www.himss.org/library/interoperability-standards/what-is-interoperability
[12] Health Informatics on FHIR: How HL7's New API is Transforming Healthcare | Mark L. Braunstein | Springer. (n.d.). Health Informatics on FHIR: How HL7's New API is Transforming Healthcare | Mark L. Braunstein | Springer.
[13] Heart, T., Ben-Assuli, O., & Shabtai, I. (2017). A review of PHR, EMR and EHR integration_ A more personalized healthcare and public health policy. Health Policy and Technology, 6(1), 20–25. http://doi.org/10.1016/j.hlpt.2016.08.002
[14] Edwards, A., Hollin, I., Barry, J., & Kachnowski, S. (2010). Barriers to cross-institutional health information exchange: a literature review. Journal of Healthcare Information Management: JHIM, 24(3), 22–34.
[15] El-Sappagh, S., Ali, F., El-Masri, S., Kim, K., Ali, A., & Kwak, K.-S. (2018). Mobile Health Technologies for Diabetes Mellitus: Current State and Future Challenges. IEEE Access, PP(99), 1–1. http://doi.org/10.1109/ACCESS.2018.2881001
[16] http://www.hl7.org/about/index.cfm?ref=nav
[17] http://www.hl7.org/implement/standards/product_brief.cfm?product_id=185
[18] Mandel, J. C., Kreda, D. A., Mandl, K. D., Kohane, I. S., & Ramoni, R. B. (2016). SMART on FHIR: a standards-based, interoperable apps platform for electronic health records. Journal of the American Medical Informatics Association, 23(5), 899–908. http://doi.org/10.1093/jamia/ocv189
[19] http://www.hl7.org/implement/standards/product_brief.cfm?product_id=7
[20] http://www.hl7.org/implement/standards/product_brief.cfm?product_id=491
[21] http://docs.smarthealthit.org
[22] https://www2.openehr.org/what_is_openehr#
[23] https://www.directtrust.org/direct-project-makes-secure-messaging-available-to-patients/
[24] Halamka, J. D. and Ekblaw, A. (2017). The potential for blockchain to transform electronic health records. Harvard Business Review, 3.
[25] Azaria, A., Ekblaw, A., Vieira, T., and Lippman, A. (2016). Medrec: Using blockchain for

medical data access and permission management. In Open and Big Data (OBD), International Conference on, pages 25–30. IEEE.

[26] Ekblaw, A., Azaria, A., Halamka, J. D., and Lippman, A. (2016). A case study for blockchain in healthcare:"medrec" prototype for electronic health records and medical research data. In IEEE Open & Big Data Conference, volume 13, page 13.

[27] https://www.businesswire.com/news/home/20180402005181/en/Humana-MultiPlan-Optum-Quest-Diagnostics-UnitedHealthcare-Launch?elq_mid=10941&elq_cid=558581

[28] http://www.who.int/goe/publications/goe_mhealth_web.pdf

[29] https://www.alivecor.com

[30] Halcox, J. P. J., Wareham, K., Cardew, A., Gilmore, M., Barry, J. P., Phillips, C., & Gravenor, M. B. (2017). Assessment of Remote Heart Rhythm Sampling Using the AliveCor Heart Monitor to Screen for Atrial Fibrillation. Circulation, 136(19), 1784–1794. http://doi.org/10.1161/CIRCULATIONAHA.117.030583

[31] https://www.diasyst.com

[32] Phillips, L. S., Branch, W. T. J., Cook, C. B., Doyle, J. P., & Barnes, C. S. (2001). Clinical Inertia. Annals of Internal Medicine, 135(9), 825–34.

[33] Phillips, L. S., Barb, D., Yong, C., Tomolo, A. M., Jackson, S. L., Olson, D. E., et al. (2015). Translating What Works. Journal of Diabetes Science and Technology, 9(4), 857–864. http://doi.org/10.1177/1932296815576000

Computer technologies in shipping and a new tendency in ship's officers' education and training

B C Belev[1] and S I Daskalov[1]

[1]Nikola Vaptsarov Naval Academy, Faculty of Navigation, 73 V. Drumev street, 9026 Varna, Bulgaria

bl.belev@naval-acad.bg

Abstract. This article presents the results of the authors' work on a project of the European program Erasmus+ "Diversification of seafarers' employability paths through collaborative development of competences and certification" – DivSea. In this project, the Nikola Vaptsarov Naval Academy is actively involved in partnerships with related maritime education institutions from Romania, Cyprus and the Netherlands. The aim of the project is introducing qualifications required by the marine labour market into the European education and training institutes to ensure transparency and recognition of skills and qualifications to facilitate learning, employability and labour mobility. Transferring the aim to the shipping, we propose to use modern ship communication equipment and computer technology to enable seafarers to access up-to-date training and qualifications at any point in the World. The created and developed training programs by the participants in the project fill the gap created by the rapid development of shipping over the last twenty years and the changes in maritime legislation in this regard. The opportunities for ship Internet delivery at every point of the world ocean are used. These communication technologies enable the training of marine professionals to become an ongoing process.

1. Introduction

The ambition of the International Maritime Organization IMO to reduce the risk in shipping worldwide and to promote safety at sea made ship owners turn to modern information technologies and their application on board ships. This interest also corresponds to the idea of reducing the number of ship crewmembers, more and more activities being assigned to the so called intelligent schemes. Starting from the beginning of 90-th in the previous century with Capt. Hedestrom and Capt. Gylden publication [1], for example, today the maritime society discuss the ideas of unmanned and autonomous ships. The authors on above suggest the use of an Integrated Bridge System (IBS), which carries out a number of activities in ship handling and observation of the situation around her. According to them, such an organization could be established on the bridge that only one person, the officer on watch, would be involved in handling of the ship, One Man Bridge Operation (OMBO). Today both IBS and OMBO are a fact and their realization was made possible by the implementation of state-of-the-art electronic data-processing equipment on the bridge.
Since this time the mankind made big steps towards today's unmanned and autonomous ships. Over the last four years, there has been increased activity among the research community in the field of autonomous vessels. At this stage, publications mainly analyze the current state of shipping and the restrictions on the use of autonomous maritime transport on international voyages [2, 3, 4, 5]. Other large-scale publications present the problems faced by large consortia such as "Mitsui O.S.K. Lines,

Ltd. - Nippon Yūsen Kaisha", "Rolls Royce - Kongsberg", "Yara", "DNV GL". The authors mainly cite the definitions of the concept of "ship" presented in the IMO conventions and resolutions, as well as the role of the crew and its physical presence on the ship. Autonomy criteria have been proposed and adopted, according to which the separation of ships and their belonging to the relevant level has been made [6, 7].

According to J. Witt [8], the following three aims are achieved through the concept of the "intelligent control":

- reduction of the number of crewmembers involved in the systems' monitoring and control;
- maximum fuel consumption efficiency due to the precise keeping of the ship on the preset course;
- precise control of the movement of the ship in heavy traffic and near dangers to navigation.

All these publications point out the importance of modern information technologies in the implementation of the above concept, as well as the necessary computer equipment. An adequate education and training related to the new reality in the shipping is also needed. Therefore, deck officers should possess the knowledge and skills necessary for the competent operation and usage of the equipment. In this respect the problem has a bearing on the education, training and post-graduate qualification of marine officers in the field of information technologies.

The DivSea project team developed a blended learning system and analyzed the opportunity of implementation for vocational education and training. The system was developed for the marine sector and comprises modules created for soft and technical skills' improvement and assessment. A survey was conducted to identify the set of skills required be the end-users. Following the survey, focus groups with marine professionals were organized to identify methods for skills recognition. The results concluded in an E-platform for Blended Learning constructed to allow participants to improve their knowledge and gradually self-assess the level of achievements. The platform is available on the project website all around the World.

2. Usage of computer systems and technologies on board merchant ships and new tendency in vocational education

In his publication in 2017 [9] P. Borkowski presents an original design of an expert system, whose function is to automatically stabilize ship's course. This is one of the examples how the technologies are placed on board of merchant vessels.

Many researchers are attempting to use different mathematical algorithms that adapt to shipping conditions. Modern information and computing technologies make it possible to implement any scientific idea. Arjid, Dharmender and Jarial [10] have analyzed a number of algorithms and the possibilities for their application in solving data clustering problems. Their study presents an in-depth study of swarm intelligence techniques and is an example of a research approach to the algorithms used in other scientific fields.

Two types of shipboard computer networks are used:

- a shipboard local computer network (LAN) which is not connected to other networks;
- an extended shipboard computer network, connected to a shore-based computer or computer network. Those networks connected to the Internet are classed in this latter type.

In shipping computers are used not only for administrative purposes but also for communication, navigation, control and monitoring of the ship power plant, control of the cargo handling operations and other operations. On modern ships information about the state of the power plant is fed from the control computer to special devices in the accommodation spaces of the responsible personnel, to the navigating bridge or to other specialized spaces. The Norwegian company Ulstein Bergen has even developed a special information system for post-repair monitoring of the power plant from a shore-based office [11]. The system has been designed as a component of the program, called "Information Technologies in Ship Operation".

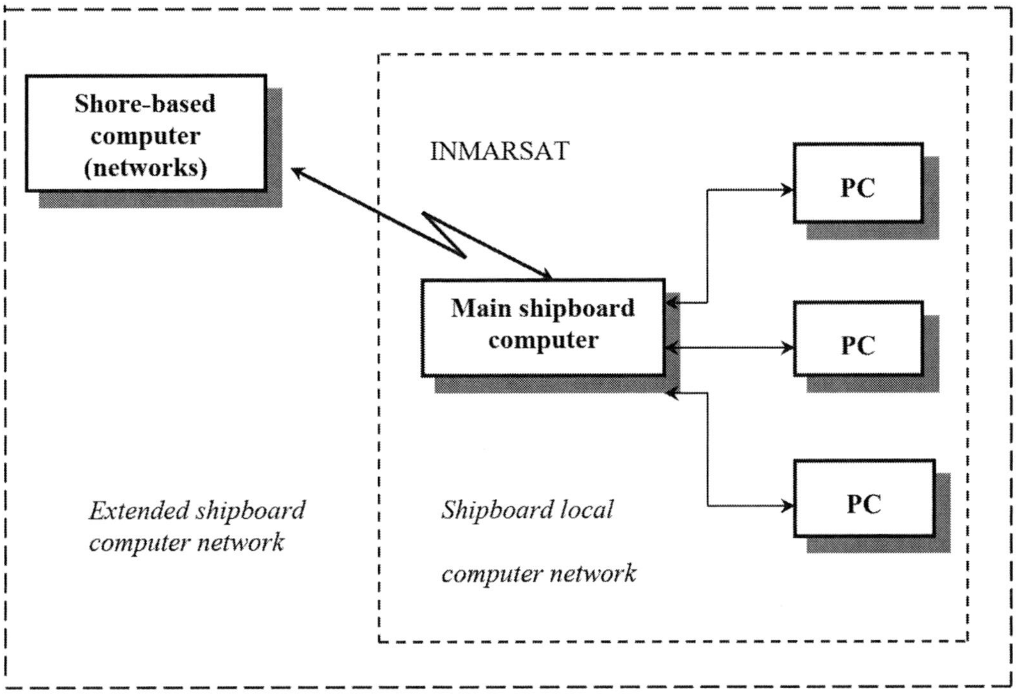

Figure 1. Principle scheme of an extended shipboard computer network

In his post in a specialized marine website Julius Patrick Apud describes very well information technology picture on board of modern merchant ships. He says that modern ships contain many different equipment and machinery that are controlled by computers. Systems such as the main engine, fuel delivery, electrical power, accommodation climate control, among others, need constant monitoring. Temperature changes, shafts revolutions per minute, fuel and oil flow rates and consumption, and other parameters need to be observed, recorded and analyzed. By using computers, the task of record-keeping and analytics is easier, which in turn improves performance [12].

The modern ship's bridge is a system of computers that performs all the tasks related to the safety of navigation and the observation of underwater and surface environment conditions around the ship (fig. 2)

Figure 2. Integrated Bridge System

Maritime education is an integral part of the global strategy for the development of world shipping. The seemingly peaceful area of the International Maritime Organization's agenda faces the challenges of the new era - the gradual realization of the idea of building and using autonomous ships. This fact makes the use of computer technology and the training of seafarers to use them even more relevant.

Training institutions are concerned about providing the most appropriate curricula that will respond to the trainees' needs. For this purpose, researchers care about updating the teaching curricula to offer teaching materials focused on learning outcomes. Cooperation with business/industry brings value by ensuring relevance of methods and technologies used in training, as specified in the European Commission's staff working document, "VET for better Skills, Growth and Jobs" [13].

According to the European Commission publication "Rethinking Education Initiative" [14], in order to build skills for the 21st century, efforts are needed to develop soft skills to enhance employability. Soft skills acquisition seems to be essential for flexibility and security of current and further employees. The right skills offer people the potential and confidence to carry out good quality jobs. All the conclusions above are fully related to the shipping and required adequate reaction from the training institutions. This approach allowed DivSea researchers to select the most relevant topics and to develop career guidance tools and learning materials for soft skills and for maritime technical skills.

Researchers collaborated to adopt a unitary system for developing the learning materials. The courses and assessment tools have been integrated in an E-Platform for blended learning available at http://divsea.cmu-edu.eu/ [15].

3. E-Platform for blended learning as a part of DIVSEA project

The IMO STCW 78/95 convention generally states the requirements for the training of ship officers to use modern information technologies. The syllabi of the maritime schools and universities, training ship officers and engineers, include subjects related to them. The syllabi of the Nikola Vaptsarov Naval Academy (NVNA) for the specialties Navigation and Ship Engineering contains the subject Information Technologies that is divided into two parts. The main purpose of the training is acquiring skills for using the computer as a universal means of data processing, for using office programs for text processing, spreadsheets, graphics processing, etc. The study of computer networks and work in the Internet was introduced long time ago.

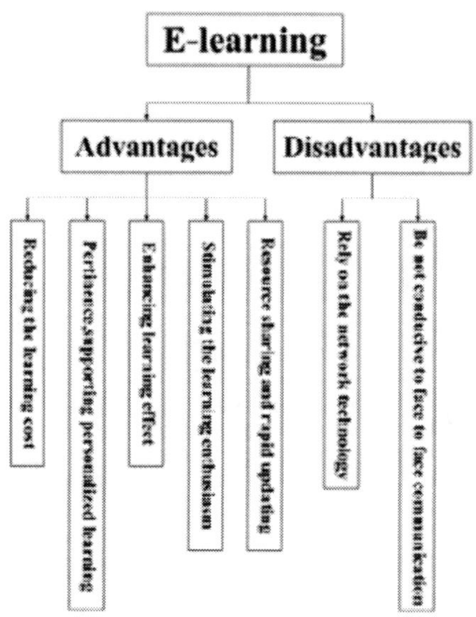

Figure 3. The advantages and disadvantages of E-learning

NVNA deliver to the students distance learning module. It's a kind of e-learning process, which is very popular within the last decade. In the study "The Application of E-learning in Maritime Education and Training in China" X. Chen, X. Bai and Y. Xiao describe well the process in China, which is a model for basic education, higher education, vocational education and so on [16]. Comparing advantages and disadvantages of e-learning and distance maritime education with traditional one, the authors says that advantages are obviously more (fig. 3):

- provides students with abundant learning resources;
- improves the flexibility of learning;
- reduce the learning cost;
- meet different levels, age, and geographic location of the students' individualized learning needs.

A small team professors from NVNA took part in a European Commission under Grant Agreement — 2016-1-RO01- KA202-024663, ERASMUS+ Strategic Partnership project "Diversification of seafarers' employability paths through collaborative development of competences and certification". Main objective of the project is to improve the employability skills and work opportunities for marine workers and to reduce the existing gap between the educational world and labour market by diversification of career paths and skills recognition [17]. The specific objectives are:

1. To build a network to continuously identify the gaps between marine labour market needs and educational field, for establishing unified system for transparency of skills and qualification by cooperation between representatives from both labour market and education.

2. To promote the continuous professional development of educational staff and marine workers in information computer technology methodologies for enhancing diversifications of career paths, by developing innovative practices for improving soft skills through exchanging good practices and collaborative short training periods.

3. To create teaching curricula focused on areas with skill shortages, targeting new professional marine qualifications in accordance with the National Qualification Frameworks, and to define competence standards orientated towards learning outcomes, supported by development of learning resources widely available as online open education resource.

4. Facilitate the recognition and validation of work-related skills and learning outcomes by referencing them to European Qualification Framework and using European Union instruments.

The authors are part of this team and participated very active in creating of e-learning platform for this new for seafarers training courses. Dissemination of project results included creation of website based on Moodle platform (fig. 4), conference in NVNA, pilot training course in Constanta Maritime University.

Figure 4. Project website http://divsea.cmu-edu.eu/

The following training courses were created in performance of DivSea project:

- Effective Communication;
- Leadership;

- Teamwork;
- Problem solving;
- High Voltage Safety and Maintenance Course;
- DP Maintenance course;
- Cargo Surveyor;
- Marine survey;
- Energy Efficiency;
- Ballast water management treating systems.

All of them are established in three versions – English, Romanian and Bulgarian language version. The interactive education, training and examination is in English language (fig. 5):

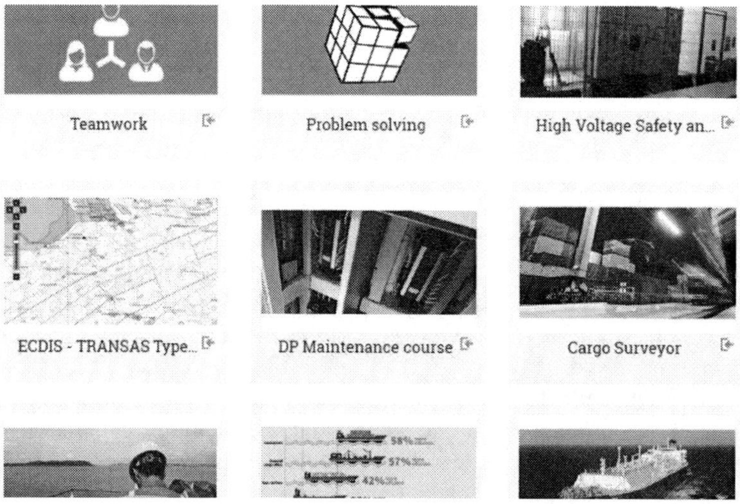

Figure 5. DivSea website Home page.

The education environment, described above is very suitable for distance learning and DivSea project aims are focused on it. The developed E-platform for blended learning is a complex learning environment intended for individual assessment of skills and expert guidance to improve and certify the beneficiaries' competences. Main feature of the platform is that it is available for onboard training also.

The dynamic change of the information technologies, used on ships, requires solving of at least two problems.

The first one is related to the necessity of constant updating of the syllabi not only in the area of information technologies but also in some specialized subjects. It is necessary to study the latest and most perspective computer software for use on board ships.

The second problem is related to the necessity of constant supplementing and enlarging the scope of the ship officers' training in the area of information technologies.

E-learning platform is not only sources of information. It also allow the participants in various courses to maintain interactive communication with the instructors and take part in peer discussions. The training information is provided in the form of multi-media products. Moreover, it offers links to sites of organizations and institutions like IMO, United Kingdom Hydrographic Office, United State Coast Guard, World Maritime University, etc. through them any user can get quick and updated information about the latest requirements of current international documents and regulations for special vocational areas.

4. Conclusion

The E-platform for blended learning comprises a set of tools for guiding users in their future career and sections designed for improving their skills and competencies. It includes information not only about the project but also about the way in which they can create an account, participate in the different courses and even get their training certificates free.

The users can use the platform for preparing for a new career also. All skills have been selected based on the results of the international cooperation between marine labour market, trainees and VET institutions carried out during the first stage of the project, that of the survey. The experts created teaching curricula and assessment materials for recognition of achievements.

The developed model can be implemented for vocational education and training in various sectors, other than the maritime sector. The efforts to involve the main factors as employees, employers and trainers offer a clear identification of needs that will impact to the relevance of the final result represented by the E-platform.

5. References

[1] Captain Hedestrom H and Captain Gylden S, Safer Navigation in the '90s – Integrated Bridge Systems, Journal of Navigation, vol. 45, 1992, p.p. 369 – 383

[2] UNCTAD, Review of Maritime Transport 2017, United Nation Publication, ISBN 978-92-1-112922-9, pp. 130

[3] Standard of Training, Certification and Watchkeeping Convention (STCW), IMO, London, 2017

[4] Analysis of regulatory barriers to the use of autonomous ships, Final report, Danish Maritime Authority, 2017, https://www.dma.dk/Documents/Publikationer/Analysis%20of%20Regulatory%20Barriers%20to%20the%20Use%20of%20Autonomous%20Ships.pdf (Accessed 12.12.2018)

[5] Decetelaere P, The legal challenges of unmanned vessels, Master Dissertation, University Gent, 2017, pp. 113

[6] Lloyds Register, Tackling an evolving threat, https://www.lr.org/en/cyber-security/, (Accessed 05.12.2018)

[7] Johns M, Seafarers and digital disruption, Hamburg School of Business Administration, 2018, pp. 32

[8] Witt J, Sutton R and Miller K, Recent Technological Advances in the Control and Guidance if Ships, Journal of Navigation, vol. 47, 1994, p.p. 236 – 258

[9] Borkowski P, Inference Engine in an Intelligent Ship Course-Keeping System, Computational Intelligence and Neuroscience, Volume 2017, Article ID 2561383, 2017, https://www.hindawi.com/journals/cin/2017/2561383/ (Accessed 01.03.2019)

[10] Kumar A, Kumar D and Jarial S, A Review on Artificial Bee Colony Algorithms and Their Applications to Data Clustering, Cybernetics and Information Technologies, Volume 17, No 3, Sofia, 2017, pp. 3 – 28

[11] Zolkiewski J and Morazzani C, Relationships, Information Technology and the Shipping Industry – An Exploratory Investigation, https://www.impgroup.org/uploads /papers/4613.pdf (Accessed 01.03.2019)

[12] Apud J, Information Technology Applications in the Maritime Industry, 2016, http://maritimereview.ph/2016/11/01/information-technology-applications-in-the-maritime-industry/, (Accessed 12.02.2019)

[13] European Commission (2012), VET for better Skills, Growth and Jobs, Staff working document. https://eur-lex.europa.eu/LexUriServ/LexUriServ.do?uri=SWD:2012:0375:FIN:EN:PDF, (Accessed 01.03.2019)

[14] European Commission (2012). Rethinking Education: Investing in skills for better socio-economic outcomes. Communication from the Commission to the European parliament, the council, the European economic and social committee and the committee of the regions. http://www.cedefop.europa.eu/files/com669_en.pdf. (Accessed 26.10.2018)

[15] Acomi N, Varsami C and Makrides G (2017). Diversification of Seafarers' Employability Paths through Collaborative Development of Competences and Certification. E-platform for Blended Learning Recommendations for recognition and validation of workrelated skills and learning

outcomes, Erasmus+ Strategic Partnership: 2016-1-RO01-KA202-024663, http://divsea.cmu-edu.eu/, (Accessed 21.03.2019)
[16] Chen X, Bai X and Xiao Y, The Application of E-learning in Maritime Education and Training in China, International Journal on Marine Navigation and Safety of Sea Transportation TransNav, vol. 11, No 2, 2017, p.p. 349-354
[17] Acomi N (2018). Developing a System for Recognition of Work-Related Skills and Learning Outcomes. Revista Romaneasca pentru Educatie Multidimensionala, 10(2), 102-111. https://doi.org/10.18662/rrem/49 (Accessed 26.01.2019)

IOP Conf. Series: Materials Science and Engineering **618** (2019) 012035 doi:10.1088/1757-899X/618/1/012035

CONCEPTUAL FRAME MODEL FOR THE PRESENTATION OF THE CONCEPTS AND RULES IN NATURAL LANGUAGE INTERFACE FOR DATABASE

M Zhekova[1] and G Totkov[2]

[1] PHD student, Department of Computer Science, University of Plovdiv „Paisii Hilendarski", Tzar Asen str.24, Plovdiv, Bulgaria

[2] Prof., Department of Computer Science, University of Plovdiv „Paisii Hilendarski", Tzar Asen str.24, Plovdiv, Bulgaria

m_jekova@uft-plovdiv.bg , totkov@uni-plovdiv.bg

Abstract: The article presents a frame model for conceptual description of a subject area by presenting a detailed structure of multiple objects and concepts, knowledge about them, hierarchies of types, rules, specifying classes and relations between them. Include all possible words and word combinations that formalize a subject area. The model itself represents knowledge and reflects the conceptual basis of the organization of the memory of a person, as well as its flexibility and visibility. When creating an appropriate database of accumulating frame models (AFM), frame questions, and at a later stage a frame model for automatic question generation, it is necessary to build an appropriate structure that allows for both flexibility and easier building of the databases of data, as well as efficient retrieval of different types of information when summarizing the results. The publication is part of a study devoted to the application of frame models for the presentation of knowledge and processes in e-learning.

1. Introduction

With native language processing, interfaces can be built to allow users to interact with data more intuitively using their native language. This paper provides a methodology for creating a question-response information system using a natural-language interface to gain knowledge for the benefit of users through a web-based system.

Natural Language Interfaces (NLI) are a kind of computer-human interfaces, where language forms such as verbs, phrases, and clauses act as parameters in the formation of queries to a data mining database. Retrieving database information requires in-depth knowledge of technical languages such as Structured Query Language [10]. By contrast, database NLI would allow questions to be asked in the native language of the user. Therefore, natural language interface information systems would be more suited to casual users as the user would not need to learn a language through which to communicate with the system.

This paper aims to compile a database query model by manipulating the queries introduced by the user, then differentiating and parameterizing the parts of the trait and reaching the rules and relationships that lead them. Steps in my work includes create frame models, define parameters for these models and then we use models and semantic defined parameters to produce user outputs. NLI is not just a question-response dialog based on a given knowledge base but also access to structured data modeled and designed syntactically and semantically classified. For this purpose, it is necessary to precisely define the scope of the subject area, correctly examine the peculiarities of the lexical units and phrases, to construct a system for their linguistic description and classification and to create a methodology for their recognition and tagging with grammatical and syntactic information. A major challenge will be to solve the problem of natural language comprehension, how to interpret the query data, and how the user's information is transformed into a question.

The purpose of the development is to construct a conceptual map in the form of a linguistically correct and argued theoretical frame model for presenting a database of concepts, relationships and relationships by describing the individual language units and rules, and then able to derive this information and identify all SQL elements included in frames, slots, attributes, aggregates, values from the input entry queried using pre-created rules for the purpose. In this way, a new database is being built - a computer model full of content extracted from various sources. And once we have the knowledge base from the limited subject area, we can derive the data from all these sources.

2. Related work

One of the tasks of computer linguistics is the creation of software programs and applications that, with the help of the knowledge formalizations realized so far, facilitate the access to the accumulated resources and the huge amount of information and work with it. Like the systems that generate text from documents, question-response systems rely on complex techniques for natural - linguistic analysis.

Here are some systems that are oriented to a multi-contextual knowledge base that describes both concepts and roles that correspond to objects in the subject area, as well as rules for constructing a query to the database and for constructing concepts.

- An information system that processes the user's question determines the type and focus of the question, the type of response, and then uses that information to create the query is a system based on retrieving the information. The PLINIUS system is designed to automatically extract knowledge from texts in chemistry [12]. Several sets of concepts such as a chemical element, integer, group, etc. are defined in it and rules for constructing the other concepts. About 150 concepts and 6 rules are described in the ontology describing the system [13]. It described using frame constructions.
- Knowledge-based systems generate a semantic description of the question that is then used to search in structured resources. They are suitable for limited domains such as tourism, medicine and transport, where there are no large volumes of pre-recorded documents. Knowledge based approaches are based on the fact that no new rules need to be added for each new domain or language of the application. Apple's SIRI is such a system. It contains information, messages are sent, events are added to a calendar instructions are received. The application uses machine learning technology to make it "smarter" and more capable of predicting and understanding the questions and requests in a natural language by the user.
- In hybrid systems, like IBM's WATSON, while the issue is semantically analyzed, responses are generated using information retrieval methods.

3. Frame model description

In artificial intelligence systems, the presentation of the database in the form of a conceptual model is close to the way the concepts are stored in the human brain.

In the development we present a method for creating a natural-language interface used for the purpose of semantic search in structured data and content through a system of questions and answers.

The description of the proposed model is presented in figure 1, which is a University database created with MySQL for a university.

Classifying, maintaining and storing information as well as defining relations between datas aims to facilitate access to the database as well as to facilitate communication between the user and the computer.

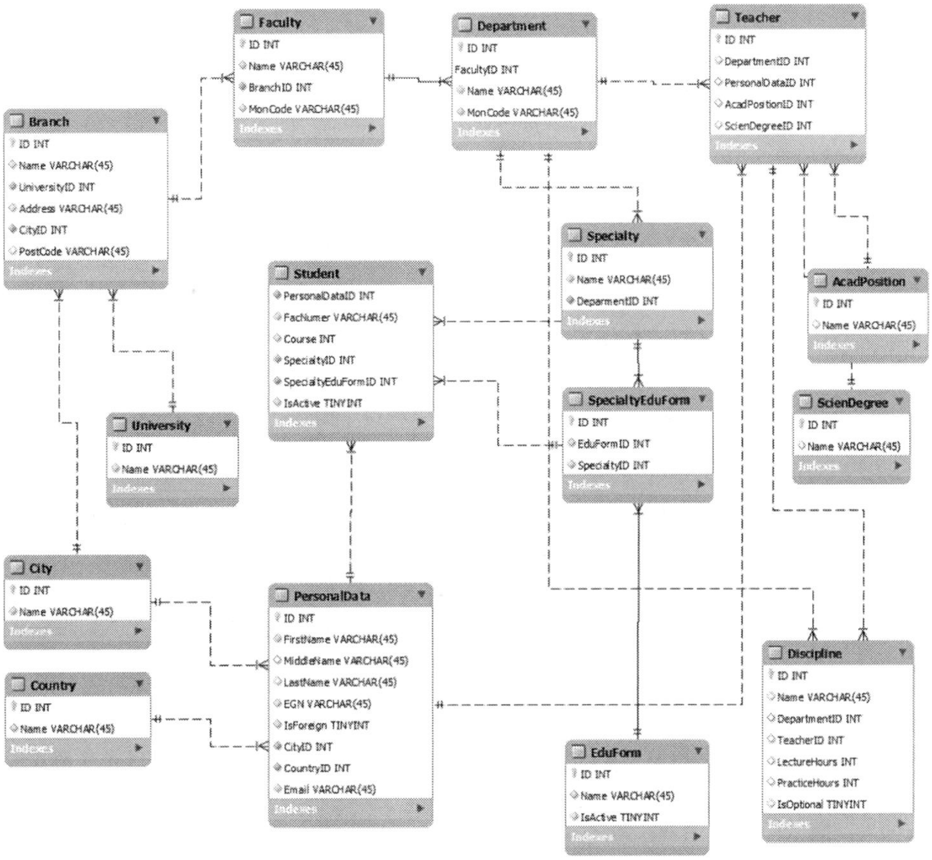

Figure 1. University database.

Several correctly normalized tables have been created in the University database - Branch, Faculty, Specialty, Discipline, Student,... Obtaining a response from this database, however, requires SQL knowledge to be learned and assimilated.

If we look at the example: User wants to view information such as - the names and codes of the courses studied by a student which faculty number equals 16810235411, whose specialty is "Business Information Technologies". From the Specialty table of the University database we use the following SQL query:

```
SELECT discipline_name, discipline_code
FROM discipline
WHERE specialty_name = “Бизнес Информационни технологии”
and student_facNumber = “16810235411”
```

Native language processing and SQL query to a Specialty Table in University database will result in the same output, and the question is not a SQL query, but the user's natural language.

```
What are the name of the discipline and code of the discipline which specialty
name equals "Business IT" and faculty number of student equals
"16810235411"?
```

In this database we can ask following questions:
-Who are the lectors working in the Department of Computer Science? and
-Who are the lectors reading lectures in the Department of Computer Science?

4. Scope of the Frame model

Scope of the proposed frame model is:

- For the purpose of testing, a University database (figure 1) has been created. For a relational database management, knowledge of software for working with MySQL databases is required.
- Language for working with the system for facilitating a wider range of users has been chosen to be Bulgarian.
- User input is in the form of questions - Who? What? What? Where? When?
- Is used a detailed structure of multiple objects and concepts, knowledge about them, hierarchies of types, rules, specifying classes and relationships between them to describe the corresponding element. Is included all possible words and word combinations that formalize a subject area. The list is updated with words and phrases specific to the specific domain area.
- The question divides into lexical units or phrases from which it is composed, giving each unit a unique number (Tokenization).
- To remove unnecessary words from the user's question. In the table 3 are specific domain area words not bearing important information, which will be updated periodically.
- Identify all SQL items included in tables, attributes, aggregates, values from the input entry queries, using pre-created rules for the purpose.
- Create an SQL template using follow template:

```
ID of slot,
ID of table name,
ID of and, ID of aggregate,
ID of interval, ID of value
```

- Create a SQL query using SQL Elements.
- For the avoidance of ambiguity, frames have been assigned a frame ID field, frame name, attribute ID, and attribute name to store all the attributes belonging to a frame.

5. Frame as a concept

The frame, according to [1], is used to represent the minimum structured information that defines unity phenomena, events, situations or processes. The term was introduced in the artificial intelligence by M. Minsky [2] as a means of presenting knowledge. According to [7], the frame model is used to represent relatively independent, logically distinct units of knowledge with the possibility of multiple use in different situations (including retrieval and aggregation of data).

Each frame has a unique name and structure consisting of stacked elements called slots. Each slot has a unique name in the frame, it contains certain information, and it carries classification information about the syntactic function and the belonging of the language unit located in it to part of the speech. The content of each slot can be a numeric or text value, predicate dependence, function, or other frame. A group of frames connected to each other forms a system [3]. The structure of the frame is dynamic, ie. allows adding new slots at any time [6].

6. Frame model description

The article presents a frame model for conceptual description of a subject area by presenting concepts and relations with their concrete representatives - prototypes and specimens. The model itself represents knowledge and reflects the conceptual basis of the organization of the memory of a person, as well as its flexibility and visibility. When creating an appropriate database of accumulating frame models (AFM), frame questions, and at a later stage a frame model for automatic question generation, it is necessary to build an appropriate structure that allows for both flexibility and easier building of the databases, as well as efficient retrieval of different types of information when summarizing the results. In the presented computer model, the task is to synthesize text to answer a question asked by a system user.

The accumulation frame model(the set of interconnected frames) is both a form, method and means of learning, logical and consecutively ordered knowledge structures composed of generalized structural models(blank frame-prototype) and their specific, filled-in representatives (frame-specimen). Both the descriptive and basic knowledge for each domain area are found in the model described above.

Table 1. A common structure of the frame model with the n-slot.

Frame name:
slot 1 name (content of slot1);
slot 2 name (content of slot2);
slot 3 name (content of slot3);
......
slot N name (content of slotN);

Frame models, composed of slots describing concepts and surations of reality, are presented within the article. Each slot is described by one or more attributes that represent knowledge describing the element. Attributes or so-called veneers serve to describe both static knowledge - values and dynamic knowledge through so-called "attached procedures" (methods) or "demons" [6]. Demons are a special type of procedures that start automatically when an event occurs. The value of a slot is of a certain type of data, which may be elementary or compiled.

The purpose of classifying data in frames is to maintain and store the data, to define the links between them and to facilitate access to them.

6.1. Description of lexical unit

According to [4], a lexical unit is the ordered three - lemma, lexical meaning, examples. In Table 2 is presented a frame-prototype - Concept with a description of the slots, defined exhaustively with the terms of computer linguistics and their content.

The frame model shown in Table 2 describes a separate lexical unit, the so- frame-prototype with a detailed description of its slots and content. In this way, the terms used in the restricted domain area are described.

In the conceptual model, elements of the system under consideration involve different concepts with their connections described by facts and rules from the respective domain area. Structurally, it is a systematically bound concept with a different level of abstraction and a prototype of the future database.

Table 2. Frame- prototype – Concept.

Name	(name / full name of frame);
EN	(name of concept in English)
Description	(lexical description, meaning)
Sunonims	(synonymous order of the frame name);
Domain	(limited subject area);
Question Word	(questions that can be asked over the name of the frame)
Part Of Speech	(grammar category)
Semantic class	(belonging to a semantic class)
Semantic role	(semantic function / role)
Sub Frame Of	(inherit from)
Is Argument In	(participates as an argument in a predicate structure)

IOP Conf. Series: Materials Science and Engineering **618** (2019) 012035 doi:10.1088/1757-899X/618/1/012035

We will present the database as a set of frame models. From the frame-prototype, by filling the slots with specific values, a frame-specimen is obtained.

7. Frame model architecture

The natural-language interface for database deals with translating user queries from its native language to the corresponding SQL query to retrieve the relational database data via a web based information system. The result is displayed on the user screen. This web-based system is initially developed for Bulgarian, and other languages will be discussed later. An algorithm for mapping a natural language, implemented in Bulgarian, has been developed effectively to convert an SQL query to create appropriate responses. The algorithm is conceived with C #, Apache, MySQL and tested successfully. The system architecture is presented in figure 1, which figure the process layout involved in transforming the user query into the NL query into a syntactic SQL query that is created by relational database management system and receives database responses.

Table 3. List of words that are not involved in the SQL query.

are| is | which| that| for | in front of | this | that | like a| in | those | on | towards | throw | will | under| another | over | must | however | anyway | yet | though | still | else | some | even | only | indeed

Table 4. Frame – prototype – Person.

Person:		
Name	(text value – first name, last name)	Who?
ID_card	(integer)	What?
City	(text value)	Where?
Address	(text value)	Where?
Post code	(integer value)	What?
Sex	(text value)	What?
Nationality	(text value)	What?
Sub Frame Of	(Person)	
Is Argument In	(live_in {Person, Address}, live_in_city{Person, City}, have_ID{Person, ID}, sex{Person, Sex}, nationality{Person, Nationality})	

The term Student is described by the criteria that we can outline a student.

Table 5. Frame – prototype – Student.

Student:		
Frame name	(Student)	
Student name	(Person.Name)	Who?
Faculty number	(text value)	What?
Semester	(integer value)	Which?
Evaluate	(function)	How much?
Edu form	(редовно/задочно/платен)	What education form?
Edu degree	(bachelor/master)	What education degree?
University	(text value)	Which university?
Specialty	(text value)	What specialty?
Diploma number	(text value)	What diploma number?
Personal skills	(text value)	What personal skills?
Role	(text value)	What role?
Internship	(text value)	What internships?
Sub Frame Of	(Person)	
Is Argument In	(study {Student, Specialty}, study{Student, Discipline}, is_in {Student, Degree}, live_in{Student, City}, ID{Student, ID_card})	

All values from Person frame - Person.Name, Person.Address, Person.Nationality .... are inherited from Student frame because Student is a Person type. Student is a Person and all attributes of Person are attributes of the Student.

When the user asks if there are more criteria in the question, if the search is more parameters, it is convenient to arrange in a cequence of parameters / parameters here are the names of the slots /.

Table 76. Rules for representing some different frame in University database.

Students	table Student	Rule for representing table Student
Faculties	table Faculties	Rule for representing table Faculties
Specialties	table Specialties	Rule for representing table Specialties
Disciplines	table Disciplines	Rule for representing table Disciplines
Teachers	table Teachers	Rule for representing table Teachers
Internship	table Internship	Rule for representing table Internship

Table 7. Rules presenting different attributes of frame Student.

Student name	attribute Student Name	Rule for representing attributes
Student faculty	attribute Fac Number	Rule for representing Fac Number
Semester	attribute Semester	Rule for representing Semester
Education degree	attribute Edu Degree	Rule for representing Edu Degree
University	attribute Uni Name	Rule for representing Uni Name
Personal Data	attribute PersonalData	Rule for representing Personal Data
Specialty	attribute Specialty Name	Rule for representing Specialty Name

Table 8. Rules for representing an 'and', 'as well as', 'or' and 'nor' concept.

and	s and	rule for representing And
as well as	s and	rule for representing And
also	s and	rule for representing And
in addition	s and	rule for representing And
or	s or	rule for representing OR
not	s not	rule for representing Not
not include	s not	rule for representing Not

Table 9. Rules for the aggregate functions.

most	aggregate max	rule for representing max()
maximum	aggregate max	rule for representing max()
highest	aggregate max	rule for representing max()
maximum number of	aggregate max	rule for representing max()
highest number of	aggregate max	rule for representing max()
smallest	aggregate min	rule for representing min()
minimum	aggregate min	rule for representing min()
smallest number of	aggregate min	rule for representing min()

Table 10. Rules for the interval 'equal' concept.

equal	interval =	rule for interval equal
exactly	interval =	rule for interval equal
match with	interval ==	rule for interval absolutely equal
less than	interval <	rule for interval less
bigger than	interval >	rule for interval greater
less than or equal to	Interval <=	rule for interval less or equal
greater than or equal to	interval >=	rule for interval greater or equal
different from	interval <>	rule for interval different

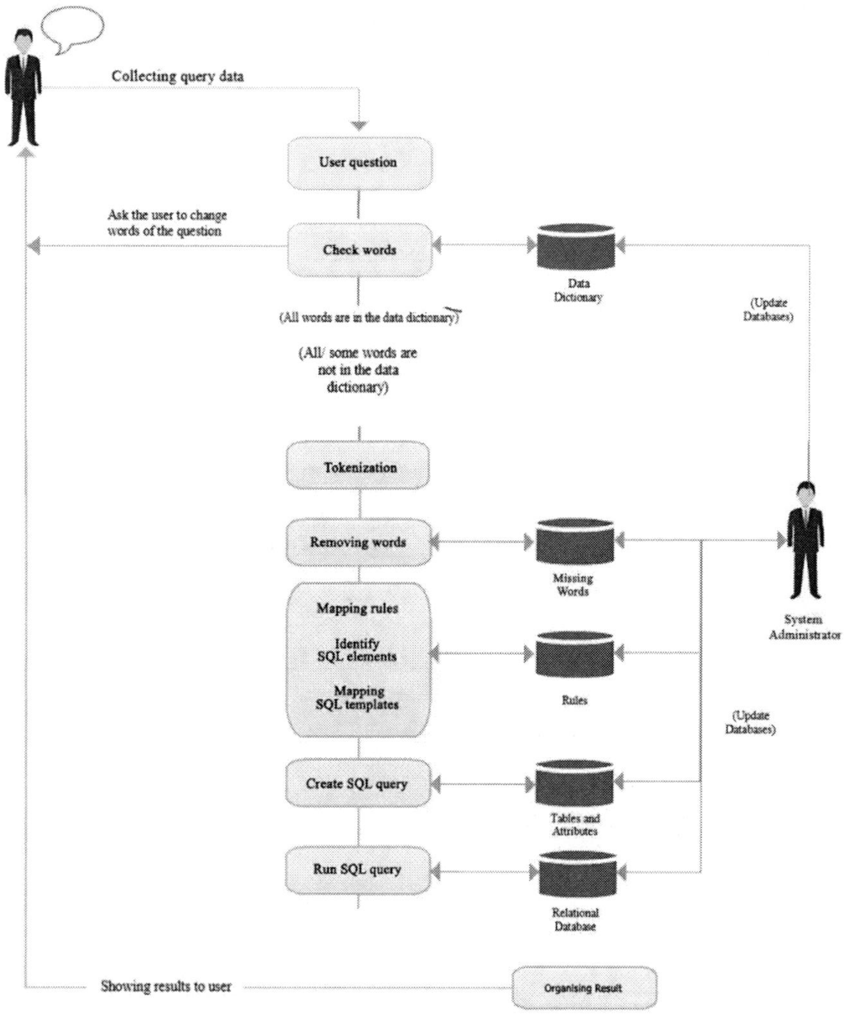

Figure 2. Architecture of a web system with a native-language interface to connect to a database.

8. Algorithm used in natural language system

The following steps describe the algorithm for the proposed natural-language system.

- Checking for an input value and if it exists, the value is removed from questional string.
- Check all the words in the query string wheather they exist in the resource dictionary / dictionary of the domain area.
- Tokenization
  - Split the questional query of tokens
  - The ID number of each identified token is given
- Remove unnecessary words / words that are not essential / from the string.
- Make a comparison by removing each last word from the question string. If compliance with an existing rule is found, the corresponding word / phrase / slogan is removed from the query string and the action is repeated until the string is run out until the query string becomes null.
- The arrays store the IDs / IDs of the identified SQL elements.

IOP Conf. Series: Materials Science and Engineering **618** (2019) 012035 doi:10.1088/1757-899X/618/1/012035

- An SQL template string is constructed considering the IDs of SQL elements available.
- Mapping the new SQL string with SQL templates.
- Generate SQL query using SQL Elements.
- The system returns an answer.
- Show the result in the user interface.

In the computer analysis of the user query first, it is divided into words and punctuation (tokenization) and then proceeds to semantic analysis at different levels where each token is defined by grammatical characteristics, belonging to syntactic categories, parts of speech, semantic classes and dictionary meaning.

In the abstract structure of the Bulgarian language, the sentence exists in the form of a scheme called a construction of sentence [9]. The sentence construct includes certain positions corresponding to given syntactic roles and arranged in a particular scheme.

In the proposed theoretical frame model, the criteria and features used to describe each slot of the frame provide information about:

- Semantic role / relationship - The purpose of semantic roles is to summarize similar semantic properties of classes of arguments. Semantic (theta) roles are reviewed by J. Penchev (1999), who distinguishes the following basic relations: agent, actor, experimental, elemental, object, patient, addressee, topic, locative, temporativ.
- Selective Restrictions - This classification divides the nouns into abstract and concrete, animate and inanimate, common and own.
- Part of the speech / Syntactic function - subject, verb, object, direct supplement.
- Part of a syntactic category - Latin alphabets are used to denote the parts of the speech to which the final elements of the sentence belong: Noun - noun, Adj - adjective, Verb - verb, P - preposition, Adv - adverb, Conj - conjunction, Pron - pronoun, Num - numerical. S is the category - sentence.

9. Conclusion

In computer systems for artificial intelligence, the presentation of the database in the form of a conceptual model is close to the way the concepts are stored in the human brain. Natural language generation is a task for generating conclusions - understandable answers, natural language texts from a computer software application, such as a database or logical form, such as a frame. Frames are also the sources from which the data will be derived in the proposed theoretical model. In the study is presented model for classifying concepts, relations and rules in the domain area. The model will be explored experimentally and analyzed in subsequent publications.

The proposed methodology is used for a question-answer information system in order to leverage knowledge for the benefit of users in solving organizational communication problems. The dependencies between words or phrases in the user query are only identified with the links that exist between concepts and rules in the subject area knowledge base. The approach strongly depends on the qualitative selection of knowledge about objects, rules and relationships between them.

Future developments will add to the existing conceptual frame model and syntactic semantic analyzer. The natural language interface for dialogue with an information system has been developed as a start in Bulgarian, with the issue of developing other languages as a task for further research.

Acknowledgements
The paper is partly supported within the project MU17-FF-023"Accumulative frame models for extraction and aggregation of data for knowledge and processes in learning" of the Scientific Research Fund at the University of Plovdiv.

IOP Conf. Series: Materials Science and Engineering **618** (2019) 012035 doi:10.1088/1757-899X/618/1/012035

References

[1] Zhekova M, Totkov G and Panayotova T, Acumulative frame models for generating test units in e-learning systems, Scientific works of the Union of scientists in Bulgaria - Plovdiv, Series C. Technique and Technology, Volume X, 2017

[2] Minsky M, A Framework for Representing Knowledge, MIT, Cambridge, 1974

Yordanova S and Stefanova K, Retrieving knowledge from unstructured data through consumer opinion analysis, issue 1, 2017

[3] Shivacheva G and Totkov G, Accumulative frame models in programming training, Oktober 2017

[4] Koeva Sv, Bulgarian FrameNet, Semantic-syntactic dictionary of the Bulgarian language, print edition, Sofia 2008

[5] Nestorova P, Semantic and syntactic peculiarities of predicates in the Bulgarian language, print edition, Plovdiv 2008

[6] Doneva R, Gaftandzhieva S and Totkov G, Frame Models in Programming Training, XI National Conference "Education and Research in the Information Society", 2018

[7] Doneva R, Gaftandzhieva S and Totkov G, Frame representations in e-learning applications and new developments, International Journal on Information Technologies & Security, No 2 (vol. 10), 2018

Penchev Y, Construction of the Bulgarian Sentence, Sofia 1984

[8] Penchev Y, Bulgarian syntax. Manage and Connect, Plovdiv 1993

[9] Kolev K, Course project in linguistics

[10] Rukshan A, Prashanthi R and Sinnathamby M, Natural Language Web Interface for Database (NLWIDB), Proceedings of the Third International Symposium, SEUSL: 6-7 July 2013, Oluvil, Sri Lanka

Barkalova P, Bulgarian syntax - familiar and unknown, 1997 Sofia, Abagar

Stoyanova Iv, Automatic recognition and tagging of composite lexical units in the Bulgarian language, 2012 Sofia

[11] Popov D, Bulgarian vocabulary, ed. "Science and art", 1994

Fellbaum C, WordNet - An Electronic Lexical Database, MIT Press, 1998

[12] Van der Vet P, Speel P and Mars N, Tne PLINIUS Ontology of ceramic materials, April 1995

[13] Terzieva T and Totkov G, An ontological approach to the learning process

Ruppenhofer J, Ellsworth M, Petruck M R L, Johnson C R, Scheffszyk J, FrameNet II: extended theory and practice, Technical Report, 2005

Smart shopping system

Dilyana Budakova, Lyudmil Dakovski

Technical University of Sofia, Plovdiv Branch, Bulgaria
European Polytechnical University, Pernik, Bulgaria

dilyana_budakova@yahoo.com, l.dakovski@gmail.com

Abstract. The paper describes the design of an intelligent system for smart shopping. The proposed prototype of the system incorporates speech enabled intelligent agent, beacon-based technology, holographic technology, picture exchange communication system, and machine learning techniques. Common use of these technologies in united intelligent systems enables people with different communication possibilities to navigate in large buildings. Some initial user opinions for the system are presented.

1. Introduction

In large and unfamiliar indoor spaces, such as airports, shopping centers, stadiums, hotels, office buildings, people may have difficulties with finding the desired destination. Many categories of people – the elderly, the children, the people with visual or hearing impairment, with difficulties in communication etc. – need specialized ways of communication. This paper presents an experimental prototype of a Intelligent Navigation System (INS), which unites possibilities of speech enabled intelligent agent, beacon-based technology, holographic technology, picture exchange communication, and machine learning techniques.

Beacons are used to mark the location of objects and navigate people in indoor spaces [1, 6, 7, 20]. They work on the principle of lighthouses by emitting signals at short intervals based on Bluetooth Low Energy (BLE) technology. The distance to the Beacon can be defined depending on the signal strength [19]. In addition to emitting advertising or other types of announcements, it is also possible to locate beacons [1,7,20].

Holograms are made of light and sound, appear in the around space and reply to gestures, voice and gaze commands [18]. A hologram can be placed and integrated in the real world or can tag along with user as an active part of user's world halping for navigation in indoor spaces.

Another possible solution to the problem of orientating people in indoor spaces are the embodied conversational information kiosks [3, 4]. These systems use the information they have both about their own location and about the layout of the building and give instructions to the users how to find the desired place in the building.

The information kiosks are a collection of different technologies such as video processing form face detection, speaker-independent speech recognition, array microphone for noise cancellation, a database system, and a dynamic question answering system [4, 21].

The Picture Exchange Communication System (PECS) [2,5] allows people with little or no communication abilities to communicate using pictures. People using PECS are taught to approach another person and give them a picture of a desired item in exchange for that item. [11,12] Screen readers [14,15,17] and text-to-speech (TTS) systems [13,16] enable blind and vision impaired people to use blind and vision impaired people to use computers and provide the key to education and employment.

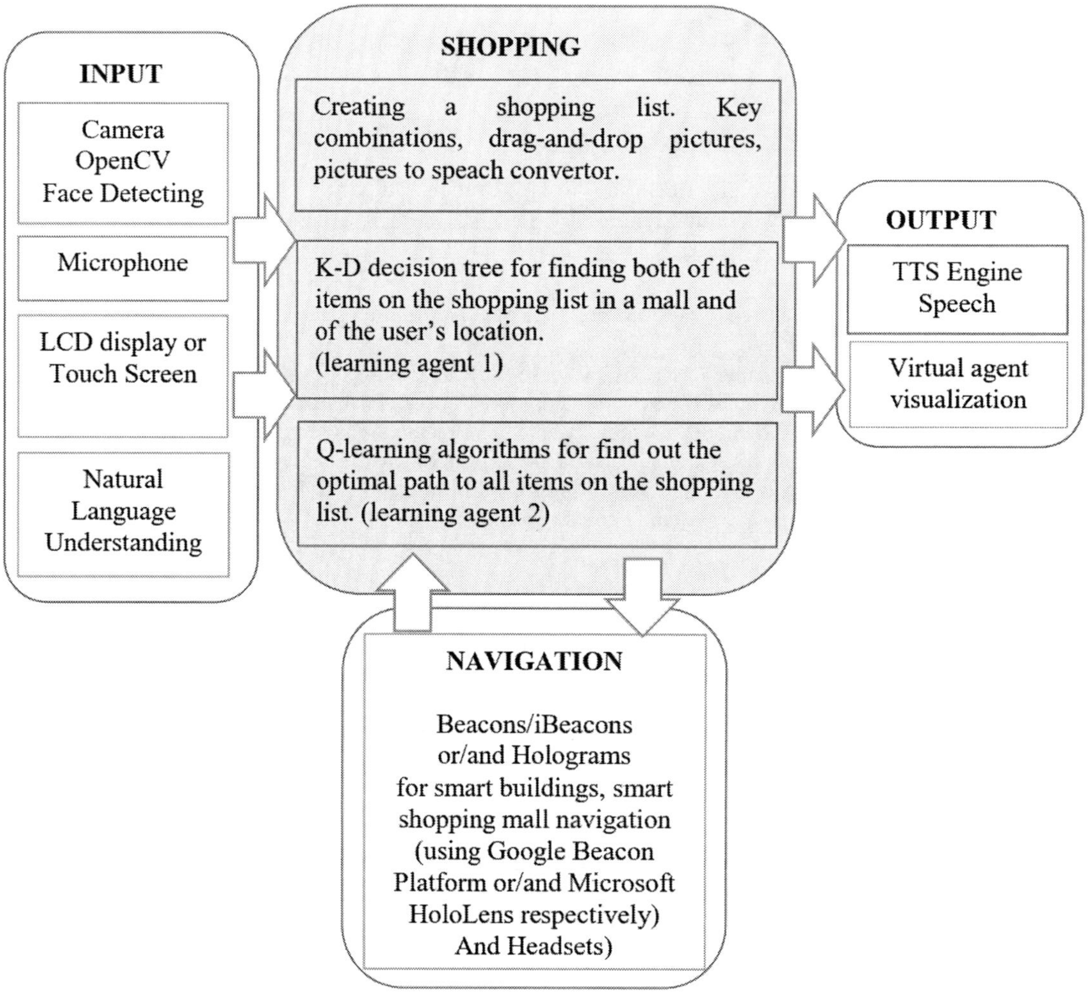

Figure 1. Block-diagram of INS

2. Design OF INS

The modulus of the system prototype are given in figure 1. The design system consists of four main blocks: input, output, shopping, and navigation.

The technologies, used in the input block, are face detection and speech recognition. The equipment comprises a camera, a microphone, a keyboard, a mouse, and a touch screen.

The output block uses speech synthesis and cartoon virtual character visualization for giving information to the user.

The block shopping includes: drag and drop pictures for creating the shopping list; pictures to speach convertor; algorithm for machine learning like decision k-d tree [10] for finding a shop, an item or a user's location; q-learning algorithm for finding the optimal path for shopping.

IOP Conf. Series: Materials Science and Engineering **618** (2019) 012036 doi:10.1088/1757-899X/618/1/012036

The navigation block consists of a sensor network of beacons and/or holograms with corresponding technologies.

The agent starts a dialogue after recognizing a human face in front of the camera. The drag-and-drop technique or key combinations are used to create the shopping list from a range of item pictures. When key combinations are used, the name of the item is pronounced, which facilitates the blind people. The agent reads the shopping list and prepares the optimal path from the entrance of the mall to the checkouts through all the shops, from which the user would like to buy something. A decision k-d tree of the location of the various shops in the shopping mall is built. The optimal path for shopping is found by means of the implemented Q-learning algorithm on individual smartphone. A simple tracker that relies on face detection is used. The general object detection algorithm consisting of a cascade of boosted classifiers proposed by Viola and Jones [8] is used to detect faces. For video processing C# wrapper for the Intel OpenCV library [9] is used.

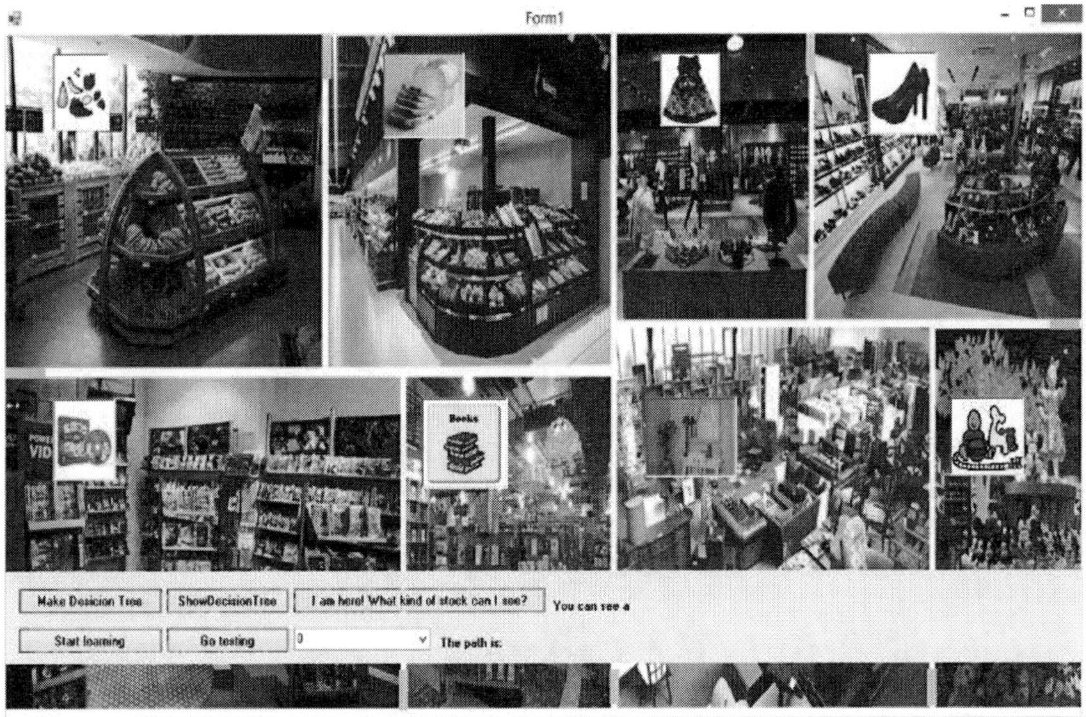

Figure 2. Exemplary placement of shops in a shopping mall

The virtual agent is visualized as a cartoon character. A number of emoticons are chosen to express the agent's emotions. The agent can execute a limited number of commands, given in a natural language.

The user has option to move the emoticon to a desired by him/her position on the picture map of the shopping center. The position of the emoticon is compared to the position of the considered axis of comparison. Depending on the result, the next axis of comparison of the decision tree is chosen. When there is no next axis of comparison in the already built decision tree, the name of the object-shop or object-commodity, located in the area, defined by the axis of comparison of the decision tree, appears on the screen (figure 2)

The axes of comparison are defined based on the location of the shops and the goods in the shopping center in two-dimensional space. The algorithm can be applied for an arbitrary shopping center. The location of each shop is pointed by placing the corresponding picture by the drag-and-drop technique on its position on the map of the shopping center.

3. Empirical survey

To investigate people's mindset towards the use of beacons, the use of holograms, drag-drop pictures, pictures to speech, TTS and Speech Synthesis an empirical study was conducted. The survey's purpose was to explore the interest of end customers in beacon-based services, holographic technology, PECS, TTS and Speech Synthesis technologies and the willingness to use them.

As a base we use [7] but append some questions about new technologies.

With this end in view, we designed a questionnaire with the following tree sections.

In the figure 3a and 3b, participants were asked to assess the usefulness of typical applications, which were based on already existing scenarios by using beacon-based or holographic realizations:

•General situations, such as navigation on airports, coupons in stores, information on exhibits in museums etc.

•Specific retail store types, e.g. supermarkets, outdoor stores, furniture shops.

•Applications in a super-market, e.g. personal welcome message, navigation to products on the shopping list, information on products, special offers, and electronic payment at the checkout.

•Applications in a stadium, e.g. offers for seat category upgrade, navigation to wardrobe/restrooms, special offers for drinks and snacks.

In the figure 3c, participants were asked, whether they (1) own a smart-phone, (2) know Bluetooth, (3) know Bluetooth Low Energy, (4) know Holographic computer, (5) know Holographic technologies, (6) know beacons, (7) have used beacons before, (8) have used holograms before, (9) are familiar with PECS, (10) are familiar with TTS and speech recognizing technologies.

This helps to understand, whether consumers are aware of beacons. Then, they were given a short introduction of the beacon technology, holographic technology, PECS, TTS and speech recognizing technologies, as a preparation for the remaining questions.

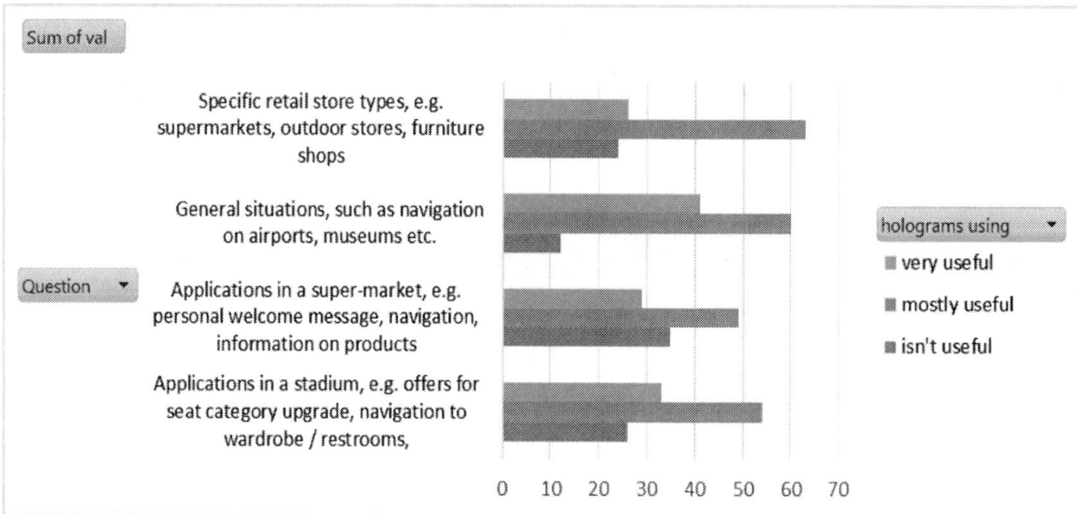

Figure 3(a). Prefered technologies

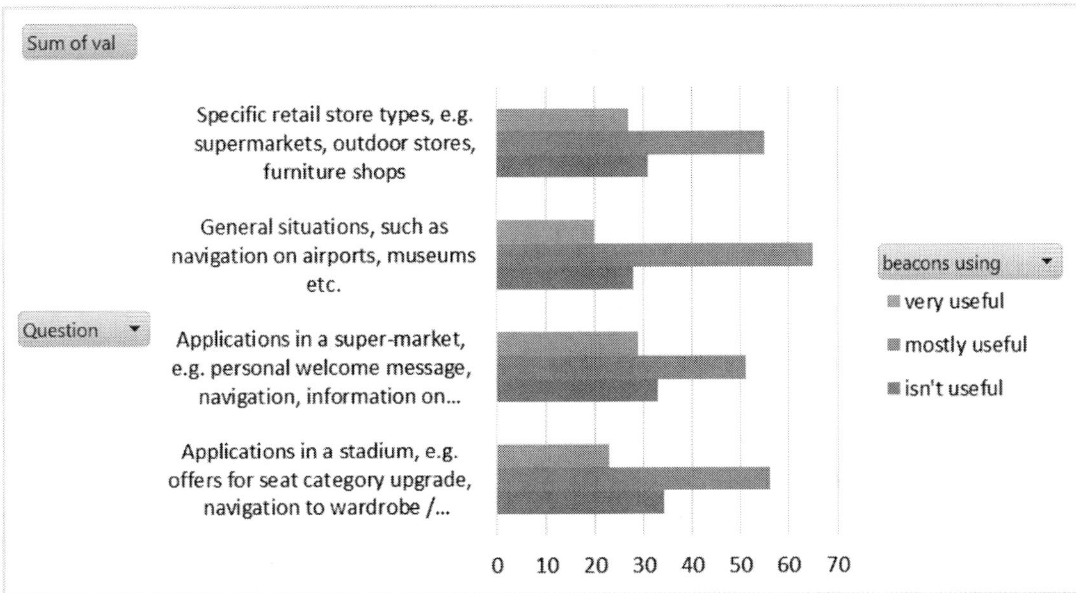

Figure 3(b). Prefered technologies

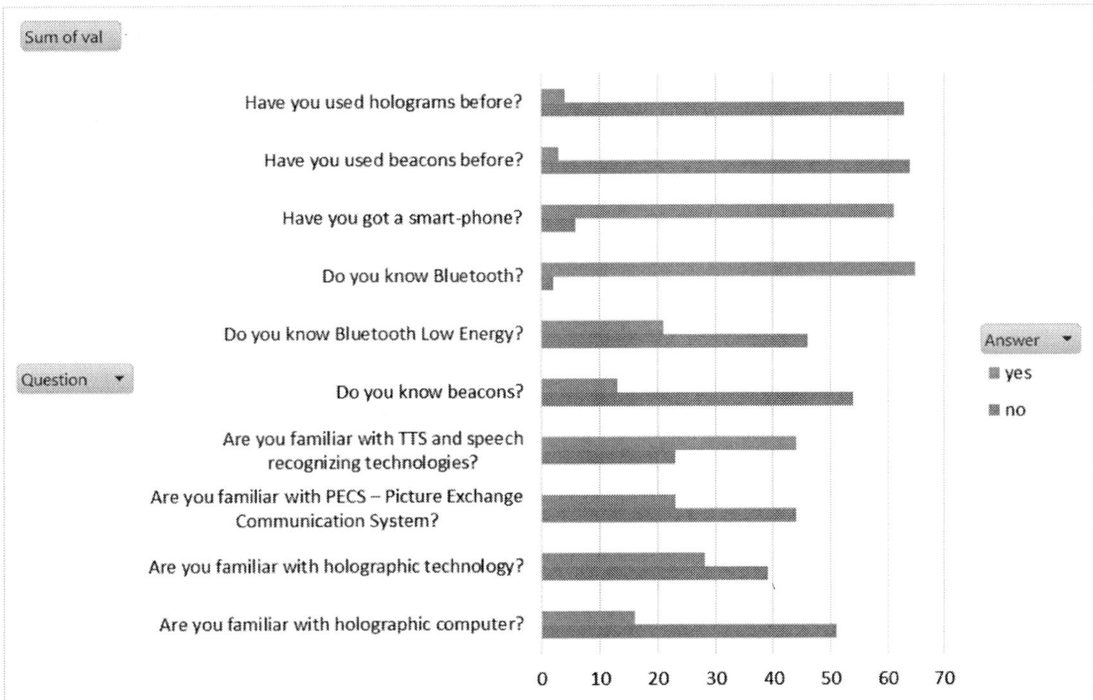

Figure 3(c). Prefered technologies

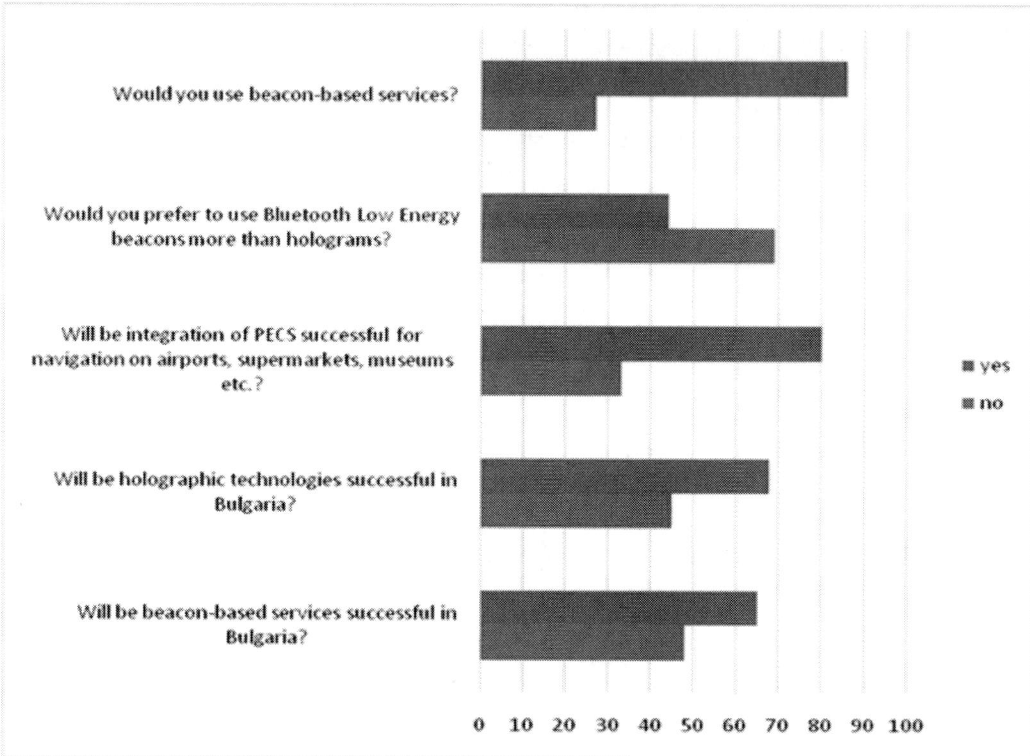

Figure 4. Readiness for useing the new technologies

In the figure 4 participants were asked whether they were willing to use beacon-based services, holographic services and if the integration of PECS will be successful in navigation applications. Participants were also asked to make an assessment whether beacon-based services and holographic services will be successful in Bulgaria.

The survey was conducted at the university. The total number of 115 students were offered the questionnaire. All of the participants were between the ages of 18 to 23 years old.

Beacon-based technology and holographic technology are little known and the services based on them are not used widely yet, but the respondents declared willingness and readiness to use them.

Five blind people aged 45-65 also took part in the survey. These respondents were not familiar with the described technologies and had not used them before. However, they do know and use in their daily routine Internet, Skype, smartphones, e-mail, all TTS programs and desktop reading programs. They showed great enthusiasm and willingness to get acquainted with beacon-based services and holographic services for navigation in buildings.

4. Conclusion

The paper describes the design of an intelligent navigation system prototype for smart shopping. The system differs from other intelligent systems by the combination of machine learning techniques, beacon-based navigation and/or holograph-based navigation in the mall, the integration of Picture Exchange Communication System in it and by its language understanding and speech synthesis capabilities, drag-and-drop techniques and keyboard button combinations enabled access. These allow various ways of accessing the system, which makes it appropriate for people with insufficiency of communication abilities. The empirical survey conducted with a limited number of users showed their positive mindset for using such navigation systems in indoor spaces.

The route planning in large buildings can use 3D geometrical and topological models.

For clients who are blind, precise 3D localization is relevant, and the pathfinding is a task that can be solved through a multi-parameter optimization.

Acknowledgments

This work was supported by the European Regional Development Fund within the OP "Science and Education for Smart Growth 2014-2020", Project Competence Centre "Smart Mechatronic, Eco-And Energy Saving Systems And Technologies", № BG05M2OP001-1.002-0023.

References

[1] Ahmetovic D., C. Gleason, C. Ruan, K. Kitani, H. Takagi, and C. Asakawa. "Navcog: A navigational cognitive assistant for the blind." In International Conference on Human Computer Interaction with Mobile Devices and Services. ACM, 2016.

[2] Bondy A. and L. Frost, PECS - Picture Exchange Communication System https://pecsusa.com/apps/

[3] Cassell J., T. Stocky, T. Bickmore, Y. Gao, Y. Nakano, K. Ryokai, D. Tversky, C. Vaucelle, and H. Vilhjalmsson, "MACK: Media lab Autonomous Conversational Kiosk," Imagina'02, Monte Carlo, 2002

[4] McCauley L. and S. D'Mello, MIKI: A speech enabled intelligent kiosk, IVA 2006, LNAI 4133, pp.132-144, 2006

[5] Lord C., M. Rutter, P.C. Dilavore, and S. Risi. ADOS - Autism Diagnostic Observation Schedule. Western Psychological Services, 2002.

[6] Seyed A. Ch., Vinod N., Kaushik S., IBeaconMap: Automated Indoor Space Representation for Beacon-Based Wayfinding, Human-Computer Interaction arXiv:1802.05735v1 [cs.HC], USA, 2018

[7] Thamm A., J. Anke, S. Haugk, D. Radic, "Towards the Omni-Channel: Beacon-based Services in Retail", Business Information Systems – 19th International Conference, BIS 2016, Leipzig, Germany, July 6-8 2016, Proceedings, pp. 181-192

[8] Viola P. and M. Jones, "Rapid object detection using a boosted cascade of simple features." Presented at CVPR, 2001

[9] The Intel Open Source Computer Vision Library" vol. 2006: Intrel Corp., 2006

[10] Winston P. H., Artificial Intelligence, Addison-Wesley Publishing Company ISBN-13: 978-0201533774 ISBN-10: 0201533774

[11] https://www.nationalautismresources.com/the-picture-exchange-communication-system-pecs/

[12] https://pecs-unitedkingdom.com/pecs/

[13] SpeechLab2 http://www.bacl.org

[14] JAWS,http://www.freedomscientific.com/Products/Blindness/JAWS

[15] Vocalizer Android/Dariya/Bulgarian voice https://play.google.com/store/apps/details?i d=es.codefactory.vocalizertts&hl=bg

[16] Innoetics TTS Reader Female Bulgarian Voice (IRINA) https://www.innoetics.com/

[17] NVDA https://www.nvaccess.org/

[18] Microsoft HoloLens holographic computer – https://www.microsoft.com/bg-bg/hololens https://developer.microsoft.com/en-us/ windows/mixed-reality/

[19] Google Beacon Platform (2016) https://www.youtube.com/watch?v=0QeY9FueMow

[20] My Dream Companion project – YGA; http://www.yga.org.tr/en/my-dream-companion

[21] Tomorrow's digital signage, today with 3d holographic kiosks, Intel IoT Digital kiosks, Solution brief, 2018.

IOP Conf. Series: Materials Science and Engineering **618** (2019) 012037 doi:10.1088/1757-899X/618/1/012037

Legal data extraction and possible applications

K Stoykov[1] and S Chelebieva

[1]Technical university of Sofia, 8 "Kl. Ohridski" Street, Sofia, Bulgaria

[1]k.stoykov@hotmail.com

[2]stefka.chelebieva1202@gmail.com

Abstract. The functioning of the modern society provokes creation of significant number of documents, most of them in text form. Many of them are legally relevant and complex in nature, which makes their understanding difficult. They are written in nature language for people, which is hard for computer processing and hence hard for further analysis. Documents represent an abundance of data in unstructured form. They can be handled using NLP methodology. Different models, trained for specific use cases are used, depending on a type of legal document. Automation of number of processes can save time and speed up work on document processing. This, of course, does not rule out the need for highly qualified lawyers. The goal is to ease their work in terms of processing large amount of documents. Increased productivity should be beneficial for both natural and legal persons when working with textual and legal issues.

1. Introduction

The presence of many data leads to a major delay in processing by humans. Automated processing is many times faster, but the data must be extremely well-structured, that is, computer easily understands SQL, but a much more difficult the context of a text [1]. The innovations in hardware in recent years has led to a boom in machine learning algorithms, including an understanding of natural language. These approaches allow a much faster analysis of texts, including legal ones. Legal documents are considered to be difficult to understand and apply [2]. However, different approaches are used to solve legal problems in specific areas [3]. The goal of the article is to show the connection between legal processes and how their analysis in large amounts can be useful and saves time [3]. The processing of natural language is directly related to the preliminary preparation of models in the given language [4], which is not very well covered in Bulgarian [5]. In order to achieve the goal of the article, different types of documents will be considered, what algorithms and approaches can be used to retrieve data. Also, tools that will be involved in processing [5, 6, 7] will also be considered.

2. Legal context

The law is divided into different branches - the common and the classic ones are: civil law, criminal law and administrative law, which in turn are divided into branches. According to the public relations governed by public law, the bond, the copyright, the hereditary, the family are branches of civil law, [8] according to the structural division of the positive law can be distinguished - material and procedural, private and public, domestic and international law. Each of them has its peculiarities and specifics, but there is a common, the normative basis of which they are subordinate, the documents and in general all the texts of a legal nature are in natural language. Large amounts of data are generated from and in the form of various documents - contracts, powers of attorney, declarations, orders, applications, requests, decisions, letters, notices, receipts, notifications, etc. Though each one of them to have an abstract structure, they are treated as unstructured information from the point of view of processing them as computer data. Different documents, depending on their type, have elements that have a specific location. The personal data of the parties to the contracts are always mentioned at the beginning of the documents, not as an annex or between the paragraphs itself, this is an example of an abstract structure.

The availability of a large set of data which must be processed is a good sign to involve AI methods for task solving and process acceleration. Below, the article lists various tools that can be used to solve different tasks, such as the use of NLP to understand text. The introduction of such a methodology is not only because it is modern or because of some major change over the past one or two years, but rather because the legal sphere is ready to accept software solutions that use technology to understand natural language and machine learning [3]. These adaptation processes are not the same throughout the world, but have begun and cover more and more countries.

3. Applications

The article will focus on examining three main areas where natural language understanding can be applied [3]. The first is dealing with contracts. They can be a variety of types - contract for purchase, rent, order, donation, replacement, etc. NLP methods could retrieve contract information and verify that it meets the requirements [9]. In some cases, this can helpful to finalize the contract more quickly and avoid mistakes. The other strand that will be considered is the legal study - analyzing law cases (judgments, definitions and orders), thus predictive systems can be trained. The last direction that will be addressed in the article is intelligent interfaces. We can imagine them as web-based platforms that are the kind of question-response. They can be used for both lawyer and lawyer training.

3.1. Contracts

The term "contract" is used with different meanings, it is legally embedded in the Law on Obligations and Contracts (the Law on Obligations and Contracts), in foreign literature it is considered as a legal relationship and as a document. [10]. The contract is a phenomenon common to the different branches of law - real estate, family (marriage), labor (individual and collective labor contracts). Administrative law is familiar with the administrative contract [11]. One of the sources of international law is the international treaty, which can be bilateral or multilateral. Many companies, institutions, governments, and so on. must handle contracts for a wide range of tasks. An example of this may be a process that verifies whether a clause in the contract is contrary to another or to a mandatory rule of law. As another example, it is possible to examine whether the legal document includes clauses that are binding on the company. Both cases only illustrate the current volume of work. Lawyers have to check many contracts many times until a final consensus is reached on both sides. Some of these views can be automated. This involves understanding and interpreting text [12]. Depending on the type of document, it must contain certain elements and have a specific structure. Processes of classification and verification of elements and structure can be automated. Another key task is to specify what the clauses are (required - price, goods, property, behavior, etc.) of the parties to the contract and what would happen and what would be due in case of non-execution of the contract – penalties, interest, etc. This can also be delegated to stand-alone systems that check the understanding document.

Similar contract processing involves reading with understanding using NLP and similar approaches like word-embeddings. For the training of such a system, a certain amount of contracts will be needed. Also other set of contacts is necessary to verify the correctness of the model obtained. The task of such a system is to define the elements of a text, to extract the structure and the parts of the contract. Including to emphasize how much the document differs from the standard of this kind.

An important part of the processing of such information is the need to anonymize personal data. We do not want our model to make decisions on a name-based basis and, above all, such data should be well protected and not publicly available.

3.2. Legal study
Apart from the analysis of a document, AI can also be used with a wider range [3]. This range includes knowledge and prediction systems.

Knowledge-based systems are software based on law, time-based regulatory changes, judgments and motivations in it, opinion of experts in the field, lawyers' opinions, scientific developments, etc. Such systems can analyze a problem very well thanks to the understanding of natural language and provide a response to a case. Although such systems seem like a search engine in a large database, their great advantage is the understanding of the language and the results they provide are not just based on keywords but also on the meaning itself.

Such systems work in a given language in most cases English is used. Though it is not part of the current article, a system that combines good practices across different legal system could be a very good research field. Such combination will significantly increase training data set and most of all, will help to transfer good practices from one legal system to another if local law permits it. This study remains for future development.

The systems that "predict" include an analysis of court judgments that have already entered into force and the production of a result for a new case. We are far from the moment when such systems can replace people in the decision-making process, but they can speed up the whole process of dealing with a case. This approach is extremely useful when we have analogous case studies and can quickly show us the relevant case law. This trained model will have an extremely large set of cases to work on and the result obtained can be as accurate as possible. The main purpose is to reduce the manual search of different judgments, motives, opinions, etc., which process takes a very long time.

3.3. Intelligent interface
In addition to the direct application for document processing and casework, the possibility of such software being used for both student and non-lawyer training in answering legal questions should not be underestimated. The training process involves examining and analyzing good practice and case studies. The typical process involves searching in documents and textbooks to find the right text. An interface that responds to user queries will be very useful in the learning process and will speed it up and, above all, increase quality because the information that will be offered will be as accurate, up to date, comprehensive as possible in more understandable language. By the same logic, such an interface would also be useful to non-jurists in solving trivial problems, which would again save a huge amount of time globally. This type of software will resemble an intelligent search engine that has to work very well with a legal context.

An example of such an interface in the field of learning is one that could answer the question: "What kind of documents are need during car police check?" "Could a pedestrian be guilty in case of an accident on a pedestrian crossing? "," What documents are needed in particular case - when issuing new identity documents - ID card or passport?". Such a platform would save a great amount of time and bring confidence and security to citizens, whether lowers or not.

4. Data extraction
The extraction of knowledge can be defined in many different ways. Even the name can be defined as the discovery of knowledge in the data [13]. It is so vast discipline that none of the definitions describe the entire process that takes place during the acquisition of knowledge. In general, the process involves data processing in order to extract useful information. It can be represented as 7 consecutive steps that are performed until the desired results are found [13].

1. Clear data - data must be cleared from noise before it can be processed [13].
2. Data Integration - Checks whether multiple sources can be combined to achieve as much knowledge as possible [13].
3. Selection of data - check how much data is correct for the task we are trying to solve [13].
4. Transformation of the data - the data shall be prepared in a suitable type for processing [13].

5. Knowledge extraction - a major part of the process where intelligent processing systems are involved. This part is fundamental to the current article, because the NLP method is just right for it [13].
6. Model Evaluation - Identify the really interesting models (templates) found in the data [13].
7. Presenting knowledge in a man-readable form [13].

5. NLP

Natural language processing (NLP) is a set of approaches that are used to process text data. The entire processing text being worked on is called a corpus. The use cases of such data processing are many - check grammar and spelling, knowledge of the next word in speech, retrieve user-based query information (intelligent interfaces, including the legal ones that were discussed in the article), categorization text, summary, etc. [14].

Data mining is a basic method that will also be used to process legal documents, and ongoing work will show several specific approaches and algorithms that can be used [15].

A word bag [15] - one of the simplest yet effective methods for presenting machine learning data. This approach ignores the order of words in sentences, the sentences and structures themselves. It loses part of the information but gives a general idea of the content of the text. Three things are taken into account:

1. Tokenization - splitting the document into words
2. Define the words and number them
3. Count how many times a word occurs.

Not all words have the same weight and a list of words that are not involved in the processing should be made - pronouns, pronouns, etc. There are similar lists of words, mainly in English, but it is not a problem to do so in other languages as well.

This method can be improved by not losing the word order. In order to do this, the word bag is not built with just one word, but with a set of consecutive words. The approach is called n-gram model, n being the number of words used to compose the word bag [15].

TF-IDF is another, very powerful method for extracting a common knowledge of text where, apart from counting words, this number is also compared to all other words in the corpus. [16]

The above methods give a general idea of the content of the text, but the legal documents will need a deeper understanding. To achieve such an understanding, methods need to be improved. One such improvement can be made in the first step of the bag with words - tokenization. In the texts there are always words that are in different forms (verb tenses, singular or plural, congressional) which do not contribute much to the ultimate understanding of the text, but rather saturate the pattern. During processing, these words can be replaced with their roots, thus reducing the number of words being worked on [15].

Also, to improve tokenization, not only the roots of the words, but also the synonyms can be used, i.e. how close the words to each other are in terms of meaning. This can be achieved by using another approach in advance - presenting the words as vectors and searching for the proximity of the vectors [4].

Another very good method of text processing is the use of recurrent neural networks. Unlike other methods that analyze a test, neural networks can generate text that makes them very convenient for use in intelligent interfaces [15].

6. Tools

To achieve the above tasks, NLP is used. As it has become clear, this is an approach to understanding text written in natural language. Based on it, various tools have been developed to facilitate the development of client-end software that can solve a given legal task with understanding.

Standford-parser [4] – one of the most commonly used NLP tools. Provides tools for solving subtasks from the above-described - basic parts of words, parts of speech, normalized teachings, numbers, etc. This way you can find how the different words are connected to one another. The library is based on the statistical model and works fast. It is mainly used for English.

Apache OpenNLP [6] – a very good text-processing library in natural language. It is very useful for solving some of the tasks in the 3 main directions that are discussed in the article. Maintains basic algorithms such as tokenization, segmentation of sentences, determination of the parts of the speech. These techniques are needed to create more sophisticated text-processing systems.

TensorFlow [17] – an open-source Google library for creating and training neural networks. Everything, working with this library, is presented as tensors of different rows.

SpaCy [18] – a statistical tool for extracting data from large arrays. Works fast with large text processing as input. As part of the functionalities it supports, it is syntax-based sentence separation, good integration with machine learning systems that includes TensorFlow.

Parse-IT – part of the digital law platform [19]. The tool adopts normative text as input and finds logical structures. Provides a fine tuning capability.

The tools mentioned here are fundamental to text processing in natural language, most commonly used is the first tool. Using only once of them could not give a direct answer and does not solve directly the tasks listed above in the article. Rather, these tools are the alphabet used to write a book, a book that solves legal problems with AI and machine learning. Instruments are mainly designed to work in English, which opens the next obstacle to their broad application, namely how easily they can be adapted to the local language of the countries. Which will be more appropriate - whether reworking the tools and re-engineering the models in a given language or automatically translating a text into English and vice versa. The answer to this question is beyond the scope of the article and is rather a reason to develop a study in this area.

7. Summary

Automated approaches, based on natural language processing, combined with machine learning can be used to process the legal documents. They cannot replace experienced lawyers, but they can do a lot about initial analysis - whether the contract is correct and includes the necessary clauses, what kind of document we have and whether it is structured correctly, automatic structure correction, intelligence interfaces and platforms, which take decision what lay is applicable in particular case. Even if it is not 100% correct, this type of systems could be an "initial" light when solving a problem and thus would save time for lawyers themselves, as well as for citizens, that is, processes can happen more quickly, more efficiently, and more accurately.

8. Conclusion

The article far cannot cover all aspects of the application of intelligent systems in law. The legal system will change over the coming years, and that should be for good. In some countries this has already begun, and in others it is about to start [5]. New companies engaged in AI and machine learning in law appear constantly [3], each solving different problems. The sphere is extremely difficult and complex, but it should start from somewhere and move forward.

The growing acceptance of AI in world law is before us. In the technological world, the new and modern today will be old tomorrow. We are not at a time when we have to think if technology can be useful, but rather try to accept it as quickly as possible and improve it so that we can make the most of it.

References

[1] Christian H¨anig, Martin Schierle, Daniel Trabold, 2010, Comparison of Structured vs. Unstructured Datafor Industrial Quality Analysis
[2] Risto Hiltunen, 2012, The Grammar And Structure Of Legal Texts
[3] Richard Tromans, 2016, Legal AI
[4] Douwe Osinga, Deep learning cookbook. 2018
[5] Neural Reasoning For Legal Text Understanding (Guido Boella Adebayo Kolawole John, Luigi Di Caro), In Proceedings of the 29th International Conference on Legal Knowledge and Information Systems (JURIX2016), 2016

IOP Conf. Series: Materials Science and Engineering **618** (2019) 012037 doi:10.1088/1757-899X/618/1/012037

[6] Andor, Daniel, et al. "Globally normalized transition-based neural networks." arXiv preprint arXiv:1603.06042 (2016)
[7] Milagro Teruel, Cristian Cardellino, Fernando Cardellino, Laura Alonso Alemany, Serena Villata, 2017, Legal text processing within the MIREL project
[8] Pavlova, M., Civil law - common part, 2nd edition, Sofi-R (Павлова, М. Гражданско право - обща част - Второ преработено и допълнено издание - Софи-Р), 2002, 38-44.
[9] Ilias Chalkidis, Achilleas Michos, Ion Androutsopoulos, 2017, Extracting Contract Elements
[10] Calamari, J.,J. Perillo. Contracts.Saint Paul, Minn.: West Publishing Co.,1987. p.3.
[11] Ruschev, I., Administrative and Private Law, Market and law (Русчев, И. Административният и частноправният договор. - Пазар и право), 2000, №1, 25-32.
[12] Adrien Bibal, Benoît Frénay, 2016, Interpretability of Machine Learning Models and Representations: an Introduction
[13] Jiawei Han, Micheline Kamber, Jian Pei, Data mining: Concepts and techniques, 2012
[14] Dr. Mariana Neves, Natural Language processing, SoSe 2016
[15] Andreas C. Müller and Sarah Guido, Introductin to machine learming with python, 2017
[16] Stoykov, K., Knowledge Extraction and Machine Learning, International Conference "Automatics and Informatics'2018", 4-6 Oct. 2018, Sofia, pp. 117 – 120, ISSN 1313-1869.
[17] Floris Bex, Joeri Peters and Bas Testerink: A.I. for Online Criminal Complaints: From Natural Dialogues to Structured Scenarios, in prof. of Artificial Intelligente for Justice workshop, collocated at the 22nd European Conference on Artificial Intelligence
[18] C. Bartolini, G. Lenzini and L. Robaldo: Towards legal compliance by correlating Standards and Laws with a semi-automated methodology, in proc. of the 28th Annual Benelux Conference on Artificial Intelligence. Amsterdam, 2016.
[19] Deerwester, Scott, et al. "Indexing by latent semantic analysis." Journal of the American society for information science 41.6 (1990): 391.

IOP Conf. Series: Materials Science and Engineering **618** (2019) 012038 doi:10.1088/1757-899X/618/1/012038

Combination of Virtual Private Network and Wireless Sensor Network: Protection Against The Interference Problem of IOT

Pramathesh Majumdar [1], Rachna Jain[2], Ayushi Pandey[3], Rana Majumdar[4], Sunil Kumar Khatri[5]

[1,3,4]ASET, Amity University Uttar Pradesh, Noida, India, pramathesh@outlook.com
[2,5]Amity Institute of Information Technology, Amity University Uttar Pradesh, Noida, Indiajain.rac16@gmail

Abstract. As the era of the Internet of Things (IoT) is emerging rapidly, the interference problem is also increasing at the same pace. In IoT based connections, as we know that devices share a common communication channel to interact with each other, the interference problem arises as all the devices are connected. Once an attacker gains access to any of the devices in an IoT enabled network, the whole IoT network becomes susceptible to breach. The biggest example of it is the IoT enabled CCTV hacking. In this work, we are proposing the usage of Virtual Private Network (VPN) and Wireless Sensor Network (WSN) to create a virtual environment and to protect each device from IP address breach that is connected to an IoT network. Using VPN, each IoT device will gain their own virtual environment which will enable it for using a dynamic IP address to connect to another IoT device's virtual environment. Using WSN, the virtual environments will be able to identify any signal interference that can cause any security breach.

1. Introduction
The main and foremost basic behind the technologies of any field in today's age is how devices can talk to each other. Keeping this in mind, several inventions were and are being made to enable devices to communicate. Internet of Things (IoT) is the filed which makes this attempt possible. But, as it shows the way how to make the devices talk to each other, it introduces some new kind threats as well. As we know, in IoT all the devices are interconnected; hence the chance of security breach increases drastically as a hacked device may lead to security shut down of all the other connected devices.
While a relatively few connected gadgets will use wireline communications, the bulk of these devices ranging in applications from farming to medical to toasters to pets will connect wirelessly over Bluetooth, Wi-Fi, or cellular. Some will be designed to limit EMI emissions to the proper channels and reject out-of-band power, but many won't.
Cyber-criminals are trying to find ways to breach the IoT security. As many devices are interconnected, the chances of vulnerability increase at the same pace. For e.g., back in 2016, the largest IoT attack of the history was launched on
Service provider "Dyn" using an IoT botnet. This lead to huge portions of the internet going down, including Twitter, the Guardian, Netflix, Reddit, and CNN.By utilizing thousands of insecure associated gadgets, hackers can create DDoS assaults that can handicap the foundation, frameworks,

and lifestyle. Or then again, assailants can specifically misuse a gadget and utilize it as a portal to further levels of a system where they assemble important private information.
The following figure shows the increased threat to data privacy of connected devices in IoT (based on the study done by ISACA [13]).

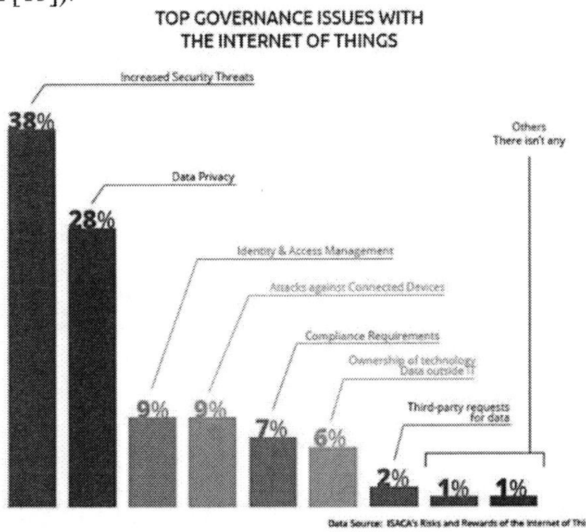

Figure 1. An Increased threat to data privacy in IoT[13]

In this work, we are introducing the concept of combining Virtual Private Network (VPN) with Wireless Sensor Network (WSN) to secure each interconnected device in a network.
Here, we assume each device as a node (n1, n2... nX). Every node will be assigned a VPN setup which will provide the device a designated way to interact with other nodes' VPN setup. Using wireless sensors network, we will be able to track the number of attempts that took place for a security breach. This record will then be used to further improve the VPN setup according to the depth of the vulnerability level.

2. Related work

At this moment, the most crucial detriments in IoT is the shortcoming of this meta-system to different sorts of ambushes. Obliged limit and dealing with restricting in IoT contraptions influences the execution of sensible to confirmation parts incredibly troublesome IoT gadgets are slanted to ambushes, their lacking level of security makes them astoundingly fragile concentrations for being vulnerable [2].
The rapid development of new applications using smartphones in the world caused all users of the IoT community to be faced with one major challenge of security in the form of side channel attacks against highly intensive 3D printing systems. The smartphone formulated Intellectual property (IP) of side channel attacks investigate against 3D printer in the physical domain through reconstructed G-code file through primitive operations [3].
Specifically, hackers make use of these devices in coordination to make botnets and make Distributed Denial of Service (DDoS) strikes against a pariah. Such disturbance can turn an IoT orchestrate improving the assignment of an industry or an organization into a security chance itself. Henceforth, the deal of these structures by malicious experts doesn't simply harm these substances, yet what's more the general ICT condition [3].
The unfortunate use of covering channels among neighboring APIs may provoke extended essentialness usage because of the ensuing re-transmissions. Covering channels among neighboring APs can decrease the execution of preliminary Wi-Fi based arranging. The correspondence goes for an API is considered as the cell. By distributing non-covering channels among neighboring cells, the block issue between channels among the API's can be settled [4-6].

The attack basically spreads by first debasing devices, for instance, webcams, DVRs, and switches that run some variation of BusyBox. It by then finds the definitive capabilities of other IoT contraptions by strategies for a mammoth drive, contingent upon a little word reference of potential username–watchword sets. It causes DDoS against a game plan of target servers by continually inducing to weakly composed IoT devices [7-8].
Most remote frameworks are demonstrated as layer 2 (L2), or data interface tradition. Starting late, IoT and huge data dealing with have propelled the use of remote sensor frameworks to interface and send data to server cultivate applications using the Internet. To do in that capacity, the execution of an IP stack on the remote center point, or the section of the IP and remote L2 mastermind, has been proposed.
They are made to allow the application of the IP framework to get to L2 remote framework center points [9].
In the other work, IoT was deciphered using an IP just to endure data made by non-IP center points. Thusly, we display another novel sort of VPN for partner PANs to server ranches, which has three features. The first is that it is a traverse framework instead of an entryway. The remote sensor orchestrates data interface diagrams are transported direct end-to-end through the IP spine [10-12].

3. Interference problem in iot: overview

The Internet of Things is the most recent in a long queue of progressively vital advancements, and the quantity of associated protests inside this web is developing at an exponential rate. Gartner anticipated that the quantity of "things" being used in 2019 would grow 22 percent and by 2020, the IT investigate firm anticipated that there would be upwards of 20.8 billion associated protests being used. Organizations and purchasers alike are finding the advantages of the IoT to surprise, and numerous are amped up for its future.
In any case, sadly, IoT gadgets still experience the ill effects of fundamental security vulnerabilities and it is accurately this absence of security that makes them so appealing to programmers. Be that as it may, it's not only a secret word issue any longer. Assailants comprehend that makers and clients are awakenings to the issue of passwords on IoT gadgets, as are looking for more intricate approaches to get to them. As this pattern proceeds, and programmers turn out to be progressively innovative while hunting down new gadgets and approaches to enroll them, there is extremely no restriction to the size and size of future DDoS assaults driven by IoT botnets.
All things considered, any gadget that has an Internet association and a processor can be abused. In a perfect world, all gadgets ought to be compelled to experience a type of system design before being utilized, as opposed to being exploitable from a default position.
Computerized Enterprises can shield their systems from DDoS assaults fuelled by IoT-driven botnets by sending continuous, the robotized arrangement at the system edge, which can quickly distinguish and alleviate DDoS action and kill dangers from entering a system. Similarly, as with all DDoS dangers, clear permeability is a significant advance in distinguishing and protecting against assaults.
Ransom Denial of Service (RDoS) assaults additionally returned in Q3 2017, the report found, as this strategy permits cybercriminals to blackmail cash from their casualties. In these assaults, the criminal will normally make an impression on the casualty requesting a payment, frequently extending from five to 200 bitcoins, as indicated by a current Kaspersky Lab report. On the off chance that the casualty declines to pay, the assailants debilitate to sort out a DDoS assault on one of the casualty's imperative online assets.
As IoT botnets keep on rising, we may soon observe programmers put on more sensational RDoS presentations to exhibit the quality of their digital capability, with the goal that their future requests for payoff should be considered more important," Stephenson said. "Paying the payment is infrequently the best guard, as it just urges these requests to spread like fierce blaze.

IOP Conf. Series: Materials Science and Engineering **618** (2019) 012038 doi:10.1088/1757-899X/618/1/012038

4. Proposed model

As mentioned previously that rapid development of new applications using smartphones in the world caused all users of the IoT community to be faced with one major challenge of security in the form of side channel attacks against highly intensive 3D printing systems .In our work, we are offering the concept to combine the Virtual Private Network (VPN) with the concept of Wireless Intrusion Detection System(WIDS) to combat such issues easily As billions of gadgets are added to the Internet of Things, organizations require an approach to increase one of a kind bit of knowledge from their IoT information.

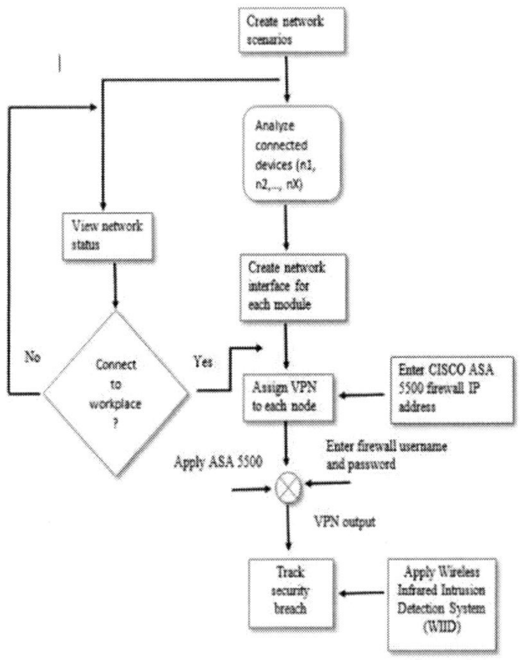

Figure 2. Flowchart of the proposed model

1. (WIID) to secure each node that is interconnected to each other in an IoT network. Here we are assuming every device as interconnected nodes which represent a module of the IoT network. The nodes are presented as n1, n2, nX, where X is the maximum number of nodes.
2. The steps that are followed in our proposed model are as follows:
3. Step 1: Create network scenarios
4. Devices are interconnected using an IoT network to create a network scenario. This is the combination of all the devices that communicate to each other via a physical or a wireless network and acts as a framework.
5. Configure the IP interfaces and IP addresses, and also include a determined default course for the framework. Moreover, we can arrange any framework wide system administrations, for example, naming and index administrations. The accompanying cases accept that we are utilizing the settled mode for arrange setup.
6. Step 2: Analyze connected devices
7. IoT Connection Service empowers organizations to reveal new information-driven bits of knowledge that can change their generation procedure, make one of a kind client encounters, increment operational effectiveness and enhance ecological stewardship.
8. Each device needs to be analyzed for the IP address configuration in order to set up a VPN. The analysis is done using the network framework created earlier with the help of NS2 tool.
9. Step 3: Create network interface for each module

10. For communicating one VPN with other VPNs, there have to be some dedicated interfaces to the VPNs. These interfaces help in talking with the other networks and work as a gateway. The following code was generated to create a network interface –
11. Create-network-interface
12. [--description <effective-point>]
13. [--testing | --no-testing]
14. [--groups <effective-point>]
15. [--ipv6-address-count <effective-point >]
16. [--ipv6-addresses < effective-point >]
17. [--private-ip-address < effective-point >]
18. [--private-ip-addresses < effective-point >]
19. [--secondary-private-ip-address-count <value>]
20. --subnet-id < effective-point> [--client-input-json < effective-point>]
21. [--generate-client-skeleton < effective-point >] --desc (double)
22. Explanation of the interface –
23. ---testings | --no-testing (Boolean)
24. Checks whether you have the required assents for the action, without making the request, and gives a botch response. In case you have the required assents, the slip-up response is Testing Operation. Else, it is Unauthorized Operation.
25. --groups (list)
26. Step 4: Assign VPN to each network

- Once the interfaces are created for each node, the VPNs are configured for the nodes to create the private_network_modules. These processes are done using the following steps – Enter into Network and sharing center of the system.
- Manage the network connections and change the adapter settings.
- Now enter into file option and click on a new incoming connection.
- Now mark which nodes have to be given VPN connections and click next.
- Choose which protocols need to be started. Generally, TCP/IPv4 needs to be started to access the network.
- Once done, click on close.

27. Now the properties of the new nodes have to be well-defined and for that, IP address ranges have to be setup.

- Enter incoming connections section in the network connection window.
- Select the TCP/IPv4 option.
- Now enter the required IP addresses in a manner that they should not conflict with the DHCP ranges.
- Now click OK.

28. Step 5: Enter CISCO ASA 5500 firewall IP address
29. Today's remote-access VPN deployments want the ability to safely and effortlessly outspread corporate network access beyond managed desktops to different users devices, while protecting endpoints and key corporate resources from ever-evolving threats. In our work, we prefer utilizing the CISCO ASA 5500 design benchmarks to setup the earth for the VPN firewall for its intrusion prevention inspection and web threat prevention capability.
30. The ASA programming depends on Linux. It runs a solitary Executable and Linkable Format program called lina. This timetables forms inside instead of utilizing the Linux offices. In the boot grouping, a boot loader called ROMMON begins, stacks a Linux portion, which at that point stacks the lina_monitor, which at that point loads lina.
31. The ROMMON likewise has a charge line that can be utilized to stack or select other programming pictures and arrangements. The names of firmware records incorporate a form indicator, - smp implies it is for a symmetrical multiprocessor (and 64 bit design), and diverse parts additionally demonstrate if 3DES or AES is bolstered or not.

32. The ASA programming has a comparable interface to the Cisco IOS programming on switches. There is a charge line interface (CLI) that can be utilized to question work or arrange the gadget. In config mode, the design proclamations are entered. The design is at first in memory as a running-config yet would regularly be spared to streak memory.
33. Step 6: Apply Wireless Infrared Intrusion Detection System (WIID)
34. The WIID framework is an infrared-based border security checking framework that comprises of multi-beam infrared sensors coordinated with double band (2.4 GHz and Global System for portable correspondence) remote specialized gadget. These are put on the assigned edge.
35. Remote infrared sensors can work in such the double radio band, which thus ensures that there is positively no loss of data. Multibeam remote infrared sensors likewise distinguish break viably progressively to create spot caution and remote alarm.
36. The brutal idea of nature/territory around the border makes it troublesome for fiber optic sensors and weight sensor strips to work viably and create fitting alarms. Subsequently, there emerges a huge window for false caution age.
37. On account of ultrasonic sensors, they don't bolster a wide range for identification of a hindrance or interloper. In spite of the fact that, the working rule of photoelectric sensors is to some degree like that of infrared (IR) sensors, their working precision is much lower contrasted with infrared sensors and their arrangements.
38. The WIID System produces visual alarms at the local security office with the end goal of crisis notice. Messages are additionally sent by means of SMS to the no-obligation security officers and watching staffs with respect to the episode.
39. The framework produces spot alarms through hooter and concentrates light so as to caution the closest security staff and after that guide them to the area of occasion at the suitable time. This framework additionally incorporates a reconnaissance camera framework that gives data about the idea of the risk.
40. There is negligible human intercession required, in this manner influencing it to fall flat evidence to human mistakes and makes funds in the enlisting of security workforce.

5. Conclusion and future scope

This model provides a systematic way to further provide security to each device connected to an IoT network by protecting them from external unauthorized interference using VPN technology. Wireless sensor technology is used to track the security breach attempts to further improvise the security level using the records gathered from the tracking system i.e. WIID. This work can be further extended to the next level using advanced VPN system using IPv6 configuration to deploy extra security level in the IoT framework.

References

[1]. C. Kolias, G. Kambourakis, A. Stavrou, and J. Voas, "DDoS in the IoT: Mirai and other botnets", Computer, vol. 50, no. 7, 2017, pp.80-84.
[2]. M. Özçelik, N. Chalabianloo, and G. Gür, G., "IEEE International Conference on Software-Defined Edge Defense Against IoT-Based DDoS", Computer and Information Technology (CIT), 2017, pp. 308-313.
[3]. M. Bilal, "A Review of Internet of Things Architecture, Technologies and Analysis Smartphone-based Attacks Against 3D printers. arXiv preprint arXiv:1708.04560", 2017
[4]. K. Wiklundh, and P. Stenumgaard, "International Symposium on EMC challenges for the internet of things", Electromagnetic Compatibility-EMC EUROPE, 2017, pp. 1-6.
[5]. A. Bakshi, L. Chen, K. Srinivasan, C.E. Koksal, and A. Eryilmaz, A. "EMIT: An efficient MAC paradigm for the Internet of Things. IEEE International Conference on", Computer Communications, INFOCOM, 2016, pp. 1-9.
[6]. J. A. Stankovic, "Research directions for the internet of things", IEEE Internet of Things Journal,

vol. 1, no. 1, 2014, pp.3-9.
[7]. N. Khalil, M. R. Abid, D. Benhaddou, and M. Gerndt, " IEEE Ninth International Conference on Wireless sensors networks for Internet of Things", Intelligent Sensors, Sensor Networks and Information Processing (ISSNIP), 2014, pp. 1-6.
[8]. A. P. Mynster, and P. T. Jensen, "International Symposium on EMC for the IoT", Electromagnetic Compatibility-EMC EUROPE, 2016, pp. 144-149.
[9]. A. Tsitsigkos, F. Entezami, T.A. Ramrekha, C. Politis, and E. A. Panaousis, "A case study of internet of things using wireless sensor networks and smartphones", Proceedings of the Wireless World Research Forum (WWRF) Meeting: Technologies and Visions for a Sustainable Wireless Internet, Athens, Greece, 2012, pp. 23-25.
[10]. H. Zin, C. Kim, M. Wu, and S. Kim, "Avoidance of channel interference in polygonal IoT networks", Concurrency and Computation: Practice and Experience, vol. 29, no. 11, 2017
[11]. H. Hata, "A bridging VPN for connecting wireless sensor networks to data centers", Journal of Reliable Intelligent Environments, vol. 3, no. 4, pp.211-219.
[12]. S. K. Dhar, S. S. Bhunia, and N. Mukherjee, "Fourth International Conference of Interference aware Scheduling of Sensors in IoT enabled Health-care monitoring system", Emerging Applications of Information Technology (EAIT), 2014, pp. 152-157.

IOP Conf. Series: Materials Science and Engineering **618** (2019) 012039 doi:10.1088/1757-899X/618/1/012039

Substantiation of multinomial classification using ensemble learning approach

Lipika Goel[1], D Damodaran[2], Mayank Sharma[1] and Sunil Kumar Khatri[1]

[1]Amity University Uttar Pradesh, Noida, India
[2]Centre for Reliability, Thiruvanmiyur, Chennai India

lipika.bose@gmail.com, damodarand@gmail.com, msharma22@amity.edu, sunilkkhatri@gmail.com

Abstract. The software defect prediction is the most operative research domain in software engineering as it enhances its reliability. The availability of defect data related to different projects leads to cross project defect prediction an open issue. This paper instanced on multiclass/multinomial classification of defect prediction on different categories of cross projects. The ensemble learning statistical models – Random forest and Gradient Boosting are used for classification. An empirical study is carried out to determine the predictive performance of the within project and cross project prediction models. Depending on the number of defects, class level information is classified into one of three defined multiclass. The homogeneous set of object oriented metrics is used for training the model. Furthermore, k-fold cross validation is done to evaluate the training accuracy of the statistical models. Major outcome of the paper concludes that multinomial/multiclass classification is applicable on cross project data and has comparable results to within project defect data with statistical significance.

1. Introduction

Identification of the defects prone classes before actual testing reduces the testing cost and efforts. It also leads to a more focused testing thereby enhancing the probability of fault free software. Much research work has been carried out in within projects data. The availability of many different projects data leads to the motivation of cross project defect prediction. In the past decade much research has focused on binary classification of CPDP. The multinomial classification in CPDP is still an open area to be focused.

The multinomial classification provides detailed information of severity of defect prone classes. It classifies the class level defect data into different groups depending on the total number of defects in each class. The Table1. summarizes the different categories of defect prediction.

The main research question addressed in this paper is Whether multinomial classification of CPDP is comparable to WPDP? This paper is divided into the following sections: Section 2 gives an overview of the literature survey. Section 3 summarizes the datasets, metrics and the models used. Section 4 states the performance measures used followed by Section 5 stating the methodology used in prediction. Section 6 tabulates the results. The conclusion is stated in the Section 7 of this paper.

Table 1. Categories of defect prediction.

Category	Description
Strict CPDP	No information of the testing data remains in the training dataset.
Mixed CPDP	Older version of the target data is added to the training dataset.
Mixed CPDP+ 10% training data	Few percentage of testing data is also included in training dataset.
Pairwise CPDP	Pairwise training and testing is carried out.
WPDP	Certain percentage of the same project is used as training and the remaining percentage as testing.

2. Background

Turhan et al [1] in 2009 used K-NN to minimize the data distribution difference followed by Naïve Bayes classifier for defect prediction. T. Zimmermann et al[2] in 2009 gave a new dimension to defect prediction, by proposing CPDP using Logistic Regression model.Ma et al. [3] in 2012 presented a novel approach Transfer Naïve Bayes for CPDP.Jaechang Nam et al.[4] in 2013 applied a method of Transfer Component Analysis for CPDP.Jaechang Nam et al.[4] in 2013 applied a method of Transfer Component Analysis for CPDP.Herbold [5] in 2013 proposed distance-based strategies (EM Clustering & NN Filtering) for selecting the training dataset. Nam et al. [9]and X. Y. Jing et al.[10] in 2015 provided solutions for HCPDP (heterogeneous). also projected a solution to heterogeneous CPDP. DuksanRyuet al.[28] in 2016 gave a solution to the class imbalance problem by creating a novel model of SVM with cognitive boosting.The combined approach of Genetic and Ensemble learning gave good results.Zhang et al.[29] in 2016 used unsupervised classifier for heterogeneous CPDP.Xin Xia et al. [30] in 2016 proposed a Hybrid Model Reconstruction Approach. Steffen Herbold et al.[31] (2017) did a comparative study to benchmark cross project defect prediction approaches. He concluded that CPDP has still not reached to a level where it can be used for the application in practice.

3. Prerequisite knowledge

3.1. Dataset & Metrics

We have taken the well-defined datasets from the publicly available PROMISE repository (www. Promisedata.org) for NASA related projects and from http://openscience.us. We have taken class level information of each dataset.

The description of the dataset is given in Table 2.

Table 2. Description of the dataset.

Project	Language	Total no. of classes	Defect prone classes
Ant 1.7	Java	745	168
Camel 1.0	Java	339	13
Camel 1.2	Java	608	216
Camel 1.4	Java	872	145
Camel 1.6	Java	965	189
Ivy1.1	Java	111	63
Ivy1.4	Java	241	17
Ivy2.0	Java	352	40
Jedit 3.2	Java	272	47
Jedit 4.0	Java	306	75
Jedit 4.1	Java	312	79
Jedit 4.2	Java	367	47
Jedit 4.3	Java	492	11
Prop1	Java	18471	2738
Prop2	Java	23041	2431

Prop3	Java	10274	1180

In this work we have focused on homogeneous CPDP for multinomial classification. The CK metrics [15] are selected and are used for software defect prediction.

The CK Metrics used in the analysis are as follows[22]:

- DIT(Depth Of Inheritance)
- WMC(Weighted Methods Per Class)
- CBO(Coupling Between Objects)
- NOC(Number of children)
- LCOM(Lack Of Cohesion Of Methods)
- RFC(Response For Class)
- LOC (Lines of Code)

3.2. Ensemble Learning Models

An ensemble contains a combination of base/ weak learners to improve the performance of the model. The prediction accuracy of an ensemble is much higher than the base learners. Base learning algorithms can be neural network, decision tree or any other machine learning algorithm. The majority voting for classification and weighted averaging for the regression are the common strategies of combining base learners. The popular and effective ensemble methods are Bagging, Boosting and Stacking. The figure1. gives a diagrammatic representation of ensemble approach.

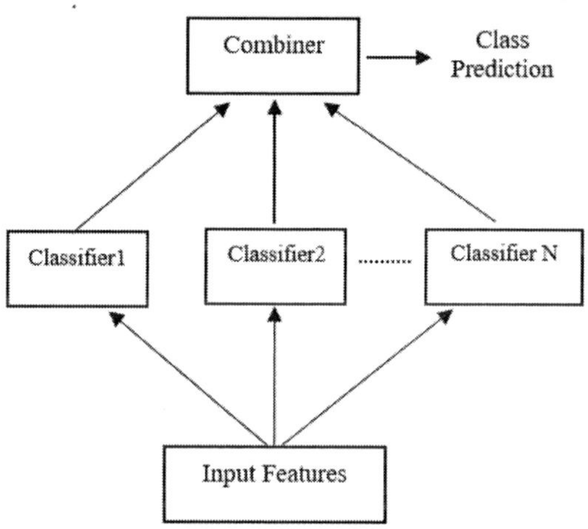

Figure 1. Ensemble Learning Approach.

The ensemble learning models used in this experiment are Random Forest and Gradient Boosting. Random forest model is a bagging technique whereas Gradient Boosting is a boosting approach[21].

Random Forest: Random Forest is a solution to most of the data science problems. It is a flexible machine learning model which is competent to perform both classification and regression. It handles missing values, dimensionality reduction methods, outliers and thereby having better results. It is an ensemble of weak models (decision trees). To classify an object each tree votes, the forest chooses the classification which has the highest number of votes. For regression it computes the average of the output produced by each tree. The major advantage of Random forest is avoiding the problem of overfitting.

Given a training set: -

$X = x_1, x_2, \ldots. x_n$ with responses $y_1, y_2, \ldots. y_n$. bagging it B times- selecting random samples.

For $b = 1,2,3,....N$ We will train a regression or classification tree on X_b, Y_b. After training the prediction of the sample x can be done either by averaging the prediction f all the regression trees or taking the majority votes in the case f classification trees.

$$F = 1/N \sum f_b(x') \quad \text{for} \quad b = 1,2,3,....N$$

The standard deviation of the predictions from different regression trees gives the estimate of uncertainty of the prediction.

$$\sigma = \sum (f_b(x') - F)^2 /(N-1)$$

Gradient Boosting: Gradient Boosting is an ensemble learning technique which has sequential predictors. In this technique the subsequent predictors enhance its learning by the faults of the previous predictors. It involves three elements:

A loss function to be optimized.

A weak learner to make predictions.

Model to add weak learners.

Gradient boosting uses decision trees as weak learners. A tree is parameterized and its parameters are modified and then it is added to the model to minimize the loss.

```
For i=1…..n
F0(x) = arg min ∑ L(yi , γ )
   γm = argmin ∑ L (yi , F m-1(xi) + γ hm(xi) )
   Fm(x) = Fm-1(x) + γmhm(x).
```

Where (xi, yi) is the training set L(y,F(x)) is a differential loss function, M is the no. of iterations, hm(x) is the base learner, γmis the multiplier.

4. Performance measures

The Table 3 summarizes the performance measures considered for multinomial classification.

Table 3. Performance measures.

Performance Measure	**Description**
Precision	It is the proportion of cases that are correctly identified to belong to class A among all the cases which classifier claims to belong to A.
Recall	It is the proportion of cases that are correctly identified to belong to class A among all the cases that truely to belong to A.
F-measure	(2*Precision* Recall)/(Precision + Recall)
Accuracy	It is the ratio of all correctly identified and classified cases.

5. Proposed methodology

We have conducted the experiment to classify our data of cross and within project into multi-classes. The three classes defined are as follows:

- Class 0: no bugs.
- Class 1: bugs are greater than 0 and less than 5.
- Class 2: bugs are greater than and equal to 5.

The technique used for learning and classification is Ensemble Learning approach. The models used are:

- Gradient Boosting
- Random Forest

The training datasets belong to different categories of CPDP. They are:

- Strict CPDP

- Mixed CPDP.
- Mixed with target Data.
- Pairwise CPDP.

Training and testing is also performed on within project dataset. Furthermore, we have carried out k-fold cross validation to evaluate the training accuracy of the data.

6. Results & Discussion

In this section we present results of the experiment conducted. The results are summarized in the below table. The performance evaluation measures used are accuracy, recall, precision, recall and f1-score. The k-fold cross validation is performed to evaluate the training accuracy of the datasets.

RQ: Whether multinomial classification of CPDP is comparable to WPDP?

From the Table 4 following observations are noted:

- The testing accuracy of CPDP was as high as 0.91 in the case of Strict CPDP. In the case of Mixed CPDP average accuracy was 0.88, 0.85 in the case of Mixed CPDP+10% WP data and 0.82 in pairwise CPDP. These values are comparable to the highest accuracy value of 0.93 achieved in the case of WPDP.
- Similarly, the value of precision was as high as 0.91 in the case of Strict CPDP. Values of 0.88 0.85, 0.76 was achieved in the case of Mixed CPDP, Mixed CPDP+10% WP data and Pairwise CPDP respectively. These values are comparable to the highest value of 0.91 achieved in the case of WPDP.
- On analyzing the value of Recall, the highest value was 0.93 for WPDP. The recall values of Strict, Mixed CPDP, Mixed CPDP+10% WP data and Pairwise CPDP was as high as 0.91, 0.88, 0.85 and 0.76 respectively. These values are thereby comparable to the recall value of WPDP.
- The F1 score in the case of Strict CPDP was 0.9 whereas for Mixed CPDP and Mixed CPDP+10% WP data the values were 0.83, 0.86 respectively. These values are also comparable to the value of 0.92 achieved in the case of WPDP.

The above stated observations infer the comparable performance of multinomial classification of CPDP to WPDP.

Table 4. Multinomial class level classification of different categories of CPDP.

Category	Training	Testing	Classifier	Class	Accuracy Training	Accuracy Testing	Precision	Recall	F1 score
Strict CPDP	jedit-3.2 , jedit-4.0, jedit-4.1, jedit- 4.2, jedit-4.3, prop1	camel-1.4	Random Forest	0			0.85	0.9	0.87
				1	0.89	0.77	0.21	0.17	0.19
				2			0.5	0.06	0.11
				AVG/TOTAL	0.89	0.77	0.75	0.77	0.76
			Gradient Boosting	0			0.86	0.86	0.86
				1	0.88	0.75	0.23	0.26	0.24
				2			0.25	0.06	0.1
				AVG/TOTAL	0.88	0.75	0.76	0.76	0.76
	ant-1.7, camel-1.6, jedit-4.0	ivy-1.4	Random Forest	0			0.96	0.85	0.9
				1	0.93	0.81	0.14	0.31	0.2
				2			0.21	0.05	0.12
				AVG/TOTAL	0.93	0.81	0.91	0.83	0.86
			Gradient Boosting	0			0.94	0.97	0.95
				1	0.84	0.91	0.17	0.06	0.09
				2			0.23	0.07	0.12
				AVG/TOTAL	0.84	0.91	0.89	0.91	0.9
Mixed CPDP	ivy-1.1, camel-1.4, jedit-3.2,jedit-4.0,prop2	jedit-4.1	Random Forest	0			0.87	0.82	0.85
				1	0.92	0.74	0.44	0.47	0.46
				2			0.38	0.73	0.5
				AVG/TOTAL	0.92	0.74	0.76	0.74	0.75
			Gradient Boosting	0			0.79	0.99	0.88
				1	0.91	0.77	0.6	0.09	0.15
				2			0.6	0.55	0.57
				AVG/TOTAL	0.91	0.77	0.74	0.78	0.71
	ant-1.7, camel-1.6, jedit-4.0,prop2	prop3	Random Forest	0			0.89	1	0.94
				1	0.92	0.88	0.27	0.01	0.02
				2			0.22	0.51	0.6
				AVG/TOTAL	0.92	0.88	0.81	0.88	0.83

IOP Conf. Series: Materials Science and Engineering **618** (2019) 012039 doi:10.1088/1757-899X/618/1/012039

Table 4. Multinomial class level classification of different categories of CPDP.

Category	Training	Testing	Classifier	Class	Accuracy Training	Accuracy Testing	Precision	Recall	F1 score
Mixed CPDP +10% WP data	prop1, camel-1.2, jedit-3.2 ,ivy1.1, (few labeled data from ivy-2.0)	ivy-2.0	Gradient Boosting	0			0.89	1	0.94
				1	0.91	0.88	0.42	0.01	0.01
				2			0.2	0.5	0.61
				AVG/TOTAL	0.91	0.88	0.83	0.88	0.83
			Random Forest	0			0.86	0.8	0.83
				1	0.88	0.72	0.39	0.46	0.42
				2			0.27	0.5	0.35
				AVG/TOTAL	0.88	0.72	0.75	0.73	0.74
			Gradient Boosting	0			0.83	0.94	0.89
				1	0.88	0.79	0.55	0.26	0.35
				2			0.38	0.5	0.43
				AVG/TOTAL	0.88	0.79	0.77	0.8	0.77
	ant-1.7, camel-1.6, jedit-4.0,prop2, (few labeled data from prop3)	prop3	Random Forest	0			0.93	0.9	0.91
				1	0.92	0.85	0.66	0.69	0.68
				2			0.36	0.75	0.49
				AVG/TOTAL	0.92	0.85	0.87	0.85	0.86
			Gradient Boosting	0			0.86	0.96	0.9
				1	0.91	0.83	0.66	0.39	0.49
				2			0.83	0.42	0.56
				AVG/TOTAL	0.91	0.83	0.82	0.83	0.81
Pairwise CPDP	jedit-4.0	prop2	Random Forest	0			0.77	0.99	0.87
				1	0.92	0.75	0.2	0.03	0.05
				2			1	0.09	0.17
				AVG/TOTAL	0.92	0.75	0.66	0.75	0.67
			Gradient Boosting	0			0.76	1	0.87
				1	0.91	0.76	1	0.02	0.03
				2			1	0.18	0.31
				AVG/TOTAL	0.91	0.76	0.82	0.76	0.67
WPDP	prop2	prop2	Random Forest	0			0.96	0.95	0.95
				1	0.92	0.91	0.19	0.21	0.2
				2			0.04	0.06	0.05
				AVG/TOTAL	0.92	0.91	0.91	0.91	0.91
			Gradient Boosting	0			0.95	0.98	0.97
				1	0.91	0.93	0.27	0.11	0.15
				2			0.07	0.25	0.11
				AVG/TOTAL	0.91	0.93	0.91	0.93	0.92
	ivy-2.0	ivy-2.0	Random Forest	0			0.84	0.91	0.87
				1	0.93	0.78	0.49	0.36	0.42
				2			0.01	0.04	0.05
				AVG/TOTAL	0.93	0.78	0.75	0.78	0.77
			Gradient Boosting	0			0.82	0.93	0.87
				1	0.99	0.78	0.51	0.29	0.37
				2			0.01	0.04	0.02
				AVG/TOTAL	0.99	0.78	0.74	0.78	0.75

7. Conclusion & Future scope

The software reliability is directly proportional to probability of error free software. Software defect prediction leads to its more focused and rigorous testing. In this paper we analyzed the performance in multinomial classification of cross and within project defect prediction. The datasets were collected from the NASA projects. CK based object oriented metrics was effective for CPDP. We first labeled our class information in three different levels depending on the number of defects in each class.The experiment was conducted by training the model on different categories of cross project defect prediction.K-fold cross validation was done to estimate the predictive performance of the models. Ensemble learning approach validated the classification results. The results indicate that multinomial classification on CPDP is feasible and comparable to WPDP. There are many verticals on which the work can be extended. Multinomial classification of Heterogeneous CPDP can be focused. Proper training dataset selection and class imbalance to optimize the performance of defect prediction are still open issues in CPDP.

References

[1] B. Turhan, T. Menzies, A. B. Bener, and J. Di Stefano, "On the relative value of cross-company and within-company data for defect prediction," Empirical Software Engineering, 14(5):540-578, 2009.

[2] T. Zimmermann, N. Nagappan, H. Gall, E. Giger, and B. Murphy, "Cross-project defect prediction," In Proceedings of the 7th Joint Meeting of the European Software Engineering Conference and the ACM SIGSOFT Symposium on the Foundations of Software Engineering (ESEC/FSE), pages 91-100, 2009. O

[3] Y. Ma, G. Luo, X. Zeng, and A. Chen, "Transfer learning for cross-company software defect prediction. Information and Software Technology", 54(3):248-256, 2012. O

[4] J. Nam, S. J. Pan, and S. Kim, "Transfer defect learning," In Proceedings of the 35th International Conference on Software Engineering (ICSE), pages 382-391, 2013. O

[5] S. Herbold, "Training data selection for cross-project defect prediction" In Proceedings of the 9th International Conference on Predictive Models in Software Engineering (PROMISE), pages 6-16, 2013. O

[6] F. Peters, T. Menzies, and A. Marcus, "Better cross company defect prediction" In 10th IEEE Working Conference on Mining Software Repositories (MSR), pages 409-418, 2013. O

[7] Panichella, R. Oliveto, and A. De Lucia, "Cross-project defect prediction models: L'Union fait la force," In Software Evolution Week-IEEE Conference on Software Maintenance, Reengineering and Reverse Engineering (CSMR-WCRE), pages 164-173, 2014.

[8] D. Ryu, , O. Choi, , & J. Baik, "Value-cognitive boosting with a support vector machine for crossproject defect prediction," 2016 Empirical Software Engineering. doi:10.1007/s10664-014-9346-4.

[9] J. Nam, and S. Kim, "Heterogeneous Defect Prediction" Proceeding ESEC/FSE 2015 Proceedings of the 2015 10th Joint Meeting on Foundations of Software Engineering (pp. 508-519). New York, NY, USA: ACM. doi:10.1145/2786805.2786814 D

[10] S.J. Pan, Q. Yang, "A survey on transfer learning, Technical Report" HKUST-CS 08- 08, Department of Computer Science and Engineering, Hong Kong University of Science and Technology, 2008.

[11] S. J. Pan and Q. Yang. "A survey on transfer learning". IEEE Trans. on Knowl. and Data Eng., 22:1345–1359, October 2010.

[12] S. Kim, H. Zhang , R. Wu, & L. Gong, "Dealing with noise in defect prediction". 2011 33rd International Conference on Software Engineering (ICSE) (pp. 481 - 490). Honolulu, HI: IEEE. doi:10.1145/1985793.1985859

[13] M. D'Ambros, M. Lanza, and R. Robbes. "Evaluating defect prediction approaches: A benchmark and an extensive comparison". Empirical Software Engg., 17(4-5):531–577, Aug. 2012.

[14] M. H. Halstead. "Elements of Software Science (Operating and Programming Systems Series)". Elsevier Science Inc., New York, NY, USA, 1977.

[15] S. R. Chidamber and C. F. Kemerer. "A metrics suite for object oriented design". IEEE Trans. Softw. Eng., 20:476–493, June 1994. D

[16] T. McCabe. "A complexity measure. Software Engineering", IEEE Transactions on, SE-2(4):308–320, Dec 1976.

[17] . J. Pan and Q. Yang. "A survey on transfer learning". IEEE Trans. on Knowl. and Data Eng., 22:1345–1359, October 2010. D

[18] X. Y. Jing, F. Wu, X. Dong, F. Qi, and B. Xu, — "Heterogeneous cross company defect prediction by unified metric representation and cca- based transfer learning,"I In Proceedings of the 10th Joint Meeting on Foundations of Software Engineering, 2015, pp. 496–507 D

[19] G. Canfora, A. De Lucia, M. Di Penta, R. Oliveto, A. Panichella, and S. Panichella. "Multi-objective cross-project defect prediction". In IEEE 6th International Conference on Software Testing, Verification and Validation (ICST), pages 252-261, 2013.

[20] Z. He, F. Shu, Y. Yang, M. Li, and Q. Wang. "An investigation on the feasibility of cross-project defect prediction." Automated Software Engineering, 19(2):167-199, 2012.
[21] W. Dai, Q. Yang, G. Xue, & Y. Yu, "Boosting for transfer learning. In Proceedings of the 24th international conference on Machine learning" 2007 (pp. 193–200). http://dl.acm.org/citation.cfm?id= 1273521. Accessed February 25, 2014.
[22] Singh, P., Verma, S., & Vyas, O."Cross Company and within Company Fault Prediction using Object Oriented Metrics". 2013, International Journal of Computer Applications, 5-11.
[23] K. Dejaeger, "Toward Comprehensible Software Fault Prediction Models Using Bayesian Network Classifiers,". IEEE Transactions on Software Engineering, 39(2), 2013, PP. 237–257.
[24] K. O. Elish, and M. O. Elish,"Predicting defect-prone software modules using support vector machines," Journal of Systems and Software, 81(5), 2008, pp. 649–660. doi:10.1016/j.jss.2007.07.040.
[25] O. Mizuno, Y. Hirata, "A Cross-Project Evaluation of Text-Based Fault-Prone Module Prediction,". Proceeding IWESEP '14 Proceedings of the 2014 6th International Workshop on Empirical Software Engineering in Practice (pp. 43-48). Washington, DC, USA: IEEE Computer Society.
[26] B. Turhan, , A. Mısırl, & A. Bener, "Empirical evaluation of the effects of mixed project data on learning defect predictors. Information and Software Technology,", 2013, 55(6), 1101–1118. doi:10.1016/j.infsof.2012.10.003
[27] P. He, B. Li, & Y. Ma, "Towards Cross-Project Defect Prediction with Imbalanced Feature Sets,". Cornell University Library, 2014.
[28] D. Ryu, J.-I.Jang , & J. Baik, "A transfer cost-sensitive boosting approach for cross-project defect prediction,". Software Quality Journal, 2015, PP.1-38. doi:10.1007/s11219-015-9287-1
[29] Zhang, Feng, et al. "Cross-project defect prediction using a connectivity-based unsupervised classifier." Proceedings of the 38th International Conference on Software Engineering. ACM, 2016.
[30] Xia, Xin, et al. "Hydra: Massively compositional model for cross-project defect prediction." IEEE Transactions on software Engineering 42.10 (2016): 977-998.
[31] Steffen Herbold et al. "A Comparative Study to Benchmark Cross-project Defect Prediction Approaches". Ieee Transactions On Software Engineering, 2017.

IOP Conf. Series: Materials Science and Engineering **618** (2019) 012040 doi:10.1088/1757-899X/618/1/012040

Algorithm for assessing the quality of service in special-purpose networks during operation

N Kireeva[1], **V Zastupov**[1]

[1]Povolzhskiy State University of Telecommunications and Informatics (PSUTI)
23 Tolstoy street, Samara, 443010,Russia

[1]zeppelinSN@yandex.ru, [2]vladimir.zastupov@mail.ru

Abstract. This article is to show an algorithm for assessing the quality of service in special-purpose networks, regardless of the type of traffic being transmitted and the network architecture. This method allows to obtain a detailed assessment of the quality of service, as well as to adjust the values of the characteristics of the network during the evaluation. The values used in the calculations are entered in the database to aid in the design of new networks.

1. Introduction

Based on the tasks performed by special-purpose networks (SPN), they are subject to increased requirements regarding reliability, security, timeliness and quality of service (QoS) in the process of information transfer. When assessing the QoS in SPN, it is often resorted to the methods of general-access telecommunications networks with similar traffic. The existing methods [1-4] of calculating the QoS in specialized networks differ in the choice of characteristics for estimating this parameter, which leads to a number of assumptions and a decrease in the accuracy of the estimation, as well as ignoring the influence of unused network properties on the quality of service indicator [5-8].

The purpose of this article is to develop a method for assessing the QoS that would be suitable for all SPN, regardless of their architecture and the type of traffic being transmitted, as well as using a full set of characteristics that somehow affect the QoS [9-11].

At the initial stage, it is necessary to configure the network so that during its operation the maximum possible load is provided, and also to prepare and record the initial data. After setting up the parameters of the SPN, an algorithm for calculating and evaluating the QoS parameters is started, consisting of several steps [12, 14].

In the framework of this study, characteristics are selected that allow an accurate assessment of the quality of network service, the methods and procedure for calculating them are determined, an algorithm is created for assessing the quality of service in special-purpose networks.

2. Estimation of reliability of a special-purpose network

First of all, the reliability of the used SPN (P_R). In this case, the reliability parameter is a property of communication that characterizes the ability to provide the required accuracy of message reproduction at delivery points [8]. Reliability is a collection of event probabilities. In this method, the probabilities of such events are used to evaluate this parameter: the transmitted message is accepted / not accepted (P_A / P_{NA}) by the user, received with an error (P_{ER}), or non-transferable data is received (P_{NT}).

$$P_R = \{P_A; P_{NA}; P_{ER}; P_{NT}\} \tag{1}$$

Comparison of the given probabilities with the established requirements and, if necessary, their normalization will be considered as an estimate of the QoS of the SPN for reliability.

3. Evaluation of network architecture parameters

At this stage, emphasis is placed on the architecture of building a SPN, the number of nodes and branches used, the number of transmission paths, and the directions of communication [13,15]. Based on these properties, the distributed load is calculated on the branches of the network (Z_i).

$$Z_i = Z + 0,6742\sqrt{Z} \tag{2}$$

where Z – average load;
i – branch ordinal number.
Next, the probability of loss and load on the direction of communication is determined:

$$P_n = \sum_{j=0}^{n} P_j \cdot E_j(Z_j) \tag{3}$$

where n – number of channels in the direction of communication;
j – channel sequence number;
P_j – probability of presence of effective channels in the direction of communication;
$E_j(Z_j)$ – probability of loss in service with load Z on the communication channel.

It is also necessary to calculate the survivability of the network [5-6] (W_{SN}) – this is a characteristic of the network's ability to ensure the establishment of connections and the transmission of messages between sources and consumers of information in the event of failure of its elements or areas without standardizing the QoS.

$$W_{SN} = 1 - \prod_{l=1}^{I}\left(1 - \prod_{k=1}^{n} W_k \prod_{i=1}^{L} W_i\right) \tag{4}$$

where I – number of independent communication directions;
n – number of switching centers in transit;
W_k – probability of survival of the switching center;
L – the number of branches in the path;
W_i – probability of branch survival.
Ultimately, the characteristics calculated by the formulas (2), (3) and (4) are compared with the required values to the investigated special-purpose network.

4. Assessment of the general system of quality indicator

At this stage, the QoS is evaluated with respect to information exchange and management, as well as the systems corresponding to them. It forms a general quality indicator system (GQIS):

$$GQIS = [Y_{IE}, Y_C, Y_{SIE}, Y_{SC}]^T \tag{5}$$

where Y_{IE} – quality information exchange (IE) system;
Y_C – quality management system;
Y_{SIE} – system of the quality indicator of the IE system;
Y_{SC} – quality system of the management system.
Elements from which the GQIS is formed, in turn, depend on the following parameters:

IOP Conf. Series: Materials Science and Engineering **618** (2019) 012040 doi:10.1088/1757-899X/618/1/012040

$$Y_{IE} = \left[t_{ds0}(r), P(t_{ds0}(r) \le t_{ds\,dm}(r)), P_{d_{IE}}, P_{s_{IE}}, C_{IE} \right]^T \quad (6)$$

where $t_{ds0}(r), P(t_{ds0}(r) \le t_{ds\,dm}(r))$ – average time and probability of timely delivery of data of different priorities $r = \overline{1,R}$;

$P_{d_{IE}}$ – authenticity of IE;

$P_{s_{IE}}$ – information security indicator while providing IE;

C_{IE} – vector of resources for IE.

$$Y_C = \left[T_{cc}, P(t_{dc} \le t_{dsc\,dm}), P_{s_c}, C_C \right]^T \quad (7)$$

where T_{cc} – average duration of the communication parameter control cycle;

$P(t_{dc} \le t_{dsc\,dm})$ – timely delivery of management messages;

P_{s_c} – an indicator of information security while providing management;

C_C – resource cost vector for management provision.

$$Y_{SIE} = [V_{SIE}, H_{SIE}, C_{SIE}]^T \quad (8)$$

where V_{SIE} – vector of IE system capacity [7];

H_{SIE} – vector of the stability indicator of the IE system;

C_{SIE} – vector of resource costs for the construction and operation of the IE system.

$$Y_{SC} = [V_{SC}, H_{SC}, C_{SC}]^T \quad (9)$$

where V_{SC} – vector of control system capacity;

H_{SC} – control stability index vector;

C_{SC} – vector of resource costs for the construction and operation of a control system.

As an allowed value of the GQIS as a generalized probabilistic indicator, we take the value equal to 0.8.

5. Algorithm

This algorithm of the method for assessing the quality of service in a special-purpose network is shown in the Fig 1.

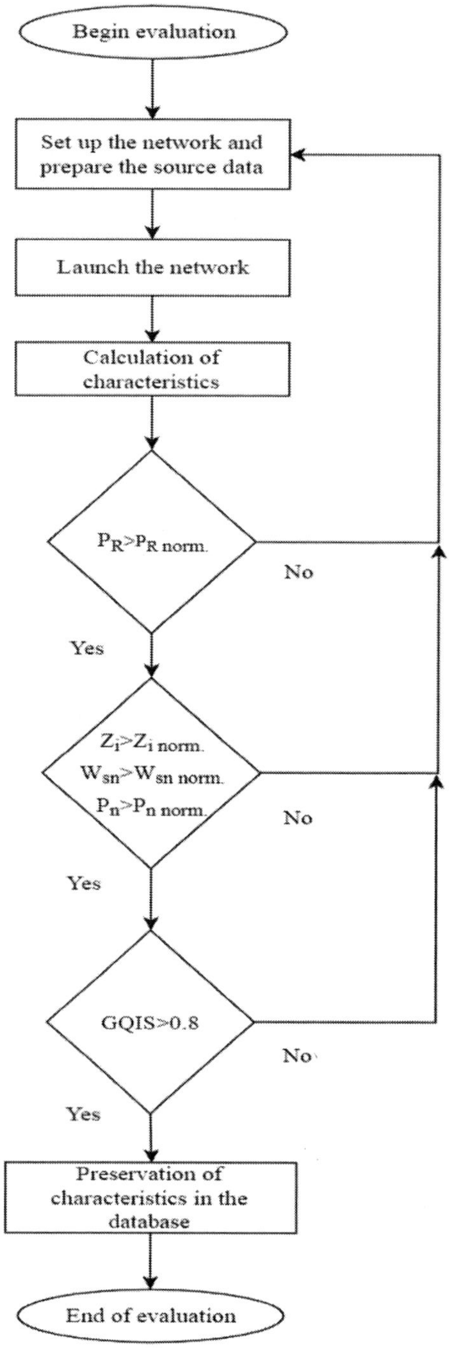

Figure 1. Algorithm for assessing the quality of service in special-purpose networks

At the initial stage, data is collected on a special purpose network (type, data transmission rate, length, number of communication branches, modulation type, etc.). After all the necessary information for the calculations is provided, the simulation modeling of the future network is launched. If the parameters amenable to preliminary calculation meet the above standards, it is possible to proceed to the organization of a special-purpose network. If any parameter (s) do not meet the specified requirements, an adjustment is made and the choice of the optimal elimination of the nonconformity. To save resources and time, successful network models are stored in a database for use as standards.

6. Conclusion

This method covers such basic segments of a special-purpose network as: a radio relay communication line, a fiber-optic communication line, a telephone communication line, a satellite communication line. Ultimately, a holistic method was developed for assessing the quality of service in special-purpose networks, covering a large area of calculations and parameters used, which increases its accuracy and efficiency.

References

[1] D.I. Mogilevich, S.V. Druzhinin, O.K. Klimovich "Methodology for assessing the quality of service in a special purpose telecommunications network with connection establishment", Information processing system, no. 7 (65), 2007, pp. 43-47.
[2] E.N. Chapurin "Methodology for assessing the quality of service provision in a special purpose infocommunication network based on traffic prioritization mechanisms", Modern problems of science and education, no. 1-1, 2015, pp. 18-26.
[3] P.V. Kolgotin, "Estimation of channel parameters and development of measuring technologies in special purpose communication networks", Young scientist, 2011. no. 10(33), pp. 34-39.
[4] M. Louta, A. Michalas, V. Loumos, Quality of Service Management in IP networks, Contel. 2003.
[5] L.P. Shcherbina, O.G. Khilko, Reliability and survivability of switched communication networks, L.:VAS, 1977.
[6] Romanov O.I. Yu.V.Gresko, O.K. Klimovich "Probabilistic model of estimation of telecommunication network survival", Collection of scientific works of VITI NTUU "KPI", 2003, no. 4, pp. 98-111.
[7] G.G. Yanovsky "Modern Problems of Science in the Field of Telecommunications (Evolution and Convergence)", Electronic Journal of St. Petersburg: SPbGUT name prof. M.A. Bonch-Bruevich, 2008, 163 p.
[8] S. I. Makarenko "Descriptive Model of a Special Purpose Communication Network", Systems of Control, Communication and Security, 2017, no. 2, pp. 113-164.
[9] Y. Klochkov, and A. Gazizulina, "Improvement of methodology of evaluation of efficiency of the etallurgical complex procebes development", Key Engineering Materials, vol. 684, 2016, pp. 453-460. 10.4028/www.scientific.net/KEM.684.453
[10] Y. Klochkov, A. Gazizulina, and N. Golovin, "Assessment of organization development speed based on the analysis of standards efficiency", Proceedings - 2nd International Symposium on Stochastic Models in Reliability Engineering, Life Science, and Operations Management, SMRLO 2016, 2016, pp. 530-532. 10.1109/SMRLO.2016.93
[11] M. Akhin, S. Kolton, and V. Itsykson, "Random model sampling: Making craig interpolation work when it should not", Automatic Control and Computer Sciences, vol. 49, no. 7, 2015, pp. 413-419. doi:10.3103/S0146411615070020
[12] M. Khludova, "Resource allocation policies for smart energy efficiency in data centers", Lecture Notes in Computer Science (including subseries Lecture Notes in Artificial Intelligence and Lecture Notes in Bioinformatics), vol. 8638, 2014, pp. 16-28
[13] Kochovski, P., Drobintsev, P. D., & Stankovski, V. (2019). Formal quality of service assurances, ranking and verification of cloud deployment options with a probabilistic model checking method. Information and Software Technology, 109, 14-25. doi: 10.1016/j.infsof.2019.01.003
[14] Musaev, A. A., & Serdyukov, Y. P. (2016). The uncertainty relation between the signals as a criterion for evaluation of the physical data rate. *SPIIRAS Proceedings, 2*(45), 58-74. doi:10.15622/SP.45.4
[15] Kalinin, M. O., & Pavlenko, E. Y. (2015). Increasing the fault tolerance and availability of software defined networks using network equipment control based on multiobjective optimization by service quality parameters. *Automatic Control and Computer Sciences, 49*(8), 673-678. doi:10.3103/S014641161508026X

Assessment of the Internet of Things Projects on the Real Estate Market

S Pupentsova[1], V Leventsov[1], M Livintsova[1], N Alexeeva[1], S Vodianova[2]
[1]Peter the Great St. Petersburg Polytechnic University, St. Petersburg, Russia
[2] J'son & Partners Consulting, Moscow, Russia

Abstract. This paper discusses the concept of the Internet of Things. It analyzes the existing definitions, and introduces a new definition of this concept. In this research study, the experience of the Internet of Things in the world has been analyzed, the possibilities of deflationary development of the economy have been identified with the use of the Internet of Things. The paper presents directions for important qualitative changes in economy and for organizational and technological transformations of production, which may be induced by the development of the Internet of Things. The "smart home" system market has been examined. Its potential has been identified for Russian companies that work in the field of information and telecommunication technology connected with the development of the Internet of Things. To justify the efficiency of money investments into the "smart home" system and services, the algorithm of cost estimation for such systems, which can be used in making a decision about investing, is proposed. The authors conducted the review of the typical projects offered with the "smart home" system. The power law dependence of the unit cost of the system installation from the floor space of the object has been received. The internal rate of return on investment for the introduction of the "smart home" system has been identified at the level of 50%.

1. Introduction

The Internet of Things (IoT) and the Internet technology (IT) as a whole, as well as other technologies of public use (for example, the electricity), make a transformational impact on various sectors of the national economy and the entire economy.

The result of the IoT technology is the reduced prime cost of transactions carried out by enterprises, growth of labor productivity, increase in efficient use of resources by enterprises and organizations. From the macroeconomic point of view, active development of the Internet services and introduction of IT mean transition to deflationary development of efficiency for all participants in the value chain, which in the IoT model is called “IoT ecosystem”, and includes end-users. This is unprecedented case in the history of the world economy. When the resources of extensive economic growth slow down due to growing production of new goods and services, better efficiency of production and marketing processes become the key focus of development.

For the public companies, increasing labor productivity, a higher level of resource utilization and reduced prime cost of transactions due to introduction of the IoT are accompanied by a growth in their shareholder value.

The research of the introduction of the IoT technologies and evaluation of its efficiency is a priority now. In this paper, an algorithm is proposed for evaluating the efficiency of introduction of the Internet of Things using an example of the “smart home” system.

2. Object and method of research

The concept of the Internet of Things (IoT) arose as a phenomenon of mass connection of various gadgets and equipment to the Internet, the growth of their functionality and autonomy of use. For the first time the term "Internet of Things" appeared in 1999 in the company "Procter & Gamble" as an idea that wireless sensors connected to each other can be built in everyday things [1].

The official definition of the Internet of Things is proposed by ITU-T Recommendations Y.2060 (2012), "Overview of the Internet of Things", where the Internet of Things is defined as "a global infrastructure for the information society, enabling advanced services by interconnecting (physical and virtual) things based on existing and evolving interoperable information and communication technologies" [2].

"The Internet of Things as a new stage in the evolutionary development of the Internet, characterized by the fact that heterogeneous networks and a multitude of sensors have to unite for interaction under the management of unified standards, while stressing that it should not become technology for the sake of technology and, therefore, the task of the info-communication industry is to demonstrate its value for every person. The objects which have an incorporated sensor connected to the Internet are called "smart things" [3].

In reference [4], the following definition of the Internet of Things is presented: "The concept of a computer network of physical objects ("things") with built-in technologies for interaction with each other or with the external environment and considering organization of such networks as a phenomenon capable of restoring economic and social processes, excluding from the part of actions and operations the necessity of human participation".

The following definitions of this term can be found [5]:

- "a single network which connects the objects of the real world, which surround us, and virtual objects;
- closer integration of the real and virtual worlds, where communication occurs between people and devices;
- an innovative direction in the trade policy focused on creation of new goods and/or on giving goods new qualities based on their interaction with each other or with the external environment".

In the publication [6] the Internet of Things is defined as "a network of interconnected "smart" devices, creation and access to which will provide greater awareness of all phases of the company's activities. In reference [7], it is a popular concept for development of computer networks, including technical devices equipped with technologies for interacting both with each other and with the environment without human participation".

In Rob Van Kranenburg's opinion [5] "the Internet of Things is a "four-layer pie", which combines four levels: the level is associated with the identification of each object; the level provides services to meet the needs of the consumer (a network of their own "things", for example – "smart home"); the level is connected with the urbanization. For example, the conception of a "smart city", where all information for residents of the city is structured with reference to a specific residential block, a particular house; the level is a sensory planet".

Forrester Consulting notes that 85% of executives agree with the definition: "The Internet of Things is a network of interconnected "smart" devices, creation and access to which will provide greater awareness of all stages of the company's activities" [6].

In reference [8], the Internet of Things (IoT) is understood as the concept of constructing computer networks from physical objects (otherwise - "things") capable of interacting with the environment or with each other.

Gartner Company interprets the term "Internet of Things" as "a network of physical objects that have incorporated technology allowing these objects to measure the parameters of their own state or the state of the environment, to use and transfer this information" [9].

Thus, amongst modern researchers, there is no single point of view on the existing phenomenon. The term of the Internet of Things is understood as an idea, concept, network, infrastructure, direction, levels, integration and many other things. As this area of economic activity is developing, researchers move away from the notion of an idea and concept, emphasizing that it is a matter of interconnection of things, practically without participation of man. At the same time, a new market segment is being formed, which, on the basis of new information opportunities, sets new requirements for goods and services in consumer and industrial markets.

Thus, the following definition of the Internet of Things (IoT) can be given: a system of integrated computer networks and connected physical objects (things) with built-in sensors and software for data collection and exchange, with the possibility of remote control and management in automated mode, creating not only new products and services but a whole infrastructure based on greater awareness.

3. The use of the internet of things

The analyzed experience of the IoT introduction in the world shows that the transition to the concept of the IoT happens due to the formation of cross-industrial open (horizontally and vertically) production and service ecosystems which unite many different information management systems of different enterprises and involve many different devices.

This approach allows implementing complex end-to-end business projects in the virtual space, with them being able to perform in an automated mode optimization administration (end-to-end engineering) of various types of resources through the entire product supply and value chain, starting from an idea, design, engineering and finishing with production, operation and utilization.

From the macroeconomic point of view, the increase in the efficiency of processes in the supply-consumer chain means a transition from an inflationary development consisting in shifting growing costs (a growth in the supplier's revenue is the growth of consumer costs) to the "next in the chain", and from the end user back to the producers (employers) through demands for bigger salaries – to deflationary development, based on better efficiency of all participants in the IoT ecosystem, including end users. A separate segment in the growth of national economies is not consumers of Internet services, but producers and providers of Internet services, products and solutions that capture traditional industry niches and reform them based on cloud technologies. Typical examples are online media, e-commerce and online taxi booking.

Increasing efficiency in the world economy leads to lower prices and, in prospect, less consumption of energy carriers and raw materials, whose supplies to the world market determine Russia's role in the global division of labor at the present time, which negatively affects the Russian economy. On the other hand, low labor productivity in Russia (figure 1) has an additional negative impact on the opportunities and resources for the development of the Russian economy. It negatively affects the volume of domestic production and consumption (the capacity of domestic markets). In figure 1, the data for productivity are taken according to [10].

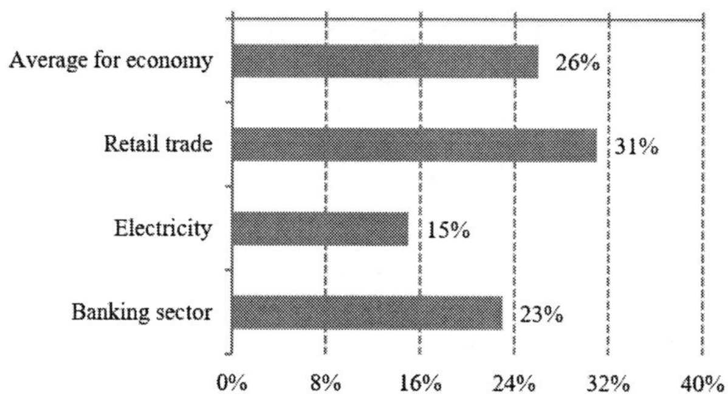

Figure 1. Labor productivity in the Russian Federation, in % from the USA (100%)

E.g., the five-fold difference in labor productivity in Russia and the USA determines the five-fold difference in housing provision, which largely determines the birth rate.

A number of factors affect the overall level of labor productivity in Russia. On the one hand, this is a level of equipment of Russian enterprises with modern means of production, including IT and Internet technologies. On the other hand, it is a relatively low level of development of modern high-tech industries, including high-tech engineering, electronics, IT and Internet services, which have a high level of labor productivity in comparison with traditional and less technologically advanced industries [22]. Also, the level of labor productivity is affected by the issues of overall organization of business management and human resources management.

Thus, there is a kind of vicious circle: the Russian economy critically needs accelerated modernization to improve the quality of life in the country and the level of its competitiveness. One of the elements of modernization should be measures to improve the efficiency of organization of labor (figure 1) and creation of new markets and high-tech industries, including those on the grounds of new business models based on IoT [23]. However, low labor productivity, a high level of costs in the Russian economy and a relatively low level of development of modern high-tech industries are the main reasons for the low volume of domestic demand and inaccessibility of investment resources. This, in turn, leads to the impossibility of implementing measures to increase labor productivity and create new markets for services and industries in Russia.

For quantitative growth of the Internet of Things and organizational and technological transformation of production, important qualitative changes must happen in the economy, such as:

- the possibility of obtaining valuable information on the nature of the product and equipment usage for all participants in the production cycle as embedded devices are becoming increasingly common. These previously unavailable data were the basis for new business models and additional earnings from the supply of new services, such as, for example, the life cycle contract for industrial equipment, contract manufacturing as a service, transportation as a service, security as a service and others;
- virtualization of production functions is accompanied by the formation of "shared economy" characterized by significantly higher efficiency and productivity due to increasing the use of available resources, changing the functionality of devices without making changes to physical objects, by changing the technologies for managing them;
- modeling of technological processes, end-to-end engineering and, as a result, optimization of the value chain at all stages of the life cycle of the product in real time, allow producing a count or small-batch product at minimal cost for the customer and with a profit for the manufacturer, which in the traditional production is only possible in case of mass production;
- reference architecture, standardized networks and a rental model, instead of paying the full cost of ownership, make the joint production infrastructure accessible for small and medium businesses, which facilitates their efforts in managing production, makes it possible to respond faster to changing market requirements and reduction in the product life cycle, and entails development and appearance of new applications and services;
- analysis of data about the user, his production facilities (machines, buildings, equipment) and the nature of consumption open possibilities for the provider to improve the client's experience, create better convenience of use, better solutions and reduce customer costs, which leads to the increased satisfaction and loyalty from work with the given supplier;
- the functioning of various sectors of the economy will be continuously complicated under the influence of technology development and will be increasingly implemented through automatic decision making by the machines themselves based on the analysis of a large amount of data from the connected devices, which will lead to a gradual decrease in the role of production personnel, including well-qualified personnel. Qualitative vocational education will be required, including engineering, special learning programs for employees and trainings.

With the development of the IoT market in Russia, the cost-effectiveness factor in the introduction of new "smart" technologies is becoming more understandable and obvious to the consumer. Both commercial consumers and households are beginning to be interested in the possibility of increasing the efficiency of resource use through IoT solutions and services, in saving without compromising quality. This will contribute to the development of mass IoT segments, where the motive for end users to use IoT solutions and services will be market incentives, rather than compulsory government programs [11, 12].

This demand is now being formed in the "smart home" segment, which is one of the IoT technologies, including solutions for creating intelligent security services and intelligent services for optimizing the use of resources by households.

The use of smart house systems by households is in an embryonic state in Russia, the level of their penetration is extremely low – less than 0.1% of the total number of apartments and individual houses, according to J'son & Partners Consulting. However, there is a need for such systems, since about half of the commissioned housing in Russia falls on individual housing (data from Rosstat), where "smart home" systems, unlike apartment blocks, can be extremely efficient.

Nevertheless, it is already possible to talk about mass implementation of certain functions of "Smart Home/Building" systems in terms of security and fire alarms, video surveillance and monitoring of resource consumption (electricity, water, heat). Figure 2 below shows the dynamics of the number of connected IoT / M2M devices in the territory of Russia at immovable objects based on actual data for 2010-2014 with a forecast up to 2018. Figure 2 shows the number of devices introduced according to [10].

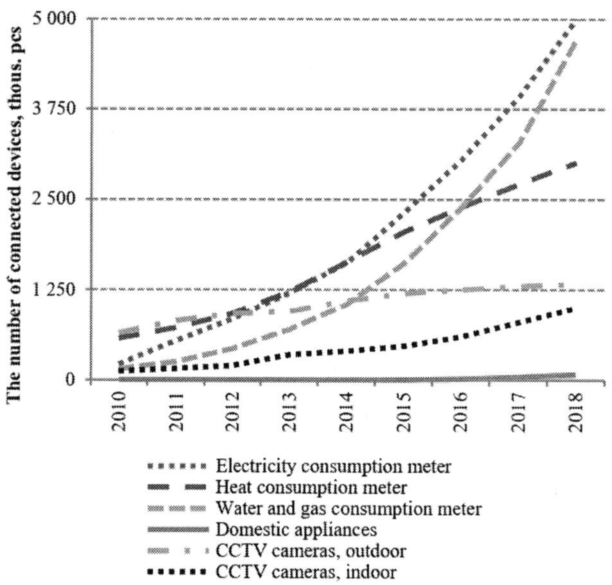

Figure 2. Installed and connected IoT/M2M devices at immovable objects.

Besides, the market of "Smart Home" systems and services has a significant growth potential in Russia due to the segment of connected home appliances (See Figure 2). Now, in those models where this feature is present, it is generally not used, inasmuch as in Russia the systems that are able to receive and process the data are absent. On the other hand, the appearance in the Russian Federation of services analogous to Amazon Dash Replenishment Service, which can reasonably be expected in the next 2-3 years, will lead to an explosive growth of connected home appliances in Russia starting from the year 2018.

As mentioned above, the segment of IoT-devices for accounting consumption of resources, which is closely connected to the segment of connected home appliances, is already developing in Russia, but specifically under the influence of governmental programs on installation of "smart meter" and connection of earlier installed meters to communication networks.

Russian companies that work in the sphere of info-communicational technologies, related to IoT system's development, have ample opportunities for development on this market. All those below can be referred to as such companies: communication service providers and data-center service providers; system integrators and software and app engineers; developers and manufacturers of electronic devices; developers and providers of integrated solutions to service platforms on the Russian IoT market.

4. "Smart home" system and estimation of its cost

The term "smart home" is now associated with individual apartments, offices and detached houses. The offered technical solutions, examples of "smart" home introduction are mainly intended for an object's management by the very residents ("inside" the object) [18].

Based on the data [13, 18] in order to manage residential objects, devices can be used, which carry out the following functions:

1) air quality monitoring (sensors of temperature, humidity, pressure, air particles);

2) water quality monitoring (acidity, redox potential, amount of oxygen, conductivity, dissolved ions);

3) level control of waste, noise (ultrasound, noise, vibration, temperature, humidity);

4) detection of infiltrations, water leakages and presence of open doors and windows;

5) management of electricity and water in containers and tanks (current, ultrasound, water availability, water level);

6) management of lighting inside an apartment and on the facade, adjustment of brightness, traffic control for automatic management of lighting in passageways;

7) management of the "smart home" system by an app on smartphone (iPad, iPhone, Android, Windows);

8) detection of available places on a parking lot;

9) control of video surveillance, audio and document management at the object with a record on the server, and etc.

Installation of all of the above systems at one object in the company "Smart Home KNX Showroom" [13] is referred to the package of the "Lux +" system. The package "Lux" includes 1-5 elements of the "smart home" system from the list above. Let us consider the possibility of evaluating the efficiency of "smart home" systems.

The review of the standard projects proposed in the "smart home" system in St. Petersburg allowed the authors to derive a power law dependence of the specific cost of installation on the floor area of the object (see figure 3). Figure 3 takes the cost of the introduced "smart home" system, package "Lux +" for the standard projects of the company "Smart Home KNX Showroom" [13].

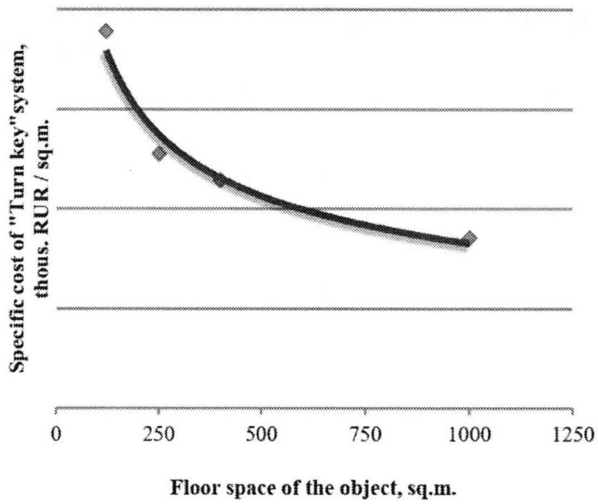

Figure 3. Dependence of the specific cost of a "smart home" system in standard projects on the floor space of an object.

Thus, for an apartment with a floor space of 180 sq. m., the cost of installation of the "smart home" system, which includes the whole list of the mentioned package of services "Lux +", will be 20 thous. RUR. for a sq. m. of the total area or about 3.6 m. RUR for an object.

Installation and usage of the "smart home" systems allow increasing of the life quality of residents, which should be estimated by the latter in monetary value [19]. An additional factor, which increases earnings from the installation of "smart home" systems, is the economy of utility services. Nevertheless, installation and maintenance of such systems require additional capital and maintenance costs [20].

The calculation algorithm presented below is based on the principle of contribution of one or another production factor (resource) into the value of the object which takes into account that the economic value of a specific element in the content of the object is defined by the change of the value of the whole object when the stated element is added or removed from it [12, p. 206].

At the moment such systems can be used in Russia by residents of elite and business class apartments. Therefore, in the selection of analogical objects, it is necessary to constrict the used sample to the objects of these classes.

Since the right for a real estate object can be held on the basis of ownership or rent, two possible calculation algorithms are presented below. The choice of an algorithm will be justified by the availability of information on the properties offered on the market that can be used as analogs. The best option is to apply both calculation algorithms with subsequent analysis of the obtained results and their harmonization.

1. The algorithm of calculations in case there is information on the prices of offers for property rights can be as follows:

1.1. Selection of elite or business class apartments in order to define the price of their offer (or without negotiated discount — transaction price).

1.2. Determination of the specific cost of 1 sq. m. of the transaction price of the apartments in the analyzed segment in order to eliminate the influence of the apartment floor space factor on the calculations.

1.3. Clarification of the location characteristics of the apartment, view characteristics, specifics of the transfer of ownership rights and purchase financing, the date of the offer, the internal condition of the apartment, rights for additional premises transferred with the apartment, for example, parking spaces for their further unification.

1.4. Calculation of the average values of unified prices for elite class apartments with installed "smart home" systems.

1.5. Calculation of the average values of unified prices for elite class apartments without "smart home" systems.

1.6. The difference between the last two values will give information on the cost of "smart home" systems.

2. *The algorithm of calculations in the presence of information on the offer to rent apartments can be as follows*:

2.1. Selection of elite or business class apartments for the purpose of determining the value of their rent under the conditions of full lease

2.2. Determination of the rental rate for 1 sq. m. of elite class apartments in order to exclude the influence of the apartment floor space factor on the calculations.

2.3. Clarification of the location characteristics of the apartment, view characteristics, specific conditions of the lease contract, the date of the offer, the internal condition of the apartment, the rights for additional premises transferred with the apartment, for example, parking spaces for their further unification.

2.4. Calculation of the average value of unified rental rates for elite class apartments with installed "smart home" systems.

2.5. Calculation of the average value of unified rental rates for elite class apartments without "smart home" systems.

2.6. Determination of the average annual cost of the utility payments and operating costs to eliminate the influence of the seasonality factor for apartments with installed "smart home" systems.

2.7. Determination of the average annual cost of utility payments and operating costs to eliminate the influence of the seasonality factor for apartments without "smart home" systems.

2.8. Determination of the cost for installation of smart home systems.

2.9. Determination of the discount rate for alternative projects.

2.10. Determination of the discounting period.

2.11. Determination of the present value of the increment of value due to the "smart home" systems, reduced by the cost of installation of these systems. To implement the presented algorithm, the calculation is carried out according to the following formula:

$$V^{SH} = -E_0 + \sum_{i=0}^{n-1} \frac{\left[\left(A_i^{SH} - A_i\right) + \left(E_i - E_i^{SH}\right)\right] \times S^{SH}}{\left(1 + \frac{Y_o}{q}\right)^{i \times q}} \qquad (1)$$

where V^{SH} — cost of the "smart home" system installed at the property object;

E_0 — initial costs for installation of the "smart home" system;

n — projection period in years;

IOP Conf. Series: Materials Science and Engineering **618** (2019) 012041 doi:10.1088/1757-899X/618/1/012041

A_i^{SH} — monthly rental rate per square meter of the floor space of the property with a "smart home" system;

A_i — monthly rental rate per square meter of the floor space of the property without a "smart home" system;

E_i — monthly service charges and maintenance costs per square meter of the property without a "smart home" system;

E_i^{SH} — monthly utility bills and maintenance costs per square meter of the property with a "smart home" system;

S^{SH} —_the floor space of the property with a "smart house" system;

Y_0 — the annual real general rate of return on the entire invested capital determined on a project with alternative risks;

q — the regularity of rental charges paid during the year, in this case it equals 12.

Consider the implementation of the proposed algorithm through an example. To implement the presented algorithm for estimation with transferable property rights, two apartments located in the prestigious district of St. Petersburg were chosen. The characteristics of these apartments are very similar and differ by the presence of a "smart home" system. The characteristics of these apartments and their offer prices are presented in Table 1.

The analysis revealed that in addition to the "smart home" system, in the sale of apartment 1 the right to a place in the parking lot is being transferred too [21]. Further studies of the real estate market in the surrounding area showed that the cost of the parking place is approximately 3.6 million rubles, which was taken into account in the calculations presented in Table 2.

Table 1. Characteristics of the analogues objects with transferable ownership rights

Parameter	**Data for objects**	
	Apartment 1	***Apartment 2***
"Smart home" system	Yes	No
Price of the offer, m. RUR	182	174
Floor space of the apartment, sq. m.	186	190
Date of the offer	April, 2017	April, 2017
Accepted negotiated discount	0.2	0.2
Transferable rights	Ownership, without encumbrance	Ownership, without encumbrance
Financing conditions	One-time payment, mortgage is possible	One-time payment, mortgage is possible
View characteristics	Yard and street	Yard and street
Floor / Number of storeys	4 / 9	5 / 8
The condition of interior decoration	excellent	excellent
Furniture	Built-in kitchen	Built-in kitchen
Additional elements of service	With a parking place in the parking lot of the house	Absent

Table 2. Calculation of the cost of the "smart home" system by using analogues objects with transferable property rights

Parameter	Data for objects	
	Apartment 1	***Apartment 2***
Price of the offer, m. RUR	182	174
Price of the transaction, m. RUR	145.6	139.2
Floor space of the apartment, sq. m.	186	190
Price of the place in the parking lot, m. RUR	3.6	-
The cost of the apartment, adjusted for the place in the parking lot, m. RUR.	142	139.2
Cost of the apartment, RUR / sq. m.	0.76	0.73
Difference in the cost of the apartment, RUR / sq. m	0.03	
Cost of the “Smart home” system, m. RUR	5.7	

Thus, the cost of the "smart home" system with the usage analogues objects with the transferable property rights is determined in the amount of 5.7 million rubles.

To implement the presented calculation algorithm with transferable tenancy rights, two apartments were chosen located in a prestigious district of St. Petersburg, with the most similar characteristics and which differ in terms of the "smart home" system. The characteristics of these apartments and the size of their rent are presented in Table 3.

Table 3. Characteristics of analogues objects with transferable lease rights

Parameter	Data for objects	
	Apartment 1	***Apartment 2***
“Smart home” system	Yes	No
Price of the offer, m. RUR	540	315
Floor space of the apartment, sq. m.	181	173
Date of the offer	April, 2017	April, 2017
Transferable rights	Rent (use and possession), without the right of residence registration	Rent (use and possession), without the right of residence registration
Financing conditions	Monthly advance payments, deposit in the amount of 3 rent payments	Monthly advance payments, deposit in the amount of 3 rent payments
View characteristics	Yard and street	Yard and street
Floor / Number of storeys	8 / 9	5 / 7
The condition of interior decoration	Excellent	Excellent
Furniture	Fully furnished	Fully furnished
Additional elements of service	Absent	Absent

IOP Conf. Series: Materials Science and Engineering **618** (2019) 012041 doi:10.1088/1757-899X/618/1/012041

When performing the calculations, it was taken into account that the nominal overall return on all invested capital determined for an alternative risk project is 25% per year in accordance with the stage of exit from a venture project and in accordance with the classification presented in [15]. The real rate of return is determined by Fisher's formula given the projected rate of increment in rental rates and utility payments in St. Petersburg by 5% per year, on average.

Based on the experience of the authors, the projected period was adopted as equal to 6 years. It is the period of the economic effect [16, p. 172]. In the calculations, it was considered that the rent was received monthly, at the beginning of each month. The results of calculations are presented in Table 4.

Thus, the cost of the "smart home" system with the usage of analogues objects with transferable tenancy rights is determined in the amount of 5.8 million rubles. The calculations made allow us to conclude that the cost of the smart home system is 5.7 – 5.8 million rubles.

Then, the entrepreneurial profit of the project for introducing the "smart home" system will amount to 2 million rubles. As shown in [17, p. 205], the amount of the contribution of any of the technological or organizational innovations in increasing the value of a business depends on the financial efficiency of the innovation. The internal rate of return (IRR) was chosen as a relative criterion of efficiency. The IRR was calculated for the projected time of net earnings receipt from the implementation of the "smart home" system, equal to six years. Note that the internal rate of return on the investment project for the introduction of the "smart home" system will be 50% of the project.

Table 4. Calculation of the cost of the "smart home" system using analogues objects with the transferable rent rights

Parameter	**Data for objects**	
	Apartment 1	*Apartment 2*
“Smart home” system	Yes	No
Cost of the rent, thous. RUR / month	540	315
Floor space of the apartment, sq. m.	181	173
Cost of the rent, thous. RUR / sq. m. / year	35.8	21.85
Average utility bills, thous. RUR / month.	21	27
Average utility bills, thous, RUR / sq. m. / annum	1.39	1.87
The installation cost of the "smart home" systems, m. RUR	3.6	-
Nominal discount rate, % p.a.	25%	
Real discount rate, % p.a.	19%	
Period of the discount, year	6	
Annuity payments, m. RUR / year	2.6	
The present value of incremental cost due to "smart home" systems, m. RUR.	9.4	
The present value of incremental cost due to "smart home" systems, reduced by the installation cost of these systems, m. RUR	5.8	

5. Conclusion

The paper analyzes the term of the Internet of things, on the basis of which a new definition of this concept is given.

The analysis of the world experience of the IoT introduction is presented and a conclusion is made about the possibility of deflationary development of economic processes due to the growth in efficiency of all participants of the IoT ecosystem.

A conclusion is made about the possibility to modernize the Russian economy by creating new markets and high-tech industries, including on the grounds of new business models based on IoT, which will lead, among other things, to improved quality of population's life.

One of the segments of IoT technologies was considered the "smart home" system. As a result of the analysis, it was revealed that the specific feature of the Russian market is an extremely low level of penetration of intelligent control systems for managing engineering systems in buildings and facilities.

An algorithm for calculating the cost of "smart house" technology is proposed, which allows investors to make informed decisions about investing money in the Internet of Things. The review of standard projects of the "smart house" system offered in St. Petersburg was conducted and a power law dependence was obtained for the unit cost of installation on the floor space of the object. The internal rate of return on the investment project for the introduction of the smart house system will be 50%.

Acknowledgment
The research carried out with the financial support of the grant from the Program Competitiveness Enhancement of Peter the Great St. Petersburg Polytechnic University, Project 5-100-2020.

References

[1] O.Y. Andreeva, Y.K. Batueva (2014) “Leaders of Innovation: Consumers of the Internet of Things”. Shumperter readings. Vol. 1. Pp. 89-94.

[2] M. Samsonov, A.Roslyakov, S. Vanyashin (2013) “From the Internet of People to the Internet of Things” , IKS. No. 5, p. 62.

[3] L.M. Martynov, S.I. Syzdaltsev (2013) “Management Development Vector in Competitive Conditions of Information and Communication Business Environment. Contemporary scientific thought”., No. 6, pp. 112-119.

[4] M.B. Yanenko, M.E. Yanenko, “Communication Marketing in Information Economy: Development Problems and Prospects of the Internet of Things”, Bulletin of Yaroslav-the-Wise Novgorod State University, 2014, No. 82, pp. 77-81.

[5] Internet of Things of Rob Van Kranenburg. Lecture in the context of Futurdesignlab [Electronic resource]. URL: http://design-union.ru/authors/theory/1490-internet-of-things-futurodesignlab (date of visit 09.04.17).

[6] Shabanov S. The Internet of Things: Myth or Reality? [Electronic resource]. URL: http://www.logistikaprim.ru/unpublish/internet-veshchei-mif-ili-realnost (date of visit 09.04.17).

[7] O.Y. Nikiforov, “Basic Technologies of the Internet of Things”, The Symbol of Science, 2015, No. 9-1, pp. 105-107

[8] M.A. Elizarov, “Development Prospects and Problems of the Internet of Things”, Nauka-Rastudent.ru, 2015, No. 12, p. 35.

[9] M. Draganov, O. Andreev, “Innovation City Marketing – Development Prospects of Concepts”, Information Technologies and Systems: Management, Economics, Transport, Law, 2015, No. 2 (16), pp. 143-149.

[10] Shekhovtsev M. Technology Shifts, the Internet of Things and Digital Economy. Report of J’son & Partners Consulting. Moscow, 2016 [Electronic resource]. URL: http://json.tv/ict_telecom_analytics_view/tehnologicheskie-sdvigi-internet-veschey-i-

tsifrovaya-ekonomika-20161209060802 (date of visit 09.04.17).

[11] Bril, A., Kalinina, O., & Ilin, I. (2017). Small innovative company's valuation within venture capital financing of projects in the construction industry. Paper presented at the MATEC Web of Conferences, , 106 doi:10.1051/matecconf/201710608010.

[12] Iliashenko, O. Y., & Shirokova, S. V. (2014). Application of database technology to improve the efficiency of inventory management for small businesses. WSEAS Transactions on Business and Economics, 11, 810-818.

[13] Price-Lists for Standard Projects of the company "Smart home KNX Showroom» in SPb: http://pro-smarthome.ru/

[14] Ozerov E.S., Pupentsova S.V. Value and Investment Potential Management in Real Estate. SPb. : Publishing House of the Polytechnic University, 2016, 602 p.

[15] Rosnano. Practical guidelines on assessing assets in terms of projects implemented with participation of the State Corporation "Russian Corporation of Nanotechnologies", part. 1. M., 2010.

[16] Bukharin N.A, Alekseeva N.S. Estimated Value of an Enterprise (Business). Assessment of Intangible Assets and Intellectual Property: Course Book. SPb.: Publishing House of the Polytechnic University, 2011, 301 p.

[17] Bukharin N.A., Ozerov E.S., Pupentsova S.V., Shabrova O.A. Estimation and Management of a Business's Value: Course Book. SPb.: Publishing House of the Polytechnic University, 2011, 238 p.

[18] A.A. Popov, "Creation of an Information System for Managing an Apartment Block Based on the Internet of Things", Bulletin of Plekhanov Russian University of Economics, 2015, No.2 (20), pp. 69-83

[19] Mokhireva, A., Loginova, P., Glebova, E., & Bespalov, V. (2018). Hardware base and experimental potential of a smart building prototype. Paper presented at the MATEC Web of Conferences

[20] Vavilov, D., Melezhik, A., & Platonov, I. (2015). Reference model for smart home user behavior analysis software module. Paper presented at the IEEE International Conference on Consumer Electronics - Berlin, ICCE-Berlin

[21] Nikolova, L. V., Kuporov, J. J., & Rodionov, D. G. (2015). Risk management of innovation projects in the context of globalization. International Journal of Economics and Financial Issues, 5(3S), 68-72.

[22] Gluhov, V., Leventsov, V., Radaev, A., & Nikolaevskiy, N. (2018). Analytical modeling of development and implementation of telecommunication technologies doi:10.1007/978-3-030-01168-0_39

[23] Glukhov, V. V., Ilin, I. V., & Lepekhin, A. A. (2018). Towards business optimization and development of telecommunication companies: Tools analysis and their adaptation opportunities doi:10.1007/978-3-030-01168-0_42

IOP Conf. Series: Materials Science and Engineering **618** (2019) 012042 doi:10.1088/1757-899X/618/1/012042

The artificial intelligence methods testing in case of engineering diagnostics system creation of the synchronous machines

V Polishchuk [1], K Baratova [2]

[1] Samara State Technical University, Samara, Russia, e-mail: polischuk_vi@mail.ru
[2] Polzunov Altai State Technical University, Altai Russia, e-mail: baratova.altai@list.ru

Abstract. Due to the lack of the exact mathematical processes description for internal faults diagnosis of synchronous machine rotor winding. To solve the problem of concurrent processing of the indirect diagnostic signs complex connected with the concrete type of multiple disabilities. It is the author's opinion that such problem is necessary to solve by artificial intelligence systems accepted in the theory and practice. The architecture of intelligent diagnostic system and the technical condition forecast of the synchronous machine rotor winding on the basis of the fuzzy logic mathematical tool are offered. Diagnosing reliability of and selectivity support determination of synchronous machine rotor winding defect category is reached by complex conjugation of separation of diagnostic information sensitive methods on the basis of intellectual digital signal processing methods. It is proved experimentally that the fuzzy logic using provided diagnosing reliability of synchronous machine rotor winding turn-to-turn short-circuit at the 1,5% level of rotor pole winding Visualization of all making decision stages about the existence and defect type was made in Fuzzy Logic Toolbox software package.

1. Introduction

The high performance system synchronous machines engineering diagnostics creation considering features of their construction and meeting the modern demand for technical and economic performances. It effectively works autonomous and as a part of automatic process control systems. It represents an urgent scientific and technical problem. The purpose of this paper consists of the allocation intellectual method development of selective diagnostic information about the turn-to-turn short-circuits on the basis of the fuzzy logic mathematical tool.

Using sensitive data to turn-to-turn short-circuit methods and the fuzzy logic mathematical tool to adjust the fuzzy inference algorithm for the flaw detection of the rotor winding turn-to-turn short-circuits. For this purpose the informative indirect indicators of turn-to-turn short-circuit in the alternator rotor, according to authors it is necessary to use regulating performance deflection and rotational frequency vibration level.

2. The research methods and techniques

Fuzzy logic using in the performance of a problem of selective flaw detection is based on interrelation between the defect and diagnostic signs allocated from the set of observable parameters. At the same time it is necessary to consider that each component is comprised of the element of fuzziness and also owing to features of obtaining primary information (a hindrance, a measurement error, etc.) and the conclusion could not be absolute ("accurate") [1].

Various parameters deviations and defects arising in of diagnostics object can influence the allocated diagnostic sign at the same time. At the same time defect influence on each of diagnostic signs can vary from any strong influence to total absence. For definition of extent of this influence in diagnostic system, the linguistic values of the validity presented to the table 1 are used.

Table 1. The linguistic values

The symbolic notation	Interpretation
NB (Negative Big)	Existence of steady diagnostic signs of considerable defect development
NM (Negative Middle)	Steady diagnostic defect signs existence
NS (Negative Small)	Existence of diagnostic signs of accident in a primitive state of development
PS (Positive Small)	The deviation of diagnostic signs at the admissible level
PB (Positive Big)	All parameters at the standard level

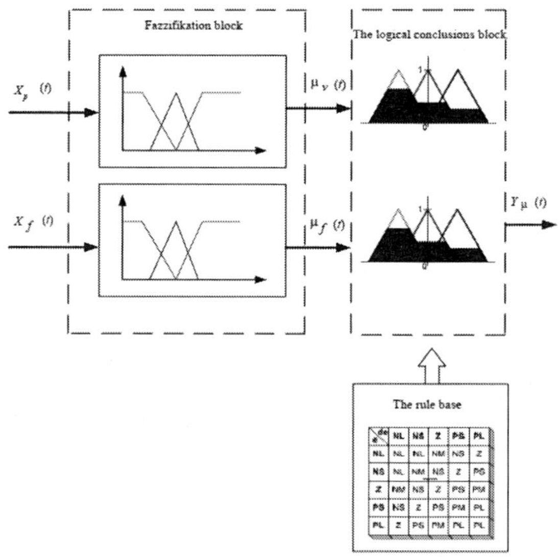

Figure 1. Turn-to-turn short-circuit diagnostic expert system of synchronous machine rotor winding on the basis of fuzzy logic.

It is necessary to set linguistic variables to use linguistic truth values. The linguistic variable is the variable which accepts values in the form of words and phrases of the natural language just like the algebraic variable accepts numerical values. The linguistic variable separate value is named as linguistic term. Term is a membership function of some set defined in the specified interval [2]. Each term of the linguistic variable can be considered as some fuzzy subset determined by the basic variable. For example, for the description of vibration existence we use Xf "Vibration", for the description of regulating performance deflection level of alternator we use Xp "Regulating performance deflection", for the description of turn-to-turn short-circuit existence we use "Turn-to-turn short-circuit". Let's say that these variables can accept values according to table 1:

Intelligent diagnostic system work of technical condition of alternator rotor on the basis of fuzzy logic is composed of two stages Pic. 1 [2]:

1) Fazzifikation is a transformation of input absolute values in linguistic value;

2) The logical conclusion with previously made knowledge base using.

Input terms are formed by the expert on the basis of the knowledge and sensitivity to defect of diagnostic signs.

The allocation method sensitivity of a diagnostic sign of turn-to-turn short-circuit in the alternator rotor winding based on the analysis of regulating performance deflection in 2,75 ... 3% of winding [3,6,9,11] has experimentally been proved. From there the following key values for the term "Regulating performance deflection" have been defined:

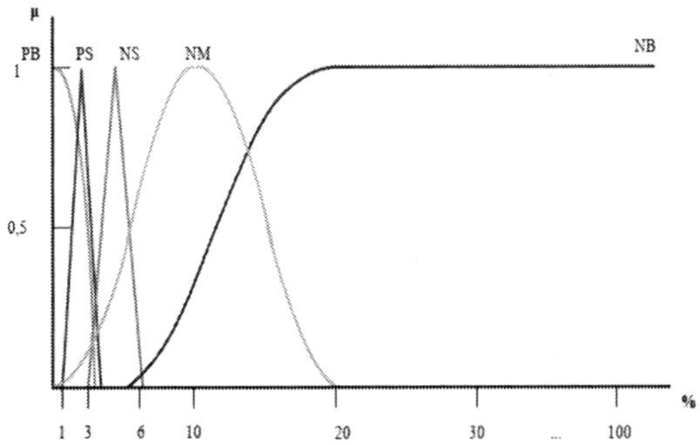

Figure 2. Input terms: "Regulating performance deflection"

When determining the term "Vibration" we recognize that this sign is confirming and has no accurate limit values. For vibration there is a standard value. The upward bias or downward bias needs to be perceived as emergence of changes in the machine, which can be a rotor winding defect. On that basis terms for "Vibration" have been defined:

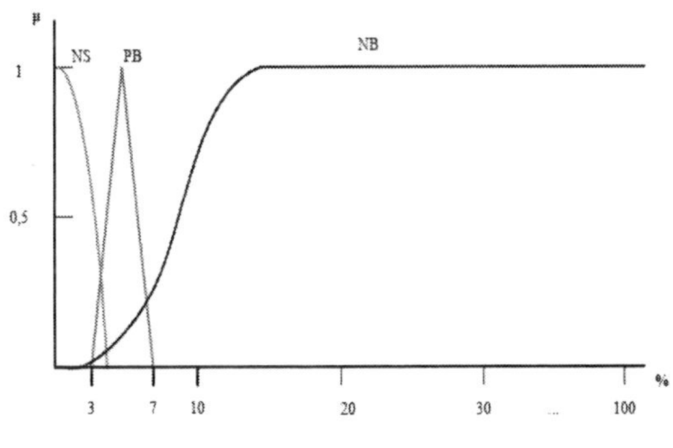

Figure 3. Input terms "Vibration"

For output variables it is also necessary to create terms. Terms for the output variable "Turn-to-turn short-circuit" proceeding from five linguistic values of a defect status, the table 1, we set the following terms:

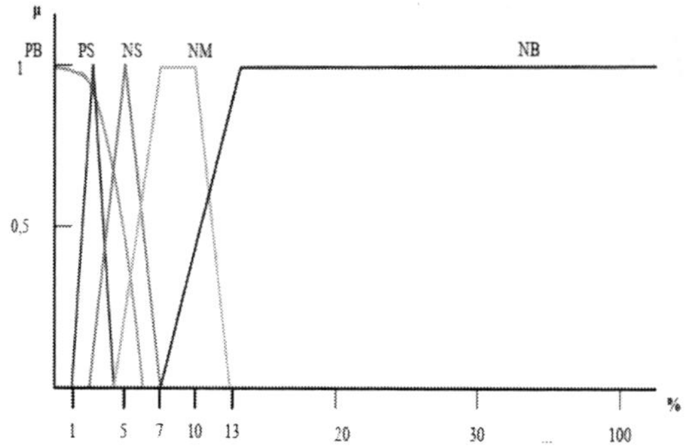

Figure 4. Output terms "Turn-to-turn short-circuit"

The interrelation between the diagnostic signs and defects is formed in reliance on rule base. Receiving the fuzzy inference is made on a specific algorithm. Basis for making operation of the fuzzy logical inference is the rule base containing fuzzy statements like "IF" - "THAT" and membership function to linguistic terms.

For the flaw detection "Turn-to-turn short-circuit" the rule base has been created, the table 2. [4,5].

Besides the rule base can be presented in the form of the structured text:

RULE_1: IF "Regulating performance deflection = PB" And "Vibration = PB" THAT "Turn-to-turn short-circuit = PB" (F1);

RULE_2: IF "Regulating performance deflection = PS" And "Vibration = PB" THAT "Turn-to-turn short-circuit = PS" (F2);

RULE_15: IF "Regulating performance deflection = PB" And "Vibration = NB" THAT "Turn-to-turn short-circuit = PS" (F15).

R.1 IF Xp is PB And Xf is PB THAT Y is PB (F1);

R.2 IF Xp is PS And Xf is PB THAT Y is PS (F2);

R.3 IF Xp is PB And Xf is NB THAT Y is PS (F15).

where F_i - reliability expert assessment of the existing rule. This coefficient defines the importance of the rule or confidence in truth degree of the decision received from fuzzy rules. [10]

Table 2. The rule base

№	Rejection of the regulating performance	Vibtation	Turn-to-turn short-circuit
1	PB	PB	PB
2	PS	PB	PS
3	NS	PB	PS
4	NM	PB	PS
5	NB	PB	PS
6	PB	NS	PB
7	PS	NS	PS
8	NS	NS	NS
9	NM	NS	NM
10	NB	NS	NB
11	PS	NB	PS
12	NS	NB	NS
13	NM	NB	NM
14	NB	NB	NB
15	PB	NB	PS

Rules from 1 to 15 have the compound conclusion. As the logic connective is presented by operator "And", that membership function of the statement in the conclusion is $\mu' = \min\{\mu(x_p), \mu(x_f)\}$.

Then all indistinct sets assigned for each term of each output linguistic variable integrate together and the single indistinct set, value for each removed linguistic variable [7 - 11] is created.

3. Experimental corroboration

In a generator charge coil the GAB-4-T/230 the tee off 4% of winding loops has been created. At their short circuit the diagnostic system has given result in the form of the fuzzy conclusion NS (Negative small, the existence of accident diagnostic signs in a primitive state of development, table 1) which is to say that steady sign of turn-to-turn short-circuit exists. Increase in vibration at 50% eccentricity was observed, but the system hasn't issued the conclusion about the existence of turn-to-turn short-circuit that means that the system works logically [12 – 21].

4. Conclusion

The diagnostic system of rotor equipment technical condition in this paper was developed, on the basis of a fuzzy logic which is capable to separate selectively diagnostic signs of the eccentricity and synchronous generator rotor winding turn-to-turn short-circuits.

References

[1]. Rutkovsky, Leshek., "Methods and technologies of artificial intelligence" transl. from Polish. / L. Rutkovsky. – Moscow: Goryachaya liniya-Telekom, 2010. – 520 p.: il. – Bibliogr.: p. 503-514. – Index: p. 515-517. – ISBN 978-5-9912-0105-6.

[2]. R. R. Yager, "Fuzzy sets and theory of opportunities. The last achievements", Radio and communication, 1986. – 408 p.

[3]. T.I. Rozum, V. I. Polishchuk "Identification method of the turn-to-turn short circuits in the synchronous generator winding", Fundamental'nye issledovaniya. – 2013. – No. 8 (part 5). – P. 1061-1065.

[4]. A.V. Leonenkov "Fuzzy modeling in MATLAB and fuzzyTECH" – SPb.: BHV-St. Petersburg, 2005. – 719 p.: il. – Decisions master. – Bibliogr.: p. 717-719. – ISBN 5-94157-087-2.

[5]. S. D Shtovba. "Design of fuzzy systems by means of MATLAB". – M.: Goryachaya liniya. – Telecom, 2007. – 288 p.

[6]. G. K. Voronovsky, K. V. Makhotilo, S. A. Sergeyev "Problems and prospects using theartificial neural networks in power system. Part I. Modeling", Problems of the general power system. – 2006. –№ 14. – P. 50-61.

[7]. L.A. Zadeh "Fuzzy algorithms", Information and Control. – 1968. – V. 12(2). – Pp. 94-102

[8]. N. N. Lebedev "Special functions and their applications": manual– 3rd prod., – SPb.: Lan, 2010. – 359 p.: il. – Textbooks for higher education institutions. Special literature. – Bibliogr.: p.355-358. – ISBN 978-5-8114-1023-1

[9]. V.I. Polischuk, V.L. Sergeev, "Adaptive Identification Method of a Signal from Stray Magnetic Field Sensor for Turbogenerator Diagnostics', Journal of Siberian Federal University,Mathematics and Physics 2015 8(2), pp 201-207, 2015.

[10]. D.A. Gnetova, V.I. Polishchuk "The research of fuzzy features of the SG rotor equipment defect", Ehlektroehnergetika glazami molodezhi 2016, VII Materials of the International youth scientific and technical conference 19 –23 September , 2016 , Part 3, pp 194-197, 2016V

[11]. V.I. Polishchuk, V.L. Sergeev, V.T. Kalayda "Models and Algorithms of Non-Stationary Signal Identification in Conditions of Uncertainty", XII International IEEE Siberian Conference on Control and Communications (SIBCON-2016), May 12–14, 2016, Moscow, Russia.D

IOP Conf. Series: Materials Science and Engineering **618** (2019) 012042 doi:10.1088/1757-899X/618/1/012042

[12]. Ageev, E. V., & Benderskaya, E. N. (2017). Natural computing: From natural phenomena to practical tasks. Paper presented at the Proceedings of 2017 20th IEEE International Conference on Soft Computing and Measurements, SCM 2017, 420-421. doi:10.1109/SCM.2017.7970604

[13]. Belenko, V., Krundyshev, V., & Kalinin, M. (2018). Synthetic datasets generation for intrusion detection in VANET. Paper presented at the ACM International Conference Proceeding Series, doi:10.1145/3264437.3264479

[14]. Dubolazov, V., & Somov, A. (2018). Interval approach of time series forecasting by neural networks for the decision support system. Paper presented at the MATEC Web of Conferences, , 170 doi:10.1051/matecconf/201817001014

[15]. Fisher, W. D., Camp, T. K., & Krzhizhanovskaya, V. V. (2016). Crack detection in earth dam and levee passive seismic data using support vector machines. Paper presented at the Procedia Computer Science, , 80 577-586. doi:10.1016/j.procs.2016.05.339

[16]. Korshunov, G. I., Nazarevich, S. A., & Smirnov, V. A. (2018). Fuzzy classification of technical condition at life cycle stages of responsible appointment systems. Paper presented at the CEUR Workshop Proceedings, , 2258 427-437.

[17]. Popov, A. V., Sayarkin, K. S., & Zhilenkov, A. A. (2018). Analysis of perspective models of artificial neural networks for control of robotic objects. Paper presented at the Proceedings of the 2018 IEEE Conference of Russian Young Researchers in Electrical and Electronic Engineering, ElConRus 2018, , 2018-January 958-961. doi:10.1109/EIConRus.2018.8317248

[18]. Tsvetkova, N. A., Tukkel, I. L., & Ablyazov, V. I. (2017). Simulation modeling the spread of innovations. Paper presented at the Proceedings of 2017 20th IEEE International Conference on Soft Computing and Measurements, SCM 2017, 675-677. doi:10.1109/SCM.2017.7970686

[19]. Utkin, L. V., Podolskaja, A. V., & Zaborovsky, V. S. (2018). A robust interval autoencoder. Paper presented at the Proceedings - 2017 International Conference on Control, Artificial Intelligence, Robotics and Optimization, ICCAIRO 2017, , 2018-January 115-120. doi:10.1109/ICCAIRO.2017.32

[20]. Utkin, L. V., & Ryabinin, M. A. (2018). A deep forest for transductive transfer learning by using a consensus measure doi:10.1007/978-3-319-71746-3_17

[21]. Zuev, D. V., Garbaruk, V. V., Khodakovsky, V. A., Fedorchuk, P. E., Blagoveshchenskaya, E. A., & Kunets, D. S. (2016). On neural network online learning algorithm. Paper presented at the Proceedings of the 19th International Conference on Soft Computing and Measurements, SCM 2016, 279-280. doi:10.1109/SCM.2016.7519753

The application of brain activity as a method for assessment of the human thermophysiological comfort and performance in cold indoor environment

R A Angelova[1, 2], E Georgieva[1], D G Markov[2], N Kehayova[2], I Simova[2], P Stankov[2], R Velichkova[2]

[1]Department of Textiles, Technical University of Sofia, 8 Kliment Ohridski Blvd., 1000, Sofia, Bulgaria
[2]Centre for Research and Design in Human Comfort, Energy and Environment, Technical University of Sofia, 8 Kliment Ohridski Blvd., 1000, Sofia, Bulgaria

joy_angels@abv.bg

Abstract. The paper presents a pilot study on the influence of cold indoor environment on the activity of the brain while solving mental tasks. The Neuro Sky Mind WaveTM technology was used to record EEGs during the cold exposure in two subzero environments. Tsai-Partington test was used to stimulate the brain activity during the state of thermophysiological comfort and during the cold exposure. The results obtained showed differences in the type of the emitted brain weaves and the total brain power used during the same activity in the cold.

1. Introduction

The human body has an extremely sophisticated thermoregulatory system that reacts to the temperature of the environment to keep the temperature of the internal organs around 37 °C. Without the protection of clothing and textile items, however, the thermoregulatory system could not cope with body protection at temperatures below 18 °C [1]: depending on the time of the exposition, metabolic and metabolism the low temperatures will lead to lighter or heavier cold-related health problems. Therefore, clothing is a significant factor for all human activities in the cold, being a protective barrier between the environment and the body and an active participant in the processes of heat and mass exchange between them [2, 3].

Despite the importance of the cold environment for the health, productivity and comfort of a substantial part of the population on earth, there are no standardized values of the required thermophysiological properties of low-temperature protective clothing for heat, air and moisture transfer [4]. An important reason for this is the interdependence of these properties as well as their dependence on the specific structural, geometric and mass characteristics of the textile articles and their combinations used in the garments [5]. The vast variety of options does not allow for standardized requirements [6]. Textiles and clothing, on the other hand, are only one aspect of thermophysiological comfort in a cold environment that cannot be considered in a complex way without an assessment of the environmental parameters (temperature, humidity, air velocity), metabolic activity and duration of exposure.

The thermophysiological reactions of the body in cold environment are initiated by the hypothalamus: the gland in the brain, which is responsible for constant comparison between the signals, coming from the blood, washing this part of the brain, and the signals, coming from the cold thermoreceptors in the skin. Therefore, the hypothesis that the brain waves measurement can give a piece of additional information on the processes inside the brain that control the thermoregulation of the human body, sounds logical.

The recorded electric potentials in the brain or electroencephalograms (EEGs) allow estimating the synaptic action related to behavior and cognition [7]. Thus, an EEG provides genuine "window on the mind," at large scales [8].

Neuro Sky Mind WaveTM is state-of-the-art technology for a recording of brain waves, created in 2009 and subject to improvements over the 2012-2013 period [9]. It has been already in use in a series of research: for emotional responses [10], the assessment of the level of meditation [11], the connection between the brain activity and adult and child reading [12], and the performance of individuals [13].

This paper presents a pilot study on the human thermophysiological comfort in the cold indoor environment via non-invasive measurements of the brain weaves. A single point EEG was used to evaluate the brain activity in an artificial cold chamber with two subzero temperatures (-1 °C and -11 °C). The brain activity and brain power were measured during completion of a Tsai-Partington test and assessed in conditions of thermal comfort, at the begging and the end of the cold exposure. The results obtained showed the measurable effect of the cold environment on the power and the type of brain activity.

2. Protocol

The experiments were performed in an artificial cold chamber (3x2x2.20m) with two subzero air temperatures (-1 °C and -11 °C). The air temperature and relative humidity in the chamber were continuously recorded during the experiment. A clothing ensemble with insulation of 1.53 clo was used [5]. Hands and heads remained uncovered. During the exposure, a thermal infrared camera, model FLIR E6 (FLIR Systems Inc., Wilsonville, OR, US) was used for remote recording of the surface temperature of the body (clothed and uncovered).

The volunteers in the experiment were three young female subjects with the following data for their height, body mass, and age: 1.60 m (SD 0.006), 52.7 kg (SD 3.05), 29.6 years (SD 3.05). According to the questionnaires none of them had diseases that require systemic use of drugs, including hormonal pills. Neuro Sky Mind WaveTM [9] was used to record single point EEGs during the tests, performed in the cold. The device explored the electrical signals generated by brain activity. The sensor was placed on the forehead (with a textile strap). An additional sensor (part of the device) was attached to the ear for measuring the heart rate.

The study was performed in two phases: Phase one (ambient temperature of -11 °C), and Phase two (ambient temperature of -1 °C) organized 48 h after the first phase. The duration of the exposure was 30 ± 5 min. Before and after the exposure each participant rested in an air-conditioned laboratory (22 °C, 65% RH) for 65 ± 5 min.

Brain wave recording was done during the completion of the Tsai-Partington test [14], used to assess the level of "alertness" and the performance. It refers to the sequential connection of the numbers 1 to 30, randomly placed in the volume of sheet A4, as shown in Figure 1. The number 1 is in the center. and during the connection of the numbers, the pen should not be lifted from the sheet. The test and the brain waves recoding lasts 30 seconds.

3. Results and discussions

Figure 2 presents an exemplary picture and the corresponding thermogram of one of the volunteers during the cold exposure: Tsai-Partington test accomplishment with the Neuro Sky Mind WaveTM device on the head. The signals obtained during the brain wave recording related to the electrical

IOP Conf. Series: Materials Science and Engineering **618** (2019) 012043 doi:10.1088/1757-899X/618/1/012043

potential (potential energy) of the brain waves (μV2) and the power, defined as energy per second ($\mu V^2/s$). The raw signal of the EEG was processed afterwards by additional software [13, 14].

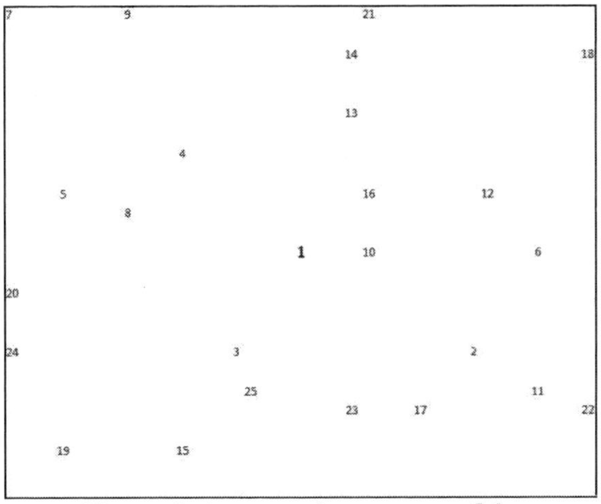

Figure 1. An exemplary pattern of the Tsai-Partington test

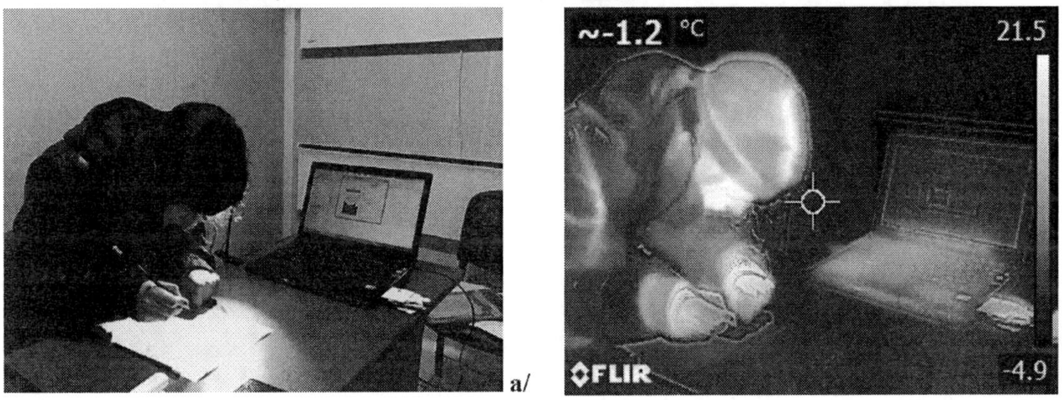

a/ b/

Figure 2. Tests and brain waves recording during the cold exposure: a) picture; b) thermogram.

Brain waves of different frequencies that correspond to a variety type of activity of the cortex were measured. The correspondence between the type of the brain wave and the human physiological reactions are summarized as follows [13]:

- Delta brain waves: 0.5-2.75 Hz; appear in states of deep sleep (no dreams) and unconsciousness, associated with subconscious thoughts and information.
- Theta brain waves: 3.5-6.75 Hz; appear in states of mild sleep or creative activity, related to intuition, fantasy, imagination.
- Alpha brain waves - low and high frequency:
  - Low-frequency alpha (7.5-9.25 Hz); appear in tranquility in a conscious state.
  - High-frequency alpha waves (10-11.75 Hz); appear in waking states of relaxation.
- Beta brain waves - low and high frequency:
  - Low-frequency (13-16.75 Hz); appear in conditions of a motor and sensory rhythm, attention, tranquility but with the ability to concentrate.
  - High-frequency (18-29.75 Hz); appear in concentration or agitation.
- Gamma brain waves: 31-49.75 Hz; appear in states of high activity.

All types of brain waves occur simultaneously but with different amplitudes. The cumulative value of brain waves represents the total value of brain power.

3.1 Results from the acclimatization

The preliminary preparation for the cold exposure of the subjects includes acclimatization in an air-conditioned laboratory (22 °C, 65% RH). The participants solved a Tsai-Partington test. Figure 3 shows the EGG of Individual 1 in the two phases of the measurements.

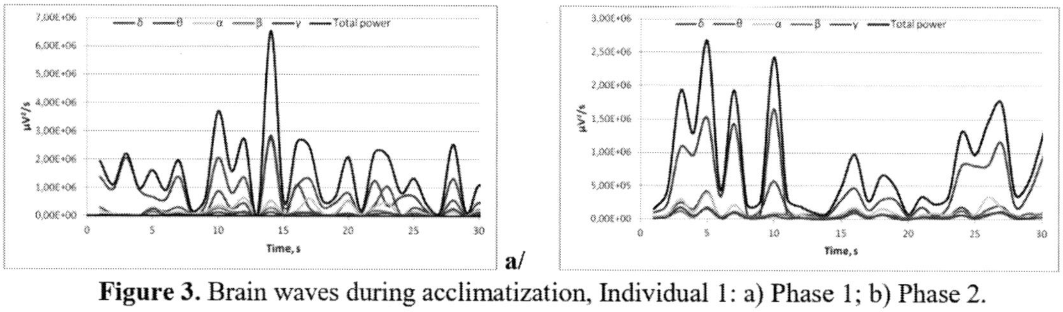

Figure 3. Brain waves during acclimatization, Individual 1: a) Phase 1; b) Phase 2.

The results obtained present short-time modulations about background levels, which is believed to be generated largely by synaptic action fields [7]. The analysis of the graphs shows that the total energy from the brain activity is determined mainly by the dominating delta waves. The same was found for the other two participants. This result coincides with the conclusions in [15-16]. The peaks appearing in Figure 3, as well as the following figures for solving the Tsai-Partington test correspond to finding a specific test digit. If the detection the next digit in the test was easier, a smaller amount of brain energy was consumed. For example, 13 peaks were recorded in Fig. 3a and the participant in the test succeeded to connect 13 consecutive numbers.

3.2 Results from the study in the cold environment

The Tsai-Partington test was completed twice: approx. 7 min and 21 min after the beginning of the cold exposure, conditionally considered as "at the beginning" and "at the end" of the exposure.

Figures 4-6 present the EGGs at the beginning and the end of the cold exposure of the three participants in Phase 1 (-11 ° C; 1.53 clo). Unlike the reference EGGs in the acclimatization (a comfortable environment), both delta and theta waves brightly expressed, together with the appearance of alpha, beta and gamma waves.

To compare the results of Figures 4-6, the mean values of the total energy, emitted from the brain cortex of each participant during acclimatization (reference value) and at the beginning and the end of the cold exposure, were calculated. The results in Fig. 7 clearly show that the brain power during the Tsai-Partington test was lowest in thermal comfort conditions during the acclimatization. Solving the same test in the cold environment, Individual 1 increased with 32% her brain energy, Individual 2 – with 29%, and in Individual 3 – with 192%. It should be noted that the reference value of the total brain energy for Individual 3 had the lowest value during acclimatization.

The protective ability of the clothing ensemble during the cold exposure decreases with time. The sedentary position of the participants, performing the tests, inhibited the heat production due to muscles work. Therefore, the cold load increased with the length of the exposure, and for two of the participants, there was a difference in the total brain power at the beginning and the end of the stay in the cold.

The change in the mean value of the total brain power is most striking for Individual 2: at the end of the exposure, the test was performed at a 48% increment of the total brain power compared with the total brain power at the beginning of the exposure. For Individual 1, there was a decrement of 11%. For Individual 3 there was a slight increment of 1%.

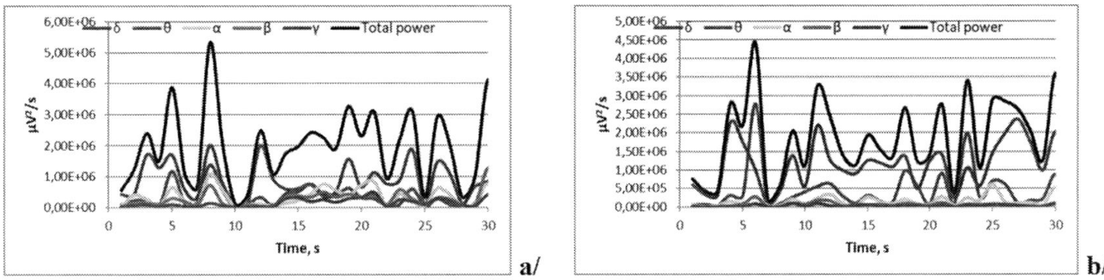

Figure 4. Brain waves during cold exposure (-11 °C; 1.53 clo), Individual 1: a) beginning of exposure; b) end of exposure.

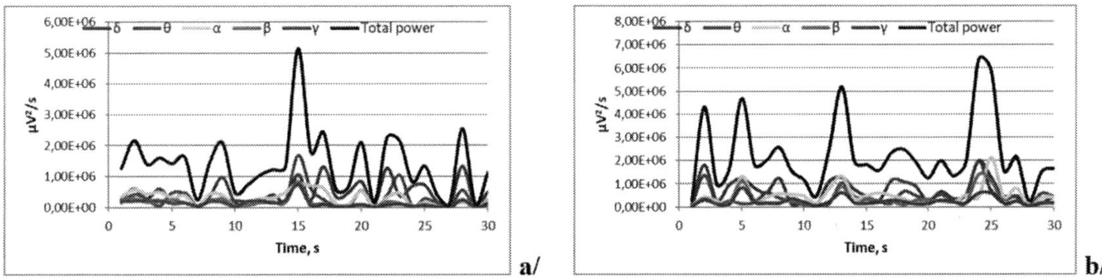

Figure 5. Brain waves during cold exposure (-11 °C; 1.53 clo), Individual 2: a) beginning of exposure; b) end of exposure.

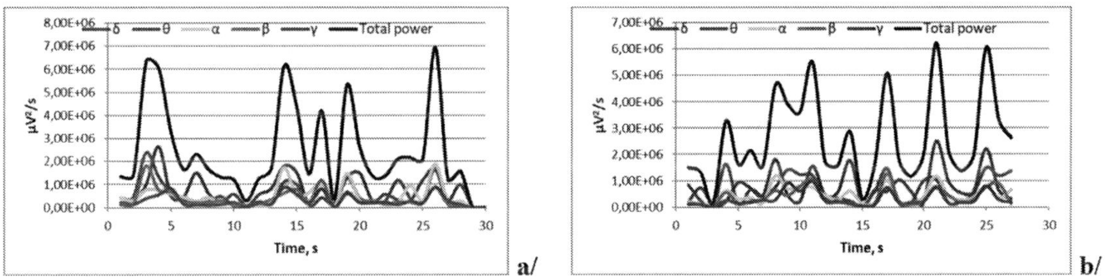

Figure 6. Brain waves during cold exposure (-11 °C; 1.53 clo), Individual 3: a) beginning of exposure; b) end of exposure.

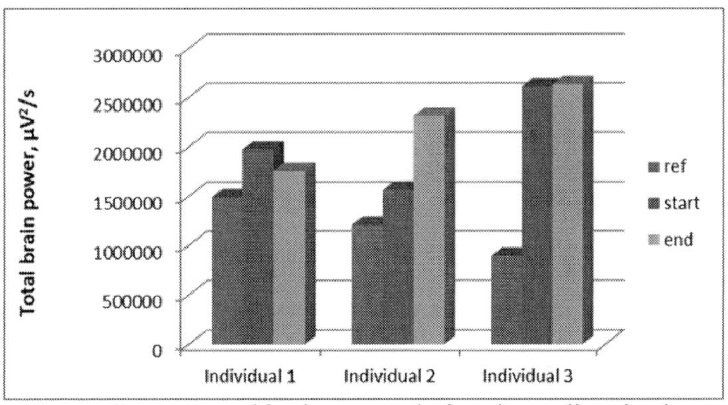

Figure 7. Mean total brain power during the acclimatization, at the beginning and at the end of the cold exposure (-11 ° C, 1.53 clo)

4. Conclusion

As a result of the present study, it can be concluded that the cold environment influences brain activity and type of brain waves. The total brain power required to perform a particular mental task in a cold environment is higher than the energy needed to complete the same task in the conditions of thermophysiological comfort. Additional research is required in order to assess the effect of the duration of the exposure in the cold environment on the total brain power, especially in the direction of increasing the exposure interval and number of participants.

Acknowledgments

The publication is supported by Bulgarian Science Fund of the Ministry of Education and Science ДН17 /12/12.12.2017.

References

[1] Havenith G 2002 The interaction of clothing and thermoregulation *Exogenous Dermatology* **1** (5) pp 221-230

[2] Morrissey M and Rossi R M 2013 Clothing systems for outdoor activities *Textile Progress* **45** (2-3) pp 145–181

[3] Celcar D 2014 A subjective evaluation of the thermal comfort of the clothing evaluated in cold environment *Glasnik Hemicara, Technologa i Ekologa Republike Srpske* **5**(10)

[4] Angelova R A 2017 Chapter Working in cold environment: Clothing and Thermophysiological comfort In *Occupational Health* InTechOpen Croatia

[5] Angelova R A et al 2016 Heat and mass transfer through outerwear clothing for protection from cold: influence of geometrical, structural and mass characteristics of the textile layers. *Textile Research Journal* **87**(9) pp 1060-1070

[6] Holmer I 2009 Evaluation of cold workplaces: an overview of standards for assessment of cold stress *Industrial Health* **47** (3) pp 228–234

[7] Nunez P L and Srinivasan R 2006 *Electric fields of the brain: the neurophysics of EEG* 2nd edition Oxford University Press NewYork

[8] Nunez P L and Srinivasan R 2006 A theoretical basis for standing and traveling brain waves measured with human EEG with implications for an integrated consciousness *Clinical neurophysiology* **117**(11) pp 2424-2435

[9] Introduction of NeuroSky MindWave, Retrieved from https://www.mindtecstore.com/mediafiles/Sonstiges/Shop/MWM2/MWM2%20introduction.pdf

[10] Crowley K et al 2010 Evaluating a brain-computer interface to categorise human emotional response *10th IEEE Int Conf on Advanced Learning Technologies* pp 276-278

[11] Guðmundsdóttir K 2011 Improving players'control over the NeuroSky brain-computer interface *School of Computer Science* Reykjavik Iniversity

[12] Mostow J et al 2011 Toward exploiting EEG input in a reading tutor, AIED'11 Proc of 15th Int Conf on Artificial Intelligence in Education

[13] Simova I et al 2015 On the occupants' performance measured by a single-point electroencephalography *Proc of Int Conf of the Faculty of Power engineering, Sozopol vol 2*

[14] Markov D et al 2014 On the influence of indoor temperature on occupant's performance *Ruse University Science Works* **53** pp 159-164

[15] Basar E et al 2001 Gamma, alpha, delta, and theta oscillations govern cognitive processes *Int. J. Psychophysiol.* **39** pp 241–248

[16] Schurmann M et al 2001 Delta responses and cognitive processing: single-trial evaluations of human visual P300 *Int. J. Psychophysiol.* **39** pp 229–239

IOP Conf. Series: Materials Science and Engineering **618** (2019) 012044 doi:10.1088/1757-899X/618/1/012044

Kinematic study of the articulated trucks operating layout of turn for articulated vehicles

Y P Sofronov[1], Y P Stoyanova[2], N E Kopralev[1] and G D Todorov[1]

[1]Faculty of Industrial Technology, Laboratory „CAD/CAM/CAE in Industry",
Technical University of Sofia, Sofia, Bulgaria,

[2]Faculty of Industrial Technology, Department "Theory of mechanisms and machines"
University of Sofia, Sofia, Bulgaria,

E-mails: ysofronov@tu-sofia.bg, yast@tu-sofia.com, nkopralev@3clab.com, gdt@tu-sofia.bg

Abstract. In this article, a kinematic analysis of articulated forklift truck turning has been made. Articulated steering systems separate the vehicle into a front and back sections, with a pivot (articulated) point directly joining them in the middle. This allows them to be rotated one compared to the other. In most cases, the turning of the vehicle is accomplished when force is applied to front section, via hydraulic cylinders thus creating torque in the articulated joint or directly applying torque via electric/ hydraulic motor. This allows such vehicles to have greater manoeuvrability and flexibility compared to the conventional forklift trucks making them preferable in the modern warehouses, which allows reduction in its isle width and as such increase in the its overall capacity which is of utmost importance. The minimum width of aisle and junction allowing articulated forklift truck with set dimensions to properly pass are been evaluated.

1. Introduction

According to the dictionary, articulated means jointed or segmented. This definition agrees well with the common impression of an articulated vehicle. Such a vehicle consists of two or more body or frame units jointed together. These units or sections should be an integral part of the vehicle. The joints may have one, two or three degrees of freedom (yaw, pitch and roll) [1, 2].

In comparison to equipment with articulated frame steering, equipment with single-axle steering and identical undercarriage geometry has the significant disadvantage that it is considerably less flexible and maneuverable and thus possesses lower performance capability [3].

With the articulated forklift, the operator can see the load being handled well than a reach truck operator can see his load. This is because here the operator will pick and place loads at a 90° angle to the machine. Visibility from the side is unobstructed when compared to a reach truck operator trying to see through the mast and reach mechanism. In fact, reach trucks operators can often be seen leaning outside of the operator's compartment to get a better view of the forks or the load being handled [4].

IOP Conf. Series: Materials Science and Engineering **618** (2019) 012044 doi:10.1088/1757-899X/618/1/012044

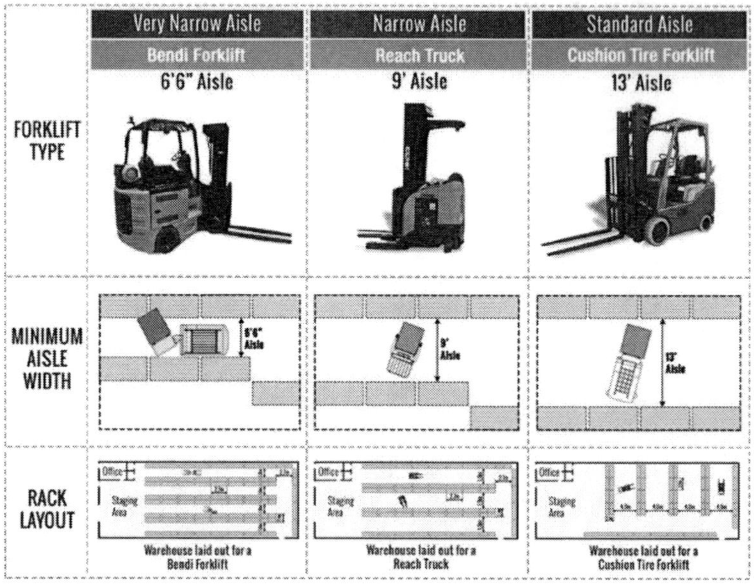

Figure 1. Comparison between articulated-, reach trucks and counterbalance forklift trucks.

Outside of the aisle, the operator's visibility can be improved in much the same way as it is in the aisle by picking and placing loads at an angle instead of straight ahead, whenever possible. The reach truck operator is always working from a position behind the mast and reach mechanism where visibility is obstructed [4].

Another big advantage over reach trucks, depending on the application, is that the articulated truck can be configured to load and unload trailers. This trailer to rack capability eliminates the expense of additional equipment and personnel needed to operate that equipment [4].

A reach truck requires staging areas for temporary storage between the dock and the rack. Because the articulated truck can work trailer to rack and back, staging areas are not required and additional pallet positions are created. Reach trucks are rarely able to load and unload trailers because of their small front wheels and low ground clearance [4].

Aim of this study is determine what should be the narrowest aisle in which the articulated forklift, with defined geometry, could operate. Usually in technical specification this is presented as steady state geometry without taking into account envelop of turning. Geometric modeling and calculations [5] provide insight into the details of how products interact with the environment, and allow new products to be evaluated in the computer, even before prototypes have been built. This is driven by the Virtual Prototyping (VP) ability to cut down developmental costs by minimizing physical testing which can result in considerably reduced design cycle time and design costs [6].

2. Exanimated model

The simplified geometric model of an existing articulated forklift truck in straight position is shown in figure 2. The geometric parameters (in meters) of the devised model are shown on Table 1.

Process of moving of such vehicles consists of two independent variables [7] and nonholonomic relations should be used to present it. The first one is the movement of the front side where traction motors are installed and second one is steering angle produced by operator with steering wheel. Steering wheel with transfer mechanisms produces torque around pivot point.

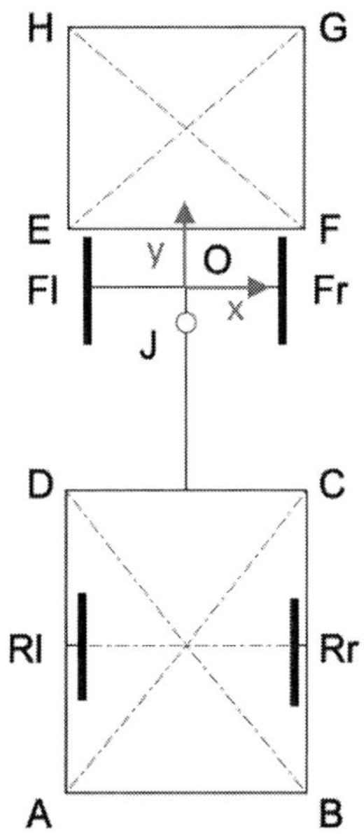

Figure 2. Defined pallet and chassis by border points.

Point	X	Y
A	-0,614	-2,514
B	0,614	-2,514
C	0,614	-1,007
D	-0,614	-1,007
E	-0,6	0,2885
F	0,6	0,2885
G	0,6	1,2885
H	-0,6	1,2885
Fl	-0,49	0
Fr	0,49	0
Rl	-0,54	-1,78
Rr	0,54	-1,78

Table 1. Point coordinates

Combinations of these variables produce trajectories which could be calculated using MS Excel spreadsheet [8]. The rear of the articulated truck is a trailing link attached to the articulated joint of the vehicle and has two degrees of freedom: it could roll in the direction of its centre-line, while also rotating about the centre of the rear axle group. Determining the trajectories [9] traced out by the border points on the vehicle is based on being able to calculate how the position of the centre of the rear axle varied as the articulated joint moves. Where the front of the forklift travelled in a straight line, the centre of the rear axle is following a curve. This type of trailing-link motion is modelled by a succession of small increments of the movement of the whole machine.

The analysis used here did not take into account the dynamic effects and finite displacements of the truck's reference points were used to calculate each new position respectively.

3. Analytical methodology

Modelling of the trajectory was done using following methods.

Rotation

This subroutine allows a point to be rotated about another point and makes use of trigonometric principles in a straightforward way.

IOP Conf. Series: Materials Science and Engineering **618** (2019) 012044 doi:10.1088/1757-899X/618/1/012044

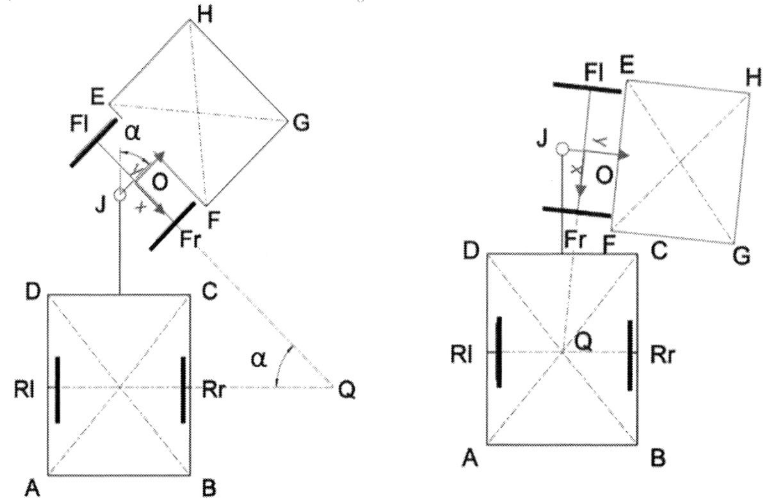

Figure 3. Defined pallet and chassis by border points.

Trail

This subroutine has been written to calculate the new position of the forklift's rear end when the front is moved through a distance in a straight line. The underlying principle is that, for an infinitesimal displacement of the head in any direction, the new position of the rear will lie on a straight line that joins the original position of itself to the new position of the front.

Rotation by Radius

For the articulated truck – or indeed for any non-articulated, front-axle-steered vehicle – there will be a specific turning radius associated with any particular position of the steering wheel (or amount of steering). The turning centre (or centre of rotation) lies on the centre-line of the rear. The turning centre also lies on the axis of each of the steered front wheels.

Subroutine Rotation by Radius calculates the new position of a reference point that is rotated about a centre with a given radius and through a given trajectory distance. It also calculates the coordinates of the centre of rotation and the direction in which the trajectory is pointing at the end of the travel

Steer

Subroutine Steer builds on the functionality of subroutine RotationByRadius, which it calls, and allows two inverse turning radii to be specified: one at the beginning and the other at the end of the travel. Within the subroutine it is assumed that the inverse turning radius varies linearly with the distance travelled: this is a fair approximation if the steering wheel is being turned at a steady rate with respect to distance travelled

SteerAndTrail

Subroutine SteerAndTrail, like subroutine Steer, builds on the functionality of subroutine RotationByRadius, which it calls, and allows beginning and ending inverse turning radii to be specified. However, SteerAndTrail calculates the new position and direction of the rear at the end of the movement by using the functionality of Trailing, which it also calls.

IOP Conf. Series: Materials Science and Engineering **618** (2019) 012044 doi:10.1088/1757-899X/618/1/012044

4. Results from performed analysis

For the examined problem here, a spreadsheet was used to determine the minimum isle width that would be required for an articulated truck to properly and safely operate in. The situation that is observed is the one that is most intriguing for the warehouse planner – a 90o corner for the forklift to steer from one isle to another perpendicular to it.

Using the built-in function MAX the X and Y coordinates of all the points of the articulated truck including the standard pallet with dimensions 1200x1000mm were traced in time of the truck's steering action, thus creating contours whose envelop can be used to determine the minimum isle width.

Diagram showing the paths envelop of the truck in proper scale is shown on figure 4. The calculated dimensions for the isle's intercrossing are shown in table 2.

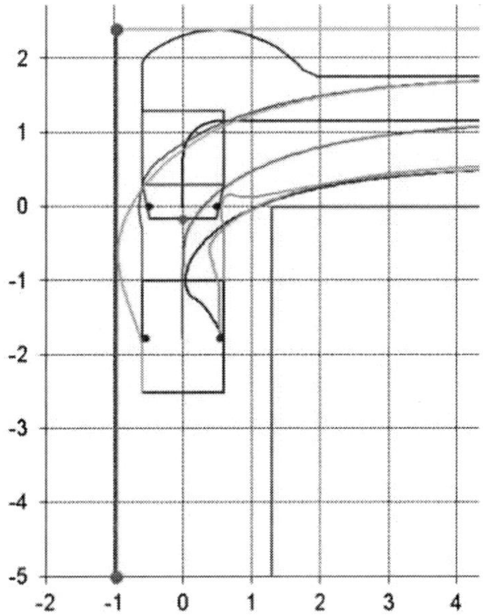

Figure 4. Generated trajectories of the reference points

Table 2. Coordinates of the calculated isle cross-section

Point	X	Y
Corner	1.3	-6
Right most	1.3	0
Bottom most	10	0
Left most	-0.952	-6
Top most	-0.952	2.379
	10	2.379
Incoming Aisle	2.252	-
Outgoing Aisle	2.379	-

5. Conclusion

This paper demonstrates the use of a spreadsheet application to solve engineering problems relating to the turning or steering of articulated forklift trucks. Analytical methodology is developed to study the turn envelop using steering kinematics for articulated forklift truck and the width of narrowest aisle in defined dimensions of the truck have been calculated. It is found that by adjusting the articulated joint position, the wheelbase and the distance between wheel groups contact centres, the distance between the isles in a warehouse can be significantly decreased – increasing the warehouse capacity and reducing the cost for pallet for the storage. The developed methodology could be used to optimize geometric parameters for vehicles with similar kind of steering kinematics and applications.

Acknowledgments
This work was supported by the European Regional Development Fund within the Operational Programme "Science and Education for Smart Growth 2014 - 2020" under the Project CoE "National center of mechatronics and clean technologies "BG05M2OP001-1.001-0008- C01.

References
[1] D-H Wu, "A theoretical study of the yaw/roll motions of a multiple steering articulated vehicle," in Proceedings of the Institution of Mechanical Engineers, Part D: Journal of Automobile Engineering, Great Britain, 2001, pp. 1257-1265.
[2] A J P Miège and D Cebon, "Active roll control of an experimental articulated vehicle," in Proceedings of the Institution of Mechanical Engineers, Part D: Journal of Automobile Engineering, Great Britain, 2005, pp. 791-806.I.S. Jacobs and C.P. Bean.
[3] Piotr A. Dudzinski, "Design Characteristics of Steering Systems For Mobile Wheeled Earthmoving Equipment," *Journal of Terramechanics*, vol. 26, no. 1, pp. 25-82, 1989.
[4] http://narrowaisleinc.com/
[5] Todorov G., K.Kamberov, B.Romanov, Ts.Ivanov, Safety and Reliability Assesment of Forklift Cabine Based on Virtual Prototype, Journal of Proceedings of the International Conference on Manufacturing Systems – ICMaS, Vol. 4, ISSN 1842-3183, Bucharest, Romania, 2009, pp 347-352
[6] Kamberov K., Zlatev B., Todorov T., Design development of a car fan shroud based on virtual prototypes, FABULOUS 2019 - 4th EAI International Conference on Future Access Enablers of Ubiquitous and Intelligent Infrastructures, Sofia, Bulgaria, March 28-29, 2019
[7] Todorov G., Kamberov K., G. Kyurkchiev, Parametric optimisation of flywheel design, Journal of the Balkan Tribological Association 24(3), pp. 390-399, 2018
[8] Jim McGovern, "Turning an Articulated Truck on a Spreadsheet", School of Mechanical and Design Engineering, Dublin Institute of Technology.
[9] Malakov, I., Zaharinov, V., Tzenov, V.: Size Ranges Optimization. Procedia Engineering, Vol. 100 (January), ISSN 1877-7058, pp. 791-800, (2015).

IOP Conf. Series: Materials Science and Engineering **618** (2019) 012045 doi:10.1088/1757-899X/618/1/012045

Analysis of some problems in the theoretical wagon strength studies due to the imperfection of the European legislation

V Stoilov[1], **S Slavchev**[1], **V Maznichki**[1, 2], **S Purgic**[1]

[1] Department of Railway Engineering, Technical University Sofia, Kliment Ohridski blvd. 8, Sofia 1000, Bulgaria

[2] v.maznichki@tu-sofia.bg

Abstract. In this paper the strength calculations of a flat wagon are presented. These calculations are carried out in accordance with European standards, but in practice, often, calculations and tests are carried out by a shortened procedure not in accordance with the standards. On the other hand, the procedures prescribed in the standards do not always cover all aspects of the actual operating conditions. Based on the satisfactory results of the calculations, the notified body often admits the wagon into service. Under real working conditions it is established that the structural integrity of the wagon underframe is impaired due to loads not clearly reflected in the standard and are therefore not taken into account either in the calculations or when the wagon is admitted into service. Some activities have been done in order to reinforce the problem areas and a repeated extended strength analysis has been carried out with the loads not described in the normative documents. Subsequently, after the reinforcing measures had been taken, the wagon was admitted back into operation. A proposal for specification of such problems in the European standard EN 12663-2: 2010 is given.

1. Introduction

The strength calculations of the wagons and their components, such as the underframe, wagon body, bogies, axles, wheels, etc. are carried out according to methodologies and calculation procedures regulated in the relevant European standards, Technical Specifications for Interoperability (TSIs) and/or UIC leaflets. Today, to a large extent, these normative documents cover the calculation methods, but most of these methods after processing have been adapted in European standards. For example, in TSI - Rolling Stock Freight Wagons [1] have mainly left over general requirements for wagon structures. Prior to the introduction of a wagon in operation, whether new or modified, the strength calculations must be made. The new wagon structures or their components shall then be tested. For wagons, the methodologies and procedures for strength analysis and testing are defined in the European standard EN 12663-2: 2010 - Railway applications - Structural requirements of railway vehicle bodies - Part 2: Freight wagons [2]. In addition to the load cases it also defines the minimum requirements for wagon body strength and associated equipment such as: roof, side and front walls, doors, stanchions, fasteners and other equipment. It provides data on materials, identifies their use, and presents principles and methods that are used to validate the structure by strength analysis and testing.

In practice, calculations and tests are often carried out in shortened procedure that is not in accordance with the standard. Usually this is due to the client's desire or it is approved by the notified body. On the other hand, the procedures prescribed in the standards do not always cover all aspects of the actual operating conditions. In this particular case, this concerns the impact of additional equipment - the stanchions - on the underframe strength of the wagon series Smnps 194. In EN 12663-2: 2010, load cases for strength calculations and strength requirements on wagon underframe and on stanchions and their fastenings are clearly described. But nowhere is exactly stated that the impact of stanchion load on the wagon underframe should be taken into account. The first part of the paper presents the subject of the study - a flat also wagon equipped with stanchions, as well as the results of the stress-strain analysis of the wagon underframe and the side stanchions. Based on the results of these calculations and other validation documents, the notified body has commissioned the wagon into operation. In the course of the operation, it is established that the structural integrity of the wagon underframe is impaired due to loads not taken into account either in the calculations or when the wagon is admitted into service as they are not specified in the standard. These problems occur in the areas of stanchions connections with the wagon underframe and are described in the second part of the paper. Once problems and their causes have been identified, measures have been taken to reinforce the problematic areas. This would not have happened if the standard specified that it was also necessary to check the impact of the equipment on the underframe strength. It should be noted that during testing the stanchions are tested as a standalone object subjected to the regulated loads and not in a combination with wagon underframe [7].

2. Subject of the study and initial results

The subject of the study is a wagon series Smrps 194, which is a modification of the wagon series Smnps 193. The three-dimensional wagon model is shown in Figure 1.

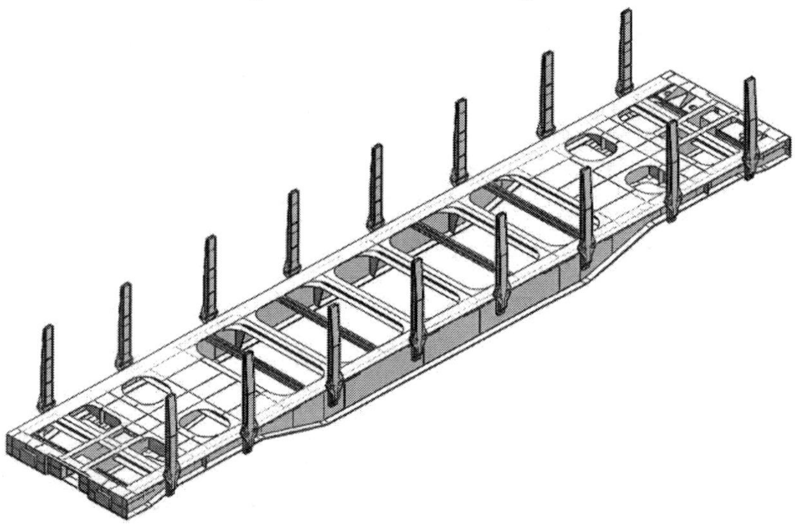

Figure 1. Three-dimensional model of wagon series Smnps 194.

The main parameters of the wagon are: gauge G1, number of axles 4, axle load 22,5 tons, payload capacity about 68 tons, own weight 22 tons, base 11,7 m, wagon length without buffers 15,5 m. The underframe of the wagon is made of steel S355J2 according to EN 10025. The theoretical studies are made by the finite element method. The software product SolidWorks is used for calculations. For the strength analysis spatial computational models were developed by using 3D solids elements. The approximation of the solution has been examined, with the optimized model being made up of 344 495 notches and 177 678 finite elements. The maximum size of the finite elements is 80 mm and the minimum (in the zones with expressed stress concentrators) is 10 mm.

2.1. Load cases

In accordance with international standards, each newly designed or modified wagon is object of theoretical (by calculation) and each newly designed wagon is object of experimental studies in order to determine stresses and deformations. Load cases and conditions are specified in European standard EN 12663-2:2010 [2], TSI subsystem "Rolling stock – Freight wagons" [1] and leaflet 577 [3] of the

IOP Conf. Series: Materials Science and Engineering **618** (2019) 012045 doi:10.1088/1757-899X/618/1/012045

International Union of Railways (UIC). According to documents cited loads are divided into the following groups: longitudinal static loads (HLC), vertical static loads (VLC), exceptional static loads (LLC) and superposition of static loads (CHLC). Additionally, three load cases (SLC) as defined in subsection 7.10 of European standard EN 12663-2:2010 [2] are applied to high strength stanchions. An overview of all load cases is shown in table 1.

Table 1. Load cases applied to wagon construction.

Load case	Description
HLC 1	Compressive force at buffer axis height 2000 kN (Table 2 of [2])
HLC 2	Compressive force 50 mm below buffer centre line 1500 kN (Table 3 of [2])
HLC 3	Compressive force applied diagonally at buffer level 400 kN (Table 4 of [2])
HLC 4	Tensile force at coupler axis level 1500 kN (Table 5 of [2])
VLC 1	Vertical load 30t distributed on area with width 2 m and length a-a = 1.3 m
VLC 2	Vertical load 39t distributed on area with width 2 m and length b-b = 5.0 m
VLC 3	Vertical load 60t distributed on area with width 2 m and length c-c = 7.8 m
VLC 4	Vertical load 68t distributed on area with width 2 m and length d-d = 9.4 m
VLC 5	Vertical load 68t distributed on area with width 2 m and length e-e =11.7 m
VLC 6	Vertical load 68t distributed on area with width 2 m and length f-f =13.5 m
VLC 7	Vertical load 30t distributed on whole width of wagon and length a-a = 1.3 m
VLC 8	Vertical load 39t distributed on whole width of wagon and length b-b = 5.0 m
VLC 9	Vertical load 60t distributed on whole width of wagon and length c-c = 7.8 m
VLC 10	Vertical load 68t distributed on whole width of wagon and length d-d = 9.4 m
VLC 11	Vertical load 68t distributed on whole width of wagon and length e-e =11.7 m
VLC 12	Vertical load 68t distributed on whole width of wagon and length f-f =13.5 m
LLC 1	Lifting at one end of the vehicle with most unfavourable vertical load (Table 7 of [2])
LLC 2	Lifting the whole vehicle at 4 lifting positions with most unfavourable vertical load (Table 8 of [2])
LLC 3	Lifting the whole vehicle at 4 lifting positions with one lifting point displaced 10 mm vertically with most unfavourable vertical load (Table 8 of [2])
CHLC 1	Compressive force at buffer axis height 2000 kN and most unfavourable vertical load
CHLC 2	Compressive force 50 mm below buffer centre line 1500 kN and most unfavourable vertical load
CHLC 3	Compressive force at buffer axis height 2000 kN and vertical load from own mass
CHLC 4	Tensile force at coupler axis level 1500 kN and most unfavourable vertical load
CHLC 5	Tensile force at coupler axis level 1500 kN and vertical load from own mass
SLC 1	Force of 35 kN at 500 mm acting from inside the wagon from the centre of the borehole (Subsection 7.10. of [2])
SLC 2	Moment of 42 kNm in transverse direction (Subsection 7.10. of [2])
SLC 3	Moment of 15 kNm in longitudinal direction (Subsection 7.10. of [2])

Methods for the application of loads from HLC, VLC, LLC and CHLC load cases groups are described in detail in [4, 5, 6]. The loads acting on the stanchions (SLC group load cases) are applied in accordance with customer requirements. The stanchions used fully correspond to those of the prototype wagon - Smnps 193. They underwent experimental testing by the Bulgarian National Research Institute for Transport (NRTI) in 2013 and were approved by the notified body. The loads and the methodology as used in the tests are the same as applied in the calculation model [7]. Loads for the SLC1, SLC2 and SLC3 load cases are applied as shown in Figures 2, 3 and 4, respectively. The displacements are limited to the three bolts connecting the stanchions to the wagon underframe.

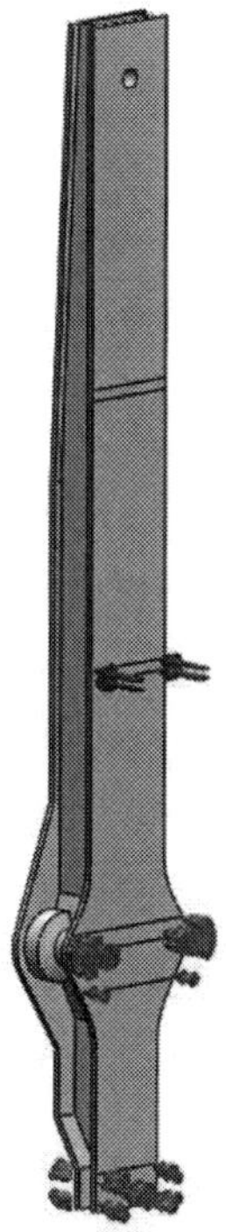

Figure 2. Load application for load case SLC 1.

Figure 3. Load application for load case 2.

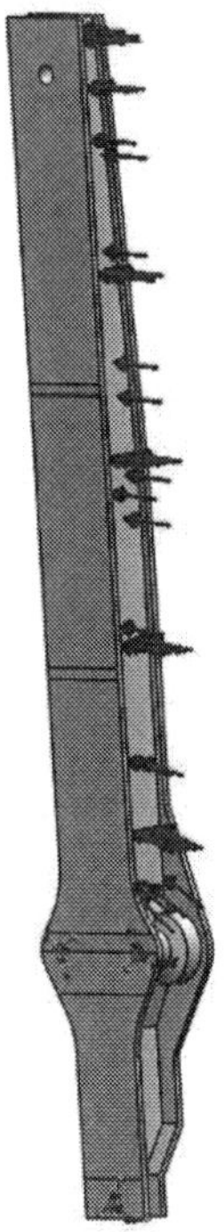

Figure 4. Load application for load case SLC 3.

2.2. Initial calculation results

The study on static strength of wagon underframe and side stanchions (load cases from Table 1) was conducted in accordance with the data in subsection 2.1.

The strength evaluation of the wagon underframe and side stanchions is made using the security factors S as given in equation (1):

$$S = \frac{\sigma_{\lim}}{\sigma_{Mises}} \geq 1,0 \tag{1}$$

where $\sigma_{\lim}$ is the yield limit of the material. For steel S355J2 the parameter is Re = 355 MPa. The stress σ_{Mises} represents the maximum stress received under the corresponding load case.

After conducting the strength analysis, the following conclusions are defined:

- For all tested load cases, the resulting stresses are lower than permissible stresses.
- There are a small number of load cases in which the stresses are close to the permissible stresses, but they are lower. They are characterized by the following features: they have a

pronounced local character and there is a very rapid decrease of the values with an increased distance from the analyzed zone.
- The overall stress level is relatively low, which is a guarantee of high strength and reliability of the wagon.
- The wagon underframe and side stanchions are dimensioned correctly with respect to static load cases.

The results of the strength analysis of the wagon underframe and the side stanchions and the above conclusions have provided grounds for the notified body to admit the wagon series Smnps 194 into service.

3. Analysis of problem areas

Two years after the wagon was taken into service, it was found out that deformations in the connection zone between side stanchions and the wagon underframe were unacceptable. Particularly great distortion of the structural strength of the underframe is noted in the side stanchion connections with the side sill in the area of the weld connecting the two vertical side sill sheets. This area is shown in Figure 5.

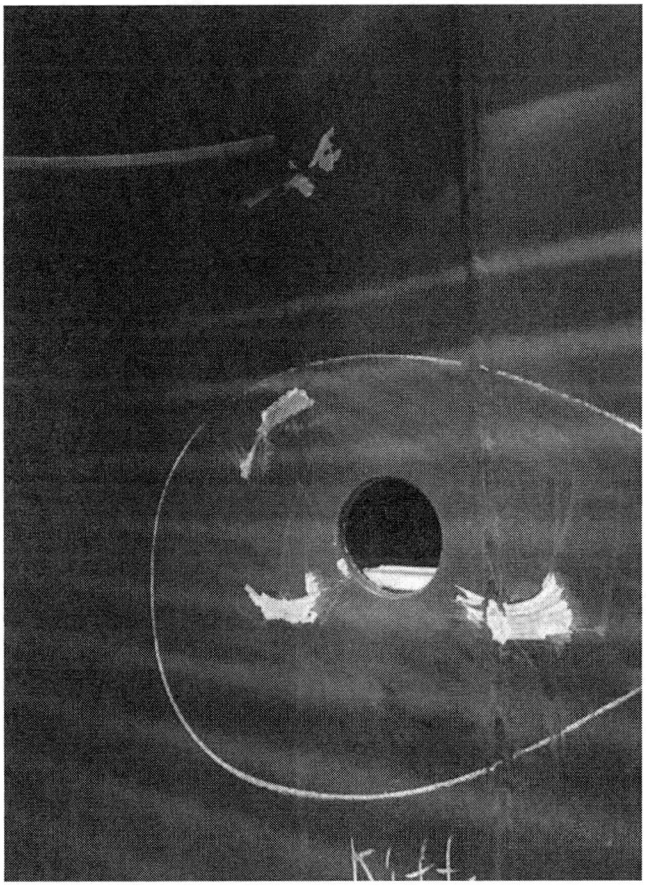

Figure 5. Deformation area at the side sill.

After analysis of the problem zones it was found that the deformations were due to the pressure applied by the stanchion transmitted through its fastening on the wagon underframe at the side sill. The strength analysis of the connection between the stanchion fastening and the wagon underframe has not been requested by the customer and the notified body, due to the fact that the on prototype wagon series Smnps 193 any strength problems have not be identified. Also this is not exactly specified in EN 12663-2:2010.

In order to determine the cause for deformations in the problem zone, it was analyzed by modeling the impact of the load from stanchion fastening on the underframe. Figure 6 shows the simulation model of the connection, and figure 7 shows the load application on the stanchions fitted on the underframe.

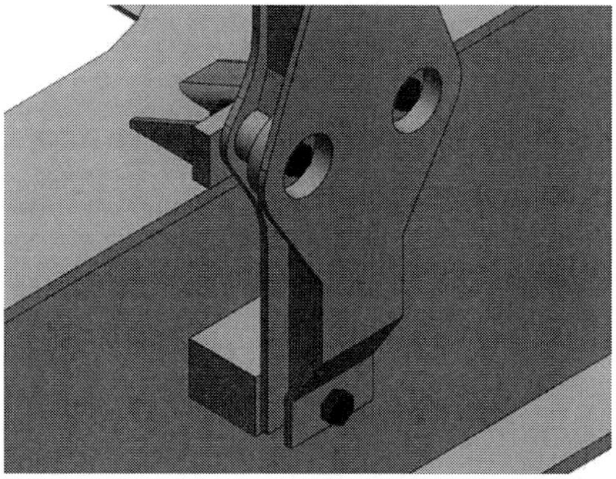

Figure 6. Model of the stanchion fastening.

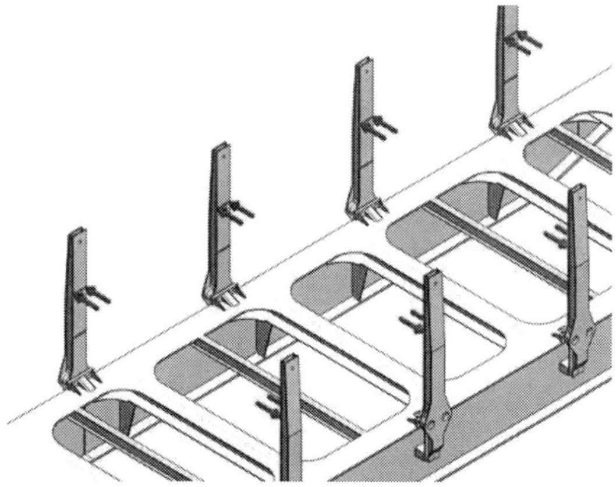

Figure 7. Load application on stanchions fitted to the wagon underframe.

The stress results obtained in the most unfavorable load case SLC 2 are shown in figure 8. Those stresses do not occur in the connection itself or in the stanchion, namely they occur in the side sill of the underframe.

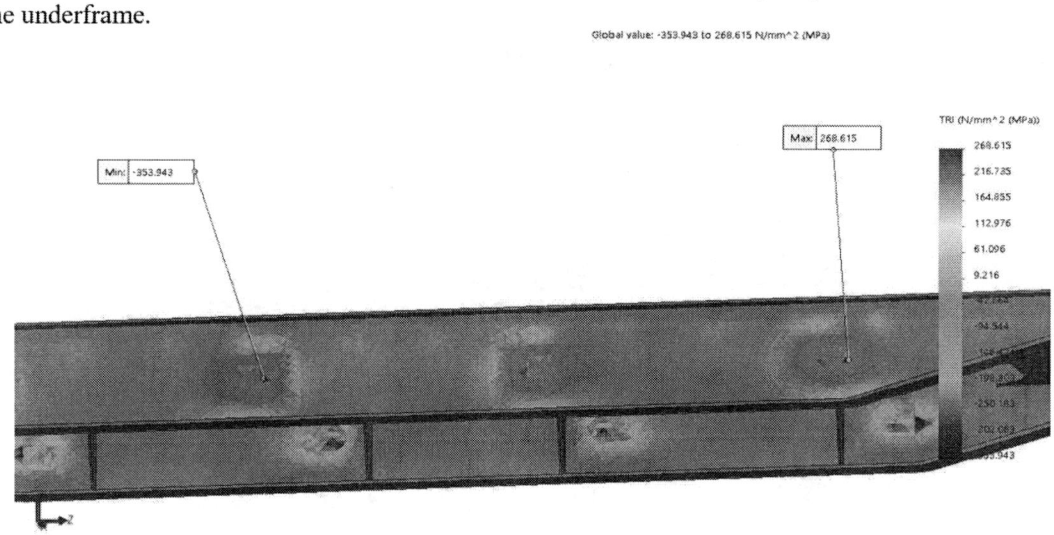

Figure 8. Calculated stress results for load case SLC 2.

Extreme stress values are: - 353,943 MPa and +268,615 MPa. They are very close to the stress values (-337 MPa) registered in the wagon partial tests conducted in Germany by DAkkS Laboratory. Considering that the maximum stress is obtained in the weld area (permissible stresses are σ_{lim}=355/1,1=323 MPa) and the minimum stresses are in parent material (permissible stresses are σ_{lim}=355MPa), then the safety factors are:

$$s_{max}=355/353.943=1{,}003 \text{ and} \quad\quad s_{min}=323/268.615=1{,}202. \quad (2)$$

Equations (2) show that the safety factors are higher than 1, but with minimal increase in load force (e.g. due to impact in the stanchion during loading and unloading operations) deformations in side sill may occur. The analysis shows that it is necessary to propose appropriate measures for reinforcement of areas in which the stanchions are connected to the wagon's side sill.

4. Reinforcement measures and results from extended calculations

Following the analysis made in section 3, it is clear that the stanchion fastenings should be subject to constructive changes as they could cause deformations in the side sills. The authors of the calculations propose to reinforce the problem zone as shown in figure 9. In figure 9 can be seen that the reinforced fastening has a larger contact area with the vertical web of the side sill. In this way, the pressure load from the stanchion acting on the underframe is redistributed more evenly over a larger area. This leads to a reduction of the maximum stresses in the problem area to 315.8 MPa, which is below the permissible 355 MPa. The maximum stresses are displaced from the underframe side sill to the stanchion body, as shown in figure 10.

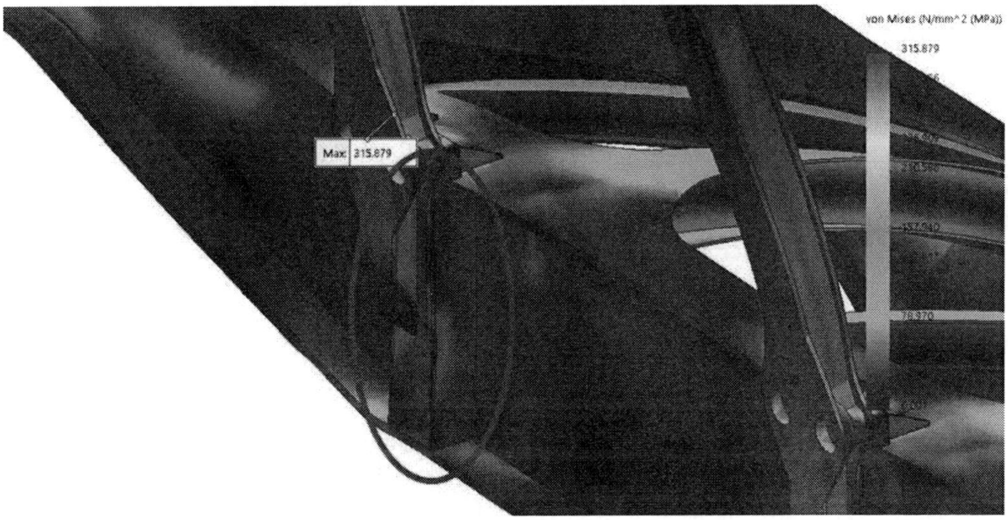

Figure 9. New fastening of the side stanchion.

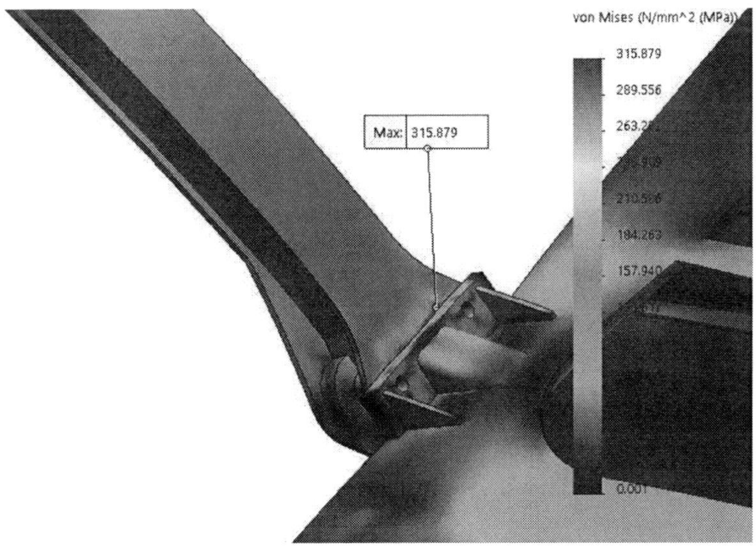

Figure 10. Maximal stresses in side stanchion for load case SLC2.

IOP Conf. Series: Materials Science and Engineering **618** (2019) 012045 doi:10.1088/1757-899X/618/1/012045

5. Conclusion

When carrying out theoretical and experimental strength studies of wagon structures, it is often the case that they are carried out by a shortened procedure, which is not in full accordance with the regulations in the relevant standards. This is usually due to the fact that a given wagon component is already realized in another similar construction and has not shown insufficient strength.

However, the calculation procedures prescribed in the standards do not always cover all the possible impacts of individual wagon components on the overall construction. This report show how this can result in a mismatch with real operating conditions and that the consequences can lead to higher costs, even though all aspects of the calculation procedures prescribed in the normative documents are respected. It follows from this that the tests and theoretical studies of wagon strength prescribed in the normative documents should be expanded or more clearly formulated in order to approximate as much as possible to the real conditions. In this way is it possible to avoid problems similar to those described in this report.

Regarding this study, the authors propose in EN 12663-2: 2010 to add a text "the strength analysis of the side stanchions with the prescribed loads should be carried out by taking into account their impact on the wagon underframe".

Acknowledgements

1. The authors would like to thank the Research and Development Sector at the Technical University of Sofia for the financial support.
2. This work was supported by the European Regional Development Fund within the Operational Programme "Science and Education for Smart Growth 2014 - 2020" under the Project CoE "National center of mechatronics and clean technologies "BG05M2OP001-1.001-0008-C01.

References

[1] EU Agency for railways 2013 *Technical Specifications for Interoperability – Rolling stock Freight wagons* (Brussels: European Railway Agency)

[2] European standard EN 12663-2:2010 *Railway applications - Structural requirements of railway vehicle bodies - Part 2: Freight wagons* (Brussels: European committee for standardization)

[3] UIC Leaflet 577 2012 *Wagon stresses* (Paris: International Union of Railways (UIC))

[4] Slavchev S, Georgieva K, Stoilov V and Purgic S 2015 Analysis of the results of theoretical and experimental studies of freight wagon Fals *Facta Universitatis, Series: Mechanical Engineering* **13(2)** 91

[5] Slavchev S, Stoilov V and Purgic S 2015 Static strength analysis of the body of a wagon series Zans *J. Balkan Tribol. Assoc.* **21(1)** 49

[6] Stoilov V, Slavchev S and Purgic S 2016 Static strength study of body of universal flat wagon *Proc. 17th Int. Sci.-Exp. Conf. on Railways (Niš)* vol 1 (Niš: University of Niš Press) p 91

[7] Bulgarian National Transport Research Institute 2013 *Report "Testing of wagon series Smnps 194 for commissioning"*

Methodology for assessment of material fatigue in the area of welded joints of railway bogies by calculation

S Slavchev[1], V Maznichki[1, 3], V Stoilov[1], S Enev[2] and S Purgich[1]

[1]Department of railway Engineering, Faculty of transport, Technical University of Sofia, Kliment Ohridski Blvd. 8, 1000 Sofia, Bulgaria
[2]Faculty of automatics, Technical University of Sofia, Kliment Ohridski Blvd. 8, 1000 Sofia, Bulgaria

[3]v.maznichki@tu-sofia.bg

Abstract. The current work presents the problem of assessment of the material fatigue in the welding area of the rail bogie frame type Y25LS-K. During the assessment process of the bogie frame dynamic strength, it was found that the results from the theoretical assessment differ from the experimental ones. An analysis of the reasons for these differences was made. A few approaches for the theoretical assessment are presented in order to match the conditions for application of the prescribed loads in calculations as close as possible to the test conditions. Proposed methodology allows the commissioning of bogies into service only based on a theoretical assessment of material fatigue in the welding areas of the bogie frames. The methodology is verified in real theoretical and experimental studies of the certified design of bogie Y25LS-K, produced by Transvagon AD, Burgas. The methodology can be used by researchers, performing a stress-strain analysis of the bogie frame as well as by the notified bodies responsible for the admission of those products into service.

1. Introduction

Railway wagon bogies are one of the most responsible units of the wagon structure. The safety of the wagon, dynamics of the vehicle, cargo consistency, passenger comfort and many others depend on the reliability of the bogies structure. This demands the development and use of precision methods for assessing the strength of the bogie frame. Especially important is the problem of the fatigue phenomenon in the welded joints. This problem is particularly relevant in the procedure for the admission of the bogie based only on calculation and without testing.

The current work proposes a method for assessment of the material fatigue in these areas theoretically, which fully corresponds to the current normative basis. It clarifies some unclearly formulated texts and allows avoiding unnecessary and inadmissible interpretations by both researchers and notified bodies.

2. Normative documents analysis

Required steps for the admission of railway wagons into service are:

- Development of design and technological documentation;
- Performing a stress-strain analysis;
- Prototype testing.

A certificate for bogie production is issued after satisfactory test results have been obtained from the prototype testing. European Union and UIC normative documents allow bogie admission based only on a stress-strain analysis. The necessary conditions are:

- the manufacturer has received a certificate for bogie with similar structural, technological and operational characteristics;
- complete volume of theoretical research with verified calculations and simulation models is done.

The normative documents related to the tests and the theoretical studies for freight bogies (including the analysis of material fatigue in the welded joints of the structure) are EN 15827:2011, EN 13749:2011, UIC 510-3, UIC 615-4, TSI, ERRI B 12/RP 17/ RP 60, EN 12663-1:2010, EN 12663-2:2010, DVS 1612:2014 and Eurocode 3 [1-11]. Their analysis shows that load forces, application locations, combinations of load cases (load groups) are clearly and explicit defined. Complete information on load cases is given in table 1and figure 1 below.

Table 1. Load cases [1,2,3,13]

Load case	Force						Twist	Braking force	
	Vertical			Transverse	Longitudinal				
[kN]	Force on sidebearer F_{z2}	Force on pivot F_{zc}	Force on sidebearer F_{z1}	F_y			g^+	Fbz	Fbx
1	-	F_z	-	-	-	-	-	-	-
2	-	$(1+\beta)F_z$	-	-	-	-	-	-	-
3	-	$(1-\beta)F_z$	-	-	-	-	-	-	-
4	-	$(1-\alpha)(1+\beta)F_z$	$\alpha(1+\beta)F_z$	$0{,}2(F_z+m^+g)$	-	-	-	-	-
5	$\alpha(1+\beta)F_z$	$(1-\alpha)(1+\beta)F_z$	-	$-0{,}2(F_z+m^+g)$	-	-	-	-	-
6	-	$(1-\alpha)(1+\beta)F_z$	$\alpha(1+\beta)F_z$	$0{,}2(F_z+m^+g)$	-	-	$+5\ ^{o}/_{oo}$	-	-
7	-	$(1-\alpha)(1+\beta)F_z$	$\alpha(1+\beta)F_z$	$0{,}2(F_z+m^+g)$	-	-	$-5\ ^{o}/_{oo}$	-	-
8	$\alpha(1+\beta)F_z$	$(1-\alpha)(1+\beta)F_z$	-	$-0{,}2(F_z+m^+g)$	-	-	$+5\ ^{o}/_{oo}$	-	-
9	$\alpha(1+\beta)F_z$	$(1-\alpha)(1+\beta)F_z$	-	$-0{,}2(F_z+m^+g)$	-	-	$-5\ ^{o}/_{oo}$	-	-
10	-	$(1-\alpha)(1-\beta)F_z$	$\alpha(1-\beta)F_z$	$0{,}2(F_z+m^+g)$	-	-	-	-	-
11	$\alpha(1-\beta)F_z$	$(1-\alpha)(1-\beta)F_z$	-	$-0{,}2(F_z+m^+g)$	-	-	-	-	-
12	-	$(1-\alpha)(1-\beta)F_z$	$\alpha(1-\beta)F_z$	$0{,}2(F_z+m^+g)$	-	-	$+5\ ^{o}/_{oo}$	-	-
13	-	$(1-\alpha)(1-\beta)F_z$	$\alpha(1-\beta)F_z$	$0{,}2(F_z+m^+g)$	-	-	$-5\ ^{o}/_{oo}$	-	-
14	$\alpha(1-\beta)F_z$	$(1-\alpha)(1-\beta)F_z$	-	$-0{,}2(F_z+m^+g)$	-	-	$+5\ ^{o}/_{oo}$	-	-
15	$\alpha(1-\beta)F_z$	$(1-\alpha)(1-\beta)F_z$	-	$-0{,}2(F_z+m^+g)$	-		$-5\ ^{o}/_{oo}$	-	-
16	-	Fz	-	-	$0{,}05x(F_z+m^+g)$	$-0{,}05x(F_z+m^+g)$	-	-	-
17	-	Fz	-	-	$-0{,}05x(F_z+m^+g)$	$0{,}05x(F_z+m^+g)$	-	-	-
18	-	$1{,}2F_z$	-	-	-	-	-	F_{bz}	F_{bx}
19	-	$1{,}2F_z$	-	-	-	-	-	$-F_{bz}$	$-F_{bx}$

The markings used in the table are in full accordance with [1-4].

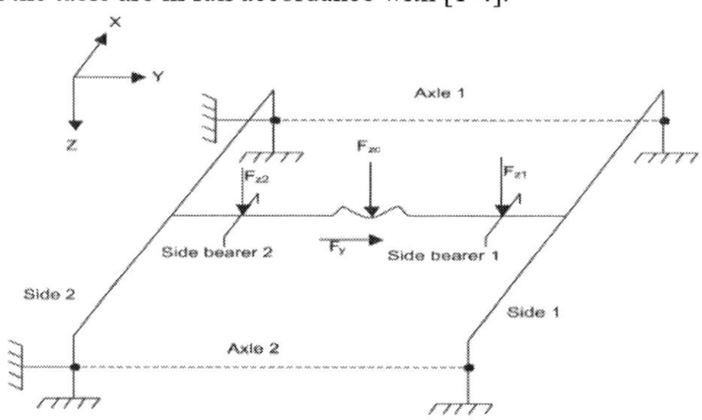

Figure 1. Centre pivot bogie loading arrangement [4].

The data from table 1 and figure 1 allow the boundary conditions for the theoretical and experimental tests of bogie frame strength for static loading modes to be precisely determined.

The normative documents analysis concerning the study of the material fatigue in the welding area in the theoretical and experimental way shows some differences.
The reason is that the theoretical assessment is done by quantitative indicators (tolerable stresses) and the experimental one - by qualitative (presence of cracks).
In particular, the assessment of material fatigue according to [1, 2, 3] is theoretically based on current European or national standards. For the purposes of this study, the German National Welding Standard DVS 1612:2014[10] was used. It is preferred by most European manufacturers of railway vehicles and sophisticated welded products due to the very good results regarding the reliability of the welds. The standard is based on Moore-Kommers-Jasper (MKJ) Curves diagram [6, 7, 10]. Depending on the type of welding structure, the welded joint type, the test methodology and the type of load, a diagram is selected to determine the allowable stresses. In general, for the study of sophisticated welded products at a complex load, the assessment is carried out on four successive criteria, three on stress components and one on complex stress state, equations (1-4).

$$\frac{\sigma_{\parallel}}{\sigma_{\parallel,lim}} \leq 1 \quad (1) \qquad \frac{\sigma_{\perp}}{\sigma_{\perp,lim}} \leq 1 \quad (2) \qquad \frac{\tau}{\tau_{lim}} \leq 1 \quad (3)$$

$$\left(\frac{\sigma_{\parallel}}{\sigma_{\parallel,lim}}\right)^2 + \left(\frac{\sigma_{\perp}}{\sigma_{\perp,lim}}\right)^2 - \frac{\sigma_{\parallel}\sigma_{\perp}}{\left|\sigma_{\parallel,lim}\sigma_{\perp,lim}\right|} + \left(\frac{\tau}{\tau_{lim}}\right)^2 \leq 1,1 \quad (4)$$

where: $\sigma_{\parallel}$ are the normal stresses longitudinally to the weld seam;
$\sigma_{\perp}$ - normal stresses transverse to the weld seam;
τ - tangential stresses longitudinally to the weld seam;
σ_{lim} and τ_{lim} - permissible stresses. [10,12]

The permissible stresses $\sigma_{lim}(R_\sigma)$ are a function of the asymmetry coefficient R_σ in the cyclic loading from equation (5), which is calculated according to equation (6).

$$R_\sigma = \frac{\sigma_{\min}}{\sigma_{\max}} \quad (5) \qquad \sigma_{lim}(R_\sigma) = 150\left(1,04^{-x}\right)\left[\frac{2(1-0,3R_\sigma)}{1,3(1-R_\sigma)}\right] \quad (6)$$

where: x is the exponent depending on the selected curve corresponding to the welded joint and load;
σ_{min} - the minimal stress in a welded joint;
σ_{max} – the maximal stress in a welded joint.

The experimental test method is described in details in [1-4]. According to it, the nominal value of the applied forces corresponds completely to the data in Table 1. The test runs in 3 stages, the total number of cycles being $10\text{x}10^6$ [1-4, 14, 17-23].
The first stage covers $6\text{x}10^6$ cycles, where the load cases have nominal load values, which the structure encounters in normal operation. The stage ends with an intermediate qualitative and quantitative assessment. The qualitative assessment is mandatory and requires complete control of the welded joints. The criterion is the absence of cracks. The quantitative assessment is most often done with Goodman-Smith diagrams [16-22]. It is allowed to exceed the permissible stresses in a limited number of controlled points. The condition for these is that these areas have to be controlled with caution in the next two stages. The second stage includes $2\text{x}10^6$ cycles. Vertical forces in this stage increase by 20%. The stage ends with an intermediate qualitative assessment, where the welds are controlled. Cracks are not allowed. The third stage includes $2\text{x}10^6$ cycles, with vertical forces including 40% overload over the nominal ones. The assessment criterion is qualitative and cracks are allowed if they do not violate the integrity of the structure.
Further analysis of normative documents shows the following differences between the theoretical and experimental fatigue studies of the material:

1. In a fatigue test of the material:

IOP Conf. Series: Materials Science and Engineering **618** (2019) 012046 doi:10.1088/1757-899X/618/1/012046

Longitudinal loads are not applied on the bogie frame. This excludes the load cases with numbers 16 and 17 from the table 1.

- Annex G of EN 13749: 2011 [2] allows the braking forces not to be applied on the structure in the case that their influence in the static tests is negligible. This text allows the testers to make a decision whether to include or exclude the load cases 18 and 19 of Table 1.
- According to Annex G of EN 13749: 2011 [2], the operating loads are applied in combination with the same phase and frequency and so to simulate the movement of the bogie in the left or right curve. The direction of the curve is being reversed every 10 to 20 cycles. With the force action thus formulated, it follows that when the pivot bearing is considered to be unloaded, the transverse force F_y becomes zero (Figure 2).

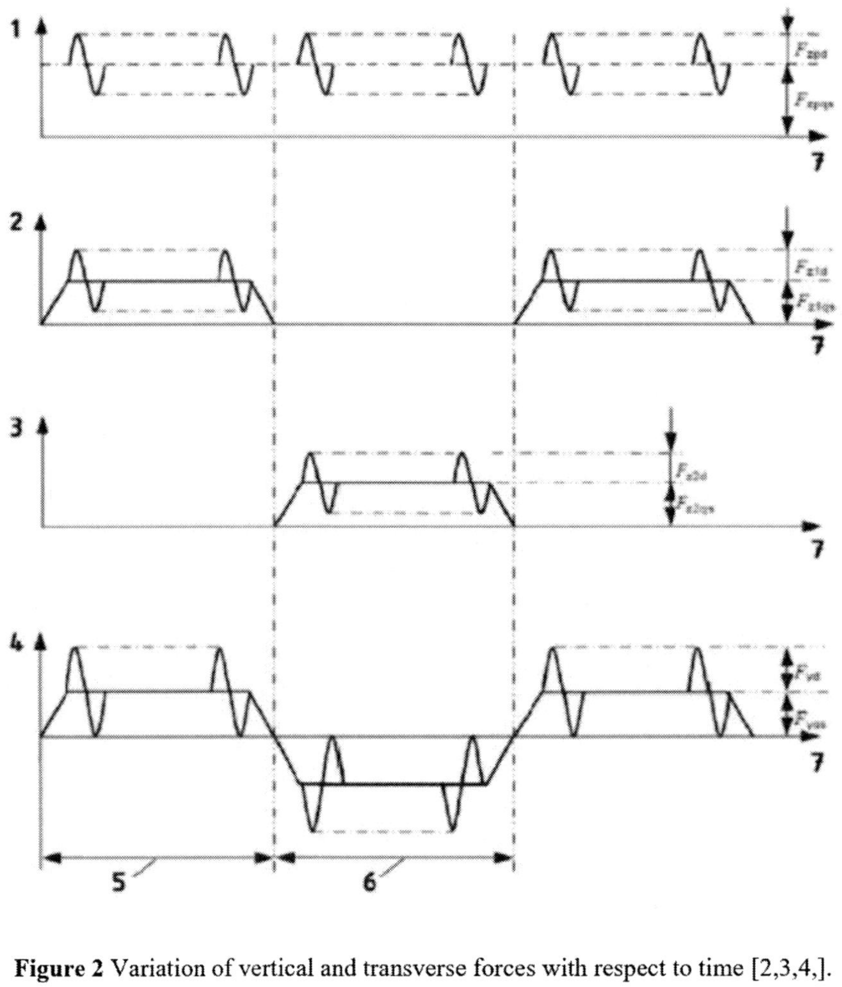

Figure 2 Variation of vertical and transverse forces with respect to time [2,3,4,].

2. In the theoretical studies of the material fatigue it is appropriate the applied forces to have a normal service value, without simulating overload by 20 or 40%, similar to the second and third stages of the test study. The reason is purely theoretical. The method for theoretical study was created on the premise that if the conditions of equations (1 - 4) are met, it is ensured that the material is indestructible in 10^7 cycles. If 20% or 40% overloads are applied, then that would mean that the bogie frame would be significantly oversized because it would not show any fatigue process at these higher values.

IOP Conf. Series: Materials Science and Engineering **618** (2019) 012046 doi:10.1088/1757-899X/618/1/012046

The differences mentioned above give grounds for proposing a detailed methodology for theoretical investigation of material fatigue.

3. Methodology for fatigue assessment in the welded joints

The theoretical evaluation of the weld seams of the structure is done in the following order:

- A simulation model for stress-strain analysis is developed. In the case that the procedure specified by the notify body provides admission without testing, then the model should be as close as possible to a verified version for a similar object by criteria: type of finite elements, minimum and maximum dimensions of the finite elements, number of elements, method of introducing boundary conditions and etc.
- Simulations are performed using all 19 regulated load cases in accordance with table. 1.
- A method of material fatigue testing is selected - MKJ diagrams, Goodman-Smith diagrams, ERRI or other method, acceptable according to European or national standards.
- Material fatigue assessment in accordance to the methodology of the chosen method is performed.
- In the case that all nodes or elements fulfil the material fatigue criteria, further calculations are not necessary and the calculation team proposes the bogie frame for admission without testing or sending it to tests.
- If a small number of nodes or elements do not fulfil the material fatigue criteria, the simulation is repeated using the first 15 load cases in accordance with table 1. Reason for this gives Annex G of [2].
- A reassessment of material fatigue is performed in accordance with the methodology of the chosen method.
- In the case that all nodes or elements fulfil the material fatigue criteria, further calculations are not necessary and the calculation team proposes the bogie frame for admission without testing or sending it to tests.
- If a small number of nodes or elements do not fulfil the material fatigue criteria, the simulation is repeated using the first 15 load cases in accordance with table 1 under assumption that the transverse force is changed by $F_y=0\text{-}F_{yd}$. This is justified by the above-mentioned analysis of the normative documents.
- A third assessment of the material fatigue is made in accordance with the methodology of the chosen method.
- In the case that all nodes or elements fulfil the material fatigue criteria, further calculations are not necessary and the calculation team proposes the bogie frame for admission without testing or sending it to tests.
- If there are still a small number of nodes or elements that do not fulfil the material fatigue criteria, the calculation team concludes that structure must be modified or suggestions are proposed for the test of bogie frame.

Figure 3 illustrates the proposed methodology.

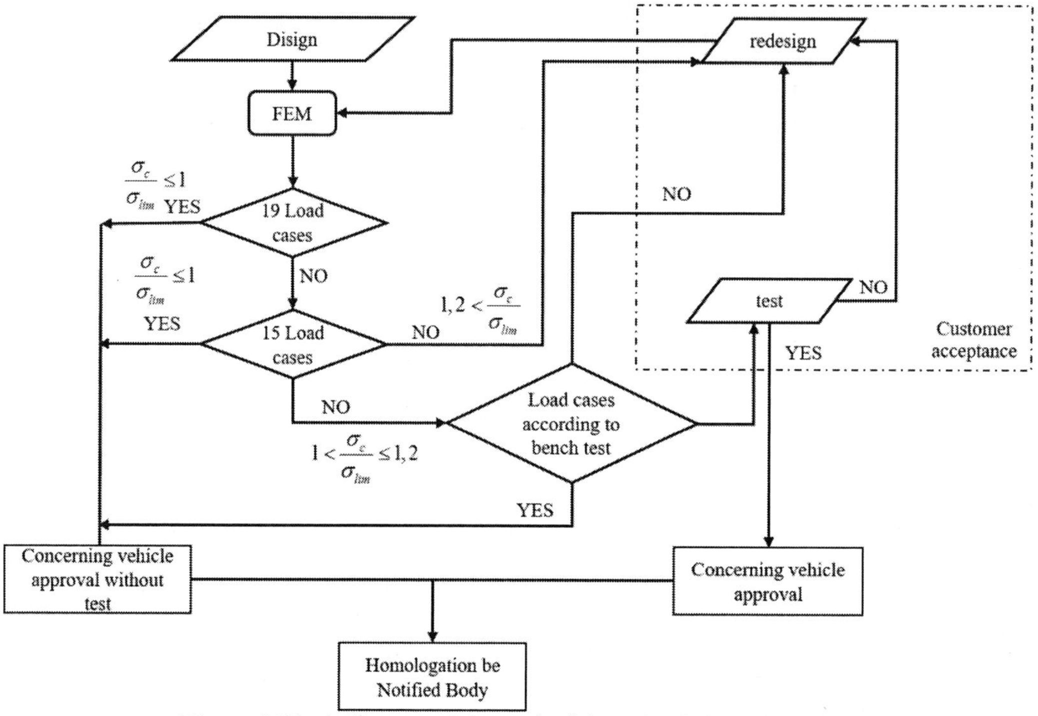

Figure 3 Block diagram of the methodology for fatigue assessment.

4. Application of the proposed methodology

The method shown in section 3 was applied during the theoretical and experimental studies of the Y25LS-K bogie manufactured by Transvagon AD, Burgas. For this purpose, the calculation model was verified using the data from the static test of the bogie. The performed comparative analysis gives very good match of the values of the theoretical and experimental stresses [15].

4.1. Assessment of material fatigue in the bogie frame according to DVS 1612:2014

Table 2 presents part of the results of the frame assessment when examining all 19 load cases of Table 1. Because of the limited volume of the report, only data for the first ten nodes with insufficient dynamic strength are shown.

Table 2. The results of fatigue assessment including all 19 load cases

Nr. Node	σ⊥min	σ⊥max	R	σ⊥zul	case min	case max	σǁmin	σǁmax	R	σǁzul	case min	case max	τmin	τmax	R	τzul	case min	case max	(4)
1012487	88,7	-128,4	-0,7	111,3	16	17	-58,8	64,6	-0,9	103,9	17	16	3,1	49,8	0,1	117,3	3	17	2,615
1012486	88,5	-119,6	-0,7	109,5	16	17	-56,1	67,1	-0,8	106,2	17	16	1,6	43,7	0,0	114,6	3	17	2,427
1012622	71,4	-120,7	-0,6	115,3	17	16	-52,0	54,8	-0,9	102,8	16	17	1,1	44,8	0,0	113,3	3	16	2,095
1012488	74,1	-119,0	-0,6	114,0	16	17	-54,2	56,2	-1,0	102,3	17	16	4,8	44,1	0,1	122,3	3	17	2,094
1012150	64,1	-113,1	-0,6	116,4	17	16	-51,7	52,2	-1,0	101,6	16	17	3,0	44,6	0,1	117,8	3	16	1,850
966820	69,9	-107,0	-0,7	112,8	16	17	-48,3	54,6	-0,9	104,7	17	16	3,6	42,9	0,1	119,5	3	17	1,796
966821	70,5	-100,1	-0,7	110,8	16	17	-45,4	56,5	-0,8	107,3	17	16	2,1	37,5	0,1	116,5	3	17	1,673
1341113	70,6	-115,3	-0,6	114,5	15	13	-15,0	46,0	-0,3	129,0	13	8	3,3	44,8	0,1	118,4	8	4	1,643
1341112	66,7	-114,5	-0,6	115,7	15	13	-13,8	45,3	-0,3	130,4	13	8	2,8	46,1	0,1	117,1	9	4	1,598
966938	54,7	-107,4	-0,5	119,1	17	16	-45,3	45,3	-1,0	101,4	16	17	2,9	41,6	0,1	117,9	3	16	1,542

The analysis of the results show that from all nodes located in the welds area only 43 show insufficient dynamic strength according to formula (4).That means that the bogie frame can be send for testing or the implementation of the methodology presented in section 3 can continue.

4.2. Assessment of material fatigue in the bogie frame according to DVS 1612: 2014 without the braking force load cases

The results from second fatigue assessment are shown in table 3. The assessment includes all load cases except the load cases, which simulate braking of the vehicle (the last four load cases from table 1). The reason for that exclusion are the results from static tests of bogie frame.

Table 3. The results of fatigue assessment including first 15 load cases presented in table 1

Nr. Node	$\sigma_{\perp min}$	$\sigma_{\perp max}$	R	$\sigma_{\perp zul}$	case min	case max	$\sigma_{\parallel min}$	$\sigma_{\parallel max}$	R	$\sigma_{\parallel zul}$	case min	case max	τ_{min}	τ_{max}	R	τ_{zul}	case min	case max	(4)
1341113	70,6	-115,3	-0,6	114,5	15	13	-15,0	46,0	-0,3	129,0	13	8	3,3	44,8	0,1	118,4	8	4	1,643
1341112	66,7	-114,5	-0,6	115,7	15	13	-13,8	45,3	-0,3	130,4	13	8	2,8	46,1	0,1	117,1	9	4	1,598
1485026	55,9	-110,1	-0,5	119,1	13	8	-11,5	39,0	-0,3	131,0	15	4	6,9	46,1	0,1	127,4	10	8	1,349
1485047	48,2	-109,3	-0,4	122,5	13	8	-8,3	40,2	-0,2	137,3	15	4	4,1	44,1	0,1	120,6	12	8	1,277
1341151	62,9	-89,6	-0,7	110,9	13	15	-15,6	57,7	-0,3	132,7	15	4	1,8	31,8	0,1	116,7	10	9	1,268
1341133	51,8	-102,0	-0,5	119,1	15	4	-6,6	42,3	-0,2	141,1	13	8	4,6	42,8	0,1	122,3	9	7	1,202
1341150	56,9	-85,6	-0,7	112,3	13	15	-15,0	58,4	-0,3	133,5	15	4	2,0	33,1	0,1	117,1	10	9	1,184
1341117	49,7	-102,4	-0,5	120,2	15	4	-4,6	41,3	-0,1	145,1	13	8	5,1	44,8	0,1	123,1	14	6	1,181
1341155	63,6	-82,4	-0,8	108,3	13	15	-10,3	65,4	-0,2	141,0	15	4	7,9	21,4	0,4	165,1	12	5	1,163
1485027	55,0	-99,2	-0,6	117,0	13	15	-10,5	38,5	-0,3	132,5	15	6	5,4	40,6	0,1	125,2	6	8	1,155

The result analysis shown that the only 13 nodes show insufficient dynamic strength according to formula (4). That means that the bogie frame can be send for testing or the implementation of the methodology presented in section 3 can continue.

4.3. Assessment of fatigue material in the bogie frame according to DVS 1612: 2014 by reducing the transverse force F_y

Table 4 presents the results of material fatigue study with the first 15 load cases. The transverse force F_y is reduced.

Table 4. The results of fatigue assessment including first 15 load cases and reduced transverse force

Nr. Node	$\sigma_{\perp min}$	$\sigma_{\perp max}$	R	$\sigma_{\perp zul}$	case min	case max	$\sigma_{\parallel min}$	$\sigma_{\parallel max}$	R	$\sigma_{\parallel zul}$	case min	case max	τ_{min}	τ_{max}	R	τ_{zul}	case min	case max	(4)
1341112	32,8	-107,5	-0,3	130,4	5	1	10,6	45,3	0,2	189,4	12	5	2,8	46,1	0,1	117,1	6	1	1,089
1341113	34,3	-106,4	-0,3	129,3	5	1	10,8	46,0	0,2	189,6	12	5	3,3	44,8	0,1	118,4	5	1	1,079
1485026	20,7	-110,1	-0,2	138,7	4	5	10,1	39,0	0,3	194,3	10	1	7,1	46,1	0,2	127,9	3	5	0,960
1485047	22,5	-109,3	-0,2	137,3	1	5	10,8	40,2	0,3	195,8	10	1	7,1	44,1	0,2	128,7	1	5	0,957
1341111	23,5	-105,0	-0,2	136,0	5	4	11,4	40,2	0,3	199,4	11	2	7,9	48,8	0,2	129,0	6	4	0,936
1341133	26,9	-102,0	-0,3	133,1	5	1	11,8	42,3	0,3	198,3	8	5	4,6	42,8	0,1	122,3	6	4	0,919
1341117	25,3	-102,4	-0,2	134,3	6	1	11,7	41,3	0,3	199,0	12	5	6,1	44,8	0,1	125,7	5	3	0,910
1485042	19,0	-104,1	-0,2	139,1	3	6	11,5	38,6	0,3	202,3	7	1	9,5	50,0	0,2	132,6	4	6	0,882
1341127	23,9	-101,7	-0,2	135,2	5	1	11,5	40,8	0,3	198,7	11	6	5,6	43,0	0,1	125,1	6	4	0,882
1485022	27,3	-98,3	-0,3	132,2	4	5	12,6	44,3	0,3	199,3	10	3	9,7	42,6	0,2	138,5	10	5	0,863

The analysis of the theoretical data shows that there are no nodes with insufficient dynamic strength. This fully corresponds to the results of the experimental tests of the bogie.

5. Conclusion

Current work presents a critical analysis of the normative documents concerning the fatigue phenomenon of the material in the welding areas of the railway bogie frame. Discrepancies are found between the theoretical and experimental assessment conditions. Based on them, a detailed methodology for theoretical study and assessment of the material fatigue is proposed, which brings it closer to the conditions of tests.

The methodology was applied in the theoretical studies of Y25LS-K bogie manufactured in TRANSVAGON AD, Burgas. The prototype was tested in 2018 in the Czech laboratory VUZ [19]. Comparative analysis showed very good compatibility of results.

Acknowledgements
1. The authors would like to thank the Research and Development Sector at the Technical University of Sofia for the financial support.
2. This work was supported by the European Regional Develop-ment Fund within the Operational Programme "Science and Education for Smart Growth 2014 - 2020" under the Project CoE "National centre of mechatronics and clean technologies "BG05M2OP001-1.001-0008-C01.

References

[1] EN 15827:2011, *Railway Applications - Requirements for Bogies and Running Gears*, 2011
[2] European standard EN 13749:2011, *Railway Applications - Wheelsets and Bogies - Method of Specifying the Structural Requirements of Bogie Frames*
[3] UIC Leaflet 510-3 *1989 Wagons - Strength Testing Of 2 And 3-axle Bogies On Test Rig*
[4] *TSI-Rolling Stock - Freight Wagons*. 2006
[5] UIC Leaflet 615-4:2003 *Motive Power Units - Bogies And Running Gear - Bogie Frame StructureStrength Tests, 2nd edition*
[6] ERRI B 12/RP 17, 8th Edition 1997 *Programme of tests to be carried out on wagons with steel underframe and body structure (suitable for being fitted with the automatic buffing and draw coupler) and on their cast steel frame bogie*
[7] ERRI B 12/RP 60, 2th Edition 2001 *Tests to demonstrate the strength of railway vehicles. Regulations for proof stress and maximum permissible stresses*
[8] EN 12663-2:2010 *Railway Applications - Structural Requirements of Railway Vehicle Bodies - Part2: Freight Wagons*
[9] EN 12663-1:2010 *Railway applications - Structural Requirements of Railway Vehicle Bodies - Part 1: Locomotives and Passenger Rolling Stock*
[10] DVS 1612:2014
[11] EN 1993-1-9:2005, *Eurocode 3: Design of Steel Structures - Part 1-9: Fatigue*, 2007
[12] Radaj D 1990 *Design and analysis of fatigue resistant welded structures*. Cambridge Abington.
[13] Slavchev S, Maznichki V, Stoilov V, Enev S and Purgic S 2018 Comparative analysis of fatigue strength of an y25ls-k bogie frame by methods of UIC AND DVS 1612 *XVIII Scientific-Expert Conference on Railways RAILCON`**18*** pp. 1-4, Nis,
[14] Bracciali A 2010 Analisi critica delle procedure di calcolo FEM dei telai carrello e delle casse dei rotabili ferroviari, ***XXXIX** convegno AIAS*, Maratea
[15] Slavchev S, Maznichki V, Stoilov V and Purgic S. 2017 Comparative analysis of the result of thestatic strength calculations and strength tests of a bogie frame Y25Ls-K. *Journal MATEC Web of Conferences* Vol. **133** Sozopol
[16] Norman E Dowling, Katakam Siva Prasad and Narayanasamy R 2012 *Mechanical Behavior of Materials,Engineering Methods for Deformation, Fracture and Fatigue* England
[17] Yongyang Zhang, Pingbo Wu, Ye Song 2015 Strength Test and Modal Analysis for a Standardized High-Speed EMU Motor Bogie Frame *4th International Conference on Sensors, Measurement and Intelligent Materials* (ICSMIM **2015**) pp 1128-1132
[18] Jung-Won Seo, Hyun-Moo Hur, Hyun-Kyu Jun, Seok-Jin Kwon, and Dong-Hyeong Lee FatigueDesign Evaluation of Railway Bogie with Full-Scale Fatigue Test, *Advances in MaterialsScience and Engineering*, Volume **2017** pp.11
[19] VUZ Plc, 2018 *Test report Strength tests of bogies and their components: Non-destructive testing – Magnetic particle test (MT) - Bogie frame Y25Ls-K* Czech
[20] Slavchev S and Stoilov V 2011 Methodology for assessment of stress in the application of "static method UIC" for the Study of fatigue when using numerical methods *„BulTrans **2011**"* pp 267-270 Sozopol.
[21] Herbert J Sutherland Optimized Goodman Diagram For The Analysis Of Fiberglass Composites Used In Wind Turbine Blades, DOI: 10.2514/6.2005-196
[22] San Romaán J L, Álvarez-Caldas C and Quesada A 2005 Structural validation of railway bogies and wagons using finite elements tools Volume: **219 (3)** pp 139-150
[23] Mancini G and Cera A 2013 Design of Railway Bogie in Compliance with new EN 13749 European Standard *Engineering Failure Analysis* vol. **31(9)** pp 412–420

IOP Conf. Series: Materials Science and Engineering **618** (2019) 012047 doi:10.1088/1757-899X/618/1/012047

Deposition and tribomechanical study of nanolaminate Ti/TiN/AlTiSiN/(AlTiSiN/TiAlSiN)n/AlTiSiN hard coating

Velko Rupetsov[1], Lilyana Kolaklieva[2], Vassiliy Chitanov[2], Milko Angelov[3], Raycho Raychev[4], Zlatko Zlatanov[4] and Chavdar Pashinski[2,4*]

[1]Faculty of Physics and Technology, Plovdiv University "Paisii Hilendarski", Plovdiv, Bulgaria
[2]Central Laboratory of Applied Physics, Bulgarian Academy of Sciences, Plovdiv, Bulgaria
[3]Milko Angelov Consulting Co. Ltd., Plovdiv, Bulgaria
[4]Department of Mechanics, Technical University - Sofia, Br. Plovdiv, Plovdiv, Bulgaria

* Email: pashinski@yahoo.com

Abstract. The application of Physical Vapor Deposition (PVD) hard coatings is one time-tested practice in the modern industry. Two of the main tendencies that promise a progress in this area are the combination of more chemical elements in one coating and the creation of nanolaminate structures. Since many combinations are possible in the design of the coatings, it could be expected they would have many different properties which are difficult to predict. In an attempt to work simultaneously in both of these directions, one nanolaminate Ti/TiN/AlTiSiN/(AlTiSiN/TiAlSiN)n/AlTiSiN coating was developed and its mechanical properties were tested. The results obtained give a reason to believe that it could be widely practically applied.

1. Introduction

The hard coatings (with a hardness of 20 GPa or more) increase the life span of the cutting tools, reduce the breaks for tool replacement and subsequent machine adjustment, allow heavy work conditions without coolant lubricants, increase productivity, etc. [1]. On the other hand, depending on the parameters of the working process (operation modes, raw material, type of machining, etc.), a choice of proper coating should be made for each particular situation [1, 2].

With the progress of PVD technologies, new possibilities have been found to combine more chemical elements in one coating, as well as a precise deposition of thin layers with a thickness of several nm. At present, the creation of nanolaminate (superlattice) coatings is a notably promising trend for development of new coatings for cutting tools.

The coatings AlTiN and AlTiSiN kind have long been exploited and are somewhat classic for application on cutting tools [1, 3-6]. It could be said that such monolayer, multilayer and nanocomposite coatings are well studied. However, because of the many possible combinations and the relatively new appearance of the nanolaminates, there is still a wide field to research them.

In this work one similar nanolaminate coating: Ti/TiN/AlTiSiN/(AlTiSiN/TiAlSiN)n/AlTiSiN is presented which possess good properties allowing its practical application for cutting tools.

2. Experimental details

Test samples of widely used steel were made and the mentioned above coating was deposited. Its tribomechanical properties were then investigated.

2.1. Preparation of the samples

The used samples of tool steel 1.2419 (ХВГ) are disc-shaped: 20 mm in diameter and 6 mm in thickness. They were previously bulk annealed and ground.

The coating was deposited by Closed Field Unbalanced Magnetron Sputtering (CFUBMS) technology which was implemented in an industrial equipment HVP100RHD (Nanotech Group Ltd.).

Prior to the deposition, the samples were mechanically cleaned, treated with an alkaline detergent, washed in an ultrasonic bath using deionized water and placed in a drier. After the samples were parched, they were loaded into the vacuum chamber and fastened on a three-axial rotation carousel.

The process was started at initial pressure of $4{,}0.10^{-5}$ mbar. At the beginning, the samples were heated in Ar/H_2 (6% H_2) atmosphere up to 390 °C for 60 min and this maximum temperature was maintained the last 10 min. Next, an etching using Ar+ ions (12 min) and Ti+ ions (10 min) was done at a pressure of $2{,}5 \div 2{,}7.10^{-3}$ mbar and a pulsed bias voltage U_{bias} = -930 V/100 kHz to the samples. The coating deposition was performed at a temperature of 378÷370 °C (a slight drop to the end) and a pressure of $2{,}0 \div 2{,}2.10^{-3}$ mbar using two Ti cathodes and one Al-12 at.% Si cathode operating in a N_2 environment. The single deposition steps were performed according to a technology recipe, details of which could be learned by Nanotech Group Ltd. In particular, for the creation of the nanolaminate structure, a pulsed bias voltage U_{bias} = -54 V/100 kHz was applied, whereby the maximum ion current to the samples I_{bias} = 1,6 A was measured respectively (three working cathodes). The total duration of the entire process was 4 hours, after which the samples were cooled by Ar flowing into the chamber until a pressure rose to ca. 10 mbar. After the temperature dropped to 120 °C, the chamber was vented and the samples were unloaded.

2.2. Tribomechanical surveys

The coating thickness which is an important complex feature, was investigated by a calo tester (created by the Central Laboratory of Applied Physics, BAS). One grind ball of bearing steel with diameter of 30 mm was used to form the crater.

The roughness of the uncoated specimen and the deposited coating was evaluated by Surftest SJ-301 (Mitutoyo Corporation).

The nanohardness and the modulus of elasticity were studied by means of Compact Platform CPX (MHT/NHT) (CSM Instruments). In this study, a Berkovich diamond tip was used. To determine the hardness values, the Oliver-Pharr method (which is software embedded in the equipment) was applied. There 3 measurements were done with each of the loads: 15, 25, 50 mN and the obtained results were averaged. The bulk hardness was evaluated using durometer TK-2M (ЗИП).

The adhesion was estimated using a Micro Scratch Tester (MST) module which is included in the mentioned above equipment of CSM Instruments. A Rockwell diamond indenter possessing a 200 μm peak rounding and a sliding distance of 3 mm were applied. The coefficient of friction and penetration depth were also taken into account during this test.

The wear resistance of the coating was assessed by two methods: ball-on-flat and ball-on-disc. In the first case, a stand СИИП-1 (developed by Faculty of Physics and Technology, PU "Paisii Hilendarski") was applied. The plied counter-part was one Al_2O_3 ball with diameter of 3 mm. The sample performed 11,0 mm long linear reciprocating motions. A total of 5 tracks were made and their average width (over 5 measurements for each) was accepted to calculate the wear volume. The second method was applied by means of a stand developed by Milko Angelov Consulting Co. Ltd. The loaded counter-part was one Al_2O_3 ball with diameter of 3/16". The sample was rotated, the average diameter of the formed circular track is 12,2 mm. Just 1 track was grooved (the purpose was to use the same sample for the two methods for greater reliability, but there cannot be grooved many non-intersecting tracks, especially in the shape of a circle), but its width was averaged over 5 measurements of different places. In the both cases, the tracks were inspected and their width was measured by means of microscopic equipment: non-contact PC-based measurement system TESA VISIO 300 (Brown&Sharpe TESA) at 100x magnification (resolution: 0,001 mm).

3. Results and discussion

A structure scheme of the deposited coating and view of the calotte section are shown in Fig. 1. The general appearance of the crater leads to the conclusion that there is relatively low internal stress in the coating because of absence of destruction traces. The color of the coating is brown, its thickness is 2,25 μm and the particular layers in it are clearly distinguished.

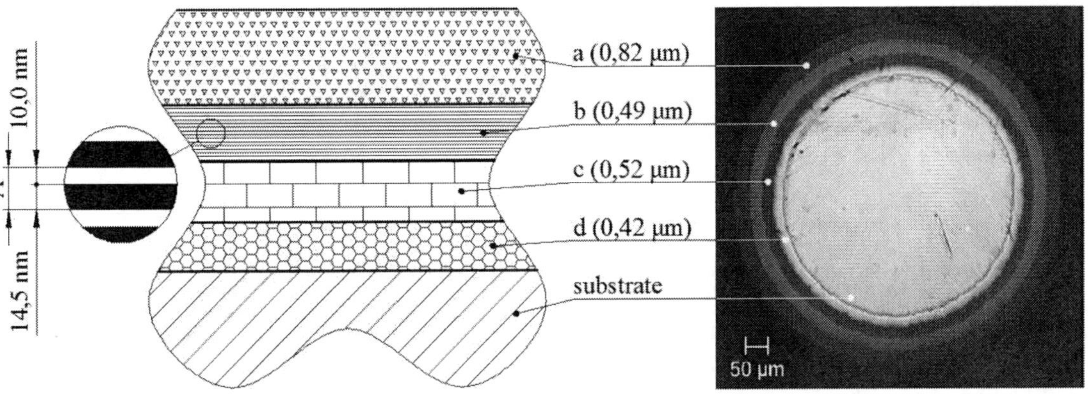

Figure 1. Structure scheme and calotte section: a) external Al-enriched AlTiSiN layer; b) nanolaminate structure (AlTiSiN/TiAlSiN)n; c) cohesive gradient AlTiSiN layer; d) adhesive Ti/TiN layers.

The outermost area is shown in Fig. 1a: external Al-enriched AlTiSiN layer (thickness of 0,82 μm) – it starts gradient with a smoothly increase of Al, but after reaching its optimum quantity (Al: 29,5 at.%, Ti: 16,5 at.%, Si: 4,0 at.%, N and impurities: the rest), the layer already grows up at constant parameters. Its function is to increase the oxidation and corrosion resistance by the formation of a passivation Al_2O_3 layer (it also has more Si which further favors this feature [4]). The nanolaminate structure (AlTiSiN/TiAlSiN)n could be seen in Fig. 1b (thickness of 0,49 μm) – it implements several functions: unloading of the stress in the coating, a barrier against oxidation and corrosion, increasing of some mechanical characteristics, etc. It possesses a bilayer period Λ of 24,5 nm (AlTiSiN – 10,0 nm and TiAlSiN – 14,5 nm) – a total of 20 repeating sublayers (in Fig. 1b they cannot be distinguished because of their small thickness, the values presented are evaluated by calculations based on the experience with other processes using this deposition equipment). The cohesive gradient AlTiSiN layer (thickness of 0,52 μm) is presented in Fig. 1c – it serves as a smooth transition to mentioned nanolaminate structure. The gradient adhesive Ti/TiN layers (thickness of 0,42 μm) are depicted in Fig. 1d – their purpose is to provide a better adhesion of the coating to the substrate and a smooth transition to the functional layers, they are typical of such coatings.

The measured roughness of the deposited coating is Ra 0,23 μm. The results of the roughness measurements of the uncoated sample (Ra 0,48 μm) and subsequently the same having a deposited coating are shown in Fig. 2. It should be noted that the vertical scales of the diagrams used for the bare substrate (Fig. 2a) and the coated sample (Fig. 2b) are different. One can see that the roughness was decreased after the deposition. It could be assumed that this is due to the intense Ar+ and Ti+ etching [7] which destroys the peaks on the surface. On the other hand, the CFUBMS method does not create droplets from the target's material which fall on the substrate and increase the roughness respectively. In a nutshell, the applied coating practically reproduces the relief of the surface which is already smoothed during the etching.

IOP Conf. Series: Materials Science and Engineering **618** (2019) 012047 doi:10.1088/1757-899X/618/1/012047

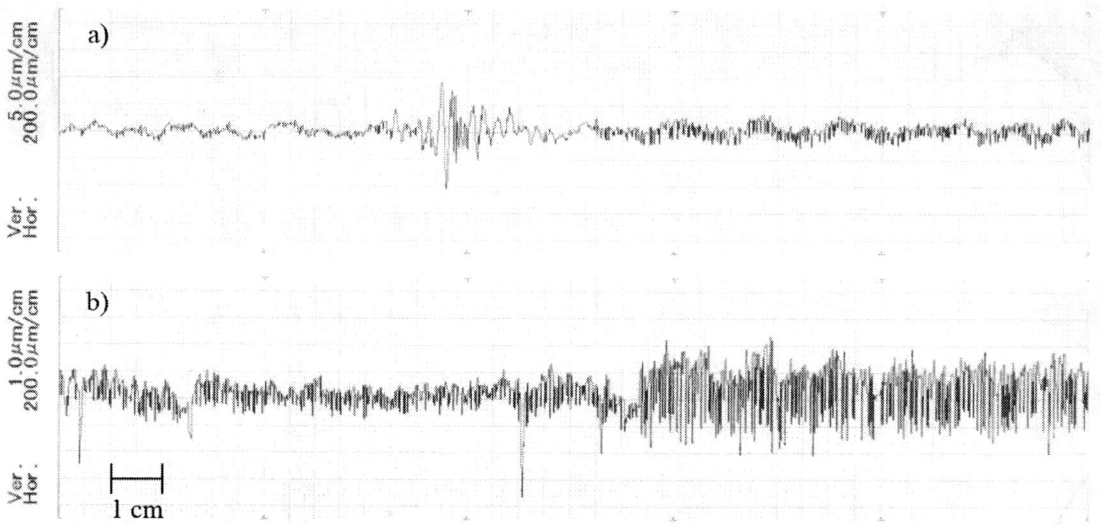

Figure 2. Roughness measurement diagrams: a) bare substrate; b) coated sample.

A plot of nanoindentation load/unload curve using a load of 25 mN (averaged over three measurements) is shown in Fig. 3. Here the penetration in the sample is 266,7 nm which represents 11,9% of its coating thickness. The curve is apparently smooth giving a prof any cracks did not occur during the indentation [8].

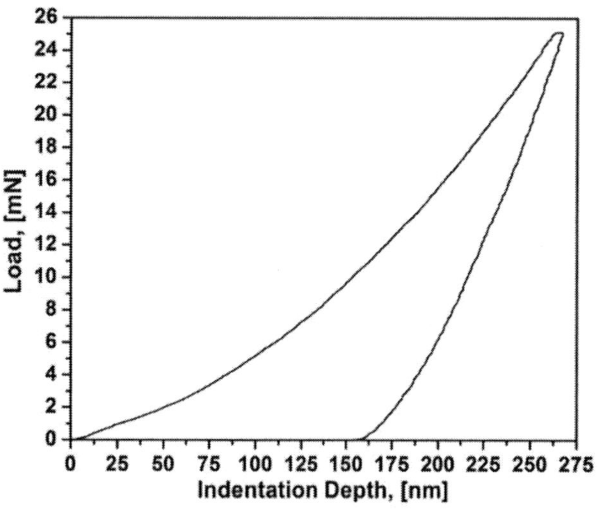

Figure 3. Nanoindentation load/unload curve using a load of 25 mN.

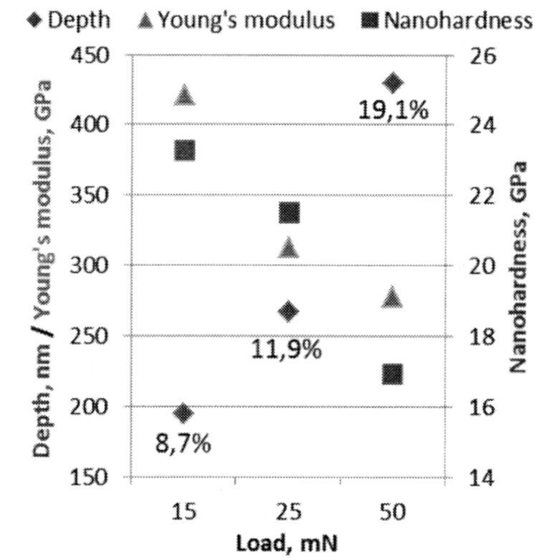

Figure 4. Depth of indentation, modulus of elasticity and nanohardness at loads of 15, 25 and 50 mN.

The indentation depth, Young's modulus and nanohardness at three loads: 15, 25, 50 mN (averaged over three measurements) are shown in Fig. 4. Below the depth is also given its percentage of the coating thickness. It could be assumed that the actual values of nanohardness: 21,5 GPa and Young's modulus: 312,7 GPa are those at a load of 25 mN. Then the penetration is relatively small (see above) hence the substrate hardness does not influence on the values. On the other hand, it is large enough preventing from Indentation Size Effects (ISEs). As for the bulk hardness of the sample, it is 53 HRC.

The scratch test is depicted in Fig. 5. The indenter is pressed by a force Fn which changes linearly from 0 to 30 N during a run x of 3 mm (Fig. 5a). The first critical load L_{C1} arises at ca. 15,0 N (first coating damages: Fig. 5b). Any delamination, i.e. - exposure of the sample, is not seen until the end of the test hence the second critical load L_{C2} is not detected. However, there are increasingly hemispherical crevices and breakages. The used equipment does not allow higher loads than 30 N.

Looking on the diagram in Fig. 5a, one can see that the values of the friction force Ft and the friction coefficient μ start to fluctuate obviously from the middle of the distance to its end. Probably, the reason is the existence of coating damages. However, Ft changes relatively linearly and smoothly. The coefficient of friction μ is almost constant, slightly increasing after the middle of the distance. This behavior of these two dimensions implies relatively low internal stress in the coating, probably because of the presence of a nanolaminate structure. The low friction coefficient μ is largely due to the use of a diamond tip. It is obvious that this coefficient increases slightly until the end of scratching when the penetration depth Pd is maximal and there are increasingly interactions with the indenter [9].

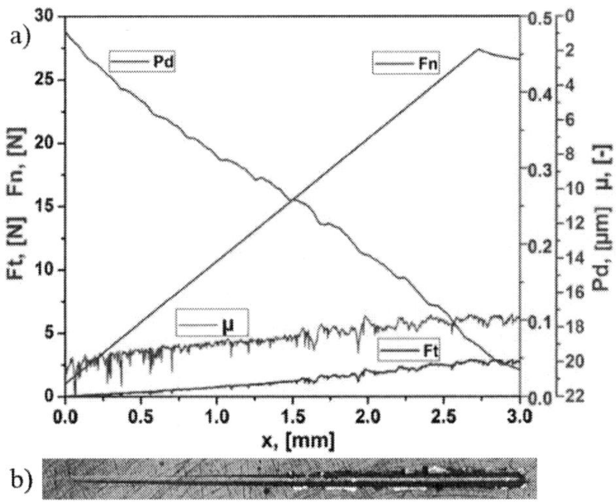

Figure 5. Scratch test: a) diagram; b) view of the trace.

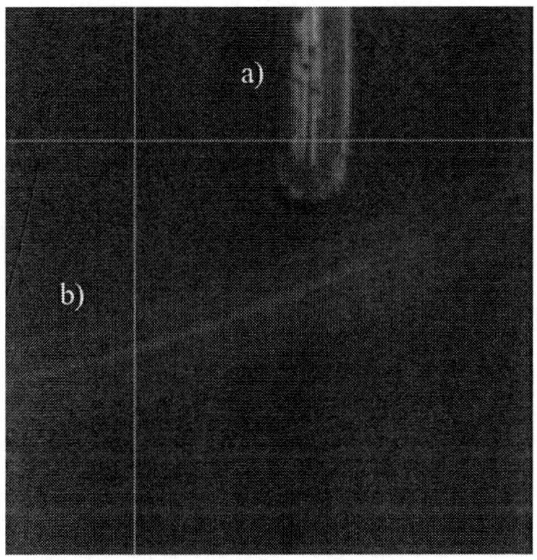

Figure 6. Tracks of the wear rate test: a) ball-on-flat; b) ball-on-disc.

The two tests of wear resistance were carried out at room temperature without use of a lubricant. Tracks of their implementation could be seen in Fig. 6. The wear rate I_w was calculated using the well-known equation:

$$I_w = V.F^{-1}.L^{-1} \left[mm^3.N^{-1}.m^{-1}\right] \quad (1)$$

where V - wear volume (mm^3), F - normal load (N), L - sliding distance (m). The wear volume was calculated geometrically after initial measurement of the track width using a microscope.

The ball-on-flat test was accomplished under the following conditions: F = 2 N, L = 50 m (2273 double strokes of 11,0 mm). Average sliding speed of the sample was 20 mm/s (of course, the moment value of sliding speed is different, since at the ends of a track the sample changes its movement direction). From a geometric point of view, the track is a relatively complex figure because of its different profile - at the edges and the middle. For the exact calculation of the wear volume, the method described in [10] was applied. A wear rate I_w of $57{,}8.10^{-6}$ mm^3/Nm was calculated.

The ball-on-disc test was implemented under the following conditions: F=5 N, L =200 m. Average sliding speed of the sample was 20 mm/s as well which was also a moment value of the speed because of an uniform sample rotation. A wear rate I_w of $26{,}0.10^{-6}$ mm^3/Nm was computed.

The values given by the two methods are of one order, but are remarkably different - about twice. As a reason, one different depth penetration (consequently – a contact with layers possessing different properties) could be excluded: it is 3,70 μm for ball-on-flat and 3,79 μm for ball-on-disc respectively. The probable reason is the difference in relative movement between the sample and counter-part. During the linear motion in two directions, it is possible to arise one increased shear stress in the track which causes a local fatigue wear, hence the track is carved faster. One scratch in the middle of the track width is noticeable in Fig. 6a which supports such an assumption. According to the studies in [11] where the same two methods under similar conditions had been used, the wear rate is about 4 times greater using ball-on-flat test. Nevertheless, the results obtained here using also the both methods are close to these known for similar coatings [5, 12].

4. Conclusions

The presented studies of the considered nanolaminate Ti/TiN/AlTiSiN/(AlTiSiN/TiAlSiN)n/AlTiSiN hard coating give a reason to assume that it is suitable for deposition on cutting tools. In the same batch, similar tools (mills) from WC-Co alloy were coated and preliminary results show that their life span is increased ca. 2,9 times compared to uncoated ones under the same operating mode. This achievement is not surprising [1, 5, 6]. However, after further experiments, even better results could probably be expected.

Acknowledgements

The authors would like to acknowledge the support of the National Programme "Young Scientists and Post-Doctoral Students" funded by the Bulgarian Ministry of Education and Science (MES).

References

[1] Veprek S and Veprek-Heijman M J.G. 2008 Industrial applications of superhard nanocomposite coatings *Surf. Coat. Technol.* **202** (21) 5063–73

[2] Xiao D-H 2009 Microstructure and properties of homogeneous and gradient Ti(C, N) coatings *J. of Materi Eng and Perform* **18** (9) 1226-9

[3] Chen L, Chang K, Du Y, Li J and Wu M 2011 A comparative research on magnetron sputtering and arc evaporation deposition of Ti–Al–N coatings *Thin Solid Films* **519** (11) 3762–7

[4] Zhu L, Hu M, Ni W and Liu Y 2012 High temperature oxidation behavior of $Ti_{0.5}Al_{0.5}N$ coating and $Ti_{0.5}Al_{0.4}Si_{0.1}N$ coating *Vacuum* **86** (12) 1795-9

[5] Warcholinski B and Gilewicz A 2011 Multilayer coatings on tools for woodworking *Wear* **271** (11-12) 2812–20

[6] Xiao B-J, Chen Y, Dai W, Kwork K-Y, Zhang T-F, Wang Q-M, Wang Ch-Y and Kim K 2017 Microstructure, mechanical properties and cutting performance of AlTiN coatings prepared via arc ion plating using the arc splitting technique *Surf. Coat. Technol.* **311** 98–103

[7] Mrochek Zh, Vershina A, Ivashchenko S, Ivanov I, Frolov I and Latushkina S 2004 *Vacuum-Plasma Coatings* (Minsk, Belarus: Tekhnoprint) pp. 77-9 (in Russian)

[8] Mendoza C, Gonzalez Z, Gordo E, Ferrari B and Castro Y 2018 Protective nature of nano-TiN coatings shaped by EPD on Ti substrates *J Eur Ceram Soc* **38** (2) 495-500

[9] Aprahamian B, Nikolova M, Zaharieva and Yankov E 2016 Protective TiN coatings deposited on copper and aluminium medium voltage busbars of electrical distribution boards *Nanoscience & Nanotechnology: Nanostructured materials application and innovation transfer* **16** (1) 44-6

[10] Rupetsov V 2014 Experimental test stand for the study of coatings resistance *Journal of Food and Packaging Science, Technique and Technologies* **1** (4) 60-5

[11] Nanovea 2017 Tribology used to compare rotative wear and linear wear *Saved from URL:* https://www.azonano.com/article.aspx?ArticleID=4708

[12] Dishliev S and Mishev G 2011 Research the wear intensity of PVD coatings on steel X46Cr13 *Proc. of the 7th Int. Conf. on Tribology (BALKANTRIB'11)* Thessaloniki, Greece, pp 31-40

IOP Conf. Series: Materials Science and Engineering **618** (2019) 012048 doi:10.1088/1757-899X/618/1/012048

System for materials testing at static loading

V Tsonev[1], N Kuzmanov[1], B Borisov[1] and K Penkov[1]

[1]Technical University – Sofia, "Strength of Materials" Department, Bulgaria

E-mail: tzonev@tu-sofia.bg

Abstract. This paper presents a newly developed system for mechanical testing at static loading. The separate modules of the system are described - tensile testing machine, grips, extensometers for axial end transverse deformations, electronic control system. The results obtained during tensile testing with the new system are given. The yield strength, the tensile strength, the elasticity modules and the Poisson's ratio of general purpose steal S235JRC are determined.

1. Introduction

The development of energy, chemical, aviation, automotive and other industries as well as many other branches of technology involves the use of new materials. That is why the problem of studying the mechanical properties of new materials is quite topical.

The engineering approach for stress-strain analysis is based on standard mechanical tests. The test piece made from the respective material is subjected to pure tension (compression), pure shearing, pure bending, etc. The axial force F, with which the test piece is loaded, is increased slowly until its destruction. The values of the force and displacement are continuously measured and recorded. The obtained results are used to determinate the desired mechanical properties.

In modern laboratories universal testing machines with servo-controlled loading mechanisms are used for conducting mechanical tests at static loading because they allow for precision loading [1,2].

This type of tensile testing machines is continuously modernized by implementing electronic devices and new software [3] which permit:

- Automatic control of the experiment;
- Reliable management of the test parameters during prolonged experiment;
- Recoding and visualization of the test parameters;
- Ease and flexibility of conducting different types of experiments using the properties of the machine software (for example, different types of cycling loading).

For measuring the deformations of the test piece most often extensometers are used. These devices convert the mechanical deformation in proportional electrical signal [4,5].

The devices are complex and very expensive. The purchase of off-the-shelf systems is ineffective. Experience shows that in most cases it is more viable to buy specific components and conduct in-house development of other components and software according to the goal of the tests. Thus, the team of scientists has the ability to maintain and improve the machinery during its service life. In addition, scientists are capable of assessing the effect of test machinery on experimental results, which is a prerequisite for more accurate interpretation of the test results.

The aim of the present research is to build a system for mechanical tests at static loading using available modules and to design and build the necessary additional devices.

2. System design

To conduct a mechanical test at static loading with measurement and recording of all appropriate values, the following modules are necessary:

- A universal testing machine used to conduct the static loading. A ZD10 tensile testing machine in non-working condition described below is available at the Strength of Materials Department at the Technical University of Sofia. The machine has to be modernized. This modernization is described below;
- Devices for deformation measurement. Extensometers built by Schenck (Germany) for measurement of longitudinal and transverse deformation are available at the Strength of Materials Department. The extensometers are described below;
- A control system for the new testing machine. It will include an electronic control unit used to control the testing machine and personal computer (PC). The system is described below.

3. Testing machine ZD10

Figure 1 shows a 3D module of the ZD10 universal tensile testing machine. The machine is designed to test materials at static loading. It comprises:

- Housing *1* of the reduction gear, attached to lead screw transmission;
- Two vertical columns *2*;
- Load cell *3*;
- Load cell housing *4*;
- Traverses – upper fixed traverse *5* and lower movable traverse *6*;
- Screw *7* with trapezoidal thread;
- Grips *8*;
- Main housing *9*. The power electronic unit, the power transformer, the DC motor, the analog motor control, the tube analog load cell amplifier, the mechanical system of the measurement scale *10* and the recording device *11* are situated in the main housing.

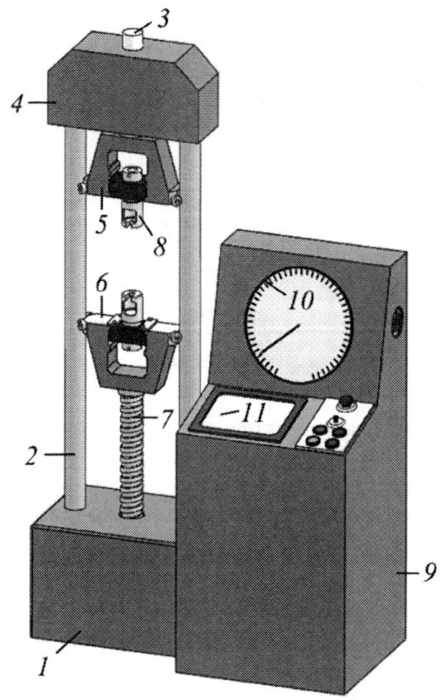

Figure 1. 3D model of the universal testing machine ZD10.

The machine is equipped with:

- Spherical grips (they allow for self-aligning of the test piece) for cylindrical test piece;
- Wedge type grips for testing flat test piece and cylindrical test piece;
- Device for compression testing;
- Device for three-point bending tests;
- Device for sheering tests.

The load cell and the electronic control unit are in non-working condition which makes the machine inoperable. On the other hand, recording of the measured values on paper is obsolete. A modernization of the tensile testing machine is required.

4. Modernization of the universal testing machine ZD10

A new strain gauge load cell is mounted. The force measuring device converts mechanical deformation into a proportional electrical signal. The elastic element of the load cell has been designed for 100 kN. The load cell is calibrated according to EN ISO 376:2011 [6].

A linear incremental encoder is mounted on the right hand-side column. It is used to measure linear displacement. The displacement measuring device generates electrical impulses whose number is proportional to the displacement. The linear encoder has a range from 0 to 800 mm.

A new servo-controlled DC motor is mounted in the main housing of the universal testing machine. The DC motor is equipped with an optical sensor (rotary encoder) with a resolution of 2000 impulses per revolution. The DC motor has a nominal speed of 2100 rpm, continuous torque of 7 Nm, a nominal voltage of 150V and a maximum current of 13 A.

5. Extensometers for measuring deformation

Figure 2 shows a 3D model of an extensometer used to measure longitudinal deformation produced by the Schenck company.

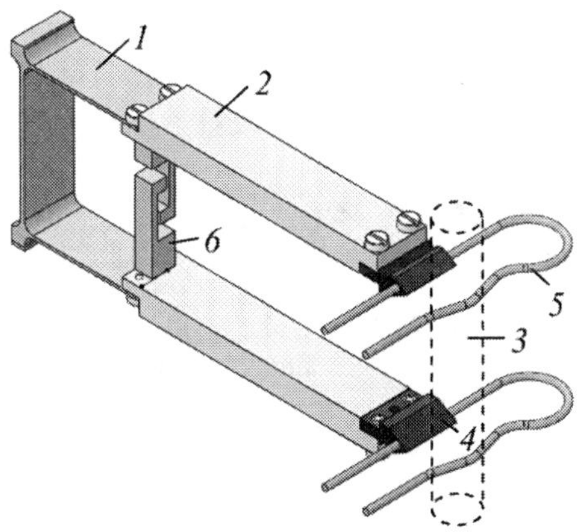

Figure 2. 3D model of an extensometer used to measure longitudinal deformation.

The elastic element *1* is attached to the arms *2*. The extensometer makes contact with test piece *3* using blades *4*, which are attached to the arms. The tight contact between the extensometer and the test piece is ensured by springs *5*. The extensometer is equipped with a number of springs suitable for test piece with different diameters. A limiter *6* is attached to the arms of the extensometer; it is used to prevent excessive deformation of the elastic element. When the test piece is deformed the arms are spread and as a result the elastic element is deformed. The gauge length of the extensometer is 25 mm. The elastic element is designed to measure deformations of ± 10 mm.

Figure 3 shows a 3D model of an extensometer used to measure transverse deformation produced by Schenck.

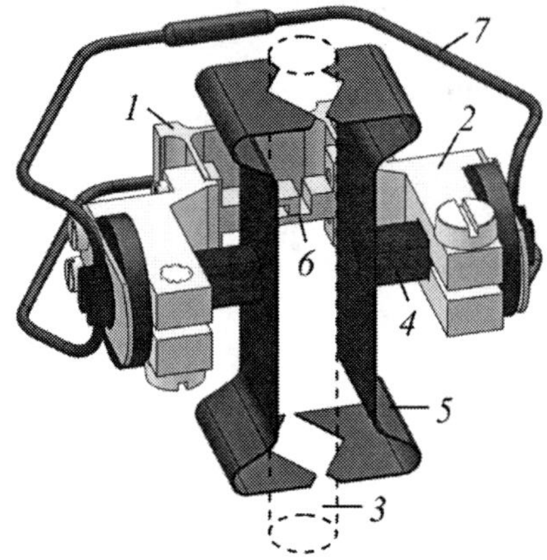

Figure 3. 3D model of an extensometer used to measure transverse deformation.

Elastic element *1* is attached to the arms *2*. The extensometer makes contact with the test piece *3* using blades *4*, which are attached to the arms. Plates 5 are attached to the blades, with are used for radial alignment of the extensometer in respect to the test piece. A limiter 6 is attached to the arms. The limiter is used to prevent excessive deformation of the elastic element. When the extensometer is mounted on the test piece the arms spread apart. As a result the elastic element is pre-deformed and when the diameter of the test piece is reduced the element strives to regain its original shape. The tight contact between the extensometer and the test piece is ensured by spring *7*. The extensometer is designed to be used with test piece with a diameter from 5 to 15 mm. The elastic element is designed to measure deformations of ± 1 mm.

The device used to measure deformation (the elastic element) converts the mechanical deformation into proportional electrical signal using strain gauges wired in full bridge. The load cell is calibrated according to EN ISO 9513:2012 [7].

6. Control system

The schematic of the used control system is shown in Figure 4. The load cell *LC*, the extensometer for measuring longitudinal deformation *E1*, the extensometer for measuring transversal deformation *E2* and the linear encoder *LE* send signal to the electronic control unit *ECU*. The *ECU* receives signal from the position limit switches *PLS1* and *PLS2*. If the limit switches are tripped, the *ECU* disengages the servo-controlled DC motor *DCM* and the reduction gearbox *RG*. To prevent external noise in the amplifier for the load cell and the linear encoder, the electronic control (*DCCU*) of the servo-controlled DC motor is mounted in separate housing from that of the *ECU*.

The *ECU* is equipped with its own memory in which the measured values can be recorded (given and measured force, longitudinal and transverse deformation and the displacement of the moving traverse). If the system works without connection to *PC*, the data is stored only in the memory of the *ECU*, which is enough for 1200 recordings. When the test is prolonged or there is a need for recording the data with a high sample rate the *ECU* is connected with a *PC*. In this case the data is transferred to PC in real time. The data is processed, visualized and recorded with specially developed software.

Using the *ECU's* keyboard, the following parameters can be adjusted:

- Calibrating values of the measuring channels;
- Testing mode – force control, displacement control or time control;
- Loading profile – trapezoidal, sinusoidal, or triangular. For each profile the number of cycles; the beginning, maximum and ending values of loading; the loading speed in kg/min or mm/min; loading time, holding time and unloading time can be set;
- The time interval for recording the measured values.

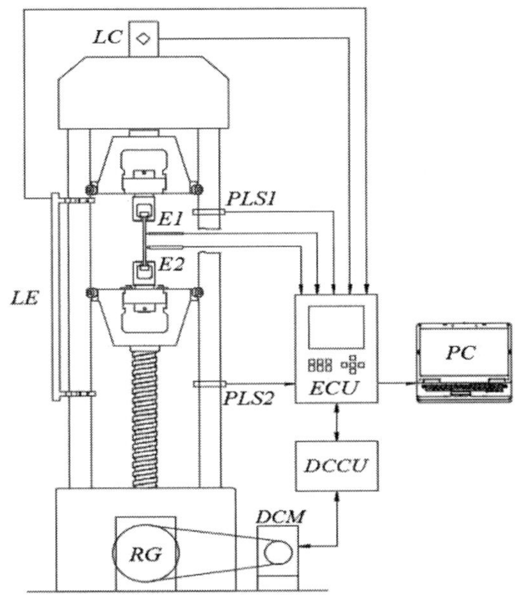

Figure 4. Schematic of the used control system.

7. Mechanical testing with the developed system

To test the newly developed system a tensile test of the test piece made from calibrated steel S235JRC is carried out. The test pieces are made from a rod with a diameter of 6 mm. The extensometers are mounted on the test piece to measure the longitudinal and transverse deformation. The general view of the test system is shown in Figure 5.

Three test piece were tested. The test pieces are subjected to pure tension while the force, longitudinal and transverse deformation are continually recoded. The testing speed was set at 1050 kg/min.

Figure 6 shows the stress-percentage extension curve of the studied steel. The rupture of the test pieces has occurred in the gauge length of the extensometer used to measure longitudinal deformation. Table 1 presents the values of the following parameters determined using the stress-strain curves:

- Tensile strength R_m – the stress corresponding to the maximum force, which the test piece can bear during testing [8];
- Proof strength, plastic extension $R_{p0,2}$ – the stress, under which the relative percentage extension is equal to 0,2%. It is determined for steel, which does not have a physical yield limit;
- Modulus of elasticity E – the tangent of the angle between the linear part of the stress-percentage extension curve and the x-axis;
- Poisson's ratio ν – the absolute value of the ratio of the transverse deformation to the longitudinal deformation in the linear part of the stress-strain curve.

The obtained results correspond to the data in the reference literature for S235JRC steel.

8. Conclusion

A modern universal system for materials testing at static loading has been developed. It can perform the following static tests – pure tension test, pure compression test, shearing test, three-point bending test. It is possible for the system to be controlled by force, by displacement or by time. A different type of loading profiles can be applied - trapezoidal, sinusoidal, or triangular. The measured values are recorded or visualized on a PC with special software.

The test results obtained with the newly developed system were given. The results show that the system can be used to determine the material properties of different materials at static loading.

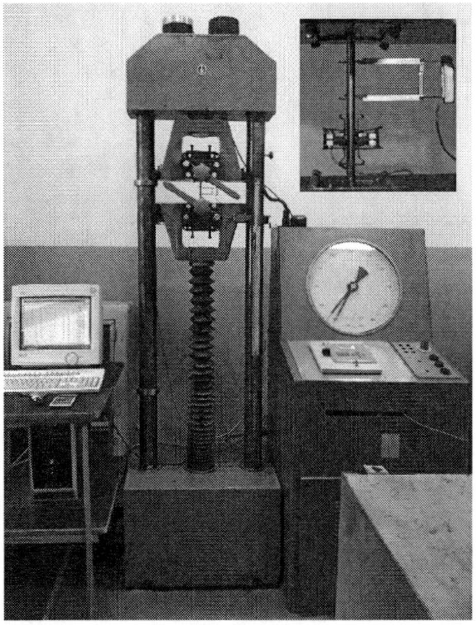

Figure 5. General view of the system for testing materials at static loading.

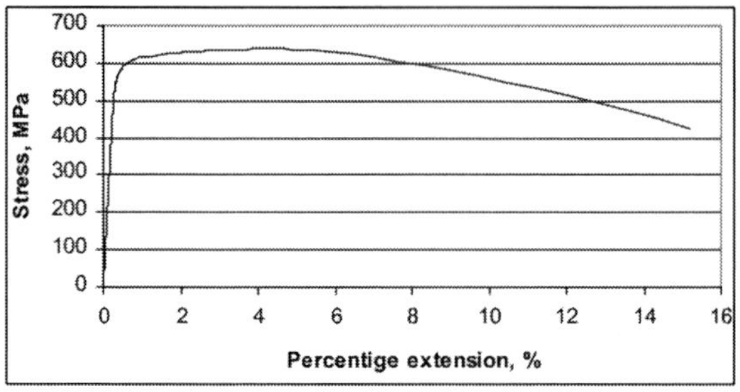

Figure 6. Stress-percentage extension curve of S235JRC steel under tension.

Table 1. Material constants of S235JR steel.

Test piece №	R_m, MPa	$R_{p0,2}$, MPa	E, GPa	ν, –
1	642	597	203	0,27
2	639	591	201	0,29
3	640	587	204	0,27
Average	640	592	203	0,28

The authors would like to thank the Research and Development Sector at the Technical University of Sofia for the financial support.

References

[1] Keppler I, Kocsis L, Oldal I and Csatár A, 2011, Determination of the discrete element model parameters of granular materials, *Journal of Hungarian Agricultural Engineering*, Volume **23**, Pages 30-32.

[2] Faltermeier A, Behr M, Rosentritt M, Reicheneder C and Müßig D, December 2007, An in vitro comparative assessment of different enamel contaminants during bracket bonding, *European Journal of Orthodontics*, Volume **29**, Issue 6, Pages 559–563.

[3] Afrose M, Masood S, Iovenitti P, Nikzad M and Sbarski I, June 2016, Effects of part build orientations on fatigue behaviour of FDM-processed PLA material, *Journal of Progress in Additive Manufacturing*, Volume **1**, Issue 1–2, Pages 21–28.

[4] Šafka J, Ackermann M, Seidl M, Běhálek L, Bobek J and Klunejko J, September 2016, Evaluation of the impact of production parameters on the final properties of the part made of nylon 12 with rapid prototyping technology (FDM), *Modern Machinery Science Journal*, Pages 956–959.

[5] Letcher T and Waytashek M, November 2014, Material Property Testing of 3D-Printed Specimen in PLA on an Entry-Level 3D Printer, *ASME 2014 International Mechanical Engineering Congress & Exposition*, Montreal, Canada.

[6] EN ISO 376:2011, Metallic materials - Calibration of force-proving instruments used for the verification of uniaxial testing machines (ISO 376:2011).

[7] EN ISO 9513:2012, Metallic materials - Calibration of extensometer systems used in uniaxial testing.

[8] EN ISO 6892-1:2016, Metallic materials - Tensile testing - Part 1: Method of test at room temperature.

CFD based design of a ventilated space

S G Mijorski[1], D G Markov[2], G T Pichurov[2], P Stankov[2], N G Ivanov[3], I S Simova[2]

[1]SoftSim Consult Ltd., Mladost-4, bl.438, 1715, Sofia, Bulgaria
[2]Centre for Research and Design in Human Comfort, Energy and Environment, Technical University of Sofia, 8 Kliment Ohridski Blvd., 1000, Sofia, Bulgaria
[3]Department of Fluid Dynamics, Combustion and Heat Transfer, Peter the Great St.Petersburg Polytechnic University, 29 Polytechnicheskaya str., St.Petersburg, 195251, Russia

peter.stankov@abv.bg

Abstract. The paper presents Computational Fluid Dynamics (CFD) based design of a densely populated classroom with mechanical ventilation. The aim is by means of numerical simulations to define the proper locations of four four-way ceiling fancoils in order to provide controlled conditions of the thermal environment in the classroom under summer conditions. CFD simulations are completed by Fluent 6.3 software package. Cooling loads through the enclosing surfaces of the room and the material properties of the modeled elements are evaluated by established and approved methods in the field of HVAC engineering.

1. Introduction

Thermal comfort of occupants is increasingly important economic factor and a very complex phenomenon [1]. The basic standards, which define the thermal comfort, are EN ISO 7730-2007 and ANSI/ASHRAE 55-2017 [2, 3]. Both standards apply the concept suggested by Fanger, [4]: a person is at thermal comfort when its whole body is at thermal balance (general thermal comfort) and there are no local discomforts, i.e. unwonted cooling/heating of some body parts. General thermal comfort is evaluated by the Predicted Mean Vote (PMV) index, which afterwards is used for evaluation of the percentage of people who will be dissatisfied with the actual thermal environment - Predicted Percentage of Dissatisfied (PPD) index. Local thermal discomforts, arising due to asymmetry in the thermal environment are radiant asymmetry, risk of draught (Draught Rating, DR), vertical temperature difference, and uncomfortable floor temperature.

Since a human body exchange heat with the surroundings at the skin by conduction, convection, radiation and evaporation, as well as within the lungs by convection and evaporation several physical parameters must be taken into account during the design of the indoor environment. With so many aspects to consider it is challenging to design a comfortable thermal environment in a densely occupied space with total volume ventilation. For this reason it is common to resort to the methods of Computational Fluid Dynamics (CFD).

The goal of this study is to design numerically the thermal environment in a classroom at the Technical University of Sofia following the ISO 7730-2007 standard, so that 24 students and a teacher are at thermal neutrality. The task is to select the positions of four four-way ceiling fancoils and the angle of the louvers at their outlets, which will ensure the occupants' thermal comfort during summer season.

CFD simulation predicted the air movement and heat exchange in the room. Fluent 6.3, [5], software package was used for the simulations. This package provides numerical algorithms for calculation and visualization of all pertinent flow parameters - velocity components, temperature and pressure. Comfort indices PMV, PPD as well as local discomfort due to draught (DR), were further programmed via a User Defined Function (UDF) following the ISO 7730-2007 standard [3]. Cooling loads through the enclosing surfaces of the room and the material properties of the modeled elements were evaluated by established and approved HVAC procedures.

At this stage of the design of the thermal environment, simplification was made assuming equally distributed flow rate through the four outlets at each fancoil and uniform velocity distribution throughout. The supply air direction was assumed identical to the inclination of the louvers.

2. Physical model

Figure 1 presents the geometry of the classroom. Although it's a box shaped room, the presence of columns, beams, window sills, and radiators makes its geometry fairly complex. A door is located on the west wall and all windows are on the east wall. Air volume of the classroom is 193.84 m^3.

Based on the preliminary HVAC calculations, it is decided to use four four-way ceiling fancoils TOSHIBA MMU-AP0092H. On the west wall of the room four air transfer grills are located. Two cases with respect to fancoil positions as illustrated on Figure 2 are identified as possible. Case 1 is with distance between the geometrical centers of fancoils along X direction δ_{xcc}=100 cm, and Case 2 is with δ_{xcc}=80 cm. For both cases Y coordinates of the geometrical centers of the fancoils are the same.

Cooling loads through all enclosing surfaces of the room are calculated based on the data about their geometry, construction materials, prescribed indoor air temperature and outdoor conditions – temperature and solar insolation. Thermal loads for all boundary elements, internal heat sources (occupants) and lighting units are defined based on Methodological guidelines for calculating the annual energy consumption, heat, humidity and energy load of buildings and individual hazardous substances [7].

3. Computational details

CFD simulations are based on steady-state Reynolds-Averaged Navier Stokes (RANS) equations, including the energy equation and Discrete Ordinate (DO) radiation model, Boussinesq approximation of buoyancy forces, the standard k-ε turbulence model and the respective boundary conditions. Numerical simulations use the pressure-based steady solver for RANS equations of Fluent 6.3, [5].

3.1. Computational grid

The computational grid was prepared in Gambit 2.4.6 mesh generator, comprising 10 cm step structured rectilinear grid, which can reduce numerical errors and guarantee faster convergence. For the purpose of the CFD simulations classroom geometry is slightly modified in order to use 10 cm structured rectangular grid. Heated dummies simulate occupants. Figure 3 shows numerical representation of 24 seated occupants and one standing upright. Figure 4 shows all elements in the computational domain for Case 1 ($\delta_{x.c.c}$=100 cm).

As shown on Figure 5a, the geometry of the fancoils is too complex. For the purpose of the study it is simplified as shown on Figure 5b.

3.2. Boundary conditions

Boundary conditions for the energy equation are presented in Table 1. The internal objects - desk, columns, and radiators participate only as obstacles, but don't exchange any heat. For simplification purposes the cooling load of lighting and ceiling is combined. This increases the total load of the ceiling from 1051 to 1251[W]. All elements are split in five different types of materials with corresponding thermal properties. The model features four pressure outlets on the west wall (the air transfer grills), four velocity outlets (the fancoils inlets), and 16 mass flow inlets (the four outlets of each fancoil). Boundary conditions on the fancoil surface are given in Table 2 and Table 3.

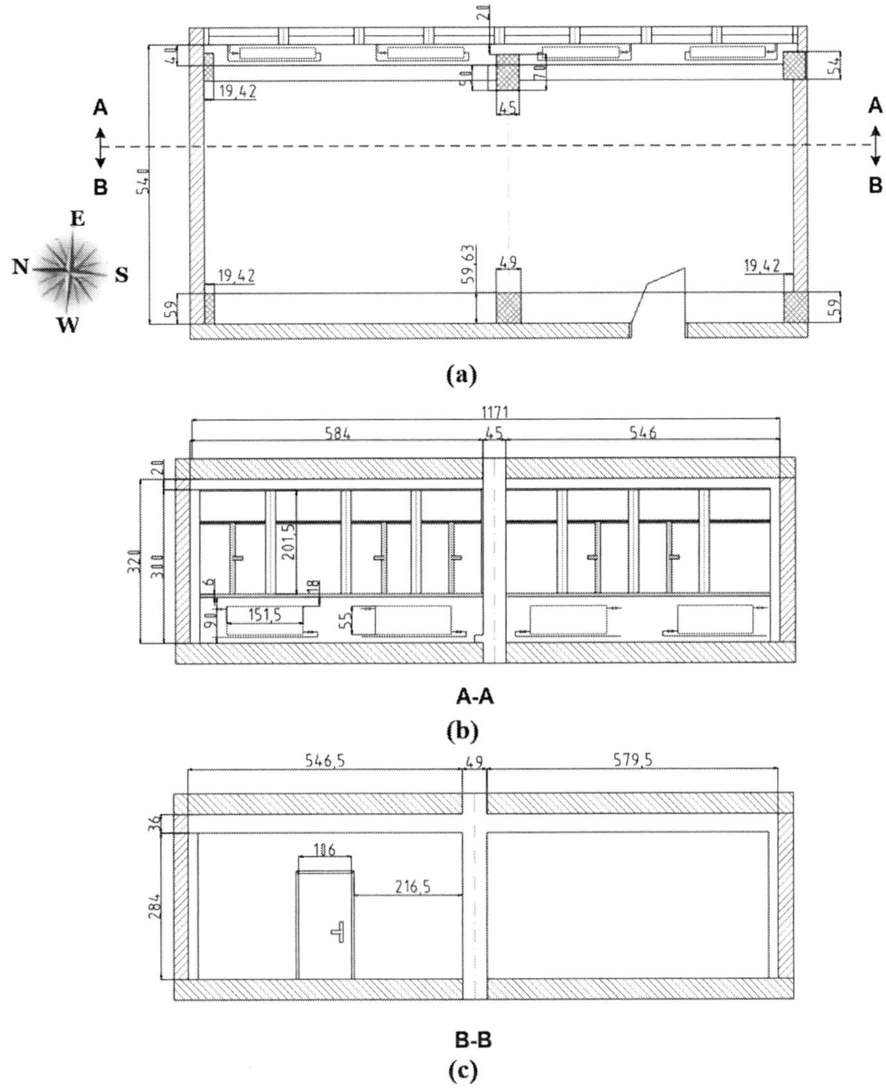

(a)

(b)

(c)

Figure 1. Classroom geometry: (a) plan; (b) section A-A; (c) section B-B..

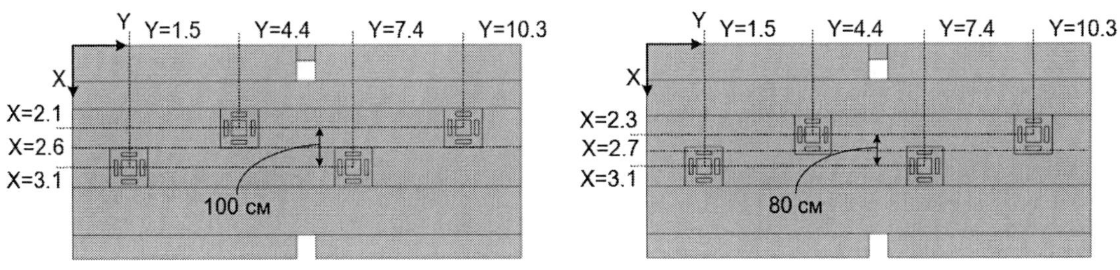

(a) (b)

Figure 2. Simulated fancoil locations: a) δ_{xcc}=100cm; b) δ_{xcc}=80cm.

IOP Conf. Series: Materials Science and Engineering **618** (2019) 012049 doi:10.1088/1757-899X/618/1/012049

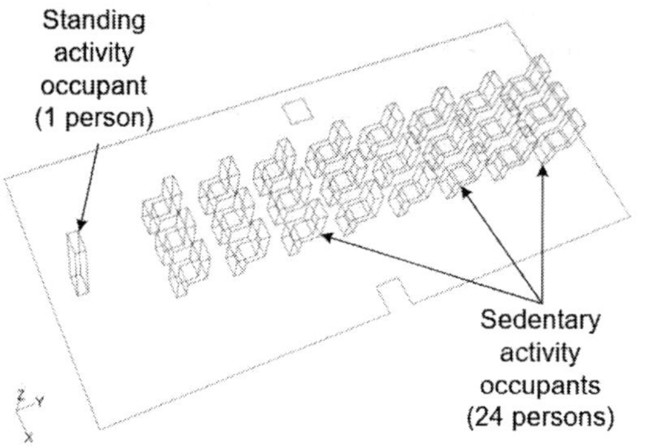

Figure 3. Numerical model of the occupants.

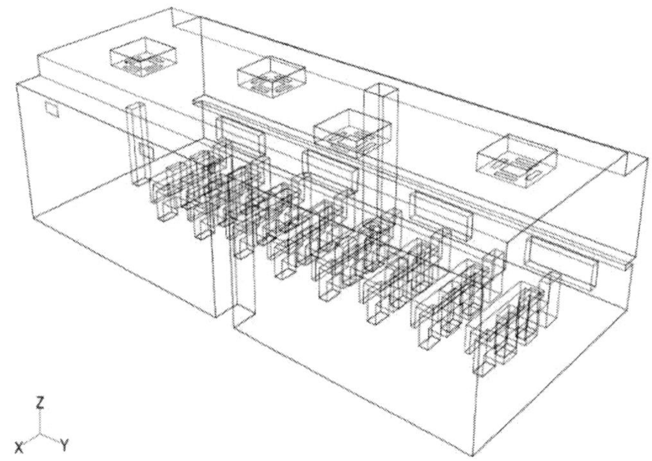

Figure 4. Numerical model of the classroom

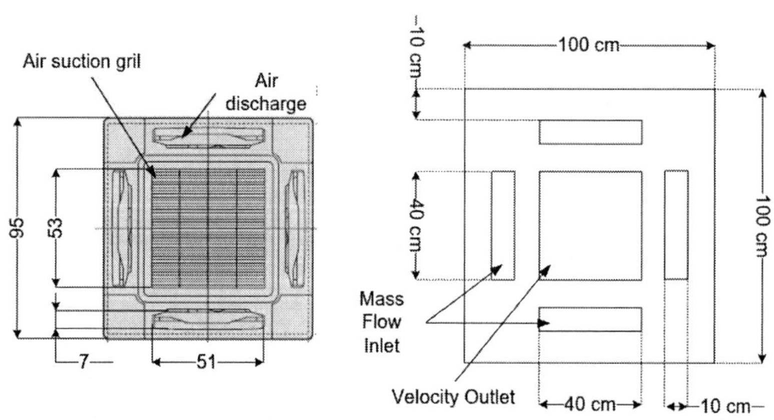

Figure 5. Fancoil geometry: (a) real; (b) numerical model.

Mass flow inlets are divided in four different groups depending on which side of the fancoil are: East, West, North, and South. This gives possibility for group setting of boundary conditions about louvers inclination, Figure 6.

Table 1. Numerical model components' boundary conditions.

Component [-]	Type of material [-]	Numerical area [m^2]	Total cooling load [W]	Heat Flux [W/m^2]
Outer wall (East)	Outer walls	17.01	39.38	2.32
Inner wall (West)	Inner walls	31.32	365.40	11.67
North wall	Inner walls	16.55	168.00	10.15
South wall	Inner walls	16.55	133.14	8.04
Windows	Glass	21.24	995.10	46.85
Floor	Floor and Ceiling	59.99	1051.00	17.52
Ceiling + lamps	Floor and Ceiling	68.61	1251.00	18.23
Sited occupants	Skin and clothing	48.00	2088.00	51.41
Standing occupant	Skin and clothing	2.12	108.98	43.50

Table 2. Fancoils – Velocity Outlets.

Recirculation elements -	**Velocity Inlet** **[m/s]**	**Turbulence Intensity** **[%]**	**Hydraulic Diameter** **[m]**	**Temperatures** **[C]**
1, 2, 3 and 4	-0.692344	10	0.2	24.5

Table 3. Fancoils – Mass Flow Inlet.

Mass Flow Inlet elements -	**Mass Flow Rate** [kg/s]	**Turbulence Intensity** [%]	**Hydraulic Diameter** [m]	**Temperatures** [C]
East, North, South and West	0.226776	10	0.5	17.7

3.3. Simulated cases

In the analysis 10 simulation cases are considered. For each of the two cases with respect to the distance between the fancoil centres along X-direction are simulated five different inclinations (α – the angle between the plane of the louver and the ceiling) of the louvers at the mass flow inlets. Cases of the inclination for both models are shown on Figure 6. The louver inclination is modelled by changing velocity profile on mass flow inlets.

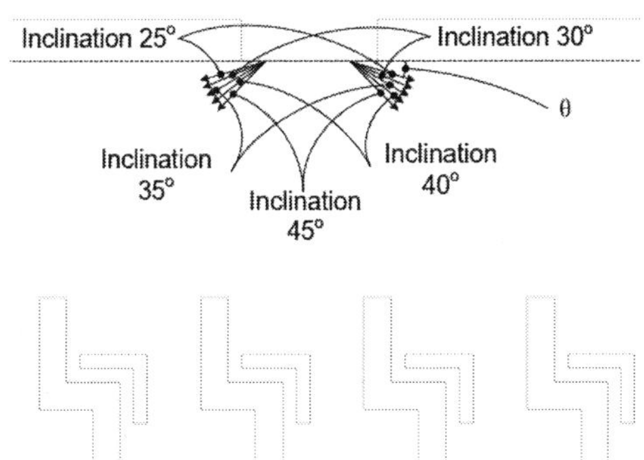

Figure 6. Louver inclination.

3.4. Operating conditions and initial guess
Operating conditions are set to: Atmospheric pressure: 94500, [Pa]; Gravity acceleration: -9.81 in Z directions, [m/s^2]; Operating temperature: 24.5, [C].

Initial guess for simulated variables are as follows: Turbulent Kinetic Energy: 0.03747055, [m^2/s^2]; Turbulent Dissipation Rate: 0.0048341, [m^2/s^3]; Temperature: 23.89264, [C].

4. Results about air velocity and temperature field
Here are presented only the results for Case 1 (δ_{xcc}=100cm) with blade angle α=30°. The variation of both air temperature and velocity magnitude are visualized in the Y-Z plane at X=2.6 m (between the fancoils – Figure 2) and in the X-Y planes at Z=0.6 m (the body centre of gravity for a seated person) and Z=1.1 m (head level of a seated person).

Air temperature variation in the classroom is visualized on Figure 7 and Figure 8 and air velocity variation on Figure 9 and Figure 10.

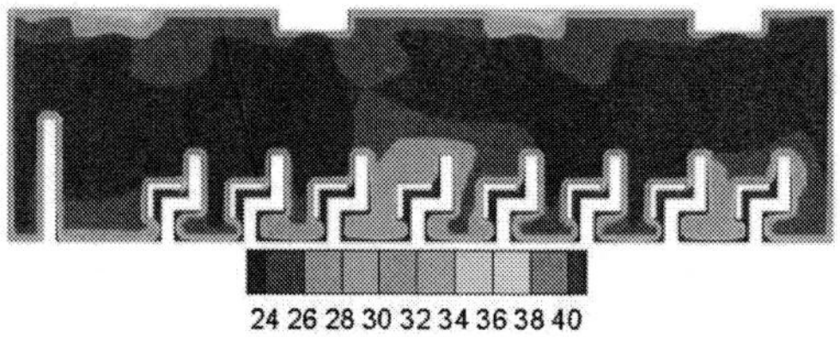

Figure 7. Contour plots of air temperature (°C) in the Y-Z plane at X=2.6 m.

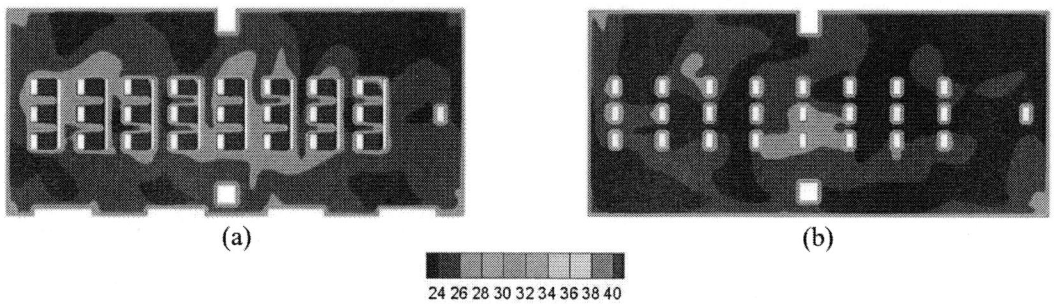

Figure 8. Contour plots of air temperature (°C) in the X-Y planes at: a) Z=0.6 m; b) Z=1.1 m.

Figure 9. Velocity magnitude (m/s) contour plots in the Y-Z plane at X=2.6 m.

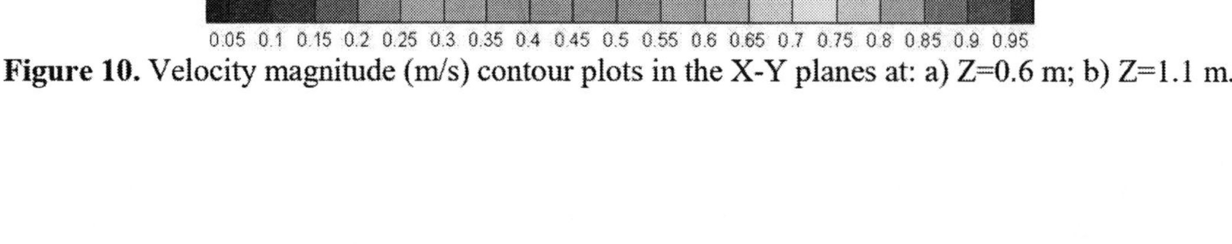

Figure 10. Velocity magnitude (m/s) contour plots in the X-Y planes at: a) Z=0.6 m; b) Z=1.1 m.

IOP Conf. Series: Materials Science and Engineering **618** (2019) 012049 doi:10.1088/1757-899X/618/1/012049

5. Results about PMV, PPD, and DR

The purpose of the simulations is to determine the basic parameters for the thermal comfort of occupants in the investigated room. The calculation of these parameters is based on the EN ISO 7730:2007 standard [3] and is described with full details in [6].

Risk of draught, which characterizes the local heating/cooling of the body by convection, depends on the local features of both the air velocity field (mean velocity and turbulence intensity) and air temperature field. In EN ISO 7730-2007 it is defined as

$$DR = (34 - t_a)(\bar{V} - 0.05)^{0.62}(0.37\bar{V}T_u + 3.14) \quad (1)$$

where DR is the percentage of people dissatisfied due to draught, %; t_a is the local air temperature, °C; $\bar{V}$ is the local mean velocity (magnitude), m/s; T_u is the local turbulence intensity, %.

Results about DR, PMV, and PPD variation in the horizontal planes Z=0.6 m and Z=1.1 m are presented on Figure 11, Figure 12 and Figure 13, respectively.

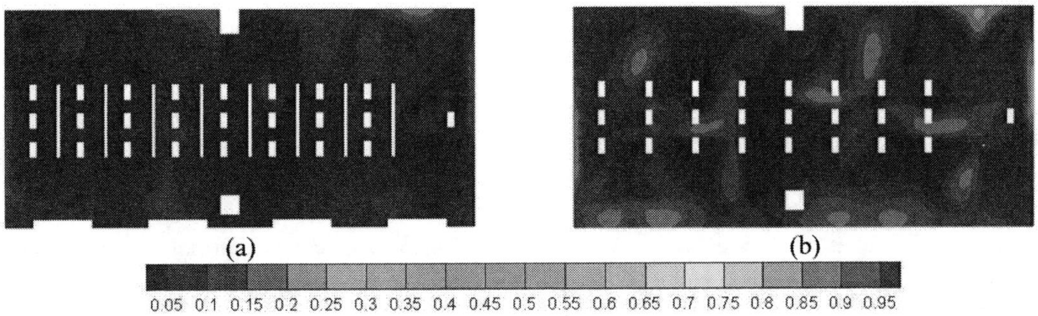

Figure 11. DR, Case 1, α=30°: a) Z=0.6 m; b) Z=1.1 m.

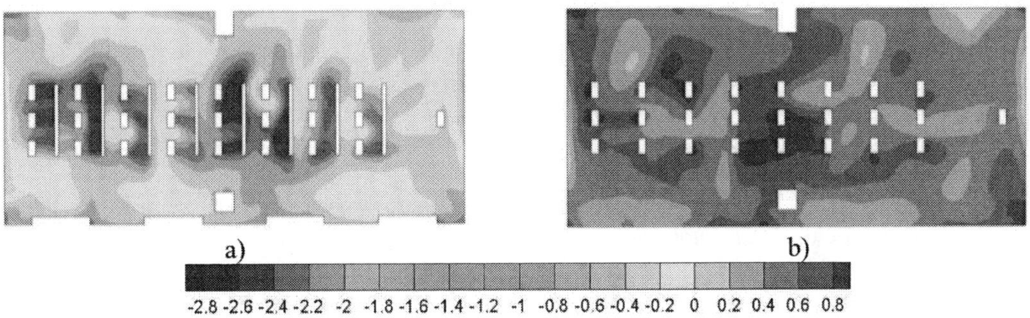

Figure 12. PMV, Case 1, α=30°: a) Z=0.6 m; b) Z=1.1 m.

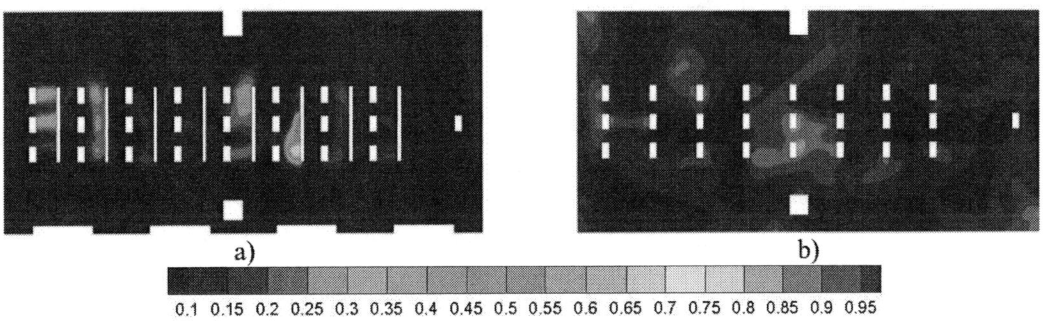

Figure 13. PPD, Case 1, α=30°: a) Z=0.6 m; b) Z=1.1 m.

6. Analysis and discussion

Based on equation 1 in order to avoid draught discomfort at head level it was decided to use four four-way ceiling fancoils, for air treatment, though only 2 are enough to remove the excess heat from the room. With four fancoils the flow rate of the recirculated air is increased but the temperature difference between the air in occupied zone and the cooled air is decreased to less than 2°C. Figure 11b shows that it is a good decision. There are a few zones only, far from the occupants, were it is possible to have 10%≤ DR ≤ 20%, which meets the requirements for category II of the EN ISO 7730-2006 standard.

EN ISO 7730-2006 standard puts requirements for PMV index at the point where the occupant's body center of gravity will be located. At all workplaces PMV ≤ 0.7, (Figure 12a). This meets the requirements for category III of the standard. The PPD distribution shows that by rearrangement of the work places it is possible to ensure that no more than 15 % of the occupants would be dissatisfied with the general thermal comfort.

The CFD simulation results show that the optimal location of the air distribution units and louvers inclination at their outlets is successfully identified.

This design was later physically implemented in the simulated room and used for further experimental studies.

7. Conclusions

The study employed the methods of CFD in the design stage of a total volume ventilation system. There is clear evidence that CFD based simulations of room air movement is the only effective tool in design optimization of HVAC systems. The simulated design was later installed and commissioned for experimental studies. It pays off to put extra coding effort and calculate thermal comfort indices that are otherwise not part of the classical CFD. The thermal comfort distribution allows not only to select a proper ventilation design, but also to allocate workplaces under optimal thermal conditions that will reduce human complaints and work absence due to illness.

Acknowledgments

The present study is supported by Bulgarian Science Fund of the Ministry of Education and Science (grant № ДНТС/Русия 02/11 from 15.06.2018) and by the Russian Foundation for Basic Research (grant № 18-58-18011).

References

[1] Angelova, R. A. 2016 Textiles and human thermophysiological comfort in the indoor environment, *Boca Raton (US)*
[2] ANSI/ASHRAE Standard 55-2017 2007 Thermal environmental conditions for human occupancy
[3] EN ISO 7730 2006 Ergonomics of the thermal environment – Analytical determination and interpretation of thermal comfort using calculation of the PMV and PPD indices and local thermal comfort criteria
[4] Fanger P O 1972 *Thermal Comfort: Analysis and Applications in Environmental Engineering*, New York: Mcgraw-Hill
[5] Fluent 6.3 2007 User's guide, *Fluent 6.3 documentation, Fluent Inc.*
[6] Markov D. 2006 Methodology For Thermal Environment Assessment in Practice, *EuroAcademy on Ventilation and Indoor Climate, ISBN – 13: 978-954-91681-4-3*
[7] Methodological guidelines for calculating the annual energy consumption, heat, humidity and energy load of buildings and individual hazardous substances 2007 *ISBN 978-954-8873-83-3, (in Bulgarian)*

Study of the influence of both a modifier P and a nanomodifier ND on the structure of a hypereutectic aluminium-silicon alloy AlSi18

Iv Panov[1], B Dochev[2], V Manolov[3], A Velikov[4], V Diakova[5], P Kuzmanov[6]

[1] Head Assistant, PhD - TU Sofia, Plovdiv Branch
[2] Assistant - TU Sofia, Plovdiv Branch
[3] Associate Professor, PhD - IMSETHC-BAS
[4] Head Assistant - IMSETHC-BAS
[5] Head Assistant, PhD - IMSETHC-BAS
[6] Head Assistant, PhD - IMSETHC-BAS

[1] specialista57@abv.bg
[2] boyan.dochev@gmail.com
[3] v.manolov@ims.bas.bg
[4] anmabg@abv.bg
[5] v_diakova@ims.bas.bg
[6] pawel_71@abv.bg

Abstract. The process of hardening a hypereutectic aluminium-silicon alloy AlSi18 in unmodified, modified and nano-modified state has been studied. Two types of modifiers have been used - a standard modifier P, introduced into the melt by means of the ligature CuP 0,4wt.%, and a nanomodifier (nanodiamonds - ND) 0,1wt.%. The influence of the various types of modifiers on the process of crystallization, as well as on the size of the primary Si crystals in the structure of the alloy and their distribution in it has also been discussed.

1. Exposition

Experiments with new types of modifiers of aluminum alloys, namely nanomodifiers (NM), have been carried out in the past years. The nanomodifiers are ultradisperse nanopowders with particles sizing 4-100 nm, having high melting temperature (~ 2273 – 3273 K depending on the composition), which are usually obtained either in result of a self-propagating high-temperature synthesis (SPHTS) [1] or by plasma-chemical synthesis (PCS) – for producing nitrides, carbides, oxides, oxi-carbides etc. [1, 2, 3]. The nanomodifier-nanodiamond (ND) is produced by detonation technology [4]. These nanoparticles, used as modifiers, have physical-chemical properties, influencing the structure of the alloys and therefore they can be used for obtaining new promising materials.

Most of the existing studies have been carried out on hypoeutectic and eutectic aluminum-silicon alloys, as well as on certain types of cast iron and steel [8, 9, 10, 11, 12, and 13]. We could not find any data about studies of the influence of nanomodifiers on the structure and the properties of hypereutectic aluminium-silicon alloys in the literature, studied by us.

Aim: The present paper aims at studying the influence of both a standard modifier – P, and a nanomodifier – ND, on the process of hardening the hypereutectic aluminum-silicon alloy AlSi18, as well as on the size of the primary Si crystals and their distribution.

2. Production and preparation of nanomodifiers

Explosive technology research, started in the Department "Space Materials Science and Nanotechnologies" at SRTI-BAS, has led to creating and patenting an industrial technology and an installation for nanodiamonds (ND) synthesis [14]. The detonation technology for nanodiamonds synthesis is based on blasting powerful explosive materials (EM) or mixtures of them with a negative oxygen balance, whereas the freed carbon transforms into a diamond. Both the pressure and the temperature in the detonation wave (P – T parameters) correspond to the thermo-dynamic stability of the diamond ($P \geq 10$ GPa; $T \geq 3000$ K). The main characteristics of the nanodiamond are given in [13].

The nanosized and hard-to-melt particles are difficult to wet (or they do not get wet at all) by liquid metals or alloys. Therefore, their homogeneous distribution when introduced into a melted metal is a very difficult task [1]. The decision to pre-coat nanoparticles has helped to solve this problem. One way to obtain active and melt-wetting NMs is the method of their mechanical-chemical treatment with pre-cladding metals in centrifugal planetary mills [2]. When introduced into the melt, the pre-cladding layer dissolves and facilitates the process of wetting, while the activated surface of the nanoparticles serves as a substrate for primary crystals formation (nucleation) [5].

Another way to prepare and activate the surface of the nanoparticles is to pre-treat them chemically until an appropriate metal coating from different metals is obtained (Ag, Ni, Cu) [6, 7].

3. Studied alloy

Hypereutectic aluminium-silicon alloy – AlSi18 – was studied. Its chemical composition is shown in table 1.

Table 1. Chemical composition of the alloy AlSi18, wt.%

Si	Fe	Cu	Mn	Mg	Cr	Ni	Zn	Pb	Al
17,55	0,120	0,025	0,047	0,001	0,001	0,005	0,102	0,01	rest

The nanomodifier ND 0,1 wt.%, used for better wetting by the melt, was Ni-coated by means of a currentless deposition method.

The used standard modifier P 0,4 wt.% was introduced into the melt by the ligature CuP10.

4. Methodology of study

The study of the hypereutectic aluminium-silicon alloy AlSi18 consisted in conducting experiments with unmodified, modified with a modifier P, and modified with a nanomodifier ND versions of the alloy.

The melting of the alloy is carried out in an electrosurgical laboratory furnace by using preliminary cleaned and dry stock materials. The melting process is carried out under a layer of roof-refining flux in an amount of 0.5 wt.% from the amount of the stock material. After melting, the melt is stirred vigorously for removing non-metallic inclusions, then the slag is removed, and the process of degassing the alloy is started. Degassing is performed at a temperature of 760 ° C by purging with argon for 3minutes, then cleaning the metal mirror and casting samples.

When a modified by P alloy is used, the modification is carried out before degassing at a temperature of 770°C. The alloy is degassed by purging with argon for 3minutes at a temperature of 760°C. The metal mirror is cleaned and the process of samples casting is started.

Typical in the preparation of the hypereutectic aluminium-silicon alloy AlSi18, modified by a nanomodifier ND, is that the modification of the alloy is carried out at a temperature of 760°C after its refining and degassing. The calculated amount of nanomodifier (ND) is mounted on the titanium stirrer of a stirring device. The nanomodifier is introduced close to the bottom of the crucible. Then mechanical stirring is started for 3 minutes at a low number of revolutions per minute – 120 – 130 min – 1, which is done to evenly disperse the nanomodifier particles throughout the melt volume.

Figure 1 presents the structure of the device for producing a sample for obtaining time-temperature curves. The thermocouple has Ni-CrNi electrodes with diameter of 0.25 mm and unprotected hot solder. Thus the thermal resistance at the border between the thermocouple and the metal is reduced, i.e., the inertness of the thermocouple decreases. The electrodes of the thermocouple are prevented from touching the melt by mounting them in a thin-walled stainless tube. The bottom is protected by a heat-insulating substrate of pre-coated sand. Thus conditions for heat exchange only through the walls of the container and through the upper free surface are ensured.

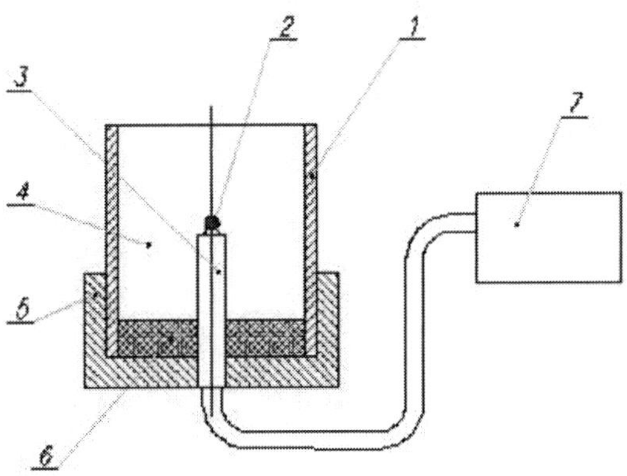

Figure 1. Device for studying the dependence of the temperature on the time of cooling
1-container, 2-thermocouple, 3-stainless tube, 4-sample, 5-base, 6-heat-insulating substrate, 7- data backup device

The temperature recording is performed using an archiver, allowing to record the discrete temperature readings for the corresponding sample over an interval of 0.2 s. A fast-running USB FLASH ARCHIVER MS DL-F2 v-2.0 is used.

5. Results

From the obtained time-temperature curve for unmodified alloy AlSi18 (figure 2) the temperature interval of alloy crystallization is defined (630°C-578°C). The characteristic change in the slope of the curve due to the solid phase separation is observed and the cooling rate decreases. The curve shows a degree of overcooling during the eutectic crystallization at 1 ° C for 5 seconds, and, consequently, the free energy of the solid phase is less than the free energy of the liquid phase, i.e. there are energy conditions for the crystallization process to occur. Due to the radiation of the hidden heat of crystallization, the temperature goes up again to the theoretical temperature of alloy crystallization (578°C) and stays at that level until the melt is completely hard. Straight liquation is observed under conditions of equilibrium cooling of the AlSi18 alloy, i.e. the hard-to-melt component pushes the fusible component to the center of the cast samples, figure 3.

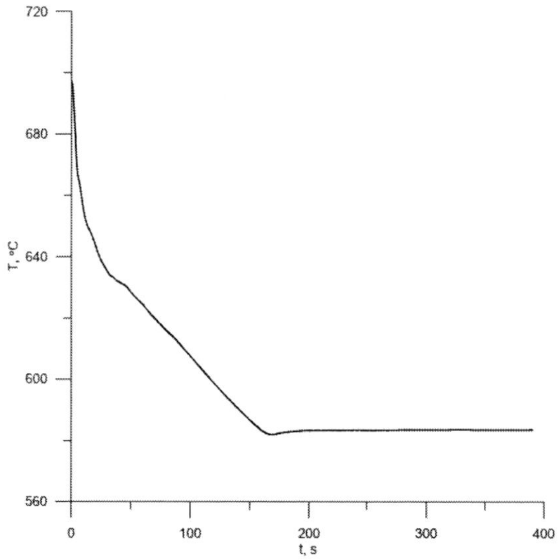

Figure 2. Time-temperature curve AlSi18

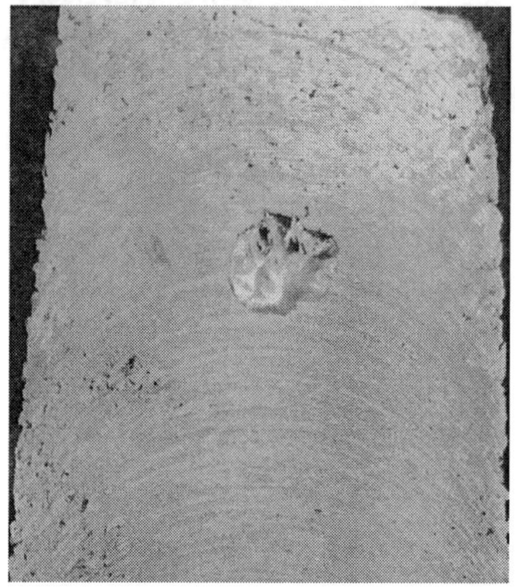

Figure 3. Distribution of SI crystals AlSi18

Figure 4 presents the time-temperature curve of a modified by a standard modifier P alloy AlSi18. Unlike the time-temperature curve of the unmodified alloy AlSi18, this time the curve does not register any level of overcooling during crystallization, which is explained by the formation of crystallization centers by the introduced into the melt modifier. A more even distribution of the primary Si crystals in the volume of the sample is observed for this type of alloy, which can be explained by the modifying effect of P, figure 5.

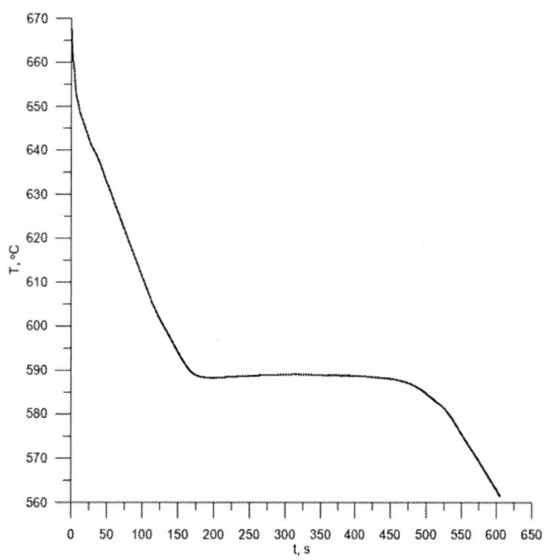

Figure 4. Time-temperature curve for AlSi18(P)

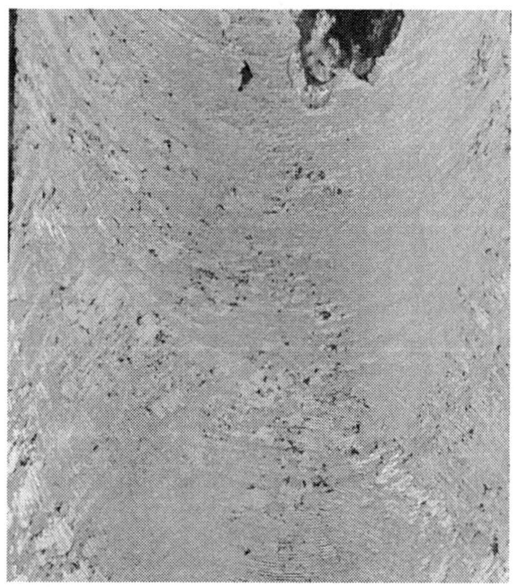

Figure 5. Distribution of SI-crystals AlSi18(P)

Figure 6 shows the time-temperature curve for a modified by a nanomodifier ND alloy. For this type of alloy a horizontal curve section at the liquidus line is observed, as well as an unregistered

IOP Conf. Series: Materials Science and Engineering **618** (2019) 012050 doi:10.1088/1757-899X/618/1/012050

degree of overcooling during crystallization at the solidus line. This is explained by the big number of crystallization centers formed by the nanomodifier ND and the intense separation of the primary Si crystals. The most even distribution of Si crystals in the volume of the sample is observed for the type of alloy, modified by a nanomodifier ND, what is explained by the greater extent of refinement of the silicon crystals and the registered area of intense silicon crystal separation at the liquidus line area of the alloy, figure 7.

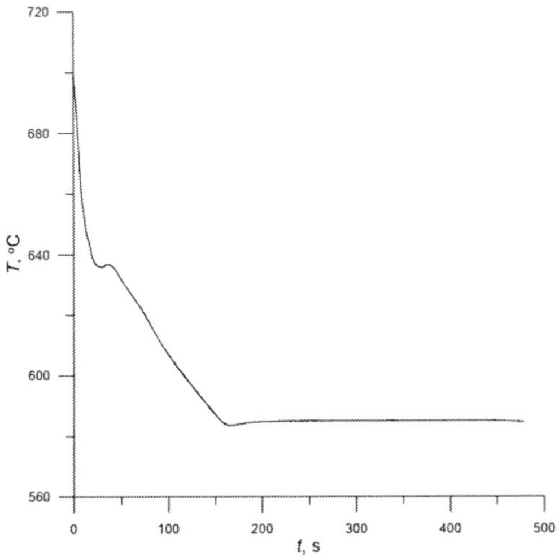

Figure 6. Time-temperature curve AlSi18ND

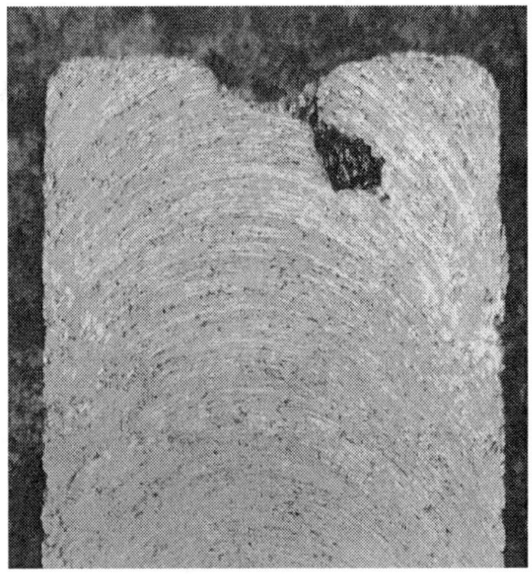

Figure 7. Distribution of Si crystals AlSi18ND

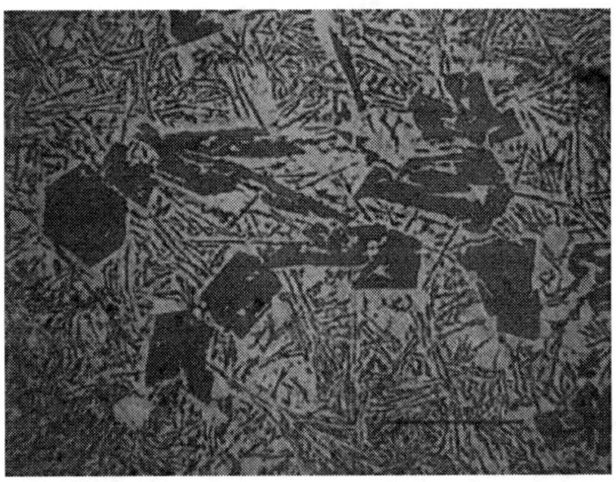

Figure 8.

From the micro-structural analysis it has been established that the primary silicon crystals are different in shape for the unmodified version of the alloy AlSi18: they are right-walled polygons, well-formed plates that resemble needles or plate-type crystals of irregular shape in the plane of the microscopic observation, figure 8. The arbitrary average diameter of the primary silicon crystals in the unmodified AlSi18 alloy ranges from 87.2 to 97.6 μm.

Figure 9.

The primary silicon crystals in the structure of the modified by P version of the alloy differ in shape and size from the ones, in the unmodified version, figure 9. Their arbitrary average diameter ranges from 59,2 to 52,1 μm. Right-walled polygonal crystals prevail and plate-type primary crystals are not observed.

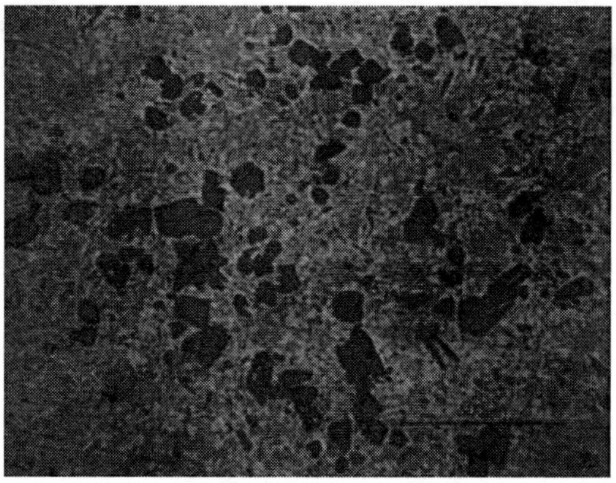

Figure 10.

Primary silicon crystals of polygonal shape predominate in the modified by a nanodiamond version of the AiSi18 alloy, while the amount of primary crystals of irregular shape is insignificant, figure 10. The arbitrary average diameter of the primary silicon crystals in a modified by a nanodiamond modifier AlSi18 alloy is 54 μm.

Table 2 illustrates the change in the value of the arbitrary average diameter of the primary Si crystals in the three versions of the AlSi18B alloy: unmodified; modified by P; and modified by ND nanomodifier.

Table 2. Arbitrary average diameter of the primary silicon crystals

Alloy	**Modifier**	**Arbitrary average diameter of the grains D [μm]**
AlSi18	-	92,4
AlSi18	P	55,7
AlSi18	ND	54

In an AlSi18 alloy, modified by phosphorus, reduction of the arbitrary average diameter of the primary Si crystals by 34,1% is observed.

In an AlSi18 alloy, modified by ND nanomodifier, reduction of the arbitrary average diameter of the primary Si crystals by 41,6% is registered.

6. Conclusions

From the conducted study it has been established that:

- The primary Si crystals in a hypereutectic aluminium-silicon alloy AlSi18 are successfully modified by means of a nanomodifier ND.
- The modification of the AlSi18 alloy by a nanomodifier ND can replace its modification by a standard P modifier.
- The replacement of the standard P modifier by a nanomodifier ND in AlSi18 alloy modification will also have ecological effect in addition.
- The conducted study confirms the positive influence of the nanomodifier ND on the size of the primary Si crystals and their distribution in the structure of the AlSi18 alloy.

Acknowledgements: The studies were supported by National Science Fund of Bulgaria, Project DN07/20/15 Dec2016.

References:

[1] Kalinina N.E., Kalinin V.T., *Nanomaterials and nanotechnologies: production, structure, application.* Monography. Dnepropetrovsk: Makovetskij, Y.V, Publishing House, 188

[2] Cherepanov A.N., managing editor Fomin V.M., 1995, Plasma-chemical synthesis of ultradisperse powders and their application in metal and alloy modification, *Low-temperature plasma,* volume 12, Novosibirsk, Science, Syberian Publishing Company RAN, 344

[3] Kalinina A.P., Cherepanov A.N., Poluboyarov V.A, and Korotaeva Z.A, A mathematical model

of the nucleation in liquid metals on ultradisperse ceramic particles, *Journal of physical chemistry,* vol. 75, №2, 283-289

[4] Cherepanov V.A., Zuev A.Y., and Tsvetkov D.S., 2011, Synthesis of ultradisperse materials and nanopowders, *Workshop on the course "Thermodynamics and structure of a solid body"*, part 2, Ekaterininburg

[5] Manolov V., Cherepanov A., Lazarova R., and Konstantinova S., March, 22-25, 2011, Properties of the aluminium alloy AlSi7Mg, modified by refractory nanopowders, Interaction of highly concentrated energy flows with materials in promising technologies and medicine, IVth all-Russan conference, Novosibirsk, 186-190, ISBN 9785-93089-034-1

[6] Kalinina A.P., Cherepanov A.N., and Borisov O.V., 1999, The effect of linear tension on the rate of heterogeneous nucleation on dispersed particles, *Metals,* №5, 40-43

[7] Popov S., Manolov V., and Cherepanov A., *Mathematical modelling of metal alloys crystallization,* http://www.nanomodifiers-ims.com/botl.htm

[8] Lazarova R, Bojanova N, Dimitrova R, Panov I and Manolov V, *Influence of Nanoparticles Introducing in the Melt of Aluminum Alloys on Castings Microstructure and Properties*, https://link.springer.com/article/10.1007/s40962-016-0033-7

[9] Lazarova R, Dimitrova R, Bojanova N, Stanev S, Velikov A and Manolov V, *Modifying of luminium alloys using nanosized powders*, Project TK 01/076 funded by the National Fund "Scientific Researches", published on http://www.nanomodifiers-ims.com/

[10] Manolov V., Cherepanov A., Lazarova R., and Konstantinova S., Influence of nanopowder inoculars on the structure and the properties of the AlSi7Mg alloy, *Foundry*, M., № **11**, 11-14, ISSN: 0024-449X, IF 0,088

[11] Stanev S., Lazarova R., Konstantinova S., and Manolov V., September, 2010, Study of the effect of nanopowder modification of AlSi7Mg alloy by thermal and metalograpic analyses, *Proceedings of the XXVIth international scientific conference "65 years Mechanical and technological Faculty at TU – Sofia* (Sozopol, Bulgaria), 277-282.

[12] Kuzmanov P, Dimitrova R, Lazarova R, Cherepanov A and Manolov V, March 2014, Investigations of the structure and mechanical properties of castings of aloy AlSi7Mg, cast iron GG15 and GG25 and steel GX120Mn12, modified by nanosized powders, *Journal of Nanoengineering and Nanosystems,* vol. **228** no. 1, pp. 11-18

[13] Gaidarova V N, Lazarova R L, Vaseva S V, Manolov V K, and Konstantinova S G, June 2010, *Microstructure and Mechanical Properties of AlSi7Mg Alloy: Effect of Added Nanodiamond,* In: Explosive Production of New Materials: Science, Technology, Business, and Innovations. Edited by A.A. Deribas and Yu.B. Scheck, Moscow, TORUS PRESS, ISBN 978-5-94588-073-3, p. 25 (Proceedings of the X International Symposium EPNM-2010 Bechichi, Montenegro)

[14] Stavrev S et al., Oct.11, 1994, *Method for production of ultradispersed diamond,* Patent US005353708A

The studies were supported by National Science Fund of Bulgaria, Project DN07/20/15 Dec2016.

IOP Conf. Series: Materials Science and Engineering **618** (2019) 012051 doi:10.1088/1757-899X/618/1/012051

Lowering the noise level in the transport flows through reduction of the traffic barrier reflected noise

I Kralov and K Nedelchev

Technical University of Sofia, Department of Mechanics

E-mail: kralov@tu-sofia.bg

Abstract. Reflected noise from traffic barriers usually leads to increasing of the noise level excited by the automobiles and trains. This effect requires the design of transport noise barriers for effective noise reduction behind them as well as to produce small level of reflected noise in source direction. This will effect in better noise environment for many passengers especially in case of busy traffic in rush hours. People affected by this effect are usually far more than the affected ones behind the barrier. One of the existed solutions is the use of expensive combined noise barriers with high degree of absorption. In this study a new passive transport noise barrier is designed, tested and analysed for a good reduction level of the reflected noise. It is done by appropriate design of the barrier profile performing source and reflected acoustic waves interference in small volumes near the barriers. The barrier construction, made by repeated, equally profiled elements, allows even the existence of acoustic bandgaps especially behind the barrier.

1. Introduction

Acoustic noise reduction is a very important process due to its direct impact on the environment and people health. There are three main ways to reduce the acoustic noise: reduction of the sound power of the source, influence on the noise radiation and noise insulation of the working/living place. In many cases application of the second method includes typical solution for reduction of noise propagation by different types of barriers, absorbers or active noise control solutions. The enormous number of recent research works and publications in this area prove the actuality of this research field [10, 11, 12, 13, 16]. Usually, expensive and sophisticated solutions have to be created in order to meet the noise level standards. The noise barriers are mainly divided into passive, active and combined. Passive solutions mainly block the spreading of sound waves and decrease the noise levels by reflecting acoustic waves [1, 2, 3, 10, 11, 12, 13, 14, 15]. On the other hand, active methods input additional energy that is used to eliminate the acoustic energy of the source by propagation of sound waves of opposite phase, by conversion of the sound energy to electricity, etc. [1, 2, 13, 17]. Combined solutions include combination of the previously mentioned ones.

Main advantages of the passive noise barriers are the relatively low cost of manufacturing, assembly and maintenance due to the use of a few materials only, the use of assemblies of equally shaped elements, etc. This is proved by the huge number of patented solutions [4, 5, 6, 7, 8, 9]. A lot of studies show good acoustic efficiency of these barriers regarding the relation of the noise level behind the barrier with respect to the source noise level [1, 2, 13].

In case of higher level of noise reduction needed, the combined types of noise barriers are used.

They usually have additional closed space for spreading the sound waves and/or additional absorption materials are used. This type of barriers is relatively more expensive for manufacturing, assembly and maintenance. Its advantage is better noise energy absorption.

In case of noise excited by the transport flows, it is very essential noise to be reduced not only behind the barrier, but also the barrier reflected noise to be reduced.

Reflected noise from traffic barriers usually leads to increasing of the noise level excited by the automobiles and trains. This effect insists transport noise barriers to be designed for effective noise reduction behind them as well as to produce small level of reflected noise in road direction. This will effect in better noise environment for many passengers especially in case of busy traffic in rush hours. People affected by this effect are usually far more than the affected ones behind the barrier. One of the existed solution is the use of the above mentioned expensive combined noise barriers of high degree of absorption.

The aim of this research study is to present a new passive transport noise barrier of high level of reduction of the reflected noise design, test and analysis. The reflected noise level is done by appropriate design of the barrier profile performing source and reflected acoustic waves interference in small volumes near the barriers. The barrier construction, made by repeated equally profiled elements, allows even the existence of acoustic bandgaps especially behind the barrier.

2. New transprot acoustic barrier of low level of reflected noise

As a result of the analysis above the following principles for the new barrier synthesis were used:

- passive noise barrier for transportation and industrial noise;
- level of noise reduction equal or better to that of the existing barriers;
- relatively low costs of manufacturing, transportation, assembly and maintenance;
- relatively high level of reflected noise reduction.

The new acoustic noise barrier construction is shown on figure 1. Noise insulating panels are made of non-metal materials. They have a half-tube like form arranged one over the other, and all their concave arc profiles have one and the same direction – to the sound source waves. The essential characteristic of the barrier construction is that the concave profile is exactly semi-cylindrical and the border area between two neighbor channels is a sharp edge. These channels are arranged into H-shaped metal columns fixed to the ground. The channels are fixed vertically to the columns through non-metal semi-disks of external diameter, equal to the internal diameter of the channels. The semi-disks are fixed to the columns by bolt or stud connection [15].

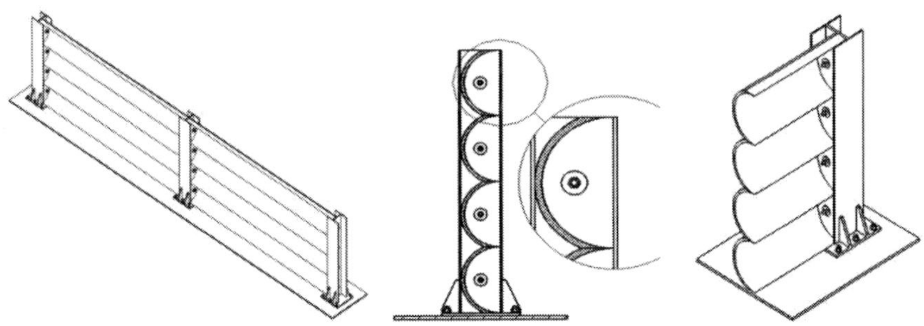

Figure 1. New acoustic barrier design [15]

The main expected effect of the new barrier is that the interference between the source and reflected sound waves is close to the barrier. In this way the noise reduction by converting sound energy into heat will have lower effect over the noise level near the source. The degree of reflection depends mainly on the barrier material, sound waves frequency and acoustic pressure level. An important parameter of the barrier construction is the absence of reflecting surface, perpendicular to the source sound waves. The exact semi-cylindrical profile of the channels focuses reflected waves to interfere with the source waves inside the profile of the channels.

To achieve better performance, geometrical parameters of the barrier can be designed for specific cases. Varying the width of the profile and the radius of the channel according to the sound power and frequency is a successful way of improving the acoustic efficiency of the barrier.

The presented construction has a lot of advantages. First, the non-metal channels are technologically easy for manufacturing. The uniform shape of elements allows big manufacturing series and reduction of production costs. Second, the elements are easy to be transported, assembled and maintained. Assembly does not require neither special machines nor expensive equipment. There is no need for the working team to possess any special skills. Third, the modular construction gives the possibility to vary easily the height of the barriers.

More advantages and areas of application of new barrier could be found in [15].

3. Simulation, test and analysis of the acoustic efficiency of new passive transport noise barrier

The main investigation of the acoustic properties of new barrier will be made by numerical simulation using a professional software for acoustics – COMSOL Multiphysics®. To verify some numerical results, a laboratory experiment using a scaled prototype of the barrier is made.

3.1. Experimental test of the scaled prototype of the new acoustic barrier

To produce some laboratory experiments a scaled prototype of the barrier is made. The prototype is shown on figure 2a and its main parameters are given in Table 1. The test is shown on figure 2b.

a) *b)*

Figure 2. a) the scaled prototype of the new acoustic barrier; b) acoustic test of the prototype

Table 1. Main parameters of the scaled prototype of the barrier under study

№	Parameter	Value
1	Diameter of the channel profile (internal), mm	76
2	Number of channels	7
3	Width of the profiled channels, mm	8
4	Material	HDPV
5	Material density, kg/m3	970

The experiments were performed in a semi-anechoic chamber. A 40 dB chamber was used. The sound system of sound generator, amplifier and specialized sound source, inside the chamber, is used to generate the noise level needed. First the noise level inside the chamber as well as at the receiving position were measured. The open side of the chamber was closed by the barrier prototype. Again the noise level inside the chamber and the noise level at the receiving position /behind the barrier/ were measured.

The acoustic measurement and analyzing system PULSE 3560B of the Bruel&Kaer equipped with the OEM software and microphone 4516A is used. Before measurements, all the standard procedures for calibration and estimation of the environmental noise influence are proceeded.

The results from the experiment are given in the table 2 and in figure 3. The following notations are used:

L_{fon} – level of the surrounding noise, dB;

L_s – noise level of the source, dB;

$L_{RP}(A)$ – A-weighted noise level at the receiving point without barrier, dB;

$L_{RP,bar}$ (A) – A-weighted noise level at the receiving point with new barrier, dB;

Eq – equivalent noise level for all the treated octaves, dB.

Table 2. Results from experimental test of the scaled prototype of the barrier under study

Octave, Hz	L_{fon} dB	L_S dB	L_{RP} (A) dB	$L_{RP,bar}$ (A) dB
63 Hz	19,7	67,9	48,2	41,8
125 Hz	22,1	68,1	54,3	47,2
250 Hz	23,0	68,2	58,9	51,2
500 Hz	19,6	68,2	62,8	49,1
1000 Hz	21,5	68,2	65,7	50,2
2000 Hz	16,7	68,2	66,4	46,6
4000 Hz	15,7	68,3	66,2	44,3
Eq	**28,3**	**76,0**	**71,8**	**56,5**

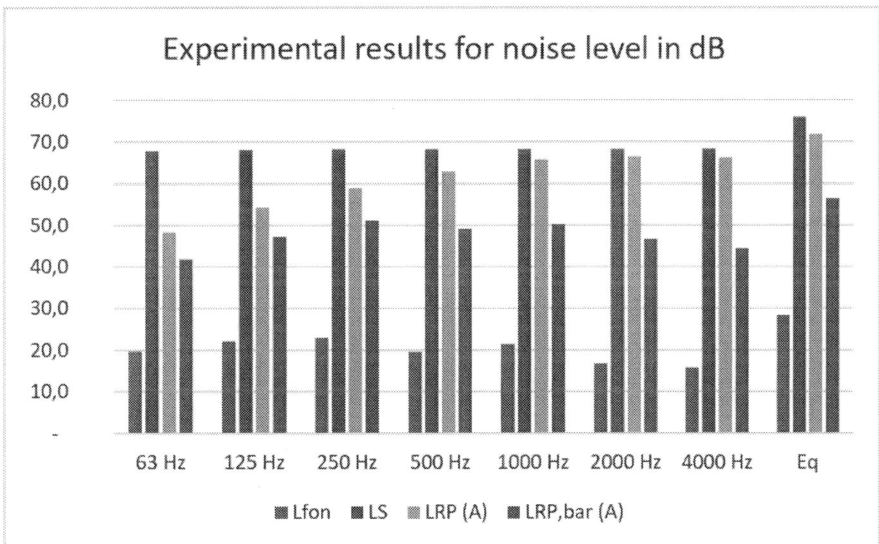

Figure 3. Results from experimental test of the scaled prototype of the barrier under study

The results show an A-weighted noise level reduction in case of use of the barrier prototype between 16 dB and 22 dB.

3.2. Numerical simulation and test of the scaled prototype of the new acoustic barrier

Numerical simulation of the same scaled barrier prototype is made in COMSOL Multiphysics®. The process of model building, model mesh, type of analysis and boundary conditions are the same as in [15]. Comparative results of the new barrier acoustic efficiency are presented in Table 3 and on figure

IOP Conf. Series: Materials Science and Engineering **618** (2019) 012051 doi:10.1088/1757-899X/618/1/012051

4. Here the following new notations are used:

$L_{RP,bar,n}$ (A) – A-weighted noise level at the receiving point with new barrier, numerically found, dB;

$L_{RP,bar,e}$ (A) – A-weighted noise level at the receiving point with new barrier, experimentally found, dB;

Table 3. Comparative results from experimental and numerical test of the scaled prototype of the barrier under study

Octave, Hz	$L_{RP,bar,n}$ (A)	$L_{RP,bar,e}$ (A)	$L_{RP,bar,n}$(A) - LRP,bar,e(A)	$L_{RP,bar,n}$(A) - LRP,bar,e(A)
	dB	dB	dB	%
63 Hz	39,6	41,8	- 2,2	-5,3%
125 Hz	48,4	47,2	1,2	2,5%
250 Hz	47,1	51,2	- 4,1	-8,0%
500 Hz	56,7	49,1	7,6	15,5%
1000 Hz	59,1	50,2	8,9	17,7%
2000 Hz	54,6	46,6	8,0	17,2%
4000 Hz	53,7	44,3	9,4	21,2%

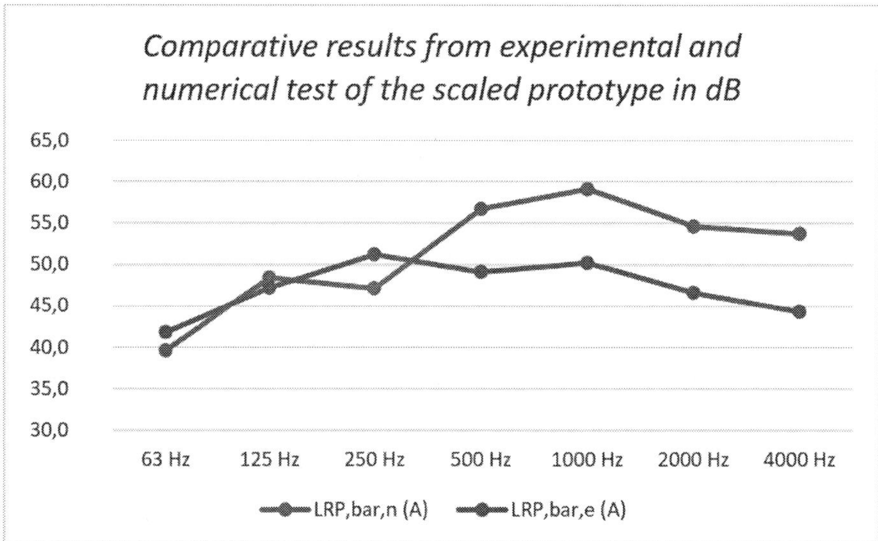

Figure 4. Comparative results from experimental and numerical test of the scaled prototype of the barrier under study

Results show relatively good relevance of the experimental and numerical results for the scaled prototype of the new barrier acoustic efficiency. Differences are less than 30% for different octave bands which is appropriate for such basic research problem and is similar to the known results in the literature. Based on this conclusion next step of investigation is done.

3.3. Numerical investigation of the acoustic efficiency of the new barrier

In this section an investigation of the noise level distribution on the both sides of the new acoustic barrier is presented in the vertical plain of a small source of spherical noise waves. The studied acoustic barrier as well as the environment around it, are modelled as a "tridimensional" 2D axisymmetric object [15]. For the present study a Frequency Acoustic Analysis - Pressure Acoustic, Frequency Domain is used. Again the COMSOL Multiphysics® is used. The process of material parameters, model building, model mesh, type of analysis and boundary conditions are the same as in [15].

To compare the influence of the reflected noise over the sound level in the area of the source (traffic flow) as well as the level of noise reduction behind the barrier, a comparative analysis is made. The results of noise distribution on both sides of new barrier and those of an often used combined barrier are compared. Second barrier consists of a perforated flat metal surface, directed to the source, mineral wool of density 170 kg/m^3 and another flat metal (aluminum) surface on the rear side. Part of the sound waves passes through the perforated screen and are absorbed form the mineral wool inside the panel of the barrier. Other part of the waves is reflected from the non-perforated part of the front panel, while relatively small amount of sound energy passes behind the barrier. This barrier has common acoustic effect: reflection and interference between source and reflected waves, and relatively good level of absorption inside the panel. Serious disadvantage here is bigger manufacturing cost due to presence of different materials, bigger transport and technological expenses etc. Schemes of the numerical tests for both type of barriers are given on figure 5. Figure 6 shows the noise distribution at 1 kHz and 2 kHz without barrier.

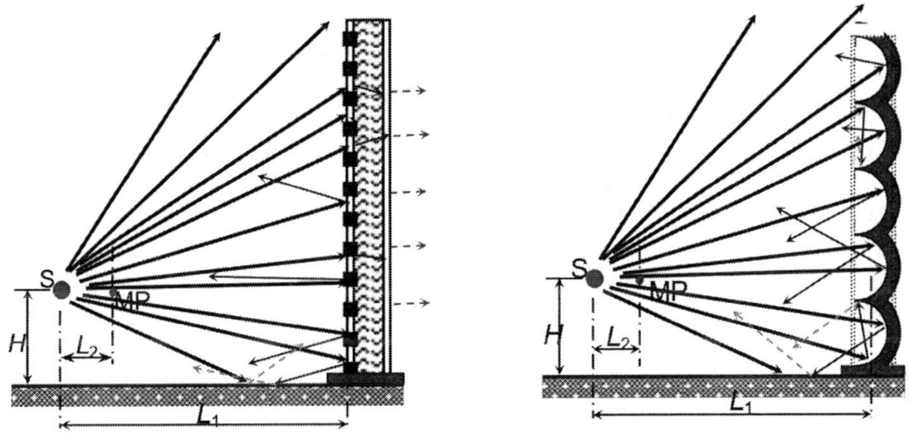

Figure 5. Schemes of the numerical tests for both type of barriers

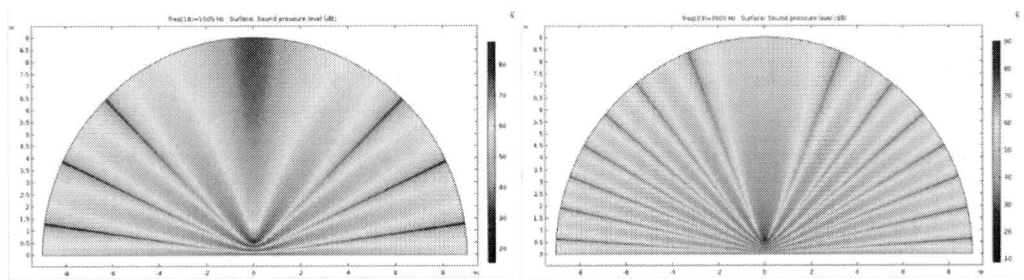

Figure 6. Source noise level distribution without barriers at frequency octave bands 1kHz and 2 kHz

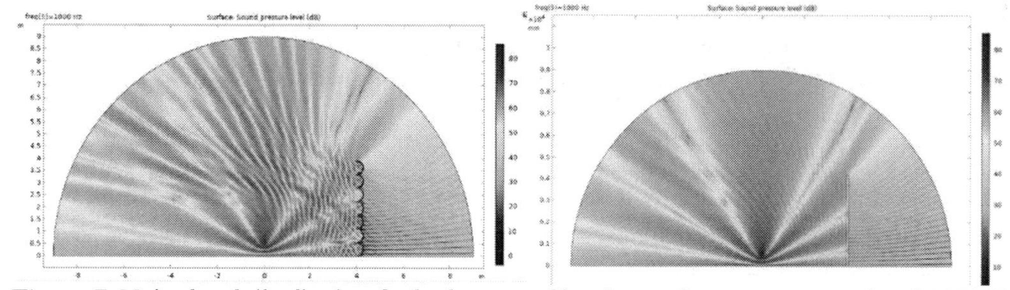

Figure 7. Noise level distribution for both types of barriers at frequency octave band 1000 Hz

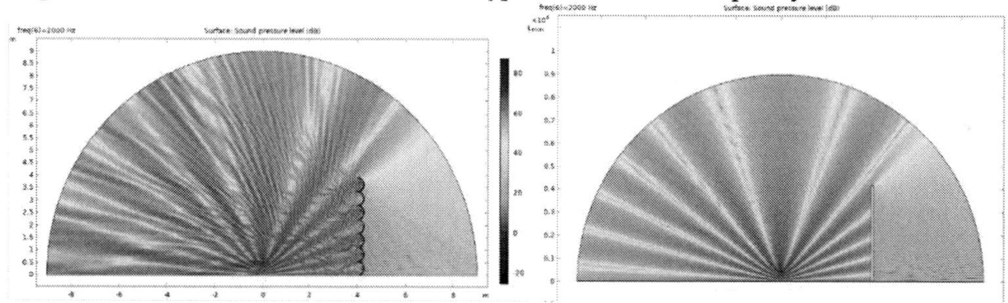

Figure 8. Noise level distribution for both types of barriers at frequency octave band 2000 Hz

Some comparative results of the both barriers noise level distribution are presented on figures 7 and 8.

The source is in the middle of the graphs and represents the transport unit noise source. The vertical plain of the source is bounded at the bottom from the road, modelled as more than 90% reflection barrier surface. On the right side at 4 m from the source, the investigated noise barrier is modelled as described above.

It could be seen from these figures that the bandgaps exist due to reflection from of the road. In case of noise barrier there is additional interference in radial and obtuse angle directions. This makes the picture much more complicated. It is obvious that the sound distribution depends on the frequency range, source noise level, distance between the source and barrier etc. To make an approximate comparison of the acoustic efficiency of both barriers, the following procedure is used. For both cases (barriers) the equivalent level of the sound pressure level is found from the measurement points (MP) of vertical lines at different distance L_2 from the source. Three lines are between the source and the barriers, while other three lines are behind the barriers. Results are presented in table 4.

Table 4. Equivalent noise level form all the points in vertical lines at given distances from the source

Leq	0,5 m	3,45 m	4 m	4,1 m	4,3 m	8 m
	SPL, dB	SPL, dB	SPL, dB	SPL, dB	SPL, dB	SPL, dB
Source without barrier	69,9	66,2	66,0	66,0	65,7	64,0
Perforated combined barrier	71,0	68,6	68,4	49,1	48,7	47,9
New type barrier	70,5	67,4	67,1	47,0	46,4	46,1

The results show that the new barrier has similar and even better acoustic efficiency both for the noise level distribution behind the barrier and for the area near the source. The results are very interesting for the distance of 4 m and 4,1 m. They show that the combined barrier excites the level of noise which is bigger than the level excited by the new barrier. This could be explained by the relatively big flat reflecting area of the front panel directly to the source area. Another interesting fact is the presence of bandgap regions behind the barriers. May be this is due to the periodic structure of the sound barrier panel shapes. This will be a new investigation task in a future work.

4. Conclusions

An investigation of the acoustic efficiency of a passive traffic noise barrier was made. Numerical simulation, verified by experiments, is used to find the noise level distribution on both sides of the barrier. The searched effect is the barrier to produce small level of reflected noise in source direction. This will effect in better noise environment for many passengers especially in case of busy traffic in rush hours which makes the number of people affected by this effect usually many times bigger than number of affected people behind the barrier. A comparative analysis with noise level distribution in case of use of expensive combined noise barriers of high degree of absorption is done.

Results and their analysis shows better acoustic efficiency of the new barrier. The barrier construction, made by repeated equally profiled elements, allows even the existence of acoustic bandgaps behind the barrier. This effect could be used for optimization of the acoustic barrier characteristics.

Acknowledgment: This work was supported by the Bulgarian Ministry of Education and Science under the National Research Programme "Low carbon energy for transport and live", approved by DCM # 577 / 17.08.2018.

References

[1] Banov S and Kralov I 2003 *Noise in Transportation Technics* (Sofia: Technical University of Sofia (in Bulgarian)) p 134
[2] Nikolov N, Benov D and Shubin I 2014 *Acoustic Design of Transport Noise Insulating Barriers*, (Sofia: ACMO Academic Press (in Bulgarian)) p 241
[3] Patent № UD 96/16230 – 30.05.1996.
[4] Patent № US № 6305492 B1 / 23.10.2001.
[5] Patent № US 7789193 B2 / 07.09.2010.
[6] Patent № AT 513615 A4 2014-06-15.
[7] Patent № 89203035.4 / 29.11.1989.
[8] Patent № DE 42 20 547 A1 / 07.01.1993.
[9] Patent № 10 2014 2017 767.7 / 10.03.2016.
[10] Kralov I, Terzieva S and Ignatov I 2011 Analysis of methods and MEMS for acoustic energy harvesting with application in railway noise reduction *Bucharest. Proc. MECAHITECH'11.* **3** 56-62
[11] Kralov I, Sinapov P, Nedelchev K and Ignatov I 2012 Friction induced rail vibrations, *AIP Conf. Proc.* **1497** 19-25.
[12] Trevor J and D'Antonio P 2009 *Acoustic Absorbers and Diffusers* (Taylor & Francis)
[13] Thompson D J 2009 *Railway Noise and Vibration* (Elsevier)
[14] Gieva E, Ruskova I, Nedelchev K and Kralov I 2018 An investigation of the influence of the geometrical parameters of a passive traffic noise barrier upon the noise reduction response, *AIP Conf. Proc.* **2048** 020020
[15] Kralov I 2017 New solution for transport and industrial noise protection through reflective noise barriers *Matec Conf. Proc.* **133** DOI: https://doi.org/10.1051/ matecconf/ 201713306001
[16] Ivanova Y, Vassilev V, Djondjorov P and Djoumaliisky S 2015 Experimental-Theoretical approach to the identification of effective sound attenuation panels from recycled materials, *J. Bul. Chem. Comm.* **42** 1–8
[17] Aleksandrova M 2018 Spray deposition of piezoelectric polymer on plastic substrate for vibrational harvesting and force sensing applications *AIMS Mat. Sci.* **5(6)** 1214-22

IOP Conf. Series: Materials Science and Engineering **618** (2019) 012052 doi:10.1088/1757-899X/618/1/012052

Some aspects of the informational support of technological processes

S Koleva

Department of Machine building Technology and Machine Tool, University of Ruse, Ruse 7000, Bulgaria

svetla@uni-ruse.bg

Abstract. The increase of the efficiency of technological processes during the various stages of product development, such as design and manufacture, is in direct connection with their informational support. In this paper are considered the state of the art of modern informational support of processes, and the methods of its realisation. On the basis of informational modelling of technological processes, are considered and analysed the capabilities of the suggested scientific approaches for the assurance of the parameters of its efficiency, including their evaluation. Established is the necessity of a new approach during the modelling of technological processes for parts manufacturing, which would enable to overcome the main problem of their development – the ambiguity of the values of the technological parameters and their influence on efficiency. In order to guarantee the efficiency of these processes, formulated are the tasks and approaches which are to be applied for developing a reliable and up-to-date informational system.

1. Introduction

Modern machining processes in production systems are characterised by the extensive use of CNC technologies; it includes not only the technological equipment, but also the design, control and management of the technological processes, at the core of which is the aim to increase their efficiency [1, 2, 4, 7, 8,]. The development of efficient technological processes include the activities of all three stages of their realisation: design, application to real manufacturing conditions, and batch production (established manufacturing process). The basis for the development of an integrated system that enables the increase of efficiency of the technological processes for machining parts is its informational support at each stage of the technological process.

2. Details of the system

The technological processes include three stages of their realisation: design, application to real manufacturing conditions, and batch production.

2.1. *Stage 1: Design of manufacturing processes*

There are two main approaches of designing manufacturing processes: the application of group technologies, and individual (expert) approach. As a result, its quality depends on the qualification level of the process engineer (expert). Both approaches are based on the method of analogies, so its efficiency depends on the correspondence between the parameters of the modelled and real processes.

The main problem of designing manufacturing processes is the uncertainty of the information connected with the characteristics of the technological system and additional means of automation, parameters of the processes, etc. The large variety of technical and technological solutions, their realisation under various limitations, such as time, economics, qualification level, results in multi-variable manufacturing processes. Finding the most optimal values of the many influential factors makes the development of efficient manufacturing processes a task with multiple optimisation parameters.

A large problem of designing efficient manufacturing processes is the uncertainty of making a decision (Figure 1), which can be grouped into 5 groups:

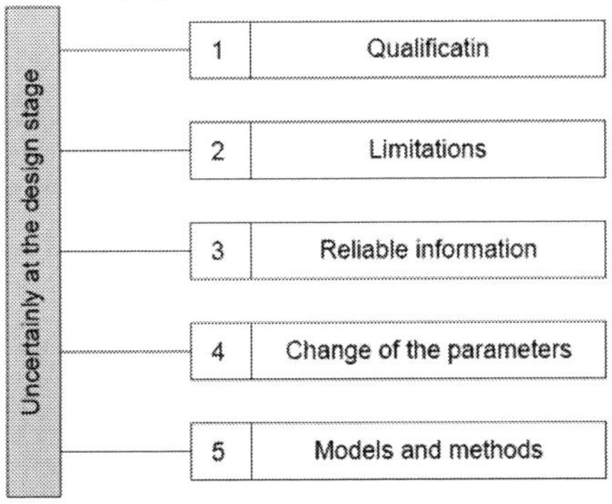

Figure 1. Groups of uncertainty in making a decision during designing manufacturing processes

Based on the notation in Figure1, uncertainty in the groups is the result of:

Group 1 – the result of insufficient experience of those making the decision (a person or team) and the knowledge of factors that influence the decision making;

Group 2 – the existence of time constraints (e.g. short time period for designing manufacturing processes, which is a condition for the inadequate consideration and evaluation of various candidate solutions), or financial constraints at decision making;

Group 3 – lack of reliable information for the values of the factors acting during the manufacturing processes, under the conditions of multi-variable occurrence.

Group 4 – the parameters of the technological equipment (machine tool, main and auxiliary equipment) change in time, which is the result of wear and tear.

Group 5 – lack of models and methods that would allow the prediction of a specific result out of the interconnected influence of the factors under the conditions of multi-variable occurence.

The lower the uncertainty during the phase of the design of the manufacturing processes, the more the real process will approach the required process under the given conditions. In order to overcome the uncertainty during the design phase, it is necessary to have informational support that provides information for:

- Regularities connected with the machined parts – here included is information for the structure of the process, clamping schemes, machining schemes, technological equipment.
- Factors and models for defining the cutting conditions and mechanical loads – the necessary information includes material characteristics, cutting tool life and tool strength parameters, stability of the technological system.
- Factors and models for defining the influence of the deviations during manufacturing – information for the geometrical accuracy of the machine tool, geometrical accuracy of the cutting tool and its setting, accuracy of installing the blank using the given scheme, deviations from the mechanical deformations, cutting tool wear rate, stability of the technological system, errors connected to accuracy control.
- Technological costs and models of defining the technological cost price – this information is connected to the general shop floor costs, costs of exploitation of the workplace, such as energy, depreciation, cutting tool costs, salaries.

The more complete and reliable the information is at this stage, the more problems could be eliminated or avoided at the stage of the design of the manufacturing process.

Figure 2 presents systematic and detailed input information, necessary for high quality design of manufacturing processes, and output information for their realisation.

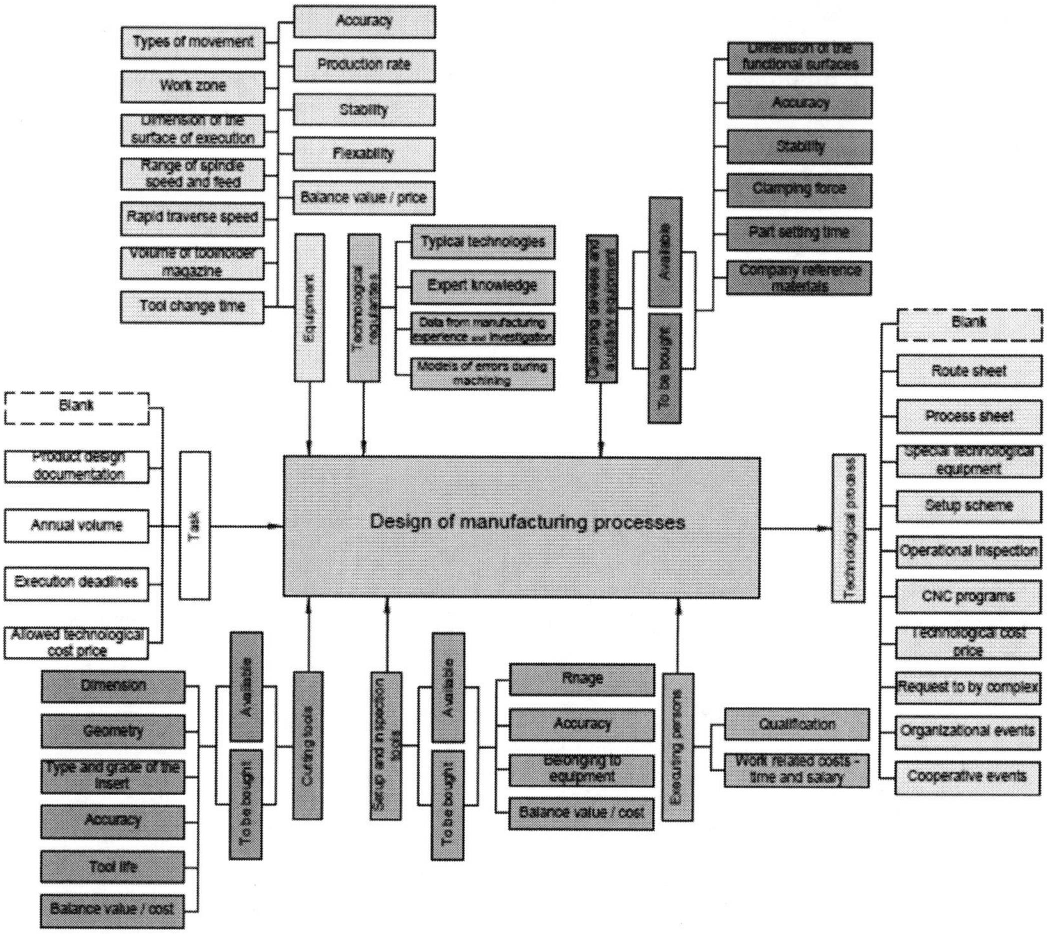

Figure 2. Scheme of the input and output information at the stage of designing manufacturing processes

2.2. *Stage 2: Implementation / preparations*

At this stage, all missed or unsuccessful solutions from the previous stage will become apparent. The activities at this stage are connected to the preparation of the equipment, setup, testing and ensuring the workability of the process in correspondence with the developed technology.

They are shown in Figure 3.

During the setup of the workpiece on the machine tool it is necessary to correctly orient it in relation to the machine and ensure its stability. In certain cases, it is necessary to perform measurements in order to establish its correct orientation. In order to eliminate scrap, it is necessary to inspect the accuracy of the surfaces of the set up blank. These deviations may be caused by wear, partial damage of the centring surfaces, or by errors of the datum surfaces of the workholding device. The results of these measurements are entered into the control program for their compensation if this is possible. It is necessary to know the limiting values of the deviations, which can or cannot be compensated for. Similar are the requirements for setting the cutting tool.

The stability of the tool is mainly influenced by the condition of the base instrumental surfaces of the machine tool, the auxiliary tools, such as pockets, bushings, and the instrumental datum surfaces of the tool itself. Preliminary data about the conditions of the base instrumental surfaces of the machine tool allow for the positioning of the tool for the finishing operation into the most accurate positions, to which some tools are particularly sensitive. During machining of rotational workpieces, for example, these are threading tools, tools for grooving and cut off, carbide drill bits, reamers, etc. On the other hand, rotating multitooth cutting tools require symmetrical positioning relative to the rotational axis.

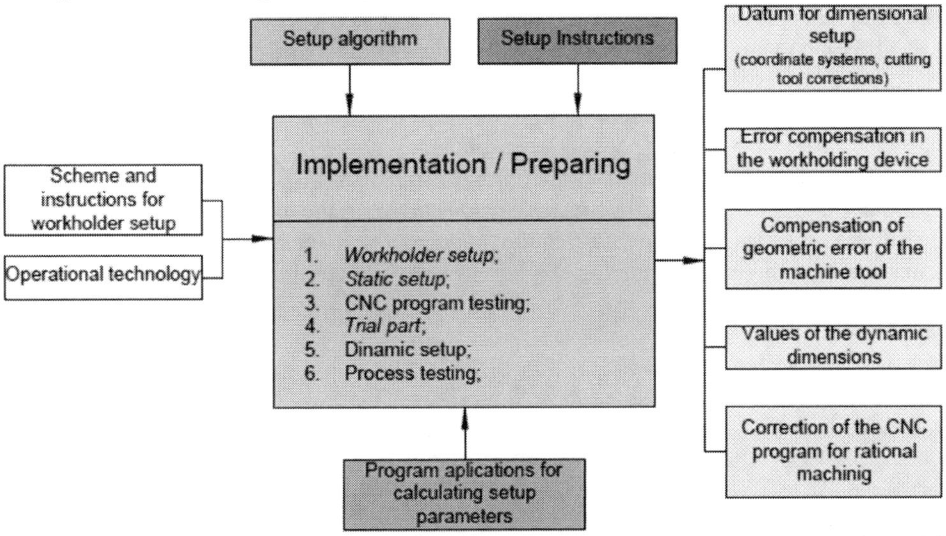

Figure 3. Scheme of the input and output information at the stage of implementation of the technological operation

High speed methods for dimensional setup, such as off-machine setup, setup using reference surfaces, automated setup on the machine tool (contact and no-contact based), define the position of the blank and cutting tool in a static position, without actual cutting. This way the dynamic errors occurring during machining of the workpieces cannot be taken into account. The only method that allows for taking into account the dynamic errors and its compensation is the method of trial cuts, if the cuts are performed on a real workpiece and under real cutting conditions (depth of cut, feed rate, cutting speed). The disadvantage of this method is increased machine down time which in general increases the manufacturing costs. Trial cuts of workpieces are performed in order to avoid scrap in cases of uncertainty of the results during machining as a result of:

- insufficient accuracy of the initial static setup; for high accuracy surfaces compensations are introduced in the direction of repairable scrap;
- for following up the load (elastic deformations), vibration resistance, during machining difficult to machine materials.

Based on the obtained results, the permanent dynamic error and the error of the initial setup are calculated.

The aim of dynamic dimensional setup is to define the dynamic error. This is necessary when machining high accuracy surfaces. It is performed based on the results of machining a small batch (3-5 parts). This way the constant errors are compensated, and an approximation of the reliability of the obtained quality parameters is performed. Testing of the process aims at more accurate definition of the dynamic error, which allows more accurate initial setup of the next batch, and especially at the approximate evaluation of the reliability of the process. It is performed during surveying and inspection of the machined batch of 10-30 parts.

At this stage occur those errors that had been made in the scheme of workpiece setup and manufacturing of the workholding device (insufficient stability and vibration resistance), incorrect cutting tool selection. Visible become all incorrect solutions that lead to difficult chip removal, improper application of cutting fluids, elimination of the flow of the material at the start and end of the machined surface, etc.

Regardless of the way the CNC program had been developed, it has to be tested on the machine tool. The aim of the test is to eliminate visible errors and collisions, corrections for more rational realisation of the machining operation (optimisation of empty passes and motions). This stage of the implementation can only be avoided if a CAM system is used for the analysis of collisions [5]. This is possible if all components of the technological system are entered into the model with their real 3D dimensions. This, however, is not always possible due to missing information and time constraints.

At this stage also become visible omissions of unforeseen locksmith or cleaning operations, connected with the conserving and packing operations.

Implementation is a stage where the machining of the first batch is performed. Another important task is elimination of mistakes and irrational decisions, making the technical documentation up-to-date including all related instructions and suggestions. During the machining of the next batch, after using the information from the implementation stage, only remain activities connected to the preparation of the technological system and its setup (described in Figure 3 in *Italic*).

2.3. *Stage 3: Batch production*

This is the stage of the established manufacturing process (Figure 4), during which the following tasks are solved:

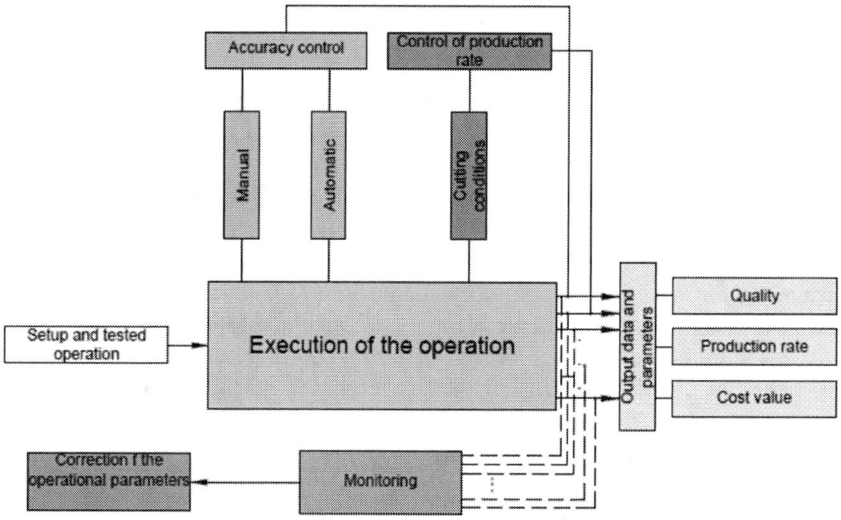

Figure 4. Scheme of the input and output information at the stage of implementation of the operation

- Maintaining the workability of the process and the quality parameters of the machined surfaces that may be altered by the disturbing factors. Some authors call this task process control [3].
- During the first, and sometimes also the next batch, data are collected and the possibilities of its optimisation during machining the next batches is analysed.

3. Conclusions

The following conclusions can be made:

1. In order to improve the quality of technological activities it is necessary to provide systematic and easily accessible information for the regularities connected to the:

1.1. development of the structure of the designed technological process;

1.2. machining scheme;

1.3. theoretical and empirical models for the definition of errors during machining, including instructions of their practical realisation;

2. In order to reduce the implementation time of a technology it is necessary to have information for the actual (at the moment of their execution) values of the technological parameters of the machine tools, workholders, cutting tools, auxiliary tools, and blanks used by the process;

3. The reduction of setup time requires the development of algorithms, instructions, and program applications for making the work of the operator easier and reducing the influence of his level of competency;
4. CNC programs alone are not sufficient to hold all information, connected to setup. In order to reduce setup time and scrap rate and avoid collisions, which are most likely at this stage, the program has to be accompanied by a scheme that can give dimensional information for the position of the work coordinate systems relative to the machine tool.
5. During testing the CNC program and its simulation testing, it is necessary to have updated 3D models of the blank, workholder, cutting tools and auxiliary tools;
6. The designed manufacturing process has to provide information for the organisation of events that can reduce machine down time to a minimum at the subsequent phases;
7. The designed manufacturing process has to provide information for the inspection of the qualitative parameters of the parts, including the inspection scheme, equipment and tools. Special attention should be paid on operational inspection if it is not automated.
8. After the implementation of the manufacturing process are formulated output data which should be saved and implemented during the preparation for machining the next batch;
9. For efficient control of the manufacturing processes at the phase of established manufacturing, it is necessary to have actual information, including real-time data for automatic control of the quality parameters of the machined parts, and also the means for its processing and interaction.

Acknowledgments
The study was supported by contract of University of Ruse “Angel Kanchev”, № BG05M2OP001-2.009-0011-C01, " Support for the development of human resources for research and innovation at the University of Ruse “Angel Kanchev”. The project is funded with support from the Operational Program " Science and Education for Smart Growth 2014 - 2020" financed by the European Social Fund of the European Union.

References
[1] B. Krupińska, D. Szewieczek Analysis of technological process on the basis of efficiency criterion, Journal of Achievements in Materials and Manufacturing Engineering, vol.17, Iss.1-2, July-August 2006
[2] B. Srinivasa Prasad, D. Siva Prasad, A. Sandeep, G. Veeraiah, Condition Monitoring of CNC Machining Using Adaptive Control, International Journal of Automation and Computing, vol.10, Iss.3, pp 202–209
[3] Geprgiev V., Salapateva S., Chetrokov I., Lengerov A., Adaptive control and active control in machine building, EKS-PRESS OOD, Gabrovo, 2012
[4] Zhilevski M., Main problems with CNC Lathes, Technical College Lovech, 2017
[5] Zhilevski M., Zhilevska M., Practical application of CAD/CAM systems in metal cutting machine tools, Unitech, Gabrovo, 2018
[6] Information support. What does it support, and what it is applied for?, https://enginclub.ru/informacionnoe-obespechenie-chto-obespechivaet-i-dlya-chego-primenyaetsya/ , 18.10.2018
[7] Maxted P., Tackling process variation in manufacturing – the benefits of modern process control techniques, IMTS conference September 2016, https://www.mapyourshow.com/MYS_Shared/imts16/handouts/Paul%20Maxted%20IMTS1 1v2.pdf , 18.10.2018
[8] Radchenko S., Information support of technological processes, http://n-t.ru/sp/lesmi/iotp.htm, 18.10.2018

IOP Conf. Series: Materials Science and Engineering **618** (2019) 012053 doi:10.1088/1757-899X/618/1/012053

Application of machine translation techniques for the automatic selection of the technological process

O Iliashenko[1], A Beseda[2], P Pelikhov[3] and T Ermilova[4]

[1]Peter the Great St.Petersburg Polytechnic University, St. Petersburg, Russia
ioy120878@gmail.com

[2, 3, 4]Department of Corporate projects Dialog IT St. Petersburg, Russia
pelikhovp@gmail.com

Abstract. This research deals with the problem of developing a design and technological documentation for a new product in terms of technological processes selection, according to the list of components from the product specification. Practical experience in the implementation of automated control systems in manufacturing enterprises of the machine-building industry and the latest technologies in the field of machine learning were used. The concept of using machine learning for automating the selection of technological processes was developed based on a list of components. As a technique of machine learning, the methodology sequence to sequence is proposed.

1. Introduction

Today, it is difficult to imagine an enterprise that does not use information technologies (IT). The interesting area is IT for the automation of processes in the manufacturing enterprise. Specialized software products have already been developed, including for product design (CAD systems), software products for technological production (CAPP systems), enterprise resource planning (ERP), advanced planning&scheduling (APS), manufacturing executing system (MES), quality assurance (QA-systems) [3]. However, the more systems and software products are used in the enterprise, the more problems arise in the organization of communications and the provision of reliable and timely information flows.

An automated production planning system based on information about the composition of products and of technological production processes makes stringent requirements to the quality and timeliness of design and technological documentation, this data is also the basis for production plans calculations. The absence or incompleteness of such information can be a serious problem for production planning, which may even require a review of the methodology and tools for calculating production plans.

According to the authors of the article, the question of timeliness and completeness of data is especially important for the engineering enterprises with design and production according to individual customer requirements, where the process of preparing the design and construction documentation is carried out practically throughout the entire production cycle. In this case, most of the composition of the product is repeated to some extent or can be attributed to a certain class of produced parts or assemblies only by the experienced employee or a production worker, but it is difficult to be identified by the available information systems. At large enterprises, where the size of the production plans consists of tens and hundreds of thousands of operations, the dimension of the task further aggravates the situation.

Bearing in mind the above, the currently known technologies for working with large data [4] and recent breakthroughs in the field of artificial intelligence [5], the authors of the research came to the following conclusion: the inadequacy and untimeliness of information about the technological process for the purposes of production planning can be compensated or completely excluded with the help of already developed and proven methods of artificial intelligence (machine translation in particular) [6,7].

2. Methodology

The first part of the work is practical and based on the implementation projects in three large machine-building holdings. The field of activity of the first company is the production of downhole equipment for the oil and gas industry. The specificity of production lies in the fact that due to different indicators inside well sites at different oil fields, every time one has to develop or refine its products. The second enterprise produces transformers for various hydroelectric power stations, which also have different characteristics, so developers have to make a large amount changes in the technological documentation. The third company is one of the leaders in the Russian market for the development and production of power and electrical equipment for the oil and gas, metallurgical and chemical industries [8]. The holding also develops and manufactures equipment for the power industry and the power grid complex.

The following methods were used for the research:

- questioning of stakeholders on the part of the customer. For this purpose, specialized questionnaires were developed for each unit. In the questionnaires, key users noted information about job responsibilities and types of documents processed. Completed questionnaires were processed by an expert in the field of automation of industrial enterprises;
- study of production documentation: technological processes, route maps, product specifications and other regulatory and reference information;
- studying software products used for the design and planning of technological processes;
- interviewing stakeholders on the part of the customer. The purpose of the interview is to cover the issues that arose after processing the questionnaire. Interview is the final touch for describing the current business process.
- Focus group discussion was used to build a future enterprise model. Focus groups discussed the problematic issues of automation, proposals on adjusting the business processes that the responsible party put forward.

In the second part of the research, a study of scientific literature devoted to the problems of neural networks took place.

The development of a new product is considered as a task. The product consists of several large assembly units, which in turn consist of smaller assembly units, etc. Each unit "explodes" until we reach the purchased materials. Thus, the specification of our future product can contain up to several thousand parts produced. It should be understood that each part has its own technological process, which can contain up to 30 operations.

During the interviewing and questioning it was revealed that during the development of new types of products, a part of the detail can be similar to those, which were previously produced, some of the details can be developed from the beginning, and some of the details may have similar characteristics. When the design drawings are prepared, it is necessary to develop the documentation of the technological processes.

Modern CAPP systems allow automatic selection of technological processes for identical (previously produced details. In this part the process is already automated, but for details that are developed from the beginning or have similar characteristics, this technique is not applicable.

During the discussions in the focus groups it was revealed that often in the developed from the beginning or modified details, the technological processes will be similar or identical for the already developed products, but the existing technology selection process will not offer the technical process, since this detail has differences from those already existing in information system.

The study of research papers devoted to such problems did not reveal a developed solution for optimizing the process of automating the selection of technological processes according to the available list of details. From the point of view of the task posed articles in the field of artificial intelligence turns out to be unexpectedly interesting, exactly the machine translation of foreign languages and machine description of images. [9]

The enormous success of the Internet search leader Google in translating texts from one language to another, (as well as other participants in the scientific and applied community in the field of machine translation) has already proved the effectiveness of machine learning methods for translation (translation of meaning) from one language to another (from one set of symbols into another)[1], [2]. and for the verbal description of the images (the translation of the meaning from one set of symbols (images and their combinations in the picture) to another (text)), as well as the verbal description of the video images (the transmission of the meaning of the pictures taking into account their changes in time).

In these quite different applied problems, in our opinion, there is a general property. The understanding of this property prompted us to the idea of its possible use for the problem described above: all these cases are similar in one fundamental property - people are trying to describe the same essence in different ways. In translation, these are two different languages. In the description of a picture and video it is a translation from the language of images, which we naturally perceive by one of our six senses, into another form of meaning transfer - words (the alphabetic display). In our case, on the one hand, we describe some output product through a set of its parts (the table is something that consists of a lid and four legs), and on the other hand, the table is something that we get when we perform a set of actions. "Operation 1: Take the table top; put it on the bench upside down. Operation 2 Take the leg of the table and screw it into one of the holes in the bottom of the table top. Operation 3-5 Repeat step 2 for the remaining table legs by placing them in the remaining free holes.

Thus, the observation described above allows us to put forward a hypothesis that by applying methods similar to machine translation techniques to the problem of auto-selection of the technological process, we can achieve the desired goal (automatic determination of the technological process of manufacturing the product according to the composition of its components).

To solve this problem, an attempt to adopt machine translation method is described in current study. Vector space models (VSMs) represent (embed) words in a continuous vector space where semantically similar words are mapped to nearby points ('are embedded nearby each other') [2]. Such a model may be interesting and can allow taking into account the cases where the similarity of the products composition takes place. Besides, it can be quite effective, when the technological processes are identical or so similar, that might be considered as identical.

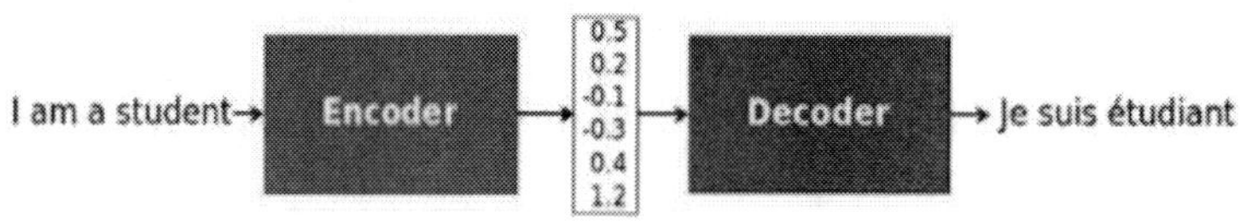

Figure 1. Seq2seq

Another model that is actively used in machine translation is a model Word2vec. There are two versions of this model CBOW and Skip-Gram [2]. Both models are based on the principle of joint occurrence of words and the possibility of predicting the probability of the appearance of a word, based on the context of its appearance. However, in the case of the existence of critical dependencies between words in the text (the sequence, the presence of related words at a considerable distance (the number of other words in the text) between them), the approaches listed above prove to be ineffective, in contrast to the one used lately and a neural-based Sequence-to-sequence (seq2seq) method.

In general, seq2seq is a sequence of two mechanisms: an encoder and a decoder and a number of auxiliary mechanisms, such as: attention mechanism [2], the mechanism for optimizing the search for the most likely answer in deciphering (beam-search).

Encoder-decoder architecture – example of a general approach for NMT. An encoder converts a source sentence into a "meaning" vector which is passed through a decoder to produce a translation [10].

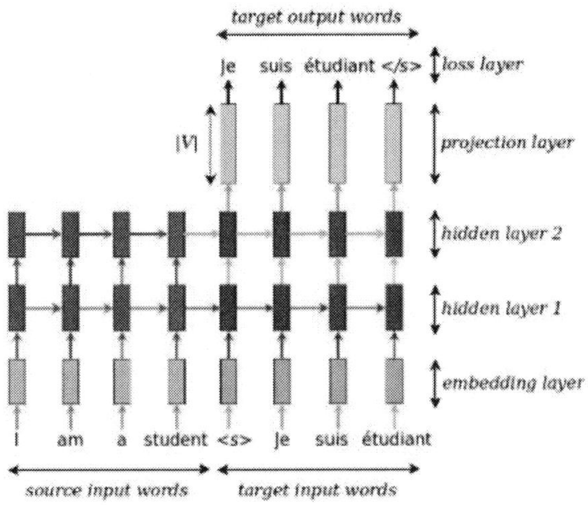

Figure 2. Seq2seq encoder.

Neural machine translation – example of a deep recurrent architecture proposed by for translating a source sentence "I am a student" into a target sentence "Je suis étudiant". Here, "<s>" marks the start of the decoding process while "</s>" tells the decoder to stop [11].

3. Results

Let's consider the concept of the proposed solution using a simple example.

In the MES of the enterprise for a few years, a sufficient large amount of data on the products produced has accumulated. At the same time, we can obtain accurate names for the parts produced, their characteristics, such as: the outer and inner diameter of the finished part, the length of the finished part, the steel grade from which it is made, the standard length, the outer and inner diameter of the workpiece from which the part is made. For convenience, these data are tabulated for three "similar" parts on each other:

Table 1. Hypothetical specification.

Parameter	**Part TL 1**	**Part TL 1-2**	**Part TD 2-1**
Outer diameter	25 mm	25 mm	24 mm
Inner diameter	20 mm	20 mm	18 mm
Length	600 mm	600 mm	600 mm
Round beam	1 unit	1 unit	1 unit
Outer diameter	27 mm	26 mm	26 mm
Inner diameter	-	10 mm	-
Length	1200 mm	1200 mm	1400 mm
Metal mark	Steel MX-10	Steel MX-10	Steel M-35

In addition, since these parts have already been produced, information about the technology of their production is available to us at the level of abstraction, which necessary for the production planning. Despite some difference in the parameters of the parts Part TL 1, Part TL 1-2, the details of the composition, the order and the time of execution of technological operations for them are identical. While for the Part TD 2-1 the composition and the order of operations are identical to the previous ones, (in this case, the main reason for the differences are the properties of the material, the part is made from - the hardness of the steel.) The resistance of the material to the reverse affects the processing time).

Table 2. Map of technological processes.

Part TL 1, Part TL 1-2			**Part TD 2-1**
Operation number	Operation	Time, min	Time, min
010	Cutting	10	15
020	Drilling	25	30
030	NC turning	20	25

040	Metal work	10	15
045	QC	5	5

Let's represent the data shown in a tabular form in a slightly different form, namely as a body of parallel texts, as it is done in solving the problems of machine translation and other similar problems [1].

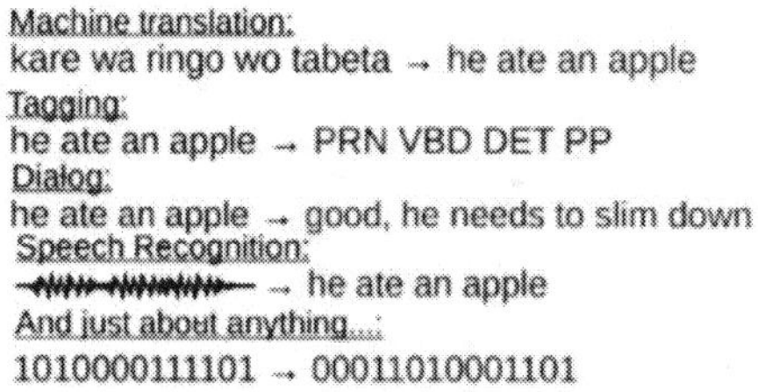

Figure 3. Data, presented as body of text.

Our example is in the form of a parallel body of text is presented in the Table 3.

Table 3. Example of a parallel body of text.

Part_TL 1 25_mm 20_mm 600_mm Round beam 27_mm 1200_mm Steel_MX-10	Cutting Drilling NC_turning Metal work QC
Part_TL 1-2 25_ mm 20_mm 600_ mm Round beam 26_mm 10_mm 1200_mm Steel MX-10	Cutting Drilling NC_turning Metal work QC
Part_TD 2-1 24_mm 18_mm 600_mm Round beam 26_mm 1400_mm Steel_M-35	Cutting Drilling NC_turning Metal work QC

As shown in the example, in most cases, the initial data for the task will be stored in a table format. In the framework of the ETL standard for machine learning tasks, we will have to perform additional actions to bring the data to the original form for the machine translation (the body of parallel texts). First, we need to present a set of procurement features in the form of a list of values with ordering, the numerical values of the measure units and their notation are represented in the form of words (by linking through the underlined). Second, we need to represent the sequence of operations, ordered by ascending number of operation in process technology, while leaving behind the numbers themselves and the execution time.

The first column of the table obtained (product description in the design view) can be interpreted as texts in the original language, the second column (product description in the technological view) as texts in the translation language.

Having performed such a transformation, we get a set of data standard for the machine translation task seq2seq, and then we can apply the technique similarly to its application to texts in natural languages (in fact our descriptions are text in natural languages, even in one natural language, but in different language subsets of the dictionary representation)

Proceeding from what has been said above, the authors of the article hypothesize that the task of automating a process based on a known composition of a assembly or a set of characteristics of a part can be logically reduced to the task of translating text from one language to another or a derivative of the task of text description of the image [12]. Accordingly, the same decision methods can be applied to the task of auto-selection of technical processes as for machine translation or textual description of the image, such as, machine translation algorithms of the seq2seq type as described in [1] and software, implementing such algorithms [11-14].

4. Discussion

According to the data available to us, the hypothesis about the similarity of the problem of auto-selection of technical processes and the task of machine translation of texts has not been proposed before and is proposed by us based on our understanding of machine translation technologies and automated text description of images, as well as it is based on our experience in automation of manufacturing enterprises based on PLM functionality, ERP, APS, MES.

In this regard, we are very interested in the opinion of the expert community on the formulation of the problem and the proposed method for solving it, it would be especially interesting to hear about the results of the practical implementation of the ideas put forward by us, for example, by popular platform Tensorflow [8] and its specialized library tf-seq2seq [9].

In addition, we see the possibility of predicting the operation time, after determining the composition of operations. Time of operation can be predicted by ML methods - standard classification approaches.

Table 4. Example of automated time calculation

Part TL 1		
Operation number	**Operation**	**Time, min**
010	Cutting	10
020	Drilling	25
030	NC turning	20
040	Metal work	10
045	QC	5

5. Conclusion

In the current research, the task of auto-selection of technological processes during the development of a new product was formulated, advanced machine translation methods were considered, and an assumption was made about the logical similarity of the problem of auto-selection of the technological process and the translation of the text, as well as the proposal for the use of machine translation technologies for solving the problem of auto-selection of technological processes.

In the case of confirmation of the formulated hypothesis and solution of the problem by machine translation methods, it is possible to expect a significant reduction in the labor intensity of the development of technological processes, which may result in a positive economic effect (increase in labor productivity and a reduction in the production cycle of the order).

Acknowledgment

The research carried out with the financial support of the grant from the Program Competitiveness Enhancement of Peter the Great St.Petersburg Polytechnic University, Project 5-100-2020

References

[1] Graham Neubig Language Technologies Institute, Carnegie Mellon University "Neural Machine Translation and Sequence-to-sequence Models: A Tutorial"

[2] Denny Britz, Anna Goldie , Minh-Thang Luong, Quoc Le "Massive Exploration of Neural Machine Translation Architectures".

[3] Aladwani, A. M. (2001). "Change management strategies for successful ERP implementation" Business Process management journal, 7(3), 266-275

[4] Saint John Walkerb. "Big Data: A Revolution That Will Transform How We Live, Work, and Think".

[5] Quoc V. Le & Mike Schuster, Research Scientists, Google Brain Team "A Neural Network for Machine Translation, at Production Scale".

[6] Klochkov, Y., & Gazizulina, A. (2016). Application of the method of performance evaluation of the production proceb design using abociative design doi:10.4028/www.scientific.net/KEM.684.448

[7] Rudskoy, A. I., Kodzhaspirov, G. E., Kitaeva, D. A., Rudaev, Y. I., & Kliber, J. (2015). To the problem of the cylindrical product with a bottom forming process optimization under superplasticity condition. Materials Physics and Mechanics, 22(2), 191-199

[8] Ilin, I. V., Izotov, A. V., Shirokova, S. V., Rostova, O. V., & Levina, A. I. (2017). Method of decision making support for it market analysis. Paper presented at the Proceedings of 2017 20th IEEE International Conference on Soft Computing and Measurements, SCM 2017, 812-814. doi:10.1109/SCM.2017.7970732

[9] Top 10 machine learning frameworks. Retrieved from http://bigdata-madesimple.com/top-10-machine-learning-frameworks/.

[10] Seq2seq. Overview. Retrieved from https://google.github.io/seq2seq/

[11] Levina, A., Ilin, I., & Esedulaev, R. (2018). IT requirements integration in high-rise construction design projects. Paper presented at the E3S Web of Conferences, , 33

doi:10.1051/e3sconf/20183303009.
[12] Neural Machine Translation (seq2seq) Tutorial. Retrieved from https://www.tensorflow.org/tutorials/seq2seq
[13] Fisher, W. D., Camp, T. K., & Krzhizhanovskaya, V. V. (2017). Anomaly detection in earth dam and levee passive seismic data using support vector machines and automatic feature selection. Journal of Computational Science, 20, 143-153. doi:10.1016/j.jocs.2016.11.016
[14] Leonov, S., Vasilyev, A., Makovetskii, A., Kuznetsov, V., & Diaz-Escobar, J. (2018). An algorithm for selecting face features using deep learning techniques based on autoencoders. Paper presented at the Proceedings of SPIE - the International Society for Optical Engineering, , 10752 doi:10.1117/12.2321068

Qualitative analysis of oscillating magnetomechanical system

T S Todorov[1*], R P Mitrev[2], B N Tudjarov[3] and R F Nikolov[4]

[1] Department of Theory of Mechanisms and Machines, Technical University of Sofia, Sofia 1000, Bulgaria
[2] Department of Logistics Engineering, Technical University of Sofia, Sofia 1000, Bulgaria
[3] Department of Computer Systems, Technical University of Sofia, Sofia 1000, Bulgaria
[4] Department of Precision Engineering and Measurement Instruments, Technical University of Sofia, Sofia 1000, Bulgaria

*tst@tu-sofia.bg

Abstract. The paper deals with mathematical modelling and qualitative analysis of a dynamic system of the type mass-spring-permanent magnet. The mass of the system is suspended on the free end of a cantilever beam and oscillates in a magnetic field created by a permanent magnet. The the magnetic force is obtained using the finite element method and approximated by a function similar to the Coulomb's Law. The dynamical model of the system is built on the basis of the Newton 2nd law. By means of qualitative analysis of a nonlinear differential equation, the bifurcation zones are determined. The phase portraits of the system are investigated depending on its bifurcation behaviour. Based on the performed analysis a method for dynamical synthesis of system parameters is proposed. Besides, some typical properties of the dynamical system are determined.

1. Introduction

In the last three decades, the microelectromechanical systems (MEMS) have attracted increasing interest in the engineering community. Niarchos [1] describes the magnetic MEMS as a new class of conventional MEMS devices with considerable potential for modern science and its practical applications. The extensive investigations in this field have naturally led to creating a novel type of models associated with the scaling laws, with changing the force dominations, time responses and system performance.

Electromechanical and magnetomechanical systems have sophisticated characteristics and the description of their behaviour is based on solving nonlinear ordinary differential equations [2]. In the general case, it relates to an analytical or numerical investigation of complex dynamical processes which are described by partial differential equations [3]. The selection of the parameters of such systems additionally can be included in the tasks of the dynamical synthesis, as far as such a selection can guarantee some requested properties of the real-life design. Dynamic models of the magnetic MEMS containing oscillating cantilever beams with permanent magnets are typically used to investigate the functioning of such devices as sensors, actuators and micro energy harvesters [4]. The studies of the practical engineering applications containing vibrating cantilever beam are extremely diverse. Primarily, they differ by the design of the structure based on the governing physical laws, defined goals of the study and complexity of the used mathematical models. During the years a number of researchers [5,6] explored the various aspects of the behavior of the cantilever beam with a tip mass. The paper [7] proves the possibility for use of an electromagnetic actuator to control the vibrations of a cantilever beam with a tip mass by different control laws. The author [8] presents a method for enhancement of the power output of energy harvester by applying a simple repulsive magnetic force to a piezoelectric cantilever beam. The complex nature of the MEMS structures uncovers a wide field for application of optimization methods to improve the characteristics of the structures. An example is a recent research [9] proposing a multi-objective optimization model of an electromagnetic actuator applied as a fast tool servo. Despite the large volume of the conducted research studies, the precise description of the complex dynamical behavior of the magnetomechanical systems [10] is still a challenging problem that needs to be addressed through additional research. Some initial investigations of the authors are presented in the paper [11].

The aim of the present paper is the investigation of a nonlinear oscillating system of the type mass-spring-permanent magnet by the use of qualitative analysis methods and determination of system parameters that provide a predefined system behaviour.

2. Magnetic force determination

The essential elements of the considered system are shown in figure 1(a). A ferromagnetic mass 2 is fixed to the free end of a cantilever beam 1 and oscillates in the force field created by a permanent magnet 3. When the cantilever is in undeformed position between the faces of the ferromagnetic mass and magnet, an initial air gap h is established. A linear generalized coordinate y is selected for describing the mass position. Zero position of this coordinate corresponds to the horizontal position of the cantilever beam.

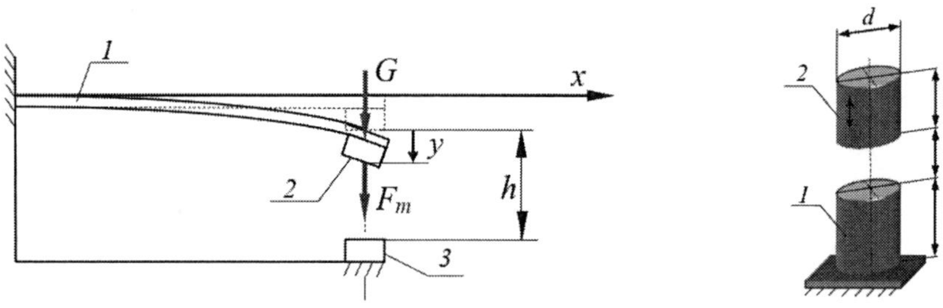

(a) (b)

Figure 1. An oscillating system with a permanent magnet.

To qualitatively investigate the dynamic behavior of the considered oscillating system, it is necessary to determine the force dependence on the position. This function is worked out for particular cases and it depends on dimensions, shape and material properties of the system elements.

A precise determination of the magnetic force is carried out by investigation of the magnetic field by using the Finite Elements Method. The system under investigation is shown in Figure 2. It consists of a cylindrical permanent magnet 1 and a ferromagnetic cylinder 2, disposed over the magnet. All dimensions are denoted in figure 1(b). The magnetic field is analyzed as an axi-symmetrical one. The task of analysis is formulated by Poisson's equation with respect to the magnetic vector potential in the cylindrical coordinate system. The investigated area is determined as a sufficiently large buffer zone (of the order of 20-multiple dimension of the system) with imposed Dirichlet's boundary conditions. The problem is numerically solved by the software product FEMM (Finite Element Method Magnetics) and calculations are automated by programs in Lua Script® language. For determining the magnetic force, the built-in Maxwell stress tensor is used. The number of the mesh nodes varies for the different tasks between 18000 and 20000. Investigations are carried out for two types of permanent magnets (BaFerrite и NdFeB) with two diameters - 8 и 13 mm. Results for the magnetic field distribution and for the attractive force between the magnet and ferromagnetic cylinder are obtained by a change of the air gap from 1 to 5 mm. Magnetic field lines for BaFerrite and NdFeB magnets with diameter 8 mm for air gaps 1 and 5 mm are visualized in figures 2(a) and 2(b). As one can see, there is no fundamental difference in the distribution field patterns. However, a significant difference in terms of force is established - for the NdFeB magnet the mean magnetic force is more than 60 times bigger than the same force for BaFerrite.

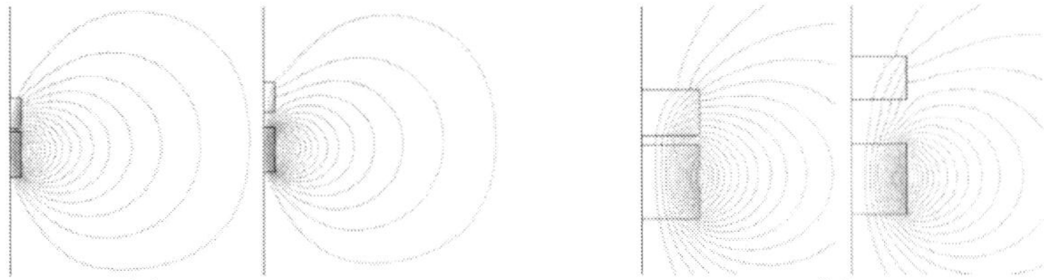

(a) (b)

Figure 2. Magnetic field lines distribution for δ=1 mm и δ=5 mm: a) BaFerrite d = 8mm; h_m = 15mm; h_c = 10 mm; b) NdFeB d = 8 mm; h_m = 15 mm; h_c = 10 mm.

The obtained by the numerical experiment force data are approximated by the relation:

$$F_m = \frac{\kappa}{(\delta_0 + \delta)^2} \tag{1}$$

which is close to the Coulomb's low [12] when δ_0 is small. Analogically to this law, here by κ is denoted the imaginary magnetic mass and δ_0 is an imaginary initial gap. The determined regression coefficients and maximal deviations of the experimental data from the computed values are shown in table 1.

Table 1. Regression coefficients for the three types of magnets.

Material dimensions d/ h_m/ h_c	$\kappa \times 10^{-7}$ [Nm^2]	δ_0 [m]	Max. deviation $\times 10^{-6}$ [N]
BaFerrite 8/15/10	6.2681702	0.001 2348	0.00086
NdFeB 8/15/10	35.921246	0.001 0328	0.207
BaFerrite 13/8/5	1.4354750	0.001 5858	0.000124

3. Mathematical model of the oscillating system

A mathematical model of the system is derived, taking into account the following more important simplifying assumptions: the influence of the distributions of all kind of loads is neglected; the system is considered as a single degree of freedom system with lumped mass and lumped stiffness; it is assumed that the mass moves rectilinearly and its rotation is ignored; all kinds of resistive forces are neglected.

The latter assumption is highly plausible in case of oscillations in a vacuum combined with low energy dissipation materials, used in some microdevices as accelerometers, energy harvesters etc.

Taking into account the adopted assumptions, notation δ=h-y, the elastic force F_e=-cy and the weight force G=mg, according to the Newton 2nd law the motion of the mass at the free end of the cantilever beam is described by the following nonlinear differential equation:

$$\ddot{y} + k^2 y = g + \frac{\kappa}{m(\delta_0 + h - y)^2} \tag{2}$$

where k^2=c/m is the natural frequency of the system, c is the spring stiffness coefficient, m is the mass, and g is the Earth gravity acceleration. For the stroke constraint of the cantilever beam it follows the constraint $y \leq h$.

After substituting $h_0=\delta_0+h$, introducing the dimensionless time $t_{new}=kt$ and change of the variable $y=h_0-\xi$ the differential equation (2) takes the following simplified form:

$$\ddot{\xi}+\xi-\alpha+\frac{\beta}{\xi^2}=0 \tag{3}$$

where $\alpha=h_0-g/k^2$ and $\beta=\kappa/\left(mk^2\right)$. The equation (3) is a nonlinear autonomous differential equation which could be solved in terms of elliptic integrals. The change of the variable leads to the following constraint:

$$\xi\geq\delta_0 \tag{4}$$

By the energy transformation $\ddot{\xi}=\frac{1}{2}\frac{d\dot{\xi}^2}{d\xi}$ and after separating of the variables the first integral of equation (3) is obtained

$$\frac{1}{2}\dot{\xi}^2=\alpha\xi-\frac{1}{2}\xi^2+\frac{\beta}{\xi}+c \tag{5}$$

By the expression (5) after substitution of different values of the integration constant c the phase trajectories in the plane $\left(\xi,\dot{\xi}\right)$ of the oscillating system can be plotted [13].

4. Determination of the bifurcation zones

The bifurcation zones are determined by the values of the parameters α and β leading to the change of the nature of the system phase states. The function:

$$f=\alpha\xi-\frac{1}{2}\xi^2+\frac{\beta}{\xi} \tag{6}$$

is considered for determination of the bifurcation values. For this purpose (6) is presented as a sum of two functions:

$$f=f_1+f_2 \tag{7}$$

where

$$f_1=\alpha\xi-\frac{1}{2}\xi^2,\; f_2=\frac{\beta}{\xi} \tag{8}$$

From equation (3) one can see that β is a positive nonzero parameter, but the parameter α theoretically can accept negative and zero values. The negative values of α are possible when simultaneously the air gap h and the imaginary gap δ_0 are small and the natural frequency k is low. Such values are relatively rarely used in real-life designs.

It is apparent that the function f_2 is discontinuous for $\xi=0$ and is strictly monotone decreasing. The function f_1 is a parabola with a zero root and a second root equal to $2a$. Two cases of the sum function f are possible. The first one corresponds to a strictly monotone decreasing function for $\xi>0$ and to a function with one extremum for $\xi<0$. The second case corresponds to a function possessing two extrema for $\xi>0$ and to a function with one extremum for $\xi<0$. At the first case, for $\xi>0$ there will not be a singular point in the phase trajectories. At the second case, for $\xi>0$ two singular points will occur corresponding to the stable and unstable equilibrium points. The limit case between the two cases corresponds to the occurrence of an inflection point in the function f leading to the appearance of a cusp point in the phase trajectories. From this limit case the condition of bifurcation is expressed as simultaneously occurring zeroes of the first and second derivatives of f :

$$\alpha-\xi-\frac{\beta}{\xi^2}=0,\; \frac{2\beta}{\xi^3}-1=0 \tag{9}$$

therefore

$$\beta=\frac{1}{2}\xi^3,\; \alpha=\frac{3}{2}\xi \tag{10}$$

which represents the parametric equation of the bifurcation curve. After isolating of the parameter ξ, the equation $\beta=f(\alpha)$ of the bifurcation curve is obtained:

$$\beta = \frac{4}{27}\alpha^3 \tag{11}$$

From the above mentioned it naturally follows that the bifurcation curve divides the plane into two zones with different behaviour of the dynamical system. For values of α and β over the bifurcation curve, i.e. when

$$\beta > \frac{4}{27}\alpha^3 \tag{12}$$

conditions for oscillations do not exist, and the mass will accomplish only a displacement with a variable acceleration. The closer to the magnet is the mass, the higher is its velocity. If α and β have values under the bifurcation curve, according to the initial conditions both oscillatory and non-periodic one-way motions are possible.

5. Topological analysis

The study of the number and type of the equilibrium points will facilitate the study of the dynamic behaviour of the system and defining the zones allowing the realization of stable oscillations around a certain position. The equilibrium or singular points [14] of the oscillating system follow from the system of algebraic equations

$$\ddot{\xi} = 0, \ \dot{\xi} = 0 \tag{13}$$

Taking into account equation (3) it follows

$$\xi^3 - \alpha\xi^2 + \beta = 0 \tag{14}$$

The number of the equilibrium points of the system is equal to the number of the real roots of equation (14) which discriminant is:

$$D = 4\alpha^3\beta - 27\beta^2 \tag{15}$$

and according to its values, three cases are considered.

Case I: For $D>0$ the equation has three distinct real roots and the following condition is satisfied:

$$\beta < \frac{4}{27}\alpha^3 \tag{16}$$

and the value of the largest positive root is:

$$\xi_1 = \frac{1}{3}\left(\alpha + \frac{\alpha^2}{\sqrt[3]{A}} + \sqrt[3]{A}\right) \tag{17}$$

where $A = \left(\alpha^3 + \frac{3}{2}\left(\sqrt{3\beta\left(27\beta - 4\alpha^3\right)} - 9\beta\right)\right)$.

Case II: For $D=0$ the equation has three real roots and a repeated root:

$$\xi_1 = -\alpha/3, \ \xi_2 = \xi_3 = 2\alpha/3 \tag{18}$$

This is achieved when

$$\beta = \frac{4}{27}\alpha^3 \tag{19}$$

Case III: For $D<0$ the equation has one real root and two complex conjugate roots. This gives the following condition:

$$\beta > \frac{4}{27}\alpha^3 \tag{20}$$

By the substitutions $\xi = z_1$ and $\dot{\xi} = z_2$, equation (3) is converted to a system of two first order equations:

$$\begin{cases} \dot{z}_1 = z_2 \\ \dot{z}_2 = \alpha - z_1 - \beta / z_1^2 \end{cases} \tag{21}$$

To find the equilibrium points for the system, we solve

$$\begin{cases} \dot{z}_1 = z_2 = 0 \\ \dot{z}_2 = \alpha - z_1 - \beta / z_1^2 = 0 \end{cases} \tag{22}$$

simultaneously and a set of i solutions $\left(z_1^*, z_2^*\right)_i$ is obtained. According to Hartman's theorem [15], the classification of the equilibrium points is performed by the use of the Jacobian matrix $\mathbf{J}$ with elements $j_{11} = j_{22} = 0$, $j_{12} = 1$, $j_{21} = 2\beta / z_1^3 - 1$. Evaluation of the Jacobian matrix $\mathbf{J}_{\left(z_1^*, z_2^*\right)_i}$ at the equilibrium points and computation of its eigenvalues allows determining the equilibrium point type. The obtained numerical results for the three discussed cases are shown in table 2 together with the corresponding phase portraits with denoted equilibrium points.

Table 2. Number and classification of the equilibrium points.

Equilibrium point	Coordinates $\left(z_1^*, z_2^*\right)_i$	Jacobian $\mathbf{J}_{\left(z_1^*, z_2^*\right)_i}$	Eigenvalues	Phase portrait
Case I: $\beta < \frac{4}{27}\alpha^3$, $\alpha = 0.01$, $\beta = 0.5 \times 10^{-7}$				
p.A Stable center	$(0.00944, 0)_1$	$\begin{bmatrix} 0 & 1 \\ -0.88108 & 0 \end{bmatrix}$	$\lambda_{1,2} = \pm 0.93866i$	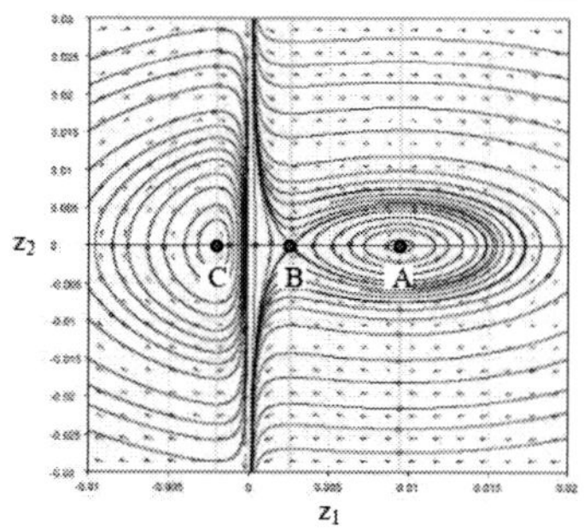
p.B Unstable saddle	$(0.00259, 0)_2$	$\begin{bmatrix} 0 & 1 \\ 4.6945 & 0 \end{bmatrix}$	$\lambda_{1,2} = \pm 2.1667$	
p.C Stable center	$(-0.00204, 0)_3$	$\begin{bmatrix} 0 & 1 \\ -12.813 & 0 \end{bmatrix}$	$\lambda_{1,2} = \pm 3.5796i$	
Case II: $\beta = \frac{4}{27}\alpha^3$, $\alpha = 0.01$, $\beta = 1.4815 \times 10^{-7}$				
p.A Degenerate, Unstable	$(0.00666, 0)_1$	$\begin{bmatrix} 0 & 1 \\ 0 & 0 \end{bmatrix}$	$\lambda_{1,2} = 0$	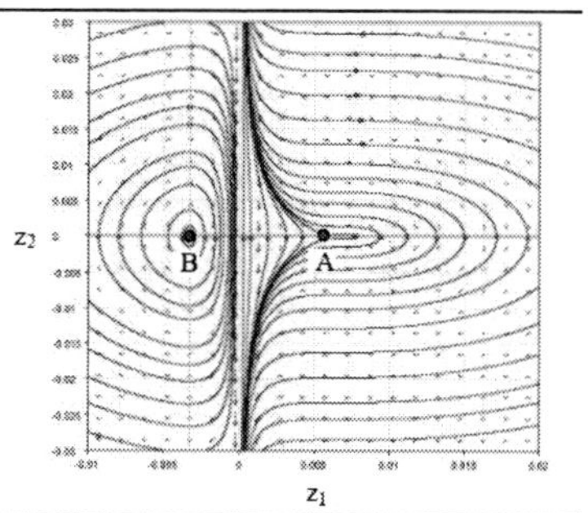
p.B Stable center	$(-0.00333, 0)_2$	$\begin{bmatrix} 0 & 1 \\ -9 & 0 \end{bmatrix}$	$\lambda_{1,2} = \pm 3i$	
Case III: $\beta > \frac{4}{27}\alpha^3$, $\alpha = 0.01$, $\beta = 5 \times 10^{-7}$				
p.A Stable center	$(-0.0056, 0)_1$	$\begin{bmatrix} 0 & 1 \\ -6.5386 & 0 \end{bmatrix}$	$\lambda_{1,2} = \pm 2.5571i$	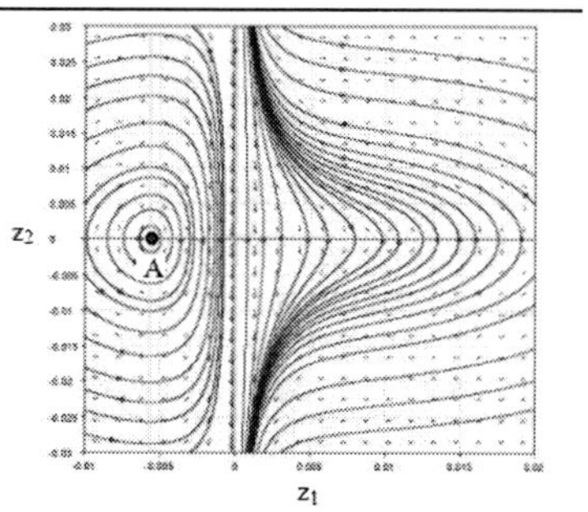

For all cases when the eigenvalues are pure imaginary complex numbers, the equilibrium point is center or focal. Applying global theorem due to Poincare [16] one determines that the equilibrium points are of type center.

As was already established, a possibility for stable vibrations in the vicinity of a value of ξ, satisfying constraint (4) exists only in Case I – p.A. This is feasible if the initial conditions of motion are within the area enclosed by the separatrix curve with the equation:

$$z_2 = \pm\sqrt{\frac{2\beta + 2c_{sep}z_1 + 2\alpha z_1^2 - z_1^3}{z_1}} \tag{23}$$

which follows directly from equation (5). In equation (23) by c_{sep} is denoted the energy constant, corresponding to the separatrix curve - the phase trajectory, passing through the saddle point *B*. For initial conditions outside of the area enclosed by the separatrix curve, only non-periodical single displacements with variable mass velocities and accelerations are possible.

The phase portrait for Case II corresponds to the bifurcation value (19). In this case, there are two equilibrium points. The unstable point *A* satisfies the constraint (4) and it is a result of merging of two equilibrium points – unstable saddle and stable center. The type of the phase lines shows that for all possible initial conditions the mass approaches the axis z_1=0 with increasing velocity.

The phase portrait for Case III shows that there is one equilibrium point and it doesn't satisfy constraint (4). The shape of the trajectories in the right half plane shows that in this case, the periodic motion is impossible.

6. Dynamical synthesis of an oscillating system

The aim of the dynamical synthesis of the considered magnetomechanical system is expressed here in finding such parameters combination, which guarantees periodic oscillations. For convenience the condition for stable oscillations (16) is rewritten in the following form:

$$\alpha > 3\sqrt[3]{\frac{\beta}{4}} \tag{24}$$

The value of the positive parameter $\beta = \kappa/(mk^2)$ is completely determined by predefined values of *k* and *m* or their combination $c=mk^2$. Taking into account that $\alpha = h_0 - g/k^2$ and using (24) one is able to determine the size of the air gap *h*:

$$h > 3\sqrt[3]{\frac{\kappa}{4mk^2}} + \frac{g}{k^2} - \delta_0 \tag{25}$$

The obtained numerical values are sufficient for choosing the rest of the geometrical parameters and material properties of the designed system but they do not guarantee oscillations with the prescribed natural frequency *k*. The reason for that is the presence of the imaginary magnetic mass which changes the natural frequency of the vibrating system.

The magnet force (1) is linearized in the vicinity of the equilibrium point with coordinate y_1, corresponding to the value of the largest positive root (17):

$$F_m \approx \frac{\kappa}{(h_0 - y_1)^2} + \frac{2\kappa(y - y_1)}{(h_0 - y_1)^3} \tag{26}$$

After inserting (26) in (2) and rearranging some terms one obtains the linearized differential equation describing the motion of the mass in the vicinity of the equilibrium point:

$$\ddot{y} + \left(k^2 - \frac{2\kappa}{m(h_0 - y_1)^3}\right)y = g + \frac{\kappa}{m(h_0 - y_1)^2} - \frac{2\kappa y_1}{m(h_0 - y_1)^3} \tag{27}$$

As one can see the term

$$k_1 = \sqrt{k^2 - \frac{2\kappa}{m(h_0 - y_1)^3}} \tag{28}$$

represents the natural frequency of the linearized magnetomechanical system and its value is less than the value of the prescribed natural frequency k of the mechanical system. Equation (28) points out the possibility for tuning the system resonant frequency by adjusting the magnetic system parameters [17]. For larger amplitudes due to the lack of isohronity of the system the frequency depends on the initial conditions.

7. Conclusion

Due to the iterative nature of the design process, in most cases, many iterations are needed to carefully design a device satisfying predefined requirements. On the basis of the presented qualitative analysis of the considered nonlinear oscillating system, a preliminarily dynamical synthesis is reasonably achieved and thus the design process is considerably facilitated. The developed mathematical model of the system under study is investigated by the use of the qualitative analysis methods of dynamical systems. Such type of analysis is the only mean of studying similar systems in the cases when the differential equation is nonlinear and cannot be solved exactly, or it leads to solutions which cannot be analyzed. The bifurcation zones corresponding to different behaviour of the system are determined. The study of the system phase portraits shows that only for a certain ratio of the parameters it is possible to have stable oscillation around an equilibrium point. In addition, some typical properties of the considered dynamical system are determined.

Acknowledgements

The authors would like to thank the Research and Development Sector at the Technical University of Sofia for the financial support.

References

[1] Niarchos D 2003 Magnetic MEMS: Key issues and some applications *Sensors Actuators A* **109** 166-73.
[2] Timoshenko S and Woinovski-Krieger S 1959 Theory of Plates and Shells (New York: McGraw-Hill) pp 63-76
[3] Nguyen V, Zehn M and Marinković D 2016 An efficient co-rotational fem formulation using a projector matrix *Facta Universitatis Series: Mechanical Engineering* **14** 227 – 240
[4] Liu S, Davidson A and Lin Q 2004 Simulation studies on nonlinear dynamics and chaos in a MEMS cantilever control system *J. Micromech. Microeng.* **14** 1064
[5] Schafer B E and Holzach H 1985 Experimental researches on flexible beam modal control *J. Guid. Control Dyn.* **8** 597–604.
[6] Ju M S, Chung Y Y and Tsuei Y G 1993 Development of electromagnetic actuators for structure control *Journal of Control Systems and Technology* **1** 235–246.
[7] Fung R F, Liu Y T, Wang C C 2005 Dynamic model of an electromagnetic actuator for vibration control of a cantilever beam with a tip mass *J. Sound Vib.* **288** 957-980.
[8] Lin J T 2010 The magnetic coupling of a piezoelectric cantilever for enhanced energy harvesting efficiency *Smart Mater. Struct.* **19** 045012
[9] Nie Y, Du Y and Xu Z 2017 Optimization Design of Electromagnetic Actuator Applied as Fast Tool Servo *Actuators* **6**, 25.
[10] Kim P, Bae S and Seok J 2012 Resonant behaviors of a nonlinear cantilever beam with tip mass subject to an axial force and electrostatic excitation *Int J Mech Sci* **64** 232-57
[11] Todorov T and Nikolov R 2008 Dynamical modeling of mechanical oscillations in magnetic field *RECENT* **9** 96-9.
[12] Walker J, Halliday D and Resnick R 2014 *Fundamentals of Physics* (New Jersey: Wiley) p 614.
[13] Manevitch L I, Kovaleva A S and Manevitch E L 2010 Limiting Phase Trajectories and Resonance Energy Transfer in a System of Two Coupled Oscillators *Math Probl Eng* **2010**

760479
[14] Khalil H K 2002 *Nonlinear Systems* (New Jersey: Prentice Hall) pp 35-88
[15] Lynch S 2010 *Dynamical Systems with Applications Using Maple* (Boston: Birkhauser)
[16] Ens R 2011 *It's a Nonlinear World* (New York:Springer) p 384
[17] Challa V R, Prasad M G and Fisher F T 2011 Towards an autonomous self-tuning vibration energy harvesting device for wireless sensor network applications *Smart Mater. Struct.* **20** 025004

IOP Conf. Series: Materials Science and Engineering **618** (2019) 012055 doi:10.1088/1757-899X/618/1/012055

A method for determining the constant DC/AC motor power range and main technological spindle unit parameters

S Guergov

Manufacturing Engineering & Machine Tools Department,
Technical University - Sofia, Bulgaria

sguergov@tu-sofia.bg

Abstract. The article offers an original method for determining the constant DC/AC motor power range and basic technological parameters of the main spindle unit for multifunctional machine tools at the design stage. The method allows them to be determined, both in terms of the strength characteristics of the gearbox (or the strength characteristics of the system "machine-equipment-tool-work piece") and in order to obtain the maximum spindle torque needed to ensure maximum machine productivity. The graphical interpretation of the method can be used both for designing new and for operating existing machines.

1. Introduction

Modern CNC machine tools (MT) are characterized by a higher level of energy costs than universal MT [1-5]. Moreover, the spare higher capacity lead to kinematic changes, increasing the mass of the machine, and hence increasing the operating costs and decreasing the wear-resistance of the machine as a whole.

On the other hand, it is known from many studies, the full power of the machines often is not used. In many other examples, some of main parameters (maximum spindle torque, maximum cutting power, etc.) are insufficient, which limits the cutting modes. Therefore, the choice of the input power and core technical parameters of the main spindle unit of the machine tools must be a production necessity and not a consequence of the market conjuncture. In the modern CNC machine tools are used stepless DC or AC motors.

That means that the correct calculation of the input power and basic technical characteristics of the machine must be carried out at the design stage. For the strength dimensioning of machine parts and units, it is necessary to determine the nominal or maximum load, as well as the calculation mode, which can be expressed by torque, load force, power or rotational speed [5,6].

In most cases the basic parameters of a newly designed machine tool are determined from the designer by analogy to other machines. Then they are complemented in the design process. In design, the basic parameters of the machine tools must be determined by analysing the family of parts for which the machine is designed.

The rated power output of the main motor and other technical parameters - the speed range, torque, the range of part's diameters, cutting speeds, etc., determine the production and economic capabilities of the machine. The correct and accurate determination of these parameters at the design stage should be based on the analysis of the physic-mechanical properties and geometric parameters of the parts family, blanks and tools related to the realization of particular technological process/processes. In this analysis, group technology methods are used, with the family of details divided into groups and a detail-type representative (complex work-piece/phantom detail) is synthesized for each group.

In the modern machine tools stepless DC or AC motors are used. The diagrams of the torque (M) and the power (P), depending on the spindle frequency n are shown in figure1.

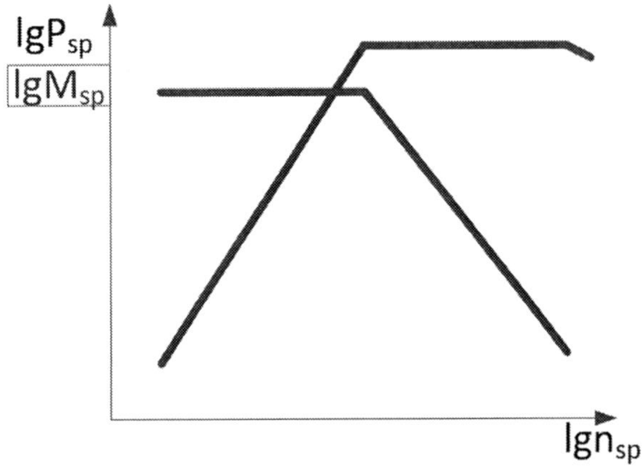

Figure 1. ($M_{sp} - n_{sp}$) Diagram

This paper presents an original method for determining the constant power range of the main DC/AC motor and the basic technological parameters of the main spindle unit for multifunctional machine tools at the design stage. In principle, the suggested method can be used to design of servo MT's drives/transmissions.

2. Methodology and research results

The function ($M_{sp} - n_{sp}$) (figure 1), expressed by the spindle power (P_{sp}) in the double logarithmic coordinate system ($\lg n_{sp} - \lg M_{sp}$) is a straight line equation (1). The angle between it and the abscissa is 45^0 (figure 2).

$$\frac{lg M_{sp}}{lg\left(\frac{1020.30P_{sp}}{\pi}\right)} + \frac{lg n_{sp}}{lg\left(\frac{1020.30P_{sp}}{\pi}\right)} = 1 \quad (1)$$

For the different cutting power values, a family of straight lines is produced. On an axis perpendicular to them, the corresponding bands can be seen (figure 2).

For the purpose of the study, the spindle power diagrams (P_{sp}) and spindle torque (M_{sp}), depending on the rotational spindle speed (n) from figure 1 can be represented as shown in figure2.

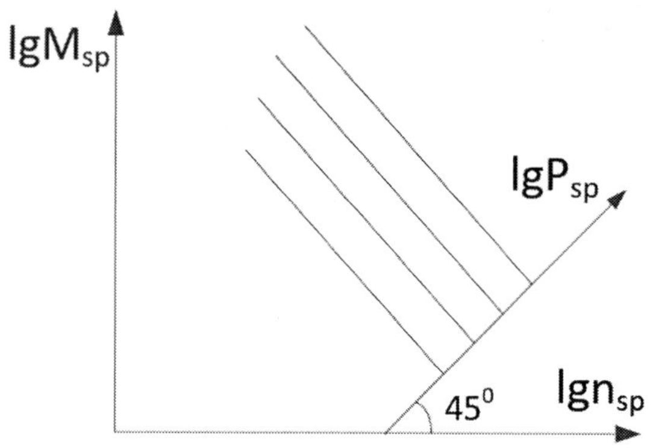

Figure 2. (P_{sp}-n_{sp}) Diagram

From the merging of the diagrams of figure 1 and figure 2, the diagram shown in figure 3 is obtained, from which both the momentary and the power values of the spindle can be taken into account.

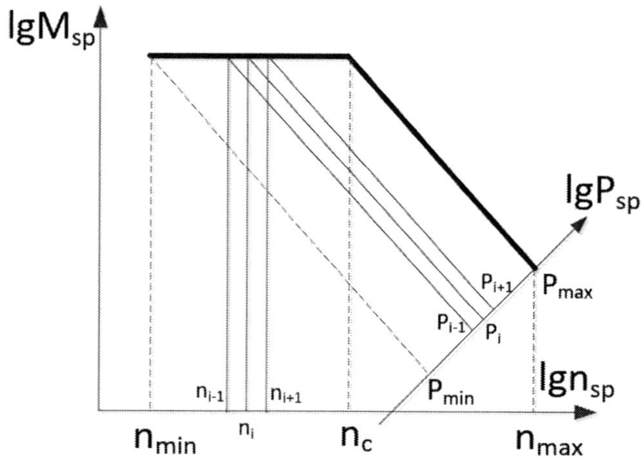

Figure 3. (M_{sp}–P_{sp}-n_{sp}) Diagram

From figure 3, the equivalent power equation (R_e) can be written. It is considered as a random probability interval within the range (P_{min}-P_{max}). This interval corresponds to the interval (n_{min}-n_{max}) and is equal to the mathematical expectation:

$$P_e = \lim \sum_{i=1}^{n} P_i P_i^* , \tag{2}$$

where n are the cutting passes (the size of the spindle speed variation range); P_i - the equally probable random magnitude of the power $n = n_{\min} \div n_c$ within the probability range P_i^*.

For P_i and P_i^* can be written:

$$P_i = \frac{n_i}{n_c} P_{\max}, \tag{3}$$

$$P_i^* = \frac{n_{i+1} - n_i}{n_c - n_{\min}} = \frac{\Delta n}{n_c - n_{\min}}. \tag{4}$$

After substitution of (3) and (4) in (2) for the equivalent power, the equation is obtained:

$$P_e = \int_{n_{\min}}^{n_c} \frac{P_{\max}}{n_c(n_c - n_{\min})} n dn = \frac{P_{\max}(n_c + n_{\min})}{2n_c}. \qquad 5)$$

Equivalent power (P_e^*) can also be calculated from the cutting conditions of the details set that are intended to be machined or using the group technology of the cutting conditions of the complex work-piece using the equation:

$$P_e^* = \sqrt{\frac{\sum_{i=1}^{n} P_i^2 t_i}{t_i}}, \tag{6}$$

where t_i is the time for processing the ith pass; P_{max} - the maximum value of the cutting power required.

After substitution of (6) into (5) for the value of the calculation frequency (n_c) is obtained the following equation:

$$n_c = \frac{n_{\min} P_{\max}}{2P_e^* \lambda - P_{\max}}, \tag{7}$$

where λ is the gearbox's efficiency.

The calculation frequency (n_c) is using for strength calculating the components of the gearbox and spindle unit. Depending on the required power for different details sets (P_e^*) the point n_c is approaching or moving away from $n_{\min}$. Thus it limits the torque M_{sp} and creates prerequisites for optimizing the strength dimensioning of the transmission.

For the rated output of the motor is obtained the following equation:

$$Pm = \frac{P_e^*}{\lambda} k, \tag{8}$$

where k is a reliability factor.

After applying P_e^* for P_m the diagram shown in figure 4 is obtained.

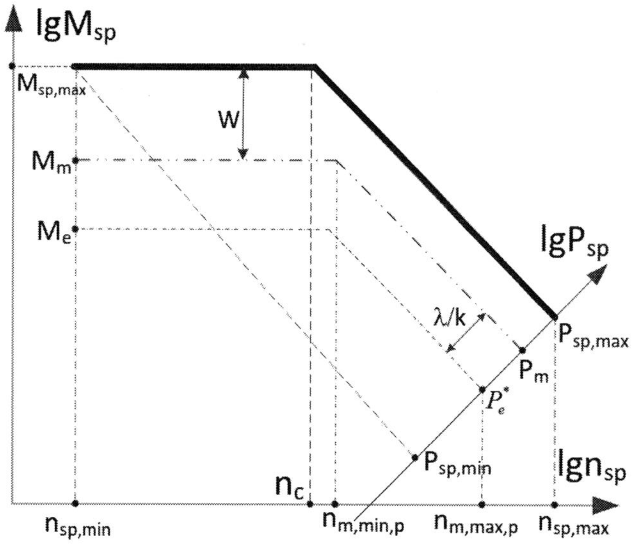

Figure 4. (M_{sp}–P_{sp}- P_e^* -P_m-n_{sp}) Diagram

3. Example

From the calculation of the cutting conditions for the processing of details set to be processed, it is obtained (figure 5):

Spindle speeds range (n_{min} – n_{max}): 45-3980 min^{-1}; t_i = 0,2 ÷ 2,0 min; λ =0,82; $P_{sp,max}$= 14,5 kW; $P_{sp,min}$ = 2,1 kW.

From (6) follows P_e^*= 11,6 kW; от (8) P_m = 14,1 kW. Choose from a catalog is selecting P_m=14,5kW, where $n_{m,max,p}$ = 4000min^{-1}, $n_{m,min,p}$ = 1150 min^{-1}, $M_{m,max}$ = 120 Nm).

From (7) is obtained for n_c = 144 min^{-1}.

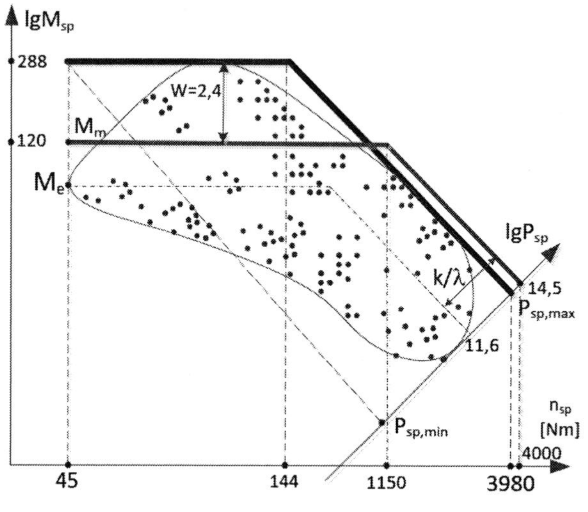

Figure 5. (M_{sp}–P_{sp}-P_m-n_{sp}) Exsample diagram

IOP Conf. Series: Materials Science and Engineering **618** (2019) 012055 doi:10.1088/1757-899X/618/1/012055

4. Conclusion

1) An original method for determining the constant DC/AC motor power range and main technical parameters of the main spindle unit for multifunctional machine tools at the design stage is proposed.
2) The method to be determined allows, both in terms of the strength characteristics of the gearbox (or the strength characteristics of the system "machine-equipment-tool-work piece") and in order to obtain the maximum spindle torque needed to ensure maximum machine productivity.
3) The method allows the determination of the basic technological parameters in order to obtain the maximum necessary spindle torque. This approach must be basic in order to obtain the maximum spindle torque needed to ensure maximum machine productivity.
4) The diagram of the type shown in figure 4 can be used both for designing new machines and for operating existing ones to quickly check for the ability and efficiency of using a particular machine to process certain details.

Acknowledgement
The authors would like to thank the Research and Development Sector at the Technical University of Sofia for the financial support.

References

[1] Defilippi A, Ippolito R and Micheletti G F 1981 NC machine tools as electric users, CIRP *Annals - Manufacturing Technology* 30(**1**) pp 323-326
[2] Dimitrov K, Dzhedzhev K and Mitsev T 2018 Enhancing Smart-home Environments Using Infrared Arrays, *9th National Conference with International Participation, ELECTRONICA–* Proceedings 8439575
[3] Nikolova N and Nikolov E 2006. Energy saving algorithms and control system, IFAC Proceedings Volumes, *International IFAC Workshop on Energy Saving Control in Plants and Buildings* ESC 2006 Bansko Bulgaria 2-5 October 2006 Code 85982, Vol. **1**, Issue PART 1, pp 141-146
[4] Holkup T, Vyroubal J and Smolik J 2013 Improving energy efficiency of machine tools, G. Seliger (Ed.), Proceedings of the *11th Global Conference on Sustainable Manufacturing - Innovative Solutions* ISBN 978-3-7983-2609-5 Universitätsverlag der TU Berlin, pp 125-130
[5] Skoczynski W,Maczka J, Wasiak Z,Roszkowski A and Pres P 2013 Assesment of Energy Consumption by Machine Tools, Acta Technica Corviniensis – Bulletin of Engineering Tome **VI**, ISSN 2067-3809, pp 103-108
[6] R. Crowder R 2006 *Electric Drives and Electromechanical Systems*, Butterworth-Heinemann is an imprint of Elsevier Linacre House Jordan Hill Oxford OX2 8DP UK ISBN–13: 978-0-7506-6740-1 p. 292

Green time period optimization at signalized intersection in Town of Pernik

D N Saliev[1] and I R Andreev[1]

[1]Technical University of Sofia, Faculty of Transport, Department of Combustion Engines, Automobile Engineering and Transport, Bulgaria

durhan_saliev@tu-sofia.bg

Abstract. Green time period optimization is essential issue about delay time reduction at signalized intersection. The features of traffic flows are variables in rush and other period of day. Optimal values of green time are that guarantee minimum of delay time. It can be accepted that the values that guarantee that in rush period is can be use in others periods of day. The main issue is being considered in paper is green time period optimization at signalized intersection for minimum delay time according to traffic flows features in rush period. That include comparison between three options of movement diagram and two option of phases for the chosen intersection in Town of Pernik. The application of result will make a lot of benefits both for the traveling and for the environment.

1. Introduction

In the last years in urbans number of vehicles have been increasing bringing seriously problems at social, environmental and economic levels. The main problem is increase delay time at intersections at greater rate. That influence of travel time of travelers (drivers and passengers), travel cost, fuel consumption, etc. That problems are traditionally solved by optimizing to minimizing total delays for vehicles. In case that approach is not efficient requires decision related to new road capacity, intelligent traffic system in new approach of traffic management, intersection reconstruction in new shape and more levels.

In this paper optimizing by minimizing total delay time for vehicle are presented for isolated intersection. Approaches used are described in [1] and [2].

2. Prerequisites and means for solving the problem

One of the busiest intersections on the territory of the town of Pernik is the four-approach junction of “Yuri Gagarin” Boulevard, “St. St. Cyril i Methodii” Street and “Pyrvi Mai” in the quarter of Dimova Mahala. Intersection of traffic flows is regulated by traffic lights. The transport link between the districts "Dimova Mahala", "Tvardi livadi", "Beli breg" and "Humni dol" is at the crossroads. A scheme of the junction is shown in Figure 1. After observation, it was found that during the evening peak hour of the day, the loading of cars at the crossroads in question was high. For some of the streams, a queue length of cars that interferes with the normal operation of the junction is formed.

The intersection in question consists of four approaches. Approach 1 (from “Yuri Gagarin” Boulevard) is a single canvas and separate traffic lanes for straight and left, each lane of 3,75 meters. Approach 2 on which the traffic road (from “Pyrvi Mai”) is a two-way roadway and consists of two traffic lanes of 3.75 m width and a total gauge of 7,5 m, which in the area of the junction do not increase their number. Approaches 3 and 4 (from “St. St. Cyril i Methodii” Street) are with one canvas and separate traffic lanes for straight, right and separate traffic lane for the turning left flows, each strip of 3,75 meters.

For the purpose of the study, the approaches of the junction were designated as shown in Figure 2.

Using described algorithms in [1] and [2] requires input data wherefore were examined the following parameters for features of traffic flows crossing the intersection: queues length for all approaches; traffic light cycle and phases' durations; intensity of traffic flows; saturation flows for all traffic lanes. Technical devices of traffic research, described in [3] and [4], were used for collection of some of these data.

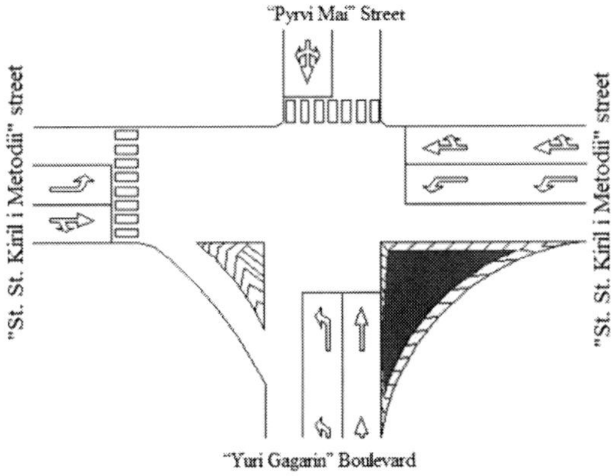

Figure 1. Intersection scheme.

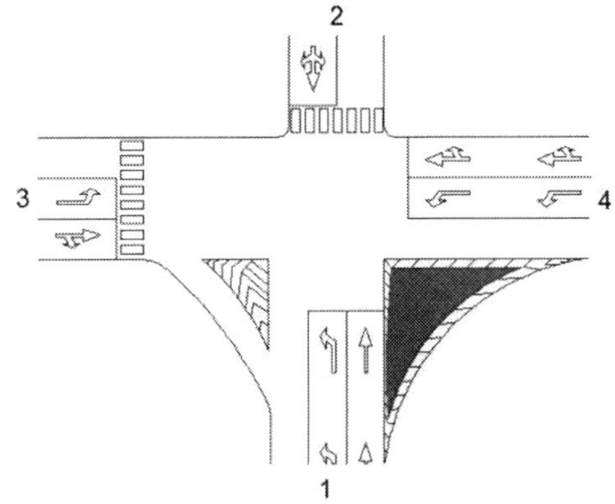

Figure 2. Intersection's approaches numbering.

The results of survey are shown in Table 1, Table 2, Table 3, Table 4, Figure 3, and Figure 4.

Table 1. Average queue length.

Flow №	Average queue lengths Q_{av} (PCU)
Flow 1-2	4,40
Flow 1-3	16,10
Flow 2-3, 2-1, 2-4	5,13
Flow 3-1, 3-4	86,26
Flow 3-2	3,27
Flow 4-2, 4-3	26,07
Flow 4-1	23,40

Table 2. Intensity by approaches at intersection.

Approach №	Intensity I_a^f (PCU/s)
Approach 1	0,31
Approach 2	0,09
Approach 3	0,34
Approach 4	0,34

Table 3. Saturation flows for traffic lanes.

Flow №	Saturation flow I_p^f (PCU/s)
Flow 1-2	0,41
Flow 1-3	0,19
Flow 2-3, 2-1, 2-4	0,40
Flow 3-1, 3-4	0,52
Flow 3-2	0,41
Flow 4-2, 4-3	0,47
Flow 4-1	0,46

The traffic flows passes in three phases (Figure 3). Cycle length is 75 s. Intergreen time (t_{ti}) for all of phases is 5 s. Green time (t_{gi}) duration is 18 s for first phase, 22 s for second phase and 20 s for the third phase (Figure 4).

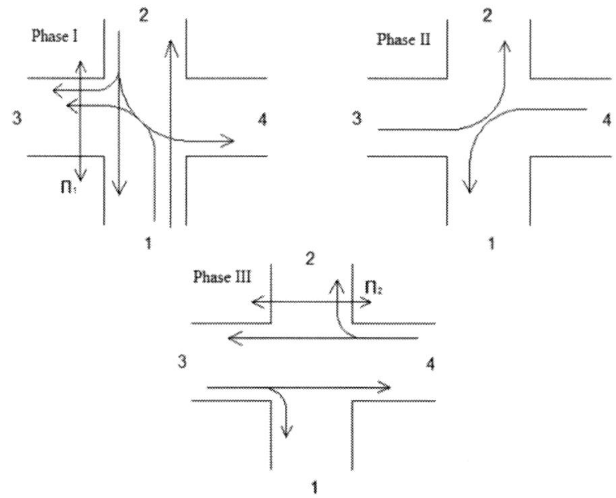

Figure 3. Phase signal diagram.

Table 4. Phase time duration.

Phase №	Duration (s)
Phase I	23
Phase II	27
Phase III	25

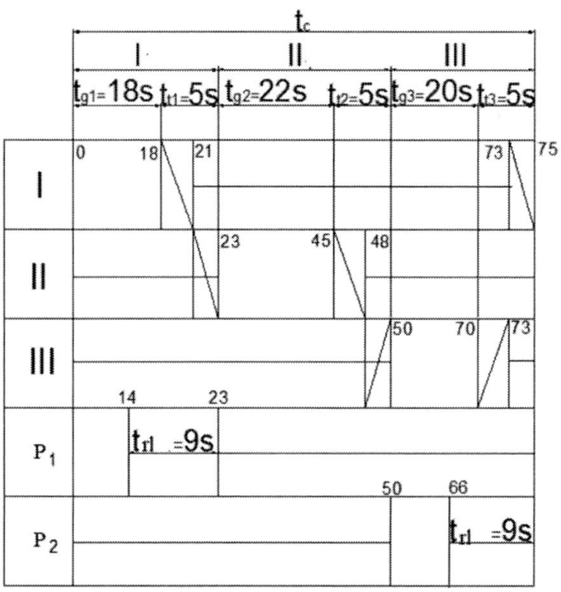

Figure 4. Time-Space diagram.

After the observation and measurements, it is clear that at the intersection there are problems with the accumulation of the most cars under flow 3-1, 3-4 running right and to the right from the busiest direction "St. St. Kiril and Metodii" boulevard. After it, there is a flow of 4-2, 4-3, and a flow of 4-1. One way to solve this problem is by optimizing the green signal times of each phase by the method described in [2]. The goal is to optimize the green signal duration of each phase, so minimizing total delay of all the junction flows.

3. Results and discussion

They were appreciated three different variants of number of phases and phase signal diagrams – optimizing the green times of current time-spacing diagram; calculation of the delay time at the junction for passing the vehicles in four phases and calculation of delay time at the junction when passing vehicles in three different phases than existing ones.

3.1. Optimizing the green times of current time-spacing diagram

The maximum duration of the traffic light cycle shall be determined in accordance with the requirements set out in [7], namely 90 seconds when the three-phase traffic is omitted. The minimum cycle time is determined according to the minimum green signal time. It takes 8 seconds in view of the traffic flow parameters for features and the possibility of pedestrian crossing in the individual phases. The cycle change step should be taken to be the minimum duration of one phase when passing by road vehicles as defined in [7]. According to this, the cycle will change from 49 to 87 seconds at an interval of 8 seconds while preserving the intergreen times for each phase.

Using the optimize algorithms in [1], we need to combine traffic flows by gathering the queue lengths, the incoming car's intensity and the saturation flows by each phase. For phase I we collect queue lengths, intensity and saturation flows of traffic flow 1-2, flow 1-3 and flows 2-3, 2-1, 2-4. For phase II of traffic flows 3-2 and 4-1. For phase III of traffic flows 3-2 and 4-1. The resulting queue lengths values are shown in Table 5, vehicle Intensity in Table 6, and saturation flow in Table 7.

The results of optimizing the duration of the traffic light phases at different cycle time values are shown in Table 8 and Figure 5.

Table 5. Average queue length.

Phase №	Average queue lengths Q_{av} (PCU)
Phase I	25,63
Phase II	26,67
Phase III	112,33

Table 6. Intensity by phases.

Phase №	Intensity I_a^f (PCU/s)
Phase I	0,39
Phase II	0,68
Phase III	0,68

Table 7. Saturation flows by phases.

Phase №	Saturation flow I_p^f (PCU/s)
Phase I	0,60
Phase II	0,87
Phase III	0,98

Table 8. Results of optimizing the duration of the traffic light phases at different cycle time values.

t_c (s)	Phase's duration (s)			T_{wmin} (h)
	Phase I	Phase II	Phase III	
39	8	8	8	15,80
47	8	8	16	12,40
55	8	10	22	12,52
63	10	12	26	12,67
71	14	14	28	13,15
79	14	16	34	13,55
87	16	20	36	14,07

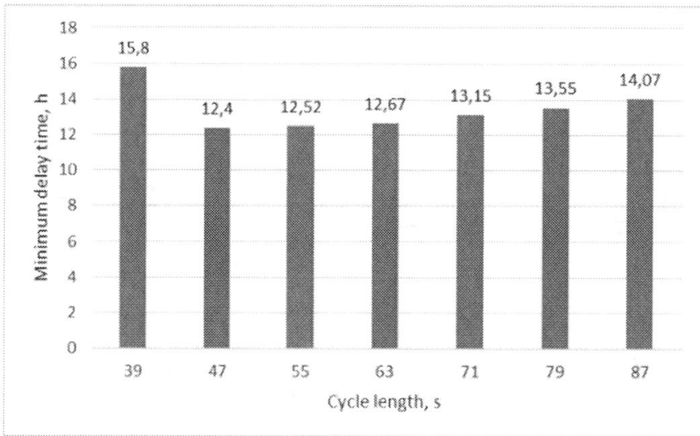

Figure 5. Delay time at different cycle time values.

It can be seen that with any change in cycle time with the specified step, the delay time tends to decrease as the minimum is reached at 47 second cycle time. After this step each successive increase in waiting time. With the corresponding green signal timing distribution, the minimum waiting time for all traffic flows at the junction is 12,40 hours for crossing the junction of the vehicles in the evening peak period.

After optimizing the green phase times, the waiting time decreases by 3,29 hours from the current 15,69 hours.

3.2. Calculation of the delay time at the junction for passing the vehicles in four phases

The junction configuration and traffic flow intensity values provide a basis for identifying four trafficking phases, aiming at a safe passage of vehicles from each approach.

In this case we use an algorithm describe in [2] for calculating cycle length and green tame duration. After that we gone optimized that value with an algorithm presented in [1].

Determined phases are that shown in Figure 6. To provide an extra time reserve when crossing the conflict zones between cars of different phases, intermediate times are foreseen. It is the duration between the end of the green period of one traffic stream and the start of the green period of a following released conflicting or partially conflicting traffic stream. In this way, it is ensured that the time taken for the last vehicle to arrive from the junction flows to pass the conflict zone with the outgoing flows from the next phase before the first vehicle reaches it. A schematic of the conflict zones in the defined phase signal diagram is shown in Figure 7. For each of them the duration of the intermediate time is determined in use the German method reached with the Japan researchers [5]. Values for clearance speed and entrance time we accept the presented in [6].

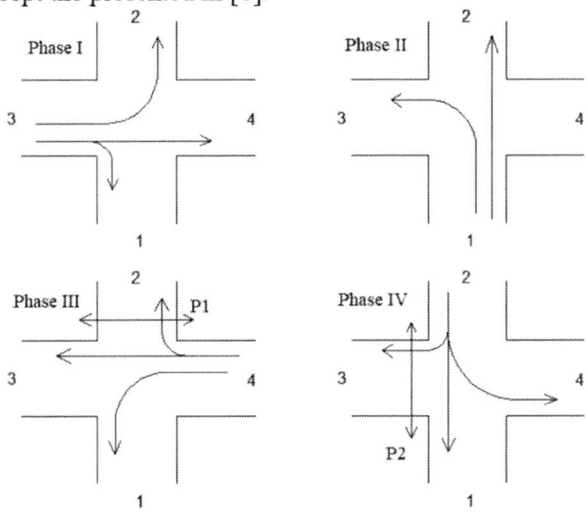

Figure 6. Phase signal diagram.

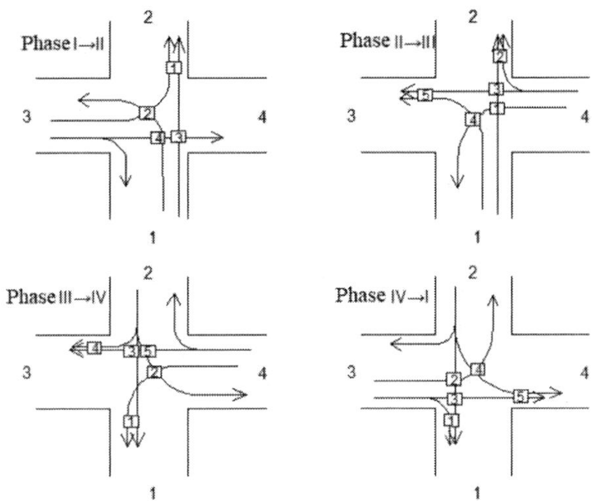

Figure 7. Schematic of the conflict zones in the defined phase signal diagram.

The intergreen time adjacent to the corresponding phase is the largest calculated value for the relevant conflict zones. It needs to be an integer in seconds. Ensuring the resulting values requires rounding to the larger integer. The adjacent intergreen times to the relevant phase are: 7 s for phase I; 5 s for phases II and III and 3 s for phase IV.

According to algorithm described in [2] it needed to calculate saturation flows for each phases. This is done for flows that are only passed through the appropriate phase. For phase I, these are flows 3-1, 3-2 and 3-4, for phase II these are flows 1-2 and 1-3, for phase III these are flows 4-1, 4-2 and 4-3, for phase IV, these are flows 2-1, 2-3 and 2-4. Result about saturation flows are: 1,0 PCU/s for phase I, II and III; 0,5 PCU/s for phase IV.

Needed green time for all traffic flows cross the junction through the evening rush period is 4212 seconds. That shows the time is not enough for vehicles passing in this period. Decision in this case is intersection reconstruction for change the shape and increase capacity.

For aim of optimization without reconstruction we accept for cycle length the maximum permissible traffic light cycle time determined in accordance with the requirements set out in [7], namely 120 seconds when a four-phase is omitted.

In this acceptance the green time period are calculated as follows: for phase I - 31 s; for phase II – 27 s; for phase III – 31 s and for phase IV – 11 s.

For calculating delay time according algorithm describe in [1] we needed to collect values for queue length for flows in each phases. Collected values for Q_{av} are: for phase I – 89,53 PCU; for phase II – 20,5 PCU; for phase III – 49,47 PCU and for phase IV – 5,13 PCU.

Phase's duration and total delay time are shown in Table 9.

Table 9. Results of phase's duration and total delay time.

t_c (s)	Phase's duration (s)				T_{wmin} (h)
	Phase I	Phase II	Phase III	Phase IV	
120	38	32	36	14	9,48

Four phases variant application decreases total delay time about 40 %.

3.3. Calculation of delay time at the junction when passing vehicles in three different phases than existing ones

To determine the number and type of phases of flows across the intersection, we chose the busiest legs to separate them in separate phases, and those with lower metrics to unify them in one phase. Legs 3 and 4 will pass into two separate phases, the intensity of these two approaches is 0,34 PCU/s. Legs 1 and 2 will pass through a common phase, their total intensity is 0,39 PCU/s.

The phases are as follows: continuing movement to the right, straight and left streams of approach 3 form Phase I, there is no leakage of pedestrian movement; continuing to the straight, right and left flows of approach 4 form a phase II in which the pedestrian movement of approach 2, parallel to the leg 3 and 4, is also omitted; continuing movement straight, right, left flows of approach 1 and 2 form a phase III in which the pedestrian movement of leg 3 is parallel to legs 1 and 2. The phase signal diagram is shown in Figure 8. A schematic of the conflict zones in the defined phase signal diagram is shown in Figure 9. For each of them the duration of the intergreen time is determined according to algorithm presented in [2]. The values are as follows: for phase I – 9 s; for phase II – 5 s; for phase III – 3 s.

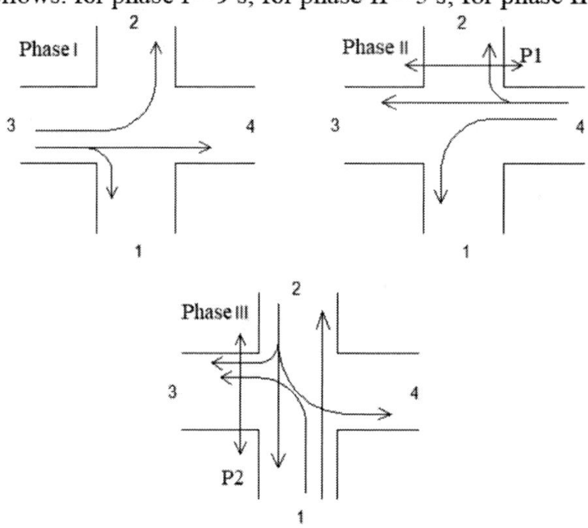

Figure 8. Phase signal diagram.

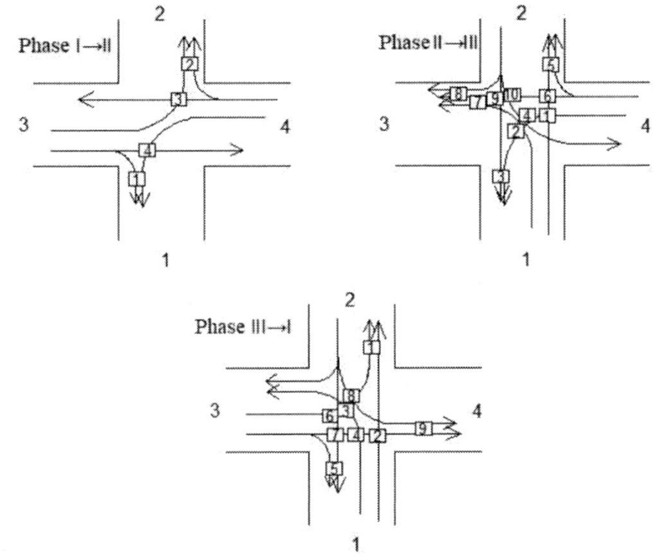

Figure 9. Schematic of the conflict zones in the defined phase signal diagram.

Calculating saturation flows for each phases is once again follow the algorithm presented in [2]. This is done for flows that are only passed through the appropriate phase. For phase I, these are flows 3-1, 3-2 and 3-4, for phase II these are flows 4-1, 4-2, 4-3 for phase III, these are flows 1-2, 1-3, and 2-1, 2-3, 2-4. The calculated values for saturation flows are: 1,0 PCU/s for phases I and II, 1,5 PCU/s for phase III.

Needed green time in this case is 3384 s. Determined value for cycle length is 283 s. This is inconsistent with the requirements set out in the [7] – 90 seconds about three phases signal diagram. This again highlights the necessity of re-engineering the junction to increase capacity.

Following the same aim - optimizing without reconstruction – we accepted for cycle length duration 90 s in accordance with [7]. In this case calculated green time period for each phases is: 27 s for phases I and II, 19 s for phase III.

For calculating total delay time we again collect the values about queue lengths for flows pass in separated phases. Collected values for Qav are as follows: for phase I – 89,53 PCU; for phase II – 49,47 PCU; for phase III – 25,63 PCU.

Phases duration and total delay time is shown in Table 10.

Table 10. Results of phase's duration and total delay time.

t_c (s)	Phase's duration (s)			T_{wmin} (h)
	Phase I	Phase II	Phase III	
90	36	32	22	9,35

Three different phases than existing ones variant decrease total delay time above 40 %.

3.4. Comparison of the results obtained

The results of the phase optimization and the duration of the traffic light phases at the crossing considered at different cycle time values are shown in Table 11 and Figure 10. It can be seen that with each of the optimizations made, the total delay time tends to decrease, as the minimum is achieved with three-phase adjustment different phases than existing ones and a 90 second cycle time. With the corresponding green signal timing, the minimum delay time for all flows at the junction is 9,35 hours.

Table 11. Comparison of the result obtained.

Variants	t_c *(s)*	$T_{w\min}$ (h)
Current	75	15,69
Optimized current	47	12,40
Four phases	120	9,48
Three phases different for existing once	90	9,35

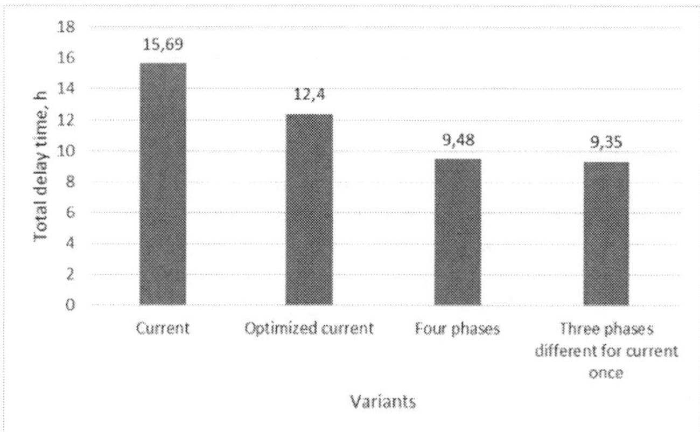

Figure 10. Result of optimization of green time period at signalized intersection in Town of Pernik.

4. Conclusions

The basic aim in this paper - optimization of green time period at signalized intersection in Town of Pernik – have been achieved. Examined variants of traffic oversight are shown different results but implementation of which one different than existing one will be better for decrease negative of reached traffic flows in evening rush period at this intersection. The best result are given in three phases dissimilar than current. This variant for passes flows through the crossroad will decrease the total delay time above 40 %. This is expected to decrease travel time, fuel consumption and the harmful components in the exhaust gases almost such values.

For more benefits and availability of resources of Municipality of Pernik that intersections must be re-engineering for change the shape (mean roundabout) or make the one more level for flows of "St. St Ciril i Metodii" street direction.

5. Acknowledgements

The author/s would like to thank the Research and Development Sector at the Technical University of Sofia for the financial support.

References

[1] Saliev D 2013 Modeling of Road Traffic in Case of Dangerous Occurrences *Dissertation* (Sofia: Technical university of Sofia Press)

[2] Saliev D 2017 *Machines, Technologies, Materials – International Scientific Journal* **Issue 1** ISSN Print: 1313-0226 ISSN Web: 1314-507X Year XI (Sofia) p 33

[3] Hristov V, Savova-Mratsenkova M, Damyanov I, Mladenov G and Palagachev G 2017 *Proc. Conf. TechCo – Lovech 2017* ISSN 2535-079X (Lovech: Gabrovo University press) p11

IOP Conf. Series: Materials Science and Engineering **618** (2019) 012056 doi:10.1088/1757-899X/618/1/012056

[4] HristovV, Damyanov I, Savova-Mratsenkova M, Palagachev G and Mladenov G, 2017 *Proc. Conf. TechCo – Lovech 2017* ISSN 2535-079X (Lovech: Gabrovo University press) p17

[5] Tang K, Ono T., Tanaka Sh and Kuwahara M 2011 *Asian Transport Studies* **Volume 1** Issue 4, (Japan) p 412

[6] Hebenstreit C, Fellendorf M, etc. 2014 ***Final*** STREP ICT Call 8FP7-ICT-2011-8 *COLOMBO Deliverable 2.2 – Policy definition and dynamic policy selection algorithms* p46

[7] Ministry of regional development and public works 2004 *Ordinance No 17 of 23 July 2001 for regulation of road traffic with light signals* (Sofia)

Method for experimental determination of the coefficients of stiffness and damping of rubber insulators

L P Kunchev and Z A Georgiev

Technical University – Sofia, Faculty of Transport, Department of Combustion Engines, Automobile Engineering and Transport, 8 Kliment Ohridski Blvd., 1000 Sofia, Bulgaria

zlaty.georgiev@gmail.com

Abstract. Rubber insulators are used in many dynamic systems to reduce the impact of vibrations. They are located in the mount system of the engine, gearbox, suspension, et cetera. Two of their characteristics are the dynamic coefficients of stiffness and damping. These characteristics are influenced by the frequency of vibration, the preload in the mounts and the hardness of the rubber. The characteristics of the rubber insulators in the suspension elements also influence the handling of the vehicle. In the work are investigated rubber insulators from different suspensions in the frequency range 70 - 220 Hz. This frequency range contains the resonant frequencies of the radial type pneumatic tyre due its interference with the road surface. The purpose of the work is to develop a method for obtaining the dynamic coefficients of stiffness and damping of the rubber insulators.

1. Introduction and overview of the problem

Rubber insulators used in the mounting of the suspension influence the magnitude and frequency of high frequency oscillations [1]. In the modeling and optimization of suspension behavior, the values of the stiffness and damping coefficients of the insulators are essential. The dynamic coefficients of stiffness and damping of the rubber mounts have non-linear characteristics and they depend on the frequency of the oscillations, the preload on them and the hardness of the rubber. These coefficients are different from those in static measurements. In the modeling of the behaviour of the dynamic system operating under a fixed operating mode (constant frequency), it is possible to use constant values of the coefficients of stiffness and damping [2]. In the case of road disturbance to the tyre and the suspension elements, the frequency of excitation depends on the speed of motion. The dynamic system operates in a wide frequency range and it is necessary to study the values of these coefficients in each resonance of the tyre and the suspension elements. The resonant frequencies of the radial type pneumatic tyre, due to its design characteristics, occur in the 80 - 250 Hz frequency range [3, 4]. The absorption of these oscillations in this frequency range is mainly carried out by the rubber insulators located at the mounts of the suspension. Above 50 Hz, regardless of its type, the shock absorber does not affect the oscillation absorption [5]. In modeling the dynamic behaviour of the suspension, the coefficients of stiffness and damping of the rubber mounts have to be set and predetermined for the respective resonance frequencies of the tyre and the suspension elements [6]. In [7], the commonly used models of rubber insulators are considered and the authors offer their own model that describes its non-linear dynamic behavior. In [8], the authors investigate the structure-borne vibrations generated by the engine and their impact on the vibro-acoustic comfort in the vehicle cabin. For the numerical method, it is used the coefficient of stiffness, which is experimentally obtained. The results show that the value of stiffness coefficient increases with the increase in frequency. In work [9] are obtained the forces before and after the rubber insulators of the diesel generator. The forces are presented as functions of the frequency of oscillation, the angle of rotation of the crankshaft and the load on the diesel unit. Suspension elements are responsible for the vehicle handling. Because of that, the optimization of the stiffness values of the rubber insulators, in order to improve comfort, is made within narrow limits [10].

This work presents a method for obtaining the dynamic coefficients of stiffness and damping of rubber insulators. Object of the study are rubber mounts for different types of suspension: the rubber insulators in the lower arm (position 2 in figure 1(a)) and in the strut rod (position 1 in figure 1(a)) for an independent front torsion spring suspension; the two rubber insulators in the lower control arm (position 3 and 4 in figure 1(b)) for front MacPherson strut suspension; the rubber insulators in the shock absorber mounts (position 6 in figure 1(c)) and that in the trailing arm (position 5 in figure 1(c)) for twist-beam rear suspension.

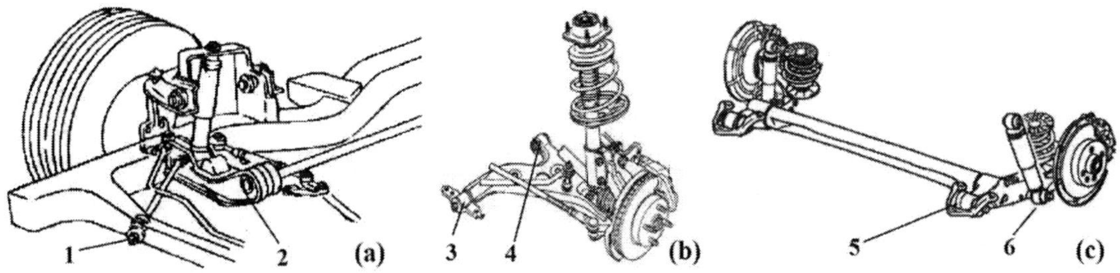

Figure 1. Rubber mounts for different types of suspension: (a) - torsion spring suspension; (b) - MacPherson strut suspension; (c) - twist-beam rear suspension [11].

2. Description of the method

The method for experimentally determining the coefficients of stiffness and damping of rubber insulators is based on a comparative method of two dynamic systems presented in figure 2. In the first dynamic system (A), the mass m_b and coefficients of stiffness c_b and damping β_b are known. To create the second system (B), three new parameters are added, from which the mass m_v is known and the coefficients of damping β_v and stiffness c_v are unknown. The first system (A), which is called "spring – mass", is a simplified model of the free end of a cantilever beam. The second system (B), which is called "spring – mass – insulator", is created by installing fixed rubber insulator to the free end of the beam. The unknown coefficients β_v and c_v are respectively the dynamic coefficients of damping and stiffness of the rubber insulator.

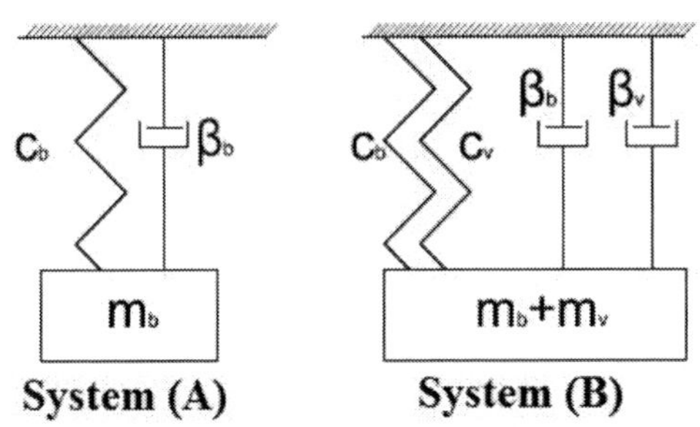

Figure 2. Dynamic system (A) "spring – mass" and dynamic system (B) "spring – mass – insulator".

By using the relationship between the described parameters in the case of free damped oscillations, the unknown coefficients β_v and c_v can be determined. For free damped oscillations, the following can be written [12]:

$$\tau = \frac{2\pi}{\Omega} = \frac{2\pi}{\sqrt{\omega^2 - n^2}}, \quad (1)$$

where τ is the period of the free damped oscillations;

Ω – the frequency of the free damped oscillations;

$\omega = \sqrt{\frac{c}{m}}$ – the natural frequency of the dynamic system;

$n = \frac{\beta}{2m}$ – coefficient that parameterizes the strength of the damping.

For the specific dynamic systems (A) and (B) $n<\omega$, so in this case the damped oscillations are periodic. Figure 3 is the time-domain response of the dynamic systems after initial perturbation. The continuous line record (figure 3) corresponds to the free damped oscillations of system (A) and the dashed line record corresponds to the free damped oscillations of system (B). Changing the masses of the systems also changes their frequency of oscillation.

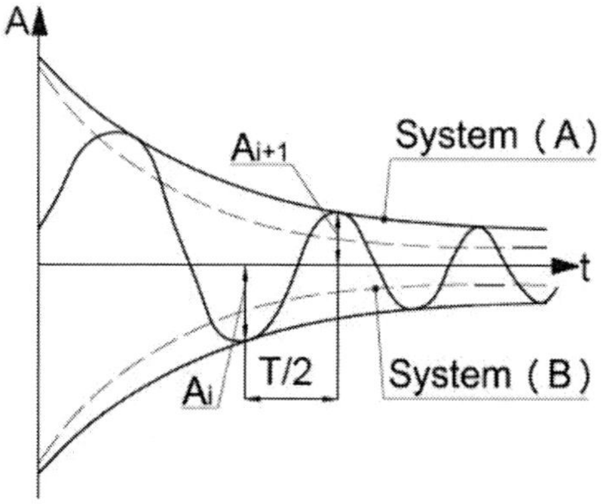

Figure 3. Time-domain response of dynamic systems (A) and (B).

For each system, the logarithmic decrement is determined. The logarithmic decrement is defined as the natural logarithm of the ratio of any two successive amplitudes:

$$\delta = \ln\frac{A_i}{A_{i+1}} = \frac{n\tau}{2} = \frac{\pi n}{\sqrt{\omega^2 - n^2}} \tag{2}$$

The logarithmic decrement characterizes damping in the dynamic systems. Possessing the time-domain responses of the two systems (A) and (B), the dynamic coefficients of damping and stiffness of the rubber insulator can be determined from equations (1) and (2). The coefficients of damping β_v and stiffness c_v are respectively:

$$\beta_v = \frac{4m_\Sigma\delta_\Sigma}{\tau_\Sigma} - \frac{4m_b\delta_b}{\tau_b} \tag{3}$$

$$c_v = \frac{4m_\Sigma}{\tau_\Sigma^2}\left(\delta_\Sigma^2 + \pi^2\right) - \frac{4m_b}{\tau_b^2}\left(\delta_b^2 + \pi^2\right), \tag{4}$$

where m_b and m_Σ are respectively the mass of the system (A) "spring – mass" and the mass of the system (B) "spring – mass – insulator" (figure 2);

τ_b and τ_Σ – respectively the periods of oscillation of system (A) and system (B);

δ_b and δ_Σ – respectively the logarithmic decrement of system (A) and system (B).

3. Method of conducting the experiment

3.1. Description of the experimental setup

A flexible steel beam 6 is welded to the upper metal plate of housing 9 (figure 4). The housing is a metal cylinder, the volume of which is filled with reinforced concrete. The natural frequency of the housing must be at least five times higher than that of the steel beam. Together with the correcting weights 5, the beam provides the possibility of varying the frequency of the oscillation of the test object in the frequency range 70 – 220 Hz. The test object (rubber insulator) 4 is attached to the bracket 3, which can be moved vertically by the loading device 2 in the support 1. The loading device 2 presses the rubber insulator to the free end of the beam with the required preload force.

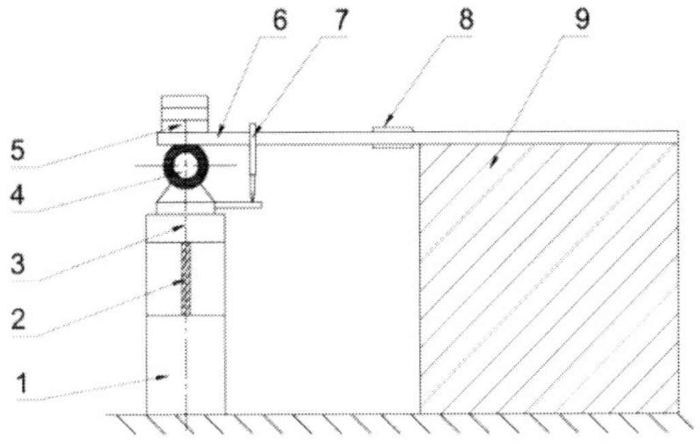

Figure 4. Experimental setup for obtaining oscillograms of free damped vibrations (1 - support, 2 - loading device, 3 - bracket, 4 - test object, 5 - correcting weights, 6 - steel beam, 7 - displacement transducer, 8 - strain gage, 9 - housing.

The excitation of the free damped oscillations in the system is performed by a single impact on the beam. The vibrations of the beam are measured by a strain gage 8 or an accelerometer (figure 4). The deflection of the free end of the beam is measured by a displacement transducer 7. In the layout of the measuring system are also included analog-to-digital converters for voltage inputs and for measuring with strain gages (positions 4 and 5 in figure 5). The obtained oscillograms are recorded with a computer. The requirements for conducting the experiment are: room temperature 20±5 °C; accuracy of the measuring equipment less than 1%; the mass (housing), to which the elastic beam is mounted, needs to has at least five times higher natural frequency than that of the beam; the elastic beam at different loads needs to has frequency of oscillation in the range of 70-220 Hz.

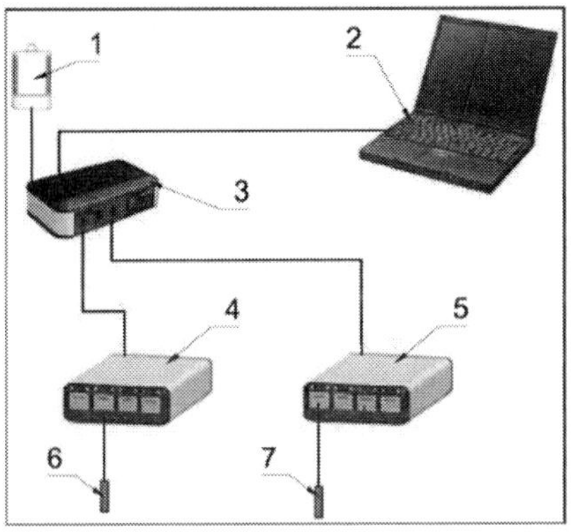

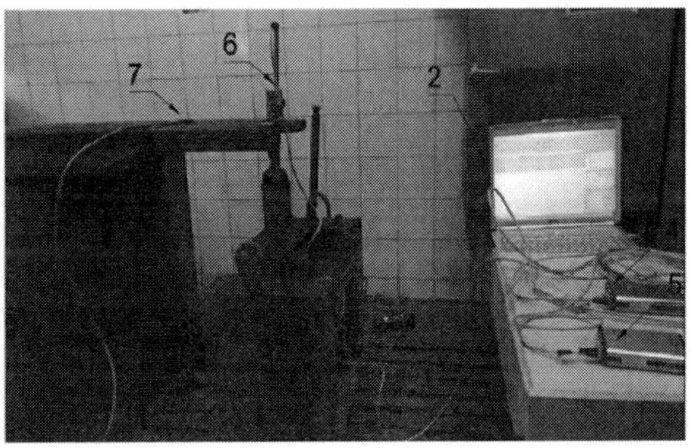

Figure 5. Layout of the measuring system (1 - power supply, 2 - read out (PC), 3 - active USB HUB, 4 - analog-to-digital converter DQ401, 5 - analog-to-digital converter DQ430, 6 - inductive displacement transducer, 7 - half bridge strain gage).

3.2. Sequence of conducting the measurements

Firstly are conducted the measurements for free damped oscillations of the cantilever steel beam – the system (A) "spring – mass" (figure 2). To change the frequency of the oscillations, different weights are mounted at the free end of the beam. The beam is excited by initial impact. The recording device saves the time-domain responses.

Secondly are conducted the measurements for free damped oscillations of the steel beam with rubber insulator mounted at its free end – the system (B) "spring - mass - insulator" (figure 2). The investigated rubber insulator is pressed against the beam with varying preload force. To change the frequency of the oscillations, different weights are mounted at the end of the beam. The beam is excited by initial impact. The recording device saves the time-domain responses.

By conducting series of experiments in the desired frequency range 70 - 220 Hz, a set of oscillograms are obtained both for the "spring – mass" system and the "spring - mass - insulator" system.

3.3. Sequence of processing the experimentally obtained oscillograms

Comparisons are made between the two systems with the same period of oscillation $\tau_\Sigma = \tau_b = \tau$. From the oscillograms for the both systems are obtained the period τ and the values of two successive amplitudes A_i and A_{i+1} (the peak stresses in the beam). The logarithmic decrements δ_Σ and δ_b of the two dynamic systems are calculated. At last the coefficients of damping β_v and stiffness c_v of the rubber insulator are determined.

4. Results

The results of the study show that the dynamic characteristics of the rubber insulators are non-linear. Figures 6 and 7 show the coefficients of damping β_v and stiffness c_v for the front and rear rubber insulators in a lower control arm (position 3 and 4 in figure 1(b)) of MacPherson strut suspension. The dynamic coefficients are presented as functions of the frequency and the preload.

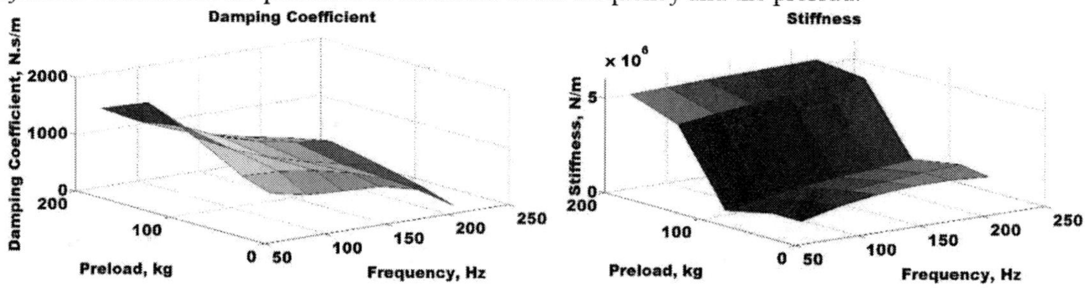

Figure 6. Coefficients of damping and stiffness for front rubber insulator in a lower control arm (position 4 in figure 1(b)) of a MacPherson strut suspension.

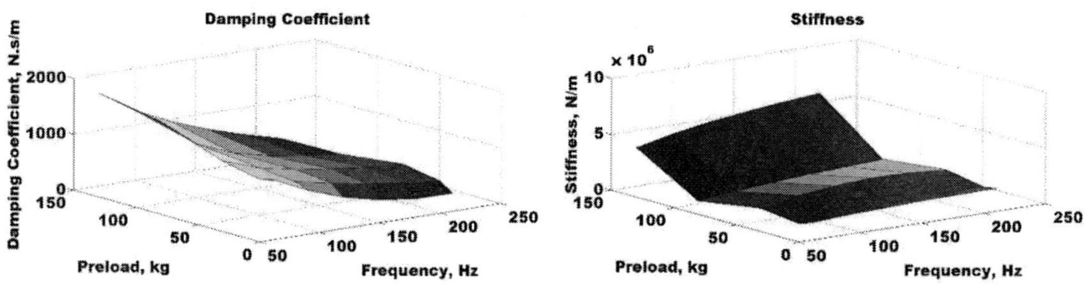

Figure 7. Coefficients of damping and stiffness for rear rubber insulator in a lower control arm (position 3 in figure 1(b)) of a MacPherson strut suspension.

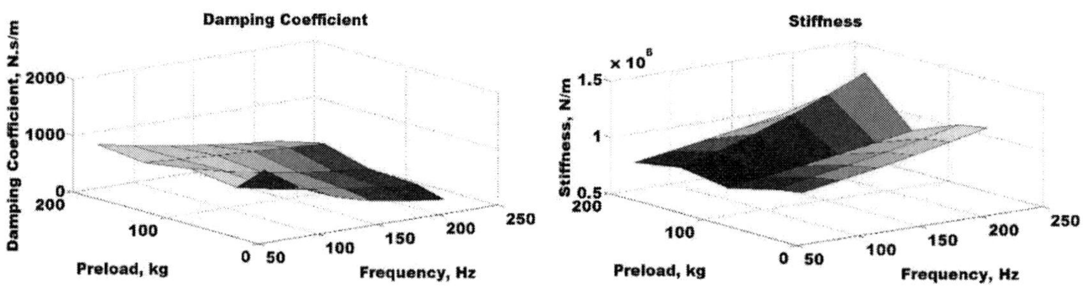

Figure 8. Coefficients of damping and stiffness for rubber insulator in a strut rod (position 1 in figure 1(a)) of a torsion spring suspension.

From the results it is noticed that the coefficient of stiffness increases with increasing the frequency and the preload force. It is observed that the coefficient of damping decreases with increasing the frequency. Also there is slight increase of the damping coefficient with increasing the preload force. Figure 8 shows the coefficients of damping β_v and stiffness c_v for the rubber insulator in a strut rod (position 2 in figure 1(a)) of a torsion spring suspension. Figure 9 shows the damping and stiffness coefficients for the rubber insulator in a trailing arm (position 5 in figure 1(c)) for twist-beam rear suspension. Since these dynamic systems have limited power, the increase in frequency itself leads to a reduction in the deformation of the rubber, which is one reason for reducing the damping coefficient.

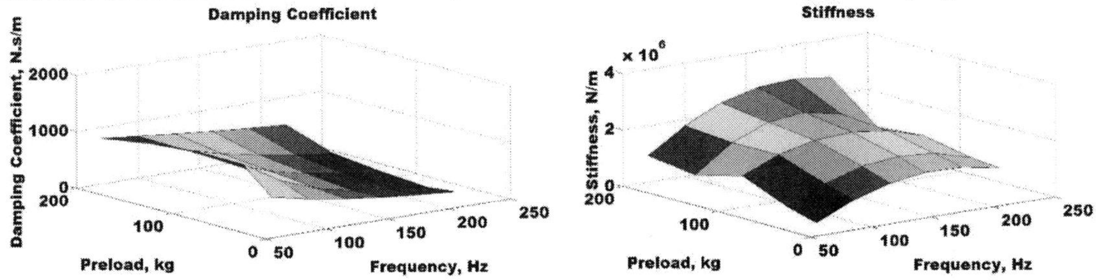

Figure 9. Coefficients of damping and stiffness for rubber insulator in a trailing arm (position 5 in figure 1(c)) for twist-beam rear suspension.

The tested rubber elements have different hardness, and those with harder rubber have a higher damping and stiffness coefficients. For measuring the hardness, it is used Shore durometer Type A. The damping coefficient is also influenced by the volume of the rubber layer. Increasing the volume of the rubber material leads to higher damping coefficient (figure 10). For suspension components that are responsible for the handling of the vehicle, the rubber insulators have higher hardness (Shore A 70) than the others (Shore A 50).

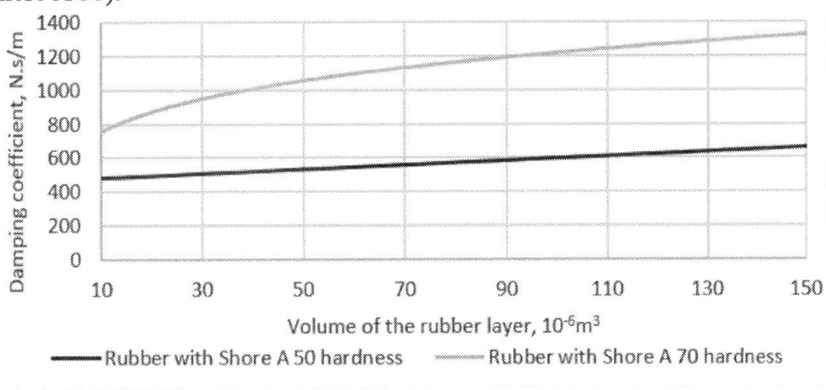

Figure 10. Coefficient of damping as a function of the volume of the rubber layer in the insulators for frequency 100 Hz and preload force 50 kg.

5. Conclusions

The proposed method provides the following opportunities:

- Possibility to obtain the damping coefficient of the rubber insulators as a function of the frequency and the preload force;
- Possibility to obtain the stiffness coefficient of the rubber insulators as a function of the frequency and the preload force;
- Possibility to estimate the damping coefficient of rubber insulators depending on the volume of the rubber layer and the hardness of the rubber;
- Possibility to study the dynamic coefficients of stiffness and damping for other rubber insulators such as the mounts of the engine, gearbox, et cetera.

IOP Conf. Series: Materials Science and Engineering **618** (2019) 012057 doi:10.1088/1757-899X/618/1/012057

Acknowledgments
This work is part of a research project № 182ПД0009-04 in support of PhD students funded by the Research Centre at the Technical University – Sofia.

References
[1] Petkov P T 2015 Studying the vibro-insulators parameters impact of automobile suspension on high vibrations transmissions (in Bulgarian) *Mech. Trans. Comm. Acad. J.* **13** № 1202 pp 23-29
[2] Ivanov A I and Petkov P T 2017 Determination of insolation parameters of machine aggregate "electric engine-compressor" (in Bulgarian) *Mech. Trans. Comm. Acad. J.* **15** № 1494 pp 1-6
[3] Société de Technologie Michelin 2002 *The tyre – Mechanical and acoustic comfort* (France)
[4] Kindt P, de Coninck F, Sas P and Desmet W 2007 Analysis of tire/road noise caused by road impact excitations SAE Technical Paper 2007-01-2248
[5] Pavlov N and Sokolov E 2017 Vibration characteristics of quarter car semi-active suspension model – numerical simulations and indoor testing *Trans Motauto World* **1** pp 11-16
[6] Kunchev L, Pavlov N and Yanachkov G 2010 Mechanic-mathematical modeling of independent suspension as take account elasticity of silent blocks *International Scientific Journal Mach. Tech. Mat.* Issue 3 pp 30-35
[7] Lijun Z, Zengliang Y and Zhuoping Y 2010 Novel empirical model of rubber bushing in automotive suspension system *Proceedings of ISMA 2010* pp 4259-74
[8] van der Linden P J G and Floetke H 2004 Comparing inverse force identification and the mount-stiffness force identification methods for noise contribution analysis *Proceedings of ISMA 2004* pp 2971-85
[9] Nikolov V and Ambarev K Mar-2014 Experimental study of the forces above and under the vibration insulators of diesel generator in transitional regimes *IJRET* **Vol. 03** Iss.03 pp 338-43
[10] Fichera G, Scionti M and Petrone F 2004 Optimization of the mount system of a vehicle's front subframe *Proceedings of ISMA 2004* pp 3395-3409
[11] Suspension design figures, available at https://www.monroe.com.au/trade-corner/tech-info/suspension-systems/spring-design.html (online 03.04.2019) http://www.autoprospect.ru/ (online 03.04.2019)
[12] Nashif A D, Jones D I G and Henderson J P 1985 *Vibration Damping* (John Wiley & Sons)

IOP Conf. Series: Materials Science and Engineering **618** (2019) 012058 doi:10.1088/1757-899X/618/1/012058

Determining the values of the Coefficient of Restitution in the meanwhile of a crash between two vehicles

Milena Savova-Mratsenkova, Georgi Djonev

Department of Engines, Automotive Machinery and Transport, Technical University of Sofia, Bulgaria

savova@tu-sofia.bg

Abstract. The current scientific research aims to present two of the methods, which are used worldwide by the automobile expert practice when investigating the velocity of motor vehicles. These methods are based on theorems from mechanics. Basic quantitative characteristic of energy loss during a two-body impact is the introduced in the impact theory – coefficient of restitution (COR). In reference to automobiles, the values of the coefficient depend on the type of the elements that have crashed as well as on the relative velocity in the meanwhile of the crash. The article presents an approach for determining the values of the above mentioned coefficient when investigating frontal or side collisions of vehicles in a movable barrier, based on multiple regression analysis. The conducted research uses real automobile expertise data from the case law.

1. Introduction

The intensive development of road transport in the past years not only helps to satisfy the constantly increasing need of the society for the use of motor vehicles, but also causes a number of negative consequences as a result of the occurrence of car accidents with either injured or killed casualties. In addition to this, the problems, which are connected with traffic safety and the ways of their solving are subjects of thorough studies and analysis. A fundamental part of them is the process of restoring the whole mechanism of the traffic accident. That process is based on research, analysis and drawing conclusions on the grounds of which the causes and the events that led to the occurrence of the problems in the first place are defined. Furthermore, these actions support the objective description of the facts and the mechanism of occurrence of accidents. Moreover, they are key elements in determining the guiltiness of the participants and in a larger scale they help the process of taking appropriate preventive measures for improving the traffic safety.

A huge part of the conducted up to this moment worldwide researches in the area are based on statistical processing of data about real road traffic accidents [7, 8, 9, 10, 14, 15, 16, 17]. It is unequivocal, however, that the collisions with the most severe consequences of all are the ones that occur between two vehicles. For the study and analysis of vehicle-to-vehicle impacts, knowledge in the field of mechanics and physico-mathematical modeling is required [1, 2, 3, 4, 5, 8, 10, 12]. The main difficulty in the expert study of the above mentioned collisions is determining the redistribution of the participating vehicles' kinetic energy before the impact into energy of deformation during the impact and residual kinetic energy of the vehicles after the impact [2, 3, 8, 9, 10, 11].

IOP Conf. Series: Materials Science and Engineering **618** (2019) 012058 doi:10.1088/1757-899X/618/1/012058

The mechanical interaction between the motor vehicles in the meantime of a collision is extremely complex and due to that it is investigated either in laboratory conditions or by using computer simulation modeling [7, 8, 13].

It is to be noted that the automobile is an intricate, mechanical, nonhomogeneous system of elements, each of which has different sturdiness and elasticity. Therefore, the complex structure of the motor vehicles nowadays is what hinders the physico-mathematical modeling of road traffic accidents [1, 4, 5, 8].

As a main characteristic of the energy loss during a vehicle-to-vehicle collision in most of the conducted researches in the particular scientific field is considered the introduced in the Theory of impact – coefficient of restitution, which mainly depends on the relative velocities of the vehicles before the impact, as well as on the type of the elements that have collided [9, 11, 12].

When analysing traffic safety and investigating the causes and severity of the traffic accidents a regression analysis may be used, with the help of which the complex interconnections between the studied parametres can be described [6, 13, 14, 15, 16, 16].

2. Exhibition

When conducting an expert analysis of a traffic accident by using the provided pre-trial data, which indicates the location of the impact and the location of the vehicles after the accident, the mechanism of the investigated incident is restored by determining the angles of motion of the vehicles and the values after the impact. During the calculation of the speed values at the moment of the impact different methods can be used. Among those methods that are most frequently used are:

2.1. „Momentum 360"

The velocities of the vehicles $\vec{V}_i$ (before the impact) and $\vec{U}_i$ (after the impact) are connected with a relation, based on the Law of storing the amount of traffic $\vec{Q}$, which is expressed by the following vector equation (1):

$$\vec{Q}= m_1.\vec{V}_1 + m_2.\vec{V}_2 = m_1.\vec{U}_1 + m_2.\vec{U}_2 \qquad (1)$$

where m_1 and m_2 are the vehicles' masses; $\vec{V}_1$ and $\vec{V}_2$ - the velocities of vehicles' mass centres before the impact; $\vec{U}_1$ and $\vec{U}_2$ - the velocities of vehicles' mass centres after the impact.

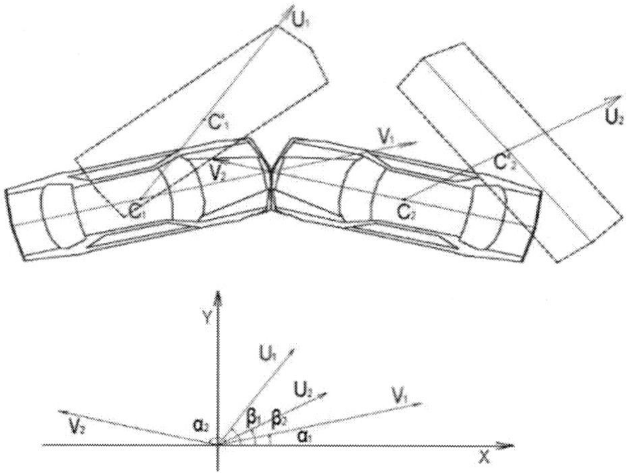

Figure 1: Angles between the vectors of the velocities before (V) and after the impact (U) and the X axis

If we design vector equation (1) on the coordinate axes of an arbitrarily chosen coordinate system, an algebraic system, composed of two equations, (2) is obtained. After we solve the algebraic system we get the values of the vehicles' velocities before the impact:

$$m_1V_1cos\alpha_1 + m_2V_2cos\alpha_2 = m_1U_1cos\beta_1 + m_2U_2cos\beta_2$$
$$m_1V_1sin\alpha_1 + m_2V_2sin\alpha_2 = m_1U_1sin\beta_1 + m_2U_2sin\beta_2 \quad (2)$$

where α_1 and α_2 are the angles of the vehicles' direction of movement before the impact, while β_1 and β_2 – the angles of the vehicles' direction of movement after the impact.

The values of the angles of vehicles' direction of movement before and after the impact shall be determined in accordance with the direction of the X axis of the selected coordinate system, which is directed along the axis of the traffic lane.

2.2. *Theory of impact* – when this method is applied a coefficient of restitution k [1, 3] is used. The coefficient of restitution is a quantitative characteristic of the loss of energy in the meantime of an impact between two vehicles. When an expert analysis of a two-vehicle traffic accident is conducted, the coefficient of restitution is presented as a ratio of the relative velocity's projection between the contact points after the impact on the axis of the directrix of the impact impulse.

When the coordinate system is rotated at an angle α_S (the angle between the vector of the impact impulse and the X axis) as a normal axis (n) is assumed the axis of the impact impulse, while the tangential axis (τ) is its perpendicular. Therefore, the formulas for determining the coefficient of restitution k and the coefficient of momentary friction λ, which are defined by the projections of the velocities on the axes of the new coordinate system, are (3):

$$k = \pm\frac{U_{n2}-U_{n1}}{V_{n1}-V_{n2}}; \qquad 1-\lambda = \frac{U_{\tau2}-U_{\tau1}}{V_{\tau1}-V_{\tau2}} \approx 1. \quad (3)$$

In a case of a *perfectly elastic collision k =1*, the whole kinetic energy of the bodies before the impact is further transformed into kinetic energy after the impact, hence there is no loss of kinetic energy.

However, in a case of a *perfectly inelastic collision k=0* the whole kinetic energy before the impact is further transformed into energy of deformation, thermal and acoustic energy.

For every distinct material object, the coefficient of restitution alters in the range: $0 < k < 1$.

When taking into account the axis n of the coordinate system, the scalar equation is as follows (4):

$$[(-\cos(\alpha_1-\alpha_s) - sin(\alpha_1-\alpha_s))V_1 + (\cos(\alpha_2-\alpha_s) + sin(\alpha_2-\alpha_s))V_2](\pm k) =$$
$$= [(-\cos(\beta_2-\alpha_s) - \sin((\beta_2-\alpha_s))U_2 + (\cos(\beta_1-\alpha_s) + \sin(\beta_1-\alpha_s))U_1]. \quad (4)$$

The value of the coefficient of restitution (k) can be estimated from the last equation, if the values of all other elements are known and if the restrictive conditions are observed (5):

$$S = m_1.\Delta V_1 = m_2.\Delta V_2\,,\ N.s; \quad 1-\lambda \approx 1. \quad (5)$$

The change of the velocity of the mass centre in each of the vehicles for the time duration of the crash is represented by the vector $\Delta\vec{V}_i$, which has the following form (6):

$$\Delta\vec{V}_i = \vec{U}_i - \vec{V}_i \quad (6)$$

The coefficient of restitution depends basically on the type of the impact, the construction of the colliding elements from the vehicles and the relative velocities of the vehicles themselves before the impact.

Modern automobiles are complex mechanical systems with various shape and different internal structure, which is defined by numerous details with various sturdiness, as well as plastic and elastic qualities, dictated by passive safety requirements. For this reason, the reliable data about the values of today vehicles' coefficient of restitution are very limited and are mainly from studies, conducted by separate car manufacturers.

IOP Conf. Series: Materials Science and Engineering **618** (2019) 012058 doi:10.1088/1757-899X/618/1/012058

The authors of the current scientific paper have set themselves the goal of developing an approach for determining the values of the coefficient of restitution in the meantime of frontal or side collisions between vehicles. This new approach is based on multiple regression analysis.

The present research uses data from 14 real judicial automotive expert reports, describing actual traffic accidents with either frontal or side collisions between the vehicles. In table 1 and 2 are shown the values of the automobile parametres, which have been used for the development of the currently presented by the authors regression model.

Table 1.

№	U_1 m/s	U_2 m/s	m_1 kg	m_2 kg	$m_1.U_1$ kg.m/s	$m_2.U_2$ kg.m/s	β_1 ^{0}	β_2^0	α_1^0	α_2^0
1	9,3	9,91	1250	1420	11625,0	14072,2	339	186	0	188
2	13,88	13,14	2010	1530	27898,8	20104,2	353	11	355	40
3	6,79	4,36	1350	1650	9166,5	7194,0	345	2	0	213
4	1,72	3,64	1150	1275	1978,0	4641,0	250	235	0	200
5	10,65	24,93	1360	1600	14484,0	39888,0	287	151	4	176
6	9,01	8,79	2000	1250	18020,0	10987,5	214	355	210	350
7	17,96	9,82	1160	1300	20833,6	12766,0	1	153	10	180
8	7,86	6,23	1600	1070	12576,0	6666,1	348	10	355	98
9	6,89	7,95	1320	1840	9094,8	14628,0	356	309	15	270
10	14,69	16,31	1470	1400	21594,3	22834,0	20	2,5	84	7
11	8,58	12,53	1440	1500	12355,2	18795,0	27	344	1	0
12	5,18	3,88	1380	1460	7148,4	5664,8	336	347	0	228
13	9,19	14,15	970	1315	8914,3	18607,2	11	0,6	3	31
14	13,66	8,7	2500	1270	34150,0	11049,0	343	80	0	179

Table 2.

№	U_1 m/s	U_2 m/s	$m_1U_1+m_2U_2$ kg.m/s	$(\beta_1-\beta_2)^0$	α_S^0	k
1	9,3	9,91	25697,2	153	9,06	0,311
2	13,88	13,14	4800,3	18	359.6	0,047
3	6,79	4,36	16360,5	17	13.29	0,107
4	1,72	3,64	6619,0	15	7,18	0,072
5	10,65	24,93	54372,0	136	38,42	0,143
6	9,01	8,79	29007,5	141	295,0	0,661
7	17,96	9,82	33599,6	152	22,34	0,482
8	7,86	6,23	19242,1	22	9,76	0,093
9	6,89	7,95	23722,8	47	31,08	0,240
10	14,69	16,31	44428,3	17,5	12,54	0,052
11	8,58	12,53	31150,2	43	17,25	0,102
12	5,18	3,88	12813,2	11	17,39	0,038
13	9,19	14,15	27521,5	10,4	7.00	0,184
14	13,66	8,7	45199,0	97	13,45	0,136

In table 2 [β1-β2] is equal to the absolute value of the difference in the angles β1 and β2.

IOP Conf. Series: Materials Science and Engineering **618** (2019) 012058 doi:10.1088/1757-899X/618/1/012058

In the conducted research are determined the values of the coefficient of restitution in case of a frontal or side collision between two vehicles, i.e. in case of an impact between two objects in motion. The above mentioned values of the coefficient of restitution are not the same with the ones in a case of an impact of a vehicle with a fixed barrier (with a distinct material object which is not in motion) [1].

On the basis of the acquired statistics data, with the help of correlation and regressive analysis, is studied the dependence of the coefficient of restitution on some of the other factors, which are shown in the tables.

The correlation analysis allows to be defined the dependencies between random variables. The change in the value of one of these random variables leads to a change in the mathematical expectation of another. The next stage in the analysis of the collected statistics data includes creating a regression equation, which aims to describe the connection between the values of the dependent variable k (the coefficient of restitution) and the chosen five independent variables [($m_1.U_1 + m_2.U_2$); (β_1-β_2); α_S; U_1; U_2] (7):

$$k = f[(m_1.U_1 + m_2.U_2),\ (\beta_1 - \beta_2),\ \alpha_S,\ U_1,\ U_2, \varepsilon]. \qquad (7)$$

In the equation ε represents the random residual components of the model.

The process of creating a regression equation is connected with the necessity of solving a series of interrelated tasks: establishing the shape of the model, determining the coefficients of the regression equation, checking the model for relevance and adequacy.

In order for the set tasks to be solved a software product for statistics processing - „Statgraphics Centurion" has been used. A polynomial model of multiple linear regression has been obtained. The said polynomial model represents the correlation between the dependent parameter *k* and the five independent variables.

In table 3 are shown the values of the coefficients in the derived model. These values are determined by the Least-squares method, which minimizes the sum of the squared residuals. So as to exist a statistically significant correlation between the factors, it is essential that P-Value < 0,05 at 95,0% or higher confidence level.

Table 3.

Parameter	*Estimate*	*Error*	*Statistic*	*P-Value*
$m_1.U_1+m_2.U_2$	-0,0000142227	0,0000053958	-2,63588	0,0271
$\beta_1 - \beta_2$	0,00228072	0,00046788	4,87465	0,0009
α_S	0,000865039	0,00024815	3,48598	0,0069
U_1	0,0286416	0,00990357	2,89205	0,0178
U_2	0,0132796	0,00910193	1,45899	0,1786

R-squared = 92,4718 percent
R-squared (adjusted for d.f.) = 89,126 percent
Standard Error of Est. = 0,0893627
Mean absolute error = 0,0573742
Durbin-Watson statistic = 2,15845
Lag 1 residual autocorrelation = -0,132544

The multiple regression equation, representing the obtained model is as follows (8):

$$k = -0{,}0000142227(m_1.U_1 + m_2.U_2) + 0{,}00228072(\beta_1 - \beta_2) +$$
$$+0{,}000865039.\alpha_S + 0{,}0286416.U_1 + 0{,}0132796.U_2 \qquad (8)$$

The R-Squared statistic indicates that the model as fitted explains 92,4718% of the variability in *k*. The standard error of the estimate shows the standard deviation of the residuals to be 0,0893627. The mean absolute error (MAE) of 0,0573742 is the average value of the residuals.

On figure 1 and figure 2 are shown the matches between the observed and the predicted values of *k*.

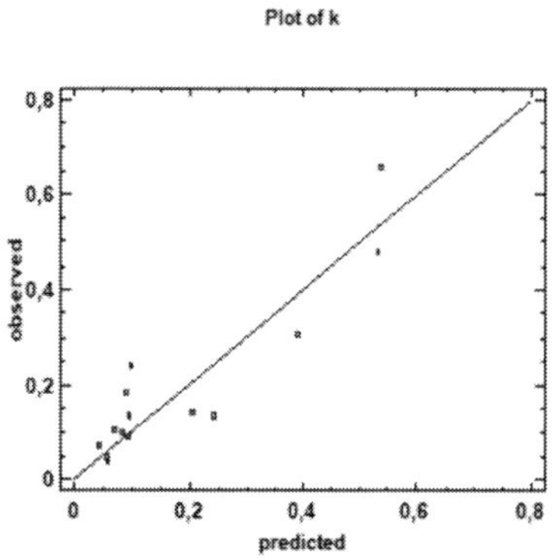

Figure 2: Plot of *k*

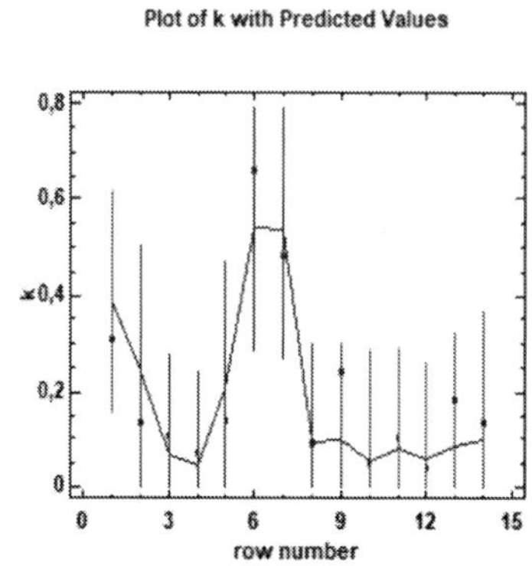

Figure 3: Plot of *k* with predicted values

Table 4 contains information about the predicted values of *k*, which have been generated by the fitted model.

Table 4.

№	*Observed Value*	*Fitted Value*	*Residual*	*Studentized Residual*	*Stnd. Error CL for Forecast*	*Lower 95,0% CL for Forecast*
1	0,311	0,38927	-0,078274	-1,07146	0,102698	0,15696
2	0,136	0,24143	-0,10543	-3,44107	0,117906	-0,02535
3	0,107	0,069955	0,037045	0,408831	0,0923063	-0,13886
4	0,072	0,043884	0,028117	0,300998	0,0901522	-0,160065
5	0,143	0,206193	-0,06319	-1,49528	0,118068	-0,060896
6	0,661	0,538994	0,122006	3,01269	0,1134	0,282464
7	0,482	0,532929	-0,05093	-1,00577	0,115776	0,271025
8	0,093	0,0928	0,000200	0,002203	0,0929675	-0,117508
9	0,24	0,099592	0,140408	1,79688	0,0913971	-0,107163
10	0,052	0,056208	-0,004208	-0,0552	0,103945	-0,178934
11	0,102	0,082093	0,0199069	0,222952	0,0940257	-0,130608
12	0,038	0,057782	-0,0197817	-0,21324	0,090986	-0,148043
13	0,184	0,089468	0,0945317	1,39705	0,104444	-0,146802
14	0,136	0,096792	0,0392078	1,06639	0,120826	-0,176536

When assessing the relevance of the regression equation it is necessary for the residuals to be analysed. For the model to be adequate, the residuals must be independent and normally distributed.

The hypothesis that in the studied model the residuals are normally distributed is then further checked.

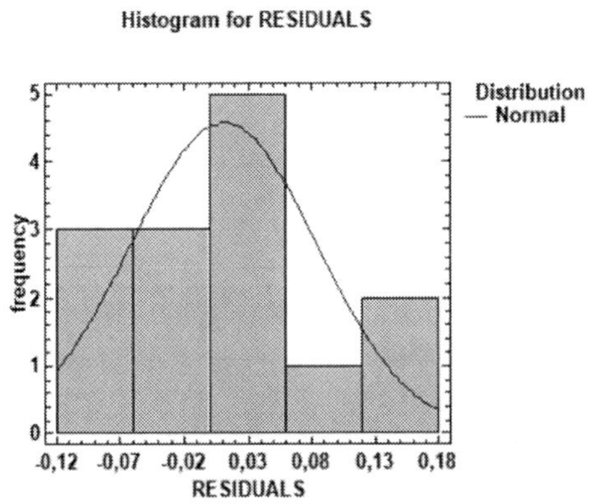

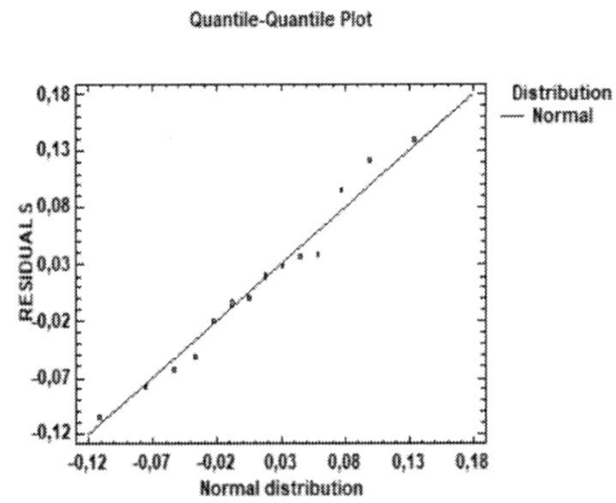

Figure 4: Histogram for residuals **Figure 5:** Normal distribution of the residuals

The parametres of the estimated normal distribution are mean = 0,0114005 and standard deviation = 0,073407.

In order for the hypothesis to be checked the test of Kolmogorov-Smirnov has been used. The results from the test are shown in table 5:

Table 5. Kolmogorov-Smirnov Test

	Normal
DPLUS	0,138127
DMINUS	0,0855679
DN	0,138127
P-Value	0,952154

The acquired results from the test give no argument for rejection of the residuals' normal distribution.

Therefore, the obtained multiple regression equation is relevant and can be used for further forecasting of the coefficient of restitution's values.

3 Conclusion

The present scientific article presents two of the methods, which are used worldwide by the automobile expert practice when investigating the velocities of motor vehicles during an impact between them with the help of theorems from mechanics. Each road accident is characterized by its own specificity, which furthermore requires the choice of an adequate method when conducting an expert analysis.

Basic quantitative characteristic of the energy loss in the meantime of a two-body collision is the introduced in the Theory of impact – coefficient of restitution. When automobiles are to be concerned, the values of the coefficient of restitution mainly depend on the type of the elements that have crashed, as well as on the relative velocities of the vehicles. The authors of the present article have created a specifically developed for the purpose model for forecasting the values of the coefficient of restitution, based on multiple regression analysis.

For the achieving of the goal that has been set, a software product for statistical processing – 'Statgraphics Centurion' has been used. With the help of that software product a polynomial model of multiple linear regression, describing the correlation between the parameter k and the five selected independent variables - ($m_1.U_1+m_2.U_2$); ($\beta_1-\beta_2$); α_S; U_1; U_2 is obtained.

During the conducted research real automobile expertise data has been used, from which the values of the independent variables in the acquired model have been extracted.

In order for the multiple regression equation to be obtained the following interrelated tasks have been solved: the shape of the model has been established, the coefficients of the regression equation have been determined, the model has been checked for significance and adequacy.

Moreover, the estimated through the acquired multiple regression equation value of the coefficient of restitution may be used in further analysis for determining the velocities of the vehicles at the moment of the impact with the help of a research method, based on the Theory of impact.

References

[1] Karapetkov St., Auto-technical expertise. Sofia, Technical University, 2005.

[2] Savova-Mratsenkova M., Djonev G., Influence of the angle of the directory on the accuracy of the measurement when determining vehicles' velocities in case of an impact between them. *Conference BulTrans, Sozopol*, 11-13 September 2017.

[3] Djonev G., Savova-Mratsenkova M., Methods for determining the velocity of vehicles in case of an impact between them. Conference BulTrans, Sozopol, 11-13 September 2017.

[4] Pisarev, A. and others. Course in Theoretical Mechanics - Part II. Sofia, Technika, 1988.

[5] Ilarionov V., Examination of an accident, Moscow, "Transport", (1989).

[6] Kachaunov T., Modelling and Optimization of Transport Processes, VTU - T. Kableshkov, 2005.

[7] Hoxha G., Shala A., Likaj R., Pedestrian crash model for vehicle speed calculation at road accident. *International Journal of Civil Engineering and Technology*, Volume **8**, Issue 9, September 2017, pp. 1093–1099, ISSN: 0976-6308, https://pdfs.semanticscholar.org/8eac/c99fba42ae9d81a2ce77fcd800240f5a8293.pdf

[8] Scott W., Bain C., Manoogian S., Cormier J., Simulation Model for Low-Speed Bumper-toBumper Crashes. *SAE International*, 04.12.2010, ISSN 0148-7191, http://www.brconline.com/files/consultants/101/simulation_model_for_low-speed_bumper-to-bumper_crashes.pdf

[9] Rose N., Fenton S., Beauchamp G., Restitution Modeling for Crush Analysis: Theory and Validation. *SAE International* 2006-01-0908, 2006, http://kineticorp.com/wp-content/uploads/2017/05/2006-01-0908-restitution-crush-analysis-theory-and-validation.pdf

[10] Rose N., Fenton S., Beauchamp G., Analysis of Vehicle-to-Ground Impacts during a Rollover with an Impulse-Momentum Impact Model. *SAE International* 2008-01-0178, http://kineticorp.com/wp-content/uploads/2016/08/2008-01-0178-Analysis-of-Vehicle-to-Ground-Impacts-During-a-Rollover-with-an-Impulse-Momentum-Impact-Model.pdf

[11] Batista M., On the Mutual Coefficient of Restitution in Two Car Collinear Collisions. 01.2006, https://arxiv.org/vc/physics/papers/0601/0601168v1.pdf

[12] Tseveennamjil B., Hudák A., Rievaj V., Determining the speed of vehicles before and after Crash. Number III, Volume **VI**, July 2011, http://pernerscontacts.upce.cz/22_2011/Tseveennamjil.pdf

[13] Doris E., The Analysis of Injury Mechanisms from Frontal Car Accidents. *University of Iceland Reykjavik, January 2017*, https://skemman.is/bitstream/1946/26755/1/The%20Analysis%20of%20Injury%20Mechanisms%20from%20Frontal%20Car%20Accidents.pdf

[14] Agyemang B., Abledu G., Semevoh R., Regression Analysis of Road Traffic Accidents and Population Growth in Ghana. *International Journal of Business and Social Research (IJBSR)*, Volume -**3**, No.-10, October, 2013, https://290-757-2-PB.pdf

[15] Gupta H., Rokade S., Development of crash prediction model using multiple regression analysis. *International Journal of Current Engineering and Scientific Research, 2017*, http://troindia.in/journal/ijcesr/vol4iss6/82-86.pdf

[16] Awel O., Adarabioyo M., Multivariate Regression Techniques for Analyzing AutoCrash Variables in Nigeria. *Journal of Natural Sciences Research*, ISSN 2225-0921 (Online) Vol.**1**, No.1, 2011, https://www.researchgate.net/profile/Olushina_Awe/publication/266165185_Multivariate_Regression_Techniques_for_Analyzing_Auto-Crash_Variables_in_Nigeria/ links/55b38e4608aed621de00f818/Multivariate-Regression-Techniques-for-Analyzing-

Auto-Crash-Variables-in-Nigeria.pdf
[17] Desai M., Patel A., *Road Accidents Study Based On Regression Model: A Case Study of Ahmedabad City*. National Conference on Recent Trends in Engineering & Technology, 13-14 May 2011, http://bvmengineering.ac.in/misc/docs/published-20papers/civilstruct/Civil/101027.pdf

IOP Conf. Series: Materials Science and Engineering **618** (2019) 012059 doi:10.1088/1757-899X/618/1/012059

A multi-criteria approach for evaluating the urban transport technologies by using SIMUS method

Svetla D Stoilova

Department of Railway Engineering, Technical University – Sofia, Bulgaria

E-mail: stoilova@tu-sofia.bg

Abstract. The study purpose a methodology based on Sequential Interactive Modelling for Urban Systems (SIMUS) method for ranking the urban transport technologies. The methodology of research includes three steps. The first step defines quantitative and qualitative criteria for the assessment of urban transport modes. Fourteen criteria of four groups, as ecological, economical, technological and social for evaluating the transport alternatives in urban environment have been defined. The second step includes a selection of alternatives of transportation by different transport means for a given route. In this research is investigated transportation by metro, buses and cars. The SIMUS method is applied in the third step for ranking the variants of transportation. The method solves successive scenarios formulated as linear programming. This step involves the ranking of alternatives and choice of optimal type of transportation. The methodology is approved for study the alternative mode of transportation for parallel routes from Sofia's transport network. It was found that the carriage by metro is the best transport technology for investigated route. The methodology and defined criteria could be applied to study the transport alternatives for different routes of Sofia's transport network, and also to assess the modes of transport in urban area of Sofia or other city.

1. Introduction

The urban transport system combines different modes of transport, which develop according to the city's territory, population size, socio-economic development and other factors. In major metropolis, the main types of public transport are metropolitan, bus, tram, trolleybus. The factors that are important for passengers to choose the type of transportation are the accessibility, time travel, freight rates, prices, etc. Undoubtedly, the electrified rail transport as metropolitan, light rail, tramway and trolleybus are environmental friendly. The bus transport and passenger car transport have a big part of urban transportations. They are the main source of air pollution. The metropolitan, as ecological and speed rail transport, occupies an important place in public transport to meet the needs of travel in major cities. The increase in environmental pollution due to exhaust from transport activities necessitate to seek efficient transport technologies that improve the quality of life in big cities in terms of air pollution.

The AHP method has been used in [1] to assess the importance of various parameters affecting the mode choice made by urban transport users e.g. time, fare, comfort, reliability and accessibility, etc. The research aims to understand the major influences on user behaviour in an environment where all modes are available for use. In [2] authors compared ANP method and AHP method to study the choice of different alternatives for urban transport. The capital cost, capacity, environmental benefit and socio-economic benefit was applied to rank the alternatives. The AHP method has been used in [3] for ranking light rail, a dual mode of tram and bus. The AHP method has been applied for selecting the best mode of transportation in Lagos State metropolis, [4]. The alternative modes of transportation were evaluated with nine decision criteria: transportation cost, environmental effect, capacity, safety, comfort, accessibility, reliability, number of interchanges required and journey time. In [5] is presented a model based on AHP method for public transport management. The following criteria have been defined: safety of the participants, ecology, availability, capacity of the network of traffic routes, and economic indicator. The ELECTRE III/IV and AHP methods taking into account two

IOP Conf. Series: Materials Science and Engineering **618** (2019) 012059 doi:10.1088/1757-899X/618/1/012059

models of decision maker preference: passenger and operator have been used in [6] to study six scenarios of transportation. A set of six criteria was proposed: waiting time; riding time; accessibility to the stop; comfort of travel; cost of rolling stock purchase; cost of building, maintenance and renewal of route. The ELECTRE III method has been applied in [7] to assess four variants of the mass transit system in Czestochowa. Two methods – TOPSIS and SAW were used and compared in [8] to determine the best transport development alternative based on traffic parameters. In [9] the authors investigated alternative-fuel buses for public transportation using multi-criteria analysis. The main criteria that were used are social, economic, technological, and transportation. Sub-criteria for each main group have been defined.

The both compensatory and non-compensatory approaches including lexicographic, SAW method, TOPSIS method and Concordance Analysis have been applied in [10] to assess the three urban public transport modes. In [11] the following criteria to assess transport alternatives for urban development have been studied: accessibility measurements, impact of each stop on the total travel time, catchment areas for population and jobs, existing urban functions in the surroundings of the stops, potential for urban development of housing, activities and equipment, potential for connection with other modes, and cost of construction.

An integrated multi-criteria approach is proposed in [12] for evaluation of public transportation systems based on Delphi method, group AHP method and ROMETHEE method. The criteria of selection of transport types that have been used are: travel cost, travel time, waiting time, suitability, accessibility and safety.

In [13] has been formulated ten clusters form the basis for the overall evaluation of the transport system - comfort, vehicles, customer service, information, reliability, security, payment, entertainment, accessibility, and terminals infrastructure. In [14] the relationships between attribute-based satisfaction, overall satisfaction and customer characteristics are studied by means of constructing three linear models. The 15 criteria have been studied. In [15] was elaborated model based on two statistical methods (factor analysis and ordered logit modelling) assessed the quality implications of the variability of the users' perceived satisfaction with respect to the public transit systems. The research conducted in [16] uses logit model to investigate the passenger flows by rail and buses in Sofia. Two types of factors are used: price and quality of service (travel time, frequency, comfort, etc.). In [17] is elaborated multi-criteria model based on rank-order centroid method. The groups of criteria that were used are: engineering designers and developers, economic and financial, environmental communities, service operators and users. In [18] two criteria have been introduced for assessment the different means of transportation: traffic performance index and weighted traffic performance indices.

It can be summarized that the transport costs, travel speed; frequency; prices of the tickets; safety, comfort are the main criteria for assessment of different alternative modes of transportation that have been studied. The main types of methods that have been used can be summarized as follow: different multi-criteria methods to determine the weights of defined criteria and rank the alternatives; statistical methods to investigate the clusters for evaluation of the transport system. The experts' assessment is used to determine the weights of criteria, which leads to the increase of subjectivism in decision making.

The SIMUS (Sequential Interactive Modelling for Urban Systems) is a hybrid method based on Liner programming, Weighed Sum Method (WSM) and Outranking and aid to solve decision-making problems with multiple objectives. SIMUS was developed by Nolberto Munier, and tested in many and diverse type of projects, [19]. The weights of criteria are not used to make decision. The application of SIMUS method does not use subjective assessments of weights of criteria by experts.

The aim of the study is to define the criteria and to develop a methodology for assessment the passenger transport modes in urban area by using SIMUS method for make decision, and taking into account of the both requirements of the carrier and the expectations of the passengers, and also environmental protection.

IOP Conf. Series: Materials Science and Engineering **618** (2019) 012059 doi:10.1088/1757-899X/618/1/012059

2. Methodology

The methodology for evaluating the effectiveness of urban transport includes the following steps:

- Step 1: Definition of the criteria to evaluate the transport technology of carriage. The choice of mode of transport depends on a set of criteria, as ecological, economical, technological, and social. In this study the following criteria have been defined: C1 - Carbon dioxide (CO_2), g/pass.km; C2 - Carbon monoxide (CO), g/pass.km; C3 - Nitrogen oxides (NOx), g.passkm^{-1}; C4 - Non-methane hydrocarbons (NC), g.pass.km^{-1}; C5 - Particulate matter (PM), g.pass.km^{-1};C6 - Transport costs for fuel (electric energy), BGN; C7 - Ticket price, BGN.pass.$^{-1}$; C8 - Direct operating costs, BGN.pass.$^{-1}$; C9 - Time travel, min.; C10 - Frequency of shipments, number of transport means.h^{-1}; C11 - Comfort. This means the possibility of a seating place during a trip, and a comfortable trip of the standing; C12 - Security. This means safety with regard to road accidents; C13 - Reliability. This means the accuracy performance of the timetable; C14 - Stability. This mean independence from meteorological conditions. The first five criteria are connected to the ecological impact of transport means. The criteria from C2 to C5 present the polluting emissions. The criteria from C6 to C8 are economic indices. C9 and C10 are technological criteria. The criteria from C11 to C14 are social indices. The values of criteria from C11 to C14 is set wit "1" if the answer for performance of criteria is "yes", and "0" if the answer for performance of criteria is "no".
- Step 2: Determination of the variants of transportation by different transport means for a given route. In this research is investigated urban transportation by metro, buses and cars.
- Step 3: Determination of the optimal variant of transportation. The study uses the SIMUS method as a tool for ranking the variants of transportation. The method is based on linear programming. It consists of following steps:

- Organization of the decision matrix of criteria and alternatives. The matrix consists two parts. The first Left Hand Side (LHS) presents the values of criteria for the alternatives. The second part Right Hand Side (RHS) establishes the limits to each criterion. The matrix consist also the type of optimization for each criterion, and the type of operator. The value of RHS could be set by decision maker or determined as equal of maximum normalized value of the row when the objective of criterion is of maximum. In the case of minimum of objective function the value of RHS is equal of minimum normalized value of the row.
- Determination the normalized values of LHS part of decision matrix. The normalization can be made by different method as total sum in row, maximum value in row, Euclidean formula and min-max. The choice of normalization system does not affect the results of ranking the alternatives.
- Determination the Efficient Results Matrix (ERM). This matrix is formed according the procedure of SIMUS method which uses simplex algorithm of linear programming. The number of linear optimization models is equal to the number of criteria. The first optimization model is compiled by using the first criterion as the objective function which is removed from decision matrix. The restrictive conditions of the model are the others rows of the matrix. The procedure is repeated with the other criteria. It is necessary the all criteria to be used as objective functions. The ERM matrix is formed based on results of all linear optimization models. The elements of ERM show the score of each alternative for each optimization model.
- Ranking the alternatives using ERM. The procedure consists the following steps: normalization of the ERM; determination the sum of all elements in each column (SC); determination the number of participation of each alternative in each column of normalized ERM named Participation Factor (PF); determination the normalized values of PF (NPF) by dividing each value of PF to the total number of criteria; determination the values of the criterion for ranking - the multiplication of the NPF by the SC for each alternative. The ranking of the alternatives is made according descending order of the criterion. The maximal values of criterion show the best alternative.
- Ranking the alternatives using Determination Project Dominance Matrix (PDM). The number of column and the rows in PDM has equal to the number of alternatives. The elements of PDM are determined successively starting from the highest value in the first row by calculating the differences between values in the same row of normalized ERM. The procedure is repeated with all the values.

IOP Conf. Series: Materials Science and Engineering **618** (2019) 012059 doi:10.1088/1757-899X/618/1/012059

The row sum and column sum of matrix PDM are determined. The ranking of the alternatives is made according criterion net dominance, which presents a difference between row and column values for the same alternative. The ranking is made according descending order of the criterion. The maximal values of criterion show the best alternative.

3. Results and discussion

The methodology is approved for study the alternative mode of transportation for parallel routes from Sofia's transport scheme. The section Sofia University St. Kliment Ohridski metro station - G.M. Dimitrov metro station has been investigated. The following alternatives of transportation have been studied: Alternative 1- metro; Alternative 2- bus; Alternative 3 - car.

This area is served by metro and on a parallel route runs bus 280. There are also cars that moved in this section.

The following conditions have been taken into account when determining the criteria's values:

- The morning peak period from 8.00 to 9.00 o'clock is examined. For a peak period, filling factor for metro trains and buses was adopted equal of 0.9. For carriage it is assumed that one or two passengers travel in one car.
- The European Environmental Protection Agency's model for estimating the emissions from a vehicle based on the total fuel consumption or engine power is used for determining the parameters of emissions from road transport. The Euro 5 standard is applied.
- The Carbon dioxide (CO2) emissions for electricity generation have been accounted for metro transport. The CO2 emissions are 0.460 $t.MWh^{-1}$ according data from National Statistical Institute of Bulgaria. The CO2 emissions is assumed to be $130g.km^{-1}$ for a car, and $170g.km^{-1}$ for a bus. When determining the pollutants emissions, a power of 180 kW for an urban bus was adopted with an average power of 80% of the bus power.
- In the study, it is assumed that in the morning peak period the metro trains are 81-740/741 series (produced from 2005 to 2013). The relative power consumption for this series is $7.02kWh.km^{-1}$. The electricity price is 282.46 $BGN.MWh^{-1}$.
- It is suggested the following: the number of cars with gasoline and diesel engines for the studied period is equal; the average fuel consumption for a car is 12 $l.100km^{-1}$; the average fuel consumption for a bus is 35 $l.100km^{-1}$ for urban carriage; the cost for average fuel consumption is 3.06 $BGN.km^{-1}$.
- When determining the direct operating costs for the metro, the costs of running trains (electricity) and wage costs of drivers are taken into account. These costs are 4.80 $BGN.trainkm^{-1}$ according data from Sofia metro. When determining the running costs of a car, it is assumed that the cost of oil, tires, maintenance, insurance, taxes, are 40% of the fuel costs, i.e. according to the above mentioned average fuel consumption for a car $12l.100km^{-1}$, the operating costs are 0.34 $BGN.km^{-1}$.
- The number of metro trains and buses of line 280 for morning peak period is determined according the data given by Urban Mobility Centre. The cars' frequency is 10 minutes.
- The ticket price for the metro and bus transport is calculated according a discount for 10 trips and it is 1.20 $BGN.ticket^{-1}$. When traveling by car, the ticket price is equal to the fuel consumption per passenger.
- The operating costs and the value of harmful emissions for different mode of transport are compared by passenger-kilometer.

Table 1 presents the parameters of investigated alternatives. Table 2 shows the LHS data of decision matrix and contains the values of criteria. Table 3 show the normalized values of LHS data and thresholds. The sum in row of LHS data is applied for normalization. The part named "Thresholds" show the type of optimization for each criterion, the respective operator, and the value of RHS that establishes the limits to each criterion. The value of RHS column is determined according the normalized values of each row.

Table 1. Parameters of alternatives.

Parameters	metro	bus	car
Length, km	4.5	5.94	5.53
Number of seats (sitting and standing)	344	95	5
Number of passengers in a vehicle in peak period, pass.vehicle^{-1}	310	85	2

Table 2. Decision matrix- LHS Data.

Criteria	LHS Data Alternative 1	2	3
C 1	10.50	1.98	65.00
C 2	0.00	0.13	0.38
C 3	0.00	0.04	0.06
C 4	0.00	0.17	0.06
C 5	0.00	0.002	0.003
C 6	1.48	4.15	1.32
C 7	1.20	1.20	0.66
C 8	0.02	0.04	0.17
C 9	8.00	17.00	14.00
C 10	3.00	8.57	10.00
C 11	0.00	0.00	1.00
C 12	1.00	0.00	0.00
C 13	1.00	0.00	0.00
C 14	1.00	0.00	0.00

Table 3. Normalized LHS Matrix and RHS.

Criteria	Alternative 1	2	3	Thresholds Action	Operator	RHS
C 1	0.14	0.03	0.84	min	≥	0.03
C 2	0.00	0.25	0.75	min	≥	0.00
C 3	0.00	0.39	0.61	min	≥	0.00
C 4	0.00	0.74	0.26	min	≥	0.00
C 5	0.00	0.40	0.60	min	≥	0.00
C 6	0.21	0.60	0.19	min	≥	0.19
C 7	0.39	0.39	0.22	min	≥	0.22
C 8	0.07	0.16	0.77	min	≥	0.07
C 9	0.21	0.44	0.36	min	≥	0.21
C 10	0.14	0.40	0.46	max	≤	0.46
C 11	0.00	0.00	1.00	max	≤	1.00
C 12	1.00	0.00	0.00	max	≤	1.00
C 13	1.00	0.00	0.00	max	≤	1.00
C 14	1.00	0.00	0.00	max	≤	1.00

Table 4 presents the linear optimization models with objective function and restrictive conditions based on SIMUS method for the first and for the second criterion. Initially, the first criterion has been used as the objective function. The restrictive conditions are presented by the others rows of normalized matrix. In table 4 x_j presents the score of each alternative; $j = 1, \ldots, m$ are the number of variants.

Table 4. Linear models.

Liner model for first criterion		Liner model for second criterion	
Objective function:		Objective function:	
$0.14.x_1 + 0.03.x_2 + 0.84.x_3 \rightarrow min$	(1)	$0.25.x_2 + 0.75.x_3 \rightarrow min$	(13)
Restrictive conditions:		Restrictive conditions:	
$0.25.x2 + 0.75.x3 \geq 0$	(2)	$0.14.x1 + 0.03.x2 + 0.84.x3 \geq 0.03$	(14)
$0.39.x2 + 0.61.x3 \geq 0$	(3)	$0.39.x2 + 0.61.x3 \geq 0$	(15)
$0.74.x2 + 0.26.x3 \geq 0$	(4)	$0.74.x2 + 0.26.x3 \geq 0$	(16)
$0.40.x2 + 0.60.x3 \geq 0$	(5)	$0.40.x2 + 0.60.x3 \geq 0$	(17)
$0.21.x1 + 0.60.x2 + 0.19.x3 \geq 0.19$	(6)	$0.21.x1 + 0.60.x2 + 0.19.x3 \geq 0.19$	(18)
$0.39.x1 + 0.39.x2 + 0.22.x3 \geq 0.22$	(7)	$0.39.x1 + 0.39.x2 + 0.22.x3 \geq 0.22$	(19)
$0.07.x1 + 0.16.x2 + 0.77.x3 \geq 0.07$	(8)	$0.07.x1 + 0.16.x2 + 0.77.x3 \geq 0.07$	(20)
$0.21.x1 + 0.44.x2 + 0.36.x3 \geq 0.21$	(9)	$0.21.x1 + 0.44.x2 + 0.36.x3 \geq 0.21$	(21)
$0.14.x1 + 0.40.x2 + 0.46.x3 \leq 0.46$	(10)	$0.14.x1 + 0.40.x2 + 0.46.x3 \leq 0.46$	(22)
$x3 \leq 1$	(11)	$x1 \leq 1$	(23)
$x1 \leq 1$	(12)	$x3 \leq 1$	(24)

The results for optimization by the first criterion are recorded in the first row of Efficient Results Matrix (ERM); the results for optimization by the second criterion are recorded in the second row of ERM matrix, Table 5. The procedure is applied consistently for all criteria. The Software SIMUS System which uses Simplex method of linear optimization has been used to make the research.

Figure 1 and Table 5 show the results of Efficient Results Matrix. The values of normalized ERM is presented in Table 6. The results presents that alternative 1 (transportation by metro) has the score "1" for eight of the criteria when they are applied as objective functions. This means that this alternative is the best according these criteria.

Table 5. ERM Matrix.

Criteria	Alternative			Objective function
	1	2	3	
C 1	0.00	0.55	0.00	0.01
C 2	1.00	0.00	0.00	0.00
C 3	1.00	0.00	0.00	0.00
C 4	1.00	0.00	0.00	0.00
C 5	1.00	0.00	0.00	0.00
C 6	0.34	0.00	0.38	0.14
C 7	0.00	0.22	0.30	0.15
C 8	1.00	0.00	0.00	0.07
C 9	0.35	0.19	0.02	0.16
C 10	0.00	0.00	0.00	0.00
C 11	0.00	0.00	1.00	1.00
C 12	1.00	0.00	0.00	1.00
C 13	1.00	0.00	0.00	1.00
C 14	1.00	0.00	0.00	1.00

Table 6. ERM Normalized.

Criteria	Alternative		
	1	2	3
C 1	0.00	1.00	0.00
C 2	1.00	0.00	0.00
C 3	1.00	0.00	0.00
C 4	1.00	0.00	0.00
C 5	1.00	0.00	0.00
C 6	0.48	0.00	0.52
C 7	0.00	0.42	0.58
C 8	1.00	0.00	0.00
C 9	0.63	0.33	0.03
C 10	0.00	0.00	0.00
C 11	0.00	0.00	1.00
C 12	1.00	0.00	0.00
C 13	1.00	0.00	0.00
C 14	1.00	0.00	0.00
Sum of Column (SC)	9.11	1.76	2.13
Participation Factor (PF)	10	3	4
Normalized PF (NPF)	0.71	0.21	0.29
Final Result (SC x NPF)	6.51	0.38	0.61
ERM Ranking			
Alternative 1 - Alternative 3 - Alternative 2			

The alternative 1 is the best also by criterion C9 (Time travel) when it are given as objective function. The results of optimization by criterion C11 (Comfort) and C12 (Security) shows that the alternative 1 and alternative 3 are the best. When the first criterion is applied as objective function, the alternative 2 have a highest score. The third alternative is the best when criterion C6 (Transport costs for fuel or electric energy) and criterion C7 (Time travel) are used separately as an objective function. The optimization by the criterion C10 (Frequency of shipments) shows that the scores alternatives are equal.

Table 7. Project Dominance Matrix (PDM).

Subordinated variants - alternatives - options Dominant Alternative	1	2	3	Row sum of dominant alternative	Net dominance
1	0	8.8	8.6	17.4	14.3
2	1.4	0	1.3	2.7	-7.7
3	1.6	1.7	0	3.3	-6.6
Column Sum of subordinated Alternatives	3.0	10.5	9.9		
PDM Ranking	Alternative 1 - Alternative 3 - Alternative 2				

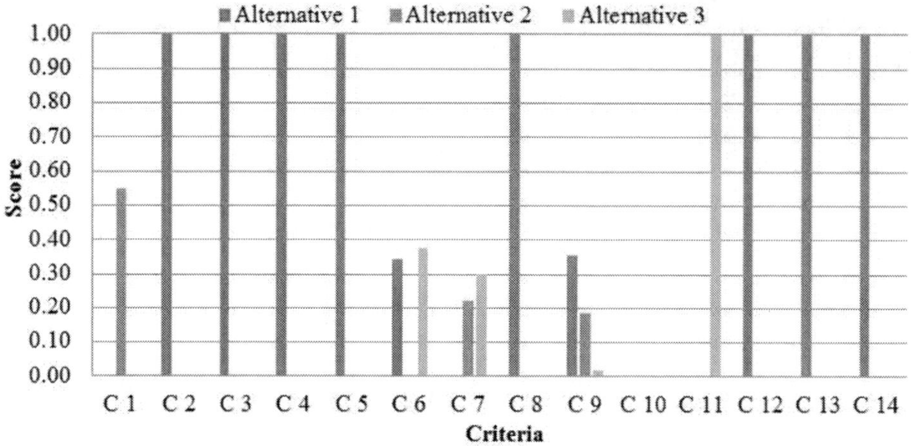

Figure 1. Score of the criteria by alternatives.

Table 6 presents the results of ERM Ranking. It can be seen in Fig. 1 that the alternative 1 is the best by ten criteria, alternative 2 is the best by three criteria, and alternative 4 satisfy four criteria. This is also given in Participation Factor (PF) in Table 6. The maximal score of Sum of Column (SC) has alternative 1(transportation by metro). The Normalized Participation Factor (NPF) has maximal value also for alternative 1. The total number of criteria is 14. This number is taken into account in the calculation of NPF. The ERM ranking is as follow: Alternative 1 - Alternative 3 - Alternative 2.

Table 7 show the PDM Ranking. The criterion of ranking - net dominance is shown in the last column of the table. The first alternative has the big dominance value of 17.4. The second alternative has the big subordinate value of 10.5. The results show that the alternative 1 has greatest value of the Net dominance and it is a best. The last row of the table presents the PDM ranking. It can be seen that the both ranking by ERM and PDM are equal.

4. Conclusion

1. Fourteen criteria that are connected to four groups, as ecological, economical, technological and social for assessment the transport alternatives in urban environment have been defined.
2. In this study has been elaborated a methodology which uses linear optimization method and SIMUS procedure for assessment alternative transport mode in urban area. The solution for each criterion, as objective function of linear optimization model has been determined.
3. The model has been experimented for parallel routes of Sofia's transport network. It was found that the carriage by metro is the best transport technology for investigated route.
4. The alternative – transportation by metro is the best for nine criteria when they are given as objective function, i.e. this includes the following criteria: minimum of carbon monoxide (CO), nitrogen oxides (NOx), non-methane hydrocarbons (NC), particulate matter (PM), direct operating costs, time travel, and maximum of security, reliability, and stability.
5. The methodology and defined criteria could be used for study the transport alternatives for different routes of Sofia's transport network, and also to assess the modes of transport in urban area of Sofia or other cities.

Acknowledgments
This research has been supported by contract №КП-06-Н27/12 of 11.12.2018 "Modelling and elaboration of complex system for selection of transport technology in transport network" funded by the National Science Fund of the Ministry of Education and Science of Bulgaria.

References

[1] Kumar C, Mangaraj B. K and Vijayaraghavan T. A. S 2015 Mode choice for urban work-based daily trips: a multi-criteria decision making model using the Analytical Hierarchical Process *WIT Transactions on The Built Environment* **146** 627-638

[2] Mintesnot G and Takano S 2007 Multi-criteria decision making for public transportation development projects using Analytic network process (ANP) *Journal of the Eastern Asia Society for Transportation Studies* **7** 223-237

[3] Lee D-J 2018 A multi-criteria approach for prioritizing advanced public transport modes (APTM) considering urban types in Korea *Transportation Research Part A: Policy and Practice* **111** 148-161

[4] Odeyalel S O, Alamu O J and Odeyale E O 2014 Performance evaluation and selection of best mode of transportation in Lagos state metropolis *International Journal for Traffic and Transport Engineering* **4(1)** 76 - 89

[5] Kavran Z, Štefančić1 G and Presečki A 2007 Multicriteria analysis and public transport management *WIT Transactions on The Built Environment* **96** 85-90

[6] Dudek M, Solecka K and Richter M 2018 A multi-criteria appraisal of the selection of means of urban passenger transport using the ELECTRE and AHP methods *Technical transactions Civil engineering* 79-93

[7] Zak J 2011 The methodology of multiple criteria decision making/aiding in public transportation *J. Adv. Transp.* **45** 1–20

[8] Jakimavičiu M and Burinskienė M 2013 Multiple criteria assessment of a new tram line development scenario in Vilnius city Public transport system *Transport* **28(4)** 431–437

[9] Tzeng G-H, Lin C-W and Opricovic S 2005 Multi-criteria analysis of alternative-fuel buses for public transportation *Energy Policy* **33** 1373–1383

[10] Keyvan-Ekbatania M and Catsa O 2015 Multi-Criteria Appraisal of Multi-Modal Urban Public Transport Systems *Proc. of 18th Euro Working Group on Transportation EWGT (Delft)* (Netherlands) 1-11

[11] L'Hostis A, Soulas C and Vulturescu B 2015 A Multi-criteria approach for choosing a new public transport system linked to urban development: a method developed in the Bahn.Ville project for a tram-train scenario in the Saint-Étienne region *Recherche Transports Sécurité*, NecPlus 17-25

[12] Nassereddinea M and Eskandarib H 2017 An integrated MCDM approach to evaluate public transportation systems in Tehran *Transportation Research Part A: Policy and Practice* **106** 427-439

[13] Barbosa B, Ferreira M, Nickel E M, Cruz J A, Forcellini F A, Garcia J, Salgueirinho J B and Guerra O A 2017 Multi-criteria analysis model to evaluate transport systems: An application in Florianópolis, Brazil *Transportation Research Part A* **96** 1–13

[14] Mouwen A 2015 Drivers of customer satisfaction with public transport services. *Transportation Research Part A* **78** 1–20

[15] Tyrinopoulos Y and Antoniou C 2008 Public transit user satisfaction: Variability and policy implications *Transport Policy* **15** 260–272

[16] Todorova M 2015 Choice of passenger transport mode using Logit model *Proc. Int. Conf. of 23nd International Symposium EURO - Zel 2015 "Recent Challenges for European Railways" (Žilina)* (Slovak Republic) pp 216-223

[17] Sirikijpanichkul A, Winyoopadit S and Jenpanitsub A 2016 A multi-actor multi-criteria transit system selection model: A case study of Bangkok feeder system *Transportation Research Procedia* **25** 3736–3755

[18] Brunner H, Hirz M, Hirschberg W and Fallast K 2018 Evaluation of various means of transport for urban areas *Energy, Sustainability and Society* **8** 9

[19] Munier N 2016 A new approach to the rank reversal phenomenon in MSDM with the SIMUS method *Multiple criteria decision making* **11** 137-152

IOP Conf. Series: Materials Science and Engineering **618** (2019) 012060 doi:10.1088/1757-899X/618/1/012060

Selecting a location for establishing a rail-road intermodal terminal by using a hybrid SWOT/MCDM model

Svetla D Stoilova and Svetoslav V Martinov

Department of Railway Engineering, Technical University – Sofia, Bulgaria

E-mail: stoilova@tu-sofia.bg

Abstract. The study proposes a methodology based on the combination of Strengths-Weaknesses-Opportunities-Threats (SWOT) analysis and Multi-Criteria Decision Making (MCDM) methods for selecting the most appropriate location for establishing an intermodal terminal. Analytic Hierarchy Process (AHP) and Preference Ranking Organization Method for Enrichment of Evaluations (PROMETHEE) are the MCDM methods used in the paper. The methodology contains four steps. In the first step we determine the alternatives variants of the location of intermodal terminal. In the second step the SWOT analysis have been applied to identify strengths, weaknesses, opportunities, and threats related to the surveyed system. The sub-criteria for each SWOT group have been defined. A total of 33 criteria are included in the study. Third step includes determination of weights of main criteria and sub-criteria by applying AHP method. The results of weights given in this step are used in the four step to ranking the alternative variants of the intermodal terminal location by using PROMETHEE method. The methodology has been approved for choosing the location of intermodal terminal in the city of Ruse. Five alternative locations of intermodal terminal have been studied. The results show that main importance of SWOT group has the strengths group criteria (0.471). It was found that the internal factors including the strengths and weaknesses have more important compared to the external factors including the opportunities and threats when studying the location of intermodal terminal. It was found that the main importance of all sub-criteria have the quantity of manipulated containers (9%), connectivity to road infrastructure (6%), rail connection with a port (5%) and average useful length of railway tracks used for loading/unloading activities (5%). The study shows that location that is situated in the place of the suspended Ruse-East railway yard is the best alternative for intermodal terminal location.

1. Introduction

The intermodal transport is a key element of the freight transport system. The essence of the intermodal transport is the organized carriage of a large number of intermodal transport units by rail, inland waterways and/or sea transport over long distances. The first and the last stage of the intermodal transport chains being carried out by road transport that has an auxiliary role. It is a possibility to being used and combined the competitive advantages of different modes of transport to achieve a cost-effective and an environment-friendly intermodal transport chain that is accomplished just in time.

An important efficiency factor of the intermodal transport system is the location of intermodal terminals. The terminals are the main element of this system. It is necessary to set up a network of interoperable intermodal terminals [1, 2] to achieve an efficient, competitive and sustainable intermodal transport system. The intermodal transport development is accompanied by research studies directed to different fields of transport science as route selection [3], cost analysis [4], decision-making [5], intermodal terminal location problem [1, 6, 7, 8], transport policy [2], etc.

Selecting a location for establishing a logistic center, incl. an intermodal terminal, depends of different conditions. To take into account the diversity of the conditions in selecting a location for establishing an intermodal terminal different decision-making techniques are used [5]. Widely used methods of selecting a location are MCDM models. Their use is mainly in decision making models when it is necessary to make an optimal decision among multiple alternatives taking into account different, often conflicting criteria [9]. The variety of the criteria allows to be classified into different groups and subgroups. A literature review of the criteria and the sub-criteria used in selecting a location of establishing intermodal terminals and logistic centers has been made in [5, 10, 11]. Three groups of approaches for solving intermodal terminal location problems have been defined in [10]. The authors have written the MCDM methods are considerably less used methods for solving location problems [10].

Widespread used decision criteria for selection of a logistic center location [5] or an intermodal terminal [10] are connected with: costs, natural resources, proximity to transport infrastructure, land use, size of land and suitability to enlargement, proximity to industrial zones, macro-economic criteria, environmental impact, etc. The most commonly used by frequency criteria for selection of a location are: cost-related criteria [10, 12, 13], connectivity and accessibility to railway network, roads and highways or port infrastructure [10, 14], availability of land and expansion possibilities [13, 15], capabilities of environment, environmental impacts [10, 14], macroeconomic indicators, accessibility to labor and distance to residential areas [14], utilities, transportation, logistics, social or investment attractiveness and benefits [12, 14], etc. All these criteria can be summarized in the following groups: location, environmental impact, technical criteria, economic and social factors, etc. The economic criteria are the most frequency used type criteria for decision making analysis in the logistics. The intermodal actors who used cost related criteria in decision making are intensive users of intermodal services in comparison of actors who used both types of criteria - quality and cost criteria [16]. The models consist a multiple criteria with different coefficients of importance their weights is necessary to be determined. One of the most commonly used method for determination of the criteria weights` is AHP method [17, 18].

The variety of the most preferred MCDM techniques used for selection a location for establishing a logistic center have been described in literature [5, 11, 19]. In some research papers a combination between different MCDM methods have been proposed. In [20] a conceptual model based on a combination of the fuzzy-analytical hierarchy process and artificial neural networks methods has been proposed by author to select the most appropriate location. In [21] authors have proposed a method for selecting of a logistic strategy by cooperating of SWOT analysis and AHP model. In addition, a combination of both methods - SWOT analysis and AHP has been used in other research fields and decision-making activities [22, 23, 24]. It can be summarized that the multi-criteria analysis is a suitable method for studying the location of a logistics center.

This paper aims to propose an approach for selection the most appropriate location for the intermodal terminal by using multi-criteria methods and taking into account multiple factors relevant for transport. The object of the research is the intermodal terminal in the city of Ruse.

2. Methodology

This paper proposes a new hybrid methodology based on the combination of SWOT analysis, AHP and PROMETHEE methods for selecting the most appropriate location for establishing a rail-road intermodal terminal.

The methodology includes the following steps:

- Step 1: Selection of alternatives of the location of intermodal terminal.
- Step 2: Definition of quantitative and qualitative criteria for the assessment of alternatives. This step of the model uses SWOT analysis as strategic planning technique to identify strengths, weaknesses, opportunities, and threats related to investigated system. It also groups these main criteria

into two categories: internal factors including the strengths and weaknesses and external factors including the opportunities and threats presented by the environment external to the investigated system. The sub-criteria for each main criterion of SWOT group are defined for the evaluation of the alternatives. In this research the following sub-criteria are proposed:

➢ Internal strengths (IS): S1 - Appropriate terrain conditions, coef.; S2 - Part of the main TEN-T network, coef.; S3 - State land ownership, coef.; S4 - Location near the urban area, coef.; S5 - Presence / Using of existing infrastructure, coef.; S6 - Level of access to railway infrastructure. It is measured by the number of railway lines with which the terminal is connected, number; S7 - Length of the longest arrival-departure railway track, m; S8 - Average useful length of railway tracks used for loading/unloading, m; S9 - Connectivity to road infrastructure, m; S10 - Rail connection with a port, m; S11 - Quantity of manipulated twenty-foot equivalent units (TEU), TEU/year.

➢ Internal weaknesses (IW): W1 - High investment costs, million EUR; W2 - High operating costs, 1000 EUR/year; W3 - High net present value (HNPV) of total costs, million EUR; W4 - Size of the site, 1000 m^2; W5 - Change of land use, coef.; W6 - Lack of direct link between railway tracks used for loading/unloading and arrival-departure railway yard, coef.; W7 - Lack of arrival-departure railway yard in the terminal, coef.; W8 - Crossing of communal infrastructure, coef.; W9 - Impact on protected areas, coef.; W10 - Lack of opportunity for further development, coef.

➢ External opportunities (EO): O1 - Possibilities for upgrading the associated infrastructure as part of TEN-T network, coef.; O2 - Opportunity for two-sided connectivity of arrival-departure railway tracks and railway tracks used for loading/unloading activities, coef.; O3 - Development of freight village, coef.; O4 - Development of additional services, coef.; O5 - Attracting of local cargo traffic, coef.; O6 - Proximity to inland waterway and equipped cargo port (use of the terminal as a Dry port), coef.; O7 - Proximity to sea-borne and equipped cargo port (use of the terminal as a Dry port), coef.

➢ External threats (ET): T1 - Competitive terminals, coef.; T2 - Increase of the percentage of road transport, coef.; T3 - Delay at the border crossings, coef.; T4 - Inappropriate conditions for transport along the river, coef.; T5 - Risk of bad meteorological conditions, coef.

Criteria S6, S7, S8, S9, S10, S11, W1, W2, W3 and W4 are quantitative and values need to be set for them. The other criteria are qualitative. They have to be set with "1" and "0". The value is "1" if the answer is "yes", and "0", otherwise.

- Step 3: Determination the weights of main SWOT criteria and sub-criteria by applying AHP method. This step uses the AHP to derive the relative priorities of each sub-criterion within the SWOT groups. The global criteria weight is obtained by multiplying the factors local weights by the specific group weight.
- Step 4: Ranking the alternative variants of the intermodal transport terminal location by using PROMETHEE method. Choice of an optimal variant.

2.1. Determination the weights of criteria by AHP Method

The AHP method is one of the most popular methods of multi-criteria analysis of decision-making. The weights of the criteria are determined by comparing the pairs of the criteria on the basis of the fundamental scale for assessing the criteria (Saaty`s scale). Table 1 presents Saaty`s scale for pair-wise comparison [25, 26].

Table 1. Saaty`s scale for pair-wise comparison.

Explanation	Intensity of importance	Reciprocal values
Equal importance	1	1
Moderate importance	3	1/3
Strong importance	5	1/5
Very strong importance	7	1/7
Extreme importance	9	1/9
Average intermediate values between two close judgments	2; 4; 6; 8	1/2; 1/4; 1/6; 1/8

The evaluation square matrix A (n, n) is formed based on the pairwise comparison on n criteria using Saaty's scale. For the matrix elements are used the following relationships:

$$a_{ii} = 1;\ a_{ij} \neq 0;\ a_{ji} \neq 0; a_{ji} = \left(a_{ij}\right)^{-1} \quad (1)$$

The weights of criteria are determined according the following relationship:

$$A.W = \lambda_{max}.W \quad (2)$$

where $W = \{w_1, \ldots, w_n\}^T$ is the normalized right eigenvector; λ_{max} - the largest eigenvalue of the evaluation square matrix A.

$$\lambda_{max} = \sum_{i=1}^{n}\left[\left(\sum_{j=1}^{n} a_{ij}\right).w_i\right] \quad (3)$$

In AHP method mathematically are validated the results using the consistency ratio CR.

$$CR = CI.RI^{-1} \leq 0.1 \quad (4)$$

where CI is the consistency index; RI is a random index.

The random matrix is given in [26]. The consistency index is:

$$CI = (\lambda_{max} - n).(n-1)^{-1} \quad (5)$$

where λ_{max} is the maximum eigenvalue of the priority matrix; n - the number of elements in the matrix.

Generally, if $CR \leq 0.10$ the consistency of the decision-maker is considered satisfactory. But if CR exceeds 0.10, some revisions of judgements may be required. In order to control the results of the methods, the consistency ratio is used to estimate directly the consistency of pairwise comparisons. The application of consistency index of experts and the conduction of sensitivity analysis of solutions verifies the results.

2.2. Ranking the alternatives variants of intermodal terminal location by using PROMETHEE method

A suitable approach to prioritize the alternatives is the outranking method of multi-criteria analysis - PROMETHEE, where the decision matrix and the weights of criteria are set by the decision maker. In the developed methodology, the weights are determined in advance using the AHP method. For each criterion it is necessary to be set the following:

- The type of optimization - maximum or minimum;
- The preference function which characterizes the difference for a criterion between the evaluations obtained by two possible decisions into a preference degree ranking from 0 to 1. Six basic preference functions have been proposed in this method - usual criterion; quasi criterion; criterion with linear preference; level criterion; criterion with linear preference and indifference area; Gaussian criterion.

The explanation and mathematical calculation steps of the PROMETHEE method are summarised below [27]:

- Step 1: This step computes, for each pair of possible decisions and for each criterion, the value of the preference degree.
- Step 2: This step consists of aggregating the preference degrees of all criteria for each pair of possible decisions.
- Step 3: This step includes the computing of the outranking flows. For each possible decision the positive outranking flow $\varphi^+(a)$ and the negative outranking flow $\varphi^-(a)$ are computed. The positive outranking flow expresses how much each alternative is outranking all the others. The negative outranking flow expresses how much each alternative is outranked by all the others.
- Step 4: In this step the net outranking flows $\varphi(a_j)$ of a_j in the alternatives set m of a possible decision are determined as a difference between $\varphi^+(a_j)$ and $\varphi^-(a_j)$.

$$\varphi(a_j) = \varphi^+(a_j) - \varphi^-(a_j) \quad (6)$$

For net outranking flow, the following conditions are valid:

$$\varphi(a_j) \in [-1; 1] \quad (7)$$

$$\sum_{j=1}^{m} \varphi(a_j) = 0 \quad (8)$$

The optimal alternative is determined by the maximum value of net outranking flows, which corresponds to the alternative with highest priority:

$$max_{a_j \in S} \varphi_j(a_j) \quad (9)$$

where $\varphi_j(a_j)$ are the net outranking flows for alternative a_j; $j = 1, \dots, m$ is the number of alternatives.

3. Results and discussion

The methodology is approved for choosing the location of intermodal terminal in the city of Ruse. It is a part of comprehensive TEN-T network. Figure 1 shows the investigated locations.

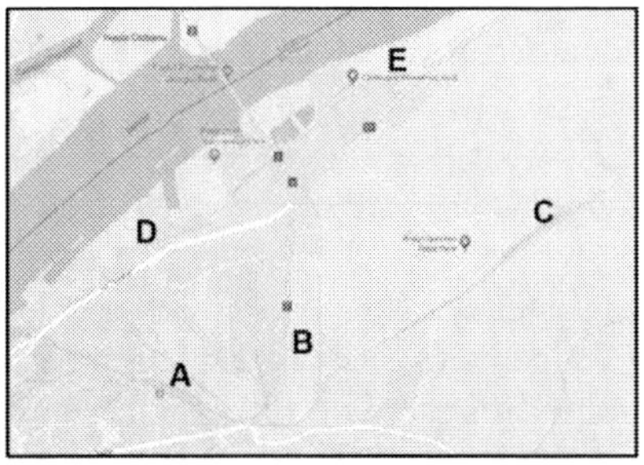

Figure 1. The alternative locations of intermodal terminal in Ruse.

3.1. *Alternative places for location of the intermodal terminal*

This research uses as alternatives for location of the intermodal terminal the places specified in [28]. The alternatives are as follow: Alternative A - The location is situated near to Ruse railway yard to the north of it; Alternative B - The location is situated in a place on site intended for the construction of Ruse technical railway station; Alternative C - The location is situated in the place of the suspended Ruse-East railway yard; Alternative D - The location is situated to the west of the Ruse-East port; Alternative E - The location is at East of Ruse free zone.

The values of the quantitative criteria for each alternative have been determined by using information given in [28]. Table 2 presents the values of the sub-criteria.

Table 2. Sub-criteria values of Decision matrix.

IS	Type	A	B	C	D	E	IW	Type	A	B	C	D	E
S1	max	1	1	1	1	1	W1	min	21.7	22.2	21.8	24.8	27.9
S2	max	1	1	1	1	1	W2	min	399.0	419.4	423.7	468.2	554.2
S3	max	0	0	1	0	0	W3	min	29.1	29.8	29.6	33.1	37.3
S4	max	1	1	1	1	1	W4	max	109.3	109.8	123.4	96.0	112.1
S5	max	0	0	1	0	0	W5	min	0	0	0	0	1
S6	max	5	5	5	3	4	W6	min	1	1	0	0	1
S7	max	760	750	774	636	750	W7	min	1	1	0	1	1
S8	max	470	530	530	470	530	W8	min	1	1	0	1	0
S9	min	1050	220	100	580	850	W9	min	0	0	0	0	1
S10	min	5800	3500	5400	10000	10000	W10	min	0	1	1	0	0
S11	max	30.6	30.7	34.6	26.9	31.4							
EO	Type	A	B	C	D	E	ET	Type	A	B	C	D	E
O1	max	1	1	1	1	1	T1	min	1	1	1	1	1
O2	max	0	1	1	0	0	T2	min	1	1	1	1	1
O3	max	0	1	1	0	0	T3	min	1	1	1	1	1
O4	max	1	1	1	1	1	T4	min	1	1	1	1	1
O5	max	1	1	1	1	1	T5	min	1	1	1	1	1
O6	max	1	1	1	1	1							
O7	max	1	1	1	1	1							

3.2. Determination the weights of criteria

This research uses 10 experts with experience in intermodal transport (academics specialists in freight transportation and managers of transport companies). The experts were asked to perform pairwise comparisons of all main criteria of SWOT group and sub-criteria. Table 3 shows the results of expert assessment for main SWOT group. Table 4, table 5 and table 6 presents expert assessment for sub-criteria. The first and the last column of the tables presents the name of criteria, others columns show the score of each expert. The value of the score is integer if the criterion that is written it the first column of the table is equal or more important than the criterion in the last column in the same row. Otherwise the reciprocal value is recorded, according Saaty's scale.

Table 3. Expert`s assessment for the main SWOT group.

	Expert										
SWOT	1	2	3	4	5	6	7	8	9	10	SWOT
EO	1/2	1/5	1/2	1/3	1/4	2	3	2	2	2	ET
EO	1/4	1/6	1/3	1/2	1/2	1/3	1/4	1/4	1/4	1/3	IS
EO	1/3	1/3	1/3	1	1/4	1/3	1/3	2	1/5	1/3	IW
ET	1/4	1/5	1/3	1/2	1/2	1/3	1/4	1/4	1/5	1/3	IS
ET	1/3	2	1/2	3	1	1/3	1/4	1	1/5	1/3	IW
IS	4	3	2	3	2	3	2	4	1/2	3	IW

Table 4. Expert assessment for sub-criteria for strengths (IS).

IS	1	2	3	4	5	6	7	8	9	10	IS	IS	1	2	3	4	5	6	7	8	9	10	IS
S1	4	2	1/2	1/2	2	2	1/2	3	1/2	2	S2	S4	1/3	1/2	1/2	1/2	1/2	1/3	1/3	1/4	1/4	1/3	S7
S1	1/5	2	2	3	2	1	1/3	3	1/3	1	S3	S4	1/4	1/3	1/3	1/2	1/2	1/3	1/4	2	1/4	1/3	S8
S1	1/4	3	3	3	2	2	2	2	2	2	S4	S4	1/4	1/3	1/3	1/2	1/2	1/3	1/4	1/4	1/5	1/3	S9
S1	1/4	2	2	2	2	1	1/3	1/2	1/4	1	S5	S4	1/2	1/2	1/2	1/2	1/2	1/3	1/3	1/4	1/4	1/3	S10
S1	1/5	2	2	2	2	1/2	1/4	1/5	2	1/2	S6	S4	1/5	1/4	1/3	1/3	1/2	1/3	1/3	1/5	1/3	1/3	S11
S1	1/4	2	2	3	2	2	2	3	1	2	S7	S5	1/3	1/2	1/2	1/2	1/2	1/3	1/4	3	1/3	1/3	S6
S1	1/4	1/3	1/3	1/2	1/2	2	2	3	1	2	S8	S5	1/3	1/2	1/2	1/2	1/2	2	1/2	2	2	2	S7
S1	1/5	1/3	1/3	1/2	1/2	1/2	1/3	1/3	1/3	1/2	S9	S5	1/4	1/3	1/3	1/2	1/3	1	1/2	2	4	1	S8
S1	1/3	1/2	1/2	1	1/2	1/2	1/4	1/4	1/3	1/2	S10	S5	1/3	1/2	1/2	1/2	1/2	1/3	1/2	3	1/3	1/3	S9
S1	1/4	1/3	1/2	1/2	1/2	1/3	1/3	1/4	1/3	1/3	S11	S5	2	1/2	1/2	1/2	1/2	1/3	1/3	2	1/4	1/3	S10
S2	1/4	2	3	3	3	3	4	3	1	3	S3	S5	1/5	1/3	1/3	1/3	1/3	1/3	1/3	1/3	1/3	1/3	S11
S2	1/4	2	2	3	4	2	3	4	4	2	S4	S6	1/2	1/2	1/2	1/2	1/2	3	3	3	3	3	S7
S2	1/4	1/2	2	2	1	1	1/2	1	1/3	1	S5	S6	1/5	1/3	1/3	1/2	1/2	3	2	2	3	3	S8
S2	1/4	2	2	2	1/2	1/2	1/3	1/3	1	1/2	S6	S6	1/2	1/2	1/2	1/2	1/2	1	1	1	1	1	S9
S2	1/4	2	2	2	1/2	1/2	1/2	1	2	1/2	S7	S6	3	2	1	1	2	1	1	1	1	1	S10
S2	1/4	1/3	1/2	1/2	1/2	2	1/2	1	2	2	S8	S6	1/4	1/3	1/3	1/2	1/2	1/3	1/3	1/3	1/3	1/3	S11
S2	1/4	1/3	1/3	2	1/2	1/3	1/3	1/4	1/2	1/3	S9	S7	1	1/2	1/2	1/2	1/2	2	1	1	2	2	S8
S2	1/2	1/2	1/2	1/2	1/2	1/3	1/3	1/4	1/2	1/3	S10	S7	3	1/2	1/2	1/2	1/2	1/3	1/2	1/2	1/3	1/3	S9
S2	1/5	1/3	1/3	2	1/2	1/3	1	1/2	1	1/3	S11	S7	3	1/2	1/2	1/2	1/2	1/3	1/2	1/2	1/3	1/3	S10
S3	3	2	2	1/2	1	3	3	1	4	3	S4	S7	1/2	1/3	1/2	1/2	1/2	1/4	1/3	1/4	1/4	1/4	S11
S3	2	1/2	1/2	1/2	1/3	1	1	1/4	1	1	S5	S8	2	2	2	2	2	1/3	1/4	1/4	1/3	1/3	S9
S3	3	1/2	1/2	1/2	1/2	1/2	1	1/3	1/4	1/2	S6	S8	4	3	3	2	2	1/3	1/3	1/3	1/3	1/3	S10
S3	1/3	1/3	1/3	1/2	1/3	1	1	1/2	2	1	S7	S8	1/3	1/2	1/3	1/3	2	1/4	1/3	1/3	1/4	1/4	S11
S3	1/3	1/4	1/4	1/2	1/3	1	1	1/2	2	1	S8	S9	4	3	1	1	3	1	1	1	1	1	S10
S3	1/3	1/2	1/2	1/2	1/3	1/3	1/3	1/3	1/2	1/3	S9	S9	1/4	1/2	1/2	1/2	1/2	1/3	1/2	1/2	1	1/3	S11
S3	2	2	2	2	2	1/3	1/3	1/3	1/2	1/3	S10	S10	1/5	1/3	1/2	1/2	1/2	1/3	1/2	1/2	1	1/3	S11
S3	1/5	1/4	1/4	1/4	1/2	1/3	1/3	1/2	1/3	1/3	S11												
S4	1/3	1/2	1/2	1/2	1/2	1/3	1/5	1/4	1/3	1/3	S5												
S4	1/4	1/3	1/2	1/2	1/3	1/4	1/3	1/4	1/4	1/4	S6												

The AHP method has been applied separately for each expert's assessment. Thus, 10 variants of results were obtained. Table 7 presents the value of CR for experts` assessments of main SWOT group, and for sub-criteria. The super decision software have been used to make research [29]. It can be seen that the equation (4) is satisfied for all cases. This shows that each expert's assessment is correct.

Table 5. Expert assessment for sub-criteria for weaknesses (IW).

IW	1	2	3	4	5	6	7	8	9	10	IW
W1	1/3	1/2	1/2	1/2	1	2	1/2	1	1	1/2	W2
W1	1/3	1/3	1	1/2	1	3	1/3	1	3	1/3	W3
W1	3	3	3	2	3	1/2	3	2	3	3	W4
W1	3	3	3	2	4	1	2	2	2	2	W5
W1	3	2	2	2	2	3	3	3	3	3	W6
W1	5	3	3	3	4	3	3	3	3	3	W7
W1	2	2	2	2	3	2	1/2	1	2	1/2	W8
W1	1/3	1/2	1	1	1	2	1/2	1	2	1/2	W9
W1	3	4	4	2	5	3	3	3	3	3	W10
W2	1/3	1/2	1/2	1/2	1	1/3	1/2	1/3	1/3	1/2	W3
W2	3	2	2	2	2	2	1/2	3	1	1/2	W4
W2	4	3	3	3	3	2	2	3	3	2	W5
W2	4	3	3	3	3	2	3	3	1/2	3	W6
W2	4	5	5	3	3	2	3	4	2	3	W7
W2	5	5	5	3	1	2	1	2	3	1	W8
W2	1/3	1/3	1	2	1/2	2	1	1	2	1	W9
W2	3	4	4	3	3	2	1	3	2	1	W10
W3	4	3	3	2	2	2	1/2	3	2	1/2	W4
W3	3	3	3	2	3	2	1/2	1/2	3	1/2	W5
W3	4	4	4	3	3	2	2	1	3	2	W6
W3	4	5	5	3	4	2	2	1	3	2	W7
W3	3	3	3	3	2	2	2	1/2	1	2	W8
W3	1/3	1/2	1	1	1/2	2	1	1	2	1	W9
W3	4	5	4	4	5	2	2	2	2	2	W10

IW	1	2	3	4	5	6	7	8	9	10	IW
W4	2	3	3	3	3	1	2	1	1	2	W5
W4	1/2	3	2	2	3	2	1/2	2	2	1/2	W6
W4	3	3	3	3	3	1	1	1	1	1	W7
W4	3	3	3	3	3	1/2	1/2	2	1/2	1/2	W8
W4	1/3	1/3	1/2	1/2	1/2	1/3	1/2	1	1/3	1/2	W9
W4	2	2	2	2	2	2	2	1/2	2	2	W10
W5	1/3	1/3	1/2	1/2	1/2	1	1	1/3	1	1	W6
W5	1/2	1/2	1/2	1/2	1/2	1	1	2	1	1	W7
W5	1/2	1/2	1/2	1/2	1/2	1	1	1	1	1	W8
W5	1/4	1/3	1/2	1/3	1/2	1/2	1/2	1	1/2	1/2	W9
W5	1/3	1/2	1/2	1/3	1/2	2	1	3	2	1	W10
W6	3	3	3	3	1	2	2	2	2	2	W7
W6	2	1/2	1/2	1/3	1/2	1/2	1/2	1/2	1/2	1/2	W8
W6	1/4	1/3	1/2	1/2	1/2	1/2	1/2	1/2	1/2	1/2	W9
W6	2	1/3	1/2	1/3	1/2	2	2	2	2	2	W10
W7	3	1/3	1/3	1/3	1/2	1/2	1/2	1/2	1/2	1/2	W8
W7	1/4	1/3	1/2	1/2	1/2	1/2	1/2	1/2	1/2	1/2	W9
W7	1/3	1/2	1/2	1/3	1/2	2	2	2	2	2	W10
W8	1/4	1/3	1/2	1/2	1/2	1/2	1	1	1/2	1	W9
W8	1/3	1/2	1/2	1/3	1/2	2	2	2	2	2	W10
W9	4	3	3	2	1	3	2	2	3	2	W10

Table 6. Expert assessment for sub-criteria for opportunities (EO) and threats (ET).

EO	1	2	3	4	5	6	7	8	9	10	EO
O1	1/3	2	2	3	1/2	2	1/2	1	1	1	O2
O1	1/3	2	2	2	1/2	1/2	1/3	3	3	3	O3
O1	1/3	1/2	1/2	2	1/2	2	3	4	1	4	O4
O1	1/2	2	1/3	2	1/2	1/3	1/5	1/3	1/4	1/3	O5
O1	1/3	1/4	1/3	1/2	1/2	1/2	1	2	1	2	O6
O1	1/3	1/3	1/3	3	1	1/2	1	2	1	2	O7
O2	1/2	1/2	1/2	1/2	1/2	1/3	2	3	1/3	3	O3
O2	1/3	1/4	1/4	1	1/2	3	1/2	2	1/4	2	O4
O2	3	3	1/3	1/2	1/2	1/3	1/3	1/3	1/3	1/3	O5
O2	1/3	1/4	1/3	1/2	1/2	1/2	1/2	1/2	1/2	1/2	O6
O2	1/3	1/3	1/3	1/2	1/3	1/2	1/2	1/2	1/2	1/2	O7
O3	1/3	1/3	1/3	1/2	1/3	2	2	1	3	1	O4
O3	2	1/2	1/2	1/2	2	1/3	1/3	1/4	1/4	1/4	O5
O3	1/4	1/4	1/3	1/2	1/2	1/2	1/2	1	1	1	O6
O3	2	1/3	1/2	1/2	1	1/2	1/2	1	1	1	O7
O4	3	3	1/2	1/2	1/2	1/3	1/4	1/3	1/3	1/3	O5
O4	1/3	1/3	1/3	1/2	1/2	1/2	1/3	1/2	1/2	1/2	O6
O4	2	1/3	1/3	1/2	1/2	1/2	1/3	1/2	1/2	1/2	O7
O5	1/3	1/5	1/3	1	1/3	1	2	2	2	2	O6
O5	1/3	1/3	3	3	1/2	1	2	2	2	2	O7
O6	5	5	2	2	2	1	1	1	1	1	O7

ET	1	2	3	4	5	6	7	8	9	10	ET
T1	2	2	2	1/2	1/2	1/2	3	1	1/3	1/3	T2
T1	4	2	2	2	3	1	2	3	1	1	T3
T1	4	1/3	2	2	1	2	2	3	3	3	T4
T1	5	2	3	4	2	2	2	2	2	2	T5
T2	3	3	3	3	2	3	1/3	1/3	4	4	T3
T2	4	1/2	1/2	1/2	1/2	2	1/2	1	3	3	T4
T2	5	2	3	3	2	1	1/2	1	1	1	T5
T3	1/3	1/2	1/2	1/2	1/2	3	2	3	2	2	T4
T3	3	3	3	4	1	1	1	1	1	1	T5
T4	4	4	4	4	2	1	1	1/2	1	1	T5

Table 7. Consistency ratio (CR).

Expert	1	2	3	4	5	6	7	8	9	10
SWOT group	0.08	0.09	0.03	0.04	0.09	0.08	0.06	0.02	0.03	0.08
IS sub-criteria	0.10	0.07	0.07	0.07	0.07	0.05	0.09	0.09	0.06	0.05
EO sub-criteria	0.09	0.10	0.08	0.05	0.09	0.05	0.09	0.05	0.09	0.05
IW sub-criteria	0.09	0.08	0.06	0.07	0.06	0.07	0.10	0.10	0.07	0.10
ET sub-criteria	0.08	0.08	0.06	0.10	0.05	0.08	0.02	0.09	0.09	0.09

The results of expert's assessments have been summarized. The local weights for each group and the global weights of criteria have been determined as average values of individual scores. The local weights are the normalised weights that show the weight of each sub-criterion in the respective group of the main criterion. The global weights show the priority of all sub-criteria taking into account the weights of main criteria. Table 8 presents the values of both types of weights.

Table 8. Local and Global weights of criteria and sub-criteria.

SWOT group	Weight	Criteria	Local weight	Global weight	SWOT group	Weight	Criteria	Local weight	Global weight
IS	0.471	S1	0.070	0.033	IW	0.252	W1	0.142	0.036
		S2	0.072	0.034			W2	0.144	0.036
		S3	0.056	0.026			W3	0.165	0.042
		S4	0.034	0.016			W4	0.091	0.023
		S5	0.070	0.033			W5	0.058	0.015
		S6	0.099	0.047			W6	0.066	0.017
		S7	0.073	0.034			W7	0.048	0.012
		S8	0.102	0.048			W8	0.081	0.021
		S9	0.125	0.059			W9	0.146	0.037
		S10	0.106	0.050			W10	0.059	0.015
		S11	0.193	0.091					
EO	0.122	O1	0.123	0.015	ET	0.155	T1	0.276	0.043
		O2	0.086	0.010			T2	0.256	0.040
		O3	0.102	0.012			T3	0.156	0.024
		O4	0.108	0.013			T4	0.190	0.029
		O5	0.210	0.026			T5	0.122	0.019
		O6	0.220	0.027					
		O7	0.151	0.018					

It was found that the main importance of SWOT group has the strengths group criteria. In the second position are the weaknesses group criteria. It can be concluded that the internal factors including the strengths (IS) and weaknesses (IW) have more important compared to the external factors including the opportunities (EO) and threats (ET). The weights of the main criteria opportunities and threats are close.

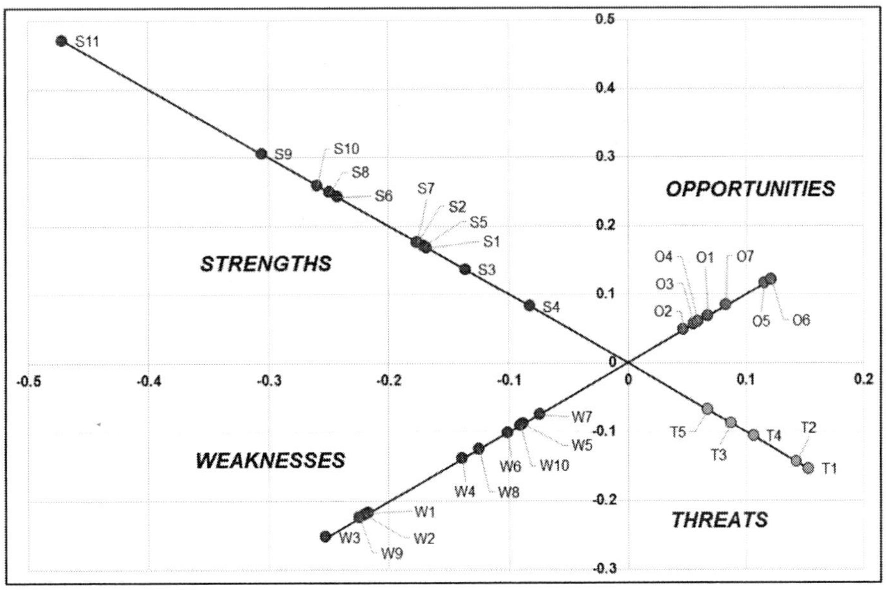

Figure 2. Overall priority score.

IOP Conf. Series: Materials Science and Engineering **618** (2019) 012060 doi:10.1088/1757-899X/618/1/012060

Figure 2 presents the overall priority scores. The lengths of the lines in the different sectors demonstrate that the strengths and weaknesses predominate in comparison to external factors of opportunities and threats. The length of line in each quadrant represents the overall importance of each group. The points in the lines show the overall priority of individual factors. The results show that the main importance have the following sub-criteria: S11 (Quantity of manipulated TEU), S9 (Connectivity to road infrastructure and S10 (Rail connection with a port) of SWOT group – strengths; W3 (HPV of total costs), W9 (Impact on protected areas), W2 (High operating costs) and W1 High investment costs) of SWOT group weaknesses; O6 (Proximity to inland waterway and equipped cargo port), O5 (Attracting of local cargo traffic) and O7 (Proximity to sea-borne and equipped cargo port) of SWOT group opportunities; T1 (Competitive terminals) and T2 (Increase of the percentage of road transport) of SWOT group threats.

The Pearson correlation coefficient as a measure of the linear correlation between two variables has been determined to assess the difference of value of weights given individual expert's assessments and its average value. Table 9 shows the values of Pearson's correlation coefficient. The values which are larger than 0.7 show presence of a very strong relationship, the values which are between 0.4 and 0.69 present strong relationship. The first column presents the number of experts, the columns indicated with IS-EW show the values of the Pearson's correlation coefficient for local weights in each SWOT group; the column for main weights presents the values of the Pearson's correlation coefficient for the weights of the main SWOT group; the last column in the table presents the values of the Pearson's correlation coefficient for the weights of all sub-criteria. It can be summarise that the results of weights given by individual expert's assessments correlate with average values of weights. For example the value 0.76 in the column IS presents the Pearson's correlation coefficient between the weights given by expert's assessment of the first expert for sub-criteria in IS group and the average value of weights given by all experts. It shows proximity in expert assessments and verifies they results.

Table 9. Values of the Pearson`s correlation coefficient by SWOT groups.

	For local weights				For main weights	For global weights
Expert	IS	IO	ET	EW		
1	0.76	0.42	0.88	0.88	0.99	0.84
2	0.87	0.58	0.94	0.94	0.93	0.85
3	0.86	0.94	0.97	0.97	0.99	0.90
4	0.73	0.72	0.91	0.91	0.65	0.58
5	0.76	0.65	0.94	0.94	0.87	0.69
6	0.92	0.85	0.91	0.91	0.98	0.89
7	0.90	0.76	0.90	0.90	0.95	0.82
8	0.78	0.58	0.82	0.82	0.88	0.72
9	0.78	0.67	0.88	0.88	0.57	0.46
10	0.92	0.58	0.90	0.90	0.98	0.86

3.3. Ranking the alternatives by using PROMETHEE method

After determining the weights of the criteria the PROMETHEE method is applied for ranking the investigated location. The weights of the criteria determined by means of AHP method are used in the PROMETHEE method to estimate the alternatives. The Visual PROMETHEE software has been used to make research [30]. The type of optimization is set in table 2. A usual preference functions have been set for the dichotomous variable, which values are 0 or 1. For others variables have been linear preference.

Figure 3 shows the ranking of the investigated location of intermodal terminal using all global criteria. In the first part of the figure is presented the ranking according values of net outranking flows;

the second part shows the global weights of the sub-criteria. It can be seen that the location C which is situated in the place of the suspended Ruse-East railway yard is the best alternative.

Figure 4 presents the ranking by using separately the local weights of each SWOT group. The Location C is the best alternative also when ranking by using strengths group criteria (IS), and when ranking by weaknesses group criteria (IW). The location B is the best alternative when using opportunities group criteria (EO). The net outranking flows for location B and location C have closes values. The value of net outranking flows for all alternatives is equal to zero when applying threats (ET) group criteria. This is due to the equal values of the threats (ET) group criteria for each of the alternatives that can be seen in table 2.

The sensitivity analysis was made for optimal alternative. It was found that the criteria S10 (Rail connection with a port), W2 (High operating costs) and W10 (Lack of opportunity for further development) have small stability intervals of weights: S10 [0-38.96%]; W2 [0-33.13%]; W10 [0-27.22%]. The others criteria have large stability intervals of weights [0-100%].

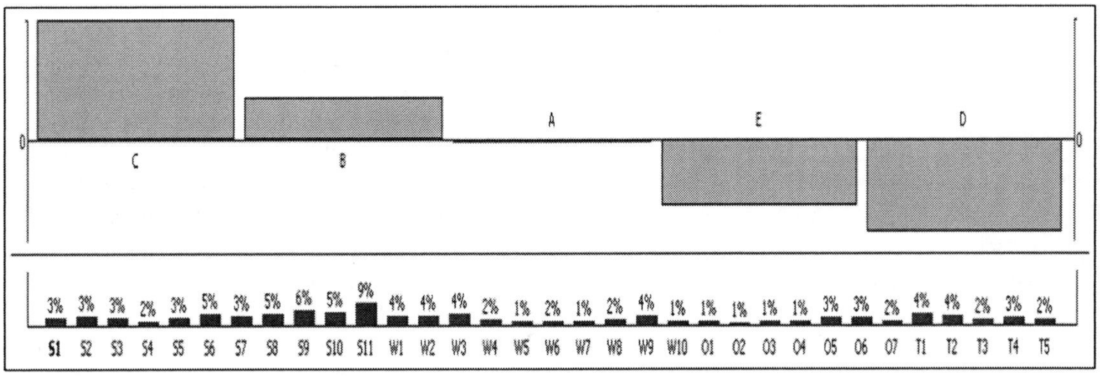

Figure 3. Ranking the alternatives locations in Visual PROMETHEE using global weights.

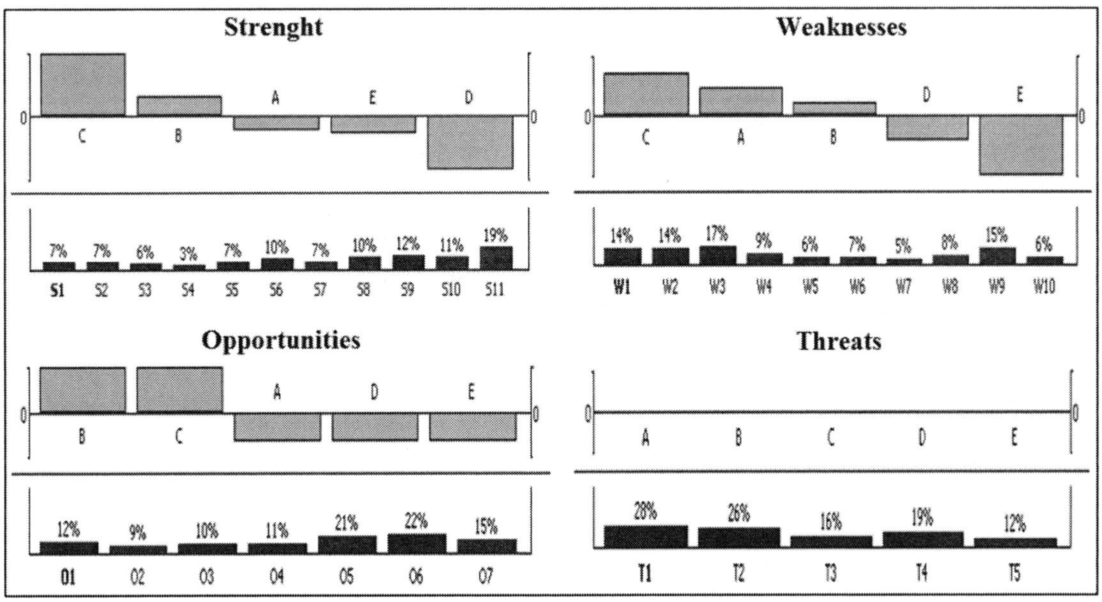

Figure 4. Ranking the alternatives locations in Visual PROMETHEE using local weights.

The results of this study in regard to the best location of intermodal terminal in the city of Ruse are identical with the recommendations given in [28].

4. Conclusion

This paper presents a combination of SWOT analysis, AHP and PROMETEE multi-criteria methods into a hybrid model for selecting the most appropriate location for the intermodal terminal. The proposed methodology that combines SWOT analysis with AHP and PROMETHEE methods is the main contributions of this paper. For the first time in this study is deal with the SWOT analysis for the intermodal terminal location problem. The papers that use SWOT analysis and AHP method in different research area do not continue the study to determine the optimal alternative.

The SWOT analysis has been used to define the main factors as strengths, weaknesses, opportunities and threats for investigated system. The sub-criteria for each main criterion of SWOT group are defined for the evaluation of the alternatives. The weights of main SWOT criteria and sub-criteria have been determined by applying AHP method. It was found that the main importance of SWOT group has the strengths group criteria (0.471). The results show that the internal factors including the strengths and weaknesses have more important compared to the external factors including the opportunities and threats when studying the location of intermodal terminal. It was found that the main importance of all sub-criteria have the quantity of manipulated TEU (9%), connectivity to road infrastructure (6%), rail connection with a port (5%) and average useful length of railway tracks used for loading/unloading (5%). The important criteria in strengths group are quantity of manipulated TEU (19%), connectivity to road infrastructure (12%) and rail connection with a port (11%); average useful length of railway tracks used for loading/unloading (10%). The important criteria in weaknesses group are HPV of total costs (17%), impact on protected areas (15%), high investment costs (14%) and high operating costs (14%). The main criteria in opportunities group are proximity to inland waterway and equipped cargo port (22%), attracting of local cargo traffic (21%) and proximity to sea-borne and equipped cargo port (15%). The main criteria in threats group are competitive terminals (28%) and increase of the percentage of road transport (26%). The study produces a ranking of alternative locations of intermodal terminal of Ruse using the proposed hybrid model. The research showed that location that is situated in the place of the suspended Ruse-East railway yard is the best alternative for intermodal terminal position. The elaborated methodology in this study could be used for investigations the optimal location of others intermodal terminals. The proposed model could also be applied for solving various problems in transport and different area.

Acknowledgments

This research has been supported by contract №КП-06-Н27/12 of 11.12.2018 "Modelling and elaboration of complex system for selection of transport technology in transport network" funded by the National Science Fund of the Ministry of Education and Science of Bulgaria.

References

[1] Limbourg S and Jourquin B 2009 Optimal rail-road container terminal locations on the European network *Transportation Research* E. **45** 551–63
[2] Macharis C, Caris A, Jourquin B and Pekin E 2011 A decision support framework for intermodal transport policy *Eur. Transp. Res. Rev.* **3** 167–78
[3] Chang T-S 2008 Best routes selection in international intermodal networks *Computers&Operations Research* **35** 2877–91
[4] Hanssen T-E S, Mathisen T A and Jørgensen F 2012 Generalized transport costs in intermodal freight transport *Social and Behavioral Sciences* **54** 189-200
[5] Uyanik C, Tuzkaya G and Oğuztimur S 2018 A literature survey on logistics centers' location selection problem *Sigma J Eng & Nat Sci* **36** (1) 141-60
[6] Sörensen K and Vanovermeire C 2013 Bi-objective optimization of the intermodal terminal location problem as a policy-support tool *Computers in Industry* **64** 128–35
[7] Sörensen K, Vanovermeire C and Busschaert S 2012 Efficient metaheuristics to solve the intermodal terminal location problem *Computers & Operations Research* **39** 2079–90
[8] Lin C-C and Lin S-W 2016 Two-stage approach to the intermodal terminal location problem

Computers & Operations Research **67** 113–19
[9] Odu G and Charles-Owaba O 2013 Review of multi-criteria optimization methods – Theory and applications, *IOSR Journal of Engineering* **3**(10) 01-14
[10] Zečević S, Tadić S and Krstić M 2017 Intermodal Transport Terminal Location Selection Using a Novel Hybrid MCDM Model *International Journal of Uncertainty, Fuzziness and Knowledge-Based Systems* **25** No. 6 853-76
[11] Özceylan E, Erbaş M, Tolon M, Kabak M and Durğut T 2016 Evaluation of freight villages: A GIS-based multi-criteria decision analysis *Computers in Industry* **76** 38-52
[12] Żak J and Węgliński S 2014 The selection of the logistics center location based on MCDM/A methodology *Transportation Research Procedia* **3** 555–64
[13] Destiny Ugo P 2015 A Multi-Criteria Decision Making for Location Selection in the Niger Delta Using Fuzzy TOPSIS Approach *Int. J. Manag. Bus. Res.* **5** (3) 215-24
[14] Uysal H T and Yavuz K 2014 Selection of Logistics Centre Location via ELECTRE Method: A Case Study in Turkey *International Journal of Business and Social Science* **5**, No. 9 276-89
[15] Stević Ž, Vesković S, Vasiljevićc M and Tepićd G 2015 The Selection of The Logistics Center Location Using AHP Method *Proc. 2nd Logistics International Conference (Belgrade)* (Serbia: University of Belgrade) pp 86-91
[16] Bontekoning Y, Macharis C and Trip J 2004 Is a new applied transportation research field emerging? - A review of intermodal rail–truck freight transport literature *Transportation Research* A **38** 1–34
[17] Saaty T 1980 *The Analytic Hierarchy Process* (New York: McGraw-Hill)
[18] Saaty R 1987 The analytic hierarchy process - what it is and how it is used *Mathematical Modelling* **9** No. 3–5 161-76
[19] Aruldoss M, Miranda Lakshmi T and Prasanna Venkatesan V 2013 A Survey on Multi Criteria Decision Making Methods and Its Applications *American Journal of Information Systems* **1** No. 1 31-43
[20] Kayikci Y 2010 A conceptual model for intermodal freight logistics centre location decisions *Procedia Social and Behavioral Sciences* **2** 6297–311
[21] Wang X, Wu P and Huang T 2017 Analysis of Logistic Strategy based on SWOT and AHP Methods *Advances in Engineering Research* **134** 70-74
[22] Görener A, Toker K and Uluçay K 2012 Application of Combined SWOT and AHP: A Case Study for a Manufacturing Firm *Procedia - Social and Behavioral Sciences* **58** 1525–34
[23] Kurttila M, Pesonen M, Kangas J and Kajanus M 2000 Utilizing the analytic hierarchy process AHP in SWOT analysis - a hybrid method and its application to a forest-certification case *Forest Policy and Economics* **1** 41-52
[24] Mehmood F, Hassannezhad M and Abbas T 2014 Analytical investigation of mobile NFC adaption with SWOT-AHP approach: A case of Italian Telecom *Procedia Technology* **12** 535-41
[25] Saaty T 2000 *Fundamentals of Decision Making and Priority Theory with the Analytic Hierarchy Process* (New York: RWS Publications) 1st ed. p 477
[26] Saaty T 2004 Fundamentals of the Analytic network process – Dependence and feedback in decision-making with a single network *Journal of Systems Science and Systems Engineering* **13** Issue 2 129-57
[27] Brans J-P and Mareschal B 2005 Multiple Criteria Decision Analysis: State of the Art Surveys *International Series in Operations Research & Management Science book series (ISOR)* **78** (New York: Springer) Chapter Promethee Methods pp 163-86
[28] Martinov S 2018 Increasing the accuracy of evaluation in selecting a location for establishing an intermodal terminal *MATEC Web of Conferences* **234** 06004
[29] https://www.superdecisions.com/ (Available online on 27.02.2019)
[30] http://visualpromethee.com/ (Available online on 27.02.2019)

IOP Conf. Series: Materials Science and Engineering **618** (2019) 012061 doi:10.1088/1757-899X/618/1/012061

Determining the velocity and the duration of the travel by the method of recording the registration numbers with the help of an integrated system of mobile LPR cameras

I S Damianov[1], G D Mladenov[1], M G Savova-Maratsenkova[1], V D Hristov[2], G D Palagachev[1]

[1]Technical University of Sofia, Faculty of Transport, Department of Combustion Engines, Automobile Engineering and Transport, Bulgaria
[2] Technical University of Sofia, Department "Automation of Electric Drives"

idamyanov@tu-sofia.bg

Abstract. This paper presents research a determining the velocity and the duration of the travel via developed intelligent system of License Plate Recognition cameras and by using the method of recording the State Control Number (DK number) with the aim of improving the traffic organization and safety, as well as reducing the harmful impact and improving the ecological indicators in the research area. The intelligent system for recording and recognition of registration numbers provides the opportunity for statistical analysis to determine traffic performance and to take specific measures so as to improve the organization and increase the traffic safety. It is based on MySQL database and PHP scripts to solve the analysis tasks.

1. Introduction

The research on transport flows is a complex task and requires significant resources [1-4]. This is due to the use of existing methods for the examination of traffic indicators [5,6]. Some of them require the installation of dedicated technical means to report traffic [7,8]. Others require multiple observers who need to be instructed in the specifics of the tasks they will be performing [9,10]. Some methods allow for the use of video capture, requiring the installation of multiple cameras for the relevant surveyed areas of the roadway [11].

For the insurance of safe, economic and ecologically friendly traffic flow it is important the basic traffic indicators to be defined and assessed. The analysis, the assessment and the optimization of the transport traffic are based on the results that are obtained, their common relation and the knowledge about the tolerable limits, characterizing the demeanour of the traffic flow.

Software for modeling the traffic movement is based on input data for its metrics obtained by some of the known methods [12].

The established system uses License Plate Recognition cameras to register the State Control Number (DK number) of each vehicle and allows continuous video surveillance of transport flows. Functional potential of the developed algorithm allows to determine traffic flow indicators and to analyze the results obtained [13, 14].

Technology includes:

1. Traffic streams capture with the License Plate Recognition Registrar.
2. Processing the Received Data from the License Plate Recognition Registrar to determine traffic flow indicators.
3. Analysis of processed results.
4. Generating database on the nature of traffic flows.

2. Prerequisites and ways to solve the problem

The aim of the present research paper is to determine the velocity and the time duration of travel of the traffic flows, which pass through a particular section of the road. An intelligent system of License Plate Recognition cameras [15] is used, with the help of which an automatized video monitoring is done and an opportunity for further statistical processing is ensured. This allows for the necessary indicators, characterizing the traffic flow in the researched area of the road, to be determined. The analysis of the obtained data is used for organization and management of the traffic flow, reducing the congestions, improving the traffic safety and reducing the harmful impact of the traffic on the environment. The necessary indicators for the assessment of the road traffic are:

- modification of the traffic in time;
- density of the traffic flows;
- intensity of the traffic flows;
- composition of traffic flows, ets.

Besides from the State Control Number (DK number), the used License Plate Recognition cameras also report other indicators like the country from which the vehicle comes, the direction in which it travels, date, hour, etc. The acquired data is saved in the form of an Excel file and photographic material.

The information from the video registrars (exact time at which the vehicle is captured on the camera, license plate of the vehicle, etc.) are imported in a specifically developed database, which allows the data to be statistically processed according to the set parametres for the specific research problem.

During the analysis of the acquired data can be determined the basic indicators like the time duration of the travel between the places where the cameras are located and the velocities of the vehicles. Histograms can also be made.

The generated database gives the opportunity for collecting information, with which thorough analysis and research about the traffic flow's indicators can be conducted.

3. Results and discussion

The subject of the present research is to determine the velocity and the duration of the travel through a particular road section of the urban street network [16].

The aim of the research is to examine and analyse the traffic flows, as well as to assess the opportunity for optimization and increase of the traffic safety and the environmental impact [17, 18].

The study has been conducted on a road section on Kliment Ohridski Boulevard during the time of the morning rush hour from 8:00 to 9:30. The values of the examined indicators were determined with the help of two License Plate Recognition cameras to register DK - numbers. The cameras were positioned at predetermined locations (counting points) and spacing from one another shown in Figure 1.

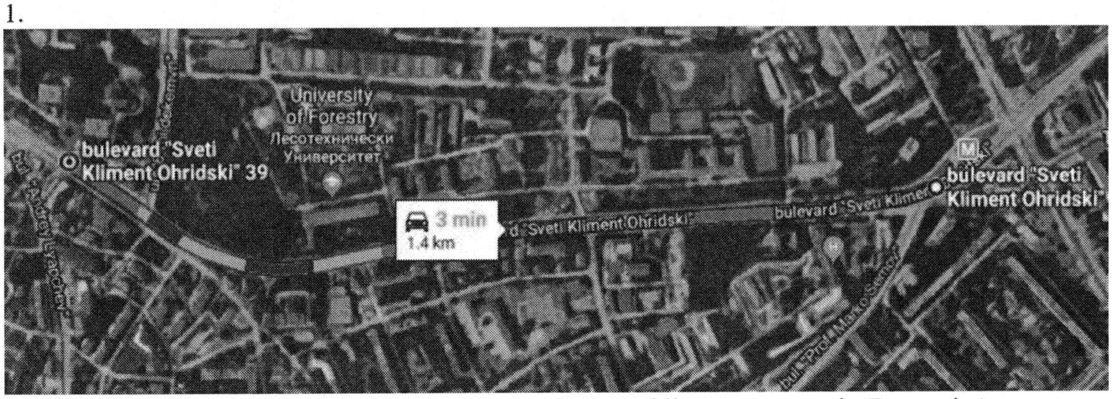

Figure. 1. Position of License Plate Recognition Mobile Cameras on the Research Area.

The tasks of the examination are defined as follows:

- intensity of the traffic flow at the counting points;
- number of vehicles which pass through the places where camera 1 and camera 2 are located and vice versa;
- time duration of the travel between the two cameras;
- velocity with which the vehicles pass through the section between the cameras.

In order for the traffic indicators to be defined, in both directions of the 1.4 km section that was examined the position of the cameras and the direction of shooting change. Thus, four data sets are obtained from each of the cameras, which are then merged in to two sets, one for each of the respective directions of the road:

- Direction I - from camera 1 to camera 2 - direction from G. M. Dimitrov Metro Station to Andrei Lyapchev Boulevard, for the time from 8:00 to 8:44;
- Direction II - from camera 2 to camera 1 - direction from Andrei Lyapchev Boulevard to G.M. Dimitrov Metro Station for the time from 8:45 to 9:30.

The time of the survey were determined by preliminary tests to determine the busy hourly periods in July 2018.

The extracted information in the Excel format is imported into the finishing database [19]. The information is presented in appropriate columns and tables.

The analysis of the data shows that the number of vehicles that have been captured via camera 1 is 251 and the number of vehicles that have been captured by camera 2 is 858 (Table 1). On the other hand, the total amount of vehicles that have passed through the lens of both of the License Plate Recognition cameras is 91. Therefore, only 36% of the vehicles pass subsequently through camera1 and camera 2. Moreover, the intensity of the traffic flow in the place where camera 2 is positioned is approximately 3.5 times higher than the intensity of the traffic flow that has been captured by camera 1. The vehicles that have been captured by the lens of camera 1 make up a little more than 10% of the total amount of the vehicles that have been detected by both of the cameras.

Table 1. Traffic flow passing through direction I.

camera 1	camera 2	Vehicles that have been captured on both of the cameras
251	858	91

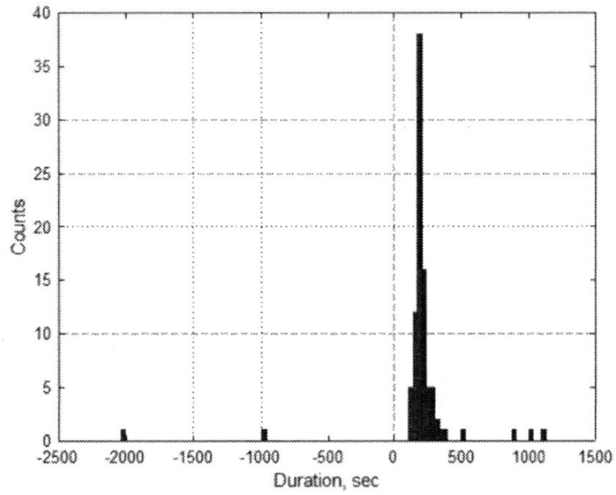

Figure. 2. Histogram of the time duration of the travel in direction 1.

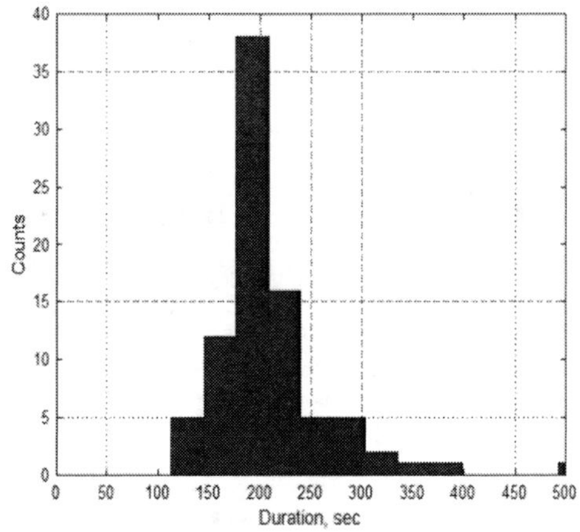

Figure 3. Histogram of the time duration of the travel (detailed).

The results from the analysis of the data about the time duration of the travel in direction 1 are shown on figure 2 and figure 3. Figure 2 and figure 3 show histograms of the time difference between the records of matched registration numbers from camera 1 and camera 2. Figure 3 presents the time distribution of the travel of all vehicles, while figure 4 shows the data distribution only in the range of 100 to 500 seconds.

However, in the records there are also exceptions. Two of the durations exceed 900 seconds (which is an indication of a stopover recorded time durations of the traveling vehicles are shorter (34 and 16 minutes), while other two of the recorded time).

The results from the analysis of the data show that the approximate time duration of the travel through that particular road section is around 3 minutes. The velocity of the vehicles is measured when the distance between the cameras (1.4km) is divided by the time duration of the travel. The distribution of the velocities of the vehicles is shown on figure 4. The approximate velocity that has been measured is around 24 km/h. It is to be taken into account that the shortest recorded time for passing through the section in between the cameras is 118 seconds. That travel time refers only to two of the recorded vehicles. It is estimated that these two vehicles have reached an average velocity of 42,71 km/h, which probably is due to the non-compliance with the speed limits. The traffic law offenders in the surveyed area are around 2-3% of the total number of vehicles that have passed between the cameras. That is an indication of a high rate of compliance with the restrictions and control of road transport.

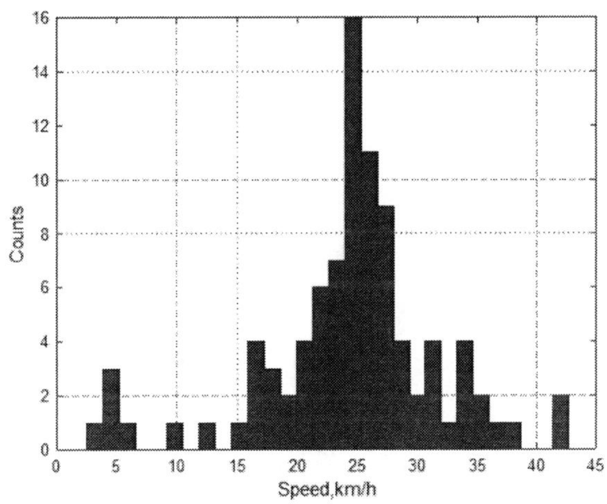

Figure 4. Histogram, showing the velocity of the vehicles in direction 1.

Analogically, the analysis of the data, acquired from the cameras in direction 2, shows that the number of vehicles that have been captured via camera 2 is 707, while the number of the vehicles that have been captured via camera 1 is 579 (table 2). The total number of the vehicles that have subsequently passed through camera 2 and camera 1 is 158. On the basis of that information it can be concluded that only 22% of the vehicles which have been captured by camera 2 have also been captured by camera 1. Moreover, the intensity of the traffic flow in the place where camera 2 is located is 1,2 times higher than the intensity in the place of camera 1. The vehicles that have passed through camera 2 are around 27% from the total amount of vehicles which have been captured by camera 1.

IOP Conf. Series: Materials Science and Engineering **618** (2019) 012061 doi:10.1088/1757-899X/618/1/012061

Table 2. . Traffic flow in direction II – from camera 2 to camera 1.

camera 1	camera 2	Vehicles that have been captured on both of the cameras
707	579	158

The results from the analysis of the data about the time duration of the travel in direction 2 are shown on figure 5 and figure 6. Figure 5 and figure 6 present histograms of the time difference between the records of the recurrent registration numbers from camera 2 and camera 1. Figure 5 shows the distribution of the time duration of the travel for all vehicles, while figure 6 presents the data of the distribution only in the range from 100 to 500 seconds.

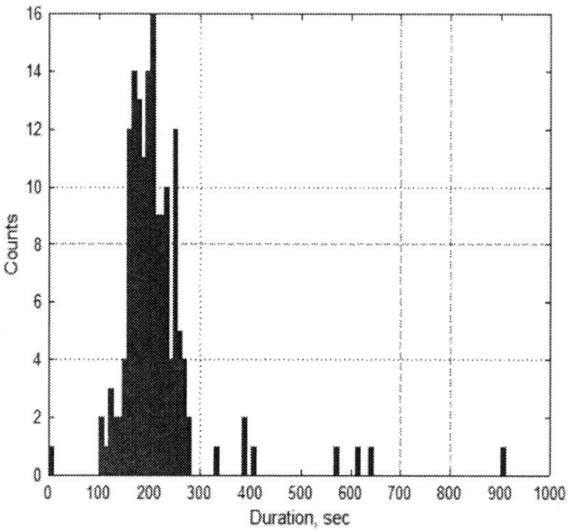

Figure. 5. Histogram of the time duration of the travel in direction 2 (main).

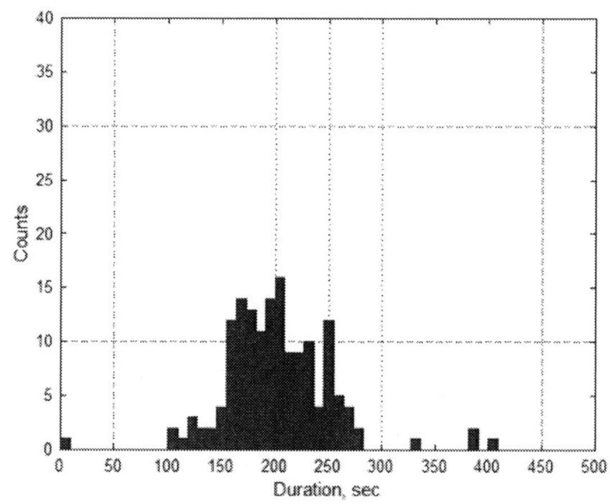

Figure 6. Histogram of the time duration of the travel in direction 2 (detailed).

The results from the analysis of the data show the average time duration of the travel in the studied section of the road is about 3 minutes. The velocity of the vehicles is analogically estimated as the velocity of the vehicles in direction 1. The distribution of the velocity of the vehicles in direction 2 is shown on figure 7. The average velocity of the traffic flow is around 25 km/h.

Unlike in direction 1, the minimal time duration of the travel in direction 2 is only 106 seconds, while the average velocity is 47,54 km/h. A total number of 6 vehicles during the conducted research have been traveling with a velocity above 40 km/h, which is probably due to non-compliance with the speed limits. The traffic law offenders in direction 2 are 3.79% from the total number of vehicles.

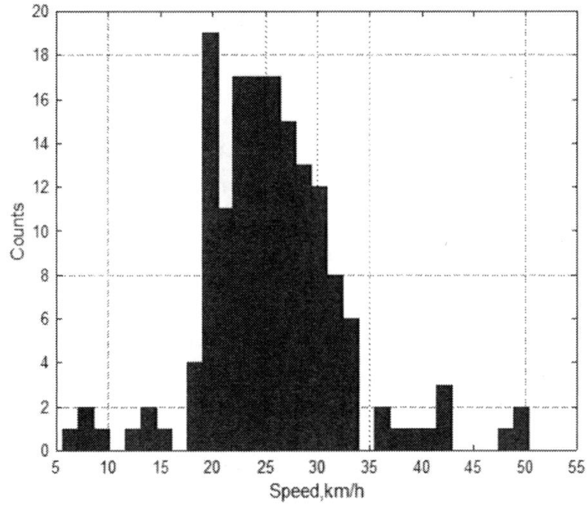

Figure 7. Histogram, showing the velocity of vehicles in direction 2.

4. Conclusion

The developed intelligent system for recording and recognition of registration numbers provides the opportunity for statistical analysis to determine traffic performance and to take specific measures so as to improve the organization and increase the traffic safety.

The research and analysis show the duration of the travel and the velocity of the vehicles, passing through a particular section of Kliment Ohridski Blvd in July 2018. The data acquired from the research will be used for the management and optimization of the transport flow, which will increase the safety and reduce the impact on the environment in the researched area.

References

[1] W. McShane, R. Roess. *Traffic engineering*. Prentice-Hall. Englewood Cliffs, N.J, (1990).
[2] *Highway Capacity Manual. TRB*, National Research Council, Washington, 2010.
[3] Kerner B., *Introduction to modern traffic flow theory and control*, Berlin, Springer, 2009.
[4] A.D. May, *Traffic Flow Fundamentals*, Prentice Hall, Englewood Cliffs (1990).
[5] *Traffic Detector Handbook*: Third Edition, US Department of Transportation, October 2006.
[6] *Traffic Monitoring Guide*, Federal Highway Administration, Washington, D.C., 2001.
[7] *A Summary of Vehicle Detection and Surveillance Technologies use in Intelligent Transportation Systems, The Vehicle Detector Clearinghouse*, updated August 2007, Available online: https://www.fhwa.dot.gov/policyinformation/pubs/vdstits2007/vdstits2007.pdf.
[8] Leduc G., *Road Traffic Data: Collection Methods and Applications*, European Communities, 2008.
[9] Guerrero-Ibáñez J., Zeadally S., Contreras-Castillo J., *Sensor Technologies for Intelligent Transportation Systems* , Sensors 2018, 18, 1212; doi:10.3390/s18041212 www.mdpi.com/journal/sensors
[10] D. N. Saliev, *Modeling of road traffic in case of dangerous occurrences,* Dissertation, A dissertation submitted in Technical University - Sofia, 2013.
[11] 11.UNECE - *Handbook on Statistics on Road Traffic, Methodology and Experience*, United Nations Economic Commission for Europe (UNECE) – Transport Division, WP6 Task Force on Road Traffic, Geneva, June 2007
[12] Krile, R., Todt, F., and Schroeder, J. 2015. *Assessing Roadway Traffic Count Duration and*

Frequency Impacts on Annual Average Daily Traffic Estimation: Assessing Accuracy Issues Related to Short-Term Count Durations. Federal Highway Administration. FHWA-PL-16-008.

[13] L.A. Klein, M.K. Mills, D.P. Gibson, *Traffic Detector Handbook*, vol. I (third ed.), Turner-Fairbank Highway Research Center, McLean, VA (2006)

[14] A. A. Shafie, M. H. Ali, Fadhlan Hafiz, Roslizar M. Ali, *"Smart Video Surveillance System For Vehicle Detection And Traffic Flow Control"*, Journal of Engineering Science and Technology Vol. 6, No. 4 (2011) 469 – 480

[15] http://www.hikvision.com/UploadFile/image/2015012013544861055.pdf

[16] F. Ryan, Olariu S., *On a Variant of the Mobile Observer Method*, IEEE Transactions on Intelligent Transportation Systems (Volume: 18 , Issue: 2 , Feb. 2017), pp 441 – 449.

[17] Kellner F., *Exploring the impact of traffic congestion on CO2 emissions in freight distribution networks*, Springer Berlin Heidelberg, 2016, Print ISSN 1865-035X, Online ISSN - 1865-0368.

[18] Guerrero-Ibáñez, J.A.; Flore-Cortés, C.; Zeadally, S. Vehicular ad Hoc Networks (VANETs): Architecture, *Protocols and Applications. In Next-Generation Wireless Technologies 4G and Beyond, 1st ed.*; Chilamkurti, N., Chaouchi, H., Zeadally, S., Eds.; Springer: London, UK, 2013; pp. 49–70, ISBN 978-1-4471-5163-0.

[19] Yordanov R., Miletiev R. Damyanov I., Iontchev E., *Measurement and analysis of the road traffic* - XXVII International Scientific Conference Electronics - ET2018, September 13 - 15, 2018, Sozopol, Bulgaria.

IOP Conf. Series: Materials Science and Engineering **618** (2019) 012062 doi:10.1088/1757-899X/618/1/012062

Optimal Control Synthesis of Semi Active Vehicle Suspension

I Angelov

Technical University of Sofia, 8 Kl. Ohridski Blvd, Sofia 1000, Bulgaria

E-mail: ivvoangelov@gmail.com

Abstract. The paper discusses the problems related with the synthesis of a linear quadratic (LQ) regulator of a semi active vehicle suspension. The synthesis combines the control through output variables with such by the input excitation i.e. classical LQ regulator with Compensator. The physical nonlinearities of the controllable magneto-rheological semi-active damper are taken into an account with the including in the feedback control circuit an inverse damper model. . A reduced suspension model is used to illustrate the synthesis (quarter car model).

1. Introduction

The quality criteria of the vibroisolation usually lay contradictory requirements for the vehicle's suspension synthesis. The ensuring enough ride comfort as generally requires a softer suspension, while the better vehicle stability and respectively better road-tire friction require a harder suspension. The compromise solutions in the conventional passive implementations are with limited possibilities. The usage of active or semi-active elements in vehicle suspensions gives considerable bigger potential in the solution of this problem. The semi-active vibroisolation is widely used because of its reliability, low price, and long exploitation period. It is known that due to the phase shifting, during some intervals of the oscillation time period, the damper directs energy into the object of vibroisolation instead of dissipating it. The main idea of the semi-active control is the reduction of the damping coefficient in these time period, which will decrease the amount of the energy transferred to the object [1÷4].

There are many types of the control laws [5÷8], but the main problem consists in the fact that, when the control is targeted to the comfort criterion the stability makes worse and vice versa. For this reason the synthesis of the optimal control requires a compromise solution satisfying both the quality factors simultaneously [9, 10, 11]. Generally the control based on a linear quadratic regulator - LQR is consider as an optimal and minimizes an integral of a time function formed by scaled variables describe the quality criteria [12÷16].

2. Dynamic model of the object

2.1. Quarter car model

On figure 1 is shown the quarter car suspension model (related to one of the wheels) with semi-active suspension, achieved by damper with controlled rheology. This model is enough adequate in case of constructive selection of the suspension performance, securing approximate independence of the motion of the front and the rear axis of the vehicle. At the same time due to its simplicity it is very appropriate for a real-time control. The results can be easily adapted for more complex models.

IOP Conf. Series: Materials Science and Engineering **618** (2019) 012062 doi:10.1088/1757-899X/618/1/012062

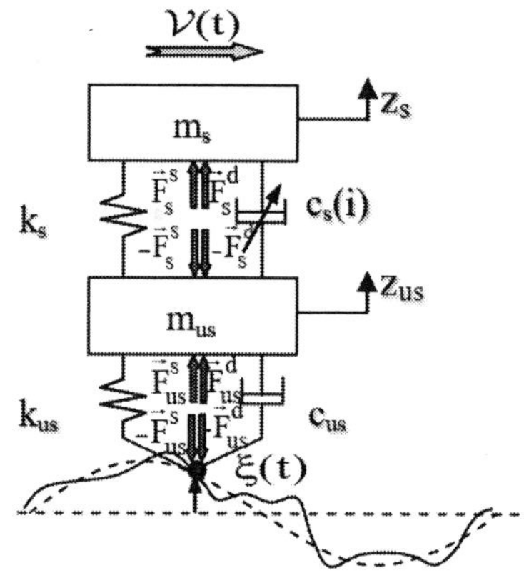

Figure1. Quarter car model

Parameters of the model are:

z_{us} and z_s - vibro-displacements in the vertical plane of the sprung and unsprung masses.

m_{us}, m_s - sprung and unsprung masses;

k_{us}, k_s, c_{us}, $c_s(i)$ - tire and suspension spring and damping coefficients;

$$s(t)=\begin{cases} 0.5a_0t^2 & , \text{ за } t<V_{max}/a_0 \\ V_{max}t-\dfrac{V_{max}^2}{2a_0}, & \text{ за } t\geq V_{max}/a_0 \end{cases}$$ - motion law, a_0 - acceleration, V_{max} - maximal velocity;

$\xi(t)=\xi_h(s)+\xi_s(s)$ - kinematical excitation from the road roughness, presented as a sum of determined and stochastic component:

$\xi_h = H_h\sin(2\pi s/L)$ models the road roughness as a harmonic function with amplitude H_h (m) and wave length L (m),

$\xi_s(s)$ kinematical excitation modeling the road roughness according ISO 8608 [19];

$\vec{F}_{us}$, $\vec{F}_s$ - forces transmitted through the tire to the unsprung mass and respectively from it to the sprung mass: $F_{us}^s = k_{us}(\xi - z_{us})$, $F_{us}^d = c_{us}(\dot{\xi}-\dot{z}_{us})$; $F_s^s = k_s(z_{us}-z_s)$, $F_s^d = f(i, z_{us}, z_s, \dot{z}_{us}, \dot{z}_s)$ where the indexes 's' and 'd' show respectively the spring and damping components and 'i' is the control signal.

2.2. Model of the semi-active damper

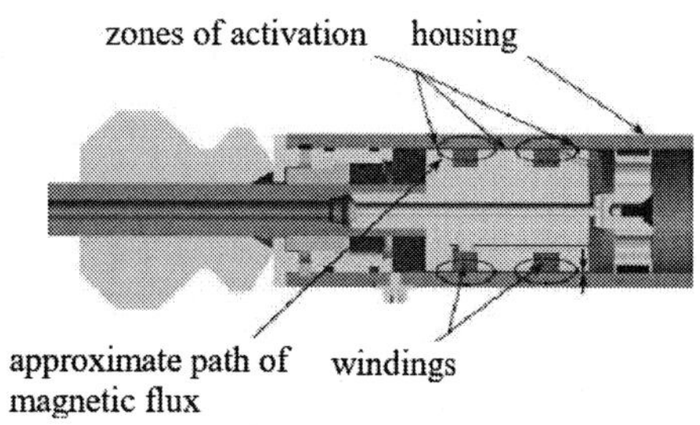

Figure 2. Scheme of MR damper

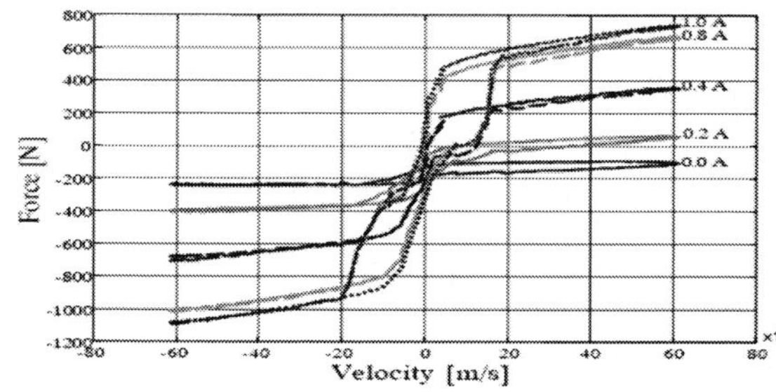

Figure 3. Performance of MR damper

The essence of the semi-active suspension consists in damper with variable rheology. The damping force depends on the kinematical performance on its deformation, and from the control signal.

The magneto-rheological (MR) dampers – figure 2 are more appropriate solution for the considered application. At them by feeding a control current signal is creates magnetic field which forms domains from the ferrite micro particles, contained in silicon based suspension. In this way the damping coefficient becomes a function of the control signal and varying in definite boundaries.

The experimental performance, that show the dependency of the force vs. deformation velocity for different values of the control current, for MR damper type RD-1005-3 manufactured by „LORD Corporation"-USA are shown on the figure 3.

[20] proposes a dynamic model of controllable MR damper based on the approximation of the hysteresis suggested by Bouc-Wen [21, 22]. Its mechanical analogue is shown on figure 4.

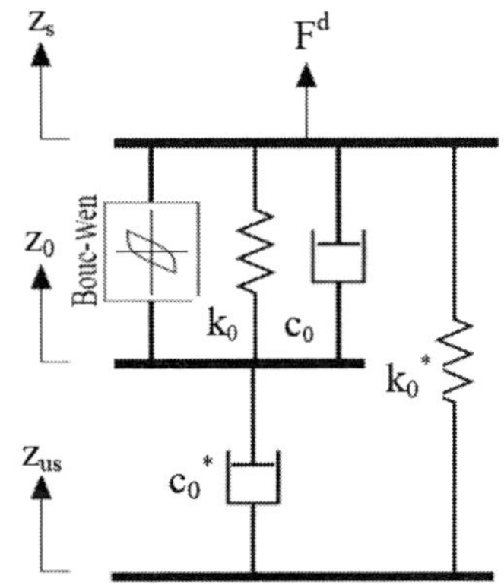

Figure 4. Bouck-Wen-Spenser mechanical model of MR damper

The damper force is calculated by the following equation:

$$F^d = k_0\left(z_0 - z_s\right) + c_0\left(\dot{z}_0 - \dot{z}_s\right) - k_h\zeta + k_0^*(z_{us} - z_s + \delta_0) = k_0^*(z_{us} - z_s + \delta_0) - c_0^*\left(\dot{z}_0 - \dot{z}_{us}\right), \quad (1)$$

where k_h (N/m) is the elastic coefficient connected with the value on which the angle of the characteristic is changed and δ_0 (m) takes into account the hydraulic accumulator.

From the force balance upon the level z_0 is obtained:

$$\dot{z}_0 = \left[c_0^*\dot{z}_{us} + c_0\dot{z}_s - k_0\left(z_0 - z_s\right) + k_h\zeta\right] / \left(c_0 + c_0^*\right). \quad (2)$$

The evolutionary variable is given by the dependency:

$$\dot{\zeta} = -\gamma\left|\dot{z}_0 - \dot{z}_s\right|\zeta\left|\zeta\right|^{\chi-1} + \beta\left(\dot{z}_0 - \dot{z}_s\right)\left|\zeta\right|^{\chi} - v\left(\dot{z}_0 - \dot{z}_s\right) = \left\{\left[\beta\text{-}sign\left(\left(\dot{z}_0 - \dot{z}_s\right)\zeta\right)\gamma\right]\left|\zeta\right|^{\chi} - v\right\}\left(\dot{z}_0 - \dot{z}_s\right), \quad (3)$$

where γ (m), β (m), v and χ are coefficients forming the hysteresis.

The values of k_h, c_0 and c_0^* as functions of the control current are given by the first order approximations:

$$k_h = k_h^c + k_h^v i', \; c_0 = c_0^c + c_0^v i', \; c_0^* = c_0^{*c} + c_0^{*v} i'. \quad (4)$$

Here i' is function taking into account the process of rheological equilibrium after the aperiodic dependency from first order:

$$\dot{i}' = \left(i - i'\right) / \tau_r, \; \tau_r \text{ – time-sample of the process, } i \in \left[0, i_{max}\right]. \quad (5)$$

2.3. *Inverse model of MR damper*

The MR damper has essential nonlinear characteristics and physical limitations. For the control purpose an inverse dynamical model is required. For a particular state of the system the model assures corresponding optimal damper force, calculated by the controller. It is obvious that due to the physical limitations, the optimum force will not be generated in some situations and generally the control will be quasi-optimal.

Due to the complexity of the problem, a neural network model is used [12, 23, 24]. Its structure is shown on figure 5.

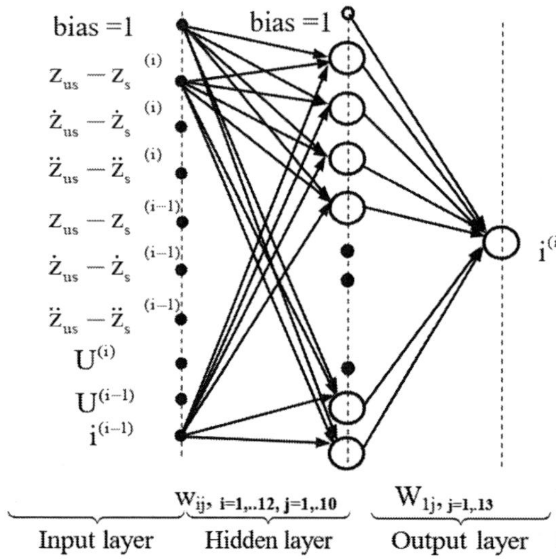

Figure 5. Inverse neural networks model of MR damper

In the input layer the value of the deformation of the damper and its first and second derivatives with respect to time are entered, as well as the control signal U (the damper force). The output layer represents a linear activation function and determines the control current, which generates the current damper force U_i.

The hidden layer consists of 12 neurons with non-linear logical sigmoid activation functions $g_i(x) = \dfrac{\theta}{1-e^{\sigma x}} - 1$, where $\theta = 2$, and σ is positive constant. The determination of the weight coefficients for the hidden layer w_{ij}, i=1,..12, j=1,..10 and the output layer W_{1j}, j=1,..13, is done by the Levenberg-Margquardt method and the Jacobian is calculated by back propagation method [25].

2.4. State space presentation

$$\begin{aligned} \dot{X} &= AX + B_\Lambda \Lambda + B_U U,\ X(0)=X_0 \\ Y &= CX + D_\Lambda \Lambda + D_U U \end{aligned} \tag{6}$$

where:

- $X = \begin{bmatrix} z_{us} & z_s & \dot{z}_{us} & \dot{z}_s \end{bmatrix}^T$ - state vector; $\Lambda = [\xi, \dot{\xi}]^T$ - input exitation; $U = F_s^d$ - the control;
- $Y = \begin{bmatrix} z_{us} & z_s & -F_{us} & \ddot{z}_s \end{bmatrix}^T$ - output vector which consists of characteristics that define the quality of the suspension – the motions z_{US} and z_S, the dynamical component of the normal force $-\vec{F}_{us}$, which determines the friction tire-road and the acceleration of the sprung mass $\ddot{z}_s$ which determines the ride comfort;
- matrices of the state A, the control B_U and the input B_Λ are :

$$A = \begin{bmatrix} 0_{2x2} & I_2 \\ -m^{-1}k & -m^{-1}c \end{bmatrix},\ B_U = \begin{bmatrix} 0_{2x1} \\ m^{-1}d_U \end{bmatrix},\ B_\Lambda = \begin{bmatrix} 0_{2x2} \\ m^{-1}d_\Lambda \end{bmatrix};$$

- and the matrices of the output variables are:

$$C=\begin{bmatrix} I_2 & & 0_{2x2} & \\ k_{us} & 0 & c_{us} & 0 \\ k_s/m_s & -k_s/m_s & 0 & 0 \end{bmatrix},\ D_U=\begin{bmatrix} 0_{3x1} \\ 1/m_s \end{bmatrix},\ D_\Lambda=\begin{bmatrix} 0_{2x2} & \\ -k_{us} & -c_{us} \\ 0 & 0 \end{bmatrix};$$

- mass, spring and damping matrixes and matrix of the excitation and control:

$$m=\begin{bmatrix} m_{us} & 0 \\ 0 & m_s \end{bmatrix},\ k=\begin{bmatrix} k_{us}+k_s & -k_s \\ -k_s & k_s \end{bmatrix},\ c=\begin{bmatrix} c_{us} & 0 \\ 0 & 0 \end{bmatrix},\ d_\Lambda=\begin{bmatrix} k_{us} & c_{us} \\ 0 & 0 \end{bmatrix},\ d_U=\begin{bmatrix} -1 \\ 1 \end{bmatrix};$$

^{T} – transpose symbol, *0* and *I* are zero and eye matrix with the shown dimensions.

3. Linear quadratic regulator synthesis [12, 13,15, 18]

The synthesis of the control aims minimizing of a square function of the characteristics Y, proving the quality of the suspension and from the control U representing the MR damper force:

$$J=X^T(T_r)FX(T_r)+\int_0^{T_r}\left[\left(CX+D_UU+D_\Lambda\Lambda\right)^TQ\left(CX+D_UU+D_\Lambda\Lambda\right)+U^TRU\right]dt \quad (7)$$

where: T_r is the time interval for the control process,
$Q,F\in\mathbb{R}^{4x4}$ are positive semi-defined
$R\in\mathbb{R}$ is positive defined (in this particular case - constant) weight matrices.

There it is accepted that $F=0_{4x4}$, and Q diagonal matrix.
After opening the brackets is obtained:

$$J=\int_0^{T_r}\left[\left(X^TQ_UX+2X^TN_{XU}U+U^TR_UU\right)+\left(2X^TN_{X\Lambda}\Lambda+2U^TN_{U\Lambda}\Lambda+\Lambda^TR_\Lambda\Lambda\right)\right]dt \quad (8)$$

where $N_{X\Lambda}=C^TQD_\Lambda\in\mathbb{R}^{4x2}$, $N_{U\Lambda}={D_U}^TQD_\Lambda\in\mathbb{R}^{1x2}$, $R_\Lambda={D_\Lambda}^TQD_\Lambda\in\mathbb{R}^{2x2}$. (9)

The Hamiltonian and the conjugated system are:

$$\bar{H}=\left(X^TQ_UX+2X^TN_{XU}U+U^TR_UU\right)+\left(2X^TN_{X\Lambda}\Lambda+2U^TN_{U\Lambda}\Lambda+\Lambda^TR_\Lambda\Lambda\right)+P^T\left(AX+B_UU+B_\Lambda\Lambda\right),$$

$$\dot{P}=-\frac{\partial\bar{H}}{\partial X}=-\left(2Q_UX+2N_{XU}U+A^TP+2N_{X\Lambda}\Lambda\right). \quad (10)$$

From the Maximum principles $\left.\frac{\partial\bar{H}}{\partial U}\right|_{U=U_{extr}}=0\Rightarrow U_{extr}\underset{\frac{\partial^2\bar{H}}{\partial U^2}=2R_U>0}{=}U_{opt}=-{R_U}^{-1}\left[{N_{XU}}^TX+{B_U}^TP+N_{U\Lambda}\Lambda\right]$

(11)

For the conjugated system is searches solution in the form:

$$P=2\left(KX-\Psi\right)\Rightarrow\dot{P}=2\left(\dot{K}X+K\dot{X}-\dot{\Psi}\right). \quad (12)$$

Combining (10)÷(12) are obtained two Riccati equations (for K=const and Ψ =const):

$$\left|\begin{array}{l}\tilde{K}A + A^T\tilde{K} + Q_U - \left(\tilde{K}B_U + N_{XU}\right)R_U^{-1}\left(N_{XU}^T + B_U^T\tilde{K}\right) = 0 \\ \tilde{\Psi} = -\left[A^T - \left(\tilde{K}B_U + N_{XU}\right)R_U^{-1}B_U^T\right]^{-1}\left[\left(\tilde{K}B_U + N_{XU}\right)R_U^{-1}N_{U\Lambda} - \left(\tilde{K}B_\Lambda + N_{X\Lambda}\right)\right]\Lambda = 0\end{array}\right. . \quad (13)$$

Matrix $\tilde{K}$ is defined by (13) based on the numerical computation of the vectors of Hamiltonian matrix [26].

Combining (11), (12) and second equation of (13) is obtained the optimal control:

$$\left|\begin{array}{l}U_{opt} = -G_X X - G_\Lambda \Lambda \\ G_X = R_U^{-1}\left(N_{XU}^T + B_U^T K\right) \\ G_\Lambda = R_U^{-1}\left\{N_{U\Lambda} + B_U^T\left[A^T - G_X^T B_U^T\right]^{-1}\left[G_X^T N_{U\Lambda} - \left(\tilde{K}B_\Lambda + N_{X\Lambda}\right)\right]\right\}\end{array}\right. . \quad (14)$$

The kinematic excitation due to the road roughness may be identified using the accelerations of the sprung and unsprung masses obtained from accelerometers:

$$\tilde{\xi} = w_{us}(s)\tilde{\ddot{z}}_{us} + w_{us}(s)\tilde{\ddot{z}}_s, \quad (15)$$

where transfer functions are: $w_{us}(s) = \dfrac{m_{us}s^2 + c_{us}s + k_{us}}{s^2\left(c_{us}s + k_{us}\right)}, w_{us}(s) = \dfrac{m_s}{c_{us}s + k_{us}}$ and with ~ is denoted the Laplace transform image.

On figure 6 is given the structural scheme of the control.

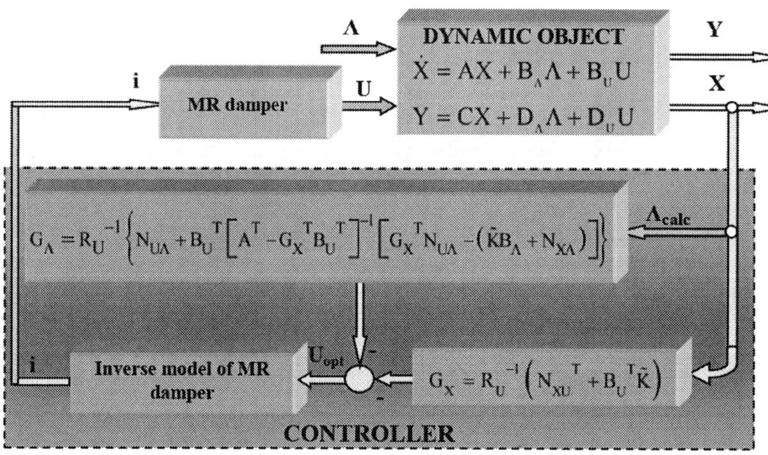

Figure 6. Structural scheme of the control

4. Numerical example

4.1. Description of the task

- It is consider a model with the following characteristics:

  m_1=40 kg; m_2=350 kg; k_1=1.6e^5 N/m; k_2=0.19e^{5}N/m; c_1=110 Ns/m.

- The parameters of the MR damper model are given in table 1

IOP Conf. Series: Materials Science and Engineering **618** (2019) 012062 doi:10.1088/1757-899X/618/1/012062

Table 1. Parameters of the MR damper model

c_0^c, Ns/m	c_0^V, Ns/mA	k_0, N/m	k_h^c, N/m	k_h^V, N/mA	c_0^{*c}, Ns/m	c_0^{*V}, Ns/mA	k_0^*, N/m
780	2230	1400	14500	71500	28000	500	540
χ	τ_r, s	γ, m^{-2}	β, m^{-2}	δ	δ_0, m	i_{max}, A	
2	0,0053	0,5 10^6	0,8 10^6	170	0	1,2	

- The parameters of the motion law are: a_0=1.1 m/s^2, V_{max} =130km/h, T_k=25s.
- The kinematic excitation due to the road roughness as function of the time is shown on figure 7.

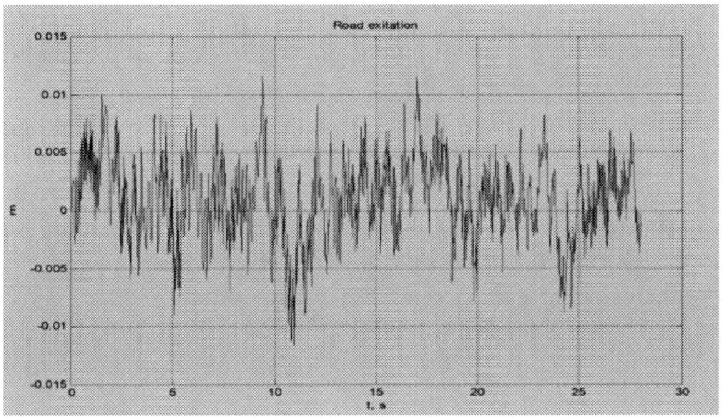

Figure 7. Kinematic excitation vs. time

- The weighted matrix in the object function are

Q=diag([1.34e4, 0.258e4, 2.462e-6, 3.453]), R=1.38e-6.

4.2. Results

Characteristics	Passive suspension b_2=1505Ns/m		With semi active LQR control	
	Z_{us}	Z_{us}	Z_{us}	Z_{us}
$\int_0^{T_k} Z(t)dt$, ms	0,346	0,23	0,378	0,0658
Max($[\lvert Z\rvert,\lvert \dot{Z}\rvert,\lvert \ddot{Z}\rvert]$),[m,m/s,m/s^2]	[0,027;1,03;79,3]	[0,036;0,27;4,67]	[0,031;1,32;104,5]	[0,005;0,12;7,84]
RMS($[Z,\dot{Z},\ddot{Z}]$)[m,m/s,m/s^2]	[0,011;0,29;16,4]	[0,01;0,082;1,5]	[0,012;0,34;24,1]	[0,022;0,04;1,66]
RMS(F_{us}^d), N	797		768,9	
J	**0,01602**		**0,01322**	

Acknowledgements: This work was supported by OP "Science and education for smart growth " – centers of excellence TU Sofia „National center of Mechatronics and clean technologies", BG05M2OP001-1.001-0008-C01 Sect. Design, synthesis and testing of noise and vibration protection

References

[1] Karnopp D, 1983 Active Damping in Road Vehicle Suspension Systems *Vehicle System Dynamics* **12** pp 291-316
[2] Karnopp D, 1990 Design principles for vibration control systems using semi-active dampers *Transactions of the ASME - J. Dynamic Systems, Measurement, and Control* **112** pp 448-455
[3] Yi K., Wargelin M., Hedrick K. 1992 Semi-active suspensions to reduce road damage: theoretical design and implementation, in *Heavy vehicles and roads: technology, safety and policy* Thomas Telford London pp 122-129
[4] Genov J. et al. 2008 Analysis and Control Laws of Semi-Active Vehicle Suspension System

with Rending an account of the Physical Characteristics of Magneto-Rheological Dampers *Proc. of the 13th Int. Conf. on Applied Mechanics and Mechanical Engineering (AMME-13) 27-29 May 2008 M.T.C. Cairo Egypt* pp 1333-1348

[5] Ahmadian M. 1997 Hybrid Semiactive Control to Secondary Suspension Applications *Proc. of Six ASME symposium on Advanced Automotive Technologies Texas 1997* pp 1-8

[6] Guglielmino E. et al. 2008 *Semi-active Suspension Control - Improved Vehicle Ride and Road Friendliness* Springer-Verlag London Ltd. p 302

[7] Savaresi S. et al. 2010 *Semi-Active Suspension Control Design for Vehicles* Butterworth-Heinemann is an imprint of Elsevier Ltd. Oxford UK p 232

[8] Preumont A. 2011 *Vibration Control of Active Structures an Introduction* Springer p 453

[9] Venkov G., Tashkov S., Genov J. 2007 Optimal Multicriterial Synthesis of Vehicle Suspension *Proc. of the 32-st Int. Conf. Applications of Mathematics in Engineering and Economics'32, Sozopol, Bulgaria, 2006* Softtrade, Sofia pp 217-221.

[10] Genov J., Tashkov S. 2015 Multi-Objective Synthesis of Vehicle Semi-Active Suspension Control Represented In A Vertical Plane Part 1 Model and Algorithm for Multi-Objective Synthesis *Proc. of Sc. Conf. on Aeronautics, Automotive and Railway Eng. and Technologies BulTrans-2015,* TU-Sofia pp 77-83

[11] Mastinu G., Gobbi M., Miano C. 2006 *Optimal Design of Complex Mechanical Systems. With Applications to Vehicle Engineering*, Springer-Verlag Berlin Heidelberg p 392

[12] Genov J., Tashkov S., Venkov G., 2009 On the Synthesis of Optimal and Quazi-optimal Control of Semi-Active insulation System *35th Int. Conf. Appl. of Math. in Eng. and Economics - AMEE-2009*, AIP Conference Proceedings **1184** pp 32-38.

[13] Genov J. et al. 2009. On the Synthesis of Optimal and Quazi-Optimal Control of Semi-Active Vibro-insulation System *11th National Congress on Theoretical and Applied Mechanics, 2-5 Sept. 2009, Borovets Bulgaria* pp 1-12

[14] Nguyen T.-A. 2006 *Application of Optimization Methods to Controller Design for Active Suspensions* Doktor-Ingenieurs genehmigte Dissertation TU Cottbus p 152;

[15] Genov J., Tashkov S. 2012 Quasi Optimal Control of Magneto-rheological Damper. *Proc. of Int. Conf. and Exh. on New Actuators and Drive Systems ACTUATOR 12, Bremen, Germany, 18-20 June 2012* pp 612-615

[16] Pareto V. 1896 *Cours D'économie politique Tome Premier* F. Rouge Libr. Édit. Lausanne p 438

[17] Pareto V. 1897 *Cours D'économie politique Tome Second* F. Rouge Libr. Édit. Lausanne p 430

[18] Genov J., Tashkov S. 2015 Multi-Objective Synthesis of Vehicle Semi-Active Suspension Control Represented in a Vertical Plane Part 2 Synthesis of Linear Quadratic Regulator *Proc. of Sc. Conf. on Aeronautics, Automotive and Railway Eng. and Technologies BulTrans-2015* pp 84-88

[19] International Standard ISO 8608 1995 *Mechanical vibration and shock – Road Surface Profiles Reporting of Measured Data*, Int. Org. of Standardization Central Secr., Switzerland, p 30

[20] Spenser Jr., B. et al. 1997 Phenomenological Model of a Magnetorheological Damper *Journal of Engineering Mechanics* **123** pp 230-238

[21] Bouc R. 1967 Forced vibration of mechanical systems with hysteresis. *Proceedings of the Fourth Conference on Non-linear oscillation, Prague, Czechoslovakia* p 315

[22] Wen Y. 1976 Method for random vibration of hysteretic systems *J. Eng. Mech.* **102** pp 249-263

[23] Kim B., Roschke P. 1999 Linearization of Magnetorheological Behaviour Using a Neural Network *Proc. of the American Control Conf., San Diego, June 1999* AACC pp 4501-4505

[24] Gilev B., Venkov G., Genov J. 2006 Neural Network Prediction of Hysteresis, *Proc. of the 31-st Int. Conf. Appl. of Math. in Eng. and Economics'31, Sozopol' 2005*, Softtrade pp 172-175

[25] Hinton G. 2013 *Computation in Neural Networks Lecture course CSC2535*, Dept. of Computer Science, University of Toronto, Canada

[26] Arnold W., Laub A. 1984 Generalized Eigen problem Algorithms and Software for Algebraic Riccati Equations, *Proc. IEEE*, **72**, No.12, pp 1746-1754

IOP Conf. Series: Materials Science and Engineering **618** (2019) 012063 doi:10.1088/1757-899X/618/1/012063

Comparative analysis of conventional and optimal methods for detail part design for light aircraft

A O Marinosyan[1] and H P Panayotov[2]

[1,2]Technical University – Sofia, Plovdiv Branch, Bulgaria

[1] sasho01@yahoo.com, [2] hristian@tu-plovdiv.bg

Abstract. The present paper evaluates the pros and cons of three different methods of detail design of parts for light aircraft: a conventional (heuristic) method, topology optimization and morphogenesis. This is done through applying these methods to solve the same design problem and analyze the results theoretically by means of finite element analysis under certain external loads. The theoretical results are compared to experimental using Digital Image Correlation (DIC) technique on real parts with the same loads applied. As a case study a control surface A-frame hinge similar to the lower rudder hinge of M-540 aircraft is considered.

1. Introduction

In the present study three different methods are used for light aircraft detail part design. In particular as a case study a typical control surface A-frame hinge is considered. Specifically the lower rudder hinge of M-540 light aircraft (Fig. 1, left) is examined. For light aircraft detail design usually heuristic and rational design is used, based on designer's experience rather than optimal design methods. Hence the aim of the study is a comparative analysis of conventional (non-optimal) and optimal design methods. Two different types of topology optimization methods are employed opposed to conventional design. The goal is to minimize the mass of the parts under same external loads mostly due to optimal material distribution. Additional constraints are imposed such as minimal displacements. In many cases topology optimization yields geometry that resembles natural shapes. Additive manufacturing such as 3D-printing is considered widely as a main method of production of optimal parts. The chosen design methods, both conventional and optimal are:

- Conventional. It is based on experience of skilled designers, and works on trial and error. Design organizations have accumulated that knowledge and distribute it to its employees as in-house design manuals. They provide allowables and guidelines for the design of different kinds of parts, as fittings, formers, ribs etc. Most times the proposed design has to be justified by simple stress calculations [3].

- Topology optimization. It may be divided into optimization problem formulation and running an optimization algorithm which suggests areas of material to be removed [2]. The solution requires additional initial and boundary conditions to be defined such as fixture, loading, material definition etc.

IOP Conf. Series: Materials Science and Engineering **618** (2019) 012063 doi:10.1088/1757-899X/618/1/012063

- Morphogenesis. It is basically another name of topological optimization, but the algorithm is capable of building structure by adding material within the design domain under pre-defined initial and boundary conditions [3],[4].

2. Detail design methods and part generation

For the abovementioned case study – A-frame part the used material is aluminum alloy with Young's modulus 7e+010 Pa, Poisson's ratio 0.346, density 2710 kg/m^3, yield strength 9.5e+007 Pa. This is a basic alloy, chosen to make possible both conventional and additive manufacturing of complex shaped parts. One load case is evaluated with two simultaneously applied perpendicular, in-plane forces at bearing hole. One with 1800N of magnitude, in plane, perpendicular to flight direction and one 180N of magnitude, in plane, parallel to the flight direction (Fig. 1, right.).

2.1. The *conventional part* was designed by heuristic approach and modeled using CATIA V5 CAD software by Dassault Systèmes®. The rationale behind the design is a strut-like structure for direct transfer of loads from the bearing hole to the attachment points – bolt holes. There is also a stabilizing connection between the struts, thus the name A-frame. The part has 8mm thickness allowance and was supposed to be fabricated by water jet cutting from 8mm thick raw plate (Fig. 1, right).

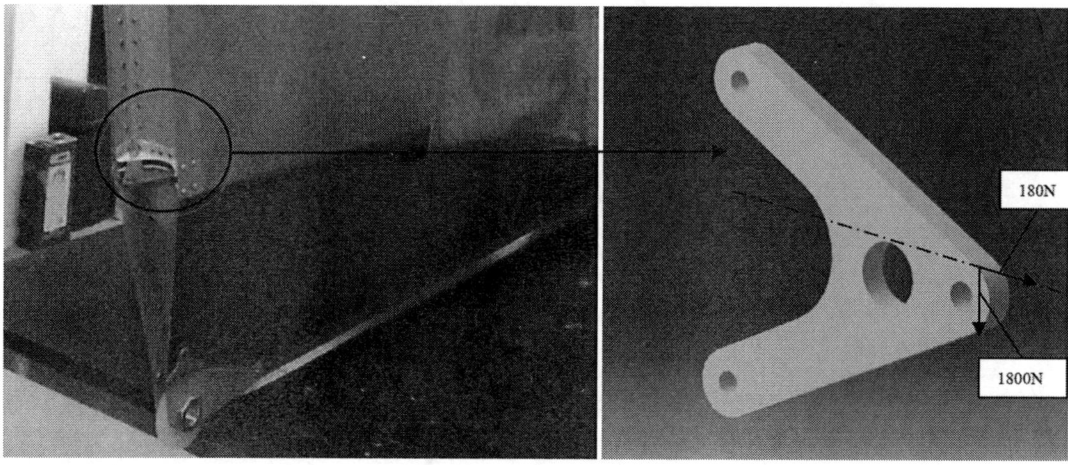

Figure 1. A-frame, generated by conventional approach

2.2. *Topology optimization* using Solid Works 2018 Topology Study software optimization module by Dassault Systèmes® was used to generate optimal part design. The algorithm removes under stressed material at certain external loads. The initial part has a triangular shape and allows the software to define the elements to be removed. The part is fixed to the attachment holes and the loads described are set (Fig. 2, left). The solver goal is Minimal mass with Displacement Constraint. In other words the Solid Works Topology study will try to find the stiffest structure possible given a certain amount of material removal. The final optimal part is generated and shown in the most right of Figure 2.

2.3. *Morphogenesis* using "Top 3D" MATLAB code by K. Liu. Topology optimization is a computational material distribution method for synthesizing structures without any preconceived shape. This freedom provides topology optimization with the ability to find innovative, high-performance structural layouts, which has attracted the interest of applied mathematicians and engineering designers. A topology optimization problem can be defined as a binary programming problem in which the objective is to find the distribution of material in a prescribed area or volume referred to as the *design domain*. A classical formulation, referred to as the *binary compliance problem*, is to find the "black and white" layout (i.e., solids and voids) that minimizes the work done by external forces (or compliance) subject to a volume constraint.

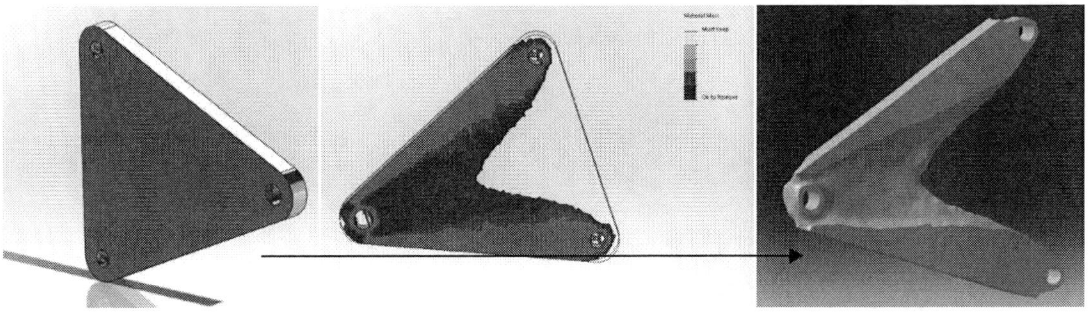

Figure 2. Topology-optimized A-frame generated in Solid Works

This code implements Sequential quadratic programming, Method of moving asymptote, and Optimality Criterion methods to the solution of the minimum compliance topology optimization problems. [4] The software requires definition of material properties and boundary conditions – fixtures and loads. Since there is not an option to define bolt fixture, the part is fixed at distributed zoned near the attachments, and the load is concentrated. The resulting part geometry form the solver (Fig. 3) can be exported in standard CAD-exchange formats. An obvious disadvantage of that product is low resolution of generated geometry.

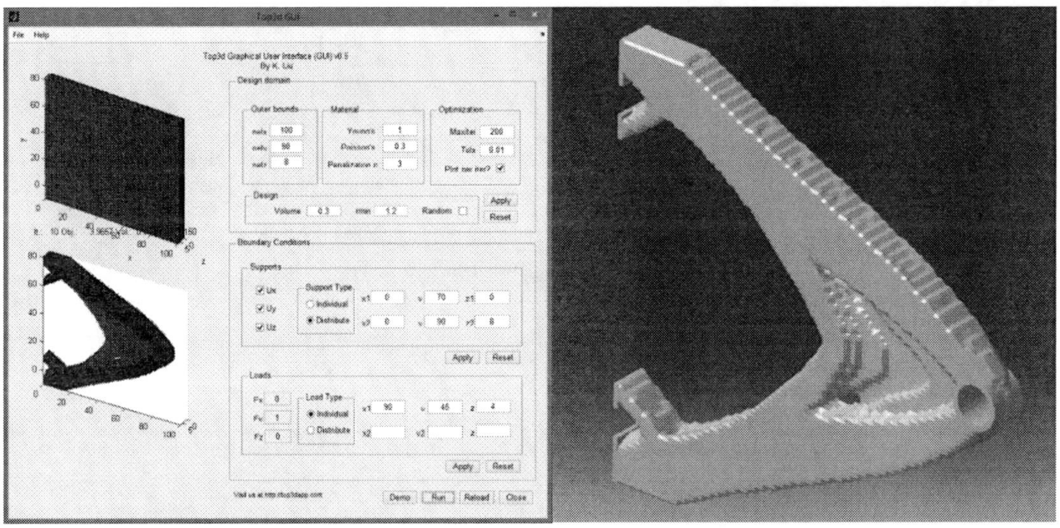

Figure 3. Morpho-generated A-frame by Top3D MATLAB optimizer

3. Finite element analysis

All parts are analysed purely theoretically in the Generative Part Structural Analysis module of CATIA V5 software tool using Finite Element Method (FEM) analysis. Since the topology optimization solvers do the synthesis of the parts it is appropriate to check the strength and stiffness of the designed parts using FEM analysis. The fixtures and loads shown on Fig. 1 are applied. The stress and displacement distributions are shown in Fig. 4 for all three parts.

The conventional part has the lowest maximum stress magnitude and the smallest maximum displacements. It is obvious from the stress distribution (Fig 4a) that there are highly loaded zones of the part and zones with low stresses. One could go for another design iteration and manually remove the low-stressed zones.

IOP Conf. Series: Materials Science and Engineering **618** (2019) 012063 doi:10.1088/1757-899X/618/1/012063

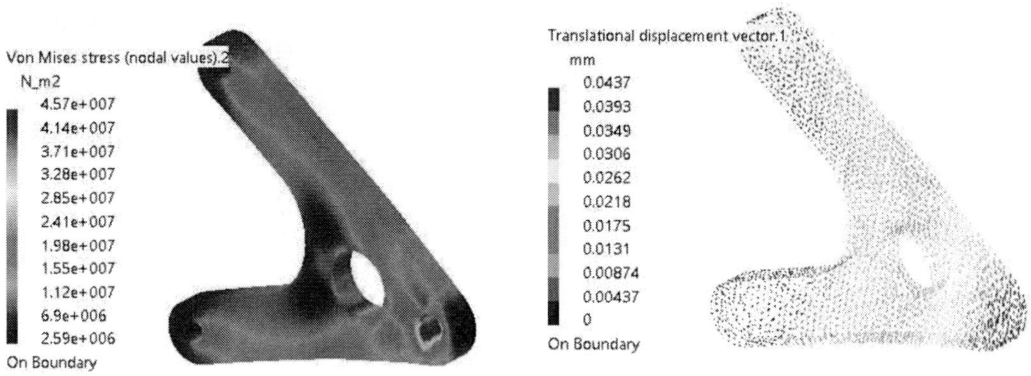

a) Conventional part

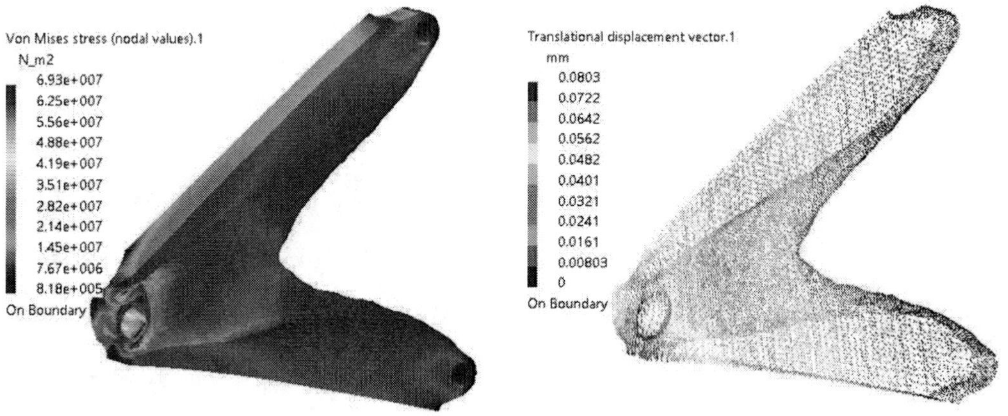

b) Topology-optimized part

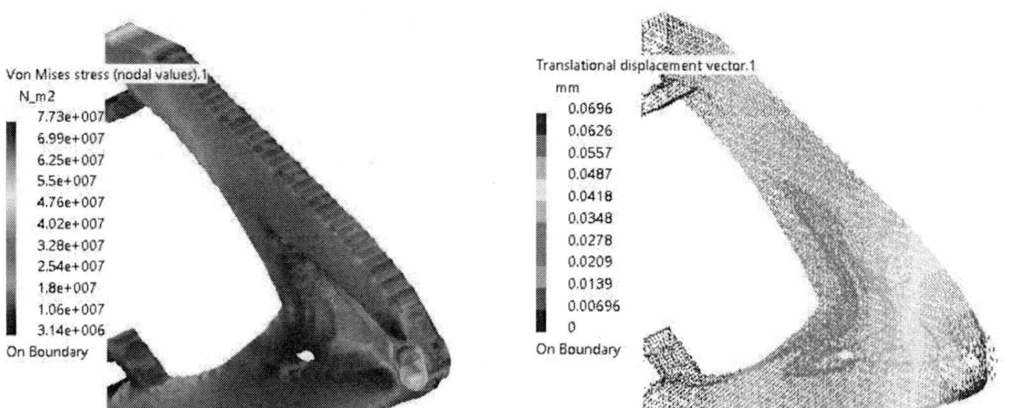

c) Morpho-generated part

Figure 4. Equivalent (von Mises) stresses (left) and displacements (right)

The optimal part, generated by Solid Works 2018 Topology Study application has greater magnitude of maximum stress and displacements than conventional part (Fig 4b). However it can be seen from the stress distributions that the maximal stresses are more evenly distributed at the "legs" of the A-frame. That is as a result of the optimizer. Nevertheless the stresses are concentrated at the loading mount.

The morpho-generated part has the biggest magnitude of maximum stress, but the maximum displacements are lower than Solid Works generated part (Fig. 4c). According to the colormap and the stress distribution it can be seen that the morpho-generated part has the most uniform stress distribution at satisfactory low displacements.

4. Experimental Digital Image Correlation (DIC) analysis

Digital Image Correlation (DIC) analysis is an experimental contactless, non-intrusive technique to determine displacement and strain distribution on deformed bodies. Two-dimensional (2-D) digital image correlation, which is used with a single camera, can measure only in-plane displacement/strain fields on plane objects. To overcome the drawback of 2-D digital image correlation, a 3-D digital image correlation technique, which combines digital image correlation with stereo vision and can measure the 3-D displacement field and surface strain field of 3-D object. The digital image correlation method uses the random speckle pattern to match the corresponding points precisely on two images [6]. The correlation of the corresponding points on the reference (non-deformed) and deformed image defines the displacement. In the present study 2D-DIC is used based on GOM Correlate software.

To determine experimentally the deformations of the designed parts a specially designed test stand is produced (Fig. 5). All three parts, conventional and optimal, are produced by milling and are shown on the right hand side in Fig. 6. The loads are generated by a hydraulic jack and read by a load cell. The resulting load is indicated on a monitor. The load case is the one described in Fig. 1. The loaded part is inclined at a certain angle ($\approx 6°$) to imitate both perpendicular forces at bearing hole.

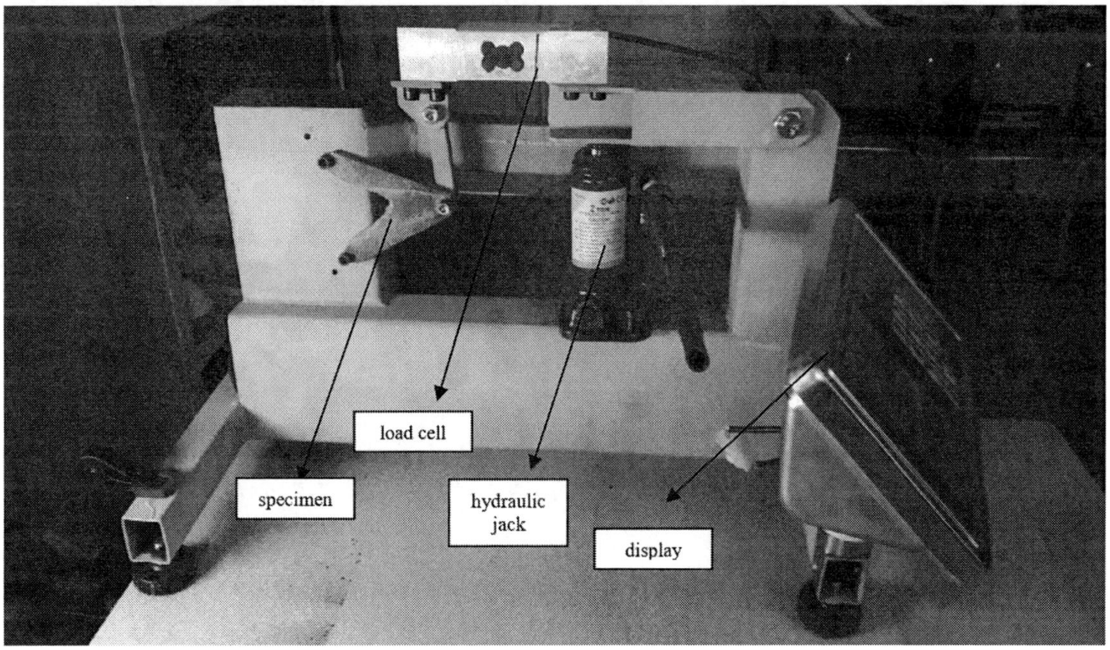

Figure 5. Experimental test stand

Figure 6 shows the experimental results using 2D-DIC technique on the designed parts.

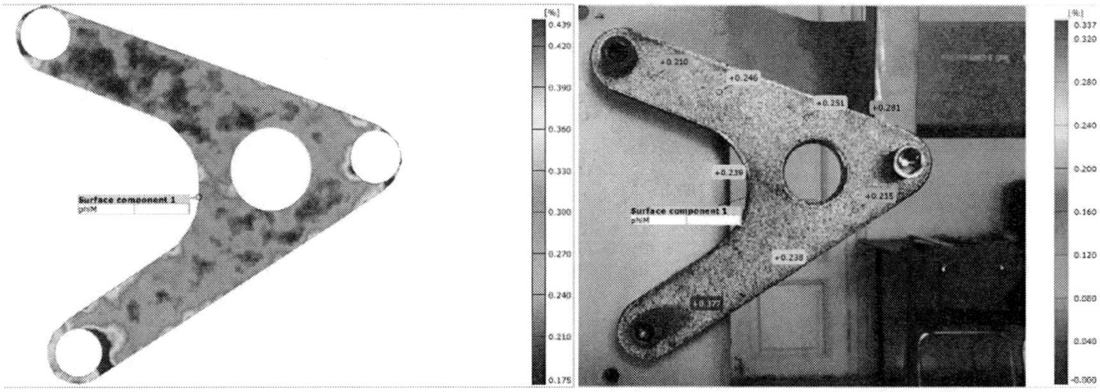

a) Conventional part

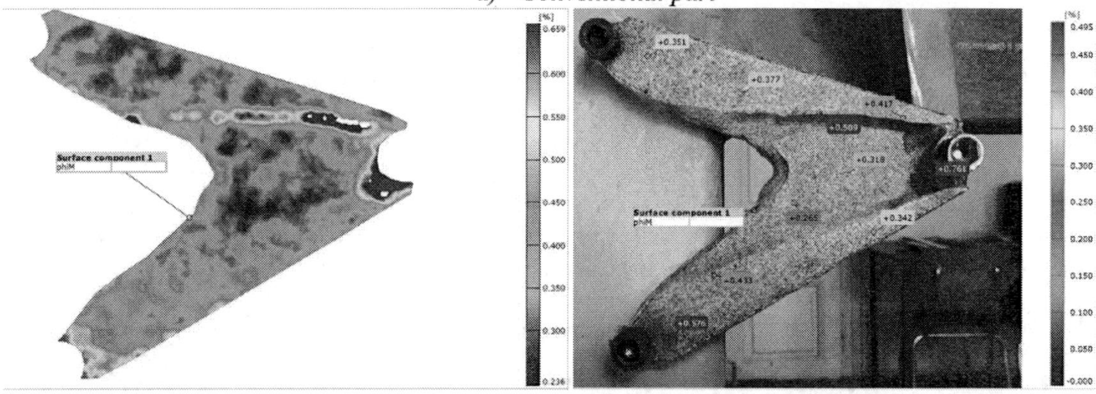

a) Topology-optimized part

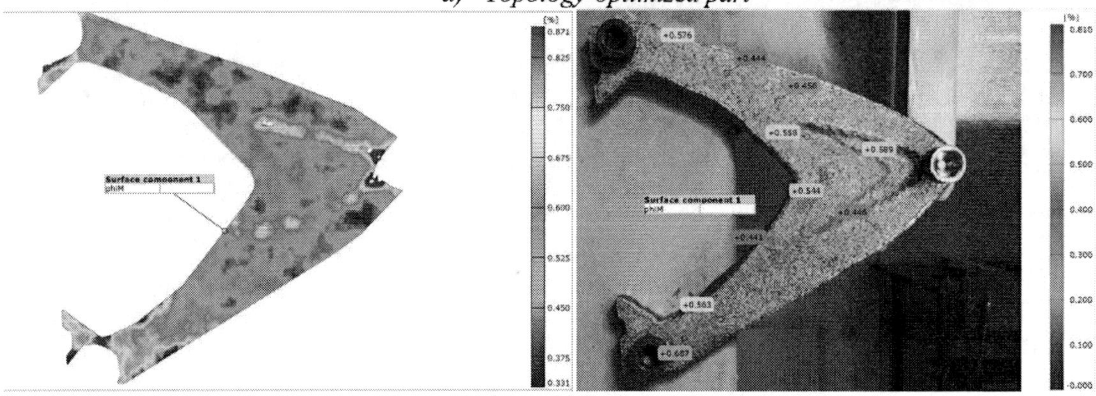

a) Morpho-generated part

Figure 6. Equivalent (von Mises) strains. Without (left) and with (right) original image

The displayed results are the equivalent (von Mises) strains [5]:

$$\varepsilon_{eq} = \frac{2}{3}\sqrt{\frac{3(e_{xx}^2+e_{yy}^2+e_{zz}^2)}{2} + \frac{3(\gamma_{xy}^2+\gamma_{yz}^2+\gamma_{zx}^2)}{4}}. \tag{1}$$

With the deviatoric strains:

$$
\begin{aligned}
e_{xx} &= +\frac{2}{3}\varepsilon_{xx} - \frac{1}{3}\varepsilon_{yy} - \frac{1}{3}\varepsilon_{zz} \\
e_{yy} &= -\frac{1}{3}\varepsilon_{xx} + \frac{2}{3}\varepsilon_{yy} - \frac{1}{3}\varepsilon_{zz} \\
e_{zz} &= -\frac{1}{3}\varepsilon_{xx} - \frac{1}{3}\varepsilon_{yy} + \frac{2}{3}\varepsilon_{zz}
\end{aligned}
\tag{2}
$$

The tensorial shear strain components of the infinitesimal strain tensor can then be expressed using the engineering strain definition, γ, as $\gamma_{ij} = 2\varepsilon_{ij}$. Finally for the strain tensor:

$$
\boldsymbol{\varepsilon} = \begin{bmatrix} \varepsilon_{xx} & \varepsilon_{xy} & \varepsilon_{xz} \\ \varepsilon_{xy} & \varepsilon_{yy} & \varepsilon_{yz} \\ \varepsilon_{zx} & \varepsilon_{yz} & \varepsilon_{zz} \end{bmatrix} = \begin{bmatrix} \varepsilon_{xx} & \frac{1}{2}\gamma_{xy} & \frac{1}{2}\gamma_{xz} \\ \frac{1}{2}\gamma_{xy} & \varepsilon_{yy} & \frac{1}{2}\gamma_{yz} \\ \frac{1}{2}\gamma_{zx} & \frac{1}{2}\gamma_{yz} & \varepsilon_{zz} \end{bmatrix}. \tag{3}
$$

Results from Fig. 6 show similar loading and deformation to the simulation results (Fig. 4). The von Misses strains are connected to von Mises stresses by the elasticity moduli of the material. Again the maximum values of von Mises strains have the smallest magnitude for the conventional part, they are bigger for the topology-optimized part and the biggest value is for the morpho-generated part. The strain distribution on the part surface is more evenly distributed for the optimized parts. Concentration and increase of the equivalent strains are observed at the attachment and bearing holes. This confirms the simulation results.

5. Discussion and future work

Conventionally designed part has 77 grams of mass and 46 MPa peak von Mises stress. Topology-optimized part has 69 grams of mass, or 10% reduction over conventional and 70 MPa peak von Mises stress. The mass of morpho-generated part is 57 grams, or 26% reduction over conventional and 78 MPa peak von Mises stress.The reduced mass is of course important advantage of optimization based methods. They could provide lighter airframes with significant fuel savings during the lifespan of the airplane. The increased stress however, could mean shorter fatigue life and reduced overall damage tolerance. It may be preferable to use topology optimized parts at non fatigue critical locations where multiple load paths are present and assure adequate periodic inspections. Another concern for optimized parts is the complexity of production, which is a major factor in parts price.

The significance of theoretical results is somehow verified by real 2D-DIC deformation tests of the produced parts. However so far only initial analysis is made and further investigations are needed to accurately confirm simulation results. The proposed methods could be used in the future work for optimal design of more complex primary structure elements such as wing spar, ribs etc. [7]

Acknowledgements

The authors would like to thank Scientific and Research Sector at TU-Sofia, Bulgaria, contract **№182PD0006-24** for funding their investigations.

This work was supported by the European Regional Development Fund within the OP "Science and Education for Smart Growth 2014-2020", Project Competence Centre "Smart Mechatronic, Eco- And Energy Saving Systems And Technologies", № BG05M2OP001-1.002-0023.

References

[1] Bendsøe MP, Sigmund O 2003 Topology optimization: theory, method and applications, Springer Verlag, ISBN 978-3-540-42992-0

[2] Doursat, R. 2008 Programmable architectures that are complex and self-organized: From morphogenesis to engineering. 11th International Conference on the Simulation and

Synthesis of Living Systems (ALIFE XI), August 5-8, 2008, University of Southampton, Winchester, UK, pp. 181-188

[3] Aage N, Andreassen E, Lazarov BS, Sigmund O, 2017 Giga-voxel computational morphogenesis for structural design, *Nature*, vol **550**, no. 7674, pp. 84-86. DOI: 10.1038/nature23911

[4] Liu K, Tovar A 2014 An efficient 3D topology optimization code written in Matlab, Structural and Multidisciplinary Optimization **50**(6), ISSN: 1615-147X (Print) 1615-1488 (Online), Springer Verlag, DOI: 10.1007/s00158-014-1107-x

[5] Equivalent Von Mises Strain https://dianafea.com/manuals/d944/Analys/node405.html

[6] Zheng-Zong Tang, Jin Liang et al. 2010 Three-dimentional digital image correlation system for deformation measurement in experimental mechanics, Optical Engineering **49**(10), 103601

[7] Marinov II 2010 The spar wing optimal design, Proceedings of scientific conference of National Military University, Dolna Mitropoliya, ISBN 978-954-713-094-4, pp. 36-38

Numerical study of heat transfer between impinging gas jets and solid surfaces

Viktor Stoyanov[1], Valyo Nikolov[2] and Manuel Mirabal Garcia[3]

[1, 2] Technical University of Sofia, Plovdiv Branch,
Department of Transport and Aviation Equipment and Technologies,
25 Tsanko Dyustabanov str., 4000 Plovdiv, Bulgaria

[3] The Autonomous University of San Luis Potosí, Institute of Physics,
64 Alvaro Obregon, Centro, 78300 San Luis, San Luis Potosí, Mexico

E-mail: viktor.stoyanov@gmx.de [1], vnikolov@tu-plovdiv.bg [2], mirabal@ifisica.uaslp.mx [3]

Abstract. When two spaces of different gas pressure are connected through a nozzle forming and spreading a gas jet with certain energetic characteristics is observed in the region of the lower pressure. If a solid body is located in the region of the jet spreading, then this case is defined as an impinging gas jet. The aim of this work is to consider a specific case of this heat transfer method using 3D virtual simulation models known as Computational Fluid Dynamics - CFD. The simulations were performed in the commercial solver Star-*CCM+* version 6.04.014 using two turbulence models: Shear Stress Transport - *SST* and v^2f Low-Reynolds Number (Re) k-ε model. The obtained results are represented by a dimensionless Nusselt number (Nu) and its local validation determines the accuracy of the solver when using a digital twin describing a real measurement set-up with known empirical measurements in digital format.

1. Theoretical prerequisites

The first laboratory tests were performed in the 1950s by K Peri [5]. His studies describ the relationship between the temperature and velocity of an impinging jet and the experimental measurements which determine the temperature value on the surface of a dense solid body around which a jet stream flows. Further background analyses and studies were conducted later by H Martin [4], S Beltaos [10] and L W Florschuetz et al. [3], and their documentation refers to the 70s and 80s of the last century. H Martin [4] also performed fundamental research and further developed the gas flow physics knowledge by proposing division of the impinging gas jet structure into three distinct flow regions: the free jet region, the stagnation flow region and the wall jet region [4]. The model he proposed illustrates the airflow rate profile of an impinging gas jet which is directly related to the heat transfer coefficient in the region of contact between the impinging jet and the surface of a heated dense solid body. The three main flow regions are shown in Fig. 1 [4].

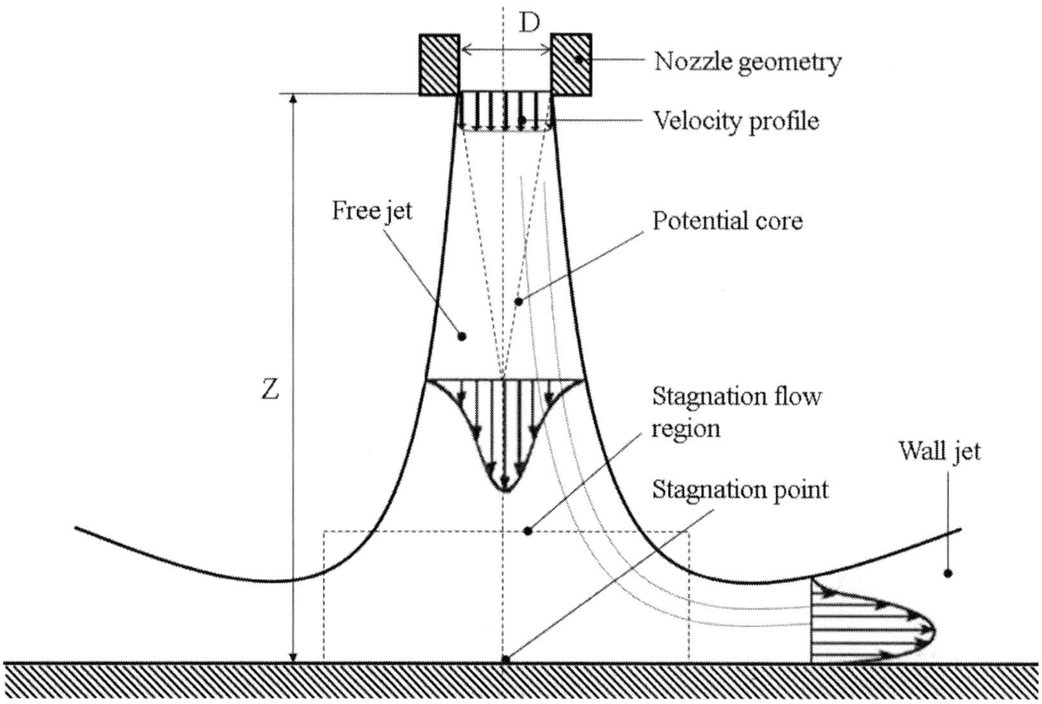

Figure 1. Schematic representation of a vertical impinging gas jet [4].

1.1. The Free Jet Region.

In this region spreading with increasing gas jet rate is observed against the fluid from the surrounding space. In the region of the potential core which is found at a distance up to 4 – 6 times the nozzle diameter no changes are observed in the speed and temperature of the impinging gas jet.

1.2. The Stagnation Flow Region

This region is characterized by the presence of a stagnation point. The pressure has a local maximum value at the stagnation point.

1.3. The Wall Jet Region

For the wall jet region formation of a laminar boundary layer is characteristic which turns into a turbulent jet at a distance 1-2 times the diameter of the nozzle [4].

Accuracy evaluation of the numerical analysis against the results of the experimental measurements of the heat transfer process can be carried out using the dimensionless *Nu*. The Nu is a universal way of expressing the efficiency of a heat transfer process and have been used in a number of scientific works as a dimensionless criterion for comparing heat transfer processes [9].

Due to the increasing complexity of the tasks solved with the help of virtual twins, in-depth analyzes are necessary to evaluate the accuracy of different software programs used for modeling the impinging gas jets. Despite the accumulated extensive practical experience in using impinging gas jets in our daily live, significant difference often occur between the basic research results in this direction and the digital twins used. In this regard a number of software program validations have been carried out in recent years. K Thiel in [6] compared three *CFD* programs, *Fluent, CFX* from *ANSYS* and *Hybrid.* In this case, experimentally obtained values NACA0012 were studied. The used meshing tools were: *ICEM CFD, GAMBIT* and *CENTAUR*. The results show that the processed validation cases from each program can be partly solved in different ways. S Spring [6,12] examined the correlation between the simulation values of heat transfer obtained by *CFX-5.7.1* and the selected turbulence models using a single impinging gas jet for validation on a flat plate and selection of two *Z/D* distances of 6 and 12 at different *Re*. *Z* defines the distance between the nozzle exit and a heat conducting solid plate, and *D* defines the inner diameter of the nozzle. F Ahmet et al. [12] examined the influence of the nozzle geometric parameters on the way of the impinging jet spreading using *CFD* simulations. A more complex experimental case was considered by P S Penumadu and A G Rao [8] which involvs the interaction of 225 impinging gas jets passing through nozzles with a diameter of 400 μm. They limited the overall dimensions of the region under investigation by using a digital twin characterized by symmetrical boundary conditions. The results of this study provide an insight into the physical processes of impinging gas jet spreading and their mutual influence.

2. Selection of test cases

The results accuracy evaluation of an impinging gas jet numerical study of solid surfaces was performed by direct comparison of the dimensionless *Nu* values obtained by experimental and simulation way. The *Nu* was selected in this case as a criterion for assessing accuracy of the simulation results because it adequately represents the correlation between geometric and energy parameters of the heat transfer process. In the experiment under consideration, the parameters describing the heat transfer process efficiency depend entirely on the nozzle geometry, formed impinging jet, temperature of the solid body around which the gas flows, the type of medium used for the experiment, and its initial characteristics. The case corresponds to the experimental set-up at Re = 23000, Z/D = 2. Characteristic of this experiment is that besides the initial maximum *Nu* in the stagnation flow region a secondary maximum is observed at a distance *2D* corresponding to twice the nozzle diameter for all experiments performed at Z/D = 2 [1-2] [7]. The studies of S Spring [11] and F Ahmet [12] and the corresponding local *Nu* obtained experimentally [1-2] were used for verifying the authenticity of *CFD* code of *STAR-CCM+*.

3. Computational details

To perform a numerical analysis of the accuracy of *STAR-CCM+ 6.04.014,* a digital twin of the experimental model was used shown in point 2 [1-2].

3.1. Domain and boundary conditions

During the simulation the boundary and initial conditions of the digital twin did not undergo change. Reading of the local *Nu* was performed by reaching a quasi-stable equilibrium state of iterations. Two turbulence models were validated - *SST* and v^2f. The models were selected after preliminary analysis of the survey results obtained with other software products such as *Fluent*, *CFX* [6] and *Fluent* [7-8] [13-14] which show satisfactory accuracy against other popular turbulence models such as k-ε and k-ω for the respective experimental set-up.

The digital twin from Fig. 2 is characterized by the following Domain and boundary conditions: a circular cross section of the air nozzle D = 26 mm, impinging jet velocity $u_j = V_0$ at Re = 23000 and kinematic viscosity at jet temperature T_j = 293,15 K.

The working medium in this system was air at an initial state corresponding to a temperature T_u = 293,15 K and a static pressure p_u = 101325 Pa. Non-permeable thermally conductive solid body was used which is located parallel to the nozzle exit cross section at a distance Z = *2D*. The initial temperature of the body was $T_j = T_u$ which was subjected to a constant external thermal influence of a specific wall heat flux q_w =300 W/m^2. Additionally, modification of the numerical domain was done with the purpose of using hardware resources efficiently and accelerating the time of obtaining the results of the digital twin investigation while maintaining the accuracy of the results. This was possible because of the axial symmetry of the system relative to the axis of the nozzle.

IOP Conf. Series: Materials Science and Engineering **618** (2019) 012064 doi:10.1088/1757-899X/618/1/012064

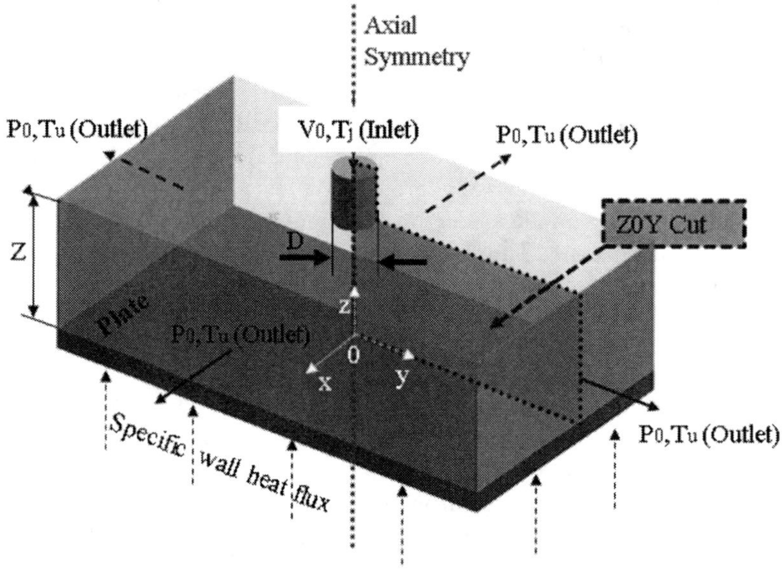

Figure 2. Numerical modeling and boundary conditions of the digital twin.

The two-dimensional domain XOZ *Section* shown in Fig. 2 is extrapolated from the basic digital twin and has the shape shown in Fig. 3. Further modification of the domain from Fig. 3 is fulfilled for better stability of the iterations in the numerical examination process. The result is a three-dimensional domain with a thickness D_S corresponding to the thickness of one cell of the numeral domain (Fig. 4).

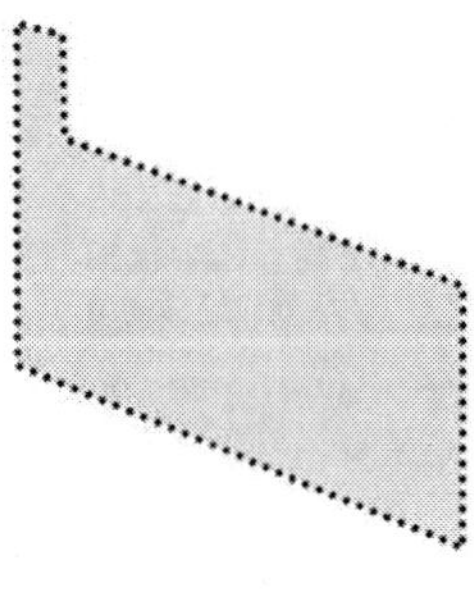

Figure 3. Two-dimensional view of the domain ZOY Section.

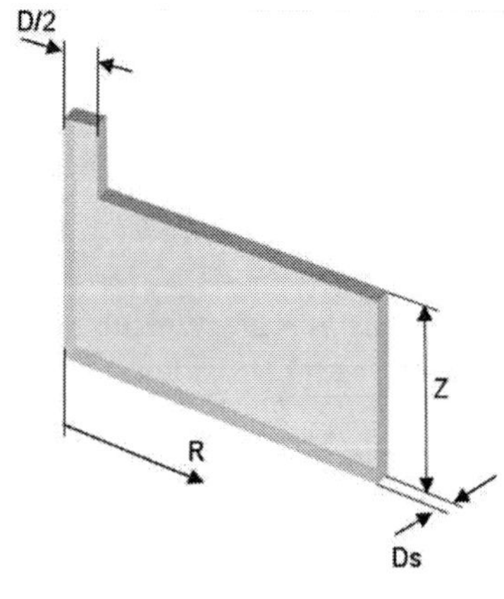

Figure 4. Three-dimensional view of the numerical domain.

The boundary areas of the digital twin in the computing domain are defined according to their physical characteristics, respectively Wall, Axial Symmetry, Wall (Heat Flux), Inlet, Stagnation Inlet and Pressure Outlet (Fig. 5).

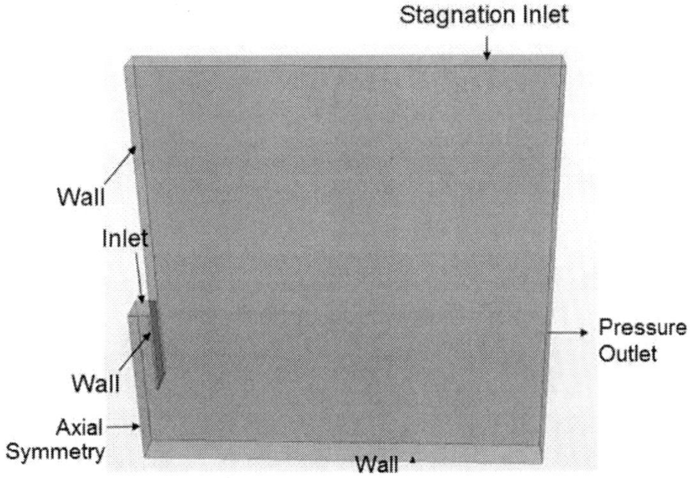

Figure 5. Boundary areas of the digital twin.

3.2. Grid generator

The grid is generated in the commercial solver *STAR-CCM+* version 6.04.014. Trimmed cells with base length of 2 mm are used. A prismatic layer of 6.6 mm thickness in 200 layers with a coefficient of expansion of 1,03 and a Dimensionless wall distance $y^+ = 0{,}065$ is defined near the wall.

Fig. 6 shows the computing domain divided into several zones. Additionally, smaller areas such as "Jet Nozzle Region", "Jet Region", "Nozzle Extrude" and "Wall Region" are integrated to the regions with large areas such as "Basis" or "Near Region". These areas are linked to each other through a refined grid which improves the results quality.

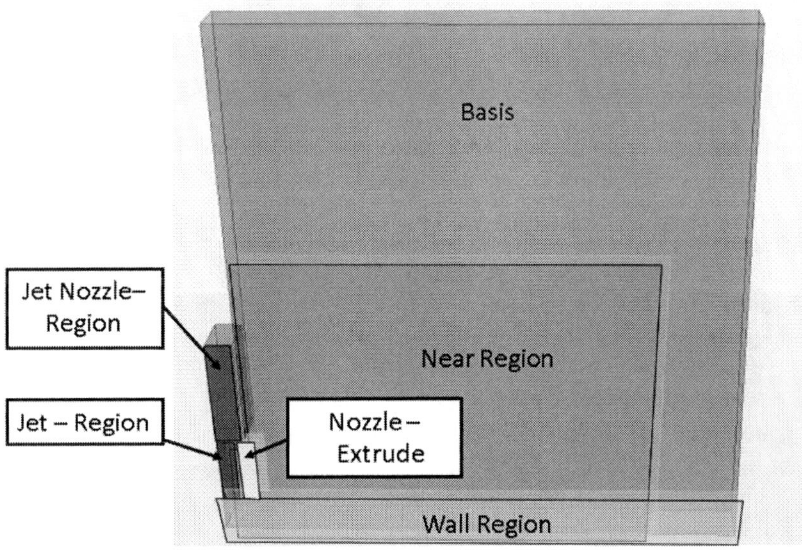

Figure 6. The location of the additional meshed areas.

Table 1 presents the ratio of the cell size in all areas relative to the basic thickness of the cell expressed as a percentage.

Table 1. The ratio of the cell size in the different areas relative to the basic thickness of the cell of 2 mm.

Area	Ratio (%)
Basis	100
Near Region	25
Jet Nozzle-Region	30
Jet-Region	20
Nozzle-Extrude	10
Wall Region	10

After analysis of the grid for independence of the result from its size, a choice was made which was used for subsequent analyzes of a modified digital twin with the number of cells 153912 (Fig. 7).

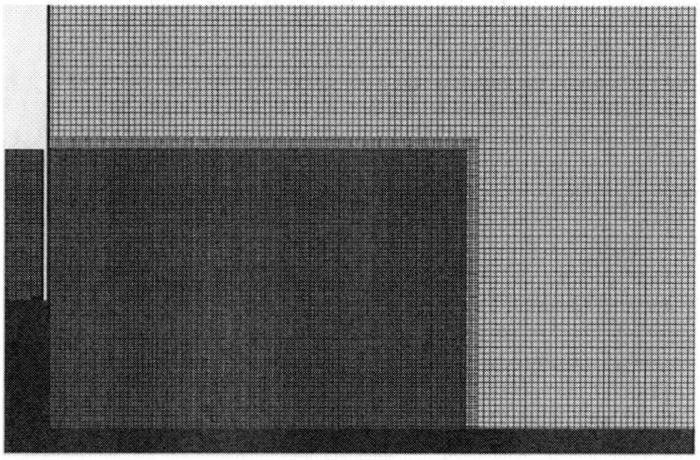

Figure 7. Three-dimensional view on a grid for the study of heat transfer between impinging gas jets and solid surfaces.

4. Numerical prediction and evaluation of turbulence models

Two pre-selected turbulence models (*v2f* and *SST*) were used to perform the simulation results analysis which are integrated into the *STAR-CCM +* software. In both cases the studies of the turbulence models were performed in completely identical domain and boundary conditions of the digital twin. The obtained results were validated with the experimentally determined values described in [1-2] whose numerical values are shown in Fig. 8.

At the stagnation point a very good match is observed between the measured and the numerical values for *Nu* of the model v^2f. In this area, the maximum deviations - more than 20 % between the *SST* model and the experimental values are observed.

The numerical prediction of the flow and heat transfer of the two turbulence models shows a deviation up to 18 % after the first maximum at a distance R/D = 1,3 and reaching a secondary maximum at R/D = 2,5. The model v^2f provides values that are closer to the maximum obtained in the experiment against SST. The v^2f model provides a well-expressed secondary maximum against *SST*. The average distribution of the Nu is characterized by a difference of less than 9%.

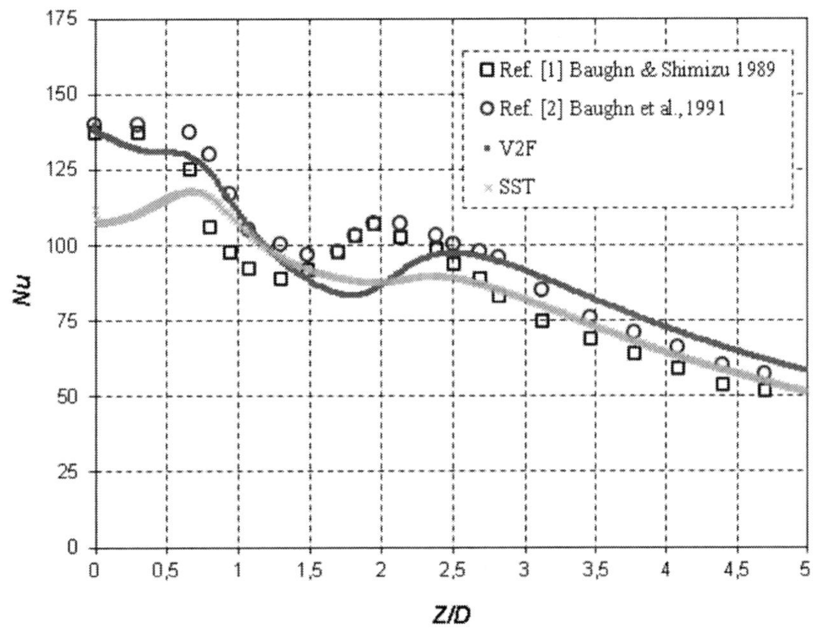

Figure 8. Local *Nu* distributions predicted by *v2f* и *SST* turbulence models by *Re* = 23000 and *Z/D* = 2 in comparison to experimental data from [1-2].

For the spreading profile of the velocity magnitude (Fig. 9) a non-homogeneous transition is observed in the contact region between the air jet and the ambient air for both turbulence models. This may be due to the transition between two meshed areas with different cell sizes, despite the used cell independence analyzes.

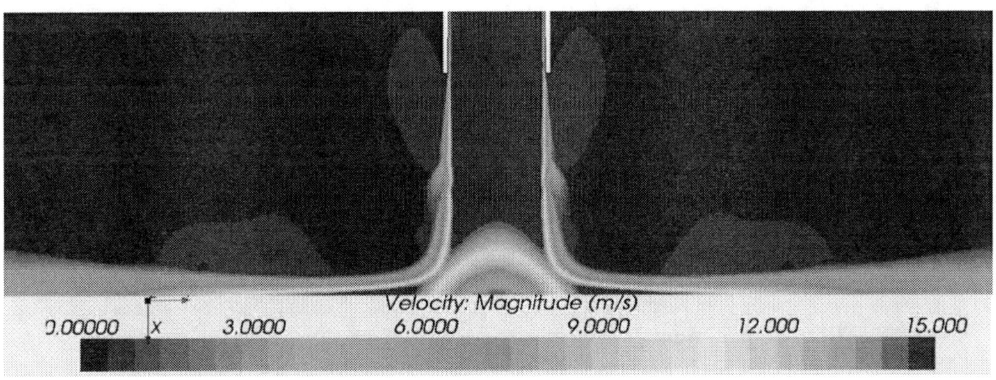

Figure 9. Visualization of the magnitude velocity distribution (*SST* turbulence model).

5. Conclusion

The two turbulence models show satisfactory comparison and reproduction of the Nu values on the considered metal plate surface despite the deviations from the secondary maximum location. Forming a primary and secondary maximum can be observed in both models, and in v^2f the obtained numerical values for the first maximum coincide completely with the performed experiments. The maximum difference between the local results obtained for the *SST*-model is below 21 % if these extremes are located for that model in the area of the first maximum, and for *v2f* in the area of the second maximum those differences do not exceed 10 %. The v^2f model achieves good consistency at the stagnation point where the difference is less than 0,5 %. At *R/D* = 2,5 to 4, the *SST*-model also achieves results with a difference of less than 1% compared to the measured values.

Reproduction of the primary and secondary maximum is achieved in both turbulence models in the considered application area. The v^2f model shows a mean deviation of the numerical results against *Nu*'s average number across the whole range which is less than 7,9 %. For the *SST*-model this value is 8,8 %. Both models can provide satisfactory heat transfer performance in case of impinging gas jet at *Re* = 23000.

Acknowledgments

This work was supported by the European Regional Development Fund within the OP "Science and Education for Smart Growth 2014-2020", Project Competence Centre "Smart Mechatronic, Eco- And Energy Saving Systems And Technologies", № BG05M2OP001-1.002-0023.

The authors would like to thank the Research and Development Sector at the Technical University of Sofia for the financial support.

References

[1] Baughin J W and Shimizu S 1989 Heat transfer measurements from a surface with uniform heat flux and an impinging jet *J. Heat Transfer*, **111** 1096–1098

[2] Baughn J W, Hechanova A E and Yan X 1991 An experimental study of entrainment effects on the heat transfer from a flat surface to a heated circular impinging jet *J. Heat Transfer* **113** 1023–1025

[3] Florschuetz W, Truman C R, Metzger D E 1981 Streamwise flow and heat transfer distributions for jet array impingement with crossflow *J. Heat Transfer* **103(2)** 337-342

[4] Martin H 1977 Heat and mass transfer between impinging gas jets and solid surfaces *Advances in Heat Transfer Academic Press New York* **13** 1–60

[5] Perry K 1954 Heat transfer by convection from a hot gas jet to a plane surface *Proc. Inst. Mech. Eng.* **168** 775-784

[6] Thiel K 2006 Vergleich des Lösungsverhalten der CFD - Programme: FluentTM, CFXTM und Hybrid *Graduate Thesis on the Institute of Aerospace Thermodynamics University of Stuttgart* 24-47

[7] Zu Y Q, Yan Y Y, Maltson J D 2007 CFD prediction for confined impingement jet heat transfer using different turbulent models *Proc. of ASME Turbo Expo 2009 Power for Land, Sea and Air Orlando Florida USA* GT2009-59488

[8] Penumadu P S, Rao A G 2017 Numerical investigations of heat transfer and pressure drop characteristics in multiple jet impingement system *J. Appl. Thermal Engineering* **110** 1511–1524

[9] Spring S 2010 Numerical Prediction of Jet Impingement Heat Transfer *PhD thesis Institute of Aerospace Thermodynamics on the University of Stuttgart* **ISBN 978-3-86853-811-3** 35-54

[10] Beltaos S 1976 Oblique impingement of circular turbulent jets. *J. Hydraulic Research* **14** 17–36

[11] Spring S, Weigand B, Krebs W, Hase M 2006 CFD Heat Predictions of a Single Circular Jet im Pinging with Grossflow. *AIAA ASME Joint Thermophysics and Heat Transfer Conference San Francisco California USA* **9** 3589

[12] Ahmet F B, Tucholke R, Weigand B, Meier K 2010 Numerical investigation of heat trans-fer and pressure drop characteristics for different hole geometries of a turbine casing im-pingement cooling system *14th International Heat Transfer Conference Washington DC USA IHTC* **14-22817** 99-212

[13] Penchev S, Seyzinski D, Stanchev J 2006 Modeling the flow in test section of wind tunnel UT-1 *Journal of the Technical University at Plovdiv "Fundamental Sciences and Applications" Plovdiv Bulgaria* **13(8)** 82-86

[14] Panayotov H, Hafizoğlu S 2014 Airfoil aerodynamics investigation using automated CFD analysis *6*th *Sc. Conf. BulTrans 2014 Sozopol Bulgaria Proc.* 86-90

IOP Conf. Series: Materials Science and Engineering **618** (2019) 012065 doi:10.1088/1757-899X/618/1/012065

Study on the critical speeds of car's movement with low and ultra-low profile tires in a turn

S Taneva[1], K Ambarev[1] and D Katsov[1]

[1]Technical University of Sofia, Plovdiv Branch, Bulgaria

E-mail: s.taneva@tu-plovdiv.bg

Abstract. This work proposes an improved algorithm for determining the critical speeds of car's movement with low and ultra-low profile tires in a turn, where sliding of its axles occurs and overturning. The critical speeds are determined by the characteristics of the lateral slipping calculated for their critical normal loading. On the basis of the algorithm an interactive programming system in the MATLAB environment is developed. Graphic dependencies for the radius of the turn are presented, taking into account the slip of the tires with different speeds of a car Honda Civic Aerodeck 1.5iLS with low profile tires of size185/60R14. Results have been obtained on critical speeds against sliding in the front and rear axles and against the car's overturning. Critical speed against overturning are determined in the absence of normal reactions for the inner steered wheels in a turn. The critical speeds of sliding are found to be lower than the overturning, and thus appear to be decisive.

1. Introduction

When the car is moving in a turn, stability may be disrupted by effects of different nature as a result of which lateral overturning or sliding can occur [1,2,3,4]. Critical speeds of movement against overturning or sliding are studied at settled turn, and where the vehicle's theory [5,6,7,8] does not account the influence of tire lateral slipping. Some authors explore the critical speed of elastic tire sliding in the front axle only [9]. This study [10] suggests dependencies for the critical speeds against overturning and sliding of the front and rear axle of a car, but it does not account the changes in the resistance to lateral slipping coefficient $k_{\delta 0}$ of the elastic front and rear wheels as well as the change in grip coefficient of lateral direction φ_y.

The purpose of the present work is to determine through the improvement interactive programming system presented in [11], the critical speeds against sliding of the front and rear axle and overturning of cars with low and ultralow profile tires, taking into account the redistribution of wheel load and its impact on the lateral slipping coefficient $k_{\delta 0i}$ with the proposed [11,12,13] dependencies.

2. Algorithm for determining the critical speeds of a car's movement in a turn

For critical speeds of movement in a turn on a horizontal road, without taking into account the elasticity of the tires [5,6], it is recommended to use the dependencies:

- against overturning

$$V_k = \sqrt{gR\frac{B}{2h}} \tag{1}$$

- against sliding

$$V_{k\varphi} = \sqrt{gR\varphi_y} \tag{2}$$

where g is acceleration of gravity;

φ_y - the grip coefficient of lateral direction;

R - radius of the turn;

B - the track width of the vehicle;

h - the height of the centre of gravity.

In the case of a steady motion of a car in a turn, with taking into account the lateral slip of tires, for the radius of the turn R_δ, on the basis of [1,2,3,5,6,7,10] and figure 1, the following dependency is applied

$$R_\delta = \frac{L}{tg(\theta_{sr} - \delta_1) + tg\delta_2} \quad . \tag{3}$$

To determine the distance R_c from the centre of the turn to the centre of gravity of the vehicle according to [11] the dependence is

$$R_c = \sqrt{R_\delta^2 - (b - R_\delta tg\delta_2)^2} \quad . \tag{4}$$

The values in formulas (3) and (4) , similarly to [5,6] are shown on figure 1.

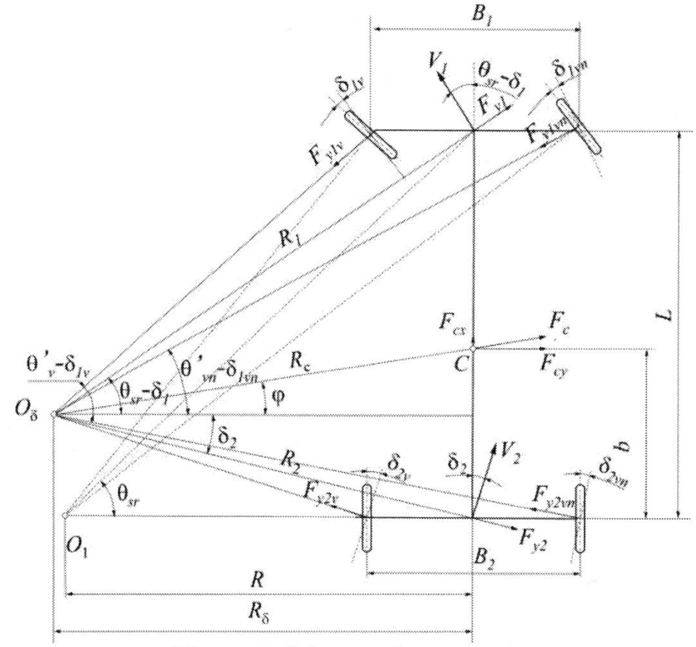

Figure 1. Scheme of a car in a turn.

According to [11] for the centrifugal force the following can be given by

$$F_c = \frac{GV_2^2}{gR_\delta^2} R_c \cos^2 \delta_2 \quad . \tag{5}$$

From figure 1 for the forces and reactions acting on the front and rear axles, vector triangles, shown in figure 2 can be constructed. Using a sinus theorem and transforming the trigonometric expressions for the summed lateral forces acting on the front and the rear axle, the following dependencies are obtained

$$F_{y1} = \frac{GbV_2^2 \cos^2 \delta_2}{gLR_\delta \cos(\theta_{sr} - \delta_1)} \tag{6}$$

$$F_{y2} = \frac{G(L-b)V_2^2 \cos^2 \delta_2}{gLR_\delta} , \tag{7}$$

where L is the wheelbase of the vehicle;

G - the weight of the vehicle;

V_2 - the speed of the rear axle, determine the speed of motion of the vehicle in a turn;

b - the longitudinal coordinate of the centre of gravity of the vehicle;

θ_{sr} - the average angle of the steered wheels.

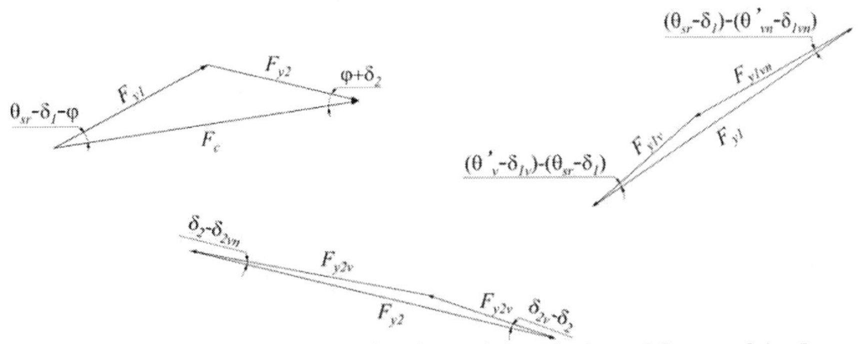

Figure 2. Calculation scheme for determining the lateral forces of the front and rear axles and the lateral reactions of the wheels.

In the case of a car moving in a turn on a horizontal road, for the critical speeds against overturning and sliding with accounting the slipping, the following dependencies can be written

$$V_k = \frac{1}{2\cos\delta_2}\sqrt{gR_\delta \frac{B_1 + B_2}{h}} \tag{8}$$

$$V_{k\varphi} = \sqrt{gR_\delta \varphi_y} \quad . \tag{9}$$

In this work, the critical speeds against sliding of the front $V_{1k\varphi}$ and the rear $V_{2k\varphi}$ axle and against overturning are defined by improvement interactive system based on the proposed in [11] for movement of a car with low and ultra-low profile tires in a turn. Initially, the minimum angles of turning of the steered wheels are set $(\theta_v^{'}, \theta_{vn}^{'})$ and the average angle of the steered wheels θ_{sr} is calculated. For each value of θ_{sr}, the speed of the car V_2 changes a small step until the critical speed is reached of the front axle $V_{1k\varphi}$ or the $V_{2k\varphi}$ rear axle. The calculation starts with the initial values of the slipping - $\delta_1 = \delta_2 = \delta_{1v} = \delta_{1vn} = \delta_{2v} = \delta_{2vn} = 0$ and by [11,12,13] the wheel loads are determined F_{z1v}, F_{z1vn}, F_{z2v}, F_{z2vn}.

By the proposed [11,12,13] dependencies and the improved graphical method of K. Enke separately for the front and the rear axle, with the obtained values of the normal loads, the characteristics of the lateral slipping are constructed, taking into account the changes in the coefficient of resistance against lateral slipping $k_{\delta 0i}$, according to [12,13]. The characteristics of the outer wheels are displaced, respectively, at a distance F_{y1} for the front and F_{y2} for the rear axle by a mirror image [11]. The intersection point of the front axle determines the slipping δ_1, while the intersection point of the rear axle - δ_2.

With the resulting values of δ_1 and δ_2 by the formula (3) and (4) R_δ and R_c are calculated. With a small value of deviation of the new R_c^{new} from the old R_c^{old} value, the iteration process stops. This has the following form:

$$\left|100\left(\frac{R_c^{new}}{R_c^{old}}-1\right)\right|<\Delta \qquad (10)$$

where Δ is the set accuracy of approximation, Δ=0,1%.
If the inequality (10) is not met with the obtained values of δ_1, δ_2 and R_δ the new wheel loads are determined F_{z1v}, F_{z1vn}, F_{z2v}, F_{z2vn} and the lateral forces F_{y1} and F_{y2}. The characteristics of slipping are constructed $\delta_{1v}=f(F_{y1v})$, $\delta_{1vn}=f(F_{y1vn})$, $\delta_{2v}=f(F_{y2v})$ and $\delta_{2vn}=f(F_{y2vn})$ by the method of K. Enke [11]. The new values of R_δ, δ_1, δ_2 and R_c^{new} are determined, until the inequality is fulfilled (10). The radius of the turn is specified R_δ, the slipping of the front and rear axle at the set value of θ_{sr}. Then the speed is increased and the calculations are repeated until the inequality is met (10). The speed increase is performed until, as shown in figure 3 there is no intersection point on the characteristics of lateral slip for inner and outer wheels for the front or rear axle. If there is no intersection between the slips δ_{1v} and δ_{1vn} the front axle is sliding and the critical speed is determined $V_{1k\varphi}$, while in the absence of the intersection point between the slips δ_{2v} and δ_{2vn} the rear axle is sliding - the critical speed is $V_{2k\varphi}$.

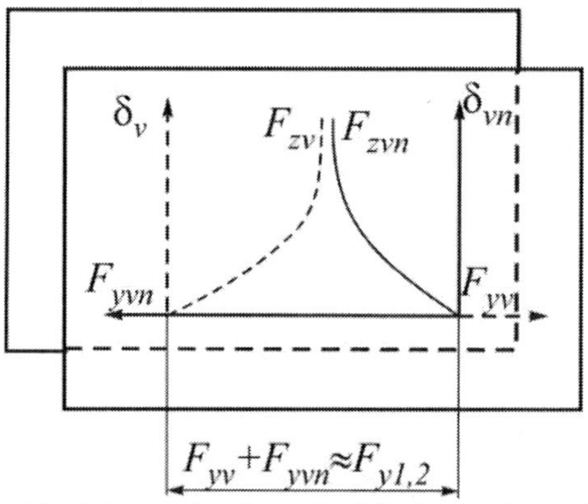

Figure 3. Graphical determination of critical speeds against sliding of the front and rear axle.

3. Interactive system for determining the critical speeds of car's movement in a turn

Figure 4 presents a block-scheme of the improved program, operating on the basis of the presented in [11] and presently algorithm. The user's communication with the created program is by the dialogue mode, i.e. question-answer. Initially, the geometric and mass parameters of the test vehicle are entered by the user: the initial vehicle speed value - V, the initial values for the angles of rotation of the inner θ_v and outer θ_{vn} wheels around the steering axis and the step $\Delta\theta_{sr}$, with which the average angle of the steered wheels changes. This allows at different speeds of movement to specify the accuracy in determining the critical speeds against sliding and overturning. The interactive system performs a quick and accurate calculation of the parameters of the turn. The program can be used to determine which of the two axles is first sliding and what the value of the critical speed against sliding is. It is checked at what speed the vehicle will overturn, by monitoring the value of θ_{sr}, in which the vertical reactions (F_{z1v}, F_{z2v}) of the inner wheels become zero or negative.

IOP Conf. Series: Materials Science and Engineering **618** (2019) 012065 doi:10.1088/1757-899X/618/1/012065

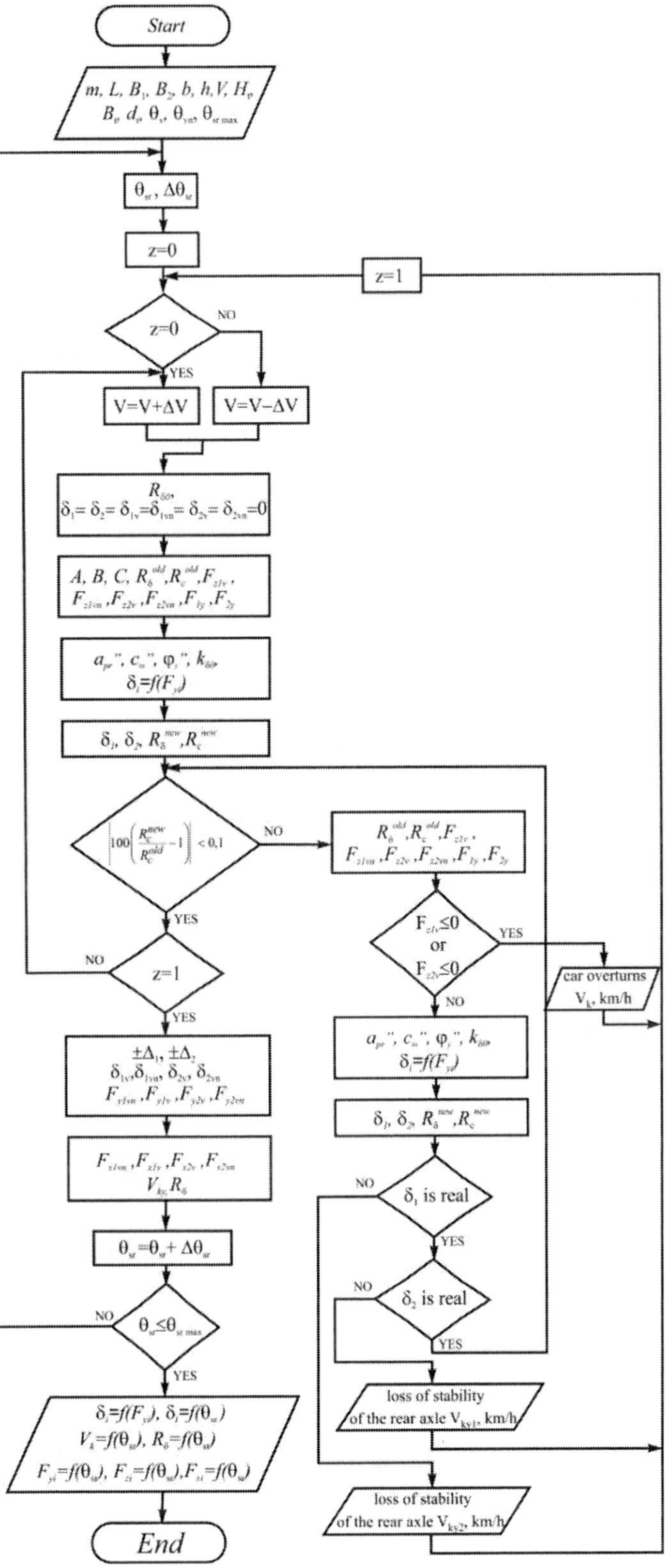

Figure 4. Block-scheme of the interactive system.

The resistance is determined at the smallest of the three speeds V_k, $V_{1k\varphi}$, $V_{2k\varphi}$.

4. Results

The subject of the study is a passenger car Honda Civic Aerodeck 1.5iLS with the following parameters: mass of the car $m = 1665\ kg$, track width of the front $B_1 = 1,475\ m$ and rear $B_2 = 1,470\ m$ axle, wheelbase $L = 2,62\ m$, maximal angel of rotation of the inner $\theta_v = 36^\circ$ and outer $\theta_{vn} = 30^\circ$ wheels around the steered axes, the coordinates of the mass centre of the vehicle, respectively $b = 1,286\ m$, $h = 0,50\ m$, low profile tires of size 185/60R14 and maximum speed $V = 191\ km/h$.
The graphical determination of the slip of the front δ_1 and rear δ_2 axles at average angle of the steered wheels $\theta_{sr} = 1,9^\circ$ and at driving speed $V = 85\ km/h$ are presented on figure 5 and 6.

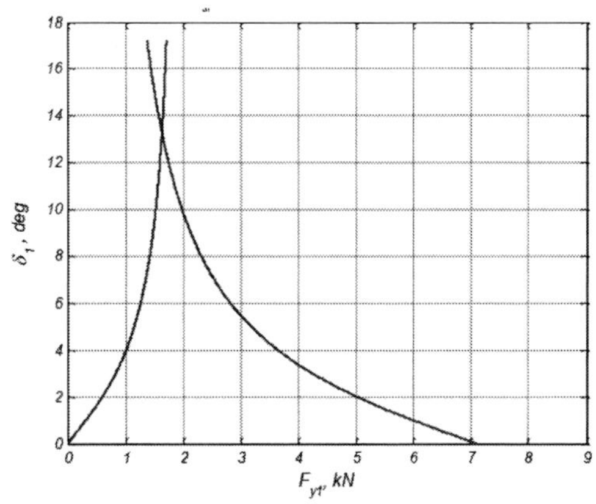

Figure 5. Determination of the slip δ_1.

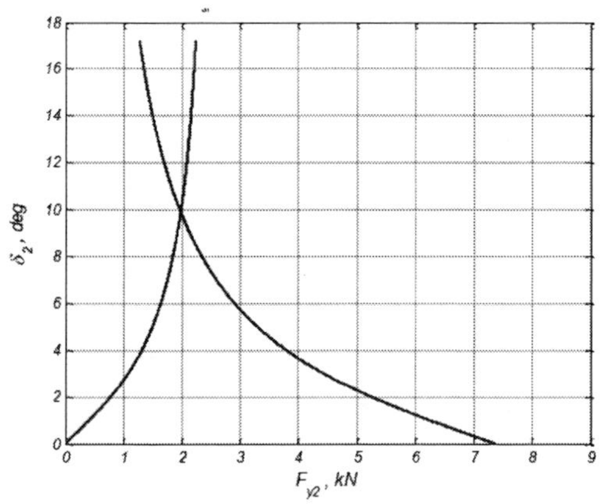

Figure 6. Determination of the slip δ_2.

Determination of the critical speeds against sliding of the front and rear axles at average angle of turning of the steered wheels $\theta_{sr} = 1,9^\circ$ are presented in figure 7 and 8. The obtained critical speed is $V = 87\ km/h$.

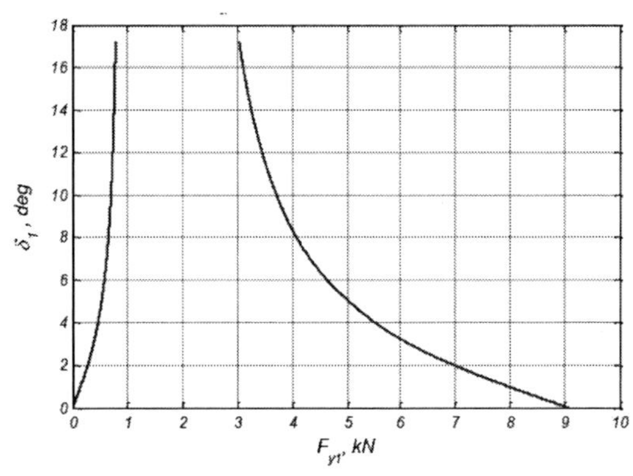

Figure 7. Determination of the critical speed against sliding of the front axle.

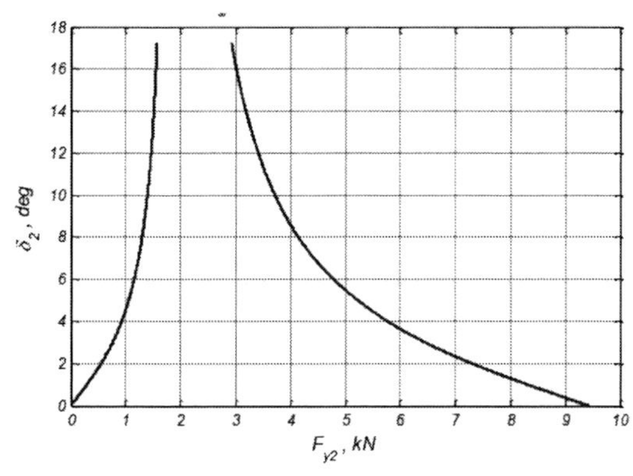

Figure 8. Determination of critical speeds against sliding of the rear axle.

Figure 9 shows the dependence of the radius of the turn R_δ of a car on the average angle of turning of the steered wheels in the plane of the road at driving velocity of 20, 50, 80, 110, 140 and 165 *km/h*.

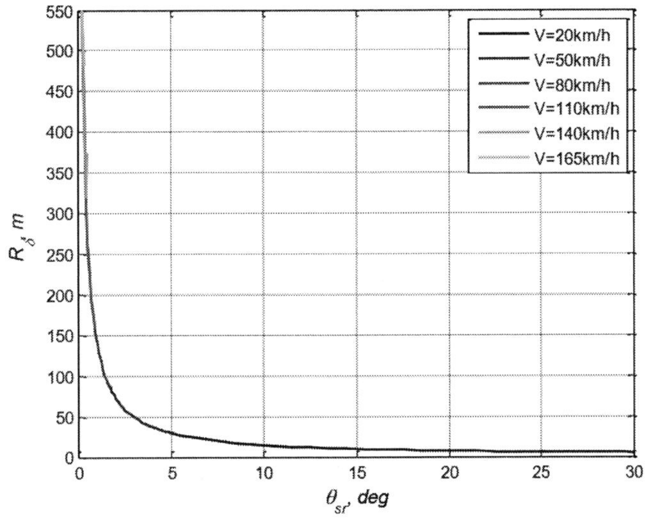

Figure 9. Dependence of the radius of the turn of a car on the average angle of the steered wheels at different driving velocity.

The analysis in figure 9 shows that by increasing vehicle speed in a turn, the angle of the steered wheels decreases and the radius increases significantly. By increasing speed respectively from 20 to 50, 80, 110, 140 to 165 km/h, the program limits the maximum angle of turning of the steered wheels. From $\theta_{sr} = 32^{\circ}$, they respectively are reduced to 5,95°, 2,15°, 1,15°, 0,55°, 0,35°, which indicates that the interactive programming system quickly and accurately determines the minimum value of the radius of the turn R_{δ}, taking into account the slipping.

Figure 10 and 11 present the results for changes of the vertical reactions of the inner wheels due to the driving speed. Critical speeds against overturning of the front and rear axles at average angle of the steered wheels $\theta_{sr} = 3{,}2^{\circ}$ are defined.

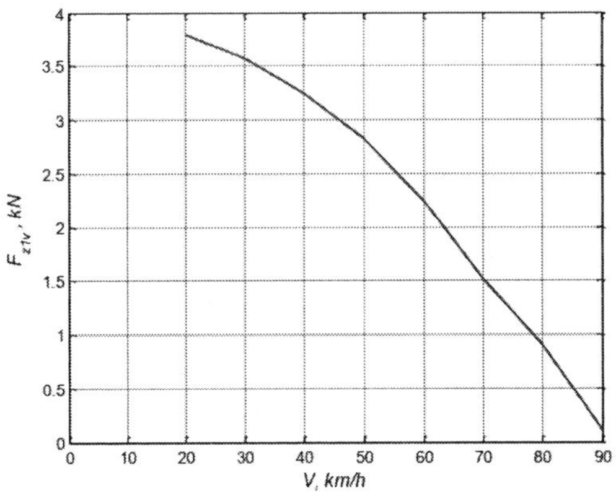

Figure 10. Vertical reaction of inner wheel of the front axle.

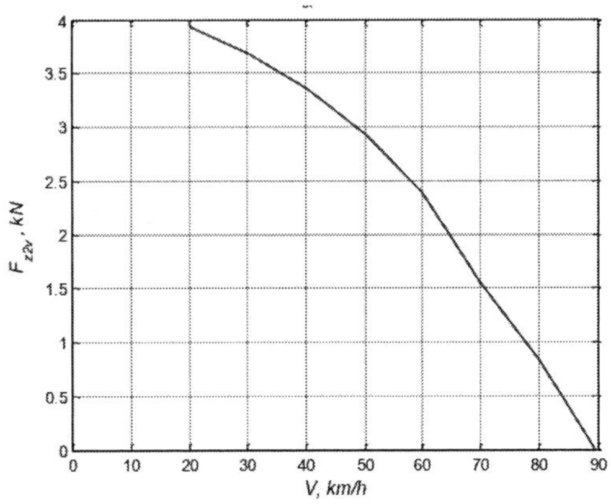

Figure 11. Vertical reaction of inner wheel of the rear axle.

Figure 12 shows the changes of the critical speeds of the car, taking into account the lateral slip of low profile tires.

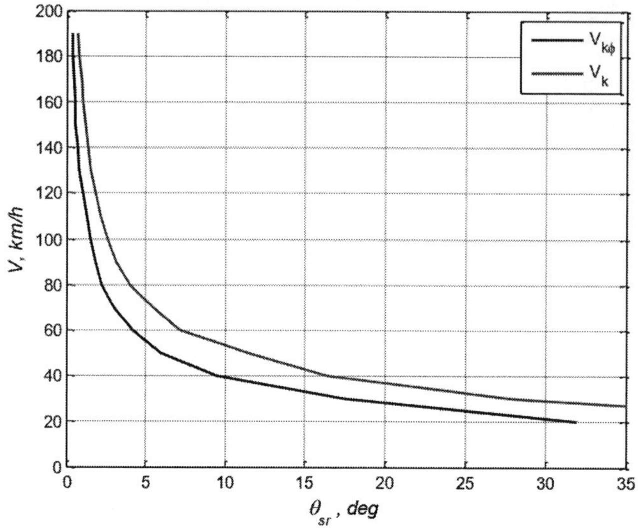

Figure 12. Changing critical speed sliding and overturning from the average angle of the steered wheels of the car.

The analysis in figure 12 shows that by increasing the average angle of the steered wheels θ_{sr} (respectively the inner and outer wheel in respect to the turn), sliding occurs at lower speeds. The critical speeds against sliding $V_{k\varphi}$ for the two axles of the car are practically the same, due to the virtually symmetrical position of the centre of gravity, and they are presented with one line. The critical speed against overturning is above against sliding, which means that the sliding occurs before the car is overturned.

5. Conclusion

From the expose and the results for the radius of the turn at different speeds and angles of the steered wheels of a car, as well as for the critical speeds against sliding and overturning, the following conclusions are made:

- The proposed improved interactive programming system to study the critical speed of a motion of a car with low and ultra-low profile tires in a turn allows their determination to be made in the design stage of the car.
- The critical speeds against sliding of the front and rear axle for studied car are practically equal.
- When a car with low-profile tires turns, the critical speeds against overturning are higher than critical speeds against sliding, and depending on the condition of the road sliding of the front or rear axle may occur.

Acknowledgments

The authors would like to thank the Research and Development Sector at the Technical University of Sofia for the financial support.

Our research, the results of which are presented in this article have been carried out under the Programme Young Scientists and Postdoctoral Students of the Ministry of Education and Science, Bulgaria.

IOP Conf. Series: Materials Science and Engineering **618** (2019) 012065 doi:10.1088/1757-899X/618/1/012065

References

[1] Hlebarski D. A., Katsov D. A., 2004 Influence of the non-independent suspension with two longitudinal arms and a panhard rod of a passenger car on its motion at settled turn. MOTAUTO'04, Proceedings, "Automobile Engineering, Industrial Trucks and Road-Building Technique. Automobile Transport", Plovdiv, Bulgaria, Volume 3, pp. 51-55, ISBN 954-9322-07-6
[2] Katsov D. A., Hlebarski D. A, 2004 Dynamics of the turn of wheel vehicle with account of the deformations of the tires and the influence of the suspension. International Journal of Computational and Numerical Analysis and Applications, Vol. 6, No. 2, pp.97-106, ISSN 1311-6789
[3] Katsov D. A., Hlebarski D. A, 2005 Kinematics of the turn of wheel vehicle with account of the deformations of the tires and the influence of the suspension. International Journal of Computational and Numerical Analysis and Applications, Vol. 25, No. 2, pp. 203-214, ISSN 1311-8080
[4] Nikolov V, 2013 A simulation study of the controllability of a car under the influence of a sudden gust of crosswind. International virtual journal for science, technics and innovations for industry "Machines, Technologies and Materials", Scientific technical union of Mechanical engineering, Sofia, Bulgaria, Year 7, Issue 7, pp. 55-59, ISSN 1313-0226
[5] Lyubenov S, Angelov B, Evtimov I, 2004 Vehicles and Tractors, Performance properties, Ruse, ISBN 954-712-240-1
[6] Falkevich B.S, 1963 Theory on a car, M. Mashgiz
[7] Pacejka H. B, 2006 Tyre and Vehicle Dynamics, Second edition, Amsterdam, Boston, Heidelberg, London, New York, Oxford, Paris, San Diego, San Francisco, Singapore, Sydney, Tokyo, ISBN - 13: 980-0-7506-6918-4
[8] Rill G, 2012 Road vehicle dynamics: Fundamental and Modeling, Taylor & Francis Group, LLC, ISBN – 13: 978-1-4398-9744-7 (eBook - PDF)
[9] Krasikov S. M. Ilarionov A. V, 1969 The condition of handling of the car on turning, Vehicle industry, vol.3
[10] Lyubenov S, Katsov D. 1986 Automobile critical speed at turns, Proceedings of the University of Ruse, Series 2, pp.179-184
[11] Taneva S, Ambarev K, Katsov D, 2019 Interactive system for study of car's movement with low and ultra-low profile tires in a turn, Proceedings of the 12th International Scientific and Practical Conference, Environment. Technology. Resources. Rezekne, Latvia (preprint)
[12] Katsov D, Taneva S, 2015 On obtaining the lateral slip characteristics of low and superlow profile tires of cars, Journal of Machine Building and Machine Science, Varna, Bulgaria, Year. X, No. 2, Vol. 24, pp. 33-38, ISSN 1312-8612
[13] Katsov D, Taneva S, 2015 Investigation of the lateral slip and grip of low and superlow profile tires of cars, Journal of Machine Building and Machine Science, Varna, Bulgaria, Year. X, No. 2, Vol. 24, pp. 39-45, ISSN 1312-8612

IOP Conf. Series: Materials Science and Engineering **618** (2019) 012066 doi:10.1088/1757-899X/618/1/012066

MULTI-CRITERIA SYNTHESIS OF FREQUENCY-MODULATED DISCRETE CONTROL OF SEMI-ACTIVE VEHICLE SUSPENSION - PART 1 ANALYSIS AND CONTROL STRATEGIES

J Genov

Technical University of Sofia, 8 Kl. Ohridski Blvd, Sofia 1000, Bulgaria

E-mail: j_genov@mail.bg

Abstract. The paper discusses the contradiction influence of system damping coefficient in the different excitation's frequency diapasons on the raid comfort and the vehicles stability on road. It is analyzing the discrete "on-off" control laws for the semi-active suspension included in a model reduced to the vehicle wheel and accounting the vertical dynamics of the vehicle so-called "Quarter car model". Based on analysis of the frequency characteristics it is propose a frequency modulation of the control realized by a signals filtration.

1. Introduction

The damping characteristics of the suspension of vehicles do not affect unambiguously on the quality of the vibroisolation. Due to the phase shifting, during certain time parts of the oscillation period; the damper directs energy into the object of vibroisolation instead of dissipating it. The amount of this induced energy depends on the damping coefficient and the frequency spectrum of the external excitation defining the phase shifting. Generally the vehicle`s comfort is associated with low values of the damping coefficient, while the stability of the vehicle with high values [1÷7]. Results close to the optimal could be achieved by usage of semi-active damper which rheology changes through a control electromagnetic field.

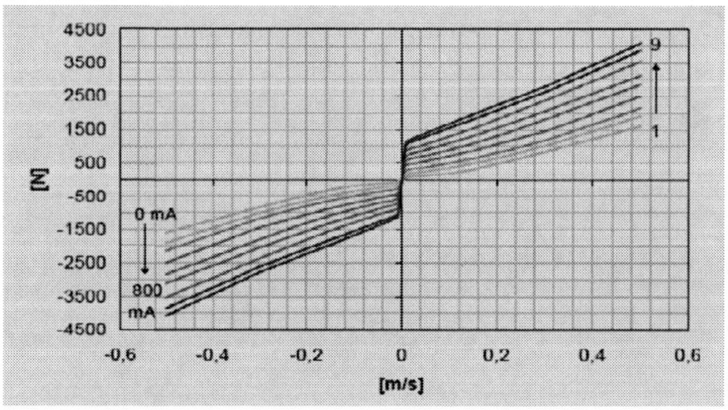

Figure 1. Characteristic of a semi-active damper [8]

On figure 1 is shown a sample characteristic of the damping force as a function of the deformation velocity, for different values of the control current.

The basic idea for the control of the semi-active damper is that when the energy is directed into the isolation object, to be switch to the lower value of damping coefficient, respectively, when the damper dissipates energy, to be realized the maximum value of the damping. This must be realized under the condition for a compromise between ride comfort and road holding because when the control is targeted to the comfort criterion the stability makes worse and vice versa. In this sense and also in a case of presence of a poly-harmonic excitation, this "on-off" approach does not give the best possible results. Additional, the sudden stiffing and softening of the suspension, leads to effects of internal impacts in the system.

2. Damping force analysis

On figure 2 is shown a one-dimensional linear suspension model for kinematic excitation. The shown dimensions are:

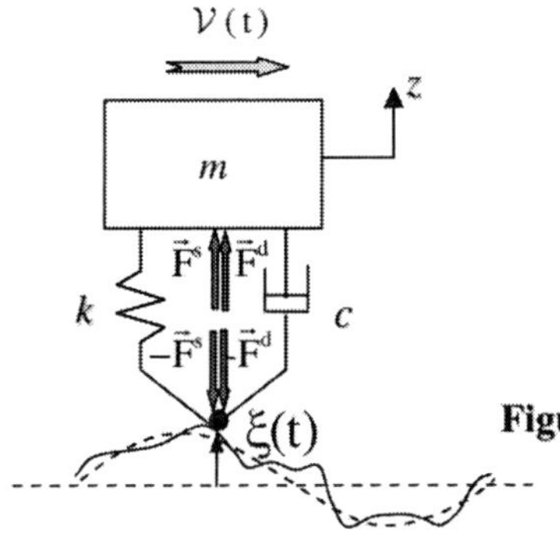

Figure2. One degree of freedom model

Parameters of the model are:
m - sprung mass; *k* - coefficient of elasticity; *c* - damping coefficient;
z(*t*),m - vertical displacement of the sprung mass considered from the static equilibrium position;
$\xi(t)$ kinematic excitation from road roughness.

The Phase - frequency characteristic between vibration velocity $\dot{z}$ and damping force $F^d = c\left(\dot{\xi} - \dot{z}\right)$ is:

$$\Phi_{\dot{z}F^d}(\omega) = \arctan\left(2n\omega / \omega_n^2\right) - \pi , \tag{1}$$

where: ω_n – natural frequency, ω – kinematic excitation frequency, $n=c/(2m)$ – damping.

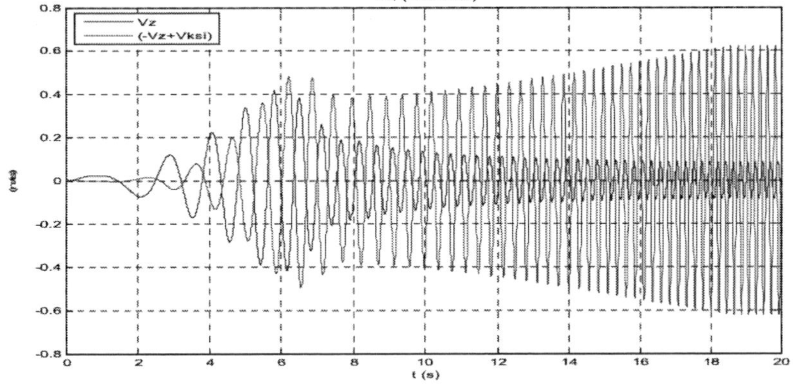

Figure 3. Sprung mass velocity and deformation velocity of damper.

On figure 3 is shown the result of the simulation $\dot{z}$ vs. $\left(\dot{\xi}-\dot{z}\right)\sim F^{d}$, for linearly increasing frequency over time.

The time corresponding to the phase shift as function of the frequency is:

$$t_0(\omega)=\Phi_{\dot{z}F^d}(\omega)/\omega\,, \tag{2}$$

and percentage part of the oscillation period during which the damper force and the mass velocity are in the same direction is calculated by the dependence:

$$100\left(2t_0(\omega)-\tau\right)/\tau=100\left(\Phi_{\dot{z}F^d}(\omega)/\pi-1\right)\%\,, \tag{3}$$

where $\tau=2\pi/\omega$ is the period of oscillation.

Graphics of the time intervals when the damper directs the energy to the mass, in percentage relation to the period of oscillation as functions of the frequency, and for different values of the damping, are shown in figure 4.

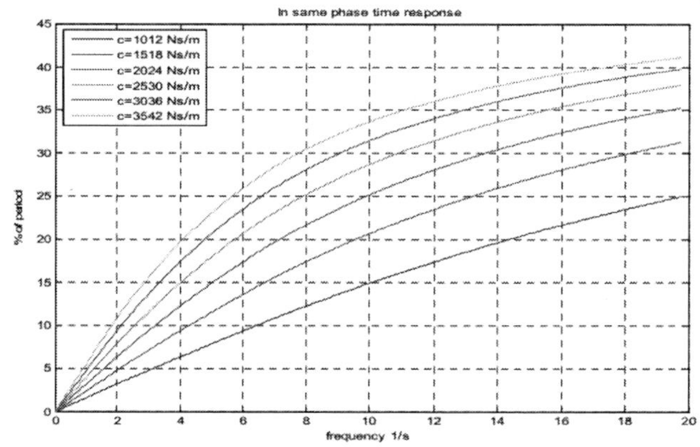

Figure 4. Percentage relation of time interval of importing energy to the period of oscillation

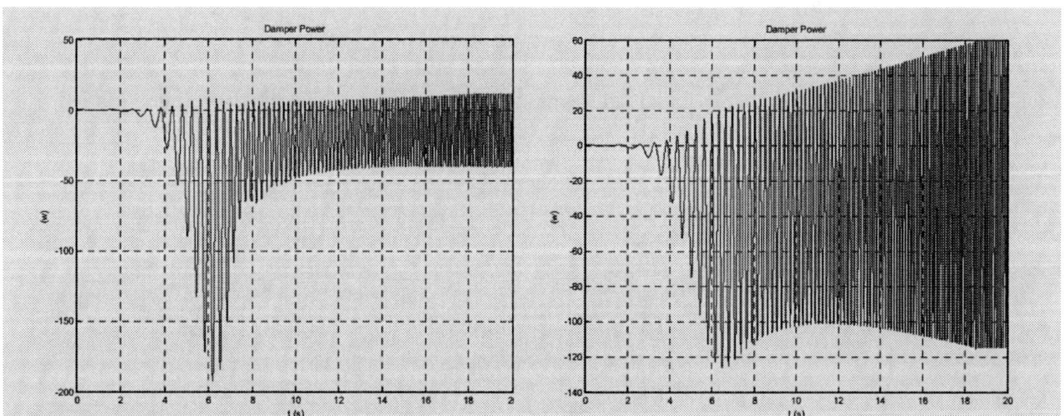

a) b)

Figure 5. Damper power for linear rising in the time frequencies for: a) n=1012 and b) n=2024 Ns/m

The analysis of the results shows that in the under-resonance and resonance frequencies the phase shifting between the force in the damper and the direction of the velocity of the mass is close to 180^0, i.e. the damper predominantly dissipates the vibrational energy. In the after-resonance region the relative duration of the intervals, in which the damper introduces energy, increases. In that as much greater the damping ratio is, these intervals are more prolonged while at the same time a higher amount of energy is introduced through them - figure 5.

In the "Quarter car" model figure 6 the influence of the damping coefficient is even more complicated.

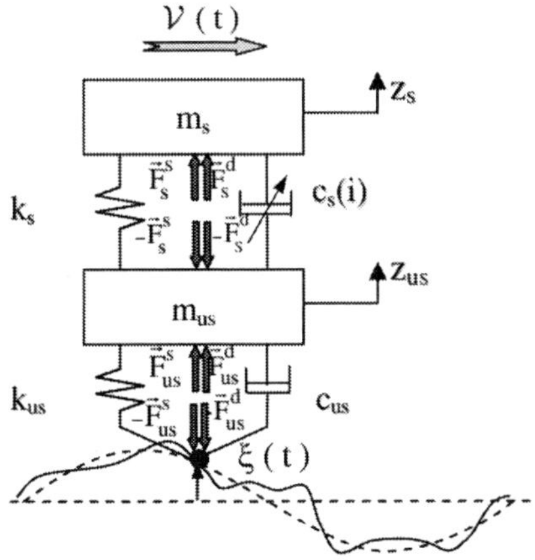

Figure6. Quarter car model

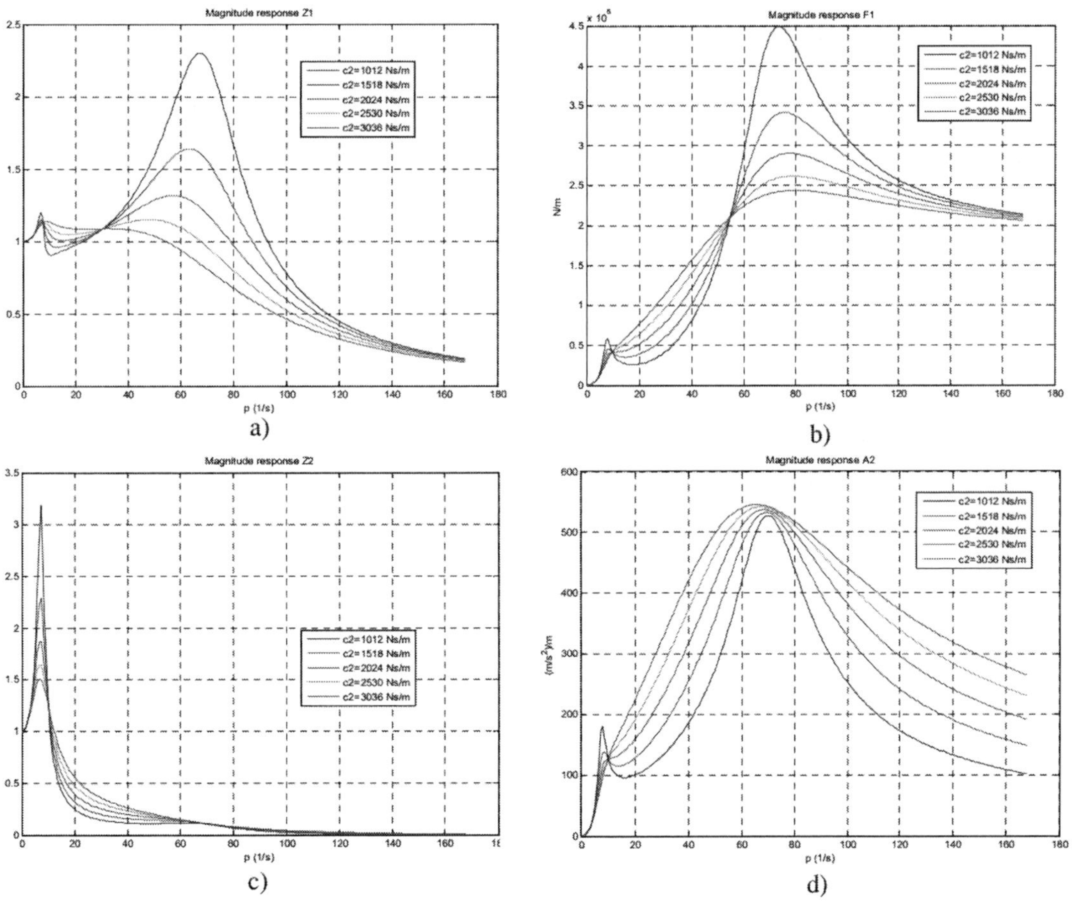

Figure 7. Magnitude responses of a) displacement of the unsprung mass b) dynamic force between the tire and road pavement c) displacement of the sprung mass d) acceleration of the sprung mass

IOP Conf. Series: Materials Science and Engineering **618** (2019) 012066 doi:10.1088/1757-899X/618/1/012066

On figure 7a) and on 7b) are given families of magnitude response for the vibro-displacement of the unsprung mass – z_{us} and the normal dynamic force between the tire and road pavement, $F_{us}^{dyn} = k_{us}\left(-z_{us}+\xi\right)+c_{us}\left(-\dot{z}_{us}+\dot{\xi}\right)$, related with the vehicle stability, and on figure 7c) and 7d) the vibro-displacement - z_s and acceleration $\ddot{z}_s$ of the sprung mass both related with the ride comfort.

It is obvious from upper graphics that in each frequency range, determined by the damping invariant frequencies, the influence of the damping is changed. Furthermore, the invariant frequencies are different for each of the magnitude characteristics.

3. Essence and laws of semi-active control system

In the semi-active dampers, the base silicon fluid contains ferrite particles. Under the influence of an electromagnetic field, they form colonial structures, which lead to considerable rise of the viscosity and respectively of the damping. In this way the damping may to varying in a wide range. A linearized model of this variation may be presented in the following form:

$$c = c^{\ell} + \left(c^{u} - c^{\ell}\right)U \tag{4}$$

where c^{ℓ} and c^{u} are respectively the lower and upper limits of the variation of the damping coefficient obtained for values 0 and 1 of the control signal U.

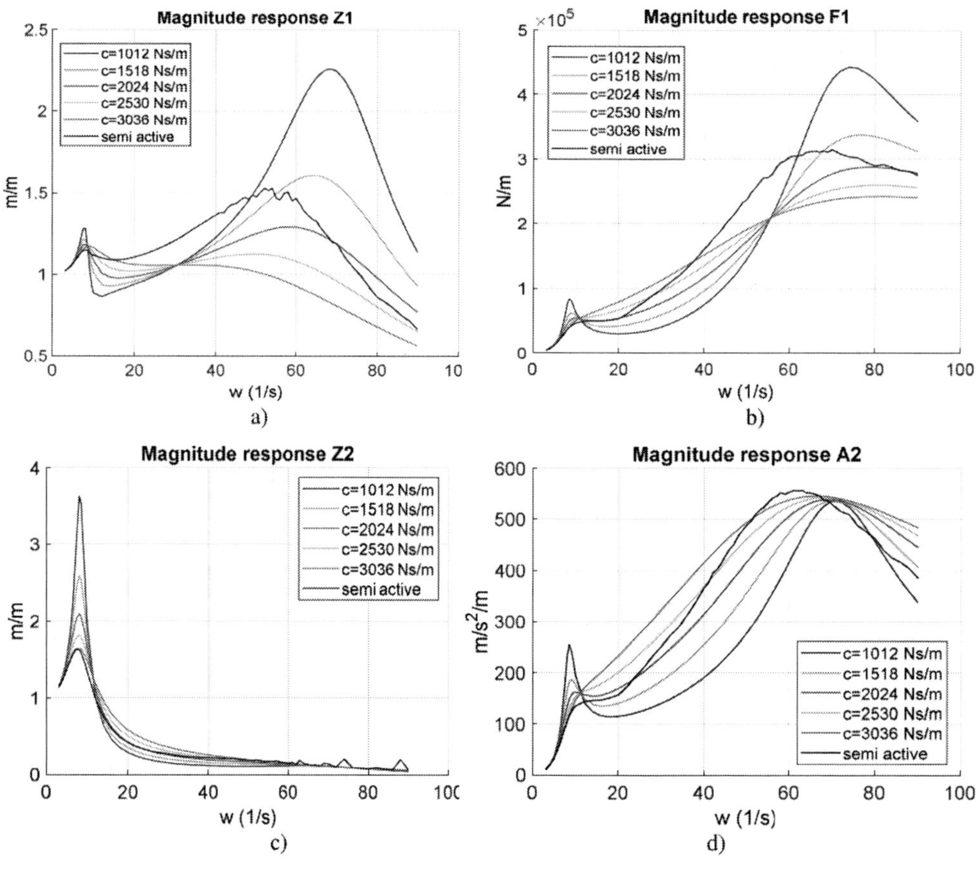

Figure 8. Magnitude responses of a) displacement of the unsprung mass b) dynamic force between the tire and road pavement c) displacement of the sprung mass d) acceleration of the sprung mass

IOP Conf. Series: Materials Science and Engineering **618** (2019) 012066 doi:10.1088/1757-899X/618/1/012066

The classical approach for vibroisolation of the sprung mass is a control that sets maximum damping when the damping force is in the opposite direction to the velocity and minimal value otherwise (so called "sky hook" type control):

$$c_s = c_s^{\ell} + \left(c_s^{u} - c_s^{\ell}\right)U_s,\ U_s = \begin{cases} 1, & \left(\dot{z}_{us} - \dot{z}_s\right)\dot{z}_s \le 0 \\ 0, & \left(\dot{z}_{us} - \dot{z}_s\right)\dot{z}_s > 0 \end{cases} \tag{5}$$

The results of this control compared with those from the passive suspension are shown of figure 8. It is seen that this control law gives good results between the two invariant frequencies.

For another type sky hook control:

$$c_s = c_s^{\ell} + \left(c_s^{u} - c_s^{\ell}\right)U_s,\ U_s = \begin{cases} 1, & \left(\dot{z}_{us} - \dot{z}_s\right)\ddot{z}_s \le 0 \\ 0, & \left(\dot{z}_{us} - \dot{z}_s\right)\ddot{z}_s > 0 \end{cases}, \tag{6}$$

the obtained results are given on figure 9.

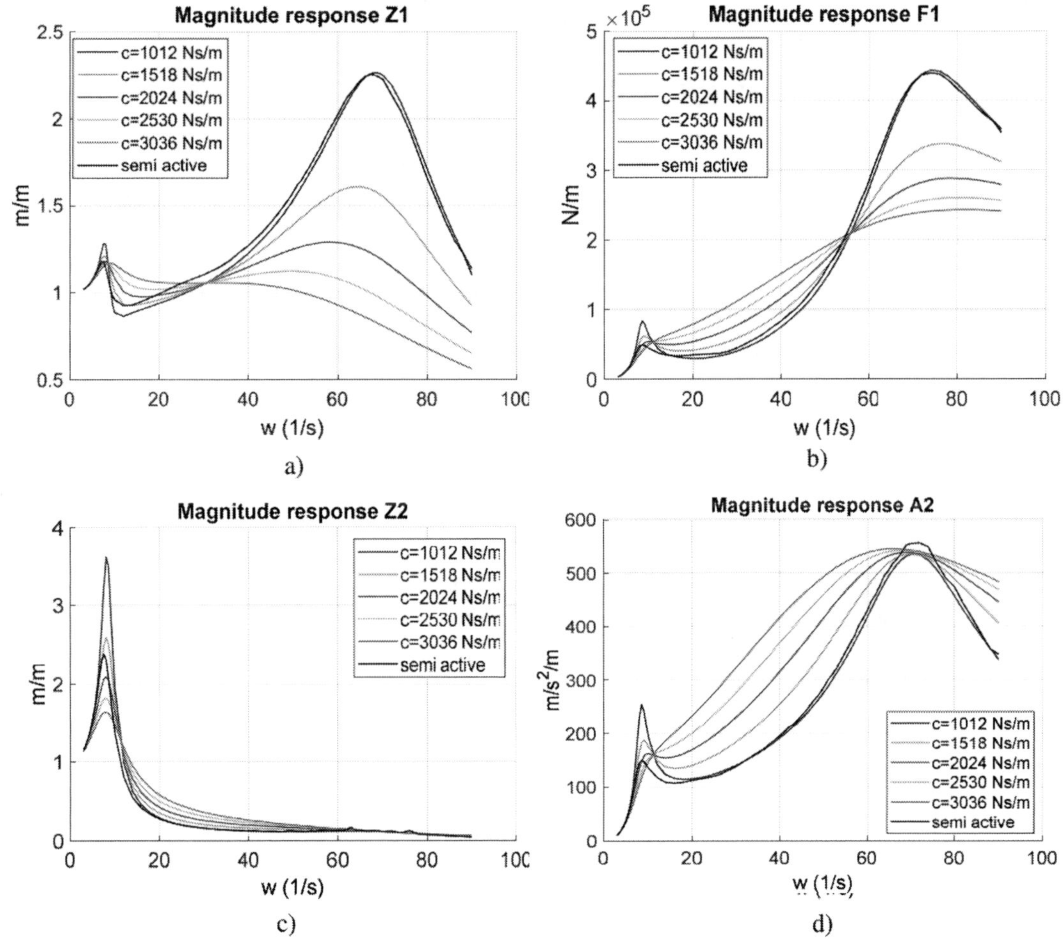

Figure 9. Magnitude responses for control of the type (6)

Here, the effect is exactly the opposite of the control type (5).

From the result analysis follows that the combination of the two control systems leads to maximum results of vibroisolation of the sprung mass. This can be accomplished by using combination filters – low-pass filter for control system (5) and band-pass filter for (6):

$$c_s = c_s^{\ell} + \left(c_s^{u} - c_s^{\ell}\right)U_s,$$

$$U_s = \begin{cases} 0 \quad , & [q>0]\cap[p>0], \quad q = F_s^d\, \ddot{z}_s,\ p = F_s^d\, \dot{z}_s \\ U_s^v = \dfrac{\left|z_s^{\ell p}\right|}{\left|z_s^{\ell p}\right| + \left|z_s^{bp}\right|}, & [q\le 0]\cap[p>0], \quad U_s^a = \dfrac{\left|z_s^{bp}\right|}{\left|z_s^{\ell p}\right| + \left|z_s^{bp}\right|}, \quad [q>0]\cap[p\le 0] \\ U_s^v + U_s^a \quad , & [q\le 0]\cap[p\le 0] \end{cases} \tag{7}$$

The frequency characteristics of the low-pass and band-pass filter are shown on figure 10.

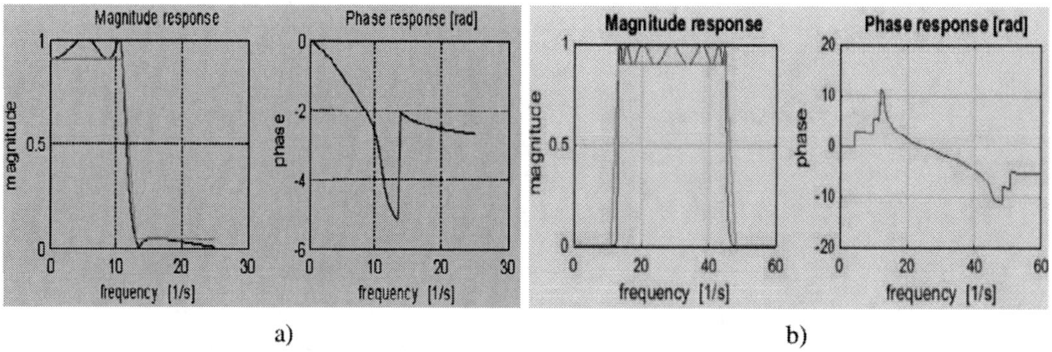

Figure 10. Frequency characteristics of modulating filters a) low-pas b)band-pass

The obtained results are shown on figure 11.

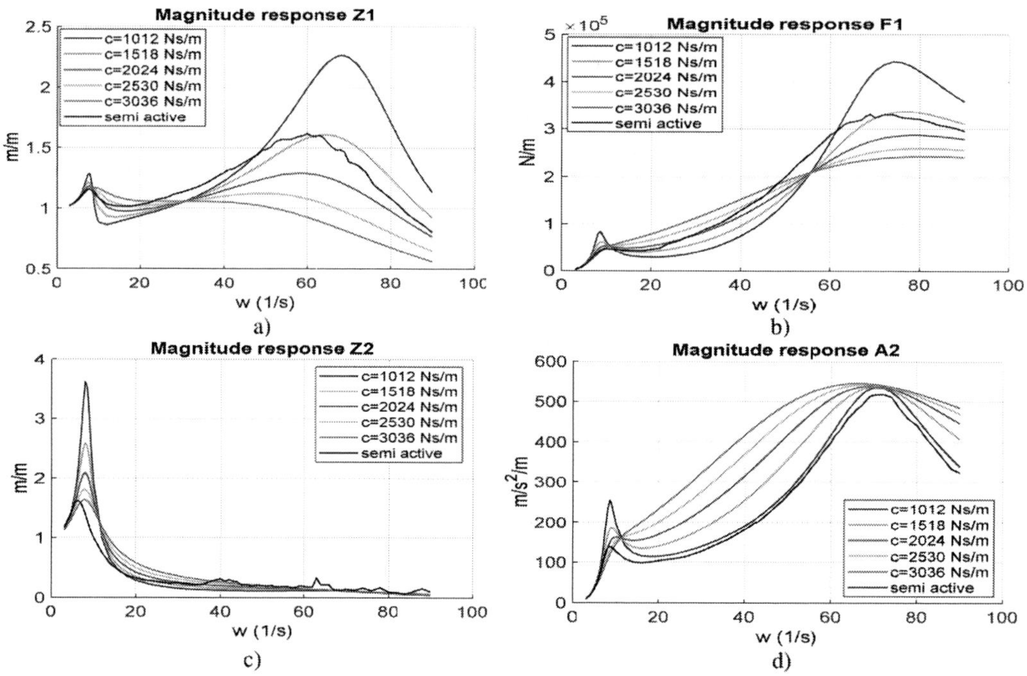

Figure 11. Magnitude responses for control of type (7)

This control realizes the best vibroisolation of the sprung mass, but the unsprung mass dynamics is increased. In order to reduce the dynamics of the unsprung mass, respectively the stability of the vehicle on the road, the following control is considered (as called "Ground hook" control type):

$$c_s = c_s^{\ell} + \left(c_s^{u} - c_s^{\ell}\right) U_{us},\ U_{us} = \begin{cases} 1, & -\left(\dot{z}_{us} - \dot{z}_s\right)\dot{z}_{us} \le 0 \\ 0, & -\left(\dot{z}_{us} - \dot{z}_s\right)\dot{z}_{us} > 0 \end{cases}. \tag{8}$$

The obtained results are shown on figure 12. Inefficiency of the control system is established between the 1st and 2nd invariant frequency for z_s and F_s.

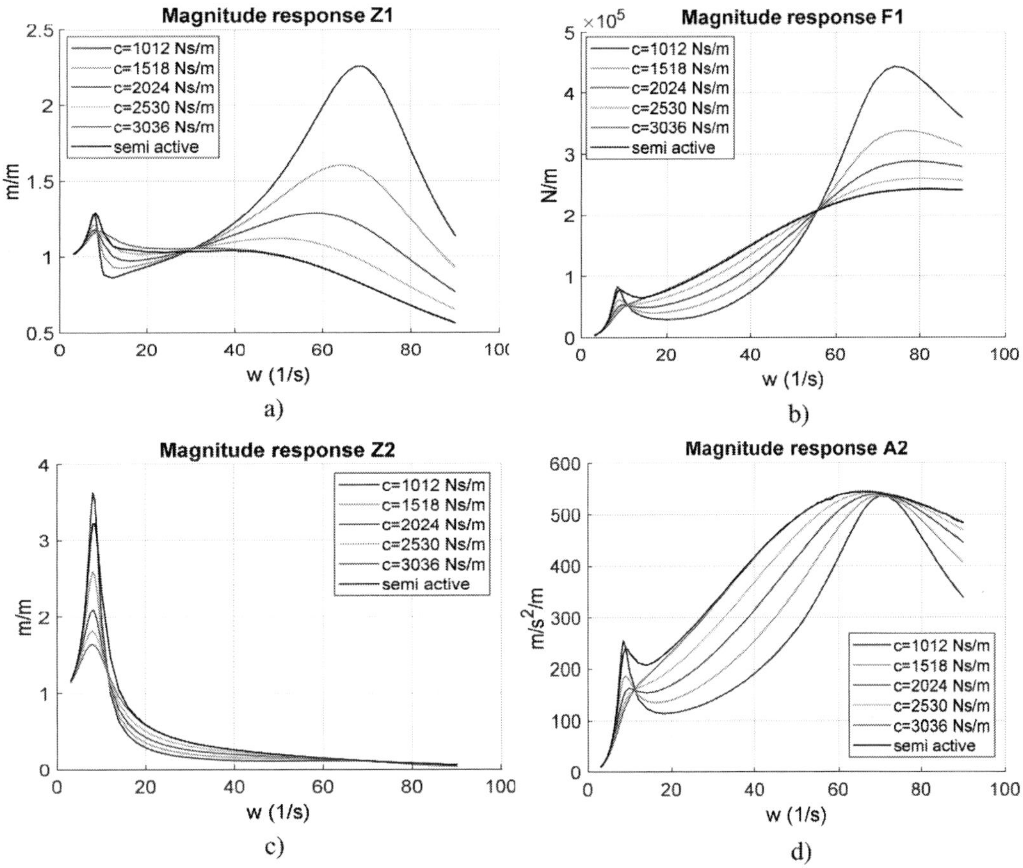

Figure 12. Magnitude responses for control of type (8)

The results are worse between the invariant frequencies.

The control similar to (6) due to the phase shift is not effective and it is consider the another form:

$$c_s = c_s^{\ell} + \left(c_s^{u} - c_s^{\ell}\right) U_{us},\ U_{us} = \begin{cases} 1, & -\left(\dot{z}_{us} - \dot{z}_s\right) z_{us} \le 0 \\ 0, & -\left(\dot{z}_{us} - \dot{z}_s\right) z_{us} > 0 \end{cases}. \tag{9}$$

The results are shown of figure 13. Those results are good between the invariant frequencies, but outside of them they are worse than for the control of type (8) and both control laws may by combined as same way as in the control (7).

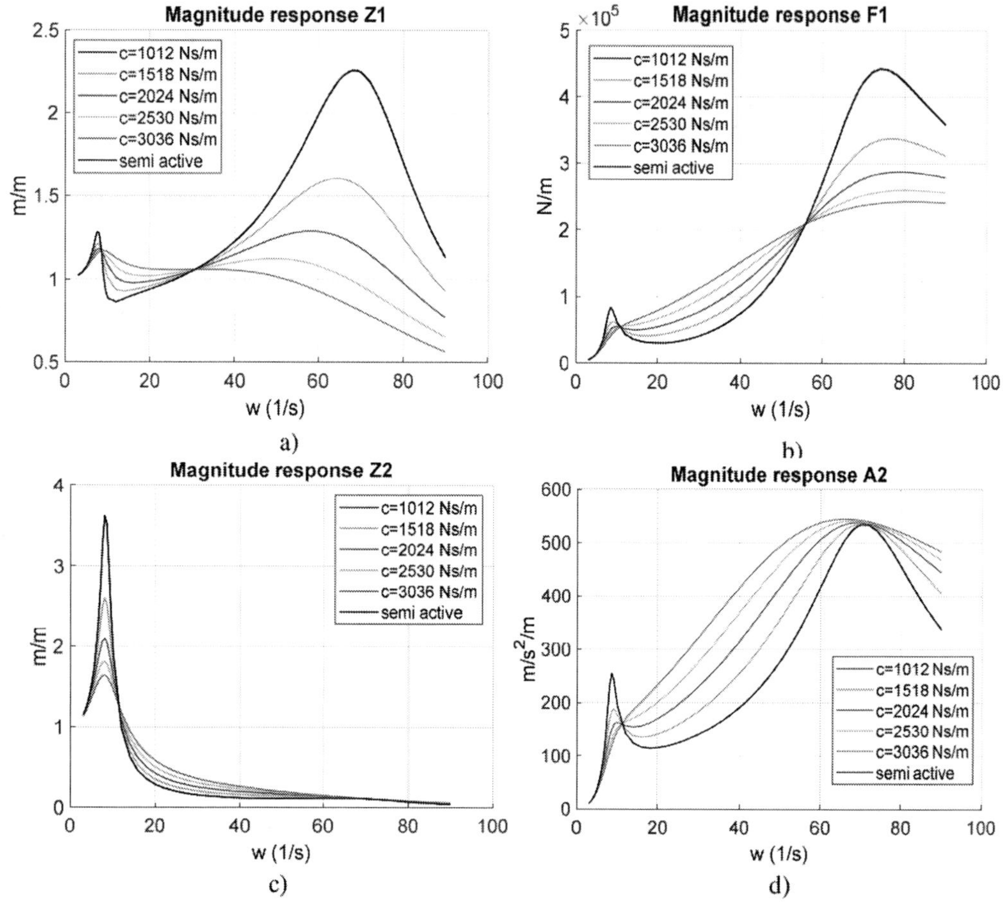

Figure 13. Magnitude responses for control of type (9)

Acknowledgements: This work was supported by OP "Science and education for smart growth " – centers of excellence TU Sofia „National center of Mechatronics and clean technologies", BG05M2OP001-1.001-0008-C01 Sect. Design, synthesis and testing of noise and vibration protection

References

[1] Dahlberg T., 1978, Ride Comfort and Road Holding of a 2-DOF Vehicle Travelling on a Randomly Profiled Road, *Journal of Sound and Vibration* **58** (2), pp.179-187

[2] Nikolov V. 2003 Simulation Model of Forced Oscillations of a Mechanical System with Four Degrees of Freedom *Mechanics of Machines* 2003 **5** pp.126-129

[3] Bala Raju A., Shruti R., Venkatachalam R., 2014 Effect of Damping on Comfort Level of a Fully Independent Suspension System of an Automobile, *Int. Jour. of Innovative Research in Science, Engineering and Technology* **3** Issue 6 2014, pp. 13604-9;

[4] Bachev V., Nikolov V., Angelov I. 2014 A Study of the Natural Frequencies of the Free Undamped Related Oscillations of a Car *Int. Jour. Engineering and Automation Problems* National Institute of Aviation Technologies Moscow Russia, **2** 2014 pp.53-58

[5] Guiggiani M., 2014 *The Science of Vehicle Dynamics. Handling, Braking and Ride of Road and Race Cars*, Springer Dordrecht Heidelberg New York London, p.356

[6] Kralov I., Nedelchev K., 2016 Mass and Elasticity Synthesis of the Support of a Generator for Vibration Energy Harvesting, *Journal of the Balkan Tribological Association*, **22**, 2016 №3 A-I, pp.3213-3227

[7] Kralov I., 2017 New Solution for Transport and Industrial Noise Protection through Reflective Noise Barriers, BulTrans2017, Sozopol, MATEC Web of Conferences **133**, 06001(2017) p.5

[8] Lord Corporation 2006 *MR Damper RD-1005-3 Product Bulletin, MR Solutions* Customer Service

IOP Conf. Series: Materials Science and Engineering **618** (2019) 012067 doi:10.1088/1757-899X/618/1/012067

MULTI-CRITERIA SYNTHESIS OF FREQUENCY-MODULATED DISCRETE CONTROL OF SEMI-ACTIVE VEHICLE SUSPENSION - PART 2 MULTI-OBJECTIVE SYNTHESIS

J Genov

Technical University of Sofia, 8 Kl. Ohridski Blvd, Sofia 1000, Bulgaria

E-mail: j_genov@mail.bg

Abstract. The paper discusses obtaining of a control law of semi-active vehicle suspension realizing compromise between requirements for ride comfort and vehicle stability on road pavement. The problem is solved as multi-objective synthesis based on a selection from the Pareto set of the optimal compromise solution by using the Game theory's strategy.

1. Introduction

The first part of the publication discusses control systems for vehicle semi-active suspension. Each of the considered control laws achieved very good characteristics with respect to one of the two main criteria: ride comfort and stability on road, but not simultaneously for the both. Achieving high quality on one of the criterions usually leads to a severe deterioration of the quality of the other criterion. This is due of the contradiction nature and characteristics of these criteria and therefore the main synthesis problem is to determine the optimal compromise.

2. Implementation of a compromise frequency-modulated control

From the analysis of the results in the first part of paper may be makes the next conclusions for the realization of the compromise control of semi-active suspension:

- For the frequency diapason up to the first invariant frequency, the control must to realize a high damping ratio;
- In the first half of the frequency range, between the first and second resonance frequency – approximately between the first and second invariant frequencies relation to the unsprung mass vibro-displacement the control must setting a minimal damping ration;
- For the second half of the above mentioned interval, optimal control is as in expression (10) from the first part of publication;
- In the zone of the second resonance and afterwards the control must realize a maximal damping coefficient for better vehicle stability.

Based on these conclusions, the frequency-modulation control is achieved by means of a four filters for each of range, respectively: low-pass filter for the first frequency range, two band-pass filters for the second and the third range and high-pass filter for the fourth.

For the considered primer the filters characteristics are given in table 1.

IOP Conf. Series: Materials Science and Engineering **618** (2019) 012067 doi:10.1088/1757-899X/618/1/012067

Table 1. Filters characteristics

	LP	BP1	BP2	HP
band pass frequencies s^{-1}	7	11÷40	47÷65.4	65.4
band stop frequencies s^{-1}	7.8	7.6, 44.4	44, 68.4	68.4
band pass magnitude	[0.9÷1]	[0.9÷1]	[0.9÷1]	[0.9÷1]
band stop magnitude	< 1e-5	< 1e-5	< 1e-5	<< 1e-5
filter's order	12	24	21	16
approximation	elliptic	elliptic	elliptic	elliptic

If the signals passed by each filter are:

$$z_{us}^{\ell p}(t),\ z_{us}^{bp_1},\ z_{us}^{bp_2}(t),\ z_{us}^{hp}(t)\ \Rightarrow\ z_{us}^{fsum}(t)=z_{us}^{\ell p}(t)+\ z_{us}^{bp_1}(t)+z_{us}^{bp_2}(t)+z_{us}^{hp}(t), \tag{1}$$

The control can be defined as:

$$\left| \begin{array}{l} c_s = c_s^{\ell} + \left(c_s^{u} - c_s^{\ell}\right) U_{us}, \\ U_{us} = \dfrac{\left|z_{us}^{\ell p}(t)\right| + k_{bp_2}\left|z_{us}^{bp_2}(t)\right| + \left|z_{us}^{hp}(t)\right|}{\left|z_{us}^{\ell p}(t)\right| + \left|z_{us}^{bp_1}(t)\right| + \left|z_{us}^{bp_2}(t)\right| + \left|z_{us}^{hp}(t)\right|},\quad k_{bp_2} = \begin{cases} 1, & -F_s z_{us} \le 0\ . \\ 0, & -F_s z_{us} > 0 \end{cases} \end{array} \right. \tag{2}$$

The obtained results are shown on figure 1.

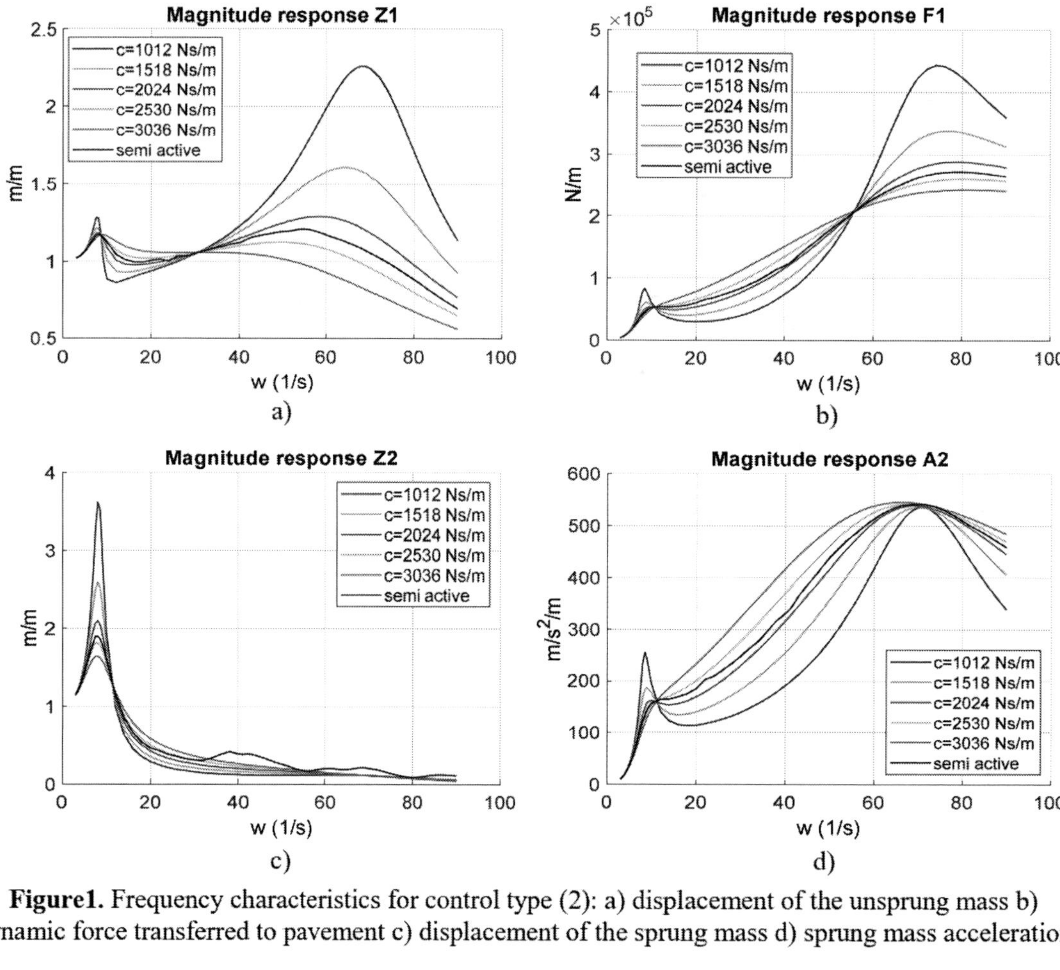

Figure1. Frequency characteristics for control type (2): a) displacement of the unsprung mass b) dynamic force transferred to pavement c) displacement of the sprung mass d) sprung mass acceleration

The control uses the feedback from unsprung displacement. For this reason the obtaining results are very good for the stability factors zus (z1) and Fus (F1). They are good and for the ride comfort's factors but for obtaining better results let's combine this control law with control law (7) from the first part of the publication:

$$c_s = c_s^{\ell} + \left(c_s^{u} - c_s^{\ell}\right)U_s,$$
$$U_s = \begin{cases} 0 \quad , \quad [q>0]\cap[p>0], \quad q = F_s^d\,\ddot{z}_s,\ p = F_s^d\,\dot{z}_s \\ U_s^v = \dfrac{\left|z_s^{\ell p}\right|}{\left|z_s^{\ell p}\right| + \left|z_s^{bp}\right|}, \quad [q\le 0]\cap[p>0],\ U_s^a = \dfrac{\left|z_s^{bp}\right|}{\left|z_s^{\ell p}\right| + \left|z_s^{bp}\right|}, \quad [q>0]\cap[p\le 0] \\ U_s^v + U_s^a \quad , \quad [q\le 0]\cap[p\le 0] \end{cases} . \tag{3}$$

This reconciliation, which implements the optimal compromise between the dynamics of the overgrown and the unredeemed masses, is done by introducing a compromise factor $\alpha \in [0,1]$ through which multi-criteria control law is represented in the form of:

$$\left| \begin{array}{l} c_s = c_s^{\ell} + \left(c_s^{u} - c_s^{\ell}\right)U, \qquad U = \alpha U_s + (1-\alpha)U_{us},\ \alpha \in [0 \div 1] \\ U_s = \begin{cases} 0 \quad , \quad [q>0]\cap[p>0] \\ U_s^v = \dfrac{\left|z_s^{\ell p}\right|}{\left|z_s^{\ell p}\right| + \left|z_s^{bp}\right|}, \quad [q\le 0]\cap[p>0] \\ U_s^a = \dfrac{\left|z_s^{bp}\right|}{\left|z_s^{\ell p}\right| + \left|z_s^{bp}\right|}, \quad [q>0]\cap[p\le 0] \\ U_s^v + U_s^a \quad , \quad [q\le 0]\cap[p\le 0] \\ q = F_{\text{сьп},2}\,\ddot{z}_s,\ p = F_{\text{сьп},2}\,\dot{z}_s \end{cases} \\ U_{us} = \dfrac{\left|z_{us}^{\ell p}(\mathrm{t})\right| + k_{bp_2}\left|z_{us}^{bp_2}(\mathrm{t})\right| + \left|z_{us}^{hp}(\mathrm{t})\right|}{\left|z_{us}^{\ell p}(\mathrm{t})\right| + \left|z_{us}^{bp_1}(\mathrm{t})\right| + \left|z_{us}^{bp_2}(\mathrm{t})\right| + \left|z_{us}^{hp}(\mathrm{t})\right|}, \\ k_{bp_2} = \begin{cases} 1, & -F_s^d z_{us} \le 0 \\ 0, & -F_s^d z_{us} > 0 \end{cases} \end{array} \right. \tag{4}$$

3. Description of optimization algorithm

The optimizing parameters are the spring coefficient of the suspension and the compromise factor:

$$\mathcal{X} = [k_s, \alpha] \in D_x . \tag{5}$$

The vector criterion consists of four object function represented the areas of the considered magnitude responses:

$$\mathrm{J}(\mathcal{X}) = \left[\mathrm{J}_1(\mathcal{X}),\ \mathrm{J}_2(\mathcal{X}),\ \mathrm{J}_3(\mathcal{X}),\ \mathrm{J}_4(\mathcal{X})\right]^{\mathrm{T}}, \tag{6}$$

where: $$\left| \begin{array}{l} J_1(\mathcal{X}) = \sum_{k=1}^{N} \left| W_{z_{us}\xi}(\omega_k) \right|_k, \ J_3(\mathcal{X}) = \sum_{k=1}^{N} \left| W_{\ddot{z}_s\xi}(\omega_k) \right|_k \\ J_2(\mathcal{X}) = \sum_{k=1}^{N} \left| W_{z_s\xi}(\omega_k) \right|_k, \ J_4(\mathcal{X}) = \sum_{k=1}^{N} \left| W_{F_{us}^{Dyn}\xi}(\omega_k) \right|_k \end{array} \right., \ \omega_k = k\Delta\omega; \quad (7)$$

$W_{z_{us}\xi}(\omega_k)$, $W_{z_s\xi}(\omega_k)$, $W_{\ddot{z}_s\xi}(\omega_k)$, $W_{F_s^d\xi}(\omega_k)$ are the discrete magnitude responses between input kinematic excitation ξ and respectively z_{us}, z_s, $\ddot{z}_s$, F_{us}^{Dyn}.

The calculation of the magnitude characteristics is based on setting of a harmonic kinematic excitation with frequency ω_k and then using the least mean squares method, an approximation the obtained numerical solution with a harmonic with the same frequency and in such way is obtaining the ratio between the amplitudes of the output magnitude and input excitation.

The multiobjective solution is obtains by the next algorithm:

- A single criteria optimization is executed for each of the criterions and are defined the optimum values:

$$J_i^{opt} = J_i\left(\mathcal{X}_i^{opt}\right), \ i=1,4; \quad (8)$$

- it is defined the maximal value of each criterion in the optimum of others:

$$J_i^{max} = \max_{j=1..4 \cap j \neq i} \left\{ J_i\left(\mathcal{X}_j^{opt}\right) \right\}, \ i=1,4; \quad (9)$$

- it is formed a cooperative game schema by scaling of the functions as:

$$J_i^*(\mathcal{X}) = \frac{J_i(\mathcal{X}) - J_i^{opt}}{J_i^{max} - J_i^{opt}}, \ i=1,..4; \quad (10)$$

- by the scaled criterions is formed the weighted criterion:

$$\mathcal{L}(\mathcal{X}) = \sum_{i=1}^{n} \lambda_i J_i^*(\mathcal{X}), \quad (11)$$

where the weighted vector is $\lambda = \left\{ \lambda_i \left| \begin{array}{l} \lambda_i \in [0,1) \\ \sum_{i=1}^{4} \lambda_i = 1 \end{array} \right. \right\}$;

- it is realized the hierarchical optimization on two levels
  - On the upper level is optimizes a super criteria with respect to weighted vector

$$S = \min_{\lambda} \max_{i=1,..4} \left(J_i^*\left(\mathcal{X}^p\right) \right), \quad (12)$$

  - On the low level Pareto solution is obtains $\mathcal{X}^p$ by minimizing the weighted criterion:

$$\mathcal{X}^p \Leftarrow \min_{\mathcal{X}} \mathcal{L}(\mathcal{X}). \quad (13)$$

This makes it possible to determine Nash's stationary solution (the optimal compromise) without identifying the entire Pareto set.

4. Numerical example

The algorithm is illustrated by the next example

Set of possible solutions: $D_x = \begin{Bmatrix} k_s \in [12000 \div 28000], N/m \\ \alpha \in [0 \div 1] \end{Bmatrix}$

The model`s parameters are: $m_{us}=40\ kg$, $m_s=320\ kg$, $k_{us}=180000\ N/m$, $c_{us}=100\ Ns/m$.

The range of variation of the damping coefficient of the semi-active damper is: $c_s \in \left[c_s^{\,l}, c_s^{\,u}\right]$,

$c_s^{\,u} = 3542\ Ns/m$, $c_s^{\,\ell} = 1012\ Ns/m$.

Number of the discrete frequencies $N = 90$, $\Delta\omega = 1s^{-1}$.

Results from the single criteria optimization:

Table 2. Results from single criteria optimization

J	Values of optimizing parameters		Values of the object functions:			
	k_s N/m	α -	J_1^{ω}	J_2^{ω}	J_3^{ω} s^{-2}	J_4^{ω} N/m
J_1	27000	0.0625	85.63	30.38	2.7922e^4	1.0973e^7
J_2	12000	0	90.2	25.02	2.6740e^4	1.1418e^7
J_3	13000	0.9375	124.07	29.55	2.3927e^4	1.65e^7
J_4	27000	0.0625	85.63	30.38	2.7922e^4	1.0973e^7

Cooperative game construction and criterion transformation:

Table 3. Game boundaries

	J_1^{ω}	J_2^{ω}	J_3^{ω} s^{-2}	J_4^{ω} N/m
J^{opt}	85.63	25.02	2.3927e^4	1.0973e^7
J^{max}	124.07	30.38	2.7922e^4	1.6500e^7

Table 4. Results from multiobjective optimization

Multiobjective optimization	Optimizing parameters		Values of the object functions::							
	k_s N/m	α -	J_1^{ω} s^{-1}		J_2^{ω} s^{-1}		J_3^{ω} s^{-3}		J_4^{ω} N/ms	
$\min\limits_{x\in\Theta} \max\limits_{i=1..4} (J_i(x))$	14250	0.4219	0.35	99	0.20	26.1	0.05	2.41e^4	0.21	1.25e^7

The obtained results are shown on figure 2. They are better for ride comfort characteristics while those for vehicle stability are very slightly degraded in some frequency diapasons in comparison with the passive implementation of the suspension.

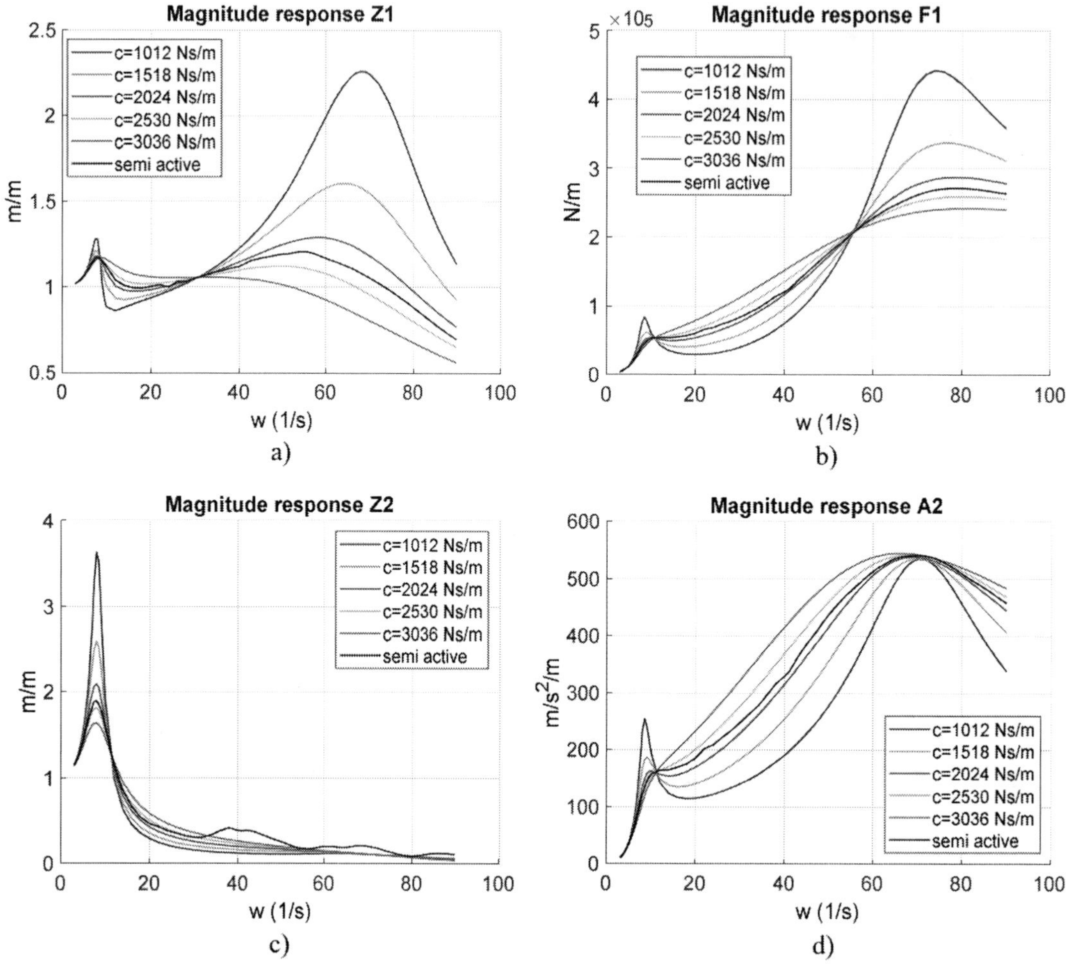

Figure 2. Magnitude responses of the multiobjective solution

5. Conclusion

Based on the analysis of the amplitude and phase-frequency characteristics, approaches for semi-active suspension control systems of a vehicle, represented by a "Quarter car" model are considered. The innovative idea in the study is the combination of the advantages of different control laws, based on frequency modulation, achieved via the signals filtration. The classical discrete "on-off" control is transformed into a continuously acting. This type of a control system gives much better performance in terms of ride comfort and stability of the vehicle.

Acknowledgements

This work was supported by OP "Science and education for smart growth " –centers of excellence TU Sofia „National center of Mechatronics and clean technologies", BG05M2OP001-1.001-0008-C01 Sect. Design, synthesis and testing of noise and vibration protection

References

[1] Nash J. 1951 Non-cooperative Games, *Annals of Mathematics* **54** 2 (September) pp. 286–295

[2] Podinovsky V. Nogin V. 1982 *Pareto-optimal solutions of multicriteria tasks*, (in Russion) M. Scince p. 256

[3] Venkov G. 1991 *Basics of Automated Design - Systems for Optimal Design* (in Bulgarian), Offset Traffic, Sofia, p.212;

[4] Genov J. Tashkov S. Venkov G. 2009 On the Synthesis of Optimal and Quazi-optimal Control of Semi-Active insulation System, *35th International Conference AMEE-2009, American Institute of Physics Conference Proceedings 1184*, pp. 32-38

IOP Conf. Series: Materials Science and Engineering **618** (2019) 012068 doi:10.1088/1757-899X/618/1/012068

Shock-wave structures of prospective combined ramjet engine

M V Chernyshov[1], K E Murzina[1], S A Matveev[1], V V Yakovlev[2]

[1] Baltic State Technical University "VOENMEH"
[2] Peter the Great St Petersburg Polytechnic University

Abstract. A concept of the air jet engine of mixed type, which combines the operating principles of detonation engine and traditional ramjet, is proposed. A part of inflow that has the smallest losses of the stagnation pressure at forming shocks in the inlet is burned in the combustion chamber of that device. Another part of the inflow, that has the smaller stagnation pressure, has the significantly higher temperature at the same time. Due to this fact, it can be effectively used as in the detonation engine.

1. Introduction

Stability and safety of supersonic flight directly depend on efficiency and constructive simplicity of jet engine. Straight-through ramjet thruster seems to be one of the most optimal decisions of the problem. Realized properly, it provides necessary thrust at relatively low drag force coefficient at supersonic flight speeds.

Jet engines of the traditional design, working on chemical fuels, have reached the limit of its efficiency, in fact, up to nowadays. For the significant growth of thrust and increase of efficiency, fundamentally new types of jet engines are to be worked out.

A concept of the straight-through detonation ramjet was discussed in scientific literature in several last decades (Dunlap R., Brehm R. L., & Nichols J. A., 1958), (Cambier J. L., Adelman H., & Menees G. P., 1990), (Choi J. Y., Jeung I. S., & Yoon Y., 1998). Despite the theoretical simplicity of the device, its functioning in practice is difficult because of formation of complex shock-wave structures with different temperatures of fuel/air mixture behind their elements. As a result, only part of fuel that has the larger temperature detonates downstream the forming shock waves (Bulat P. V., Denissenko P. V., & Volkov K. N., 2016), (Bulat P. V. & Denissenko P. V., 2016). In particular, flow temperature is much higher downstream the Mach stem, than behind a system of incident shock and the reflected one, at simple Mach reflection. Consequently, as a rule, only part of the fuel/air mixture that situates downstream Mach stem really detonates.

In its turn, as the studies (Oswatitsch K., 1956), (Hermann R., 1956), (Petrov G. I., 1992) of supersonic air inlets shows, full pressure losses in the stream behind the incident shock and the reflected one is significantly smaller than downstream the Mach stem. That part of the flow is preferable for the succeeding application in a combustion chamber of the "classical" ramjet.

The essential distinctions (at least, in several times, but sometimes – by a factor of ten) in gas stagnation pressures and temperatures downstream the triple-shock configurations (TCs) of various types (for example, behind the TCs at Mach reflection) were discussed in (Uskov V. N. & Chernyshov M.V., 2006),

(Chernyshov M. V. & Kapralova A. S., 2017). The extremal properties of shock-wave systems (i.e.. sequences of shocks and expansion waves) and structures (i.e. sets of shocks and waves crossing in common point) were applied earlier (Omel'chenko A. V. & Uskov V. N., 1995), (Omel'chenko A. V. & Uskov V. N., 1996), (Omel'chenko A. V. & Uskov V. N., 1997), (Omel'chenko A. V. & Uskov V. N., 1999) to optimization of shapes of streamlined bodies (Silnikov M. V. & Chernyshov M. V., 2015), supersonic inlets (Oswatitsch K., 1956), (Hermann R., 1956), (Petrov G. I., 1992), in jet technologies and other engineering application of gas dynamics.

Based on the marked special features of occurring shock-wave structures, we propose the concept of the combined air breather, with detonation of air/fuel mixture downstream the Mach stems of the triple configurations, and with its "ordinary" combustion downstream the system of oblique reflected shocks.

In this study, we briefly analyse the relations of temperatures and full pressures downstream the shock-wave structure of Mach shock reflection in the narrowing channel (for example, in supersonic inlet). It is shown that the relations can reach very high quantities, and it serves as a theoretical ground for the outworked design of the prospective air breather.

2. Model and methods

2.1. Flow structure and basic relations on gasodynamic discontinuities

As a model problem that characterizes the operating principle of the prospective device, we consider the supersonic (with Mach number M of the undisturbed flow) stream of the inviscid perfect gas with ratio γ of the specific heats (for definiteness, $\gamma = 1.4$ in all following computational examples) in the plane narrowing channel between two wedges (see Figure 1).

The strength J_1 (relation of static pressures downstream the shock and upstream it) of the incident oblique shock 1 that forms at the entrance into the channel depends on the flow deflection angle β_1 at the front edge of the wedge:

$$\tan\beta_1 = \sqrt{\frac{(1+\varepsilon)M^2 - \varepsilon - J_1}{J_1+\varepsilon}} \cdot \frac{(1-\varepsilon)(J_1-1)}{(1+\varepsilon)M^2 - (1-\varepsilon)(J_1-1)} \,. \tag{1}$$

Here $\varepsilon = (\gamma-1)/(\gamma+1)$.

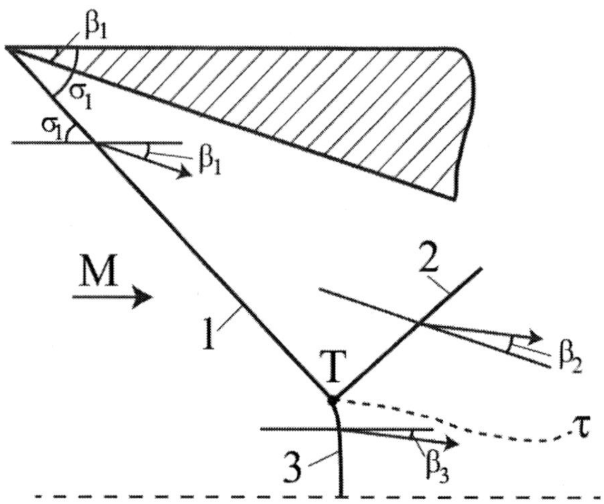

Figure 1. General layout of the flow

At rather sufficient strength J_1 of the incident shock, its irregular (Mach) reflection from the flow symmetry plane occurs with formation of the reflected shock 2 (with strength J_2) and main shock (Mach stem) 3 (with strength J_3). The strengths of the shocks 1-3 and the corresponding flow deflection angles β_i $(i = 1..3)$ are constrained with compatibility conditions at the slipstream τ that emanates from their common (triple) point T:

$$J_1 J_2 = J_3, \; \beta_1 + \beta_2 = \beta_3 . \qquad (2)$$

Flow deflection angles β_2 and β_3 at the shocks 2 and 3 are connected with strengths of the corresponding shocks and flow Mach numbers upstream them with the relations that are similar to (1). Flow Mach number M_i behind any shock i depends on shock strength J_i and flow Mach number M_j before the shock as it follows:

$$M_i = \sqrt{[(J_i + \varepsilon)M_j^2 - (1-\varepsilon)(J_i^2 - 1)] / [J_i(1 + \varepsilon J_i)]} . \qquad (3)$$

2.2. Single Mach reflection, domain of existence

To calculate the appearing TC (i.e., to solve the system (1-3) of the relations on gasodynamic discontinuities), it is necessary to specify Mach number M of the undisturbed flow and the strength J_1 of the incident shock. The range of the acceptable strengths of the shocks widens unlimitedly at large Mach numbers Therefore the TCs are being analyzed traditionally at the plane $(M; \sigma_1)$ Here σ_1 is inclination angle of the shock 1 to flow velocity vector upstream it (see Figure 1) that is bound with shock strength as in follows:

$$J_1 = (1+\varepsilon)M^2 \sin^2 \sigma_1 - \varepsilon .$$

IOP Conf. Series: Materials Science and Engineering **618** (2019) 012068 doi:10.1088/1757-899X/618/1/012068

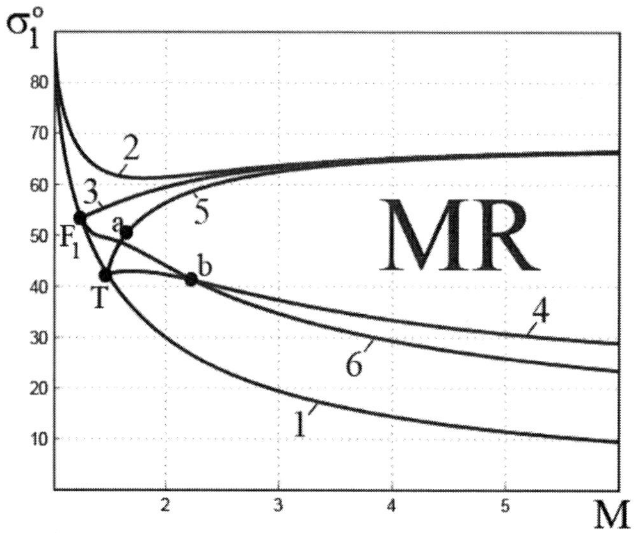

Figure 2. Triple-shock configuration existence domain

The curve 1 in Figure 2 corresponds to the degeneration of the shock 1 into the weak (acoustic) discontinuity ($\sigma_1 = \arcsin(1/M)$, $J_1 \to 1$) and serves as a lower border of TC existence domain. Curve 2 corresponds to critical flow velocity downstream the shock 1 ($M_1 = 1$). Curve 3 is the upper border of TC existence domain. It starts from the point F_1, where $M_{F_1} = \sqrt{(2+2\sqrt{\varepsilon})/(1+2\sqrt{\varepsilon})} = 1.245$ at the curve 1, and obeys the following relation:

$$M = \sqrt{[A(B-C)-1+\varepsilon]/\varepsilon},$$

$$A = (1+\varepsilon J_1)/[(1+\varepsilon)(J_1(1-3\varepsilon)-4\varepsilon^2)],\ B = J_1(1-2\varepsilon-\varepsilon^2)-2\varepsilon^2,\ C = 2\varepsilon\sqrt{\varepsilon(1+\varepsilon J_1)(J_1+\varepsilon)}.$$

Depending on flow deflection directions at the shocks 1-3, triple configuration of the first type (TC-1, $\beta_1\beta_2 < 0$, $\beta_1\beta_3 < 0$), the second one (TC-2, $\beta_1\beta_2 < 0$, $\beta_1\beta_3 > 0$), and the third one (TK-2, $\beta_1\beta_2 > 0$, $\beta_1\beta_3 > 0$) can be separated. As a rule, TC-1 and TC-3 result in special cases of the interaction of contra-directional and co-directional shocks, correspondingly TC-2 results in general case of Mach reflection.

Curve 4 at Figure 2 borders the domains of TC-1 and TC-2 existence. It corresponds to the stationary Mach configuration with normal Mach stem ($\beta_3 = 0$). Corresponding value $J_1 = J_N(M)$ of the strength of the shock 1 can be determined from the equation:

$$\sum_{n=0}^{3} E_n J_N^n = 0,$$

$$E_3 = 1-\varepsilon,\ E_2 = -[(1+\varepsilon-\varepsilon^2+\varepsilon^3)M^2 + (1-\varepsilon)(1-\varepsilon+\varepsilon^2)],$$

$$E_1 = \varepsilon[(1+\varepsilon)M^2+1-\varepsilon]\cdot[(1-\varepsilon)M^2-2+\varepsilon],\ E_0 = (1-\varepsilon)(M^2-1)((1+\varepsilon)M^2-\varepsilon).$$

According to theoretical and experimental results generalized of (Ben-Dor G., 2007), (Vasil'ev E. I., 1999), TC-1 does not appear at Mach reflection in steady flows (Mach reflection at $J_1 < J_N(M)$ is so-called von Neumann reflection with Mach stem that is convex upstream the flow). Therefore, according to well-known von Neumann criterion (Ben-Dor G., 2007), (Vasil'ev E. I., 1999), (Adrianov A. L., Starykh A. L., & Uskov V. N., 1995), (Hornung H. G., 1986) of Mach/regular shock reflection transition, curve 4 is the lower border of single Mach reflection existence domain.

Curve 5, that corresponds to Mach reflection with normal reflected shock ($\beta_2 = 0$), serves as the upper border of TC-2 existence domain at (M, σ_1) diagram. Corresponding triple configuration (TCT-2-3, (Uskov V. N. & Chernyshov M.V., 2006)) is transitional between TC-2 and TC-3. Mach reflection with formation of TC-3 (so-called Guderley reflection (Ben-Dor G., 2007) or Vasil'ev one (Vasil'ev E. I., 1999) realizes very seldom in practice (detached shock formation before the narrowing channel is much more likely). Therefore the curve 5, which obeys the equation

$$M^4 - rM^2 + (J_1 - 1)(J_1 + 2 - \varepsilon)/(1 - \varepsilon) = 0,$$

$$r = (J_1 - 1)(J_1 + 2 - \varepsilon)/(J_1 + \varepsilon) + (J_1 + \varepsilon)/(1 + \varepsilon) + (1 + \varepsilon J_1)^2 / \left[(1 - \varepsilon)(J_1 + \varepsilon)^2\right],$$

serves as the upper border of single Mach reflection existence.

Both curves 4 and 5 (lower and upper borders of single Mach reflection domain) come out from the point T ($M_T = \sqrt{(2 - \varepsilon)/(1 - \varepsilon)} = 1.483$) at curve 1. Thereby, the triple-shock configurations of the second type considered here exist only at $M > M_T$.

2.3. Target functions

Distinction of temperatures and stagnation pressures behind the shocks 2 and 3 (in corresponding streams separated by the slipstream) can be characterized by functions:

$$I_T = T_3/T_2 = \Theta_3/(\Theta_1\Theta_2) = E_3/(E_1E_2), \quad I_{p_0} = p_{02}/p_{03} = I_1I_2/I_3 = [E_3/(E_1E_2)]^{(1+\varepsilon)/2\varepsilon}. \quad (4)$$

Here $E_i = (1 + \varepsilon J_i)/(J_i + \varepsilon)$ is the relation of gas densities before the i-th shock and downstream it according to Hugoniot shock adiabat, $\Theta_i = E_i J_i$ is the relation of the static temperatures (T), $I_i = (J_i E_i^\gamma)^{(1-\varepsilon)/2\varepsilon}$ is the stagnation pressure (p_0) recovery factor.

The following functions characterize the distinction of flow parameters downstream TCs and in the undisturbed flow:

$$I_{T_3} = T_3/T = \Theta_3, \quad I_{T_2} = T_2/T = \Theta_1\Theta_2, \quad I_{p_{03}} = p_{03}/p_0 = I_3, \quad I_{p_{02}} = p_{02}/p_0 = I_1I_2. \quad (5)$$

Consequently, high growth of temperature at the shock 3 and relatively high full pressure recovery behind the shock 2 (comparing with the undisturbed flow or the flow at the opposite side of the slipstream) can be estimated through the study of the objective functions (4-5) at the domain MR between the curves 4 and 5 in Figure 2.

3. Results and discussion

3.1. Theoretical results

Curve 6 (Figure 2) indicates the triple configurations with extreme values of both the full pressure relation and the static temperature relation across the slipstream at any Mach number of the undisturbed flow. At small Mach numbers ($M > M_{F_1}$), such extreme TCs belong to the third type and correspond to special cases of the intersection of the shock of the same direction. Starting from $M = M_a = 1.596$, extreme TCs correspond to Mach reflections (TC-2 are extreme). At last, at $M > M_b$ (here $M_b = \sqrt{4-3\varepsilon+\varepsilon^2}/(1-\varepsilon) = 2.254$), triple-shock configuration of the first type are extreme both on relations of temperatures and stagnation pressures.

The extreme values of the functions I_T and I_{p_0}, that can be reached at any Mach number for the TC of absolutely voluntary type, are shown by curves 1a and 1b in Figure 3. Those "optimal" values strive to finite limits (Uskov V. N. & Chernyshov M.V., 2006) at large Mach numbers:

$$I_T \to 1/\varepsilon = 6.0, \qquad I_{p_0} \to \varepsilon^{-\frac{1+\varepsilon}{2\varepsilon}} = 529.1 \tag{6}$$

Similar values of the functions I_T and I_{p_0}, that are reached at the lower border of single Mach reflection domain (corresponding to stationary Mach configuration, or to "mechanical equilibrium" von Neumann reflection transition criterion), are shown by curves 2a and 2b in Figure 3. Those relations of flow parameters behind the stationary Mach configuration also strive asymptotically to some finite limits (Uskov V. N. & Chernyshov M.V., 2006):

$$I_T \to -\left[1+2\varepsilon-2\varepsilon^3+\varepsilon^4+(1-\varepsilon)\sqrt{D}\right]/\left[2\varepsilon(2-\varepsilon)\right] = 3.363,$$

$$I_{p_0} \to \left[\frac{1+2\varepsilon-2\varepsilon^3+\varepsilon^4+(1-\varepsilon)\sqrt{D}}{2\varepsilon(2-\varepsilon)}\right]^{\frac{1+\varepsilon}{2\varepsilon}} = 69.72.$$

Here $D = (1+\varepsilon)^2 - \varepsilon(1-\varepsilon)\left[2(1+\varepsilon)(2-\varepsilon) - \varepsilon^3(1-\varepsilon)\right]$. Those limits are large, but they accede significantly to the limits (6) for voluntary TCs

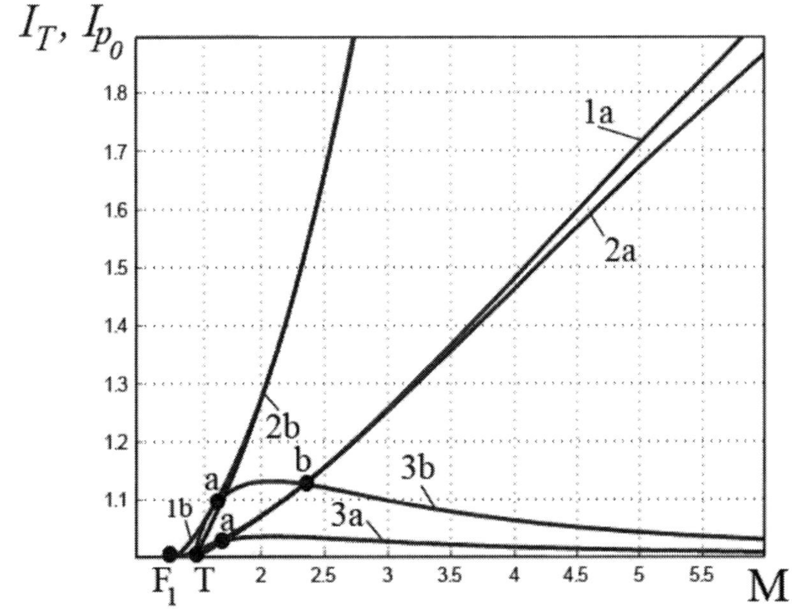

Figure 3. Relations of flow parameters downstream the triple point

However, comparing the curves 1a and 1b, 2a and 2b, we see that the considered relations of flow parameters calculated for stationary Mach configurations differ from the "optimal" ones insufficiently at small and moderate Mach numbers.

Curves 3a and 3b correspond to relations of temperatures and full pressures at the slipstream behind the transitional configuration TCT-2-3 (at the upper border of single Mach reflection existence domain). Indicated values coincide with the "optimal" ones only at $M = M_a$, and they are much smaller that the "optimal" values at moderate and large Mach numbers. As it is shown at (Uskov V. N. & Chernyshov M.V., 2006), those values asymptotically strive to unity at $M \to \infty$.

The relations (5) of flow parameters downstream some elements of TCs at Mach reflection and in the undisturbed flow require additional studies.

3.2. Numerical simulations

Flow in supersonic inlet of the prospective combined jet engine was simulated numerically A planar supersonic stream of the perfect gas flows into a narrowing channel between two wedges (see Figures 4 and 5). Mach number of the undisturbed flow $M = 5$ at normal atmospheric static pressure and temperature, ratio of gas specific heats $\gamma = 1.4$ Standard $k - \varepsilon$ turbulence model was used and academic version of ANSYS Fluent program code was applied to flow field calculations.

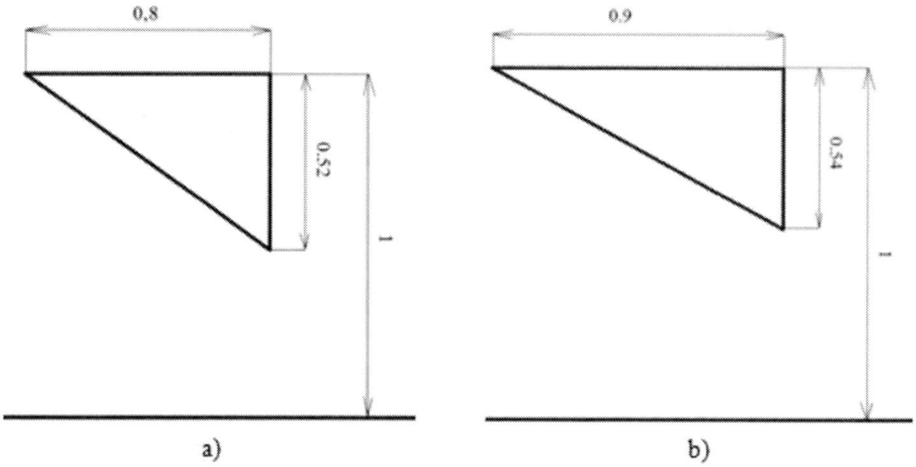

Figure 4. Geometrical parameters of idealized supersonic inlet: a) with Mach stem; b) with detached shock

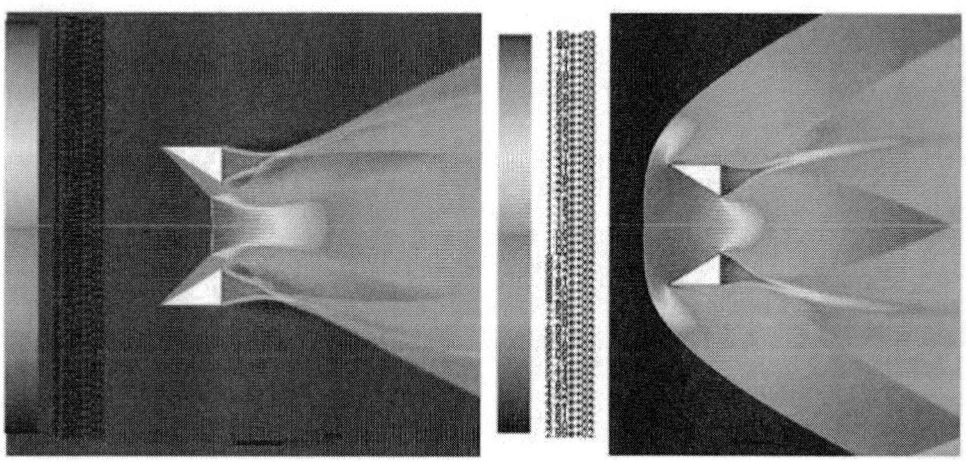

Figure 5. Spatial distribution of temperature at the flow with Mach reflection (left) and at the flow with detached shock (right)

It occurred that rather small variation of geometrical parameters of the supersonic inlet (compare Figures 4a and 4b; all linear dimensions are given in meters) can lead to crucial variation of flow field and corresponding shock-wave structure. Regime with detached bow shock (Figure 5, right) leads to crucial increase of drug force and decrease of stagnation pressure comparing to the regime with Mach reflection (Figure 5, left) which was theoretically analyzed above. So, as we can see, not only optimal flow parameters of the flow with Mach stem formation, but also stability of this flow should be studied for the flight safety problem decision.

4. Conclusion

Relations of temperatures and full pressures in the streams behind the triple point of Mach reflection can reach very large values. This is the reason why high-temperature stream of air-fuel mixture behind the Mach stem (disc) is much more applicable to detonation engine, and high-pressure stream behind the system of reflected oblique shock is more useful in the combustion chamber of the ordinary ramjet. Derived relations theoretically justify such design of the prospective engine.

To realize proposed combined (hybrid) engine in practice, it is necessary to separate the flows downstream the triple point of Mach reflection. It means that the algorithm to determine the situation of triple point (or the Mach stem height) is necessary to be worked out. Analytical methods for shock-wave structures of supersonic flows in overexpanded jets and narrowing channels (see, for examples, studies (Omel'chenko A. V., Uskov V. N., & Chernyshev M. V., 2003), (Chanetz B., et al., 2003), (Silnikov M. V., Chernyshov M. V., & Uskov V. N., 2014), other works of V N Uskov and his colleagues) can be adopted for the solution of this problem.

Acknowledgement

This work was financially supported by the Ministry of Education and Science of the Russian Federation (agreement No. 14.577.21.0277, a unique identifier of applied scientific research RFMEFI57717X0277) and the grant from the Program Competitiveness Enhancement of Peter the Great St.Petersburg Polytechnic University, Project 5-100-2020.

References

[1] Adrianov A, Starykh A & Uskov V 1995 Interaction of Stationary Gasodynamic Discontinuities Nauka, Novosibirsk 180 p.
[2] Ben-Dor G 2007 Shock Wave Reflection Phenomena Springer Berlin, New York 342 p.
[3] Bulat P, Denissenko P & Volkov K 2016 Trends in the development of detonation engines for high-speed aerospace aircrafts and the problem of triple configurations of shock waves Part I Research of detonation engines Scientific and Technical Jour 16 (1), 1-21 (in Russian)
[4] Bulat P, Denissenko P (2016) Trends in the development of detonation engines for high-speed aerospace aircrafts and the problem of triple configurations of shock waves Part II Research of counterpropagating shock waves and triple shock wave configurations Scientific and Technical Journal of Information Technologies, Mechanics and Optics 16 (2), 199-223 (in Russian)
[5] Cambier J, Adelman H & Menees G 1990 Numerical simulations of an oblique detonation wave engine Journal of Propulsion and Power 6 (3), 315-323
[6] Chanetz B, Chernyshev M, Durand A, Khotyanovsky D, Kudryavtsev A, Omelchenko A & Ivanov M 2003 Analytical, numerical, and experimental investigation of shock wave reflection transition induced by variation of distance between wedges Proceedings of West East High Speed Flow Fields Conference CIMNE, Barcelona, 274-281
[7] Chernyshov M, Kapralova A 2017 Triple configurations of pursuit shock waves in conditions of ambiguity of the solution Vestnik Kontserna VKO "Almaz-Antey" 4, 46-52 (in Russian)
[8] Choi J, Jeung I & Yoon Y 1998 Numerical Study of Scram Accelerator Starting Characteristics AIAA Journal 36 (6), 1029-1038
[9] Dunlap R, Brehm R & Nichols J 1958 A Preliminary Study of the Application of Steady-State Detonative Combustion to a Reaction Engine Journal of Jet Propulsion 28 (7), 451-456
[10] Hermann R 1956 Supersonic inlet diffusers and introduction to internal aerodynamics Minneapolis: Minneapolis-Honeywell Regulator Company, Minneapolis, 378 p.
[11] Hornung H 1986 Regular and Mach reflection of shock waves Annual Review of Fluid Mechanics 18, 33-58

[12] Omel'chenko A, Uskov V & Chernyshev M 2003 An Approximate Analytical Model of Flow in the First Barrel of an Overexpanded Jet Technical Physics Letters 29 (3), 243-245
[13] Omel'chenko A, Uskov V 1995 Optimal Shock Wave Systems Fluid Dynamics 30 (6), 905-911
[14] Omel'chenko A, Uskov V 1996 Optimal Shock-Wave Systems Under Constraints on the Total Flow Turning Angle Fluid Dynamics 31 (4), 597-603
[15] Omel'chenko A, Uskov V (1997) An optimal shock-expansion system in a steady gas flow Journal of Applied Mechanics and Technical Physics 38 (2), 204-210
[16] Omel'chenko A, Uskov V (1999) Optimum overtaking compression shocks with restrictions imposed on the total flow-deflection angle Journal of Applied Mechanics and Technical Physics 40 (4), 638-646
[17] Oswatitsch K (1956) Gas Dynamics New York: Academic Press, 610 p.
[18] Petrov G I (1992) Aeromechanics of High Velocities and Space Research: Selected Works Nauka, Moscow, 306 p.
[19] Silnikov M, Chernyshov M 2015 The interaction of Prandtl-Meyer wave and quasi-one-dimensional flow region Acta Astronautica 109, 248-253
[20] Silnikov M, Chernyshov M & Uskov V 2014 Two-dimensional over-expanded jet flow parameters in supersonic nozzle lip vicinity Acta Astronautica 97, 38-41
[21] Uskov V, Chernyshov M 2006 Special and extreme triple shock-wave configurations Journal of Applied Mechanics and Technical Physics 47 (4), 492-504
[22] Vasil'ev E 1999 Numerical simulation of weak shock diffraction over a wedge under the von Neumann paradox conditions Computational Mathematics and Mathematical Physics 39 (8), 1335-1345.
[23] Zybina O, Gravit M, Stein Y 2017 Influence of carbon additives on operational properties of the intumescent coatings for the fire protection of building constructions IOP Conference Series: Earth and Environmental Science 90 (1) DOI: 10.1088/1755-1315/90/1/012227
[24] Pykhtin K, Simankina T, Sharmanov V, Kopytova A 2017 Risk-based approach in valuation of workplace injury rate for transportation and construction industry IOP Conference Series: Earth and Environmental Science 90 (1) DOI: 10.1088/1755-1315/90/1/012065
[25] Korshunov G, Polyakov S, Shunmin L 2017 Assurance of reliability and safety in liquid hydrocarbons marine transportation and storing IOP Conference Series: Earth and Environmental Science, 87 (6) DOI: 10.1088/1755-1315/87/6/062009
[26] Burlow V, Grachev M, Shlygina S. 2017 Adoption of management decisions in the context of the uncertainty of the emergence of threats Proceedings of 2017 20th IEEE International Conference on Soft Computing and Measurements SCM 2017 pp. 107-108. DOI: 10.1109/SCM.2017.7970510

IOP Conf. Series: Materials Science and Engineering **618** (2019) 012069 doi:10.1088/1757-899X/618/1/012069

Stochastic algorithms in multimodal 3PL segment for the digital environment

S Krasnov, E Zotova, S Sergeev, A Krasnov, M Draganov

Peter the Great St.Petersburg Polytechnic University, St.Petersburg, Russia

hsm.krasnov@gmail.com; sergeev2@yandex.ru; zea0284@gmail.com; alex073ul@mail.ru

Abstract. This paper discusses the problem of creation of control algorithms when working in the Third Party Logistic concept. It considers the process of supply management when several means of transport and an intermediate distribution center are used. It is assumed that goods are processed in the Machine-to-Machine mode by means of machine-readable codes. The controlling server processes the data and makes a decision either to use one or another mode of transport, or to store the cargo temporarily at the distribution center. An option is provided when an emergent stock of a limited y is placed directly at the end point so that the impact of market uncertainty in demand can be mitigated. The methods of mathematical modeling in the time of uncertainty of the operational environment are applied. The algorithm is based on the stochastic programming method. The practical focus of the work lies in orienting the results on multimodal logistics, which makes the model scalable. The computations, which use the algorithms developed in the paper, provide commercial network managers with a convenient tool for planning current activities taking into account market uncertainty in all operations, when dealing both with consumers and suppliers throughout the business chain, including distribution centers as independent nodes of machine-to-machine interaction. The implementation of the proposed models not only will bring economic benefits and reduce the global traffic, but will also help improve the environment and give a significant social effect.

1. Introduction

Globalization provides a wide area for business opportunities. At the same time, the coverage of the potential consumers is not limited spatially and geographically, and among the variety of markets the manager has to choose the best options from an economic point of view [1]. As work in any segment of the economy takes place in conditions of demand uncertainty, it is necessary to involve the mathematical tools [2], the criterion, as a rule, is the functionality formed from the profit maximization condition.

In the transition to the Machine-to-Machine interaction concept [3], a number of problems arise, associated with both the identification of items (goods in any container), as well as addressing the process of their movement. The widespread use of intelligent automation tools such as laser and radio frequency scanners, robotic warehouse complexes, inter-operational autonomous transport is based on the use of machine-readable codes [4]. For example, to solve the identification problem, it is possible to use the entire set of tools, ranging from linear barcode and 2D QR (Quick Response Code) to RFID (Radio Frequency Identification) tag recognition tools.

The practicability of application is primarily subject to the conditions of the item identification throughout the track, as well as restrictions on the equipment used in the transportation process. According to the accepted standards EAN/UCC and ITF [5], the maximum allowable encoding has from 10208 to 23648 characters. The size of the data array depends on the possibility of error correction, starting from the lower L level (allows recovery of up to 7% of information loss) with intermediate M and Q levels and ending with the upper H level (restores up to 30% of the information lost due to the ambient environment and other adverse factors).

The current level of machine-to-machine interaction is organized on the basis of cloud Internet solutions, so as control objects logistics algorithms [6] use both single units of cargo and large quantities formed in blocks. In this case, combined schemes based on RFID tags containing addresses of individual units and requiring access to the forwarding server are used. Such identification allows to switch to M2M algorithms during cross-docking and pick-by-line selection procedures [7], which are present in the vast majority of logistics schemes.

The difference in these technologies lies in the fact that cross-docking does not imply changing the machine labeling during processing by 3PL (Third Party Logistic) operator. Its advantage is a high speed SKU (Stock Keeping Unit) processing of a goods items. This minimizes the time spent within the DC (distribution center). Pick-by-line technology includes SKU processing, the restructuring of unit loads, which implies their deconsolidation. Thus, the goods are not delivered to the stock, but to the unloading zone. Accordingly, each of the smaller units must have its own machine-readable label with complete information.

This type of interaction covers exceptionally large data arrays, millions of records, each has 10-15 thousand characters, since today the global trend is to consolidate commercial structures in the network under a single brand, for conduction of unified distribution and marketing policy. In addition, it is in the paradigm of commercial networks and on the basis of SCM (supply chain management) technologies [8] distribution centers formed as the nodes, and the concept of third-party logistics emerged. Today, 3PL providers are actively working in the segment of commercial and, primarily, retail chains.

Due to the flexibility of M2M technologies and alternative transportation modes, as well as the expansion of functionality, a higher level of service such as 4PL and 5PL has formed in the world. Such forms are more common in other segments of the economy, for example, in P2P (Peer-to-Peer or Partner-to-Partner) networks, online auctions, etc. In this work, they can be considered as a prospect of applying the obtained results.

2. Statement of a problem

Consider the problem of digital technologies usage in the concept of Machine-to-Machine interaction and machine-readable standard codes with corrective properties in order to switch to the end-to-end automation in the most common segment of multimodal transport of the 3PL concept.

The conducted analysis of the world experience in the digital tracking application showed that the main effect of the implementation can be achieved only by applying the capabilities of formalized algorithms based on the mathematical theory of optimal processes [9], as it is much more effective than a simple reduction in the forwarding costs. The practical orientation of this study is the focus of the results on multimodal logistics and the creation of a scalable model. This will allow to develop the results in the areas of the following, higher levels of third-party logistics.

Let us introduce the formal descriptions necessary for building a mathematical model. We define that the planning of the multimodal transport operator (MTO) is carried out in conditions of market uncertainty, which is expressed in a set of random variables c_j, where $i = 1, 2, ..., m$. This includes quantitative values of supply requests, volumes of raw materials, components, traffic, consolidated supply and their known distribution functions, based on statistical analysis of market data. Such factors of the market environment as demand, prices, personnel, risks affect corresponding figures of the mathematical expectation $m_j = E(c_j)$ and dispersion D_j. The only limitation is the Lindeberg condition: for any $\tau > 0$, the following must be true

$$\lim_{x \to 0} \frac{1}{B_n^2} \sum_{k=1}^{n} \int_{|x-m_k|>\tau \cdot B_n} (x-m_k)^2 c_k(t)dt = 0 \quad (1)$$

where $B_n = \sqrt{\sum_{k=1}^{n} D_k}$, - $c_k(t)$ distribution density, t -time, x -parameter.

Moreover, the dependence of the values c_j, as a rule, are seasonal [10]. For this purpose, use the following expression obtained in [11]: $m_j = A_{0j} + \sum_k A_{kj} Cos(2\pi \frac{1}{12}(k+\theta_j))$, where k is a planning period number, A_{0j}, A_j, θ_j are Fourier expansion coefficient.

Next, we introduce control variables x_j: $x_j \geq 0$; $j = 1,2,...,n$ which define both the traffic volume by individual modes of transport, and the choice of routes, including the impact of capacity, and data related to the indicators of distribution centers used as route nodes.

The task is to determine the variables x_j, $j = 1,2,...,n$ in the conditions of stochastic nature of information on the actual values taken c_j $j = 1,2,...,n$.

Next, we introduce the deterministic values a_{ij} $i = 1,2,...,m$; $j = 1,2,...,n$, as well as the parameters b_i, defining the constraints of the model. This includes such values as carrying capacity of the vehicles, the timetable of the railway, port, storage capacity, the performance of stevedores, the operating costs of warehouses, cost of rent and energy.

We formulate the problem in terms of separable functions. It is necessary to optimize the functional $\sum_{j=1}^{n} c_j x_j$, with condition $\sum_{j=1}^{n} a_{ij} x_j = b_i$ $i = 1,2,...,m$.

The simplest interpretation of one set of constraints, for example, is deadweight of the consolidated transport pool.

By analyzing the search for the optimization of the quality functional, it can be separately noted that when only the x_j, $j = 1,2,...,n$ values are determined in the absence of complete information about the values c_j, the required solution of the optimum of the mathematical expectation $E\left[\sum_{j=1}^{n} c_j x_j\right]$ under these constraints is reduced to the solution of the equivalent conventional deterministic linear programming problem. For this purpose, based on the properties of random functions, the condition functional is rewritten as $\sum_{j=1}^{n} E\left[c_j\right] x_j$. At this, the expected values of the corresponding coefficients of the original problem are chosen as the coefficients in the expression for the objective function.

However, in cases where even some of the values x_j require information for a more accurate determination, the problem of stochastic linear programming can not be solved by such a simple method.

One of the methods of solving [12] can be the application of mathematical game theory, discussed in detail in [13], but it is necessary to take into account additional factors for the multidimensional model. First of all, it should be noted that an extended planning horizon is always considered and the problem becomes multistage, and, with a large length of the optimization vector, it is necessary to form iterative procedures that are convenient for programming. In such conditions, it is possible to propose another method of obtaining the desired result, we outline it below.

3. Development of practical algorithm for the limited dimension problem

Let the subdivision of the commercial network be engaged in planning of multimodal transportation and utilization of distribution centers for a well-known long time horizon.

The total volume of cargo required for business activity in the planning period is limited by the value T (the dimension is arbitrary, for example, in tons or pallets).

At the beginning of each of the periods that comprise the planned horizon, the enterprise should calculate the following controlled variables:

x_1 is the amount of goods and cargo transported by one mode of transport (e.g. rail cars);

x_2 is the amount of goods and cargo transported by second mode of transport (as a rule, road transport);

x_3 is a goods and cargo processed at distribution centers.

The dimension of x_1, x_2, x_3 is arbitrary, for example, in tons or pallets/ In order to implement the delivery program x_1, units $a_1 x_1$ of the first type of transport will be required for the planned period, respectively, for delivery x_2, units $a_2 x_2$ of the second transport will be required for the planned period. We introduce the designation D_1 and D_2 for the quantitative characteristics of the maximum [13] fleet of each of the transport modes.

Next, we introduce the price list for the unit load of the first mode of transport equal to r_1, and r_2 for the second mode. At the same time, the costs for each transportation are equal $r_j(a_j x_j)$ for each track j $(j = 1,2)$. In the model we should consider the associated costs for the maintenance of units of transport, stevedoring, insurance, etc. Its amount depends directly on, for example, conditions of deliveries. The amount of such costs varies significantly for different terms of delivery. Let's denote them by servicing units of each modes of transport j $(j = 1,2)$, equal to e_j. Thus, the value $e_j x_j$ is the total associated costs for mode j. To form an optimization criterion, we calculate the cost value Q:

$$Q = (a_1 r_1 + e_1)x_1 + (a_2 r_2 + e_2)x_2 + c_3 x_3 \qquad (2)$$

where c_3 is handling costs per unit of cargo from x_3. Then, by introducing the notation: $c_j = r_j a_j + e_j$ $(j = 1,2)$, we rewrite the cost optimization criterion in a form of minimum search:

$$\min_j \left(Q = \sum_{j=1}^{3} c_j x_j \right). \qquad (3)$$

Note, what with deterministic a_j $(j = 1,2)$, c_j $(j = 1,2,3)$ and D_j, the problem can become trivial [14] and be solved by linear programming method. In commercial applications, it is necessary to minimize Q taking into account the limitations:

$$\sum_{j=1}^{3} x_i = T, \; a_1 x_1 + s_1 = D_1, \; a_2 x_2 + s_2 = D_2, \; x_j \geq 0, s_i \geq 0, \qquad (4)$$

where s_i is the deficit of corresponding units of transport.

Due to the conditions of market uncertainty described above, in practice enterprise managers will be able to have real data on costs r_1, r_2 (and, accordingly, values c_1, c_2), only a certain time lag after making a decision on the values x_1, x_2, x_3.

In addition , there are values a_1, a_2 in a range of random variables, as well as potential levels D_1, D_2 of the transport pool. Therefore, in practice the s_1, s_2 indices will only be known once the exact values a_j and D_j are obtained. It follows that under conditions of market uncertainty [15] the a_j and D_j coefficients are random and the structure of the model constraints is not completely defined.

For completeness of the optimization problem, it is necessary to take into account that the multimodal transport operator delivers an additional amount of cargo that is not specified in the contract at a higher price. This formally corresponds to the excess of $a_j x_j$ over D_j.

In order to take this factor into account, we add (t_1, t_2) to the left side of the balance ratios (4), and ($f_1 t_1 + f_2 t_2$) to the formula (2) of the objective function. This is interpreted by the $t_i \geq 0$ values as the excess of volumes over the contracted ones for transport j, and ($f_i t_i$) introduces the appreciation factor f_j for the delivery mode of t_i cargo units.

Finally, we formulate the optimization problem as a search for the minimum cost of the extended criterion Q^*. Let us introduce the following formal description:

$$Q^* = \sum_{j=1}^{3} c_j x_j + f_1 t_1 + f_2 t_2, \text{ under constraints}$$

$$a_1 x_1 + s_1 - t_1 = D_1; \quad a_2 x_2 + s_2 - t_2 = D_2; \quad \sum_{j=1}^{3} x_i = T \qquad (5)$$

The coefficients of appreciation in the directions f_i have a random nature, their values at the beginning of the planning period and at the time of determining the indicators for x_j are known only at a level of the probability distribution function. In addition, real data t_i, and, accordingly, s_i, are calculated after the values of random variables that define this mathematical model are determined.

4. Development and programming of machine-to-machine interaction

The solution of this problem can be reduced to an iterative algorithm of machine-to-machine interaction. Here, data about cargo parameters such as dimensions, weight, destination, time limits and restrictions on the transportation mode is written in QR code and computer automatically reads it. This serves as the source data for the calculation. The degree of market uncertainty σ is used as a parameter of the normal (Gaussian) distribution, expressed as a percentage of the expected value. The computation consists of four basic steps outlined below.

1. Using simplex method on PC, we calculate values x_1, x_2, x_3 on the basis of certain data only relative to the values e_1, e_2 and c_3.
2. From market monitoring, Internet survey data, we obtain an analytics by random values c_1, c_2, f_1, f_2, a_1, a_2, D_1, D_2 and the degree of their correlation with the values x_1, x_2, x_3.
3. Using PC, we calculate values c_1, c_2, t_1, t_2.
4. We carry out the calculation of x_1, x_2, x_3, using the condition of minimization of the functional Q^*.

Since when conducting real business, risk management is a must, it becomes possible to determine the final number W of states characterized by a set of indicators (f_1, f_2, a_1, a_2, D_1, D_2) and the their probabilities vector $\overline{P}$ of the same length. To program the algorithm, we reduce these states to a Ω matrix of $W \times 6$ size. If $W = 3$, then the matrix and vector have the form:

$$\Omega = \begin{Vmatrix} f_{11}, f_{12} & a_{11}, a_{12} & D_{11}, D_{12} \\ f_{21}, f_{22} & a_{21}, a_{22} & D_{21}, D_{22} \\ f_{31}, f_{32} & a_{31}, a_{32} & D_{31}, D_{32} \end{Vmatrix}, \ \overline{P} = (p_1, p_2, p_3), \ \sum_i p_i = 1 \qquad (6)$$

In this case, the calculation is carried out using mathematical expectations. We obtain: $E\left[a_j\right] = \sum_{q=1}^{3} p_q a_{qj}$, and knowing $E\left[r_j\right]$ and e_j, define $E\left[c_j\right] = E\left[r_j\right] \cdot E\left[a_j\right] + e_j$ $j = 1,2$. Finally, the formulation of the model is as follows:

Minimize the expression

$$E[c_1]x_1 + E[c_2]x_2 + c_3 x_3 + p_1 [f_{11} t_{11} + f_{12} t_{12}] +$$
$$+ p_2 [f_{21} t_{21} + f_{22} t_{22}] + p_3 [f_{31} t_{31} + f_{32} t_{32}] \qquad (7)$$

under following limitations:

$$\sum_{j=1}^{3} x_i = T;$$

$$a_{11}x_1 + s_{11} - t_{11} = D_{11};\ a_{12}x_2 + s_{12} - t_{12} = D_{12};\ a_{21}x_1 + s_{21} - t_{21} = D_{21};\ a_{22}x_2 + s_{22} - t_{22} = D_{22};\quad (8)$$
$$a_{31}x_1 + s_{31} - t_{31} = D_{31};\ a_{32}x_2 + s_{32} - t_{32} = D_{32}.$$

The calculations are carried out having regard to market uncertainty, which is measured by a single parameter σ, the standard deviation of the normal distribution, expressed as a percentage of the expected value.

.

5. Simulation on a computer

The proposed technique allows you to program an iterative algorithm on a computer. The equations obtained in this work were programmed in the Mathcad environment. For this purpose, the calculations outlined by steps 1 ... 4 in section IV of this paper are made in sequence. Data sets, having different limitations by the number of vehicles and the position of the distribution center, were used as an example. The initial data sets D_1 and D_2 and matrix (6) Ω to be used in the calculation have the values summarized in Table 1 and Table 2.

Table 1. The values of the initial data for calculation limitations.

Intermediate DC		Emergent DC	
option 1	$D_1 \le 12$	option 3	$D_1 \le 12$
option 2	$D_2 \le 20$	option 4	$D_2 \le 20$

Table 2. Values of matrix Ω elements to be used in the calculation.

20 pct	10 pct	70 rbl/pcs	90 rbl/pcs	40 pcs	40 pcs
15 pct	15 pct	110 rbl/pcs	80 rbl/pcs	60 pcs.	30 pcs
10 pct	20 pct	70 rbl/pcs	90 rbl/pcs	30 pcs	60 pcs

According to the model, it was assumed that decisions are made on the entire track without the participation of managers, which is equivalent to machine-to-machine interaction. However, this assumes that all the information needed for the solution is contained in the machine-readable code.

In the calculation process the computer makes from 1000 to 5000 iterations to find a solution. The calculation stops when the specified accuracy level is reached.

Figures 1-4 show the results of computer simulation of planned activities of multimodal transport operator using the machine-to-machine interaction. The value of the standard deviation of demand expressed as a percentage of the average value is shown on the X-axis. The calculation problem is formulated for two means of transport, with the park having limitations by number D_1 and D_2. There is also an option of an emergent DC to smooth out fluctuations in demand. This corresponds to two limitation options for the transport pool and the position of the distribution center in the supply chain.

Note that the Y-axis shows the initial values of the amount of cargo moved by two means of transport and located in DC, corresponding to the deterministic problem.

Figure 1 and Figure show the obtained results and we can see that the dynamics of using the intermediate DC changes as soon as one of the means of transport is fully loaded. In addition, it follows from the results of the calculation that the influence of the second means of transport is smaller than the first one. This can be explained by the smoothing of the load due to the intermediate DC. The calculations in Figure 3 and Figure 4 show the strong influence of the limitations by the number of park units on the operation of the emergent DC. This fact should be taken into account when multimodal logistics is designed. This is due to the fact that the emergent warehouse is located at the end point and, accordingly, in direct proximity to the customers. As a rule, the costs of its maintenance are many times higher than those of an intermediate DC.

The analysis of the obtained data leads to the conclusion that decisions are strongly dependent not only on the degree of market uncertainty σ , but also on the role of configuration and technologies used in the delivery process.

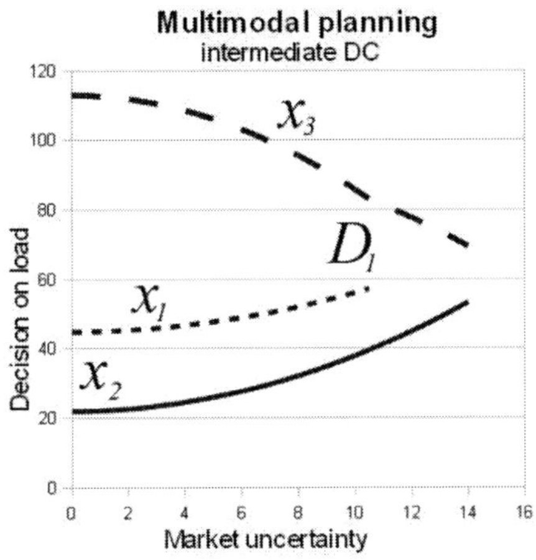

Figure 1. The plan under limitation on D_1.

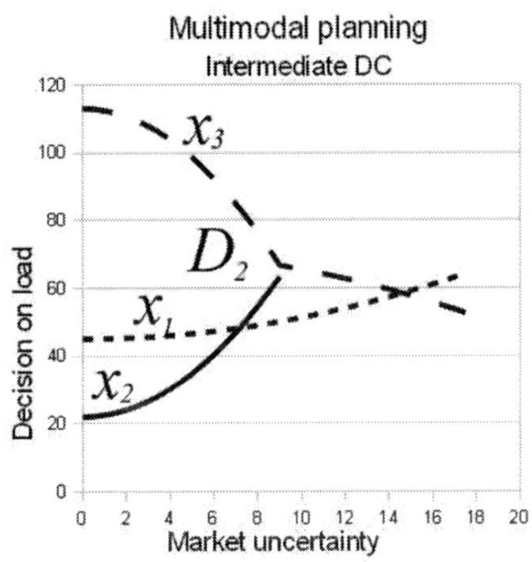

Figure 2. The plan under limitation on D_2.

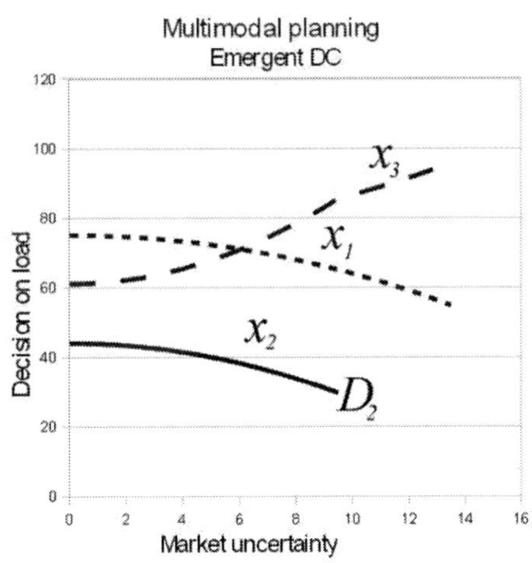

Figure 3. The plan under limitation on D_2

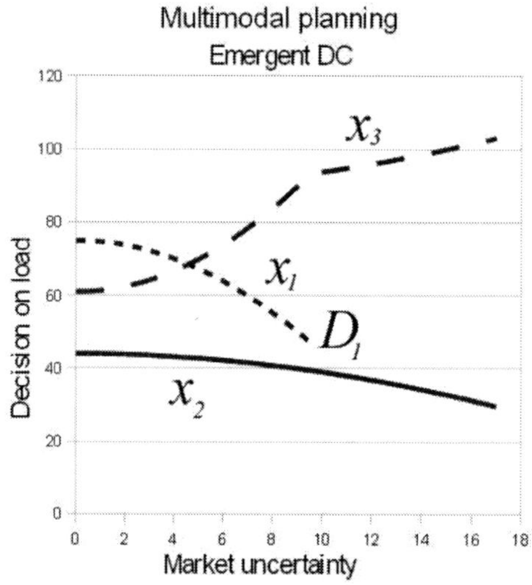

Figure 4. The plan under limitation on D_1

So, if the distribution center is located between transition from one transport to another (figures 1-2), its utilization is reduced, but if it is located is in the area of the final recipient (figures 3-4), then it plays the role of the emergent stock that helps to smooth the uneven demand.

The formulas for three levels of the set of indicators obtained for a particular example are actually scalable and allow to write the algorithm in a matrix form. The dimension of the model is not limited, not only parameters related to multimodal transportation can be used as the control variables, but also related processes such as cross-docking, pick-by-line, which are included in the overall machine-to-machine interaction chain. We can separately consider joint optimization when taking into account the Private Label segment [16], which accounts for up to 25% of the turnover of retail chains. A distinctive feature for the part of the variables is that they will be determined for this position, but for their distribution parameters it is necessary to specify zero dispersion and use the same expressions.

Also, you can include in the parameters such important indicators for cross-border commerce [17] as customs clearance, border control, including delays in crossing state borders and change of transport operator. At the same time, it is necessary to control the volume of machine-readable codes on the transported loads so that it contains the information necessary for the algorithm.

6. Conclusion

The proposed mathematical model allows one to choose the optimal load ratio when several means of transport and intermediate storage are used. Also, a distinctive feature of the model is the consideration of the stochastic nature of the demand for goods, which reflects market uncertainty. In addition, the model takes into account the location of the warehouse for intermediate storage, either between transportation stages, as in the cross-docking technology, or at the final unloading point in a form of a consolidation warehouse. The main application area of these results is the automation of decision-making in planning [18] the activities of the multimodal transport operator. The advantage of this method is the possibility of using machine-to-machine interaction technologies, including cloud-based ones. Due to the obtained algorithm, it is possible to automate some procedures used on the controllers of automated warehouses, palletized robots, and interoperation automatic transport. Such intelligent systems will make it possible not only to exclude human forwarders from the technological process, but also to improve economic performance by reducing costs, and choosing the best inter-transport interaction. Their usage towards higher levels of third-party logistics is promising. Since additional conditions can be introduced in the calculation process, it will allow us to observe cargo handling modes and to take into account their limitations on storage periods,. Implementing the subsystems [19] in the software applications used by managers of network commercial enterprises, in a form of an expert system, will provide a tool of competitive advantage when current activities are planned.

Acknowledgment
The research carried out with the financial support of the grant from the Program Competitiveness Enhancement of Peter the Great St.Petersburg Polytechnic University, Project 5-100-2020.

References
[1] Ilin I V, Izotov A V, Shirokova S V, Rostova O V and Levina A I 2017 Method of decision making support for it market analysis. *Paper presented at the Proceedings of 2017 20th IEEE International Conference on Soft Computing and Measurements*, SCM 2017, 812-814. doi:10.1109/SCM.2017.7970732
[2] Alexandrova I and Zhabko A 2018 A new LKF approach to stability analysis of linear systems with uncertain delays *Automatica*, 2018, vol. 91, pp. 173-178
[3] Glukhov V V and Balashova E 2015 Operations strategies in info-communication companies *Lecture Notes in Computer Science* (including subseries Lecture Notes in Artificial Intelligence and Lecture Notes in Bioinformatics), 2015, No. 9247, pp. 554-558
[4] Kravets O J, Barkalov S A, Butyrina N A, Sekerin V D and Gorokhova A E 2018 Processes of multidimensional classification of scoring objects with heterogeneous features based on the neural networks modeling *International Journal of Pure and Applied Mathematics*. 2018. vol. 119. no. 7a. C. 875-879
[5] Krasyuk I A, Bakharev V V, Kozlova N A and Mirzoeva D D 2017 Staffing in the sphere of trade: the main issues and prospects of solution *Proceedings of 2017 IEEE 6th Forum Strategic Partnership of Universities and Enterprises of Hi-Tech Branches* (Science. Education. Innovations), SPUE 2017 6. 2018. pp. 48-50.
[6] Tsatsulin A N, Babkin A V and Babkina N I 2016 Analysis of the structural components and measurement of the effects of cost inflation in the industry with the help of the index method *Proceedings of the 28th International Business Information Management Association Conference* Vision 2020: Innovation Management, Development Sustainability, and Competitive Economic Growth, pp. 1559-1573
[7] Yanenko M 2016 Cost-Based Brand Management *International Business Managem*ent Year: 2016. Volume: 10. Issue: 26. Page No.: 5991-5995.
[8] Babkin A V and Kudryavtseva T J 2015 Identification and analysis of instrument industry cluster on the territory of the Russian Federation *Modern Applied Science,* 9 (1), pp. 109-118. DOI: 10.5539/mas.v9n1p109
[9] Shchukin T 2014 Neuronet: the next generation communication environment *Technowars* - 2014. № 5. p. 66-85
[10] Borisoglebskaya L N and Sergeev S M 2018 Model of assessment of the degree of interest in business interaction with the university *Journal of Applied Economic Sciences* 2018, vol. XII, Iss. 8 (54) Winter 2017, pp. 2423-2448
[11] Aleksandrov A and Zhabko A 2014 Delay-independent stability of homogeneous systems *Applied Mathematics Letters*. 2014. vol. 34. no. 1. pp. 43-50.
[12] Babkin A V, Plotnikov V A and Muraveva S V 2015 Integrated industrial structures in the economy of Russia: Organizational forms and typology *Proceedings of the 25th International Business Information Management Association Conference* - Innovation Vision 2020: From Regional Development Sustainability to Global Economic Growth, IBIMA 2015, pp. 1286-1293
[13] Podvalny S L and Provotorov V V 2015 The questions of controllability of a parabolic systems with distributed parameters on the graph // International Conference "*Stability and Control Processes*" in Memory of V.I. Zubov (SCP) 2015. pp. 117-119

[14] Provotorov V V 2015 Boundary control of a parabolic system with distributed parameters on a graph in the class of summable functions *Automation and Remote Control*. 2015. vol. 76. no. 2. C. 318-322.

[15] Karvanen Juha, Rantanen Ari and Luoma Lasse 2014 Survey data and Bayesian analysis: a cost-efficient way to estimate customer equity *Quantitative Marketing and Economics* 12 (3) pp. 305–329

[16] Krasnov S V, Sergeev S M, Mukhanova N V and Grushkin A N Methodical forming business competencies for private label *Reliability, Infocom Technologies and Optimization (Trends and Future Directions)* 6th International Conference ICRITO 2017, p. 569-574

[17] Kapustina I V, Kirillova T V, Ilyina O V, Razzhivin O A and Smelov P A 2017 Features of Economic Costs of Trading Enterprise: Theory and Practice *International Journal of Applied Business and Economic Research* vol. 15. №11. 2017. pp.1-10

[18] Lavrova D, Pechenkinc A and Gluhov V 2015 Applying correlation analysis methods to control flow violation detection in the internet of things *Source of the Document Automatic Control and Computer Sciences*, 49 (8), pp. 735-740. DOI: 10.3103/S0146411615080283

[19] Kagermann H, Wahlster W and Helbig J 2013 Recommendations for implementing the strategic initiative Industrie 4.0 *Final report of the Industrie 4.0 Working Group* 2013

IOP Conf. Series: Materials Science and Engineering **618** (2019) 012070 doi:10.1088/1757-899X/618/1/012070

Business-Science-Education: a Collaboration for Competitive and Sustainable Growth of the Wine Industry

M Angelova, P Georgiev, G Dimitrova and D Pastarmadzhieva

Faculty of Economics and Social Sciences, University of Plovdiv Paisii Hilendarski, 4000 Plovdiv, 24 Tzar Asen str, Bulgaria

mina.angelova@uni-plovdiv.bg

Abstract. Optimization of the collaboration "business-science-education" is a prerequisite for the increase of the competitiveness of the enterprises. The article introduces a scientific solution of a practical problem through the creation of a model for informal education of specialists in the wine industry. The presented models are a part of the Virtual Educational Space (VES) as a smart space, context-dependent, based on a script and a controlled infrastructure. The results of the approbation of the model prove its significance for the practical processes and the increase of the competency of the specialists in connection to the implementation of innovation. Particularly to the wine industry, the suggested model for informal education represents a specific innovation, which approbates the known "good practices" in the strategy "Lifelong Learning".

1. Introduction

The competitiveness is an economic and managerial category, which in conditions of an increasing globalization, appears as a main force for the development of both each individual business units and national economies. The category is examined as a multi-aspect and polyvalent concept, in which evaluation is a leading priority for each management level. The tendency in the contemporary world forces the enterprises to be proactive, future-oriented and striving for foreseeing future changes. This is connected with the search and formation of new means for competitive behaviour so as to position themselves strategically on the market and generate their sustainable development.

To a huge extent the competitiveness is formed by the innovation activity of the business enterprises, and a huge impact on this activity has the national innovation policy. A possible path for its application is by the use of science and knowledge in the practical activity and the creation of innovation economy and national policy that correspond to the requirements of the EU. With the adoption of the Lisbon strategy (year 2000) has been emphasized on the scientific researches and innovation as key competitive advantages to which the member states of the EU have to aim as goals.

In 2010 the strategy is changed into "Europe 2020", which has "3 mutually fortifying priorities" – for a smart (economy of knowledge), sustainable (competitive economy), and integrated (economy with high levels of employment) development. One of its main goals is the investments in research and development activities (R&D) with a relative amount of 3% of EUs GDP. The latter is also a main goal of the functioning in the boundaries of the EU the Competitiveness Council, which functions in 4 main areas for "increasing of the competitiveness and enhancement of the growth of the EU", one of which is exactly innovation and scientific researches.

As a result from **a preliminary research** by a team of experts, regarding the realization of a project in the development of a conceptual model for implementation of innovation and increase the competitiveness of industrial enterprises № KP-06-M25/5/17.12.2018, financed by the National Scientific Researches Fund, 2018-2020, defined the need for increasing research activity so as to ensure opportunities for informal education, which has to be targeted to specialists in the wine industry enterprises. The obtained results show that there is a need of an appropriate form of education for the enhancement of the levels of innovation in SMEs in the wine industry. The managerial staff is not well aware of the opportunities and does not have a clue of those, which can be integrated in their enterprises. The collaboration between the experts in the project and team members of wine enterprises formed prerequisites for the creation of a unified information archive of knowledge or Virtual Educational Space (VES).

The goal of the research is to present the relation between the real-world problems of the business, the expertise of the scholars, directed to the solving of the problems and the opportunities of informal education of the specialists.

2. Triangle of Knowledge "Science-Business-Education"

A main challenge for Bulgaria is to achieve the goals set by the adopted in 2011 national strategy for development of scientific researches 2020 [1], which aim is to "enhance the development of science in Bulgaria so as to convert this experience into a factor for the development of the economy, based on knowledge and innovation activity"[1].

The aim is at encouraging the interaction of the "**triangle of knowledge**", which includes an interconnected system "science-business-education". The latter corresponds to the idea of approbation of the scientific researches and realization of the educational product in the practical activity with intent to form a unified scientific-manufacturing process and a competitive economy. The "triangle of knowledge" is accepted as a "main constructive element, which creates, generates and transforms knowledge from science to the business and its use in the real economy. "[2].

In the current research is presented an example of collaboration between experts from science and representatives of the business sphere. Science attempts to provide a solution to a real-world problem for business, determined by a preliminary empirical study. The developed model for informal education with the help of the formation of smart agents and ontologies will provide guidelines to the representatives of the business and science spheres for the level of acquired education by the specialists.

2.1. Correlation Science-Business

The correlation between science and business have been evaluated in the annual report for global competitiveness (The Global Competitiveness) of the World Economic Forum. The same is accepted as one of the fullest, most complete and representative sources for evaluation of the competitiveness of nations. The data for economic growth is presented via the Global Index for Competitiveness (The Global Competitiveness Report), which shows the prosperity of each individual country. The report is prepared on the basis of 12 points (columns), divided into 3 groups, which measure the macro- and microeconomic aspects of the competitiveness, encompassing 111 indicators that take into account all specifications of each of the individual countries included in it. From 1999 onwards Bulgaria is included in the Report via the Centre for Economic Growth as a partner of the World Economic Forum.

Pursuant to the data in The Global Competitiveness Report, the positions of Bulgaria in the indicator for cooperation in Research and Development activity for the period 2013-2018 are presented in Table 1.

For the first three years of the examined period unstable positions can be seen for our country regarding the indicator for cooperation in Research and Development activity. According to analysts, the reason roots in the ineffective national administration, corruption practices, difficult access to financing and other.

For the last 3 years, however, a rapid stable growth can be observed, as in 2018 our country reaches 41st place out of 140 countries. The indicated can be considered as a significant progress in the sphere.

Table 1. The position of Bulgaria regarding cooperation in R&D activity, for the period 2013-2018.

The Global Competitiveness Report	University-industry collaboration in R&D, Rang*
The Global Competitiveness Report 2013–2014	117/148
The Global Competitiveness Report 2014–2015	113/144
The Global Competitiveness Report 2015–2016	112/140
The Global Competitiveness Report 2016–2017	74/138
The Global Competitiveness Report 2017–2018	74/137
The Global Competitiveness Report 2018	41/140

2.2. Relationship Business-Education

The aim of this research is to find an appropriate model for informal education of specialists from the wine industry, which would help in the integration of innovation. Similar models have been developed by the authors, as the idea will be adapted and developed to the specific sphere of work. The presented models are part of VES [3] as a smart space, context-dependent, based on a script and a controlled infrastructure.

The online education is getting more and more used in last years. The team of experts, representatives of science, is aimed at the development of a conceptual model for the integration of innovation and increasing of competitiveness of the industrial enterprises by using a VES. The current conceptual example is applied to the formalized representation of opportunities for increasing and integration of innovation in the wine industry.

The conceptual model for informal education of specialists from the wine industry (figure 1) appears as an agent-oriented space [4], which is built in 5 modules:

Module 1: Education. Lifelong Learning;
Module 2: Examination of user's knowledge level;
Module 3: Generation of knowledge for the wine industry via surveys;
Module 4: Measuring of innovation, as a model for measuring is yet to be done;
Module 5: Consultations and recommendations.

BLA – Basic Learning Agent

TA – Test Agent

UKA – Update Knowledge Agent – responsible for the gathering of knowledge from surveys and filling of ontologies LO and ICO, if new knowledge is present. Archives the results from the conducted surveys.

IMA – Innovative Measurement Agent – responsible for the measuring of the level of the imported innovation and recommendations for the importing of appropriate innovation for a specific user.

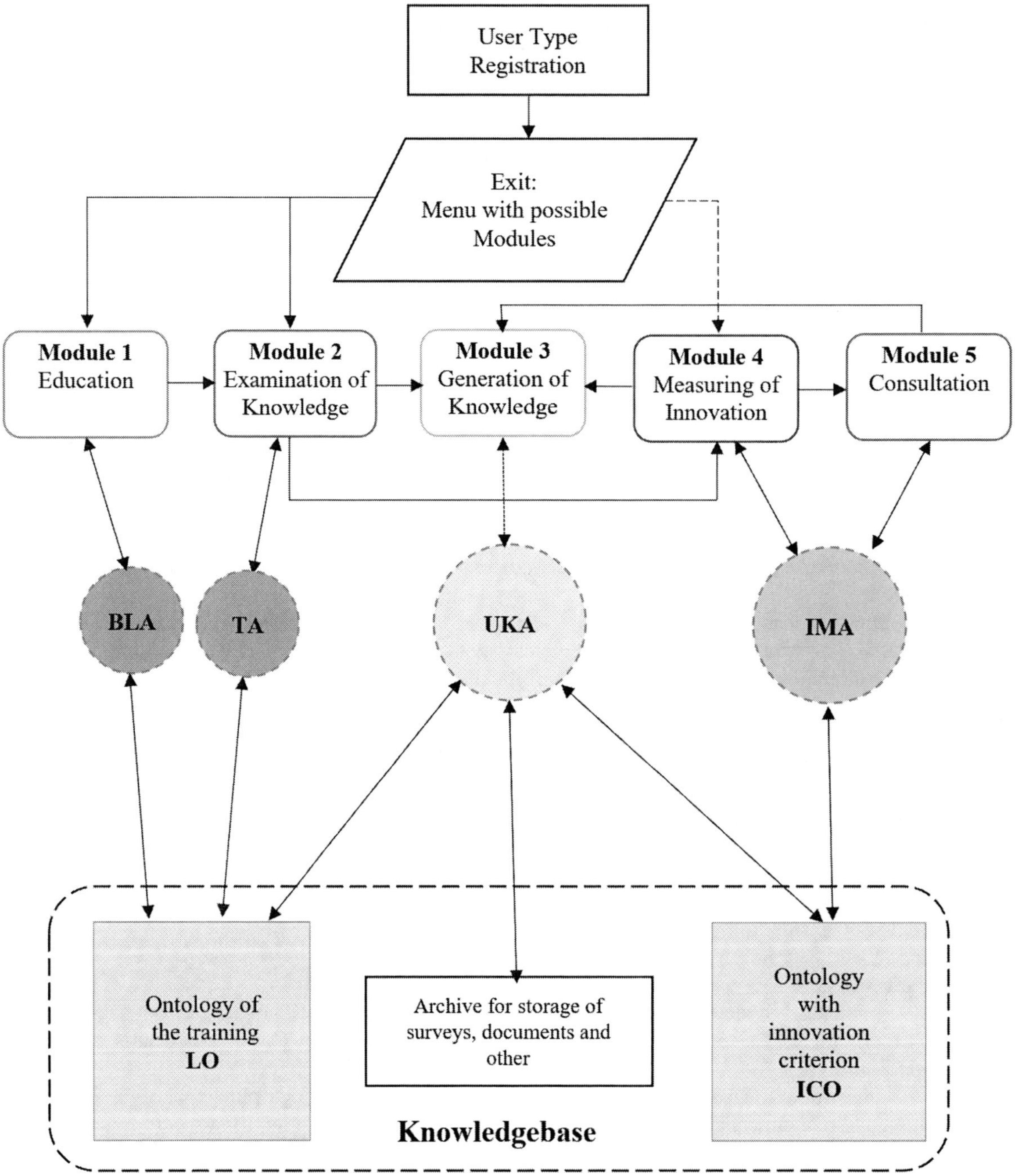

Figure 1. A conceptual model for informal learning

The Knowledgebase is built up of two ontologies and one archive: LO – Learning Ontology – for preparing of training and examination of knowledge. ICO – Innovation Criteria Ontology

Archive – storage of surveys, documents and other.

3. Approbation of the Science-Business-Education Collaboration through an Informal Learning Model

Opportunities for using innovative products in companies depend on many factors such as company security, crediting, and senior managers' capabilities to develop and deploy new products and technologies, and more. With the sustainable growth of the economy, the trend of improving the entrepreneurship and business environment in Bulgaria continues. Specifically for the wine industry, the proposed informal learning model is a specific innovation, which exemplifies the "good practices" set out in the Lifelong Learning Strategy. Observations show that the qualification level of the Bulgarian labor force has a steady trend of decrease, even of degradation. Thus, the increasing human resource quality is a priority strategic problem for our country [8].

Lifelong learning can be defined as any purposeful learning activity that serves to continually improve people's knowledge, abilities and competence. It is necessary to create an interest for both employers and employees to constantly increase their qualifications through legal regulation, rigor and control of the occupation of qualified candidates, which ensures the achievement of the expected results of the Lifelong Learning strategy. In order to successfully implement the strategy in an organization, it is necessary to create opportunities for certification of results of distant, informal and self-dependent learning, of skills acquired in practice and recognition of all learning achievements - knowledge, skills and competences. This will enable employees to combine the performance of their duties with learning. The essence of the model is represented by five modules that create opportunities for increasing the competence of the specialists.

3.1. Module 1 - Education

Module 1 provides informal training for wine industry specialists to learn about innovative techniques. Informal learning is purposeful and organized, but its successful completion does not lead to the acquisition of a level of education (class, stage) or degree of professional qualification (including qualification as part of a profession). The most common forms of informal learning are courses, private lessons, seminars, workshops with training, workplace training, occupational safety briefings, etc. All of these require the employee to be separated from his / her employment position, which is associated with financial loss for the enterprise. Successful completion of non-formal learning optimizes the process of learning, testing, counseling, etc., with no need for absenteeism.

Possible topics and sections of the training are related to **innovation management; use of financial instruments; GDPR and security; opportunities for developing wine tourism**.

In *Module 1* the informal learning is expected to be done using agency-oriented architectures and knowledge-based ontologies. The intelligent agent, the BLA (basic learning agent), provides information on a specific subject to predefined criteria. Preliminary criteria are taken as a limiting condition once a questionnaire is completed to determine the knowledge and interests of the consumer. These conditions in the questionnaire are foreseen in the next version of the informal learning model. The current version provides an agent solution that monitors the provision of structured content by learning ontology (LO). The ontology is divided into two sub-ontologies KLO and TLO. Where KLO (knowledge learning ontology) contains knowledge about all types of innovations, and TLO (test learning ontology) contains user knowledge verification questions and is used by TA (test agent) in Module 2.

In consideration with the examinated agent architecture, it is important to note some definitions that will later be considered in the research. There are various definitions of agents in the literature, but only some of them are presented here.

- „An agent is an identity that can perceive the environment by means of sensors and effects that have influence through effectors“ [5].
- „Autonomous agents are computing systems that inhabit some complex dynamic environment, perceive and act autonomously in that environment, and acting in this way reach a multitude of goals or actions for which they are created“ [6].
- „Intelligent agents continuously perform three functions:

- Adoption of dynamic conditions in the environment;
- Action to influence the conditions in the environment;
- Reflections on interpretation of perceptions, solving problems, making conclusions and determining impacts on the environment“ [7].

BLA agent for management the Knowledge Learning Ontology – KLO

- **Beliefs**
  - o Data ontology;
  - o User responses processed by the agent to determine restrictive conditions.
- **Desires**
  - o Reading of ontology;
  - o Retrieving data from ontology against restrictive conditions.

KLO – It is envisaged to develop a model covering the types of innovation and their aspects. One idea of building the KLO ontology might look like this:

Types of innovation -> sub- type -> definitions -> concepts

Or represented in OWL (Web Ontology Language): Class -> subclass ->annotations ->individuals

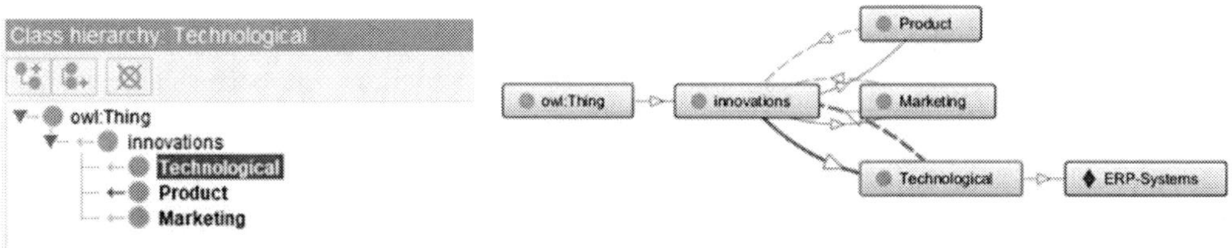

Figure 2. Web Ontology Language

3.2. Module 2 - Verification of knowledge

Knowledge verification is possible even if *Module 1* is omitted, as shown in Figure 1, only if minimal knowledge is covered. Otherwise, the user is required to return to the initial position of *Module 1*. Analogously to the above models, a test agent model - TA and ontology of exams - TLO are applied.

Agent running an ontology of knowledge issues:

- **Beliefs:**
  - o Ontology with questions, sequence of questions;
  - o List of questions already asked;
  - o User responses;
  - o List of questions to be asked.
- **Desires**
  - o Reading ontology, retrieving question lists and answers from it;
  - o Asking a next question;
  - o Receiving a response from the user.

A model for ontology of knowledge verification is to be developed. Since the two ontologies are to be linked and form the Learning Ontology (LO), the model will be refined and presented in the next team's work (see Figure 3).

3.3. Module 3 - Generation of Knowledge

It is planned to build a survey to generate knowledge about innovative and useful practices in micro and small wine enterprises.

The Update Knowledge Agent (UKA) is responsible for gathering knowledge, by querying and completing LO and ICO ontologies if new knowledge is available. Retains polls as an archive in the repository. At first glance, the task of solving is simple in collecting certain data in the form of a poll and storing them in the repository (database) as an archive. This is an easily solved and fully automated task.

On the other hand, the system's ability to detect new knowledge and to update ontologies with it would be most valuable. The latter is a serious challenge and requires significant efforts, but the authors are determined to offer a solution.

The goal is to develop a model for detecting new system intelligence by the intelligent agent UKA, then outputting a message to an expert to validate the data and assess whether or not where to revise the ontologies LO and ICO.

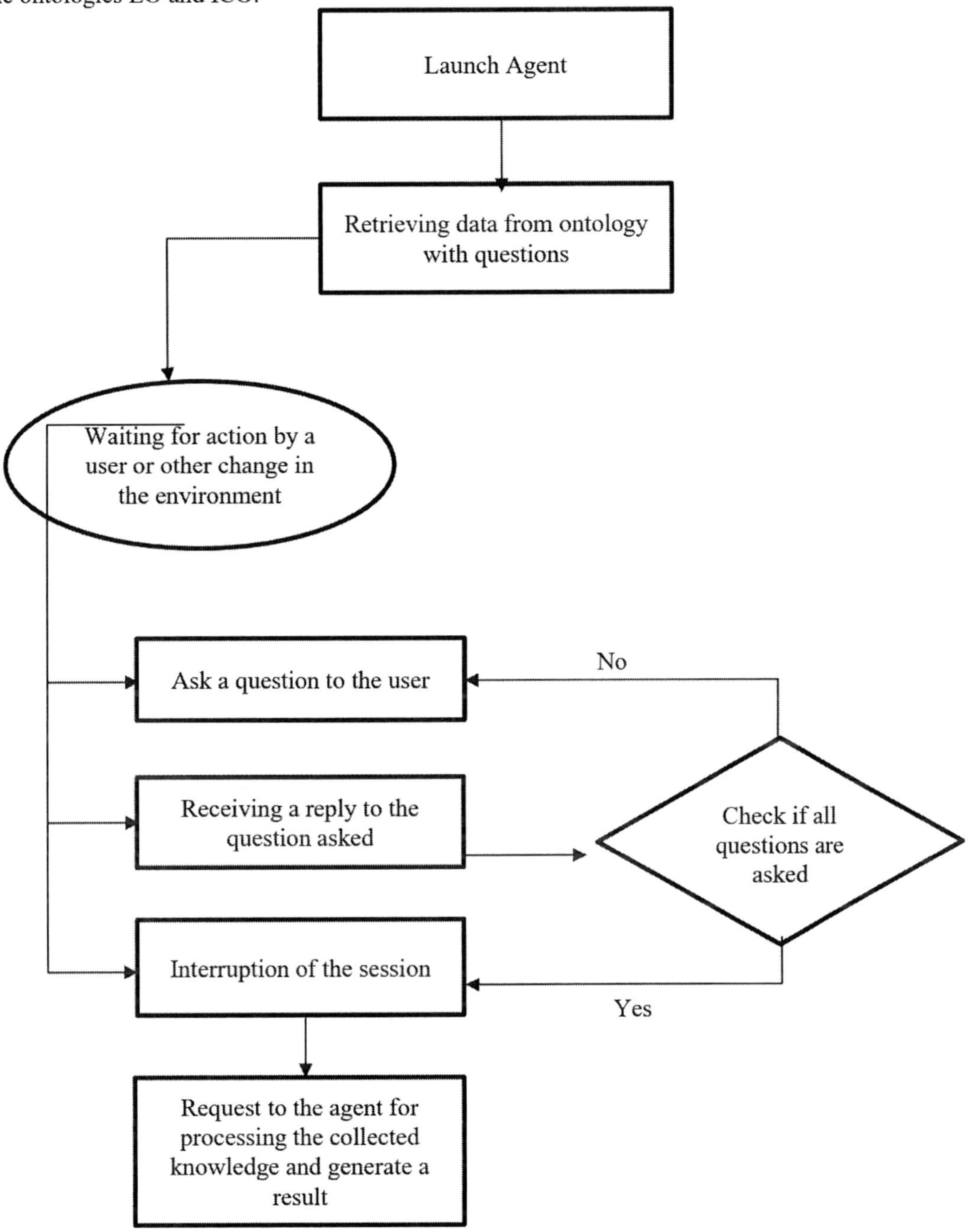

Figure 3. A Module for Knowledge Verification

3.4. Module 4 – Innovativeness Measurement

Measuring the innovativeness of a micro- or small enterprise in the wine industry is the biggest challenge for the authors. First, on the basis of a study, a definition of the term "innovation" is going to be applied. Secondly, in consideration with the accepted concept follows a presentation in a formal model. The aim of the upcoming research is to find the exact criteria that can assess the innovativeness in the specific field. In parallel, a questionnaire is developed through the Innovative Measurement Agent (IMA) responsible for measuring the level of innovations introduced and recommendations for introducing appropriate innovations for a particular user in *Module 5*. The investigation is also focused in creating an appropriate Innovation Criterion Ontology (ICO).

3.5. Module 5 – Consulting and Recommendations
In the last module (stage) is foreseen that the system had already generated the necessary information about the knowledge level of the particular specialist and the lack of other appropriate knowledge, with the possibility to offer recommendations to the user. The goal is to show the level of innovation in *Module 4* and to highlight the weaknesses and the possibility of introducing innovations into *Module 5* that other enterprises have successfully introduced and used.

Conclusion
The fundamental objective of the research is to collaborate on the three-way science-business-education relationship by developing and adapting a model for informal learning of wine industry specialists. The models presented are part of the Virtual Education Space as an intelligent space, context-sensitive, script-based, and controlled infrastructure. The results of the preliminary study confirm the lack of such a model and the possibility of forcing research efforts. The results of the methodology's testing prove its significance for the practice and potential for transforming human resources into human capital for organizations.

Acknowledgments
This article is a part of a project № KP-06-M25/5/17.12.18, financed by National Science Fund, Ministry of Education and Science.

References
[1] National Strategy for Development Research 2020 - http://www.strategy.bg/StrategicDocuments/View.aspx?lang=bg-BG&Id=708 , 02.2019.
[2] Dimitrova G and Dakova M 2016 Cooperation between business, science and education - a factor for sustainable development and competitiveness, Int. business Conf. "*Science and business for intelligent future*", University of Varna, (available in Bulgarian)
[3] Valkanov V, Stoyanov S, Valkanova V 2015 Building a virtual education space, *WMSCI 2015* - 19th World Multi-Conf. on Systemics, Cybernetics and Informatics, Proc., 1, pp. 322-326
[4] Georgiev P, Valkanov V, Sandalski M, Stoyanov S 2016 Intelligent environment for manager's training, Proc. Int. Scientific Conf. "*High Technologies. Business. Society 2016*", ISSN: 1310-3946 (available in Bulgarian)
[5] Stoyanov S, Popchev I, Ganchev I and O'Droma M 2005 From CBT to e-Learning, *Information Technologies and Control*, vol. 4, pp. 2-10, ISSN 1312-2622.
[6] Stoyanov S, Ganchev I, Popchev I, O'Droma M and Sgurev V 2010 An Approach for the Development of a Context-Aware and Adaptive eLearning Middleware, *J. Intelligent Systems: From Theory to Practice*, vol. SCI 299, Berlin, ISBN: 978-3-642-13428-9, pp. 519-535.
[7] Stoyanov S, Ganchev I, Popchev I and O'Droma M 2008 An Approach for the Development of InfoStation-Based eLearning Architecture, *Compt. Rend. Acad. Bulg. Sci.*, vol. 62, no. 9, pp. 1189-1198
[8] Mihova T, Anguelov K, Ferdov A 2018 Specificity of the Training of The Employees in High-Technological Enterprises, IX Nat. Conf. with Int. Participation (ELECTRONICA), Sofia, IEEE Catalog Number:CFP18P58-POD,ISBN:978-1-5386-5802-4

Application of Business Crisis Forecasting Models in Bulgaria

T Gigova[1], V Nikolova-Alexieva[2] and K Valeva[3]

[1] Faculty of Mechanical Engineering, Department of Industrial Business, Technical University-Branch Plovdiv, Technical University - Sofia, Plovdiv Branch, 25 Tsanko Diustabanov Str., Plovdiv, 4000, Bulgaria
[2] Faculty of Economics, Department of Industrial Business and Entrepreneurship, University of Food Technologies, 4000 Plovdiv, Bulgaria
[3] Faculty of Economics, Department of Industrial Business and Entrepreneurship, University of Food Technologies, 4000 Plovdiv, Bulgaria

E-mail: t_gigova@abv.bg, valentina_nikolova@abv.bg, tomika888@yahoo.com

Abstract. Every business organization faces a crisis at a certain point in its development. Crisis phenomena for overcoming preventive measures can lead to serious economic and social problems for the organization or its bankruptcy. The crisis is a widely held view and its indications affect different aspects of the financial, social, spirit life of the nation. It's important to accent not only to the consequences of its manifestation but also to the opportunities for overcoming. This study focuses to analyze the specific models for assessing the dangers of crisis in Bulgarian enterprises. The report aims to systemize the methods and models for assessing the crisis in the enterprises from different sectors of the Bulgarian economy and to define the specific influential factors on them. A questionnaire was used to collect data from a sample of 312 entreprises which were selected through stratified random sampling method. Collected data were analyzed using descriptive and inferential statistics with the aid of Statistical Package for Social Sciences (SPSS). The study revealed that in the food industry is used Altman's five-factor Z-score while in the enterprises from the financial sector, telecommunications and the IT sector the Springate model is used. Tourism businesses, those in the education sector, and chemical and health enterprises prefer to predict crises through Toffler's model. The results show that crises are somewhat the driving force behind the creation of innovation and are a prerequisite for the emergence of modern forms of entrepreneurship so necessary to stabilize the economic system.

1. Introduction

Crisis phenomena for overcoming preventive measures can lead to serious economic and social problems for the organization or its bankruptcy. Business crises occur at different intervals, provoked by a number of factors, characterized by different scales and forms of manifestation, and have the ability to repeat, i.e. have a cyclical character.

The development of the world economy, including the Bulgarian economy [1], [2], [3], is still under the impact of the global economic crisis. Since 2009, they have had a strong impact on the dynamics of the development of the industrial sector and its companies and enterprises in Bulgaria [4], [5]. Objective was the deterioration of most business performance indicators, a sharp decline in innovation and foreign investment in our economy due to loss of interest and incentives, the high degree of uncertainty and risk in its development [6], [7]. These crisis circumstances increase the interest of researchers and scientists in revealing the genesis, prerequisites and trends in the global economic and financial crisis in the national economies and its impact, especially in the industrial sector, businesses and enterprises [3],[4].

The main purpose of the study is to analyze the theory in the sphere of anti-crisis management to systematise the indicators, models and methods for crisis forecasting in the Bulgarian enterprises from the different branches of the economy.

The objective is decomposed into the following research tasks:

First, revealing the peculiarities and causes of company crises, the types of crises and the factors for their development.

Secondly, on the basis of a critical analysis of research, to develop the principles, approaches and concepts of anti-crisis management of industrial enterprises.

The object of research is 312 enterprises, representing all sectors of the Bulgarian economy.

The subject matter of the study is the specific models for anti-crisis management, in the context of the real opportunities for introducing strategic innovations as a factor for achieving efficiency and competitiveness in the long-term perspective .

The research supports **the main research hypotheses,** namely that the concept of crisis can not be unified, because of the many areas in society, which is manifested. The systematization of factors generating crisis development in enterprises shows that external factors have an impact on the formation of their competitive environment, and internal factors affect their management, but regardless of whether they are internal or external they have a constructive or negative influence on their development.

In the present study, the following **methods were used:** analysis and synthesis; induction and deduction; analogy and comparisons; quantitative methods, by statistical processing of SPSS (one-dimensional, two-dimensional distributions, χ-square and correlation coefficient).

Data and information from NSI, SAPI, Agrostatistics Department at The Ministry of Agriculture and Food, Marketing and Market Research Agency ICAP, EUROSTAT and others were collected, analyzed and synthesized. Studies, strategies, outcomes and guidelines of work from national and regional programs have been used. In the development of the analysis, results of the Association of Meat Processors in Bulgaria and the Association of Industrial Capital in Bulgaria surveys, as well as similar surveys and relevant information sources on the subject of analysis were used.

Literature Review

Every business organization faces a crisis at a certain point in its development. The etymology of the term "crisis" is of Greek origin. It is associated with a phenomenon or a set of phenomena, both personal and social, bearing negative changes. Personally, the concept of crisis is identifiable as a stressful state due to health, financial or other reasons. In the public context, the concept is associated with an unstable and dangerous situation caused by economic, political or international problems. The crisis, according to the glossary of foreign words [8], means "acute change, turning, decisive moment, verdict, decision on a question in a questionable situation. "In ancient times, the Greek philosopher Hippocrates considers the crisis limited only to the human organism [9]. In the field of international relations, the "crisis" is related to the emergence and development of conflicts characterized by a threat to basic social values (independence, freedom, etc.). In the economy, as a common feature of economic crises, "there is a shortage of resources for the effective realization of the assigned functions of enterprises, corporations, economic systems, the national economy of the country or the world as a whole" [10].

1.1. The Concept of Business Crisis

Business crises are a common phenomenon and an integral part of their emergence, economic development and liquidation. They start with the onset of the first symptoms of the crisis and continue until its final positive or negative resolution.

Depending on the behavior and performance of a company, the onset of the crisis develops in a typical direction, going through the following phases - strategic crisis, revenue crisis, liquidity crisis and bankruptcy [10]. (see figure 1)

Strategic crisis. Also known as the hidden insolvency or latency stage. The strategic crisis does not threaten the enterprise in the present, but in the future. In this state, the enterprise only deals with invisible disproportions. Prerequisites for the emergence of a strategic crisis are changes in consumer behaviour, population structure, market developments, legislation and politics.

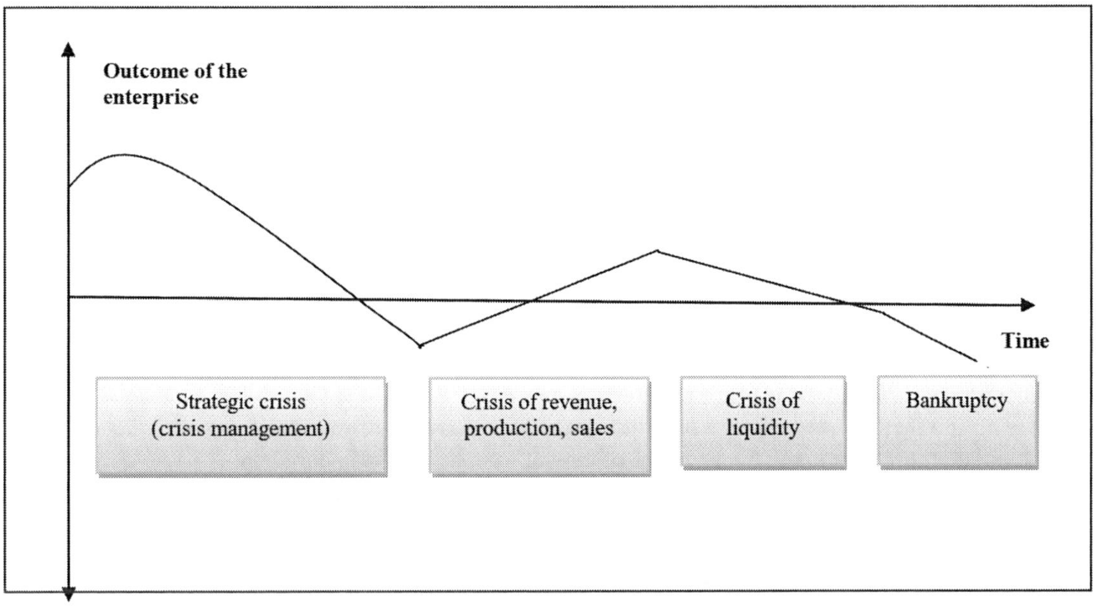

Figure 1. Outcome of the enterprise as a symptom of the crisis.
Source: Iliev Y., Anti-Crisis Management of Industrial Firms, Avangard Prima, S., 2011, p. 48

From the economic point of view, the first symptoms of the strategic crisis are the decline in business activity, the reduction in sales volumes, the reduction of the prices of the products due to over production or loss of quality, the reduction of production capacity, the loss of markets and, the market position of the enterprise.

Depending on the reasons for the emergence of a strategic crisis, there are different possibilities for reaction - by changing the location of the company, attracting new customers, diversifying own products, merging companies.

Revenue crisis, also known as a stage of apparent insolvency, a stage of financial instability or a period of crash. An enterprise is in a revenue crisis when its profits or turnover begin to decline and debt increases. The stage is characterized by the company's inability to perform its cash obligations in full and in full. This is a period in which it begins to experience difficulties for timely payment.

Crisis of liquidity. Most often, this is the phase in which a crisis is being talked about, but this is not yet the final stage. The stage is characterized by the fact that the company can not cover its obligations in a timely manner. Creditors are not ready to negotiate, suppliers are only delivering on advance payment. Unless rapid action is taken at this stage, the path to bankruptcy is inevitable.

Bankruptcy (insolvency). Bankruptcy characterizes a situation in which an entity is unable to continue to operate in its present form and scale. Bankruptcy has two outcomes:

• *positive* - reorganization of the enterprise through a rehabilitation plan, transformation or business combination with other enterprises and restoration of its activity to a qualitatively new level or in a new legal-organizational form;

• *negative* - termination of the enterprise's activity, liquidation and its complete deletion as a business entity.

According to Argenty J., the most difficult from a technical point of view is the problem of identifying the initial symptoms of the financial crisis due to the creative keeping of companies' accounts [9]. It distinguishes the causes of bankruptcy from symptoms according to three indicators:

- ***financial*** - are established using the techniques of financial analysis;
- ***non-financial*** - stemming from management, marketing, planning, motivation of staff, etc .;
- ***terminal*** - regardless of which stage of the crisis process they are identified, they characterize the final phase of the financial crisis - bankruptcy.

The financial crisis can also be triggered by a suboptimal structure of capital, assets, investment, unbalanced time and money flows, impaired financial sustainability, and so on [4], [7]. It creates the greatest potential threats to the successful operation of the enterprise, and may lead to insolvency [2], [3].

The stage of bankruptcy has far-reaching negative consequences and generates a destructive potential, which is expressed in the fact that:

- Failing enterprises create serious risks for successful businesses and their partner commercial banks, causing them material damage in the process of business interaction.
- Failing enterprises complicate the formation of the revenue side of the state budget and extra-budgetary funds, slowing economic and social development programs and worsening the macroeconomic indicators of the national economy.
- Using the inefficiently allocated credit resources, bankruptcy firms have a downward impact on the overall profit rate and the turnover of capital in their entrepreneurial sphere.
- Failing companies cut jobs and reduce the number of people employed in public production, where social tensions escalate and the social crisis worsens in society.

The exact establishment of the stage of the crisis process in the enterprise requires a reliable information base that has important practical relevance in determining the real scale and expected consequences, choice of strategy and countermeasures.

1.2. Models for predicting and assessing the dangers of a crisis

The ability to predict an enterprise's insolvency crisis before it arises is important for both its management and creditors. Business insolvency research dates back to the late 1930s in the United States. At the beginning of the twentieth century a number of researchers were dealing with bankruptcy problems. Thus, for example, analyzing 183 companies experiencing financial difficulties between 1921 and 1930, suggest that the most reliable indicator of forthcoming bankruptcy is to reduce the ratio between net working capital and the sum of assets [11].

On the basis of a comparative study of the indicators of 20 bankrupt companies in the period 1920-1929 and the indicators of 19 successful companies during the same period concluded that the two ratios - between profit and equity and between equity and debt are the most reliable indicators to predict the risk of bankruptcy. S. Mervin based on data from 939 companies over a ten-year period, shows the possibility of predicting bankruptcy for 4-5 years until it arrives, through the dynamics of three ratios - to cover interest payments, the ratio between net working capital and the sum of assets, and the relationship between own capital and debt [10].

Table 1 .The coefficient of W. Beaver

Indicator	formula	Meaning metric		
		For successful companies	5 years before bankruptcy	1 year before bankruptcy
Bever coefficient	Net profit + Amortization / Liabilities	0,4 - 0,45	0,17	- 0,15

The first attempt to develop a model for predicting insolvency by using financial ratios is by W.Beaver [12]. It compares indicators of 79 insolvent companies with the same indicators in 79 stable enterprises, and finds that five years before the collapse, the difference between the two sets of indicators are significant and there are signs of the on-going crisis. Beaver's model is a ratio that is calculated as the ratio between the amount of cash flow and the total amount of the entity's liabilities. Table 1 presents this coefficient (See Table 1).

In 1972, British economist R. Liss developed a business model in the UK, using indicators of liquidity, profitability and financial independence of the business organization [9], [10]. The model of R. Lys represents the following formula:

$$Z = 0.063 * X1 + 0.092 * X2 + 0.057 * X3 + 0.001 * X4, \quad (1)$$

where:

X1 is the Turnover / Asset Amount;
X2 - Profit before Interest and Taxes / Amount of Assets;
X3 - Retained earnings / Amount of assets;
X4 - Equity / Debt.

Interpretation of results:

$Z < 0.037$ - high probability of bankruptcy
$Z > 0.037$ - the probability of bankruptcy is small

The Springate model was built by Gordon Spring in 1978 [13]. The model was built by testing 40 asset companies with an average of $ 2.5 million. In the process of setting up 19 financial ratios considered to be representative in the final version only four quotes remain on the model. Its general appearance is as follows:

$$Z = 1.03 * X1 + 3.07 * X2 + 0.66 * X3 + 0.4 * X4, \quad (2)$$

where:

X1 is the Net Working Capital / Asset Amount;
X2 - Profit before interest and taxes / Amount of assets;
X3 - Accounting Profit / Current Liabilities;
X4 - Net Sales Income / Amount of Assets.

At Z < 0.862 it is assumed that the company is in poor financial condition and financially difficult; her assessment is "crash".

The advantage of the model are that the model has a good level of predictability. ***Deficiencies*** of the model are: There is no sectoral and regional one differentiation, and it is appropriate for use in advanced economies.

The Taffler model for predicting the probability of bankruptcy was proposed in 1977 and is a fourfold forecasting model [14]. It is recommended to take into account the current trends in business and the impact of prospective technologies on the structure of financial indicators. The model is expressed by the formula:

$$Z = 0.53 * X1 + 0.13 * X2 + 0.18 * X3 + 0.16 * X4, \quad (3)$$

where:

X1 is Profit before Interest & Tax / Current liabilities;
X2 - Current Assets / Total Liabilities;
X3- Current liabilities / Amount of assets;
X4 - Net Sales Income / Amount of Assets.

Interpretation of results:

$Z > 0,3$ - low probability of bankruptcy;
$Z < 0.2$ - a high probability of bankruptcy.

The advantage of the model is that model parameters are calculated on the basis of the financial statement, which allows it to be used in external analysis. ***Deficiencies*** of the model are: it does not account for sectoral and regional specifics of the enterprise and only at a negative value of Z it can be claimed that the enterprise is close to bankruptcy.

A **significant group of models** for predicting the probability of bankruptcy are the Z-models developed by E. Altman [15], [16] often referred to as the "father" of modern theory of bankruptcy forecasting. Altman develops two-factor, five-factor, and seven-factor Z models. The two-factor model has the following features:

$$Z = - 0.3877 - 1.0736 * K1 + 0.0579 * K2, \quad (4)$$

where:

K1 is a current liquidity ratio;

K2 - the financial dependency ratio, which represents the relationship between the debt and the balance sheet total of the asset (liability).

Interpretation of results:

Z> 0 - the probability of bankruptcy is over 50% and increases with the increase of Z;

Z = 0 - the probability of bankruptcy is 50%.

The model is not adequately explained in the literature, and thus is associated with its low popularity and limited use. Its ***advantage*** is the ability to be used in a limited amount of information about the enterprise. The ***drawback*** of the model is that it does not provide high precision on bankruptcy forecast. It also does not include performance metrics for the resources used.

The most popular is the **five-factor model** for predicting corporate bankruptcy, known as ***Altman's Z-Score***. The essence of the model is to select relatively independent coefficients and determine their respective weights. When developing the model, E. Altman analyzed 66 companies, half of which went bankrupt in the period 1946-1965, and the other half worked successfully. He has also investigated 22 analytical ratios that can be used to predict the risk of bankruptcy. From these metrics, he selects the five most important for the prognosis and integrates them into the multifactorial regression equation. Variables are classified into the following groups: liquidity, yield, indebtedness, solvency and efficiency. These indicators include Altman in the original five-factor model, which has the following appearance:

$$Z = 1,2 * x1 + 1,4 * x2 + 3,3 * x3 + 0,6 * x4 + 0,999 * x5, \quad (5)$$

where:

x1 is the ratio of working capital / amount of all assets. Characterizes the share of working capital in assets.

x2 - the ratio of undistributed profit / total of all assets. Characterizes the profitability of assets on the basis of undistributed profits.

x3 - the ratio of profit before interest and taxes / amount of all assets. Shows the profitability of assets based on operating profit.

x4 - equity value / carrying amount of attracted capital. Characterizes the funding ratio.

x5 - sales / total income ratio of all assets. Shows the return on all assets.

The coefficient obtained by applying the model allows to determine the probability of an enterprise bankruptcy. For this purpose, Altman formed the following intervals with different probability of bankruptcy:

Z <1.23 - financially troubled enterprise, potential bankruptcy (crisis)

1.23 <Z <2.90 - uncertainty, area of uncertainty, high probability of bankruptcy in the near 2-3 years (pre-crisis state)

Z> 2.9-secure enterprise, there is no risk of bankruptcy (steady state)

Therefore, the result assuming that the undertaking under study is experiencing financial difficulties is Z <2.9, and the result at which the enterprise is classified as actually going bankrupt is Z <1.23.

Advantages of the model are that the indicators in the model characterize different aspects of the business. ***Deficiencies*** of the model are: the model is developed on the basis of data from a developed economy. Components of the formula are related to various crises in the enterprise - X1 - with a crisis in management; X4 - with the financial crisis; X2, X3 and X5 - with an economic crisis in the enterprise.

1.3.Exploration

The survey of anti-crisis management models used was conducted between 1.09.2018 and 15.12.2018 among companies of various industries and sizes. The survey is based on open and closed questionnaire. A questionnaire was used to collect data from a sample of 312 entreprises which were selected through stratified random sampling method.

Responding companies are of varying size as staff, with the smallest companies with 1 to 10 employees and more than 500 employees being distributed in a balanced manner (see fig. 2)

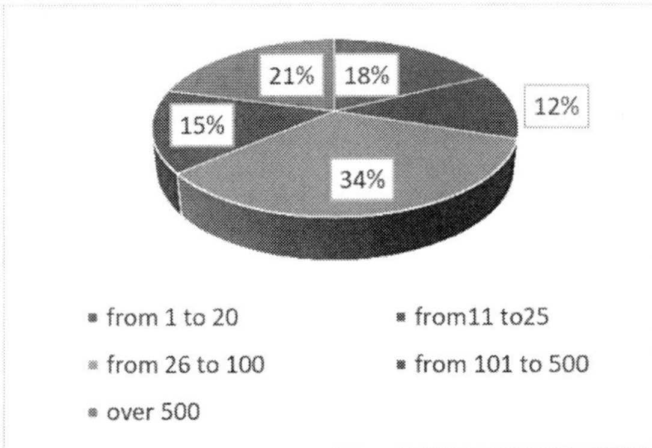

Figure 2. Number of employees

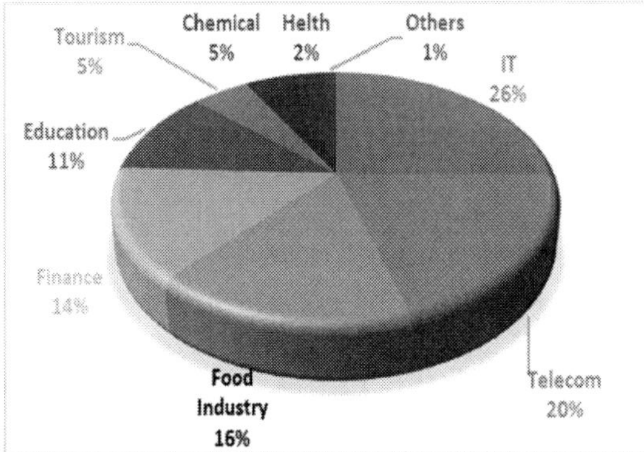

Figure 3. Distribution of respondents by industry

About 10% of respondents say they have not introduced anti-crisis management. Logically, the smallest companies with a turnover of less than BGN 500,000 have the least use of similar systems, with only 12% of these companies responding with "yes". An interesting result is that among the companies with sales revenues of over BGN 500,000 there is no relation between the size of the company and the system. Even from companies with over BGN 10 mln turnover for 2018, 68% use anti-crisis management.

Figure 4. Methods of anti-crisis management used by the industrial sectors

25% of the respondents are representatives of the IT sector, 20% are from Telecommunications, 16% from the food industry, 14% from the financial sector, 11% from education, 5% from tourism, 5% from chemical industry, 2 from medicine % and others - 1% (see fig. 3).

Research has shown that the most preferred method of anti-crunching in the food industry is Altman's Z-Score model (see fig.4). Businesses in the tourism sector and education prefer to use Taffler model. Companies in the financial sector, telecommunications and the IT sector anticipate crises with the help of the Springate model. The other two models- the model of R. Lys and the model of W. Beaver are applied significantly less frequently by some companies from the IT sector, telecommunications, health and finance.

Conclusion

From the interpretation of the different strategic options to overcome crises in enterprises, two options for their anti-crisis management can be summarized. According to the first, anti-crisis management is achieved by realizing cost savings of the enterprise and selling surplus assets. According to the second, anti-crisis management is realized through the production and sale of competitive products. Anti-crisis management, aimed at increasing sales and improving economic, financial and social status, is preferable to anti-crisis management leading to cost reductions and the sale of available assets, so the two options are not opposed, as the first one can achieve quick effect and to provide the resource for realization of the second.

The limited time resource for crisis management requires the use of anti-crisis management models that predict a crisis situation that can greatly contribute to the detailed planning of the enterprise's operations and hence the preparation of preliminary decision-making options.

Acknowledgements

The authors would like to thank the Research and Development Sector at the Technical University of Sofia for the financial support.

References

[1] Tepavicharova, M., M Sikudy, (2016), "The role of the social innovation in the development of the social entrepreneurshipand the addition of value in society", *First Conference on Social entrepreneurship*, Academic Publishing Agricultural University, Plovdiv, Bulgaria, ISBN 978-954-517-249-6, pp. 148 – 156,

[2] Londev A.,(2015) „Fiscal Stability of the Economy", *Yearbook of the Higher School of Security and Economics*, Plovdiv, vol 12, ISSN 1313-8472, pp. 197-206

[3] Londev A.,(2017), "Financial fraud and tax speculation, *Politics and Security Journal* , HSSE, Plovdiv, Vol.1, ISSN 2535-0358, pp. 131-137

[4] Ganchev D. and Naidenov S.,(2018), Bulgarian Industry and Technology Challenges, Eleventh International Scientific Applied Conference "Digital Economy and Blockchain Technology"- Varna, ISBN 978-619-7026-28-3; pp. 203-209

[5] Kasabadzhakova, S.,(2014),"Life Cycle Assessment of Innovative Product", International Journal of Science and Innovation, ISSN 1314-6907, pp.245

[6] Tepavicharova, M., L. Boykova, (2016), "Social media and transparency in public sector", In: KNOWLEDGE, International Journal Scientific papers, Vol.15., ISSN 1857-92, p. 629 – 633

[7] Ganchev D.,(2016), "Prospects for Economic Growth of the Bulgarian Economy", International Scientific Conference "Economic Welfare through Sharing Knowledge", 80th Anniversary of "D. A. Tsenov" ,November 9-10, Svishtov, vol. 1, ISBN 978-954-23-1185-0; pp. 217-222, 2016

[8] Hauschild J, (1998) "Diagnosis of the Crisis by Balance Analysis", *Journal of Economic Perspectives,* Vol. 6, No. 3, pp.155–185

[9] Argenty J., Corporate collapse, *McGraw-Hill*, UK, 1976, p.146

[10] Iliev, Y. et al.,(2011), Anti-Crisis Management of Industrial Firms, pp.64

[11] Fitzpatrick, P. J. (1932). "A Comparison of the Ratios of Successful Industrial Enterprises With Those of Failed Companies".*The Certified Public Accountant*, October, pp.1598-1604.

[12] Beaver, W. H. (1968). Alternative Accounting Measures as Predictors of Failure. The Accounting Review, 43(1), 113-122

[13] Springate, Gordon L.V.,(1978) "Predicting the Possibility of Failure in a Canadian Firm".

Unpublished M.B.A. Research Project, Simon Fraser University, pp. 234
[14] Taffler, R.J. (1977), "Finding those companies in danger using discriminant analysis and financial ratio data: a comparative based study city business school", *City University Business School*, London, Working Paper No. 3; p.50-54
[15] Altman, E, Haldeman, R G, Narayanan, P, (1977) "ZETA analysis: a new model to identify bankruptcy risk of corporations", Journal of Banking & Finance vol. 45, pp.345-358
[16] Altman, E. I., (1968): "Financial Ratios, Discriminant Analysis and the Prediction of Corporate Bankruptcy", *Journal of Finance* vol. 23 pp. 589-609

IOP Conf. Series: Materials Science and Engineering **618** (2019) 012072 doi:10.1088/1757-899X/618/1/012072

Determining the Nature of Competition in Bulgarian Food Industry

T Gigova[1] and K Valeva [2]

[1] Faculty of Mechanical Engineering, Department of Industrial Business, Technical University-Branch Plovdiv, Technical University - Sofia, Plovdiv Branch, 25 Tsanko Diustabanov Str., Plovdiv, 4000, Bulgaria

[2] Faculty of Economics, Department of Industrial Business and Entrepreneurship, University of Food Technologies, 26 "Maritza" Bld, 4000 Plovdiv, Bulgaria

E-mail: t_gigova@abv.bg, tomika888@yahoo.com

Abstract. The highly competitive environment for the food industry after the accession of Bulgaria to the European Union puts companies under conditions where the focus is on the quality and products'safety. Both the competitiveness theory and the topic of food safety have been widely discussed in recent years, and research in this area is growing, but there is a lack of in-depth information in the Bulgarian literature on the relationship between competitiveness ↔ food safety systems and the open question of how strategies food safety would increase corporate competitiveness. Examining this relationship, it is possible to uncover those factors stemming from the requirements of food safety systems, the improvement of which can help the food industry companies to increase their competitiveness. The subject of the study is the different approaches and methods of assessment and analysis outlining the opportunities for increasing the competitiveness of enterprises from the food industry through the implementation of food safety systems. The report aims to reveal the interrelationship between competitiveness and food safety systems and the implementation of methodologies to increase their competitiveness. The study and assessment of factors to improve the competitiveness of food business enterprises through the implementation of food safety systems is based primarily on expert judgment as well as on marketing and diagnostic analysis. Emphasis is placed on factors that arise from the requirements of food safety systems and a few extras that are important for the formation of strong competitive advantages.

1. Introduction

Competition is a driving force and an incentive to increase the efficiency of business operations. It is an integral part of the market economy and should be considered in relation to the market as a whole system [1]. Since competition is a competition between different business entities, it allows for the most efficient use of resources and the most successful organization of business to achieve the ultimate goal of making more profits [2]. Competition can be defined as a driving force for the development of society. It provokes the use of the best of abilities and knowledge, leads to increased human wealth and diversity, leads to rational behavior on the part of sellers and buyers, and rational use of resources [3],[4]. According to Joseph Schumpeter, in the economic system, the new combinations make their way, defeating the old ones, and for him this process is the basis of competition. Enhancing the ability of enterprises to meet standards more stringently is likely to create new forms of competitive advantage [5]. In this regard, research has been carried out on some sectors - Spring and Isaac emphasize the fundamental importance of food safety to the competitiveness of the meat industry in Australia, Canada, the UK and the United States. Issues related to increasing business competitiveness are particularly relevant as they affect in particular the food industry and, more generally, they are also linked to national competitiveness [6], [7].

Under increasingly sophisticated market relationships, the success of the food industry is driven by their flexibility and adaptability to the demands of their environment, as well as the prudent management of intra-company processes focused on the quality and safety of production.

The **research is focused on** the food industry - a sector with long traditions, occupying a major share of the Bulgarian industry. In addition, food productions fall into a risk group from a safety perspective and it is appropriate to target food companies and their food safety systems as an area that can be developed and used as a source of competitive advantage;

The main purpose of the study is to reveal the interrelationship between the competitiveness of the food safety systems and to implement a methodology for increasing the competitiveness of the enterprises in the food industry.

The objective is decomposed into the following research tasks:

First, to review the current state of the food industry and to develop a methodology algorithm for increasing the competitiveness of enterprises in the food industry through food safety systems.

Second, to approve the established methodology in Bulgarian food industries selected on the basis of relevant criteria and to provide recommendations for increasing the competitiveness of the sector through the implementation of food safety systems.

The object of research is 198 enterprises, representing food sector of the Bulgarian economy.

The subject of the study is the different approaches and methods of assessment and analysis outlining the possibilities for increasing the competitiveness of the food enterprises by applying the food safety systems.

The research supports **the main research hypotheses,** namely that the food safety management systems are tied to company competitiveness and competitive advantages and a methodology can be applied to reveal opportunities for increasing the competitiveness of food business companies by systems of food safety.

In the present study, the following **methods were used:** methods for collecting primary empirical information, questionnaire, deep interview method, expert assessment were applied. Methods of analysis and synthesis, comparison and analogies are also used; dialectical, logical, causal, economic-statistical and comparative analysis.

Studies, strategies, outcomes and guidelines of work from national and regional programs have been used. In the development of the analysis, results of the Association of Meat Processors in Bulgaria and the Association of Industrial Capital in Bulgaria surveys, as well as similar surveys and relevant information sources on the subject of analysis were used.

Literature Review

The term competition "concurrentia" derives from the Latin word "co cure", which means I run to the goal [. In daily vocabulary, it is used in the sense of rivalry, competition, and race between people, companies, countries and regions to achieve a particular goal. There are different definitions of competition in economic literature. Porter describes it as a "competition between companies in which they use national, regional, and global strategies" [1]. According to Ribov, "competition is an interaction between business entities, each of which strives to realize its interests, regardless of the other subjects." According to Ganchev, [3] "competition, each pursues one's own interest, but at the same time the interests of society are realized". Vladimirova defines it as "a rivalry between economic agents to achieve the goals set (the most efficient use of resources, maximizing profits, achieving a high market share, etc.), which takes place in the conditions of interaction between them [2].

The diversity of competitive relations in the economy is divided into **three levels**:

- *Micro level* - for individual types of products, productions or companies;

• *Meter level* - for a given industry, corporate unions and conglomerate type companies;
• *Macro level* - for a country.

The study of competition takes place through the so-called competition mechanism, which consists of the following elements:

• *Sources of competitive advantage*. There are two types - lower costs and differentiation of goods and services.

• *Subject to competition* - these are: consumer demand (domestic and international), factors of production (labor, land, capital and information) and power.

• *Subject of competition* - here are individuals (physical or legal) leading to competition for the objects of competition, ie. competing companies that produce or sell similar products and services.

• *The subject of competition* is goods, services or products.

• *Competitive struggle methods* - applied by subjects to the competition process for sites.

To survive on a highly competitive market, subjects can apply different methods of competitive struggle - to reduce prices, improve quality, expand service, and more [7]. Unlike competition, which is characteristic of the market, competitiveness is a characteristic of the particular economic entity / system. In other words, competitiveness can be defined as the ability to compete [8].

There is no complete and accurate definition of competitiveness. It is so comprehensive that different definitions are not contradictory but rather complementary. The existing definitions of competitiveness differ depending on the level they are relevant to. According to some of them, competitiveness is linked to the production and distribution of manufactured products [9]. Other definitions are based on the understanding of the efficiency of the production process and the potential for innovation and technological excellence. According to Michael Porter, competitiveness refers to the way in which the available resources are used [1]. The link between the notions of competition and competitiveness can be explained in the following way: competition and competitive relations predetermine the competitive advantages that are a factor of competitiveness.enterprises, corporations, economic systems, the national economy of the country or the world as a whole" [10].

1.1. Form the model of perfect and imperfect competition

The neo-classical theory of perfect competition is based on unrestricted competition between business entities. The main moments in it were developed by Adam Smith, and A. Marshall, L. Walras, E. Djevins and Pareto are involved in the development of the model [1],[2]. The main points in A. Smit's theory of perfect competition are:

- Defines the notion of competition as rivalry, rising prices when supply is reduced and prices fall in surplus supply.
- Defines the main principle of competition - "The principle of the invisible hand". Competition forces market players to follow the instructions of the "invisible hand" and is a mechanism to automatically achieve market equilibrium.
- Develops the theoretical foundations of a flexible competition mechanism that balances the industry's profit rate, leading to an optimal allocation of resources across the various sectors of the economy.
- Defines the basic conditions for effective competition, including the availability of a large number of sellers and buyers, the mobility of the resources used, the availability and availability of information on the level of demand, supply and prices, which makes it possible to make optimal choices.
- Defines competition as a means of solving contradictions between private and public interests.

Perfect competition - characterized by a large number of players (sellers and buyers) on the market. Each of them has enough full market information and none of them can control market demand, delivery of goods and their price.

Perfect competition is growing into its antipode - the monopoly, that is, in imperfect competition. It appears in a market where individual producers (sellers) can have some impact on the price.

In the monopoly theoretical model there is only one producer or consumer who has no competitors on the market, there are no substitutable goods or services [8]. There are different circumstances in which a firm may become a monopolist. In this sense, it is a closed, natural or open monopoly.

- *Closed monopoly* - is typical for cases where the company is protected by competition by means of legal restrictions (by the Copyright Agency).
- *Natural monopoly* - exists when there is a company producing a unique product on the market and there are some limitations on the factors of production (unique natural resources).
- *Open monopoly* - for a while, a company is the only supplier of a particular product without any specific protection from competition, as in the closed and natural monopoly, but in the future a competitor may appear.

When examining market conditions, the authors of the theory of *imperfect competition* focus their attention on the situations of *polypole, oligopoly and monopoly*

- ✓ *Polypol* is a market with many participants, but they are mostly small companies;
- ✓ The *oligopoly* is a market with a small number of participants, but medium-sized companies;
- ✓ The *monopoly* is a market with only one bidder on the demand or supply side, but it is a big company.

In order to regulate competition, Bulgaria adopt and enforce anti-monopoly legislation.

There are quantitative methods defining market conditions for determining the nature of competition. Widespread indicators are two: the concentration ratio of leading (usually four) firms and a Herfindal-Hirschman index.

Concentration ratio of the four largest companies - represents the sum of the relative shares of the sales value of the four largest companies. Determined by the formula:

$$CR4 = MS1 + MS2 + MS3 + MS4 \qquad (1)$$

where:

CR4 is the concentration factor (concentration ratio);
MS1, 2, 3, 4 – market share (as a percentage) of the four largest companies.

The index may range from zero in pure competition to 100% in monopoly. An indicator value above 40% is indicative of oligopolistic competition, and below 40% for monopolistic competition.

Herfindahl-Hirschmann Index - represents the sum of squares of market shares of major competitors. Determined by the formula:

$$HHI = \sum_{i=1}^{N}(MS)_i^2 \qquad (2)$$

where:

HHI is the Herfindal-Hirschman index;
MS - market share in percent;
N - the number of competitors.

The higher the index, the more monopolized the relevant market and:

HHI ≤ 1000 means a competitive market;
1000 ≤ HHI ≤ 1500 - Slightly monopolized market;
1500 ≤ HHI ≤ 2500 - moderately monopolized market;
HHI ≥ 2500 - high concentration of monopoly power;
HHI = 10,000 - full monopoly of 1 company.

1.2. Functions and importance of competition

The **first function** is a counterpart. It finds expression in the impact of competition in two directions:

- ➢ Competition is a universal tool for comparing the effectiveness of different companies to discover and stimulate the most economically active [5], [7].

- Such a match meets the interests of the consumer, who chooses the trader giving him the most advantageous terms.

The **second function** is selective. It involves removing market participants who work economically ineffective. As a result, overall efficiency of production and trade is increasing.

The **third function** is related to its disciplining power in the market economy. The availability of alternative sellers and buyers in the market to some extent forces companies to comply with consumer wishes and to seek more efficient ways to carry out their business. Companies that fail to provide consumers with quality goods and services at competitive prices will suffer losses and will eventually go bankrupt.

The **fourth feature** focuses on the dynamics. Firms operating under strong competition find new ideas, develop new offers on the market, new channels for realization, new forms of service, and make profits as innovators. At the same time, they take the risk that new ideas and developments will not be accepted by the market, which would bring them losses. This is how dynamic changes are made, in which the old methods and methods give way to the new ones.

1.3. Competition and efficiency

Competition is a driving force and an incentive to increase the efficiency of business operations. It is an integral part of the market economy and should be considered in relation to the market as a whole system [3], [9].

Since competition involves a struggle between different business entities, it enables the most efficient use of resources and the most successful organization of the business to achieve the ultimate goal of gaining more profits [6], [10].

Competition is closely related to efficiency as:

- Involves actions whereby the business seeks to offer higher quality and more efficient products and services to achieve higher turnover and hence greater profit than other manufacturers and traders.
- Competition stimulates efficiency by introducing innovation. Scarce public resources and limited demand necessitate the introduction of new ideas, methods, tools, knowledge and experience that, under the influence of competition, provide businesses with high performance and success in competitive struggle.
- The market raises the drive for greater profit and maximizing profits. It forces manufacturers to implement the latest advances in science and technology, creating conditions for increasing production efficiency. Universal competition is a prerequisite for thriving the one who is quicker and best adapted to changes in consumer demand and most effectively introduces the latest advances in scientific and technological progress by reducing production and labor costs.

1.4. Analysis of the company's competitiveness

The research and analysis of the competitiveness of a company is carried out at three levels - national (macro level), sectoral and company level. It can take place both in the order and in the opposite direction: company, industry, national level. The relationship of the analysis levels with the analysis tools is presented in Fig. 1.

At national level, the Porter Diamond tool is most commonly used to analyze competitiveness. The four main tips in the "diamond" are: *Factors of production; Demand conditions; Connected and supporting industries. Company strategy, Structure and competition.*

The elements are: **Competition between existing companies**; **The threat of new firms entering**; The emergence of **substitute products; Contractual power of suppliers; The contractual strength of customers** is directly reflected not only on the particular firm, but also on the competition in the industry, which may cause a direct clash between different companies. As a result of the sectoral level analysis, we need to get an evaluation of the competitive advantage over the rivals.

At company level, the competitive advantage can be determined by the following methods:

Competitive triangle - is a comparison with the leading competitor. It is often used also the abbreviation three K (client, company, competitor). The competitive triangle reflects the fact that the consumer perceives the value offered by a particular company and its main competitor.

Value chain. It uses the system approach to form the market advantage of the company. It divides company activities into two types: primary and secondary.

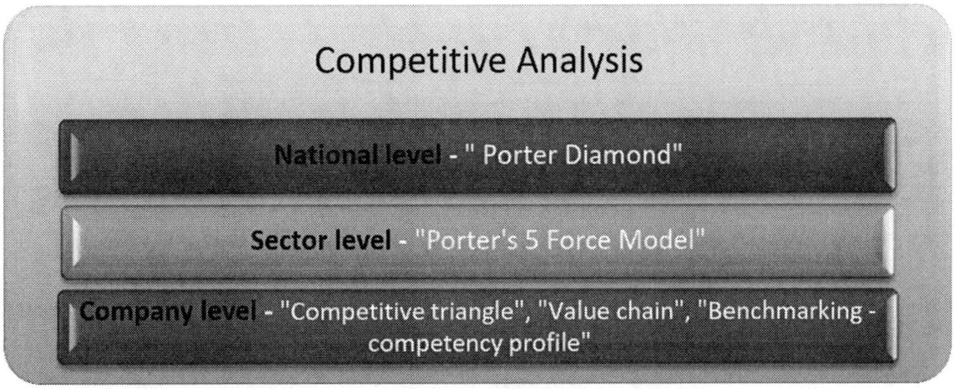

Figure 1. Relationship between Analysis Levels and Competitive Analysis Methods

At the industry level, the *"Porter's Five Force Model"* is the most widespread [2]. Its main

Benchmarking method - drawing up a profile of competence. The method allows for ongoing monitoring of the main rivalries and comparison with customer requirements, not the establishment of the results after the occurrence of the unfavorable events or at the end of a certain period [3],[9].

In the current conditions of increasing competition between rivals, the idea of the dynamic nature of competitive advantage and hyper-competition is becoming more and more popular [4],[10]. These models emphasize the speed and aggressiveness of the action taken and the countermeasures in the particular market, resulting in competitive advantages quickly becoming eroded.

Hyper-competition. The model is based on the following prerequisites. Any competitive advantage may be copied or undermined by other market participants. Once overcome, it becomes a burden. Blindly following the idea of maintaining the market advantage can be a fatal delusion because the pursuit of perfecting the advantage diverts attention from the formation of a new competitive advantage. Overtaking of the immediate competitors is not once established forever. It represents a number of small steps, each of which allows to undermine, destroy the advantage of the opponent or build the next competitive advantage before eroding the present. In this direction, the following strategies can be implemented:

1. **The strategy of undermining the market advantage** includes two elements: *consumer satisfaction* and *strategic prediction (prediction)*. They help to explore in more detail the requirements of current users or to open new customers whose needs are not satisfied. This strategy can also be used by the market leader, and in this case, it will be possible for him to preserve his position [.

2. **The tactic of erosion of the market advantage** includes three types of possible actions: *changing the game's rules* - the conditions of competition by opening new ways of servicing consumers; *disclosure of strategic intentions* - to drive action and track competitors' responses; *simultaneous and consistent attacks* on rivals to confuse and block their efforts to maintain the market advantage. These actions shape the direction or nature of the response of competitors.

3. **The possibilities of undermining the competitor's market advantage** are two: *speed and surprise*. Positioning the product and brand with the ability to reposition it as quickly and in a way that will surprise the competition.

Exploration

The survey of was conducted between September of 2018 and February of 2019 among companies of various food' industry subsectors and sizes. The survey is based on open and closed questionnaire distributed by email. A questionnaire was used to collect data from a sample of 198 entreprises which were selected through stratified random sampling method from all subsectors of Bulgarian food industry.

Respondents are representatives of all major subsectors of the food industry in Bulgaria and their relative shares are shown in figure 2.

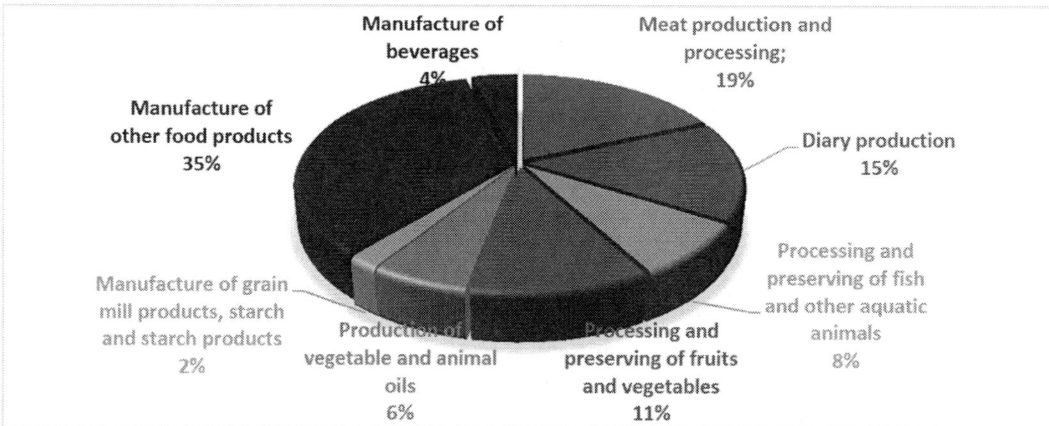

Figure 2. Share of respondents from food subsectors in Bulgaria

It appears that the largest share is the respondents representing the subsector - production of other food products - 35%; followed by representatives of the meat processing industry - 19%, the dairy sector - 15%, followed by preserved plums and vegetables - 11%, canned fish - 8%; the production of vegetable oil and animal oil - 6%, manufacturer of beverages are 4%, and with the smallest share being the respondents-bakers and producers of starch products - 2%.

Table 1. *Determining the nature of competition in Bulgarian food industry*

Manufacture of food products	Widespread indicators		The nature of competition
	the concentration ratio of leading firms	*Herfindal-Hirschman index*	
1. Meat production and processing	48 %	HHI < 1 000	Competitive market
2. Diary production	62 %	1 000 < HHI< 1 500	Slightly monopolized market
3. Processing and preserving of fruits and vegetables	79%	1500 < HHI < 2 500	Moderately monopolized market
4. Production of vegetable and animal oils	82%	1 500< HHI <2 500	Moderately monopolized market
5.Processing and preserving of fish and other aquatic animals	99%	HHI =10 000	Full monopoly
6. Manufacture of other food production	45 %	HHI < 1 000	Competitive market
7. Manufacture of grain mill products and starch products	%	HHI > 2 500	High concentration of monopoly power
8. Manufacture of beverages	52%	HHI < 1 000	Competitive market

We applied the quantitative method to measure market conditions for determining the nature of competition in Bulgarian food industry (see Table 1).

The analysis of the survey results shows that the meat industry companies in Bulgaria are operating under conditions of monopolistic competition. Dairy firms are operating under a slightly monopolized market, with the concentration ratio of leading firms CP> 40%, which is indicative of working in oligopolistic competition. Companies in the beverage sector in Bulgaria, as well as those from the sub-sector of other food products, operate under conditions of monopolistic competition. Companies in the canned fruits and vegetables sector and the production of vegetable and animal oil work in a moderately monopolized market. The companies from the subsector of grain mill manufacturing and starch products works in the condition of high concentration of monopoly power.

2. Conclusion

The most important sectors (meat processing, production of other food products and beverage production) are identified as well as those with the best prospects from an economic point of view (canned fruits and vegetables, vegetable and animal oil, dairy products). An analysis of the industrial structure and the market conditions in which they work is carried out. As a result of the survey, it has been established that the food industry in Bulgaria is experiencing difficulties in marketing and innovation activity, offering products that meet specific consumer needs, using modern methods of analysis, applying management practices, building on voluntary standards. The conclusion is that the main subsectors of the food industry in Bulgaria are operating in the conditions of monopolistic competition, which requires the adoption of concrete strategic decisions for affirming their competitive advantages and expanding the market positions.

Acknowledgements

The authors would like to thank the Research and Development Sector at the Technical University of Sofia for the financial support.

References

[1] Porter M.,(2004) „The Competitive Advantage of Nations", Ed. Classics and Style
[2] Vladimirova, J. (2012), "Competition and Competitiveness of Commercial Companies", Avangard Prima. Sofia
[3] Ganchev D. and Naidenov S.,(2018), Bulgarian Industry and Technology Challenges, Eleventh International Scientific Applied Conference "Digital Economy and Blockchain Technology"- Varna, ISBN 978-619-7026-28-3; pp. 203-209
[4] Mihova, T., Anguelov, K., (2018), Research on interactions among universities and high-technological enterprises at qualification of employees, IX National conference with international participation ELECTRONICA, Sofia
[5] Tepavicharova, M., (2016), "Possibilities of benchmarking for measuring the effectiveness of the human resources management in the organization", In: KNOWLEDGE, International Journal Scientific papers, Vol.15.2, p. 549 – 554, ISSN 1857-92
[6] Londev A.,(2015) „Fiscal Stability of the Economy", *Yearbook of the Higher School of Security and Economics*, Plovdiv, vol 12, ISSN 1313-8472, pp. 197-206
[7] Angelova, M., Pastarmadzhieva, D. et al (2017) "Innovative model for business financing in the area of wine production"; International scientific conference „Applied Modeling in Economics, Finance and Social Sciences" 27-31.08.2017, Hisar, Bulgaria, bhttp://dataconferences.org/page/publications
[8] Zlatanova-Pazheva, E., Survey of food preferences of children and teenagers in Southern Bulgaria, Scientific works of University of Food Technologies-Plovdiv, 2016, Volume 63, Issue 2, p. 321-327
[9] Ganchev D.,(2016), "Prospects for Economic Growth of the Bulgarian Economy", International Scientific Conference "Economic Welfare through Sharing Knowledge", 80th Anniversary of "D. A. Tsenov", November 9-10, Svishtov, vol. 1, ISBN 978-954-23-1185-0; pp. 217-222
[10] Kulova, I.,(2016), "Challenges in front of pricing policy of the companies in bulgaria at the beginning of 21st century", Міжнародної науково-практичної конференції: «Досягнення науки в 2016 році», Центр наукових публікацій, Київ, ISSN: 5836 – 4978, pp. 126-128

IOP Conf. Series: Materials Science and Engineering **618** (2019) 012073 doi:10.1088/1757-899X/618/1/012073

Features of industrial management in the field of ecology

Oleg Burgonov[1], Dmitrii Kruglov[1], and Elena Ivleva[1]

[1] Saint-Petersburg University of Management Technologies and Economics, 190103 Saint-Petersburg, RussiaDepartment

burgonov@list.ru, kdvspb@list.ru, ivlevaes@mail.ru

Abstract. The processes of global reorganization make managers pay attention to environmental safety and regulation of economic systems. In modern conditions, it is possible to solve environmental problems only by accelerating the development of economic activity. An entrepreneur who seeks to improve competitiveness and becomes a leader, both at the international and national levels. This happens when the company's management implements environmental programs related to financial activities and trade.

1. Introduction

Development of technologies and techniques concentrates the changes which after reaching a certain threshold will become an impetus for transition to a new technological paradigm. Emergence of the new paradigms will result in development of the economy. Leading European and global companies focus their efforts on moving from the 4th to the 5th level of process maturity [1]. As a result, new business lines and forms of management, as well as new industries are created. New terms appear in the economic science: green entrepreneurship, eco-entrepreneurship, ecopreneurship, eco-friendly products, green market, green business. The latter is about environmentally friendly business results (product) and environmentally friendly technologies of its production.

In parallel with origination of scientific management, new forms of organization of social labor will be prepared on the basis of relevant theoretical studies. Such studies will form methodological and theoretical bases of the next stage of science development. Use of the general scientific methods helps to summarize that we are at the threshold of a new form of industrial management. The management system itself becomes the most important intangible asset. Currently the researchers put forward an idea that the success of business in XXI century will be determined not only by achievements in development of production technologies, but also by the level of development of industrial organizations management. Business tries to improve its management using new methodology and tools [2].

The study supports the thesis that using the ecological concept of industrial management in the manufacturing enterprises is a strategic approach that makes businesses more competitive and socially responsible.

The purpose of the study is to identify the dependencies and relationships between various environmental factors and the level of development of industrial enterprises, as well as their impact on innovation activity, efficiency and competitiveness.

IOP Conf. Series: Materials Science and Engineering **618** (2019) 012073 doi:10.1088/1757-899X/618/1/012073

2. Setting the problem

Global restructuring processes force managers to pay attention to environmental safety and regulation of economic systems. In modern conditions the only way to solve environmental problems is to accelerate the corporate development. The entrepreneur strives to improve the level of competitiveness and becomes a leader, both internationally and nationally. It happens when the company's management implements environmental programs related to financial activities and trade. Entrepreneurship is not only creation of a product, but also creation of new values and ideas. However, if solutions to the problems are not known, the management needs consulting services on environmental management:

- about the existing regulations in the field of environmental management;
- for organization of training system for employees on environmental standards;
- for creation of an image of "environmentally friendly enterprise";
- on conducting a production policy aimed at transition to new types of raw materials, business processes, technologies.

It means that solving the initial problems will help small and medium-sized businesses to find additional ways for making a profit. The ecological acceptability of different activities should be checked during the examination. In the process of carrying out of such an analysis of the environmental component developed by the International Bank for Reconstruction and Development, it is possible to assess the environmental impact of investment projects planned for implementation. As market conditions change in the field of investment projects involving the use of low-waste technologies, it is possible to predict an increase in the role of consulting services. Companies operating in the field of technically complex products are not sufficiently studied. If you need a detailed analysis of the decision-making mechanism within the existing enterprise management system, the environmental problems can be considered by the marketing department. In this regard we see a formation of environmentally oriented market space and management.

Most prognoses related to the environment show that there is a beginning of global irreversible processes that may lead to disastrous results. Such scientists as D. Meadows and D. Forrester from the Club of Rome in 1972 in their work "The Limits to Growth" criticized the negative trends of human development [3, 4]. Pollution of the environment is an important indicator of civilization development, which will grow if effective measures are not taken. The development paradigm has four components: improvement of the environment, efficient use of resources, social and economic development. It should be emphasized that the measures taken to address the environmental problems at the state level are insufficient, often requiring a regional level, for example, the EU [5,6].

In addition to the 1996 concept of sustainable development, there are other documents in the field of environmental development. The Government of the Russian Federation, realizing the importance of movement in this direction, has made certain commitments in the form of "Fundamentals of the State Policy for the Period Until 2030" and the State Program "Environmental Protection for the Period Until 2020". Total amount of funds planned for "Environmental Protection for the Period Until 2020" program is 335.8 billion rubles. The main direction of creation of these documents is the balance of interests between the preservation of the environment and the interests of economic development. The purpose of the state policy in the field of ecological development is addressing the issues providing growth of nonpolluting economy [7]. Achievement of this goal is only possible through creation of an effective management system. The environmental policy is aimed at meeting the requirements of organizations related to environmental protection, which can lead to significant costs, for example, investments to bring agricultural enterprises in line with the standards of organizations for protection of animals.

Environmental management is usually associated with investments in environmental activities. Russian statistics traditionally take into account the costs of environmental objectives in three areas: for protection of atmospheric air, water resources, lands. We see an increase in costs for all rows of the table (table 1.) [8].

Table 1. Investments in fixed assets aimed at environmental protection and rational use of natural resources (in actual prices, million rubles; 1992 – billion rubles).

Years	Total	Atmospheric air	Water resources	Land
1992	53	9.2	33	7.1
2000	22339	7948	8251	3520
2005	58738	19839	26143	9206
2010	89094	28127	48025	9340
2015	151767	40120	78941	15703

3. Basic results

In Russia, environmental entrepreneurship as a segment of the national market appeared in 1988-1989. In 1991, there were more than 10 thousand enterprises that declared provision of environmental services, but only 20% of them actually worked in environmental field. In year 2000 their number increased, when reorganization of state environmental management bodies diminished the clout of this activity. Currently, there is a pronounced trend of activity in this economic sector.

Current types of environmental entrepreneurship in Russia are: wind power, solar power, hydropower, bio- and geothermal energy, production of energy-saving equipment, recycling and reuse of waste, etc.

In parallel, there are two types of environmentally oriented businesses: new business and established business which is able to shift to green technologies. The first is business, which is initially based on the idea of environmental entrepreneurship and is a new direction created in accordance with the idea of innovative development, being an accurate representation of sustainable development and environmental safety principles [9, p.102].

A survey of 47 respondents (managers) helped to identify the main trends and problems of development of environmental entrepreneurship in the Russian Federation. According to the largest number of respondents (79%) the main obstacle for them is lack of support from the state. The three main problems include the lack of a regulatory framework (21%), a tax and credit support (30%), the need to provide state order for the products of environmental entrepreneurship (49%). At the same time, most of the respondents believe that the environmental business activities will make a real contribution to the environmental improvement.

The need for state regulation of environmental activities is primarily caused by the fact that in this case the use of only market mechanisms is significantly complicated for a number of reasons. First of all, it is rather difficult for individual subjects of environmental entrepreneurship to concentrate the large-scale resources for the implementation of the necessary investments in environment-oriented production facilities. It requires significant capital investment, market forecasting, staff training, etc.

Secondly, many enterprises producing environmental products and technologies do not become efficient immediately. Therefore, isolated, without state support, environmental production facility can suffer significant losses or even go bankrupt. External incentives or guarantees of full or partial recovery are required.

Thirdly, the economic effect of the use of environmentally-friendly and environmental restoration technologies and industries is manifested in various types and many areas of activity. Therefore, it is often difficult to assess the full effect of activity at the level of an environmental business entity, because some portion of associated effects are outside the scope of its activities. In addition, in the current realities, environmental products are not in demand, as environmental protection has not become one of the main objectives of corporate business strategies.

Therefore, it is necessary to provide measures for increase in demand for environmental products through development of real programs with real funding sources, necessary infrastructure and legal framework in order not to become dependent on foreign environmental technologies in the near future and not to lose the environmental products market.

The lack of financing for modernization of production facilities in view of the environmental situation, once again confirm the need for greening of production processes by the way of creation of better management system. The environmental management system is focused on achievement of results at a low cost. In this regard, organizational activities are a priority. However, it requires highly skilled workers.

"Strategic management" concept provides two degrees of readiness of the businesses to the solution of environmental safety problems: environmental and ecologized management options. Ecologized management is a management of an economic entity by adapting the existing infrastructure to the requirements of international and national regulations in the field of resource saving. This situation is typical for the majority of Russian enterprises. Environmental management - is the management of environmentally safe production facilities, providing a balance between economic and environmental performance throughout the life cycle of the production complex and its products. An example is PAO Gazprom, which plays a key role in the economical development. Gazprom is developing its long-term development strategy. It takes into account not only the objectives of the company's shareholders and the economic interests of the state, but also the role of improvement of the environmental situation [10].

Considering participation of Russia in a community of countries with developed economies, it is necessary to say about compliance of these States with common international norms and rules. A company that has implemented environmentally-oriented tools in its management system (ISO 14000) will get advantages by gaining economic benefits and achieving significant competitive advantages [11]. Organizations that are members of EMAS (Environmental Management and Audit System) together with implementation of environmental policy will have the right to use a special logo. This attribute will increase confidence in the environmental activities of the company and increase the chances in a competitive struggle in the international market.

A number of Russian companies already meet the requirements of ISO 14000, EMAS and bring their products to international markets. Such enterprises include Sokol Aircraft Plant in Nizhny Novgorod, Engine-Building Plant in Rybinsk, Aluminum Plant in Krasnoyarsk, etc. Formation of new categories and recognition of such qualifications as manager, is a feature of economic development, determined by the technical achievements of our time. At the same time, the digitalization of society makes profound changes in the organization of production. Digital technologies helps to promptly identify errors of individual responsible persons and managers in production organization and management, to identify their incompetence, to open a prospect of improving the level of the managers. Unfortunately, freeing a person from hard physical labor, digital technologies place heavy demands on physiological capabilities and mental activity.

In this regard the concept of environmental management becomes particularly relevant - environmental management is a management of activities through implementation of measures aimed at safety of working conditions of the managers. The stated above elements of environmentally oriented management are interrelated with the environmental idea (figure 1.).

The process of environmental management should start with determination of the interests of the economic entity for which the management system is developed. Uncoupled from interests of economic entities, it makes no sense to consider environmental risks and consequences of the entrepreneurial activity.

Then the purpose of the subject-oriented environmental management is to maximize the equity capital taking into account the interests of other partners, the society and nature in the form of limitations on a number of actions. When building an environmental management system, the interests of owners and non-owners of the business entity should be taken into account, first of all, they include the life and health of all people in direct contact with it.

Along with the financial consequences of the environmental problems that reduce business profits by additional compensation payments, it is also necessary to take into account the reputational and ethical categories up to stopping and closing of the production facilities. The balanced social development should be based on the mechanism of biological balance of the environmental and individual safety. This approach requires fundamental changes, which are based on the environmental focus of all human activities.

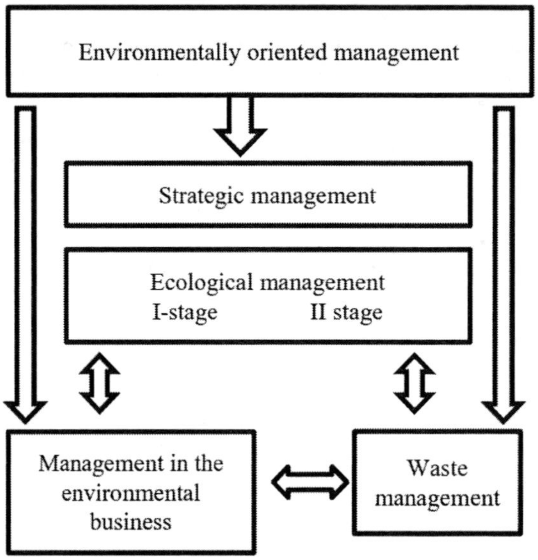

Figure 1. Elements of environmentally oriented management.

Currently, we need to solve the following problems for the development of this sector of the economy. First, it is a lack of regulations that would clearly identify this type of activity and its results. Secondly, it is necessary to determine the goals and instruments of state regulation in this field. Thirdly, development of a system of elements for environmental business management mechanism, which effective functioning will help to achieve the goals. Fourth, implementation of continuous monitoring of the management system functioning.

A variety of objectives aimed at provision of balanced operation of the constituent elements of the management mechanism is based on three levels: operational, tactical and strategic. At the operational level, they create conditions for activities in the field of environmental entrepreneurship, contributing to an increase in production, sales of environmental products in the external and domestic markets and increase its competitiveness through the growth of innovative activity of environmental businesses. The tactical level of the goals is aimed at creation of conditions for effective functioning of the mechanism of environmental entrepreneurship management, contributing to the development of market relations and diversity of ownership in this field of activity, as well as full promotion of innovative activity.

The strategic level of the goals should be aimed at creation of investment attractiveness of the environmental sector of the national economy, contributing to creation of a favorable economic, environmental and social environment.

The following indicators are proposed as indicators of achievement of the goals:

Level I:

- solvency of environmental companies;
- share of innovative products in the revenue;
- budget revenue growth;

Level II:

- growth of infrastructure support tools;
- simplification of conditions for starting and doing business;

Level III:

- average profitability of environmental entrepreneurship;
- investment attractiveness in this field;
- growth of social parameters of the region;
- reducing environmental risks.

Statistics confirm a possibility of identifying the factor of "environmental innovation" both within the structure of organizations engaged in environmental innovation, and in the special costs associated with environmental innovations [8].

Table 2. Environmental innovations.

Parameters \ Years	2012	2013	2014
Share of companies with environmental innovations	22.4	12.9	13.6
Costs associated with environmental innovations, mln. rub.	27768	15098	20914
Share of organizations using the pollution control system	12.1	10.8	11.1

4. Conclusion

The report attempts to point out some important factors affecting the environment on the functioning of industrial enterprises. The results of the survey show that there are still a small number of Russian companies that, mainly with state participation, take advantage of environmental management as part of strategic management. In this regard, it is necessary to green the industry and use the opportunities provided by the environmental management system.

References

[1] Mihova T.; Alexieva-Nikolova V., (2018), "Conceptual impact model of process management on the meat industry enterprises in Bulgaria" International Scientific Conference on Engineering, Technologies and Systems TECHSYS 2018, Technical University-Sofia, Plovdiv branch 17-19 May 2018, Plovdiv, Bulgaria

[2] Mihova T.; Alexieva-Nikolova V., (2016), "Principles and instruments of lean methodology" International Scientific Conference on Engineering, Technologies and Systems TECHSYS 2016, Technical University – Sofia, Plovdiv branch 26-28 May 2016, Plovdiv, Bulgaria

[3] J. Forrester, World dynamics. M.: AST, 2006, 384 p.

[4] Meadows D. The Limits to Growth. 30 years later. M.: Akademkniga, 2007, 342 p

[5] Bankova I., (2016), "Potentials for the development of transport infrastructure of bulgaria in the context of sustainable development" International Scientific Conference on Engineering, Technologies and Systems TECHSYS 2016, Technical University – Sofia, Plovdiv branch 26 – 28 May 2016, Plovdiv, Bulgaria

[6] Petkova-Kozovska M., (2017), "Problems and solutions of waste management in global and national aspect" International Scientific Conference on Engineering, Technologies and Systems TECHSYS 2017, Technical University – Sofia, Plovdiv branch 18 – 20 May 2017, Plovdiv, Bulgaria

[7] Fundamentals of the state policy in the field of environmental development of the Russian Federation for the period till 2030, http://kremlin.ru/events/president/news/ 15177.

[8] Russia in figures 2016, data from www.gks.ru

[9] Kazantseva A. N. Role of environmental business in development of eco-products market / / Economics and management, 2017, No.3, p. 100-105

[10] Ignatov V. G., Kokin, A.V., Sidorov P. T. Environmental management at the enterprise: how to make money in terms of greening the economy. - Rostov-on-don: Publishing house of the Northern Caucasus Higher School Scientific Center, 2008.

[11] Pashkov E. V., Fomin G. S., Krasnyi D. V. ISO 14000 international standards. Fundamentals of environmental management. M.: Publishing and Printing Complex "Publishing standards", 2012, p. 464.

Digital business models in industrial enterprises

T Mihova[1] and K Chukalov[2]

[1]Technical University-Sofia, branch Plovdiv, Faculty of Mechanical Engineering, Department of Industrial Management

expert2009@abv.bg

Abstract. The report examines the reasons for turning business models into digital ones and reveals their nature and technological trends. The authors try to point out what are the first concrete steps in the company practice for digital transformation of business models. Based on our own research, the level of digital maturity of small and medium-sized industrial enterprises in Bulgaria

1 Introduction

We are witnessing how the use of the terms "digitization", "digital models", "digital transformation" is increasing on a daily basis. In the scientific and media space, a number of publications emerge about the essence of these concepts, roundtables and various forums are being held on this issue. Surveys from leading consultancy companies show, that in the era of the Fourth Industrial Revolution, when more and more sectors are forced to change their business models, Europe is lagging behind in the field of digital transformation. The consulting company PricewaterhouseCoopers (PwC) predicts that digitization and adaptation of smart technology in business can lead to a 14% increase in global GDP by 2030 [1].

It is a challenge for us to point out in this report what are the specific actions on the way of digital transformation of business models. For this purpose, we conducted a survey in small and medium-sized industrial enterprises on the practice of digitizing existing business models, on the basis of which we drew conclusions about the digital maturity of the surveyed representatives of the small and medium business.

2 Basic views of business model and digital business model

Before clarifying the essence of the digital business model, we will look at what the business model is. The term "business model" has become particularly popular since the second half of the 20th century, as a result of the big changes in the creation and supply of products by the companies, thanks to the new technologies and the e-commerce opportunities. The idea of a business model arises from the need to understand and explain the new approaches to doing business through the Internet [2]. In the specialized literature, there are a number of definitions of the validity of the concept in question. According to some authors, the business model shows how earnings can be increased from a particular economic activity (how to make money out of it) [3]. Others mention only the components of the business model, namely: value, market segment, value chain, cost structure and potential profit, network value, competitive strategy. In his monograph, Daniel Pavlov makes a detailed overview of the different concepts of the essence of the business model and explores the links of the business model with other management tools, among which: business idea, business strategy, business plan, business project and production program. [4]. We support his position that the most comprehensive definition of a business model is that of A. Osterwalder, who links the concepts of "business" („purchase and sale of products and services " и „making money ") and „model" („the presentation of something through a simple description, which can be used to make an assessment ") individually, to arrive at a precise definition of the business model. Thus, the business model serves as a scheme, which shows how an organization works – he transforms the company's strategic goals into a simplified model, which clearly shows the functions of the business and can be used as a plan on which the it can be structured [5]. If we summarize, the business model shows the way from idea to realization (Figure 1). The development of a business model begins after the business idea is generated and ends with the transition to the development of a business plan. In this way, we can justify the key role of the business model to show the way of the idea of economic activity towards the real environment in which it is to be realized.

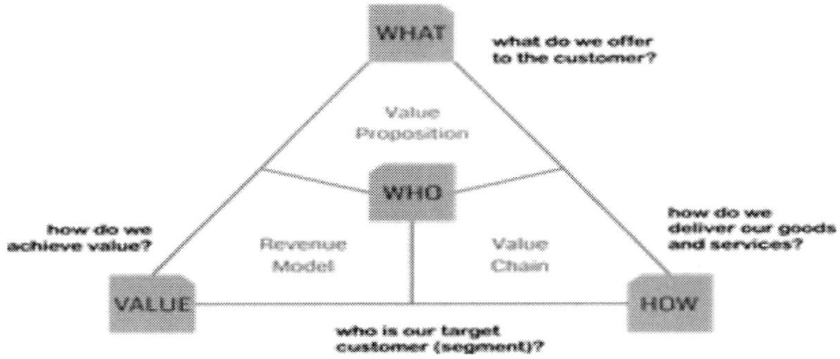

Figure 1. Business model

To try to answer the question of what a "digital" business model is, it is necessary first of all to identify some of the reasons for the digital transformation. Systematically, they are mostly related to the growing customer expectations, search for speed and flexibility, the need to know the customers and the engagement of the companies with them, increasing digital maturity at all levels. Digital transformation involves a rediscovery of the business as a whole, including hisstrategy, leadership models, organizational processes and culture. Digitization will affect the relations with the different stakeholders [6].

All this can serve as a basis for summarizing that digital business models are those where digital services play an important role in creating value for the product / service or even underpin the business itself.

The analysis of publications on this topic enables us to highlight those with a practical focus, which give specific directions for turning the business models into digital. After extensive research inside companies across a wide range of industries, MIT research scientist Stephanie Woerner and coauthor Peter Weill have developed a powerful yet simple framework to guide organizational thinking about which digital business models could be right for your company and where profits can be made [7].

Other publications state a specific model for digital transformation and digital enterprise building, as well as recommendations for successful digital transformation [8].

It is essential to identify the technological trends in the digital transformation of the business models.

3 Technological trends

A number of articles on the dynamic development of technology are published in the media and science space [9]. For us, distinguishing with an emphasis on the practical orientation of the subject are the project materials DIGITRANS, which in a systematic way indicate the current technological trends in digital transformation [10].

3.1 Internet of Things (IoT)

The Internet of Things describes the vision of the physical and digital worlds becoming one as a result of smart objects being connected through the Internet. Depending on the perspective, the focus can be put on the connected things themselves, the protocols or the opportunities and semantic challenges. The idea goes back to the early 90s with the vision of Ubiquitous Computing and has been developed further over the years with the term of Pervasive Computing as the precursor to the Internet of Things. Simply put, IoT is the concept of connecting any device to the Internet (or a local network). It includes smart devices like phones or tablets as well as original dumb devices like lights, door locks or HVAC that are made smart by enhancing them with intelligence and network connectivity. The IoT trend is driven by decreasing costs of technologies and the sky-rocketing penetration of smart devices (Figure 2).

Figure 2. Internet of Things (IoT)

3.2 Blockchain

The blockchain technology describes an algorithm and distributed data structures for secure data transfers without a central administration (e.g. electronic cash transfers). It was originally designed for the crypto-currency Bitcoin and the blockchain concept was mainly driven by the rejection movement against money and bank-controlled payments regulated by governments. The original vision of the Bitcoin developers was to enable people to spend money without friction, intermediaries, regulations or the necessity to know or trust third parties (Figure 3).

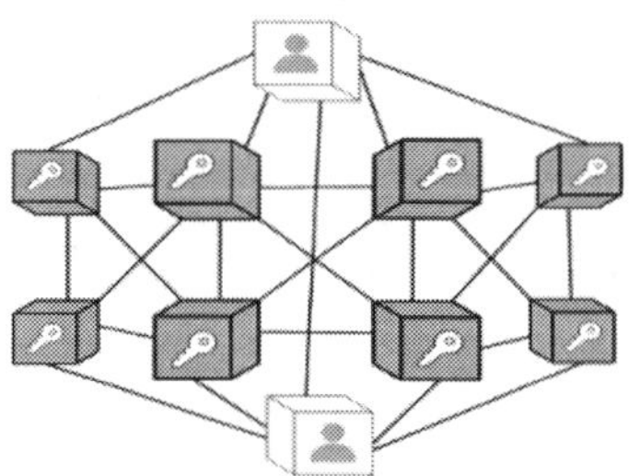

Figure 3. Block chain

3.3 Big Data

When searching the Internet for a definition of Big Data, one will inevitably stumble across the V-Model describing Big Data with its three dimensions: volume, velocity and variety. These dimensions make it difficult to handle with present technologies. Big Data is mostly unstructured data that exceeds the processing capacity of conventional database systems. It is too big, moves too fast or does not fit the structures of existing database architectures (Figure 4).

Figure 4. Big Data

3.4 Machine Learning

The objective followed with Machine Learning is to enable machines to imitate and adapt human-like behaviour. This is done by teaching machines to learn from experience. Machine Learning algorithms contain methods to learn from data without relying on predetermined equations and thus memorizing it. The aim is to find natural patterns and get insights from data and ideally make predictions for data-driven decision making. Examples for the use of Machine Learning are personal assistants like Siri, medical diagnosis or personalized shopping suggestions and ads (Figure 5).

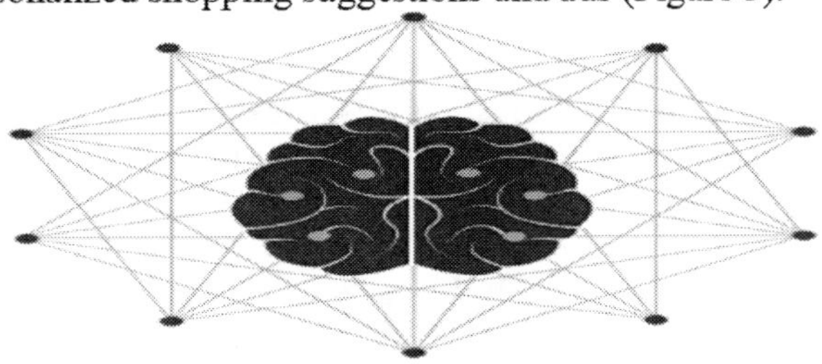
Figure 5. Machine Learning

In summary, these are the current technological trends in synthesized form in the digital transformation of the business models.

4 Methodology and results of the study

4.1 Methodology of the study

In previous publications, we mainly studied the attitudes and readiness of industrial enterprises to digitally transform existing business models. Now, our goal is to analyze the concrete steps in the practice of digital business models.

The survey was conducted in 50 small and 50 medium-sized industrial enterprises in Bulgaria in the period October 2018 - January 2019. Businesses are selected according to criteria from a previous survey – informed and attentive to the first steps of digitization of business models. The methods used in the survey are polls and interviews with business managers. The used model is "Digital Maturity Business Model", Developed by Researchers of Digital Management and Marketing at the University of Salford, UK, which determines the level of digital maturity of an enterprise [11]. Polls include issues related to specific activities undertaken with regard to digital business models and expected results. Participants in the survey are managers of three levels of government: operationally, secondary and higher, covering 328 people. The distribution is as follows – 194 are representatives of the middle business and 134 are managers of small businesses. Enterprises are from the sectors "Manufacture of computer and communication equipment, electronic and optical products", "Manufacture of electrical equipment", "Manufacture of machinery and equipment", "Food Manufacturing", "Manufacture of beverages" and "Manufacture of paper, paperboard and products of paper and paperboard".

4.2 Results of the study

The analyses of the results of the study, in accordance with the Digital Maturity Business Model used, are as follows:

1. 6% of the participants from the small and 4% of the representatives of the medium enterprises, are still at Level 0, where there is no digital presence – enterprises do not have a web site and presence in digital networks.

2. 72% of small respondents and 76% of medium-sized enterprises are at Level 1, which means they have a presence in the digital world – businesses provide digital information to partners and customers.

3. 20% of small and medium-sized enterprises surveyed are at Level 2, for which it is characteristic that companies have an electronic commitment. They have two-way communication, E-business processes, E-supply management, human resources and e-commerce.

4.3 Conclusions from the study

From the brief analysis of the survey results, it is logical to conclude that still a small part of small and medium enterprises have digitized the existing business models. The main reasons that survey participants point to the low level of digital maturity are related to limited financial resources for the use of cutting-edge technologies, limited capacity to manage them and lack of well-prepared teams. The expected results of the digital transformation of small and medium business representatives are related to facilitating communication with clients and partners, attract more customers, creating businesses with greater engagement with their clients and partners, improving productivity and generating higher revenue.

4.4 Recommendations

We will offer a practice-based methodology for a digital business model from the DIGITRANS project, which consists of the phases of Preparation, Innovation, Transformation [10].

During the Innovation phase, new ideas are being developed to create a digital business model. The innovation phase includes two sub-phases: Analysis - where the current situation, the situation of customers and competitors are analysed; and Design – where new digital ideas are created based on the results of the analysis. With the completion of the Innovation phase, we go through the whole cycle of developing a new digital business model – generate an idea, creating and testing the first prototype of the idea and identifying in the channel of business models what added value creates the new idea for a digital business model. Also, in this phase, different methods and tools are used to develop innovative ideas for digital business models in day-to-day work.

In the transformation phase, the idea of a digital business model, developed during the innovation phase is is realized and implemented. This requires a comprehensive approach, which is aimed at further steps to develop a digital business model and reflects the overall strategy as well as the processes and change in employee attitudes, to cope with the challenges of digital transformation. In particular, this includes the following steps:

1. Digital roadmap
2. Involve the team
3. New styles of work
4. Continuous improvement

In the media, as well as in specialized literature, there are a number of recommendations for transforming existing business models into digital ones. This should be done after preparation, including in-depth knowledge of digital transformation and analysis of the specific features of each company.

5. Conclusion

In the report, we are attempting to showcase the essence of the digital business models, current technological trends and above all to see what the level of digital transformation of industrial enterprises is. Undoubtedly, the majority of businesses are convinced of their benefits for their development and in recent years they have been trying to respond to challenges by introducing new models and methods of organizational work, cultural and technological level. At this stage, businesses still have a relatively low level of digital maturity, due to the insufficiently well-recognized need for digital business models and the inadequate preparation for digital transformation. New technologies, like the Internet of Things, calculation of large volumes of data (Big Data), artificial intelligence (AI) and machine self-learning will fundamentally alter the production paradigm and the business as a whole. The understanding and widespread use of new technologies not only will change how customers perceive what "value" is but will also affect the way people work in the organization.

Acknowledgements
The author/s would like to thank the Research and Development Sector at the Technical University of Sofia for the financial support.

References

[1] Europe lags behind digital transformation, https://www.capital.bg
[2] Lambert S, A 2008 Conceptual Framework for Business Model Research, published on the official website of Maribor University, http://ecom.fov.unimb.si/proceedings.nsf/0/1e893ee544d680fec12574810042ac2d/$FILE/22Lambert.pdf, c.c. 1-13
[3] Lang J., and the Cambridge Entrepreneurship Center 2002The High-Tech Entrepreneur's Handbook Pearson Education Limited, UK. p.12
[4] Pavlov D 2011 The Business Model - an element of a system of management tools, Publishing „A Group", ISBN 978-954-8039- 10-9
[5] Osterwalder A 2004 The Business Model Ontology – a Proposition in a Design Science Approach, dissertation work, published on the official website of the University of Lausanne, f. 11-23, available at: http://www.hec.unil.ch/aosterwa/PhD/Osterwalder_PhD_BM_Ontology.pdf
[6] Shatarova D, Zlatanova-Pazheva E 2017 Human resources in the digital age – trends and challenges, Scientific Session "*Days of Science 2017*" of the USB - Plovdiv
[7.] Weill P, Woerner S 2018 What's your digital business model ? Harvard business review press, Boston, Massachusetts 02163, ISBN: 978-1-63369-270-1
[8] Alexieva V, Vyleva K 2018 Digital transformation of Bulgarian enterprises in the conditions of industry 4.0 Scientific papers at VUSI, international scientific conference "*Security and Economy in the Uncertain World - Dilemmas and Challenges*"
[9] Gigova T 2015 Dynamics of technological development Journal of the Technical University – Sofia Plovdiv branch, "*Fundamental Sciences and Applications*" Vol. **21**, 311-314, ISSN 1310-8271
[10] *Project DIGITRANS*, https://digitrans.me/
[11] Digital Business Maturity Model: The future of #DigitalBusiness, https://blogs.salford.ac.uk/

IOP Conf. Series: Materials Science and Engineering **618** (2019) 012075 doi:10.1088/1757-899X/618/1/012075

Investigation of the specifics of the motivation profile in the high-tech enterprises

T Mihova[1], K Chukalov[2], A Ferdov[3]

[1]Technical University-Sofia, branch Plovdiv, Faculty of Mechanical Engineering, Department of Industrial Management

expert2009@abv.bg

Abstract. The report justifies the need to investigate the motivation profile of high-tech enterprises in order to use adequate motivational tactics and techniques in order to have satisfied and committed workers. Different views are given, as well as the author's position on the nature of the motivation profile. Based on own research, conclusions are drawn on the relevance of the motivational factors, the level of satisfaction and commitment from these factors. Specific guidelines and recommendations are given on how to use the motivation potential of the considered factors.

1. Introduction

In today's days, many scientific publications are devoted to motivating workers in enterprises, motivational factors and techniques, a number of successes have been achieved theoretically on mechanisms for increasing work motivation. In-depth studies of the management of the motivation process have been made and its impact on human capital, in line with the European policies for the development of human capital in Bulgaria [1,2]. Shown are the reasons why business practice lags behind scientific methods, methodologies and mechanisms to motivate workers [3].

With the dynamically developing high technology and growing competitiveness of enterprises, motivation issues are of utmost importance for their effective management. High-tech enterprises require highly qualified human resources with specific competencies [4]. In order to make full use of their potential, it is important to know how to motivate the workers most effectively to become satisfied and committed to corporate goals. Company practice has shown that in some cases, notwithstanding management skills for the use of appropriate motivational tactics and techniques, as well as work satisfaction, workers are not very much involved in the labor process. This thesis is the basis of our research – to understand the motivational potential of the factors that will make the workers more engaged in the work tasks in the high-tech enterprises. In this connection, it is a challenge for us to examine the specifics of the motivation profile in high-tech enterprises, as a continuation of a previous study, but with greater scope and enriched methodology, through the prism of satisfaction and commitment to work [5].

2. Basic views of motivational profile

Before we give some insights into the motivation profile, we'd like to look at what high-tech businesses are, what distinguishes them from other businesses. International practice perceives these high-tech industries and services, within which direct costs for scientific, research and development (SRD) are above 8 % of the sales of the relevant sector. According to the Organization for Economic Cooperation and Development (OECD), aerospace, computer and office equipment, radio and television equipment and pharmaceuticals are all in high-tech production. High-tech services are telecommunications, computer and information services, financial services and services related to research and development. As a medium-high technology is assumed the production of research equipment, electrical industry, chemical industry, the production of transport facilities [6]. The peculiarities of high - tech enterprises are related to dynamic changes in technology and products;

significant investment in science, research and development (10 times more than in low-tech, as a percentage of sales revenue); global competition in production, financing and consumption [7]. Logically, the specificity described leads to the need for specific competences of human capital, some of which are: ability to work in an interdisciplinary environment, ability to develop a network, capacity to analyze and understand complex IT tools, organizational culture and managerial skills [4]. All this also determines the specificity of the motivation profile of the employees in the high-tech enterprises. After explaining the specifics of the surveyed enterprises, we will also look at the nature of the motivational profile.

The review of the scientific literature enables us to get acquainted with a number of definitions of the nature of the motivational profile, some of which we have mentioned in a previous publication [5]. Some authors define it as "a complete, comprehensive image of the level of staff satisfaction (or workers in the group) according to the impact of the main motivational factors ", and others as "an analytical image of the extent to which the company meets the different needs of its employees" [8,9].

We are supporters of Tsvetan Davidkov's technology in building the motivational profile, which implies two lines of assessment of the factors of the work situation: by degree of importance and by level of satisfaction. As a rule, for each factor of labor motivation the two grades are compared and this comparison is the basis for description, explanation, prediction of the motivational situation [10]. Our contribution to this methodology is expressed in the assessment and the level of commitment, since it is precisely engaged workers that are needed by high-tech enterprises. After pointing out some views about the motivation profile and our position will you familiarize yourself with the methodology and the results of the study.

3 Methodology and results of the study

3.1 Methodology of the study

The study was conducted in eighty seven high-tech enterprises in Bulgaria between February 2018 and February 2019. Of these, twenty-two are large enterprises, and sixty-five are small and medium-sized enterprises (SME). Twenty-six enterprises have foreign management (Germany, France, Austria, Sweden, Denmark, Italy, Belgium and others.) Participants in the survey are 4350 workers, elected on the principle of education and training. 50% of them (2175) have higher engineering education, as majors that prevail are „Mechanical engineering and technology ", „Mechanical engineering and instrumentation ", „Transport equipment and technologies ", „Electrical engineering ", и „Electronics". We categorically designate these participants in the Group A study. The other half have secondary vocational education in the fields „Technical sciences, manufacturing and construction " and „Information and communication technologies ". These are Group B participants. The method is borrowed from Tsvetan Davidkov, being supplemented and enriched with determination of the degree and commitment in the participants' labor process [10].

The study covers six stages: develop surveys; conduct the poll; ranging the motivational factors; determining the degree of satisfaction; determining the degree of engagement; an indication of the motivation potential of each factor. The survey maps the questions related to the assessment of the significance of 15 motivational factors, with an evaluation scale of 1 to 7 as follows: 1 – most significant; 2 – significant to a great extent; 3 - moderately significant; 4 – significant to a lesser extent; 5- rather insignificant; 6 – insignificant; 7 – absolutely insignificant. The choice of motivational factors is the result of previous research experience of the authors and from preferences demanded by high-tech enterprises. Participants evaluate the extent to which each motivational factor is important for their better performance, as against each motivation factor they mark the corresponding rank in the range from 1 to 7. Similarly, the satisfaction and commitment to work is examined by each factor. As a result, the average value of the indicator is calculated and factoring is performed. The study is conducted for Group A and Group B participants. The main idea is to compare the importance of motivational factors with their actual state, measured by the satisfaction and commitment of the participants in the survey. The logic on which the thesis is based is that the strong significance of the motivation factor and the low level of satisfaction with the factor, as well as the low level of commitment are a sign of a factor with high motivation potential. Accordingly, the antithesis can be formulated as follows: the low significance of the factor and a high level of satisfaction and commitment to the motivation factor imply its insufficient motivational potential.

3.2 Results of the study

As a result of the survey we received the following results for Group A participants (Table 1)

Table 1.

Motivation factor	Average value of significance (avg.)	Rank 1	Satisfaction (avg.)	Rank 2 satisfaction	engagement (avg.)	Rank 3 engagement	Difference (Rank1-Rank 2-Rank 3
Workplace Security	3,26	11	1,86	1	1,88	1	10/10
Conditions and safety at work	3,42	12	2,28	2	2,65	2	10/10
Organization of working hours	4,86	15	2,36	3	3,86	6	12/9
Organization of Work	2,92	10	2,48	4	3,64	5	6/5
Remuneration	1,52	5	4,62	11	4,80	11	-6/-6
Benefits	1,58	6	4,28	10	4,42	8	-4/-2
Teamwork	3,58	13	2,86	5	2,80	3	8/10
Communication with colleagues and supervisors	2,85	9	3,16	6	3,88	7	3/2
Self-sufficiency of work	4,26	14	3,26	7	4,48	9	7/5
Personal responsibility	2,68	8	3,84	9	3,62	4	-1/4
Recognition of personal Abilities	2,42	7	3,66	8	5,22	13	-1/-6
Interesting and challenging Tasks	1,32	2	5,42	13	5,86	15	-11/-13
Opportunity for initiative and expression	1,38	3	5,68	14	4,82	12	-11/-9
Ability to upgrade the qualification	1,46	4	4,86	12	4,66	10	-8/-6
Opportunity for development and self-realization	1,26	1	5,90	15	5,64	14	-14/-13

Ranking in a column titled Rank 1 is based on the degree of importance of motivational factors for the participants in the survey. Estimates close to 1.00 indicate the high degree of relevance of the respondent factor, those that are close to 7 indicate the low significance of the relevant motivation factor. Ranking in the Rank 2 column results from the stated level of satisfaction with the specific motivation factors for the respondents, and in Rank 3 is the ranking based on the engagement of participants in the work process. Estimates close to 1.00 show a high level of satisfaction and commitment, and close to 7.00 - a low level of satisfaction and commitment. The last column of the table is the difference between the ranges of significance of the factors and the ranks of satisfaction and commitment (rank significance – rank satisfaction / rank engagement). According to the logic of the research procedure, motivational factors with a negative difference with a higher absolute value, have significant motivational potential.

In the same way, the results for Group B were also obtained.

3.3 Conclusions from the study

By analyzing the results obtained in the table, we can draw the following conclusions for Group A participants - highly qualified specialists with higher engineering education:

1. The first five most significant motivational factors are „Opportunity for development and self-realization ", „Interesting and challenging tasks ", „Opportunity for initiative and expression ", „Ability to upgrade the qualification " and „Labor remuneration ".
2. The lowest level of satisfaction, the participants get from „Opportunity for development and self-realization ", „Opportunity for initiative and expression ", „Interesting and challenging tasks ", „Ability to upgrade the qualification " and „Labor remuneration ".
3. The lowest level of engagement creates „Interesting and challenging tasks ", „Opportunity for development and self-realization "," „Opportunity for initiative and expression " and „Labor remuneration ".

The conclusions that we can make after analyzing the results for Group B participants – workers with secondary education are as follows:

1. The first five most significant motivational factors are „Ability to upgrade the qualification ", „Labor remuneration ", „Organization of working hours ", „Communication with colleagues and supervisors ".
2. The lowest level of satisfaction, participants get from „Recognition of personal abilities ", „Labor remuneration ", „Benefits ", „Ability to upgrade the qualification ", „Organization of working hours ".
3. The lowest level of engagement, the participants get from „Labor remuneration ", „Benefits ", „Recognition of personal abilities ", „Ability to upgrade the qualification ", „Communication with colleagues and supervisors ".

3.4 Recommendations

The analysis of the results obtained in the last column of the table gives us the opportunity to highlight the factors with the greatest motivational potential for the two groups of participants. For Group A are the following: „Opportunity for development and self-realization ", „Opportunity for initiative and expression ", „Interesting and challenging tasks ", „Ability to upgrade the qualification " and „Labor remuneration ". For Group B – are the following: „Labor remuneration ", „Benefits ", „Recognition of personal abilities ", „Ability to upgrade the qualification ", „Communication with colleagues and supervisors ". It is noticeable that the differences between the factors that satisfy and engage the surveyed are small.

What is the practical result of the fact that on the basis of the survey we have identified the factors with great motivational potential? Based on these factors, management of high-tech enterprises can design motivational policies, programs, tactics, techniques, approaches, actions and more. That is what our recommendation is – all motivational actions should be tailored to an in-depth study of the specifics of the motivation profile of workers, through the prism of their satisfaction and commitment to the work process.

4 Conclusion

In the report, we attempted to look at the specifics of high-tech enterprises and, on this basis, to draw the need for a specific approach to building the motivation system in these enterprises. For this purpose, we used the motivational profile as a tool. Based on a study of the significance of motivational factors, the degree of satisfaction and engagement of workers in high-tech enterprises, we have been able to draw conclusions about the motivation potential of the factors. Thus, we can

IOP Conf. Series: Materials Science and Engineering **618** (2019) 012075 doi:10.1088/1757-899X/618/1/012075

recommend HR managers, to periodically examine the motivation potential of the factors, which they use in order to have engaged employees and which is more important to be able to keep them in this dynamically changing business environment.

Acknowledgements
The author/s would like to thank the Research and Development Sector at the Technical University of Sofia for the financial support.

References
[1] Anglenova M 2018 Managing the (de)motivational process and its impact on human capital, *Yearbook of Faculty of Economic and Social Sciences* (FESS), University of Plovdiv"
[2] Teneva At 2010 European policies for the development of human capital in Bulgaria, Magazine *KSI Knowledge society*, Vol .**3**, Number 3, ISSN 1313-4787
[3] Koleva St, Semyonov P 2013 Establishment of individual and group motivation impact on employees in the business organizations *Economic and social alternatives*, edition **4**
[4] Dimov D Competencies of The Personnel in High-Tech Enterprises https://e-university.tu-sofia.bg
[5] Mihova T, Nikolova-Alexieva V, Ferdov Al 2018 Human Resources Motivation in the Bulgarian High-Tech Industrial Enterprises, *Scientific Researches of the Union of Scientists in Bulgaria – Plovdiv*, Series B. Natural Sciences and the Humanities, Vol. **XVIII**, ISSN-1311-9192 (Print), ISSN-2534-9376 (On- line)
[6] National Strategy for the Development of High Technology in the Republic of Bulgaria (1999)
[7] Koleva V 2006 Managing team motivation and effective behavior, *Management and Sustainable Development* 3-4/2006(15)
[8] Iliev Y 1993 Motivation of the staff, Sofia: „*Lyuren* ", 40
[9] Djonev S, Dimitrov P 2015 Organizational counseling (Vol. **3**). Sofia: *NGB Consulting*, 144
[10] Davidkov T 2018 Magazine „ Rhetoric and communications ", edition **36**

Features in management and monitoring of production process in manufacturing industry using integrated information technologies solutions

Pavel Vitliemov[1]

[1]University of Ruse, Department of Management and Business Development, Bulgaria

pvitliemov@uni-ruse.bg

Abstract. This paper gives a brief overview of the conditions, approaches and prospects of production control and monitoring of manufacturing industry through implementing information systems for optimizing the manufacturing process and to provide an answer to the question of the specificities for successful management of machine manufacturing enterprise through integrated software solutions. New integrated IT solutions based on the newest business management trends and information technologies that includes tools for management of the production and materials, as well as inventory planning, allocation of production resources, process analysis and production organization leads to improvement of the efficiency of business management and production management processes in manufacturing.

1. Introduction

Every business, notwithstanding its scale and phase of development, needs information. The process of management decision making is not always supported by the required information. This requires that more attention is paid on the activities of collection; processing, storing and using information adequate to the management needs (11).

For the purpose of helping the managers use their best abilities for high quality decision making, the required information should be provided promptly, at the adequate place and in the appropriate form. The fast development of business management software in the recent years contributed to the increase of the efficiency of many enterprises. The use of specialized software intended for the process control in production companies with different organizational and territorial structures contributes to the flexible management of the technical preparation of manufacture, the manufacture itself and the processes of material supplies and product sales (17).

These complex integrated systems have many advantages. First, they provide to the management complete overview on the corporate resources and processes running in the company, at any time, via a single interface. Second, they contribute to the optimization of resources and allow minimizing stock and more efficient personnel management (20).

Integrated information systems attracted considerable attention in the business world at the burn of the millennium. Despite the sizable investments involved, these software packages were a breakthrough, where they were installed by a vast majority manufacturing firms (18). After the initial excitement, however, an increasing number of managers have started to complain about the shortcomings of these systems. The main critique is that integrated information systems impede making changes to new product development and introduction process of an organization in dynamic business environment. Over the past couple of decades, much focus has been placed on creating models for how products are "defined and documented" in information systems.

The purpose of this paper is to make a short analysis of the terms and conditions, approaches and prospects related to the successful management of machine manufacturing by implementation of complex integrated information systems for process control.

2. Literature review

Integrated information systems are modular software packages that integrate a firm's business functions around a common database and standardized processes that are configured to fit the needs of the user organizations (22,25). Substantial research efforts have been directed to this special category of enterprise software. The notorious failures of some of the early implementations (see examples in, 23) gave rise to much research on the typical pitfalls and success factors of implementing information systems (for reviews and recent examples, see, e.g., 2,25,26,29,31). The variability in the outcomes of the implementations also motivated broad research on integrated information systems overall performance effects (12,14) and end users' assimilation of the implemented systems (1,3,24,27,30).

Over time, the studies on integrated information systems have also started to cover more specialized tools, such as customer relationship management, manufacturing planning, advanced production scheduling, supply chain management, and sourcing software (1,21). This trend has been motivated by practical considerations, as firms have increasingly implemented standalone software (4,6,9). Although most firms do not have plans to completely abandon their integrated information systems, many have implemented standalone tools to replace some functionality of their information systems (7,10).

Because integrated information systems' inflexibility is the most often cited reason why managers consider replacing them with other solutions (10), this study analyses the terms and conditions, approaches and prospects related to the successful management of machine manufacturing by implementation of such systems.

The connection between Product Development and Manufacturing logically falls within the New Product Development and Introduction (NPDI) process of an organization. A common language connecting the functions within the NPDI processes is needed and, once established, it can be used within and across the integrated information systems (13, 15). These systems will eliminate the need to manually transform product information, starting with the development processes, into manufacturing processes that are consistent across regions or manufacturing facilities. They will enable increased efficiency, productivity and quality of products as well as agility across the enterprise. The following benefits are a few examples that illustrate the magnitude of the opportunity:

- Reduction in search time for data – Estimates of time spent looking for data in the NPDI process range from 20 percent to 50 percent because data is not centralized, is not in a standard format and often is not electronically available to all parties.
- Reduced time to create master recipes
- Shorten manufacturing planning and site sourcing processes – Planning processes typically require site level knowledge of what materials can be used and what equipment constraints can be met.

3. Specificities of production preparation management

Management of machine manufacturing is the most complex case of production management. As a rule, machine manufacturing industry produces items with complex hierarchic structure. These items are usually multi-variant on all levels of production. In difference to the chemical industry where the product has its process route from the raw material to the end product, in machine manufacturing every assembly, part and semi-finished product have individual process routes. In other words, machine manufacturing industry produces complex multi-variant products based on complex multi-variant process routes.

IOP Conf. Series: Materials Science and Engineering **618** (2019) 012076 doi:10.1088/1757-899X/618/1/012076

The issue becomes more complicated due to the fact that each complex structure changes in the course of time, the design specifications and the process routes change too. Some changes are related only to the documentation and the future manufacture, other refer only to the manufacture at the given moment. Some changes affect the whole item, other - only certain components. Some changes enter into force immediately, other - from a specified moment on.

Therefore manufacture should be managed under conditions of constantly changing normative base.

The practical implementation requires real-time updating of the normative base. At any time it must be guaranteed that the participants in the production involved in the technological process have up-to-date information and that the up-to-date information about such technological process is kept only in the process control system (8).

The experience of the leading companies developing business management software in the recent years shows that the production management by means of information system is only possible when the normative base for the process engineers is created and maintained in the system itself (fig.1). Only if the process engineers create and update the specifications, technologies, the consumption rates of materials with the resources of the system, it may be expected that the production will be managed by means of up-to-date normative base.

The efficient management of machine manufacturing requires huge and various information about items, semi-finished products, auxiliary materials, tools, equipment, operations, workplaces, consumption rates, notices of amendments. Usually this whole information is scattered in numerous flow charts, operational charts and other process documents (19).

Data must be attached to the relevant product structure. Information about the auxiliary materials, instruments and equipment, kept in the process documents, shall be automatically attached to the product structure (fig. 2). Any real-time amendment of that information must update the relevant product structure. At any time the production worker, by opening the content of the product structure must see in the system guaranteed up-to-date information. He/ she should be sure that the up-to-date information about such product structure is kept in the process control system of the industrial enterprise and only in it.

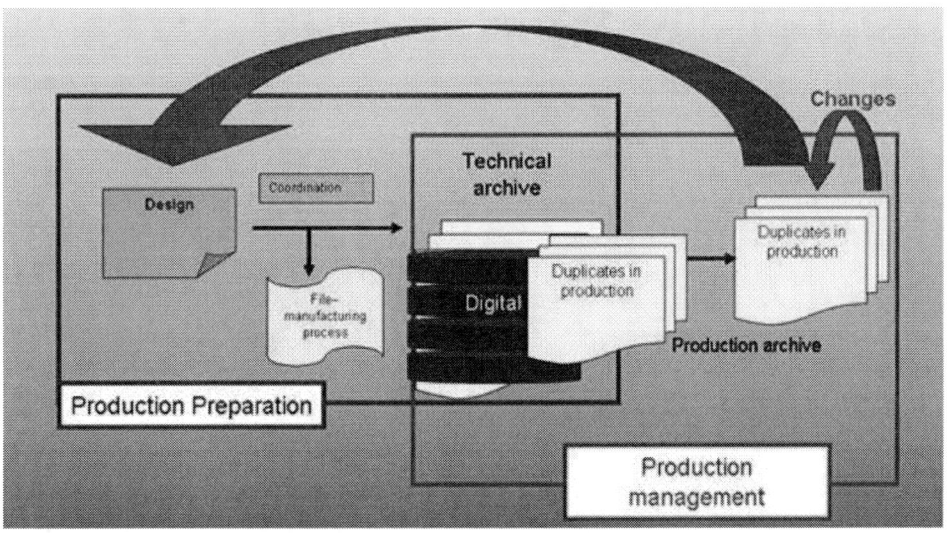

Figure 1. Model of information system implementing the interaction between the product preparation and the production management (28)

The key for the successful implementation of production management system is to ensure the work of the process engineers within the system as an integral part thereof. The system must contain the whole panoply of resources required for the work of the process engineers:

IOP Conf. Series: Materials Science and Engineering **618** (2019) 012076 doi:10.1088/1757-899X/618/1/012076

- To maintain the following processes:
  - Design development - development of basic non-graphic design information: product structure, design specifications, group specifications;
  - Forwarding of the tree structure to the production;
  - Change of the product structure in result of production practices and at the clients' request;
  - Function available in the product structure enabling the description of the sequence of assembly, processing, disassembly.
- To generate documents:
  - Flow chart;
  - Operational card;
  - Technical control card;
  - Equipment card;
  - Order card;
  - Instruments card;
  - Registration of amendments card.
- To maintain functions increasing the efficiency of work of the process engineers:
  - Standard flow technological processes;
  - Group flow technological processes;
  - Library of standard operations;
  - Real-time order for equipment design;
  - Calculator of consumption rates for materials and labour expenses;
  - Control on specific operations.
- To be integrated with CAD/CAM systems – integration means capability for importing the results from the work of the CAD/CAM systems in the control system.

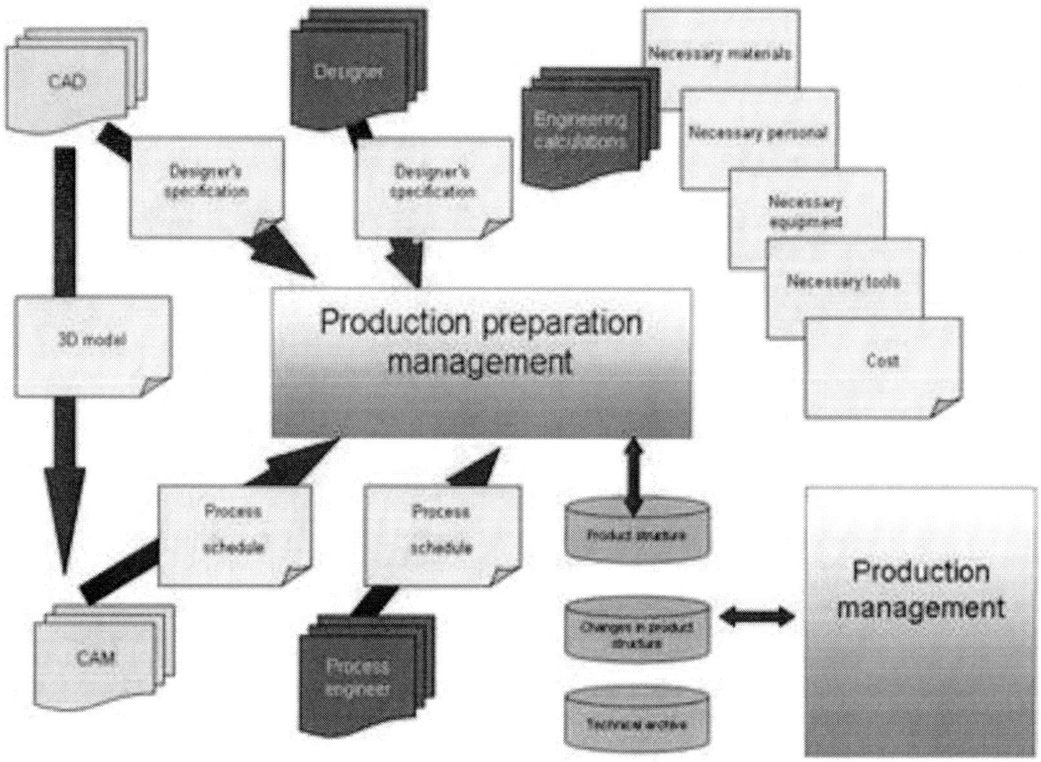

Figure 2. Model of information system unifying the information split into many schedule and operation maps and other processing documents (28)

The production control via automated systems requires that the process tree of the product represents a design tree with added elements of the technological processes - work in progress, auxiliary materials, instruments, equipment. Such additional information changes the design tree of the product to such extent that the changes become ambiguous. Generally machine manufacturing may not create unambiguous link between the design and the process tree of the product. Therefore within the production control system independent process tree for the product should be maintained independently.

4. Manufacturing supervision in machine manufacturing enterprise

The management with plan orders underlies the manufacturing supervision. Depending on the operative information (terms of the orders, provision of materials for the orders, the situation in the workshops), the production supervisor makes decisions for the issuing of plan orders (5). For each order he/ she is able to see the current deficit of materials and semi-finished products, the estimated provision as at the time of work commencement, the possible redirections of deficit materials, the loading and the resource equipment as a function of time.

The supervisor shall issue the order, which actual execution shall start within the next 3-5 days. Upon confirmation, plan orders will become workshop orders and shall become compulsory for execution in the workshop.

The plan orders and workshop orders for a particular workshop represent the plan of the enterprise (fig.3). The workshop actually may execute only workshop orders. Thus the management strategy specifying what should be produced and when should be produced with accuracy of 3-5 days is determined on corporate level by taking into account the interests and the policies of the enterprise in general.

The practice shows (16) that the implementation of information systems for production management in machine manufacturing enterprises results in reduction of the work in progress by about 25% and in increase of the productivity by up to 10%.

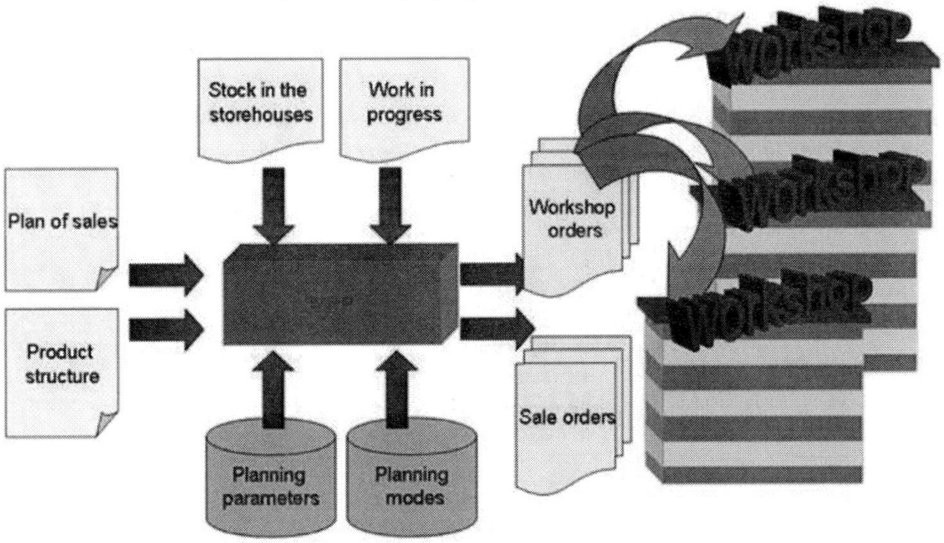

Figure 3. Workshop plan and it's updating (TECHNOCLASS, 2018)

Until the implementation of such system, the production was always "overloaded", there was no objective idea in what phase of implementation was the order, the work-in-progress was uncontrollable because it was driven by the interests of the workshop and not by the interests of the enterprise.

Conclusion

Increasing complexity continues to be one of the biggest challenges facing manufacturing today that operates in an environment of change and uncertainty. In practice, the reconfiguration work is generally considered onerous, as one must pay attention to the numerous interactions between the integrated features that are used across the organisation, a fact that has led many firms to avoid changes as much as possible. Moreover, all changes must fit the data structures and logic of the software, limiting the options for how processes can be redesigned.

This study has provided an answer to the question of the specificities for successful management of machine manufacturing enterprise through integrated software solutions, which are the following:

- the management of machine manufacturing starts by technological preparation of production, and reasonably it is within the management system of the machine manufacturing and it is impossible to separate it as a process and to expect success of management;
- the machine manufacturing schedule is made every day on the basis of the MRP standard and depending on the type of production it is rescheduled on a daily basis (or at least once a week) and for that reason the workshop orders are issued for those parts, which production should start within the next three to five days.
- as the operative planning resolves tasks connected with workshop resources management, the APS standard may be used for its implementation;
- supervision by integrated software systems has the following advantages:
  - real-time maintenance of exhaustive information for the production resources;
  - functionality for selection of appropriate resources for implementation of the process operations on the basis of data about the process capabilities, ensured technical documentation, installed special equipment, available instruments;
  - functionality for assessment of the production tasks with a view to the availability of staff with the required qualification and authorities;
  - functionality for assessment of the production tasks with a view to the availability of the required semi-finished products, basic and auxiliary materials.

References

[1] Bendoly, E., Cotteleer, M.J., (2008); Understanding behavioral sources of process variation following enterprise system deployment, Journal of Operation Management, 26 (1): 23–44.
[2] Berente, N., Yoo, Y., (2012); Institutional contradictions and loose coupling: post implementation of NASA's enterprise information system, Information System Research 23 (2): 376–396.
[3] Boudreau, M.-C., Robey, D., (2005); Enacting integrated information technology: a human agency perspective, Organization Science 16 (1): 3–18.
[4] Buttle F. (2009); Customer Relationship Management: Concepts and technologies, Elsevier Ltd., 2nd Edition.
[5] Camuffo A., Gerli F. (2005); The competent production supervisor: A model for effective performance; Industrial Performance Center Working Paper Series; Massachusetts Institute of Technology.
[6] Deloitte, (2010); Best-of-Breed Enterprise Applications. Deloitte Development LLC,Hermitage, TN.
[7] Drexl A.; Kimss A. (1998); Beyond Manufacturing Resource Planning (MRP II): Advanced Models and Methods for Production Planning, Springer-Verlag.
[8] Etera-soft (2014); Etera-soft product website, retrieved from http://www.etera-soft.com in April, 2014.
[9] Fauscette, M., (2013); Maintaining ERP Systems: The Cost of Change. IDC, Framingham,MA.
[10] Ganly, D., Montgomery, N., (2012); Hype Cycle for ERP 2012. Gartner, Stamford, CT.
[11] Günther, H.-O., van Beek, P. (Eds.) (2003); Advanced Planning and Scheduling Solutions in Process Industry, Berlin: Springer.

[12] Harris, J.G., Davenport, T.H., (2006); New Growth from Enterprise Systems: Achieving High Performance through Distinctive Capabilities. Accenture Institute for High Performance Business, Wellesley, MA.
[13] H. Lan, et al., A web-based manufacturing service system for rapid product development, Computers in Industry 54 (1) (2003, May) 51–67.
[14] Hendricks, K.B., Singhal, V.R., Stratman, J.K., (2007); The impact of enterprise systems on corporate performance: a study of ERP, SCM, and CRM system implementations, Journal of Operations Management 25 (1): 65–82.
[15] MESA International white paper #49, (2014), Enterprise Recipe Management.
[16] Mihaylov T., Vitliemov, P. (2014); Management of processes in manufacturing industry with MES system, Ruse: Primaks.
[17] Nedyalkov, A. (2010); Investigation and improvement of information system for operation management at small and medium enterprises, Doctorate Thesis, University of Ruse, Bulgaria.
[18] Olhager, J., Selldin, E., (2003); Enterprise resource planning survey of Swedish manufacturing firms, European Journal Operation Research 146 (2): 365–373.
[19] Open Data (2014); Information about advanced OPERA IT solution for the manufacturing industry, Retrieved from http://www.opendatasrl.it in September, 2014.
[20] Peltonen H. (2000); Concepts and an Implementation for Product Data Management, Doctorate Thesis, Helsinki University of Technology.
[21] Rai, A., Hornyak, R., (2013); The impact of sourcing enterprise system use and workprocess interdependence on sourcing professionals' job outcomes, Journal of Operation Management 31 (6): 474–488.
[22] Ranganathan, C., Brown, C.V., (2006); ERP investments and the market value of firms:toward an understanding of influential ERP project variables; Information System Research 17 (2): 145–161.
[23] Robey, D., Ross, J.W., Boudreau, M.-C., (2002); Learning to implement enterprise systems: an exploratory study of the dialectics of change, Journal of Management Information System19 (1): 17–46.
[24] Saeed, K.A., Abdinnour, S., Lengnick-Hall, M.L., Lengnick-Hall, C.A., (2010); Examining the impact of pre-implementation expectations on post-implementation use of enterprise systems: a longitudinal study, Decision Sciences 41 (4): 659–688.
[25] Sasidharan, S., Santhanam, R., Brass, D.J., Sambamurthy, V., (2012); The effects of social network structure on enterprise systems success: a longitudinal multilevel analysis, Information System Research 23 (3): 658–678.
[26] Seddon, P.B., Calvert, C., Yang, S., (2010); A multi-project model of key factors affecting organizational benefits from enterprise systems, MIS Quart, 34 (2): 305–328.
[27] Stadtler H., Kilger C. (2010); Supply Chain Management and Advanced Planning: Concepts, Models, Software and Case Studies; Springer.
[28] TECHNOCLASS (2016); TECHNOCLASS ERP&ME System, Retrieved from http://techno-class.com in July 2018.
[29] Terry, L, Neff, S. (1992); Material Requirements Planning: Integrating Material Requirement Planning and Modern Business, NY: McGraw-Hill, Inc.
[30] Xue, Y., Liang, H., Wu, L., (2011); Punishment, justice, and compliance in mandatory IT settings, Information System Research 22 (2): 400–414.
[31] Yeh, C.-H., Xu, Y., (2013); Managing critical success strategies for an enterprise resource planning project, European Journal of Operation Research 230 (3): 604–614.

IOP Conf. Series: Materials Science and Engineering **618** (2019) 012077 doi:10.1088/1757-899X/618/1/012077

The application of game and psychological techniques when surveying children's opinion

E Zlatanova-Pazheva

Technical University of Sofia, Branch Plovdiv, Industrial management Department, 4000 Plovdiv, Bulgaria

elyzlatanova@abv.bg

Abstract. The article explores the game as a method that can be used to conduct surveys with children using the possibilities of Internet in an entertaining manner. An empirical survey of children's opinion based on the method of games has been presented. The description of the method by which a study is conducted is in terms of the psychological techniques that have been used, mainly the psychology of colours.

1. Introduction

The constantly changing consumer wishes, tastes and emotions are a serious challenge for companies seeking sustainable development [1]. The key success factor is the timely, and even the anticipated, discovery of consumer's emotional changeability. The possibility to influence this volatility will provide one of the premises for market success [2, 3].

The interest in children as an important market segment is related to the fact that children are:

• an important market of consumers who spend their own money on their own wishes and needs,

• an influential market driving their parents' money for their own benefit,

• a future market for all goods and services, which if cultivated will now provide a steady stream of new consumers.

Over the last decades, the children market has become significant as young people have greater disposable income than previous generations. Due to the fact that the purchasing power of young people continues to grow, marketers, manufacturers and advertisers are significantly interested in developing effective methods to reach the young market [4]. In order to accomplish this task it is necessary to know the behavior of the child consumer.

Surveying the child's opinion is a great challenge. This is due to several reasons:

First, the specificity of this age group implies and requires special attention to the choice of a method of study,

Second, the necessity of taking into account the mental, cognitive and age-related development of children in forming the questions that will be addressed to them;

Third, the challenge of keeping their attention during the study.

There are different research methods for collecting primary data. Some of the most commonly used in practice are Interview, Observation, Experimentation, Questioning [5]. The most commonly used in practice is the questioning, because it allows the opinion of a large number of respondents to be explored [6, 7]. The types of questioning are: personal questioning, mail questioning, questioning at the points of sale and Internet questioning [7]. A number of studies provide information on a successful survey through questioning among children [8, 9, 10, 11].

The report discusses an online survey method through questioning. The goal of the article is to present a method for conducting surveys with children, which uses the possibilities of Internet and applies game and psychological techniques.

2. The game as a method for conducting online surveys with children

Rumen Stamatov has defined the game as one of the main forms by which the child realizes his presence in the being [12]. Gili Freedman and Mary Flanagangames define game as a situation with rules that require individuals to make judgments or decisions [13].

Adam Lobel claims that the play is integral to human development. He states that play helps fulfill basic needs, providing a sense of accomplishment with success and mastery, a sense of autonomy with its lack of constraints, and a sense of relatedness with its ability to bring people together in pursuit of a common goal [14].

Greg Costikyan points out that the game interacts with the players, and if not it's not a game [15]. Each game has a beginning and an end, there is an absolute line in it. The game must create joy for the player and to be interesting. In other case the player will stop playing it. When we are talking for playing a game we could think for something that is not serious, because it's an entertainment [15]. But Rumen Stamatov disprove this by saying that "knowing that this is just a game does not exclude the seriousness and dedication to the game, the playing child is completely conquered by the game, and the game is not only unserious, the game rises to the serious, and the serious - to the game" [12].

Ana Grudeva examines the "serious games", by saying that they can be used and applied in different aspect of e-learning for children. She defines "serious games" as interactions between the educational goal and the technologies provided by video games [16].

There are different classification of games. Gili Freedman and Mary Flanagangames point out four types of games: physical games (sports), behavioral economics games, board games, and digital games [13]. Ana Grudeva uses this classification: class games, frame games, "serious games", and educational simulations [16].

Gili Freedman and Mary Flanagangames states that games can be used to collect data (i.e., games can be used as instruments), but they can also be used as a way of altering a participant's cognitive state, emotional experience, or social situation [13]. Games allow social and personality psychologists to understand how individuals think about and interact with the world around them. Not only do games have a rich history of being used in social and personality psychology, they also hold the promise of pushing the field forward. Racial biases, affective forecasting, competitiveness, empathy, and power are just a few of the topics that have been successfully explored using game methods. Whether games are used as measures, manipulations, or contexts in laboratory experiments and field studies, their incorporation into psychological science has allowed for research that may not otherwise have been conducted. More broadly, games are an important and growing paradigm in social and personality psychology, and knowledge of how to effectively craft them will allow researchers to stay on the cutting edge [13].

Recently, multiple reviews have highlighted the impact of games on psychological states and behaviors however, reviews have not yet considered the ways in which games can be used as research methods in their own right. Unlike some methods in the psychology methods toolbox, games can be used as measures, manipulations, and experimental contexts for many different populations: they are a universal form of engagement and interaction with far-reaching applications [13].

3. An empirical survey of children's opinion using a method based on the games

Children easily lose interest and get bored. The way of examining a child's opinion should be interesting, fun and different. We consider this to be the most appropriate way for children not only to lose interest in completing the questionnaire but also to feel time-consuming. The opportunities offered by the online environment create the necessary prerequisites for achieving this intention.

A method that has been used for surveying the consumer behavior of children based on the games will be discussed.

In this survey, two age groups are defined - children (7-9 years) and teenagers (12-14 years). A web-based application based on the game as a method for studying children was created. It is based on various visual stimulants and the psychology of colors.

Rumen Stamatov notes that the child plays because the game is a pleasure and that is his freedom. The game brings joy, pleasure, self-assurance. This is the basis of the method we have developed. Rumen Stamatov defines the play as "a voluntary action, limited in time and space, requiring compliance with voluntarily accepted binding rules, which are inherently self-centered and accompanied by a sense of tension and joy from the awareness of" something different "by everyday life" [12].

The game is played with rules. Even when rules are denied, this is the reason for introducing new rules. The rules set the laws governing the temporary world created by the game. They are indisputable. Violation of the rules destroys the world of the game, it ceases to exist, the "illusion" of the game, the inner freedom and self-expression, the unconscious "it is not quite real". In this study, we came from this concept.

Taking into account the different periods of development of the two groups, bringing with them a different stage in cognitive and mental development, different interests and needs, we have sought to adapt the used visual stimuli. One aspect of visual stimulation can be found in the actual view of the web-based application, and the other in the way of asking each question.

The client view of the site is organized as a Border Layout. The border layout is organized in five regions – north item, south item, west item, east item and center item (figure 1). The north item in the developed site includes a brown shelf and the menu. The south item includes the same brown shelf, but with different foods and the contact information. The west and the east items include the hosts – a boy and a girl, respectively. All of the mentioned items are permanent. The center item is with dynamic content and shows the interviewers and the questions (figure 2).

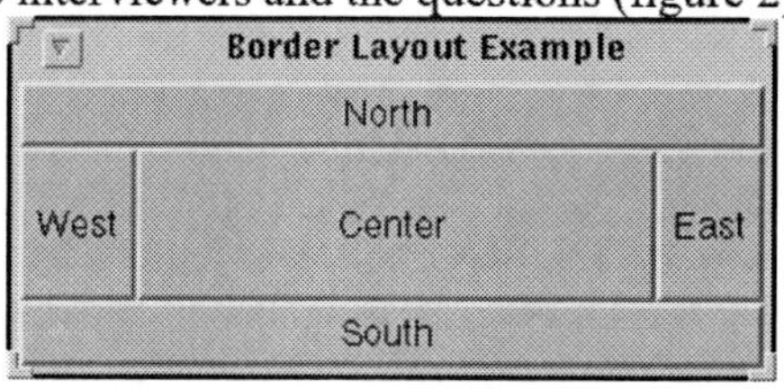

Figure 1. Border Layout Example

- Border Layout

When creating the view, the psychology of colors is used, which is also a basic marketing technique. The large number of studies in the field of color psychology show that each color has its own psychological significance and character. In this direction, it is precisely found that the colors contain a certain emotional associativity [17]. Because of the power of influence on people's will and actions, colors are a powerful psychological weapon and a means of attracting attention, interest, and ultimately decision-making [18]. Each color has its own emotional language, which brings feeling and emotions. This is because every nuance in the color range is associated with positive or negative emotions, and this gives it the power to evoke and remind different emotions [19].

The main colors used in the view are:

Green is the color of nature and life. The green is related to healing, prosperity, reincarnation, harmony, tranquility and generosity. Popular in most cultures and is the color of nature and fertility. It symbolizes ecology, growth, rebirth, and the natural environment. It is the quietest color for the human eye, showing stability and endurance, hope, nobility, generosity, freshness. Green has a great healing power. Green means safety, so it is often used in the advertising of medicines and medical products. This color embodies security. Except as a basic color in the view the green was also used for two of the children interviewers (figure 4a).

Brown is a serious, earthly color that suggests stability, structure and support. The color of security, protection and material wealth it is associated with the protection and support of the family, with a clear sense of duty and responsibility. The brown is friendly and easy to access. It is associated with the overall, natural and organic production - all that is related to agriculture, open nature and agriculture. The brown soothes us. The brown does not look for attention. It prefers to remain in the background, allowing other colors to shine. It has strength and maturity, making it stable. Brown is the most dominant color on the planet, along with the green. It soothes and stabilizes while the green balances and refreshes.

- Colors in which the "virtual hosts" are dressed

Two "virtual hosts" has been created – a girl, dressed in pink and a boy, dressed in blue (figure 2, figure 3).

The blue is the color of the sky and the sea. It is connected with depth and stability, godliness and sincerity. It has a soothing effect, it is associated with peace and tranquility. Studies show that is very well received among male audiences [20]. It is also a receding color, expressing infinity and coolness. It creates a sense of calm and security and is a great choice in the design of travel agencies.

The pink personifies compassion, caring, and love. It is associated with unconditional love and understanding. Pink is feminine and romantic, intimate, lively, rational. This color is intuitive and shows its tenderness and kindness with its empathy and sensitivity. In color psychology, pink brings hope. Positive color, filling with warm and soothing feelings. The pink personifies the sweetness and innocence of the child in each of us. This is the color of uncomplicated emotions, experience and naivety.

Figure 2. View of a question from the child questionnaire

Figure 3. View of a question from the teen questionnaire

Often certain colors are associated with a given gender. In this regard, a number of psychological studies have been carried out, which show that women prefer more red color, and men prefer the more recent blue color [21]. These are the colors that we can define mainly for feminine and respectively male colors. A typical example in this regard is the pink color for women. This color is most commonly used in women's ads [21]. We associate with women all bright and cheerful colors - talking about cheerfulness and frivolity. Men are often associated with most cold colors that speak of crude power and mysticism.

- Center Item of the Border Layout

In the rest, we used another concept. In the children section, it is related to asking a question from an interviewer who is a computer-assembled fruit, vegetable or other food that aims to provoke interest and joy in the children and create a sense of tranquility. In this way we expect them to trust the well-meaning and smiling "interviewer" (figure 2).

The same is true for the teenagers, with the difference that their "interviewers" are for slightly older children, so we used computer-minded and well-meaning emoticons (figure 3). Examples of both concepts are shown in figure 2 and figure 3.

❖ Colors of the "virtual interviewers"

Besides those that have already been discussed are also used:

Yellow is the color of the sun. The yellow color is the lightest basic color, it is expansive - it creates a sense of movement and often dominates other colors. This cheerful color is always followed by the eye, and in its brightest shades it can be more aggressive than the red color. Raises the spirit, stimulates communication, symbol of intelligence and attention to detail. Yellow is the color of sunlight. Its use produces pleasant feelings. It is connected with joy, happiness, intellect and energy, creates a warming effect, generates cheerfulness and stimulates mental activity (figure 4 b, figure 5).

Red is the color of fire, love and life. It is associated with energy, strength, power, determination, passion. As the most attractive color, it has a very high visibility and is therefore often used to indicate danger - high voltage, traffic signs, traffic lights and fire-fighting facilities. It is used as a symbol of courage and can therefore be seen in many national flags. This color is often associated with energy, so it can be used to promote energy drinks, games, cars, sports related subjects, and high physical activity. In painting and design, the red is perceived with the following characteristics: strength, cheerfulness and dynamics [22]. This is the most intrusive color of all and instantly eye-catching (figure 4 c).

Orange is the color of joy. For the human eye, this color is very hot and creates a feeling of warmth. It is connected with joy, sun and tropics. Orange represents enthusiasm, charm, happiness, creativity, determination, attraction, success, encouragement and stimulation. However, orange is not as aggressive as red. It is very well accepted by young people. This color embodies the sense of humor, optimism and friendship. It stimulates trust, tolerance, and sense of community. As a citrus color, the orange color is associated with healthy food and stimulates appetite. It is very effective in promoting sales of food and toys (figure 4 d).

White is associated with light, goodness, innocence and modesty. It is considered the color of perfection, it can be birth, purity and simplicity. The white color is the opposite of the black color. It usually has positive perception and means safety. In the world of fashion, in the furniture and interior design - white and black are extremely fashionable colors. In the advertising, the white color is associated with coolness and cleanliness, so it is common in bathroom and bedroom stores, dry cleaning, in cleaning service ads. White is also used as a symbol of simplicity in high-tech products. The color of snow promotes cleansing of thoughts and mental clarity. The white recreates the idea of hope and expectations for healing and restoration, which is why it is the color of hospitals and medical aprons for a long time. White is often used as a symbol of safety in the promotion of medical products and equipment (figure 4 e).

Purple is the color of spiritual power. Purple colors are associated with fun, festivity and solemnity. Unusual and and magical, the purple color is often associated with royal peoples. It symbolizes power, nobility, luxury and ambition, suggesting a sense of wealth, wastefulness, self-confidence and independence. This magical color is suitable for use in graphic or interior design when it is necessary to bring a sense of fiction, mystery and magic [20] (figure 4 f).

Figure 4. The children interviewers

Multi-color uses color combinations consisting of more than two colors. They rely on multi-color color psychology. The message they want to convey is related to emotions such as happiness, joy, lightness in life (figure 4 g).

Figure 5. Examples for the teen interviewers

In conclusion of the presented method we will bring facts showing the success of the conducted research. The return of the received responses is presented in figure 6.

Less than 30% did not participate in the study in both age groups.

Only 13.6% of the children and 8.6% of the teenagers did not fill the questionnaires to the end, so their answers did not participate in the statistical processing of the results.

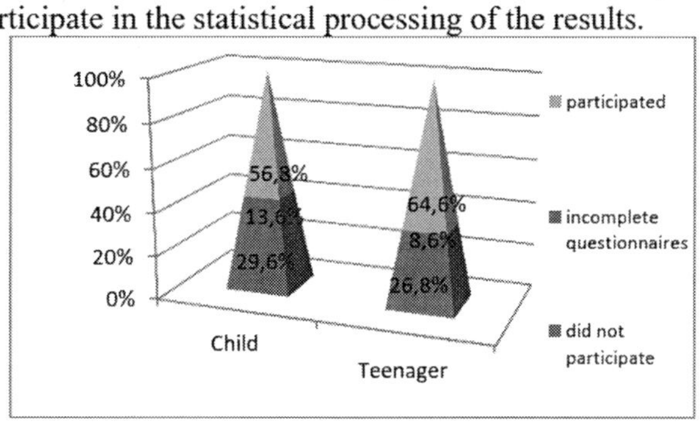

Figure 6. Distribution of participation in the study of sampled students

4. Conclusion

This report presented a method of surveying children's views, combining the capabilities of modern technologies and allowing interest in a fun and entertaining way. The modern generation of children has grown up with new technologies. Using those possibilities to communicate with them is an effective way to conduct a survey.

Acknowledgements
The author/s would like to thank the Research and Development Sector at the Technical University of Sofia for the financial support."

References

[1] Bichurova I 2015 Strategicheski instrumentarium v podkrepa na ustoychivoto turistichesko razvitie i konkurentosposobnost, *Nauchni trudove na Sayuza na uchenite Plovdiv, Seriya A. Obshtestveni nauki, izkustva i kultura* **Tom I** 123 – 126

[2] Nikolova – Alexieva V, Mihova T and Gigova T 2015 Problems in the marketing of bulgarian tourism *Научный журнал "Научно-образовательный потенциал молодежи в решении актуальных проблем XXI века" Красноярскии государственъи аграрнъй университет Ачинскии филиал* **V**

[3] Daylianova H and Mihova T 2015 Communication and the Internet *Scientific works of the Union of Scientists in Bulgaria-Plovdiv* series A. Public sciences, art and culture Vol. **II**

[4] Zlatanova-Pazheva E 2017 Izsledvane hranitelnoto povedenie na detsa i tiyneydzhari v zavisimost ot tipa naseleno myasto, v koeto uchat *Sbornik dokladi ot Shesta Mezhdunarodna Nauchna Konferentsiya „Tehnika, Tehnologii i Sistemi" TECHSYS 18 - 20 may Plovdiv* 347-352

[5] Uzunova Yu 2003 Prilozhni marketingovi izsledvaniya (izd. Romina)

[6] Genova Zh 2009 Sotsiologiya za studenti po sotsialni i stopanski nauki, administratsiya i upravlenie, Vtoro preraboteno i dopalneno izdanie (Universitetsko izdanie „Paisiy Hilendarski", Plovdiv)

[7] Zhelev S 2002 Marketingovi izsledvaniya (kratak kurs) (izd. Trakiya-M, Sofiya)

[8] Cooke L and Wardle J 2005 Age and gender differences in children's food preferences, *British Journal of nutrition* **93** 741-746

[9] Pirouznia M 2001 The influence of nutrition knowledge on eating behavior – the role of grade level *Nutrition and food science* **31** Number 2 62-66 (MCB University Press)

[10] Umeh K and Crabtree L 2006 Is fruit and vegetable intake in children a rationalistic choice?, *British Food Journal* **108** No. 10 859-874 (Emerald Group Publishing Limited 0007-070X)

[11] Cammarota G, Graziani M.P, Laurino C and Pellicano M.P 2008 Gender differences in the eating behaviour of Italian schoolchildren, *La Rivista di Scienza dell'Alimentazione*, numero **2**, anno 37

[12] Stamatov R 2000 Detska psihologiya (izd. Hermes, Plovdiv)

[13] Freedman G and Flanagangames M 2017 From dictators to avatars: Furthering social and personality psychology through game methods, *Soc Personal Psychol Compass* (John Wiley & Sons Ltd)

[14] Lobel A 2016 Game on: The relation between gaming & emotion regulation development, (Ridderprint)

[15] Costikyan G 2002 I Have No Words & I Must Design: Toward a Critical Vocabulary for Games *Proceedings of Computer Games and Digital Cultures Conference* (ed. Frans Mäyrä, Tampere University Press)

[16] Grudeva A 2014 Uchene vav virtualniya svyat na serioznite igri *Spisanie na Sofiyskiya universitet za obrazovatelni izsledvaniya*

[17] Pavlova A 2010 Znakovata sistema na tsvetovete *Nauchno elektronno spisanie Medii i*

obshtestveni komunikatsii (izd. UNSS, br. 7) razdel "Reklama", Available from: http://www.media-journal.info/
[18] http://www.eprints.ugd.edu.mk./
[19] Stoyanov H 2015 Znachenieto na tsvetovete v marketinga – kakvo e vazdeystvieto im varhu potrebitelite (publikuvano v http://www.lifehack.bg/)
[20] http://www.travel-studio.bg/ Tsvetovete i tyahnoto psihologichesko vazdeystvie
[21] Kaftandzhiev H 2013 Harmoniya v reklamnata komunikatsiya (Mnogo prezaredena) (Universitetsko izdatelstvo „Sv. Kliment Ohridski", Sofiya)
[22] http://www.franchising.bg/ Kakvo kazvat tsvetovete za Vashiya brand?

IOP Conf. Series: Materials Science and Engineering **618** (2019) 012078 doi:10.1088/1757-899X/618/1/012078

Critical Thinking About Urban Studies Linked with Thermodynamic Terms

Seda H. Bostancı

Asscoc.Prof.Dr. Tekirdağ Namık Kemal University, Political Science and Public Administration, Süleymanpaşa, Tekirdağ, Turkey

shbostanci@nku.edu.tr

Abstract. The interaction of urban science with psychology, biology, ecology, and technology creates a rich diversity. Thermodynamics and information theory are among the methods of analysis used to understand urban systems. These methods have a variety of options, and their decision-making processes are complex. Definitions related to the concept of city vary depending on the approach of social science disciplines such as urban science, sociology, geography, economy and public administration. Geographic Information System (GIS) enables the creation of interactive maps of the urban system's transformation processes. Thus, city maps have gained qualifications for the examination of the possible changes and future scenarios. The relationship between city and thermodynamics has been developed with implementations such as energy, sustainability, system approach, spatial analysis, aesthetic evaluation methodologies, urban growth, and urban models. The thermodynamic interpretation of implementations provides concrete data such as heat map of the cities, and energy potential. The interpretation of entropy at the intersection of thermodynamics and information theory gives different perspectives to the concepts such as coding, redefining aesthetics and urban growth. The studies which relates urban issues and entropy concept are very different in approach and content from each other. This situation has also created a ground for meaning shifts. The aim of this study is to critically discuss the methods, concepts, and theories of urban studies related to the thermodynamic terms. As a method, the studies on the approaches of the concepts of thermodynamics in urban planning and urban design will be evaluated with a literature review.

Introduction

The system approach, diversity, network relations, complexity are among the prominent adjectives in the definitions of a city structure. Since the emergence of the urban planning discipline, the redundancy of options and possibilities has led urban analysis to the search for integrated models. In this process, the concepts of thermodynamics and entropy have been integrated into the academic researches of the profession by an analysis of the theory and implementation characteristics. There are basically two different relationships between the city and thermodynamics. The first is related to the three basic laws of thermodynamics. Heat map and energy usage of cities can be given as examples. These features can be measured in a very tangible way and have a scientific nature. The second is associated with entropy in information theory including its re-formulation with a different interdisciplinary interpretation, depending on the second law of thermodynamics. It includes the calculation of the aesthetics and informatics-based information depending on various criteria at the plan (two dimensional) and skyline (three dimensional) scales in the urban form by using the formula developed by Shannon (1948) [1].

A source of information can be considered as a mathematical model allowing the symbols to occur randomly in a row. These symbols can be electrical quantities, computer data, image sequences, or waveforms. All these symbols are the alphabet of the model that is the result of probabilistic processes. This model includes terms that are measurable, computable and has a time concept. The calculation of these measurements is related to probability spaces and random variables. These concepts can be defined as nonnegativity (equation 1), normalization (equation 2) and countable additivity (equation 3).

$$P(F) \geq 0 \quad (1) \qquad P(\Omega) = 1 \quad (2) \qquad P(\cup_{i=1}^{\infty} F_i) = \sum_{i=1}^{\infty} P(F_i) \quad (3)$$

Here is the probability space, consists of (Ω, P, B) and a probability measure P that assign a real number P(F) to every member F of the field. It is possible to expand this theorem and to apply to different areas [2].

The cities are dynamic structures, and they change according to time in terms of visual and construction techniques. These changes can be evaluated as a random process and the perception of a dynamic system. Two models can be used here. The first model is a series of random variables that are very basic. The second model is assumed that it consists of an abstract dynamic system formed by a space of probability and a transformation over space. Both models can be related to each other by taking into account a time shift as a conversion. In the mathematical description of the physical world with laws of thermodynamics, there are statistics and probability theory. Meier (1962) revealed the concept as the communication theory of urban growth [3]. Wilson (1969), in transport planning, developed entropy-maximizing spatial interaction models [4]. Batty and Mackie (1972) dealt with spatial entropy and spatial interaction models in urban systems [5]. Based on these, the spatial infrastructure of statistical physics and information theory has been developed [5,6]. There is a need to make an abstraction with coincidence and symbols in order to associate information theory with urban systems. In this way, the definition of a specific information-based entropy can be created from the definition of physical entropy. When the process is reduced to the problem of measuring the information with abstraction, the measurement of the aesthetic value with entropy is performed. The greatest support for aesthetic theory in this field was provided by the entropy formula developed by Shannon (1948) [1]. This process can be described as the aesthetic interpretation of entropy. After this stage, spatial implementations of the entropy method begin. The information entropy approach is used in areas such as urban growth, architectural facade features, urban skylines, art, and aesthetic evaluation. On the basis of each of these areas, message items and information concepts follow a similar process, but the interface models integrated with the implementations differ. In order to better interpret the subject, it has been decided that the implementations of entropy in the field of evaluation of aesthetics and urban skyline will be considered as a constraint in the information theory part of the study.

The general criticism of the studies in this area is the acquisition of partially modified data based on the interpretation of the architectural, aesthetic and environmental psychology parameters. There is a conflict in the interpretation and handling of these data.

The definition of thermodynamics in heat and energy and its meaning in information theory and a binary grouping related to the urban structure will be performed in this study.

One of the biggest challenges in urban planning studies is to be able to relate two different definitions of entropy with thermodynamic laws and its formula in information theory with urban studies. Figure 1 includes a sketch outline prepared for this relationship.

In this schematic explanation, entropy relations and the city can be seen in two different aspects. These relations are different from each other. In one side looking at entropy from the thermodynamic standpoint, urban entropy is a kind of waste (energy, heat, pollution) related to dense urbanization and urban decay. On the other hand, from the information theory viewpoint, the entropy is a kind of measure of criteria research which is related to urban aesthetics in terms of urban aesthetic values. From this approach, Table 1 is prepared to define the urban system in the context of entropy. Comprehensive information on these definitions can be found in the next section.

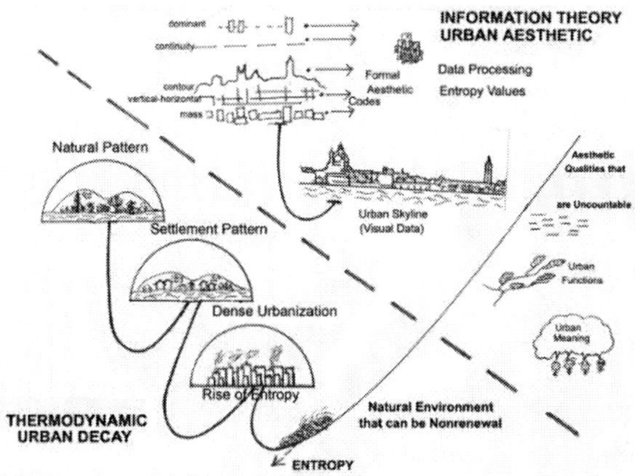

Figure 1. The binary character of the city and entropy: informational interpretation of entropy measure of urban aesthetics and thermodynamic definition of urban decay.

Table 1. Entropy concepts and their relations with urban system.

	Entropy	
Thermodynamic Laws		**Statistics and Probability Theory**
Physical Entropy		Entropy Related to Information Theory
Urban Relations -Efficient use of energy and sustainability -Heat map and thermodynamic structure of the city -Strategy for designing and management of a low-entropy city -Urban metabolism concept for understanding self-organization of open thermodynamic systems		**Urban Relations** - Spatial organization as the communication theory of urban growth -Understanding urban design aesthetic criteria in the coding system, by the effects of information theory -Understanding facade aesthetic dimensions of architectural styles of the buildings -Measuring aesthetic quality of urban skylines

The main objective of the study is to draw attention to dual structure of thermodynamics and information entropy and their discrete effects to urban studies.

Thermodynamics and Urban Studies

Thermodynamics: Heat-Energy and the City Structure

Carnot, the founder of thermodynamics, found the right expression for the maximum efficiency of any heat machine operating between two given temperatures [7]. From a thermodynamic viewpoint, resources are transformed from the usable energy state into an unavailable form, and this form is defined as entropy in Thermo dynamics developed by Clausius (equation 4).

$$\Delta S \geq 0 \tag{4}$$

Within the semi-open system of the world, sources such as sun and wind are renewable energy sources. With these features, renewable energy sources have the potential to turn into clean energy, as they offer an alternative to the situation of risk of increased entropy. Its expression in thermodynamics is that the energy cycle is a partially reversible process.

Thus, if the energy-based entropy map is prepared for the city, the natural resources and mineral reserves are determined, and the energy potential of them is shown on this map when the city is first established. So, the entropy accumulation of the city from past to present and what it means for the city's future can be defined. The relationship among thermodynamics, energy and urban structure is closely related to the concept of sustainability. Fistola (2012) describes the effects of water pollution-producing activities on the soil, electromagnetic pollution, air pollution as urban entropy producing activities. Thus, it is seen that strategies should be developed to reduce urban entropy for a sustainable future [8]. In a similar approach, Fistola and La Rocca (2014) defined urban entropy as a fundamental obstacle to creating a sustainable and intelligent urban system [9]. There are several studies suggesting the reduction of urban entropy for the ecological survival of the urban system. Pelorosso et.al. (2017) defines internal entropy as socio-ecological and structural complexity and wastes as external entropy [10]. These studies show the direct meaning and implementations of entropy in thermodynamics laws.

Gandy (2006) interpreted urban entropy in a different way. "The growing presence of nature within former industrial landscapes can be conceived as a kind of urban entropy whereby the distinction between human artifice and ecological succession becomes progressively blurred" [11]. This interpretation differs from other urban entropy definitions. This information shows that there are differences in the definition of urban entropy in the thermodynamic context. Mohajeri et.al. (2016) has addressed urban entropy as a flexible indicator for urban mobility and sustainability. Measurements have been made on the solar potential of urban entropy [12]. From a thermodynamic point of view, this study is important as an urban study in which the entropy measurements are implemented in real terms and used for solar energy measurements.

The concept of low entropy city is a common denominator of urban metabolism and sustainability approaches. Organic structures of cities are human population and ecosystems. These two organic structures are seen as urban metabolism. In terms of thermodynamics, the metabolism of cities is a growing and evolving open system [10]. In the approach of urban metabolism, there is an understanding that the city has an organic system, not an autonomous system, and that this system needs external material and energy change. Identification of appropriate entropy indicators that can spatially assess the effectiveness of green interventions at different urban scales is necessary for implementations. This makes entropy suitable for use as an effective tool in sustainability studies [13].

According to Bristow and Kennedy (2013) thermodynamics has two main effects of cities and these are; "(1) the implications of the second law of thermodynamics on the efficient use of energy; and (2) the self-organization of open thermodynamic systems" [14]. Thus, it is seen that the real meaning of thermodynamics is discussed in the definition of two different processes in urban studies. Different implementations and researches can be performed with various combinations of these approaches.

The Second Law of Thermodynamics in Information Theory: Information Aesthetics in Urban Design

Entropy as the second law of thermodynamics, redefined in the information theory, has developed according to statistics and probability theory. In relation to the random processes, the logarithmic expression used by Boltzmann (1896) [15, 16], who defines the definition of statistical entropy and improves its formula, is similar to the modern concept of information theory. This equation is expressed below (equation 5).

$$S = k \log W \tag{5}$$

$$\Delta S = S_2 - S_1 = K_B \ln \frac{\Omega_2}{\Omega_1} \tag{6}$$

Schrödinger (1944) defined the negative entropy as a contribution to the statistical entropy described by Boltzmann [17, 18]. Statistical entropy is expressed by the above equations. Here; when the initial state is set to 1, S1=KBlnΩ1. Similarly, the equation S2 can also be written. In this case, ΔS is given in equation 6. (Here S is state function). Equation (7) is obtained when equation (3) is written taking into account the gas molecules (V1, V1), Avogadro number (N) and Boltzmann constant (kB).

$$\Delta S = S_2 - S_1 = N\, k_B \ln \frac{V_2}{V_1} \tag{7}$$

When the equation (7) is arranged, equation (5) is obtained. In equation (8), Boltzmann equation is considered equal to R/NA. Where R is the gas constant NA is the number of Avagadro. As the entropy changes with temperature, the temperature is assumed to be constant.

$$\Delta S = nR \ln \frac{V_2}{V_1} \tag{8}$$

$$\Delta S \geq \frac{q_{rrev}}{T} \tag{9}$$

It is understood that the entropy from the equations can be defined according to the second law of thermodynamics in an irreversible and reversible process. In this sense, entropy becomes meaningful when the entropy of the environment and the system is also taken into consideration. Consequently, the concept of entropy ΔS is expressed by equation (9). Where q (heat) is T (temperature). An entropic amount is often used to determine whether a given reaction takes place or not. This situation can be used in the development of the environment and the city [19, 20, 21]. Shannon is the person who defines the entropy concept of physical processes within the scope of information theory, he defines the statistical entropy theory as the mean information (knowledge) and constitutes it as the basis of modern communication theory [1]. The mathematical structure equation (10) of communication theory created by Shannon (1948) is known as information theory nowadays and has been a pioneer in the development of communication technologies.

$$Hs = \sum pi\ log_2\, pi \tag{10}$$

A different approach in historical development was made by Prigogine (1989), who defined entropy on the behavior of open expansionary structures [22, 23]. Considering the meaning of entropy within other mathematical structures, Renyi (1961), who demonstrated the relationship between chaotic structures and fractals, made a new definition of the fractal dimensions of entropy [24, 25].

There are recent studies on the relationship between statistical entropy with topological structures and their participation in learning structure (algorithms) [26, 27, 28]. Interpretation of entropy in interdisciplinary studies in the fields of social sciences, economics, and psychology has become a subject of criticism over time by Shannon. Shannon (1956), who published an article on this subject, stated that information theory, which started as a technical tool for communication engineering, attracted great interest in the scientific and popular press. He points out that it is possible to make real progress in the theory of communication by only maintaining a purely scientific attitude [29]. Loewenstein (2013) states that information theory has deficits in meaning when describing the message concept. He states that Shannon formula provides data about how much of a message has knowledge of but does not provide data about the meaning value of the information content [30]. This approach is similar to Shannon's criticism. Arnheim (1974) mentions that there are some misunderstandings and confusion in the use of entropy in the field of art and aesthetics. He draws attention to the disadvantages of limiting with risky and reductionist approaches and from moving away from instrumentalized meaning such as a kind of the wrong definition of entropy such as entropy is the relationship between order and disorder [31]. Sahyun (2013) mentions that in the image processing studies, entropy approach is limited in aesthetic evaluation processes and they should be supported with psychometric studies [32]. Gauvrit et.al. (2017) use in the field of aesthetic and environmental psychology, Shannon entropy as a measure of complexity as a measure of the fact that creates a meaning shift, instead of a function of density (or symbol redundancy) emphasized that it would be more accurate [33]. Bostancı (2008), in her study on the aesthetic evaluation of urban skylines, described entropy as value range in the formal aesthetic evaluation data instead of as an expression of entropy increase or decrease [27, 34].

$$H = -\sum pi \, log_2 \, (pi + \varepsilon) \tag{11}$$

Bostancı (2008) by adding the epsilon to Shannon's formula, with equation (11) this modified formula, tries to find a solution for the uncertainty in the information entropy studies on urban aesthetics. Thus, an adaptation has been made that allows the probability values that cannot be measured in the aesthetic criteria of urban skylines to be calculated without going to infinity [27]. The given examples show that like the creator of the theory, the experts from interdisciplinary fields searching of this theory also try to identify the risks such as meaning shift, disconnection of the main theory. With its various debates, information aesthetics has reached a holistic dimension where art and aesthetic analyses have been predicted by measurability of aesthetic features of the cities, presented chronologically in Table 2.

Table 2. Conceptual development of information aesthetics.

Formation and Conceptual Development of Infrastructure of Information Aesthetics
Birkhoff (1933) [35,27] *Mathematical formula of formal aesthetics*: Aesthetic value = Order/Complexity (Analytical aesthetic approach)
Bense (1956) [37] created an approach to *information aesthetics* by the roots of Information Theory: Shannon (1948) [1] + Semiotic: Peirce (1934) [36] + Analytical Aesthetics Birkhoff (1933) [35] = Bense (1956) [37]
Gunzenhäuser (1962) created *a formula for information theory measurement of aesthetic value* as R: Redundancy, H: Entropy R/H: 1/H – 1/ Hmax [38, 27].
Nees (1969) from Bense's Stuttgart School developed the concept of *aesthetic complexity* by investigating Bense's theories related to philosophy, mathematics, and aesthetics [39, 40].
Moles (1966) wrote a book on *information theory and aesthetic perception* [41].
Berleyne (1971) questioned *the relationship between behavioral theory and aesthetics related to information theory* [42].
Kiemle (1967) *first used the information aesthetics for architectural facade analysis* with reference to Bense's work [43, 44].
Buttlar et al. (1973) conducted studies on architectural facade analysis and compared the late nineteenth-century buildings with modern buildings [45, 46].

After these studies, especially in the field of computerized visualization, information entropy in the measurements of cities and architectural facades entropy concept transformed into a tool that makes aesthetics measurable to a certain extent. Its implementations about the evaluation of architectural facades, [46, 47]; the numerical evaluation of form characteristics of urban skyline [48, 49, 50] are among the pioneering and inspiring researches in this field. Studies have gained new dimensions from facade analysis, including the aesthetic evaluation of urban skylines [34]. In this area, there are studies involving GIS modeling of urban skylines [51]. Stamps (2010) conveyed his experiments on the relationship between form and aesthetics which he developed due to information theory to virtual reality models of spaces. Stamps' entropy models are tested with experimental aesthetic approaches and become a model of an integrated aesthetic solution [52]. Gauvrit et.al. (2017) used entropy and algorithmic information theory-based studies of spatial aesthetics in relation to experimental aesthetic subjects to evaluate perceived beauty in random texture patterns [33]. Thus, the field of information aesthetics has managed to include the concepts of probability and mathematics next to the philosophical and psychological meanings of aesthetics with computer-aided innovative solutions, different aesthetics evolution methods and integrated approaches from the 1933s to the present.

Conclusion

Measurement of the energy potential of the city and changes in heat map according to years show great consistency with the formula, $\Delta S \geq 0$. In addition, there are more researches carried on information entropy in urban design and planning. The main problem of this field is the need for the development of a coding system in order Shannon's formula ($H_S = -\sum p_i \log_2 p_i$) to function. In the fiction of this code-probability relationship, the other studies have to follow this process after the pioneering studies. This brings out the risk of the increase in repetitive scientific publications. As Loewenstein (2013) [30], Arnheim (1974) [31] and Gauvrit et.al. (2017) [33] who are experts from different disciplines point out, the research questions how much these studies can be effective on aesthetics and the sensory system of human being. In order to come to a conclusion, many researchers state that these studies are a helpful method in urban studies. Generally, entropy is a component of the formula in the composition of different methods. Experimental aesthetics is the most common integrated method analysis in this field. In the future studies, all urban studies on the concepts of thermodynamics by using meta-data analysis method will be examined considering their scientific or theoretical background, concepts, method approaches, results, suggestions for solution, differences in interpreting and understanding, deficiencies and illusions, as a result, the concepts of thermodynamics for the city will become clear. The realization of such a work is possible with a series of international workshops where scientists from different disciplines in this field come together. Consequently, the studies dealing with urban entropy in a subject of heat and energy have acquired scientific content that brings urban planning and engineering together as a result of using thermodynamic concepts. On the other hand, despite all critical and controversial issues, the information aesthetics discussed in the evaluation of urban aesthetics has gone through the all stages (assumptions, a search for a method, historical content, criticism and updatability) that must be accepted as a theory in social sciences and has become a method approach which has a theoretical background in the urban aesthetics. In the academical studies in this field, descriptive and critical studies are needed as much as practices.

References

[1] Shannon, C. E. (1948). A mathematical theory of communication. Bell system technical journal, 27(3), 379-423.
[2] Gray, R.M. (1990), Entropy and Information Theory, Springer Science and Business Media LCC.
[3] Meier, R.L., 1962. A Communications Theory of Urban Growth, M.I.T. Press, Cambridge.
[4] Wilson, A. G. (1969). The use of entropy maximising models, in the theory of trip distribution, mode split and route split. Journal of Transport Economics and Policy, 108-126.
[5] Batty, M., and Mackie, S. (1972). The calibration of gravity, entropy, and related models of spatial interaction. Environment and Planning A, 4(2), 205-233.
[6] Batty, M. (1974). Spatial entropy. Geographical analysis, 6(1), 1-31.
[7] Hocaoğlu, D. (2008) Termodinamiğin İkinci Kanunu ve Entropy, Ders Notları.
[8] Fistola, R. (2012). Urban entropy vs sustainability: a new town planning perspective. The Sustainable City VII: Urban Regeneration and Sustainability; WIT Press: Ashurst, UK, 155, 1195.
[9] Fistola, R., and La Rocca, R. A. (2014). The Sustainable City and the Smart City: measuring urban entropy first. WIT Transactions on Ecology and the Environment, 191, 537-548.
[10] Pelorosso, R., Gobattoni, F., and Leone, A. (2017). The low-entropy city: A thermodynamic approach to reconnect urban systems with nature. Landscape and Urban Planning, 168, 22-30.
[11] Gandy, M. (2006). Urban nature and the ecological imaginary. In In the nature of cities, 78-89.
[12] Mohajeri, N., Upadhyay, G., Gudmundsson, A., Assouline, D., Kämpf, J., and Scartezzini, J. L. (2016). Effects of urban compactness on solar energy potential. Renewable Energy, 93, 469-482.
[13] Pelorosso, R., Gobattoni, F., Ripa, M. N., and Leone, A. (2018). Second law of thermodynamics and urban green infrastructure–A knowledge synthesis to address spatial planning

strategies. TeMA Journal of Land Use, Mobility and Environment, 11(1), 27-50.
[14] Bristow, D. N., and Kennedy, C. A. (2013). Maximizing the use of energy in cities using an open systems network approach. Ecological modelling, 250, 155-164.
[15] Boltzmann, L. (1896). Entgegnung auf die wärmetheoretischen Betrachtungen des Hrn. E. Zermelo. Annalen der physik, 293(4), 773-784.
[16] Klein, M. J. (1973). The development of Boltzmann's statistical ideas. In The Boltzmann Equation (pp. 53-106). Springer, Vienna.
[17] Schrödinger, E. (1944) What is Life? Cambridge University Press, Cambridge.
[18] Ho, M.W. (1994). What is (Schrödinger's) negentropy. Modern Trends in BioThermoKinetics, 3(1994), 50-61.
[19] Boltzmann, L. (1896/1964). Lectures on Gas Theory, translated by S.G. Brush, University of California Press, Berkeley.
[20] Borgnakke, C., Sonntag., R.E. (2009). Fundamentals of Thermodynamics, seventh edition, Wiley, pp.127-154.
[21] Chemistry, (2018) Statistical Entropy, https://chem.libretexts.org/Bookshelves/Physical_and_Theoretical_Chemistry_Textbook_Maps/Supplemental_Modules_(Physical_and_Theoretical_Chemistry), 6.3.2019.
[22] Prigogine, I. (1989). What is entropy ? Naturwissenschaften, 76(1), 1-8.
[23] Progogine, I., and Stengers, I. (1984). Order out of chaos: Man's new dialogue with nature. Fontana, London.
[24] Rényi, A. (1961) "On measures of entropy and information", *Proc. of the 4th Berkeley Symp. on Math. Statistics and Prob.*, vol. 1, pp. 547-561.
[25] Renner, R., & Wolf, S. (2004, June). Smooth Rényi entropy and applications. In International Symposium onInformation Theory, 2004. ISIT 2004. Proceedings. (p. 233). IEEE.
[26] Cambel, A. B. (1993). Applied chaos theory: A paradigm for complexity. Elsevier.
[27] Bostancı, S.H. (2008). Kent Siluetlerinin Entropi Yaklaşımı ile Değerlendirmesi (Doctoral dissertation, Fen Bilimleri Enstitüsü), Istanbul Technical University, Istanbul.
[28] Guillen, M. (2012). Five equations that changed the world: the power and poetry of mathematics. Hachette Books.
[29] Shannon, C. E. (1956). The bandwagon. IRE Transactions on Information Theory, 2(1), 3.
[30] Loewenstein, W. (2013). Physics in mind: a quantum view of the brain. Basic Books (AZ).
[31] Arnheim, R. (1974). *Entropy and art: An essay on disorder and order*. University of California.
[32] Sahyun, M. R. (2013, March). Aesthetics and entropy II: a critical examination. In Human Vision and Electronic Imaging XVIII (Vol. 8651, p. 86511K). International Society for Optics and Photonics.
[33] Gauvrit, N., Soler-Toscano, F., & Guida, A. (2017). A preference for some types of complexity comment on "perceived beauty of random texture patterns: A preference for complexity". Acta psychologica, 174, 48-53.
[34] Bostanci, S. H., and Ocakçi, M. (2011). Innovative approach to aesthetic evaluation based on entropy. European Planning Studies, 19(4), 705-723.
[35] Birkhoff, G. D. (1933). Aesthetic measure (p. 209). Cambridge, Mass.
[36] Peirce, C. S. (1934). Collected papers of Charles Sanders Peirce (Vol. 5). Harvard University.
[37] Bense, M. (1956). Ästhetische Information [Aesthetic information]. Krefeld, Germany: Agis.
[38] Gunzenhäuser, R. (1962). Ästhetisches Mass und ästhetische Information: Einführung in die Theorie GD Birkhoffs und die Redundanztheorie ästhetischer Prozesse. Schnelle.
[39] Nees, G. (1969) Generative Computergraphik, Siemens AG, Berlin.
[40] Klütsch, C. (2007). Computer Graphic—Aesthetic Experiments between Two Cultures. Leonardo, 40(5), 421-425.
[41] Moles, A. (1966). Information theory and esthetic perception, trans. JE Cohen, University of Illinois Press.
[42] Berlyne, D.E. (1971). Aesthetics and psychology, Appleton- Century-Crofts, New York.

[43] Kiemle, M. (1967). Ästhetische Probleme der Architektur unter dem Aspekt der Informationsästhetik.
[44] Bonsiepe, G. (1968). A method of quantifying order in typographic design. Visible Language, 2(3), 203-220.
[45] Buttlar, A. von, Seling, H., Wetzig, A. "The value of historic architecture" in How to Play the Environment Game, (Ed. Crosby, T.), Penguin Books, 74-81.
[46] Krampen, M., 1979. Meaning in the Urban Environment, Pion Limited, London.
[47] Minai, A. T. 1989. Design as Aesthetic Communication, Peter Lang.
[48] Salingaros, N.A., 2000. Complexity and Urban Coherence, Journal of Urban Design, 5, 291-316.
[49] Stamps III, A. E. (2003). Advances in visual diversity and entropy. Environment and Planning B: Planning and Design, 30(3), 449-463.
[50] Stamps III, A., Nasar, J. L., and Hanyu, K. (2005). Using pre-construction validation to regulate urban skylines. Journal of the American Planning Association, 71(1), 73-91.
[51] Stamps III, A. E. (2010). Slines, entropy, and environmental exploration. Environment and Planning B: Planning and Design, 37(4), 704-722.
[52] Akdağ, S. G., and Bostancı, S. H. (2013). The impacts of prestige projects on the skyline of Istanbul. International Journal of Architectural Research: ArchNet-IJAR, 7(2), 269-281.

Plug-in for creation of gamified courses in the e-learning environment moodle

M Gachkova[1] and E Somova[2]

University of Plovdiv "Paisii Hilendarski", 24 Tzar Assen st., 4000 Plovdiv, Bulgaria

[1]mariq.gachkova@gmail.com; [2]eledel@uni-plovdiv.bg

Abstract. The article examines the challenges associated with designing and delivering effective training. The authors propose the use of serious games in learning (in particular gamification) to achieve higher motivation in learners than traditional e-learning. Pedagogical approaches, theories and pedagogical models, suitable for the design and construction of gamified e-courses are presented and analysed. A plugin is presented for creating gamified e-courses in a Moodle e-learning environment. The plug-in can be used to automatically transform a standard e-course into a gamified one or the initial creation of a gamified course. It is shown how standard Moodle environment elements can be used in game contexts.

1. Introduction

The accelerated development of technologies has led to a drastic change in the image of the modern learner. The passive trainee of the past has become an active, searching and demanding learner who wants fast, attractive, quality and effective education based on the latest technologies and tools.

This change is naturally leading to, if not a shift, then at least an adaptation of the pedagogical methods, approaches and strategies, used in contemporary education.

According to Garrison [1], effective learning requires a large repertoire of techniques or "recipes" that poses the challenges of designing and delivering satisfying learning content. E-learning technologies provide different kinds of opportunities for studying and assessment, which cannot be ignored. Therefore, it is imperative to reconsider the purpose, approach and nature of the learning activities used [1] in contemporary learning.

While blended learning and e-learning approaches have great potential to improve the learning process compared to the traditional learning environment, many pedagogical models are not designed (intended) to be suited to such environments [2].

Pedagogical models, besides the effectiveness of learning, also address other aspects. Similar to architectural models, pedagogical models should not prescribe a specific way of realization, but rather provide a science-based framework for pedagogical design [3]. Designing a learning environment includes a cascade of design solutions that are made at different levels: from the "strategic level" (ex., the type of learning objectives) to the "application level" (ex. screen design of the environment) [4].

Games have a strong motivating mechanism, so they are increasingly used to motivate learners to achieve more effective learning (through so-called serious games).

Known serious games categories [5] are:

- Game-based learning (GBL) – uses video and electronic games for achieving educational goals;
- Gamification of learning – integrates game elements and techniques with the e-learning process;

- Organizational-dynamic games – they teach and reflect the dynamics of organizations at 3 levels: individual behaviour, group behaviour and culture dynamic;
- Simulation games – games, used for the acquisition or training of different skills, teaching effective behaviour in the context of simulated conditions or situations;
- Edutainment – presents content, designated simultaneously for education and enjoyment (the term is a combination of the words education and entertainment).

The generation, currently entering the higher education system, grew up with access to the Internet and early use of computers, mobile devices and gaming devices. This group has a different pattern of behaviour in media consumption, communication, and therefore different expectations in the educational environments. Passive use of educational content and traditional "face-to-face" and "distance learning" pedagogical methods are no longer sufficient for the new generation learners [6].

The article presents a study of learning approaches for designing game-based e-courses. The authors rely on the game-based methodology to meet the new requirements and expectations of learners in contemporary learning.

Section 1 of the article discusses learning approaches, theories and pedagogical models suitable for designing and constructing game-based e-courses with common concepts and components with gamification in learning: Keller's ARCS model for enhancing learners motivation, Malone's theory for Inner Motivation Based Learning, Motivational Active Learning (MAL), Colb's Modeling Process Based on "Learning Style," and Bloom's Revised Taxonomy for Assessing Cognitive Outcomes.

Section 2 introduces the authors' plug-in for creating game-based courses in Moodle. The paper ends (Section 3) with a conclusion summarizing the contributions of the authors as well as with some ideas for future works.

2. Approaches and models for designing game-based e-courses

The key concept in gamification of learning is motivation. According to Kapp [7], two types of motivation are distinguished, namely **intrinsic** (internal) or **extrinsic** (external). Intrinsic motivation originates from the learner, while extrinsic motivation is driven by external factors (for example, the teacher sets a high score in fulfilling predefined learning objectives).

Games are a powerful motivation tool that is used in many learning methods and approaches. The authors make a correlation between the learning motivation and the motivation to play.

2.1. ARCS model

In the 1980s, educational psychologist John Keller presented the four-component ARCS educational model, which is an instructional design approach. The ARCS model [8] examines the following components: **Attention, Relevance, Confidence, and Satisfaction.**

Attention can be acquired in two ways: achieving through perception - by using a surprise or an unexpected event to gain interest; or achieving by asking - by stimulating curiosity through asking challenging questions or placing difficult problems that need to be addressed. Methods of attracting learners' attention include the use of:

- Active participation - use of game strategies, role plays or other practical methods to attract the learners' attention to participate actively;
- Variability - using a variety of methods for presenting learning materials (e.g. videos, short lectures, mini-discussion groups) while taking into account the individual differences in the learners' studying styles;
- Humour - use of a small amount of humour;
- Mismatch and conflict - use of the "devil's advocate" approach, in which those who contradict the previous experience are presented;
- Experience - using visual stimuli, illustrating teaching materials with examples from the real world, etc.;
- Inquiry - Asking questions and problems that learners must solve (e.g. brainstorming activities).

In order to achieve **relevance**, it is recommended to use terminology presented with concrete examples familiar to the learners. The possible strategies outlined by Keller are:

- Target orientation – learner orientation about the importance of a learning objective, describing both the importance of achieving a goal and how it will help the learner now and in the future;
- Conformity - compliance of the learning motivation with the students' motivation;
- Familiarity - shows how new knowledge is related to the learners' existing knowledge.

Methods for achieving **confidence** in students include the use of:

- Helping learners understand the probability of success – learners' motivation will decrease if they think their goals are unachievable or the cost to achieve them (time or effort) is too high;
- Setting goals and prerequisites – with clear student performance requirements and assessment criteria, the probability of success can be assessed more accurately;
- Meaningful student success opportunities – explain to the learner what the meaning of achieving the specific learning objectives are;
- Learners' growth - making small steps for growth during the learning process;
- Feedback - providing feedback and opportunities to acquire achievements / successes in the course for which feedback is received;
- Learning control – provide learners the ability to control learning and evaluation and make them believe that their success is a direct result of the effort they've made.

Learners' **satisfaction** can be achieved by:

- Rewarding - receiving rewards for an achievement;
- Usefulness – opportunity for learners to feel the usefulness of the learning by using newly acquired knowledge in a real environment or a real task;
- Satisfaction – giving learners feedback about their results and ensuring results are enhanced;
- Avoiding underestimation - avoiding overly easy jobs.

2.2. Malone's theory of learning through intrinsic motivation

Thomas Malone [8] explores the motivating power of games by analysing predominantly the elements of entertainment. As a result, he presents a model of game motivation analysis that has three key elements: **challenge, fantasy,** and **curiosity.**

The challenge depends on goals set with unspecified results (there is no certainty that the final goal will be achieved). A variety of difficulty levels, sets of level crossing objectives, hidden information detection, or game-based challenges can be used. According to Malone, the learning objectives should be leaner relevant and easily attainable. To enhance the challenge, it is desirable to provide feedback on learner's progress in the course of achieving the learning objectives.

Malone defines **fantasy** as an environment that "induces mental images of things that are not present within the real experience of the learner." Using fantasy can make learning environments more interesting and more educational, because fantasy has both cognitive and emotional advantages. Fantasies in games mostly satisfy the emotional needs of gamers. Fantasy is defined as **extrinsic** (external) and **intrinsic** (inner).

Learning environments can trigger the **curiosity** of the learner if they provide an optimal level of information complexity and an exciting environment. According to Malone, curiosity is sensory (attracting attention by reflecting changes in light, sound or other sensory stimuli of the environment) and cognitive (provoked by the prospect of changing cognitive structures to a higher level, which can be achieved by making students believe that their knowledge is incomplete or inconsistent, thus motivating them to learn more). To engage the curiosity of learners, it is proposed to use surprising and constructive feedback as well as hidden resources in the learning environment.

2.3. Motivational Active Learning (MAL)

Pirker & Guetl [2] presents a strategy game-based approach used to design Motivational Active Learning (MAL) in traditional, mixed and distance learning. According to Pirker & Guetl [2], the approach has a great potential for implementation in both primary and secondary education and for raising students' interest and commitment.

The typical learning material for MAL is divided into small pieces of information (called theoretical units). Learners are introduced to the key concepts with the theoretical units. Then they use the acquired knowledge to solve small problems and tasks. In addition, learners are committed to apply the theoretical knowledge in researches or discussions. Most of the offered assignments are group ones (with group size of 2-4 learners). Before and after each learning material, students take an individual test and compare their knowledge before and after the material.

2.4. Kolb's Experiential Learning Model

The Kolb's experiential learning model examines one aspect of learning – the learning style (the way students most effectively perceive, process, store, and reproduce information) that is completely independent of the other components. Learning style, according to Tennant [9], represents a relatively stable characteristic of one's personality and expresses preference for the ways of perception and processing of the information, coming from the environment.

Kolb [10] introduces the notion of a "learning cycle" in which four phases of the learning process, requiring different skills, are distinguished:

- Feeling – gaining experience from personal experiences;
- Watching - based on a specific experience;
- Thinking - collecting observations and transforming them into theoretical models;
- Doing – based on what is already learned, new ideas and solutions are created, which are then tested in practice.

Kolb distinguishes four learning styles, resulting from combination between each of the two phases in the learning cycle [11]:

- The dreamer - specific experience, observation and reflection;
- The thinker – observation, reflection and abstract thinking;
- The decision maker - abstract thinking and active experimentation;
- The performer - active experimentation and specific experience.

2.5. A learning model based on Bloom's revised taxonomy

Gloria, Bellotti, & Berta [12] define the Revised Bloom Taxonomy model as the most popular cognitive approach used in serious games (including their assessment).

Based on the Bloom Taxonomy for Classification of Educational Goals [13], in 2001 the Bloom's revised Taxonomy was introduced, which gives a more dynamic concept of classification. Taxonomies distinguish six cognitive levels of learner knowledge. Bloom defines the definition of educational goals for different cognitive levels through certain actions (by verbs) and objects (by nouns), while the revised taxonomy uses actions - categories and subcategories (through verbs and gerunds).

Anderson and Cratul [13] further develop the idea by identifying 19 specific cognitive subtasks, which complement the six basic categories. According to Gloria, Bellotti, & Berta [12] the two pedagogical models of Kolb and Bloom are complementary and can be easily applied together.

3. Plug-in for creating gamified e-courses

The attempts have been made to gamify e-courses in Moodle [14], which show an increase in learners' motivation and their success. Gachkova [14] describes an exemplary game-based course in which gamification is accomplished using standard Moodle game elements (points, badges, time and conditional constraints), and standard non-gaming Moodle elements, showing how they can be used in a gaming context (section, group, and course description - for storyline description).

To implement more motivating gamified courses that follow and apply the well-known pedagogical approaches and models, described in Section 2, the authors have created a Moodle plug-in called "Gamified course view" ("course format" type). When designing learning courses, the authors will be able to apply the models and approaches from Section 2 in practice.

The plug-in allows users to design and create a game-based view of their e-learning course. Previewing a sample course through this plug-in is presented in Figure 1. The course design is accomplished on game levels that can be unlocked or locked, depending on whether the course author has set entry requirements for a given level and whether these requirements are satisfied by the learner at the time of the learning. Figure 1 shows that level 2 has already been reached (is visible) after meeting its entry requirements, consisting of completing given Level 1 activities (with certain results). It also shows that the upcoming levels 3 and 5 are available, but for level 4 the entry constraints have not been reached yet and is inactive for opening.

With this plug-in, an already existing standard e-course can be automatically converted to a gamified course, ensuring effortless transformation of existing standard e-courses. The second way to use this plug-in is to create a new gamified course by including the desired learning resources and activities in an empty course from this format. To support the course-makers, a level template is offered in which all the specific game elements (see Table 1) are included with the appropriate support that can be applied in the course.

Table 1. Game elements in the level template of the plugin.

Gamified element	Possible Moodle Element	Additional Conditions
Story/History	Label, Page, File	–
Game Rules	Label, Page, File	–
Challenge	Page, File, Folder, URL, Book, Lesson, Assignment, Choice, Quiz, Glossary, Workshop, Wiki, Database, Forum, Chat, External tool, Survey	Bonus points (not required)
Hidden Treasure	Page, File, Folder, URL, Book, Lesson, Glossary, Forum, External tool	Conditional entry restrictions, Visible disabled element
Reward	Page, File, Folder, URL, Book, Lesson, Glossary, Forum, External tool	Conditional entry restrictions, Visible disabled element
Combo	Label, Page, File, Folder, Glossary, Database, Book, Lesson, Chat, External tool	Conditional entry restrictions, Visible disabled element, Time constraint
Badge	Badge	Receiving conditions
Socializing	Forum, Chat	–

The gamified course is built from several game levels (similar to standard Moodle sections), each of them containing learning resources and activities, made using game elements. The plug-in utilizes standard Moodle elements in a game context (Table 1). The proposed game elements in the template can be deleted (if they are unnecessary), added (to increase the identical items), edited and set (to be described and filled with the appropriate content), so that the user creates his own specific level (with a concrete methodology) as part of the course.

A different view of the gamified course (see Figure 1), compared to Moodle's standard course, is realized to increase awareness and confidence of the learner on the progress of the course and its level of progress over the course. The course is visualized as a wizard (with level sequence). Learners has the opportunity to see the resources and activities for learning presented in a game context.

Moodle's default course settings have added additional features like using / changing the default level names, changing the section zero position, changing the list style for the levels, colouring the high level in the course, colouring the current level with specific colour, etc.

The plug-in is implemented with multilingual support in Bulgarian and English. The functionality for automatic update of the plugin version is added.

The plug-in was created on the base of PHP, HTML, JavaScript and CSS.

To fully implement game concepts in the Moodle e-learning environment, a second plug-in called the "Block gamification dashboard" is being developed, in which completely new gaming elements that are not standard for Moodle. The new elements that are being developed are: Leader Board, adding a setting for achieving next level based on algorithm, activity course log by avatars, etc.

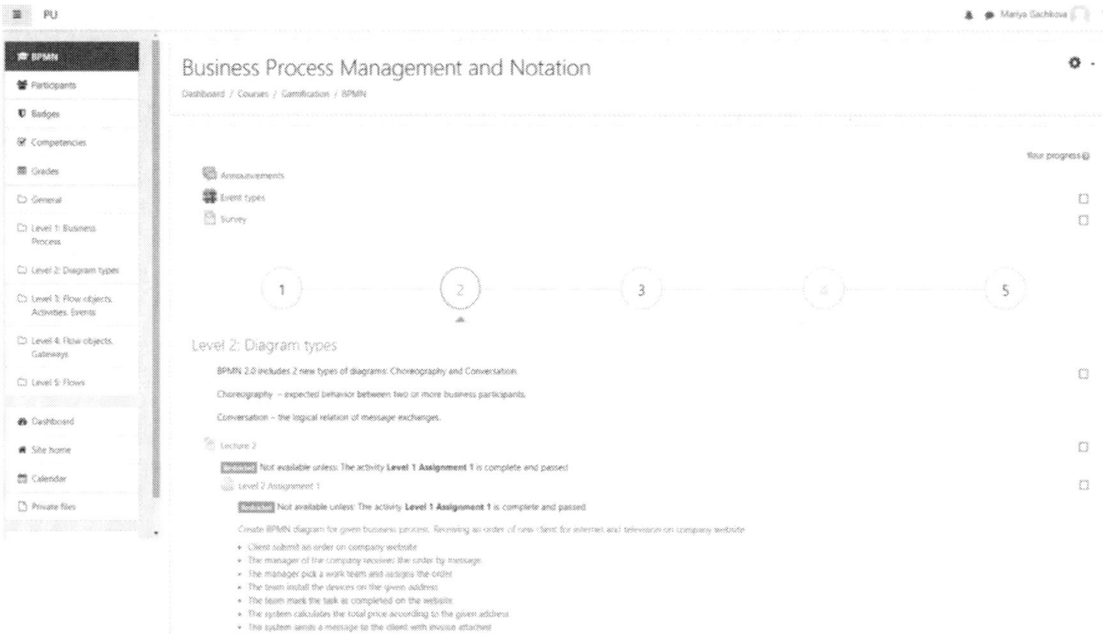

Figure 1. Visualization of a gamified course by the plug-in

4. Conclusion

The reviewed pedagogical models and learning approaches are built upon similar theories and are partly complementary. Active experimentation and repeatability in Kolb's learning cycle, the attracting of attention in the ARCS model, the challenge, fantasy and curiosity elements of Malone's theory are just one part of the examples of similarity with the concept of gamification and their application for enhancing the intrinsic and extrinsic motivation of learners.

Games or even the application of certain game elements give better attitudes towards learning than traditional teaching methods and increase motivation in different learner groups and learning situations. E-courses designed with the help of gamification provide effective training for different learners in areas of higher declarative knowledge and memory levels [15]. They are perfectly suited for the new generation (the so-called Y generation), which enters the educational environment and grew up with access to the Internet and early use of computers, mobile devices and gaming devices.

A plug-in is being developed to support the education of students at the "Paisii Hilendarski" University of Plovdiv, which offers game-based learning in the Moodle e-learning environment, using the Moodle elements in a game context. The next goal of the authors is to offer a second plug-in with more game elements that are not specific to Moodle.

References

[1] Tennant M. 1988 Psychology and Adult Learning.

[2] Pirker J, Guetl C. 2016 Engaging Students in Digital Learning, on Motivational Active Learning in Blended and Virtual Learning Scenarios.

[3] Niegemann HM, Hoscheid-Mauel D, Molz M, Eckhardt A, Schnotz W, Hessel S. 2004 Reconstructing Instructional Design Models. Developing a Heuristic Advisory System as Support for Instructional Design Research.

[4] Niegemann H. and H. S. &. H. S. 2018 Pedagogical design patterns for e-learning: A new approach to instructional design theory.

[5] Malone TW. 1981 Toward a Theory of Intrinsically Motivating Instruction, *Cognitive Science*, **volume 5**, N° 4, pp. 333-369.

[6] Kolb A. Y. и Kolb D. A. 2005 Academy of Management Learning & Education Learning Styles, Acad. of Man. Learning and Education, **volume 4**, N° 2.
[7] Kapp K. 2012 The Gamification of Learning and Instruction: Game-based Methods and Strategies for Training and Education, *Pfeiffer*.
[8] Keller JM. 2010 Motivational Design for Learning and Performance. SSBM;
[9] Greenaway TP. Experiential learning articles + critiques of David Kolb's theory Available: http://www.reviewing.co.uk/research/experiential.learning.htm. [last accessed: 20.02.19].
[10] Gloria A. D., Bellotti F. и Berta R., *2014* Serious Games for education and training, *IJSG,* **volume 1**, N° 1.
[11] Garrison DR. 2011 E-Learning in the 21st Century: A Framework for Research and Practice, 2nd edition. Available: https://dl.acm.org/citation.cfm?id=1996369. [Last accessed 8 12 2018].
[12] Gachkova M and Somova E. 2016 Game-based approach in e-Learning, *on IX National Conference ERIS.*
[13] Anderson LW, Bloom BS and Krathwohl DR. 2001 A taxonomy for learning, teaching, and assessing: a revision of Bloom's taxonomy of educational objectives.
[14] Chang V, Guetl C. 2010 Generation Y Learning in the 21st Century: Integration of Virtual Worlds and Cloud Computing Services, Global Learn, **volume 2010**, N° 1, pp. 1888-1897, 2010.
[15] Gachkova M, Somova E and Takev M. 2018 Learning and Assessment Based on Gamified e-Course in Moodle. *Mathematics and Informatics;* 61:444–54.

The role of information and communication technologies as a part of business intelligence in improving the wealth of nations

S Evseeva[1], O Kalchenko[2], K Plis[3], O Evseeva[4]

[1,2,3,4]Peter the Great St. Petersburg Polytechnic University
Polytechnicheskaya, 29 St.Petersburg, 195251, Russia

[1]evseeva_sa@spbstu.ru, [2]kalchenko_oa@spbstu.ru, [3]k.plis@igms.info, [4]evseeva@spbstu.ru

Abstract. The paper is devoted to the task of determining the role of ICT as a part of business intelligence in improving wealth as the resultant factor in the development of the country and society. The authors consider the development of the wealth concept in a historical retrospective. The role of information and technologies as a part of business intelligence in the development of countries has been studied and generalized, digital transformation has been identified as the leading factor in ensuring the country's economic growth. The paper presents data on the development of the digital economy for the period from 2010 to 2016 in developed countries and analyzes the state of the digitalization process in Russia. The authors put forward and test the hypothesis about the impact of ICT as a part of business intelligence on the Human Development Index, which determines the level of the country's wealth according to the current concept adopted in the world community. The results of the research in Russia showed that there is a close relationship between the HDI as a result of the wealth indicator and the majority of indicators characterizing the virtual - digital environment - the Internet, computers, global virtual networks and mobile communications. In conclusion, the paper presents promising directions for further deeper analysis of revealing the degree of influence of each of the identified indicators, as well as the possibility of improving the country's wealth based on the identified role of ICT as a part of business intelligence.

1. Introduction

The continuing complexity of public structures and relationships, which are increasingly based on modern digital technologies, causing exponential growth of data flows, highlights the issue of the formation of the digital economy. The importance of ongoing processes has made it possible to raise the question of the formation of a new type of economy, where relations over the production, processing, storage, transfer and use of increasing amounts of data become dominant. Data become the basis of economic analysis, which studies the patterns of functioning of modern socio-economic systems.

Information and communications technologies (ICTs) are a very fast growing new technological area. ICTs, by their performance and potential, offer numerous options to drive wealth and development. Information systems research can make an important contribution to knowledge at the nexus of information, organizations, and the natural environment; to the development of innovative strategy [1]. At present, human capital and information resources have become the main factors in the development of modern society and the world economy as a whole. The information resource and its knowledge and information are part of the accumulated and functioning human capital, it is its base and foundation. At the same time, the information resource is also an independent factor of development, subject to analysis and study from the point of view of the theory and practice of economic growth and development, the formation of civil society, provision, general, economic and information security, etc.

All of the above has a direct impact on the wealth of nations, changing and improving it. And it determines the relevance of the chosen research topic and aim which is to investigate the role of information and communications technology (ICT) in improving the wealth of nations.

Research methods used: system analysis, synthesis, comparison and generalization, deduction and induction, correlation analysis.

The initial data are the statistical data from Federal State Statistics Service, report of Higher School of Economics, State Corporation «Rostec», Human development data, International Telecommunication Union for 2010-2016.

2. The investigation of ICT's role in the wealth of nations

Throughout human history, the concept of wealth and ways of achieving it has different meanings. Economics was initially studying ways and means of achieving wealth. Since the middle of the XVIII century, the economy had become political, and the aspirations of people were aimed at finding wealth. The wealth of the nation began to be measured in stock keeping units (SKUs), in which retailers assess the types of goods they have in stocks. About the SKUs of different countries is stated in the book "The Origin of Wealth" of Eric Beinhocker, the founder of McKinsey Global Institute ideas, which was released in 2006. At the Voltaire time, material wealth and individual freedom were two sides of the same coin, they conditioned each other and together formed the basis of cultural progress [2]. According to Adam Smith, only the motive of profit makes economic development possible. Incentives should be shaped so that they are also among those who create this wealth. For Jean-Jacques Rousseau wealth was a harmonious life in nature and with nature. According to the International Monetary Fund 2011 study [3] developing countries in the early phase of their catch – up race need inequality - social injustice to quickly create a basis for balanced growth. The capitalist dynamic that Adam Smith praised for ensuring the provision of the population and improving its wealth, in fact - according to Karl Marx - leads to more and more deep crises and, finally, to the final collapse. For John Stuart Mill the wealth of a nation consists not only in possibly a more efficient production of goods. Economist Milton Friedman believes that internal and external security are the responsibilities of the state, and everything else is superfluous. According to John Maynard Keynes, if citizens save, then the state should just spend money, because the general economy leads to a general loss of wealth - paradox of thrift. According to Joseph Schumpeter, profits - and thereby growth and wealth – arise through innovation, credit and entrepreneurial spirit [2].

The essence of wealth has changed dramatically in people's minds after the first and the second industrial revolutions. At the end of the last century, the world community recognized that the existing model of world development, based on economic growth and material wealth, had exhausted itself and a system of new views was required. In this regard, the international community, with the active participation of the World Bank and the United Nations, has developed a concept of human development, according to which human potential is the main driving force of economic development and national wealth. For an integrated assessment of the level and quality of life of the population, an indicator was developed - the Human Development Index (HDI).

At the same time information as a factor of production is extremely necessary in modern conditions of functioning. It ensures the promptness of the decisions made, helps develop the entrepreneurial ability and increase the efficiency of the production process. The availability of information reduces the impact of such an external factor as uncertainty. The theory of sustainable development is one of the complex concepts, it is designed to solve the problem of uncertainty in the development of the macroeconomic system in the long-term period of time. At the moment, this theory is rapidly developing and leaves open many theoretical questions about the essence of sustainable development and applied problems of the formation of a model of sustainable development in specific conditions of developed and developing countries.

IOP Conf. Series: Materials Science and Engineering **618** (2019) 012080 doi:10.1088/1757-899X/618/1/012080

Initially, the category of "sustainable development" was proposed as a response to the environmental challenges of our time and was understood in the sense of the conscious direction of development for the long-term provision of mankind with natural resources, provided that the environment is not destroyed. However, gradually the views of scientists on this issue have become broader, and therefore there have been incentives to talk about the direction of economic development to meet the broad social needs, as well as to maintain the ability to meet the similar needs of all future generations [4]. Thus, the theory of sustainable development accumulates the theories of economic growth and development and is potentially capable of becoming a general theory of quantitative and qualitative economic development.

The intensive economic development of many developed countries and their increasing involvement in the process of globalization led to the emergence of such terms as "information society", "knowledge society". Russia seeks to keep up with the level of the more developed countries. Therefore, the development path chosen by the government includes, first of all, the development of a socially-oriented society, or a social state.

At present, the goal of a social state is to ensure decent living conditions and prosperity for all citizens. This means that the state is obliged to create all conditions for self-development of the individual, excluding factors of social dependency. In reality, not only the state, but also the citizens themselves, on the basis of guaranteed freedom and property relations, should be responsible for the material provision of citizens.

Differences between the industrial and information state of society are enormous. The information society [5] has overcome the traditional forms of commodity exchange and private property, based primarily on the unequal distribution of material wealth.

At present, the main conditions determining the wealth of an individual are his upbringing and education, respectively, the main public institutions that shape future wealth are the family and the school. No market mechanism can provide the consumer with a better life than he deserves in terms of his level of knowledge and degree of cultural education. Based on the received level of knowledge, a person forms a certain mode of consumption, where the result should be the education of reasonable consumption. With the emergence of the need for creative activity, other needs are formed, and the pleasure associated with biological instincts must go to the second place, because the perception itself must change. Approaches to determining the essence and level of wealth are not the same, because the possibilities and conditions for achieving it are different [6].

When the biological instincts of people go to the background, then the economy, formed on the basis of information, itself will be the source of growth. The development of the information economy contributes to economic growth [7], to the improvement of the well-being of citizens and the quality of life of the population.

Information is largely a public good, and the act of recognizing a public good is to use it in one form or another. In the production of goods and services, creativity is of the utmost importance. The result of creative work is getting information or knowledge. If the industrial era was characterized by the production and consumption of a tangible product, only the material product that will contribute to the development of the individual will be valued in the information society [8].

In modern society, the functioning of the entire economic system will cause a change in the main characteristics of the dominant benefits in society. With the development of the information society, material wealth will no longer dominate as much as at present. In the information society, wealth theory is changing: it is creative work that will form a noneconomic wealth based on the fact that self-affirmation, self-education is the perfect benefit for the activity of a person as an individual.

Many factors affect on the standard of living. If the standard of living is taken as an economic category, then it will be influenced by such factors as real wages, real incomes of the population, minimum wage level, subsistence level, and if we take a generalized indicator, then - the GDP per capita, final consumption expenditure, etc. If the standard of living is closely related to the quality of life of the population, then it will be influenced by the development of the infrastructure, the health of the population, the environmental state of the environment, the availability of people's leisure time, etc.

Previously, the level of socio-economic development of society was not measured by the total amount of time for the implementation of people's creative activity. With the development of the information economy, wealth can be viewed as a broad, capacious concept, which is a complex socio-economic phenomenon that unites and incorporates various characteristics of the style, image and quality of life of the population, which need to be approached from a level viewpoint, taking into account, moreover, "higher activity" in society. At the same time, aggregate income is no longer regarded as the main factor. It is presented as a necessary subfactor for the development of higher activity, which contributes to raising the level of education and health of the individual.

Raising the standard of living of the population is a priority in the development of the state as a whole, since a high standard of living contributes to the reduction of social and economic tensions in the country and to an increase in people's satisfaction. High level and quality of life allow an individual to develop his spiritual and cultural human potential more actively, to reveal himself as a person and all the time he strives for perfection [8].

There are new social elevators, geographic horizons of opportunities. Thanks to more comfortable cities, effective public institutions and affordable public services the conditions of everyday life of citizens are improved. States oriented on innovation and research, like a magnet attract qualified personnel as a key resource of digital economies.

According to the forecast digitalization will provide GDP growth from 19 to 34% by 2025.

Digital transformation is one of the main factors of the world economic growth [9]. According to estimates of the Global Institute McKinsey, In China, up to a 22% increase in GDP by 2025 can be due to Internet technologies. In the US, the expected increase in value created by digital technologies, may amount to 1.6-2.2 trillion US dollars by 2025.

Using ICT in all key activities - trade, transport, finance, industry, public services, education, public health and the public sector - not only radically changes the people's lives, but transforms economic structures. So, according to the forecasts of the American consulting company BCG, by 2035 the volume of the digital economy in the world will reach $ 16 trillion. Today in Europe, the share of the digital economy in GDP exceeds 5%, in the USA - 6%. The contribution of digital technology to the UK's GDP reaches 12%. In China, the level of digitalization is higher than the level of the economy as a whole - the Internet accounts for more than a third of the country's GDP. The digital economy of China is considered to be a group of "beginning leaders".

The share of the digital economy in the GDP of developed countries increased from 4.3% to 5.5% between 2010 and 2016 [9], and in developing countries GDP from 3.6% to 4.9%. In the G20 countries, this indicator has grown over five years from 4.1% to 5.3%. The world leader in the share of the digital economy in GDP is Great Britain - 12.4% (table I).

According to a study by analysts at International Data Corporation, published in 2016, the total world cost of digital transformation technology will grow by 16.8% annually and reach 2.1 trillion US dollars by 2019.

According to the forecasts of the consulting company Accenture, the use of digital technologies should add 1.36 trillion US dollars in 2020, or 2.3% of GDP in the total GDP of dozens of the world's leading economies. The GDP of developed countries will grow by 1.8%, and the GDP of developing countries - by 3.4% on the basis of "digital economy".

Table 1. The groth of the digital economy share in the gdp of the g20 countries from 2010 to 2016 (the share in the digital economy in gdp) [9]

Country	**Years**	
	2010	***2016***
the United Kingdom,	8,3%	12,4%
the Republic of Korea	7,3%	8,0%
China	5,5%	6,9%
India	4,1%	5,6%
Japan	4,7%	5,6%
the United States,	4,7%	5,4%

Country	Years	
	2010	*2016*
Mexico	2,5%	4,2%
Germany	3,0%	4,0%
Saudi Arabia	2,2%	3,8%
Australia	3,4%	3,7%
Canada	3,0%	3,6%
Italy	2,1%	3,5%
France	2,9%	3,4%
Argentina	2,0%	3,3%
the Russian Federation	1,9%	2,8%
South Africa	1,9%	2,5%
Brazil	2,2%	2,4%
Turkey	1,6%	2,3%
Indonesia	1,3%	1,5%

In Russia, the share of the digital economy in GDP is 2.8%, or 75 billion US dollars (according to BCG). The bulk - $ 63 billion - falls on the consumption sphere (Internet commerce, services, online search and offline purchases). If in 2010 the share of Internet trading in all sales was 1.7% ($ 12 billion), then in 2016 it grew to 3.2% ($ 43 billion). Export of IT-technologies amounted to 7 billion US dollars.

The intersectoral effect of digitalization has increased 5.5-fold since 2010: from 5 to 27.7 trillion rubles. This effect is obtained from the introduction of electronic trading platforms, the growth of transactions on bank cards, the growth of ROPO segments and online advertising [17].

At the same time, Russia lags behind the countries-leaders of digitalization for 5-8 years. If the current growth rates of the digital economy of Russia remain at the same level, by 2020, due to the high rate of global changes and innovations, this gap will be 15-20 years. In recent years the state of infrastructure in Russia has improved, primarily in terms of penetration of the wired Internet (70.4% of the total population). There are also serious achievements in the availability of broadband and mobile Internet, in the proliferation of smartphones.

Today, the digital economy is named one of the priority areas of the Strategy for Scientific and Technological Development of Russia. Full-fledged sequential digitalization of the Russian economy will become a platform for a qualitative change in its structure and long-term opportunities.

Many specialists are inclined to pin great hopes on the "Digital" economy, believing that their implementation will significantly increase labor productivity, which in turn will lead to economic growth. This statement is true only with large reservations [18].

The emergence and introduction the new technologies (industrial, financial, managerial, social and any other) can lead to a huge number of positive effects and consequences for the economy: increase the labor productivity; increase capitalization; improving the quality of life; the formation of new markets; increasing the efficiency of resource utilization (assets, capital, competencies); increase the competitiveness; increased security [9].

So far, while the process of globalization has not been completed, all of the above effects are due to the fact that the country that first mastered the new technology, was profiting in international markets. In this regard it is necessary to recognize that the development of technologies and the "Digital" economy is an important direction for Russia economic development [19].

The digitalization of the Russian economy can become a source of long-term economic growth and a rise in the wealth of nations. We consider the impact of information and communication technologies on the wealth of nations. As an index characterizing the the wealth of nations, the human development index (HDI) [11] was chosen. As indicators characterizing the development of ICT, the indicators proposed by E.V. Popov, K.A. Semyachkov, V.L. Simonov [12] were chosen. A number of indicators were replaced by those close to them because of the absence of statistical information. And instead of a set of indicators that characterize ICT costs, only one indicator was used. This is due to the fact that only the specific weights of costs that are interconnected are openly available, which can distort the results of the analysis because of their mutual influence on each other [20-22]. Most of these indicators are key in assessing the development of information and communication technologies and are used by international organizations [13]. The analysis is based on the data for 2003-2015 (table II). To test the hypothesis on the available data, the correlation coefficients [14] of the HDI and ICT development indicators were calculated (table III).

Table 2. Main indicators of wealth and ICT for analysis [11, 15, 16, 17]

Year	HDI	Gross added value, bln. rub.	Costs of organizations on ICT, mln. rub.
2010	0,785	1354	515648
2011	0,792	1534	603007
2012	0,799	1780	842674
2013	0,803	1845	1245705
2014	0,805	2149	1174913
2015	0,804	2262	1184184

Table 3. Correlation coefficient between HDI and ICT's indicators

Indicator	Correlation coefficient
Fixed-telephone subscriptions per 100 inhabitants	0,038
The share of organizations using servers in the total number of organizations surveyed	0,759
The share of organizations using local computer networks in the total number of organizations surveyed	0,921
The share of organizations using e-mail in the total number of organizations surveyed	0,961
The share of organizations using global information networks in the total number of organizations surveyed	0,974
The share of organizations that had a website on the Internet, in the total number of organizations surveyed	0,977
Number of personal computers per 100 employees - total, pcs.	0,988
Volume of communication services, million rubles	0,986
Number of subscriber units of mobile radiotelephone (cellular) communication per 1000 population, units	0,964
Number of PCs with Internet access per 100 employees	0,98
Gross added value, bln. rub.	0,927
Costs of organizations on ICT, mln. rub.	0,951

Table 4. Main indicators of wealth and ict for analysis [11, 15, 16, 17]

Year	HDI	Fixed-telephone subscriptions per 100 inhabitants	The share of organizations using servers in the total number of organizations surveyed	The share of organizations using local computer networks in the total number of organizations surveyed	The share of organizations using e-mail in the total number of organizations surveyed	The share of organizations using global information networks in the total number of organizations surveyed	The share of organizations that had a website on the Internet, in the total number of organizations surveyed	Number of personal computers per 100 employees - total, pcs.	Volume of communication services, million rubles	Number of subscriber units of mobile radiotelephone (cellular) communication per 1000 population, units	Number of PCs with Internet access per 100 employees
2003	0,741	24,97	8,3	45,8	48,6	50,1	13,5	18	396 500,6	246,7	4
2004	0,748	26,73	8,4	49,7	53,2	57,3	14,4	20	541 187,9	496,0	5
2005	0,754	27,92	9,3	52,4	56,0	54,3	14,8	23	659 909,5	862,6	7
2006	0,761	30,63	11,3	57,0	63,6	62,5	21,1	26	833 168,0	1 081,1	9
2007	0,769	31,59	13,0	56,4	69,1	68,7	19,8	29	1 035 949,6	1 199,3	11
2008	0,776	31,83	14,5	59,3	74,4	74,7	22,8	32	1 221 499,8	1 397,8	13
2009	0,773	31,71	16	60,5	78,5	79,3	24,1	35	1 274 257,3	1 613,8	15
2010	0,785	31,38	18,2	68,4	81,9	83,4	28,5	36	1 355 549,9	1 663,7	18
2011	0,792	30,82	19,7	71,3	83,1	85,6	33,0	39	1 424 869,3	1 790,3	21
2012	0,799	29,40	18,9	71,7	85,2	87,5	37,8	43	1 530 986,4	1 826,9	24
2013	0,803	28,19	19,7	73,4	86,5	88,7	41,3	44	1 608 804,8	1 933,3	26
2014	0,805	26,58	26,6	67,2	84,2	89,8	40,3	47	1 650 601,9	1 908,4	29
2015	0,804	24,71	47,7	63,5	84,0	89,0	42,6	49	1 672 028,0	1 937,8	31

Table III shows a very strong positive correlation on the Chaddock scale [16] between the HDI and the following indicators: the share of organizations using local computer networks in the total number of organizations surveyed, the share of organizations using e-mail in the total number of organizations surveyed, the share of organizations using global information networks in the total number of organizations surveyed, the share of organizations that had a website on the Internet, in the total number of organizations surveyed, number of personal computers per 100 employees, volume of communication services, number of subscriber units of mobile radiotelephone (cellular) communication per 1000 population, number of PCs with Internet access per 100 employees, gross added value, costs of organizations on ICT. Strong positive relations between HDI and the share of organizations using servers in the total number of organizations surveyed are revealed (correlation coefficient $r = 0.76$). A very weak connection between HDI and fixed-telephone subscriptions per 100 inhabitants (correlation coefficient $r = 0.04$) is stated.

3. Conclusions

As a result of the study, a number of conclusions can be drawn. The development of modern society takes place within the framework of a new era - digitalization. Digitalization transforms the social paradigm of people's lives. It opens up unprecedented opportunities for acquiring new knowledge, expanding an outlook, development of new professions and improvement of professional skill.

The digital economy is no longer a myth, but a reality that is based on the use of ICT. The digital economy breaks the habitual models of the industrial markets. It increases the competitiveness of their participants. Thus, digitalization determines the growth prospects of companies, industries and national economies generally. The emergence of digital players has already changed the face of entire industries - tourism, telecommunications, printing, passenger transportation, in particular taxi services. Number of rental options through the website of Airbnb, founded in 2008, already exceeds the general offer of the rooms of the three largest world hotel chains, which work for many decades. The welfare of the country and society as a whole is based on successful economic development, which, in its turn, at the moment is essentially dependent on the level of development of information and communication technologies (ICT).

The result of the study was the confirmation of hypotheses about the connection between the level of the country's welfare and a number of ICT indicators. Based on the correlation analysis, statistical models were constructed showing the relationship between the HDI as a resultant indicator reflecting the level of the country's welfare and indicators characterizing the development of ICT. It can be concluded that the development of information and communication technologies as a factor of economic growth positively affects the HDI in particular, and the level of wealth of nations as a whole. Fixed-telephone subscriptions per 100 inhabitants are rather weakly related to HDI – it reflects the transition to a digital economy and the increasing virtualization of activities through the use of modern ICTs.

As limitations it should be mentioned that we took only example of Russia and based only on data in open access. It could be one of the prospective – to analyze the impact of each of factors of ICT in different countries and do comparative analysis. Also as perspective directions for further research, it is necessary to note the increase in the sample of time series and the analysis of the HDI influence by the three main elements that make up its composition: life expectancy, literacy level and living standards of the population.

The development and implementation of models for assessing the impact of ICT on the development of the economy of research objects at the level of countries and individual regions creates possibilities to formulate priorities in the choice of factors and measures to improve their competitiveness and raise the level of its welfare.

Acknowledgment

The research carried out with the financial support of the grant from the Program Competitiveness Enhancement of Peter the Great St.Petersburg Polytechnic University, Project 5-100-2020.

References

[1]. O.A. Kalchenko "The role of human capital in aspects of innovations for sustainable socio-economic development." ICTM 2016, 2016: 44.
[2]. Von Wallwitz, Georg. "Mr. Smith und das Paradies.", 2013.
[3]. A.G. Berg, J.D. Ostry "Inequality and Unstable Growth: Two Sides of the Same Coin?", IMF staff discussion note, IMF, 2011, 29 p.
[4]. Thomas Taro Lennerfors, Per Fors, Jolanda van Roijen "ICT and environmental sustainability in a changing society: The view of ecological World Systems Theory", Information Technology & People, Vol.28 Issue 4, 2015, pp. 758-774.
[5]. Ilin, I. V., Iliashenko, O. Y., & Levina, A. I. (2016). Application of service-oriented approach to business process reengineering. Paper presented at the Proceedings of the 28th International Business Information Management Association Conference - Vision 2020: Innovation Management, Development Sustainability, and Competitive Economic Growth, 768-781.
[6]. A.R. Bril, O.V. Kalinina, I.V. Ilin " Economic Analysis of Projects in the Improvement of the HR Management System of Enterprises", Proceedings of the 29th International Business Information Management Association Conference – Sustainable Economic Growth, Education Excellence, and Innovation Management through Vision 2020, 2017, pp. 2268-2277.
[7]. G. Trentin "Technology and Knowledge Flow", 1st Edition, Chandos Publishing, 2011, 196 p.
[8]. In the forefront of the digital economy. [Online]. Available: http://ar2016.rostec.ru/digital/
[9]. Human Development Data (1990-2015). [Online]. Available:http://hdr.undp.org/en/data
[10]. E.V. Popov, K.A. Semyachkov, V.L. Simonova "Assesing the impact of information and communication technologies on innovative activity of regions". Finance and Credit. 46, 2016, pp. 46-60.
[11]. Global Information Technology Report 2013. Growth and Jobs in a Hyperconnected World. [Online]. Available: http://www.ifap.ru/library/book533.pdf
[12]. B. Mirkin, "Core Concepts in Data Analysis: Summarization, Correlation and Visualization" , Springer Science & Business Media, 2011, 390 p.
[13]. Digital economy indicators. [Online]. Available: https://www.hse.ru/primarydata/ice2017
[14]. Global and regional ICT data. [Online]. Available: https://www.itu.int/en/ITU-D/Statistics/Pages/stat/default.aspx
[15]. Russian Federation Federal state statistics servece. [Online]. Available: http://www.gks.ru
[16]. R. E. Chaddock, "Principles and Methods of Statistics "(1st Edition), Houghton Miffin Co., The Riverside Press, Cambridge.
[17]. Didenko, N., Skripnuk, D., Mirolyubova, O., & Radion, M. (2017). Analysis of rural areas development of the region using the ADL-model. Paper presented at the Research for Rural Development, , 2 142-147. doi:10.22616/rrd.23.2017.061
[18]. Rodionov, D. G., Kudryavtseva, T. J., & Skhvediani, A. E. (2018). Human development and income inequality as factors of regional economic growth. European Research Studies Journal, 21(Special Issue 2), 323-337.
[19]. Shabunina, T. V., Shchelkina, S. P., & Rodionov, D. G. (2017). An innovative approach to the transformation of eco-economic space of a region based on the green economy principles. Academy of Strategic Management Journal, 16(Specialissue1), 176-185.
[20]. Kałkowska, J., & Kozlov, A. V. (2016). Decision making process for the knowledge-based enterprise: Fuzzy sets theory application to strategic management doi:10.1007/978-3-319-28564-1_14
[21]. Silkina, G. Y. (2017). Information and communication technologies in ensuring of innovative development. Paper presented at the Proceedings of the 29th International Business Information Management Association Conference - Education Excellence and Innovation Management through Vision 2020: From Regional Development Sustainability to Global Economic Growth, 1165-1176.
[22]. Victorova, N. G., Yablokov, D. Y., Yevstigneev, E. N., & Valebnikova, N. V. (2019). The analysis of the impact of infotelecommunication factors on the russian tax administration. Paper presented at the Proceedings of the 32nd International Business Information Management Association Conference, IBIMA 2018 - Vision 2020: Sustainable Economic Development and Application of Innovation Management from Regional Expansion to Global Growth, 3160-3169.

Digital Entrepreneurship: Doing Business for Smart and Sustainable Bio-Based Economy

V Nikolova-Alexieva[1] and M Angelova[2]

[1] Faculty of Economics, Department of Industrial Business and Entrepreneurship, University of Food Technologies, 4000 Plovdiv, Bulgaria
[2] Faculty of Economics and Social Sciences, Department of Management and Quantitative Methods in Economics, University of Plovdiv Paisii Hilendarski, 4000 Plovdiv, 24 Tzar Asen str, Bulgaria

mina.angelova@uni-plovdiv.bg

Abstract. Bio-based products can have a lower environmental and climate footprint than products based on non-renewable raw materials. There is a high potential for the production and use of renewable raw materials for industrial applications, notably the use of lignocellulose from agricultural and forestry residues, industrial crops, by- products/co-products and waste. New low-carbon, resource-efficient and sustainable value chains can be set up using this agricultural and forestry biomass to produce bio-chemicals, biomaterials and bioenergy. These chains will contribute to the diversification of the rural economy and the reinforcement of the industrial base. It will create rural growth and employment, and will help to meet energy and climate policy targets for 2030. This paper aims to study the role of digital entrepreneurship as an essential factor to increase the creativity and innovativeness for sustainable growth of Bulgarian bio-based economy. The results show that socially responsible behavior of the digital enterprises can provide added value based on a favorable reputation that they create.

1. Introduction

The theme significance takes a central position in discussions, researches and organisational activities connected with digital entrepreneurship. This has an essential role in the Bulgarian economic environment as well as the national strategies for economic growth. All statements lead to the necessity of investigating the opportunities for transformation of bio-based organizations in different way of doing business in digital era. The digital economy in the European Union is growing much faster than the rest of the economy, which will lead to the creation of more jobs and the continuation of the digital transformation of the European economy. Bio-based economy as a part of circular economy covers all sectors and systems that use biological resources. It is one of the largest and most important sectors of the EU and includes agriculture and forestry, fisheries, agro-food, biomass and bio-based products. Its annual turnover is about 2 trillion euros, and it employs about 18 million people. Bio-economy is also a key area for stimulating growth in rural and coastal areas.

1.1. Literature Review

Nowadays, the largest global firms are no longer those operating in the industrial sectors but those belonging to the digital economy. The concepts entrepreneur spirit and entrepreneur ecosystem in Bulgaria are mentioned a lot but barely in the last few years can be observed the standing out of pronounces examples and good practices. The Republic of Bulgaria is situated in Southeastern Europe, located in the east part of the Balkan Peninsula. Its uniqueness is hidden in the balanced combination

IOP Conf. Series: Materials Science and Engineering **618** (2019) 012081 doi:10.1088/1757-899X/618/1/012081

of incredible nature, black sea strip of wonderful beaches, beautiful mountains for summer and winter tourism, rivers and so on. Besides all these natural resources, Bulgaria is a country with poorly developed economy and is far behind the economics of other countries in the European Union [1]. In terms of Global Competitiveness Index, Bulgaria also lags behind the majority of the EU member states [2] but data of The Global Competitiveness Report 2018 presents a positive tendency and the country takes 41st place of 140 countries. Bulgaria still poorly utilizes the advantages of the global process for raising the national competitive power, realizing an economic structural reorganization and modernization of the technical manufacture [3].

During the period of planned economy, large state-owned industrial enterprises using mass production methods and relatively inflexible production processes and producing for geographically restricted markets, dominated the economies in Central and Eastern Europe [4]. Transition to a market economy involves profound economic changes and sometimes – but not necessarily – political change as well [5] in all Eastern European countries. The transition economies have lower rates of entrepreneurship than are observed in most developed and developing market economies [6].

Bulgaria was from those nations that had little or no experience of a market economy because communist planning and industrialisation were contemporaneous. In line with Estin and Mickiewicz [6] as a result, laws and market supporting institutions had to be developed from scratch. Bulgaria started the transition process in 1989 and was one of the first transition countries to adopt a new constitution. The preparation of the accession of Bulgaria to the European Union exercised a positive influence on the environment for enterprise development [2]. In 2007, after fulfilling economic and political criteria, Bulgaria joined the European Union. The transition created many opportunities for entrepreneurship in transition countries and entrepreneurship became an important factor for the transition from centrally-planned to market economy [4].

In summary, during the last decade, the Bulgarian economy has achieved macroeconomic stability and growth. Various measures were implemented in order to improve the environment for doing business especially for SMEs [2]. Entrepreneurial Orientation (EO) may be seen as an important organizational resource for international involvement. The governmental restrictions of economic freedom appear to impact entrepreneurial activity differently depending on the particular freedom restricted by government and the entrepreneur's motive for engaging in entrepreneurial action [7]. Entrepreneurs exploit new opportunities and are associated with disturbing the market equilibrium. They often revolutionize industries overturning long-established technologies, business models and dominant companies [8]. In order to do so, they innovate and take risks. Entrepreneur is a person who acts in hazardous circumstances, or in other words, a person who buys products with a known price, to sell them with unknown price in the future [9]. Drucker's [10] concept of management by objectives and self-control states the need to manage business by balancing a variety of needs and goals, rather than subordinating an institution to a single value.

1.2. The Concept 'Digital Entrepreneurship'

In the last ten years, empirical evidence suggests the rise of a new category of Entrepreneurship: **Digital Entrepreneurship**, as a relevant socio-economic and technological phenomenon, which can be considered as the joining of traditional entrepreneurship with an emphasis on leveraging new digital technologies in novel ways, such as social, mobile, analytics, cloud and cyber-solutions, all in order to shift the traditional way of creating and doing business in the digital era [11].

Digital entrepreneurship can be defined as embracing "new ventures and the transformation of existing business by creating and using novel digital technologies. Digital enterprises are characterised by a high intensity of utilisation of new digital technologies (particularly social, mobile, analytics and cloud solutions) to improve business operations, invent new (digital) business models, sharpen business intelligence, and engage with customers and stakeholders through new (digital) channels" [12]. Researchers argue that a nation's digital entrepreneurial capacity depends largely on digital entrepreneurial behaviour, culture, and strategies as well as a supportive innovation ecosystem in which governments, industry, business, educational institutions and NGOs (non-government organizations) work together [13].

The concept 'Digital Entrepreneurship' has been used by some researchers and policy makers but its definition remains quite elusive. The research is in its infancy and the scholarship of this field is poorly developed. Several review articles on entrepreneurship identify other gaps in understanding the use of digital technologies by entrepreneurs [14]. A matter of interest represents the elaborated conceptual model to study digital entrepreneurship by Zhao and Collier [13]. They selected as the theoretical foundations three theories, given the social and networked nature of digital entrepreneurship: social network theory; social capital theory; and institutional theory. By using these theories, their model helps explore how social networks at individual, institutional and societal levels, and social capital, online and offline, affect digital opportunity identification and exploration as well as entrepreneurial outcomes.

Other researchers propose conceptual model of the implementation of digital entrepreneurship in the family business [15]. Basly and Hammouda's model describes the main aspects that might be affected by digital entrepreneurship in family businesses, explains the aims of the necessary digital transformation and provides recommendations for that a conducive context and necessary conditions could be implemented in order to succeed in this organizational change process.

Table 1. *A Model of Entrepreneurial Culture*

Dimension	*Definition*	*Key Characteristics/Attributes*	*Opposite Characteristics/Attributes*
Organizational Enthusiasm	A pattern of values, assumptions, and practices demonstrating an excitement and passion for the organization, its goals, and purposes.	✦ enthusiasm for accomplishing organizational goals and the organization's mission ✦ understanding the organizational vision ✦ passion for the work ✦ sense of unity of purpose	• failing to know or understand what success for the organization is or how it can/will be achieved • failing to believe or support the organization's direction or purpose
Stakeholder Alignment	A pattern of values, assumptions, and practices demonstrating the importance of building and maintaining mutually beneficial relationships with key stakeholders, such as customers, suppliers, partners, and investors	✦ a focus on building and maintaining relationships with key stakeholders ✦ appreciating how stakeholders impact the business ✦ emphasizing strong relationships based on loyalty and fairness	• viewing stakeholders as a means to an end • viewing stakeholder relationships as necessarily zero-sum • viewing relationships as only valuable if benefits outweigh costs in the short term
Learning & Development Support	A pattern of values, assumptions, and practices demonstrating individual and group self-improvement, learning, and professional development.	✦ optimism toward improvement ✦ valuing efforts to change, learn, and improve one's self and others ✦ encouraging employee professional development	• organizational members lack ambition, initiative, or an interest in improvement • resistance to acquiring new knowledge or skills
Opportunity Driven Change	A pattern of values, assumptions, and practices demonstrating a willingness to change in order to identify/develop opportunities and execute on them.	✦ pursuing innovation and change to better execute on existing opportunities ✦ pursuing innovation and change as a means for creating new opportunities	• reluctance to try new things or experiment • hesitancy in the face of uncertainty • disinterest in hearing new ideas or approaches to existing problems
Cohesiveness	A pattern of values, assumptions, and practices demonstrating organizational members' bond and commitment to each other and the organization, irrespective of circumstance.	✦ organizational members take responsibility for their actions ✦ members feel like the organization can do things on its own (internally) ✦ trust that everyone gives best effort ✦ members feel like they can rely on others to participate	• organizational members shirk responsibility or hide in the group • members lack confidence in the abilities of one another • members are secretive and uncooperative

Source: Own interpretation

1.3. A Model of Entrepreneurial Culture Applied in Digital Entrepreneurship

An example of digital entrepreneurs are all those pro-active people who create, maintain and take on the financial risks of an internet business. They are the new breed of Internet entrepreneurs who have started businesses such as Google, Yahoo, MySpace, Facebook, YouTube, Second Life, LinkedIn, eBay and others. The measurement challenge of digital entrepreneurship lays in the pervasive nature of the phenomenon itself that cannot be captured by count-based measures of individual-level entrepreneurial action. Therefore it becomes important to monitor the conditions which set the business context of entrepreneurs in the different EU countries [16]. The European Index of Digital Entrepreneurship Systems (EIDES) addresses the measurement challenge by appraising the

IOP Conf. Series: Materials Science and Engineering **618** (2019) 012081 doi:10.1088/1757-899X/618/1/012081

framework and systemic conditions for **stand-up, start-up, and scale-up** activities in the EU28 countries. Furthermore, the EIDES index also attempts to disentangle the digital component of the just-mentioned entrepreneurial conditions and stages of development.

Numbers of authors explore corporate entrepreneurs who had deployed the technology in the international speed in the digital entrepreneurial era successfully and identified the entrepreneurship opportunities in the same arena [17]. The development of digital technologies leads to changes in existing business models and the creation of new ones, the introduction of new products and services, and the increase in the efficiency of business processes and therefore making enterprises more competitive. M. Turuk claims that 'enterprises that fail to adapt adequately to the digital environment will inevitably face a decline in competitiveness, and some of them may face the very survival on the market' [17]. His research examines the use of digital technologies in business and argues the importance of using them for the growth and development of an enterprise, and to emphasize the importance of digitization for local development and the strengthening of the national economy as a whole. Therefore, it's significant the role of applied entrepreneurial culture at a national level that is a base for digital entrepreneurship development.

A preliminary research of the authors of this article connected with entrepreneurial culture addresses the manner in which the facets of **Entrepreneurial Culture** combine to give the construct meaning. To reiterate, an additive relationship is appropriate when the effect of each facet on the focal construct is independent of the effects of the other facets. This proposed five dimension framework aims to comprehensively describe Entrepreneurial Culture as a cultural subtype. It aims to accurately describe entrepreneurial culture as an Organizational Culture centrally concerned with opportunities. We argue that these facets are necessary and jointly sufficient for the meaning of the construct. This implies that each facet must have some non-zero level of all attributes to possess an Entrepreneurial Culture. The proposed model can be applied successfully in elaboration a digital entrepreneurship. However, the training of human resources in the era of high technology digital enterprises is a leading purpose, which is a factor in ensuring the sustainable development and growth of the Bulgarian economy [18].

2. The Concept of "Bio-based economy"

Bio-economy covers all sectors and systems that use biological resources. It is one of the largest and most important sectors of the EU and includes agriculture and forestry, fisheries, agro-food, biomass and bio-based products. Its annual turnover is about 2 trillion euros, and it employs about 18 million people. Bio-economy is also a key area for stimulating growth in rural and coastal areas. The new bio-economy strategy fits in with the Commission's efforts to further boost jobs, growth and investment. It aims to improve and expand the sustainable use of renewable sources to overcome global challenges such as climate change and sustainable development.

Biological resources and ecosystems in the world are limited, so efforts and an innovative approach are needed to feed the population and provide clean water and energy. Bio-economy can turn algae into fuel, recycle plastics, produce furniture or garbage from waste, and create organic fertilizers from industrial waste products. It has the potential to create 1 million new green jobs by 2030.

There is a need for a change in the system, the way we produce, consume and dispose of goods. By developing its bio-economy - the renewable segment of the circular economy – EU countries can find new ways of providing food, goods and energy without depleting the limited biological resources of the planet. Moreover, not only environmental protection and climate change make them reconsider their economic paradigm and modernize their production patterns: there is also a significant potential for new green jobs, especially in rural and coastal areas. The EU has to pave the way for the transformation of waste, residues and unnecessary items into high value products, environmentally friendly chemicals, feed and textiles. Research and innovation play a key role in accelerating the building of a green European economy and achieving the goals of the United Nations for sustainable development. This strengthening of the bio-economy can make a significant contribution to achieving a wide range of EU objectives, including mitigation of climate change, circular economy and resource efficiency, environmental protection, job creation, growth and revenue. Achieving a sustainable circular bio-economy requires concerted work by the public and private sectors. Major sectors of the bio-economy are *agriculture and forestry, fish and aquaculture, bio-energy and biofuels, food industry, bio-based products and processes.*

3. Digital Marketing and its Implementation over the Design for Digital Entrepreneurship

Globalization and Information Technology (IT) are the cause of changing business models [18]. The unpredictability and dynamics of today's business requires connecting all participants in business processes and an immediate response to the market. Business needs to be done in real time. Many agricultural and bio-economic companies, based on customer loyalty, improve their quality of e-services [9] [18]. The need to introduce digital marketing has become more and more tangible over the years and, alongside traditional marketing tools, Internet and digital technologies are increasingly being used to reach marketing goals. Digital marketing uses a wide range of information technologies to achieve greater value through effective segmentation in the market, geared towards differentiation and positioning, effective planning and distribution, product and service pricing and product pricing, and sales that will satisfy consumers as individuals and achieve the organization's goals. Digital marketing is also closely matched with e-commerce. For companies in the bioeconomy sector and the production of biobased products, the internet can be a valuable tool on the market to complement commercial activities [19]. The agricultural and food sector is not fully competitive in the use of new information technologies. For this reason, information technologies, especially the Internet, have the potential to promote the economic characteristics of the agricultural sector. Digital marketing in agribusiness brings many benefits, such as reduced agency costs (wholesale and retail), reduced time consumption, collection, selection and processing of information, resulting in improved supply chain management, and to expanding market participation and / or expanding new markets.

3.1. Results and Discussion

The **main goal** of this research is to identify the Internet and the digital marketing involvement in the bio-based economy sectors in Bulgaria and explore the attitudes of the employees in these sectors. The data on the use of digital marketing and digital entrepreneurship in Bulgarian companies from the bio-economy sectors were collected through an online survey conducted in December 2018. 123 companies representing key sectors of the Bulgarian bio-economy participated in the survey. The questionnaire includes general business information, respondent characteristics, web sales and web site between business entities and Internet business views. Single-band (frequency and distribution) and binary (chi-square) data analysis methods were used to analyze the collected data. Data is analyzed using the SPSS statistical package.

The total number of survey respondents included a total of 70 limited liability companies, 26 family agricultural companies, 13 joint stock companies, 10 craft enterprises and 4 cooperative societies. Their share of the total number of respondents is shown on figure 1. Of the surveyed, 92.5% are private, and only 7.5% are public companies (See fig.2).

The majority of enterprises surveyed are manufacturing (77.5%). Commercial enterprises are 26.1% and those in the service sector 20.4%. The majority of respondents are micro-enterprises with up to 10 employees (51.4%), respondents with up to 50 employees are 22.6%, 19.7% are medium-sized enterprises with 51 to 250 employees and only 6.3% companies are large with more than 251 employees.

The largest number is the number of respondents, producers of organic food - 34, 27 are the companies from the bioenergy sector, 17 are the agrarian farms, followed by companies, producers of heat and electricity - 14, 9 are the companies from the fish industry and aquaculture and at least the representatives of the forestry industry - 6. The distribution of respondents by sectors of the bio economy is shown in figure 3.

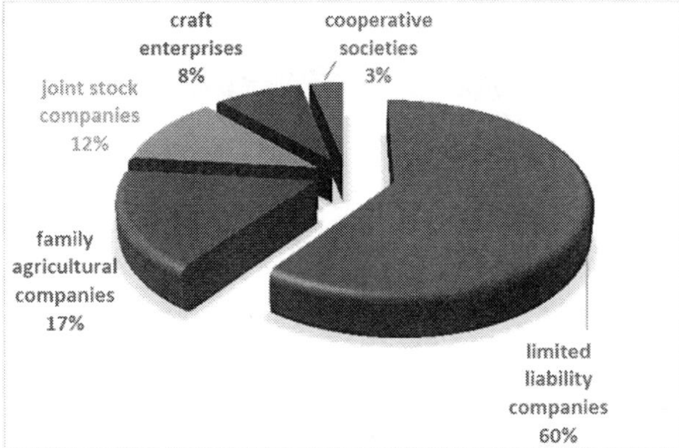

Figure 1. Legal Status of Responding Companies

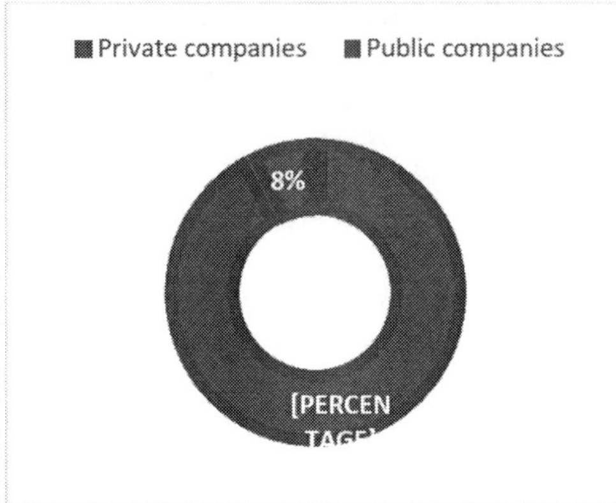

Figure 2. Distribution by type of ownership

The majority of respondents have an annual turnover of more than 50 000 000 BGN (47%), followed by those with an annual turnover between 10 000 000 BGN and 50 000 000 BGN (25%). Businesses with an annual turnover between 850,000 BGN and 10,000,000 BGN make up 15% of the respondents, while those with an annual turnover between 50,000 BGN and 850,000 BGN account for 11% of the respondents. The share of companies with annual turnover between 5 000 BGN and 10 000 BGN (2%) is the lowest. More men were involved in the study, compared to 58% and 42%, respectively. The respondents aged between 35 and 50 are dominated, and the lowest is the number of participants over the age of 50. Almost 74 % of respondents have secondary or high education and 12% of respondents with a doctorate degree. The respondents with less than 10 years of service experience in the current business unit dominate (78.8%) while only 3.0% of the respondents in the current business entity work for more than 20 years. Even 68% of the surveyed businesses have their own website, which is a smaller proportion compared to the results of the research conducted in Greece, where 78% of the surveyed businesses have a website and in Spain, where even 93% of them have a website. Among those who do not have a website, most of them plan to make their own website in near future, while 21% of the respondents do not intend to make a website.

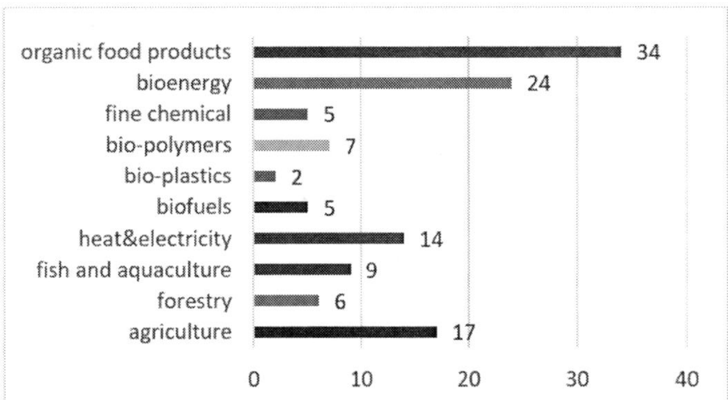

Figure 3. Distribution by sector of bio economy

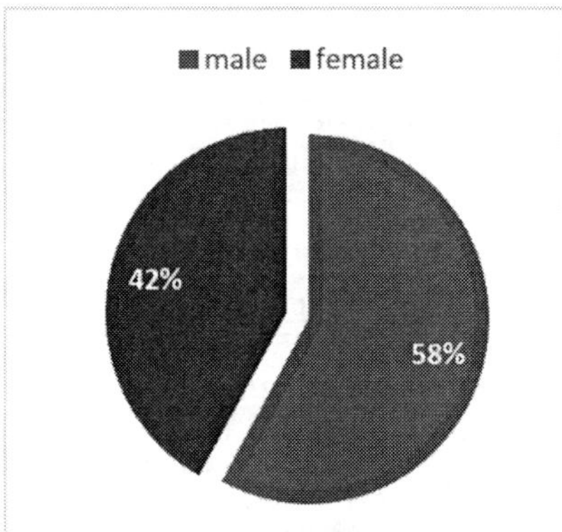

Figure 4. Distribution by sex

The conclusions that can be made are that companies in different sectors of the bio economy with higher annual turnover are more likely to get a website than those with lower turnover. There is a correlation between company incomes and their presence in the Internet space. However, there is no link to the website with other business characteristics (number of employees, business level, company seat, business sector), ie ($p \geq 0{,}05$). Out of all 83 respondents who have a web page, more than half (60%) believe that the website's creation moderately impacts its recognition on clients and their potential partners, 22% said it had a significant impact on recognizability of business and 8% consider that the impact of the webpage is insignificant. Only 10% of the respondents believe that the impact of the website is very important to create visibility of their business among customers and potential partners. The webpages of bio-based businesses with the broadest coverage are general information about their business (62%) and the catalog of their products (53%), while online trading and electronic payments are neglected. (See table 2).

Table 2. *Profile of Companies in the bio economy sectors on the indicators for digital marketing*

Web site content	**N**	**%**
Business profile – general information	76	62
Links to other useful information	7	6
Product catalog	65	53
Methods of electronic payment	6	4,9
Online retail	15	12
Information about physical stores or locations where products can be purchased	56	46
Internal employee communication	5	4,1
Price list	12	9,8
Online wholesale	4	3,2
Sections with special facilities for specific customers	10	8
Professional advice for customers	27	22
Link to professional associations	9	7,3

Source: Own data

About 85% of respondents believe their websites cannot fully meet the needs of advertising. Only 14.3% of respondents sell their products online, while the remaining 85.7% do not trust e-sales. The survey results show that the annual turnover of the bio-based companies is related to the online sale of products ($p \leq 0.05$). Companies with low annual earnings often offer an opportunity for e-sales compared to businesses with high annual earnings. The number of employees is also related to electronic sales ($p \leq 0.05$). Companies with fewer employees often take electronic sales. An important reason for the untapping of e-sales is the unwillingness of customers to use this method due to the risk of billing or theft. The high price for such a service is considered to be less important. Some of the other reasons why e-sales are not used are: people's ignorance, a small number of products, the inadequacy of products for such a form of sales, lack of time and staff.

Table 3. *Attitude to digital business*

Item	**N**	**min**	**max**	**Mean**	**Standard deviation**
Organic food and bio-based products can't be sold via Internet	121	1	5	2,83	1,134
Digital business is a necessary precondition for successful business in the future	123	2	5	4,13	0,723
Information on more complex bio-based products is difficult to present on the Internet	121	1	5	2,56	0,999,
Digital business will significantly reduce the importance of intermediaries in our bioeconomy sector in the next three years	123	2	5	3,24	0,923
Internet will not have a very high impact on the business of my company in the future	122	1	5	2,98	1,134
Competitors are pushing for the need to use digital business	123	1	5	3,87	0, 798

Source: Own data

Respondents believe that e-commerce is essential for business development in the future (average of 4.13 on a scale of 5 degrees, where 1 means that it disagrees and 5 means completely agree). They consider that competition requires the use of electronic sales (average 3.87). Respondents believe that information on sophisticated products can hardly be represented on the Internet (average 2.56) and that they can not sell organic foods and bio-based products online. Also, the surveyed enterprises believe that the Internet has a major impact on the business.

Conclusion
As a global medium the Internet has a major impact on today's business world. Today we receive all the necessary product information via the Internet. From the digital marketing in the marketing strategic objectives the biggest part belongs to the creation of a website, which is the first step to attract new customers and keep the old ones. The results showed that the companies from main bioeconomy sectors are aware of the advantages and disadvantages of the digital marketing, and the use of e-marketing is to be a target of this kind of trade. The basic limit, in the opinion of the respondents, is the ignorance of the business partners where occurs the need for education of the subjects of agriculture and related industries as well as within the consumers about the opportunities and advantages of the e-commerce.

This report is funded from the National Science Program "Healthy Foods for a Strong Bioeconomy and Quality of Life" of the Ministry of Education and Science, approved by decision of the Council of Ministers №577 / 17.08.2018 , contract No 68 / NSP under Work Package 4.2 "Regional ecosystems for bio-economy"

References
[1] Davidkov T and Yordanova D 2015 Enhancing SME internationalization in a transition economy: the role of internal factors, *Int. J. of Latest Trends in Finance & Economic Sciences*, Vol. 5, No. 3, pp.945–956
[2] Angelova M and Nikolova-Alexieva V 2018 Business climate and preconditions for reviving the Bulgarian industry, *Int. Scientific J. 'Machines, Technologies, Materials'*, Vol. 1, No. 2, pp.69–72, Print ISSN: 1313-0226, Web ISSN: 1314-507X
[3] McMillan J and Woodruff C 2002 The central role of entrepreneurs in transition economies, *J. of Economic Perspectives*, Vol. 16, No. 3, pp.153–170
[4] Dana L-P and Ramadani V 2015 *Family businesses in transition economies. Management, succession and internationalization*, Springer Int. Publ. Switzerland. p.363, ISBN 978-3-319-14209-8 (eBook)
[5] Estrin S and Mickiewicz T 2010 Entrepreneurship in Transition Economies: the role of Institutions and generational change, *IZA Discussion Paper No. 4805*
[6] McMullen J, Bagby D and Palich L 2008 Economic freedom and the motivation to engage in entrepreneurial action, *Entrepreneurship Theory and Practice*, Vol. 32, No. 5, pp.875–895
[7] Širec K and Močnik D 2017 The competitiveness of established entrepreneurs in Balkan countries, *World Review of Entrepreneurship, Management and Sust. Development*, Vol. 13, Nos. 2/3, pp.141–158, Inderscience Enterprises
[8] Hisrich R, Peters P and Shepherd D A 2017 *Entrepreneurship*, 10th ed., Irwin McGraw-Hill, New York, ISBN: 978-0-07-811284-3[online] http://www.wuperbooks.com/uploads/ December 20, 2017]
[9] Drucker P F 2001 *The Practice of Management*, p.471, Klasika i stil, Sofia (in Bulgarian)
[10] https://www.journals.elsevier.com/technological-forecasting-and-social-change/call-for-papers/digital-entrepreneurship-creating-and-doing-business-in-the-digital-era, 07.03.2019
[11] European Commission "Fuelling Digital Entrepreneurship in Europe"
[12] Zhao F and Collier A 2016 Digital Entrepreneurship: Research and Practice. 9th Annual Conf. of the EuroMed Academy of *Business Innovation, Entrepreneurship and Digital Ecosystems*, pp. 2173 - 2182, ISBN: 978-9963-711-43-7
[13] Kiss A N, Danis W M and Cavusgil S T 2012 International entrepreneurship research in emerging economies: A critical review and research agenda, *Journal of Business Venturing* (27:2), 266-290
[14] Basly S and Hammouda A 2018 Are family and digital entrepreneurship incompatible? A model of the implementation of digital entrepreneurship in the family business, Conf.: *Digital Innovation and Finance Conference*, Lyon
[15] Autio E, Szerb L, Komlósiand É and Tiszberger M 2018 *The European Index of Digital Entrepreneurship Systems*, DOI: 10.2760/39256, Publisher: Publications Office of the European Union, ISBN: 978-92-79-91303-7
[16] Venkatesh K A and Pushkala N 2018 Digital entrepreneurship: the technology deployment in internationalization speed in the digital entrepreneurship era, *Int. J. on Recent Trends in Business and Tourism*, vol 2(4), pp. 39-42
[17] Turuk M 2018 The importance of digital entrepreneurship in economic development, 7th Int. Scientific Symp.: *Economy of Eastern Croatia - Vision and Growth*, Osijek, Croatia
[18] Mihova T, Anguelov K, Ferdov A 2018 Specificity of the training of the employees in high-technological enterprises, IX Nat. Conf. with Int. Participation (*ELECTRONICA*), Sofia, IEEE Catalog Number: CFP18P58-POD, ISBN:978-1-5386-5802-4
[19] Anguelov K, Ferdov A and Mihova T 2018 Research on interactions among universities and high-technological enterprises at qualification of employees, IX Nat. Conf. with Int. Participation (*ELECTRONICA*), Sofia, IEEE Catalog Number: CFP18P58-POD, ISBN:978-1-5386-5802-4

Technological innovations and total factor productivity in dairy industry

Asen Konarev[1, 2], Snezhinka Konstantinova[2] and Gergana V – Georgieva[2]

[1] Technologies and Innovations Corporation AD, Plovdiv, Bulgaria
[2] University of Food Technologies, Plovdiv, Bulgaria

E-mail: akonarev@abv.bg, sks_ko@abv.bg, geriveselinova@yahoo.com

Abstract. This study analyses the main factors influencing the development of companies in dairy industry. Particular attention is paid to technological innovations by exploring two groups of companies: applying membrane-cell technology and without applying it. The study has successfully used tools of production functions to determine the total factor productivity in microeconomics. It is determined the impact of technological innovation on overall factor productivity and corporate growth.

1. Introduction

The dairy industry is an important subsector of the food and drink industry with great social and economic impact. In recent years on progress of subsector has an adverse effect a set of macroeconomic factors: over-supply of milk in world market, reduction in Chinese imports, embargoes from the Russian Federation and others. These factors formed the tendency of significant reduction of redeemed price of milk in Bulgarian market. Undoubtedly this causes the compression of the raw material base of the sub-sector and reduces production. For the past 20 years, the number of dairies has fallen more than 2 times (from around 450 in 2000 to around 200 in 2017).

Under these conditions, managers are oriented more on technological process as a decisive factor for efficient use of scarce raw materials, improving quality and expanding product range dairies to avoid massive competitive pressure.

This will allow not only to increase exports through a specific price and specific niche positioning but also to increase sales in the country (Bulgaria is a net importer of milk and dairy products worth nearly USD 100 million per year).

2. Methods and methodology

Corporate growth is the growth of the core business unit of sub-sector – dairy sector. It is known that corporate growth can be measured by various parameters Konarev and Konstantinova [1]. In this case, the amount of the company's net sales revenue will be used as a growth indicator. The main model for research of corporate growth of dairies is as follows:

IOP Conf. Series: Materials Science and Engineering **618** (2019) 012082 doi:10.1088/1757-899X/618/1/012082

$$Cg = F\ (C, L, T_{fp})$$

, where
C_g is corporate growth;
C – capital;
L – labor;
T_{fp} – total factor productivity.

Using this model, corporate growth is assumed to be the result of three main factors:
a) capital – additional investment costs;
b) labor – an increase in staff costs;
c) total factor productivity.

Total factor productivity is a complex factor – it includes all other factors beyond labor and capital.

It is believed that the technological process – implementation of new technologies, new machinery, new combination of resources, new forms of organization, new methods of marketing, etc. is the most influential factor in this group of factors.

However, included in the total factor productivity are factors stimulating corporate growth, as well as factors restraining it such as reduction in the amount of raw material, decrease in market prices, loss of markets, increased competitive pressures, embargoes, oversupply on the world market and others.

In order to examine the dynamics of corporate growth in dairy industry, a random sample of 10 enterprises, divided into two groups:

Group I – companies applying membrane-cell technologies in production process;

Group II – companies that do not apply these technologies.

The analysis period is 2007 – 2016.

The choice of 2007 as the beginning of a period was driven by our desire to analyze corporate growth after the global economic crisis of 2008. The following model was used to determine total factor productivity Konstantinova and Konarev [2]:

$$\frac{A_t^i - A_{t-n}^i}{A_{t-n}^i} = \frac{Y_t^i - Y_{t-n}^i}{Y_{t-n}^i} - Cd_i \frac{K_t^i - K_{t-n}^i}{K_{t-n}^i} - La_i \frac{L_t^i - L_{t-n}^i}{L_{t-n}^i} \tag{1}$$

, where
A_t^i – total factor productivity of the i-th dairy company in year *t* (year 2016);
Y_t^i – net sales of i-th company in t-th year;
Cd_i – capital intensity of i-th company's production for the relevant period (2007 – 2016);
L_t^i – total amount of labor costs of i-th company for the year *t*;
La_i – labor intensity of i-th company's production for the same period;.
N – number of years of the analyzed period.

This model represents a modification of R. Solow's model [3] for economic growth analysis at national level:

$$Yt = At\,.\,F\ (Kt, Lt), \tag{2}$$

, where
Yt – the aggregate production of the country;
Kt – amount of physical capital used in production;
Lt – the amount of labor inputs;
At – total factor productivity.

This model was also used by other authors in Bulgaria and abroad Ganev, Todorov *et al.* [4, 5, 6] for national analysis. The motivation of the authors to modify the model (2) in a model (1) is aspiration to use this toolkit for successful research at the level of the individual business unit – the dairy company.

3. Results and discussion

Primary data from companies enables Table 1 to be compiled.

Table 1. Increase of the main indicators of the dairy companies for the period 2007 – 2016 (in thousands of BGN)

Companies	Revenue (ΔY)	Capital (ΔK)	Labor (ΔL)
Group I: using membrane technology			
1. OMK AD	32 202	26 629	6 712
2. Dimitar Madzhyarov EOOD	14 008	8 308	726
3. Polidey-2 OOD	12 412	4 547	527
4. Rodopeya Belev EOOD	5 331	3 761	561
5. Kostovi EMK OOD	887	632	136
Total for the group	68 840	46 883	8 665
Group II: without application of membrane technology			
1. Shipka 99 AD	10 241	9 627	530
2. Elite 95 EOOD	10 042	24 767	596
3. Bour Chvor EOOD	13 512	8 687	1 824
4. Milk Group EOOD	-	-	-
5. Rojen – Milk OOD	70	623	77
Total for the group	34 498	43 702	3 027

The analysis shows that only one of the ten companies analyzed did not achieve corporate growth – Milk Group EOOD. It is necessary to point out that labor is measured through wage and social security costs, the capital – through the balance sheet of the total amount of equity and debt capital, and revenue – through net sales revenue.

After defining the basic parameters of the model (1), including the capital and labor productivity of each of the companies surveyed for analyzed period is compiled Table 2:

Table 2. Contribution of capital and labor to corporate growth

Companies	Revenue (ΔY)	Capital (ΔK)	Labor (ΔL)
Group I: using membrane technology			
1. OMK AD	0.9517	0.7480	0.2179
2. Dimitar Madzhyarov	0.7011	0.3678	0.0456

EOOD			
3. Polidey-2 OOD	2.3419	0.6599	0.0958
4. Rodopeya Belev EOOD	1.9527	1.2821	0.2991
5. Kostovi EMK OOD	0.9989	0.8210	0.2148
Total for the group	1.0285	0.6584	0.1587
Group II: without application of membrane technology			
1. Shipka 99 AD	1.5163	1.3857	0.0798
2. Elite 95 EOOD	1.2666	7.0428	0.0914
3. Bour Chvor EOOD	2.8392	2.1124	0.4644
4. Milk Group EOOD	-	-	-
5. Rojen – Milk OOD	2. 3289	1.5617	0.1927
Total for the group	1.7474	3.0002	0.2338

The data from Table 2 make it possible to determine the contribution of total factor productivity to corporate growth, using the model of authors presented in modification (1):

(a) For a first group of milk-processing companies applying membrane-cell technology in their production activities:

$\Delta A/A = 1.0285 - 0.6584 - 0.1587 = 0.2114$;

b) For a second group of companies that do not apply membrane-cell technology in their business:

$\Delta A/A = 1.7474 - 3.0002 - 0.2338 = - 1.4866$.

The results for the first group of companies indicate that analyzed nearly 10-year period (2007 – 2016) corporate growth to the whole group is 102.85%. It was formed in the following way:

- "capital" factor – 65.84%;
- labor factor – 15.87%;
- total factor productivity – 21.14%.

The second group of companies achieved higher corporate growth – 174.74%. However, this growth is extremely capital-intensive, which is confirmed by the empirical parameters of the model for the influence of the individual factors, as follows:

- "capital" factor – 300.02%;
- labor factor – 23.38%;
- total factor productivity – 148.66%.

As mentioned total factor productivity is a complex factor, which include, as factors that stimulate growth and factors that retain it. In this case, large investments in the second group of dairies have supported corporate growth.

First and second group of companies worked under the same economic conditions during the analyzed period. In principle, it can be assumed; that on them are equally influences both stimulating and restraining factors that form the overall factor productivity.

The only difference that exists between them is the application of membrane cell technology in production. This gives the authors reason to conclude that the application of membrane-cell technology has had a positive impact on the corporate growth of dairy companies.

Technological progress through the introduction of these technologies in the first group (companies) has not only blocked a number of growth factors but has increased this growth by

21.14%. The opposite is the result for a second group of companies, that to overcome the restraining factors and implement growth were forced to use additional significant capital that is out of reach for most dairies.

4. Conclusions

The presented study gives the following main conclusions:

1. Membrane cell technologies stimulate corporate growth in the dairy industry.
2. Business model that managers should implement to boost the development of milk processing companies has accelerated deployment of process, including membrane-cell technology.
3. Factoring analysis of corporate growth over the survey period (2007 – 2016) shows that capital expenditures are the main factor of growth. The impact of labor, measured through employers' wage and social security costs, is relatively low. These results are quite disturbing, considering limited capital and labor resources, and high expectations of technological progress.
4. The three-factor model proposed by the authors can be successfully used in the theory and practice of analyzing, planning and forecasting the development of dairy companies.

References

[1] Konarev A and Konstantinova S 2017 *Analysis of the economic activity of industrial companies.* (Plovdiv: Publishing house KSI) p 227
[2] Konstantinova S and Konarev A 2017 *Corporate growth and total factor productivity in industrial companies* vol 15 (1) (Stara Zagora: Trakia Journal of Sciences) pp 191–194
[3] Solow R M 1957 *Technical Change and the Aggregate Production Function* vol 39 (3) (Review of Economics and Statistics), pp 312–320
[4] Ganev K 2005 *Measuring Total Factor Productivity: Growth Accounting for Bulgaria* (Discussion papers of the BNB, DP/48/2005) pp 8–9
[5] Todorov I 2015 *The Economic Growth of Bulgaria in the Period of the Currency Board.* (Sofia: Economic Alternatives) issue 3
[6] *The Conference Board* 2015 (Productivity Brief) - www.conference-board.org
[7] U S Department of Labor *Preliminary Multifactor Productivity Trends* 2014
- www.bls.gov/news.release/prod3.nr0.htm

IOP Conf. Series: Materials Science and Engineering **618** (2019) 012083 doi:10.1088/1757-899X/618/1/012083

Development of QFD methodology

Y Klochkov[1], A Gazizulina[1], M Ostapenko[2]
[1]Peter the Great St. Petersburg Polytechnic University, St.Petersburg, Russia

[2] Tyumen industrial University, Tyumen, Russia

y.kloch@gmail.com, albinagazizulina@gmail.com

Abstract. The production of elongated shaped charges is particularly dangerous (or in the language of ISO 9001 "special") and difficult from the point of view of quality control, so it is necessary to pay special attention to the dynamics of customer requirements, the level of development of competitors in order to remain a competitive organization. The proposed work addresses the issues of determining the ways of product development based on the analysis of the dynamics of consumer requirements and the level of competition, a mechanism is shown to identify bottlenecks of products based on the results of the organization's resource audits.

1. Introduction

A promising method for managing the quality of production of elongated shaped charges (ESC) is the method of structuring the quality function (QFD) over the life cycle stages of production [7].

In this paper, we propose approaches to improving the QFD methodology based on development (ESC), which will solve the following tasks:

I. Constantly monitor the level of demand,
II. Determine the relationship between customer requirements and characteristics (ESC),
III. Analyze the quality level of competitors' products,
IV. Determine the direction of development (ESC), based on an analysis of customer requirements and/or the position of the organization on the market,
V. Identify bottlenecks in the production process (ESC) associated with the level of quality of existing resources in the enterprise,
VI. Form requirements for products of outsourcer and / or supplier.

Fundamentals of QFD (in some sources referred to as "the deployment of quality functions"), the reader can find in almost all textbooks on quality management issues [2, 16, 17-21]. In this paper, we propose an improved model of its application that takes into account modern trends in quality management systems (one of which is the involvement of high-class specialists by outsourcing part of the production process to outsourcers) [1]. The tasks of tracking the dynamics of demand and determining the relationship between customer requirements and product characteristics have already been resolved (tasks 1 and 2) [3, 4]. Naturally, some of the characteristics of the manufacturer's products (ESC) are purchased, then it becomes necessary to build a Quality House of the supplier (outsourcer) [22, 23]. When applying QFD analysis technology from the constructed Quality House, the relationship between customer requirements and product characteristics is known [5]. From here you can calculate the degree of influence of outsourcer products on customer satisfaction:

$$B_{out} = \frac{R \cdot \sum I_{\text{аут}}}{R \cdot \sum I} = \frac{\sum I_{\text{аут}}}{\sum I}, \quad (1)$$

Where B_{out} – the degree of influence of outsourcer products on the fulfillment of customer requirements;

R – the absolute weight of the requirement;

$\sum I$ – the relationship between demand and product characteristics (▲ - 1; ○ - 3; ● - 9);

$\sum I_{\text{out}}$ – the relationship between the user's demand and the characteristics of the outsourcer's products.

The sum of the degrees of influence of the producer and outsourcer on the satisfaction of the consumer is equal to unity, then:

$$P_{out} = 1 - B_{out}, \quad (2)$$

where P_{out}– the degree of influence of the producer's products on the fulfillment of the customer's requirements;

B_{out} – the degree of influence of outsourcer products on the fulfillment of customer requirements.

Assume that, as a result of interaction with the consumer, the level of fulfillment of the requirement [6], which is equal to Q, was determined, then the loss level is equal to:

$$Y_{loss} = 1 - Q. \quad (3)$$

Then it can be argued that the losses (expressed in fractions) through the fault of the outsourcer were:

$$P_{out} = B_{out} \cdot Y_{loss}, \quad (4)$$

where B_{out} - the influence degree of outsourcer products on the fulfillment of customer requirements;

y_{losses} - level of loss or degree of customer satisfaction.

2. Development of the QFD-analysis method

This assumption is acceptable if the links between requirements and characteristics are constantly updated, which will make it possible to assert that the losses in the level of satisfaction are distributed by the consumer in such a way that the dissatisfaction with the corresponding characteristic increases the degree of its importance.

Accordingly, the satisfaction of the consumer with the output of the outsourcer will be:

$$YD_{out} = 1 - P_{out}. \quad (5)$$

Knowing the relationship between the products of the outsourcer and the manufacturer's products (by analyzing the "roof" of the Quality House), it is possible to determine the degree of the organization's dependence on the activity of the outsourcer and the weight of the outsourcer's products in meeting the customer's requirements [3, 15].

The development of the QFD-analysis method is possible by clarifying the absolute and relative weights of the product characteristics. When analyzing the consumer evaluation of products produced by the enterprise and its competitors, it is possible to carry out an accurate calculation of the absolute value of the product characteristic, knowing the degree of realization of the consumer's demand (column K table 1) and the weight of its importance (column B table 1).

The calculation of the absolute value of the importance of the product characteristics taking into account the consumer's opinion will be carried out as follows:

$$B_{\text{char.coun.}} = \sum(I_x \cdot R_x) + \sum[\frac{P_{laggx}}{T_x}(I_x \cdot R_x)], \quad (6)$$

where I_x – the relationship between demand and product characteristics (▲ - 1; ○ - 3; ● - 9);

R_x -the absolute weight of the requirement;

P_{lagg_x} – lagging the consumer's assessment of the extent to which the organization's requirement is being met from the target value;

T_x – the target value of the degree of realization of the customer's demand.

Calculation of the absolute value of the importance of the product characteristics taking into account the situation on the market will be performed as follows:

$$B_{\text{char.coun.}} = \sum(I_x \cdot R_x) + \sum[\frac{C_{laggx}}{P_x}(I_x \cdot R_x)], \quad (7)$$

where I_x – the relationship between demand and product characteristics (▲ - 1; ○ - 3; ● - 9);

R_x – the absolute weight of claim;

C_{lagg_x} – lagging of the consumer's assessment of the degree of realization of the requirement of the organization in question from the maximum valuation in the market;

P_x – the maximum value of the consumer's assessment of the degree to which the requirement is met in the market.

Table 1. Development of the QFD-analysis method.

	Customer Requirements	Absolute weight requirement	Product Features					
			1	…	*n*	*1*	…	*m*
A	*B*	*C*	*D*	*E*	*F*	*G*	*H*	*I*
1	Req1	X	▲				●	
2	Req2	Y	○		▲			○
3	Req3	Z		○		○		
4	Req4	P	●			▲		
…	…	…	…	…	…	…	…	…

	Customer Requirements	Absolute weight requirement	Evaluation of the implementation requirements Competitors			
			I	II	III	…
A	*B*	*C*	*J*	*K*	*L*	*M*
1	Req1	X	I_{req1}	II_{req1}	III_{req1}	…
2	Req2	Y	I_{req2}	II_{req2}	III_{req2}	…
3	Req3	Z	I_{req3}	II_{req3}	III_{req3}	…
4	Req4	P	I_{req4}	II_{req4}	III_{req4}	…
…	…	…	…	…	…	…

	Customer Requirements	Absolute weight requirement	target	backlog	maximum	losses
A	*B*	*C*	*N*	*O*	*P*	*Q*
1	Req1	X	The maximum possible estimate	N-J	max(K,L,M)	P-J
2	Req2	Y		N-J	max(K,L,M)	P-J
3	Req3	Z		N-J	max(K,L,M)	P-J
4	Req4	P		N-J	max(K,L,M)	P-J
…	…	…	…	…	…	…

Based on the results of the construction of the Quality House, the level of delay in the degree of realization of the consumer's demand from the maximum score becomes known (column P Table 1) [5]. Suppose that this lag is related to the work of a particular resource, which participates in the production of relevant characteristics. Then, by determining (for example, based on the expert method or regression analysis) the degree of participation ("the share of resources used" in table 2) of each resource in the production of the relevant characteristic, we calculate the level of claims and / or the quality level of the resource based on the consumer's opinion ("consumer votes").

Table 2. The relationship of characteristics with resources

	Product Features					
	1	...	n	1	...	m
resources	D	E	F	G	H	I
	Share of resources used					
Res.1	0,5		0,2	0,8	0,3	0,4
Res.2		1		0,2	0,2	0,6
Res.3	0,5		0,8		0,5	
Σ	1	1	1	1	1	1

We will carry out calculations for resource Res.3 (Table 2). This resource is involved in the production of three characteristics (columns D, F, H), then the resource affects the fulfillment of the customer's requirements, respectively, Req1, Req2, Req4 - through the characteristic on the column "D"; Req 2 - on the column "F"; Req 1 - on the column "H". We will assume that the estimates of the implementation of requirements were distributed as follows:

Req1 - 10 points out of 10 possible;
Req2 - 8 points out of 10 possible;
Req4 - 9 points out of 10 possible.

Let us determine the degree of participation of the resource in the realization of the requirement through characteristics (the matrix of communication requirements and characteristics). Reg2 is related to the characteristics in the columns "D", "F", "H", respectively, to "3", "1" and "3" points, Res.3 resource is involved only in the formation of "D" and "F" ", Therefore, then the degree of influence of the characteristics on the demand will be respectively 3/7; 1/7; and 3/7. The degree of participation of the resource in the production of these characteristics according to Table 2 0.5; 0.8 and 0, respectively. Then the level of quality of the resource "Res.3" in the implementation of the requirement of Req2 is:

$$1-(0{,}5\cdot\frac{3}{7}+0{,}8\cdot\frac{1}{7})\cdot\left(\frac{10-8}{10}\right)=0{,}934$$

or in alphabetical order:

$$\text{Yqr} = 1-\Sigma(\text{degres}\bullet\frac{I_x}{\sum I_x})\bullet\left(\frac{P_{\iota aggx}}{T_x}\right), \tag{8}$$

where degrees is the degree of resource participation ("share of resources used" in Table 2).

Having calculated all the values of Yqr, it is possible to determine the average value and root-mean-square deviation of the quality level of the resource (and these calculations were obtained by analyzing the "consumer's voice"). Using the same approach, it is possible, knowing the levels of resource quality (established, for example, based on the results of audits, self-assessment, analysis of the level of defectiveness, etc.), to determine delays in the degree of realization of customers' requirements.

3. Result

As shown, the development of the method requires additional analyzes and calculations, and, consequently, the decision to improve production processes and / or design. Therefore, it is necessary to assess the adequacy of the constructed Quality House. Such an assessment is proposed as follows:

When the columns "J", "K", "L" are received, you can calculate the probability of product sales and compare this value with the sales figures;

Calculating the mean and standard deviation of Y_{qr}, you can compare these data with the actual level of quality of resources (established, for example, based on the results of audits, self-assessment, analysis of the level of defectiveness, etc.). Coincidence of the results of calculations will indicate the adequacy of the application of quality management methods.

But it should be remembered that the consumer's opinion differs from the manufacturer's opinion.

Acknowledgments
The paper is based on the financial support of the Council on Grants of the President of the Russian Federation (Project No. MK-3229.2018.8 "Improving the Standardization Management Method"). Peter the Great St. Petersburg Polytechnic University and the grant from the Program Competitiveness Enhancement of Peter the Great St.Petersburg Polytechnic University, Project 5-100-2020.

References
[1] A. Ashok Kumar, V. K. Tewari, J. Maiti, and B. Nare, "Design and selection of agricultural machinery using a quality function deployment technique", *Agricultural Engineering International: CIGR Journal*, vol. 17, no. 4, 2015, pp. 91-99.
[2] C. Chen, and C. Huang, "The synergy of QFD and TRIZ for solving EMC problems in electrical products - A case study for the notebook PC", *Journal of Industrial and Production Engineering*, vol. 32, no. 5, 2015, pp. 311-330. 10.1080/21681015.2015.1046507
[3] Chen, W. -., & Chiang, Y. -. (2015). A study of establishing the requirement assessment and improvement strategy to the product design based on customer-oriented. Journal of Quality, 22(6), 559-580. 10.6220/joq.2015.22(6).04
[4] H. Cheng, H. Liu, and T. Liao, "An integrated product design and selection approach using grey QFD and grey TOPSIS methods", *International Journal of Information and Management Sciences*, vol. 26, no. 1, 2015, pp. 29-50. 10.6186/IJIMS.2015.26.1.3
[5] J. Chun, J.Cho, "QFD model based on a suitability assessment for the reduction of design changes in unsatisfactory quality", *Journal of Asian Architecture and Building Engineering*, vol. 14, no. 1, 2015, 113-120. 10.3130/jaabe.14.113
[6] F. Franceschini, and D. Maisano, "Prioritization of QFD customer requirements based on the law of comparative judgments", *Quality Engineering*, vol. 27, no. 4, pp. 437-449. 10.1080/08982112.2015.1036292
[7] J. Jin, P. Ji, Y. Liu, and S. C. Johnson Lim, "Translating online customer opinions into engineering characteristics in QFD: A probabilistic language analysis approach", *Engineering Applications of Artificial Intelligence*, no. 41, 2015, pp. 115-127. 10.1016/j.engappai.2015.02.006
[8] C. K. M. Lee, C. T. Y. Ru, C. L.Yeung, K. L. Choy, and W. H. Ip, "Analyze the healthcare service requirement using fuzzy QFD",. *Computers in Industry*, vol. 74, 2015, pp. 1-15. 10.1016/j.compind.2015.08.005
[9] M.-.C. Li, C.-.B.Yeh, A. Abbas, and H. Kuo, "The application of FMEA and QFD to the improvement of packaging processes for transistors", *Journal of Quality*, vol. 22, no. 5, 2015, pp. 441-460. 10.6220/joq.2015.22(5).05
[10] X. G. Luo, C. K. Kwong, J. F. Tang, and F. Q. Sun, "QFD-based product planning with consumer choice analysis", *IEEE Transactions on Systems, Man, and Cybernetics: Systems*, vol. 45, no. 3, pp. 454-461. 10.1109/TSMC.2014.2347916
[11] C. Saricam, F. Kalaoglu, and A. Aksoy, "Internal benchmarking methodology and its implementation on apparel retail industry", *Tekstil Ve Konfeksiyon*, vol. 25, no. 4, 2015, pp. 352-358.
[12] E. Vezzetti, M. Alemanni, and J. Macheda, "Supporting product development in the textile industry through the use of a product lifecycle management approach: A preliminary set of guidelines", *International Journal of Advanced Manufacturing Technology*, vol. 79, no. 9-12, 2015, pp. 1493-1504. 10.1007/s00170-015-6926-4
[13] F. Wang, H. Li, A. Liu, and X. Zhang, "Hybrid customer requirements rating method for customer-oriented product design using QFD", *Journal of Systems Engineering and Electronics*, vol. 26, no. 3, 2015, pp. 533-543. 10.1109/JSEE.2015.00061
[14] X. Zhang, S. Tong, H. Eres, K. Wang, and M. Kossmann, "Towards avoiding the hidden traps

in QFD during requirements establishment", *Journal of Systems Science and Systems Engineering*, vol. 24, no. 3, 2015, pp. 316-336. 10.1007/s11518-015-5275-z

[15] V.G. Kartashevskii, N.V. Kireeva, M.A. Buranova, and L.R. Chupakhina, "Approximation of distributions in the problems of the analysis of self-similar traffic", *Third International Scientific-Practical Conference Problems of Infocommunications Science and Technology PIC S and T 2016 - Conference Proceedings*, 2016, pp. 105 – 108.

[16] V.G. Kartashevskii, N.V. Kireeva, M.A. Buranova, and L.R. Chupakhina, "Simulation and analysis of generalized queue system with arbitrary distributions of system parameters", *Infokommunikacionnye tehnologii*, no. 3, 2015, pp. 252-258

[17] V. Kartashevskiy, N. Kireeva, M. Buranova, and L. Chupakhina, "Study of queuing system G/G/1 with an arbitrary distribution of time parameter system", *2015 2nd International Scientific-Practical Conference Problems of Infocommunications Science and Technology, PIC S and T 2015 - Conference Proceedings*, 2015, pp. 145-148. doi:10.1109/INFOCOMMST.2015.7357297

[18] Zharov, V. S., & Kozlov, A. V. Management of technological development of enterprises on the basis of a life cycle model. Paper presented at the *Proceedings of the 2018 International Conference "Quality Management, Transport and Information Security, Information Technologies", IT and QM and IS 2018, 2018,* 181-184. doi:10.1109/ITMQIS.2018.8525109

[19] Kudryavtsev, D., Grigoriev, L., & Koryshev, I. (2014). Applying quality function deployment method for business architecture alignment. Paper presented at the Proceedings of the 8th European Conference on Information Management and Evaluation, ECIME 2014, 118-127.

[20] Lukichev, S., & Romanovich, M. The quality management system as a key factor for sustainable development of the construction companies. Paper presented at the *Procedia Engineering,2016, 165*, 1717-1721. doi:10.1016/j.proeng.2016.11.914

[21] Alexandr, V. B., Gileva, A. T., Margarita, P. G., & Gorshenina, E. M. (2018). Enterprise's innovative infrastructure development model based on quality function deployment method. Paper presented at the *Proceedings of the 2018 International Conference "Quality Management, Transport and Information Security, Information Technologies", IT and QM and IS 2018,* 27-34. doi:10.1109/ITMQIS.2018.8524924

[22] Aytasova, A. S., Erofeeva, K. Y., & Solopova, N. A. (2019). Improvement of the "research and development " and "development R&D" process maps for enterprises of the nuclear industry. Paper presented at the Proceedings of the 2019 IEEE Conference of Russian Young Researchers in Electrical and Electronic Engineering, ElConRus 2019, 1361-1364. doi:10.1109/EIConRus.2019.8657201 Retrieved from www.scopus.com

[23] Kudryavtsev, D., Grigoriev, L., & Bobrikov, S. (2014). Strategy-focused and value-oriented capabilities: Methodology for linking capabilities with goals and measures. Paper presented at the CEUR Workshop Proceedings, , 1182 Retrieved from www.scopus.com

IOP Conf. Series: Materials Science and Engineering **618** (2019) 012084 doi:10.1088/1757-899X/618/1/012084

Data Management for Solving the Problem of Sequential Shaping Simulation

V Mikheev[1], A Grechnikova[1], D Agafonova[1], Y Klochkov[2]
[1] Department of Metal Forming, Samara University, Samara, Russia
[2] Peter the Great St. Petersburg Polytechnic University, St. Petersburg, Russia

klochkov_yus@spbstu.ru

Abstract. The publication presents the results of a study of the process of investing in the innovation of the Russian manufacturing industry (sample 2009-2017). Factors of economic efficiency of investment projects of the industry are revealed. It is proved that the efficiency of investment is determined by external factors, primarily by the level of economic concentration (CR3) of the market of innovative products. The coordination of the scientific result with the industry consolidation dynamics model (A.T. Kearney) is presented.

1. Introduction

Sheathing of aircrafts is the most large-sized and thin sheeted parts. The sheath itself is subject to increased demands on the quality of the enclosing surface associated with the shape of the shell of double curvature. The most characteristic property of this type of parts is their low stiffness [1,2]. Thin sheet shell under the influence of its own weight changes its shape and can not be a carrier of dimensions for linking the units and parts of the aircraft. At the same time, their bearing capacity and strength are determined by the thickness of the sheet preform. Significant overall dimensions of the skins, as well as their relatively small number on the aircraft, determined special requirements for data management in the construction of the model of sequential shaping and to automated means of their production [3].

The hood refers to the forming operations associated with an increase in the area of the sheet blank and subsequent localization of the tension in the dangerous section of the part, as a result the sheet stock quickly becomes thinner and collapses, which is the cause of high rejection and large technological waste [4, 6]. The reason for the localization of the stretching of the sheet blank is, first of all, the unevenness of the deformation due to the complex shape of the shell and the influence of frictional forces.

Many manufacturers of aircraft traditionally collect aircraft with a large amount of refinement of parts. This introduces new requirements to the tightening processes and leads to the need to develop a new method for geometric modeling of surfaces of various shapes, as well as a new method for forming an equal thickness shell of double curvature.

2. Principles for the formation of a shell surface model of more complex geometric shapes

In reality, the geometric model of elliptic shells of double curvature describes the relationships between the aerodynamic properties of the skin contours in the aircraft's reference frame and the characteristics of the geometric shape of the shell. The data of this model in the form of a description of the guiding and forming surfaces are used for graphical representation of the design object, as well as for the development of technological equipment and programs for their processing on machines with numerical control (CNC) [5,7].

A method is proposed for constructing a mathematical model of the surface of enclosing shells, which can be designed as a document containing directions of the guiding and generating surfaces in a certain coordinate system [9]. In addition, to determine the geometric parameters of the shell-forming shells, the direction of the skin is compatible with one of the orthogonal planes of symmetry F1, which determines the longitudinal shaping contour of the section of the shell surface passing through the point O. The second plane of symmetry F2 is orthogonal to the first determines the position of the central transverse contour of the section of the shell surface, also passing through the point 0. The point O in this case is considered as an elliptic point and represents the pole of the surface The shell of a double curvature of a biconvex form [10,12].

To determine the geometric parameters of a double-curvature shell, it is necessary to build it into a 3D model on a computer. The direction of the cover is compatible with one of the orthogonal planes of symmetry F1, which defines the longitudinal shaping contour of the section of the shell surface passing through the point O [13]. The second plane of symmetry F2 orthogonal to the first determines the position of the central cross section of the shell surface also passing through the point O. Figure 1 shows the surface of a biconcave double curvature shell having a ratio of the length (L) to the width (B) of the selected sheet blank is less than one and the double angle Pulling the grip of the punch work piece in a direction stitched equal to 1800.

As a result, conditions are created for the geometric linking of the conjugate surfaces of the envelope-forming shells and the conical punches according to the symmetry principle, determined by the symmetry planes and. Then, under the mathematical model of the surface of shells, we mean a set of algorithms and numerical data that are necessary and sufficient for unambiguous determination of differential-geometric parameters of characteristic sections and points of the analysed surface [14-16].

The accuracy of the formation of the contours is achieved through the use of a computer, which in the calculation process selects an analytic contour curve and compares it with the original data set. Automation of calculations and transformation of geometric parameters into digital information allow applying the processing of tightening punches on CNC machines [17].

However, the achieved accuracy of processing the tightening punch does not solve the problem entirely, it is necessary to provide a regulated tolerance for thickness variation when the sheet blank is tightened. All this introduces new requirements to the technology of production of encapsulating shells of aircraft and leads to the need to develop new technical and technological solutions that ensure the production of shell-forming shells with minimal variability [4, 5].

The initial stage in modelling the sequential forming scheme of the cover is the preparation of geometric models of the elements participating in the process, the sheet blank, the punch, and the part to be obtained - double curvature shells. The double curvature shell has a rather complex geometric shape, which is difficult to describe with the standard ANSYS / LS-DYNA surface modelling tools [18-20].

For this purpose, it is advisable to use a specialized CAD system, such as Compass-3D. The model of the part is the prototype of the punch model, taking into account the dimensions of the straight sections of the latter. Also, given that one of the features of LS-DYNA is the need to break down into finite elements and absolutely rigid bodies, the punch was presented in the form of a surface (shell) directly in contact with the work piece. When constructing the model of the work piece, the dimensions of the clamping areas - sections of the sheet billet, which are in the process of shaping in the clamps of the tension press, are also taken into account. An important part of the sequential forming scheme of the tightening is the movement of the clamps of the tensioning press along the necessary trajectory [21]. All motion trajectories are approximated by a sufficient number of points, which we will call steps.

For the convenience of modelling the process throughout its length, a step-by-step numbering of steps is used, which breaks the entire process into 84 steps.

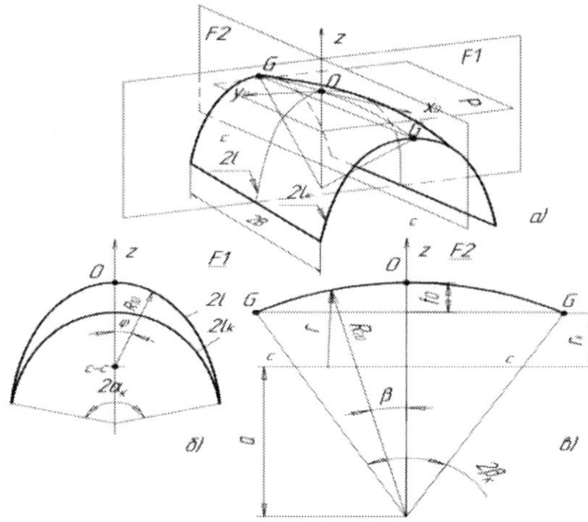

Figure 1. The shell of a biconvex form of significant double curvature (an element of the cladding of the outer surface of a motor nacelle engine)

The compass-3D geometry created earlier in the surface modelling module of the Compass-3D system is imported into the Ansys environment using the IGES format. It is also necessary to take into account that the dimension of the imported model is millimetres, while the dimension used in the Ansys environment is meters. To reconcile the dimension, the geometry must be scaled based on the ratio 1: 1000, representing it, thus, in meters. The finite element mesh is constructed using the constructed surfaces. When the physical model was divided into finite elements, an ordered partition into quadrangular elements was used [22]. The longitudinal and transverse lines of the curvatures of the surfaces of the shell, punch, and also the work piece itself were broken into 100 segments each, which corresponds to a breakdown into 10,000 elements.

3. Numerical experiments

In the course of the work, a large number of numerical experiments were carried out, as a result of which an optimal kinematic model of the sequential formation pattern of the skin was obtained and a series of final numerical experiments were performed in which various external and internal parameters varied. External parameters include the coefficients of friction of the various facing of the punch and the coefficients of the anisotropy of the properties of the sheet material.

For the initial analysis of the results of numerical experiments, the values of the effective deformation and shell thickness were considered along the sequence of the sequential circuit at the characteristic points and in the corresponding elements (Figure 2):

- element A with the serial number 4991 is a characteristic point on the edge of the part;
- element B with serial number 4951 is a characteristic point at the pole of the shell;
- element C with the serial number 9551 is a characteristic point on the descent of the shell.

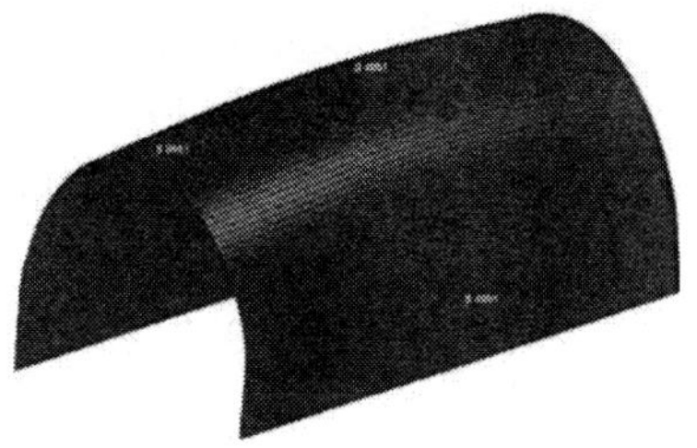

Figure 2. Arrangement of elements in characteristic points

Internal parameters include the kinematic angular values associated with the angle of coverage of the tightening punch and the angle of extension of the shell between the cladding operations created by the movement of clamps of the blank. The choice of angular parameter values is realized due to understanding of several reasons:

• The localization of deforming forces in the vicinity of the pole of the tight punch at the first tightening operation due to the redistribution of the load from the flat sections of the punch is taken into account;

• The features of the formation of the central part of the shell at the initial stage of the first hanging operation are taken into account, where the localization of the deforming forces in the vicinity of the pole of the tight punch is preserved.

• the features of the double curvature shell are taken into account, due to the fact that when the surface of the surface is bent it remains unchanged, as does its Gaussian curvature, although the principal curvatures vary inversely.

Figure 3 shows a "symmetric" palette of distributions of the values of deformations and thickness along the average surface of the shells. This indicates, first of all, the possibility of controlling the process of shaping the shell and its thinning. With regard to the quantitative evaluation of the obtained values of deformations and thickness, for this purpose there are characteristic elements A, B, C

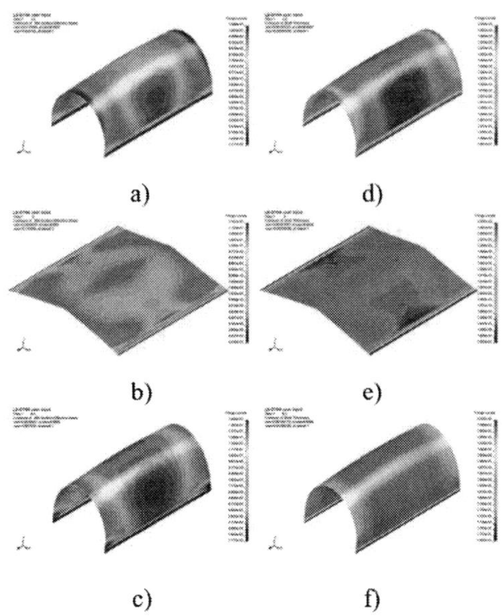

Figure 3. Distributions of the data of the numerical experiment in the following stages: after the first skinning by an angle of 900 (a, d), after the part is bent through the angle 120 and its

extension (b, e), at the end of the molding operation (c, f); a, b, c - distribution of effective deformations along the middle shell surface; d, e, f is the shell thickness distribution

Full-scale quantitative information is carried by the graphs of the change in effective deformations along the average surface of the shell and the thickness of the shell at its characteristic points, depending on the time vector of the combined operations according to the sequential scheme (Figure 4).

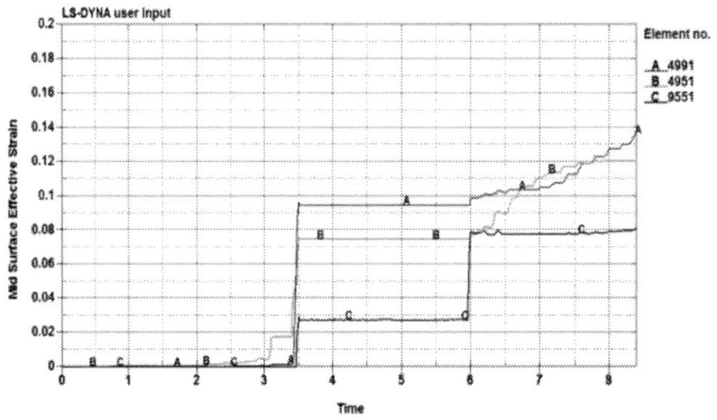

a)

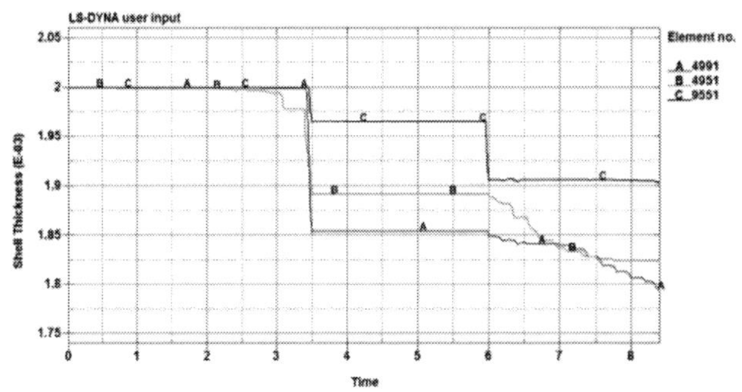

b)

Figure 4. Graphs of the change in effective deformations along the average shell surface (a) and shell thickness (b) at its characteristic points

In quantitative terms, this numerical experiment is the best. This is due to the fact that the values of the shell thicknesses at the end of the circuit take the following values: A - 1.903 mm; B - 1,823 mm; C - 1.795 mm. Correspondingly, the transverse thickness difference (between points A and B) is 0.080 mm, and the longitudinal difference in thickness (between B and C) is 0.028 mm.

4. Model of the analysis of quality indicators

To analyse all values of the thickness of the shell, the Minitab software complex was designed to process statistical data. Minitab is often used in combination with the Six Sigma technique. To build the diagrams in the Minitab software, we used the procedure for analyzing the reproducibility for normal playback (Capability Analysis Normal). In the construction, the tolerance for the thickness of the work piece mm was set. On the abscissa axis, the thickness values of the shell are plotted, and the frequency of their recurrence along the ordinate axis. When constructing a histogram, the program automatically determined the number and width of columns with a base equal to the width of the interval along the thickness. The height of the column corresponds to the frequency of the data on the thickness in this interval. Then on the histogram the upper and lower limit values of the norm were determined, and the program automatically has constructed a curve of distributions of data on frequency. As a result, you can easily understand the type of histogram distribution.

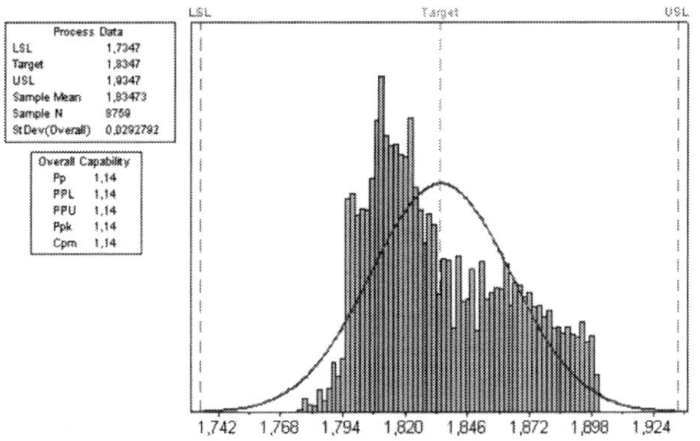

Figure 5. Six Sigma diagram for the given numerical experiment

The analysis of the Six Sigma diagrams includes the calculations of the following statistical parameters: the average thickness value over the shell surface, the standard deviation of the thickness and the efficiency factor. In addition, for each numerical experiment, the lower and upper tolerance limits were defined, the so-called control norms and. With the known numerical value of the coefficient of validity, an analysis was made of the reproducibility of the shaping process by the skin. For this, 5 intervals were considered. In particular, six numerical experiments were detected in the range, of which only two were performed without corrugation, including the one given, which is the best.

5. Conclusion

Information technology without analysing the quality of the information provided will not be able to solve the problem of reducing the cost of production of aircraft. The authors, having experience in constructing a model of the surface of shells of complex geometric shapes, should note that errors occur not only related to the calculation of curvature and loads, but also to the data management model. Therefore, in this paper, the principles of constructing planes of complex geometric shapes are considered in detail.

Acknowledgment

The research carried out with the financial support of the grant from the Program Competitiveness Enhancement of Peter the Great St. Petersburg Polytechnic University, Project 5-100-2020.

References

[1] V. A. Glouschenkov, F. V. Grechnikov, and B. S. Malyshev, "Pulse-magnetic processing technology when making parts and units of aerospace engineering", Journal De Physique.IV : JP, 7(3), 1997, pp. C3-45-C3-48.

[2] F. V. Grechnikov, E. A. Noskova, O. G. Savel'eva, "A study of the anisotropy of properties of sheet semiproducts from alloy AMg10", *Metal Science and Heat Treatment,* vol. 51, no. 7-8, 2009, pp. 326-329. doi:10.1007/s11041-009-9163-3

[3] A. A. Vasilyev, N. G.Kolbasnikov, A. I. Rudskoy, D. F Sokolov, and S. F. Sokolov, "Kinetics of structure formation in the heating of cold-rolled automotive steel sheet", *Steel in Translation,* vol. 47, no. 12, 2017, pp. 830-838. doi:10.3103/S0967091217120130

[4] J. A. Baimova, R. T. Murzaev, and A. I. Rudskoy, "Discrete breathers in graphane in thermal equilibrium", *Physics Letters, Section A: General, Atomic and Solid State Physics,* vol. *381*, no. 36, 2017, pp. 3049-3053. doi:10.1016/j.physleta.2017.07.027

[5] R. G. Burkovsky, I. Bronwald, D. Andronikova, B. Wehinger, M. Krisch, J. Jacobs, D. Gambetti, K. Roleder, A. Majchrowski, A.V. Filimonov, A.I. Rudskoy, S.B. Vakhrushev, and A. K. Tagantsev, "Critical scattering and incommensurate phase transition in

antiferroelectric PbZrO3 under pressure", *Scientific Reports,* vol. 7, 2017. DOI:10.1038/srep41512

[6] G. Kodzhaspirov, and A. Rudskoy, "The effect of thermomechanical processing temperature-strain-time parameters on the mesostructure formation", *Materials Science Forum,* vol. 879, pp. 2407-2412. DOI: 10.4028/www.scientific.net/MSF.879.2407

[7] C. Apostolopoulos, A. Drakakaki, A. Apostolopoulos, T. Matikas, A. I. Rudskoi, and G. Kodzhaspirov, "Characteristic defects-corrosion damage and mechanical behavior of dual phase rebar", *Materials Physics and Mechanics*,vol. 30, no. 1, 2017, pp. 1-19.

[8] V. A.Glushchenkov, F. V. Grechnikov, D. G. Chernikov, and A. A. Lazareva "Formation of electrocontact plugs of anodes of metallurgical production", *Russian Journal of Non-Ferrous Metals, vol. 53, no.* 1, 2012, pp. 20-25. doi:10.3103/S1067821212010099

[9] R. A. R. Izamshah, J. P. T Mo,., and S. "Ding, Finite element analysis of machining thin-wall parts", *Key Engineering Materials,* vol.458, 2011, pp. 283-288. DOI: 10.4028/www.scientific.net/KEM.458.283

[10] G. A. Politova, N. Y.Pankratov, P. Y. Vanina, A. V. Filimonov, A. I. Rudskoy, G. S. Burkhanov, A. S. Ilyushin, I. S. Tereshina, (2017). "Magnetocaloric effect and magnetostrictive deformation in tb-dy-gd-co-al with laves phase structure", *Journal of Magnetism and Magnetic Materials,* 2017. DOI:10.1016/j.jmmm.2017.11.016

[11] F. V. Grechnikov, and A. I. Khaimovich, Development of the requirements template for the information support system in the context of developing new materials involving big data. Paper presented at the *CEUR Workshop Proceedings,* vol. 1490, 2015, pp. 364-375. doi:10.18287/1613-0073-2015-1490-364-375

[12] N. V. Andreeva, A. A. Naberezhnov, M. V. Tomkovich, B. Nacke, V. Kichigin, A. I. Rudskoy, and A. V. Filimonov, "Surface morphology and structure of double-phase magnetic alkali borosilicate glasses", *Metal Science and Heat Treatment,* vol. *58*, no.7-8, 2016, pp. 479-482. DOI:10.1007/s11041-016-0039-z

[13] R. Leishman, J. Ferrin, T. McLain, J. MacDonald, S. Quebe, and R. Beard, R. Utilizing an improved rotorcraft dynamic model in state estimation. Paper presented at the *IEEE International Conference on Intelligent Robots and Systems,* 2011, pp. 5173-5178. doi:10.1109/IROS.2011.6048591

[14] A. I. Rudskoi, G. E.Kodzhaspirov, and E. I. Kamelin, "Simulation and prediction of the development of dynamic recrystallization during the deformation of low-alloy low-carbon steel blanks", *Russian Metallurgy (Metally),*vol. 2016, no. 10, 2016, pp. 956-959. doi:10.1134/S0036029516100177

[15] N. V. Andreeva, A. V. Filimonov, A. I. Rudskoi, G. S. Burkhanov, I. S.Tereshina, G. A Politova,. and I. A. Pelevin, "A study of nanostructure magnetosolid Nd–Ho–Fe–Co–B materials via atomic force microscopy and magnetic force microscopy", *Physics of the Solid State,* vol. *58*, no. 9, 2016, pp. 1862-1869. doi:10.1134/S1063783416090079

[16] I. I. Musabirov, I. M. Safarov, M. I.Nagimov, I. Z. Sharipov, V. V. Koledov, A. V. Mashirov, A.I. Rudskoi, R. R. Mulyukov, "Fine-grained structure and properties of a Ni2MnIn alloy after a settling plastic deformation", *Physics of the Solid State*, vol.58, no. 8, pp. 1605-1610. doi:10.1134/S1063783416080217

[17] D. S. Lisovenko, J. A.Baimova, L. K. Rysaeva, , V. A. Gorodtsov, A. I. Rudskoy, and S. V. Dmitriev, "Equilibrium diamond-like carbon nanostructures with cubic anisotropy: Elastic properties", *Physica Status Solidi (B) Basic Research,* vol. *253*, no. 7, 2016, pp. 1295-1302. doi:10.1002/pssb.201600049

[18] V. Imayev, R. Gaisin, A. Rudskoy, T. Nazarova, R. Shaimardanov, and R. Imayev, "Extraordinary superplastic properties of hot worked ti-45Al-8Nb-0.2C alloy", *Journal of Alloys and Compounds,* vol. *663*, pp. 217-224, 2016. doi:10.1016/j.jallcom.2015.11.228

[19] A. I. Rudskoy, , S. Y. Kondrat'ev, and Y. A. Sokolov, "New approach to synthesis of powder

and composite materials by electron beam. part 1. technological features of the process", *Metal Science and Heat Treatment,* vol. *58*, no. 1, 2016, pp. 27-32. doi:10.1007/s11041-016-9959-x

[20] A. I. Rudskoy, G. E. Kodzhaspirov, J. Kliber, and C. Apostolopoulos, "Advanced metallic materials and processes", *Materials Physics and Mechanics,* vol. *25*, no.1, 2016, pp. 1-8.

[21] F. V. Grechnikov, E. G. Dem'yanenko, and I. P. Popov, Development of the manufacturing process of aluminum alloys with high strength and conductance. *Russian Journal of Non-Ferrous Metals,*vol. *56*, no. 1, 2015, pp. 15-19. doi:10.3103/S1067821215010083

[22] A. I. Rudskoy, T. S. Kol'Tsova, T. V. Larionova, A. N. Smirnov, E. S. Vasil'Eva, and A. G. Nasibulin, "Gas-phase synthesis and control of structure and thickness of graphene layers on copper substrates", *Metal Science and Heat Treatment,* vol. *58*, no. 1, 2016, pp. 40-45. doi:10.1007/s11041-016-9962-2

IOP Conf. Series: Materials Science and Engineering **618** (2019) 012085 doi:10.1088/1757-899X/618/1/012085

Filling a Knowledge Base for Expert System in Information Security

N Kireeva[1], I Pozdnyak[1], A Gazizulina[2]

[1]Povolzhskiy State University of Telecommunications and Informatics, Information Security Department, Samara, Russia
[2]Peter the Great St.Petersburg Polytechnic University, St.Petersburg, Russia
Corresponding author's e-mail: albinagazizulina@gmail.com

Abstract. This article describes a study of parameters of a comprehensive assessment of the organization's information security. The formation of a knowledge base based on international and national standards and various industry requirements is proposed. Based on the standards, methods for filling the knowledge base for the expert system and a model for the functioning of this system are proposed. As a result, with a help of a specific user interaction algorithm with the output mechanism, it is possible to segregate the industry component in information security assessment process that will allow more accurate identification of risks and making recommendations for protection.

1. Introduction

Modern studies of information systems show that at present, any organization is engaged in storing or processing information, in addition, one way or another it deals with personal data, the protection of which is also assigned to the organization itself. The number of service are according by the Internet increased significantly. The analysis of the organizations using the Internet network in Russia in the per cent of the total number of the organizations shows that in 2005-2017 years the number of organizations using the Internet increased in 1.7 times. It is 88,9% (figure 1, figure 2). Proportion of the Internet users in Russia is 81% general public. Including 65% go online every day. At the same time, about 37%, according to data for 2017, use the Internet to purchase goods and services.

An important component in monitoring the international and national of information security in an organization is an audit, which consists in obtaining objective qualitative and quantitative assessments of the current state of security of an organization in accordance with certain criteria and safety indicators. As a rule, an audit requires the verifier to have in-depth knowledge of the legal framework in the field of protection, as well as the specifics of the data stored and processed in a particular information system owned by the organization [1-3, 7].

Existing audit systems are based on expert systems. Expert systems (ES), the most typical implementation of artificial intelligence, are now used in various areas for risk assessment. An important feature of expert systems is their ability to learn, which is associated with a continuous process of knowledge detection and data mining. That is why the first component of the expert system should be the knowledge base [4-6].

According to experts, in the near future, expert systems will play a leading role in all phases of design, development, production, distribution, sales, support and provision of services. Their technology, having received commercial distribution, will provide a revolutionary breakthrough in the integration of applications from ready intelligently interacting modules.

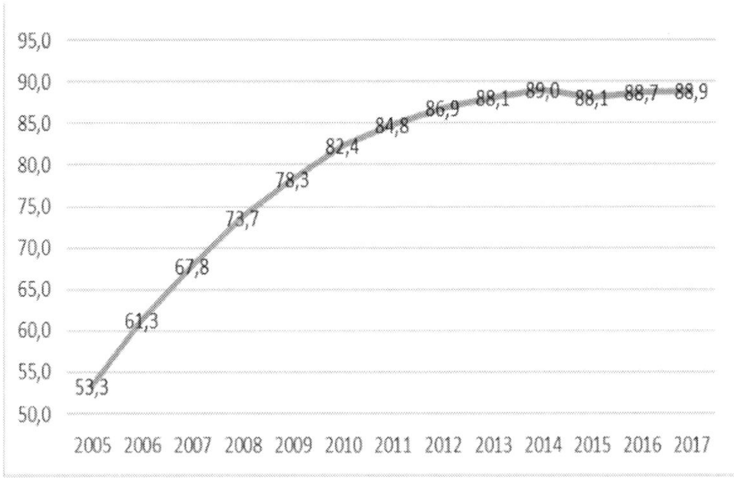

Figure 1. The unit weight of organizations that used the Internet for the period from 2005 to 2017

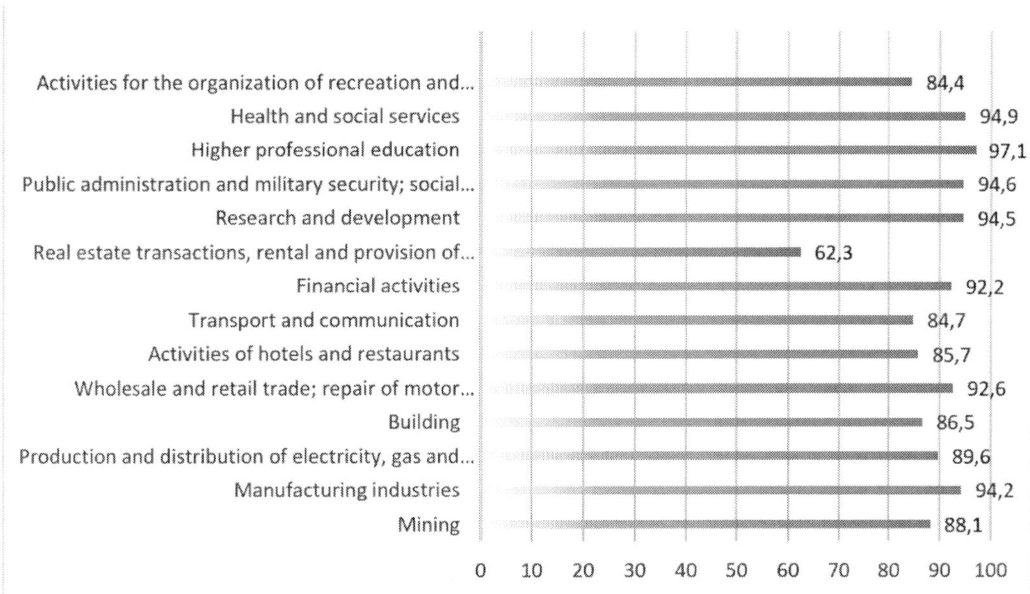

Figure 2. Organizations by economic activity using the Internet

Among specialized knowledge-based systems, real-time expert systems, or dynamic expert systems, are most significant. They account for 70 percent of this market [8, 11-13].

One of the most important requirements for expert systems are:

Ensuring the filling of the knowledge base of varying degrees of complexity with minimal time and labor.

Providing problem and subject-oriented system configuration.

Ensuring the creation and support of interfaces for various categories of users.

Ensuring the level of information security (by user categories) and the prevention of unauthorized access.

Often while building the knowledge base for the expert system, the industry components remain unknown and unappreciated, which determines the relevance and practical significance of studying the problem stated in this article [14 – 16].

In connection with the dynamic development of information technologies and the widespread use of information systems of various architectures and difficulties in the activities of modern small enterprises, there is the problem of a comprehensive assessment system for information security. At the same time, a team of specialists is often needed to solve this problem.

The goal of the presented work is to create an Expert system, within which it is possible to conduct a comprehensive information security (IS) assessment for both a typical information system and an enterprise system for a particular industry (gas, energy, telecommunications, banking, etc.). It is planned to develop an algorithm for interaction between the user and the information system, which will allow, if necessary, to identify the sectoral component in the process of information security assessment, as well as to describe in detail the methodology for filling the expert system's internal knowledge base.

2. Knowledge Base Analysis

As part of the process of comprehensive information security assessment of an information system, the creation of an expert system is possible, justified, relevant and under necessary condition. There are several reasons for this:

the possibility of making a subjective decision inherent in a person is reduced to zero;

it is possible to divide one large task into several subtasks;

the use of a human expert is problematic and expensive (especially in small enterprises), and the use of ES allows you to reduce the number of specialists and experts to analyze and evaluate IS;

the complexity of managing, analyzing and information security assessment leads to the fact that a person needs a lot of time in order to reach the level of an expert.

At the moment, expert systems are mostly focused very narrowly and do not always give the desired effect.

The scheme of functioning of the Expert systems in its general form is shown in figure 3.

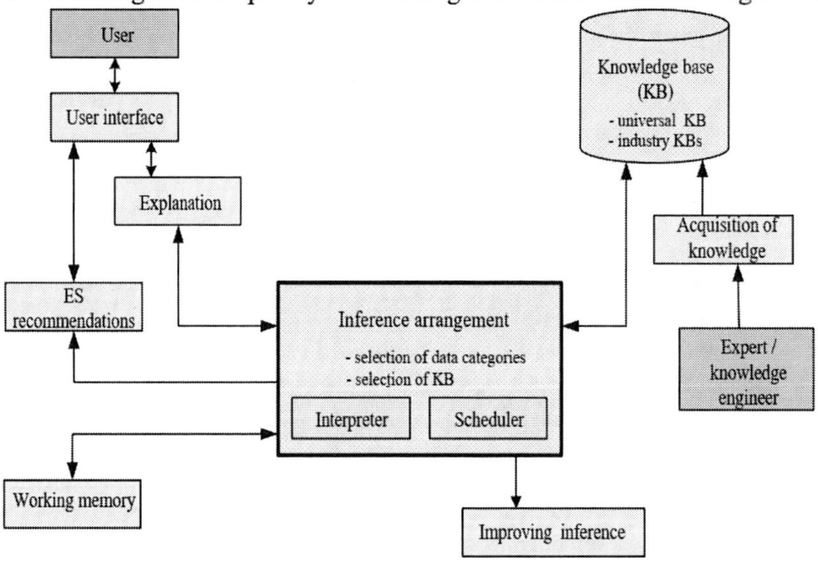

Figure 3. The interaction of elements of Expert systems

The main elements of any ES is the Knowledge base (KB) and the logical inference unit. The Knowledge base must be sufficiently demonstrative and at the same time allow the construction of structural models of an information system, models of threats and vulnerabilities associated with individual components of the information system. This will allow to identify those elements and objects of the information system, the risk and damage from the violation of security of which is the most critical.

IOP Conf. Series: Materials Science and Engineering **618** (2019) 012085 doi:10.1088/1757-899X/618/1/012085

Today, when building an Expert systems knowledge base, the process of acquiring the necessary knowledge is the most laborious. In doing so, the knowledge engineer decides what method of knowledge engineering will be used, and how he will interact with the expert.
Conversations with an expert. In this group of methods distinguish observational and intuitive approaches.
With an observational approach, they follow the work of the expert, trying not to do anything that could affect the work of the expert in solving the problem. The observation is followed by the clarification stage, at which the knowledge engineer together with the expert analyzes the record of the expert's stage of work, performed by the observer. This approach is also called protocol analysis.
In an intuitive approach, in one case, the expert acts as the developer of a model of his behavior when solving problems, in the second, as a knowledge engineer.
The expert choose is the principal question at the stage of formation of the knowledge base. The expert personality can be considered as a dynamic model. We will use the personality dynamic model described by Milovanov V.P. [9] like the expert, who will be a developer of the behavior model in solving problems, and as a knowledge engineer.
Let x is the expert x information (the expert x problems); y is the information for the expert from engineer. The expert gives the opinion (assessment) to the engineer y. We will have the next model:

$$\begin{cases} \dfrac{dx}{dt} = \alpha_1 y + \alpha_2 xy - \alpha_3 x^2 y, \\ \dfrac{dy}{dt} = \beta_1 x - \beta_2 xy. \end{cases} \tag{1}$$

Here $\alpha_1 y$ is the engineer y knowledge, agreed with the expert x (α1 - relevancy of expert x assessment), and $\beta_1 x$ – is the expert x knowledge, agreed with the engineer y, $-\beta_2 xy$ is the engineer y expectations, forwarded to the expert x, $\alpha_2 xy$ is a part shows the expert x listens the engineer y and forms agreed answers on the proposals of the engineer y; $-\alpha_3 x^2 y$ shows the expert x listens the engineer y proposals, forms agreed answers on the proposals of the engineer y (multiplier x^2), forms the actions to bring in knowledge base.
The equilibrium position of the model (1) is:

$$\bar{x} = \frac{\alpha_2 \pm \sqrt{\alpha_2^2 + 4\alpha_3\alpha_1}}{2\alpha_3}, \qquad \bar{y} = \frac{\beta_1}{\beta_2}. \tag{2}$$

The values of σ and Δ are as follows:

$$\sigma = \alpha_2 \bar{y} - 2\alpha_3 \overline{xy} - \beta_2 \bar{x}, \qquad \Delta = \beta_2 \overline{xy}\left(2\alpha_3 \bar{x} - \alpha_2\right) \tag{3}$$

When 2$\alpha_3 \bar{x} > \alpha_2$ and $2\alpha_3 \overline{xy} + \beta_2 \bar{x}$ there is a bifurcation and an expert arises, whose opinions are reliable.
We now calculate the Lyapunov dimension α:

IOP Conf. Series: Materials Science and Engineering **618** (2019) 012085 doi:10.1088/1757-899X/618/1/012085

$$\begin{aligned} \alpha = -\frac{\pi}{4\left(\alpha_1+\alpha_2\bar{x}-\alpha_3\overline{x^2}\right)}\Big\{ & \left(\alpha_2\bar{y}-2\alpha_3\overline{xy}\right)\left(\beta_1-\beta_2\bar{y}\right)\left(\alpha_2-2\alpha_3\bar{x}\right)^2+\left(\alpha_2\bar{y}-2\alpha_3\overline{xy}\right) \\ & \left(\alpha_1+\alpha_2\bar{x}-\alpha_3\bar{x}^2\right)\left(\beta_2{}^2+\alpha_3\beta_2\bar{y}\right)-2\left(\alpha_2\bar{y}-2\alpha_3\overline{xy}\right)\left(\alpha_1+\alpha_2\bar{x}-\alpha_3\bar{x}^2\right)\left(-\alpha_3\bar{y}\right)+ \\ & +\alpha_3\overline{y}\left(\alpha_2-2\alpha_3\bar{x}\right)\left[\left(\alpha_1+\alpha_2\bar{x}-\alpha 3\bar{x}^2\right)\left(\beta_1-\beta_2\bar{y}\right)-2\left(\alpha_2\bar{y}-2\alpha_3\overline{xy}\right)^2\right]- \\ & -\left[\left(\alpha_2\bar{y}-2\alpha_3\overline{xy}\right)^2+\left(\alpha_1+\alpha_2\bar{x}-\alpha_3\bar{x}^2\right)\left(\beta_1-\beta_2\bar{y}\right)\right]2\left(-\alpha_2\bar{y}+2\alpha_3\overline{xy}\right)\alpha_3\Big\} \end{aligned} \tag{4}$$

Because, obviously, the value α can be both positive and negative, it means that expert x gives consistent or inconsistent estimates.

The position of the equilibrium of the value of the opinion (assessment) of the expert and the engineer will fall into the knowledge base ($C_1C_2 = 1$).

3. Assessment Filling Knowledge Base

In our work, we will build a knowledge base on the basis of international and national standards in the field of information security and various industry documents (International and national standards, Industry standards, Working documentation, Technical requirements, Industry recommendations). In addition to this, a knowledge engineer will prepare a series of typical tasks for discussion with an expert.

To begin, consider the standard ISO/IEC 27002 “Information technology – Security techniques – Code of practice for information security management”. Section 6.1 “Internal organization” consider information security management issues in an organization. To create the Knowledge base, we use subsections 6.1.1 - 6.1.8. Each contains questions necessary for consideration. For example, does the organization have an information security council (clause 6.1.2), are information security responsibilities distributed (clause 6.1.3), is there an independent audit (audit) of information security (clause 6.1.8) and etc. Questions of section 6.2 “External Parties” (touches issues of supporting security of organization’s information processing tools and information assets when accessed by third parties) will be included in the knowledge base. Application of Section 7 “Asset Management” in the knowledge base helps to determine the assets in relation to which the information security assessment will be carried out, as well as the appropriate protection of the organization’s selected assets. Should not be overlooked and organizational measures relating to the actions of staff. To do this, apply Section 8 “Human Resources Security”. This includes taking security issues into official duties when hiring personnel (clause 8.1.1), and user training (clause 8.2.2), and issues related to the transformation or change of employment of employees (clause 8.3). Section 9 “Physical and Environmental Security” contains the necessary information for the developed knowledge base on protected areas, equipment safety inside these areas, as well as general measures to control IS. Section 10 “Communications and Operations Management” will help fill the knowledge base with information about:

- operational procedures and responsibilities,
- management of third party services,
- planning and acceptance of systems,
- protection against malicious and mobile programs,
- reservations
- network security management,
- handling media,
- exchange of information,

- e-commerce service,
- monitoring.

Section 11 "Access Control" will allow filling the knowledge base of the ES with knowledge of:

- access control business requirements
- user access management,
- user responsibilities
- network access control,
- control of access to the operated system,
- control access to information and application programs
- mobile computing and remote work.

Issues related to Section 12 "Information Systems Acquisition, Development and Maintenance", Section 13 "Information Security Incident Management", Section 14 "Business Continuity Management" and Section 15 "Compliance", will also be fully included when creating the knowledge base. More fully with these sections can be found in ISO/IEC 27002.

Next, we consider the standard ISO/IEC 27001 "Information technology. Security techniques. Information security management systems. Requirements". Appendix A of this standard contains the objectives and management measures that are closely related to ISO/IEC 27002. Therefore, when filling the knowledge base of the developed expert system, care should be taken to prevent the repetition of similar situations. For example, section A.5 of the appendix corresponds to section 5 "Security policy" (ISO/IEC 27002), section A.12 of the appendix - with section 12 "Acquisition, development and maintenance of information systems" (ISO/IEC 27002). An exception is Section A.13, Information Security Incident Management. Section 13.2 "Information Security Incident Management and Necessary Improvement" (ISO/IEC 27002) partially complies with it.

Despite this, the objectives and management measures from Appendix A (ISO/IEC 27001) are mandatory for inclusion in the knowledge base. If they do not override the measures and methods of managing information security systems (ISO/IEC 27002), then they are their complement. The list of management measures contained in these two standards is not exhaustive and can be supplemented by a knowledge engineer and expert.

The following standard to which attention should be paid when filling the ES knowledge base is ISO/IEC 27005 « Information technology - Security techniques - Information security risk management». The knowledge base should include the types of assets for which risks will be assessed:

- Business processes (or subprocesses) and business activities;
- Information;
- Hardware;
- Data processing equipment (active);
- Mobile equipment;
- Stationary equipment;
- Peripheral processing equipment.
- Data carrier (passive)
- Electronic media.
- Software
- Operating system.
- Software maintenance, maintenance or administration.
- Batch software or standard software.
- Business application, etc.
- Standard business application.
- Specific business application.
- Network
- Wednesday and support.

- Passive or active repeaters.
- Communication interface, etc.

Also in the knowledge base must make the concept of "asset value". At the same time, it should be determined by the owners and users of assets in the process of using the expert system. In addition, the concepts of "threat", "source of threat", "object of threat", "probability of threat realization", "vulnerability", "degree of vulnerability", and "protective measures" should be introduced. A list of typical types of threats is presented in Appendix C of the standard, examples of common vulnerabilities are given in Appendix D of ISO/IEC 27005.

In total, four national standards in the field of information security formed the basis of the knowledge base of the developed expert system. They contain the most important information (knowledge), reflecting a comprehensive information security assessment of the information system of any enterprise.

In addition, it is necessary to consider and include in the KB data of various industry documents. Since there are a huge number of industries, it is not possible to include all possible standards and recommendations. However, the knowledge base should be filled with documents from the most sought-after industries. For example, the banking sector. For him, there are a number of standards that will help fill the BS with industry knowledge: STO BR IBBS-1.0-2008, STO BR IBBS-1.1-2007, STO BR IBBS-1.2-2009, RS BR IB IBBS-2.0-2007, RC BR IB IBS-2.2-2009.

One of the main standards is the Bank of Russia standard: "Ensuring the information security of organizations of the banking system of the Russian Federation. General provisions "(STO BR IBBS-1.0-2008). Section 7 of this document will allow defining information security requirements for assigning and distribution roles and ensuring confidence in personnel, information security requirements for automated banking systems at life cycle stages, as well as requirements for ensuring information security in access control and registration, in the use of Internet resources, in the use of cryptographic information protection, banking payment and information technology processes. Section 8 "Information Security Management System of Organizations of the Banking System of the Russian Federation" examines in detail the information security management, which will fully fill the knowledge base with information on the banking industry.

During the filling of the Knowledge Base with the help of industry standards it is necessary to remember that situations, questions and knowledge can already be entered into the database. In this case, you should check the nuances associated with a particular industry.

4. Conclusion

The proposed algorithm for filling the knowledge base, as well as building an expert system, allows for a comprehensive assessment of the information security of an enterprise taking into account the industry component, which is a necessary and important condition during an information security audit. The interaction of the Expert system user with the output mechanism will be carried out using a specific algorithm that, if necessary, will allow to highlight the component for the industry in the information security assessment process. The result of this interaction will be the recommendations issued by the system on the protection of the selected asset, and the main risks will be identified. Based on the results of the expert audit, preparation of administrative documents is being carried out, which determine recommendations for building or upgrading the existing information protection system. In addition, this algorithm for filling the expert system knowledge base will reduce the financial and time costs for conducting an audit.

Acknowledgment

The research carried out with the financial support of the grant from the Program Competitiveness Enhancement of Peter the Great St. Petersburg Polytechnic University.

The section «Knowledge Base Analysis» is based on the financial support of the Council on Grants of the President of the Russian Federation (Project No. MK-3229.2018.8 "Improving the Standardization Management Method"). Peter the Great St. Petersburg Polytechnic University.

References

[1]. Bataev, A. V. (2017). Analysis of the use of cloud services and assessment the possibilities of introducing in Russian financial institutions. Paper presented at the Proceedings of the 29th International Business Information Management Association Conference - From Regional Development Sustainability to Global Economic Growth, 25-35.

[2]. Demidenko, D. S., Malinin, A. M., & Litvinenko, A. N. (2017). A new classification of the risks of the quality processes. Paper presented at the Proceedings of the 30th International Business Information Management Association Conference, IBIMA 2017 - Vision 2020: Sustainable Economic Development, Innovation Management, and Global Growth, 2017-January 2892-2897.

[3]. Didenko, N. I., Skripnuk, D. F., & Mirolyubova, O. V. (2017). Big data and the global economy. Paper presented at the Proceedings of 2017 10th International Conference Management of Large-Scale System Development, MLSD 2017, doi:10.1109/MLSD.2017.8109611

[4]. Klochkova, E. S., Bolsunovskaya, M. V., & Shirokova, S. V. (2019). The significance of humanities for engineering education. Paper presented at the Proceedings of 2018 17th Russian Scientific and Practical Conference on Planning and Teaching Engineering Staff for the Industrial and Economic Complex of the Region, PTES 2018, 265-268. doi:10.1109/PTES.2018.8604199

[5]. Didenko, N. I., Skripnuk, D. F., Kikkas, K. N., Sevashkin, V., Romashkin, G., & Kulik, S. V. (2018). Innovative and technological potential of the region and its impact on the social sector development. Paper presented at the International Conference on Information Networking, 2018-January 611-615. doi:10.1109/ICOIN.2018.8343191

[6]. Didenko, N., Skripnuk, D., Mirolyubova, O., & Radion, M. (2017). Analysis of rural areas development of the region using the ADL-model. Paper presented at the Research for Rural Development, 2 142-147. doi:10.22616/rrd.23.2017.061

[7]. Glukhov, V. V., Ilin, I. V., & Iliashenko, O. J. (2016). Improving the efficiency of architectural solutions based on cloud services integration doi:10.1007/978-3-319-46301-8_43

[8]. Gorodetskiy, A. E., & Tarasova, I. L. (2019). Situational control a group of robots based on SEMS doi:10.1007/978-3-319-99759-9_2

[9]. Milovanov V.P. (2005) Sinergetika i samoorganizatsiya. Obshchaya i sotsial'naya psikhologiya /USSR, p.208

[10]. Ilin, I. V., Iliashenko, O. Y., & Levina, A. I. (2016). Application of service-oriented approach to business process reengineering. Paper presented at the Proceedings of the 28th International Business Information Management Association Conference, 768-781.

[11]. Klochkov, Y., Klochkova, E., Antipova, O., Kiyatkina, E., Vasilieva, I., & Knyazkina, E. (2016). Model of database design in the conditions of limited resources. Paper presented at the 2016 5th International Conference on Reliability, Infocom Technologies and Optimization, ICRITO 2016: Trends and Future Directions, 64-66. doi:10.1109/ICRITO.2016.7784927 Retrieved from www.scopus.com

[12]. Kostin, G. A., Pokrovskaia, N. N., & Ababkova, M. U. (2017). Master-chain as an intellectual governing system for producing and transfer of knowledge. Paper presented at the Proceedings of 2017 IEEE 2nd International Conference on Control in Technical Systems, CTS 2017, 71-74. doi:10.1109/CTSYS.2017.8109491

[13]. Kovalenko, I. I., Sokolitsyn, A. S., & Sokolitsyna, N. A. (2018). The enterprise's automated management stability system taking into account its life cycle stage. Paper presented at the Proceedings of the 2018 International Conference "Quality Management, Transport and Information Security, Information Technologies", IT and QM and IS 2018, 357-360. doi:10.1109/ITMQIS.2018.8524966

[14]. Kozlovsky, V., Klochkov, Y., Ostapenko, M., Ushanova, N., & Antipov, D. (2016). Conformity assessment of car quality databases exemplified on the case of the russian car manufacturer 'auto VAZ'. Paper presented at the 2016 5th International Conference on Reliability, Infocom Technologies and Optimization, ICRITO 2016: Trends and Future Directions, 57-60. doi:10.1109/ICRITO.2016.7784925

[15]. Naidenova, X., Parkhomenko, V., & Curbatov, S. (2017). A neural-network like logical-combinatorial structure of data and constructing concept lattices. Paper presented at the CEUR Workshop Proceedings, 1921 62-76.
[16]. Pechenkin, A., & Demidov, R. (2018). Applying deep learning and vector representation for software vulnerabilities detection. Paper presented at the ACM International Conference Proceeding Series, doi:10.1145/3264437.3264489

Analytical model of the Interrelation between Enterprise's Activity and its Financial Stability

A Sokolitsyn[1], M. Ivanov[2], N Sokolitsyna[3] and I Babkin[4]

[1] Peter the Great Saint-Petersburg Polytechnic University, Saint Petersburg, Russia
alex.sokol1951@yandex.ru

[2] Peter the Great Saint-Petersburg Polytechnic University, Saint Petersburg, Russia
ivanov_fuit@rambler.ru

[3] Peter the Great Saint-Petersburg Polytechnic University, Saint Petersburg, Russia
natasokoli@yandex.ru

[4] Peter the Great Saint-Petersburg Polytechnic University, Saint Petersburg, Russia
ivan.babkin@spbstu.ru

Abstract. To inquire aims of diversification as the most important Russian enterprises resultatives increase mechanism, the article considers scheme of enterprise functioning in diversification motives and aims interconnection and executes analyzing firm activity character change with its diversification and character of financial impact on its diversification. Analyzing particular diversification aims has allowed considering threats and negative factors, obligate to taking into account in forming enterprise activity diversification strategy to taking into account its specialties. Forming enterprise activity and financial sustainability basic assessing figures ordered graph and considering their calculation formulas has allowed modeling this graph. Practical realizing modeling enterprise activity diversification and financial sustainability basic figures ordered for a research object – St-Petersburg oil-gas machine-making enterprise taking into account presented graphic interpretation of its financial condition shows this research practical sustainability and opportunity to spread into other industries. The elaborated graph is the basis to form economically-mathematical models and algorithms of defining several kinds of enterprise's credits. Besides that, it is one of effective instruments of complex economic analysis, which is the base of backgrounding potential enterprise development destinations with statistic modeling enterprise production process parameters. Therefore, elaborated graph practical realization helps to sufficient increasing effectiveness of managing enterprise.

1. Introduction

Diversification is one of the most important mechanisms for strategic development of an enterprise related to ensuring competitiveness and increasing the efficiency of its operations. Diversification is not limited to introducing diversity in the range of products or services provided. It is considered in a broader sense as an expansion of the sphere of activity of enterprises in various branches of the national economy. Analysis of Russian enterprises' activities shows that many of them diversify operations, pursuing the goals of introducing innovations in the production program and the enterprise management system in order to improve their position on the market, to respond promptly and adequately to changes in the market situation, and to increase the operational efficiency. At the same time, the relationship between the diversification of the enterprise's operations and its financial stability is traced, subject to the management decisions on choosing the goals and directions for development of the enterprise. Problems of strategic development of the enterprise on the basis of diversification and provision of its financial stability have attracted various researchers' attention for many years. But, unfortunately, the modern Russian economy has some specific factors that do not allow copying foreign experience in this sphere in Russia. For this reason Russian researchers consider this problem given both the specifics of the economy as a whole and industry specific features (Ushich (2016); Zhukova (2014); Kuzmenko and Zotkov (2015); Rodionov and Kudryavtseva (2016); Klochkov (2018)). In this regard, the paper is aimed to form mechanism of diversification, taking into account the main goals, forms, methods, economic advantages and threats of diversification as a strategic development direction of the enterprise. To achieve the set aim, a scheme of the company's functioning has been formed in relation to the diversification objectives and goals, on the basis of which an ordered graph of the main diversification indicators of enterprise operations and its financial stability has been developed. The perceived interrelation between the basic diversification indicators of the enterprise operations and its financial stability while carrying out their qualitative estimation allows modelling and estimating various management policies of the enterprise and choosing the best one (Sokolitsyn, et al., (2016); Sokolitsyn, et al., (2016); Rudskaya, et al., (2017, 2018); Dvas (2018)).

2. Main part

In the current economic situation, the majority of both foreign and Russian companies are considering diversification of their operations as a strategic approach to improve their effectiveness in the conditions of uncertainty of the modern market economy. In this connection there is scientific and practical necessity to analyze the literature on the problems of improving the diversification mechanism given the main objectives, forms, methods, economic advantages and threats of diversification as a strategic development approach of the company. It is necessary to research interdependent diversification factors as a strategic direction for development and sustainable effective functioning of enterprises in the short and long term perspective, procedures to enhance managerial decision-making on business diversification strategies taking into account their specifics. For example, Kuzmin, et al., (2015) considered the concept of a project approach to diversifying the enterprise's activities, which proposed statistical and dynamic economic-mathematical models aimed at optimization of enterprise diversified development. Based on empirical analysis A. Lee, et al., (2016) showed how diversification impacts the sustainability of Korean construction companies and found out that in order to efficiently generate diversification strategies in Korea's economic environment, construction firms need to diversify the life cycle of a project, focusing on more profitable projects, engineering and consulting in a short-term period. Alternatively, they should enter an asset management market to get long-term operating profits. The importance of the optimal choice of direction for the trade activity of the enterprise as one of the most important components of enterprise diversification is justified in foreign literature in more detail than in Russian literature (Shpak, et al., (2016)). Three models for diversifying the company's trading activities are proposed. The first model is based on the evaluation of the sales channels potential. By means of the second model, the optimal quantitative distribution of products between sales channels is determined, which ensures high profitability of each unit in the assortment and the entire enterprise. The third model is the optimal distribution of products between sales channels, given the cooperation experience between enterprises and sales channels in the past, which ensures a minimum risk and sufficient profitability of each sales channel. It is noted that the results obtained by the proposed models are compared with the enterprise data to identify the advantages of each model.

The results of the research presented in the paper by scientist Selçuk (2015) are of particular interest for the development of the Russian economy. The paper analyses the relationship between diversification and profits in developing countries and shows that diversified firms in emergency markets make more profits than single-segment firms in the same industries. It is proved by the existence of a premium for diversification. In addition, the results of tests conducted are quite interesting for Russian corporations, which revealed the relationship between diversification and success at the firm level, rather than at the level of an individual enterprise (Mackey, et al., (2017)). The papers (Barnes, et al., (2015) and Yang, et al., (2017)) study what is the influence degree of enterprises' operations diversification on their performance given the industry specifics. Barnes, et al., (2015) assessed the impact of diversification on the results of agricultural business by determining the economic viability of firms in Scotland and Sweden. They have shown that farms which open up additional businesses outside traditional agriculture, diversify and get revenue from two or more agricultural enterprises are economically stronger than narrowly-specialized farms. Yang, et al., analyzed the impact of hotel business diversification on its performance. Thus, the analyzed published sources show the complexity and wide possibilities of enterprise operations diversification for improving the effectiveness of the enterprise's straegic development.

Investigation of the diversification goals as the most important mechanism for increasing the efficiency of Russian firms

At the present time various prerequisites for operations diversification have been formed at enterprises of the national industry. Therefore, diversification in Russia will be carried out simultaneously at different stages, in contrast to the world practice, where this phenomenon has evolved naturally. Determining the diversification goals is an important aspect in considering this economic phenomenon. The goals of production diversification depend directly on the financial condition, possibilities of the enterprise and attractiveness of industries for firms. An analysis of the results of some foreign and national research studies of this problem makes it possible to identify certain objectives and goals that most often serve as incentives for expanding the entrepreneurial activity of an enterprise (figure 1).

As a result of diversification, there is a constant floating and the distribution of capital between different spheres, depending on the decrease or increase in the rate of profit, the fluctuations in profitability in various industry segments are smoothed out, and the company becomes more stable. Table 1 shows the interrelated but multidirectional processes that arise from operations of the company which pursues a diversification strategy aimed to enhance stability.

Table 1. Change in the company's operations after diversification

traditional production	**new production (diversification)**
• a drop in the rate of return; • capital extraction.	• high rate of return; • capital investment.
decrease in the volume of traditional production	increase in the volume of production of new products
negative economies of scale	Positive economies of scale
reduction in supply	increase in supply
increase in price of the goods	decrease in price of the goods
increase in profits	decrease in profits
settle of losses	settle of profits

In general, we can distinguish two main trends of the company's development: strengthening of traditional business (monobusiness) or creating a diversified enterprise (diversification). When considering the problem of diversification, it is reasonable to analyze the financial impact of this growth strategy on the enterprise development. The impact of finances on development directions is very significant and usually requires a vast expenditure of financial resources with a high risk associated with obtaining an appropriate return on investment (table 2).

Table 2. Characteristics OF THE FINANCIAL IMPACT ON THE DIVERSIFICATION OF FIRMS

Financial impact	**Related diversification**	**Nonrelated diversification**
type of financial return	moderate return	high return
distribution of financial return over time	mid to long term period	long-term period
necessary financial resources	medium amount of financial resources	large amount of financial resources
level of risk in the strategy	Moderate	high
possibility of financial synergy	High	high

Due to the fact that the diversification of the firm has a certain dualism, the management team when choosing a development direction of should consider both the positive and negative consequences of this phenomenon in each individual case. Often when considering the issues of diversification, such important factors as problems of developing a common corporate strategy, specifics of business interaction are not taken into account. It is the lack of a circumspect consistent development strategy of the firm which must be seen as the main factor of many failures of the diversification strategy. By analyzing a number of specific goals that entrepreneurs seek to achieve through diversification, their common threats to a business have been identified:

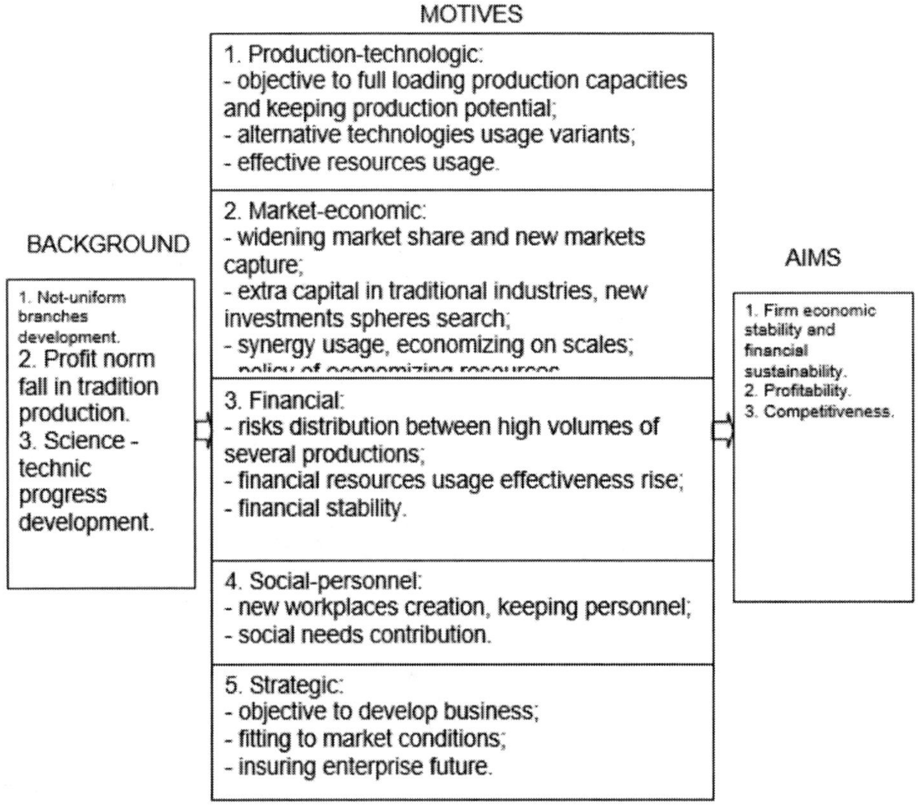

Figure 1. The scheme of enterprise operation in interrelation of the diversification incentives and goals

1. Increasing business stability. Don't-put-all-your-eggs-in-one-basket approach is used in economic practice. This principle is one of the most common arguments of diversified Russian companies. But basically this approach is reasonable for portfolio investors who reduce the risk of capital investment in the stock market. At the same time, to implement the diversification strategy at a single firm level is not so simple. Due to a real deficit of professional managers in Russia it is still possible to have benefits when one management team administers several diverse businesses. So sometimes the benefits from the professionalism of the management team can exceed the drawbacks from dispersion of resources. This strategy loses its effectiveness when the company achieves a certain size. It is often too expensive for Russian firms to develop several different directions.

2. Increasing the size of the business. You cannot estimate a company solely by its size. There are more important criteria like the amount of profit and indicators of operating profitability. In Russia, companies often increase their business only because they do not know how to spend money. This indicates that there is no strategy for the company-wide development, which is a mandatory condition for diversification efficiency.

3. Use of available resources. Russian practice proves the fact that when the low cost of a purchased business is the determining criterion for its acquisition, serious problems usually arise in adapting this business to the activities of the company as a whole. So the new owner often has to change the corporate strategy dramatically, which may be more expensive in the long run.

4. Increasing the efficiency through synergy.

While the scale effect prevails over the losses from the complication of the controlled object, the total sum of these effects has a positive sign and the company's value really increases as the range of activities expands. But from a certain threshold value, the growth rates of the scale effect slow down, and the losses rise at an increasing rate. As a result, cash flows decrease, and the enterprise devalues, contrarily to the projected growth of its value. Thus, we can talk about the existence of some optimal diversification level, which maximizes business efficiency.

Also, we must not forget about financial constraints on the rate at which an optimal level of diversification is achieved. As practice shows, underestimation of the fact that not every diversification is efficient and cost-effective, drove many enterprises with an unreasonably high growth to the brink of bankruptcy. It is important to understand that business diversification is not a universal tool for solving all economic problems.

As a result, it is possible to formulate the following negative factors, which must be taken into account when forming an expansion strategy:

1. inefficient use of funds;
2. red tape;
3. decrease in motivation and quality of work;
4. dispersion of the management team's attention;
5. brand dilution;
6. lack of information

Strategic directions of diversification policy and the results of economic activity depend on the financial situation, business characteristics and trends of the company's evolution. But diversification of the enterprise as a way of development has a strong impact on its financial condition. Thus, there is a certain relationship between diversification of the company and its financial stability.

Modeling an ordered graph of the enterprise's diversification main indicators and its financial stability

Analysis of financial indicators allows identifying critical points in the company activities, to define the causes of problems and to plan actions to address them. At the same time, it is important not only to assess the current financial condition of the enterprise, but also to project the obtained data on the nearest or more distant future, that is, to predict the expected parameters of financial condition on the basis of financial analysis.

In order to strengthen its position and survive in a competitive struggle, the company has both to increase its revenue and profit, and expand its market niche. This is achieved by increasing its share in the total volume of market transactions both by means of increasing the volume of output and through diversification of products. In competitive environment, the company in an effort to ensure rapid growth, is forced to implement more profitable, but more risky projects. Therefore, in order to maintain financial stability, it is important to find a "golden mean", that is, to choose a growth model that would give the company an opportunity to develop and avoid bankruptcy. Growth rates consistent with such a model are considered sustainable.

It is important to understand that for an unambiguous, elaborated and exhaustive analysis of the enterprise financial condition it is not enough to assess only the common used financial indicators according to the relevant criteria. The evaluation criteria should be adjusted based on statistical data, given the industry specifics and characteristics of a particular business. The main indicators of the enterprise's operations diversification and financial stability were estimated for the given study object St-Petersburg oil-gas machine-making enterprise LLC SCB Gazstroymashina, on the basis of the modelled ordered graph and taking into account the industry specifics. This enterprise makes science-constructive elaborations and manufacturing equipment for oil-gas industry enterprises and energetic complex objects. Conditionally current enterprise activity can be divided into following key destinations: manufacturing equipment in own production complex; elaborating and manufacturing equipment on customer requirements; consulting and recommendations on effective providing oil and gas industries enterprise with technic means.

Firm production usage sphere is wide. It is demanded by both big enterprises, executing extorting, transforming and transporting oil and gas, and engineering-exploitation infrastructure organizations, executing direct giving natural raw to consumers and responsible for pipeline quality. Biggest product consumers are Gazprom OJC, Russian VINCs and building organizations. Gazprom OJC share in the enterprise revenue exceeds 60%. SCB Gazstroymashina specialists actively participate R&D works of gas holding.

Company activity analysis shows that Gazstroymashina has started process of widening activity on production diversification basis. The enterprise started more active serial manufacturing a particular product. Firstly it refers to rising manufacturing goods for conducting pipelines, for example, centrators. The enterprise has increased manufacturing products with sufficiently shorter production cycle, but not abandoning orientation on high-capital and technologic projects on projecting new equipment. Such development way on business diversification base last years is popular for Russian oil-gas enterprises. Widening made product nomenclature and also choosing new destinations of functioning in crisys situations strengthen firms economic position.

Assessing financial sustainability of researched enterprise functioning and development on base of researching enterprise functioning in diversification motives and aims interconnection, analyzing changes of character of enterprise activity in its diversification and financial impact on its diversification uses ordered graph of basic Gazstroymashina activity and financial sustainability figures. This graph is constructed on statistic data base on 20 quarters in 2012 – 2016 for enterprise – research object.

Calculation of the estimated indicators complex used in the modelled ordered graph of the main indicators of enterprise's operations diversification and financial stability taking into account researched enterprise specific is presented in table 3.

Table 3. Calculation of the estimated indicators complex used in the modeling of the ordered graph for researched enterprise

Indicator name, legend	Calculation formula
Assets turnover (CA)	$C_A = R/A$, where R – revenue for the reporting period; A – average balance sheet assets of the enterprise for the reporting period.
Current assets turnover (CCC)	$C_{CC} = R/CC$, where R – revenue in the reporting period; CC – average balance sheet current assets of the enterprise for the reporting period.
Non-current assets turnover (CFC)	$C_{FC} = R/FC$, where R – revenue in the reporting period; FC – average balance sheet non-current assets of the enterprise for the reporting period.
Payable accounts turnover (CCI)	$C_{CI} = R/CI$, where R – revenue for the reporting period; CI – account payable for the reporting period.
Receivable accounts turnover (CDI)	$C_{DI} = R/DI$, where R – revenue in the reporting period; DI – accounts receivable for the reporting period.
Ratio of equity and borrowed funds (CLOM)	$C_{LOM} = (LM + CI)/CR$, where LM – borrowed funds; CI – accounts payable; CR – own funds (capital and reserves).
Ratio of equity capital provision (COM)	$C_{OM} = CR - FC/CC$, where CR - own funds (capital and reserves); FC– noncurrent assets, CC – current assets.
Current liquidity ratio (CTL)	$C_{TL} = CC/(CI + LM)$, where CC – current assets; CI – accounts payable; LM – borrowed funds.
Ratio of internal growth rate (CIGT)	$C_{IGT} = SE/CR = DN \times R_{OM}$, where SE - accumulation fund; CR - owned funds (capital and reserves); DN – rate of net profit distribution to the accumulation fund; ROM – return on equity.
Return on commercial output	$R_{CP} = BP/C \times 100$,

IOP Conf. Series: Materials Science and Engineering **618** (2019) 012086 doi:10.1088/1757-899X/618/1/012086

Indicator name, legend	Calculation formula
(RCP)	where BP – balance profit; C – value of commercial output
Return on equity (ROM)	$R_{OM} = BP/CR \times 100,$ where BP – balance profit; CR – owned funds (capital and reserves).
Return on assets (RA)	$R_A = BP/A \times 100,$ where BP – balance profit; A – average balance sheet assets of the enterprise for the reporting period.
Return on sale (RS)	$R_S = BP/R \times 100,$ where BP – balance profit; R – revenue for the reporting period.

The main indicators complex of the enterprise's operations diversification and financial stability were estimated for the given study object taking into account the researched enterprise industry specifics (table 4).

Hence, it is reasonable to build a multidimensional model, which is a graphic interpretation of the financial condition of the enterprise - the study object (figure 2). On certain financial parameters it defines areas of feasible and crisis activity on the basis of the general recommended values for oil and gas industry enterprises (Sokolitsyn, et al., 2016).

Table 4. Calculation of the basic estimated indicators complex used in the modelled ordered graph for the study object – Saint-Petersburg machine building enterprise of the oil and gas complex.

№	Indicator	Balance sheet data for the reporting period / Estimated values for the same period, thousand rubles.
1	Non-current assets (FC)	664
2	Current assets (CC), including: Receivable accounts (DI) Stock and costs (C)	6980 1261 734
3	Capital and reserves (CR)	3890
4	Short-term liabilities (STP), including: Borrowed funds (LM) Accounts payable (CI)	3754 0 3754
5	Assets (A)	7644
6	Revenue (R)	11611
7	Prime cost (C)	10450
8	Balance profit (BP)	1161
9	Net profit (NP)	882
10	Return on commercial output (R_{CP})	11.11%
11	Return on equity (R_{OM})	29.84%
12	Return on assets (R_A)	15.18%
13	Return on sale (R_S)	9.99 %
14	Assets turnover (C_A)	1.52
15	Current assets turnover (C_{CC})	1.66
16	Non-current assets turnover (C_{FC})	17.48
17	Payable accounts turnover (C_{CI})	3.09
18	Receivable accounts turnover (C_{DI})	9.20
19	Ratio of equity and borrowed funds (C_{LOM})	0.96
20	Ratio of equity capital provision (C_{OM})	0.46
21	Current liquidity ratio (C_{CL})	1.86
22	Ratio of internal growth rate (C_{IGT})	5.67%

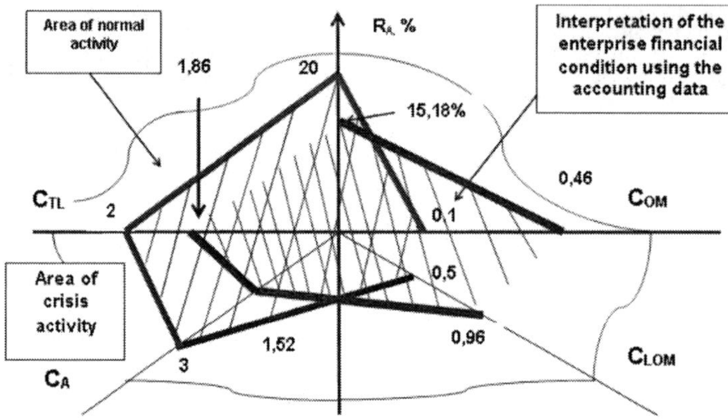

Figure 2. Graphic interpretation of financial condition of the given enterprise.

Legend: $\mathbf{R_A}$ - Return on assets; $\mathbf{C_{TL}}$ - Current liquidity ratio; $\mathbf{C_{OM}}$ - Ratio of equity capital provision; $\mathbf{C_A}$ - Assets turnover; $\mathbf{C_{LOM}}$ - Ratio of equity and borrowed funds.

Taking into account the given facts, the ordered interrelation between basic indicators of diversification operations and financial stability of the enterprise can be represented by the following graph (figure 3), where the central indicators of the company's diversification activity and financial stability are represented by the volume of revenue and the coefficient of internal growth rate. As can be seen from the graph, if an enterprise plans to change the revenue volume, for example, by increasing it through diversification of products, it will affect its entire system of financial stability indicators. Conversely, if it aims to change the internal rate of growth by changing the amount of reinvested net profit, it will affect the indicators of financial stability and diversification operations.

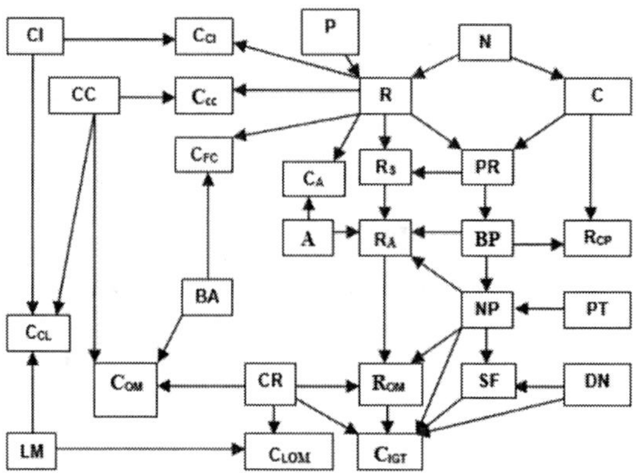

Figure 3. The ordered graph of main indicators of enterprise activity diversification and its financial stability.

Legend: P – product price; N – volume of sales in natural units; R – revenue; C – prime cost; PR, BP, NP –profit on sales, balance and net profit, respectively; DN – rate of net profit distribution to the accumulation fund (AF); PT – income tax amount; A, CC and FC – total, current and non-current assets, respectively; CR - capital and reserves; LM – borrowed assets; CI - payable accounts; C_A, C_{CC}, C_{FC}, C_{CI} – assets turnover, current assets turnover, non-current assets turnover, payable accounts turnover, respectively; C_{LOM}, C_{OM}, C_{CL}, C_{IGT} – ratio of equity and borrowed funds, ratio of equity capital provision, current liquidity ratio, ratio of internal growth rate, respectively; R_{OM}, R_{CP}, R_A, R_s – return on equity, return on commercial output, return on assets and return on sale, respectively.

As can be seen from inquiry results (figure 2 and table 4), considered oil-gas industry has sufficient difference in both average and individual figures from normative and recommended numbers in several economic sources, which shows necessary to take into account industrial and specific features of researched enterprise.

As can be seen from table 4, formed basic assessment figures complex is sufficient to assess financial sustainability of enterprise – research object with taking into account its specific conditions.

Inquiry results shows that forming the ordered graph of main indicators of enterprise activity diversification and its financial stability be executed on sufficient statistic base of researched enterprise taking into account its industrial and specific conditions of both functioning and development.

Conclusions

Results of researching enterprise functioning in diversification motives and aims interconnection, analyzing changes of enterprise activity character in its diversification and character of financial impact on its diversification taking into account its industry, scale and financial sustainability had allowed sufficiently define both industrial and specific features of researched enterprise activity aims, motives and character taking into account haven financial resources.

The perceived relationship between the indicators of product diversification and financial stability provided they are qualitatively evaluated allows modelling and evaluating various strategies of the enterprise development and choosing the best one. Thus, when analyzing and modeling the enterprise's operations, it is necessary to take into account the specifics of their functioning, the specifics of the industry and products, as these factors have a significant impact on the efficiency of companies' operations.

Results of analyzing figures, characterizing its long-term financial sustainability, allows defining figures' contents, background financial sustainability factors and assessment criteria taking into account oil-gas industry specific. Besides that, they allow finding and economically backgrounding oil-gas industry enterprises activity diversification basic destinations, differing with taking into account their development condition, oil-gas equipment market conjuncture and industry specific. Executed inquires allow interconnect enterprise diversification motives and aims with elaborating new scheme of enterprise functioning and development, taking into account diversification background. Elaborated graph allows define factors and indicators interconnection of enterprise diversification and financial stability and is the base to elaborate economically-mathematical models and algorithms of defining several kinds credits for enterprise diversification, and also one of effective instruments of complex economic analysis, which is the base for backgrounding potential most effective enterprise diversification destinations with statistic modeling enterprise production process parameters and resource costs. At the same time, it is important to understand that business expansion should be circumspect, balanced and be carried out on the basis of a specifically formed strategy of enterprise diversification, in compliance with its specifics and threats to its effective development.

Acknowledgements

The research carried out with the financial support of the grant from the Program Competitiveness Enhancement of Peter the Great St.Petersburg Polytechnic University, Project 5-100-2020.

References

[1] Barnes, A.P., et al., 2015, The influence of diversification on long-term viability of the agricultural sector. Land Use Policy, 49, 404-412

[2] Kuzmenko, S.V., Zotkov, O.M., 2015, Enterprise production activity diversification and entrepreneurial activity effectiveness. In Youngs and science: VI Russian science-technical students, postgraduate and young scientists conference collection, Publisher: Siberia federal university, Krasnoyarsk: Russia

[3] Kuzmin, O.Y., Feshchur, R.V., and Drymalovska, K.V., 2015, The concept and the instruments of project approach to diversification of enterprise activities. Actual Problems of Economics, 174 (12), 415-422

[4] Lee, S., et al., 2016, Impact of Business Portfolio Diversification on Construction Company Insolvency in Korea. Journal of Management in Engineering, 32 (3), article number 05016003

[5] Mackey, T.B., Barney, J.B. and Dotson, J.P., 2017, Corporate diversification and the value of individual firms: a Bayesian approach. Strategic Management Journal, 38 (2), 322-341

[6] Rodionov, D.G., Kudryavtseva, T.J., 2016, Factors of the effective development of the St. Petersburg instrument engineering cluster. International Journal of Economics and Financial

Issues, 6 (2), 298-306
[7] Selçuk, E.A., 2015, Corporate diversification and firm value: evidence from emerging markets. International Journal of Emerging Markets, 10 (3), 294-310
[8] Shpak, N., Kyrylych, T. and Greblikaite, J., 2016, Diversification models of sales activity for steady development of an enterprise, Sustainability (Switzerland), 8 (4), article number 393
[9] Sokolitsyn A.S., et al., 2016. Corporative industrial firms enterprises activity diversification management and providing financial sustainability mechanisms improvement, Polytechnic university press, St-Petersburg
[10] Sokolitsyn, A.S., et al., 2016. Corporative industrial structures activity management and providing financial sustainability mechanisms improvement, Polytechnic university press, St-Petersburg
[11] Ushich, K.Y., 2016, Financial viability assessment for defining enterprise diversification destinations. J: Economics, sociology and law, 7(1), 49-54
[12] Yang, Y., Cao, Y. and Yang, L.-T.G., 2017, Product diversification and property performance in the urban lodging market: The relationship and its moderators. Tourism Management, 59, 363-375.
[13] Zhukova, O., 2014, Interrelation between figures of enterprise activity diversification and its financial sustainability. International science-inquiry journal, 9(28), 75-77.
[14] Klochkov, Y., Klochkova, E., Volgina, A., Dementiev, S. (2016). Human factor in quality function deployment. Source of the Document Proceedings - 2nd International Symposium on Stochastic Models in Reliability Engineering, Life Science, and Operations Management, SMRLO 2016, 7433156, pp. 466-468. DOI: 10.1109/SMRLO.2016.81
[15] V. E. Shchepinin, V. A. Leventsov, B. F. Zabelin, E. A. Konnikov and E. O. Kasianenko, "The content aspect of the tendency to reflect the actual result of management," 2017 6th International Conference on Reliability, Infocom Technologies and Optimization (Trends and Future Directions) (ICRITO), Noida, India, 2017, pp. 657-662. doi: 10.1109/ICRITO.2017.8342509
[16] Klochkov, Y., Klochkova, E., Didenko, N., Frolova, E., & Vlasova, N. (2018). Development of methodology for assessing risk of loss of a consumer through the fault of an outsourcer. Paper presented at the 2017 International Conference on Infocom Technologies and Unmanned Systems: Trends and Future Directions, ICTUS 2017, , 2018-January 719-724. doi:10.1109/ICTUS.2017.8286101
[17] Kudryavtseva, T., Rodionov, D., Kravchenko, V., Maryta, V. (2016). Special economic zones as an instrument of industrial policy pharmaceutical clusters in Russia. Proceedings of the 28th International Business Information Management Association Conference - Vision 2020: Innovation Management, Development Sustainability, and Competitive Economic Growth, p.1008-1018.
[18] Plotnikova, E.V., Pavlova, D.I. (2016). Assessment of housing stock and its influence on Sustainable regional development management. Proceedings of the 28th International Business Information Management Association Conference - Vision 2020: Innovation Management, Development Sustainability, and Competitive Economic Growth, p. 2639-2648
[19] Rudskaya, I. A., & Rodionov, D. G. (2018). Comprehensive evaluation of russian regional innovation system performance using a two-stage econometric model. *Espacios, 39*(4)
[20] Dvas, G. V., & Dubolazova, Y. A. (2018). Risk assessment and risk management of innovative activity of the enterprise. Paper presented at the Proceedings of the 31st International Business Information Management Association Conference, IBIMA 2018: Innovation Management and Education Excellence through Vision 2020, 5650-5653.
[21] Rudskaia, I., Rodionov, D., & Degtereva, V. (2017). Assessment of the effectiveness of regional innovation systems in russia. Paper presented at the Proceedings of the 29th International Business Information Management Association Conference - Education Excellence and Innovation Management through Vision 2020: From Regional Development Sustainability to Global Economic Growth, 3437-3449.

IOP Conf. Series: Materials Science and Engineering **618** (2019) 012087 doi:10.1088/1757-899X/618/1/012087

Functional hybrid decision-making system for difficult formalized productive and economic problems

S Korjagin[1], A Babkin[2] and P Klachek[3]

[1, 3] Engineering and Technical Institute Baltic Federal University of Immanuel Kant Kaliningrad, Russia
SKoryagin@kantiana.ru; pklachek@mail.ru

[2]Peter the Great St. Petersburg Polytechnic University, Saint-Petersburg Russia
al-vas@mail.ru

Abstract. A steady trend of the development of modern processes in human activities (social, economic, bioengineering, medical, technological) is a complication of the tasks, which are to be tackled on different levels of administration. This is connected with growth of amount of inconsistent information, necessary for well-grounded decision making, as well as with time saving for decision making. To improve the quality of decisions taken in these conditions, intellectual decision-making systems and artificial intelligence methods, on which the procedures for the direct preparation of solutions to various tasks are shifted, are increasingly being used. The article deals with the fundamentals of the structural organization of functional hybrid intelligent decision-making systems designed to solve difficult formalizable production and economic tasks, and suggests a new cognitive approach in the development of integration and effective management of formalized and weakly formalized knowledge in decision-making systems.

1. Introduction

A steady trend of the development of modern processes in human activities (social, economic, bioengineering, medical, technological) is a complication of the tasks, which are to be tackled on different levels of administration. This is connected with growth of amount of inconsistent information, necessary for well-grounded decision making, as well as with time saving for decision making. To improve the quality of decisions taken in these conditions, intellectual decision-making systems and artificial intelligence methods, on which the procedures for the direct preparation of solutions to various tasks are shifted, are increasingly being used. At the same time, as the accumulated experience of using intellectual decision-making systems in the production and economic sphere shows [1,2,3], the quality of the solutions they develop directly depends on how much mathematical methods prevail over the heuristic methods used to prepare solutions. From this point of view, every possible set of production and economic tasks can be roughly divided into three classes [1,4-7]: formalizable, difficult formalizable and non-formalized. At the same time, the overwhelming majority of problems arising in engineering, economics, politics, medicine, biology, business, and other fields belong to the second class, which is difficult formalizable [1,4-8]. The main features of this class of problems are [1,8-15]: a large number of factors affecting the behavior of the system, regarding which the decision is made; the heterogeneity of the data in which it operates; significant a priori uncertainty that does not allow us to determine the basic laws of its behavior (acceptable for it in the mathematical and statistical sense) using traditional methods of data analysis, but it is known that such patterns exist, and the basic composition of informative variables is roughly determined, as well as other factors detailed in [1, 8-15]. Thus, there is a need to integrate mathematical methods with heuristic methods and information technologies, which leads to the idea of hybrid systems. An important subclass of hybrid systems is hybrid intelligent systems (GIS) [1, 4-7, 10-13, 15]. A hybrid intelligent system is commonly understood as a system in which more than one method of simulating human intellectual activity is used to solve the problem [1, 4-7, 10-11]. Thus, GIS is a complex of: analytical models, expert systems, artificial neural networks, fuzzy systems, genetic algorithms, simulation statistical models [1, 4-7, 10-13].

Despite a considerable success in the field of application of hybrid intelligent systems, especially in weakly structured subject areas [1,4-7], and effective applied systems for solving difficult formalizable production and economic problems have not been created to date. The main problem is that when solving problems that are difficult to formalize, integration of two or more functional components (modeling methods) is required [1], wherefore serious problems arise (detailed in [1]) in the field of integration and control of formalizable and weakly formalized knowledge, as well as when creating application software, which requires the use of a non-standard approach [16-18].

This work is devoted to increasing the efficiency of intelligent information processing and management systems, in particular computer decision support systems, in solving difficult formalizable production and economic tasks, based on functional hybrid intelligent decision-making systems (FHDMS), which, as practice has shown [1, 4, 7], are able to successfully cope with the complexity of difficult formalizable problems and develop qualitative solutions in various subject areas [1, 4-7].

2. Materials and Methods (Model)

The main methodological elements of FHDMS are the notion of a nonhomogeneous problem-system, its homogeneous subtasks and an integrated method-system [1]. In figure 1. A non-homogeneous problem-system is presented.

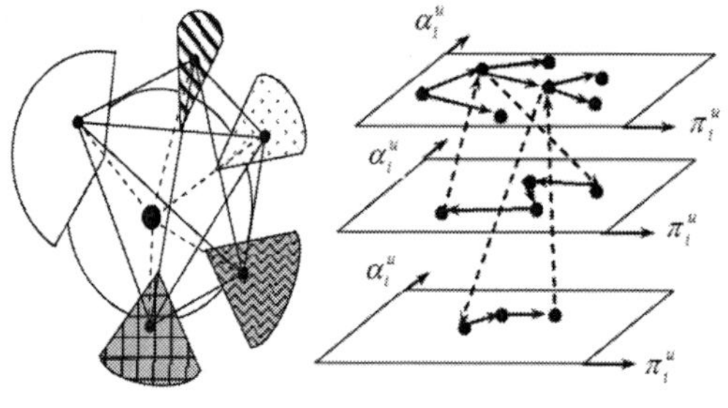

A. B.

Figure 1. Methodological basis FHDMS.

The initial, nonhomogeneous problem-system with subtasks is connected with the relations of inclusion "whole-part", which are indicated in figure 1.A. dotted line. These relations define the composition of the problem-system, which can vary within certain limits, without affecting the quality of the system. The other part of the links, already between subtasks, limits the degree of freedom between elements and does not allow one to solve some problems without others, setting the order on the cause-and-effect and time scales. A particular significance in the definition of a problem as a system has the characteristic of emergence when the properties of the system cannot be obtained by simply summing the properties of its individual elements. In other words, the problem-system must have properties that none of its component parts have in their individuality. For a nonhomogeneous problem-system, with the variability of its composition and structure, the method-system of its solution is constructed over some interrelated set of niches (autonomous operators [1, 4, 7]). Thus, FHDMS is a system that has the architecture of information exchange, input, output and functions in a nonhomogeneous state space. We call such a functioning a hybrid imitation process [1]. In figure 1.B. an example of the organization of a hybrid imitation process in FHDMS is presented, in a polymorphic regime [1]. In this mode, when modeling the problem-system, autonomous operators function simultaneously, i.e. parallel, and the user has the ability to observe changes in the state vector of FHDMS. This results in an effect well known in electronics when the same circuit is modeled as an analog device, which makes it possible to observe thermal, wave effects, and a discrete-logical device, which allows one to study diagrams of pulse shapes and amplitudes [23-26]. The relationship of element integration here arises as internal non-verbal images in the user's memory, which, comparing the dynamics of the device from different points of view, can "see" what cannot be detected in a single-model simulation.

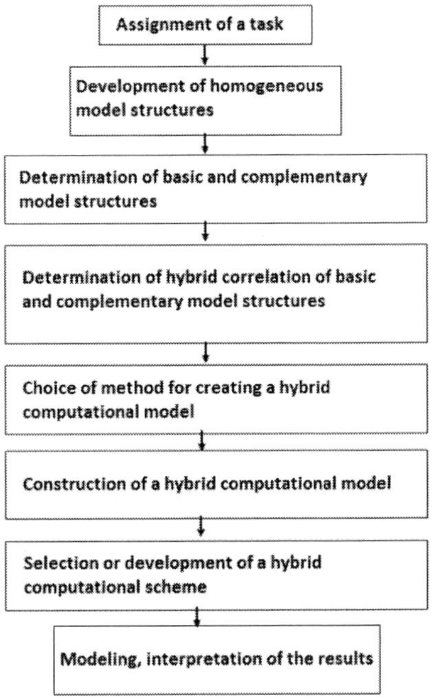

Figure 2. The universal architectural and technological scheme of modeling based on hybrid computational intelligence.

In [1], the concept of complex, weakly formalized, multicomponent economic systems, operating under conditions of uncertainty and difficult formalizable problems was introduced [1]. The set-theoretic representation of this class of complex systems is also realized in sources [4,7]. At present, the problem of developing methods and tools for hybrid modeling of weakly formalized, multicomponent economic systems [1] and their practical implementation for the construction of hybrid intelligent models of various types of component structures of complex economic systems remains extremely urgent. The hybrid intelligent model of complex, weakly formalized, multicomponent economic system is a model that has a universal character, functioning in real-time conditions and is based on the principles of intellectual activity of a human [1].

In figure 2. The universal architectural-technological scheme of the hybrid intelligent model of complex, weakly formalized, multicomponent economic system is given. The general formal scheme of the hybrid intelligent model of complex, weakly formalized, multicomponent economic system is given in [1]:

$$\dot{m}^{a}_{\Gamma BM} |^{\Gamma a} = < M^{u}, M^{h}, T_{гс}, I^{вс} > \qquad (1)$$

Where: $M^{u}=\{m_1^{u},\dots,m_{N_u}^{u}\}$ - a set of basic model structures [1]; $M^{h}=\{m_1^{h},\dots,m_{N_h}^{h}\}$ - a set of complementary model structures, ($\forall m_i^{u}\ \exists M^{h}=\{m_1^{h},\dots,m_{N_h}^{h}\}$, where $i=1,\dots,N_{\Pi}$, $\forall i\ (N_h=\mathrm{var}\,y)$, $m_i^{u}\in M^{u}$) [2,6]; $T_{гс}$ - a table of hybrid strategies [1]; $I^{гви}$ - the interpreter representing the four processes [2,6]:

$I^{\Gamma BИ}=<I^{\Gamma 1},I^{\Gamma 2},I^{\Gamma 3},I^{\Gamma 4}>$, where: $I^{\Gamma 1}$— the process of studying the hybrid correlation of basic and complementary model structures; $I^{\Gamma 2}$ - the selection process in accordance with $T_{гс}$ the method of hybrid computational intelligence [1]; $I^{\Gamma 3}$— the development process in accordance with the hybrid computational intelligence method of the hybrid computational model; $I^{\Gamma 4}$ — process of selecting or developing a hybrid computational scheme.

In figure 3. The FHDMS model for complex, weakly formalized, multicomponent economic system is proposed, which assumes the realization on the basis of (1) the model of the collective decision-making process [1] peculiar to the large class of complex, weakly formalized, multicomponent economic system.

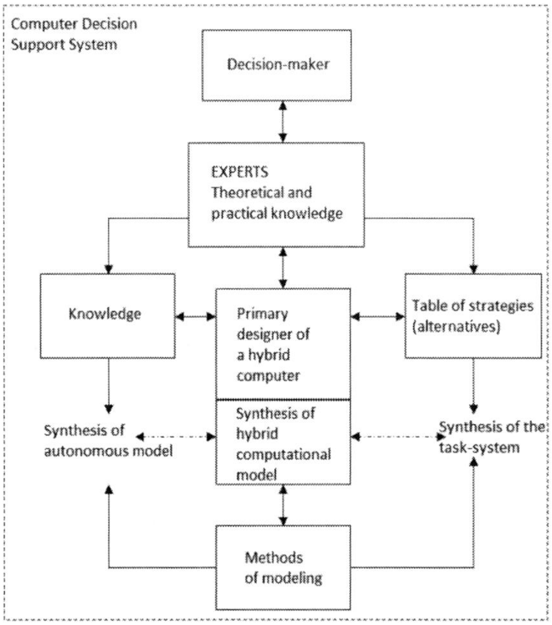

Figure 3. Model FHDMS in complex, weakly formalized, multicomponent economic system.

An essential advantage of the proposed approach was the possibility of a rapid transition to the automated development process of FHDMS based on multi-agent systems (MAS) [1,4,7,9]. In figure 4. the FHDMS architecture based on MAS is presented, which was considered in detail in [1].

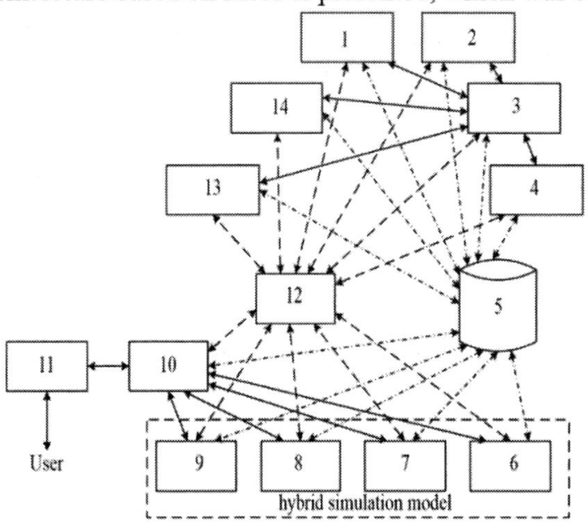

Figure 4. Architecture of FHDMS based on multi-agent systems, 1 - Logical agent, 2 - Fuzzy agent, 3 - Agent converter, 4 - Linguistic agent, 5 - System model, 6 - Search agent solutions 4, 7 - Search agent solutions 3, 8 - Search agent solutions 2, 9 - Search agent solutions 1, 10 - Interface agent, 11 - Agent decision-making, 12 - Agent middleman, 13 - Analysis agent, 14- Stochastic agent.

3. Results and Discussion

IOP Conf. Series: Materials Science and Engineering **618** (2019) 012087 doi:10.1088/1757-899X/618/1/012087

Approbation of the proposed solutions was carried out using the example of creating an intelligent control system for the technological process of drilling oil and gas [1]. One of the most promising technologies for thermal recovery of heavy oils and natural bitumen in a system of horizontal wells is the process of Steam Assisted Gravity Drainage [1]. An important significance for the implementation of this technology is the choice of optimal parameters of the coolant injected into the formation. The heat content of the steam depends on its dryness. The dependence of the radius of the zone of heating of the inter-well space is established as a function of the injection time of the steam for different values of dryness [19-21]. Thus, the fulfillment of the requirements arising from the established regularity of heat exchange between the reservoir and surrounding rocks is of great practical importance for increasing the technological efficiency of developing deposits with hard-to-recover oil reserves. Controlling the performance, as well as the quality of the steam generated by them, significantly increases the possibility of optimizing operating modes.

World experience testifies to the possibility of effective development of deposits of heavy oils and bitumen by mining, open-pit method and surface wells, in such countries as Russia, Canada, Venezuela, China, etc. [1]. In Russia, the experience of developing deposits of heavy oils and natural bitumen has been accumulated in Tatarstan, Udmurtia, the Republic of Komi and other oil-producing regions. Feasibility and economic efficiency of the development of deposits of heavy oils and natural bitumen is determined by the geological and physical conditions of their occurrence, physical and chemical characteristics, reservoir data, as well as the results of their processing.

One of the most promising technologies for thermal recovery of heavy oils and natural bitumen in a system of horizontal wells is the process of Steam Assisted Gravity Drainage (SAGD) [1], currently used in Canada, Venezuela, the USA. In addition to the classical method of SAGD technology, there are several of its various modifications, allowing for the possibility of taking into account the features of the geological and physical properties of reservoirs, and the regulation of the development of oil fields currently being implemented in Russia (Tatarstan, Komi Republic).

The essence of the technology lies in the formation of a "steam chamber" by injecting steam into the upper horizontal well end and taking out the oily liquid through the lower horizontal end of the production well. The zone of oil displacement is constantly increasing and the flow of oil spreads to the roof of the formation, and then it expands horizontally, merging with the heated zones located around the neighboring wells [22-24, 30]. Thus, at the boundary of the steam chamber, the steam condenses when the heat of the inactive oil is transferred, and the heated oil is displaced by its own weight and the condensed steam from the top to the bottom, that is, the performance of the plant is influenced by two factors - gravity drainage and pressure displacement.

To optimally form a steam chamber, it is necessary to preheat the interwell space for several months by circulating the steam in both wells. As a result, conductive heating of the bottomhole formation zone is carried out, the viscosity of the oil is reduced and the hydrodynamic connection between the wells is improved.

In accordance with (1), based on the method of knowledge genesis [1], a hybrid computational model was developed for calculating the required volume of steam, which when performing pumping at the preheating stage, takes into account the geological characteristics of the experimental section [1]:

$$Q=\left(\frac{2\pi(T_3-T_n)z\lambda_F}{(X_3-X^*)\left(\ln\frac{2\sqrt{\frac{\lambda_F}{\rho_{CK}}t^*}}{r_{CKK}}-0.29\right)}\right)^{a^3_{nf}} \tag{2}$$

Where: Q - the steam flow rate at pumping, π - the mathematical constant, T_3 - the temperature of the steam at the bottom of the well, T_n - the initial temperature of the formation, z - the length of the heated section, λ_F - the thermal conductivity of the soil, X_3 - the dryness of the steam at the bottom of the well, X^* - the dryness of the steam at the collar, ρ_{CK} - the density of the formation frame, C_r - the latent heat of vaporization, C_{CK} - specific heat of the formation frame, ρ_n - the vapor density, t^* - the time measured from the beginning of the circulation of the working agent inside the pipeline, the, $\alpha^3_{nf} =< C_r, C_{CK}, \rho_n)), Y(\Gamma, \text{mathematical degree}), KB, F^{\mu}, F^{TS}, F^{y}, I^{f} >$ - the classical fuzzy system [2].

An important indicator, both at the preheating stage and during the formation of the steam chamber, is the choice of the optimal parameters of the coolant pumped into the reservoir. Water vapor, which has a greater enthalpy (heat content) compared to hot water, provides a higher oil recovery. The change in the enthalpy of saturated water vapor depends both on temperature and pressure, and on its dryness. The higher the dryness at the same pressure and temperature, the greater the heat content, i.e., the more qualitative the steam, the faster and the more heat-efficient it is possible to realize warming of the inter-well space and the further process of realizing the Steam-Assisted Gravity Drainage.

In accordance with (1), based on the method of knowledge genesis [1], a hybrid computational model was developed for calculating the radius of heating of the inter-well space, depending on the pumping period for different values of dryness [1]:

$$r=\left(\sqrt{\frac{QC_r X_3 \rho_n t}{\pi(C_r X_3 \rho_n + C_{CK}\rho_{CK}(T_3 - T_n)}}\right)^{a^3_{rf}} \quad (3)$$

Where: r - radius of formation heating, t - time of steam pumping, m - formation porosity, h - distance between wells, β - coefficient of anisotropy of the reservoir, $\alpha^3_{nr} =< m, h, \beta)), Y(\Gamma, \text{mathematical degree}), KB, F^{\mu}, F^{TS}, F^{y}, I^{f} >$ - classical fuzzy system [1].

Currently, in collaboration with leading specialists of the Harvard School of Engineering and Applied Sciences, the development of innovative FHDMS is being carried out with the technological process of drilling oil and gas [1] (figures 4, 5).

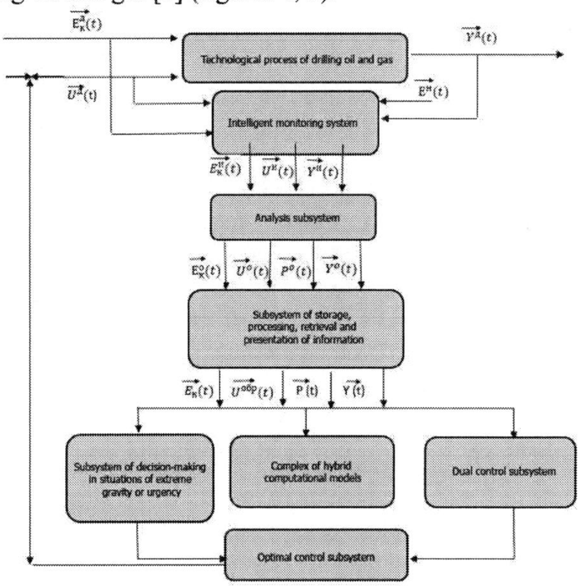

Figure 5. Architecture of innovative FHDMS in the technological process of drilling oil and gas [1].

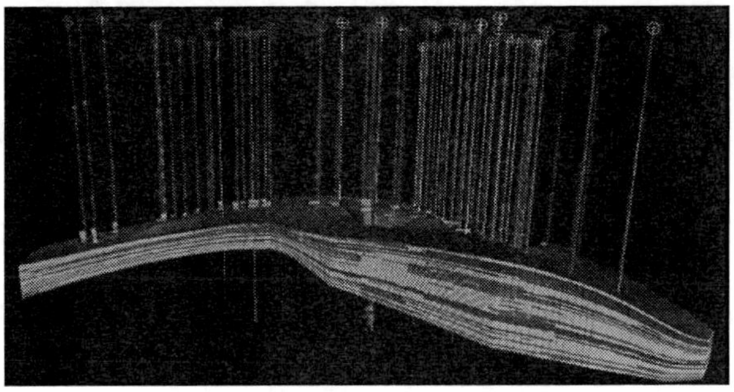

Figure 6. An example of modelling, based on (2.3), optimal volumes and modes of injection of steam into the well [1].

Preliminary tests of the FHDMS prototype used in the oil and gas drilling process at the Alpine oil field in the USA, Alaska, showed a perspective and massive opportunities of the suggested approach, (see section 2: Methods and Materials), and the developed prototype FHDMS [1].

4. Conclusion

Opening the way to a new scientific discipline - knowledge engineering, a group of scientists led by J. Weizenbaum (USA) put forward the hypothesis that a person is good at solving problems through his professional knowledge. The problems are enough to be transferred to a computer, so that it can solve practical problems. Thus, expert systems appeared. Another hypothesis can be put forward - a person solves complex problems well thanks to collective efforts in decision support systems, and for a computer to solve such problems, it is necessary to learn how to simulate in its memory the work of the group of people over a practical problem. In this way, hybrid intelligent systems were created, in particular functional hybrid intelligent systems. They are the focus of this article.

For the first time in the field of methodological support for the creation of functional hybrid intelligent decision-making systems, which as the practice has shown in [1,4-7,10-13], are capable to successfully cope with the complexity of difficult formalizable production and economic tasks and develop qualitative solutions in various subject areas [1 , 4-7,10-13]. This promising approach is conceived by the authors on the "border" of hybrid systems, synergetic artificial intelligence, neuro and psychophysiology, philosophy, cybernetics and centers around self-organization of groups of people that develop and make decisions in complex situations [15].

Currently, FHDMS has been successfully applied in the modeling of complex systems and the solution of complexes of applied production and economic problems, in agriculture, engineering, aircraft building, water ecosystems, etc. [1,4,7,9]. Work in this area is actively continuing.

Acknowledgements

The research carried out with the financial support of the grant from the Program Competitiveness Enhancement of Peter the Great St.Petersburg Polytechnic University, Project 5-100-2020.

References

[1] P. M. Klachek, K. L. Polupan, S. I. Koryagin, I. V. Liberman. Gibridnyy vychislitel'nyy intellekt. Osnovy teorii i tekhnologiy sozdaniya prikladnykh sistem. Kaliningrad: Publishing house of the Baltic Federal University of Immanuel Kant, 2018.

[2] A.V Kroshilin, A.V Babkin, S.V. Kroshilin, Peculiarities of decision-making support systems development based on fuzzy logic // Nauchno-tekhnicheskie Vedomosti of SPbGPU. Economics, 2010, vol. 2, No. 97, pp. 57-83.

[3] A.N. Tsatsulin, A.V. Babkin, N.I. Babkina, “Analysis of the structural components and measurement of the effects of cost inflation in the industry with the help of the index method”, Proceedings of the 28th International Business Information Management Association Conference - Vision 2020: Innovation Management, Development Sustainability, and Competitive Economic Growth, 2016. pp. 1559-1573.

[4] A.V. Kolesnikov, Gibridnyye intellektual'nyye sistemy: Teoriya i tekhnologiya razrabotki. SPb.: Izd-vo SPbGTU, 2001.

[5] A.V. Gavrilov, Gibridnyye intellektual'nyye sistemy. Novosibirsk: Izd-vo NGTU, 2003.
[6] N.G. Yarushkina, Osnovy teorii nechetkikh i gibridnykh sistem. M.: Finansy i statistika, 2004.
[7] A.V. Kolesnikov, I.A. Kirikov, Metodologiya i tekhnologiya resheniya slozhnykh zadach metodami funktsional'nykh gibridnykh intellektual'nykh sistem. M.: IPI RAN, 2007.
[8] I.Z. Batyrshin, A.A. Nedosekin, A.A. Stetsko, V.B. Tarasov, A.V. Yazenin, N.G. Yarushkina, Nechetkiye gibridnyye sistemy: teoriya i praktika. M.: Fizmatlit, 2007.
[9] A.V. Kolesnikov, I.A. Kirikov, S.V. Listopad, S.B. Rumovskaya, A.A. Domanitskiy, Resheniye slozhnykh zadach kommivoyazhera metodami funktsional'nykh gibridnykh intellektual'nykh sistem. M.: IPI RAN, 2011.
[10] L.R. Medsker, Hybrid Intelligent Systems. Boston: Kluwer Academic Publishers, 1995.
[11] S. Wermter, .R Sun, Hybrid Neural Systems. Heidelberg, Germany: Springer-Verlag, 2000.
[12] M. Negnevitsky, Artificial Intelligence. A guide to intelligent systems. Harlow, England: Addison-Wesley, 2005.
[13] O. Castillo, P. Mellin, Hybrid Intelligent Systems. Heidelberg, Germany: Springer-Verlag, 2006.
[14] T.A. Gavrilova, V.F. Khoroshevskiy, Bazy znaniy intellektual'nykh sistem. SPb.: Piter, 2000.
[15] M. Negnevitsky, Artificial Intelligence. A guide to intelligent systems. Harlow, England: Addison-Wesley, 2005.
[16] A. A. Rudenko, D. V. Antipov, and M.O. Iskoskov, "Ensuring business continuity of a machine building enterprise through organisation development of manufacturing system", International Journal of Applied Engineering Research, vol. 11, no. 5, 2016, pp.3211-3215.
[17] V. N. Kozlovskiy, D. V. Antipov, and A. V. Zayatrov, "Methodology for analysis and forecasting of cars in operation quality", Actual Problems of Economics, vol. 186, no. 12, 2016, pp.387-398.
[18] A. D. Novokshenov, P. A. Marchenko, A. S. Nemov, and A. I. Borovkov, "Optimizing the support of a stellar sensor", Russian Engineering Research, vol. 38, no. 1, 2018, pp. 7-12. doi:10.3103/S1068798X18010148
[19] O. V. Antonova, Y. Y. Boldyrev, A. I. Borovkov, and I. B. Voinov, "On the development of a design procedure of the hydrodynamic characteristics of hydrogenerator thrust bearings taking into account the heat exchange", Journal of Machinery Manufacture and Reliability, vol. 46, no. 6, 2017, pp. 572-578. doi:10.3103/S1052618817060036
[20] A. Nemov, A. Novokshenov, A. Lagutkina, S. Sycheva, D. Mamchits, A. Borovkov, V. Vershkov, D. Shelukhin, V. Lukyanov, "Multiphysics engineering analysis for high field side reflectometry", Fusion Engineering and Design, vol. 124, 2017, pp. 501-506. doi:10.1016/j.fusengdes.2017.05.077
[21] O. V. Antonova, A. I. Borovkov, Y. U. Boldyrev, and I. B. Voynov, "Variational problem for hydrogenerator thrust bearing", Materials Physics and Mechanics, vol. 34, no. 1, 2017, pp. 97-102. doi:10.18720/MPM.3412017_12
[22] L. N. Borisoglebskaya, and S. M. Sergeev, "Model of assessment of the degree of interest in business interaction with the university", Journal of Applied Economic Sciences, vol. 12, no. 8, 2018, pp. 2423-2448.
[23] A. Vasilyev, V. Lozhkin, D. Tarkhov, O. Lozhkina, and V. Timofeev, "Physical and mathematical modeling of pollutant emissions when burning peat", Journal of Physics: Conference Series, vol. 919, no. 1, 2017. doi:10.1088/1742-6596/919/1/012001
[24] A Vilken, V., Kalinina, O., & Dubgorn, A. (2018). Specificity of high-rise construction and real estate markets in the regional economy: An analysis of russian practice (example of st. petersburg). Paper presented at the E3S Web of Conferences, , 33 doi:10.1051/e3sconf/20183303012
[25] M. A. Poltavtseva, and A. I. Pechenkin, "Intelligent data analysis in decision support systems for penetration tests", Australian Journal of Political Science, vol. 51, no. 8, 985-991. doi:10.3103/S014641161708017X

IOP Conf. Series: Materials Science and Engineering **618** (2019) 012088 doi:10.1088/1757-899X/618/1/012088

The control model of safety management systems

V G. Burlov, O M Lepeshkin, M O Lepeshkin and F A Gomazov

Peter the Great St.Petersburg Polytechnic University, St.Petersburg, Russia

E-mail: burlovvg@mail.ru, fgomazov@gmail.com

Abstract. This article presents a control model of security management systems from destructive software and hardware effects. Considered descriptive and analytical control model of security, presents the structural and logical scheme of a control model of security management systems from destructive software and hardware effects, carrying out the search and elimination of conflicts, associated with the logical inconsistency of the structure of the control system, and conflicts associated with ongoing processes, i.e. with the fulfillment of the tasks of the regulations, operating in the system. It is proposed to realization the model, using the language of radical schemes, based on the stand for providing integrated developments. The model integrates with mathematical means of searching, predicting and eliminating conflicting schemes in the system.

1. Introduction

The impact on information is determined by the results of an evaluation of the capabilities (capacity, equipment and motivation) of external and internal violators, analysis of possible vulnerabilities of the information system, possible ways to implement threats to information safety and the consequences of violation of the properties of information safety (confidentiality, integrity, accessibility).

When determining the threats to information safety, it is necessary to take into account the structural and functional characteristics of the information system, the information technologies used and the characteristics (conditions) of the functioning of the information system. Consequently, the effectiveness of measures taken to protection information depends on the quality of the definition of threats to information safety for a specific information system in the specific conditions of its operation.

Modern management systems must respond promptly to destructive software and hardware effects, both external attacks and internal malicious effects, while preserving their accessibility, reliability and efficiency. In many ways, the purpose of the safety process is to increase network resiliency. Instead of becoming "victims" of unauthorized impacts, networks should be able to "absorb" attacks and remain operational [1-5,7].

Destructive effects on management systems can occur in many ways and can differ from each other. It can also be attacks on the infrastructure, computer network, the denial of services in order to disrupt their normal functioning [6, 8-12].

Any software or hardware is not perfect, and you can always assume, that there are vulnerabilities in it, that allow you to perform any actions, that violate the established order of resources use. In addition, reacting to unauthorized activity or attempts to hack the system in near real-time mode is almost impossible, if these functions are performed manually [13-15].

At the present stage of development of safety systems in management systems the task of developing control of security becomes more urgent, which should ensure detection of destructive effects on the communication network in a time close to real mode [2, 16-18].

To ensure compliance with the safety, requirements for management systems, they need to be represented as a complex system using object-oriented design tools.

2. Structure of the control model of the security of management systems from destructive software and hardware effects

Increasing the security of management systems (MS), it is proposed to make by realizing a process approach, that involves control the logical (regulatory) condition of the system. This can be achieved by registering the current condition of the MS and the processes, implemented by the system; simulation of software and hardware effects, or processes with simultaneous collection and processing of data on the compliance of system characteristics (system parameters) with set requirements, for the goal of identifying vulnerabilities and evaluating security; visualization of the composition, structure and process of the functioning of the system (established segments, fragments, elements); verification of the set condition and the regulations of the functioning of the elements of the system that ensure safety (their compliance with specified protection profiles and regulations) [3, 17-22].

To be able to monitor the logical (regulatory) condition of a MS necessary to develop a control model their security from destructive software and hardware effects, which allows to identify information conflicts (logical inconsistency, vulnerability of the system, violation of the integrity of the process). The tasks, associated with ensuring logical consistency of information processes, also accessibility of MS services can be solved by searching for conflicting schemes in the environment of radicals and their elimination.

2.1. Verbal control model of security management systems from destructive software and hardware effects

The basis of the construction of the MS control system should be laid formalized set of information processes, the structure of which must be developed by mapping the MS in the environment of radicals.

At this stage, the model must solve the problems of searching and eliminating conflicts, associated with the logical inconsistency of structure MS regulations.

At this stage, the model solve tasks of search and resolve conflicts, associated with to the logical inconsistency of structure regulations MS. When functioning in a MS, in the process of execution regulations arise conflicts, associated with the execution of processes (transactional tasks of regulations, operating in the system). The task model is the early detection and elimination of conflicts [4].

From these statements it follows, that the model of the MS must operate in two modes - *static and dynamic*, realization the solution of tasks of control at each stage of operation of the system.

The order of the functioning of the control model of security MS can be determined in the form of an oriented graph (figure 1), which includes:

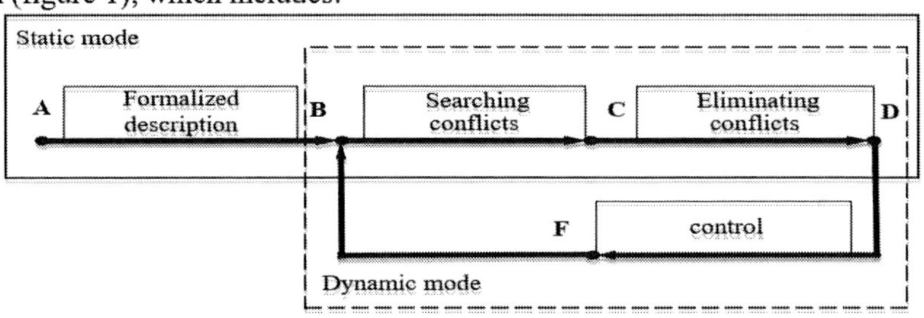

Figure 1. Oriented graph, reflecting the order of the functioning of the control model of security MS.

- *in a static mode* - the construction of the initial description of the structure of the control system in the environment of radicals, the search and elimination of conflicts - the subgraph ABCD.

- *in a dynamic mode* - the process of searching and eliminating conflicts in the process functioning of the control system in the process implementation of regulations in force in the system - the subgraph BCDF.

The solution of task of control of security MS from destructive software and hardware effects (DSHE) is achieved by identifying and eliminating conflicts, arising in the system support in decision-making in the time mode close to real information signals about detected conflict conditions (associated with the logical inconsistency of the structure of regulations and violation of the integrity of processes (committed transactional tasks of the regulations operating in the system)).

Alignment of the conformity of the current condition and regulation of the functioning of the system to a specified security profile will facilitate with a high probability of identifying conflicting schemes, associated with the logical inconsistency of the structure of regulation and the implementation of processes (regulations), operating in the system.

Thus, the structural-logical scheme of the control model of security MS from DSHE will look like the following (figure 2)

The methodology for solving the tasks of the life cycle of a complex system based on the stands for the provision of integrated development, proposed by professor Chechkin A.V., is completely suitable for solving the tasks of control of the security MS from DSHE, because:

- firstly, it includes a formalized description of the object of control;
- secondly, it includes the means of managing conflicts arising in the process of functioning of the object of control [5, 23].

The control model of MS should be implemented in the language of radical schemes and integrated with mathematical means of searching, forecasting and eliminating conflicting schemes in the system.

In addition, the method of modeling control of MS includes two stages:

- development of a formalized description of the control system of MS, which includes the structure, properties and relationships of the MS elements, also control of MS;
- development of the activation control subsystem of MS, which implements a closed management system for the control model of MS and a formalized description of control of MS with mathematical means for searching, forecasting and eliminating information conflicts in the system.

One of the private scientific tasks is the development of mathematical means of searching and eliminating conflict schemes. Therefore, the task of realization a control model in the language of radical schemes is a formalized description of control elements of MS, also the integration of the developed mathematical means of searching and eliminating information conflicts with this model [6].

Analyzing the composition of MS, it should be noted, that they are fully implemented on the basis of information technology (IT). The equipment consists of the IT elemental base, the so-called "IT products".

The evaluation of the safety of IT products is regulated by GOST R ISO / IEC 15408 "Methods and means of ensuring safety. Criteria for evaluation the safety of information technology".

The term "IT product" is not defined unambiguously, it can be understood by any type of entity, built using IT, whether it is a complete system, used by one organization or a line of ready-to-use IT products for implementation (delivery) to different customers. As an example of an IT system, can be used for example, control means to the alienation (transfer) of information from removable computer storage media, which include the following components distributed by the components of the information system or are themselves finished products:

- specialized removable computer storage media;
- initialization software;
- management software;
- software for interaction with removable computer storage media.

Since IT products can be used in different ways and in many types of operational environments, the notion of safety will vary depending on the IT product. Therefore, the end result of the evaluation can never be the conclusion that "this IT product is safe", instead it is argued that "this IT product complies with this safety specification".

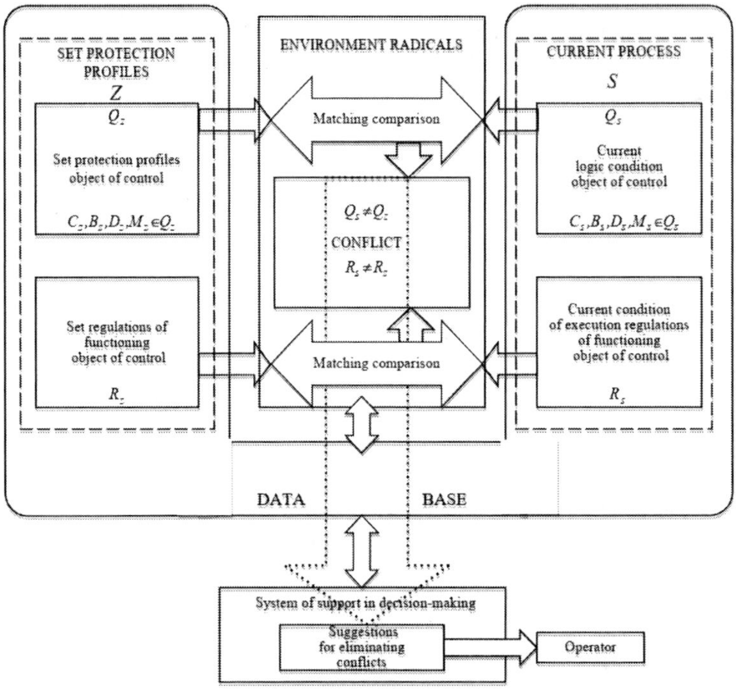

Figure 2. Structural and logical scheme of the control model of security MS from DSHE

GOST R ISO / IEC 15408 specifies standardized specifications of safety for:

- determining the specific content required to evaluation of IT product for compliance with the safety specification;
- creating conditions for comparing the safety specifications of various IT products.

GOST R ISO / IEC 15408 defines two different types of safety specifications: protection profiles (PP) and safety tasks. The difference between them is determined by their purpose in the typical development process, the evaluation of IT product and its acquisition by the customer. The protection profile, in fact, is a document in which the customer's safety requirements are stated in a formalized, standardized form.

The results of an evaluation of IT product can be used by the asset owner when deciding whether to accept the risk, associated with the exposure of assets exposed to specific threats.

PP, as a rule, are developed by the authorized federal executive body (FSTEC of Russia), large organizations, groups of organizations, departments, etc., as their development requires considerable effort.

PP contains for use as a specification of safety "Functional requirements of safety".

The goals of safety according to GOST is a short and abstract statement of the proposed solution to problem.

It is very important to understand that the goals are not the recommendations, but the necessary conditions for the functioning of the object of control (OC) according to the specification. All these goals (tasks) must be fully achieved, and the responsibility for their achievement rests with the user of the object of control or him organization (for example, the information system operator); itself the object of control will not achieve these goals. If at least one of these goals is not achieved, there may be a situation, where the object of control will functioning in an unsafe mode.

Requirements of safety are the result of converting goals s of safety for OC. They are usually presented at a more detailed level of abstraction, but they must be a full representation (goals s of safety must be fully taken into account) and be independent of any particular technical solution (implementation).

The OC can include resources in the form of electronic data carriers (such as main memory, disk space), peripherals (such as printers), and computing capabilities (such as processor time), that can be used to process and store information and are subject evaluation.

The assessment first of all confirms, that a certain set of functional safety requirements applies to OC resources. Requirements of safety define the rules, by which the OC manages the use and access to its resources and, thus, to information and services, controlled by the OC [7, 24].

In aggregate, those parts of the OC, that are aimed at the correct realization of safety requirements, are defined as safety functions of the evaluation object. Safety functions combine the functionality of all hardware, software and hardware-software of the OC, on which as directly and as indirectly responsible for ensuring safety. In fact, the mechanisms of the safety functions mechanisms are the rules, defined in the safety requirements, and the necessary opportunities are provided.

Safety requirements are selected based on the model of common functionality opportunities of OC. In this functional model defined resources, users, subjects, objects and operations. Then the safety requirements determine the safety functionalities opportunities thus, within the framework of the functional model of the object of control, the safety goals are achieved [8].

In relation to the MS, it is proposed to consider as its safety control objects IT products its constituent elements of the equipment as its constituent parts.

Therefore, for the evaluation of security control MS it is necessary to compare with set relevant protection profile goals, requirements, safety functions and resources of the object of control, the current "stationary" system-state and "dynamic" process execution regulations of object of control.

2.2. Analytical control model of security management systems from destructive software and hardware effects.

For the formation of the initial data, characterizing the conflict situation, set the mathematical notation of the sets, that make up the model of the estimated system:

$C(c_1,..., c_n)$ - a set goals of the security system, which should be achieved in the process functioning of the system;

$B(b_1,..., b_m)$ - a set of security requirements, the implementation of which is necessary to achieve the set security goals of the system;

$D(d_1,..., d_k)$ - a set of functions, the implementation of which is achieved to meet the requirements of security;

$R(a_1,..., a_j)$ - a set tasks of the regulations, achieved in the process functioning of the system;

$M(m_1,..., m_j)$ - a set of necessary resources to achieve the tasks of the regulations;

C_z, B_z, D_z, M_z, R_z - goals, requirements, functions, resources and regulations set by the PP of system;

C_s, B_s, D_s, M_s, R_s – goals, requirements, functions, resources, that determine the direct structure of the MS regulations, and regulations achieved in the system.

The first stage of compliance approvals solves the task of eliminating information conflicts, related to the structure of the MS regulations:

If $C_s=C_z$, then there are no conflict schemes in the system, and if $C_s \neq C_z$, then in the system there is a conflict, expressed in the absence of compliance of the stated goals with the goals, set by the corresponding PP.

Coordination of the "Goals-Requirements" is to eliminate information conflicts, associated with the lack of the possibility of achieving the stated goals due to non-compliance with the established safety requirements, and, as a consequence, the lack of integrity of the process of implementation of certain regulations.

If $B_s=B_z$, then there are no conflicting schemes in the system, and if $B_s \neq B_z$, then in the system there is a conflict, expressed in the absence of compliance with the requirements set by the requirements, set by the corresponding PP.

The coordination of the "Requirements -Functions" is to eliminate information conflicts, related to non-fulfillment of the requirement: for each function, the realization of which is achieved by the corresponding requirement, the functionality possibility, that implement this function must be defined.

If $D_s=D_z$, then there are no conflict schemes in the system, and if $D_s\neq D_z$, then in the system there is a conflict, expressed in the absence of compliance of the realization functions with the functions set by the corresponding PP [16-19].

The coordination of the "Functions -Resources" is to eliminate information conflicts, related to the uncertainty of resources for the realization of some function.

If $M_s=M_z$, then there are no conflicting schemes in the system, and if $M_s\neq M_z$, then in the security system there is a conflict, expressed in the discrepancy of the available resources, used to realization functions, resources set by the corresponding PP.

Let us present system set PP of the target function of a collectively sets of goals, requirements, functions and resources, necessary to ensure the security of the object of control:

$$Q_z=f(C_z,B_z,D_z,M_z).$$

And the logical condition of the system goal function collectively sets the claimed goals, defined requirements, realized the functions and the resources available in the system:

$$Q_s=f(C_s,B_s,D_s,M_s).$$

Then the assertion, that when $Q_s=Q_z$ conflict schemas do not exist.

Conversely, when $Q_s\neq Q_z$ conflict schemes are present in the system, which means, that the integrity of the logical structure of the regulations is violated, which indicates a decrease in the protection of the OC.

The regulations R_z, defined PP, is a sequence of minimal logically-formalized tasks a_z, set PP, meaningful and committed only fully:

$$R_z=a_{z(1)}+a_{z(2)}+....+a_{z(n)}=\sum_{k=1}^{n}a_{z(k)}.$$

The task a_z of the regulations R_Z, set by PP, is the use of certain PP resources M_Z, at the set time t_z:

$$a_z=M_z\cdot t_z,$$

were $a_z \in R_z$, a_z - *const*.

When performing some tasks $a_{s(i)}$ in process of implementation of regulations R_s in time $t_{s(i)}$ with using the resources $m_{s(i)}$, it must belong to a task sequence as performed in the process of implementation of the regulations R_s, i.e. $a_{s(i)} \in R_s$.

In this case, the task $a_{s(i)}$ must correspond to the set PP task $a_{z(i)}$ of regulation R_Z, i.e. $a_{z\,(i)} \in R_z$. Used in the process of making tasks $a_{s(i)}$ resources $m_{s(i)}$ and the time $t_{s(i)}$ must comply with certain resources $m_{z(i)}$ and a set time $t_{z(i)}$, also set PP.

If $a_{s\,(i)} \notin R_s$, this indicates, that there is a conflict, expressed in violation of the order of implementation of the regulations, and, as a rule, in violation of the integrity of the process of functioning of the system, which is defined by PP.

If $m_{s(i)} \neq m_{z(i)}$, this indicates a conflict condition, associated with the uncertainty of resources to perform some task $a_{s(i)}$ of regulations R_s.

If $t_{s(i)} > t_{z(i)}$, this indicates, that in set PP time the task $a_{s(i)}$ of regulations R_s will not be committed, which means availability of conflict schemas and the violation of the integrity of the process of implementation of regulations.

On the basis of this control system is presented in the form of functions of levels of agreement:

$$F_{purs.\,to\,1'st\,eq.}=f(C,B);$$

$$F_{purs.\,to\,2'nd\,eq.}=f(B,D);$$

$$F_{purs.\,to\,3'rd\,eq.}=f(D,M);$$

$$F_{purs.\,to\,1'st\,st.}=f(F_{purs.\,to\,1'st\,eq.},F_{purs.\,to\,2'nd\,eq.},F_{purs.\,to\,3'rd\,eq.}).$$

Next, made agreement the dynamic levels:

$$F_{purs.to\,1} = f(M_R);$$

$$F_{purs.to\,2} = f(t_R),$$

where M_R - the set of regulation resources, t_R - the regulation time;

$$F_{purs.to\,1'st\,st.} = f(F_{purs.to\,1}, F_{purs.to\,2}).$$

The goal function of control security of MS from destructive software and hardware effects will have the form:

$$F_{CK} = f(F_{purs.to\,1'st\,st.}, F_{purs.to\,2'nd\,st.}).$$

3. Conclusion

To create a control model of security management systems from destructive software and hardware effects proposed to use the concept of building a stand for complex development, which implements the solution of the tasks of complex system life cycle on the basis of two components: the model of the object under study and the means of conflict management of the object under study. The proposed model is based on the c agreement of the system components (goals, requirements, functions, resources).

On the basis environment radicals and a model for the agreement of components of the system proposed to develop an activating subsystem for control security MS from destructive software and hardware effects, which realizes a closed management system model control and is able to solve the tasks of detection informational conflicts and their elimination in process of functioning of MS.

Acknowledgment
The research carried out with the financial support of the grant from the Program Competitiveness Enhancement of Peter the Great St. Petersburg Polytechnic University.

References

[1] Bolodian I, Melikhov A, Tanklevskiy L 2017 Fire safety arrangement of inhabited pressurized compartments of manned spacecraft Acta Astronautica, 135 pp 92-99 DOI: 10.1016/j.actaastro.2016.10.003

[2] Bardin A, Korotkov A, Kazarnovskij V 2017 Limit of fire resistance of a steel transport structures. Simplified calculation methods *IOP Conference Series: Earth and Environmental Science, 90 (1)* DOI: 10.1088/1755-1315/90/1/012207

[3] Bolgov I, Kaverzneva T, Kolesova S, Stolyarov O, Tarkhov D 2016 Neural network model of rupture conditions for elastic material sample based on measurements at static loading under different strain rates *Journal of Physics: Conference Series* DOI: 10.1088/1742-6596/772/1/012032

[4] Bolodian I, Melikhov A, Tanklevskiy L 2017 Fire safety arrangement of inhabited pressurized compartments of manned spacecraft Acta Astronautica, 135 pp 92-99 DOI: 10.1016/j.actaastro.2016.10.003

[5] Gravit M, Gumerova E, Bardin A, Lukinov V 2018 Increase of Fire Resistance Limits of Building Structures of Oil-and-Gas Complex Under Hydrocarbon Fire *Advances in Intelligent Systems and Computing,* 692 pp 818-829 DOI: 10.1007/978-3-319-70987-1_87

[6] Zybina O, Gravit M, Stein Y 2017 Influence of carbon additives on operational properties of the intumescent coatings for the fire protection of building *constructions IOP Conference Series: Earth and Environmental Science* 90 (1) DOI: 10.1088/1755-1315/90/1/012227

[7] Korshunov G, Polyakov S, Shunmin L 2017 Assurance of reliability and safety in liquid hydrocarbons marine transportation and storing *IOP Conference Series: Earth and Environmental Science, 87 (6)* DOI: 10.1088/1755-1315/87/6/062009

[8] Pykhtin K, Simankina T, Sharmanov V, Kopytova A 2017 Risk-based approach in valuation of workplace injury rate for transportation and construction industry *IOP Conference Series: Earth and Environmental Science* 90 (1) DOI: 10.1088/1755-1315/90/1/012065

[9] Ustinov A.,. Zybina O, Tanklevsky L., Lebedev V, Andreev A. 2018 Intumescent coatings with

improved properties for high-rise construction *(E3S Web of Conferences,33 (1),* DOI: 10.1051/e3sconf/20183302039.

[10] Klochkov, Y. (2018). Conflicts between quality management methods. Paper presented at the 2017 International Conference on Infocom Technologies and Unmanned Systems: Trends and Future Directions, ICTUS 2017, , 2018-January 34-36. doi:10.1109/ICTUS.2017.8285970

[11] Klochkov, Y., Gazizulina, A., Golovin, N., Glushkova, A., & Zh, S. (2018). Information model-based forecasting of technological process state. Paper presented at the *2017 International Conference on Infocom Technologies and Unmanned Systems: Trends and Future Directions, ICTUS 2017, , 2018-January* 709-712. doi:10.1109/ICTUS.2017.8286099

[12] Bolodian I, Melikhov A, Tanklevskiy L 2017 Fire safety arrangement of inhabited pressurized compartments of manned spacecraft Acta Astronautica, 135 pp 92-99 DOI: 10.1016/j.actaastro.2016.10.003

[13] Bataev, A. V. (2017). Questions of safety of mobile banking in russian financial institutions. Paper presented at the Proceedings of the 30th International Business Information Management Association Conference, IBIMA 2017 - Vision 2020: Sustainable Economic Development, Innovation Management, and Global Growth, , 2017-January 89-99.

[14] Burlov, V., Andreev, A., & Gomazov, F. (2018). Development of a model for the management of environmental safety of the region, taking into account of the GIS capacity. Paper presented at the *MATEC Web of Conferences, , 193* doi:10.1051/matecconf/201819302038

[15] Burlov, V., Andreev, A., & Gomazov, F. (2018). Mathematical model of human decision - A methodological basis for the realization of the human factor in safety management. Paper presented at the *Procedia Computer Science, , 145* 112-117. doi:10.1016/j.procs.2018.11.018

[16] Burlov, V., Andreev, A., Gomazov, F., & Somga-Bichoga, N. (2018). System integration of security maintenance processes in knowledge management. Paper presented at the *Proceedings of the European Conference on Knowledge Management, ECKM, , 1* 112-122.

[17] Burlov, V., & Grachev, M. (2017). Development of a mathematical model of traffic safety management with account for opportunities of web technologies. Paper presented at the *Transportation Research Procedia, , 20* 100-106. doi:10.1016/j.trpro.2017.01.023

[18] Byzov, A. P., Efremov, S. V., & Gomazov, F. A. (2018). Social and economic aspects of acceptable risk of ensuring ecological safety at management of municipal waste. Paper presented at the *Proceedings of the 2018 IEEE International Conference "Management of Municipal Waste as an Important Factor of Sustainable Urban Development", WASTE 2018,* 10-13. doi:10.1109/WASTE.2018.8554106

[19] Gazizulina, A., Eskina, E., Vasilieva, I., & Valeeva, O. (2017). The reasons for the increase in self-organization in companies. *International Journal of Reliability, Quality and Safety Engineering, 24*(6) doi:10.1142/S0218539317400022

[20] Idrisova, J. I., Myasnikov, V. N., Uljanov, A. I., & Belina, N. V. (2018). Increasing the efficiency of labor protection in the enterprise. Paper presented at the *International Conference on Information Networking, , 2018-January* 586-588. doi:10.1109/ICOIN.2018.8343186

[21] Iliashenko, O. Y., Levina, A. I., & Borremans, A. D. (2017). The approach to the formal specification of static structure of the system: Mapping UML to EXPRESS-G. Paper presented at the *Proceedings of the 29th International Business Information Management Association Conference - Education Excellence and Innovation Management through Vision 2020: From Regional Development Sustainability to Global Economic Growth,* 1177-1187.

[22] Ilin, I. V., Frolov, K. V., & Lepekhin, A. A. (2017). From business processes model of the company to software development: MDA business extension. Paper presented at the *Proceedings of the 29th International Business Information Management Association Conference - Education Excellence and Innovation Management through Vision 2020: From Regional Development Sustainability to Global Economic Growth,* 1157-1164.

[23] Savoshinsky, O. P., Zakharova, A. A., & Pak, A. V. (2018). Fire safety management in transportation of municipal wastes with the use of geographic information systems. Paper presented at the *Proceedings of the 2018 IEEE International Conference "Management of Municipal Waste as an Important Factor of Sustainable Urban Development", WASTE 2018,* 55-57. doi:10.1109/WASTE.2018.8554130

[24] Varlataya, S. K., Evdokimov, V. E., & Urzov, A. Y. (2017). Methods for calculating frequency of maintenance of complex information security system based on dynamics of its reliability. Paper presented at the *IOP Conference Series: Materials Science and Engineering, 262*(1) doi:10.1088/1757-899X/262/1/012207

Factors of economic efficiency of innovative entrepreneurship in the manufacturing industry

A Alekseev[1], K Khlebnikov[2], E Fomin[3], N Fomina[4], A Salkina[3], E Konnikov[5]

[1] Department of Economics and Management of Enterprises and Production Complexes St. Petersburg State University of Economics, St. Petersburg, Russia
[2] Institute of Economics and Law Grozny State Oil Technical University named after Academician M.D. Millionshchikov, Grozny, Russia
[3] Department of Taxation and Audit Samara State Economic University, Samara, Russia
[4] JSC «United Match Company», St. Petersburg, Russia
[5] Graduate School of Economics and Technologies, Peter the Great St. Petersburg Polytechnic University, St. Petersburg, Russia

Abstract. The publication presents the results of a study of the process of investing in the innovation of the Russian manufacturing industry (sample 2009-2017). Factors of economic efficiency of investment projects of the industry are revealed. It is proved that the efficiency of investment is determined by external factors, primarily by the level of economic concentration (CR3) of the market of innovative products. The coordination of the scientific result with the industry consolidation dynamics model (A.T. Kearney) is presented.

1. Introduction

Localizing a single innovation into the project, it is advisable to reason from the position of the investment cycle with traditional indicators of (volume) effect and efficiency. Academically, the first is expressed in net present value (NPV), the second - the return on investment (ROI). The investment project is based on an assessment of the economic effect created by innovation. The questions of interpretation and assessment of the economic effects of innovations were considered by Russian and foreign scientists: Vorobyov I.A. (2014); Stolyar T.V. (2016); Rubtsova E.M. (2016); Chabanyuk O.V. (2013); Buchaeva S.A. & Hajiyev M.M. (2012); Filatov E.A. & Polyakov S.A. (2015); Kutsenko E.I. (2015); Tyukavkin N.M. (2016); Obukhova O.V. & Rachek S.V. (2011); Loktionova E.A. (2011); Kolk M. & Eagar R. (2014); Polder M. & Leeuwen G. (2010); Kuczmarski T.D. (2000); Drake M.P., Sakkab N. & Jonash R. (2006); Koskinen Y. & Maeland J. (2016); and others. Consolidating the results of the scientific discussion, we can formulate the following adopted provisions:

- The result of an innovative project is expressed by local effects of the company's economic activity: price growth, cost reduction, productivity growth, etc.;
- Local effects are determined and economically interpreted according to directions, classes of innovations: product, process, marketing, organizational;
- Econometric models of interpretation of local innovation effects in profit and profitability of projects are described.

Thus, in the scientific literature methods of planning economic effects and efficiency of investing capital in innovative projects are formed.

On the other hand, the authors note the incompleteness of the scientific discourse on the nature and factors of the economic efficiency of investing in innovation. In most scientific papers, investments in R&D are investigated. The search for interrelations between this factor and effects is very popular in economic studies due to the availability of information on expenditures on R&D and economic performance at the micro, mezo- and macro levels. Such information is contained in public reports of enterprises and state statistics. Most often, the effect of investing in R&D is defined as "productivity" (the resultant attribute): Bobillo A.M. et al. (2006), Lim E.S. et al. (2012), Drake M.P. et al. (2006). But it is necessary to understand the limitations of research results of this relationship, objectively formulated by the authors of publications. In particular (the first constraint), R&D costs range from 0.3 to 23% (OECD, 2015) of the full investment budget of the innovation project, which also includes investment in the cycle of production, operation and disposal. Formally, the investment cycle of the innovation project (Fomina N.E. & Terentyev A.V., 2012) is determined by the full amount of costs from the "investment memorandum" to "utilization". Objectively, it is possible to judge the effects and effectiveness of innovations only in relation to the realized investment cycle, the full amount of the costs of the innovation project (a posteriori). The second restriction: innovative projects do not always affect the main economic activities of the enterprise, they can have a local character. In particular, if the scale of the project is insignificant in the financial turnover of the enterprise. This statement is entirely true for medium- and low-tech industries, the level of science intensity of which is less than 7%. The third limitation is related to the popularity of the paradigm of "open innovation" and the network model of innovation process organization. Innovative projects can be organized by a number of specialized enterprises as a consortium (network, distributed form of an innovative project). Accordingly, the reflection in the management accounting of a single enterprise will not be correct from the point of view of estimating productivity growth. So, correlating the productivity of the enterprise in the framework of the main business activity and investing in R&D has many limitations and assumptions in the economic modeling of the effectiveness of investment in innovation. According to the authors, an objective assessment of the effectiveness of investing in innovation should be based on a survey of individual projects, investment cycles. In this formulation of the problem, the complexity of the statistics arises: the management reporting of investment projects aimed at implementing an innovative solution is needed. The collection of such information must be carried out "manually", interpreting the management and accounting reporting of individual innovative projects.

Nevertheless, the authors would like to focus on individual scientific studies in which an attempt to determine the nature and factors of economic efficiency of innovative projects within the framework of a full cycle of investments is made. It is solidly understood that the economic effect and effectiveness of an innovative project is a balance of three potentials (Rothwell R., 1992): innovative (scientific and technical) enterprise; innovative (meso- and macro-) infrastructure; market. At the present time, only empirical models of the interconnection of components are proposed, not supported by the econometric justification, multifactorial regression models, etc. But it is possible to single out a number of studies of single factors of investment effectiveness in innovative projects. Analysis of publications allows to conclude that the detected factors are concentrated in internal (potential, infrastructure) and external (market) environments. Internal factors were investigated by Morris, L. (2009), Tsai C.-L. & Chang H.-C. (2016) - innovative infrastructure; Bueche A. M. (1979) - the amount of intellectual capital; Vega-Jurado, J. et al. (2008) - availability of own research divisions of enterprises; Quintella, V. et al. (2017) - the level of financial risk; and others. Among external factors, Tajeddini K. et al. (2006) indicate the dependence on external marketing factors, the secondary impact of the potential and infrastructure. Caggese, A. (2012) proposed a model of assessment through an integral level of risk. Zahra S. & Das S. (1993) formulated dependence on the "level of leadership", interpreted through the market share. Nanda R. & Rhodes-Kropf M. (2013) determine the effectiveness of the growth rates of the market. The above-mentioned relationships are not universal, researchers localize the result by individual sectors, activities, countries, types of economies.

In the context of the above analysis, it seems relevant to expand the scientific discussion about the factors of economic efficiency of investing in innovation, in regional (Russia) and industry (processing industry) projections.

2. Object and method of research

IOP Conf. Series: Materials Science and Engineering **618** (2019) 012089 doi:10.1088/1757-899X/618/1/012089

Object of research: the nature and factors of economic efficiency of investing in the innovation of the national manufacturing industry.
Method of research: correlation analysis. The task of correlation analysis was set to identify the factors that affect the performance indicator - the level of return on investment (ROI) in innovative projects. The authors identified 8 internal and 4 external factors (Table 1), hypothetically determining the profitability of investments in innovative projects. The choice of hypothetical factors is determined by the following criteria: the frequency of allocation in the scientific literature; measurability; isomorphism, relative independence from the scale of projects; the possibility of comparison in projects; availability of management accounting documents (business plans, investment projects, forms 1-4 of the Russian Accounting Regulations, reports on marketing research).
The sample of the study is based on a survey of the retrospective of 92 innovative projects of manufacturing enterprises implemented in the period 2009-2017. Statistics on internal investment indicators of projects are supplemented with information on external indices of the industry obtained from business plans of innovation projects, marketing research and official statistics (in particular, FAMS). To clarify individual quantitative assessments and justify the nature of economic efficiency, in-depth interviews were conducted with the leaders of 17 innovative projects.

3. Results of research

The results of survey of the sample of innovative projects are presented as a table (1) of correlation coefficients (r) of hypothetical factors with ROI as a resultant indicator. The analysis allows to formulate the following conclusions: external factors have a relatively greater impact on the economic efficiency of investing in innovation; the «highest» level of correlation is determined with the factor of economic concentration of the industry, the market of innovation project (coefficient by the sum of 3, CR3).

Table 1. Coefficients of correlation (r) of hypothetical factors with roi iin the innovation project (resultant indicator)

Indicators, %	**r**
Internal: the average value of the indicator in the sample during the implementation of the innovative project	
The growth rate of the revenue of an industrial enterprise	0,245
Profitability of operating activities	0,446
Proportion of staff engaged in R&D	0,172
Margin coefficient	0,451
Science-intensive: costs of R&D in revenue	0,332
The share of R&D costs in the budget of the innovation project	0,227
The ratio of innovation costs to annual revenue	0,199
The share of innovation costs in total investment	0,394
External: by industry affiliation of the enterprise	
Market share by main business activity	0,447
Average market growth rate	0,571
Coefficient of economic concentration by the sum of 3 (CR3)	0,763
Science intensity of the industry (average for OECD groups)	0,311

The correlation field of the factor of economic concentration (figure 1) allows to objectively see the impact on the resultant indicator - the level of economic efficiency of innovation projects.

IOP Conf. Series: Materials Science and Engineering **618** (2019) 012089 doi:10.1088/1757-899X/618/1/012089

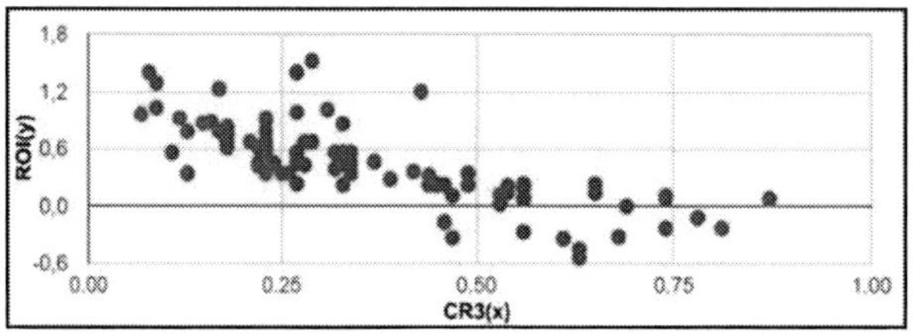

Figure 1. Correlation field: factor - coefficient of economic concentration (CR3); resultant indicator - return on investment (ROI)

The results of the research are in solidarity with the scientific conclusions about the dominance of external factors in the formation of economic efficiency of investment in innovation: Tajeddini K. et al. (2006), Caggese A. (2012), Zahra S. & Das S. (1993), Nanda R. & Rhodes-Kropf M. (2013). We should pay attention to the "average" level of correlation with the resultant indicator and other external factors. The market share of the industrial enterprise in the main economic activity (r = 0.447) characterizes the significant marketing potential of promotion and sale of innovative products. That is, the innovative product can be distributed according to the enterprise channels of sales formed within the framework of the activity basis. And the average market growth rates (r = 0.571) characterize the dynamics of demand and are clearly associated with the level of economic concentration of the industry, which is objectively shown in the model A.T. Kearney (Schuh C. et al., 2011, figure 3). The presented conclusions are also consistent with the modern methodological provisions of the theory of industrial markets (Carlton D.W. & Per-loff J.M., 2000).

In the development of a formal assessment of the results of the correlation analysis, the authors conducted in-depth interviews with the leaders of innovation projects. One of the questions was aimed at obtaining an expert assessment of the balance of influence of internal and external factors on the economic results of the innovation project. The results obtained (Table 2) show the dominance of external, market conditions in the formation of the project's effectiveness relative to the internal potential.

The level of correlation of external factors, the results of in-depth interviews and the consistency of methodology justify the first theoretical conclusion about the dominant of external factors in the formation of the economic efficiency of innovative entrepreneurship.

The observed formal relationship between the factor of economic concentration and the efficiency of investing in innovative projects (conclusion 2) requires an explanation from the position of its nature, which will justify the objectivity of the phenomenon. With reference to the correlation field (figure 1), a trend line can be detected (figure 2), which separates 2 zones (zone A and B) with the following economic logic: Markets with a small (<0.45) level of economic concentration (zone A) have a low level of intensity of competition. There are no large participants ("leaders", found by Zahra S. & Das S., 1993) in the market, competition is close to "perfect". The market of the innovative product is scarce, the demand is not satisfied, the consumption standards have not been formed. There are practically no entry barriers to the market.

Table 2. Expert estimates of the balance of influence (the survey scale) of external and internal factors on the economic result of the innovation project. interpreted according to in-depth interviews with the leaders of 17 innovative projects

Scale of assessment	**100**	**75**	**50:50**	**75**	**100**	
External						**Internal**
Results of the interpretation. Normalized values						
Industry					***Influence***	
					External	***Internal***
Composite materials					0,76	0,24
Metalworking					0,62	0,38

Woodworking	0,53	0,47
Biotechnology	0,39	0,61
Mechanical engineering	0,71	0,29
Medical devices and equipment	0,83	0,17
Light and optical instruments	0,79	0,21

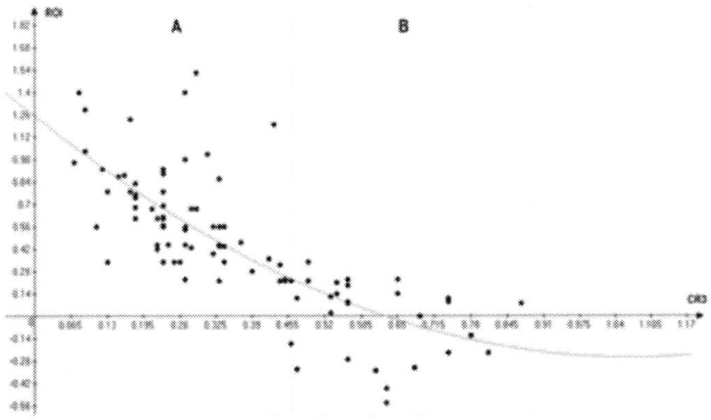

Figure 2. The trend of the influence of the level of economic concentration on the effectiveness of investing in innovation

The reverse situation in the zone of high concentration (zone B) - very rigid barriers to entry to the market formed by 3 or 4 leaders. Major players in the market (included in CR3) implement strategies for "lead-in leadership" due to "economies of scale" that do not allow new products to enter the market. This is especially evident in the project area (figure 2, zone B), which leads the innovative product to highly consolidated markets (CR3 > 0.5). More than half of the innovative projects in this consolidation zone (B) have negative profitability of investment. In in-depth interviews (3 projects with negative ROI) project managers unambiguously identified high market barriers as a reason for failure. We are talking about barriers, formed primarily by "... critically low price of market leaders ..., which does not allow to create economically efficient small lots of innovative products" (citation of the expert).

The proposed view on the nature of the connection of factors finds its opening in the consolidation model of the industry A.T. Kearney (Schuh C. et al., 2011). The well-known diagram of the consolidation dynamics of the industry (figure 3) highlights 4 stages of the industry development from the point of view of the change in the coefficient of economic concentration. The "beginning" stage characterizes the "emergence" of the industry, initiated by a new technological platform - a radical innovation. The stage of "growth" is the process of evolutionary development of a market with a low intensity of competition and a deficit of a morally new product. At the stage of "specialization", the market becomes highly consolidated, develops on the basis of mergers and acquisitions (M & A), high entry barriers are created. Innovative projects of small innovative enterprises (SIE) have practically no potential for investment scaling - they are either blocked or absorbed by leaders.

The authors combined the trend of the influence of the level of economic concentration on the effectiveness of investing in innovation (figure 2) in the industry consolidation diagram A.T. Kearney (figure 3). Overlapping gives an understanding of similar 2 zones (figure 2, A, B) predicting the economic efficiency of investing in innovative projects. The "start-grow" stage (CR3 = 15-45%) is potentially safe for investing in innovative projects due to the expected low entry barriers. And the zone of the "specialization" stage (CR3 > 45%) is determined by the highly risky zone of investing in innovation. The project may have high scientific and technical potential, be supported by an infrastructure, but high economic concentration blocks the marketing opportunities of an industrial enterprise. Of course, at the stage of "specialization" there is an exception for market leaders entering the CR3-4 assessment, their innovative projects have a marketing advantage, which provides them with the prospect of economic efficiency of investments. However, the market leaders at the stage of specialization realize only innovative projects that have a minor moral, technical and operational novelty. Incremental innovations of leaders fix the barriers of the market, which "ordinary" industrial enterprises can not overcome, even when investing in radical innovations. Therefore, the stage of "specialization" is seen as stagnating in the innovative development of the industry, investment in which, even in incremental innovations, is not economically effective. By the end of the "specialization" phase, the beginning of "equilibrium" is already a new technological platform (radical innovation), within which the potential for economic efficiency of investment is characterized by zone A - the "start" stage. So, the two highlighted zones of economic efficiency of investments (figure 2) are in solidarity with the current methodology of strategic planning, with ideas about the genesis of the development of industries.

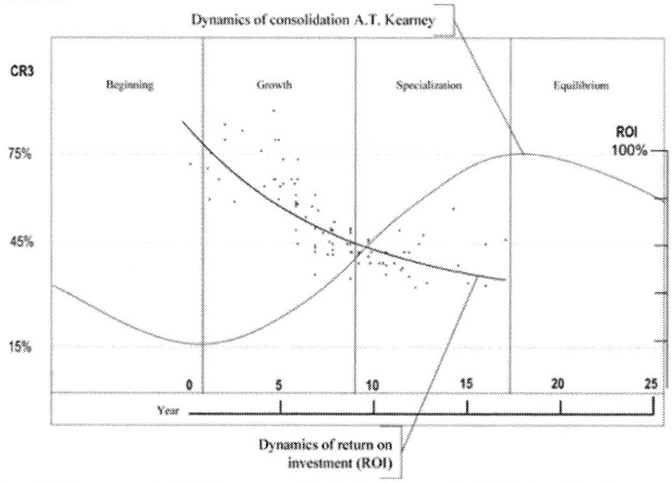

Figure 3. Combined diagram A.T. Kearney (Schuh C. et al, 2011) with the consolidation of the industry (market) and the trend of the polynomial function of efficiency in investing in innovation (figure 2)

The explanation of the genesis of economic efficiency of investing in innovative projects of the national manufacturing industry is the scientific result and the contribution of the authors to the theory of innovation, the economy of industry, and the industrial markets. The practical side of the presented research can be formulated as the basis for making a decision on investing in innovative projects, the criterion for the formation of a portfolio of technological innovations by manufacturing enterprises.

4. Conclusion

The authors state the theoretical position that in the conditions of the modern economy, the effectiveness of investing in innovation is determined not so much by the internal potential of the enterprise and its surrounding infrastructure (without denying their importance in the balance of potentials), but by external, market factors - the level of consolidation of the industry in platform of new technological innovation.

The practical side of the result can be formulated as a criterion for selecting industries (markets, types of activities) for investing in innovations by manufacturing enterprises. The formulation of the criterion: as optimal are determine the industries, where the level of economic concentration at the time of the start of sales of the innovative product is 15-45%, which characterizes the stage of "growth" in the diagram of A.T. Kearney.

The nature of the economic effectiveness of investments in innovative projects is considered to be a scientifically new position with objective limitations: regional and sectoral. The expansion of the statistical material and the sample will make it possible to clarify the following positions in the framework of further research:

- Regression analysis will allow to select a descriptive model, the trend of the relationship between return on investment and the level of economic concentration;

- Multifactorial regression analysis (Table 1) will allow to take into account a number of internal and external factors with a high and medium level of correlation;
- To consider applicability and specificity of the revealed interrelation with reference to separate branches and levels of manufacturability (science intensity);
- To expand the study to linking the dynamics of projects to the consolidation of the industry, including the identifying of the best way to estimate the concentration (coefficients HHI, CR3, CR4, etc.).

Acknowledgment
The research carried out with the financial support of the grant from the Program Competitiveness Enhancement of Peter the Great St.Petersburg Polytechnic University, Project 5-100-2020.

5. References

[1] Buchaeva SA, Hajiyev M.M. (2012) "Types of economic effects and the features of their assessment for innovation". Scientific and technical bulletin of St. Petersburg State Polytechnic University. Economic sciences, 4 (151), 225-229 (in Russian)
[2] Vorobiev I.A. (2014) "Evaluation of the effectiveness of the innovation project". The entrepreneur's guide, 21, 106-113 (in Russian)
[3] Kutsenko E.I. (2015) "Aspects of the effectiveness evaluation of an innovative project". Bulletin of Orenburg State University, 8 (183), 73-78 (in Russian)
[4] Loktionova E.A. (2011) "Methodology for assessing the economic effect of innovation in production preparation". Bulletin of Tambov University. Series: The humanities, 4 (96), 62-67 (in Russian)
[5] Obukhova O.V., Rachek S.V. (2011) "Modern methodical approaches to determining the economic efficiency of innovation-investment projects". Bulletin of the Ural State Economic University, 4 (36), 49-54 (in Russian)
[6] Rubtsova E.M. (2015) "On some features of determining the economic effect of innovation projects". Bulletin of Nizhny Novgorod University. Series: Social Sciences, 3 (39), 59-69 (in Russian)
[7] Stolyar T.V. (2016) "Evaluation of the effectiveness of innovation". Scientific research and development. Economy, 4, 4, 37-42 (in Russian)
[8] Tyukavkin N.M. (2016) "Methodical apparatus for analysis and evaluation of the efficiency of the enterprise innovation activity". Internet Journal of Science, 8, 1 (32), 12 (in Russian)
[9] Filatov E.A., Polyakov S.A. (2015) "Alternative factor analysis of the innovation efficiency assessment model". Innovation and investment, 9, 18-20 (in Russian)
[10] Fomina N.E., Terentyev A.V. (2012) "Innovation and investment cycle of industry". Issues of economics and law, 45, 148-152 (in Russian)
[11] Chabanyuk O.V. (2013) "Theoretical and methodological approaches to determining the effectiveness of innovation". Bulletin of Moscow University of the Ministry of Internal Affairs of Russia, 4, 183-185 (in Russian)
[12] Bobillo A.M., Sanz J. A. R. & Gaite F.T. (2006) "Innovation Investment, Competitiveness, and Performance in Industrial Firms". November 2006 Thunderbird International Business Review, 48(6), 867 – 890.
[13] Bueche, A. M. (1979) "Investment in innovation". The Journal of Technology Transfer, 3(2), 1–13.
[14] Caggese, A. (2012) "Entrepreneurial risk, investment, and innovation". Journal of Financial Economics, 106(2), 287–307.
[15] Carlton D.W., Perloff J.M. (2000) "Modern Industrial Organization". Addison-Wesley.
[16] Dernis H., Dosso M., Hervás F., Millot V., Squicciarini M. & Vezzani A. (2015) "World Corporate Top R&D Investors: Innovation and IP bundles". JRC and OECD common report. Luxembourg: Publications Office of the European Union. Publication of European Commission's Joint Research Centre – Institute for Prospective.

[17] Drake M.P., Sakkab N. & Jonash R. (2006) "Maximizing Return On Innovation Investment". November 2006 Research Technology Management, 49(6), 32-41.
[18] Drake, M. P., Sakkab, N., & Jonash, R. (2006) "Maximizing return on innovation investment". Research Technology Management, 6308 (November-December), 32–41.
[19] Kolk, M. & Eagar, R. (2014) "How to manage your return on investment in innovation". PRISM, (1), 66–83.
[20] Koskinen, Y. & Maeland, J. (2016) "Innovation, Competition, and Investment Timing". Review of Corporate Finance Studies, 5(2), 166–199.
[21] Kuczmarski, T.D. (2000) "Measuring your return on innovation". Marketing Management, 9(1), 24–32.
[22] Lim E.S., Lee C. & Nagaraj S. (2012) "The Relationship Between Innovation and Firm Performance: Evidence from Malaysian Manufacturing Firms". Proceedings of The Asian Business and Management Conference 2012, Osaka, Japan (ISSN: 2186-5914).
[23] Morris, L. (2009) "The Innovation Infrastructure". International Journal of Innovation Science, 1(1), 41–49.
[24] Nanda, R., & Rhodes-Kropf, M. (2013) "Investment cycles and startup innovation". Journal of Financial Economics, 110 (2), 403–418.
[25] Polder, M., & Leeuwen, G. (2010) "Product, process and organizational innovation: drivers, complementarity and productivity effects". Publications 2010s-28, (23719), 1–46.
[26] Quintella, V., Silva A.., Silva Jr.A., Embiruçu Jr.J. A. M. "Financial exposure and technology innovation investment: Measuring project results in Brazilian commodity industries". January 2017 Academia Revista Latinoamerica de Administracion, 30(4), 547-564.
[27] Rothwell, R. (1992) "Successful industrial innovation: critical factors for the 1990s". R&D Management, 22(3), 221–240.
[28] Schuh C., Kromoser R., Strohmer M. F., Pérez R.R., Triplat A. The A.T. Kearney Strategy Chessboard. USA, Illinois: Marketing & Communication A.T. Kearney inc., 2011.
[29] Tajeddini, K., Trueman, M. & Larsen, G. (2006) "Examining the Effect of Market Orientation On Innovativeness". Journal of Marketing Management, 22(5–6), 529–551.
[30] Tsai, C.-L., & Chang, H.-C. (2016) "Evaluation of critical factors for the regional innovation system within the HsinChu science-based park". Kybernetes, 45(4), 699–716.
[31] Vega-Jurado, J., Gutiérrez-Gracia, A., Fernández-de-Lucio, I., & Manjarrés-Henríquez, L. (2008) "The effect of external and internal factors on firms' product innovation". Research Policy, 37(4), 616–632.
[32] Zahra, S., & Das, S. (1993) "Innovation strategy and financial performance in manufacturing companies: An empirical study". Production and Operations Management, 2(I), 15–37.
[33] Bataev A.V. & Rodionov D.G., "Cloud computing: Evaluation use under the crisis in Russia". 2018 7th International Conference on Industrial Technology and Management (ICITM), Oxford, 2018, pp. 224-228. doi: 10.1109/ICITM.2018.8333951.
[34] Klochkov, Y., Its, A., Vasilieva, I. (2016). Development of FMEA method with the purpose of quality abebment of can stock production. Key Engineering Materials, 684, pp. 473-476. DOI: 10.4028/www.scientific.net/KEM.684.473.
[35] Rodionov, D., Nikolova, L., Velikova, M., Kalubi, R.D.M. (2016). Competitive strength of a concession integration model of labour market to education market interaction. Proceedings of the 28th International Business Information Management Association Conference - Vision 2020: Innovation Management, Development Sustainability, and Competitive Economic Growth, p. 689-699.
[36] Ozerov E. S., Pupentsova S. V., Leventsov V. A. and Dyachkov M. S., "Selecting the best use option for assets in a corporate management system," 2017 6th International Conference on Reliability, Infocom Technologies and Optimization (Trends and Future Directions) (ICRITO), Noida, India, 2017, pp. 162-170. doi: 10.1109/ICRITO.2017.8342418.

Structure of a semisimple dihedral group algebra

I Gradeva and Y Epitropov

Plovdiv University 'P. Hilendarski' – 24 Tzar Asen Str., Plovdiv, Bulgaria

E-mail: epitropov@uni-plovdiv.bg

Abstract. Let K be an arbitrary field, whose characteristic does not divide the order of the dihedral group D_{2m} of order $2m$, where m is odd. In this paper we examine the structure of the semisimple dihedral group algebra KD_{2m}. For this purpose, we find a complete system of minimal central orthogonal idempotents of the group algebra. Through it we define the minimal components of KD_{2m} and its Wedderburn decomposition. The results we get are as general as possible, i.e. without requiring the field to be finite.

1. Introduction.

Let D_{2m} be a dihedral group of order $2m$, where m is odd, determined by the conditions [1, p. 171]

$$D_{2m} = \left\langle a,b \,\middle|\, a^2 = 1, b^m = 1, m = 2s+1, aba = b^{-1} \right\rangle, \tag{1}$$

and K is such a field that the group algebra KD_{2m} of D_{2m} over K is semisimple. To describe the structure of KD_{2m} it is enough to find a complete system of minimal central orthogonal idempotents of KD_{2m} and the Wedderburn decomposition of KD_{2m} [2, Theorem 1.4.4].

For the semisimple rings the Wedderburn-Artin Theorem holds [3, Theorem 2.6.18]. It proves the existence of an isomorphism between a semisimple ring and a direct sum of some matrix algebras over the division rings. But the Theorem does not show how this isomorphism is defined, nor does it give the specific look of the direct matrix summands. This means that for each particular semisimple ring the problem of finding its Wedderburn decomposition does not have a standard solution.

For a semisimple dihedral group algebra KD_{2m} over the finite field K basic results are obtained. In Section 4.3. of [4] the primitive central idempotents of the group algebra are found and its Wedderburn decomposition is described. In [5] the central irreducible idempotents of KD_{2m} are exactly found and their number is calculated (Theorem 3.3), and Wedderburn decomposition is described in Theorem 4.4. Some results for finite semisimple dihedral group algebras can be found in [6] and [7], where group codes over dihedral groups are examined.

The above-mentioned results for semisimple group algebra KD_{2m} of a dihedral group D_{2m} have been obtained from the various authors under a substantial restriction for the field K, namely when it is finite. The contribution of our work compared to those described above is that we overcome this restriction and describe the structure of KD_{2m} in the most general case, i.e. without requiring the field K to be finite.

The structure of the paper is as follows. Section 1 contains the necessary preliminary information. In Section 2 under condition that K is a field of decomposition of D_{2m}, we find the central orthogonal idempotents of the semisimple dihedral group algebra KD_{2m} (Theorem 2) and we prove that they are minimal (Theorem 3). From this follows that the idempotents found in Theorem 3 form a complete system of minimal central orthogonal idempotents of KD_{2m}. In Theorem 4 for an arbitrary field K we find a complete system of minimal central orthogonal idempotents of the group algebra. In Section 3 we determine the minimal components of KD_{2m} and describe its Wedderburn decomposition (Theorem 6).

We will recall some basic facts about idempotents of group algebras that we will use in our work. Let K be field and G is such a finite group that the group algebra KG is semisimple. The idempotent e of the group algebra KG is minimal if KGe is a minimal left ideal of KG. The idempotent e of KG is primitive, if for an arbitrary extension $\overline{K}$ of the field K it does not decompose into a sum of two orthogonal idempotents of the group algebra $\overline{K}G$. From here follows that every primitive idempotent is minimal. (But in the general case is not true that every minimal idempotent is primitive.) The field K is a field of decomposition of the group G if every minimal idempotent of KG is primitive. For every homomorphism χ of G in the multiplicative group K^* of K the element $e_\chi = |G|^{-1}\sum_{g\in G}\chi(g^{-1})g$ is an idempotent of KG. If G is abelian and K contains a primitive n-th root of 1, where n is the exponent of G, then $\{e_\chi \mid \chi \in Hom(G,K^*)\}$ is a complete system of primitive idempotents of KG, which is an orthogonal base for KG over K [8, Lemma 4.1.1]. In the general case this is not true.

We will introduce some notations, having in mind the conditions (1), which we will use throughout this paper. Namely,

1. d is a divisor of m;
2. $C(R)$ is a centre of a ring/algebra R;
3. $Z_d = \{0,1,\dots,d-1\}$ is a ring of the remainders modulo d;
4. Z_d^* is a multiplicative group of Z_d;
5. $J = J(Z_d^*/\langle d-1\rangle)$ is a complete system of representatives of Z_d^* by subgroup $\langle d-1\rangle$ for $d\neq 1$ and $J=1$ for $d=1$;
6. η is a primitive d - th root of 1.

Let τ_s be a primitive s-th root of 1 and $Gal(K(\tau_s)/K)$ be a group of Galois of the extension $K(\tau_s)$ of the field K. Let $\theta_\lambda \in Gal(K(\tau_s)/K)$ be an automorphism, for which $\theta_\lambda(\tau_s) = \tau_s^\lambda$. Then we denote

7. $\Gamma_s = \{\lambda \mid \theta_\lambda \in Gal(K(\tau_s)/K)\}$.

Then $\Gamma_s \le Z_s^*$ and $|\Gamma_s| = |Gal(K(\tau_s)/K)|$. Furthermore,

8. $\Pi = \Pi\{\Gamma_d/\Gamma_d \cap \langle d-1\rangle\}$ is a complete system of representatives of Γ_d by subgroup $\Gamma_d \cap \langle d-1\rangle$ for $d\neq 1$ and $\Pi = 1$ for $d=1$.
9. $P = P\{Z_d^*/\Gamma_d.\langle d-1\rangle\}$ is a complete system of representatives of Z_d^* by subgroup $\Gamma_d.\langle d-1\rangle$.

2. Finding of a complete system of minimal central orthogonal idempotents of a semisimple dihedral group algebra KD_{2m}.

We denote with $C(a^ib^j)$ the class elements of the group D_{2m}, conjugated with the element a^ib^j from D_{2m}. Because $i=0$ has only two values: $i=0$ and $i=1$, then

$$C(b^j)=\{b^j,b^{-j}\},\ j=0,\dots,s,\ C(ab^j)=\{a,ab,\dots,ab^{2s}\}.$$

It is clear that the number of these conjugacy classes is $s+2$.

Let D_{2m} be a dihedral group with determining conditions (1) and K be a field of decomposition of D_{2m}. First, we will find one complete system of minimal central orthogonal idempotents of the group algebra KD_{2m}.

Let d be a fixed divisor of m. We define idempotents α_{di} and β_{dj} of the group algebra KD_{2m} the following way:

$$\alpha_{1i}=\frac{1}{2}(1+(-1)^i a),\text{ where } i=0,1 \text{ for } d=1 \text{ and } \alpha_{d0}=1 \text{ for } d\neq 1,\ \beta_{dj}=\frac{1}{m}\sum_{t=0}^{m-1}\left(\frac{b}{\eta^j}\right)^t \text{ for } j\in Z_d^*. \quad (2)$$

Let we denote

$$B_{11}=\beta_{11}=\frac{1}{m}\left(1+b+\dots+b^{2s}\right) \text{ and } B_{dj}=\beta_{dj}+\beta_{d,d-j} \text{ for } d\neq 1,\ j\in J.$$

Lemma 1. The element B_{dj} of the semisimple dihedral group algebra KD_{2m} is a central idempotent for every divisor d of m and for every $j\in J$.

Proof. Since B_{dj} is a sum of orthogonal idempotents of the group algebra KD_{2m}, then B_{dj} is an idempotent of KD_{2m}. Furthermore B_{dj} is a sum of classes conjugated elements, then B_{dj} is a central idempotent of KD_{2m}. #

Theorem 2. The elements $e_{dij}=\alpha_{di}B_{dj}$ are central orthogonal idempotents of the semisimple dihedral group algebra KD_{2m}.

Proof. Let $\alpha_{d_1i_1}$ and $B_{d_1j_1}$ be idempotents of the group algebra KD_{2m}, corresponding to a divisor d_1 of m. Then $e_{dij}e_{d_1i_1j_1}=\alpha_{di}B_{dj}\alpha_{d_1i_1}B_{d_1j_1}=\alpha_{di}\alpha_{d_1i_1}B_{dj}B_{d_1j_1}$.

1. If $d=d_1$, $i=i_1$ and $j=j_1$, then $e_{dij}^2=\alpha_{di}^2B_{dj}^2=\alpha_{di}B_{dj}=e_{dij}$, i.e. e_{dij} is an idempotent.

2. If $d\neq d_1$ or $i\neq i_1$, or $j\neq j_1$, then at least one of the equations $\alpha_{di}\alpha_{d_1i_1}=0$ or $B_{di}B_{d_1i_1}=0$ holds. Thus $e_{dij}e_{d_1i_1j_1}=0$ and those idempotents are orthogonal.

Therefore, $e_{dij}=\alpha_{di}B_{dj}$ are orthogonal idempotents of the group algebra KD_{2m}.

3. We will prove, that for fixed d,i,j the idempotent $e=e_{dij}$ is a central idempotent of KD_{2m}. First we will determine the coefficient of a^ib^j in e.

Let $d=1$. Then $e=\frac{1}{2m}(1+\varepsilon a)\left(1+b+\dots+b^{2s}\right)$, where $\varepsilon=\pm 1$. But

$$e=\frac{1}{2m}\left(1+(b+b^s)+(b^2+b^{2s-1})+\dots+(b^s+b^{s+1})+\varepsilon(a+ab+\dots+ab^{2s})\right).$$

Thus e is a finite sum of the form $e=\sum A_iC_i$, where A_i are coefficients of the field K and C_i are sums of one conjugacy class. Thus e is a central idempotent of KD_{2m}.

Let $d\neq 1$. Then $e=\beta_{dj}+\beta_{d,d-j}$. Again e is a finite sum of the form $e=\sum A_iC_i$, where A_i are coefficients of the field K and C_i are sums of one conjugacy class. Thus e is a central idempotent of KD_{2m}. #

Theorem 3. Let $e = e_{dij} = \alpha_{di} B_{dj}$ be fixed idempotent of the semisimple group algebra KD_{2m} and K be a field of decomposition of dihedral group D_{2m}. The ideal $KD_{2m}e$ of KD_{2m} is a central simple algebra.

Proof. Let d be fixed divisor of m, $j \in J$ and $i = 0,1$ for $d = 1$ and $i = 0$ for $d \neq 1$. We will prove that for a fixed idempotent $e = \alpha_{di} B_{dj}$ of the group algebra KD_{2m} the isomorphism $C(KD_{2m}e) \cong K$ holds.

Let x be an arbitrary element of the ideal $KD_{2m}e$, i.e. $x = \sum_{t=0}^{m-1}(\lambda_{0t}b^t + \lambda_{1t}ab^t)e$. If $x \in C(KD_{2m}e)$, then $xa = ax$, i.e. $\sum_{t=0}^{m-1}(\lambda_{0t}ab^t + \lambda_{1t}b^t)e = \sum_{t=0}^{m-1}(\lambda_{0t}ab^{m-t} + \lambda_{1t}b^{m-t})e$. Thus $\lambda_{kt} = \lambda_{k,m-t}$ for every $k = 0,1$ and for every $t = 0,\dots,s$. Then $x = \sum_{t=0}^{s}(\lambda_{0t}(b^t + b^{-t}) + \lambda_{1t}a(b^t + b^{-t}))e = \sum_{t=0}^{s}(\lambda_{0t} + \lambda_{1t}a)(b^t + b^{-t})e$. It is easy to check that $b^t\beta_{dj} = \frac{1}{m}\sum_{k=0}^{m-1}\frac{b^{k+t}}{\eta^{jk}} = \eta^{jt}\beta_{dj}$. We will prove, that $(b^t + b^{-t})e \in Ke$. Indeed, for $d = 1$ we have $(b^t + b^{-t})e = 2e \in Ke$. For $d \neq 1$ we have $(b^t + b^{-t})e = \frac{1}{m}(\eta^{jt} + \eta^{-jt})e \in Ke$. From this follows that if $x \in C(KD_{2m})$, then x has the form $x = (\mu_0 + \mu_1 a)e$, where $\mu_0, \mu_1 \in K$. From $x \in C(KD_{2m})$ also follows that $bx = xb$, i.e. $(\mu_0 b + \mu_1 ab^{-1})e = (\mu_0 b + \mu_1 ab)e$. This means that $\mu_1 = 0$. Thus $x = \mu_0 e \in Ke$, which shows that $C(KD_{2m}e) \subseteq Ke$.

The reverse inclusion, i.e. $Ke \subseteq C(KD_{2m}e)$ is obvious, therefore $C(KD_{2m}e) = Ke$. But $Ke \cong K$ is met. Thus, the centre of $KD_{2m}e$ is a field, therefore $KD_{2m}e$ is a central simple algebra over K. #

From the proof of Theorem 3 follows that the idempotents e_{dij} are minimal since they generate central simple algebras.

For $d = 1$ the number of these idempotents is 2 and for $d \neq 1$ their number is $\frac{\varphi(d)}{2}$. But the number of the conjugacy classes in the group D_{2m} is the same. Therefore, the previously determined e_{dij} are all central minimal orthogonal idempotents of KD_{2m}.

From the proof of Theorem 3 it also follows that for the field K to be a field of decomposition of the group D_{2m} it is sufficient that it contains all sums of the form $\eta^t + \eta^{-t}$ for $t = 0,\dots,s$.

The results in Lemma 1, Theorem 2 and Theorem 3 are obtained in the case, when K is a field of decomposition of the group D_{2m}. Now we will examine and the case, when K is not a field of decomposition of the group D_{2m}.

Let K be a field, which is not a field of decomposition of the group D_{2m} and the characteristic of K does not divide the order of D_{2m}. Let $\overline{K}$ be an extension of K, which contains the m-th roots of 1, i.e. $\overline{K}$ is field of decomposition of the group D_{2m}. We will find the minimal central orthogonal idempotents of the group algebra KD_{2m} for a fixed divisor d of m. For this is sufficient to sum the minimal central orthogonal idempotents of the group algebra $\overline{K}D_{2m}$ by conjugation over K. We obtain the following idempotents: $e_{dij} = \alpha_{di} B_{dj}$, where α_{di} and B_{11} have the same form as in (2) and for

$d \neq 1$ we have $B_{dq} = \sum_{t\in\Pi}(\beta_{d,tq} + \beta_{d,d-tq})$, where q runs P. Therefore, their total number is $2 + \sum_{\substack{d/m \\ d\neq 1}} \frac{\varphi(d)}{|\Gamma_d\langle d-1\rangle|}$.

And so we proved the following

Theorem 4. The minimal central orthogonal idempotents $e_{dtq} = \alpha_{dt} B_{dq}$ of the semisimple dihedral group algebra KD_{2m} over arbitrary field K form a complete system minimal central orthogonal idempotents and their number is $2 + \sum_{\substack{d/m \\ d\neq 1}} \frac{\varphi(d)}{|\Gamma_d\langle d-1\rangle|}$.

3. Structure of the semisimple dihedral group algebra KD_{2m}.

Let K be a field of decomposition of the group D_{2m} and $e = e_{dij} = \alpha_{di} B_{dj}$ is a fixed idempotent of the group algebra KD_{2m} for $j \in J$ and $i = 0,1$ for $d = 1$ and $i = 0$ for $d \neq 1$.

Theorem 5. If K is a field of decomposition of the dihedral group D_{2m}, then the ideal $KD_{2m}e$ of the semisimple dihedral group algebra KD_{2m} contains at most a first degree of the element be.

Proof. For $d = 1$ we have $be = e$.

Let $d \neq 1$ and $f(x) = (x - \eta^j e)(x - \eta^{-j} e)$ be a polynomial over the field Ke. We denote $e_j = \alpha_{di}\beta_{dj}$. Then $e_{d-j} = \alpha_{di}\beta_{d,d-j}$ and $e = e_j + e_{d-j}$. In the proof of Theorem 3 we saw that $b^t\beta_{dj} = \eta^{jt}\beta_{dj}$ for every positive integer t. Thus $be_j = \eta^j e_j$ and $be_j = \eta^{-j} e_{d-j}$, i.e. $(b - \eta^j)e_j = 0$ and $(b - \eta^{-j})e_{d-j} = 0$. Then $f(be) = (b - \eta^j)(b - \eta^{-j})e_j + (b - \eta^j)(b - \eta^{-j})e_{d-j} = 0$. So, we proved that be is a root of a second-degree polynomial over Ke. Therefore, the ideal $KD_{2m}e$ contains at most a first degree of the element be. #

Considering that $Ke \cong K$ is fulfilled, from Theorem 5 follows:

For $d = 1$, since $ae = \pm e$, then $\dim_K KD_{2m}e = 1$, i.e. $KD_{2m}e \cong K$.

For $d \neq 1$ the ideal $KD_{2m}e$ contains at most a first degree of the elements ae and be. Thus, the dimension of $KD_{2m}e$ over the field K does not exceed 4. And so, $KD_{2m}e$ is decomposed over K in a direct sum of two ideals with dimension 1 and $\frac{\varphi(d)}{2}$ ideals with dimension, which not exceeding 4. But then it follows that

$$2m = \dim_K KD_{2m} \leq 2 + \sum_{\substack{d/m \\ d\neq 1}} 4\frac{\varphi(d)}{2} = 2\sum_{d/m}\varphi(d) = 2m .$$

Therefore $\dim_K KD_{2m} = 4$.

From the proof of Theorem 3 it is clear for K that to be a field of decomposition of the group D_{2m}, it is sufficient that K contains all the sums of the form $\eta^j + \eta^{-j}$ for $j \in J$. If K is not a field of decomposition of the group D_{2m}, then for a fixed $q \in P$ we denote with $\overline{K}$ the field, which is an extension of K with $\eta^{qt} + \eta^{-qt}$ for some $t \in \Pi$. Then for $d = 1$ the ideal $KD_{2m}e$ is isomorphic to K and for $d \neq 1$ the ideal $KD_{2m}e$ is isomorphic to the algebra $M_2(\overline{K})$ from the matrices of order 2 over the field $\overline{K}$.

IOP Conf. Series: Materials Science and Engineering **618** (2019) 012090 doi:10.1088/1757-899X/618/1/012090

From everything so far it follows

Theorem 6. Wedderburn decomposition of the semisimple group algebra KD_{2m} of dihedral group D_{2m} over the field K is:

$$KD_{2m} \cong K \oplus K \oplus \sum_{\substack{d/m \\ d\neq 1}} \oplus M_2(\overline{K}),$$

where the number of all direct summands is $2+\sum_{\substack{d/m \\ d\neq 1}} \frac{\varphi(d)}{|\Gamma_d\langle d-1\rangle|}$.

Acknowledgments. This work is partially supported by a Project No RD-08-118/04.02.2019 of Shumen University.

References

[1] Vinberg E 2013 *Course in Algebra* (Moscow). (in Russian)
[2] Herstein I 2005 *Noncommutative rings* (Cambridge University Press).
[3] Milies C and Sehgal S 2016 *An Introduction to Group Rings* (Springer).
[4] Bakshi G, Gupta S and Passi I 2013 *J. Ramanujan Math. Soc.* **28** 141-58.
[5] Martinez F 2015 *Finite Fields Th. App.* **35** 204-14.
[6] Wong D and Ang M 2013 *Malays. J. Math. Sci.* **7** 101-16.
[7] Vedenev K and Deundyak V 2018 *Modelling and Analysis of Information Systems* **25** 232-45. (in Russian)
[8] Karpilovsky G 1990 *Unit groups of group rings* (Longman Higher Education).

Variational principle for a class of nonlocal boundary value problems for mixed type higher-order equations

G P Paskalev
Department of Mathematics, Physics, Chemistry, Technical University–Sofia, Branch Technical University-Sofia, Plovdiv Branch, 25 Tsanko Dyustabanov Str, 4000 Plovdiv, Bulgaria

E-mail: g.p.paskalev@abv.bg

Abstract. For considered class of nonlocal boundary value problems for mixed type PDE of arbitrary high order an implicit symmetrizing operator [2,4] is built. Equivalence of the problem to the problem of minimization of appropriate quadratic functional is proved. Existence and uniqueness of the generalized solution in variation sense are obtained.

1.Introduction

Let $s,m,n \geq 1$ are integer, $x = (x_1,...,x_n)$ and $D \subset R^n$ be a bounded domain with a smooth boundary ∂D $G = D \times (0,T)$, $\Gamma = \partial D \times (0,T)$. If $\nu_0 = const. > 0$, we suppose also that $\sum_{|\alpha|=|\beta|=m} a_{\alpha\beta}(x)\xi^\alpha \xi^\beta \geq \nu_0 |\xi|^2 \quad \forall x \in \overline{G}, \forall \xi \in R^{n-1}$ i.e. $M[u] = \sum_{|\alpha|=|\beta|=m} a_{\alpha\beta}(x) D_x^\alpha D_x^\beta u$ is a strong elliptic operator, where $a_{\alpha\beta}(x) \in C^\infty(\overline{D})$, $a_{\alpha\beta}(x) = a_{\beta\alpha}(x)\ \forall x \in \overline{G}\ \forall \alpha,\beta : |\alpha| = |\beta| = 1$, $k_{2s-1}(t,x), k_{2s}(t,x) \in C^\infty(\overline{G})$.

Let we have $k_{2s}(T,x) = k_{2s}(0,x) = 0 \quad \forall x \in \overline{D}, \quad k_{2s-1}(T,x) = k_{2s-1}(0,x) \quad \forall x \in \overline{D}$,

$$P[u] = k_{2s}(t,x) D_t^{2s} u + k_{2s-1}(t,x) D_t^{2s-1} u,\ L[u] \equiv P[u] + M[u] - Cu,\ C = const > 0.$$

We shall propose also that

$$k_{2s}(T,x) - D_t k_{2s}(T,x) = k_{2s}(0,x) - D_t k_{2s}(0,x) \neq 0 \quad \forall x \in \overline{D}.$$

Consider the equation

$$L[u] = f(t,x). \tag{1}$$

The last equation is a of mixed type in D and on the bottoms of the cylindrical domain D it is of parabolic type.

Let $\lambda \neq 0, |\lambda| < 1$ is a real number. Consider the following boundary value problem. To find a solution to the equation (1) in G, satisfying the boundary conditions:

$$D_x^\alpha u|_\Gamma = 0,\ |\alpha| \leq m-1, \tag{2}$$

$$D_t^i u(T,x) = \lambda D_t^i u(0,x),\ i = \overline{0,2s-2}. \tag{3}$$

The nonlocal boundary value problem (1)-(3) is considered in the papers [8,9]. In the present paper we apply for the first time the Shalov's variational method ([2,3]) to the problem with nonlocal boundary conditions (3) for mixed type equation. In the partial case when m=s=1 the problem (1)-(3) is investigated in [1].

2.Function spaces and definitions

Definition 1: *The operator L is called* $K-symmetric$ in the space X, if

$$(LV, KU)_0 = [U,V]_X \quad \forall U,V \in X.$$

Definition 2: *The operator L is called* $K-positive$ in the space X, if

$$(LU, KU)_0 = \|U\|_X^2 \ \forall U \in X.$$

Let $\widetilde{C}^\infty(\overline{G})$ be the space of infinitely smooth in $C^\infty(\overline{G})$ functions, satisfying the boundary conditions (2),(3) and $\widetilde{C}_*^\infty(\overline{G})$ is the corresponding space of infinitely smooth in $\overline{G}$ functions, satisfying the boundary conditions ad joint to (2),(3):

$$D_x^\alpha v|_\Gamma = 0,\ |\alpha| \leq m-1, \tag{4}$$

$$D_t^i v(0,x) = \lambda D_t^i v(T,x),\ i = \overline{0,2s-2}. \tag{5}$$

Define $H^{s,m}(G)$ as the closure of $\widetilde{C}^\infty(\overline{G})$ with respect to the norm

$$\|v\|_{s,m}^2 = \int_G \sum_{\frac{i}{s}+\frac{|\alpha|}{m}\le 1} (D_t^i D_x^\alpha u)^2\, dtdx. \tag{6}$$

Let $H^{s,m,*}(G)$ is the closure of the space $\widetilde{C}_*^\infty(\overline{G})$ with respect to the norm (6). The scalar product in $L_2(G)$ we shall denote by $(.,.)_0$ and the corresponding norm - by $\|.\|_0$.

Let the constant κ is defined by the equality $e^{\kappa T} = \lambda^{-2}$.

Let $N(1)=1, N(s)=2s$ for $s>1$. We suppose that for any $(t,x)\in\overline{G}$ is fulfilled the condition:

$$2k_{2s-1}(t,x) + rD_t k_{2s}(t,x) - N(s)\kappa k_{2s}(t,x) \ge \delta > 0\ \ r = 2p-1, 2p-4s+1,\quad p=\overline{0,l}.$$

If L^* is formally ad joined operator to L and, for any element $v\in H^{2s-1,m,*}(G)$ denote by Kv the solution of the following nonlocal problem (1) with boundary conditions (4),(5):

$$\mathrm{L}^*(\mathrm{Kv}) = \mathrm{v} + \sum_{\frac{i}{2s-1}+\frac{|\alpha|}{m}\le 1} (-1)^{i+|\alpha|} D_t^{2i} D_x^{2\alpha}\mathrm{v},$$

$$D_x^\alpha Kv|_\Gamma = 0,\ |\alpha|\le m-1, \qquad D_t^i Kv(0,x) = \lambda D_t^i Kv(T,x),\ i=\overline{0,2s-2}.$$

From the a priori estimates, obtained in the paper [1]:

$$\|Lu\|_{-(2s-1,m)} \ge C_1\|u\|_{2s-1,m} \quad \forall u\in\widetilde{C}^\infty(\overline{G}),$$

$$\|L^*v\|_{-(2s-1,m)} \ge C_2\|v\|_{2s-1,m} \quad \forall v\in\widetilde{C}_*^\infty(\overline{G}),$$

where C_1, C_2 are positive constants, it follows that if the constant C is sufficiently large, then the solution Kv exists and is unique.

By integration by parts, now, for any functions $u,v\in\widetilde{C}^\infty(\overline{G})$, it obtains that

$$(Lu, Kv)_0 = [u, v]_{2s-1,m} \quad \forall u, v \in \widetilde{C}^{\infty}(\overline{G}) \tag{7}$$

Changing the places of u and v, we have

$$(Lv, Ku)_0 = [u, v]_{2s-1,m} \quad \forall u, v \in \widetilde{C}^{\infty}(\overline{G}). \tag{8}$$

From (7) and (8) follows that the operator L is $K-symmetric.$

Now for $u \equiv v$ we obtain

$$(Lu, Ku)_0 = \|u\|^2_{2s-1,m} \quad \forall u \in \widetilde{C}^{\infty}(\overline{G}),$$

which means that L also is $K-positive\,operator.$

Definition 3: *The function* $u \in \widetilde{H}^{2s-1,m}(G)$ *is called a generalized solution to the problem (1)-(3) if*

$$(f, Kv)_0 = [u, v]_{2s-1,m} \quad \forall v \in \widetilde{H}^{2s-1,m}(G).$$

3. Results

Theorem 1: *The function* $u \in \widetilde{H}^{2s-1,m}(G)$ *is a generalized solution to the problem (1)-(3) if and only if* u *realizes the minimum of the following quadratic functional*

$$D[u] = \|u\|^2_{2s-1,m} - 2(f, Ku)_0 \tag{9}$$

on the space $\widetilde{H}^{2s-1,m}(G)$.

Theorem 2: *For any function* $f \in L(G)$ *there exists a unique solution t o the variational problem to minimize the quadratic functional (10) on the space* $\widetilde{H}^{2s-1,m}(G)$. *The problem (9) is correctly posed – in sense that to the small variance of* f *in* $L^2(G)$ *corresponds a small variance of the solution on the space* $\widetilde{H}^{2s-1,m}(G)$.

4. Proofs

To prove the above theorems we need to obtain the following estimate.

Lemma: *There exists a positive constant* $C_2 > 0$, *such that*

$$\|Kv\|_0 \le C_2 \|v\|_{2s-1,m} \quad \forall v \in \widetilde{C}^\infty(\overline{G}). \tag{10}$$

Proof: Since $Kv \in \widetilde{H}^{2s-1,m}(G)$, using the a priori estimate from [8], we have

$$\|Kv\|_0 \le C_3 \|L^*(Kv)\|_{-(2s-1,m)} =$$

$$C_3 \left\| \mathrm{v} + \sum_{\frac{i}{2s-1}+\frac{|\alpha|}{m}\le 1} (-1)^{i+|\alpha|} D_t^{2i} D_x^{2\alpha} \mathrm{v} \right\|_{-(2s-1,m)} = C_4 \sup_u \frac{\left|(\mathrm{v} + \sum_{\frac{i}{2s-1}+\frac{|\alpha|}{m}\le 1} (-1)^{i+|\alpha|} D_t^{2i} D_x^{2\alpha} \mathrm{v}, u)_0\right|}{\|u\|_{2s-1,m}} =$$

$$C_4 \sup_u \frac{|(v,u)_{2s-1,m}|}{\|u\|_{2s-1,m}} \le C_4 \|v\|_{2s-1,m},$$

where L^* is the corresponding formally ad joint operator to L, C_2, C_3 are some positive constants, and $\|L^*[Kv]\|_{-(2s-1,m)}$ is a negative norm of Lax.

Proof of theorem 1: Let we suppose $U_0 \in \widetilde{H}^{2s-1,m}(G)$ minimizes the quadratic functional (9). Then $D[U_0] \le D[U] \ \forall U \in \widetilde{H}^{2s-1,m}(G)$. If we pose $U = U_0 + \alpha V \ \forall V \in \widetilde{H}^{2s-1,m}(G), \forall \alpha \in R^1$, then obtain

$$D[U_0 + \alpha V] - D[U_0] = \|U_0 + \alpha V\|^2_{2s-1,,m} - \|U_0\|^2_{2s-1,m} - 2(F, KU_0 + \alpha KV - KU_0)_0 \ge 0,$$

$$\alpha^2 \|V\|^2_{2s-1,m} - 2\alpha\{(F, KV)_0 - [U_0, V]_{2s-1,m}\} \ge 0,$$

$\forall V \in \widetilde{H}^{2s-1,m}(G)\ \forall \alpha \in R^1$, from where follows

$$(F, KV)_0 - [U_0, V]_{2s-1,m} = 0 \quad \forall V \in \widetilde{H}^{2s-1,m}(G)$$

and the function U_0 is generalized solution to the problem (1)-(3).

Vice versa, if U_0 is the corresponding generalized solution, from the equalities

$$D[U]=\|U\|^2_{2s-1,m}-2(F,KU)_0=\|U\|^2_{2s-1,m}-2[U,U_0]_{2s-1,m}=\ \|U-U_0\|^2_{2s-1,m}-\|U_0\|^2_{2s-1,m}$$

follows that the element U_0 minimizes the quadratic functional (9) on the space $H^{2s-1,m}(G)$.

Proof of theorem 2: From (10) it obtains that the quadratic functional (9) is bounded from below:

$$D[U]=\|U\|^2_{2s-1,m}-2(F,KU)_0=\|U\|^2_{2s-1,m}-2C\|F\|_0\cdot\|U\|_{2s-1,m}\geq -C^2\|F\|^2_0.$$

Let we pose $d=\inf\limits_{U\in\tilde{H}^1(G)}D[U]$. Then $\exists\{U_j\}_{j=1}^{\infty}\subset\tilde{H}^{2s-1,m}(G):\lim\limits_{j\to\infty}D[U_j]=d$. For fixed $\varepsilon>0$ we can take $m,n: D[U_m]\leq d+\frac{\varepsilon^2}{4}$, $D[U_n]\leq d+\frac{\varepsilon^2}{4}$. Now from the inequality $D[U_m/2+U_n/2]\geq d$, from the parallelogram equality and from the estimate (10) we obtain

$$\|U_m-U_n\|^2_{2s-1,m}=2D[U_m]+2D[U_n]-4D[U_m/2+U_n/2]\leq 2d+\varepsilon^2/2+2d+\varepsilon^2/2-4d=\varepsilon^2.$$

Hence the sequence $\{U_j\}_{j=1}^{\infty}$ is fundamental and $\exists U_0\in\tilde{H}^{2s-1,m}(G):\|U_j-U_0\|_{2s-1,m}\xrightarrow[j\to\infty]{}0.$

Now

$$|D[U_j]-D[U_0]|\leq\left|\|U_j\|^2_{2s-1,m}-\|U_0\|^2_{2s-1,m}\right|+2|(F,KU_j-KU_0)_0|\leq$$

$$\left|\|U_j\|^2_{2s-1,m}-\|U_0\|^2_{2s-1,m}\right|+2C\|F\|_0\|U_j-U_0\|^2_{2s-1,m},$$

which implies $D[U_j]\xrightarrow[j\to\infty]{}D[U_0]\equiv d.$

The uniqueness of the solution to the considered variation problem follows immediately from the parallelogram equality. The correctness of the considered problem obtains from the estimate $\|U_2-U_1\|_{2s-1,m}\leq 2C\|F_2-F_1\|_0$, which follows from (10). The proof is complete.

5.Conclusion.

The offered variational approach is classical for elliptic partial differential equations and has long and interesting history [4,11]. $K-symmetric$ and $K-positive$ operators are notions, entered by V.M.Shalov in sixties of twentieth century [2]. He shows that all the known fundamental boundary value problems for the string equation can be solved using some minimization of quadratic functional [3]. Also, after that a sequence results for problems for parabolic type equations are obtained.

For some classes non elliptic partial differential equations with constant coefficients the above variational approach is applied in the papers [5,6]. Considered variational method is applied for multidimensional parabolic and ultraparabolic equations in [7,10]. In the last four papers we use Fourier transform to build in explicit form a symmetrizing operator and the Marcinkiewich–Michlin Theorem for the multiplicators of Fourier integral to obtain appropriate a priori estimates.

In [4] the variational approach is applied to nonlocal boundary value problem for a class of hyperbolic-parabolic equations. In present paper for the first time is built a symmetrizing operator for nonlocal boundary value problem for partial differential equation of arbitrary high order of mixed type. For such equations the symmetrizing operator is solution to the formally ad joined problem with a "special" right part.

The smoothness of the solution to the problem (1)-(3) is investigated in [9]. The investigation of the smoothness of the solution to the variational problem (9) reduces to the investigation of the properties of the corresponding symmetrizing operator.

References

[1] Karatoprakliev G D 1987 *Diff. Uravnenia* Minsk **23** 1 pp 78-84

[2] Shalov V M 1963 *Dokl. A. N. SSSR* **151** 3 pp 511-512

[3] Shalov V M 1965 *Diff. Uravnenia* Minsk **1** 10 pp 1338-1365

[4] Paskalev G P 2018 *Journal of the Technical University at Plovdiv* "Fundamental Sciences and Applications" **24** pp 317-319

[5] Paskalev G P 2017 *Journal of the Technical University at Plovdiv* "Fundamental Sciences and Applications" **23** pp 63-67

[6] Paskalev G P 2013 *Journal of the Technical University at Plovdiv* "Fundamental Sciences and Applications" **19** pp 141-145

[7] Paskalev G P 1999 *Journal of the Technical University at Plovdiv* "Fundamental Sciences and Applications" **8** pp 69-78

[8] Paskalev G P 2000 *Diff. Uravnenia* Minsk **36** 3 pp 393-399

[9] Paskalev G P 2000 *Diff. Uravnenia* Minsk **36** 6 pp 799-805

[10] Paskalev G P 1992 *Diff. Uravnenia* Minsk **28** 9 pp 1640-1641

[11] Paskalev G P 1990 *Comptes Rendus de l'Academie bulgare des Sciences* **43** 10 pp 23-25

Microscopic and surface analyses of corroded Mg alloys

Y Marcheva [1], M Loukaycheva [1], B Dimova [1] V Tsonev [1] and N Kuzmanov [1]
[1]Technical University – Sofia, Bulgaria

E-mail: ysm@tu-sofia.bg

Abstract. Using optical and SEM microscopic observations and EDX analysis, the corroded surfaces of AZ91, AM50 and AE42 Mg-alloys in 3.5%NaCl were studied at condition of open circuit potential. Different surface composition, morphology, density and characteristics of the surface films were established. The corrosion potential, recorded for 5 days in the case of AE was studied and specific behaviour was observed. EDX analysis established the presence of Mg and Al in the corrosion products of the three alloys and Ce for AE42 and formation of oxides and/or hydroxides of these elements was assumed. A mechanism of dissolution of Mg alloys at OCP condition in NaCl solutions was discussed.

1. Introduction

Magnesium has a lot of advantages [1] and modest cost towards the other metals. Compared with polymers, Mg alloys show superior mechanical properties. The use of magnesium alloys in various industries have been increasing in the past decade, and it is expected that in the coming years this trend will continue. Recently, new Mg alloys with excellent mechanical properties have been reported, [2-6] and their use for industrial purposes appears to be promising.

The most common magnesium die alloy is **AZ91**. This alloy has extremely good castability and shows high mechanical strength at ambient temperatures. Typical applications include automotive engine parts, computer parts, and parts for handheld tools. Due to the high aluminium content, the alloy has limited ductility and impact strength. To solve this problem **AM** alloys are developed. They are characterized with better ductility and impact strength. Typical parts made from this alloys include automotive instrument panel subparts, steering wheel cores, seat parts, etc. Extensive efforts have been made to develop magnesium die casting alloys with improved creep properties at temperatures exceeding 120°C. This would open up new applications in manufacturing of elevated temperature-exposed engine parts and automatic transmissions. For this purpose the **AE** types of alloys have been developed [4] by addition of rare earth (RE) elements. This is an effective way to improve the mechanical properties at elevated temperatures and the corrosion resistance to some degree of magnesium alloys [7, 8].

However, the satisfactory mechanical properties are not enough to enable a magnesium alloy to serve in a practical environment. In fact, the use of Mg alloys remains notably limited because of its high reactivity. Magnesium alloys usually have relatively low corrosion resistance, which is one of the main obstacles that impede the use of magnesium alloys in real world environments. Therefore, corrosion performance is a major issue for research and development of magnesium alloys [6].

The aim of the presented paper was to perform a comparative study of the morphology and composition of the corroded surfaces of three types of Mg alloys – AZ91, AM50 and AE42 as a part of more detailed study of the corrosion properties of these alloys.

2. Experimental

The corrosion environment was a 3.5% NaCl solution with initial pH value 6.7. The specimens were cylindrical (d = 13.5 mm). Before the immersion in the solution the specimens were subjected to the treatment as follows: mechanical cleaning with sand paper (finest grit – 2000); degreasing in an ultrasound bath with ethanol; isolating the side surfaces of the specimens with polymer material and immersion in the corrosion environment up to 120 hours (5 days). Each specimen was immersed individually in a flask with 100 mL corrosion solution. All experiments were performed at room temperature.

Open circuit potential measurements (OCP) were performed using pH/ORP/mV bench meter HI 3220 (Hanna Instruments) with automatic recording and data collection. The potentials were measured and presented versus standard silver electrode (SSE).

The surface appearance was studied by optical observations using inverted metallographic microscope Optika XDS-3MET with built-in camera for digital photography and scanning electron microscopy (SEM) TESCAN Lyra XMU, equipped with EDX spectrometer Bruker GmbH with primary energy 30 keV and detector type XFlash 5010.

3. Results and discussion

At Figure1 the corrosion potentials of the studied alloys are presented.

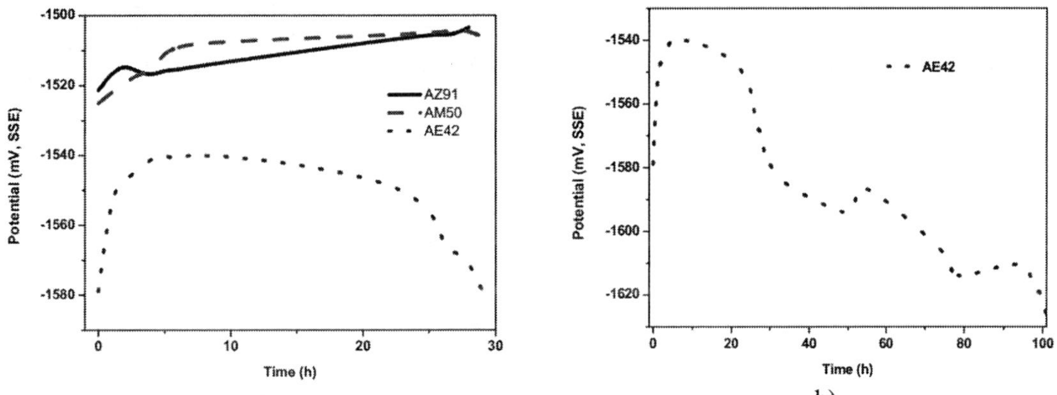

Figure 1. Open circuit potential (OCP versus SSE) as a function of time of immersion in 3.5% NaCl.

It is seen, that the potentials of the AZ and AM alloys are almost the same and reach a stable value at about 6 hours, which vary slightly up to end of the study. During the first 5 hours the corrosion potential of AE alloy has similar behaviour as the other two alloys, but has a more negative value (Figure 1a). Additionally, it can be seen that for AE42, the initial potential change in positive direction is sharper.

After 5h immersion the behaviour of the potential of AE is quite different compared to the other alloys. Here, the potential is going to the negative direction and this tendency is keeping up to the end of the experiment. The potential become more negative than the initial corrosion potential (Figure 1b).

Cain and co-workers [9] recently summarized a compilation of corrosion potentials for a number of commercial and experimental magnesium alloys. Results were published for the solution of 0.1M NaCl and the OCP was measured for 10 min prior to polarization testing. The conclusion was, that only minimal changes to the OCP of Mg-alloys were observed. This was contributed to the limited solubility of most elements in crystalline Mg and that Mg is weakly polarisable.

In our study the corrosion media was more concentrated compared to the solution, used in [9]. Data about the prolonged OCP measurement during the whole corrosion testing period were not found.

IOP Conf. Series: Materials Science and Engineering **618** (2019) 012092 doi:10.1088/1757-899X/618/1/012092

The results received in our study, for the evolution of OCP are in agreement with the published [9] for the first hours of the experiment in the case of AE. The changing of the OCP with time in negative direction for this alloy is an interesting result, which needs more attention and more detailed study.

The appearance of the corroded surface after 2 and 5 days of immersion, received by optical microscopy, is shown at Figure 2.

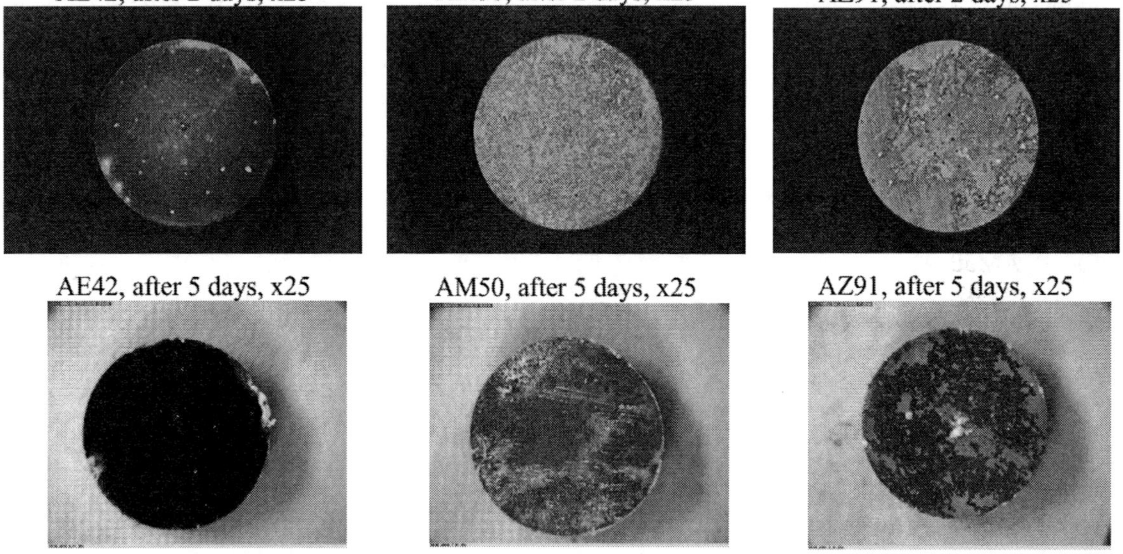

Figure 2. Corroded surface photographs of studied alloys.

In the first 2 days (48h) the surfaces of AZ and AM are still bright (AZ is rougher), without specific features, while the AE show formation of a darker film on the surface. This appearance can be related to the observed OCP change of AE and the stability of the corrosion potential of AZ and AM in this period of immersion time.

After 5 days (120h) of immersion, the surface appearance of the three alloys is very different. AZ has dark spots, and it is less uniform, AM has bright surface with white residue, while the surface of AE is covered with dense black compact film.

SEM images of the surfaces (Figure 3) confirms the formation of a dense and compact film on the AE and the presence of corrosion products on AZ and AM surfaces with structure quite different compared to AE.

Figure 3. SEM images of the corrosion product film on the surface of studied alloys.

EDX analysis of the surfaces before and after the immersion in NaCl are given in Table 1 and Table 2.

Table 1. Chemical (elemental) composition (wt. %) of the surfaces of the studied alloys before the immersion in the corrosion environment.

Alloy	Mg	Al	Zn	Cu	Fe	Mn	Si	Ce
AZ91	88.50	8.97	1.27	0.50	0.34	0.29	0.13	–
AM50	94.76	4.01	0.39	0.25	0.16	0.37	0.05	–
AE42	93.86	3.29	0.25	0.16	0.09	0.37	0.02	1.95

Table 2. Chemical (elemental) composition (wt. %) of the surfaces of the studied alloys after the immersion in the corrosion environment (3.5% NaCl) for 3 days (72h).

Alloy	Mg	Al	Zn	Cu	Fe	Mn	O	Cl	Ce
AZ91	44.95	3.33	0.89	0.36	0.21	0.27	43.71	2.84	–
AM50	44.92	1.80	0.49	0.26	0.19	0.16	46.34	5.14	–
AE42	36.28	9.82	0.45	0.52	0.13	1.40	40.42	0.06	6.53

After the immersion in the NaCl solution, in the composition of the corrosion products on the alloy surface, besides the main elements, new elements as O and Cl appear (Table 2). The ratio Mg:O is close to 1:1, which suggest, that the corrosion products on the alloy surface are composed mainly by Mg-oxides and hydroxides.

Moreover, except Mg the results show and the presence of Al in the corrosion products of the three alloys, which means that Al is dissolved during the immersion in the corrosion environment and probably the alloys form $Mg(OH)_2$ and $Al(OH)_3$ precipitates in the solution and the white residue on the specimen surface.

In the corrosion products on AE42 surface, the content of Al is much higher (about 3 times and more) than the one in the other alloy's corroded surfaces and at the same time the Mg and Cl content is the lowest. Additionally, in the AE corrosion product a considerable amount of Ce is detected. The formation of black, dense film on the surface of AE42 is probably due to formation of mixed oxides and/or hydroxides of Al, Ce and Mg. The elements in the film are uniformly spread. The negligible content of Cl on the surface could explain the most compact and dense film on AE compared to the others, bearing in mind the destructive action of chlorine ions on the passive films on the alloy surface.

It is well established [1], that the cathodic reaction during the corrosion of Mg-alloys is water reduction, often also simply termed as HE (hydrogen evolution). The overall corrosion reaction for Mg alloys is:

$$Mg + 2H_2O = Mg(OH)_2 + H_2$$

It has been shown that the Mg corrosion product exhibits enhanced catalytic activity towards the HE reaction [1]. In other words, the dark corrosion film formed during Mg corrosion at the OCP promotes the evolution of H_2 at a faster rate than uncorroded areas on the sample. There is substantial evidence that the dark film is the primary site of cathodic reaction for Mg dissolving at its OCP [10].

The morphology of the attack is directly dependent on the concentration of NaCl in the electrolyte and the amount of Fe present in the material [11]. At a fixed Fe concentration, radial disc-like propagation is exhibited in highly concentrated NaCl solutions whereas lower concentrated NaCl environments lead to a filiform-like attack. This behaviour was attributed to an ohmic effect that influences the current flow in the electrolyte between the net anodes (leading edges) and the net cathodes (dark regions), which are galvanically coupled. A solution with higher NaCl concentration, and thus higher conductivity, allows for more distant galvanic coupling and leads to a radial disk-like attack. On the other hand, solutions with lower conductivity constrain the current flux between net anodes and cathodes, leading to filiform-like tracks that maintain the distance between the galvanically coupled regions at a minimum [11].

According to [11] the introduction of rare earth (RE) elements may lead to the improved surface oxide stability of the alloy, which may help reduce the rate of dissolution of the alloy. One of the important effects of RE on the corrosion resistance of Mg is the so-called "scavenger effect", i.e., some impurity elements in Mg, such as Fe, severely deteriorates its corrosion resistance, and RE is said to cancel their influence by the formation of intermetallic compounds with the impurities [12]. Meanwhile, the corrosion resistance of Mg mainly depends on the surface oxide film, i.e., the corrosion property of Mg should be affected by RE addition since the surface oxide film changes with the RE addition.

4. Conclusions

A set of optical and SEM microscopic observations of the corroded surface show the different surface composition, morphology, density and characteristics of the surface films. The most compact film is formed on the surface of AE42

EDX analysis of the alloy surfaces were performed. The presence of Mg and Al in the corrosion products of the three alloys was established and of Ce for AE42. Formation of oxides and/or hydroxides of these elements was assumed. The negligible content of Cl on the surface of AE42 could explain the most compact and dense film on this alloy compared to the others, bearing in mind the destructive action of chlorine ions on the passive films on the alloy surface. A mechanism of dissolution of Mg at OCP condition in NaCl solutions was discussed.

The authors would like to thank the Research and Development Sector at the Technical University of Sofia for the financial support.

References

[1] Esmaily M, Svensson JE, Fajardo S, Birbilis N, Frankel GS, Virtanen S, Arrabal R, Thomas S and Johansson LG 2017 *Progress in Materials Science* **89** 92
[2] Shaw B and Wolfe R 2005 Corrosion of Magnesium and Magnesium-Base Alloys, in: *ASM Handbook Volume 13B, Corrosion: Materials,* ed S. Cramer and B. Covino (ASM International, Ohio) pp 205-227
[3] Housh S, Mikucki B and Stevenson A, Selection and application of magnesium and magnesium alloys, 1992 in: *ASM Handbook* (Vol. 2 - Properties and Selection: Nonferrous Alloys and Special-Purpose Materials), (ASM International) pp1424-1425
[4] Westengen H, Magnesium Alloys: Properties and applications, in: 2001 *Encyclopedia of Materials: Science and Technology* (Elsevier Science Ltd.) pp 4745-4753
[5] Revie R 2011 *Uhlig's Corrosion Handbook* (3rd edition, John Wiley & Sons Inc.) p 847
[6] Song G, Atrens A and StJohn D 2001 *Magnesium Technology*, ed J. N. Hryn (TMS) pp 255-262
[7] Fan Y, Wu G and Zhai Ch, 2006 *Materials Science and Engineering A*, **433** 208
[8] Birbilis N, Easton MA, Sudholz AD, Zhu SM and Gibson MA 2009 *Corrosion Science* **51** 683
[9] Cain T, Bland LG, Birbilis N, Scully JR. 2014 *Corrosion* **70** 1043
[10] Curioni M 2014 *Electrochim Acta* **120** 284
[11] Williams G, Birbilis N, McMurray HN 2015 *Faraday Discuss* **180** 313
[12] Takenaka T, Ono T, Narazaki Y, Naka Y, Kawakami M. 2007 *Electrochimica Acta* **53** 117

IOP Conf. Series: Materials Science and Engineering **618** (2019) 012093 doi:10.1088/1757-899X/618/1/012093

Green inhibitor of corrosion of aluminium alloy EN AW-2011 in acidic environment

Kalina Kamarska
Department of Mathematics, Physics, Chemistry, Technical University–Sofia, Branch Plovdiv, Bulgaria

E-mail: kamarska@tu-plovdiv.bg

Abstract. This work presents the results of corrosion inhibition of aluminium alloy EN AW-2011 in 1M HCl environment by Rosmarinus officinalis, using gravimetric and gasometric techniques. The inhibition efficiency increases with increasing the concentration. The investigation showed optimal inhibition efficiency of about 62%. The results confirmed that the adsorption of Rosmarinus officinalis reduces the corrosion rate of this alloy in the acidic medium.

1. Introduction

The first section in your paper Aluminium alloy EN AW-2011has a wide range of applications in the car industry, in some chemical processing industries and the aircraft industry. It is especially suitable for parts and structures requiring high strength and is often used to make parts that require good strength at temperatures of up to 150°C such as nuts, bolts, screws, studs, car parts and more [1]. Aluminium and its alloys are very reactive metals, predisposed to corrosion. Due to their wide applications, they frequently come in contact with acids, such as hydrochloric acid, during pickling, de-scaling or electrochemical etching of aluminium. Therefore it is very important to add inorganic and organic corrosion inhibitors in order to prevent the dissolution of metal. There are a number of methods for preventing corrosion and one of them is using inorganic corrosion inhibitors.

Inorganic corrosion inhibitors are effective but have a negative influence on the environment. Plant extracts and oils, which consist of organic substances (essential oils, saponnins, amino and organic acids) with heteroatoms oxygen, nitrogen and sulphur in functional groups, are known to evince corrosion inhibitor action [2, 3]. Green inhibitors are eco-friendly, non-toxic, highly efficient, renewable and cheap [4].

A review of the literature written on the matter revealed that plant materials such as Aloe plant extract [5], Sansaviera infasciata extract [6], aqueous [7] and ethanol extract of Rosemary [8], extract of Mentha pulegium [3], Terminalia ivorensis extract [9], acetone extract of red onion skin [10], gum exudates from pachylobus edulis in the presence of halide ions, [11], to mention but a few, have all been used to inhibit the corrosion of aluminium in acid media in recent years. In Rosmarinus officinalis, the presence of phenolic diterpenoids with antioxidant activity such as rosmarin acid, rosmanol, rosmadial, epirosmanol and carnosol. These organic compounds are generally considered to be the main corrosion inhibitors present in Rosemarinus [8].

The present work is focused on the corrosion inhibition performance of essential oil of Rosmarinus officinalis on aluminium alloy EN AW-2011 in 1M HCl solution. The study was conducted by using gravimetric and gasometric techniques.

2. Experimental

2.1. *Gravimetric technique*

The experiments were performed with cylindrical samples of aluminium alloy EN AW-2011 with an exposed area of 8.54 cm^2. Before immersion in the solution, the experimental samples were mechanically cleaned, washed in detergent for five minutes and then rinsed with distilled water and dried. The dried samples were weighed with an analytical balance Acculab ATILON with accuracy to ± 0,0001g and then immersed in a test solution of 1 M HCl for 3 hours at room temperature. The aluminium alloy was exposed to the acid solution without the inhibitor and also to acid environment containing different concentrations of essential oil of Rosmarinus officinalis (from 0,0% to 0,05%) which served as the inhibitor. After the immersion period, the samples were withdrawn from the solution, washed with distilled water, dried under a stream of air and weighed. For each of the immersion period the corrosion rate (CR) of alloy were determined using the following equation:

$$CR = \frac{(m_1 - m_2)}{S.t} \quad (1),$$

where m_1 is the weight of the sample, g; m_2 is the weight of the sample after the corrosion test, g; S is the area of the sample, m^2 and t is the test time, h.

The surface coverage (θ) and the inhibition efficiency (η %) were also calculated from the following equation:

$$\theta = \frac{CR_{blank} - CR_{inh}}{CR_{blank}} \quad (2) \text{ and}$$

$$\eta\,\% = \theta\,.\,100 \quad (3),$$

where CR_{blank} and CR_{inh} are the corrosion rate values in the absence and presence of the inhibitor, respectively.

2.2. *Gasometric technique*

The progress of the corrosion reaction was defined through volumetric measurement of the evolved hydrogen gas [3]. The materials used for constructing the working electrode were the same as those used for the gravimetric technique. From the volume of the evolved hydrogen gas, the corrosion rate was calculated using the following equation:

$$\eta\,\% = \frac{V_{blank} - V_{inh}}{V_{blank}} ,$$

where V_{bank} and V_{inh} are the volumes of hydrogen gas that evolved in the absence and presence of the inhibitor, respectively.

3. Results and discussion

3.1 Gravimetric technique

The obtained corrosion rate data of EN AW-2011 are presented in figure 1. The gravimetric results show that the inhibition efficiency increased and the corrosion rate decreased with the increase in the concentration of the inhibitor, indicating that the Rosmarinus officinalis oil acted as a good inhibitor.

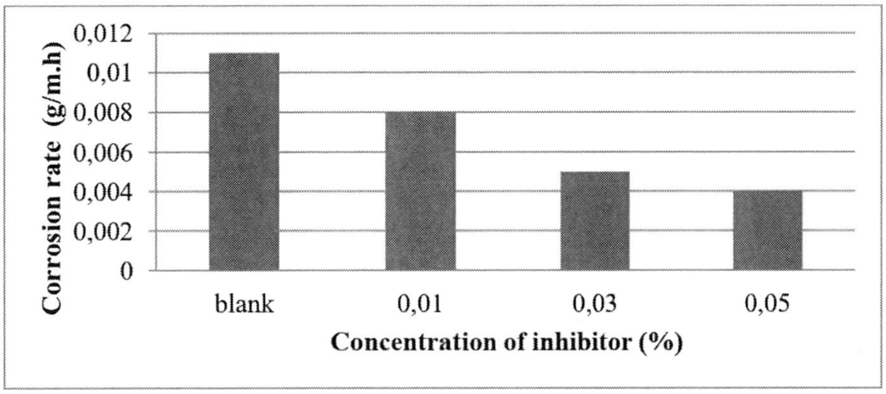

Figure 1. Corrosion rate of EN AW-2011in the absence and presence of different concentrations of Rosmarinus oil after 3 hours of immersion in 1 M HCl.

The inhibition efficiencies of the different concentrations of Rosmarinus officinalis oil as the inhibitor are shown in figure 2.

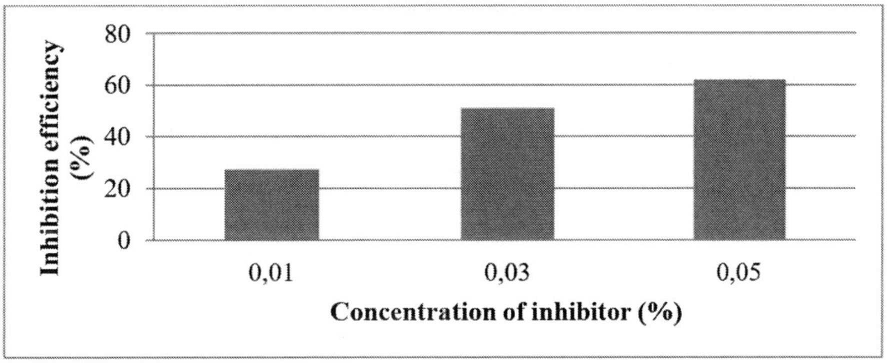

Figure 2. Inhibition efficiencies of Rosmarinus oil at various concentrations after 3 hours of immersion in 1 M HCl.

The results obtained in this work also showed that in the employed conditions the highest inhibition efficiency of 62% was obtained for 0.05 % concentration of Rosmarinus officinalis oil.

The corrosion rate and inhibition efficiencies of Rosmarinus oil at different concentrations are summarised in table 1.

Table 1. Inhibition efficiencies of various concentrations of Rosmarinus oil after 3 hours of immersion in 1 M HCl.

Concentrations of Rosmarinus officinalis, %	CR, $g/m^2.h$	Inhibition efficiencies, %
blank	0,011	-
0.01	0,008	27,3
0.03	0,005	51
0.05	0,004	62

The successful inhibitory action of oil of Rosmarinus officinalis could be explained by the presence of organic compounds such as rosmadial, rosmanol, epirosmanol, carnosol and others [12-14] in it.

3.2. Gasometric technique

The volume of hydrogen evolved during the corrosion reaction of aluminum alloy immersed in 1 M HCl solution for 3 hours is measured as a function of the concentration of Rosmarinus, and the data are represented graphically in figure. 3.

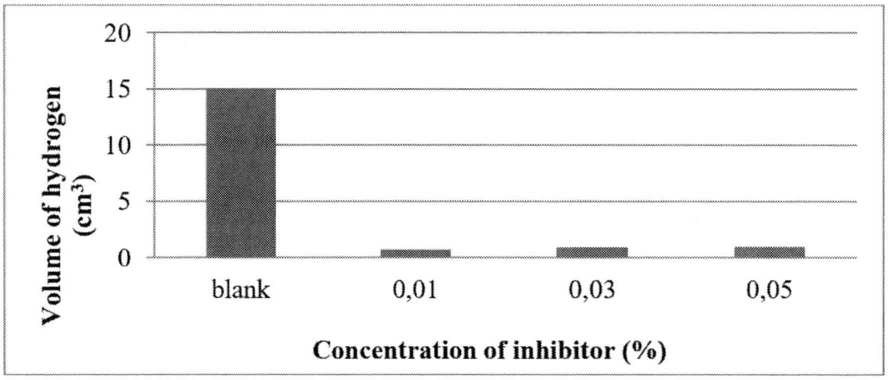

Figure 3. Volume of hydrogen evolved during the corrosion of EN AW- 2011 in the absence and presence of different concentrations of Rosmarinus officinalis oil after 3 hours of immersion in 1 M HCl.

The volume of the evolved hydrogen was observed to sensibly reduce in the presence of Rosmarinus oil and it decreased with the increase in the concentration of the inhibitor of Rosmarinus.

4. Conclusion

On the basis of this study, it can be concluded that the rate of corrosion of aluminium alloy EN AW-2011 in hydrochloric acid environment is a function of the concentration of the oil of Rosmarinus officinalis. The results obtained in this work showed that, in the conditions employed in the present study, the oil of Rosmarinus officinalis proved to be a good corrosion inhibitor for this alloy in 1 M HCl solution. The inhibition efficiency increases with increasing the concentration of Rosmarinus oil.

References

[1] Davis J R 2001 *Alloying: Understanding the Basics* (Ohio: ASM International)

[2] Prabhu Dand Padmalatha R 2013 *Int. J. ChemTech. Res.* **5**, 2690

[3] Khadraoui A, Khelifa A, Touafri L, Hamitouche H and Mehdaoui R 2013 *J. Mater. Environ. Sci.* **4** 663

[4] Sangeetha M, Rajendran S, Sathiyabama J and Krishnaveni A 2013 *Port. Electrochim. Acta* **31**, 41

[5] Al-Turkustani A, Arab S and Aldahiri R 2010 *Mod. Appl. Sci.* **4** 105

[6] Oguzie E 2007 *Corrosion Sci.* **49** 1527

[7] Radošević J, Kliškić M, Višekruna A 2001 *Kem. Ind.* **50** 537

[8] Deyab M 2015 *J. Taiwan Inst. Chem. Eng.* **58** 536

[9] Wisdom J, Kelechi1 D and Ugochukwu J 2018 *IRJET* **5** 121

[10] James, O and Akaranta O 2009 *Afr, J. Pure Appl. Chem.* **3** 262

[11] Umoren S, Obot I, Ebenso E and Okafor P 2008 *Port. Electrochim. Acta* **26** 267

[12] Inatani R, Nakatani N and Fuwa H 1983 *Agric. Biol. Chem.* **47** 521

[13] Schwarz K and Ternes W 1992 *Z. Lebensm. Unters. Forsch.* **195** 99

[14] Schwarz K, Ternes W and Schmauderer E 1992 Z. *Lebensm. Unters. Forsch.* **195** 104

IOP Conf. Series: Materials Science and Engineering **618** (2019) 012094 doi:10.1088/1757-899X/618/1/012094

Investigation of the effect of porosity on the particle deposition efficiency in the model of an open cell foam filter

O V Soloveva, R R Khusainov, E G Sheshukov and R R Yafizov

Institute of Heat Power Engineering, Kazan State Power Engineering University, Krasnoselskaya st. 51, Kazan, 420066, Russian Federation

E-mail: solovyeva.ov@kgeu.ru

Abstract. In this work, we performed experimental and numerical simulation of gas flow in the model of the open cell foam filter. We investigated the relationship between the porosity of the medium ε, and the parameter Po characterizing the number of pores per inch. We obtain a porous medium from the inverse matrix of the computational model made on a 3D printer. The calculation results of the pressure drop depending on the average filtration rate correlate well with the experimental data. Particle trajectories make it possible to determine the particle deposition efficiency in a porous layer for three values of the porosity of the medium $\varepsilon = 0.6$, $\varepsilon = 0.7$ and $\varepsilon = 0.8$. The deposition efficiency in the case of porosity $\varepsilon = 0.8$ correlate well with the data obtained by the semi-empirical formula of the authors of [1], however, for the other two values of the porous medium, the efficiency curves differ significantly, so the dependence was obtained for a material with high porosity. We define the filter quality parameter by the ratio of the particle deposition efficiency to the pressure drop value in the porous layer.

1. Introduction

Open cell foams have a very diverse use [2-10]. They are used as thermal insulation [11, 12], in heat exchange amplifiers in electrical systems, where the radiator is extremely important [13, 14], as receivers of solar radiation [15, 16], and as open cell foam matrix catalyst using instead of industrial catalysts [17]. Also, the open cell foam material is used as an active filter for aerosol particles due to the low drag coefficient and developed surface [18-21].

For the gas flow numerical simulation in the porous medium, it is necessary to make a geometric model that will correspond to the real structure of the material. The authors of [22-24] use a porous medium from simplified cells with an ordered structure. However, the real porous medium structure is not orderly, and it provides significant changes in the flow hydrodynamics. A model that represents the actual structure can be obtained using microtomography technology [15, 16]. The model obtained by the microtomography of the medium further requires long and laborious processing, cleaning from «suspended in space» elements that are not related to the general structure. The disadvantage of this method is that the resulting geometry requires lengthy and time-consuming processing. Modern technologies make it possible to print material on a 3D printer with specified parameters of a porous medium. Due to the complex shape of the porous media geometry, the majority of published works based on experimental data [25, 26].

IOP Conf. Series: Materials Science and Engineering **618** (2019) 012094 doi:10.1088/1757-899X/618/1/012094

The use of simplified models of the flow in porous structures leads to significant differences in experimental and calculated data [27-30]. There are experimental studies of heat and mass transfer in a porous medium [16, 23, 25-29].

The authors of the work [31] compare the heat exchanger performance with the open cell foam material and with finned tubes. The results show that the open cell foam material not only intensifies the heat transfer but at the same time reduces the weight and size of the heat exchanger. The work [32] presents the results of heat transfer measuring inside rectangular aluminum blocks, with a constant supply of heat to the porous medium. The authors propose an analytical model for heat exchange in a metallic porous medium. In [33], the authors investigated the effect of the open cell foam material thickness on the heat transfer properties and pressure drop in the vertical channel.

Numerical simulation of processes in porous media can overcome all experimental problems and accurately take into account the smallest geometric details. Modern computational capabilities make it possible to create a detailed geometry of the computational area, set the material properties and boundary conditions.

Previously, researchers based their numerical studies of porous media mainly on the averaged volume method and the Darcy law. In [34], the authors made a detailed numerical simulation of the flow in a metallic open cell foam material. Detailed numerical simulation is the most accurate method for studying flow in porous filters, because it is inapplicable to many averaged flow models in porous media due to a complex porous material structure.

The purpose of this work is to determine the pressure drop value in the open cell foam material, which is a model of an aerosol filter, and the particle efficiency deposition for three values of medium porosity.

2. Problem formulation

Let us make experimental studies and numerical simulation of the air flow in open cell foam material with different medium porosity ε (the ratio of the pore volume in the total body volume) and a fixed pore size $d_c = 4$ mm. The computational model of an open cell foam filter is a set of intersecting spheres randomly arranged in space. (Fig. 1).

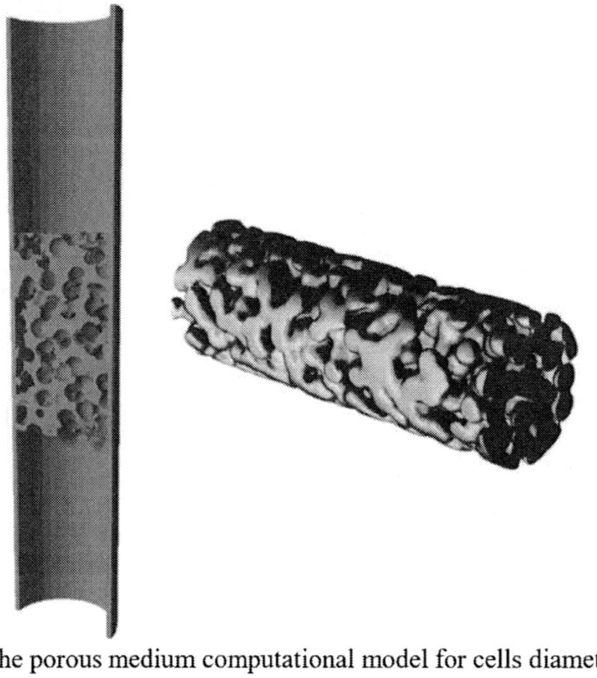

Figure 1. The porous medium computational model for cells diameter $d_c = 4$ mm.

We made calculations in the ANSYS Fluent software package (v. 19.0) for the viscous incompressible gas model by solving the Navier-Stokes equations using the finite volume method. There are two main approaches to describe gas flow in a porous medium. The first approach is based on the averaged flow models that include an adjustment for the medium porosity (and permeability), such as, the well-known Darcy model. The second approach involves the calculation of the flow in a porous medium based on the Navier-Stokes equations, taking into account the geometry details. This approach is the most accurate because it takes into account the pores shape, pore location, fiber diameter, and other parameters. In this case, taking into account the resistance of the medium occurs due to the peculiarities of the internal structure. Direct numerical simulation for a detailed picture of the flow, reflecting the vortices in each pore, was carried out using a grid of 40 million cells. Figure 2 shows the grid quality.

Figure 2. Fragment of the grid partitioning of the computational domain.

We set the condition "mass flow inlet" at the left border of the computational area corresponding to the flow inlet, and the condition "pressure outlet" at the right border. Also, the remaining boundaries were defined as "wall" automatically. We performed calculations for three values of medium porosity $\varepsilon = 0.6$, $\varepsilon = 0.7$ and $\varepsilon = 0.8$.

3. Determination of the parameter *Po* (porosity)

We selected five cross sections of the porous layer: 0 cm, 1 cm, 2 cm, 3 cm, and 4 cm from the beginning of the porous section (Table 1) to determine the parameter (counting the average number of pores per inch). The cross-section pictures of the porous medium make it possible to determine the average porosity parameter.

For medium porosity $\varepsilon = 0.6$, $\varepsilon = 0.7$, $\varepsilon = 0.8$ porosity parameters are $Po = 6$, $Po = 7$, $Po = 8$ respectively.

Table1. Pictures of the cross-section of the open cell foam model.

	$l_1 = 0$ cm	$l_2 = 1$ cm	$l_3 = 2$ cm	$l_4 = 3$ cm	$l_5 = 4$ cm
$\varepsilon = 0.6$	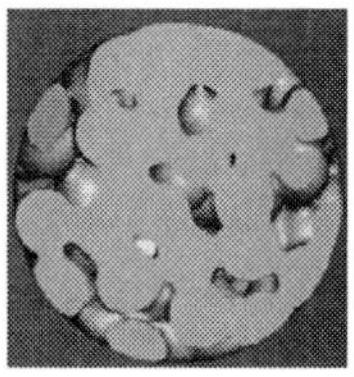	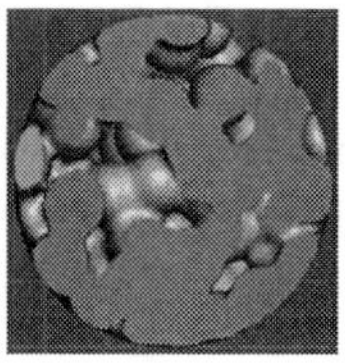	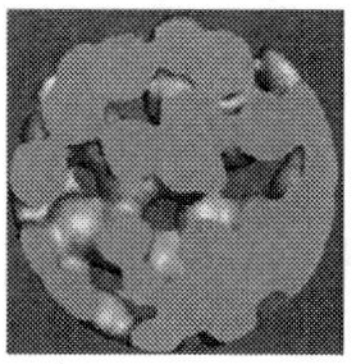	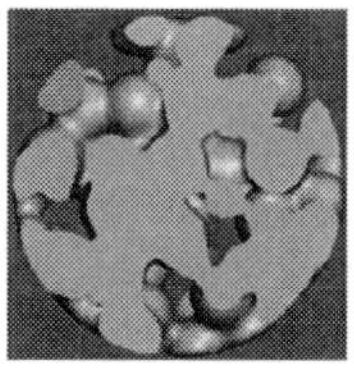	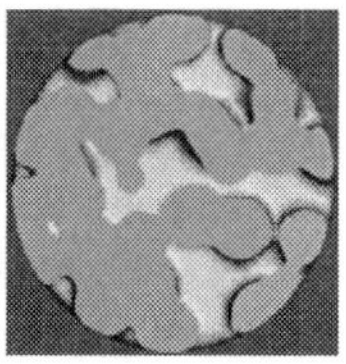
$\varepsilon = 0.7$	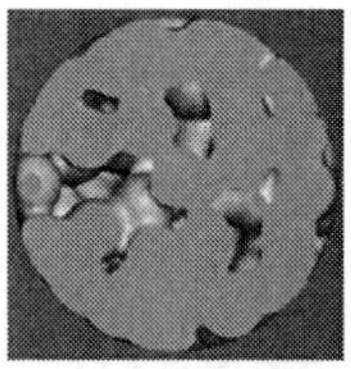	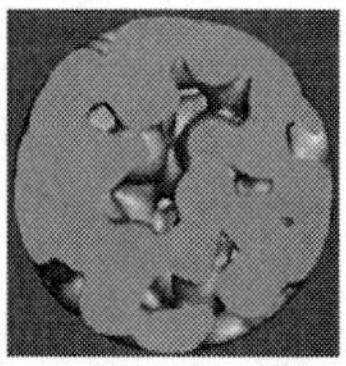	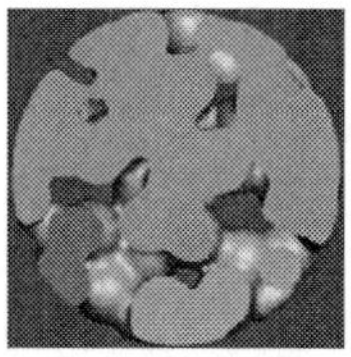	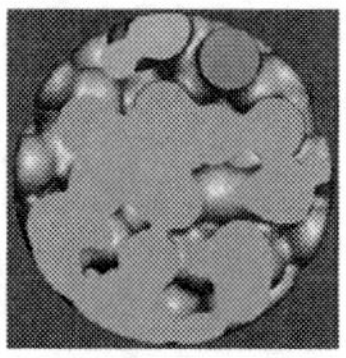	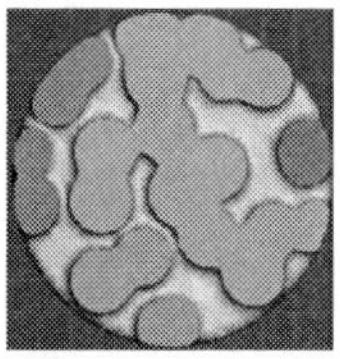
$\varepsilon = 0.8$	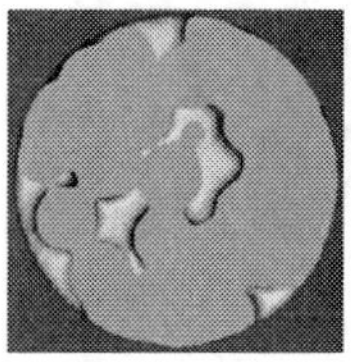	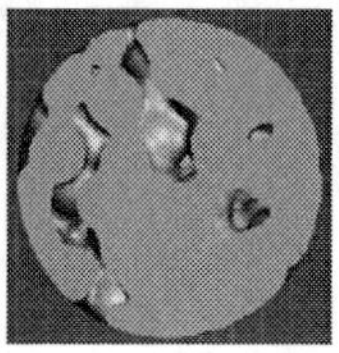	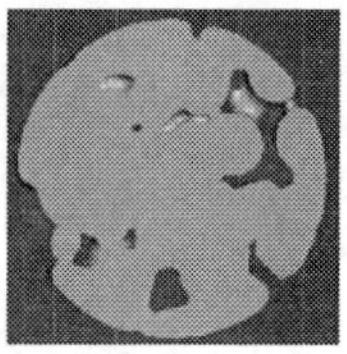	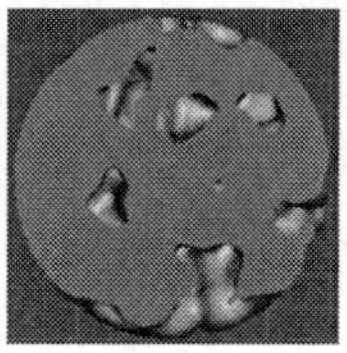	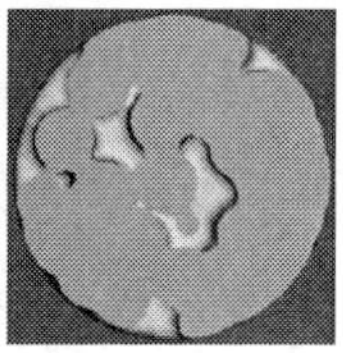

4. Results of experiment and numerical simulation

It is necessary to make sure that the model is correct in order to accurate calculation of the particle deposition efficiency. Experimental samples completely represent the computational model geometry taking into account the inverse structure (the numerical calculation is in the inner porous space). In the experiment, air is injected into the porous medium by a compressor, and the pressure drop is measured by «Testo 510 manometer», the flow rate is measured by «Testo 450 hot-wire anemometer». The pressure drop values in the porous region are determined depending on the average filtration rate. The results of experimental studies and numerical simulations correlate well. The insignificant difference in the study results at high flow velocities is related to developed turbulent flow and difficulty in «trapping» vortices in numerical calculation. It is possible that the grid partition of 40 million cells is not enough for such a complex area, but computational capabilities limit the increase of the grid elements number. The average filtration rate of gas suspension in technical devices varies from 1 to 5 m/s, the deviation in the results of numerical and experimental studies can be acceptable, and at speeds up to 5 m/s, there is a complete agreement of the obtained data.

The particle deposition efficiency is the primary parameter in a filter quality. Also, it is a ratio of the settled particles number to the started particles number:

$$E = \frac{N}{N_0}.$$

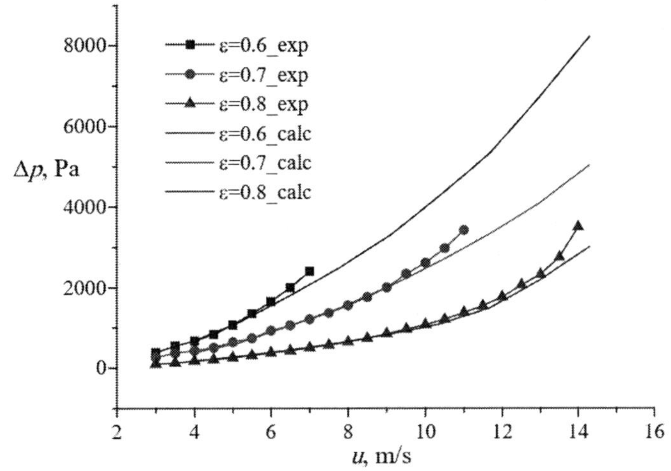

Figure 3. The pressure drop dependence on the flow rate for three values of the medium porosity: lines with symbols – experimental data, lines without symbols – calculation results.

Helmann [1] derived a semi-empirical dependence for evaluating of the particle deposition efficiency of in diffusion neglect.

$$E = 1 - P,\quad \ln(P) = -\frac{t}{d_f}\cdot\left(5.486\cdot \mathrm{St}^{2.382} + 3.891\cdot N_G^{0.88}\right)$$

where t is the filter thickness, d_f is the fiber diameter of the porous medium, St is Stokes number, N_G is the parameter of gravitational settling.

$$\mathrm{St} = \frac{\rho_p\cdot d_p^2\cdot u_0}{18\cdot\mu\cdot d_f},\quad N_G = \frac{\rho_p\cdot d_p^2\cdot g}{18\cdot\mu\cdot u_0}$$

d_p is the particle diameter, u_0 is the gas velocity, μ is viscosity.

$$d_f = 0.009633\cdot Po^{-1.216}$$

The parameters t and d_f are given in millimeters.

The correctness confirmation of the model made it possible to calculate the particle deposition efficiency in the gas flow velocity

According to the numerical calculation and the semi-empirical Hellman formula, the efficiency curves of particle deposition correlate well for the case of maximum porosity $\varepsilon = 0.8$, however, the efficiency curves diverge for the case of lower porosity because the Hellman formula works only in a certain range of environmental parameters in the performed experiment. The starting particles location is also significant. For a medium with lower porosity, a substantial area is formed in the cross-section, comparable to the pore area on which particles additionally deposited if particles start far from the pores of the filtering material. These two factors explain the discrepancy between the experimental data and the numerical calculation results for the values of medium porosity $\varepsilon = 0.6$ and $\varepsilon = 0.7$.

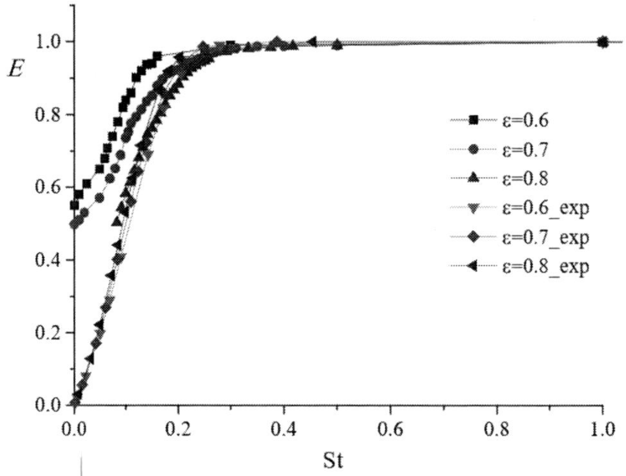

Figure 4. The dependence of the particle deposition efficiency on the Stokes number, obtained by numerical calculations and experimental studies.

5. Conclusion

In this work, we investigate the dependence of the medium porosity, which uses the mechanics of porous media and the porosity parameter Po used in industry. We carried out numerical calculations and experimental studies of the gas flow in a tube containing a porous layer, which is a filter model for various porosities of the medium. The curves of numerical calculation results for the pressure drop values correlate well with the curves of experimental data. The particle deposition efficiency was calculated depending on the Stokes number, we found that the calculations results and experimental data from other authors are entirely consistent for the porosity $\varepsilon = 0.8$, and there is a significant difference for $\varepsilon = 0.7$ and $\varepsilon = 0.6$ due to the limits of applicability of the semi-empirical formula and the particle start position.

Acknowledgements
The reported research was funded by Russian Foundation for Basic Research, grant № 19-07-01188.

References

[1] Hellmann A, Pitz M, Schmidt K, Haller F and Ripperger S 2015 *Aerosol Sci. Technol.* **49** 16
[2] Calmidi V V and Mahajan R L 1999 *J. Heat Trans.* **121** (2) 466
[3] Zhao C Y 2012 *Int. J. Heat Mass Trans.* **55** (13-14) 3618
[4] Muley A, Kiser C, Sundén B and Shah R K 2012 *Heat Trans. Eng.* **33** (1) 42
[5] Han X H, Wang Q, Park Y G, T'joen C, Sommers A and Jacobi A 2012 *Heat Trans. Eng.* **33** (12) 991
[6] Mahjoob S and Vafai K 2008 *Int. J. Heat Mass Trans.* **51** (15-16) 3701
[7] Nawaz K 2014 *Aerogel coated metal foams for dehumidification applications* (Ph.D., University of Illinois at Urbana-Champaign)
[8] Dai Z, Nawaz K, Park Y, Chen Q and Jacobi A M 2012 *Heat Trans. Eng.* **33** (1) 21
[9] Ejlali A, Ejlali A, Hooman K and Gurgenci H 2009 *Int. Com. Heat Mass Trans.* **36** (7) 674
[10] Boomsma K, Poulikakos D and Zwick F 2003 *Mech. Materials.* **35** (12) 1161
[11] Baillis D, Raynaud M and Sacadura J F 1999 *J. Thermophys. Heat Trans.* **13** (3) 292
[12] Lee S C and Cunnington G R 2000 *J. Thermophys. Heat Trans.* **14** (2) 121
[13] Hayes A M, Khan J A, Shaaban A H and Spearing I G 2008 *Int. J. Thermal Sci.* **47** (10) 1306
[14] Pavel B I and Mohamad A A 2004 *Int. J. Heat Mass Trans.* **47** (23) 4939

[15] Al-Raoush R and Alsaleh M 2007 *Powder Technol.* **176** (1) 47
[16] Beugre D, Calvo S, Dethier G, Crine M, Toye D and Marchot P 2010 *J. Comp. App. Math.* **234** (7) 2128
[17] Peng Y and Richardson J T 2004 *App. Cat. A: Gen.* **266** (2) 235
[18] Mardanov R F, Soloveva O V and Zaripov S K 2016 *IOP Conf. Ser.: Materials Sci. Eng.* **158** 012065
[19] Soloveva O V, Solovev S A, Khusainov R R, Popkova O S and Panenko D O 2018 *J. Phys.: Conf. Ser.* **944** (1) 012113
[20] Soloveva O V, Solovev S A, Khusainov R R and Shubina A S 2019 *J. Phys.: Conf. Ser.* **1158** (4) 042023
[21] Solovev S A, Soloveva O V and Popkova O S 2018 *Rus. J. Phys. Chem. A.* **92** (3) 603
[22] Perrot C, Panneton R and Olny X 2007 *J. App. Phys.* **101** (11) 113538
[23] Sullivan R M, Ghosn L J and Lerch B A 2008 Int. J. Solids Structures. **45** (6) 1754
[24] Bai M and Chung J N 2011 *Int. J. Therm. Sci.* **50** (6) 869
[25] Edouard D, Lacroix M, Huu C P and Luck F 2008 *Chem. Eng. J.* **144** (2) 299
[26] Dietrich B 2012 *Chem. Eng. Sci.* **74** 192
[27] Mahjoob S and Vafai K 2008 *Int. J. Heat Mass Trans.* **51** (15-16) 3701
[28] Fourie J G and Du J P 2002 *Chem. Eng. Sci.* **57** (14) 2781
[29] Bhattacharya A, Calmidi V V and Mahajan R L 2002 *Int. J. Heat Mass Trans.* **45** (5) 1017
[30] Huu T T, Lacroix M, Huu C P, Schweich D and Edouard D 2009 *Chem. Eng. Sci.* **64** (24) 5131
[31] Dai Z, Nawaz K, Park Y, Chen Q and Jacobi A M 2012 *Heat Trans. Eng.* **33** (1) 21
[32] Dukhan N and Chen K C 2007 *Exp. Therm. Fluid Sci.* **32** (2) 624
[33] Kamath P M, Balaji C and Venkateshan S P 2013 *Int. J. Therm. Sci.* **64** 1
[34] Chiappini D, Bella G, Festuccia A and Simoncini A 2015 *Com. Comp. Phys.* **18** (3) 707

Numerical simulation of the isoparaffins dehydrogenation in a bidisperse fluidized bed

S A Solovev[1], R R Yafizov[2] and A V Antipin[3]

[1]Institute of Digital Technologies and Economics, Kazan State Power Engineering University, Krasnoselskaya st. 51, Kazan, 420066, Russian Federation
[2]Institute of Heat Power Engineering, Kazan State Power Engineering University, Krasnoselskaya st. 51, Kazan, 420066, Russian Federation
[3]Institute of Mathematics and Mechanics, Kazan Federal University, Kremlevskaya st. 18, Kazan, 420008, Russian Federation

E-mail: solovev.sa@kgeu.ru

Abstract. We built a mathematical model and carried out a numerical simulation for the reaction of isoparaffins dehydrogenation in a fluidized-bed reactor using the example of dehydrogenating isobutane to isobutylene. Experimental data determine the parameters for the construction of the model. Numerical studies with the choice of the average diameter for the polydisperse catalyst were conducted. We performed the calculations for cases of the monodisperse and bidisperse composition of particles, carried out the analysis of the influence of the fine particles present on the reaction product yield in the numerical simulation.

1. Introduction

Fluidized bed reactors are used in the chemical and oil refining industry and the gasification of coal and biomass [1-3]. Reactors are widespread due to the high-efficiency properties of heat and mass transfer.

Mathematical modeling of fluidization generally bases on the Eulerian-Lagrangian and Eulerian-Eulerian approaches. Researchers consider the solid phase as discrete particles in the Eulerian-Lagrangian approach (for example, [4, 5]). In the Eulerian-Eulerian approach, the primary phase and the discrete phase are as continuous. To take into account the interaction of particles in the fluidized bed, by analogy with the kinetic theory of gases, an equation was added describing the change in the kinetic energy of the particles due to their collisions (for example, [6]). This approach is less demanding in terms of computational resources and requires discretization of particle characteristics. Discretization can be carried out using a multi-group approach, the essence of which is to divide the domain of definition into groups of diameters and assign them to correspond some value of the cumulative distribution function that characterizes the fraction of the group in the particle bed. The transport equations are solved for each group separately. The works [7-9] demonstrate a multi-group approach to simulate a fluidized bed. In [10], the method of taking into account polydispersity can affect the hydrodynamic characteristics of the flow obtained by numerical simulation.

We have considered the isoparaffins dehydrogenation on the example of the isobutane dehydrogenation process to isobutylene. Isobutylene is an essential monomer for petrochemical synthesis. On its basis, a wide range of products is obtained, such as synthetic rubbers, films, fibers, high-octane components of automobile gasoline. The primary process for the production of isobutylene is the isobutane dehydrogenation in a catalyst fluidized bed.

2. Problem formulation and mathematical models

It is necessary to build a model for the isoparaffins dehydrogenation in a fluidized bed reactor. The model based on the example of the isobutane dehydrogenation to isobutylene reaction in the presence of a chromium-aluminum catalyst.

We verify mathematical models with experimental data. For the fluidized bed hydrodynamic models, the experimental setup is a glass tube with an internal diameter of 2.2 cm (fig. 1). The porous material is at the bottom of the tube, which ensures a uniform flow of air supplied from the compressor. We measured the volumetric flow at the outlet of the tube using a rotameter, filmed the process of fluidization on a high-speed camera, and estimated the bed height by frames.

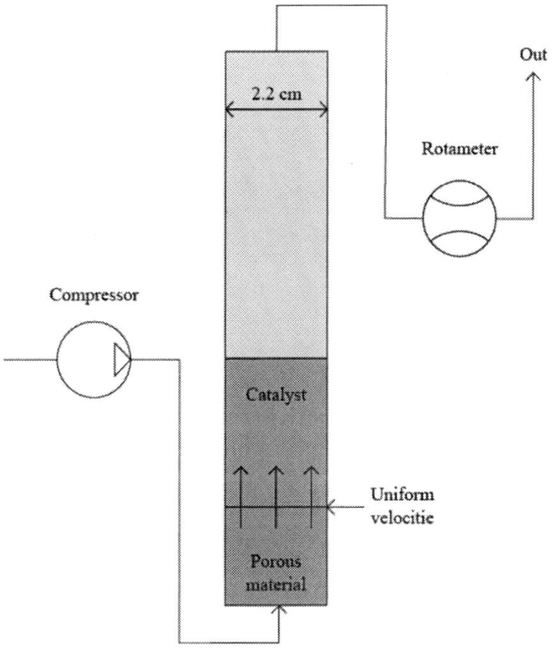

Figure 1. The scheme of the experimental setup

In order to verify the constructed mathematical model of the isobutane dehydrogenation to isobutylene in a fluidized bed reactor, we conducted experimental studies at a high-temperature reactor. The experimental setup is an insulated metal tube with an inner diameter of 3 cm. A 75-ml catalyst is poured into the tube. At the bottom of the tube is a porous material to ensure a uniform flow of raw gas (isobutane). Around the metal case are electric heaters that maintain the temperature of 570°C. We performed an analysis of contact gas at the tube outlet. For the experiment under consideration, the product yield is 45 ± 2%.

2.1. Fluidized bed mathematical model

For the Eulerian-Eulerian approach, we write down the conservation laws for mass, momentum, and energy.

The mass conservation equation for gas or solid phase:

$$\frac{\partial \alpha_i \rho_i}{\partial t} + \nabla \cdot (\alpha_i \rho_i \vec{v}_i) = 0, \tag{1}$$

where α_i is volume fraction, ρ_i is density, $\vec{v}_i$ is velocity.

The momentum conservation equation for gas phase:

$$\frac{\partial \alpha_g \rho_g \vec{v}_g}{\partial t} + \nabla \cdot \left(\alpha_g \rho_g \vec{v}_g \vec{v}_g\right) = -\alpha_g \nabla p + \nabla \cdot \bar{\bar{\tau}}_g + \alpha_g \rho_g \vec{g} + \sum_j K_{gj} \left(\vec{v}_g - \vec{v}_j\right), \quad (2)$$

where p is pressure, $\bar{\bar{\tau}}_g$ is stress tensor for the gas phase, K_{gj} is interfacial interaction coefficient between gas phase and *j*-th discrete solid phase. In equation (2), the stress tensor is

$$\bar{\bar{\tau}}_g = \alpha_g \mu_g \left(\nabla \vec{v}_g + \nabla \vec{v}_g^{\,T}\right) + \alpha_g \frac{2}{3} \mu_g \nabla \cdot \vec{v}_g \bar{\bar{I}}, \quad (3)$$

where μ_g is shear viscosity, $\bar{\bar{I}}$ is unit tensor.

The momentum conservation equation for solid phase:

$$\frac{\partial \alpha_s \rho_s \vec{v}_s}{\partial t} + \nabla \cdot \left(\alpha_s \rho_s \vec{v}_s \vec{v}_s\right) = -\alpha_s \nabla p - \nabla p_s + \nabla \cdot \bar{\bar{\tau}}_s + \alpha_s \rho_s \vec{g} + \sum_j K_{sj} \left(\vec{v}_s - \vec{v}_j\right), \quad (4)$$

where p_s is granule pressure in a solid phase, $\bar{\bar{\tau}}_s$ is stress tensor for solid phase, K_{sj} is interfacial interaction coefficient between the solid and *j*-th gas or solid phases. In equation (4), the stress tensor is:

$$\bar{\bar{\tau}}_s = \alpha_s \mu_s \left(\nabla \vec{v}_s + \nabla \vec{v}_s^{\,T}\right) + \alpha_s \left(\lambda_s - \frac{2}{3} \mu_s\right) \nabla \cdot \vec{v}_s \bar{\bar{I}}, \quad (5)$$

where μ_s, λ_s is shear and bulk viscosity.

The energy conservation equation for gas or solid phase:

$$\frac{\partial \alpha_i \rho_i h_i}{\partial t} + \nabla \cdot \left(\alpha_i \rho_i \vec{v}_i h_i\right) = \alpha_i \frac{\partial p_i}{\partial t} + \bar{\bar{\tau}}_i : \vec{v}_i + \sum_j Q_{ij}, \quad (6)$$

where h_i is enthalpy, Q_{ij} is heat transfer between the *i*-th and *j*-th phases.

The equation for the solid phase granule temperature [11]:

$$\frac{3}{2}\left[\frac{\partial}{\partial t}\left(\alpha_s \rho_s \Theta_s\right) + \nabla \cdot \left(\alpha_s \rho_s \vec{v}_s \Theta_s\right)\right] = \left(-p_s \bar{\bar{I}} + \bar{\bar{\tau}}_s\right) : \nabla \vec{v}_s + \nabla \cdot \left(k_{\Theta s} \nabla \Theta_s\right) - \gamma_{\Theta s} + \phi_{sg}, \quad (7)$$

where $k_{\Theta s}$ is the granule energy diffusion coefficient; Θ_s is solid phase granule temperature; $\gamma_{\Theta s}$ is energy dissipation from particle collisions; ϕ_{sg} is energy exchange between solid and gas phases.

The considered question's system is not closed. To close the system, use the models from [6, 13, 14]. For calculations, a dispersion turbulence $k-\varepsilon$ model was used. We solve the written system of differential equations in the ANSYS Fluent software package.

2.2. Average particle size mathematical model

We verified the mathematical model with experimental data. In the experiment, the catalyst is a polydisperse system of alumina microspherical particles belonging to the Geldart B group according to the classification. The curve of the particle size distribution function $f(D)$ is shown in fig. 2

Let us consider several ways of choosing the number and particle size for numerical simulation using the Eulerian-Eulerian approach. When conducting numerical calculations to save computational resources, as a rule, choose only one average particle diameter. One way to select the average diameter is to determine the mode on the particle size distribution curve $f(D)$

$$D_{\text{mod}} = \arg\max_D \left(f(D)\right). \quad (8)$$

To calculate the fluidized bed apparatus, it is also customary to use the mean Sauter diameter.

$$D_{32} = \int_0^\infty f(D) D^3 dD \Big/ \int_0^\infty f(D) D^2 dD. \quad (9)$$

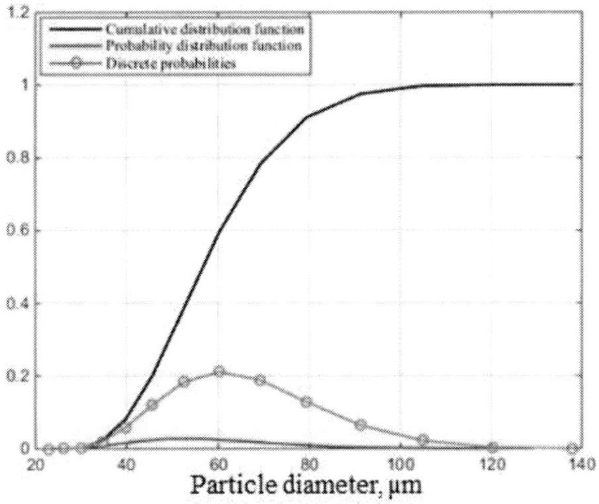

Figure 2. Particle characteristics

A detailed analysis of the effective diameter selection based on the balance of forces acting on a particle in a fluidized bed was carried out in [10]. Several fluidization regimes are considered, such as inertial, creeping, and intermediate. Resistance to gas flow for a single spherical particle [15]

$$F_D = \frac{1}{2}\rho_g v_{rel}^2 C_D A, \tag{10}$$

where $v_{rel} = v_s - v_g$ is relative gas velocity around the particle, A is relative area (area of body projection on the plane). In the case of spherical particle

$$A = \frac{1}{4}\pi D^2. \tag{11}$$

We introduce an additional effective diameter over the related area [16]. Let us consider the resistance force of a spherical particle in the form (10), taking into account the sectional area (11). Also, let us consider the expectation A_{av} for a particles mixture with a density distribution of particle diameter $f(D)$:

$$A_{av} = \frac{1}{4}\pi\int_0^\infty f(D)D^2 dD. \tag{12}$$

Equating (11) and (12) we find the effective particle diameter D_{eff} for the mixture:

$$D_{eff} = \left(\int_0^\infty f(D)D^2 dD\right)^{0.5}. \tag{13}$$

The choice of the fractions number and the particle size play an essential role in the behavior of the fluidized bed. In some cases, the addition of fine particles leads to bed expansion and an increase in the porosity of the dense phase [17-20]. However, in other cases, the addition of these particles leads to a decrease in solid-phase porosity [21, 22].

We select the fraction of coarse and fine particles for a bidisperse particles mixture. As a criterion for fine particles, we consider the terminal velocity

$$v_t = \left(\frac{4gD(\rho_p - \rho_f)}{3\rho_f C_D}\right)^{0.5}. \tag{14}$$

Let the diameter D_s of a particle under the action of gravity and resistance is in equilibrium. Then particles with a diameter less than D_s will be carried away by the flow, and particles with a larger diameter D_s will tend downwards.

For polydisperse particles, the following conditions we recommend for choosing effective diameters [16]:

$$D_{32} \le D_{eff} \le D_{mod}. \tag{15}$$

2.3. *Chemical reaction mathematical model*

The primary reaction of this process is the isobutane dehydrogenation. Under the dehydrogenation understand the chemical processes associated with the elimination of hydrogen atoms from organic compounds. Dehydrogenation reactions are reversible, strongly endothermic, occurring with an increase in volume due to hydrogen evolution.

$$C_4H_{10} \leftrightarrow C_4H_8 + H_2 - 115.1 \text{ kJ/mol}. \tag{16}$$

For the mathematical model, we define the chemical reaction rate constant. The Arrhenius equation gives the general dependence of the rate constant on temperature:

$$\frac{d\ln k}{dT} = \frac{E_a}{RT^2}, \tag{17}$$

where E_a is activation energy.

$$\ln k = -\frac{E_a}{RT} + const, \tag{18}$$

$$k = A\exp\left(\frac{-E_a}{RT}\right). \tag{19}$$

The parameters of the reaction rate constant and activation energy are obtained from experimental data carried out on high-temperature isobutane dehydrogenation reactors.

In the industrial process of isobutane dehydrogenation, along with the target reaction (16), a number of side reactions also take place, for the simulation was selected only one main reaction (16) taking place in the block.

The volume-mole concentration of the source substance, expressed in terms of the hydrodynamic characteristics of the flow, can be written as follows:

$$c = \frac{m_{C4H10}}{M_{C4H10}V_r} = \frac{\rho_{mixture} \cdot V_r \cdot Y_{C4H10}}{M_{C4H10} \cdot V_r} = \frac{\rho_{mixture} \cdot Y_{C4H10}}{M_{C4H10}}, \tag{20}$$

where m_{C4H10} is isobutane mass, M_{C4H10} is isobutane molar mass, V_r is the volume in which the reaction takes place, Y_{C4H10} is isobutane mass concentration, $\rho_{mixture}$ is density for a mixture of isobutane, isobutylene, and hydrogen.

The model of an ideal incompressible gas determine the density of the mixture:

$$\rho_{mixture} = \frac{P}{RT\sum_i \frac{Y_i}{M_i}}, \tag{21}$$

where P is the pressure of a mixture, M_i is the molar mass of i-th component, R is universal gas constant, T is the temperature of a mixture, Y_i is the mass concentration of the i-th component.

Using the law of mass action and the Arrhenius equation, we obtain expressions characterizing the mass transfer from isobutane to isobutylene and hydrogen,

$$w_{C4H10} = k \cdot \rho_{mixture} \cdot Y_{C4H10} \cdot f(\alpha), \tag{22}$$

$$w_{C4H8} = \frac{M_{C4H8}}{M_{C4H10}} \rho_{mixture} \cdot Y_{C4H10} \cdot f(\alpha), \tag{23}$$

$$w_{H2} = \frac{M_{H2}}{M_{C4H10}} \rho_{mixture} \cdot Y_{C4H10} \cdot f(\alpha). \tag{24}$$

The function expresses the dependence of the reaction rate on the amount of catalyst:

$$f(\alpha) = \frac{\alpha}{\alpha_{max}}, \tag{25}$$

where α is catalyst volume fraction, α_{max} is maximum catalyst volume fraction.

3. Results

For numerical calculations, we used the fluidized bed model with a new effective diameter. We performed the calculations for the cases of a monodisperse and bidisperse catalyst. Fig. 3 shows pictures of the catalyst particles volume fraction in a reactor model. Gas velocity is 0.1 m/s.

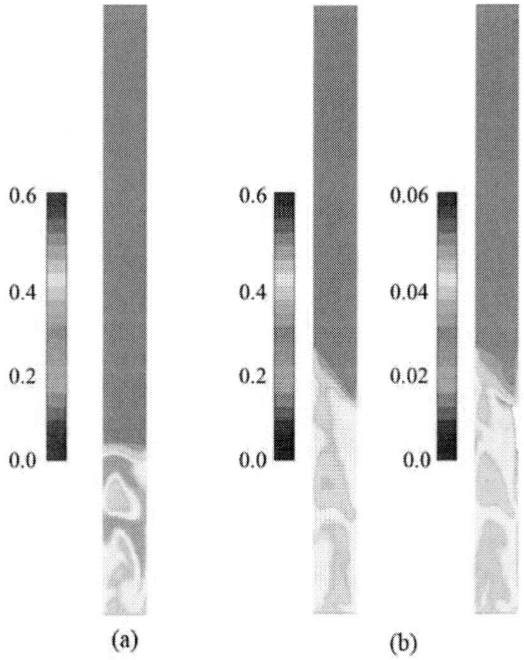

Figure 3. The catalyst particles volume fraction in the reactor tube: a - calculation of the monodisperse catalyst, b - calculation of the bidisperse catalyst.

For the case of the bidisperse catalyst, the particle bed is significantly higher, and the volume fraction is smaller. Also fig. 3b shows the field of the volume fraction of fine particles on a scale ten times smaller than the scale for coarse particles. Fine catalyst particles are not separated in the bed, and clots and rarefied zones correspond to the behavior of coarse catalyst particles.

Fig. 4a shows the gas temperature patterns in the reactor model for the calculation of a monodisperse and bidisperse catalyst. The absorption of heat in the case of the bidisperse particles is more significant than of the monodisperse particles. It indicates a higher intensity of the chemical reaction that goes with the heat absorption.

Fig. 4b shows pictures of the mass fraction of the reaction product (isobutylene) in the reactor model for calculating the monodisperse and bidisperse catalysts. The reaction product is formed in the zone where the catalyst circulates, and in the upper part of reactor model the mass fraction of isobutylene does not change.

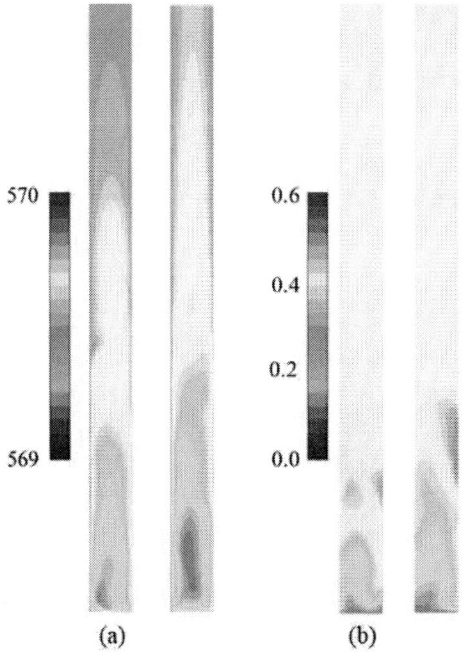

Figure 4. a - the gas temperature (°C), b - the mass fraction of the reaction product.

From the analysis of the calculations, we found that the product yield was: for the case of calculating the monodisperse catalyst - 45.43%, for the case of calculating the bidisperse catalyst - 43.69%. Thus, taking fine particles into account when modeling fluidized bed reactors can have a significant impact on the calculation results.

Acknowledgements
The reported research was funded by Russian Foundation for Basic Research and the government of the Republic of Tatarstan of the Russian Federation, grant № 18-48-160006.

References

[1] Yang W C 2003 *Handbook of Fluidization and Fluid-Particle System* (New-York: Marcel Dekker)
[2] Basu P 2006 *Combustion and Gasification in Fluidized Beds* (Boca Raton: CRC/Taylor & Francis)
[3] Sadeghbeigi R 2012 *Fluid Catalytic Cracking Handbook* (Elsevier Butterworth-Heinemann)
[4] Chen X and Wang J 2014 *Powder Technol.* **254** 94
[5] Zhou Q, Wang C, Wang H and Wang J 2016 *Int. J. Miner. Process.* **151** 40
[6] Gidaspow D 1994 *Multiphase Flow and Fluidization* (Boston: Academic Press)
[7] Shah S, Myöhänen K, Kallio S, Ritvanen J and Hyppänen T 2015 *Powder Technol.* **274** 239
[8] Solov'ev S A, Egorov A G, Lamberov A A, Egorova S R and Kataev A N 2016 *Catal. Ind.* **8(1)** 48
[9] Soloveva O V and Solovyev S A 2016 *IOP Conf. Ser.: Materials Sci. Eng.* **158(1)** 012086
[10] Loth E, O'Brien T, Syamlal M and Cantero M 2004 *Powder Technol.* **142** 209
[11] Ding J and Gidaspow D 1990 *AIChE J.* **36** 523

[12] Syamlal M 1987 *The Particle-Particle Drag Term in a Multiparticle Model of Fluidization* (Morgantown WV: EG & G Washington Analytical Service Center)
[13] Lun C K K, Savage S B and Jeffrey D J 1984 *J. Fluid Mech.* **140** 223
[14] Gunn D J 1978 *Int. J. Heat Mass Transfer*. **21** 467
[15] White F M 1991 *Viscous Fluid Flow* (New York: McGraw-Hill)
[16] Solovev S A, Solovev O V and Antipin A V 2019 *J. Phys.: Conf. Series.* **1158(4)** 042022
[17] Beetstra R, Nijenhuis J, Ellis N and van Ommen J R 2009 *AIChE J.* **55** 2013
[18] Brouwer G C, Wagner E C, van Ommen J R and Mudde R F 2012 *Chem. Eng. J.* **207-208** 711
[19] Gu Y, Ozel A and Sundaresan S 2016 *AIChE J.* **62(7)** 2271
[20] Soloveva O V, Solovev S A, Egorova S R, Lamberov A A, Antipin A V and Shamsutdinov E V 2018 *Chem. Eng. Res. Des.* **138** 212
[21] Khoe G K, Ip T L and Grace J R 1991 *Powder Technol.* **66** 127
[22] Saayman J, Ellis N and Nicol W 2013 *Powder Technol.* **245** 48

IOP Conf. Series: Materials Science and Engineering **618** (2019) 012096 doi:10.1088/1757-899X/618/1/012096

Numerical investigation of the granule size effect on the reaction product yield in a catalyst fixed bed

S A Solovev[1], R R Khusainov[2] and D R Nasretdinova[2]

[1]Institute of Digital Technologies and Economics, Kazan State Power Engineering University, Krasnoselskaya st. 51, Kazan, 420066, Russian Federation
[2]Institute of Heat Power Engineering, Kazan State Power Engineering University, Krasnoselskaya st. 51, Kazan, 420066, Russian Federation

E-mail: solovev.sa@kgeu.ru

Abstract. The research work presents the results of mathematical modeling of the ethylbenzene dehydrogenation process to styrene in a fixed catalyst bed. Calculations for the different length granule packaging are carried out. Calculated values of the reaction product yield (styrene) are in a wide range of gas velocities. There are significant differences in the yield of the product at high gas velocities for catalyst granules of different shapes. Of the considered catalyst granules forms, we got the best yield efficiency by the use of granules, which provide maximum surface area and porosity in the fixed bed catalyst.

1. Introduction

Catalytic processes with a fixed bed widely use in the chemical industry. Despite numerous developments in the field of application of structured packings catalyst [1], the random arrangement of granules is of particular interest, since such an arrangement presence in modern reactors. Accounting for the heterogeneity of the bed structure based on numerical modeling and experimental studies of chemical processes in cylindrical reactors was carried out in [2, 3]. Impressive results were obtained based on the averaged model and semi-empirical correlations.

Expanding the classical approach using the concepts of medium-radial porosity and medium-radial velocity profiles, the authors of [4-7] took into account the strong influence of the bed structural properties on heat and mass transfer as a whole. However, the use of such an average model may be incorrect if there are only a few granules in the reactor's cross section. For such cases, axial symmetry is not fulfilled, and it is impossible to describe local phenomena by the averaged model. Detailed research is possible using 3D-model [8]. The authors of [9] take into account the local features of the full three-dimensional structure of the fixed bed without simplification under the assumption of symmetry.

The primary requirement for conducting detailed numerical modeling is the creation of the porous structure geometry consisting of randomly spaced granules. To create a numerical model, use X-ray computed tomography data or magnetic resonance imaging methods as one of the suitable methods for creating a raster three-dimensional geometry image. These methods make it possible to digitize the geometry of existing samples, but do not allow the creation of environments with predetermined parameters of a porous medium. Creating the porous medium geometry based on the Monte Carlo method, or the method of a separate element ("distinct element method" - DEM) allows to create a geometry with predetermined properties (size, granule shape, porosity), and, if necessary, print out objects in 3D the printer.

Detailed modeling has shown that the use of averaged flow models in porous media is often incorrect. To determine the model correctness in the case of direct numerical simulation, conduct experimental studies of the gas movement inside a porous medium created by a computer model using three-dimensional printing.

Optimization of a fixed catalyst bed chemical reactors is often carried out by numerical simulation using computational fluid dynamics (CFD) since in most cases hydrodynamics affect the course of a chemical reaction [11-13]. CFD modeling is based on the fundamental conservation laws of mass, momentum, energy and does not depend on the types of reactors and their scales, which allows us to consider multiscale phenomena occurring in reactors [10-12, 14].

In [15, 16], a detailed numerical study of the pressure drop during the gas flow through spherical granules randomly based on CFD modeling was performed. They compared the results obtained for different geometries and flow rates with the experimentally obtained dependencies. Finally, velocity and pressure fields calculated from CFD simulations are further used to interpret the differences between CFDs and empirical dependencies to determine the pressure drop value.

In this work, we constructed a mathematical model of the ethylbenzene dehydrogenation reaction to styrene in a fixed bed catalyst. The granular bed bases on the DEM. A granule model represents cylinders of different sizes. A comparative analysis of the catalyst granules shape and size influence on the reaction product yield is carried out by numerical simulation. The calculations were carried out in a wide gas velocity range, allowing us to estimate the product yield, both for laboratory installations and industrial reactors.

2. Mathematical models

2.1. Mathematical model of packing fixed catalyst bed granules

For the Eulerian-Eulerian approach, we write down the conservation laws for mass, momentum, and
The main methods of creating geometry with elements randomly located in space are the Monte Carlo method and the DEM, which is actively used to model bulk media with non-spherical granules. In the case of the fixed bed geometry creation, the use of the Monte Carlo method may be incorrect, since it does not take into account the additional forces acting on the granule, they must be connected separately. The DEM is a numerical method for solving problems that require a description of the bulk media mechanics. Presented in [17], DEM was developed to analyze the problems of rock movement using deformable polygonal blocks and then applied to soils [18].

We considered the dynamic process in which the flow velocity depends on the physical properties of a discrete system, and assumed that the speed and acceleration are constant in each time step. The solution scheme is identical to the scheme used by the explicit finite difference method. The DEM based on the idea that the selected time step can be so small that during a single time step, the disturbances cannot spread farther from any particle than its immediate neighbors. Then the forces acting on any particle at any time step are determined solely by its interaction with the particles with which it contacts. Since the rate at which a disturbance can propagate is a function of the discrete system physical properties, it is possible to calculate the time step to satisfy this constraint. The use of an explicit (as opposed to implicit) numerical scheme allows one to simulate the nonlinear interaction of a large number of particles without excessive requirements for RAM, as well as an iterative procedure.

The calculations performed by the DEM method alternate with the application of Newton's second law to particles and the law of the displacement force at the contact area. Newton's second law is used to determine the motion of each particle arising from the contact and volume forces acting on it, while

IOP Conf. Series: Materials Science and Engineering **618** (2019) 012096 doi:10.1088/1757-899X/618/1/012096

the law of displacement force is used to update the contact forces arising from the relative motion in each contact area. In this case, there is a constant update of the position of the walls coordinate.

At the beginning of each time step, the set of contacts updates relative to the known positions of the particles and walls. Then, for each contact, the force law of displacement is applied to update the contact forces based on the relative motion between two objects in the contact and the contact model. Then the motion law is applied to each particle in order to update its speed and position based on the resultant force and moment arising from the contact forces and other forces acting on the particle. Also, the position of the walls is updated depending on the specified velocity of the walls.

Fig. 1 shows a constructing example package of cylindrical catalyst granules for a model of a laboratory reactor 3 cm in diameter. The bed height is 4.6 cm, the granule diameter is 3 mm, and the granule length is 6 mm. After falling into the flask of the reactor, the granules are in a random order. This approach imitates well the poured catalyst granules in a laboratory reactor, as well as in a large-scale industrial reactor.

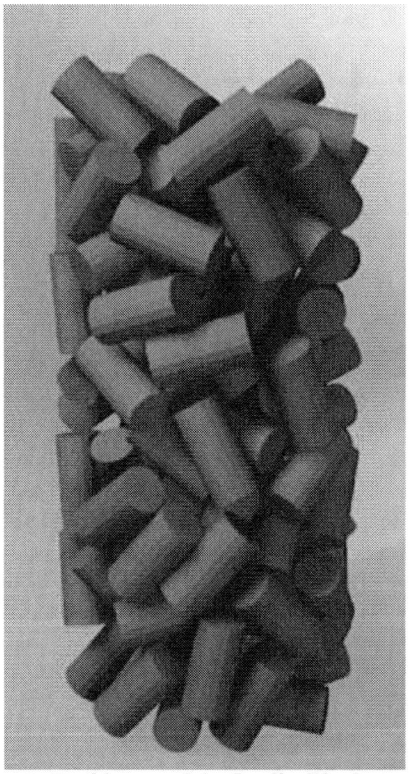

Figure 1. Packing model of cylindrical granules

2.2. Mathematical model for calculating the dynamics of the gas movement and heat and mass transfer processes

The described mathematical problem was solved by the finite volume method with a grid partition of the considered reactor region. For a multicomponent gas phase, the conservation laws of mass, momentum, and energy are satisfied. In the study of this problem, the following equations we solved.

The mass conservation equation for the gas phase:

$$\frac{\partial \rho_g}{\partial t} + \nabla \cdot (\rho \vec{v}) = 0, \tag{1}$$

where ρ is gas density, $\vec{v}$ is gas velocity.

The momentum conservation equation for the gas phase

$$\frac{\partial \rho \vec{v}}{\partial t} + \nabla \cdot (\alpha \rho \vec{v}\vec{v}) = -\nabla p + \nabla \cdot \bar{\bar{\tau}} + \rho \vec{g}, \quad (2)$$

where p is pressure, $\bar{\bar{\tau}}$ is gas phase stress tensor. In equation (2), the stress tensor

$$\bar{\bar{\tau}} = \mu\left(\nabla \vec{v} + \nabla \vec{v}^T\right) + \frac{2}{3}\mu \nabla \cdot \vec{v}\bar{\bar{I}}, \quad (3)$$

where μ is shear viscosity, $\bar{\bar{I}}$ is unit tensor.

The energy conservation equation for the gas phase

$$\frac{\partial \rho h}{\partial t} + \nabla \cdot (\rho \vec{v} h) + \nabla \cdot \vec{J} = \frac{\partial p}{\partial t} + \bar{\bar{\tau}} : \vec{v} \quad (4)$$

where h is gas enthalpy, $\vec{J}$ is diffusion flux, which arises due to gradients of concentration and temperature.

We divided the entire computing area of the reactor into finite elements of a triangular shape, the dimensions of which are sufficient to determine the characteristic factors of the phenomenon under study. In the calculations carried out in this paper, the typical number of finite elements ranged from 35,000,000 to 40,000,000 elements. Fig. 2 shows an example of a grid partition of the computational domain under consideration.

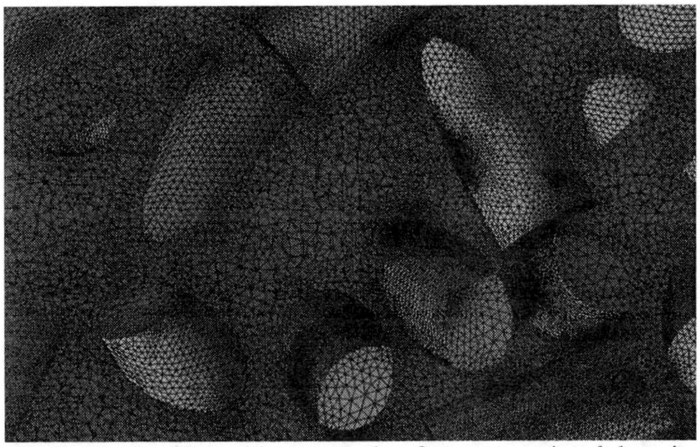

Figure 2. A grid partition example of a computational domain

To numerically solve the problem of hydrodynamics and heat and mass transfer in a catalyst fixed bed reactor for the ethylbenzene dehydrogenation to styrene, we need to specify conditions at the boundaries of the considered region. We set the boundary conditions following the mechanisms of operation of the calculated laboratory reactor and the solver used at all boundaries of the computational domain. So, we set the impenetrable wall condition on all impermeable surfaces, and a temperature equal to the temperature of the reactor heated walls on the outer wall of the area under consideration. At the site of the gas flow in the model of the reactor were set conditions for the gas mass flow. At the site in the reactor model, the conditions for external pressure outside the area under consideration we set for the gas outlet.

The considered gas phase flow is considered a multicomponent ideal incompressible gas consisting of raw material (ethylbenzene), water vapor, and the reaction products (styrene and hydrogen).

2.3. *Mathematical model of the chemical reaction kinetics of the ethylbenzene dehydrogenation to styrene*

The primary reaction of this process is the ethylbenzene dehydrogenation to styrene. Under the dehydrogenation understand the chemical processes associated with the elimination of hydrogen atoms from organic compounds. Dehydrogenation reactions are reversible, strongly endothermic, occurring with an increase in volume due to hydrogen evolution.

$$C_8H_{10} \leftrightarrow C_8H_8 + H_2 - Q. \tag{5}$$

For surface reactions, the rate of adsorption and desorption is determined both by chemical kinetics and by diffusion to and from the surface.

Consider the theoretical information from chemical kinetics used to build the model. The chemical reaction rate is usually known as the change in the amount of a substance reacting or resulting from a reaction per unit of time per unit volume.

$$\sum_i v_i B_i = \sum_j u_j A_j, \tag{6}$$

$$w = \frac{1}{V}\frac{dn}{dt} = \frac{dc}{dt},\ V = const, \tag{7}$$

where n is the number of moles, c is volume-mole or volume-molecular concentration of a substance.

In the case when the stoichiometric equation (6) of a simple reaction of type (7) is known, to determine the reaction rate should be in the form uniquely:

$$w = \frac{1}{v_i V}\frac{dn_i}{dt} = \frac{1}{u_k V}\frac{dn_k}{dt},\ V = const, \tag{8}$$

where v_i and u_k refer to the stoichiometric coefficients of the starting materials and reaction products, respectively.

The central postulate of chemical kinetics says: "The rate of chemical reaction (with constant volume) is proportional to the product of current concentrations of the initial substances, raised to some degree."

$$w = -\frac{1}{v_i}\frac{dc_i}{dt} = \frac{1}{u_j}\frac{dc_j}{dt} = k\prod_i c_i^{p_i}. \tag{9}$$

The order and the molecular reaction. The numbers p_i in equation (10) are called reaction orders by substance, and their sum is called reaction order.

$$\sum_i p_i = n. \tag{10}$$

For reactions of type (6) the molecular is

$$v = \sum_i v_i. \tag{11}$$

In the case of elementary (one-step) reactions, the molarity coincides with its order. Thus, the molarity of the considered reaction (5) and its order are equal to unity: $v = 1$ and $n = 1$.

The general dependence of the rate constant on temperature gives the Arrhenius equation:

$$\frac{d\ln k}{dT} = \frac{E_a}{RT^2}, \tag{12}$$

where E_a is activation energy.

$$\ln k = -\frac{E_a}{RT} + const, \tag{13}$$

$$k = A\exp\left(\frac{-E_a}{RT}\right). \tag{14}$$

IOP Conf. Series: Materials Science and Engineering **618** (2019) 012096 doi:10.1088/1757-899X/618/1/012096

3. Results

Consider the model of a laboratory reactor for a granular catalyst bed for the ethylbenzene dehydrogenation to styrene. For comparison, we construct two models of cylindrical granule packages: 6 mm in length and 9 mm in length. The granule diameter is 3 mm. The granules are poured into a cylindrical flask with a diameter of 2.8 cm and are 4.6 cm in height. At the same time, longer granules can form larger free areas in which gas will move without contact with the catalyst surface.

Under laboratory conditions on small-sized reactors in the reaction of ethylbenzene dehydrogenation to styrene, the average velocity of the raw materials mixture and water vapor can be in the order of 0.1 m/s. In this case, for example, in industrial radial-type reactors, the average gas velocity can be in the order of 10 m/s. Therefore, we will perform calculations for a wide range of gas velocities.

Verification of the mathematical model was carried out for laboratory experiments with an average gas velocity of about 0.1 m/s. The product yield on the experimental setup ranged from 66-70%. The results of numerical calculations are presented in fig. 3.

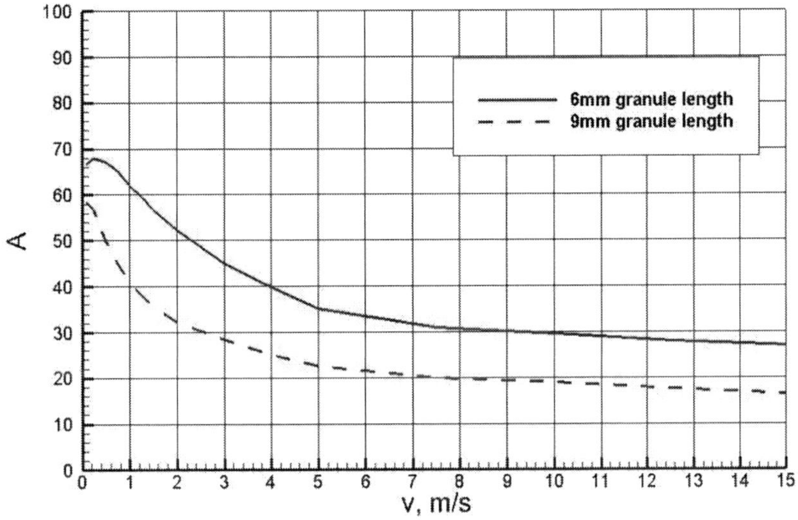

Figure 3. The calculated yield of the product depending on the gas velocity for cylindrical pellets

There is a significant decrease in product yield at high gas velocities, and this is due to the decrease in the time of gas contact with the catalyst granule surface. The use of larger granules allows for a lower yield of the dehydrogenation reaction product. Thus, the granule size and porosity of the catalyst bed can have a significant impact on the yield of the reaction product.

Fig. 4 shows the fields of mass fraction of the reaction product (styrene) for calculations at a gas velocity: (a) - 0.1 m/s, (b) - 15 m/s.

At a higher gas velocity due to the short time of gas contact with the surface, the mass fraction of the reaction product decreases significantly with increasing gas velocity.

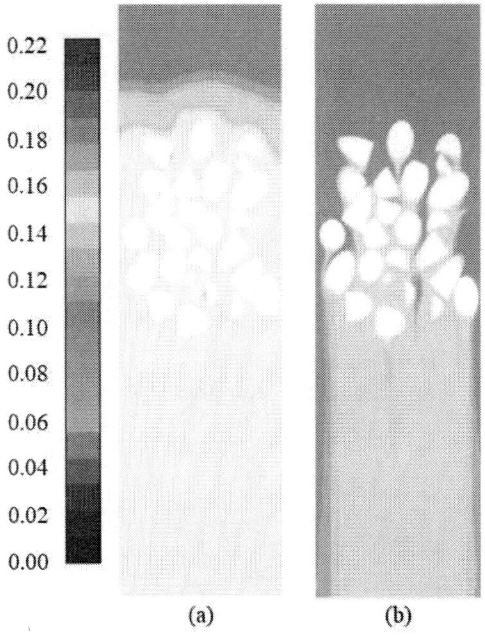

Figure 4. Mass content of styrene in the reactor near the cylindrical catalyst granules.

Acknowledgements
The reported research was funded by Russian Foundation for Basic Research and the government of the Republic of Tatarstan of the Russian Federation, grant № 18-41-160005.

References

[1] Calis H P A, Nijenhuis J P B C, Paikert B C, Dautzenberg F M and Van Den Bleek C M 2001 *Chem. Eng. Sci.* **56** (4) 1713
[2] Daszkowski T and Eigenberger G 1992 *Chem. Eng. Sci.* **47** (9-11) 2245
[3] Papageorgiou J N and Froment G F 1995 *Chem. Eng. Sci.* **50** (19) 3043
[4] Küfner R and Hofmann H 1990 *Chem. Eng. Sci.* **45** (8) 2141
[5] Bey O and Eigenberger G 1997 *Chem. Eng. Sci.* **52** (8) 1365
[6] Giese M, Rottschäfer K and Vortmeyer D 1998 *AIChE J.* **44** (2) 484
[7] Winterberg M, Tsotsas E, Krischke A and Vortmeyer D 2000 *Chem. Eng. Sci.* **55** (5) 967
[8] Dixon A G and Nijemeisland M 2001 *Ind. Eng. Chem. Res.* **40** (23) 5246
[9] Freund H, Zeiser T, Huber F, Klemm E, Brenner G, Durst F and Emig G 2003 *Chem. Eng. Sci.* **58** (3-6) 903
[10] Gunjal P R and Ranade V V 2007 *Chem. Eng. Sci.* **62** (18-20) 5512
[11] Gunjal P R, Kashid M N, Ranade V V and Chaudhari R V 2005 *Ind. Eng. Chem. Res.* **44** (16) 6278
[12] Gunjal P R, Ranade V V and Chaudhari R V 2005 *AIChE J.* **51** (2) 365
[13] Jiang Y, Khadilkar M R, Al-Dahhan M H and Dudukovic M P 2001 *Cat. Today.* **66** (2-4) 209
[14] Ranade V V 2001 *Computational Flow Modeling for Chemical Reactor Engineering* (Academic Press, London)
[15] Pavlišič A, Pohar A and Likozar B 2018 *Powder Technol.* **328** 130
[16] Della Torre A, Montenegro G, Onorati A and Tabor G 2015 *Energy Procedia.* **81** 836
[17] Cundall P A 1971 *The measurement and analysis of accelerations in rock slopes* (Ph.D. thesis, Imperial College, London)
[18] Cundall P A and Strack O D L 1979 *Geotechnique.* **29** (1) 47

Effect of temperature on the copigmentation interaction between aronia anthocyanins and sinaptic acid with different methods

I Petrova[1] and V Gandova[2]

[1] Technical University – Sofia, Bulgaria
[2] University of Food Technologies, Plovdiv, Bulgaria

E-mail: ivalinapetrova.com@abv.bg

Abstract. The effect of temperature and concentrations on the stability of anthocyanin extract from aronia as pigment and sinapic acid as copigment was evaluated. Different temperatures and copigment concentrations were used for the investigation pigment:copigment interaction. Equilibrium constant of copigmentation process was determined. Thermodynamic parameters Gibbs free energy (ΔG), enthalpy (ΔH), entropy (ΔS) and kinetic parameters activation energy (Ea) and z – factor were calculated. At high temperatures, around 80 ^{0}C destruction was observed and after that at 25 ^{0}C, restoration of the complex was observed again according the calculations. This was conformation for reversibility of copigmentation process.

1. Introduction

Anthocyanins represent one on the most important and most widely spread groups of plant pigments of the class of flavonoids. This group of pigments was responsible for the existence of most of the red, blue, and purple colors in flowers and fruits [1]. The last years pigment:copigment couple between different acids as copigment and different pigments were studied. Intermolecular copigmentation reactions between five anthocyanins and five phenolic acids acting as copigments were invesigated [2]. Thermodynamic and kinetic parameters of copigmentation proces were studied by Dangle et al. [3]. The temperature-dependence visible spectrum of the copigmented solutions yields the copigmentation enthalpy and entropy changes. The experiments were carried out at pH 3.5 and pH 5.5. Copigmentation appears to be clearly enthalpically driven (exothermic process with unfavourable entropy change) in the case of chlorogenic acid and (+)-catechin. Relaxation kinetic measurements on the malvin-rutin system give striking evidence of the strong slowing down that a good copigment causes to the anthocyanin fading process (hydration).

From the decrease in the apparent first order rate constant of hydration on copigment addition, a general method for a quick determination of the copigmentation stability constants was proposed. Changes in the visible spectra of strawberry anthocyanins were studied depending on the addition of quercetin as a copigment [4]. The investigations were provide at different temperatures between 20 ^{0}C and 50 ^{0}C. The thermodynamic parameters of the system, ΔG, ΔH, and ΔS were calculated as a function of temperature. The couple of authors investigated kinetic of pigment:copigment interactions. According the Shakeel et al. [5], the solubilities of sinapic acid (SA) were determined in eleven different neat solvents at five different temperatures (T = 298.15 K to 318.15 K) and atmospheric pressure. Experimental solubilities expressed in mole fraction of SA were correlated well with Apelblat and van't Hoff models. The results of thermodynamic analysis indicated an endothermic and entropy-driven dissolution of SA in most of the neat solvents. Gallic acid was evaluated as copigment for an anthocyanin crude extract of Cabernet Sauvignon grape skins [6]. Comparison of the stability of the grape anthocyanins and their interaction with gallic acid were investigated using kinetic parameters. The highest half life values for anthocyanin and percent color retention were reached when the samples were maintained at pH 3.0, temperature of 4 ± 1 ^{0}C and in the absence of light. Extracts from purple carrot and red flesh potatoes showed higher stability according Reyes et al. [7]. Stability to thermal degradation of extracts (pH=3) followed first order kinetics. Degradation parameters such as half life ($t_{1/2}$), degradation rates (k), D-values, z-factor and Q10-values were determined. Elderberry (Sambucus nigra L.) was considered an interesting fruit as food ingredient, due to its high content of anthocyanins that give products an attractive red colour [8]. Activation energy, degradation rates (k) and half life were obtained. Kinetic parameters calculated for elderberry juice can be used to design a thermal treatment to obtain a high retention of colour and bioactive compounds. The effect of temperature on the stability of three purified anthocyanin sources in a soft drink (pH= 3, 10 °Brix) stored at (4, 20, 30 and 50) ^{0}C for 60 days was investigated by Zozio et al. [9]. Two classical empirical approaches (Arrhenius and Ball models) were used to describe the thermal degradation kinetic of these three anthocyanins. At all temperatures, the degradation rate constant (k) for black carrot anthocyanins was less than those in açai and blackberry (0.42×10^{-2}, 0.77×10^{-2} and 1.08×10^{-2}) d^{-1}, respectively, at 30 ^{0}C). Anthocyanins in black carrot degraded less rapidly than those in açai and Andean blackberry. The activation energy (Ea) for degradation of black carrot anthocyanins was 63.2 ± 4.3 kJ mol^{-1}, and 66.3 ± 2.7 kJ mol^{-1} and 91.2 ± 0.4 kJ mol^{-1} for açai and blackberry anthocyanins, respectively, at 20–50 ^{0}C. These higher Ea of blackberry anthocyanins as compared with those of black carrot and açai imply that a small temperature increase was sufficient to degrade them more rapidly. The kinetic studies on thermal stability of anthocyanins isolated from the dry calyces of *Hibiscus sabdariffa* L. (roselle) in aqueous solutions (55–98 ^{0}C) was studied by Gradinaru et al. [10]. The rate constants for degradation were obtained from first order reaction kinetic plots.

The scope of this work is to calculate the thermodynamic and kinetic parameters at different temperatures and copigment concentrations using experimental data.

2. Experimental

2.1. Chemicals

The copigment sinapic acid was used from Sigma – Aldrich, 98 % (Germany). The reagents used for the McIlvaine buffer pH 3.4 citric acid monohydrate and disodium hydrogen phosphate dodecahydrate, were from Merck (Darmstadt, Germany). The adsorbent Amberlite XAD 16N resin was purchased from Sigma Aldrich Co. (St. Louis, MO, USA). All other reagents and solvents used were of analytical grade.

2.2. Extraction, purification and determination of anthocyanins

Aronia anthocyanins were extracted and purified as described by Shikov et al. [11]. The total monomeric anthocyanins were assessed by the pH-differential method as described in Shikov et al. [12].

2.3. Preparation of model solutions

Stock solutions of aronia extract, on the basis of the total anthocyanins, and sinapic acid were prepared in McIlvaine buffer (0.1 M, pH 3.4). Model solutions of aronia anthocyanins (1×10^{-4} M) were obtained by mixing equal volumes (5 ml) of the corresponding stock solutions and were left for equilibration (30 min at 25 ^{0}C).

2.4. Spectrophotometric measurements

Spectral changes were followed by recording the absorption spectra of model solutions from 400 to 700 nm with Halo RB–10 spectrophotometer (Dynamica Scientific, United Kingdom), using 1 cm path length cuvettes. Before measurements, samples were incubated in a thermostatic shaker water bath (NUVE, Turkey) for 30 min at 25 ^{0}C.
Before measurements the samples were thermostated (VEB MLW Prufgepate-Werk Medingen sitz Freital, Germany) at 25, 40, 60 or 80 ^{0}C, respectively, at heating and then cooling to 60, 40, 25 ^{0}C.

2.5. Modelling of thermodynamic and kinetic parameters

After spectrophotometer measurements the equilibrium constant was calculated using the following equation: $\ln[(A - Ao)/Ao] = \ln[K] + n \times \ln[C]$ (1)

where A and Ao - the absorption maximum values of the anthocyanin solution with and without added copigment; C - the molar copigment concentration; K - the equilibrium constant and n - the stoichiometric ratio of the reaction [4].

Thermodynamic parameters Gibbs free energy, enthalpy and entropy were calculated using the following equations:

$$\Delta G = -RT \ln Kp \quad (2)$$

where R - universal gas constant ($R = 8.314\ J.K^{-1}.mol^{-1}$), T - absolute temperature (K), ΔG - Gibbs free energy ($kJ\ mol^{-1}$), K – equilibrium constant.

The enthalpy was calculated by applying the Vant-Hoff equation.

$$\frac{d \ln K}{d(1/T)} = \frac{-\Delta H}{R} \quad (3)$$

ΔH - enthalpy for the copigmentation reaction ($kJ.mol^{-1}$).

Once the Gibbs free energy and the enthalpy were obtained, the entropy were determined by using classic thermodynamic equation (4):

$$\Delta S = \frac{(\Delta H - \Delta G)}{T} \quad (4)$$

ΔS - entropy for the copigmentation reaction ($kJ.K^{-1}.mol^{-1}$).

For calculation of kinetic parameters two models Arrhenius and Ball's were chosen.

– The first was a conventional chemical kinetic model, Arrhenius model [9, 13]. The linear relationship between lnk and 1/T was observed:

$$k = k_0 e^{-\frac{E_a}{RT}} \quad (5)$$

where Ea - activation energy ($kJ.mol^{-1}$); R - universal gas constant ($R = 8.314\ J.K^{-1}.mol^{-1}$), T - absolute temperature (K).

– The second model was a Ball's model [9, 14], defines a decimal reduction time which was related to temperature via a factor z.

$$D = \ln 10k \quad (6)$$

$$D = D_0 10^{-\frac{T}{z}} \quad (7)$$

where D - decimal reduction time at temperature T (s); D_0 - value of D extrapolated at 0 ^0C; z - expressed in ^0C.

The model's parameters were identified, using linear regression on the logarithmic curves of experimental data. Although the z value was estimated from Ea using the relationship:

$$z = \ln(10)\frac{RT^2}{E_a} \quad (8)$$

2.6. Statistical analysis

The results reported in the present study were the mean values of two independent determinations. The data of different samples were analyzed independently by Origin Pro 7.0 software and the standard deviations of the parameters were determined.

3. Results and discussion

In this study the model solutions were prepared with stoichiometry quantity aronia anthocyanins and variations of sinapic acid as copigment change between 1:0 to 1:50 concentration. The thermodynamic parameters of the copigmentation reaction are presented in Table 1.

Table 1. Equilibrium constants and thermodynamic parameters for the copigmentation interaction between aronia anthocyanins and sinapic acid at different temperatures

aronia:sinapic acid			
K	ΔG [kJ.mol^{-1}]	ΔH [kJ.mol^{-1}]	ΔS [kJ.mol^{-1}]
10046.16±2.21	-23.83±0.81	-28.25±0.78	-0.014±0.009
9063.58±1.47	-22.71±0.18	-27.53±0.99	-0.015±0.009
1412.14±5.66	-19.47±0.77	-27.25±1.15	-0.024±0.003
26.54±1.71	-9.07±0.65	-13.94±0.49	-0.029±0.004
1530.98±2.91	-19.69±0.42	-27.22±1.57	-0.023±0.004
6049.08±4.24	-22.65±0.57	-27.35±0.64	-0.017±0.005
9727.47±0.94	-22.75±0.84	-28.05±1.27	-0.015±0.006

The equilibrium constant exhibited different values at different temperatures. The constant (K) displayed a strong dependence on the temperature. The gradual heating from 25 to 80 ^{0}C resulted in significant decrease of the K values at high temperature, indicating thermal instability of the pigment:copigment complex. The subsequent cooling shifted the reaction toward the product formation, regenerating the complex. This thermal reversibility of the copigmentation reaction was of particular importance, implying that the anthocyanin-stabilizing ability of the added copigments would not be cancelled by the heat treatments.

At 25 ^{0}C the constant showed the highest value. At the same temperature the Gibbs energy exhibited the highest value too: -23.83±0.81 kJ mol^{-1}. This result was similar at heating system and shows that the system was most stable at 25 ^{0}C. At temperatures above 25 ^{0}C, the stability decreased and interactions between pigment and copigment slowed down. At all temperatures in two temperature regimes negative Gibbs energies were calculated. These results were connected with the spontaneous process of copigmentation leading to formation of the products. On the basis of Gibbs energy changes (Table 1) at the all temperatures it can be concluded that the process of copigmentation is possible only at temperatures to 25 ^{0}C. According the Petrova et al. [4] appears the similar Gibbs free energy in system, but in this work observed reversibility of copigmentation process.

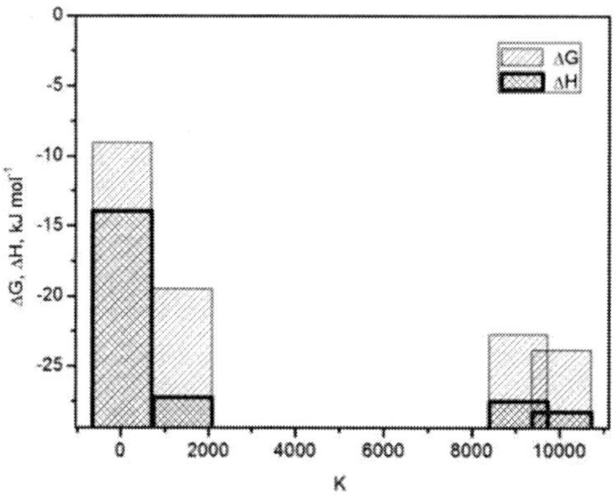

Figure. 1. Dependence between Gibbs free energy and enthalpy and equilibrium constant of the copigmentation interaction between aronia anthocyanins and sinaptic acid in model solutions upon heating system.

The enthalpy and entropy changes of the process were also negative at all temperatures. It can be concluded that such a dependence on temperature a consequence of the exothermic copigmentation and enthalpically driven (ΔH < 0) process was observed. The negative value of the entropy, ΔS, indicates that copigment formation establishes greater order in the system.

The graphical view of Gibbs free energy and enthalpy dependence on the equilibrium constant was presented in Figure 1. The stability of the system was determined by the decrease in Gibbs energy to a minimal negative value. On the figure the minimal value was observed at 25 ^{0}C. The second temperature close to the first one was 40 ^{0}C. At the other temperatures, the Gibbs energy increased to values, which was connected to the break-down of the pigment: copigment bond and destroyed of the system. Enthalpy of the process exhibit the similar results. At low temperature shown the small negative enthalpies and with increase of temperature enthalpy values increase too. This two results connected with stability of pigment:copigment system at low temperature.

The kinetic investigation of system was provide by the kinetic parameters - degradation rates (k), Activation energy (Ea) and z-factor [9]. According the experimental results and calculations, interaction between pigment and copigment described as first order reaction. This result was connected with result of Gradinaru et al. [10]. Anthocyanin degradation was connected with degradation rates constant (k). The values of constant k, activation energy and z-facor are presented in Table 2.

Table 2. Kinetic parameters for thermal investigation of aronia anthocyanin:sinapic acid interactions following the Arrhenius and Ball models.

System	t, ^{0}C	k, s^{-1}	Ea, kJ.mol^{-1}	z, ^{0}C
Aronia: sinapic acid 1:0/1:50	25	0.0241±0.0141	85.868±2.12	21.177±1.76
	40	0.0338±0.0282		
	60	0.0651±0.0062		
	80	0.0879±0.0071		
	80/60	0.0597±0.0191	59.901±3.55	29.965±2.32
	60/40	0.0411±0.0132		
	40/25	0.0301±0.0209		

The kinetic investigation of system was provide the kinetic parameters degradation rates (k), Activation energy (Ea) and z-factor using two classical kinetic models Arrhenius and Ball's. The rate constant was the smallest at 25 ^{0}C at heating and which was proof of system stability at this temperature. After heating system to 80 ^{0}C and foloweing cooling the rate constant increases and after that decreases there values. These results were associated with the reversibility of the copigmentation process in the system. Figure 2 presented dependance betwen lnk and 1/T. Linear dependance and good corelations observed at two temperature range.

The activation energies (Ea) ranged was calculated between 85.868±2.12 at heatin to 59.901±3.55 at cooling, and the z-factor ranged between 21.177±1.76 and 29.965±2.32 compared with values of Zozio et al. [9]. This values were associated with a low z-factor and with the degradation of anthocyanin rate, which was comparatively slower at high activation energy (Ea). When the temperature in the system increases, the values of z-factor increases and the activation energy Ea decreases. On cooling system Ea was have an extremely high value again and a correspondingly low z-factor. These results were expected and were related to the irreversibility of the copigmentation process on the system. According the Kopjar et al. [15] activation energies (Ea) ranged between (63 and 91) kJ mol^{-1}, and the z-factor ranged between (18 and 27) ^{0}C confirm results obtained in this work at heating and at cooling of the system.

Similar activation energy (Ea) [6] was observed in grape extract 81.34 kJ mol^{-1} too. In two determinations by thermodynamic (low negative Gibbs free energy) and kinetic (higher activation energy and less rate degradation) parameters the results confitm the reversibility of the copigmentation process and higher stability of system at low temperature.

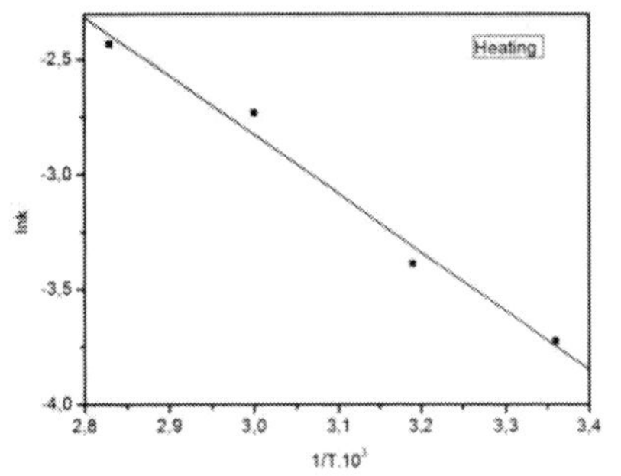

a)

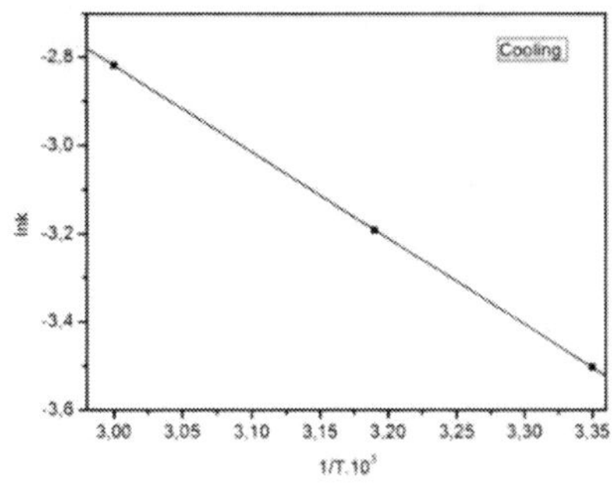

b)

Figure. 2. Correlation relationships between degradation rate constants as a function of temperature 1/T. a) presented system at heating; b) presented system at cooling.

4. Conclusions

The thermal stability of aronia anthocyanin:sinapic acid was provide first by heating system to 80 ^{0}C and then cooling to 25 ^{0}C. At the same temperatures destruction was observed and after that restoration of complex was observed accorsing the calculations. The thermodynamic and kinetic parameters confirm this results. On the basis of calculations according the thermodynamic and kinetic parameters obtained in all temperature range it can be concluded that the process of copigmentation is possible only to temperature 40 ^{0}C or lower.

Acknowledgements

The authors would like to thank the Research and Development Sector at the Technical University of Sofia for the financial support.

References

[1] Nikkhah E, Khayamy M, Heidari R, Jame R 2007 *Journal of Biological Sciences* **7 (8)** 1412
[2] Eiro M and Heinonen M 2002 *Journal of Agricultural and Food Chemistry,* **50 (25)** 7461
[3] Dangles O and Brouillard R 1992 *Can. J. Chem.* **70** 2174
[4] Petrova I, Shikov V, Gandova V, Mihalev K, Dimitrov D 2017 *Bulgarian Chemical Communications* **49 (1)** 115
[5] Shakeel F, Haq N, Raish M, Anwer MK, Al-Shdefat R 2016 *Journal of Molecular Liquids* **222** 167
[6] Falcao, A. Falcao, E. Gris, M. Bordignon-Luiz 2008 *Brazilian Journal of Food Technology* **11 (1)** 63
[7] Reyes L and Cisneros-Zevallos L 2007 *Food Chemistry* **10 (3)** 885
[8] Casatia C, Baezaa R, Sancheza V, Catalanob A, L´opezb P, Zamora M 2015 *Journal of Berry Research* **5 (4)** 29
[9] Zozio S, Pallet D, Dornier M 2011 *Fruits* **66 (3)** 203
[10] Gradinaru G, Biliaderis C, Kallithraka S, Kefalas P, Garcia-Viguera C 2003 *Food Chemistry* **83 (3)** 423
[11] Shikov V, Kammerer D, Mihalev K, Mollov P, Carle R 2008 *Journal of Agricultural and Food Chemistry* **56 (18)** 8521
[12] Shikov V, Mihalev K, Yoncheva N, Karagyozov V, Mollov P 2012 *Journal of EcoAgriTourism* **8 (2)** 231
[13] Cisse M, Vaillant F, Acosta O, Dhuique-Mayer C, Dornie M 2009 *Journal of Agricultural and Food Chemistry* **57 (14)** 6285
[14] Kopjar M, Jaksic K, Pilizota V 2012 *Journal of Food Processing and Preservation* **36 (6)** 542
[15] Kopjar M, Piližota V, Šubaric D, Babic J 2009 *Croatian Journal of Food Science and Technology* **1 (1)** 24